Problem Books in Mathematics

Series editor:

Peter Winkler
Department of Mathematics
Dartmouth College
Hanover, NH 03755
USA

For further volumes:
http://www.springer.com/series/714

Leszek Gasiński • Nikolaos S. Papageorgiou

Exercises in Analysis

Part 1

Springer

Leszek Gasiński
Faculty of Mathematics
and Computer Science
Jagiellonian University
Krakow, Poland

Nikolaos S. Papageorgiou
Department of Mathematics
National Technical University
Athens, Greece

ISSN 0941-3502
ISBN 978-3-319-35535-1 ISBN 978-3-319-06176-4 (eBook)
DOI 10.1007/978-3-319-06176-4
Springer Cham Heidelberg New York Dordrecht London

Mathematics Subject Classification (2010): 00A07, 46-XX, 30LXX, 28-XX, 60G46

Printed on acid-free paper

Springer is part of Springer Science+Business Media (www.springer.com)

Preface

The aim of this book is to review the theory of some basic topics in Analysis and accompany the theory with problems and their solutions. With the problems the reader can test his/her understanding of the theory and also discover extensions of the theory and additional results which are not so standard in the literature. The topics covered span a more or less standard advanced undergraduate and graduate curriculum in Analysis. More precisely, we focus on the following subjects:

1. Metric Spaces
2. Topological Spaces (here there is also some introductory material on Algebraic Topology)
3. Measure, Integration and Martingales (including L^p-spaces)
4. Measure and Topology (covering issues concerning the interplay between measure theory and topology)
5. Functional Analysis (with emphasis on basic Banach space theory).

Each one of the above five subjects corresponds to a different chapter. In the first part of each chapter, we present the basic theory, with all the main definitions and results. We also include comments and remarks expanding on the concepts and results, but no proofs. This review material will help the reader refresh his/her knowledge of the theory before tackling the problems. In each chapter the theory is followed by problems and their detailed solution. In each chapter, there are at least 170 problems, marked with $\star$, $\star\star$ or $\star\star\star$ according to their difficulty. Some of those problems complement the theory, while the rest check the reader's understanding of the theory. We strongly encourage the reader to put some substantial personal effort in trying to solve the problem, before looking at the solution. Otherwise, they will get no benefit from reading the book. On the other hand, a serious effort to solve the problem themselves and subsequently compare their solution with the one provided (or checking to see where they failed to come up with the right arguments to produce a solution) will help them to achieve a solid understanding of the theory.

It is not easy to provide the origin of each of the problems and of their solutions. They can be traced as problems included in the books mentioned in the literature (where they are stated without proofs), or they can be found in the problem books listed in the References, or they are standard exercises in the public domain, or they were accumulated through the years from teaching undergraduate and graduate courses on these or closely related subjects. In any case, the books mentioned in the References can be a valuable source for additional theoretical material and more problems. Our book is only the starting point (helpful we hope).

The authors express their gratitude to Springer, New York, for its highly professional assistance and above all we want to thank our editor, Mrs. Elizabeth Loew, for her strong moral support, patience and kind understanding.

Krakow, Poland — Leszek Gasiński
Athens, Greece — Nikolaos S. Papageorgiou

Contents

1 Metric Spaces 1
1.1 Introduction 1
1.1.1 Basic Definitions and Notation 1
1.1.2 Sequences and Complete Metric Spaces 4
1.1.3 Topology of Metric Spaces 4
1.1.4 Baire Theorem 8
1.1.5 Continuous and Uniformly Continuous Functions 9
1.1.6 Completion of Metric Spaces: Equivalence of Metrics 14
1.1.7 Pointwise and Uniform Convergence of Maps . 16
1.1.8 Compact Metric Spaces 17
1.1.9 Connectedness 22
1.1.10 Partitions of Unity 26
1.1.11 Products of Metric Spaces 28
1.1.12 Auxiliary Notions 31
1.2 Problems 33
1.3 Solutions 64
Bibliography 191

2 Topological Spaces **193**
2.1 Introduction 193
2.1.1 Basic Definitions and Notation 193
2.1.2 Topological Basis and Subbasis 198
2.1.3 Nets 199
2.1.4 Continuous and Semicontinuous Functions . . . 201
2.1.5 Open and Closed Maps: Homeomorphisms . . 208
2.1.6 Weak (or Initial) and Strong (or Final) Topologies 209
2.1.7 Compact Topological Spaces 213

2.1.8 Connectedness 217
2.1.9 Urysohn and Tietze Theorems 223
2.1.10 Paracompact and Baire Spaces 224
2.1.11 Polish and Souslin Sets 226
2.1.12 Michael Selection Theorem 229
2.1.13 The Space $C(X;Y)$ 230
2.1.14 Elements of Algebraic Topology I: Homotopy . 233
2.1.15 Elements of Algebraic Topology II: Homology . 245
2.2 Problems 265
2.3 Solutions 293
Bibliography 401

3 Measure, Integral and Martingales **403**
3.1 Introduction 403
3.1.1 Basic Definitions and Notation 403
3.1.2 Measures and Outer Measures 407
3.1.3 The Lebesgue Measure 412
3.1.4 Atoms and Nonatomic Measures 414
3.1.5 Product Measures 415
3.1.6 Lebesgue–Stieltjes Measures 416
3.1.7 Measurable Functions 418
3.1.8 The Lebesgue Integral 425
3.1.9 Convergence Theorems 430
3.1.10 L^p-Spaces 431
3.1.11 Multiple Integrals: Change of Variables 437
3.1.12 Uniform Integrability: Modes of Convergence . 440
3.1.13 Signed Measures 446
3.1.14 Radon–Nikodym Theorem 449
3.1.15 Maximal Function and Lyapunov Convexity Theorem 450
3.1.16 Conditional Expectation and Martingales . . . 451
3.2 Problems 460
3.3 Solutions 497
Bibliography 631

4 Measures and Topology **633**
4.1 Introduction 633
4.1.1 Borel and Baire σ-Algebras 633
4.1.2 Regular and Radon Measures 635

4.1.3 Riesz Representation Theorem for Continuous Functions . . . 636
4.1.4 Space of Probability Measures: Prohorov Theorem . . . 638
4.1.5 Polish, Souslin and Borel Spaces . . . 640
4.1.6 Measurable Multifunctions: Selection Theorems 643
4.1.7 Projection Theorems . . . 647
4.1.8 Dual of $L^p(\Omega)$ for $1 \leqslant p \leqslant \infty$ . . . 648
4.1.9 Sequences of Measures: Weak Convergence in $L^p(\Omega)$ . . . 649
4.1.10 Covering Theorems . . . 652
4.1.11 Lebesgue Differentiation Theorem . . . 654
4.1.12 Bounded Variation and Absolutely Continuous Functions . . . 658
4.1.13 Hausdorff Measures: Change of Variables . . . 672
4.1.14 Carathéodory Functions . . . 675
4.2 Problems . . . 677
4.3 Solutions . . . 710
Bibliography . . . 833

5 Functional Analysis 835
5.1 Introduction . . . 835
5.1.1 Locally Convex, Normed and Banach Spaces . 835
5.1.2 Linear Operators: Quotient Spaces—Riesz Lemma . . . 838
5.1.3 The Hahn–Banach Theorem . . . 840
5.1.4 Adjoint Operators and Annihilators . . . 842
5.1.5 The Three Basic Theorems of Linear Functional Analysis . . . 843
5.1.6 The Weak Topology . . . 846
5.1.7 The Weak* Topology . . . 848
5.1.8 Reflexive Banach Spaces . . . 850
5.1.9 Separable Banach Spaces . . . 851
5.1.10 Uniformly Convex Spaces . . . 853
5.1.11 Hilbert Spaces . . . 854
5.1.12 Unbounded Linear Operators . . . 860
5.1.13 Extremal Structure of Sets . . . 863
5.1.14 Compact Operators . . . 865
5.1.15 Spectral Theory . . . 867

5.1.16 Differentiability and the Geometry of Banach Spaces . . . 870
5.1.17 Best Approximation: Various Theorems for Banach Spaces . . . 874
5.2 Problems . . . 876
5.3 Solutions . . . 903
Bibliography . . . 1003

Other Problem Books **1005**

List of Symbols **1007**

Index **1015**

Chapter 1

Metric Spaces

1.1 Introduction

1.1.1 Basic Definitions and Notation

Definition 1.1
*A **metric space** is a pair (X, d_X) of a set X and a function $d_X : X \times X \longrightarrow \mathbb{R}$, which has the following properties:*
(a) $d_X(x, y) \geqslant 0$ *for all* $x, y \in X$ *and* $d_X(x, y) = 0$ *if and only if* $x = y$;
(b) $d_X(x, y) = d_X(y, x)$ *for all* $x, y \in X$;
(c) $d_X(x, y) \leqslant d_X(x, u) + d_X(u, y)$ *for all* $x, y, u \in X$ *(**triangle inequality**).*
The function $d_X(\cdot, \cdot)$ *is called a **metric** or **distance**.*

Remark 1.2
If (a) in the above definition is replaced by a weaker requirement:
(a)′ $d_X(x, y) \geqslant 0$ for all $x, y \in X$ and if $x = y$, then $d_X(x, y) = 0$,
then d_X is said to be a ***pseudometric*** or ***ecart*** or ***semimetric*** and (X, d_X) is a ***pseudometric space*** (or ***semimetric space***).

Example 1.3
(a) Suppose that $\{(X_k, d_{X_k})\}_{k=1}^N$ are metric spaces and set

$$X = \prod_{k=1}^N X_k$$

L. Gasiński and N.S. Papageorgiou, *Exercises in Analysis: Part 1*,
Problem Books in Mathematics, DOI 10.1007/978-3-319-06176-4_1,

and

$$\hat{d}_p(x,y) = \Big(\sum_{k=1}^{N} d_{X_k}(x_k,y_k)^p\Big)^{\frac{1}{p}} \qquad \forall\, x,y \in X,$$
$$\hat{d}_\infty(x,y) = \max\big(d_{X_k}(x_k,y_k):\ 1 \leqslant k \leqslant N\big) \qquad \forall\, x,y \in X,$$

where $1 \leqslant p < +\infty$, $x = (x_k)_{k=1}^N$, $y = (y_k)_{k=1}^N \in X$. Then $(X, \hat{d}_p)$ $(1 \leqslant p < +\infty)$ and $(X, \hat{d}_\infty)$ are metric spaces. In particular, if $X_k = \mathbb{R}$ and

$$d_{\mathbb{R}}(x,y) = |x-y| \qquad \forall\, x,y \in \mathbb{R},$$

then $d_{\mathbb{R}}$ is a metric on $\mathbb{R}$. For $N \geqslant 1$, $\hat{d}_2$ is ***the Euclidean metric*** on $\mathbb{R}^N$.

(b) Suppose now that we have an infinite family $\{(X_k, d_{X_k})\}_{k \geqslant 1}$ of metric spaces and assume that

$$\sup\{d_{X_k}(x,y):\ x,y \in X,\ k \geqslant 1\} < +\infty.$$

We set

$$X = \prod_{k \geqslant 1} X_k$$

and

$$\hat{d}_X(x,y) = \sum_{k \geqslant 1} \frac{1}{2^k} d_{X_k}(x_k,y_k) \quad \forall\, x = \{x_k\}_{k \geqslant 1},\ y = \{y_k\}_{k \geqslant 1} \in X.$$

Then $(X, \hat{d}_X)$ is a metric space.

(c) Let X be an arbitrary nonempty set and let

$$d_X^d(x,y) = \begin{cases} 0 & \text{if } x = y, \\ 1 & \text{if } x \neq y, \end{cases} \qquad \forall\, x,y \in X.$$

Then (X, d_X^d) is a metric space and d_X^d is called the ***discrete metric*** on X.

(d) Let $X = C[a,b]$ (the space of continuous functions on $[a,b]$). We set

$$d_X^1(f,g) = \int_a^b |f(t) - g(t)|\,dt \quad \text{and} \quad d_X^\infty(f,g) = \max_{a \leqslant t \leqslant b} |f(t) - g(t)| \qquad \forall\, f,g \in X.$$

Then (X, d_X^1) and (X, d_X^∞) are metric spaces. However, as we will see later (see Remark 1.82) these metric spaces are fundamentally different. We call d_X^1 the ***L^1-metric*** and d_X^∞ the ***uniform metric*** or ***supremum metric*** on $X = C[a, b]$.

(e) Let $X = \mathbb{R} \cup \{\pm\infty\}$ (extended real line) and set

$$d_X(x, y) \;=\; \left|\tan^{-1} x - \tan^{-1} y\right| \qquad \forall\, x, y \in X$$

(recall that $\tan^{-1}(\pm\infty) = \pm\frac{\pi}{2}$ and $\tan^{-1}$ is injection). Then (X, d_X) is a metric space.

Proposition 1.4
If (X, d_X) *is a metric space and*

$$\hat{d}_X(x, y) \;=\; \frac{d_X(x, y)}{1 + d_X(x, y)} \qquad \forall\, x, y \in X,$$

then $(X, \hat{d}_X)$ *is a metric space.*

Remark 1.5
Note that

$$\hat{d}_X(x, y) \;<\; 1 \qquad \forall\, x, y \in X.$$

Moreover, using this proposition, we can see that the space X of all real sequences equipped with the distance

$$d_X(x, y) \;=\; \sum_{k \geqslant 1} \frac{1}{2^k} \frac{|x_k - y_k|}{1 + |x_k - y_k|}$$

is a metric space (see Example 1.3(b)).

Definition 1.6
Let (X, d_X) *be a metric space.*
(a) *The* ***open ball centred at*** $x_0 \in X$ ***of radius*** $r > 0$ *is defined by*

$$B_r(x_0) \;=\; \left\{x \in X:\; d_X(x, x_0) < r\right\}.$$

(b) *A set* $C \subseteq X$ *is said to be* ***bounded*** *if it is contained in some open ball. The set* C *is unbounded if this is not the case.*

*(c) The **diameter** of a set $C \subseteq X$ is given by*

$$\operatorname{diam} C = \sup\{d_X(x,y) : x, y \in C\}$$

(if $C = \emptyset$, then $\operatorname{diam} C = 0$).
(d) For any $x \in X$ and $A, B \subseteq X$, we define:

$$\operatorname{dist}(x,A) = \inf\{d_X(x,a) : a \in A\}$$

and

$$\operatorname{dist}(A,B) = \inf\{d_X(a,b) : a \in A, \ b \in B\}.$$

1.1.2 Sequences and Complete Metric Spaces

Definition 1.7
Let (X, d_X) be a metric space and let $\{x_n\}_{n\geqslant 1} \subseteq X$ be a sequence.
(a) We say that the sequence $\{x_n\}_{n\geqslant 1} \subseteq X$ converges to $x \in X$ if and only if

$$d_X(x_n, x) \longrightarrow 0 \quad \text{as} \quad n \to +\infty,$$

i.e., for any $r > 0$, we can find an integer $n_0 = n_0(r) \geqslant 1$ such that

$$x_n \in B_r(x) \qquad \forall\, n \geqslant n_0.$$

The point $x \in X$ is said to be the limit of the convergent sequence $\{x_n\}_{n\geqslant 1}$.
*(b) Let (X, d_X) be a metric space. A sequence $\{x_n\}_{n\geqslant 1} \subseteq X$ is said to be a **Cauchy sequence** if for any given $\varepsilon > 0$, we can find an integer $n_0 = n_0(\varepsilon) \geqslant 1$ such that*

$$d_X(x_n, x_k) \leqslant \varepsilon \qquad \forall\, n, k \geqslant n_0.$$

*(c) A metric space (X, d_X) is said to be **complete** if every Cauchy sequence in X is convergent in X.*

1.1.3 Topology of Metric Spaces

To be able to proceed further with the study of metric spaces, we need to introduce some topological material associated with them. We will return to these concepts in a more general setting in Chap. 2.

Definition 1.8
Let (X, d_X) be a metric space.
(a) *A set $U \subseteq X$ is said to be* **open** *if for every $x \in U$ we can find $r = r(x) > 0$ such that $B_r(x) \subseteq U$.*
(b) *A set $C \subseteq X$ is said to be* **closed** *if the set $X \setminus C$ is open.*
(c) *The family of open sets $U \subseteq X$ is called the* **topology** *determined by the metric d_X* (**metric topology**).
(d) *A set $A \subseteq X$ is called a* **neighbourhood** *of a point $x \in X$ if there exists $r > 0$ such that $B_r(x) \subseteq A$.*

Proposition 1.9
If (X, d_X) is a metric space, then
(a) *$\emptyset$ and X are open sets;*
(b) *if $\{U_i\}_{i \in I}$ is any family of open sets in X, then $\bigcup_{i \in I} U_i$ is an open set too;*
(c) *if $\{U_k\}_{k=1}^N$ is any finite family of open sets in X, then $\bigcap_{k=1}^N U_k$ is an open set too.*

Using de Morgan laws and Definition 1.8(b), we also have

Proposition 1.10
If (X, d_X) is a metric space, then
(a) *$\emptyset$ and X are closed sets;*
(b) *if $\{C_i\}_{i \in I}$ is any family of closed sets in X, then $\bigcap_{i \in I} C_i$ is a closed set too;*
(c) *if $\{C_k\}_{k=1}^N$ is any finite family of closed sets in X, then $\bigcup_{k=1}^N C_k$ is a closed set too.*

We can characterize sets in terms of convergent sequences.

Proposition 1.11
If (X, d_X) is a metric space,
then $C \subseteq X$ is closed if and only if every convergent sequence $\{x_n\}_{n \geqslant 1} \subseteq C$ has limit in C.

Definition 1.12
Let (X, d_X) be a metric space and $E \subseteq X$. The ***subspace (metric) topology*** *on E induced by the metric d_X is the family*

$$\{E \cap U : U \subseteq X \text{ open}\}.$$

Definition 1.13
Let (X, d_X) be a metric space and let $E \subseteq X$.
(a) We say that $x \in E$ is an ***interior point of*** *E if there exists $r > 0$ such that $B_r(x) \subseteq E$. The set of all interior points of E is called the* ***interior of*** *E and is denoted by $\operatorname{int} E$.*
(b) We say that $x \in E$ is a ***limit*** *(or* ***cluster*** *or* ***accumulation)*** ***point*** *of E if $B_r(x) \cap (E \setminus \{x\}) \neq \emptyset$ for every $r > 0$. The union of E with the set of all its accumulation points is called the* ***closure*** *of E and is denoted by $\overline{E}$.*
(c) The ***boundary*** *of E is defined by $\partial E = \overline{E} \cap \overline{E^c}$.*
(d) We say that $x \in E$ is an isolated point of E, if there is an $r > 0$ such that $B_r(x) \cap (E \setminus \{x\}) = \emptyset$. Hence $\{x\}$ is a relatively open subset of E.
(e) We say that $E \subseteq X$ is a ***perfect set*** *if every point of E is an accumulation point.*

Theorem 1.14
Let (X, d_X) be a metric space and let $E \subseteq X$.
(a) $x \in \operatorname{int} E$ if and only if there exists $r > 0$ such that $B_r(x) \subseteq E$.
(b) $x \in \overline{E}$ if and only if for every $r > 0$, we have $B_r(x) \cap E \neq \emptyset$.
(c) $x \in \partial E$ if and only if for every $r > 0$, we have $B_r(x) \cap E \neq \emptyset$ and $B_r(x) \setminus E \neq \emptyset$.
(d) x is an accumulation point of E if and only if we can find a sequence $\{x_n\}_{n \geqslant 1} \subseteq E \setminus \{x\}$ such that $x_n \longrightarrow x$.

Proposition 1.15
Let (X, d_X) be a metric space and let $E, F \subseteq X$. If $E \subseteq F$, then $\operatorname{int} E \subseteq \operatorname{int} F$ and $\overline{E} \subseteq \overline{F}$.

Definition 1.16
For every $r > 0$ and for every $x \in X$, we define a ***closed ball***

$$\overline{B}_r(x) = \{u \in X : d_X(u, x) \leqslant r\}.$$

Remark 1.17
Any closed ball is a closed set (in the sense of Definition 1.8(b)) and we always have

$$\overline{B_r(x)} \subseteq \overline{B}_r(x) \qquad \forall\, x \in X,\ r > 0.$$

The above inclusion in general cannot be replaced by equality. To see this let the set X has at least two elements and let (X, d_X^d) be a discrete metric space (see Example 1.3(c)). If $x \in X$, then

$$\overline{B_1(x)} = \{x\} \subsetneq X = \overline{B}_1(x).$$

Proposition 1.18
If (X, d_X) is a metric space and $E \subseteq X$, then
(a) *$\operatorname{int} E$ is open and is the union of all open sets contained in E, i.e., $\operatorname{int} E$ is the largest open set contained in E (see Proposition 1.9(b)).*
(b) *$\overline{E}$ is closed and is the intersection of all closed sets which contain E, i.e., $\overline{E}$ is the smallest closed set containing E (see Proposition 1.10(b)).*

Corollary 1.19
If (X, d_X) is a metric space and $E \subseteq X$, then
(a) *E is open if and only if $E = \operatorname{int} E$.*
(b) *E is closed if and only if $E = \overline{E}$.*

Definition 1.20
Let (X, d_X) be a metric space. A set $D \subseteq X$ is ***dense in*** *$E \subseteq X$ if $E \subseteq \overline{D}$. We say that D is* ***dense*** *(or* ***everywhere dense****) if $\overline{D} = X$.*

Definition 1.21
Let (X, d_X) be a metric space and $x \in X$.
(a) *A family of open sets $\{U_i\}_{i \in I}$ all containing x is said to be a* ***local base at*** *x if for every open set U containing x, we can find $i \in I$ such that $U_i \subseteq U$.*
(b) *A family $\{U_i\}_{i \in I}$ of nonempty sets in X is a* ***base for the metric topology of*** *X if every open set in X is the union of a subfamily of $\{U_i\}_{i \in I}$.*
(c) *We say that X is* ***second countable*** *if it has a countable base for the metric topology.*

(d) *A family $\mathcal{F}$ of open sets in X is an* ***open cover of*** *X if*

$$\bigcup_{U \in \mathcal{F}} U = X.$$

An ***open subcover*** *of $\mathcal{F}$ is a subfamily of $\mathcal{F}$ which is an open cover of X.*
(e) *We say that X is* ***Lindelöf*** *if every open cover of X contains a countable subcover.*
(f) *We say that X is* ***separable*** *if it has a countable dense subset.*

Proposition 1.22
Every metric space has a countable local base at every point.

Remark 1.23
The above proposition explains the name "second countable" in Definition 1.21(c). As we shall see in Chap. 2, for a topological space the property of having a countable local base at every point is called ***first countability***. The above proposition says that every metric space is first countable. This is the reason why sequences suffice to describe the various topological notions of a metric space. However note that not every metric space is second countable.

Proposition 1.24
Let (X, d_X) be a metric space. The following three properties are equivalent:
(a) *X is second countable.*
(b) *X is Lindelöf.*
(c) *X is separable.*

1.1.4 Baire Theorem

One of the most important properties of a metric space is completeness and many fundamental results of analysis depend critically on this property. A main device through which completeness becomes a powerful tool is the so called ***Baire category theorem***.

Definition 1.25
Let (X, d_X) be a metric space. A set $E \subseteq X$ is said to be ***nowhere dense*** *if $\operatorname{int} \overline{E} = \emptyset$. A set $E \subseteq X$ is said to be* ***meager*** *or* ***of first category*** *if E can be written as a countable union of nowhere dense sets. If $E \subseteq X$ is not of the first category, then we say that it is* ***of second category***.

Theorem 1.26 *(Baire Category Theorem)*
If (X, d_X) *be a complete metric space, then*
(a) *the intersection of countably many open dense sets is dense in* X*;*
(b) *if* $X = \bigcup_{n \geqslant 1} C_n$ *with* $C_n \subseteq X$ *closed, then for at least one* $n_0 \geqslant 1$*, we have* $\operatorname{int} C_{n_0} \neq \emptyset$.

Corollary 1.27
(a) *A complete metric space is of second category.*
(b) *In a complete metric space a meager set has empty interior, i.e., its complement is dense in* X.

Theorem 1.28 *(Cantor Intersection Theorem)*
If (X, d_X) *be a metric space,*
then the following two statements are equivalent:
(a) X *is complete;*
(a) *every decreasing sequence* $\{C_n\}_{n \geqslant 1}$ *(i.e.,* $C_{n+1} \subseteq C_n$ *for all* $n \geqslant 1$*) of nonempty closed sets in* X *such that* $\operatorname{diam} C_n \searrow 0$ *as* $n \to +\infty$ *has singleton intersection.*

1.1.5 Continuous and Uniformly Continuous Functions

Definition 1.29
Let (X, d_X) *and* (Y, d_Y) *be two metric spaces. A function* $f\colon X \longrightarrow Y$ *is said to be* ***continuous at*** $x \in X$ *if for every* $\varepsilon > 0$*, there exists* $\delta = \delta(x, \varepsilon) > 0$ *such that*

$$d_X(x, y) < \delta \implies d_Y\big(f(x), f(y)\big) < \varepsilon.$$

We say that $f\colon X \longrightarrow Y$ *is* ***continuous****, if it is continuous at every* $x \in X$.

Proposition 1.30
If (X, d_X) *and* (Y, d_Y) *are two metric spaces and* $f\colon X \longrightarrow Y$ *is a function,*
then f *is continuous at* $x \in X$ *if and only if for any sequence* $x_n \longrightarrow x$ *in* X*, we have that* $f(x_n) \longrightarrow f(x)$ *in* Y.

Corollary 1.31
If X, Y, V *are three metric spaces,* $f: X \longrightarrow Y$ *is continuous at* $x \in X$ *and* $g: Y \longrightarrow V$ *is continuous at* $f(x) \in Y$,
then the composition $g \circ f: X \longrightarrow V$ *is continuous at* $x \in X$.

Proposition 1.32
If (X, d_X) *and* (Y, d_Y) *are two metric spaces and* $f: X \longrightarrow Y$ *is a function,*
then the following statements are equivalent:
(a) f *is continuous;*
(b) for every open $U \subseteq Y$, *the set* $f^{-1}(U) \subseteq X$ *is open;*
(c) for every closed $C \subseteq Y$, *the set* $f^{-1}(C) \subseteq X$ *is closed;*
(d) $f(\overline{E}) \subseteq \overline{f(E)}$ *for all* $E \subseteq X$;
(e) $\overline{f^{-1}(A)} \subseteq f^{-1}(\overline{A})$ *for all* $A \subseteq Y$.

The last proposition leads to the following definition.

Definition 1.33
Let X, Y *be two sets and let* E *be a proper subset of* X. *If* $f: E \longrightarrow Y$ *is a function, then* $\hat{f}: X \longrightarrow Y$ *is said to an* ***extension*** *of* f, *if* $\hat{f}|_E = f$. *The function* f *is called the* ***restriction*** *of* $\hat{f}$ *on* E.

Theorem 1.34
If X *and* Y *are two metric spaces,* $D \subseteq X$ *is dense and* $f: D \longrightarrow Y$ *is a function,*
then f *admits a continuous extension* $\hat{f}: X \longrightarrow Y$ *if and only if for every* $x \in X$ *the limit* $\lim\limits_{u \to x} f(u)$ *exists in* Y *and is equal to* $f(x)$ *if* $x \in D$.
When the extension exists, it is unique.

Proposition 1.35
If X *and* Y *are two metric spaces,* $\mathcal{X}$ *is a nonempty collection of mutually disjoint nonempty subsets of* X, $f: \bigcup \mathcal{X} \longrightarrow Y$ *and for each* $E \in \mathcal{X}$, *we have* $E \cap \left(\overline{\bigcup(\mathcal{X} \setminus \{E\})}\right) = \emptyset$ *and* $f|_E$ *is continuous,*
then $f: \bigcup \mathcal{X} \longrightarrow Y$ *is continuous.*

Proposition 1.36
If (X, d_X) *is a metric space,*
then the metric $d_X: X \times X \longrightarrow \mathbb{R}$ *is continuous, where in* $X \times X$ *we consider any product metric* $\hat{d}_p$ *with* $1 < p < +\infty$ *(see Example 1.3).*

Definition 1.37
Suppose that (X, d_X) *and* (Y, d_Y) *are two metric spaces and* $f: X \longrightarrow Y$ *is a function. For each* $x \in X$*, we define*

$$\omega_f(x) \stackrel{\text{def}}{=} \inf_{r>0} \operatorname{diam} f\big(B_r(x)\big).$$

The function ω_f *is called the* ***oscillation function for*** f*.*

Remark 1.38
(a) Suppose that (X, d_X) and (Y, d_Y) are two metric spaces and $f: X \longrightarrow Y$ is a function. The function f is continuous at $x \in X$ if and only if $\omega_f(x) = 0$ (see Problem 1.46).
(b) If (X, d_X) is a metric space and $f: X \longrightarrow \mathbb{R}$ is a function, then for every $x \in X$, we have $\omega_f(x) = \limsup_{u \to x} f(u) - \liminf_{u \to x} f(u)$.

Definition 1.39
Let X *and* Y *be two metric spaces and let* $f: X \longrightarrow Y$ *be a function.*
(a) *We say that* f *is an* ***open function*** *if for every open set* $U \subseteq X$*, the set* $f(U)$ *is open in* Y*.*
(b) *We say that* f *is a* ***closed function*** *if for every closed set* $C \subseteq X$*, the set* $f(C)$ *is closed in* Y*.*
(c) *We say that* f *is a* ***homeomorphism*** *if* f *is continuous, bijective and* $f^{-1}: Y \longrightarrow X$ *is continuous (i.e.,* f *is bicontinuous bijection). The metric spaces* X *and* Y *are said to be* ***homeomorphic spaces*** *if there exists a homeomorphism* $f: X \longrightarrow Y$*.*

Theorem 1.40
If X *and* Y *are two metric spaces and* $f: X \longrightarrow Y$ *is bijective, then the following statements are equivalent:*
(a) f *is a homeomorphism;*
(b) f *is continuous and open;*
(c) f *is continuous and closed;*
(d) $f(\overline{E}) = \overline{f(E)}$ *for every* $E \subseteq X$*.*

Definition 1.41
Let (X, d_X) and (Y, d_Y) be two metric spaces. A function $f\colon X \longrightarrow Y$ is an ***isometry*** *if $d_Y\big(f(x_1), f(x_2)\big) = d_X(x_1, x_2)$ for all $x_1, x_2 \in X$. The metric spaces X and Y are said to be* ***isometric****.*

Evidently an isometry is a continuous function.

Remark 1.42
Using the notions of homeomorphism and isometry, we can speak about the topological properties and the metric properties of a metric space X. Topological are those properties which are preserved by homeomorphisms, and metric are those properties which are preserved by isometrics. For example openness, closedness or being a convergent sequence in X are topological properties of X, while Cauchy sequences and completeness are metric properties. Of course the topological structure is more flexible than the metric structure.

The next theorem is a particular case of the so called ***Urysohn lemma*** (see Theorem 2.136).

Theorem 1.43
If X is a metric space and $A, C \subseteq X$ are disjoint closed sets, then there is a continuous function $f\colon X \longrightarrow [0,1]$ such that

$$f|_A \equiv 0 \quad \textit{and} \quad f|_C \equiv 1.$$

Similarly the next theorem is a particular case of the so called ***Tietze extension theorem*** (see Theorem 2.138).

Theorem 1.44
If X is a metric space, $C \subseteq X$ is a nonempty closed subset and $f\colon C \longrightarrow \mathbb{R}$ is a continuous function,
then f admits a continuous extension $\hat{f}\colon X \longrightarrow Y$ such that

$$\inf \hat{f}(X) = \inf f(C) \quad \textit{and} \quad \sup \hat{f}(X) = \sup f(C).$$

In the definition of continuity (see Definition 1.29), δ depends on x and ε. If δ can be chosen independently of $x \in X$, then we have a stronger form of continuity called uniform continuity.

Definition 1.45
Let (X, d_X) and (Y, d_Y) be two metric spaces and $f: X \longrightarrow Y$. We say that f is **uniformly continuous** *if for every $\varepsilon > 0$, there exists $\delta = \delta(\varepsilon) > 0$ such that*

$$d_X(x_1, x_2) < \delta \quad \Longrightarrow \quad d_Y(f(x_1), f(x_2)) < \varepsilon.$$

Proposition 1.46
Let X, Y, V be three metric spaces.
(a) If $f: X \longrightarrow Y$ is uniformly continuous and $E \subseteq X$, then $f|_E$ is uniformly continuous too.
(b) If $f: X \longrightarrow Y$ is uniformly continuous, then maps Cauchy sequences to Cauchy sequences.
(c) If $f: X \longrightarrow Y$ and $g: Y \longrightarrow V$ are both uniformly continuous functions, then so is $g \circ f: X \longrightarrow Y$;

Combining Theorem 1.34 with Proposition 1.46(b), we get at once the following fundamental extension theorem.

Theorem 1.47
<u>If</u> X and Y are metric spaces, Y is complete, $D \subseteq X$ is dense and $f: D \longrightarrow Y$ is a uniformly continuous function,
<u>then</u> f admits a uniformly continuous extension $\hat{f}: X \longrightarrow Y$.

An important subclass of uniformly continuous functions are the so called Lipschitz functions.

Definition 1.48
Let (X, d_X) and (Y, d_Y) be two metric spaces, $k \geqslant 0$ and let $f: X \longrightarrow Y$ be a function.
(a) We say that f is **Lipschitz continuous** *with Lipschitz constant k (or k-Lipschitz), if*

$$d_Y(f(x_1), f(x_2)) \leqslant k d_X(x_1, x_2) \qquad \forall\, x_1, x_2 \in X.$$

If $k \in [0, 1)$, then we say that f is a **contraction** *or k-***contraction***.*
If $k = 1$, then we say that f is **nonexpansive***.*
(b) We say that f is a **locally Lipschitz function** *if for all $x \in X$, we can find an open set U_x such that $x \in U_x$ and a constant $k_x > 0$ such that*

$$d_Y(f(x_1), f(x_2)) \leqslant k_x d_X(x_1, x_2) \qquad \forall\, x_1, x_2 \in U_x.$$

Theorem 1.49 *(Banach Fixed Point Theorem)*
If (X, d_X) is a complete metric space and $f: X \longrightarrow X$ is a contraction, then there exists unique $x_0 \in X$ such that $f(x_0) = x_0$ (x_0 is called ***fixed point*** *of f).*

1.1.6 Completion of Metric Spaces: Equivalence of Metrics

Definition 1.50
Let (X, d_X) be a metric space. The metric space (Y, d_Y) is said to be a ***completion*** *of (X, d_X) if (Y, d_Y) is complete and X is isometric to a dense subset of Y.*

Theorem 1.51
Every metric space (X, d_X) has a completion and any two completions of X are isometric, i.e., if (Y_1, d_{Y_1}) and (Y_2, d_{Y_2}) are completions of (X, d_X), then there exists an isometry $\varphi: Y_1 \longrightarrow Y_2$.

As we already mentioned in Remark 1.42, homeomorphic but non-isometric spaces need not have the same Cauchy sequences. To guarantee this, we need for the metrics to satisfy the following.

Definition 1.52
Let d_X^1 and d_X^2 be two metrics on X. We say that d_X^1 and d_X^2 are ***(topologically) equivalent*** *if for every sequence $\{x_n\}_{n\geqslant 1} \subseteq X$ and $x \in X$, we have*

$$d_X^1(x_n, x) \longrightarrow 0 \iff d_X^2(x_n, x) \longrightarrow 0.$$

Then the metric spaces (X, d_X^1) and (X, d_X^2) are said to be ***(topologically) equivalent metric spaces.***

Remark 1.53
By Proposition 1.30, two metrics d_X^1 and d_X^2 on X are (topologically) equivalent if and only if the identity function

$$i: (X, d_X^1) \longrightarrow (X, d_X^2)$$

is a homeomorphism. This is equivalent to saying that the metric topologies corresponding to d_X^1 and d_X^2 (see Definition 1.8(c)) are identical. This justifies the term "topologically". Continuing in this path, we say that metrics d_X^1 and d_X^2 on X are ***uniformly equivalent*** if and only if the identity functions

$$i\colon (X, d_X^1) \longrightarrow (X, d_X^2) \quad \text{and} \quad i'\colon (X, d_X^2) \longrightarrow (X, d_X^1)$$

are both uniformly continuous and we say that metrics d_X^1 and d_X^2 on X are ***Lipschitz equivalent*** if and only if both the above identity functions i and i' are Lipschitz continuous.

Proposition 1.54

Two metrics d_X^1 and d_X^2 on X are Lipschitz equivalent if and only if there exist $\vartheta > \beta > 0$ such that

$$\beta d_X^1(x_1, x_2) \leqslant d_X^2(x_1, x_2) \leqslant \vartheta d_X^1(x_1, x_2) \qquad \forall\, x_1, x_2 \in X.$$

Evidently the isometry satisfies all three equivalences (topological, uniform and Lipschitz). So, it is the strongest form of equivalence between metric spaces. Definition 1.52 and Remark 1.53 prompt us to introduce some weaker forms of equivalence of different metric spaces. In fact the first one (homeomorphic) has already been introduced in Definition 1.39(c), but for the sake of completeness we recall its definition here.

Definition 1.55

Let (X, d_X) and (Y, d_Y) be two metric spaces.

(a) *X and Y are said to be* **homeomorphic** *(or* **topologically homeomorphic***) if there exists a homeomorphism (i.e., a bicontinuous bijection) $f\colon X \longrightarrow Y$.*

(b) *X and Y are said to be* **uniformly equivalent** *if there exists a bijective function $f\colon X \longrightarrow Y$ such that both f and f^{-1} are uniformly continuous.*

(c) *X and Y are said to be* **Lipschitz equivalent** *if there exists a bijective function $f\colon X \longrightarrow Y$ such that both f and f^{-1} are Lipschitz continuous.*

Definition 1.56

A metric space (X, d_X) is said to be **topologically complete** *if there is a metric $\hat{d}_X$ which is (topologically) equivalent to d_X such that the space $(X, \hat{d}_X)$ is complete.*

Definition 1.57
Let X be a metric space.
(a) *A set $E \subseteq X$ is said to be a G_δ-set if $E = \bigcap_{n\geqslant 1} U_n$ with $U_n \subseteq X$ open for all $n \geqslant 1$.*
(b) *A set $D \subseteq X$ is said to be a F_σ-set if $D = \bigcup_{n\geqslant 1} C_n$ with $C_n \subseteq X$ closed for all $n \geqslant 1$.*

The next theorem characterizes topologically complete metric spaces.

Theorem 1.58 *(Alexandrov Theorem)*
A metric space (X, d_X) is topologically complete if and only if X is a G_δ-set in its completion.

1.1.7 Pointwise and Uniform Convergence of Maps

Next we introduce two modes of convergence for sequences of functions.

Definition 1.59
Let X be a set and (Y, d_Y) a metric space. Consider functions $f_n, : X \longrightarrow Y$ for $n \geqslant 1$ and $f : X \longrightarrow Y$.
(a) *We say that the sequence $\{f_n\}_{n\geqslant 1}$* **converges pointwise** *to f if*

$$d_Y\left(f_n(x), f(x)\right) \longrightarrow 0 \quad \text{as } n \to +\infty \qquad \forall\, x \in X.$$

We denote it by

$$f_n \longrightarrow f.$$

(b) *We say that the sequence $\{f_n\}_{n\geqslant 1}$* **converges uniformly** *to f if*

$$\sup_{x\in X} d_Y\left(f_n(x), f(x)\right) \longrightarrow 0 \quad \text{as } n \to +\infty.$$

We denote it by

$$f_n \rightrightarrows f.$$

Remark 1.60
It is clear from the above definition that uniform convergence implies pointwise convergence. The converse is not true.

Example 1.61
Continuity is not preserved by pointwise convergence. Let $X = Y = [0, 1]$ with the metric induced from $\mathbb{R}$ and consider the sequence of functions $f_n\colon [0, 1] \longrightarrow [0, 1]$, defined by

$$f_n(x) = x^n \qquad \forall\, x \in [0, 1],\ n \geqslant 1.$$

Then

$$f_n \longrightarrow f,$$

where f is the discontinuous (at $x = 1$) function

$$f(x) = \begin{cases} 0 & \text{if } x \in [0, 1), \\ 1 & \text{if } x = 1. \end{cases}$$

The sequence $\{f_n\}_{n\geqslant 1}$ does not converge uniformly.

Proposition 1.62
If (X, d_X) and (Y, d_Y) are two metric spaces and $\{f_n\colon X \longrightarrow Y\}_{n\geqslant 1}$ is a sequence of continuous functions such that

$$f_n \rightrightarrows f,$$

then $f\colon X \longrightarrow Y$ is continuous (i.e., the uniform limit of continuous functions is a continuous function).

1.1.8 Compact Metric Spaces

In Definition 1.21(d), we introduced the notion of open cover. Using it we can introduce the important notion of compactness.

Definition 1.63
We say that a metric space (X, d_X) is **compact** *if every open cover $\mathcal{F}$ of X has a finite subcover, i.e., there is a finite subfamily $\{U_k\}_{k=1}^N \subseteq \mathcal{F}$ such that $X = \bigcup_{k=1}^N U_k$. A set $E \subseteq X$ is said to be compact if it is compact for the subspace metric topology (see Definition 1.12).*

Remark 1.64
Evidently $E \subseteq X$ is compact if and only if every family of open sets in X whose union contains E has a finite subfamily whose union contains E.

Directly from De Morgan law, we obtain the following result.

Proposition 1.65
A metric space X is compact if and only if every family of closed sets with empty intersection has a finite subfamily with empty intersection.

Definition 1.66
Let X be a metric space. A family $\mathcal{F}$ of subsets of X is said to have the ***finite intersection property*** *if every finite subfamily of $\mathcal{F}$ has a nonempty intersection.*

Proposition 1.67
A metric space X is compact if and only if every family of closed subsets of X with the finite intersection property has nonempty intersection.

A direct consequence of Proposition 1.24 is the following result.

Proposition 1.68
Every compact metric space is separable.

Proposition 1.69
Let X be a metric space.
(a) Every compact subset of X is closed and bounded.
(b) Every closed subset of a compact set is compact.

Definition 1.70
Let (X, d_X) be a metric space.
(a) We say that a set $E \subseteq X$ is ***sequentially compact*** *if every sequence $\{x_n\}_{n\geqslant 1} \subseteq E$ has a subsequence $\{x_{n_k}\}_{k\geqslant 1}$ which converges to a point of E.*
(b) We say that a set $E \subseteq X$ is ***totally bounded*** *if for any $\varepsilon > 0$ there is a finite number of open balls $B_\varepsilon(x_k)$, $k = 1, \dots, N$, with centres $x_k \in X$ such that*

$$E \subseteq \bigcup_{k=1}^{N} B_\varepsilon(x_k).$$

The set $\{x_k\}_{k=1}^{N}$ of the centres of the balls is called an ***ε-net*** *of E.*

Theorem 1.71
Let X be a metric space. The following statements are equivalent:
(a) X is compact;

(b) X is sequentially compact;
(c) X is complete and totally bounded.

Definition 1.72
Suppose that X and Y are two metric spaces and $f: X \longrightarrow Y$ is a function. We say that f is **proper** *if the inverse image of any compact set in Y is compact in X.*

Remark 1.73
Even if f is continuous, f need not be proper. For example a constant $\mathbb{R}$-valued function on a noncompact metric space is continuous but not proper.

The next few results relate compactness with continuous functions.

Proposition 1.74
If X and Y are two metric spaces, $f: X \longrightarrow Y$ is a continuous function and $E \subseteq X$ is a compact set,
then $f(E) \subseteq Y$ is compact (i.e., continuous functions send compact sets to compact sets).

Theorem 1.75 *(Weierstrass Theorem)*
If X is a compact metric space and $f: X \longrightarrow \mathbb{R}$ is a continuous function,
then f attains its supremum and infimum, i.e., there exist $\underline{x}, \overline{x} \in X$ such that

$$f(\underline{x}) = \inf f(X) \quad \text{and} \quad f(\overline{x}) = \sup f(X).$$

Proposition 1.76
If X and Y are two metric spaces, X is compact and $f: X \longrightarrow Y$ is a continuous bijection,
then f is a homeomorphism.

Proposition 1.77
If X and Y are two metric spaces, X is compact and $f: X \longrightarrow Y$ is continuous,
then f is uniformly continuous.

Many important spaces in analysis (such as $\mathbb{R}^N$) are not compact, but behave locally as compact spaces.

Definition 1.78
A metric space (X, d_X) is said to be ***locally compact*** *if each point $x \in X$ has a closed ball $\overline{B}_r(x)$ which is compact. Equivalently, we can say that a metric space (X, d_X) is locally compact, if every point $x \in X$ has an open neighbourhood $U \subseteq X$ (see Definition 1.8(d)) such that $\overline{U}$ is compact (cf. Proposition 1.69).*

Theorem 1.79
<u>If</u> X is a locally compact metric space,
<u>then</u> the following statements are equivalent:
*(a) $X = \bigcup\limits_{n \geqslant 1} K_n$ with each $K_n \subseteq X$ compact (i.e., X is **σ-compact**);*
(b) X is separable;
(c) there exists an increasing sequence $\{U_n\}_{n \geqslant 1}$ of open sets in X such that $\overline{U}_n$ is compact and $\overline{U}_n \subseteq U_{n+1}$ for all $n \geqslant 1$ and we have $X = \bigcup_{n \geqslant 1} U_n$.

Proposition 1.80
Any open or closed subset of a locally compact metric space is itself locally compact. Also a locally compact metric space is open in its completion.

Definition 1.81
We say that a metric space X is a ***Baire metric space*** *if every intersection of a nonempty countable collection of open dense subsets of X is dense in X.*

For a given metric space X, let

$$C(X) = \{f : X \longrightarrow \mathbb{R} : f \text{ continuous}\}.$$

We furnish $C(X)$ with the supremum metric d^∞, defined by

$$d^\infty(f, g) = \max_{x \in X} |f(x) - g(x)|$$

(see Example 1.3(d), where $X = [a, b] \subseteq \mathbb{R}$). Then by Proposition 1.62, $(C(X), d^\infty)$ is a complete metric space.

Remark 1.82
If $X = [a, b]$ and instead of d^∞ we use metric

$$d^1(f, g) = \int_a^b |f(t) - g(t)| \, dt$$

(see Example 1.3(d)), then $\big(C(X), d^1\big)$ is not a complete metric space.

Definition 1.83
Let (X, d_X) be a metric space and let $\mathcal{Y}$ be a family of functions $f : X \longrightarrow \mathbb{R}$.
(a) *We say that $\mathcal{Y}$ is* **equicontinuous** *at $x \in X$ if for any given $\varepsilon > 0$, we can find $\delta = \delta(\varepsilon, x) > 0$ such that*

$$d_X(x, u) < \delta \implies \big|f(x) - f(u)\big| < \varepsilon \qquad \forall\, f \in \mathcal{Y}.$$

(b) *We say that $\mathcal{Y}$ is* **pointwise bounded** *if for every $x \in X$ we can find $M_x > 0$ such that*

$$\big|f(x)\big| \leqslant M_x \qquad \forall\, f \in \mathcal{Y}.$$

(c) *We say that $\mathcal{Y}$ is* **uniformly equicontinuous** *on X, if it is equicontinuous at every $x \in X$ and the $\delta > 0$ in the definition* ***(a)*** *above can be chosen independently of $x \in X$.*
(d) *We say that $\mathcal{Y}$ is* **uniformly bounded** *if there exists $M > 0$ such that*

$$\big|f(x)\big| \leqslant M \qquad \forall\, f \in \mathcal{Y},\ x \in X.$$

Theorem 1.84 *(Arzela–Ascoli Theorem)*
If X is a compact metric space and $K \subseteq C(X)$ is equicontinuous at each $x \in X$ and pointwise bounded, then
(a) *K is uniformly equicontinuous and uniformly bounded on X;*
(b) *K is relatively compact in $\big(C(X), d^\infty\big)$ (i.e., $\overline{K}^{d^\infty}$ is compact in $C(X)$);*

In fact ***(a)*** *and* ***(b)*** *are equivalent.*

1.1.9 Connectedness

After compactness, we introduce connectedness which is the other fundamental topological notion in the theory of metric spaces.

Definition 1.85
Let X be a metric space. A ***separation*** *of X is a pair of nonempty, disjoint open sets $U, V \subseteq X$ such that $X = U \cup V$. We say that X is* ***disconnected*** *if there exists a separation of X and* ***connected*** *otherwise.*

Remark 1.86
Connectedness is a property of a space, not a property of subsets as are, for example, openness and closedness. Of course, we can talk about connected subsets of X, by which we mean connected in the subspace metric topology. In this context we can always consider a separation of $E \subseteq X$ to be a pair of open sets $U, V \subseteq X$ such that $U \cap E \neq \emptyset$, $V \cap E \neq \emptyset$ and $E \subseteq U \cup V$. Note that if the pair $\{U, V\}$ is a separation for the disconnected metric space X, then U and V are also closed (such sets which are both open and closed are called ***clopen***). Finally, if $A \subseteq Y \subseteq X$, then A is connected in X if and only if it is connected in Y.

Proposition 1.87
A metric space X is connected if and only if the only subsets of X which are both open and closed (i.e., are clopen) are $\emptyset$ and X itself.

Proposition 1.88
A nonempty subset of $\mathbb{R}$ is connected if and only if it is an interval.

Proposition 1.89
If X is a metric space and $E \subseteq X$ is a connected set,
then so is every set A such that $E \subseteq A \subseteq \overline{E}$.
In particular the closure of a connected set is connected.

The well known ***Bolzano Theorem*** (or ***Intermediate Value Theorem***) from introductory calculus is a particular case of the following more general theorem.

Theorem 1.90
The continuous image of a connected metric space is connected.

Simple examples show that the intersection of two connected metric spaces need not be connected. Under certain conditions, a union of connected subsets is connected.

Theorem 1.91
If X is a metric space, $\{E_i\}_{i \in I}$ is a family of connected subsets of X and at least one of the following conditions is satisfied:
(a) $E_i \cap E_j \neq \emptyset$ *for all* $i, j \in I$;
(b) *there exists* i_0 *such that* $E_i \cap E_{i_0} \neq \emptyset$ *for all* $i \in I$,
then the set $E = \bigcup_{i \in I} E_i$ *is connected.*

Definition 1.92
Let X be a metric space and $x \in X$. The union $C(x)$ of all connected subsets of X containing x is called the ***connected component of*** x. *Evidently $C(x)$ is a maximal connected subset of X (see Theorem 1.91).*

Proposition 1.93
If X is a metric space, then
(a) *the distinct connected components of X form a partition of X;*
(b) *every connected component $C \subseteq X$ is closed;*
(c) *every nonempty connected subset of X which is both open and closed is a connected component of X;*
(d) *if $y \in C(x)$, then $C(x) = C(y)$.*

Definition 1.94
A metric space X in which the connected components are all singleton sets is said to be ***totally disconnected.***

As with compactness (see Definition 1.78), we can have a local version of the notion of connectedness.

Definition 1.95
A metric space X is said to be ***locally connected at*** x *if it has a local basis at x (see Definition 1.21(a)) consisting of connected sets. We say that X is* ***locally connected,*** *if it is locally connected at every point $x \in X$, i.e., X has a basis for the metric topology (see Definition 1.21(b)) consisting of open connected sets.*

Proposition 1.96
A metric space X is locally connected if and only if for every open set $U \subseteq X$, the connected components of U are open too.

This is another notion of connectedness of metric spaces.

Definition 1.97
A metric space X is said to be ***path-connected*** *if for every $x_1, x_2 \in X$ there is a continuous function (path) $f\colon [0,1] \longrightarrow X$ such that $f(0) = x_1$ and $f(1) = x_2$.*

Proposition 1.98
If X is a metric space and $x_0 \in X$,
then X is path-connected if and only if every $x \in X$ can be joined to x_0 by a path.

Proposition 1.99
A path-connected metric space is connected, but the converse fails in general.

Example 1.100
Consider $E \subseteq \mathbb{R}^2$, defined by $E = G \cup T_0$, where

$$G = \{(x_1, x_2) \in \mathbb{R}^2 : 0 < x_1 \leqslant \tfrac{2}{\pi}, x_2 = \sin \tfrac{1}{x}\}, \qquad T_0 = \{0\} \times [-1, 1].$$

Then E is connected but not path-connected.

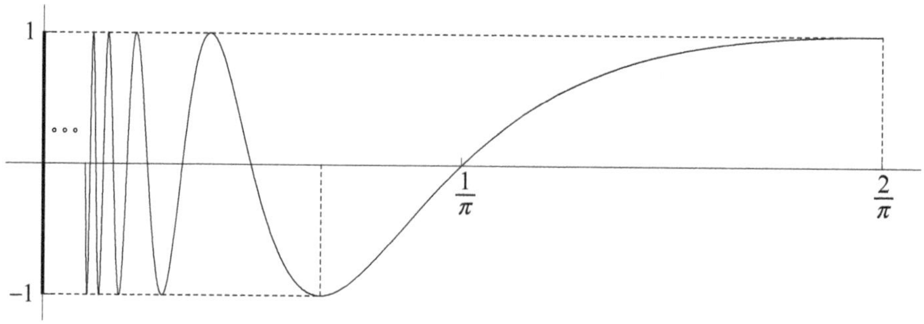

As for connected metric spaces (see Theorems 1.90 and 1.91), we have corresponding properties for path-connected spaces.

Proposition 1.101
(a) *The continuous image of a path-connected metric space is path-connected.*
(b) *If $\{E_i\}_{i\in I}$ is a family of path-connected subsets of X such that $\bigcap_{i\in I} E_i \neq \emptyset$, then the set $\bigcup_{i\in I} E_i$ is path-connected too.*

By Proposition 1.101(b) one can make the following definition.

Definition 1.102
Let X be a metric space and let $x \in X$. The ***path-connected component*** *of x in X is the maximal subset of X containing x which is path-connected.*

Definition 1.103
A metric space X is said to be ***locally path-connected at*** *x if it has a local basis at x consisting of path-connected sets. We say that X is* ***locally path-connected,*** *if it is locally path-connected at every point $x \in X$, i.e., X has a basis for the metric topology consisting of open path-connected sets.*

Proposition 1.104
<u>If</u> X is a metric space,
<u>then</u> the following two statements are equivalent:
(a) *X is locally path-connected;*
(b) *the path-connected components of X are open (hence closed too).*

Proposition 1.105
<u>If</u> X is a locally path-connected metric space,
<u>then</u> the connected components and the path-connected components of X coincide.
Moreover, X is connected if and only if it is path-connected.

Remark 1.106
Another way to view the connected and the path-connected components of a metric space X is the following. We define an equivalence relation on X, called the ***connectivity relation*** (respectively,

path-connectivity relation), by saying that $x_1 \sim x_2$ if there exists a connected subset of X containing both x_1 and x_2 (respectively, if there is a path in X from x_1 to x_2). Then $X/_\sim$ (the set of equivalence classes) are the connected (respectively, path-connected) components of X.

Theorem 1.107
If X is a connected, locally compact metric space, then X is σ-compact (see Theorem 1.79).

Definition 1.108
*Let (X, d_X) be a metric space. For $x, u \in X$ and $\varepsilon > 0$ an **ε-chain** connecting x and u is a finite set $\{c_0, \ldots, c_n\} \subseteq X$ such that $c_0 = u$, $c_n = u$ and $d_X(c_k, c_{k+1}) \leqslant \varepsilon$ for all $k \in \{0, \ldots, n-1\}$. We write $x \overset{\varepsilon}{\sim} u$, if there exists an ε-chain connecting x and u (evidently for every $\varepsilon > 0$ this is an equivalence relation). We say that X is **well-chained** if for every $\varepsilon > 0$ and all $x, u \in X$, we have $x \overset{\varepsilon}{\sim} u$.*

1.1.10 Partitions of Unity

Definition 1.109
*Let $\mathcal{X}$ be a cover of a metric space X (i.e., $\mathcal{X}$ is a family of sets such that $\bigcup\limits_{A \in \mathcal{X}} A \supseteq X$). We say that $\mathcal{X}$ is **locally finite** if for every $x \in X$, there is an open set $U \ni x$ (a neighbourhood of x) which intersects only a finite number of sets in $\mathcal{X}$.*

Definition 1.110
*Let X be a metric space and let $f\colon X \longrightarrow \mathbb{R}$ be a function. The **support** of f is defined as*

$$\operatorname{supp} f = \overline{\{x \in X :\ f(x) \neq 0\}}.$$

Definition 1.111
*A family $\{\varphi_i\}_{i \in I}$ of continuous functions from a metric space X is called a **partition of unity** if*
(a) *the family $\big\{ \{x \in X :\ \varphi_i(x) \neq 0\} \big\}_{i \in I}$ is a locally finite open cover of X;*
(b) *for all $x \in X$, we have $\sum\limits_{i \in I} \varphi_i(x) = 1$ (from **(a)** we see that for each $x \in X$, the sum has only finitely many nonzero terms).*

If $\mathcal{X}$ is an open cover of X and for each $i \in I$, there is some $U \in \mathcal{X}$ such that $\operatorname{supp}\varphi_i \subseteq U$*, then we say that the partition of unity $\{\varphi_i\}_{i\in I}$ is **subordinate** to $\mathcal{X}$.*

Since an open cover may not be locally finite, we need the notion of a refinement.

Definition 1.112
*Let $\mathcal{X}$ be a cover of the metric space X. A cover $\mathcal{X}_0$ is called a **refinement** of $\mathcal{X}$ if for every $U_0 \in \mathcal{X}_0$, we can find a set $U \in \mathcal{X}$ such that $U_0 \subseteq U$.*

Theorem 1.113
If X is a locally compact, σ-compact metric space and $\mathcal{X}$ is an open cover of X,
then $\mathcal{X}$ has a countable locally finite open refinement $\mathcal{X}_0$.
Moreover, if $U_0 \in \mathcal{X}_0$, then $\overline{U}_0$ is compact.

Proposition 1.114
If X is a σ-compact metric space and $\mathcal{X}$ is a locally finite open cover of X,
then $\mathcal{X}$ is countable.

The next result provides a crucial tool to construct a partition of unity for locally compact, σ-compact metric spaces.

Theorem 1.115 *(Shrinking Lemma)*
If X is a locally compact, σ-compact metric spaces and $\mathcal{X}$ is a locally finite open cover of X,
then for each $U \in \mathcal{X}$, we can find an open set V_U such that $\overline{V}_U \subseteq U$ and $\mathcal{X}_0 \stackrel{\text{def}}{=} \{V_U : U \in \mathcal{X}\}$ is a locally finite open cover of X.

Theorem 1.116
If X is a locally compact, σ-compact metric spaces and $\mathcal{X}$ is an open cover of X,
then there is a partition of unity subordinate to $\mathcal{X}$.

1.1.11 Products of Metric Spaces

Definition 1.117
Let (X_k, d_{X_k}), for $k = 1, \dots, N$, be metric spaces and for $x = (x_1, \dots, x_N)$, $y = (y_1, \dots, y_N) \in X = \prod_{k=1}^{N} X_k$, we define

$$d_X^\infty(x, y) = \max\{d_{X_k}(x_k, y_k) : k = 1, \dots, N\}$$

(see also Example 1.3(a)).

Proposition 1.118
(X, d_X^∞) is a metric space and it is called the product metric space of $X_1, \dots, X_k$.

Remark 1.119
Other equivalent metrics on $X = \prod_{k=1}^{N} X_k$, are

$$d_X^p(x, y) = \Big(\sum_{k=1}^{N} (d_{X_k}(x_k, y_k))^p\Big)^{\frac{1}{p}}$$

with $1 \leqslant p < +\infty$. Note that

$$d_X^\infty(x, y) \leqslant d_X^p(x, y) \leqslant d_X^1(x, y) \leqslant n d_X^\infty(x, y) \qquad \forall\, x, y \in X = \prod_{k=1}^{N} X_k.$$

Thus all the above metrics are Lipschitz equivalent (see Remark 1.53 and Proposition 1.54).

Proposition 1.120
For a given $x \in X = \prod_{k=1}^{N} X_k$ and $r > 0$, we have

$$B_r(x) = \prod_{k=1}^{N} B_r^{X_k}(x_k),$$

where $B_r^{X_k}(x_k) = \{u \in X_k : d_{X_k}(u, x_k) < r\}$.

Corollary 1.121
The family

$$\{\prod_{k=1}^{N} B_r^{X_k}(x_k) : \; x_k \in X_k, \; 1 \leqslant k \leqslant N, \; r > 0\}$$

is a basis for the product metric topology on $X = \prod_{k=1}^{N} X_k$.

Proposition 1.122
If $U_k \subseteq X_k$, *for* $k = 1, \ldots, N$, *are open subsets of* X_k,
then $U = \prod_{k=1}^{N} U_k$ *is an open subset of* $X = \prod_{k=1}^{N} X_k$.

Definition 1.123
For the given product set $X = \prod_{k=1}^{N} X_k$, *we can always define the* ***projection*** *on the nth factor*

$$p_n \colon X = \prod_{k=1}^{N} X_k \longrightarrow X_n$$

($n = 1, \ldots, N$*) by*

$$p_n(x) = x_n \qquad \forall\, x = (x_1, \ldots, x_N) \in X.$$

Proposition 1.124
If (X_k, d_{X_k}), *for* $k = 1, \ldots, N$, *are metric spaces and* $X = \prod_{k=1}^{N} X_k$ *is furnished with the* d_X^∞*-metric,*
then the projection map $p_n \colon : X = \prod_{k=1}^{N} X_k \longrightarrow X_n$ *is continuous and open.*

Remark 1.125
The projection function p_n is in general not closed.

Proposition 1.126
If (X_k, d_{X_k}), *for* $k = 1, \ldots, N$ *and* (Y, d_Y) *are all metric spaces,*
then $f \colon Y \longrightarrow (X, d_X^\infty)$ *is continuous (respectively, uniformly continuous) if and only if for every* $k = 1, \ldots, N$, *the composite function* $p_k \circ f \colon Y \longrightarrow X_k$ *is continuous (respectively, uniformly continuous).*

Proposition 1.127
If (X_k, d_{X_k}), *for* $k = 1, \ldots, N$ *are metric spaces and* $E_k \subseteq X_k$, *for* $k = 1, \ldots, N$,
then

$$\overline{\prod_{k=1}^{N} E_k}^{d_X^\infty} = \prod_{k=1}^{N} \overline{E}_k^{d_{X_k}}.$$

We can also consider infinite metric products. So, suppose that we have a sequence $\{(X_n, d_{X_n})\}_{n\geqslant 1}$ of metric space such that

$$\sup\{d_{X_n}(x,y) : \ x, y \in X, \ n \geqslant 1\} \leqslant M$$

for some $M > 0$. Then we define

$$\hat{d}_X(x,y) = \sum_{n\geqslant 1} \frac{1}{2^n} d_{X_n}(x_n, y_n) \qquad \forall\, x, y \in X = \prod_{n\geqslant 1} X_n$$

(see Example 1.3(b)).

Proposition 1.128
$(X, \hat{d}_X)$ *is a metric space.*

We consider the family

$$\mathcal{B} = \{B_r^m(x) = \prod_{n=1}^{m} B_r^{X_n}(x_n) \times \prod_{n\geqslant m+1} X_n : x \in X = \prod_{n\geqslant 1} X_n, m \in \mathbb{N}, r > 0\}.$$

Proposition 1.129
$\mathcal{B}$ *is a basis for the* $\hat{d}_X$*-metric topology on* $X = \prod_{n\geqslant 1} X_n$.

Proposition 1.130
$(X = \prod_{n\geqslant 1} X_n, \hat{d}_X)$ *is a complete (respectively, compact) metric space if and only if each* (X_n, d_{X_n}), *for* $n \geqslant 1$, *is a complete (respectively, compact) metric space.*

We conclude with a theorem, which is useful in many theoretical and applied settings.

Theorem 1.131
(a) *If* X *is any set,* (Y, d) *is a complete metric space and*

$$B_b(X;Y) = \{f\colon X \longrightarrow Y : \ f \text{ is bounded (i.e., } \operatorname{diam} f(X) < +\infty)\}$$

furnished with the supremum distance

$$d^\infty(u,v) = \sup_{x \in X} d\big(u(x), v(x)\big),$$

then $\big(B_b(X;Y), d^\infty\big)$ *is a complete metric space.*
(b) *If* (X, d_X) *and* (Y, d_Y) *are two metric spaces with* Y *being complete and*

$$C_b(X;Y) = \big\{f \in C(X;Y) : \ f \text{ is bounded}\big\},$$

then $\big(C_b(X;Y), d^\infty\big)$ *is complete.*

1.1.12 Auxiliary Notions

Let us introduce some notions that will be needed in the forthcoming problems.

Definition 1.132
(a) *For any function* $f\colon X \longrightarrow Y$*, the* ***graph*** *of* f *is the set* $\operatorname{Gr} f \subseteq X \times Y$*, defined by*

$$\operatorname{Gr} f = \big\{(x,y) \in X \times Y : \ y = f(x)\big\}.$$

(b) *For any function* $f\colon X \longrightarrow \mathbb{R}$*, the* ***epigraph*** *of* f *is the set* $\operatorname{epi} f \subseteq X \times \mathbb{R}$*, defined by*

$$\operatorname{epi} f = \big\{(x,\lambda) \in X \times \mathbb{R} : \ f(x) \leqslant \lambda\big\}.$$

Definition 1.133
Suppose that (X, d_X) *is a metric space and* $\{E_n\}_{n\geqslant 1}$ *is a sequence of subsets (possibly empty) of* X*. We define the* ***lower Kuratowski limit*** *of the sets* $\{E_n\}_{n\geqslant 1}$*, by*

$$\liminf_{n\to+\infty} E_n = \big\{x \in X : \ \text{for every } r > 0 \text{ there is an integer } n_r \geqslant 1, \ \text{such that } B_r(x) \cap E_n \neq \emptyset \text{ for all } n \geqslant n_r\big\}$$

and the ***upper Kuratowski limit*** *of sets* $\{E_n\}_{n\geqslant 1}$, *by*

$$\limsup_{n\to+\infty} E_n = \Big\{x \in X : \text{ for every } r>0, \text{ we have } B_r(x)\cap E_n \neq \emptyset, \\ \text{for an infinite number of the sets } E_n\Big\}.$$

Definition 1.134
Suppose that (X, d_X) *is a bounded metric space and* $P_f(X)$ *denotes the collection of nonempty closed subsets of* X. *For every* $A, B \in P_f(X)$, *we set*

$$h(A,B) \stackrel{\text{def}}{=} \sup\Big\{\big|\operatorname{dist}(x,A)-\operatorname{dist}(x,B)\big| : x\in X\Big\}.$$

The metric space $\big(P_f(X), h\big)$ *is called the* ***Hausdorff metric space*** *on* $P_f(X)$ *(see Problem 1.176).*

We write $A_n \xrightarrow{h} A$ *for the convergence in the Hausdorff metric space* $\big(P_f(X), h\big)$ *(i.e.,* $A_n \xrightarrow{h} A$, *if* $h(A_n, A) \longrightarrow 0$*).*

Definition 1.135
Suppose that X *is a metric space and* $f\colon X \longrightarrow \mathbb{R}$ *is a function.*
(a) *We say that* f *has a* ***local minimum*** *at* $x \in X$ *if there exists an open set* U_x *containing* x *such that* $f(x) \leqslant f(u)$ *for all* $u \in U_x$.
(b) *We say that* f *has a* ***strict local minimum*** *at* $x \in X$ *if there exists an open set* U_x *containing* x *such that* $f(x) < f(u)$ *for all* $u \in U_x \setminus \{x\}$.
(c) *We say that* f *has a* ***local maximum*** *at* $x \in X$ *if there exists an open set* U_x *containing* x *such that* $f(x) \geqslant f(u)$ *for all* $u \in U_x$.
(d) *We say that* f *has a* ***strict local maximum*** *at* $x \in X$ *if there exists an open set* U_x *containing* x *such that* $f(x) > f(u)$ *for all* $u \in U_x \setminus \{x\}$.

1.2 Problems

Problem 1.1 *
Suppose that (X, d_X) is a metric space. Show that a Cauchy sequence in X converges if and only if it has a convergent subsequence.

Problem 1.2 *
Suppose that (X, d_X) is a metric space and $\{x_n\}_{n\geqslant 1}$ is a sequence in X. Assume that subsequences $\{x_{2n}\}_{n\geqslant 1}$, $\{x_{2n+1}\}_{n\geqslant 1}$ and $\{x_{3n}\}_{n\geqslant 1}$ converge. Show that $\{x_n\}_{n\geqslant 1}$ is a convergent sequence.

Problem 1.3 *
Suppose that (X, d_X) is a metric space, $x \in X$ and $\{x_n\}_{n\geqslant 1}$ is a sequence in X such that for any subsequence $\{x_{n_k}\}_{k\geqslant 1} \subseteq \{x_n\}_{n\geqslant 1}$, we can find a further subsequence $\{x_{n_{k_l}}\}_{l\geqslant 1} \subseteq \{x_{n_k}\}_{k\geqslant 1}$ such that $\lim\limits_{l\to+\infty} x_{n_{k_l}} = x$, then $\lim\limits_{n\to+\infty} x_n = x$ (we call this property ***Urysohn criterion for convergence***).

Problem 1.4 *
Suppose that (X, d_X) is a metric space and $\{x_n\}_{n\geqslant 1} \subseteq X$ is a Cauchy sequence. Show that we can find a subsequence $\{x_{n_k}\}_{k\geqslant 1}$ of $\{x_n\}_{n\geqslant 1}$ such that

$$d_X(x_{n_k}, x_{n_m}) \leqslant \tfrac{1}{2^k} \qquad \forall\, k, m \geqslant 1,\ k \leqslant m.$$

Problem 1.5 *
Suppose that (X, d_X) is a metric space and $\{x_n\}_{n\geqslant 1}$ is a Cauchy sequence in X.
(a) Show that we can find a subsequence $\{x_{n_k}\}_{k\geqslant 1}$ of $\{x_n\}_{n\geqslant 1}$ such that

$$\sum_{k\geqslant 1} d_X(x_{n_k}, x_{n_{k+1}}) < +\infty.$$

(b) Show that, if every sequence $\{y_n\}_{n\geqslant 1}$ in X, which satisfies

$$\sum_{n\geqslant 1} d_X(y_n, y_{n+1}) < +\infty,$$

is convergent, then X is complete.

Problem 1.6 *
Let $l^1 \stackrel{\text{def}}{=} \{\hat{u} = \{u_n\}_{n\geqslant 1} : u_n \in \mathbb{R}, \sum_{n\geqslant 1} |u_n| < +\infty\}$ be equipped with the metric

$$d_1(\hat{u}, \hat{x}) = \sum_{n\geqslant 1} |u_n - x_n| \qquad \forall\, \hat{u}, \hat{x} \in l^1.$$

Also let

$$c_0 \stackrel{\text{def}}{=} \{\hat{u} \in l^1 : \text{there exists } n_0 \geqslant 1 \text{ such that } u_n = 0 \text{ for all } n \geqslant n_0\}.$$

Show that:
(a) (l^1, d_1) is a complete metric space;
(b) c_0 is dense in l^1.

Problem 1.7 **
Show that there is a function $f\colon \mathbb{R} \longrightarrow \mathbb{R}$ such that

$$\{x \in \mathbb{R} : f \text{ is discontinuous at } x\} = \mathbb{Q}.$$

Problem 1.8 *
Let (X, d_X) be a metric space. Show that X is a singleton if and only if every bounded sequence is convergent.

Problem 1.9 *
Show that every complete subspace C of a metric space X is closed.

Problem 1.10 **
Let $M \subseteq C[a, b]$ be the set of all functions $f \in C([a, b])$ which are monotone on $[c, d] \subseteq [a, b]$ with $[c, d]$ nonempty and not a singleton. Show that $\operatorname{int} M = \emptyset$. As usual on $C([a, b])$ we consider the supremum metric d^∞ (see Example 1.3(d)).

Problem 1.11 *
Suppose that X is a metric space and $E \subseteq X$. Show that x is an accumulation point of E if and only if $x \in \overline{E \setminus \{x\}}$.

Problem 1.12 *
Suppose that X is a metric space and D is a dense subset of X. For every nonempty open set $U \subseteq X$, show that $U \subseteq \overline{U \cap D}$. In particular $U \cap D$ is dense in U with the subspace metric topology.

Problem 1.13 ★★
Let $D \subseteq \mathbb{R}$ be a nonempty set and let

$$E \stackrel{\text{def}}{=} \{x \in \overline{D}: \text{ we can find } \varepsilon > 0 \text{ such that } (x, x+\varepsilon) \cap D = \emptyset\}.$$

Show that E is at most countable.

Problem 1.14 ★★
Suppose that (X, d_X) is a metric space and $\mathcal{Y}$ is a countable family of separable metric subspaces of X. Show that $X_0 = \bigcup_{Y \in \mathcal{Y}} Y$ is separable.

Problem 1.15 ★
(a) Give an example of a metric space (X, d_X) and of two balls $B_R(x) \subsetneq B_r(y)$, where $R > r$ and $x, y \in X$.
(b) Show that, if $B_R(x) \subsetneq B_r(y)$ for some $R, r > 0$ and $x, y \in X$, then $R < 2r$.

Problem 1.16 ★★
Let X be a separable metric space. Show that $X = U \cup C$, where U is open and countable, C is a closed set containing only accumulation points (i.e., C is perfect) and $U \cap C = \emptyset$. This is the so called ***Cantor–Bendixson theorem***.

Problem 1.17 ★★★
Let $\{q_n\}_{n \geqslant 1}$ be an enumeration of the rational numbers in $\mathbb{R}$. For each $x \in \mathbb{R}$, we set

$$L_x \stackrel{\text{def}}{=} \{n \in \mathbb{N}: \ q_n \leqslant x\}.$$

We introduce the function $f \colon \mathbb{R} \longrightarrow \mathbb{R}$ by

$$f(x) = \sum_{n \in L_x} \frac{1}{2^n}.$$

Clearly f is strictly increasing, it has a jump discontinuity at every rational number and

$$f(x) \longrightarrow 0 \quad \text{as } x \to -\infty \quad \text{and} \quad f(x) \longrightarrow 1 \quad \text{as } x \to +\infty.$$

Show that f is right continuous and $f|_{\mathbb{R}\setminus\mathbb{Q}}$ is continuous.

Problem 1.18 **
Let l^∞ be the space of all bounded sequences in $\mathbb{R}$ furnished with the supremum metric d^∞, defined by

$$d^\infty(x,y)=\sup\big\{|x_n-y_n| : n\geqslant 1\big\} \quad \forall\, x=\{x_n\}_{n\geqslant 1},\, y=\{y_n\}_{n\geqslant 1}\in l^\infty.$$

Show that l^∞ is not separable.

Problem 1.19 **
Show that if X is a separable metric space, then $\operatorname{card} X \leqslant \mathfrak{c}$ (where $\mathfrak{c} = \operatorname{card} 2^{\mathbb{N}}$; assuming the continuum hypothesis $\mathfrak{c}$ is the cardinal number of $\mathbb{R}$).

Problem 1.20 **
Suppose that (X, d_X) is an unbounded metric space. Show that X admits a sequence with no convergent subsequence.

Problem 1.21 **
Suppose that (X, d_X) is a metric space and $\varphi\colon [0,+\infty) \longrightarrow [0,+\infty)$ is a nontrivial, increasing concave function such that $\varphi(0) = 0$. Show that the function, defined by

$$\hat{d}_X(x,y) \;=\; \varphi\big(d_X(x,y)\big) \qquad \forall\, x,y\in X$$

is a metric on X.

Problem 1.22 **
(a) Suppose that X and Y are two metric spaces and $E, C \subseteq X$ are two nonempty open (or closed) sets such that $X = E\cup C$. Assume that $f_1\colon E \longrightarrow Y$ and $f_2\colon C \longrightarrow Y$ are both continuous and $f_1\big|_{E\cap C} = f_2\big|_{E\cap C}$. Let $g\colon X \longrightarrow Y$ be defined by

$$g(x) \;\stackrel{\text{def}}{=}\; \begin{cases} f_1(x) & \text{if} \quad x\in E, \\ f_2(x) & \text{if} \quad x\in C. \end{cases}$$

Show that g is continuous.
(b) Is the conclusion true, if we drop the hypothesis that E, C are both open (or close) subsets of X?

Problem 1.23 **
Suppose that X is a separable metric space and

$$\mathcal{Y}_o \;\stackrel{\text{def}}{=}\; \big\{U\subseteq X :\; U \text{ is open}\big\}, \quad \mathcal{Y}_c \;\stackrel{\text{def}}{=}\; \big\{C\subseteq X :\; C \text{ is closed}\big\}.$$

Show that

$$\operatorname{card} \mathcal{Y}_o = \operatorname{card} \mathcal{Y}_c \leqslant \mathfrak{c}$$

($\mathfrak{c} = \operatorname{card} 2^{\mathbb{N}}$; we assume the continuum hypothesis).

Problem 1.24 *
(a) Suppose that (X, d_X) is a metric space and $\{x_n\}_{n\geqslant 1} \subseteq X$ is a sequence. Suppose that $u_n \longrightarrow u$ in X where for all $n \geqslant 1$, u_n is an accumulation point of the sequence $\{x_n\}_{n\geqslant 1}$. Show that u is an accumulation point of the sequence $\{x_n\}_{n\geqslant 1}$.
(b) Show that the set of accumulation points of any sequence in a metric space is closed.

Problem 1.25 **
Suppose that E is a set and $B(E) \stackrel{\text{def}}{=} \{f: E \longrightarrow \mathbb{R} : f \text{ is bounded}\}$. We furnish $B(E)$ with the supremum metric

$$d^\infty(f, g) \stackrel{\text{def}}{=} \sup_{s\in E} \big|f(s) - g(s)\big|.$$

Show that $\big(B(E), d^\infty\big)$ is a complete metric space.

Problem 1.26 **
Suppose that X is a separable metric space and $f: X \longrightarrow \mathbb{R}$ is a function. Let L be the set of all strict local minimizers of f. Show that L is at most countable.

Problem 1.27 **
(a) Let (X, d_X) and (Y, d_Y) be two metric spaces, let $f: X \longrightarrow Y$ be a continuous function and let $\{C_n\}_{n\geqslant 1}$ be a sequence of subsets of X such that $\bigcap\limits_{n\geqslant 1} C_n \neq \emptyset$ and $\operatorname{diam} C_n \longrightarrow 0$. Show that $\operatorname{diam} f(C_n) \longrightarrow 0$ as $n \to +\infty$.
(b) Is it possible to drop the hypothesis that $\bigcap\limits_{n\geqslant 1} C_n \neq \emptyset$? Justify your answer.

Problem 1.28 **
Suppose that X is a metric space and $D \subseteq X$ is a nonempty and closed or open set. Show that ∂D (the boundary of D) is nowhere dense. Is this true if D is an arbitrary nonempty subset of X?

Problem 1.29 ⋆
Show that in a metric space X every nowhere dense set has a dense complement. Is the converse true? Justify your answer.

Problem 1.30 ⋆
Suppose that (X, d_X) is a metric space and let C be a nonempty, closed subset of X. Show that C in nowhere dense if and only if for every nonempty open set U, we can find a ball in $U \setminus C$.

Problem 1.31 ⋆⋆⋆
Suppose that (X, d_X) is a complete metric space and D is a nonempty perfect subset of X. Show that $\operatorname{card} D \geqslant \mathfrak{c}$ ($\mathfrak{c}$ being the cardinality of the continuum).

Problem 1.32 ⋆
Use the Baire category theorem, to show that $I = [0, 1]$ is uncountable.

Problem 1.33 ⋆
Show that a countable complete metric space has an isolated point.

Problem 1.34 ⋆⋆
Suppose that X is a complete metric space and $\{C_n\}_{n\geqslant 1}$ is a sequence of nonempty closed subsets of X such that $X = \bigcup_{n\geqslant 1} C_n$. Show that the set $\bigcup_{n\geqslant 1} \operatorname{int} C_n$ is dense in X.

Problem 1.35 ⋆⋆⋆
Suppose that (l^∞, d^∞) is as in Problem 1.18 and $C \subseteq l^\infty$ is the subset of all convergent real sequences. Show that C is nowhere dense in l^∞.

Problem 1.36 ⋆⋆
Suppose that X and Y are metric spaces and $f: X \longrightarrow Y$ is a function. Show that the following two statements are equivalent:
(a) f is uniformly continuous;
(b) for all sequences $\{u_n\}_{n\geqslant 1}, \{x_n\}_{n\geqslant 1} \subseteq X$ such that $d_X(u_n, x_n) \longrightarrow 0$, we have $d_Y\big(f(u_n), f(x_n)\big) \longrightarrow 0$.

Problem 1.37 **

(a) Let $X = C_b(\mathbb{R})$ (the space of bounded continuous functions $f\colon \mathbb{R} \longrightarrow \mathbb{R}$) be furnished with the supremum metric

$$d^\infty(f,g) = \sup_{t\in\mathbb{R}} \big|f(t) - g(t)\big| \qquad \forall\ f,g \in X.$$

For $f \in X$ and $r \in \mathbb{R}$, we set $f_r(t) \stackrel{\text{def}}{=} f(t+r)$. Then $f_r \in X$. Show that, if $f \in X$ is uniformly continuous, then $d^\infty(f_r, f) \longrightarrow 0$ as $r \to 0^+$.

(b) Does the above remain true if we replace the uniform continuity of f by continuity?

Problem 1.38 *

Suppose that X is a separable metric space, Y is a metric space and $f\colon X \longrightarrow Y$ is a continuous function. Show that $f(X)$ is separable.

Problem 1.39 **

Suppose that X and Y are two metric spaces, X is complete and $f\colon X \longrightarrow Y$ is homeomorphism which is uniformly continuous. Must Y be complete? Justify your answer.

Problem 1.40 **

Give examples that illustrate each of the following statements:

(a) A continuous and open function need not be closed (see Definitions 1.29 and 1.39).

(b) A continuous and closed function need not be open.

(c) An open and closed function need not be continuous.

(d) A continuous function need not be open nor closed.

Problem 1.41 *

Suppose that X and Y are two metric spaces and $f\colon X \longrightarrow Y$ is an injection. Show that f is open if and only if f is closed.

Problem 1.42 **

Let (X, d_X) be a complete metric space and let $f\colon X \longrightarrow X$ be a function such that $f^{(k)}$ is a contraction for some $k \geqslant 1$ (recall that $f^{(k)} = \underbrace{f \circ \ldots \circ f}_{k\text{-times}}$). Show that f has a unique fixed point.

Problem 1.43 ***
Let (X, d_X) and (Y, d_Y) be two metric spaces and let $f\colon X \longrightarrow Y$ be a function. We say that f is ***asymptotically nonexpansive*** if there exists a sequence $\{k_n\}_{n\geqslant 1} \subseteq [1, +\infty)$, with $k_n \longrightarrow 1$ as $n \to +\infty$ such that

$$d_Y\big(f^{(n)}(u), f^{(n)}(x)\big) \leqslant k_n d_X(u, x) \qquad \forall\, u, x \in X,\ n \geqslant 1.$$

Give an example of an asymptotically nonexpansive function which is not nonexpansive.

Problem 1.44 *
Suppose that M is a metric space, X is a complete metric space, $f\colon M \times X \longrightarrow X$ is a function such that
(i) for every $x \in X$, the function $r \longmapsto f(r, x)$ is continuous on M; and
(ii) for every $r \in M$, the function $x \longmapsto f(r, x)$ is a contraction with contraction constant $k < 1$ independent of $r \in M$.

Show that there exists a unique continuous function $u\colon M \longrightarrow X$ such that $u(r) = f\big(r, u(r)\big)$ for all $r \in M$.

Problem 1.45 ***
Let us consider $X = (0, +\infty)$ with the usual metric. Let $f \in C^1(X; X)$ be such that $x\big|f'(x)\big| \leqslant kf(x)$ for all $x \in X$ and some $k \in [0, 1)$. Show that f has a unique fixed point.

Problem 1.46 **
Show that f is continuous at $x \in X$ if and only if $\omega_f(x) = 0$ (where ω_f is the oscillation function for f).

Problem 1.47 **
Suppose that (X, d_X) and (Y, d_Y) are two metric spaces and $f\colon X \longrightarrow Y$ is a function. Consider the oscillation function ω_f. Show that for every $\lambda \in \mathbb{R}$, the set $\big\{x \in X : \omega_f(x) < \lambda\big\}$ is open (i.e., the function $x \longmapsto \omega_f(x)$ is upper semicontinuous; see Definition 2.46).

Problem 1.48 **
(a) Let X and Y be two metric spaces with Y complete. Suppose $f\colon X \longrightarrow Y$ is bijective and uniformly continuous and that f^{-1} is continuous. Show that X is complete.
(b) Is the result still true if f is only continuous (instead of uniformly continuous)?

Problem 1.49 **

Suppose that X is a complete metric space, $\{C_n\}_{n\geqslant 1}$ is a decreasing sequence of nonempty closed sets in X such that $\operatorname{diam} C_n \longrightarrow 0$ (see Definition 1.6(b)) and $f\colon X \longrightarrow X$ is a continuous function. Show that $f\big(\bigcap_{n\geqslant 1} C_n\big) = \bigcap_{n\geqslant 1} f(C_n)$.

Problem 1.50 **

Show that every separable metric space is homeomorphic to a subset of the Hilbert cube $\mathcal{H} = [0,1]^{\mathbb{N}}$.

Problem 1.51 **

Suppose that A and B are two dense subsets of $\mathbb{R}$ and $f\colon A \longrightarrow B$ is an increasing bijection. Show that f is a homeomorphism.

Problem 1.52 **

Let (X, d_X) and (Y, d_Y) be two metric spaces with Y being complete. Let $D \subseteq X$ be a dense subset and let $f\colon D \longrightarrow Y$ be a k-Lipschitz function. Show that there exists a unique k-Lipschitz function $\hat{f}\colon X \longrightarrow Y$ such that $\hat{f}|_D = f$.

Problem 1.53 ***

Show that every metric space is isometrically embedded into the space of bounded, uniformly continuous functions.

Problem 1.54 **

Let X be a normed space and let $r > 0$. Show that X and $B_r = \big\{x \in X : \|x\| < r\big\}$ are homeomorphic.

Problem 1.55 ***

Suppose that X is a metric space, Y is a complete metric space, $D \subseteq X$ is a nonempty set and $f\colon D \longrightarrow Y$ is a continuous function. Show that f can be extended continuously to a G_δ-subset of X containing D.

Problem 1.56 ***

Suppose that X and Y are complete metric spaces, $A \subseteq X$, $B \subseteq Y$ are nonempty sets and $f\colon A \longrightarrow B$ is a homeomorphism and surjection. Show that f can be extended to a homeomorphism between G_δ-sets containing A and B respectively.

Problem 1.57 ***
Let X be a metric space and suppose that $A \subseteq X$ is homeomorphic to a complete metric space. Show that A is a G_δ-set in X.

Problem 1.58 **
Show that $\mathbb{Q}$ is not a G_δ-set in $\mathbb{R}$ or equivalently the set of irrational numbers $\mathbb{R} \setminus \mathbb{Q}$ is not an F_σ-set.

Problem 1.59 **
Show that in a complete metric space which has no isolated points, a countable dense subset cannot be G_δ.

Problem 1.60 ***
Let X be a complete metric space. Show the following:
(a) $A \subseteq X$ has a meager complement if and only if A contains a dense G_δ-set.
(b) $A \subseteq X$ is meager if and only if A is contained in an F_σ-set whose complement is dense.

Problem 1.61 ***
(a) Suppose that (X, d_X) is a metric space such that every continuous function $f\colon X \longrightarrow \mathbb{R}$ is in fact uniformly continuous. Show that X is complete.
(b) Show that the inverse theorem is not true, i.e., there exists a complete metric space X and a continuous function $f\colon X \longrightarrow \mathbb{R}$ which is not uniformly continuous.

Problem 1.62 **
Suppose that (X, d_X) and (Y, d_Y) are two metric spaces and $f\colon X \longrightarrow Y$ is a function. Show that the set

$$C \stackrel{\text{def}}{=} \{x \in X : \ f \text{ is continuous at } x\}$$

is a G_δ-subset of X.

Problem 1.63 **
Show that there is no function $f\colon \mathbb{R} \longrightarrow \mathbb{R}$ for which

$$\{x \in \mathbb{R} : \ f \text{ is continuous at } x\} \ = \ \mathbb{Q}.$$

Problem 1.64 ***
Let X be a metric space. A function $f\colon X \longrightarrow \mathbb{R}$ is said to be ***lower semicontinuous*** if for every $\lambda \in \mathbb{R}$, the set $\{x \in X : f(x) \leqslant \lambda\}$ is closed (cf. Definition 2.46). Suppose that X is complete and $\mathcal{F}$ is a family of lower semicontinuous functions $f\colon X \longrightarrow \mathbb{R}$. Assume that for every $x \in X$, there exists $M_x > 0$ such that

$$f(x) \leqslant M_x \qquad \forall\, f \in \mathcal{F}.$$

Show that there exist a nonempty open set $U \subseteq X$ and $m_0 \geqslant 1$ such that

$$f(x) \leqslant m_0 \qquad \forall\, f \in \mathcal{F},\ x \in U.$$

Problem 1.65 **
Let $\chi_{\mathbb{Q}}$ be the characteristic function of the set of rational numbers, i.e.,

$$\chi_{\mathbb{Q}}(x) \stackrel{\text{def}}{=} \begin{cases} 1 & \text{if } x \in \mathbb{Q}, \\ 0 & \text{if } x \in \mathbb{R} \setminus \mathbb{Q}. \end{cases}$$

Show that $\chi_{\mathbb{Q}}$ is not the pointwise limit of a sequence of continuous functions on $\mathbb{R}$.

Problem 1.66 **
Let (X, d_X) be a complete metric space and consider a sequence $\{f_n\colon X \longrightarrow \mathbb{R}\}_{n \geqslant 1}$ of continuous functions such that $f_n \longrightarrow f$ (pointwise). Show that the set

$$E \stackrel{\text{def}}{=} \{x \in X : f \text{ is continuous at } x \in X\}$$

is a dense G_δ-set in X.

Problem 1.67 **
Let $f\colon \mathbb{R}^2 \longrightarrow \mathbb{R}$ be a function which has partial derivatives at every point in $\mathbb{R}^2$. Show that f is differentiable on a dense G_δ subset of $\mathbb{R}^2$.

Problem 1.68 ***
Find two metrics which are topologically but not uniformly equivalent.

Problem 1.69 **
Find two metrics which are topologically equivalent but only one defines a complete metric space.

Problem 1.70 ***

Find two metrics which are uniformly equivalent but not Lipschitz equivalent.

Problem 1.71 **

Let $X = C^1([0,1])$ and consider the following two metrics on X:

$$\begin{aligned} d_1^\infty(u,v) &= \max_{t\in[0,1]} |u(t)-v(t)| + \max_{t\in[0,1]} |u'(t)-v'(t)|, \\ d^\infty(u,v) &= \max_{t\in[0,1]} |u(t)-v(t)|. \end{aligned}$$

Show that the two metrics d_1^∞ and d^∞ are not topologically equivalent.

Problem 1.72 **

Let (X, d_X) and (Y, d_Y) be two metric spaces and let $f\colon X \longrightarrow Y$ be a continuous function. Show that there exists a topologically equivalent metric $\hat{d}_X$ on X such that the function $f\colon (X, \hat{d}_X) \longrightarrow Y$ is Lipschitz continuous.

Problem 1.73 **

Let (X, d_X) and (Y, d_Y) be two metric spaces and suppose that $\{f_n\colon X \longrightarrow Y\}_{n\geqslant 1}$ is a sequence of continuous functions. Show that there exists a topologically equivalent metric $\hat{d}_X$ on X and a topologically equivalent metric $\hat{d}_Y$ on Y such that for each $n \geqslant 1$, the function $f_n\colon (X, \hat{d}_X) \longrightarrow (Y, \hat{d}_Y)$ is Lipschitz continuous.

Problem 1.74 *

Show that the completion of a separable metric space is separable too.

Problem 1.75 *

Suppose that $D \subseteq [a,b]$ and

$$\mathcal{Y} \stackrel{\text{def}}{=} \{f \in C[a,b] : \ f(t) = 0 \text{ for all } t \in D\}.$$

Show that $\mathcal{Y}$ is a closed subset of $C[a,b]$ with the uniform metric $d^\infty_{C[a,b]}$ (see Example 1.3(d)).

Problem 1.76 **

Let $X = C([0,1];[0,1])$ be furnished with the supremum metric

$$d^\infty(f,g) = \max_{s\in[0,1]} |f(s)-g(s)|$$

(see Example 1.3(d)). Suppose that

$$E_1 \stackrel{\text{def}}{=} \{f \in X : f \text{ is injective}\},$$
$$E_2 \stackrel{\text{def}}{=} \{f \in X : f \text{ is surjective}\},$$
$$E \stackrel{\text{def}}{=} E_1 \cap E_2 = \{f \in X : f \text{ is bijective}\}.$$

Check whether E_1, E_2 and E are closed in X.

Problem 1.77 ***
Let $X = C([0,1])$ be furnished with the supremum metric

$$d^\infty(f,g) = \max_{t\in[0,1]} |f(t) - g(t)| \qquad \forall\, f,g \in X.$$

Let

$$E \stackrel{\text{def}}{=} \{f \in X : f \text{ is differentiable at at least one point in } [0,1]\}.$$

Show that E is of first category in X (see Definition 1.25).

Problem 1.78 *
Let (X, d_X) be a metric space and let $A \subseteq X$ be any set. Show that the function $f\colon X \ni x \longmapsto \operatorname{dist}(x, A) \in \mathbb{R}$ is Lipschitz continuous with Lipschitz constant 1 (so also continuous).

Problem 1.79 *
Show that in a metric space (X, d_X) every closed set is G_δ and every open set is F_σ.

Problem 1.80 **
Let $D \subseteq \mathbb{R}$ be an uncountable set. Show that D has uncountably many accumulation points (see Definition 1.13(b)).

Problem 1.81 **
Let X be a metric space and let $f\colon X \longrightarrow \mathbb{R}$ be a lower semicontinuous function. Show that for every open set $U \subseteq \mathbb{R}$, the set $f^{-1}(U)$ is F_σ.

Problem 1.82 ***
Let X be a complete metric space and let $f\colon X \longrightarrow \mathbb{R}$ be a lower semicontinuous function. Show that f is continuous on a dense G_δ-subset of X.

Problem 1.83 **

Suppose that X is a nonempty set and (Y, d_Y) is a metric space.
(a) Show that if $\{f_n \colon X \longrightarrow Y\}_{n \geqslant 1}$ is a sequence of bounded functions and $f_n \rightrightarrows f$, then f is bounded.
(b) Show that a sequence of unbounded functions $\{f_n \colon X \longrightarrow Y\}_{n \geqslant 1}$ cannot converge uniformly to a bounded function $f \colon X \longrightarrow Y$.

Problem 1.84 *

Suppose that X and Y are two metric spaces, $\{f_n \colon X \longrightarrow Y\}_{n \geqslant 1}$ is a sequence of functions continuous at x_0 and $f \colon X \longrightarrow Y$ is a function such that $f_n \rightrightarrows f$. Show that f is continuous at x_0.

Remark. Applying this problem to a sequence of continuous functions, we get the completeness of the space $C(X;Y)$ (see Proposition 1.62).

Problem 1.85 **

Let X be a metric space and $\{f_n \colon X \longrightarrow \mathbb{R}\}_{n \geqslant 1}$, $\{g_n \colon X \longrightarrow \mathbb{R}\}_{n \geqslant 1}$ two uniformly convergent sequences of functions. Is it true that the sequence $\{f_n g_n\}_{n \geqslant 1}$ is uniformly convergent too?

Problem 1.86 *

Suppose that (X, d_X) and (Y, d_Y) are two metric spaces and $\{f_n \colon X \longrightarrow Y\}_{n \geqslant 1}$ is a sequence of Lipschitz continuous functions, all with the Lipschitz constant $k > 0$ (k-Lipschitz functions). Suppose that $f_n \rightrightarrows f$. Show that $f \colon X \longrightarrow Y$ is k-Lipschitz too.

Problem 1.87 *

Suppose that (X, d_X) and (Y, d_Y) are two metric spaces and $\{f_n \colon X \longrightarrow Y\}_{n \geqslant 1}$ is a sequence of uniformly continuous functions such that $f_n \rightrightarrows f$. Show that $f \colon X \longrightarrow Y$ is uniformly continuous too.

Problem 1.88 **

Let X and Y be two metric spaces.
(a) Is it possible for a sequence $\{f_n \colon X \longrightarrow Y\}_{n \geqslant 1}$ of discontinuous functions to converge uniformly to a continuous function f?
(b) Let $f \colon X \longrightarrow X$ be a function such that $f \circ f$ is continuous. Does it imply that f is continuous?

Problem 1.89 **

Suppose that X is a metric space, Y is a complete metric space, $A \subseteq X$ is a set and $\varphi\colon A \longrightarrow Y$ is a continuous function. Show that there exists a set $A_0 \subseteq X$ and a function $\varphi_0\colon A_0 \longrightarrow Y$ such that

(a) $A \subseteq A_0$

(b) A_0 is a G_δ subset of X and

(c) φ_0 is a continuous extension of φ.

Problem 1.90 *

Let X, Y and V be three metric spaces with Y being compact. Let $f\colon X \times Y \longrightarrow V$ be a continuous function and let $v_0 \in V$. Assume that for every fixed $x \in X$, the equation $f(x, y) = v_0$ (in $y \in Y$) has a unique solution $s(x) \in Y$. Show that the function $s\colon X \longrightarrow Y$ is continuous.

Problem 1.91 ***

Suppose that (X, d_X) is a compact metric space and $f\colon X \longrightarrow X$ is a continuous function such that

$$d_X\big(f(x), f(u)\big) \geqslant d_X(x, u) \qquad \forall\, x, u \in X.$$

Show that f is an isometry.

Problem 1.92 *

Show that any uniformly continuous function $f\colon (a, b) \longrightarrow \mathbb{R}$ defined on a bounded interval (a, b) is bounded.

Problem 1.93 ***

Suppose that (X, d_X) is a compact metric space and $f\colon X \longrightarrow X$ is surjective and such that

$$d_X\big(f(x), f(u)\big) \leqslant d_X(x, u) \qquad \forall\, x, u \in X$$

(nonexpansive). Show that f is an isometry (see Definition 1.41).

Problem 1.94 **

(a) Suppose that (X, d_X) is a compact metric space and $\{K_n\}_{n\geqslant 1}$ is a decreasing sequence of closed subsets of X. We set $K \stackrel{\text{def}}{=} \bigcap_{n\geqslant 1} K_n$. Show that $\operatorname{diam} K_n \longrightarrow \operatorname{diam} K$.

(b) Show that any of two assumptions of compactness of X and the closedness of sets K_n cannot be dropped.

Problem 1.95 **

Suppose that X and Y are two compact metric spaces and consider the space $C(X;Y)$ furnished with the supremum metric d^∞. Is $\big(C(X;Y), d^\infty\big)$ a compact metric space?

Problem 1.96 **

Suppose that X is a compact metric space and $\{f_n\}_{n\geqslant 1} \subseteq C(X)$ is a sequence of functions. Show that
(a) if $f_n \rightrightarrows f$ and $x_n \longrightarrow x$ in X, then $f_n(x_n) \longrightarrow f(x)$;
(b) if $f_n(x_n) \longrightarrow f(x)$ for all sequences $\{x_n\}_{n\geqslant 1} \subseteq X$ such that $x_n \longrightarrow x$, then $f \in C(X)$ and $f_n \rightrightarrows f$.

Problem 1.97 **

(a) Suppose that U is a nonempty open set in $\mathbb{R}^N$ and $\{C_n\}_{n\geqslant 1}$ is a decreasing sequence of closed, bounded sets with $\bigcap\limits_{n\geqslant 1} C_n \subseteq U$. Show that $C_{n_0} \subseteq U$ for some $n_0 \geqslant 1$.
(b) Does the result remain true if we drop the assumption of the boundedness of the sets C_n for $n \geqslant 1$? Justify your answer?

Problem 1.98 **

Let $f\colon \mathbb{R} \longrightarrow \mathbb{R}$ be a continuous function such that $\lim\limits_{z\to\pm\infty} f(x) = 0$. Show that f is uniformly continuous.

Problem 1.99 **

Let $f\colon \mathbb{R} \longrightarrow \mathbb{R}$ be a continuous function and suppose that there exists a uniformly continuous function $h\colon \mathbb{R} \longrightarrow \mathbb{R}$ such that $\lim_{|t|\to+\infty} \big(f(t) - h(t)\big) = 0$. Show that f is uniformly continuous too.

Problem 1.100 **

(a) Let X be a metric space and let $\{x_n\}_{n\geqslant 1} \subseteq X$ be a sequence such that $x_n \longrightarrow x \in X$ as $n \to +\infty$. Let $C = \{x\} \cup \{x_n : n \geqslant 1\}$. Show that the set C is compact.
(b) Suppose that X and Y are two metric spaces and $f\colon X \longrightarrow Y$ is a function. Show that f is continuous if and only if for every compact set $C \subseteq X$, the function $f|_C$ is continuous.

Problem 1.101 **
Show that, if $f\colon X \longrightarrow Y$ is a continuous and proper function, then f is closed.

Problem 1.102 **
Suppose that X and Y are two metric spaces and $f\colon X \longrightarrow Y$ is a continuous function. Show that the following two statements are equivalent:
(a) Every sequence $\{x_n\}_{n\geqslant 1} \subseteq X$ for which $f(x_n) \longrightarrow y$ in Y has a subsequence $\{x_{n_k}\}_{k\geqslant 1}$ such that $x_{n_k} \longrightarrow x$ in X.
(b) The function f is closed and for all $y \in Y$, the set $f^{-1}(y)$ is compact in X.

Problem 1.103 **
Suppose that X and Y are two metric spaces and $f\colon X \longrightarrow Y$ is a continuous function which satisfies any of the two equivalent statements in Problem 1.102. Show that f is proper (see Definition 1.72).

Problem 1.104 **
Suppose that X is a compact metric space and $\{f_n\colon X \longrightarrow \mathbb{R}\}_{n\geqslant 1}$ is a sequence of continuous functions such that

$$f_n \searrow f \quad \text{or} \quad f_n \nearrow f \quad \text{as } n \to +\infty,$$

for some continuous function $f\colon X \longrightarrow \mathbb{R}$ (by $f_n \searrow f$ we denote the pointwise convergence $f_n \longrightarrow f$ (see Definition 1.59) of a decreasing sequence of functions, i.e., for all $n \geqslant 1$ and all $x \in X$, we have $f_{n+1}(x) \leqslant f_n(x)$; analogously by $f_n \nearrow f$ we denote the pointwise convergence $f_n \longrightarrow f$ of an increasing sequence of functions, i.e., for all $n \geqslant 1$ and all $x \in X$, we have $f_{n+1}(x) \geqslant f_n(x)$). Show that

$$f_n \rightrightarrows f$$

(see Definition 1.59). This result is known as ***Dini theorem***.

Problem 1.105 **
In the Dini theorem (see Problem 1.104) show that none of the four conditions can be dropped (namely the compactness of X, the continuity of the limit function f, the continuity of functions f_n (at least for large n) and the monotonicity of the sequence $\{f_n\}_{n\geqslant 1}$).

Problem 1.106 **

(a) Suppose that X is a compact metric space and $f\colon X \longrightarrow X$ is an isometry. Show that $f(X) = X$.

(b) Does the result remain true if instead of the compactness of the space X we assume only that it is bounded? Justify your answer.

Problem 1.107 ***

Let $f\colon \mathbb{R}^N \longrightarrow \mathbb{R}^N$ be an isometry and assume that there exists $x \in \mathbb{R}^N$ such that $f(x) = x$. Show that f is a homeomorphism.

Problem 1.108 ***

Suppose that (X, d_X) is a totally bounded metric space. Show that any sequence $\{x_n\}_{n\geqslant 1} \subseteq X$ has a Cauchy subsequence.

Problem 1.109 ***

Suppose that (X, d_X), (Y, d_Y), (V, d_V) are three metric spaces, Y is compact and $f\colon X \times Y \longrightarrow V$ is a continuous function. Show that

$$f(x, \cdot) \rightrightarrows f(x_0, \cdot) \quad \text{when } x \longrightarrow x_0 \quad \text{in } X.$$

Problem 1.110 ***

Suppose that X and Y are two metric spaces with Y being compact, $f\colon X \times Y \longrightarrow \mathbb{R}$ is a continuous function and let us set

$$m(x) \stackrel{\text{def}}{=} \inf\big\{f(x, y) :\ y \in Y\big\} \qquad \forall\, x \in X.$$

Show that $m\colon X \longrightarrow \mathbb{R}$ is continuous.

Problem 1.111 **

Suppose that X and Y are two compact metric spaces, V is a metric space and $f\colon X \times Y \longrightarrow V$ is a continuous function. For every $x \in X$, let f_x be the continuous function $Y \ni y \longmapsto f(x, y) \in V$. We furnish $C(Y; V)$ with the supremum metric

$$d^\infty(g, h) \;=\; \max_{y\in Y} d_V\big(g(y), h(y)\big) \qquad \forall\, g, h \in C(Y; V).$$

Show that the function $x \longmapsto f_x$ is continuous from X into $C(Y; V)$ and $\{f_x\}_{x\in X}$ is an equicontinuous family in $C(Y; V)$.

Problem 1.112 ***

(a) Show that the Banach fixed point theorem (see Theorem 1.49) is no more valid if instead of assuming that f is a contraction, we assume that

$$d_X(f(x), f(u)) < d_X(x, u) \qquad \forall\, x, u \in X,\ x \neq u.$$

(b) Suppose that (X, d_X) is a compact metric space and $f\colon X \longrightarrow X$ satisfies the inequality

$$d_X(f(x), f(u)) < d_X(x, u) \qquad \forall\, x, u \in X,\ x \neq u.$$

Show that f has a unique fixed point (i.e., there exists $x_0 \in X$ such that $f(x_0) = x_0$).

Problem 1.113 **

Let $X = C([0, b])$ with the supremum metric d^∞, defined by

$$d^\infty(u, v) = \max_{0 \leqslant t \leqslant b} |u(t) - v(t)| \qquad \forall\, u, v \in X.$$

Let $k\colon [0, b] \times [0, b] \longrightarrow \mathbb{R}$ be a continuous function, $\lambda \in \mathbb{R}$ and $f \in X$. Find a condition on λ so that the integral equation

$$u(t) = \lambda \int_0^b k(t, s)u(s)\, ds + f(t), \quad t \in [0, b],$$

has a solution $u \in X$.

Problem 1.114 **

Show that a locally compact metric space X is a Baire metric space.

Problem 1.115 **

Let $f\colon \mathbb{R}^N \longrightarrow \mathbb{R}$ be a continuous function such that $\lim_{\|x\| \to +\infty} f(x) = +\infty$. Show that f is bounded below and there exists $x_0 \in \mathbb{R}^N$ such that $f(x_0) = \inf_{\mathbb{R}^N} f$.

Problem 1.116 **

Let X be a metric space such that every continuous function $f\colon X \longrightarrow \mathbb{R}$ attains its supremum on X. Show that X is compact. (Compare this problem with the Weierstrass theorem; see Theorem 1.75).

Problem 1.117 ***
(a) Suppose that $D \subseteq \mathbb{R}^N$ is an F_σ-set and $f\colon \mathbb{R}^N \longrightarrow \mathbb{R}$ is a continuous function. Show that $f(D)$ is an F_σ-set too.
(b) Is the image of a G_δ-set by a continuous function necessarily a G_δ-set?

Problem 1.118 **
(a) Suppose that (X, d_X) is a metric space and K, C are two nonempty disjoint subsets of X such that K is compact and C is closed. Show that $\operatorname{dist}(K, C) > 0$.
(b) Is the above conclusion true if K is only closed but not compact?

Problem 1.119 **
Suppose that (X, d_X) is a metric space and K, C are two nonempty disjoint subsets of X such that K is compact and C is closed. Show that there exists an open set $U \subseteq X$ such that $K \subseteq U$ and $C \cap U = \emptyset$.

Problem 1.120 **
(a) Suppose that X is a compact metric space, Y is a metric space and $f\colon X \longrightarrow Y$ is a continuous function. Show that $f(\overline{D}) = \overline{f(D)}$ for all $D \subseteq X$.
(b) Show that the above is not true if X is only complete (but not compact).

Problem 1.121 **
Suppose that (X, d_X) is a metric space such that for every continuous function $f\colon X \longrightarrow \mathbb{R}$ and any $D \subseteq X$, we have $f(\overline{D}) = \overline{f(D)}$.
(a) Show that X is complete.
(b) Is X necessarily compact?

Problem 1.122 **
Suppose that (X, d_X) is a metric space.
(a) Show that X is compact if and only if for every continuous function $f\colon X \longrightarrow \mathbb{R}$ and any $D \subseteq X$, we have $f(\overline{D}) = \overline{f(D)}$.
(b) Is the above result true if we replace $\mathbb{R}$ by another metric space?

Problem 1.123 **
Suppose that (X, d_X) is a locally compact metric space and K is a nonempty compact subset of X. Show that there exists $r > 0$ such that $\overline{K_r}$ is compact, where

$$K_r \stackrel{\text{def}}{=} \{x \in X : \operatorname{dist}(x, K) < r\}.$$

Problem 1.124 **
Suppose that (X, d_X) and (Y, d_Y) are two metric spaces and $f\colon X \longrightarrow Y$ is a function.
(a) Show that, if f is continuous, then $\operatorname{Gr} f$ is closed in $X \times Y$. Give an example showing that the converse is not true.
(b) Show that when Y is compact, then f is continuous if and only if $\operatorname{Gr} f$ is closed in $X \times Y$.

Problem 1.125 ***
Show that the interval $[0, 1]$ cannot have a countable partition into nonempty closed sets.

Problem 1.126 ***
Suppose that (X, d_X) is a compact metric space and $\{C_k\}_{k=1}^{N}$ $(N > 1)$ are nonempty closed subsets of X with an empty intersection. Show that there is a number $\delta > 0$ such that every set which meets all the sets C_k for $k = 1, \ldots, N$ has diameter bigger or equal to $\delta > 0$.

Problem 1.127 **
Suppose that X is a compact metric space, $f\colon X \longrightarrow \mathbb{R}$ is a function and assume that for every $s \in \mathbb{R}$, the set $f^{-1}\big([s, +\infty)\big)$ is closed. Show that there exists $x_0 \in X$ such that $f(x_0) = \sup\limits_{\mathbb{R}} f < +\infty$.

Problem 1.128 *
Show that every metric space in which every closed ball is compact is complete and a set is complete if and only if it is closed and bounded.

Problem 1.129 **
Let $f\colon \mathbb{R}^N \longrightarrow \mathbb{R}$ be a function such that $\liminf\limits_{\|x\| \to +\infty} \frac{f(x)}{\|x\|} > 0$. Show that for every $\lambda \in \mathbb{R}$, the set $\overline{\big\{x \in \mathbb{R}^N : f(x) \leqslant \lambda\big\}}$ is compact.

Problem 1.130 **
Let $G\colon [0, 1] \times [0, 1] \longrightarrow \mathbb{R}$ be a continuous function. Consider the space $C\big([0, 1]\big)$ equipped with the supremum metric d^{∞}, i.e.,

$$d^{\infty}(f, g) = \max_{t \in [0,1]} \big|f(t) - g(t)\big| \qquad \forall\, f, g \in C\big([0, 1]\big)$$

(see Example 1.3). Let $L\colon C([0,1]) \longrightarrow C([0,1])$ be the function, defined by

$$L(f)(t) \stackrel{\text{def}}{=} \int_0^1 G(t,s)f(s)\,ds.$$

Show that, if $E \subseteq C([0,1])$ is bounded, then the set $\overline{L(E)} \subseteq C([0,1])$ is compact.

Problem 1.131 **
Is the set $\{\sin(nx) : n \geqslant 1\} \subseteq C([-\pi,\pi])$ relatively compact?

Problem 1.132 **
Suppose that d_X^1 and d_X^2 are two topologically equivalent metrics on X and $E \subseteq X$ is a nonempty closed set which is locally compact with respect to d_X^1. Show that E is locally compact with respect to d_X^2 too.

Problem 1.133 **
Let $X = \{(0,0)\} \cup \{(x, \sin\frac{1}{x}) : x > 0\}$ be a metric space with the natural metric induced by the Euclidean metric in $\mathbb{R}^2$. Show that X is not locally compact.

Problem 1.134 ***
(a) Suppose that X and Y are two metric spaces with X being locally compact and $f\colon X \longrightarrow Y$ is a continuous and open function. Show that $f(X)$ is locally compact.
(b) Is the above still true if f is only continuous (but not necessarily open)?
(c) Is the above still true if f is only open (but not necessarily continuous)?

Problem 1.135 **
Show that every locally compact metric space X is homeomorphic to a complete metric space.

Problem 1.136 **
Show that $\mathbb{Q}$ is not homeomorphic to a complete metric space.

Problem 1.137 **
Suppose that (X, d_X) is a compact metric space and $\mathcal{Y}$ is an open cover of X. Show that there is a number $\delta > 0$ such that for any $A \subseteq X$ with $\operatorname{diam} A < \delta$ (see Definition 1.6), we can find $U \in \mathcal{Y}$ such that $A \subseteq U$.

Remark. This number $\delta > 0$ is called the ***Lebesgue number of the cover*** $\mathcal{Y}$.

Problem 1.138 ***
Find a bounded function on $\mathbb{R}$ which realizes neither its maximum nor minimum on any compact nontrivial interval.

Problem 1.139 *
Show that a totally bounded metric space is separable.

Problem 1.140 ***
Let (X, d_X) be a metric space. Show that the following statements are equivalent:
(a) (X, d_X) is a compact metric space.
(b) If $\hat{d}_X$ is another metric topologically equivalent to d_X, then $(X, \hat{d}_X)$ is a complete metric space.

Problem 1.141 *
Suppose that X and Y are two metric spaces, $f: X \longrightarrow Y$ is a continuous function and $K \subseteq X$ is a compact set. Show that for every $\varepsilon > 0$, we can find $\delta = \delta(\varepsilon) > 0$ such that

$$\forall x \in K,\ u \in X: \quad \left[d_X(x,u) \leqslant \delta \implies d_Y\big(f(x), f(u)\big) \leqslant \varepsilon\right].$$

Problem 1.142 ***
Suppose that (X, d_X) is a metric space and $\mathcal{B}$ is the family of all bounded sets of X (sets with finite diameter). The ***Kuratowski measure of noncompactness*** $\alpha: \mathcal{B} \longrightarrow \mathbb{R}_+$ is defined by

$$\alpha(B) \stackrel{\text{def}}{=} \inf\big\{d > 0 : B \text{ can be covered by a finite number of sets of diameter } \leqslant d\big\} \qquad \forall\, B \in \mathcal{B}.$$

Show that
(a) If B is compact, then $\alpha(B) = 0$.
(b) If $\alpha(B) = 0$, then B is totally bounded.
(c) α is monotone, i.e., if $B \subseteq B'$, then $\alpha(B) \leqslant \alpha(B')$.
(d) $\alpha(\overline{B}) = \alpha(B)$ for all $B \in \mathcal{B}$.
(e) If (X, d_X) is a complete metric space and $\{C_n\}_{n\geqslant 1} \subseteq \mathcal{B}$ is a decreasing sequence of nonempty, closed sets and $\alpha(C_n) \longrightarrow 0$, then $\bigcap_{n\geqslant 1} C_n \neq \emptyset$ and it is compact.

Problem 1.143 **
(a) Suppose that X and Y are two metric spaces and $f\colon X \longrightarrow Y$ is an injection which maps compact sets in X into compact sets in Y. Show that f is continuous.
(b) Show that the assumption on the injectivity of f cannot be dropped.

Problem 1.144 ***
(a) Suppose that (X, d_X) is a compact metric space and $C_1, \ldots, C_m$ (with $m \geqslant 2$) are nonempty closed sets in X such that $\bigcap\limits_{k=1}^{m} C_k = \emptyset$. Show that there exist open sets $U_1, \ldots, U_m$ in X such that $C_k \subseteq U_k$ for all $k \in \{1, \ldots, m\}$ and $\bigcap\limits_{k=1}^{m} U_k = \emptyset$.
(b) Does the above result remains true if X is any metric space (not necessarily compact)? Justify your answer.

Problem 1.145 **
Suppose that X is a normed vector space (for example $X = \mathbb{R}^N$), $f\colon X \longrightarrow X$ is a locally Lipschitz function (i.e., for all $x \in X$, we can find an open set U_x such that $x \in U_x$ and $k_x > 0$ such that $\|f(u) - f(u')\| \leqslant k_x \|u - u'\|$ for all $u, u' \in U_x$; see Definition 1.48(b)) and C is a compact subset of X. Show that $f|_C$ is Lipschitz continuous, i.e., there exists $k_C > 0$ such that $\|f(u) - f(u')\| \leqslant k_C \|u - u'\|$ for all $u, u' \in C$.

Problem 1.146 *
Let X and Y be two metric spaces with Y being compact and let $C \subseteq X \times Y$ be a closed subset. Show that $\mathrm{proj}_X(C) \subseteq X$ is a closed subset (where proj_X denotes the projection map on X).

Problem 1.147 *
Let X be a metric space. We say that X has ***property S*** if for every $\varepsilon > 0$, X admits an open cover by connected sets of diameter less than $\varepsilon > 0$. Are the following statements true:
(a) If X has a dense subset with property S, then X has property S.
(b) If X has property S, then any dense subset of X has property S.

Problem 1.148 ***
Suppose that (X, d_X) is a compact metric space, $\{x_n\}_{n\geqslant 1} \subseteq X$ is a sequence such that $d_X(x_{n+1}, x_n) \longrightarrow 0$ and A is the set of accumulation points of the sequence $\{x_n\}_{n\geqslant 1} \subseteq X$. Show that A is connected.

Problem 1.149 **

(a) Let $f\colon [0,1] \longrightarrow [0,1]$ be a continuous function. Show that f has a fixed point, i.e., there exists $x \in [0,1]$ such that $f(x) = x$.
(b) Is the above true if we replace the interval $[0,1]$ by $(0,1)$?
(c) Is the above true if we replace the interval $[0,1]$ by any compact metric space (X, d_X)?

Problem 1.150 **

Let $A \subseteq \mathbb{R}$ be a nonempty and connected set. Assume that $A \subseteq \mathbb{Q}$. Show that A is a singleton.

Problem 1.151 **

Let X be a metric space. Show that the following two statements are equivalent:
(a) X is disconnected.
(b) There exists a surjective continuous function $h\colon X \longrightarrow \{0,1\}$, where $\{0,1\}$ is furnished with the discrete metric.

Problem 1.152 **

(a) Show that for $N > 1$, the set $\mathbb{R}^N \setminus \{0\}$ is connected.
(b) Show that $\mathbb{R}$ and $\mathbb{R}^N$ (with $N > 1$) cannot be homeomorphic.

Problem 1.153 **

Let $\varphi\colon \mathbb{R}^N \longrightarrow \mathbb{R}$ (with $N \geqslant 2$) be a continuous and surjective function. Show that for all $u \in \mathbb{R}$, the set $\varphi^{-1}(\{u\})$ is not bounded.

Problem 1.154 **

Are the intervals $[0,1]$ and $(0,1)$ homeomorphic? Justify your answer.

Problem 1.155 **

Let $E \overset{\text{def}}{=} \{u \in \mathbb{R}^N : \ u \text{ has at least one irrational component}\}$. Is E path-connected? Justify your answer.

Problem 1.156 **

(a) Let (X, d_X) be a compact metric space. Show that the following statements are equivalent:

> **(i)** X is locally connected;
> **(ii)** for every $\varepsilon > 0$, X is the finite union of connected sets with diameter less or equal to ε.

(b) Is the above equivalence still true if we drop the assumption on compactness of X?

Problem 1.157 **
(a) Suppose that X is a metric space and $\{A_k\}_{k=1}^m$ are connected subsets of X such that $A_k \cap A_{k+1} \neq \emptyset$ for all $k \leqslant m-1$. Show that the set $\bigcup\limits_{k=1}^{m} A_k$ is connected.

(b) Is the above result true for a countable family of sets $\{A_k\}_{k=1}^\infty$? Justify your answer.

Problem 1.158 *
Let (X, d_X) be a metric space with connected open balls (for example $X = \mathbb{R}^N$ for $N \geqslant 1$), let $C \subseteq X$ and for all $\varepsilon > 0$ let

$$C_\varepsilon \stackrel{\text{def}}{=} \{x \in X : \operatorname{dist}(x, C) < \varepsilon\}$$

(the ε-neighbourhood of C). Show that, if C is connected, then C_ε is connected too.

Problem 1.159 **
Suppose that $f \in C(\mathbb{R})$. Show that $\operatorname{epi} f$ is path-connected.

Problem 1.160 **
Let (X, d_X) be a metric space.
(a) Show that if X is connected, then X is well-chained.
(b) Is the opposite true?

Problem 1.161 **
Suppose that (X, d_X) is a compact metric space. Show that X is connected if and only if X is well-chained.

Problem 1.162 **
Suppose that X and Y are metric spaces and $f\colon X \longrightarrow Y$ is a homeomorphism. Show that f maps connected components of X to connected components of Y, i.e., $f\big(C(x)\big) = C\big(f(x)\big)$ for all $x \in X$.

Problem 1.163 *
Consider the set $E \stackrel{\text{def}}{=} \{(x, \sin\frac{1}{x}) : x \in [0,1]\} \cup \big(\{0\} \times [-1,1]\big)$. Is the set E locally connected? Justify your answer.

Problem 1.164 **
Suppose that T is an interval in $\mathbb{R}$ and $f\colon T \longrightarrow \mathbb{R}$ is a monotone function (i.e., f is either increasing or decreasing). Show that f is continuous if and only if $f(T)$ is an interval.

Problem 1.165 ***
Suppose that $T = [a, b]$ and $f\colon T \longrightarrow \mathbb{R}$ is a differentiable function. Show that f' has the Darboux property (i.e., if $\eta, \vartheta \in f'(T)$, then f' takes all values $\lambda \in [\eta, \vartheta]$).

Problem 1.166 **
Let (X, d_X) be a connected metric space with at least two elements. Show that X is uncountable.

Problem 1.167 **
Suppose that X is a metric space, $\{C_n\}_{n\geqslant 1}$ is decreasing sequence of nonempty compact and connected subsets of X. Show that the set $C = \bigcap\limits_{n\geqslant 1} C_n$ is nonempty, compact and connected too.

Problem 1.168 **
Suppose that (X, d_X) and (Y, d_Y) are two metric spaces with Y being complete and $g\colon X \times Y \longrightarrow Y$ is a function such that

$$d_Y\big(g(x, y), g(x, y')\big) \leqslant k d_Y(y, y') \qquad \forall\, x \in X,\ y, y' \in Y$$

for some $k \in (0, 1)$. Show that there exists a continuous and bounded function $f\colon X \longrightarrow Y$ such that $g\big(x, f(x)\big) = f(x)$ for all $x \in X$.

Problem 1.169 **
Suppose that (X, d_X) is a metric space and $K, C \subseteq X$ is two nonempty compact subsets. Show that there exist $a \in K$ and $c \in C$ such that $d_X(a, c) = \operatorname{dist}(K, C)$.
Moreover, if $K = C$, show that there exist $a', c' \in K$ such that $d_X(a', c') = \operatorname{diam} K$.

Problem 1.170 **
Suppose that (X, d_k), $k = 1, \ldots, m$ are metric spaces and $X = \prod\limits_{k=1}^{m} X_k$. For a sequence $\{x_n\}_{n\geqslant 1}$ in X, let $\{x_n^k\}_{n\geqslant 1}$ be its projection on X_k, $k = 1, \ldots, m$. Show that $\{x_n\}_{n\geqslant 1}$ is a Cauchy sequence in X if and only if $\{x_n^k\}_{n\geqslant 1}$ are Cauchy sequences in X for all $k = 1, \ldots, m$.

Problem 1.171 ***

Suppose that (X, d_X) is a metric space and $\{E_n\}_{n\geqslant 1}$ is a sequence of subsets (possibly empty) of X. Show that

$$\begin{aligned}
\liminf_{n\to+\infty} E_n &= \{x \in X : \lim_{n\to+\infty} \operatorname{dist}(x, E_n) = 0\} \\
&= \Big\{x \in X : \text{ there exist } x_n \in E_n \text{ such that} \\
&\qquad d_X(x_n, x) \longrightarrow 0\Big\}; \\
\limsup_{n\to+\infty} E_n &= \{x \in X : \liminf_{n\to+\infty} \operatorname{dist}(x, E_n) = 0\} \\
&= \{x \in X : \text{ there exist } x_{n_k} \in E_{n_k}, \text{ with } n_1 < n_2 < \ldots, \\
&\qquad \text{such that } d_X(x_{n_k}, x) \longrightarrow 0\} \\
&= \bigcap_{m\geqslant 1} \overline{\bigcup_{n\geqslant m} E_n}; \\
\liminf_{n\to+\infty} E_n &\subseteq \limsup_{n\to+\infty} E_n; \\
\overline{\liminf_{n\to+\infty} E_n} &= \liminf_{n\to+\infty} E_n = \liminf_{n\to+\infty} \overline{E}_n; \\
\overline{\limsup_{n\to+\infty} E_n} &= \limsup_{n\to+\infty} E_n = \limsup_{n\to+\infty} \overline{E}_n.
\end{aligned}$$

In particular both sets $\liminf\limits_{n\to+\infty} E_n$ and $\limsup\limits_{n\to+\infty} E_n$ are closed.

Problem 1.172 **

Suppose that (X, d_X) is a metric space, $\{C_n\}_{n\geqslant 1} \subseteq 2^X \setminus \{\emptyset\}$ is a sequence of sets and $x \in X$. Show that

$$\limsup_{n\to+\infty} \operatorname{dist}(x, C_n) \leqslant \operatorname{dist}\big(x, \liminf_{n\to+\infty} C_n\big).$$

Problem 1.173 **

Suppose that $\{C_n\}_{n\geqslant 1}$ is a sequence of nonempty closed sets in $\mathbb{R}^N$ and $C_n \subseteq \overline{B}_R$ for all $n \geqslant 1$. Assume that $\liminf\limits_{n\to+\infty} C_n = \limsup\limits_{n\to+\infty} C_n = C$. Show that for every $x \in \mathbb{R}^N$, we have $\operatorname{dist}(x, C_n) \longrightarrow \operatorname{dist}(x, C)$.

Problem 1.174 **

Suppose that (X, d_X) is a metric space, $P_f(X)$ is the collection of nonempty closed subsets of X, $\varphi\colon X \longrightarrow \mathbb{R}$ is a continuous function, $C \in P_f(X)$, $\{C_n\}_{n\geqslant 1} \subseteq P_f(X)$ is a sequence such that $C \subseteq \liminf\limits_{n\to+\infty} C_n$. Let $m_n = \inf_{C_n} \varphi$ and $m = \inf_C \varphi$. Show that $\limsup\limits_{n\to+\infty} m_n \leqslant m$.

Problem 1.175 ***

Suppose that X is a metric space and $\{x_{m,n}\}_{m,n\geqslant 1}$ is a "double" sequence in X such that $x_m = \lim\limits_{n\to+\infty} x_{m,n}$ and $x = \lim\limits_{m\to+\infty} x_m$. Show that there exist sequences $\{n_m\}_{m\geqslant 1}$ and $\{m_n\}_{n\geqslant 1}$ such that $x = \lim\limits_{m\to+\infty} x_{m,n_m} = \lim\limits_{n\to+\infty} x_{m_n,n}$.

Problem 1.176 ***

(a) Suppose that (X, d_X) is a bounded metric space and $P_f(X)$ denotes the collection of nonempty closed subsets of X. For any $A, B \in P_f(X)$, we set

$$h(A,B) \stackrel{\text{def}}{=} \sup_{x\in X} \big|\operatorname{dist}(x,A) - \operatorname{dist}(x,B)\big|.$$

Show that h is a metric on $P_f(X)$. It is known as the ***Hausdorff metric***; see Definition 1.134.

(b) For any $A, B \in P_f(X)$, we set

$$\overline{h}(A,B) \stackrel{\text{def}}{=} \max\big\{\sup_{a\in A}\operatorname{dist}(a,B),\ \sup_{b\in B}\operatorname{dist}(b,A)\big\}.$$

Show that

$$h(A,B) = \overline{h}(A,B) \qquad \forall\, A, B \in P_f(X).$$

(c) For any $A, B \in P_f(X)$, we set

$$\overline{\overline{h}}(A,B) \stackrel{\text{def}}{=} \inf\big\{\varepsilon > 0:\ A \subseteq B_\varepsilon \text{ and } B \subseteq A_\varepsilon\big\},$$

where

$$S_\varepsilon \stackrel{\text{def}}{=} \{x \in X:\ \operatorname{dist}(x,S) \leqslant \varepsilon\} \qquad \forall\, S \in P_f(X).$$

Show that

$$\overline{\overline{h}}(A,B) = \overline{h}(A,B) \qquad \forall\, A, B \in P_f(X).$$

So in fact, we have three equivalent definitions of Hausdorff metric on $P_f(X)$.

Problem 1.177 **

Let X be a bounded metric space and let $\big(P_f(X), h\big)$ be the Hausdorff metric space (see Definition 1.134 and Problem 1.176).

(a) Show that for any $D_1, D_2, E_1, E_2 \in P_f(X)$, we have

$$h(D_1 \cup D_2,\ E_1 \cup E_2) \leqslant \max\{h(D_1, E_1),\ h(D_2, E_2)\}.$$

(b) Show that, for any k-contraction $\eta\colon X \longrightarrow X$, we have

$$h(\eta(D), \eta(E)) \leqslant kh(D, E) \qquad \forall\ D, E \in P_f(X).$$

Problem 1.178 ***

(a) Show that, if X is a complete bounded metric space, then so is the Hausdorff metric space $(P_f(X), h)$.
(b) Show that, if X is a compact metric space, then so is the Hausdorff metric space $(P_f(X), h)$.

Problem 1.179 **

Suppose that (X, d_X) is a compact metric space and $f, g\colon X \longrightarrow X$ are two k-contractions with $k \in [0, 1)$. Let $P_f(X)$ be the collection of all nonempty closed subsets of X and let $\xi\colon P_f(X) \longrightarrow P_f(X)$ be the function, defined by

$$\xi(C) \stackrel{\text{def}}{=} f(C) \cup g(C).$$

Show that there exists a unique $C_0 \in P_f(X)$ such that $\xi(C_0) = C_0$.

Problem 1.180 **

Suppose that X is a bounded metric space and h is the Hausdorff metric on $P_f(X)$. Let

$$P_k(X) \stackrel{\text{def}}{=} \{C \subseteq X : C \text{ is nonempty and compact}\}.$$

(a) Show that $P_k(X)$ is closed in $(P_f(X), h)$.
(b) Show that, if X is a complete metric space, then $(P_k(X), h)$ is a complete metric space too.

Problem 1.181 **

Suppose that (X, d_X) is a bounded metric space and h is the Hausdorff metric on $P_f(X)$. Show that the space X is compact if and only if the space $(P_f(X), h)$ is compact.

Problem 1.182 **

Suppose that (X, d_X) is a bounded metric space, $\{C_n\}_{n\geqslant 1} \subseteq P_f(X)$ is a sequence such that $C_n \xrightarrow{h} C \in P_f(X)$ (h denotes the Hausdorff metric on $P_f(X)$), C_n is connected for every $n \geqslant 1$ and C is compact. Show that C is connected too.

Problem 1.183 **

Let X be a separable bounded metric space. Show that $(P_k(X), h)$ is a separable, complete metric space (h is the Hausdorff metric).

Problem 1.184 **

Show that the Hausdorff metric h depends on the metric d_X (i.e., it is not topological; cf. Remark 1.42).

Problem 1.185 **

Find a sequence of closed sets $\{C_n\}_{n\geqslant 1}$ in some metric space such that

$$\liminf_{n\to+\infty} C_n = \limsup_{n\to+\infty} C_n = C$$

for some closed set C, but $C_n \not\longrightarrow C$ in the Hausdorff distance.

Problem 1.186 **

Suppose that (X, d_X) is a bounded metric space, $C \in P_f(X)$, $\{C_n\}_{n\geqslant 1} \subseteq P_f(X)$ and $C_n \xrightarrow{h} C$ (i.e., converges in the sense of Hausdorff). Show that

$$\liminf_{n\to+\infty} C_n = \limsup_{n\to+\infty} C_n = C.$$

Problem 1.187 **

Let (X, d_X) be a complete bounded metric space and let $\{C_n\}_{n\geqslant 1}$ be a Cauchy sequence in $(P_k(X), h)$. Let $C \stackrel{\text{def}}{=} \overline{\bigcup_{n\geqslant 1} C_n}$. Show that $C \in P_k(X)$.

1.3 Solutions

Solution of Problem 1.1
"$\Longrightarrow$": If $\{x_n\}_{n\geqslant 1}$ is a Cauchy sequence in X (see Definition 1.7) and $x_n \longrightarrow x$, then every subsequence of $\{x_n\}_{n\geqslant 1}$ also converges to x.

"$\Longleftarrow$": Suppose that the subsequence $\{x_{n_k}\}_{k\geqslant 1}$ of $\{x_n\}_{n\geqslant 1}$ converges to $x \in X$. Then for a given $\varepsilon > 0$, we can find integer $n_0 = n_0(\varepsilon) > 0$ such that

$$d_X(x_{n_k}, x) < \tfrac{\varepsilon}{2} \quad \text{and} \quad d_X(x_n, x_m) < \tfrac{\varepsilon}{2} \qquad \forall\, n_k, n, m \geqslant n_0.$$

Then

$$d_X(x_n, x) \leqslant d_X(x_n, x_{n_k}) + d_X(x_{n_k}, x) < \varepsilon \qquad \forall\, n \geqslant n_0,$$

so $x_n \longrightarrow x$ (X, d_X).

Solution of Problem 1.2
For a given $\varepsilon > 0$, we can find an integer $n_0 = n_0(\varepsilon) \geqslant 1$ such that

$$\begin{aligned} d_X(x_n, x_m) &< \tfrac{\varepsilon}{3} && \forall\, n, m\text{-even},\ n \geqslant n_0, \\ d_X(x_n, x_m) &< \tfrac{\varepsilon}{3} && \forall\, n, m\text{-odd},\ n \geqslant n_0, \\ d_X(x_n, x_m) &< \tfrac{\varepsilon}{3} && \forall\, n, m\text{-multiples of 3},\ n \geqslant n_0. \end{aligned}$$

Let us choose $k, l \geqslant n_0$ such that they are both multiples of 3, k is even and l is odd. Then for $n, m \geqslant n_0$, we need to consider only the case when n is even and m is odd. We have

$$d_X(x_n, x_m) \leqslant d_X(x_n, x_k) + d_X(x_k, x_l) + d_X(x_l, x_m) < \varepsilon,$$

which proves that the sequence $\{x_n\}_{n\geqslant 1}$ is a Cauchy sequence in X. Because it has a convergent subsequence, it converges itself (see Problem 1.1).

Solution of Problem 1.3

Let us proceed by contradiction. So suppose that x is not the limit of the sequence $\{x_n\}_{n\geqslant 1}$. Thus, there exists $\varepsilon > 0$ such that for all $N \in \mathbb{N}$ there exists $n > N$ such that $d_X(x_n, x) \geqslant \varepsilon$. First, for $N = 1$, let us find $n_1 > 1$ such that $d_X(x_{n_1}, x) \geqslant \varepsilon$. Next, for $N = n_1$, let us find $n_2 > n_1$ such that $d_X(x_{n_2}, x) \geqslant \varepsilon$. So, proceeding by induction, we can find a subsequence $\{x_{n_k}\}_{k\geqslant 1} \subseteq \{x_n\}_{n\geqslant 1}$, with the property, that

$$d_X(x_{n_k}, x) \geqslant \varepsilon \qquad \forall\, k \geqslant 1.$$

This subsequence has no further subsequence convergent to x. So, we obtained a contradiction with our hypothesis.

Solution of Problem 1.4

Since $\{x_n\}_{n\geqslant 1} \subseteq X$ is a Cauchy sequence (see Definition 1.7), for a given $\varepsilon > 0$ we can find $N(\varepsilon) \geqslant 1$ such that

$$d_X(x_k, x_m) \;<\; \varepsilon \qquad \forall\, k, m \geqslant N(\varepsilon).$$

Let $\varepsilon = \frac{1}{2}$ and choose $n_1 \geqslant N(\frac{1}{2})$. Then

$$d_X(x_{n_1}, x_m) \;\leqslant\; \tfrac{1}{2} \qquad \forall\, m \geqslant n_1.$$

Suppose that we have produced a finite sequence $\{x_{n_i}\}_{i=1}^k$ which is strictly increasing and satisfies

$$d_X(x_{n_k}, x_m) \;\leqslant\; \tfrac{1}{2^k} \qquad \forall\, m \geqslant n_k.$$

Let $\varepsilon = \frac{1}{2^{k+1}}$ and choose integer $n_{k+1} > n_k$, $n_{k+1} > N\big(\frac{1}{2^{k+1}}\big)$. Then

$$d_X(x_{n_{k+1}}, x_m) \;\leqslant\; \tfrac{1}{2^{k+1}} \qquad \forall\, m \geqslant n_{k+1}$$

and so by induction, we have a sequence $\{x_{n_k}\}_{k\geqslant 1}$ which is a subsequence of $\{x_n\}_{n\geqslant 1}$, satisfying

$$d_X(x_{n_k}, x_{n_m}) \;\leqslant\; \tfrac{1}{2^k} \qquad \forall\, k \leqslant m.$$

Solution of Problem 1.5

(a) Since $\{x_n\}_{n\geqslant 1}$ is a Cauchy sequence, we can find $n_1 \geqslant 1$ such that

$$d_X(x_m, x_k) \leqslant \tfrac{1}{2} \qquad \forall\, m, k \geqslant n_1.$$

We set $u_1 = x_{n_1}$. Then choose $n_2 > n_1$ such that

$$d_X(x_m, x_k) \leqslant \tfrac{1}{2^2} \qquad \forall\, m, k \geqslant n_2.$$

We set $u_2 = x_{n_2}$. Inductively, we produce $\{u_m = x_{n_m}\}_{m\geqslant 1}$, a subsequence of $\{x_n\}_{n\geqslant 1}$ such that

$$d_X(u_m, u_{m+1}) \leqslant \tfrac{1}{2^m} \qquad \forall\, m \geqslant 1.$$

Hence

$$\sum_{m\geqslant 1} d_X(u_m, u_{m+1}) \leqslant \sum_{m\geqslant 1} \tfrac{1}{2^m} = 1 < +\infty.$$

(b) Next, let $\{x_n\}_{n\geqslant 1}$ be a Cauchy sequence in X. From part **(a)**, we can find a subsequence $\{u_m\}_{m\geqslant 1}$ of $\{x_n\}_{n\geqslant 1}$ such that

$$\sum_{m\geqslant 1} d_X(u_m, u_{m+1}) < +\infty.$$

Then by hypothesis $\{u_m\}_{m\geqslant 1}$ is convergent, hence so is $\{x_n\}_{n\geqslant 1}$, which proves the completeness of X.

Solution of Problem 1.6

(a) Let $\{\hat{u}^k\}_{k\geqslant 1}$ be a d_1-Cauchy sequence. Then, for a given $\varepsilon > 0$, we can find $k_0 = k_0(\varepsilon) \geqslant 1$ such that

$$d_1(\hat{u}^k, \hat{u}^m) \leqslant \varepsilon \qquad \forall\, k, m \geqslant k_0.$$

Hence

$$\sum_{n=1}^{\infty} |u_n^k - u_n^m| \leqslant \varepsilon \qquad \forall\, k, m \geqslant k_0 \tag{1.1}$$

(where $\hat{u}^k = \{u_n^k\}_{n\geqslant 1}$ and $\hat{u}^m = \{u_n^m\}_{n\geqslant 1}$). It follows that for every fixed $n \geqslant 1$, $\{u_n^k\}_{k\geqslant 1}$ is a Cauchy sequence and so we have

$$u_n^k \longrightarrow \vartheta_n \quad \text{as } k \to +\infty.$$

Let $\hat{u} = \{\vartheta_n\}_{n\geqslant 1}$. If in (1.1), we let $m \to +\infty$, then

$$\sum_{n=1}^{\infty} |u_n^k - \vartheta_n| \leqslant \varepsilon \qquad \forall\, k, \geqslant k_0, \tag{1.2}$$

so

$$\sum_{n=1}^{\infty} |\vartheta_n| \leqslant \sum_{n=1}^{\infty} |u_n^{k_0}| + \varepsilon$$

and thus $\hat{u} = \{\vartheta_n\}_{n\geqslant 1} \in l^1$. Moreover, from (1.2), it follows that

$$d_1(\hat{u}_n, \hat{u}) \longrightarrow 0$$

and so we conclude that (l^1, d_1) is a complete metric space.

(b) Evidently $c_0 \neq l^1$ (for example $\hat{u} = \left\{\frac{1}{n^r}\right\}_{n\geqslant 1}$ with $r > 1$, belongs in l^1 but not in c_0). Let $\hat{u} = \{u_n\}_{n\geqslant 1} \in l^1$. For every $k \in \mathbb{N}$, let $\hat{u}^k = \{u_n^k\}_{n\geqslant 1}$ with $u_n^k = u_n$ if $n \leqslant k$ and $u_n^k = 0$ if $n > k$. Then $\hat{u}^k \in c_0$ and

$$d_1(\hat{u}^k, \hat{u}) = \sum_{n=k+1}^{\infty} |u_n| \longrightarrow 0 \quad \text{as } k \to +\infty.$$

This proves the d_1-density of c_0 in l^1 (see Definition 1.20).

Solution of Problem 1.7

Consider the function

$$f(x) = \begin{cases} \frac{1}{q} & \text{if } \ x = \frac{p}{q},\ p, q \in \mathbb{Z} \setminus \{0\}, \text{ relatively prime}, \\ 1 & \text{if } \ x = 0, \\ 0 & \text{if } \ x \in \mathbb{R} \setminus \mathbb{Q}. \end{cases}$$

Then f is continuous at every $x \in \mathbb{R} \setminus \mathbb{Q}$ and discontinuous at every $x \in \mathbb{Q}$. To see this note that, if $x \in \mathbb{R} \setminus \mathbb{Q}$ and $x_n \longrightarrow x$, with $x_n = \frac{p_n}{q_n} \in \mathbb{Q}$, $p_n \in \mathbb{Z}$, $q_n \in \mathbb{N}$, p_n, q_n relatively prime, then $q_n \longrightarrow +\infty$ and so

$$f(x_n) = \frac{1}{q_n} \longrightarrow 0 = f(x).$$

This proves the continuity of f at every irrational point. Clearly f is discontinuous at $x = 0$. Finally let $x \in \mathbb{Q} \setminus \{0\}$. Then $x = \frac{p}{q}$ with $p, q \in$

$\mathbb{Z} \setminus \{0\}$ relatively prime. Suppose that the sequence $\{x_n\}_{n \geqslant 1} \subseteq \mathbb{R} \setminus \mathbb{Q}$ satisfies $x_n \longrightarrow x$. Then $f(x_n) = 0$ for all $n \geqslant 1$ and so they cannot converge to $f(x) = \frac{1}{q}$. Hence f is discontinuous at every rational number.

Solution of Problem 1.8

"$\Longrightarrow$": If X is a singleton, then every sequence is constant, hence convergent.

"$\Longleftarrow$": Suppose that X is not a singleton and let $x, u \in X$ be such that $x \neq u$. Let

$$x_n = \begin{cases} x & \text{if } \ n = 2k, \\ u & \text{if } \ n = 2k+1 \end{cases} \qquad \forall\, k \geqslant 1.$$

then the sequence $\{x_n\}_{n \geqslant 1}$ is bounded, but not convergent, a contradiction.

Solution of Problem 1.9

Let $x \in \overline{C}$. Then we can find a sequence $\{x_n\}_{n \geqslant 1} \subseteq C$ such that $x_n \longrightarrow x$. The sequence $\{x_n\}_{n \geqslant 1}$ being convergent in X is also a Cauchy sequence in C and the latter is complete by hypothesis. Hence $x \in C$ and so C is closed.

Solution of Problem 1.10

We proceed by contradiction. So, suppose that $\operatorname{int} M \neq \emptyset$. Thus, there are $f \in M$ and $\varepsilon > 0$ such that

$$B_{4\varepsilon}(f) \subseteq M.$$

We will construct a function $h \in B_{4\varepsilon}(f) \setminus M$. Let $x_0 \in (c, d)$. Because f is continuous, so we can find $\delta > 0$, $\delta < \frac{1}{2}(d - x_0)$ such that

$$\forall x \in [c, d] : \ |x - x_0| \leqslant 3\delta \implies \big|f(x) - f(x_0)\big| \leqslant \varepsilon.$$

We define the function $h \in C[a, b]$ is such a way that

$$h\big|_{[a, x_0]} = f\big|_{[a, x_0]} \quad \text{and} \quad h\big|_{[x_0 + 2\delta, b]} = f\big|_{[x_0 + 2\delta, b]},$$

on the interval $[x_0, x_0 + \delta]$, $\operatorname{Gr} h$ is a segment joining $\big(x_0, f(x_0)\big)$ with $\big(x_0+\delta, f(x_0)-2\varepsilon\big)$ and on the interval $[x_0+\delta, x_0+2\delta]$, $\operatorname{Gr} h$ is a segment joining $\big(x_0 + \delta, f(x_0) - 2\varepsilon\big)$ with $\big(x_0 + 2\delta, f(x_0 + 2\delta)\big)$. Evidently h is continuous, it is not monotone (as it is strictly decreasing on $[x_0, x_0+\delta]$ and strictly decreasing on $[x_0 + \delta, x_0 + 2\delta]$), so $h \notin M$ and $h \in B_{4\varepsilon}(f)$, a contradiction.

Solution of Problem 1.11

By Theorem 1.14(d), x is an accumulation point of E (see Definition 1.13(b)) if and only if we can find a sequence $\{x_n\}_{n\geqslant 1} \subseteq E \setminus \{x\}$ such that $x_n \longrightarrow x$. Hence x is an accumulation point of E if and only if $x \in \overline{E \setminus \{x\}}$ (see Proposition 1.11).

Solution of Problem 1.12

Suppose that $x \in U$. Since D is dense in X (see Definition 1.20), we have

$$B_{\frac{1}{n}}(x) \cap D \neq \emptyset \qquad \forall\, n \geqslant 1.$$

Let $x_n \in B_{\frac{1}{n}}(x) \cap D$ for $n \geqslant 1$. Evidently

$$x_n \longrightarrow x \quad \text{in } X.$$

Since U is open we can find $n_0 \geqslant 1$ such that

$$B_{\frac{1}{n}}(x) \subseteq U \qquad \forall\, n \geqslant n_0.$$

Hence

$$x_n \in U \cap D \qquad \forall\, n \geqslant n_0$$

and so $x \in \overline{U \cap D}$. Therefore $U \subseteq \overline{U \cap D}$ and so $U \cap D$ is dense in U with the subspace metric topology (see Definition 1.12).

Solution of Problem 1.13

Let $x \in E$ and choose $u_x > x$, $u_x \in \mathbb{Q}$ such that $(x, u_x) \cap D = \emptyset$. We will show that, if $x, y \in E$, $x \neq y$, then $u_x \neq u_y$. Assume without any loss of generality that $x < y$ and arguing indirectly suppose that $u_x = u_y$. Then $y \in (x, u_x)$ and so (x, u_x) is a neighbourhood of $y \in \overline{D}$, which means that $(x, u_x) \cap D \neq \emptyset$, a contradiction. Therefore we have established an injective function $E \ni x \longmapsto u_x \in \mathbb{Q}$. This implies that E is at most countable.

Solution of Problem 1.14

Let $Y \in \mathcal{Y}$ and let D_Y be a countable dense subset of Y (see Definitions 1.20 and 1.21). Let

$$D \stackrel{\text{def}}{=} \bigcup_{Y \in \mathcal{Y}} D_Y.$$

Then D is countable (being the countable union of countable sets) and we claim that D is dense in X_0. To this end let $x \in X_0$. Then $x \in Y$ for some $Y \in \mathcal{Y}$ and so $\operatorname{dist}(x, D_Y) = 0$ (see Definition 1.6(d)), hence $\operatorname{dist}(x, D) = 0$, which proves the claim.

Solution of Problem 1.15

(a) Let $X = [0, +\infty)$ with the natural metric induced from $\mathbb{R}$. Note that

$$B_4(1) \;=\; [0,5) \;\subsetneq\; [0,6) \;=\; B_3(3).$$

(b) As $B_R(x) \subsetneq B_r(y)$, so we can find $z \in B_r(y) \setminus B_R(x)$. Then

$$d_X(z, y) \;<\; r \quad \text{and} \quad d_X(z, x) \;>\; R.$$

As $x \in B_r(y)$, we have $d_X(x, y) < r$. Then

$$r \;>\; d_X(x, y) \;\geqslant\; d_X(x, z) - d_X(z, y) \;>\; R - r,$$

so $R < 2r$.

Solution of Problem 1.16
Let

$$C = \{x \in X : \text{ for every } r > 0,\, B_r(x) \text{ is uncountable}\}$$

and let $U = X \setminus C$. Suppose that $x \subset U$. Then we can find $r_x > 0$ such that the ball $B_{r_x}(x)$ is at most countable. Note that $B_{r_x}(x) \subseteq U$. Consider the family $\{B_{r_x}(x)\}_{x\in U}$. This is an open cover of U and by Proposition 1.24, it has a countable subcover $\{B_{r_{xn}}(x_n)\}_{n\geqslant 1}$. Then

$$U = \bigcup_{n\geqslant 1} B_{r_{xn}}(x_n) \quad \text{is open and countable}$$

and so C is closed and of course perfect (see Definition 1.13(e)).

Solution of Problem 1.17
First we will show that $f|_{\mathbb{R}\setminus\mathbb{Q}}$ is continuous. Let us fix an irrational number $e \in \mathbb{R} \setminus \mathbb{Q}$. Let $\varepsilon > 0$ and choose $m = m(\varepsilon) \in \mathbb{N}$ such that

$$\sum_{n\geqslant m} \frac{1}{2^n} < \varepsilon.$$

We set

$$\delta = \min\{|q_k - e| : k = 1, \ldots, m-1\}.$$

Let $x \in \mathbb{R}$ be such that $|x - e| < \delta$. If $x < e$, then $L_e \setminus L_x \subseteq \{k\}_{k\geqslant m}$ and so

$$f(x) - f(e) \leqslant \sum_{n\in L_e\setminus L_x} \frac{1}{2^n} \leqslant \sum_{n\geqslant m} \frac{1}{2^n} < \varepsilon.$$

If $x > e$, then $L_x \setminus L_e \subseteq \{k\}_{k\geqslant m}$ and so

$$f(e) - f(x) \leqslant \sum_{n\in L_x\setminus L_e} \frac{1}{2^n} \leqslant \sum_{n\geqslant m} \frac{1}{2^n} < \varepsilon.$$

Thus f is continuous at e. Since e was an arbitrary irrational number, we conclude that $f|_{\mathbb{R}\setminus\mathbb{Q}}$ is continuous.

Now we will show that f is right continuous at rational numbers. Let us fix $\varepsilon > 0$ and a rational number u. Let $\{x_k\}_{k\geqslant 1} \subseteq \mathbb{R}$ be such that

$$x_k \longrightarrow u^+ \quad \text{as } k \to +\infty.$$

Since f is strictly increasing, we have

$$f(u) \leqslant f(x_k) \qquad \forall\, k \geqslant 1.$$

Then

$$\left|f(x_k) - f(u)\right| = f(x_k) - f(u) = \sum_{n \in S_{k,u}} \frac{1}{2^n},$$

where

$$S_{k,u} \stackrel{\text{def}}{=} \left\{n \in \mathbb{N} : u < q_n \leqslant x_k\right\}.$$

Choose an integer $m_0 \geqslant 1$ such that

$$\sum_{n \geqslant m_0} \frac{1}{2^n} < \varepsilon$$

and consider the set $A \subseteq \mathbb{R}$, defined by

$$A = \bigcap_{\{n \in \mathbb{N}:\, n < m_0,\, q_n > u\}} (u, q_n).$$

Evidently $A = (u, q)$ for some $q \in \mathbb{Q}$. Moreover, every rational number in A is of the form q_n, with $n \geqslant m_0$. Because $x_k \longrightarrow u^+$, we can find an integer $k_0 \geqslant 1$ such that

$$x_k \in A \qquad \forall\, k \geqslant k_0.$$

Therefore

$$\left|f(x_k) - f(u)\right| < \varepsilon \qquad \forall\, k \geqslant k_0$$

and so f is right continuous at u.

Solution of Problem 1.18

Let D be any countable subspace of l^∞. Let $\{x^n\}_{n \geqslant 1}$ be the sequence of all elements of D. In particular, for all $n \geqslant 1$, we have $x^n = \{x_k^n\}_{k \geqslant 1} \in l^\infty$. We define a new real sequence $u = \{u_n\}_{n \geqslant 1}$ as follows

$$u_n = \begin{cases} 1 & \text{if } x_n^n < 0, \\ -1 & \text{if } x_n^n \geqslant 0. \end{cases}$$

Then $u = \{u_n\}_{n \geqslant 1} \in l^\infty$ and $\operatorname{dist}(u, D) \geqslant 1$ (see Definition 1.6(d)). So D is not dense in l^∞ (see Definition 1.20) and since D was an arbitrary countable subset of l^∞, we conclude that l^∞ is not separable (see Definition 1.21).

Solution of Problem 1.19

Let $D = \{x_n\}_{n \geqslant 1}$ be a countable dense subset of X (see Definitions 1.20 and 1.21). Consider the following collection of open balls $\{B_{\frac{1}{k}}(x_n)\}_{n,k \geqslant 1}$. This collection is countable. Let $\{\hat{B}_m\}_{m \geqslant 1}$ be an enumeration of this countable collection. For every $x \in X$, let

$$L_x = \{m \in \mathbb{N} : x \in \hat{B}_m\}.$$

Clearly the function

$$X \ni x \longmapsto L_x \in 2^{\mathbb{N}}$$

is an injection. Therefore

$$\operatorname{card} X \leqslant \operatorname{card} 2^{\mathbb{N}} = \mathfrak{c}.$$

Solution of Problem 1.20

Evidently X is infinite (as it is unbounded; see Definition 1.6(b)). Let $u \in X$ and for every $n \geqslant 1$, define

$$E_n = \{x \in X : d_X(x, u) > n\}.$$

Since X is unbounded, we have $E_n \neq \emptyset$ for all $n \geqslant 1$. Let $x_n \in E_n$ for every $n \geqslant 1$. Let $y \in X$ and let $m \geqslant 1$ be an integer such that

$$d_X(u, y) + 1 \leqslant m.$$

Then for all $n \geqslant m$, we have

$$1 \leqslant n + 1 - m \leqslant d_X(x_n, u) - d_X(u, y) \leqslant d_X(x_n, y),$$

so the sequence $\{x_n\}_{n \geqslant 1}$ does not converge to y. Since $y \in X$ was arbitrary, we infer that the sequence $\{x_n\}_{n \geqslant 1}$ has no convergent subsequence.

Solution of Problem 1.21
First we show that $\varphi^{-1}(0) = \{0\}$. Indeed, from the concavity of φ, we have that

$$\varphi(u+v) - \varphi(v) \leqslant \varphi(u) - \varphi(0) = \varphi(u) \qquad \forall\, u, v > 0.$$

So, if $\varphi(u) = 0$ for some u, then by the fact that φ is increasing, we have

$$\varphi(u+v) = \varphi(v) \qquad \forall\, v > 0.$$

Let $v = u$. Then $\varphi(2u) = 0$ and so inductively

$$\varphi(ku) = 0 \qquad \forall\, k \geqslant 0,$$

from which we conclude that $\varphi \equiv 0$, a contradiction (since φ is non-trivial). Hence $\varphi(u) > 0$ for all $u > 0$ and so $\varphi^{-1}(0) = \{0\}$. This implies that

$$\hat{d}_X(x,y) = 0 \iff x = y.$$

Also, clearly we have

$$\hat{d}_X(x,y) = \hat{d}_X(y,x) \qquad \forall\, x, y \in X.$$

Finally, for $x, y, z \in X$, by the concavity of φ, we have

$$\varphi\big(d_X(x,y)+d_X(y,z)\big)-\varphi\big(d_X(y,z)\big) \leqslant \varphi\big(d_X(x,y)\big)-\varphi(0) = \varphi\big(d_X(x,y)\big)$$

and

$$\varphi\big(d_X(x,z)\big) \leqslant \varphi\big(d_X(x,y) + d_X(y,z)\big)$$

(since φ is increasing). Therefore

$$\varphi\big(d_X(x,z)\big) \leqslant \varphi\big(d_X(x,y)\big) + \varphi\big(d_X(y,z)\big)$$

so

$$\hat{d}_X(x,z) \leqslant \hat{d}_X(x,y) + \hat{d}_X(y,z),$$

i.e., $\hat{d}_X$ satisfies the triangle inequality. We conclude that $\hat{d}_X$ is a metric on X.

Solution of Problem 1.22

(a) We assume that E and C are both open subsets of X (the proof is similar if we assume that both are closed). Let $U \subseteq Y$ be an open set. Then

$$g^{-1}(U) \cap E = f_1^{-1}(U) \quad \text{and} \quad g^{-1}(U) \cap C = f_2^{-1}(U).$$

Because f_1 and f_2 are both continuous, the set $f_1^{-1}(U)$ is open in E and the set $f_2^{-1}(U)$ is open in C. Since E and C are both open in X, it follows that the sets $g^{-1}(U) \cap E$ and $g^{-1}(U) \cap C$ are both open in X. But

$$\begin{aligned} g^{-1}(U) &= g^{-1}(U) \cap X = g^{-1}(U) \cap (E \cup C) \\ &= \big(g^{-1}(U) \cap E\big) \cup \big(g^{-1}(U) \cap C\big), \end{aligned}$$

so $g^{-1}(U)$ is also open and thus g is continuous.

(b) If we drop the hypothesis that E and C are both open (or closed), the result fails. Let

$$X = \mathbb{R}, \quad E = \mathbb{Q} \quad \text{and} \quad C = \mathbb{R} \setminus \mathbb{Q}.$$

We consider the functions $f_1 \colon E \longrightarrow \mathbb{R}$ and $f_2 \colon C \longrightarrow \mathbb{R}$, defined by

$$\begin{aligned} f_1(x) &= 1 \qquad \forall\, x \in E, \\ f_2(x) &= 0 \qquad \forall\, x \in C. \end{aligned}$$

Evidently both f_1 and f_2 are continuous. Then

$$g(x) = \begin{cases} 1 & \text{if} \quad x \in \mathbb{Q}, \\ 0 & \text{if} \quad x \in \mathbb{R} \setminus \mathbb{Q}, \end{cases}$$

and this function is discontinuous everywhere.

Solution of Problem 1.23

By Proposition 1.24, we know that X is second countable. So, let $\mathcal{B} = \{U_n\}_{n \geqslant 1}$ be a basis for the metric topology on X. Every open set in X is the countable union of elements in $\mathcal{B}$, hence $\operatorname{card} \mathcal{Y}_c \leqslant \mathfrak{c}$. Finally note that the function $\mathcal{Y}_0 \ni U \longmapsto U^c \in \mathcal{Y}_c$ is a bijection.

Solution of Problem 1.24

(a) Let A be the set of accumulation points of the sequence $\{x_n\}_{n\geqslant 1}$ (see Definition 1.13(b) and Theorem 1.14(d)). We have

$$A = \bigcap_{n\geqslant 1} \overline{\{x_k : k \geqslant n\}}.$$

The set A is closed and since $u_n \in A$ for all $n \geqslant 1$, we have $u \in A$.

(b) This comes immediately from **(a)** and Proposition 1.11.

Solution of Problem 1.25

Let $\{f_n\}_{n\geqslant 1} \subseteq B(E)$ be a d^∞-Cauchy sequence. So, for a given $\varepsilon > 0$, we can find $n_0 = n_0(\varepsilon) \geqslant 1$ such that

$$d^\infty(f_n, f_k) \leqslant \varepsilon \qquad \forall\, n, k \geqslant n_0,$$

so

$$\big|f_n(s) - f_k(s)\big| \leqslant \varepsilon \qquad \forall\, n, k \geqslant n_0,\ s \in E \tag{1.3}$$

and thus $\big\{f_n(s)\big\}_{n\geqslant 1} \subseteq \mathbb{R}$ is a Cauchy sequence for all $s \in [0,1]$.

So $f_n(s) \longrightarrow f(s)$ for all $s \in [0,1]$, for some $f(s) \in \mathbb{R}$. Then $f\colon E \longrightarrow \mathbb{R}$ is a function and we will show that $f \in B(E)$. If in (1.3) we let $k \to +\infty$, we have

$$\big|f_n(s) - f(s)\big| \leqslant \varepsilon \qquad \forall\, n \geqslant n_0,\ s \in E \tag{1.4}$$

so

$$\big|f(s)\big| \leqslant \sup_{s\in E}\big|f_n(s)\big| + \varepsilon \qquad \forall\, n \geqslant n_0,\ s \in E$$

and thus $f \in B(E)$.

From (1.4), it is clear that $d^\infty(f_n, f) \longrightarrow 0$ and so we conclude that $\big(B(E), d^\infty\big)$ is a complete metric space.

Solution of Problem 1.26

Let $x \in L$ (see Definition 1.135(b)) and let U_x be the open set containing x such that

$$f(x) < f(u) \qquad \forall\, u \in U_x \setminus \{x\}.$$

Since X is separable (see Definition 1.21), it is second countable (see Proposition 1.24) and so it has a countable basis $\mathcal{B}$. We can find at least one $B_x \in \mathcal{B}$ such that $x \in B_x \subseteq U_x$. Let $\xi\colon L \longrightarrow \mathcal{B}$ be the function which assigns to $x \in L$ this set $B_x \in \mathcal{B}$. This function is injective, since if $x, y \in L$, $x \neq y$ and $\xi(x) = \xi(y)$, then $x, y \in U_x \cap U_y$ and so $f(x) < f(y)$, $f(y) < f(x)$, a contradiction. This proves the injectivity of ξ, which in turn by the countability of the basis $\mathcal{B}$, implies that L is at most countable.

Solution of Problem 1.27

(a) Since $\operatorname{diam} C_n \longrightarrow 0$ and $\bigcap\limits_{n\geqslant 1} C_n \neq \emptyset$, we have that

$$\bigcap_{n\geqslant 1} C_n = \{x\}.$$

Since f is continuous, for a given $\varepsilon > 0$, we can find $\delta > 0$ such that

$$d_X(u,x) \leqslant \delta \quad\Longrightarrow\quad d_Y\big(f(u), f(x)\big) \leqslant \varepsilon.$$

Also, we can find $n_0 \geqslant 1$ such that

$$d_X(u,x) \leqslant \delta \qquad \forall\, n \geqslant n_0,\ u \in C_n.$$

Hence for $n \geqslant n_0$, we have that

$$f(C_n) \subseteq \overline{B}^Y_\varepsilon\big(f(x)\big) = \big\{y \in Y :\ d_Y\big(y, f(x)\big) \leqslant \varepsilon\big\},$$

so

$$\operatorname{diam} f(C_n) \leqslant 2\varepsilon \qquad \forall\, n \geqslant n_0.$$

Since $\varepsilon > 0$ was arbitrary, we conclude that

$$\operatorname{diam} f(C_n) \longrightarrow 0 \quad \text{as } n \to +\infty.$$

(b) If we drop the hypothesis that $\bigcap\limits_{n\geqslant 1} C_n \neq \emptyset$, then the conclusion of the problem fails. To see this let $X = (0,1]$ and $Y = \mathbb{R}$ with the usual metric. Let $C_n = \big(0, \frac{1}{n}\big]$ and

$$f(x) = \sin \tfrac{1}{x} \qquad \forall\, x \in X.$$

Then $\operatorname{diam} C_n \longrightarrow 0$, but for each $n \geqslant 1$, $f(C_n) = [-1, 1]$ and so

$$\operatorname{diam} f(C_n) = 2 \qquad \forall\, n \geqslant 1.$$

Solution of Problem 1.28
By definition (see Definition 1.13(c)), we have

$$\partial D = \overline{D} \cap \overline{D^c}.$$

Therefore without any loss of generality we may assume that D is closed. We need to show that $\operatorname{int} \partial D = \emptyset$ (see Definition 1.25). Arguing by contradiction, suppose that $\operatorname{int} \partial D \neq \emptyset$ and let $x \in \operatorname{int} \partial D$. Then we can find $r > 0$ such that

$$B_r(x) \subseteq \partial D = D \cap \overline{D^c} \subseteq D.$$

Hence

$$B_r(x) \cap D^c = \emptyset,$$

a contradiction to the fact that $x \in \overline{D^c}$.

For a general nonempty subset D of X, this is no longer true. Consider $X = \mathbb{R}$ and $D = \mathbb{Q}$. Then $\partial D = \mathbb{R}$.

Solution of Problem 1.29
Let $D \subseteq X$ be a nowhere dense set (see Definition 1.25). We know that $\operatorname{int} D \subseteq \operatorname{int} \overline{D} = \emptyset$ (see Proposition 1.15). Hence $(\operatorname{int} D)^c = X$ and so $\overline{D^c} = X$, which means that D^c is dense in X and $X \setminus D^c$ is nowhere dense (see Definition 1.20).

The converse is not true, as the set of rationals $\mathbb{Q}$ has a dense complement in $\mathbb{R}$ but is not nowhere dense.

Solution of Problem 1.30

"$\Longrightarrow$": Suppose that C is nowhere dense (see Definition 1.25). Then C^c is dense in X (see Definition 1.20 and Problem 1.29). So, for every nonempty open set U, we have $U \setminus C = U \cap C^c \neq \emptyset$. The set $U \cap C^c$ is open. So, it contains a ball.

"$\Longleftarrow$": Every nonempty open set intersects C^c, hence C^c is dense in X, which in turn implies that C is nowhere dense (because the set C is closed).

Solution of Problem 1.31

To prove the claim of the problem, we will construct an injection of $\{0,1\}^{\mathbb{N}}$ into D. Because D is nonempty and perfect (see Definition 1.13(e)), D is infinite. Let $u_0 \neq u_1$ in D and let us set $\varepsilon_1 = \min\left\{\frac{1}{2}, \frac{1}{3}d_X(u_0,u_1)\right\}$. We define

$$\begin{aligned} D(0) &= \left\{u \in D : \ d_X(u,u_0) \leqslant \varepsilon_1\right\}, \\ D(1) &= \left\{u \in D : \ d_X(u,u_1) \leqslant \varepsilon_1\right\}. \end{aligned}$$

These are disjoint infinite and closed subsets of D such that $\operatorname{diam} D_0 \leqslant 1$ and $\operatorname{diam} D_1 \leqslant 1$. Now, let $n \geqslant 1$ and for each n-tuple $(a_1, \ldots a_n) \in \{0,1\}^n$, suppose that we have infinitely many closed subsets $D(a_1,\ldots,a_n)$ of D with diameter less or equal to $\frac{1}{n}$ and that these sets are pairwise disjoint. We choose $u(a_1,\ldots,a_n,0) \neq u(a_1,\ldots,a_n,1)$, both in $D(a_1,\ldots,a_n)$ and we set

$$\varepsilon_{n+1} = \min\left\{\tfrac{1}{2(n+1)}, \ \tfrac{1}{3}d_X\big(u(a_1,\ldots,a_n,0), u(a_1,\ldots,a_n,1)\big)\right\}.$$

For $k = 0, 1$, we define

$$D(a_1,\ldots,a_n,k) = \left\{u \in D(a_1,\ldots,a_n) : d_X\big(u(a_1,\ldots,a_n,k),u\big) \leqslant \varepsilon_{n+1}\right\}.$$

Then the family $\left\{D(a_1,\ldots,a_{n+1}) : \ (a_1,\ldots,a_{n+1}) \in \{0,1\}^{n+1}\right\}$ consists of pairwise disjoint infinite closed subsets of D, each with diameter less or equal to $\frac{1}{n+1}$. Therefore, for every $\hat{a} = \{a_n\}_{n\geqslant 1} \in \{0,1\}^{\mathbb{N}}$, we have a decreasing sequence $\left\{D(a_1,\ldots,a_n)\right\}_{n\geqslant 1}$ of infinite closed subsets of D whose diameter tend to 0 as $n \to +\infty$. Since (X, d_X) is complete, by the Cantor intersection theorem (see Theorem 1.28), we have

$$\bigcap_{n\geqslant 1} D(a_1,\ldots,a_n) = \{u(\hat{a})\}.$$

If $\hat{a}\neq\hat{b}$, $\hat{a},\hat{b}\in\{0,1\}^{\mathbb{N}}$, then for some $n_0\in\mathbb{N}$, we have $a_{n_0}\neq b_{n_0}$ and so

$$u(\hat{a}) \in D(a_1,\ldots,a_{n_0}) \quad\text{but}\quad u(\hat{b}) \notin D(a_1,\ldots,a_{n_0}).$$

Therefore $u(\hat{a})\neq u(\hat{b})$ and so the function $\hat{a}\longrightarrow u(\hat{a})$ is a bijection from $\{0,1\}^{\mathbb{N}}$ into A. So, we conclude that $\operatorname{card} D\geqslant \operatorname{card}\{0,1\}^{\mathbb{N}}=\mathfrak{c}$.

Solution of Problem 1.32

Proceeding by contradiction, suppose that $[0,1]$ is countable and put its elements in a sequence $\{x_n\}_{n\geqslant 1}$. Let

$$C_n = \{x_n\} \qquad \forall\, n\geqslant 1.$$

Then each C_n is nowhere dense (see Definition 1.25) and

$$[0,1] = \bigcup_{n\geqslant 1} C_n.$$

But $[0,1]$ is a complete metric space, a contradiction (see Theorem 1.26).

Solution of Problem 1.33

Suppose that no point is isolated (see Definition 1.13). Then each singleton is a closed nowhere dense set and their union is the whole space, which being complete is of second category (see Definition 1.25), a contradiction (see Theorem 1.26).

Solution of Problem 1.34

Let $U\subseteq X$ be a nonempty open set. We have

$$U = \bigcup_{n\geqslant 1}(C_n\cap U)$$

and $\{C_n \cap U\}_{n\geqslant 1}$ is a sequence of sets closed in U. Since U is open and X is a complete metric space, by Theorem 1.26, we can find at least one $n_0 \geqslant 1$ such that

$$\operatorname{int}(C_{n_0} \cap U) = \operatorname{int}_U(C_{n_0} \cap U) \neq \emptyset.$$

Hence

$$\Big(\bigcup_{n\geqslant 1} \operatorname{int} C_n\Big) \cap U \neq \emptyset$$

and since U was arbitrary open set in X, we conclude that $\bigcup_{n\geqslant 1} \operatorname{int} C_n$ is dense in X (see Definition 1.20).

Solution of Problem 1.35

First we show that C is a closed subset of l^∞. To this end let $\{\hat{x}^n\}_{n\geqslant 1} \subseteq C$ be a sequence such that

$$\hat{x}^n \longrightarrow \hat{x} \in l^\infty \quad \text{in } (l^\infty, d^\infty).$$

We have

$$\hat{x}^n = \{x_m^n\}_{m\geqslant 1} \quad \text{and} \quad \hat{x} = \{x_m\}_{m\geqslant 1}.$$

Let $\varepsilon > 0$. As $\{\hat{x}^n\}_{n\geqslant 1}$ is a Cauchy sequence, there exists $k_0 = k_0(\varepsilon) \geqslant 1$ such that

$$d^\infty(\hat{x}^m, \hat{x}^k) < \tfrac{\varepsilon}{3} \qquad \forall\, m, k \geqslant k_0.$$

So, in particular

$$|x_n^m - x_n^k| < \tfrac{\varepsilon}{3} \qquad \forall\, m, k \geqslant k_0,\ n \geqslant 1.$$

Next, as $\hat{x}^n \longrightarrow \hat{x}$, we can find $n_0 = n_0(\varepsilon) \geqslant 1$ such that

$$|x_m^n - x_m| < \tfrac{\varepsilon}{3} \qquad \forall\, n \geqslant n_0,\ m \geqslant 1.$$

Then for $m \geqslant k \geqslant k_0$ and for fixed $n \geqslant n_0$, we have

$$|x_m - x_k| \leqslant |x_m - x_m^n| + |x_m^n - x_k^n| + |x_k^n - x_k| < \varepsilon.$$

But $\varepsilon > 0$ was arbitrary. Therefore $\{x_m\}_{m\geqslant 1}$ is a Cauchy sequence and so $\hat{x} \in C$. Thus C is a closed set.

To show that C is nowhere dense in l^∞ (see Definition 1.25), it suffices to show that $l^\infty \setminus C$ is dense in l^∞ (see Definition 1.20). So, let $\hat{x} \in l^\infty$ and $\varepsilon > 0$. We need to show that

$$B_\varepsilon(\hat{x}) \cap (l^\infty \setminus C) \neq \emptyset.$$

If $\hat{x} \in l^\infty \setminus C$, then since the latter is open, the above intersection is indeed nonempty. Hence the interesting case is when $\hat{x} \in C$. Then $\hat{x} = \{x_n\}_{n\geqslant 1}$ and $x_n \longrightarrow b$. We can find $n_0 = n_0(\varepsilon) \geqslant 1$ such that

$$|x_n - b| \leqslant \tfrac{\varepsilon}{2} \qquad \forall\, n \geqslant n_0.$$

Choose $\hat{u} = \{u_n\}_{n\geqslant 1} \in l^\infty$ as follows

$$u_n = \begin{cases} x_n & \text{if } n < n_0, \\ b + \frac{\varepsilon}{2} & \text{if } n \geqslant n_0 \text{ and } n \text{ is odd}, \\ b - \frac{\varepsilon}{2} & \text{if } n \geqslant n_0 \text{ and } n \text{ is even}. \end{cases}$$

Then, we have

$$d^\infty(\hat{x}, \hat{u}) = \sup_{n\geqslant 1} |x_n - u_n| < \varepsilon.$$

Therefore $\hat{u} \in B_\varepsilon(\hat{x})$. Note that

$$\limsup_{n\to+\infty} u_n = b + \tfrac{\varepsilon}{2} \quad \text{and} \quad \liminf_{n\to+\infty} u_n = b - \tfrac{\varepsilon}{2},$$

so $\hat{u} = \{u_n\}_{n\geqslant 1} \in l^\infty \setminus C$ and thus $B_\varepsilon(\hat{x}) \cap (l^\infty \setminus C) \neq \emptyset$, which proves that $l^\infty \setminus C$ is dense in l^∞.

Solution of Problem 1.36

"(a) $\Longrightarrow$ (b)": Let $\{u_n\}_{n\geqslant 1}, \{x_n\}_{n\geqslant 1} \subseteq X$ be two sequences such that $d_X(u_n, x_n) \longrightarrow 0$. Let $\varepsilon > 0$. Since f is uniformly continuous (see Definition 1.45), we can find $\delta = \delta(\varepsilon) > 0$ such that

$$d_X(u, x) \leqslant \delta \quad \Longrightarrow \quad d_Y\big(f(u), f(x)\big) \leqslant \varepsilon.$$

We can find $n_0 \geqslant 1$ such that

$$d_X(u_n, x_n) \leqslant \delta \qquad \forall\, n \geqslant n_0.$$

Hence

$$d_Y\big(f(u_n), f(x_n)\big) \leqslant \varepsilon \qquad \forall\, n \geqslant n_0$$

and so we conclude that

$$d_Y\big(f(u_n), f(x_n)\big) \longrightarrow 0.$$

"**(b)** $\Longrightarrow$ **(a)**": Suppose that f is not uniformly continuous. Then

$$\exists \varepsilon > 0\ \forall \delta > 0\ \exists u, x \in X : \ d_X(u, x) \leqslant \delta \text{ and } d_Y\big(f(u), f(x)\big) > \varepsilon.$$

So, for any $n \geqslant 1$, we can find $u_n, x_n \in X$ such that

$$d_X(u_n, x_n) \leqslant \tfrac{1}{n} \quad \text{and} \quad d_Y\big(f(u_n), f(u_n)\big) > \varepsilon \qquad \forall\, n \geqslant 1.$$

Thus we have obtained two sequences $\{u_n\}_{n\geqslant 1}, \{x_n\}_{n\geqslant 1} \subseteq X$ such that $d_X(u_n, x_n) \longrightarrow 0$, but $d_Y\big(f(u_n), f(x_n)\big) \not\longrightarrow 0$, a contradiction.

Solution of Problem 1.37

(a) Let $\varepsilon > 0$. Since f is uniformly continuous (see Definition 1.45), we can find $\delta = \delta(\varepsilon) > 0$ such that

$$|t - s| \leqslant \delta \quad \Longrightarrow \quad \big|f(t) - f(s)\big| \leqslant \varepsilon.$$

Let $|r| \leqslant \delta$. We have

$$|t + r - t| = |r| \leqslant \delta$$

and so

$$\big|f_r(t) - f(t)\big| = \big|f(t+r) - f(t)\big| \leqslant \varepsilon,$$

so

$$d^\infty(f_r, f) \leqslant \varepsilon \qquad \forall\, |r| \leqslant \delta.$$

This proves that $d^\infty(f_r, f) \longrightarrow 0$ as $r \to 0^+$.

(b) No. To see this, let

$$S_n = \sum_{k=1}^{n} \tfrac{1}{k} \qquad \forall\, n \geqslant 1.$$

We have that $S_n \longrightarrow +\infty$ (as these are the partial sums of the harmonic series). Let $f\colon \mathbb{R} \longrightarrow \mathbb{R}$ be the function, defined by

$$f(t) = \begin{cases} 0 & \text{if } t < 1, \\ \sin(n+1)\pi(t - S_n) & \text{if } t \in [S_n, S_{n+1}),\ n \geqslant 1. \end{cases}$$

The function f is continuous, but not uniformly continuous.

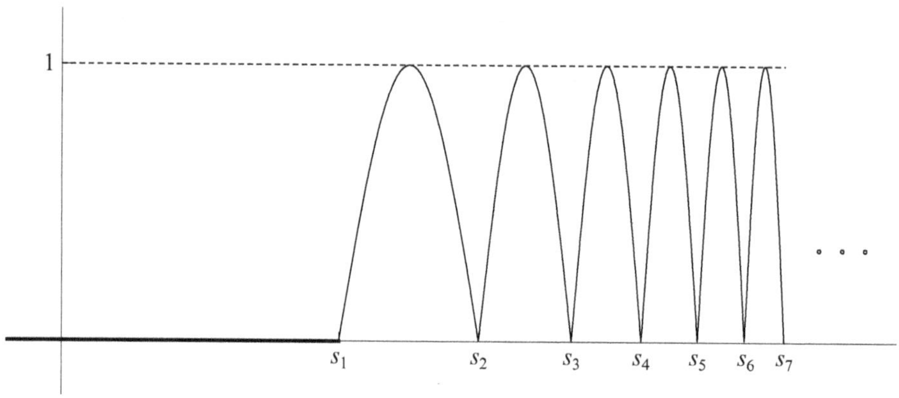

We have

$$\begin{aligned} d^\infty(f_{\frac{1}{2n}}, f) &= \sup_{t\in\mathbb{R}} \left|f_{\frac{1}{2n}}(t) - f(t)\right| \geqslant f_{\frac{1}{2n}}(S_n) - f(S_n) \\ &= \sin n\pi(S_n + \tfrac{1}{2n} - S_n) - \sin n\pi(S_n - S_n) \\ &= \sin \tfrac{\pi}{2} - \sin 0 = 1. \end{aligned}$$

So, in particular $d^\infty(f_r, f) \not\longrightarrow 0$ as $r \to 0^+$.

Solution of Problem 1.38

Let D be a countable dense subset of X (see Definitions 1.20 and 1.21). Then by the continuity of f, we have $f(X) = f(\overline{D}) \subseteq \overline{f(D)}$ (see Proposition 1.32(d)). Note that $f(D)$ is countable. Therefore $f(X)$ is separable.

Solution of Problem 1.39

No. Let $X = \mathbb{R}$, $Y = (-1,1)$ and let $f\colon X \longrightarrow Y$ be the function, defined by

$$f(x) = \begin{cases} \frac{x}{1-x} & \text{if } x \in (-\infty, 0), \\ \frac{x}{1+x} & \text{if } x \in [0, +\infty). \end{cases}$$

Then f is a bijective and

$$f^{-1}(y) = h(x) = \begin{cases} \frac{y}{1+y} & \text{if } y \in (-1, 0), \\ \frac{y}{1-y} & \text{if } y \in [0, 1). \end{cases}$$

Note that f and h are both continuous and so f is a homeomorphism. Also for $x, u \geqslant 0$:

$$\left|f(x) - f(u)\right| = \left|\tfrac{x}{1+x} - \tfrac{u}{1+u}\right| = \tfrac{|x-u|}{(1+x)(1+u)} \leqslant |x-u|,$$

for $x, u \leqslant 0$:

$$\left|f(x) - f(u)\right| = \left|\tfrac{x}{1-x} - \tfrac{u}{1-u}\right| = \tfrac{|x-u|}{(1-x)(1-u)} \leqslant |x-u|,$$

and for $u \leqslant 0 \leqslant x$:

$$\left|f(x) - f(u)\right| = \left|\tfrac{x}{1+x} - \tfrac{u}{1-u}\right| = \tfrac{|x-u-2xu|}{(1+x)(1-u)} \leqslant \tfrac{(x-u)(1+x-u)}{(1+x)(1-u)} \leqslant |x-u|.$$

This proves that f is Lipschitz continuous, hence uniformly continuous (see Definition 1.45). However, note that $X = \mathbb{R}$ is complete, but $f(X) = Y = (-1, 1)$ is not.

Solution of Problem 1.40

(a) Consider the function $f\colon \mathbb{R}^2 \longrightarrow \mathbb{R}$, defined by

$$f(u_1, u_2) = u_1.$$

Evidently f is continuous and open (as the projection function on the first coordinate), but it is not closed. Consider the closed set

$$C = \{u = (u_1, u_2) \in \mathbb{R}^2 : u_1 u_2 = 1\}$$

(a hyperbola). Then $f(C) = \mathbb{R} \setminus \{0\}$ which is not closed.

(b) Consider the function $f\colon \mathbb{R} \longrightarrow \mathbb{R}$, defined by $f(x) = c$, where $c \in \mathbb{R}$ (constant function). Then f is continuous and closed, but it is not open.

(c) Let $f\colon \mathbb{R} \longrightarrow \{0,1\}$ (where on $\{0,1\}$ we have the discrete metric) be the function, defined by

$$f(x) = \begin{cases} 1 & \text{if } x \in \mathbb{Q}, \\ 0 & \text{if } x \in \mathbb{R}\setminus\mathbb{Q}. \end{cases}$$

Then f is open and closed, but it is not continuous.

(d) Consider the function $f\colon \mathbb{R} \longrightarrow \mathbb{R}$, defined by $f(x) = e^{-x}$. Then f is continuous. However, f is not closed. Indeed let $C = [0,+\infty)$. Then C is closed in $\mathbb{R}$, but $f(C) = (0,1]$ is not closed in $\mathbb{R}$.

Also, f is not open. Let $U = (-1,1)$. Then U is open in $\mathbb{R}$, but $f(U) = \left(\frac{1}{e}, 1\right]$ is not open in $\mathbb{R}$.

Solution of Problem 1.41

"$\Longrightarrow$": Let $f\colon X \longrightarrow Y$ be an open injection (see Definition 1.39). We claim that $f^{-1}\colon f(X) \longrightarrow X$ is continuous. Let $U \subseteq X$ be a nonempty open set. Since f is open, the set $f(U)$ is open in Y, hence in $f(X)$ too (see Definition 1.12). But

$$f(U) = (f^{-1})^{-1}(U).$$

So, indeed f^{-1} is continuous (see Proposition 1.32). Then for every nonempty closed set $C \subseteq X$, we have that the set $f(C) = (f^{-1})^{-1}(C)$ is closed (due to the continuity of $f^{-1}\colon f(X) \longrightarrow X$; again see Proposition 1.32) and this proves that f is closed.

"$\Longleftarrow$": The proof is similar, reversing the roles of open and closed sets.

Solution of Problem 1.42

Suppose that

$$d_X\big(f^{(k)}(x),\ f^{(k)}(u)\big) \leqslant c d_X(x,u) \qquad \forall\, x,u \in X,$$

with $0 \leqslant c < 1$. Then by the Banach fixed point theorem (see Theorem 1.49), we can find a unique $x \in X$ such that

$$f^{(k)}(x) = x.$$

We have

$$\begin{aligned} d_X\big(f(x), x\big) &= d_X\big(f\big(f^{(k)}(x)\big),\ f^{(k)}(x)\big) \\ &= d_X\big(f^{(k)}\big(f(x)\big),\ f^{(k)}(x)\big) \leqslant c d_X\big(f(x), x\big), \end{aligned}$$

so

$$0 \leqslant (1-c) d_X\big(f(x), x\big) \leqslant 0$$

and thus

$$d_X\big(f(x), x\big) = 0,$$

i.e., $f(x) = x$.

So, $x \in X$ is a fixed point of f too.

Finally, if $f(u) = u$, then $f^{(k)}(u) = u$ and from the uniqueness of x, we have $u = x$.

Solution of Problem 1.43

Let

$$l^1 = \Big\{\hat{u} = \{u_n\}_{n \geqslant 1} :\ u_n \in \mathbb{R},\ \sum_{n \geqslant 1} |u_n| < +\infty\Big\}$$

be equipped with the metric

$$d_1(\hat{u}, \hat{x}) = \sum_{n \geqslant 1} |u_n - x_n| \qquad \forall\ \hat{u}, \hat{x} \in l^1$$

(see Problem 1.6).

Let $\{a_n\}_{n \geqslant 2} \subseteq \mathbb{R}$ be a sequence such that $a_n \in (0,1)$ for all $n \geqslant 1$ and

$$\lim_{n \to +\infty} \pi_n = \tfrac{1}{2}, \qquad \text{where } \pi_n = \prod_{k=2}^{n} a_k \text{ for } n \geqslant 1.$$

For example, one can take $a_n = \frac{(n-1)(n+1)}{n^2}$ for all $n \geqslant 2$ and then, it is easy to check that

$$\pi_n = \frac{n+1}{2n} \qquad \forall\ n \geqslant 2.$$

Let $X = B_1^{l^1}(0)$ and consider the function $f\colon X \longrightarrow X$, defined by

$$f(u_1, u_2, u_3, \ldots) = (0, u_1^2, a_2 u_2, a_3 u_3, \ldots) \qquad \forall \hat{u} = (u_1, u_2, u_3, \ldots) \in X.$$

Then

$$\begin{aligned} d_Y\big(f(\hat{u}), f(\hat{x})\big) &= |u_1^2 - x_1^2| + \sum_{n\geqslant 2} a_n |u_n - x_n| \\ &= |u_1 - x_1| \cdot |u_1 + x_1| + \sum_{n\geqslant 2} a_n |u_n - x_n| \\ &\leqslant 2\sum_{n\geqslant 1} a_n |u_n - x_n| = 2d_X(\hat{u}, \hat{x}) \qquad \forall\, \hat{u}, \hat{x} \in X, \end{aligned}$$

so f is Lipschitz continuous with Lipschitz constant 2. Note that the constant 2 cannot be improved (decreased), as taking

$$\hat{u}^n = \big(1 - \tfrac{1}{n}, 0, 0, 0 \ldots\big) \quad \text{and} \quad \hat{x}^n = \big(1 - \tfrac{2}{n}, 0, 0, 0 \ldots\big) \qquad \forall\, n \geqslant 2,$$

we get

$$d_X\big(f(\hat{u}^n), f(\hat{x}^n)\big) = \big(2 - \tfrac{3}{n}\big) \cdot \tfrac{1}{n} = \big(2 - \tfrac{3}{n}\big) d_X(\hat{u}^n, \hat{x}^n) \qquad \forall\, n \geqslant 1.$$

Thus f is not nonexpansive (see Definition 1.48).

On the other hand, for all $n \geqslant 2$ and all $\hat{u}, \hat{x} \in X$, we have

$$\begin{aligned} d_X\big(f^{(n)}(\hat{u}), f^{(n)}(\hat{x})\big) &= \prod_{k=2}^{n} a_k |u_1^2 - x_1^2| + \sum_{s\geqslant 2} \prod_{k=2}^{s+n-1} a_k |u_s - x_s| \\ &\leqslant 2\sum_{s\geqslant 1} \prod_{k=2}^{n} a_k |u_s - x_s| = 2\sum_{s\geqslant 1} \pi_n |u_s - x_s| \\ &= 2\pi_n \sum_{s\geqslant 1} |u_s - x_s| = 2\pi_n d_X(\hat{u}, \hat{x}) \end{aligned}$$

and also

$$\lim_{n\to+\infty} 2\pi_n = 1,$$

so f is asymptotically nonexpansive function.

Solution of Problem 1.44
By the Banach fixed point theorem (see Theorem 1.49), for every $r \in M$, we can find a unique $u(r) \in X$ such that

$$u(r) \;=\; f\big(r, u(r)\big).$$

It remains to show that the function $r \longmapsto u(r)$ is continuous. So, let $r_n \longrightarrow r$. Then

$$\begin{aligned} d_X\big(u(r_n),\ u(r)\big) \;&=\; d_X\big(f(r_n, u(r_n)),\ f(r, u(r))\big) \\ &\leqslant\; d_X\big(f(r_n, u(r_n)),\ f(r_n, u(r))\big) \\ &\quad\; + d_X\big(f(r_n, u(r)),\ f(r, u(r))\big) \\ &\leqslant\; k d_X\big(u(r_n), u(r)\big) + d_X\big(f(r_n, u(r)),\ f(r, u(r))\big) \end{aligned}$$

(see Definition 1.48), so

$$d_X\big(u(r_n), u(r)\big) \;\leqslant\; \tfrac{1}{1-k} d_X\big(f(r_n, u(r)),\ f(r, u(r))\big) \;\longrightarrow\; 0$$

(from hypothesis *(i)*) and hence $u(r_n) \longrightarrow u(r)$. Thus the function $r \longmapsto u(r)$ is continuous.

Solution of Problem 1.45
Let $0 < x < u$. We have

$$\begin{aligned} \big|\ln f(u) - \ln f(x)\big| \;&=\; \Big|\int_x^u \frac{f'(s)}{f(s)}\,ds\Big| \leqslant \int_x^u \frac{|f'(s)|}{f(s)}\,ds \\ &\leqslant\; \int_x^u \frac{k}{s}\,ds = k\big|\ln u - \ln x\big|. \end{aligned}$$

On X we consider the metric

$$\hat{d}_X(u, x) \;=\; \big|\ln u - \ln x\big| \qquad \forall\, u, x \in X.$$

Then $(X, \hat{d}_X)$ is a complete metric space and from the above estimate, we see that f is k-contraction for the metric $\hat{d}_X$. So, by the Banach fixed point theorem (see Theorem 1.49), f has a unique fixed point.

Solution of Problem 1.46

"$\Longrightarrow$": Suppose that f is continuous at $x \in X$. Then for a given $\varepsilon > 0$, we can find $\delta = \delta(\varepsilon, x) \in (0, \varepsilon)$ such that

$$d_Y\big(f(x), f(u)\big) \;<\; \tfrac{\varepsilon}{2} \qquad \forall\, u \in B_\delta(x).$$

Hence, we have

$$d_Y\big(f(v), f(u)\big) \;<\; \varepsilon \qquad \forall\, u, v \in B_\delta(x)$$

(triangle inequality), so

$$\operatorname{diam} f\big(B_\delta(x)\big) \;\leqslant\; \varepsilon.$$

Because $\varepsilon > 0$ was arbitrary, letting $\varepsilon \searrow 0$, we conclude that $\omega_f(x) = 0$ (see Definition 1.37).

"$\Longleftarrow$": Suppose that $\omega_f(x) = 0$. The for a given $\varepsilon > 0$, we can find $\delta_\varepsilon > 0$ such that

$$\operatorname{diam} f\big(B_\delta(x)\big) \;\leqslant\; \varepsilon \qquad \forall\, \delta \in (0, \delta_\varepsilon).$$

Thus, if $d_X(x, u) < \delta$, then

$$d_Y\big(f(x), f(u)\big) \;\leqslant\; \operatorname{diam} f\big(B_\delta(x)\big) \;\leqslant\; \varepsilon$$

and this proves that f is continuous at $x \in X$.

Solution of Problem 1.47

It suffices to show that for every $\lambda \in \mathbb{R}$ the complement

$$D_\lambda = \big\{x \in X : \; \omega_f(x) \geqslant \lambda\big\}$$

is closed (see Definition 1.37). To this end let $\{x_n\}_{n\geqslant 1} \subseteq D_\lambda$ be a sequence such that $x_n \longrightarrow x$. Then for a given $\varepsilon > 0$, we can find $n_0 \in \mathbb{N}$ such that

$$d_X(x, x_n) \;\leqslant\; \tfrac{\delta}{2} \qquad \forall\, n \geqslant n_0,$$

so

$$B_{\frac{\delta}{2}}(x_n) \;\subseteq\; B_\delta(x) \qquad \forall\, n \geqslant n_0.$$

Hence

$$\operatorname{diam} f\big(B_\delta(x)\big) \;\geqslant\; \operatorname{diam} f\big(B_{\frac{\delta}{2}}(x_n)\big) \;\geqslant\; \omega_f(x_n) \;\geqslant\; \lambda \qquad \forall\, n \geqslant n_0.$$

Since $\delta > 0$ was arbitrary, we get that $\omega_f(x) \geqslant \lambda$, so $x \in D_\lambda$. Therefore D_λ is closed in X (see Proposition 1.11).

Solution of Problem 1.48

(a) Let $\{x_n\}_{n\geqslant 1} \subseteq X$ be a Cauchy sequence. Since by hypothesis f is uniformly continuous (see Definition 1.45), the sequence $\{f(x_n)\}_{n\geqslant 1}$ is a Cauchy sequence in Y (see Proposition 1.46(b)). By the completeness of Y, we have $f(x_n) \longrightarrow y$. Then the continuity of f^{-1} implies that

$$x_n \longrightarrow f^{-1}(y).$$

Therefore the Cauchy sequence converges in X and so X is complete.

(b) No. Let $X = (0, +\infty)$, $Y = \mathbb{R}$ and $f(x) = \ln x$. Then f and f^{-1} are continuous bijections and Y is complete, but X is not complete.

Solution of Problem 1.49

We know that

$$f\Big(\bigcap_{n\geqslant 1} C_n\Big) \subseteq \bigcap_{n\geqslant 1} f(C_n).$$

So, we need to show that the opposite inclusion also holds. To this end, let $y \in \bigcap\limits_{n\geqslant 1} f(C_n)$ and set

$$E_n = C_n \cap f^{-1}(y) \qquad \forall\, n \geqslant 1.$$

As C_n is closed and $f^{-1}(y)$ is closed (because $\{y\}$ is closed in X and use Proposition 1.32), then $E_n \subseteq X$ is closed for every $n \geqslant 1$ and

$$\operatorname{diam} E_n \leqslant \operatorname{diam} C_n \longrightarrow 0 \quad \text{as } n \to +\infty.$$

Since X is complete, we can apply Theorem 1.28 and have that

$$\bigcap_{n\geqslant 1} E_n = \{x_0\},$$

hence

$$f^{-1}(y) \cap \Big(\bigcap_{n \geqslant 1} C_n\Big) = \{x_0\}$$

and so $y \in f\Big(\bigcap_{n \geqslant 1} C_n\Big)$. This proves that

$$f\Big(\bigcap_{n \geqslant 1} C_n\Big) \supseteq \bigcap_{n \geqslant 1} f(C_n)$$

and so finally

$$f\Big(\bigcap_{n \geqslant 1} C_n\Big) = \bigcap_{n \geqslant 1} f(C_n).$$

Solution of Problem 1.50

Let $\{x_n\}_{n \geqslant 1}$ be a countable dense subset of a metric space (X, d_X) (see Definitions 1.20 and 1.21). Let

$$h_m(x) = \min\big\{1, d_X(x, x_m)\big\} \qquad \forall\, m \geqslant 1.$$

We consider the function $h\colon X \longrightarrow \mathcal{H}$, defined by

$$h(x) = \big(h_m(x)\big)_{m \geqslant 1} \qquad \forall\, x \in X.$$

Since each h_m is continuous, it follows that h is continuous. Suppose that $h(x) = h(y)$ and let $x_{n_k} \longrightarrow x$ (recall that $\{x_n\}_{n \geqslant 1}$ is dense in X). Then

$$d_X(y, x_{n_k}) = d_X(x, x_{n_k}) \longrightarrow 0 \quad \text{as } k \to +\infty$$

and so $x = y$, i.e., h is injective. Finally, we need to show that h^{-1} is continuous. So, suppose that $h(y_n) \longrightarrow h(y)$. For a given $\varepsilon \in (0,1)$, choose x_m such that

$$d_X(y, x_m) < \varepsilon.$$

Because $h(y_n) \longrightarrow h(y)$, we have

$$d_X(y_n, x_m) \longrightarrow d_X(y, x_m)$$

and so

$$d_X(y_n, x_m) < \varepsilon \qquad \forall\, n \geqslant n_0,$$

for some $n_0 \geqslant 1$, hence

$$d_X(y_n, y) \leqslant 2\varepsilon \qquad \forall\, n \geqslant n_0,$$

hence $y_n \longrightarrow y$ in X and this proves the continuity of h^{-1}. Therefore h is a homeomorphism into $\mathcal{H}$.

Solution of Problem 1.51

Since f is an increasing bijection, it is strictly increasing. Let $x_0 \in A$. Since A is dense in $\mathbb{R}$ (see Definition 1.20), we have

$$C = A \cap (-\infty, x_0) \neq \emptyset \quad \text{and} \quad f|_C < f(x_0).$$

Let

$$m(x_0) = \sup\{f(x) : x \in A, x < x_0\}.$$

Evidently

$$m(x_0) \leqslant f(x_0).$$

If $m(x_0) < f(x_0)$, then exploiting the density of B in $\mathbb{R}$, we can find $u \in B$ such that

$$m(x_0) < u < f(x_0).$$

Let $x \in A$ be such that $u = f(x)$. Then $x < x_0$ and so

$$u = f(x) \leqslant m(x_0),$$

a contradiction. So, $m(x_0) = f(x_0)$.

Similarly, we show that

$$f(x_0) = \inf\{f(x) : x \in A, x > x_0\}.$$

Now, let $\varepsilon > 0$ be given. We can find $x_1, x_2 \in A$ such that $x_1 < x_0 < x_2$ and

$$f(x_0) - \varepsilon < f(x_1) \leqslant f(x_2) \leqslant f(x_0) + \varepsilon,$$

so

$$f(x_0) - \varepsilon \leqslant f(x) \leqslant f(x_0) + \varepsilon \qquad \forall\, x \in [x_1, x_2] \cap A.$$

Thus we conclude that f is continuous at x_0, hence f is continuous.

Also f^{-1} is an increasing bijection from B to A and so from the first part f^{-1} is also continuous. So, we conclude that f is homeomorphism.

Solution of Problem 1.52

Let $x \in X$ and let $\{x_n\}_{n\geqslant 1} \subseteq D$ be a sequence such that $d_X(x_n, x) \longrightarrow 0$ (recall that D is dense in X; see Definition 1.20). Note that for all $n, m \geqslant 1$, we have

$$d_Y\big(f(x_n), f(x_m)\big) \leqslant kd_X(x_n, x_m)$$

and $d_X(x_n, x_m) \longrightarrow$ as $n, m \longrightarrow +\infty$. So $\big\{f(x_n)\big\}_{n\geqslant 1} \subseteq Y$ is a Cauchy sequence. Since Y is complete, the sequence $\big\{f(x_n)\big\}_{n\geqslant 1}$ is convergent. Let us set

$$\hat{f}(x) = \lim_{n\to+\infty} f(x_n).$$

Note that due to the Lipschitz property of f, this definition is independent of the approximating sequence $\{x_n\}_{n\geqslant 1} \subseteq D$. Indeed, if $\{x_n'\}_{n\geqslant 1} \subseteq D$ is another sequence such that $d_X(x_n', x) \longrightarrow 0$, then

$$d_Y\big(f(x_n), f(x_n')\big) \leqslant kd_X(x_n, x_n') \leqslant kd_X(x_n, x) + kd_X(x, x_n') \longrightarrow 0.$$

So, we have a function $\hat{f}\colon X \longrightarrow Y$. If $x, u \in X$ and $\{x_n\}_{n\geqslant 1}, \{u_n\}_{n\geqslant 1} \subseteq D$ are two sequences such that $d_X(x_n, x) \longrightarrow 0$ and $d_X(u_n, u) \longrightarrow 0$, then

$$d_Y\big(f(x_n), f(u_n)\big) \leqslant kd_X(x_n, u_n) \qquad \forall\, n \geqslant 1,$$

so

$$d_Y\big(\hat{f}(x), \hat{f}(u)\big) \leqslant kd_X(x, u),$$

i.e., $\hat{f}$ is k-Lipschitz and $\hat{f}|_D = f$.

Solution of Problem 1.53

Let (X, d_X) be a metric space and let us fix $x_0 \in X$. For every $x \in X$, we consider the function $\xi_x\colon X \longrightarrow \mathbb{R}$, defined by

$$\xi_x(u) = d_X(x, u) - d_X(u, x_0).$$

We have

$$|\xi_x(u)-\xi_x(v)|\leqslant|d_X(x,u)-d_X(x,v)|+|d_X(u,x_0)-d_X(v,x_0)|\leqslant 2d_X(u,v),$$

so ξ_x is uniformly continuous (in fact 2-Lipschitz; see Definition 1.45).

Also, from the triangle inequality, we have

$$|\xi_x(u)| \leqslant d_X(x,x_0) \qquad \forall\, u \in X$$

and thus ξ_x is bounded. Finally, note that

$$\begin{aligned}|\xi_x(u) - \xi_v(u)| &= |d_X(x,u) - d_X(u,x_0) - d_X(v,u) + d_X(u,x_0)| \\ &= |d_X(x,u) - d_X(v,u)| \leqslant d_X(x,v) \qquad \forall\, x,v,u \in X\end{aligned}$$

and

$$|\xi_x(v) - \xi_v(v)| = d_X(x,v) \qquad \forall\, x,v \in X.$$

Therefore

$$d^\infty(\xi_x,\xi_v) = \sup_{u\in X} |\xi_x(u) - \xi_v(u)| = d_X(x,v),$$

so $x \longmapsto \xi_x$ is an isometry.

Solution of Problem 1.54

Let $f\colon X \longrightarrow X$ be defined by

$$f(x) = \tfrac{rx}{1+\|x\|}.$$

Then

$$\|f(x)\| < r \qquad \forall\, x \in X.$$

So, the range of f is in B_r. Let $u \in B_r$. Then there exists a unique $x \in X$ such that $u = f(x)$; in fact we have $x = \frac{1}{r-\|u\|}u$. So, f is a bijection from X onto B_r and $g = f^{-1}$ is given by

$$g(u) = \tfrac{u}{r-\|u\|}.$$

Evidently f and $g = f^{-1}$ are both continuous. Therefore f is a homeomorphism.

Solution of Problem 1.55

Let d_Y be the metric on Y. For every $x \in \overline{D}$, let $\omega_f(x)$ be the oscillation of f at x (see Definition 1.37). We set

$$C = \{x \in \overline{D} : \omega_f(x) = 0\}.$$

We have

$$C = \bigcap \{x \in \overline{D} : \omega_f(x) < \tfrac{1}{n}\},$$

so C is a G_δ-set in X (see Problem 1.47).

We define a continuous function $g\colon C \longrightarrow Y$, which extends f, as follows. Let $x \in C$ and let $\{x_n\}_{n\geqslant 1} \subseteq D$ be a sequence such that $x_n \longrightarrow x$. Since $\omega_f(x) = 0$, it follows that $\{f(x_n)\}_{n\geqslant 1} \subseteq Y$ is a Cauchy sequence in (Y, d_Y). The completeness of (Y, d_Y) implies that the sequence $\{f(x_n)\}_{n\geqslant 1}$ is convergent. Let

$$g(x) = \lim_{n\to+\infty} f(x_n).$$

Then it is clear that g is well defined, continuous and $g\big|_D = f$. Therefore g is the desired extension of f.

Solution of Problem 1.56

Let $h = f^{-1}$. By Problem 1.55, we can find a G_δ-set $\hat{A} \subseteq X$ containing A and a continuous extension $\hat{f}\colon \hat{A} \longrightarrow Y$ of f. Similarly, we can find a G_δ-set $\hat{B} \subseteq Y$ containing B and a continuous extension $\hat{h}\colon \hat{B} \longrightarrow X$ of h. We introduce the sets

$$\begin{aligned} G(\hat{f}) &= \operatorname{Gr}\hat{f} = \{(x,y) \in \hat{A} \times Y : y = \hat{f}(x)\}, \\ G(\hat{h}) &= \{(x,y) \in X \times \hat{B} : x = \hat{h}(y)\}. \end{aligned}$$

Let proj_X and proj_Y be the projections on X and Y respectively. We define

$$\begin{aligned} A^* &= \operatorname{proj}_X\big(G(\hat{f}) \cap G(\hat{h})\big) = \{x \in \hat{A} : (x, \hat{f}(x)) \in G(\hat{f})\}, \\ B^* &= \operatorname{proj}_Y\big(G(\hat{f}) \cap G(\hat{h})\big) = \{y \in \hat{B} : (\hat{h}(y), y) \in G(\hat{h})\}. \end{aligned}$$

The set $G(\hat{h})$ is closed in $X \times \hat{B}$ and $\hat{B}$ is a G_δ-set in Y. Hence $G(\hat{h})$ is a G_δ-set too. Since $\hat{f}$ is continuous on the set G_δ-set $\hat{A}$, it follows that A^* is a G_δ-set too. Similarly, we show that B^* is a G_δ-set. Note that $f^* = \hat{f}\big|_{A^*}$ is a homeomorphism from A^* onto B^* and $f^*\big|_A = f$.

Solution of Problem 1.57

By hypothesis there exists a complete metric space (Y, d_Y) and a homeomorphism $h\colon A \longrightarrow Y$. For $x \in \overline{A}$, we set

$$h_r(x) \;=\; \operatorname{diam} h\big(A \cap B_r(x)\big).$$

If $h_r(x) < \frac{1}{n}$, then $\operatorname{diam} h(\overline{A} \cap B_{\frac{r}{2}}(x)) \leqslant \frac{1}{n}$ and so the sets

$$C_n \;=\; \big\{x \in \overline{A}:\; h_r(x) \geqslant \tfrac{1}{n} \text{ for all } r > 0\big\}$$

are closed in $\overline{A}$, hence closed in X. The set

$$D \;=\; \overline{A} \setminus \bigcup_{n\geqslant 1} C_n$$

is a G_δ set. We will show that $A = D$. If $x \in A$, then due to the continuity of h, we have that $x \notin C_n$ for all $n \geqslant 1$ and so $x \subset D$. This shows that $A \subseteq D$.

Next, let $x \in D$. Let $\{x_n\}_{n\geqslant 1} \subseteq A$ be a sequence such that $x_n \longrightarrow x$. For every $\varepsilon > 0$, we can find an integer $k \geqslant 1$ such that $\frac{1}{k} < \varepsilon$. Since $x \notin C_k$, we can find $r_k > 0$ such that

$$h_{r_k}(x) \;<\; \tfrac{1}{k} \;<\; \varepsilon.$$

We can find $n_0 \geqslant 1$ such that $x_n \in B_{r_k}(x)$ for all $n \geqslant n_0$. The definition of h_{r_k} implies that

$$d_Y\big(h(x_n), h(x_m)\big) \;<\; \varepsilon \qquad \forall\, n, m \geqslant n_0,$$

so $\big\{h(x_n)\big\}_{n\geqslant 1}$ is a Cauchy sequence in Y.

The completeness of Y implies that $h(x_n) \longrightarrow y \in Y$. Recall that h is a homeomorphism. So, we have $x_n \longrightarrow h^{-1}(y)$. Since $x_n \longrightarrow x$, it follows that

$$x \;=\; h^{-1}(y) \;\in\; h^{-1}(Y) \;=\; A.$$

This proves that $D \subseteq A$ and so we conclude that $D = A$ and we see that A is a dense G_δ-set in X.

Solution of Problem 1.58

We argue by contradiction. So, suppose that $\mathbb{Q}$ is a G_δ-set in $\mathbb{R}$ (see Definition 1.57), so

$$\mathbb{Q} = \bigcap_{n\geqslant 1} U_n \quad \text{with} \quad U_n \subseteq \mathbb{R} \text{ open.}$$

Then $\mathbb{R} \setminus \mathbb{Q} = \bigcup_{n\geqslant 1} U_n^c$ and U_n^c is closed for every $n \geqslant 1$. Because U_n^c contains only irrationals, it can not contain a nonempty interval and so

$$\operatorname{int} U_n^c = \emptyset \qquad \forall\, n \geqslant 1,$$

which means that U_n^c is nowhere dense (see Definition 1.25). Let $\{q_n\}_{n\geqslant 1}$ be an enumeration of the rationals and set

$$C_n = U_n^c \cup \{q_n\} \qquad \forall\, n \geqslant 1.$$

Then C_n is closed and

$$\operatorname{int} C_n = \operatorname{int} U_n^c \cup \operatorname{int} \{q_n\} = \emptyset,$$

i.e., C_n is nowhere dense. Note that $\mathbb{R} = \bigcup_{n\geqslant 1} C_n$, which contradicts the Baire category theorem (see Theorem 1.26).

Solution of Problem 1.59

Let $D = \{x_n\}_{n\geqslant 1}$ be a countable dense set (see Definition 1.20) in a complete metric space X. Suppose that D is a G_δ-set (see Definition 1.57). Then we can find open sets $U_n \subseteq X$ for $n \geqslant 1$ such that

$$D = \bigcap_{n\geqslant 1} U_n.$$

Evidently, since $D \subseteq U_n$ for all $n \geqslant 1$, each U_n is open and dense in X. Let us set

$$V_n = U_n \setminus \bigcup_{k=1}^{n} \{x_k\} \qquad \forall\, n \geqslant 1.$$

Since X has no isolated points (see Definition 1.13(d)), each V_n is open and dense. Note that $\bigcap_{n\geqslant 1} V_n = \emptyset$, which contradicts the Baire category theorem (see Theorem 1.26).

Solution of Problem 1.60

(a) "$\Longrightarrow$": Since by hypothesis the complement of A is meager (see Definition 1.25), then

$$A = \Big(\bigcup_{n\geqslant 1} D_n\Big)^c, \quad \text{with } D_n \text{ being nowhere dence.}$$

Hence, we have

$$A = \bigcap_{n\geqslant 1} D_n^c \supseteq \bigcap_{n\geqslant 1} \big(\overline{D}_n\big)^c = C.$$

But the sets $\big(\overline{D}_n\big)^c$ are open and dense (see Definition 1.20 and Problem 1.29). So C is a G_δ-set (see Definition 1.57) and because X is complete, by Theorem 1.26, it is also dense.

"$\Longleftarrow$": Suppose that A contains a dense G_δ-set C. Hence

$$C = \bigcap_{n\geqslant 1} U_n, \quad \text{with } U_n \text{ open.}$$

Evidently for every $n \geqslant 1$, U_n is dense in X. Hence

$$\operatorname{int}\big(U_n^c\big) = \big(\overline{U}_n\big)^c = X^c = \emptyset,$$

so for all $n \geqslant 1$, U_n^c is a nowhere dense closed set. We have

$$A^c \subseteq C^c = \bigcup_{n\geqslant 1} U_n^c$$

and so A^c is meager.

(b) "$\Longrightarrow$": Suppose that A is meager. Then by part **(a)** we can find a dense G_δ-set C such that $C \subseteq A^c$. Therefore

$$A \subseteq C^c$$

and C^c is a F_σ-set (being the complement of a G_δ-set), whose complement C is dense.

"$\Longleftarrow$": Suppose that $A \subseteq E$ with E being an F_σ-set with dense complement. Then $E^c \subseteq A^c$, where E^c is a dense G_δ-set. Invoking part **(a)**, we conclude that $A = (A^c)^c$ is meager.

Solution of Problem 1.61

(a) Let X^* be the completion of X (see Theorem 1.51). Reasoning indirectly, suppose that $X^* \neq X$ and let $x^* \in X^* \setminus X$. Consider the continuous function $f\colon X \longrightarrow \mathbb{R}$, defined by

$$f(x) = \frac{1}{d_{X^*}(x^*, x)} \qquad \forall\, x \in X,$$

with d_{X^*} being the metric of the completion X^*.

As $x^* \in X^* \setminus X$, we see that f is continuous (see Proposition 1.36).

But f is not uniformly continuous. Indeed, assume that f is uniformly continuous (see Definition 1.45). So, for $\varepsilon = 1$, there exists $\delta > 0$ such that

$$\forall x, y \in X : \; d_X(x, y) < \delta \implies \big|f(x) - f(y)\big| < 1.$$

Because X is dense in X^* (see Definitions 1.20 and 1.50), there exists $x_0 \in X$ such that

$$d_{X^*}(x_0, x^*) < \tfrac{\delta}{2}.$$

Again using the density of X in X^*, we can choose $x_1 \in X$ such that

$$d_{X^*}(x_1, x^*) < \min\Big\{\tfrac{\delta}{2}, \frac{1}{1 + \frac{1}{d_{X^*}(x_0, x^*)}}\Big\}.$$

Then

$$d_{X^*}(x_0, x_1) \leqslant d_{X^*}(x_0, x^*) + d_{X^*}(x^*, x_1) < \tfrac{\delta}{2} + \tfrac{\delta}{2} = \delta.$$

But

$$\big|f(x_0) - f(x_1)\big| = \Big|\frac{1}{d_{X^*}(x_0, x^*)} - \frac{1}{d_{X^*}(x_1, x^*)}\Big| > 1,$$

a contradiction.

(b) Let $X = \mathbb{R}$. This a complete metric space, but the function

$$f\colon \mathbb{R} \ni x \longmapsto x^2 \in \mathbb{R}$$

is continuous and not uniformly continuous.

Solution of Problem 1.62

From Problem 1.46, we know that f is continuous at $x \in X$ if and only if $\omega_f(x) = 0$. So

$$C = \{x \in X : \omega_f(x) = 0\} = \bigcap_{n \geqslant 1} \{x \in X : \omega_f(x) < \tfrac{1}{n}\}.$$

But by Problem 1.47, for each $n \geqslant 1$, the set

$$\{x \in X : \omega_f(x) < \tfrac{1}{n}\}$$

is open. Hence C is a G_δ-set (see Definition 1.57).

Solution of Problem 1.63

Suppose that we can find a function $f\colon \mathbb{R} \longrightarrow \mathbb{R}$ for which

$$\{x \in \mathbb{R} : f \text{ is continuous at } x\} = \mathbb{Q}.$$

From Problem 1.62, we would have that $\mathbb{Q}$ is a G_δ-set (see Definition 1.57), which contradicts Problem 1.58.

Solution of Problem 1.64

For every integer $m \geqslant 1$ and every $f \in \mathcal{F}$, we consider the set

$$C_{m,f} = \{x \in X : f(x) \leqslant m\}.$$

Since f is lower semicontinuous, the set $C_{m,f}$ is closed for all $f \in \mathcal{F}$ and all $m \geqslant 1$, hence the set

$$C_m = \bigcap_{f \in F} C_{m,f}$$

is closed too. By hypothesis

$$X = \bigcup_{m \geqslant 1} C_m.$$

Since X is complete, by the Baire category theorem (see Theorem 1.26), we can find an integer $m_0 \geqslant 1$ such that $\operatorname{int} C_{m_0} \neq \emptyset$. Hence

$$f(x) \leqslant m_0 \qquad \forall\, f \in \mathcal{F},\ x \in U = \operatorname{int} C_{m_0}.$$

Solution of Problem 1.65

Arguing by contradiction, suppose that we can find a sequence of continuous functions $f_n \colon \mathbb{R} \longrightarrow \mathbb{R}$ such that

$$f_n(x) \longrightarrow \chi_{\mathbb{Q}}(x) \qquad \forall\, x \in \mathbb{R}.$$

Let

$$U_n = f_n^{-1}\big((\tfrac{1}{2}, +\infty)\big) \qquad \forall\, n \geqslant 1.$$

Then U_n is open (since f_n is continuous) for $n \geqslant 1$ and from the pointwise convergence of the sequence $\{f_n\}_{n \geq 1}$ to $\chi_{\mathbb{Q}}$, we have

$$\bigcap_{n \geqslant 1} U_n = \mathbb{Q},$$

so $\mathbb{Q}$ is a G_δ-set (see Definition 1.57), which contradicts Problem 1.58.

Solution of Problem 1.66

For a given $\varepsilon > 0$, let

$$\begin{aligned} A_m^{\varepsilon} &= \big\{x \in X : \big|f_m(x) - f(x)\big| \leqslant \varepsilon\big\} \qquad \forall\, m \geqslant 1, \\ D^{\varepsilon} &= \bigcup_{m \geqslant 1} \operatorname{int} A_m^{\varepsilon}, \\ \hat{E} &= \bigcap_{n \geqslant 1} D^{\frac{1}{n}}. \end{aligned}$$

We claim that $\hat{E} = E$. To this end let $x \in E$. Since $f_n \longrightarrow f$, we can find an integer $m \geqslant 1$ such that $|f_m(x) - f(x)| \leqslant \frac{\varepsilon}{3}$. Also the continuity of f_m (by hypothesis) and of f at $x \in E$, implies that there exists $\delta = \delta(x, \varepsilon) > 0$ such that

$$\left|f_m(u) - f_m(x)\right| \leqslant \tfrac{\varepsilon}{3} \quad \text{and} \quad \left|f(u) - f(x)\right| \leqslant \tfrac{\varepsilon}{3} \qquad \forall\, u \in \overline{B}_\delta(x).$$

Hence by the triangle inequality, we have

$$\begin{aligned} \left|f_m(u) - f(u)\right| &\leqslant \left|f_m(u) - f_m(x)\right| + \left|f_m(x) - f(x)\right| + \left|f(u) - f(x)\right| \\ &\leqslant \varepsilon \qquad \forall\, u \in \overline{B}_\delta(x), \end{aligned}$$

so $x \in \operatorname{int} A_m^\varepsilon \subseteq D^\varepsilon$. Since $\varepsilon > 0$ was arbitrary, it follows that $x \in \hat{E}$. Therefore $E \subseteq \hat{E}$.

Next let $x \in \hat{E}$. For a given $\varepsilon > 0$, let $n \geqslant \frac{3}{\varepsilon}$. We have that $x \in D^{\frac{1}{n}}$ and so there is an integer $m \geqslant 1$ such that $x \in \operatorname{int} A_m^{\frac{1}{n}}$. Then we can find $\delta > 0$ such that $\overline{B}_\delta(x) \subseteq A_m^{\frac{1}{n}}$, so

$$\left|f_m(u) - f(u)\right| \leqslant \frac{1}{n} \qquad \forall\, u \in \overline{B}_\delta(x)$$

and

$$\left|f_m(u) - f_m(x)\right| \leqslant \frac{1}{n} \qquad \forall\, u \in \overline{B}_\delta(x)$$

(from the continuity of f_m). Then, the triangle inequality implies that

$$\left|f(u) - f(x)\right| \leqslant \frac{3}{n} \leqslant \varepsilon \qquad \forall\, u \in \overline{B}_\delta(x),$$

so f is continuous at x, i.e., $x \in E$ and so $\hat{E} \subseteq E$. Therefore finally $\hat{E} = E$.

Next, for all $\varepsilon > 0$ and $m \geqslant 1$, we introduce the sets

$$G_m^\varepsilon = \left\{x \in X : \left|f_m(x) - f_{m+k}(x)\right| \leqslant \varepsilon \text{ for all integers } k \geqslant 1\right\}.$$

Clearly each G_m^ε is closed and by the pointwise convergence of the sequence $\{f_n\}_{n\geqslant 1}$ to f, we have

$$\begin{aligned} X &= \bigcup_{m\geqslant 1} G_m^\varepsilon \qquad \forall\, \varepsilon > 0, \\ G_m^\varepsilon &\subseteq A_m^\varepsilon \qquad \forall\, \varepsilon > 0,\ m \geqslant 1, \end{aligned}$$

so

$$\bigcup_{m\geqslant 1} \operatorname{int} G_m^\varepsilon \subseteq D^\varepsilon \qquad \forall\, \varepsilon > 0.$$

Hence

$$X\backslash D^\varepsilon \subseteq X\backslash \bigcup_{m\geqslant 1} \operatorname{int} G_m^\varepsilon = \bigcup_{m\geqslant 1} G_m^\varepsilon \backslash \bigcup_{m\geqslant 1} \operatorname{int} G_m^\varepsilon \subseteq \bigcup_{m\geqslant 1} (G_m^\varepsilon \backslash \operatorname{int} G_m^\varepsilon),$$

so

$$\bigcup_{k\geqslant 1} (X \setminus D^{\frac{1}{k}}) \subseteq \bigcup_{k\geqslant 1} \bigcup_{m\geqslant 1} (G_m^{\frac{1}{k}} \setminus \operatorname{int} G_m^{\frac{1}{k}})$$

and thus

$$\begin{aligned} X \setminus E &= X \setminus \hat{E} = X \setminus \bigcap_{k\geqslant 1} D^{\frac{1}{k}} = \bigcup_{n\geqslant 1} (X \setminus D^{\frac{1}{k}}) \\ &\subseteq \bigcup_{k\geqslant 1} \bigcup_{m\geqslant 1} (G_m^{\frac{1}{k}} \setminus \operatorname{int} G_m^{\frac{1}{k}}) \subseteq \bigcup_{k\geqslant 1} \bigcup_{m\geqslant 1} \partial G_m^{\frac{1}{k}}. \end{aligned}$$

But $\partial G_m^{\frac{1}{k}}$ is a nowhere dense set (see Definition 1.25). Therefore $X \setminus E$ is a subset of a meager set, hence E contains a countable intersection of open dense sets (see Definition 1.20 and Problem 1.60(a)), which is dense too, since X is complete (see Theorem 1.26). So, finally E is a dense G_δ-set in X (see Definition 1.57).

Solution of Problem 1.67

We have

$$\begin{aligned} \tfrac{\partial f}{\partial x}(x,y) &= \lim_{n\to+\infty} n\big|f\big(x+\tfrac{1}{n},y\big) - f(x,y)\big|, \\ \tfrac{\partial f}{\partial y}(x,y) &= \lim_{n\to+\infty} n\big|f\big(x,y+\tfrac{1}{n}\big) - f(x,y)\big|. \end{aligned}$$

From Problem 1.66, we know that $\frac{\partial f}{\partial x}$ is continuous an a dense G_δ set $E_1 \subseteq \mathbb{R}^2$ (see Definitions 1.57 and 1.20) and $\frac{\partial f}{\partial y}$ is continuous an o dense G_δ set $E_2 \subseteq \mathbb{R}^2$. We set $E = E_1 \cap E_2$. Then on E both partial derivatives of f exist and are continuous, hence f is differentiable on E. Clearly E is dense and G_δ.

Solution of Problem 1.68
Let $X = (0, 1]$ and let d_1 be the metric on X induced by the Euclidean metric on $\mathbb{R}$, i.e.,

$$d_1(x, u) = |x - u| \qquad \forall\, x, u \in X.$$

Also, let $d_2 \colon X \times X \longrightarrow \mathbb{R}_+$ be defined by

$$d_2(x, u) = \left|\tfrac{1}{x} - \tfrac{1}{u}\right| \qquad \forall\, x, u \in X.$$

Clearly d_2 is a metric. Note that

$$d_1(x, u) = |x - u| \leqslant \tfrac{|x-u|}{xu} = d_2(x, u) \qquad \forall\, x, u \in X.$$

Therefore the metric d_2 is stronger than d_1 (i.e., the metric topology on X generated by d_2 is richer (has more open sets) than that generated by d_1) and the identity function $i \colon (X, d_2) \longrightarrow (X, d_1)$ is continuous (in fact Lipschitz continuous with Lipschitz constant 1).

Let $V = [1, +\infty)$ and let d be the metric induced by the usual metric on $\mathbb{R}$. We consider the bijections $f \colon X \longrightarrow V$ and $h \colon V \longrightarrow X$, defined by

$$f(x) = \tfrac{1}{x} \quad \text{and} \quad h(v) = \tfrac{1}{v} \qquad \forall\, x \in X,\ v \in V.$$

Both functions are bijections and $f \colon (X, d_1) \longrightarrow (V, d)$ is continuous. Also, for all $v, y \in V$, we have

$$d_2\big(h(v), h(y)\big) = \left|\tfrac{1}{h(v)} - \tfrac{1}{h(y)}\right| = |v - y| = d(v, y)$$

and so $h \colon (V, d) \longrightarrow (X, d_2)$ is an isometry.

Note that $i = h \circ f \colon (X, d_1) \longrightarrow (X, d_2)$ is continuous and so we have proved that (X, d_1) and (X, d_2) are indeed topologically equivalent.

Next we show that the two metrics d_1 and d_2 are not uniformly equivalent. If this is not the case, i.e., the two metrics are uniformly equivalent, then $i \colon (X, d_1) \longrightarrow (X, d_2)$ is uniformly continuous (see Definition 1.45) and so we can find $\delta > 0$ such that

$$d_2(x, y) < 1 \qquad \forall\, x, y \in X, \text{ with } d_1(x, y) < \delta. \tag{1.5}$$

Let $\{x_n\}_{n \geqslant 1} \subseteq X$ be a sequence, given by $x_n = \frac{1}{n}$ for $n \geqslant 1$. Then

$$d_1(x_n, x_{n+1}) = \tfrac{1}{n(n+1)} \quad \text{and} \quad d_2(x_n, x_{n+1}) = \big|n - (n + 1)\big| = 1.$$

For our $\delta > 0$, we can find $n_0 \geqslant 1$ such that

$$d_1(x_n, x_{n+1}) \; < \; \delta \qquad \forall\, n \geqslant n_0,$$

while

$$d_2(x_n, x_{n+1}) \; = \; 1 \qquad \forall\, n \geqslant 1,$$

a contradiction with (1.5). This proves that d_1 and d_2 are not uniformly equivalent (see Remark 1.53).

Solution of Problem 1.69

Let $X = (0,1]$ and consider the metrics d_1 and d_2 by

$$d_1(x,u) \; = \; |x-u| \quad \text{and} \quad d_2(x,u) \; = \; \left|\tfrac{1}{x} - \tfrac{1}{u}\right| \qquad \forall\, x,u \in X.$$

From Problem 1.68, we know that d_1 and d_2 are topologically equivalent. Let $x_n = \frac{1}{n}$ for $n \geqslant 1$. Then $\{x_n\}_{n\geqslant 1}$ is d_1-Cauchy and

$$x_n \; \longrightarrow \; 0 \notin X.$$

So, (X, d_1) is not complete.

From the solution of Problem 1.68, we know that the function $h\colon V = [1, +\infty) \longrightarrow X$, defined by

$$h(v) \; = \; \tfrac{1}{v} \qquad \forall\, v \in V$$

is an isometry from V with the usual metric it inherits from $\mathbb{R}$ onto (X, d_2). So, h preserves Cauchy sequences. Because V is complete, it follows that (X, d_2) is complete.

Solution of Problem 1.70

Let $X = (0, +\infty)$ and let $f\colon X \longrightarrow (0,1)$ and $h\colon (0,1) \longrightarrow X$ be defined by

$$f(x) \; \stackrel{\text{def}}{=} \; \tfrac{x}{1+x} \quad \text{and} \quad h(v) \; \stackrel{\text{def}}{=} \; \tfrac{v}{1-v} \qquad \forall\, x \in X,\; v \in (0,1).$$

Both functions are continuous and $f^{-1} = h$. Hence f is a homeomorphism on $\mathbb{R}$ onto X (both spaces with the Euclidean metric topologies). Note that

$$f \in C^1(X), \quad \text{with } f'(x) \; = \; \tfrac{1}{(1+x)^2} \; > \; 0.$$

Therefore f is strictly increasing. So, for all $r, s \geqslant 0$, we have

$$\begin{aligned} f(r+s) &= \tfrac{r+s}{1+(r+s)} \leqslant \tfrac{r+s+rs}{1+r+s+rs} = \tfrac{r+s(1+r)}{(1+r)(1+s)} \\ &\leqslant \tfrac{r(1+s)+s(1+r)}{(1+r)(1+s)} \leqslant \tfrac{r}{1+r} + \tfrac{s}{1+s} = f(r) + f(s). \end{aligned}$$

Let d_1 be the usual metric on $\mathbb{R}$ and let $d_2 \colon X \times X \longrightarrow \mathbb{R}_+$ be defined by

$$d_2(x,u) \stackrel{\text{def}}{=} f\big(|x-u|\big) \qquad \forall\, x,u \in X.$$

Using the properties of f established above, we verify that d_2 is a metric on $\mathbb{R}$. Note that $f(r) \leqslant r$ for all $r \geqslant 0$ and so

$$d_2(x,u) \leqslant d_1(x,u) \qquad \forall\, x,u \in X.$$

Then the identity function $\bar{i} \colon (X, d_1) \longrightarrow (X, d_2)$ is Lipschitz continuous with Lipschitz constant 1, hence uniformly continuous too (see Definition 1.45).

We will show that also the identity function $i \colon (X, d_2) \longrightarrow (X, d_1)$ is uniformly continuous. Let $\varepsilon > 0$ be given. Let us set $\delta = \frac{\varepsilon}{1+\varepsilon}$. Then, if $d_2(x,u) \leqslant \delta$, so

$$\tfrac{|x-u|}{1+|x-u|} \leqslant \tfrac{\varepsilon}{1+\varepsilon} \quad \text{and thus} \quad d_1(x,u) = |x-u| \leqslant \varepsilon.$$

Hence, we have shown that

$$\forall \varepsilon > 0\ \exists \delta > 0\ \forall x,u \in X: \ d_2(x,u) < \delta \implies d_1(x,u) < \varepsilon,$$

so the identity function $i \colon (X, d_2) \longrightarrow (X, d_1)$ is uniformly continuous. Thus the metrics d_1 and d_2 are uniformly equivalent.

On the other hand, for every $u \geqslant 0$, we have $d_1(0,u) = u$ and $d_2(0,y) = f(y) < 1$. So there is no $k > 0$ such that

$$d_1(0,u) \leqslant k d_2(0,u) \qquad \forall\, u \in X$$

(let $u = k+1$). This implies that d_1 and d_2 are not Lipschitz equivalent (see Proposition 1.54).

Solution of Problem 1.71
Let

$$u_n(t) = \tfrac{t^n}{n} \qquad \forall\, t \in [0,1],\ n \geqslant 1$$

and let

$$u^*(t) = 0 \qquad \forall\, t \in [0,1].$$

Then

$$d^\infty(u_n, u^*) = \tfrac{1}{n} \longrightarrow 0 \quad \text{as } n \to +\infty.$$

On the other hand

$$d_1^\infty(u_n, u^*) = \tfrac{1}{n} + 1 \longrightarrow 1 \quad \text{as } n \to +\infty$$

(see Definition 1.52).

Solution of Problem 1.72
Let

$$\hat{d}_X(x,u) = d_X(x,u) + d_Y\big(f(x), f(u)\big) \qquad \forall\, x,u \in X.$$

Clearly $\hat{d}_X$ is a metric on X. Moreover, note that

$$\lim_{n\to+\infty} \hat{d}_X(x_n, x) = 0 \qquad \Longleftrightarrow \qquad \lim_{n\to+\infty} d_X(x_n, x) = 0.$$

Therefore $\hat{d}_X$ and d_X are equivalent metrics on X (see Definition 1.52). Note that

$$d_Y\big(f(x), f(u)\big) \leqslant \hat{d}_X(x,u) \qquad \forall\, x,u \in X,$$

hence the function $f\colon (X, \hat{d}_X) \longrightarrow Y$ is Lipschitz continuous (see Definition 1.48).

Solution of Problem 1.73
Let

$$\hat{d}_Y(y,v) = \frac{d_Y(y,v)}{1 + d_Y(y,v)} \qquad \forall\, y,v \in Y.$$

This is a bounded metric on Y which is topologically equivalent to d_Y (see Definition 1.52). Also on X we introduce the distance function

$$\hat{d}_X(u,x) = d_X(u,x) + \sum_{n\geqslant 1} \tfrac{1}{2^n} \hat{d}_Y\big(f_n(u), f_n(x)\big) \qquad \forall\, u,x \in X.$$

It is easy to check that $\hat{d}_X$ is a metric on X and

$$\hat{d}_X(u_n, u) \longrightarrow 0 \iff d_X(u_n, u) \longrightarrow 0,$$

so $\hat{d}_X$ and d_X are topologically equivalent.

Moreover, we have

$$\hat{d}_Y(f_n(u), f_n(x)) \leqslant 2^n \hat{d}_X(u, x) \qquad \forall\, u, x \in X,\ n \geqslant 1,$$

so for each $n \geqslant 1$, the function $f_n\colon (X, \hat{d}_X) \longrightarrow (Y, \hat{d}_Y)$ is Lipschitz continuous (see Definition 1.48).

Solution of Problem 1.74

Let Y be the completion of the metric space X and let $h\colon X \longrightarrow h(X) \subseteq Y$ be the isometry such that $\overline{h(X)} = Y$ (see Definition 1.50). Since X is separable (see Definition 1.21), it admits a countable dense set D (see Definition 1.20). Then the countable set $h(D) \subseteq Y$ satisfies

$$h(X) = h(\overline{D}) \subseteq \overline{h(D)} \subseteq \overline{h(X)} = Y$$

(see Proposition 1.32(d)), so

$$Y = \overline{h(X)} = \overline{h(D)},$$

i.e., Y is separable too.

Solution of Problem 1.75

If $D = \emptyset$, then the result is obvious, so let us suppose that $D \neq \emptyset$. Fix $t \in D$ and let

$$\mathcal{Y}_t = \{f \in C[a, b] : f(t) = 0\}.$$

Then since uniform convergence (i.e., convergence in the $d^{\infty}_{C[a,b]}$-metric; see Definition 1.59) implies pointwise convergence, we infer that $\mathcal{Y}_t$ is $d^{\infty}_{C[a,b]}$-closed. Note that $\mathcal{Y} = \bigcap\limits_{t \in D} \mathcal{Y}_t$. Hence $\mathcal{Y}$ is $d^{\infty}_{C[a,b]}$-closed.

Solution of Problem 1.76

The set E_1 is not closed in X. To see this, consider the functions $f_n \in X$ for $n \geqslant 1$, defined by

$$f_n(s) = \tfrac{s}{n} \qquad \forall\, n \geqslant 1,\ s \in [0,1]$$

Then $f_n \in E_1$ for each $n \geqslant 1$ but $f_n \xrightarrow{d^\infty} 0 \notin E_1$.

The set E_2 is closed in X. To see that, let $\{f_n\}_{n\geqslant 1} \subseteq E_2$ be a sequence and assume that

$$f_n \xrightarrow{d^\infty} f$$

(i.e., $f_n \rightrightarrows f$; see Definition 1.59). Let $t \in [0,1]$ and $n \geqslant 1$. We can find $s_n \in [0,1]$ such that $f_n(s_n) = t$. Passing to a subsequence if necessary, we may assume that $s_n \longrightarrow s$. Then $f(s_n) \longrightarrow f(s)$. We have

$$\big|t - f(s_n)\big| = \big|f_n(s_n) - f(s_n)\big| \leqslant d^\infty(f_n, f) \longrightarrow 0,$$

so $f(s) = t$. Since $t \in [0,1]$ was arbitrary, we conclude that $f \in E_2$.

Finally E is not closed. To see this, consider functions

$$f_n(s) = \begin{cases} 2\frac{n-1}{n}s & \text{if } s \in \big[0, \frac{1}{2}\big], \\ \frac{2}{n}s + \frac{n-2}{n} & \text{if } s \in \big(\frac{1}{2}, 1\big], \end{cases} \qquad \forall\, n \geqslant 2.$$

Then $f_n \in E$ and $f_n \xrightarrow{d^\infty} f$, where

$$f(s) = \begin{cases} 2s & \text{if } s \in \big[0, \frac{1}{2}\big], \\ 1 & \text{if } s \in \big(\frac{1}{2}, 1\big]. \end{cases}$$

However, it is clear that $f \notin E$ (it has a flat part on $\big[\frac{1}{2}, 1\big]$).

Solution of Problem 1.77

For integers $m, n \geqslant 1$, we define

$$C_{m,n} = \big\{f \in X : \ \exists t \in [0,1]\ \forall s \in [0,1] : \\ |s-t| \leqslant \tfrac{1}{m} \Longrightarrow \big|f(s) - f(t)\big| \leqslant n|s-t|\big\}.$$

We will show that $C_{m,n}$ is nowhere dense in X (see Definition 1.25). In fact, if $f \in C_{m,n}$, we let

$$m(f) = \max_{t\in[0,1]} \left|\frac{f(t\pm\frac{1}{m}) - f(t)}{\frac{1}{m}}\right|.$$

Let $a(x)$ be a continuous, piecewise linear function bounded by 1 and with each linear piece having slope $3m(f)$. Then, for every $\varepsilon > 0$, we have

$$d^{\infty}(f+\varepsilon a, f) \leqslant \varepsilon \quad \text{and} \quad f+\varepsilon a \notin C_{m,n}.$$

So d^{∞}-int $C_{m,n} = \emptyset$, i.e., $C_{m,n}$ is nowhere dense.

Since (X, d^{∞}) is a complete metric space (see Proposition 1.62), by the Baire category theorem (see Theorem 1.26), the set $\bigcup\limits_{m,n\geqslant 1} C_{m,n}$ is nowhere dense in X. Hence

$$D = X \setminus \bigcup_{m,n\geqslant 1} C_{m,n} \quad \text{is of second category}$$

(see Definition 1.25). But if $f \in D$, then for every $t \in [0,1]$ and every $n, m \geqslant 1$, we can find points $s \in [0,1]$ such that

$$|s-t| \leqslant \tfrac{1}{m} \quad \text{and} \quad \left|\tfrac{f(s)-f(t)}{s-t}\right| > n$$

and so f is not differentiable at any $t \in [0,1]$.

Solution of Problem 1.78

Let $x, y \in X$ and let $\varepsilon > 0$. From the definition of dist (see Definition 1.6(d)), there exists $z \in A$ such that

$$d_X(x,z) < f(x) + \varepsilon.$$

As we also have $d_X(y,z) \geqslant f(y)$, so

$$f(y) - f(x) \leqslant d_X(y,z) - d_X(x,z) + \varepsilon \leqslant d_X(x,y) + \varepsilon.$$

Because $\varepsilon > 0$ was arbitrary, we conclude that

$$f(y) - f(x) \leqslant d_X(x,y).$$

Exchanging the roles of x and y, we also have

$$f(x) - f(y) \leqslant d_X(x,y),$$

so finally

$$\big|f(x) - f(y)\big| \leqslant d_X(x,y).$$

This proves that f is Lipschitz continuous with Lipschitz constant 1.

Solution of Problem 1.79

Let $C \subseteq X$ be a closed set. Then

$$C = \big\{x \in X : \operatorname{dist}_X(x,C) = 0\big\}$$

(see Definition 1.6(d)) Hence

$$C = \bigcap_{n\geqslant 1} \{x \in X : \operatorname{dist}_X(x,C) < \tfrac{1}{n}\}.$$

The continuity of the function $x \longmapsto \operatorname{dist}_X(x,C)$ (see Problem 1.78), implies that the set

$$\big\{x \in X : \operatorname{dist}_X(x,C) < \tfrac{1}{n}\big\} \quad \text{is open}$$

(see Proposition 1.32). Therefore it follows that C is G_δ (see Definition 1.57). Because the complement of a G_δ set is an F_σ-set. Therefore the second part of the problem follows from the first.

Solution of Problem 1.80

For every $n \in \mathbb{Z}$ let $D_n = D \cap [n, n+1)$. Then $D = \bigcup\limits_{n\geqslant 1} D_n$. Since D is uncountable, at least one of the sets D_n, say D_{n_0}, is uncountable. So, in D_{n_0} we can find a sequence with distinct terms, which has a convergent subsequence. The limit of this sequence is an accumulation point of D (see Definition 1.13(b) and Theorem 1.14(d)). Now let

$$S = \big\{x \in \mathbb{R} : x \text{ is an accumulation point in } D\big\}.$$

We have just seen that $S \neq \emptyset$ and of course S is closed. Hence S^c is open and so by Problem 1.79, $S^c = \bigcup_{n \geqslant 1} C_n$ with $C_n \subseteq \mathbb{R}$ closed for every $n \geqslant 1$. If S is countable, then D must have uncountable many elements in S^c and so in some C_{n_0}. From the first part of the proof, $D \cap C_{n_0}$ has an accumulation point x. Then

$$x \in \overline{D \cap C_{n_0}} \subseteq \overline{D} \cap \overline{C}_{n_0} = \overline{D} \cap C_{n_0},$$

so $x \in C_{n_0}$ and $x \in S$. But $S \cap C_{n_0} = \emptyset$, a contradiction.

Solution of Problem 1.81

Evidently it suffices to consider $U = (a, b)$, $a, b \in \mathbb{R}$, $a < b$. We have

$$f^{-1}\big((a,b)\big) = \bigcup_{n \geqslant 1} \big[f^{-1}\big((-\infty, b - \tfrac{1}{n}]\big) \cap f^{-1}\big((a, +\infty)\big)\big].$$

But from the lower semicontinuity of f, we have that

$$f^{-1}\big((-\infty, b - \tfrac{1}{n}]\big) \text{ is closed and } f^{-1}\big((a, +\infty)\big) \text{ is open}$$

(see Problem 1.64).

Also, from Problem 1.79, we have that $f^{-1}\big((a, +\infty)\big)$ is an F_σ-set (see Definition 1.57(b)). Hence

$$f^{-1}\big((a,b)\big) \text{ is an } F_\sigma\text{-set.}$$

Solution of Problem 1.82

For every $n \geqslant 1$, we cover $\mathbb{R}$ with a countable family of open sets $\{U_k^n\}_{k \geqslant 1}$ such that

$$\operatorname{diam} U_k^n < \tfrac{1}{n} \qquad \forall\, n, k \geqslant 1$$

(see Proposition 1.24). Note that all sets $f^{-1}(U_k^n)$ are F_σ-sets (see Problem 1.81). Hence

$$f^{-1}(U_k^n) = \bigcup_{m \geqslant 1} C_{k,m}^n \quad \text{with } C_{k,m}^n \text{ closed for all } m \geqslant 1.$$

We have $X = \bigcup_{k,m \geqslant 1} C_{k,m}^n$, so

$$D_n = \bigcup_{k,m \geqslant 1} \operatorname{int} C_{k,m}^n \text{ is open dense in } X \text{ for all } n \geqslant 1$$

(see Problem 1.34). We have $D = \bigcap_{n \geqslant 1} D_n$ is a dense G_δ-set in X (see Theorem 1.26 and Definitions 1.57 and 1.20) and since the oscillation of f on each D_n is less than $\frac{1}{n}$, we infer that $f|_D$ is continuous.

Solution of Problem 1.83

(a) Since by hypothesis $f_n \rightrightarrows f$ (see Definition 1.59), we can find an integer $n_0 \geqslant 1$ such that $\sup_{x \in X} d_Y\big(f_{n_0}(x), f(x)\big) < 1$. So, for all $x_1, x_2 \in X$, we have

$$\begin{aligned} & d_Y\big(f(x_1), f(x_2)\big) \\ \leqslant\ & d_Y\big(f(x_1), f_{n_0}(x_1)\big) + d_Y\big(f_{n_0}(x_1), f_{n_0}(x_2)\big) + d_Y\big(f_{n_0}(x_2), f(x_2)\big) \\ \leqslant\ & 2 + \operatorname{diam} f_{n_0}(X) \end{aligned}$$

(see Definition 1.6(b)), hence $\operatorname{diam} f(X) < +\infty$, i.e., f is bounded too.

(b) By way of contradiction, suppose that $f_n \rightrightarrows f$ and $f\colon X \longrightarrow Y$ is bounded. Then we can find $x_0 \in X$ and $r_0 > 0$ such that $f(X) \subseteq B_{r_0}(x_0)$. Again we can find an integer $n_0 \geqslant 1$ such that

$$\sup_{x \in X} d_Y\big(f_{n_0}(x), f(x)\big) \ < \ 1.$$

Hence $f_{n_0}(X) \subseteq B_{r_0+1}(x_0)$, a contradiction to the fact that f_{n_0} is unbounded. This shows that the limit function f cannot be bounded.

Solution of Problem 1.84

Let $\varepsilon > 0$. Since $f_n \rightrightarrows f$ (see Definition 1.59), we can find $n_0 = n_0(\varepsilon) \geqslant 1$ such that

$$d_Y\big(f_{n_0}(x), f(x)\big) \ \leqslant \ \tfrac{\varepsilon}{3} \qquad \forall\, x \in X.$$

The continuity of f_{n_0} at x_0 implies that there exists $\delta > 0$ such that

$$d_Y\big(f_{n_0}(x), f_{n_0}(x_0)\big) \ \leqslant \ \tfrac{\varepsilon}{3} \qquad \forall\, x \in \overline{B}_\delta(x_0)$$

(where $\overline{B}_\delta(x_0) = \{x \in X : d_X(x, x_0) \leqslant \delta\}$). Then for every $x \in \overline{B}_\delta(x_0)$, we have

$$\begin{aligned} d_Y\big(f(x), f(x_0)\big) &\leqslant d_Y\big(f(x), f_{n_0}(x)\big) + d_Y\big(f_{n_0}(x), f_{n_0}(x_0)\big) \\ &+ d_Y\big(f_{n_0}(x_0), f(x_0)\big) \leqslant \varepsilon, \end{aligned}$$

so f is continuous at x_0.

Solution of Problem 1.85

No. Let $X = \mathbb{R}$ and for every $n \geqslant 1$, let

$$f_n(x) = x \qquad \forall\, n \geqslant 1,\ x \in \mathbb{R}, \ g_n(x) = \tfrac{1}{n} \qquad \forall\, n \geqslant 1,\ x \in \mathbb{R}.$$

Evidently $f_n \rightrightarrows f = id_{\mathbb{R}}$ and $g_n \rightrightarrows 0$ (see Definition 1.59) Then

$$(f_n g_n)(x) = \tfrac{x}{n} \qquad \forall\, n \geqslant 1,\ x \in \mathbb{R}.$$

Then $f_n g_n \longrightarrow 0$, but the above convergence is not uniform.

Solution of Problem 1.86

Since by hypothesis $f_n \rightrightarrows f$ (see Definition 1.59), for a given $\varepsilon > 0$, we can find an integer $n_0 \geqslant 1$ such that

$$d_Y\big(f_n(x), f(x)\big) < \tfrac{\varepsilon}{2} \qquad \forall\, x \in X,\ n \geqslant n_0.$$

For $x, u \in X$ and all $n \geqslant n_0$, we have

$$\begin{aligned} d_Y\big(f(x), f(u)\big) &\leqslant d_Y\big(f(x), f_n(x)\big) + d_Y\big(f_n(x), f_n(u)\big) + d_Y\big(f_n(u), f(u)\big) \\ &\leqslant \varepsilon + k d_Y(x, u) \end{aligned}$$

(see Definition 1.48). Because $\varepsilon > 0$ is arbitrary we let $\varepsilon \searrow 0$ and obtain

$$d_Y\big(f(x), f(u)\big) \leqslant k d_Y(x, u) \qquad \forall\, x, u \in X,$$

hence f is k-Lipschitz.

Solution of Problem 1.87

Because $f_n \rightrightarrows f$ (see Definition 1.59), for a given $\varepsilon > 0$, we can find $n_0 \geqslant 1$ such that

$$d_Y\big(f_n(x), f(x)\big) < \tfrac{\varepsilon}{3} \qquad \forall\, x \in X,\ n \geqslant n_0.$$

By hypothesis f_{n_0} is uniformly continuous (see Definition 1.45). So, we can find $\delta = \delta(\varepsilon) > 0$ such that

$$d_Y\big(f_{n_0}(x), f_{n_0}(u)\big) < \tfrac{\varepsilon}{3} \qquad \forall\, x, u \in X, \text{ with } d_X(x, u) < \delta.$$

Then for all $x, u \in X$ with $d_X(x, u) < \delta$, we have

$$\begin{aligned} d_Y\big(f(x), f(u)\big) &\leqslant d_Y\big(f(x), f_{n_0}(x)\big) + d_Y\big(f_{n_0}(x), f_{n_0}(u)\big) \\ &+ d_Y\big(f_{n_0}(u), f(u)\big) \\ &< \tfrac{\varepsilon}{3} + \tfrac{\varepsilon}{3} + \tfrac{\varepsilon}{3} = \varepsilon, \end{aligned}$$

so f is uniformly continuous.

Solution of Problem 1.88

(a) Yes. Let $X = Y = [0,1]$ and let $\{f_n\colon [0,1] \longrightarrow [0,1]\}_{n \geqslant 1}$ be the sequence of discontinuous functions, defined by

$$f_n(x) = \begin{cases} \frac{1}{n} & \text{if } x \in [0,1] \cap \mathbb{Q} \\ 0 & \text{if } x \in [0,1] \setminus \mathbb{Q} \end{cases} \qquad \forall\, n \geqslant 1.$$

Note that

$$f_n \rightrightarrows f \equiv 0$$

(see Definition 1.59) and of course f is continuous (in fact the functions f_n are discontinuous at every $x \in X$).

(b) No. Consider the function $f\colon \mathbb{R} \longrightarrow \mathbb{R}$, defined by

$$f(x) = \begin{cases} 1 & \text{if } x \in \mathbb{Q}, \\ -1 & \text{if } x \in \mathbb{R} \setminus \mathbb{Q}. \end{cases}$$

Then

$$f \circ f \equiv 1.$$

So, $f \circ f$ is continuous, but f is discontinuous (in fact f is discontinuous at every $x \in X$).

Solution of Problem 1.89

As in Definition 1.37, for every $x \in X$, let

$$\omega_\varphi(x) = \inf_{r>0} \operatorname{diam} \varphi\big(B_r(x) \cap A\big).$$

We know that φ is continuous at x if and only if $\omega_\varphi(x) = 0$ (see Problem 1.46). We set

$$A_0 = \big\{x \in \overline{A} : \omega_\varphi(x) = 0\big\}$$

(note that, if $x \in \overline{A}$, then $B_r(x) \cap A \neq \emptyset$). Evidently $A \subseteq A_0$. Let $x \in A_0$. Then we can find a sequence $\{x_n\}_{n\geqslant 1} \subseteq A$ such that $x_n \longrightarrow x$. Since $\omega_\varphi(x) = 0$, we see that $\big\{\varphi(x_n)\big\}_{n\geqslant 1} \subseteq Y$ is a Cauchy sequence. Since Y is a complete metric space, we can find $y \in Y$ such that

$$\varphi(x_n) \longrightarrow y \quad \text{in } Y$$

and we can easily see that this limit is independent of the choice of the approximating sequence $\{x_n\}_{n\geqslant 1}$. So, if we set $\varphi_0(x) = y$, then $\varphi_0 \colon A_0 \longrightarrow Y$ is well defined and continuous because

$$\omega_{\varphi_0}(x) = 0 \qquad \forall\, x \in A_0.$$

We need to show that A_0 is a G_δ-subset of X (see Definition 1.57(a)). Let

$$U_n = \big(x \in X : \omega_\varphi(x) < \tfrac{1}{n+1}\big).$$

From Problem 1.47, we know that the set U_n is open in X. We have

$$A_0 = \Big(\bigcap_{n\geqslant 1} U_n\Big) \cap \overline{A}.$$

But from Problem 1.79, we know that $\overline{A}$ is a G_δ-subset of X. So, we conclude that A_0 is a G_δ-set.

Solution of Problem 1.90
Let $x_n \longrightarrow x$ in X. Then

$$f(x_n, s(x_n)) = v_0 \qquad \forall\, n \geqslant 1.$$

Since Y is compact, we can find a subsequence $\{s(x_{n_k})\}_{k\geqslant 1}$ of $\{s(x_k)\}_{n\geqslant 1}$ and $y \in Y$ such that $s(x_{n_k}) \longrightarrow y$ in Y. Exploiting the continuity of f, we have

$$f(x_{n_k}, s(x_{n_k})) \longrightarrow f(x, y),$$

hence $f(x, y) = v_0$ and so by hypothesis $y = s(x)$. Since every subsequence of $\{s(x_n)\}_{n\geqslant 1}$ has a further subsequence which converges to $s(x)$, we conclude that for the original sequence, we have $s_n(x) \longrightarrow s(x)$ (see Problem 1.3) and so s is continuous.

Solution of Problem 1.91
We need to show that f is a bijection and that f^{-1} is continuous. Note that if $f(x) = f(u)$ then $d(x, u) = 0$ and so $x = u$. This proves that f is injective. Next we show that f is surjective. Arguing by contradiction, suppose that there exists $x \in X$ such that $x \notin f(X)$. Then since $f(X)$ is compact, we have

$$d_X(x, f(X)) = c > 0.$$

We have

$$c \leqslant d_X(x, f^{(k)}(x)) \leqslant d_X(f^{(n)}(x),\ f^{(n+k)}(x)) \qquad \forall\, k, n \geqslant 1$$

(recall that $f^{(m)} = f \circ \ldots \circ f$ m-times for all $m \geqslant 1$). Then the sequence $\{f^{(n)}(x)\}_{n\geqslant 1}$ has no convergent subsequence in $f(X) \subseteq X$, a contradiction. This proves the surjectivity of f, hence f is a bijection.

Let $x, u \in X$. Consider two sequences $\{f^{(n)}(x)\}_{n\geqslant 1}$ and $\{f^{(n)}(u)\}_{n\geqslant 1}$. Because X is compact, we can find a subsequence $\{n_k\}_{k\geqslant 1}$ such that both sequences $\{f^{(n_k)}(x)\}_{k\geqslant 1}$ and $\{f^{(n_k)}(u)\}_{k\geqslant 1}$ are convergent and thus they are Cauchy sequences.

Let $\varepsilon > 0$. We can find $k \geqslant 1$ such that

$$d_X(f^{(n_k)}(x), f^{(n_l)}(x)) \leqslant \varepsilon \qquad \forall\, k, l \geqslant k_0$$

and

$$d_X\big(f^{(n_k)}(u), f^{(n_l)}(u)\big) \leqslant \varepsilon \qquad \forall\, k, l \geqslant k_0.$$

Let $\overline{n} = n_{k_0+1} - n_{k_0}$. Then, using the hypothesis on f, we have

$$d_X\big(x, f^{\overline{n}}(x)\big) \leqslant d_X\big(f^{n_{k_0}}(x), f^{n_{k_0+1}}(x)\big) \leqslant \varepsilon$$

and

$$d_X\big(u, f^{\overline{n}}(u)\big) \leqslant d_X\big(f^{n_{k_0}}(u), f^{n_{k_0+1}}(u)\big) \leqslant \varepsilon$$

Hence

$$\begin{aligned} d_X\big(f(x), f(u)\big) &\leqslant d_X\big(f^{(\overline{n})}(x), f^{(\overline{n})}(u)\big) \\ &\leqslant d_X\big(f^{(\overline{n})}(x), x\big) + d_X(x, u) + d_X\big(u, f^{(\overline{n})}(u)\big) \\ &\leqslant 2\varepsilon + d_X(x, u), \end{aligned}$$

so, letting $\varepsilon \searrow 0$, we have

$$d_X\big(f(x), f(u)\big) \leqslant d_X(x, u)$$

and thus we conclude that f is an isometry (see Definition 1.41).

Solution of Problem 1.92
By Theorem 1.47, f admits a uniformly continuous extension (see Definition 1.45) $\hat{f}\colon [a, b] \longrightarrow \mathbb{R}$ and $\hat{f}\big([a, b]\big)$ is compact. So, f is bounded (see Definition 1.6(b)).

Solution of Problem 1.93
Let $x, u \in X$.Let us define two sequences

$$\begin{aligned} &x_{-1} = f(x), \quad x_0 = x \quad \text{and} \quad x_n \in f^{-1}(x_{n-1}) \qquad \forall\, n \geqslant 1, \\ &u_{-1} = f(u), \quad u_0 = u \quad \text{and} \quad u_n \in f^{-1}(u_{n-1}) \qquad \forall\, n \geqslant 1. \end{aligned}$$

Because X is compact, we can find a subsequence $\{n_k\}_{k \geqslant 1}$ such that

$$x_{n_k} \longrightarrow x \quad \text{and} \quad u_{n_k} \longrightarrow u \quad \text{as } k \to +\infty.$$

Let $\varepsilon > 0$. We can find $k_0 \geqslant 2$ such that

$$\begin{aligned} d_X(x_{n_k}, x) &\leqslant \tfrac{\varepsilon}{2} \qquad \forall\, k, l \geqslant k_0, \\ d_X(u_{n_k}, u) &\leqslant \tfrac{\varepsilon}{2} \qquad \forall\, k, l \geqslant k_0. \end{aligned}$$

Then

$$\begin{aligned} d_X(x_{k_0}, x_{n_{k_0+1}}) &\leqslant d_X(x_{k_0}, x) + d_X(x, x_{n_{k_0+1}}) \leqslant \varepsilon, \\ d_X(u_{k_0}, u_{n_{k_0+1}}) &\leqslant d_X(u_{k_0}, u) + d_X(u, u_{n_{k_0+1}}) \leqslant \varepsilon. \end{aligned}$$

Let $\overline{n} = n_{k_0+1} - n_{k_0} - 1$. Then

$$\begin{aligned} d_X(x_{\overline{n}}, x_{-1}) &\leqslant d_X(x_{n_{k_0+1}}, x_{k_0}) \leqslant \varepsilon, \\ d_X(u_{\overline{n}}, u_{-1}) &\leqslant d_X(u_{n_{k_0+1}}, u_{k_0}) \leqslant \varepsilon \end{aligned}$$

Hence

$$\begin{aligned} d_X(x, u) &= d_X(x_0, y_0) = d_X(f(x_1), f(u_1)) \leqslant d_X(x_1, u_1) \leqslant \ldots \\ &\leqslant d_X(x_{\overline{n}}, u_{\overline{n}}) \leqslant d_X(x_{\overline{n}}, x_{-1}) + d_X(x_{-1}, u_{-1}) + d_X(u_{-1}, u_{\overline{n}}) \\ &\leqslant d_X(f(x), f(u)) + 2\varepsilon. \end{aligned}$$

Because $\varepsilon > 0$ was arbitrary, we get

$$d_X(x, u) \leqslant d_X(f(x), f(u)),$$

thus using the assumption, we deduce that f is isometry.

Solution of Problem 1.94

(a) If any of the sets K_n is empty, then the result is obvious. So let us assume that all sets K_n are nonempty. Clearly the sequence $\{\operatorname{diam} K_n\}_{n\geqslant 1} \subseteq [0, +\infty)$ is decreasing and bounded below by $\operatorname{diam} K$. So

$$\operatorname{diam} K \leqslant \lim_{n\to+\infty} \operatorname{diam} K_n.$$

Because K_n is compact and the distance function d_X is continuous (see Proposition 1.36), we can find $x_n, u_n \in K_n$ for $n \geqslant 1$ such that

$$d_X(x_n, u_n) = \operatorname{diam} K_n \qquad \forall\, n \geqslant 1.$$

By the compactness of X and by passing to a suitable subsequence if necessary, we may assume that

$$x_n \longrightarrow x \quad \text{and} \quad u_n \longrightarrow u \quad \text{in } X.$$

We claim that $x, u \in K$. Note that

$$x_n, u_n \in K_m \qquad \forall\, n \geqslant m \geqslant 1.$$

Hence

$$x, u \in K_m \qquad \forall\, m \geqslant 1$$

and so $x, u \in K$. We have

$$\operatorname{diam} K \geqslant d_X(x, u) = \lim_{n \to +\infty} d_X(x_n, u_n) = \lim_{n \to +\infty} \operatorname{diam} K_n,$$

so

$$\operatorname{diam} K = \lim_{n \to +\infty} \operatorname{diam} K_n.$$

(b) Let us consider $X = \{\frac{1}{n} : n \geqslant 1\}$ with the natural metric induced from $\mathbb{R}$. Then X is not compact (as the Cauchy sequence $\{\frac{1}{n}\}_{n \geqslant 1}$ has no limit in X). Let

$$K_k = \{1\} \cup \{\tfrac{1}{n} : n \geqslant k\} \qquad \forall\, k \geqslant 1.$$

Then the sequence $\{K_n\}_{n \geqslant 1}$ is a decreasing sequence of closed subsets of X with

$$\operatorname{diam} K_k = 1 \qquad \forall\, k \geqslant 1.$$

But $K = \bigcap\limits_{k \geqslant 1} K_k = \{1\}$ and so

$$\operatorname{diam} K_k \not\longrightarrow \operatorname{diam} K = \emptyset.$$

Now, let $X = [0, 1]$ (which is compact). Let $\{q_n\}_{n \geqslant 1}$ be an enumeration of rationals in X. Let

$$K_k = \{q_n : n \geqslant k\} \qquad \forall\, k \geqslant 1.$$

Then $\{K_k\}_{k \geqslant 1}$ is a sequence of decreasing nonempty sets which are not closed, with

$$\operatorname{diam} K_k = 1 \qquad \forall\, k \geqslant 1.$$

But $K = \bigcap\limits_{k \geqslant 1} K_k = \emptyset$ and so

$$\operatorname{diam} K_k \not\longrightarrow \operatorname{diam} K = 0.$$

Solution of Problem 1.95

No. To see this let $X = Y = [0,1]$ and let $\{f_n\}_{n\geqslant 1} \subseteq C(X;Y)$ be the sequence, defined by

$$f_n(x) = \begin{cases} nx & \text{if } 0 \leqslant x \leqslant \frac{1}{n} \\ 1 & \text{if } \frac{1}{n} < x \leqslant 1 \end{cases} \qquad \forall\, n \geqslant 1.$$

Then $f_n \longrightarrow f$ (pointwise), where

$$f(x) = \begin{cases} 0 & \text{if } x = 0, \\ 1 & \text{if } x = 1. \end{cases}$$

Since the limit function is discontinuous, we infer that the convergence is not uniform (see Proposition 1.62), i.e., it is not in the d^∞-metric. Therefore $\{f_n\}_{n\geqslant 1}$ is a sequence in $\big(C(X;Y), d^\infty\big)$ with no convergent subsequence. So, we conclude that $\big(C(X;Y), d^\infty\big)$ is not compact.

Solution of Problem 1.96

(a) We have

$$\begin{aligned} \big|f_n(x_n) - f(x)\big| &\leqslant \big|f_n(x_n) - f(x_n)\big| + \big|f(x_n) - f(x)\big| \\ &\leqslant \|f_n - f\|_\infty + \big|f(x_n) - f(x)\big|. \end{aligned}$$

Since $f_n \rightrightarrows f$ (see Definition 1.59), we have $\|f_n - f\|_\infty \longrightarrow 0$. Also, since $f \in C(X)$ (see Proposition 1.62), we have that $f(x_n) \longrightarrow f(x)$. Therefore, finally we have $f_n(x_n) \longrightarrow f(x)$.

(b) First we show that if $x_n \longrightarrow x$ in X and $r_n \longrightarrow +\infty$ with $\{r_n\}_{n\geqslant 1} \subseteq \mathbb{N}$ strictly increasing, then $f_{r_n}(x_n) \longrightarrow f(x)$. To this end, let

$$u_n = \begin{cases} x_k & \text{if } n = r_k, \\ x & \text{if } n \notin \{r_k :\, k \geqslant 1\}. \end{cases}$$

Evidently $u_n \longrightarrow x$ and so $f_n(u_n) \longrightarrow f(x)$. In particular then $f_{r_k}(u_{r_k}) = f_{r_k}(x_k) \longrightarrow f(x)$.

Now suppose that f is not continuous. Then we can find $x \in X$, $x_k \longrightarrow x$ and $\varepsilon > 0$ such that

$$\big|f(x_k) - f(x)\big| > \varepsilon \qquad \forall\, k \geqslant 1.$$

The hypothesis implies that $f_n \longrightarrow f$. So, for every $k \geqslant 1$, we have $f_n(x_k) \longrightarrow f(x_k)$ as $n \to +\infty$. Then, by induction we can find a strictly increasing sequence $\{r_k\}_{k\geqslant 1} \subseteq \mathbb{N}$ such that $|f_{r_k}(x_k) - f(x)| > \varepsilon$, contradicting the first part of the solution of statement (b).

Next suppose that the sequence $\{f_n\}_{n\geqslant 1}$ does not converge uniformly to f. So, by passing to a subsequence if necessary, we may assume that for some $\varepsilon > 0$, we have $\|f_n - f\|_\infty \geqslant \varepsilon$ for all $n \geqslant 1$. The compactness of X implies that we can find a sequence $\{x_n\}_{n\geqslant 1} \subseteq X$ such that

$$\left|f_n(x_n) - f(x_n)\right| \geqslant \varepsilon \qquad \forall\, n \geqslant 1.$$

Passing to a next subsequence if necessary, we may assume that $x_n \longrightarrow x$ (recall that X is compact). Then by hypothesis, $f_n(x_n) \longrightarrow f(x)$. So, we have

$$\varepsilon \leqslant \left|f_n(x_n) - f(x_n)\right| \leqslant \left|f_n(x_n) - f(x)\right| + \left|f(x) - f(x_n)\right| \longrightarrow 0,$$

a contradiction. This proves that $f_n \rightrightarrows f$.

Solution of Problem 1.97

(a) Evidently each C_n is compact and so is $C_n \setminus U = C_n \cap U^c$. We have

$$\bigcap_{n\geqslant 1}(C_n \cap U^c) = \Big(\bigcap_{n\geqslant 1} C_n\Big) \cap U^c = U \cap U^c = \emptyset.$$

Then by Proposition 1.67 the sequence of closed sets $\{C_n \cap U^c\}_{n\geqslant 1}$ cannot have the finite intersection property (see Definition 1.66). Hence we can find an integer $n_0 \geqslant 1$ such that $\bigcap\limits_{n=1}^{n_0} C_n \cap U^c = \emptyset$. Since the sequence $\{C_n\}_{n\geqslant 1}$ is decreasing (i.e., $C_1 \supseteq C_2 \supseteq C_3 \supseteq \ldots$) we must have $C_{n_0} \cap U^c = \emptyset$, hence $C_{n_0} \subseteq U$.

(b) If the sets C_n are not bounded, the result does not hold. To see this, let $U = (0,1) \subseteq \mathbb{R}$ and let $C_n = \{\frac{1}{2}\} \cup [n, +\infty)$ for $n \geqslant 1$. Then $\{C_n\}_{n\geqslant 1}$ is a decreasing family of closed sets such that $\bigcap\limits_{n\geqslant 1} C_n = \{\frac{1}{2}\} \subseteq U$, but $C_n \nsubseteq U$ for all $n \geqslant 1$.

Solution of Problem 1.98

From the hypothesis, we know that for a given $\varepsilon > 0$, we can find $M > 0$ such that

$$|f(x)| < \tfrac{\varepsilon}{2} \qquad \forall\, x \notin [-M, M].$$

By Proposition 1.77, $f|_{[-M,M]}$ is uniformly continuous (see Definition 1.45). So, we can find δ_1 such that

$$|f(x) - f(y)| < \tfrac{\varepsilon}{2} \qquad \forall\, x, y \in [-M, M], \text{ with } |x - y| < \delta_1.$$

Since f is continuous at $\pm M$, we can find $\delta_2 = \delta_2(M, \varepsilon) > 0$ such that

$$|f(x) + f(\pm M)| < \varepsilon, \quad \text{whenever } |x - (\pm M)| < \delta_2.$$

Let $\delta = \min\{\delta_1, \delta_2\}$. Then for $x < M < y$ with $|x - y| < \delta$, we have

$$|f(x) - f(y)| \leqslant |f(x) - f(M)| + |f(M) - f(y)| < \varepsilon,$$

and for $x > -M > y$ with $|x - y| < \delta$, we have

$$|f(x) - f(y)| \leqslant |f(x) - f(-M)| + |f(-M) - f(y)| < \varepsilon,$$

which proves the uniform continuity of f.

Solution of Problem 1.99

Let $\varepsilon > 0$. By hypothesis, we can find $M > 0$ such that

$$|f(t) - h(t)| \leqslant \tfrac{\varepsilon}{3} \qquad \forall\, |t| \geqslant M.$$

Let $T_\varepsilon = \big[-M - \varepsilon, M + \varepsilon\big]$. Then $f|_{T_\varepsilon}$ is uniformly continuous (see Definition 1.45 and Proposition 1.77). So, we can find $\delta = \delta(\varepsilon) \in (0, \varepsilon)$ such that

$$|f(t) - f(s)| \leqslant \varepsilon \qquad \forall\, t, s \in T_\varepsilon,\ |t - s| \leqslant \delta.$$

Since h is uniformly continuous, we can choose $\delta = \delta(\varepsilon) \in (0, \varepsilon)$ even smaller if necessary, so that

$$|h(t) - h(s)| \leqslant \tfrac{\varepsilon}{3} \qquad \forall\, t, s \in \mathbb{R},\ |t - s| \leqslant \delta.$$

Now, if $|t|, |s| \geqslant M$ and $|t-s| \leqslant \delta$, then

$$\left|f(t)-f(s)\right| \leqslant \left|f(t)-h(t)\right| + \left|h(t)-h(s)\right| + \left|h(s)-f(s)\right| \leqslant \varepsilon.$$

Finally, if $|t| \leqslant M$ and $|s| \geqslant M$ or conversely $|t| \geqslant M$ and $|s| \leqslant M$ and $|t-s| \leqslant \delta$, then

$$|t| \leqslant |s| + \varepsilon \leqslant M + \varepsilon \qquad \forall\, t, s,\ |t| - |s| \leqslant |t-s| \leqslant \varepsilon,$$

so, we infer that $t, s \in T_\varepsilon$ and thus $\left|f(t)-f(s)\right| \leqslant \varepsilon$. This proves the uniform continuity of f.

Solution of Problem 1.100

(a) Let $\mathcal{F}$ be an open cover of C. Because $\mathcal{F}$ covers C, there exists $U_0 \in \mathcal{F}$ such that $x \in U_0$. Because the set U_0 is open and $x \in U_0$, so there exists $r > 0$ such that $B_r(x) \subseteq U_0$ (see Definition 1.8(a)). Because $x_n \longrightarrow x$, so there exists $n_0 \geqslant 1$ such that $x_n \in B_r(x) \subseteq U_0$ for all $n \geqslant n_0$ (see Definition 1.7). Now, for all $k \in \{1, \ldots, n_0 - 1\}$, we can find $U_k \in \mathcal{F}$ such that $x_k \in U_k$. So, for an arbitrary open cover $\mathcal{F}$, we have found a finite subcover $\{U_k\}_{k=0}^{n_0}$ of the set C and thus the set C is compact (see Definition 1.63).

(b) "$\Longrightarrow$": This is obvious.

"$\Longleftarrow$": Suppose that $x_n \longrightarrow x$. Then $C = \{x\} \cup \{x_n : n \geqslant 1\}$ is a compact subset of X and so by hypothesis $f|_C$ is continuous. Therefore $f(x_n) \longrightarrow f(x)$, which proves the continuity of f (see Proposition 1.30).

Solution of Problem 1.101

Let $f\colon X \longrightarrow Y$ be a continuous and proper function (see Definition 1.72). Let $C \subseteq X$ be a nonempty and closed set. We need to show that $f(C)$ is closed. So, let $\{y_n\}_{n \geqslant 1} \subseteq f(C)$ be a sequence such that

$$y_n \longrightarrow y \quad \text{in } Y.$$

Then $y_n = f(x_n)$ with $x_n \in C$ for all $n \geqslant 1$ and the set

$$K = \{y\} \cup \{y_n : n \geqslant 1\} \subseteq Y$$

is compact (see Problem 1.100(a)). By the properties of f, the set $f^{-1}(K)$ is compact and $\{x_n\}_{n \geqslant 1} \subseteq f^{-1}(K)$. So, by passing to a subsequence if necessary, we may assume that

$$x_n \longrightarrow x \quad \text{in } C$$

(since C is closed; see Proposition 1.11). Then $f(x_n) \longrightarrow f(x)$ (since f is continuous; see Proposition 1.30) and so $y = f(x) \in f(C)$, which shows that $f(C)$ is closed (see Proposition 1.11). Therefore f is a closed function (see Definition 1.39(b)).

Solution of Problem 1.102

"(a) $\Longrightarrow$ (b)": Let $C \subseteq X$ be a nonempty closed set. We need to show that $f(C)$ is closed. So, let $\{y_n\}_{n \geqslant 1} \subseteq f(C)$ be a sequence such that $y_n \longrightarrow y$ in Y. Then $y_n = f(x_n)$, with $x_n \in C$ for all $n \geqslant 1$. By hypothesis we can find a subsequence $\{x_{n_k}\}_{k \geqslant 1}$ of $\{x_n\}_{n \geqslant 1}$ such that $x_{n_k} \longrightarrow x$ in X. Then $x \in C$ (as C is closed) and because f is continuous, we have

$$f(x_{n_k}) \longrightarrow f(x) = y \in f(C),$$

which proves that $f(C)$ is closed. Also, if $\{x_n\}_{n \geqslant 1} \subseteq f^{-1}(y)$, then

$$f(x_n) = y \qquad \forall\, n \geqslant 1$$

and so by hypothesis $\{x_n\}_{n \geqslant 1}$ admits a subsequence $\{x_{n_k}\}_{k \geqslant 1}$ such that $x_{n_k} \longrightarrow x$ in X. Note that $f(x) = y$ and so $f^{-1}(y)$ is compact in X.

"(b) $\Longrightarrow$ (a)": Let $\{x_n\}_{n \geqslant 1} \subseteq X$ be a sequence such that $f(x_n) \longrightarrow y$ in Y. Let

$$C_n = \{x_k : k \geqslant n\} \quad \text{and} \quad K = f^{-1}(y).$$

Since by hypothesis f is closed, we have

$$f(\overline{C}_n) = \overline{f(C_n)}$$

(see Proposition 1.32(d)) and because $f(x_n) \longrightarrow y$, we have

$$\{y\} = \bigcap_{n\geqslant 1} \overline{f(C_n)} = \bigcap_{n\geqslant 1} f(\overline{C}_n).$$

So, if $x \in K$, then

$$f(x) = y \in \bigcap_{n\geqslant 1} f(\overline{C}_n),$$

which implies that the closed subsets $D_n = K \cap \overline{C}_n$ of K are nonempty. Note that every finite subfamily of $\{D_n\}_{n\geqslant 1}$ has a nonempty intersection. Due to the compactness of K, Proposition 1.67 implies that

$$\bigcap_{n\geqslant 1} D_n = \bigcap_{n\geqslant 1} (K \cap \overline{C}_n) = K \cap \Big(\bigcap_{n\geqslant 1} \overline{C}_n\Big) \neq \emptyset.$$

This means that $\{x_n\}_{n\geqslant 1}$ has an accumulation point in K (see Definition 1.13(b) and Theorem 1.14(d)), which is equivalent to saying that there is a subsequence $\{x_{n_k}\}_{k\geqslant 1}$ of $\{x_n\}_{n\geqslant 1}$ and $x \in K$ such that $x_{n_k} \longrightarrow x$ in X.

Solution of Problem 1.103

Let $K \subseteq Y$ be a nonempty compact set and let $\{x_n\}_{n\geqslant 1} \subseteq f^{-1}(K)$ be a sequence. Then

$$f(x_n) \in K \qquad \forall\, n \geqslant 1$$

and since K is compact, we can find a subsequence $\big\{f(x_{n_k})\big\}_{k\geqslant 1}$ of $\big\{f(x_n)\big\}_{n\geqslant 1}$ such that

$$f(x_{n_k}) \longrightarrow y \quad \text{in } Y.$$

By statement **(a)** in Problem 1.102, we can find a further subsequence $\big\{x_{n_{k_m}}\big\}_{m\geqslant 1}$ such that $x_{n_{k_m}} \longrightarrow x$ in X. This shows that $f^{-1}(K)$ is compact.

Solution of Problem 1.104
We do the case when $f_n \searrow f$ (the proof of the case $f_n \nearrow f$ being similar). We have

$$f_n - f \geqslant 0 \qquad \forall\, n \geqslant 1.$$

For a given $\varepsilon > 0$, let

$$U_n = \{x \in X : 0 \leqslant (f_n - f)(x) < \varepsilon\} \qquad \forall\, n \geqslant 1.$$

For every $n \geqslant 1$, the set U_n is open (since the function $f_n - f$ is continuous; see Proposition 1.32) and

$$\bigcup_{n \geqslant 1} U_n = X.$$

By the compactness of X, we can find an integer $m \geqslant 1$ such that

$$\bigcup_{n=1}^{m} U_n = X$$

(see Definition 1.63). Note that $\{U_n\}_{n \geqslant 1}$ is an increasing sequence (i.e., $U_n \subseteq U_{n+1}$ for $n \geqslant 1$). Hence $U_m = X$ and so

$$0 \leqslant (f_n - f)(x) < \varepsilon \qquad \forall\, n \geqslant m,\ x \in X,$$

which implies that $f_n \rightrightarrows f$ (see Definition 1.59).

Solution of Problem 1.105
First consider $X = [0, 1)$ and

$$f_n(x) = x^n \qquad \forall\, x \in [0, 1),\ n \geqslant 1.$$

Then $f_n \searrow 0$ but not uniformly (see Definition 1.59). So, the compactness of X cannot be dropped.

Next let $X = [0, 1]$ and

$$f_n(x) = x^n \qquad \forall\, x \in [0, 1],\ n \geqslant 1.$$

Then $f_n \searrow \chi_{\{1\}}$ but not uniformly. So, the continuity of the limit function cannot be dropped.

Next let $X = [0,1]$ and let $f_n\colon [0,1] \longrightarrow \mathbb{R}$, $n \geqslant 1$ be the functions, defined by

$$f_n(x) = \begin{cases} 0 & \text{if } x = 0, \\ 1 & \text{if } 0 < x < \frac{1}{n}, \\ 0 & \text{if } \frac{1}{n} \leqslant x < 1. \end{cases}$$

Then $f_n \searrow 0$, but not uniformly. So, the continuity of functions f_n (at least for large n's) cannot be dropped.

Finally, let $X = [0,1]$ and let $f_n\colon [0,1] \longrightarrow \mathbb{R}$, $n \geqslant 2$ be the functions, defined by

$$f_n(x) = \begin{cases} nx & \text{if } 0 \leqslant x < \frac{1}{n}, \\ 2 - nx & \text{if } \frac{1}{n} \leqslant x < \frac{2}{n}, \\ 0 & \text{if } \frac{2}{n} \leqslant x \leqslant 1. \end{cases}$$

Note that $\{f_n\}_{n \geqslant 2}$ is not monotone, $f_n \longrightarrow 0$ but we do not have uniform convergence (see Definition 1.59). Indeed,

$$\begin{aligned} \beta_n &= \sup_{x \in [0,1]} |f_n(x)| = \sup_{x \in [0,1]} f_n(x) \\ &= f_n\left(\tfrac{1}{n}\right) = 1 \not\longrightarrow 0 \quad \text{as } n \to +\infty. \end{aligned}$$

So, the monotonicity condition on $\{f_n\}_{n \geqslant 1}$ cannot be dropped.

Solution of Problem 1.106

(a) Proceeding by contradiction, suppose that $f(X) \neq X$. So we can find $u \in X$ such that $u \notin f(X)$. Since $f(X)$ is compact (see Definition 1.63 and Proposition 1.74), we have

$$\operatorname{dist}(u, f(X)) = \varepsilon > 0$$

(see Definition 1.6). For every integer $n \geqslant 1$, let

$$f^{(n)} = \underbrace{f \circ \ldots \circ f}_{n\text{-times}}.$$

Evidently for every $n \geqslant 1$, $f^{(n)}$ is an isometry (see Definition 1.41) and

$$f^{(n)}(X) \subseteq f(X) \qquad \forall\, n \geqslant 1.$$

Then, if $n > m \geqslant 1$, we have

$$\varepsilon = \operatorname{dist}\big(u, f(X)\big) \leqslant \operatorname{dist}\big(u, f^{(n-m)}(X)\big) \leqslant d_X\big(u, f^{(n-m)}(u)\big),$$

so

$$\varepsilon \leqslant d_X\big(f^{(m)}(u), f^{(n)}(u)\big) \qquad \forall\, n, m \geqslant 1.$$

From this it follows that the sequence $\{f^{(n)}(u)\}_{n\geqslant 1} \subseteq X$ has no convergent subsequence, a contradiction to the fact that X is compact (see Definition 1.70 and Theorem 1.71).

(b) Let $X = \mathbb{N}$ be equipped with the discrete metric $d^d_{\mathbb{N}}$ (see Example 1.3) and let

$$f : \mathbb{N} \ni x \longmapsto x + 1 \in \mathbb{N}.$$

Then $(\mathbb{N}, d^d_{\mathbb{N}})$ is bounded (see Definition 1.6(b)) but not compact, f is continuous, but $f(\mathbb{N}) \neq \mathbb{N}$.

Solution of Problem 1.107

Let $x \in \mathbb{R}^N$ be the fixed point of f, i.e., $f(x) = x$. Let $r > 0$ be arbitrary and let and $u \in \overline{B}_r(x) = \{u \in X : \|u - x\| \leqslant r\}$. Then, since f is an isometry, we have

$$\big\|f(u) - x\big\| = \big\|f(u) - f(x)\big\| = \|x - u\| \leqslant r,$$

so $f(u) \in \overline{B}_r(x)$. Thus $f : \overline{B}_r(x) \longrightarrow \overline{B}_r(x)$ and by Problem 1.106, we have that

$$f\big(\overline{B}_r(x)\big) = \overline{B}_r(x).$$

Then, Proposition 1.76, implies that $f : \overline{B}_r(x) \longrightarrow \overline{B}_r(x)$ is a homeomorphism.

Let $u \in \mathbb{R}^N$ and consider the closed ball $\overline{B}_r(x)$ with $r = \|u - x\| + 1$. Then $u \in \overline{B}_r(x)$ and from the previous part of the solution, we can find $v \in \overline{B}_r(x)$ such that

$$f(v) = u.$$

Since $u \in \mathbb{R}^N$ was arbitrary, we infer that f is surjective. Therefore, f is a homeomorphism on all of $\mathbb{R}^N$.

Solution of Problem 1.108

Let $\{x_n\}_{n\geqslant 1} \subseteq X$ be a sequence. As X is totally bounded (see Definition 1.70(b)), so we can find a finite cover of X by balls of radius $\frac{1}{2}$:

$$X = \bigcup_{i=1}^{N_1} B_{\frac{1}{2}}(a_i^1)$$

(for some $N_1 \geqslant 1$, $a_i^1 \in X$ with $i = 1, \ldots, N_1$). Because the above cover is finite, we can find a subsequence of $\{x_n\}_{n\geqslant 1}$ denoted by $\{x_n^{(1)}\}_{n\geqslant 1}$ such that all values of $\{x_n^{(1)}\}_{n\geqslant 1}$ are in one of the balls of the above cover. So, in fact

$$\operatorname{diam}\{x_n^{(1)} : n \geqslant 1\} \leqslant \tfrac{1}{2}.$$

Next we choose further subsequences proceeding by induction. Suppose that we have already chosen subsequence $\{x_n^{(k)}\}_{k\geqslant 1}$ with the property, that

$$\operatorname{diam}\{x_n^{(k)} : n \geqslant k\} \leqslant \tfrac{1}{k}.$$

Because X is totally bounded, we can find a finite cover of X by balls of radius $\frac{1}{2(k+1)}$:

$$X = \bigcup_{i=1}^{N_{k+1}} B_{\frac{1}{2(k+1)}}(a_i^{k+1})$$

(for some $N_{k+1} \geqslant 1$, $a_i^{k+1} \in X$ with $i = 1, \ldots, N_{k+1}$). Because the above cover is finite, we can find a subsequence of $\{x_n^{(k)}\}_{n\geqslant k}$ denoted by $\{x_n^{(k+1)}\}_{n\geqslant k+1}$ such that all values of $\{x_n^{(k+1)}\}_{n\geqslant k+1}$ are in one of the balls of the above cover. So, in fact

$$\operatorname{diam}\{x_n^{(k+1)} : n \geqslant k+1\} \leqslant \tfrac{1}{k+1}.$$

Finally, let us take a "diagonal subsequence" $\{x_n^{(n)}\}_{n\geqslant 1} \subseteq \{x_n\}_{n\geqslant 1}$. Note that $\{x_n^{(n)}\}_{n\geqslant 1}$ is a Cauchy sequence (see Definition 1.7). To see this, let $\varepsilon > 0$ be arbitrary and let us choose $N > \frac{1}{\varepsilon}$, $N \in \mathbb{N}$. Then for any $m > n \geqslant N$, we have

$$d_X(x_n^{(n)}, x_m^{(m)}) \leqslant \operatorname{diam}\{x_k^{(N)} : k \geqslant N\} \leqslant \frac{1}{N} < \varepsilon.$$

Solution of Problem 1.109

Let $\varepsilon > 0$. For a given $(x_0, y) \in X \times Y$, by the continuity of $f(\cdot,\cdot)$, we can find $\delta_1(y), \delta_2(y) > 0$ such that

$$\text{if} d_X(x, x_0) < \delta_1(y) \text{and } d_Y(u, y) < \delta_2(y), \quad \text{then } d_V(f(x,u), f(x_0,y)) < \frac{\varepsilon}{2}$$

(see Definition 1.29). The balls $\{B_{\delta_2(y)}(y)\}_{y \in Y}$ form an open cover of Y. The compactness of Y (see Definition 1.63) implies that, we can find a finite subcover $\{B_{\delta_2(y_n)}(y_n)\}_{n=1}^N$ of Y. Let

$$\delta_1 = min\{\delta_1(y_n) : 1 \leqslant n \leqslant N\}$$

and let $u \in Y$. Assume that

$$d_X(x, x_0) < \delta_1$$

and $n \in \{1, \ldots, N\}$ is such that

$$u \in B_{\delta_2(y_n)}(y_n).$$

Then

$$d_X(x, x_0) < \delta_1 \leqslant \delta_1(y_n) \quad \text{and} \quad d_Y(u, y_n) < \delta_2(y_n)$$

and so

$$d_V(f(x,u),\ f(x_0,y_n)) < \frac{\varepsilon}{2}.$$

It also follows that

$$d_V(f(x_0,u),\ f(x_0,y_n)) < \frac{\varepsilon}{2}.$$

Then using the triangle inequality, we conclude that

$$d_V(f(x,u),\ f(x_0,u)) < \varepsilon \qquad \forall\, u \in Y, \text{ with } d_X(x, x_0) < \delta_1,$$

so

$$f(x,\cdot) \rightrightarrows f(x_0,\cdot) \quad \text{when } x \longrightarrow x_0 \quad \text{in } X.$$

Solution of Problem 1.110

Let $\{x_n\}_{n\geqslant 1} \subseteq X$ be a sequence such that $x_n \longrightarrow x$ in X. We choose $y_n \in Y$ such that

$$f(x_n, y_n) \leqslant m(x_n) + \tfrac{1}{n} \qquad \forall\, n \geqslant 1.$$

Since Y is compact, we can find a subsequence $\{y_{n_k}\}_{k\geqslant 1}$ of $\{y_n\}_{n\geqslant 1}$ such that

$$y_{n_k} \longrightarrow y \in Y.$$

The continuity of f implies that

$$f(x_{n_k}, y_{n_k}) \longrightarrow f(x,y) \quad \text{as } k \to +\infty$$

and from the definition of m, we have

$$m(x) \leqslant f(x,y).$$

Hence

$$m(x) \leqslant f(x,y) = \lim_{k\to+\infty} f(x_{n_k}, y_{n_k}) \leqslant \liminf_{k\to+\infty} m(x_{n_k}).$$

On the other hand, from the Weierstrass theorem (see Theorem 1.75), we know that we can find $\overline{y} = y(x) \in Y$ such that

$$m(x) = f(x,\overline{y}).$$

We have

$$m(x_n) \leqslant f(x_n,\overline{y}) \qquad \forall\, n \geqslant 1$$

and so

$$\limsup_{n\to+\infty} m(x_n) \leqslant f(x,\overline{y}) = m(x).$$

It follows that $m(x_{n_k}) \longrightarrow m(x)$. Since every subsequence of $\{m(x_n)\}_{n\geqslant 1}$ has a further subsequence converging to $m(x)$, we conclude that $m(x_n) \longrightarrow m(x)$ (Urysohn criterion for convergence; see Problem 1.3). This implies that m is continuous.

Solution of Problem 1.111

Since $X \times Y$ is compact (see Proposition 1.130), f is uniformly continuous (see Definition 1.45 and Proposition 1.77). So, for a given $\varepsilon > 0$, we can find $\delta = \delta(\varepsilon) > 0$ such that

$$d_X(x_1, x_2) < \delta \text{ and } d_Y(y_1, y_2) < \delta \implies d_V\big(f(x_1, y_1), f(x_2, y_2)\big) < \varepsilon.$$

In particular keeping $y \in Y$ fixed, we have

$$d_X(x_1, x_2) < \delta \quad \implies \quad d_V\big(f(x_1, y), f(x_2, y)\big) < \varepsilon,$$

so

$$d_X(x_1, x_2) < \delta \quad \implies \quad d^\infty\big(f_{x_1}, f_{x_2}\big) \ < \ \varepsilon.$$

This proves that the function $x \longmapsto f_x$ is continuous from X into $\big(C(Y;V), d^\infty\big)$. Since X is compact, the continuity of $x \longmapsto f_x$ implies that the set $\{f_x : x \in X\} \subseteq C(Y;V)$ is compact (see Proposition 1.74). So, by the Arzela–Ascoli theorem (see Theorem 1.84), we infer that family $\{f_x\}_{x\in X}$ is equicontinuous (see Definition 1.83).

Solution of Problem 1.112

(a) For any $n \geqslant 1$, let

$$S_n \;=\; \sum_{k=1}^{n} \frac{1}{k}.$$

Let $C = \{S_k : \ k \geqslant 1\}$ be equipped with the natural metric induced from $\mathbb{R}$. Then the set C is complete (but not compact).

Let $f\colon C \ni S_n \longmapsto S_{n+1} \in C$. Then f has no fixed point and for any $n > m$, we have

$$\begin{aligned} d_C\big(f(S_n), f(S_m)\big) &= S_{n+1} - S_{m+1} = \sum_{k=m+2}^{n+1} \frac{1}{k} < \sum_{k=m+1}^{n} \frac{1}{k} \\ &= S_n - S_m = d_C(S_n, S_m). \end{aligned}$$

(b) First we show the uniqueness of the fixed point. Indeed, if we can find $x, u \in X$ such that $x \neq u$, $f(x) = x$ and $f(u) = u$, then

$$d_X(x,u) \;=\; d_X\big(f(x), f(u)\big) \;<\; d_X(x,u),$$

a contradiction. So, the fixed point (if it exists) is unique.

Next we turn our attention to the existence of a fixed point. Consider the function $\varphi\colon X \longrightarrow \mathbb{R}_+$, defined by

$$\varphi(x) \;\stackrel{\text{def}}{=}\; d_X\big(f(x), x\big).$$

Then, for all $x, u \in X$, we have

$$\begin{aligned}
\big|\varphi(x) - \varphi(u)\big| &= \big|d_X\big(f(x),x\big) - d_X\big(f(u),u\big)\big| \\
&\leqslant \big|d_X\big(f(x),x\big) - d_X\big(x,f(u)\big)\big| \\
&\quad + \big|d_X\big(x,f(u)\big) - d_X\big(f(u),u\big)\big| \\
&\leqslant d_X\big(f(x),f(u)\big) + d_X(x,u) \;<\; 2d_X(x,u),
\end{aligned}$$

so φ is Lipschitz continuous.

Since X is compact, by Theorem 1.75, we can find $x_0 \in X$ such that

$$\varphi(x_0) \;=\; \inf\big\{\varphi(x) :\; x \in X\big\}.$$

Suppose that $\varphi(x_0) > 0$. Then $f(x_0) \neq x_0$ and so

$$\varphi\big(f(x_0)\big) \;=\; d_X\big(f^{(2)}(x_0), f(x_0)\big) \;<\; d_X\big(f(x_0), x_0\big) \;=\; \varphi(x_0),$$

which contradicts the minimality of x_0. Therefore $\varphi(x_0) = 0$ and so $f(x_0) = x_0$.

Solution of Problem 1.113

Consider the (affine) function S, defined by

$$S(u)(t) \;=\; \lambda \int_0^b k(t,s)u(s)\,ds + f(t) \qquad \forall\, u \in X,\; t \in [0,b].$$

Clearly, the continuity of k implies that $S(u) \in X$ for all $u \in X$. For every $u, v \in X$, we have

$$\begin{aligned} d^\infty\big(S(u), S(v)\big) &= \max_{0 \leqslant t \leqslant b} \Big|\lambda \int_0^b k(t,s)u(s)\,ds - \lambda \int_0^b k(t,s)v(s)\,ds\Big| \\ &= \max_{0 \leqslant t \leqslant b} |\lambda| \Big| \int_0^b k(t,s)\big(u(s) - v(s)\big)\,ds\Big|. \end{aligned}$$

The continuity of k on the compact $[0, b] \times [0, b]$ implies that there exists $M > 0$ such that

$$\big|k(t,s)\big| \leqslant M \qquad \forall\ (t,s) \in [0,b] \times [0,b]$$

(from the Weierstrass theorem; see Theorem 1.75). Then

$$d^\infty\big(S(u), S(v)\big) \leqslant |\lambda| M d^\infty(u,v) b.$$

So, if $|\lambda| < \frac{1}{Mb}$, then

$$d^\infty\big(S(u), S(v)\big) \leqslant k d^\infty(u,v) \qquad \forall\ u, v \in X,$$

with $k \in (0,1)$.

Since (X, d^∞) is a complete metric space, invoking the Banach fixed point theorem (see Theorem 1.49), we can find $u \in X$ such that

$$u(t) = \lambda \int_0^b k(t,s)u(s)\,ds + f(t) \qquad \forall\ t \in [0,1].$$

Solution of Problem 1.114

Let $\{U_n\}_{n \geqslant 1}$ be open dense subsets of X (see Definitions 1.20 and 1.78). We need to show that

$$\bigcap_{n \geqslant 1} U_n \text{ is dense in } X$$

or equivalently that for every nonempty open $V \subseteq X$, we have

$$V \cap \Big(\bigcap_{n \geqslant 1} U_n\Big) \neq \emptyset.$$

Since U_1 is dense in X, we have

$$V \cap U_1 \neq \emptyset.$$

Because X is locally compact, we can find a nonempty relatively compact open set V_1 such that

$$\overline{V}_1 \subseteq V \cap U_1$$

(see Definition 1.78 and Proposition 1.80). Then due to the density of U_2, we have

$$V_1 \cap U_2 \neq \emptyset$$

and so we can find a nonempty relatively compact open set V_2 such that

$$\overline{V}_2 \subseteq V_1 \cap U_2.$$

By induction, we generate a whole sequence $\{V_n\}_{n \geqslant 1}$ of nonempty relatively compact open sets such that

$$\overline{V}_n \subseteq V_{n-1} \cap U_n \qquad \forall\, n \geqslant 1$$

(with $V_0 = V$). Note that the sequence $\{V_n\}_{n \geqslant 1}$ is decreasing and by Proposition 1.67, we have

$$\bigcap_{n \geqslant 1} \overline{V}_n \neq \emptyset.$$

Hence, if $x \in \bigcap\limits_{n \geqslant 1} \overline{V}_n$, then we have $x \in V \cap \Big(\bigcap\limits_{n \geqslant 1} U_n\Big)$, which proves that $\bigcap\limits_{n \geqslant 1} U_n$ is dense in X. Therefore X is a Baire metric space (see Definition 1.81).

Solution of Problem 1.115

For any given $\eta > 0$, we can find $M = M(\eta) > 0$ such that

$$f(x) \geqslant \eta \qquad \forall\, x \in \mathbb{R}^N,\ \|x\| \geqslant M.$$

Let $u \in \mathbb{R}^N$, $\eta = f(u)$ and $M = M(\eta) > 0$ be as above. Since $f|_{\overline{B}_M}$ is continuous, it is bounded below (see Definition 1.6(b)) and we can find $x_0 \in \overline{B}_M$ such that

$$f(x_0) = \inf_{\overline{B}_M} f \leqslant f(u)$$

[see the Weierstrass theorem (Theorem 1.75)]. If $x \in \mathbb{R}^N$ with $\|x\| \geqslant M$, then

$$f(x) \geqslant \eta = f(u) \geqslant f(x_0)$$

and so we conclude that

$$f(x_0) = \inf_{\mathbb{R}^N} f.$$

Solution of Problem 1.116

Arguing by contradiction, suppose that X is not compact. Then we can find a sequence $\{x_n\}_{n\geqslant 1} \subseteq X$ with no convergent subsequence (see Theorem 1.71). Then the set $C = \{x_n : n \geqslant 1\}$ is closed in X and let $f \colon C \longrightarrow \mathbb{R}$ be defined by $f(x_n) = n$. Then f is continuous and so by the Tietze extension theorem (see Theorem 1.44), we can find a continuous extension $\hat{f} \colon X \longrightarrow \mathbb{R}$ of f. Evidently $\hat{f}$ is a continuous function on X which does not attain its supremum, a contradiction to our hypothesis.

Solution of Problem 1.117

(a) Since by hypothesis $D \subseteq \mathbb{R}^N$ is an F_σ-set (see Definition 1.57), we have that

$$D = \bigcup_{n\geqslant 1} C_n,$$

with $C_n \subseteq \mathbb{R}^N$ being closed for all $n \geqslant 1$. Without any loss of generality, we may assume that

$$C_n \subseteq C_{n+1} \qquad \forall\, n \geqslant 1$$

(indeed we can always replace C_n by $\bigcup\limits_{k=1}^{n} C_n$). Let

$$\hat{C}_n = C_n \cap [0,n]^N \qquad \forall\, n \geqslant 1.$$

Then, for every $n \geqslant 1$, the set $\hat{C}_n$ is compact and

$$D = \bigcup_{n \geqslant 1} \hat{C}_n.$$

Then

$$f(D) = f\Big(\bigcup_{n \geqslant 1} \hat{C}_n\Big) = \bigcup_{n \geqslant 1} f(\hat{C}_n)$$

and for all $n \geqslant 1$, the set $f(\hat{C}_n)$ is compact (see Proposition 1.74) hence closed (see Proposition 1.69). Therefore $f(D)$ is an F_σ-set too.

(b) No. To show this, we will construct a continuous function $f\colon \mathbb{R} \longrightarrow \mathbb{R}$ such that $f(\mathbb{Z}) = \mathbb{Q}$. The set $\mathbb{Z}$ is a G_δ-set (since $\mathbb{Z}$ is closed in $\mathbb{R}$) but $\mathbb{Q}$ is not a G_δ-set (see Problem 1.58). Let $\{q_n\}_{n \geqslant 1} \subseteq \mathbb{Q}$ be an enumeration of $\mathbb{Q}$. Then for every $x \in \mathbb{R}$, we can find $k \in \mathbb{Z}$ such that $k < x \leqslant k+1$ and we define

$$f(x) = q_{k+1} + \big(q_{k+1} - q_k\big)(x - k - 1).$$

Then, the function $f\colon \mathbb{R} \longrightarrow \mathbb{R}$ is continuous and $f(\mathbb{Z}) = \mathbb{Q}$.

Solution of Problem 1.118

(a) We argue by contradiction. So, suppose that $\operatorname{dist}(K, C) = 0$ (see Definition 1.6). We can find a sequences $\{u_n\}_{n \geqslant 1} \subseteq K$ and $\{x_n\}_{n \geqslant 1} \subseteq C$ such that

$$d_X(u_n, x_n) \longrightarrow 0.$$

Because of the compactness of K, by passing to a suitable subsequence if necessary, we may assume that

$$u_n \longrightarrow u^* \in K.$$

Then also

$$x_n \longrightarrow u^*$$

and because C is closed, we have $u^* \in C$. Therefore $u^* \in K \cap C$, a contradiction to the hypothesis that $K \cap C = \emptyset$.
(b) No. To see this, let $X = \mathbb{R}^2$ and consider two sets

$$K = \left\{ \left(x, \tfrac{1}{x}\right) : x > 0 \right\} \quad \text{and} \quad C = \left\{ (x, 0) : x \geqslant 0 \right\}.$$

Then K and C are nonempty closed disjoint sets, but $\operatorname{dist}(K, C) = 0$.

Solution of Problem 1.119

Let $r \stackrel{\text{def}}{=} \frac{1}{2}\operatorname{dist}(K, C)$ (see Definition 1.6). From Problem 1.118(a), we know that $r > 0$. Let

$$U \stackrel{\text{def}}{=} \bigcup_{x \in K} B_r(x).$$

Then the set U is open (see Proposition 1.9(b)), $K \subseteq U$ and $C \cap U = \emptyset$ (from the choice of $r > 0$).

Solution of Problem 1.120

(a) Note that $\overline{D} \subseteq X$ is compact. So, due to the continuity of f, the set $f(\overline{D})$ is compact (see Proposition 1.74), hence closed (see Proposition 1.69(a)). Since $f(D) \subseteq f(\overline{D})$, it follows that

$$\overline{f(D)} \subseteq f(\overline{D}).$$

On the other hand the continuity of f implies that

$$f(\overline{D}) \subseteq \overline{f(D)}$$

(see Proposition 1.32(d)). So, we conclude that

$$f(\overline{D}) = \overline{f(D)}.$$

(b) The inclusion $f(\overline{D}) \subseteq \overline{f(D)}$ is always true (see Proposition 1.32(d)). The opposite inclusion can fail.

Let $X = Y = D = \mathbb{R}$ and $f(x) = e^x$. Then

$$f(\overline{D}) = (0, +\infty) \subsetneq [0, +\infty) = \overline{f(D)}.$$

Solution of Problem 1.121

(a) Let X^* be the completion of X (see Definition 1.50 and Theorem 1.51). Reasoning indirectly, suppose that $X^* \neq X$ and let $x^* \in X^* \setminus X$. Consider the function $f\colon X \longrightarrow \mathbb{R}$, defined by

$$f(x) = d_{X^*}(x^*, x) \qquad \forall\, x \in X,$$

with d_{X^*} being the metric of the completion X^*. The function f is continuous (see Proposition 1.36).

Let $D = X$. Then

$$\overline{D} = D \quad \text{and} \quad 0 \notin f(\overline{D}),$$

while $0 \in \overline{f(D)}$ (because X is dense in X^*; Definition 1.20), so

$$f(\overline{D}) \subsetneq \overline{f(D)},$$

a contradiction. Therefore $X = X^*$ and so X is complete.

(b) Yes. Let us proceed by contradiction and suppose that X is not compact. Thus, there exists a sequence $\{x_n\}_{n \geqslant 1} \subseteq X$ with no convergent subsequence. Note that the set

$$D = \{x_n : n \geqslant 1\}$$

is closed, and the function

$$f\colon D \ni x_n \longmapsto \frac{1}{n} \in \mathbb{R}$$

is continuous on D. Using Tietze extension theorem (see Theorem 1.44), we can extent f to a function continuous on X (still denoted by f). Note, that

$$\begin{aligned} f(D) = f(\overline{D}) &= \left\{\tfrac{1}{n} : n \geqslant 1\right\} \subsetneq \overline{\left\{\tfrac{1}{n} : n \geqslant 1\right\}} \\ &= \{0\} \cup \left\{\tfrac{1}{n} : n \geqslant 1\right\} = \overline{f(D)}, \end{aligned}$$

a contradiction. So X is compact.

Solution of Problem 1.122

(a) "$\Longrightarrow$": This comes from Problem 1.120(a).

"$\Longleftarrow$": This comes from Problem 1.121(b).

(b) The inclusion "$\Longrightarrow$" is always true as it was stated in Problem 1.120(a).

The inclusion "$\Longleftarrow$" need not be true if we replace $\mathbb{R}$ by another metric space. Let $Y = \{0, 1\}$ be with the discrete metric. Then any continuous function $f\colon X \longrightarrow Y$ is constant and of course the condition

$$f(\overline{D}) = \overline{f(D)}$$

is satisfied for any $D \subseteq X$. But X can be any metric space (not necessarily compact).

Solution of Problem 1.123

Since X is locally compact, for every $x \in K$, we can find $r(x) > 0$ such that the ball $\overline{B}_{r(x)}(x)$ is compact. The family $\{B_{r(x)}(x)\}_{x \in K}$ is an open cover of K. Because K is compact, we can find a finite set $\{x_k\}_{k=1}^N$ such that

$$K \subseteq \bigcup_{n=1}^{N} B_{r(x_n)}(x_n) = U.$$

The set U is open and

$$\overline{U} = \overline{\bigcup_{n=1}^{N} B_{r(x_n)}(x_n)} \subseteq \bigcup_{n=1}^{N} \overline{B}_{r(x_n)}(x_n) = \widetilde{K}$$

and $\widetilde{K}$ is compact. Note that the function $h\colon X \longrightarrow \mathbb{R}_+$, defined by

$$h(x) = \operatorname{dist}(x, U^c)$$

is continuous (see Definition 1.6 and Problem 1.78) and $h|_K > 0$ (see Problem 1.118). Invoking the Weierstrass theorem (see Theorem 1.75), we can find $x_0 \in K$ such that

$$h(x_0) = \operatorname{dist}(x_0, U^c) = \inf\{\operatorname{dist}(x, U^c) : x \in K\},$$

so $h(x_0) = r > 0$. We claim that $K_r \subseteq U$. Indeed, if $v \notin U$, then

$$\operatorname{dist}(v, K) = \inf \{d_X(v, y) : y \in K\} \geqslant \operatorname{dist}(v, U^c) = h(x_0) = r > 0.$$

So, $v \notin K_r$ and we have proved the claim. Finally note that the compactness of $\overline{U}$ implies that $\overline{K}_r$ is compact too (see Proposition 1.69).

Solution of Problem 1.124

(a) Let $\{(x_n, f(x_n))\}_{n\geqslant 1} \subseteq \operatorname{Gr} f$ (see Definition 1.132) be a sequence and assume that

$$(x_n, f(x_n)) \longrightarrow (x, y) \quad \text{in } X \times Y.$$

Due to the continuity of f, we have

$$f(x_n) \longrightarrow f(x) \quad \text{in } Y.$$

Hence $f(x) = y$ and so

$$(x, y) = (x, f(x)) \in \operatorname{Gr} f,$$

which proves that the set $\operatorname{Gr} f \subseteq X \times Y$ is closed.

To show that the converse is not in general true, let us consider the function $f : \mathbb{R} \longrightarrow \mathbb{R}$, defined by

$$f(x) = \begin{cases} \frac{1}{x} & \text{if } x \neq 0, \\ 0 & \text{if } x = 0. \end{cases}$$

Then $\operatorname{Gr} f$ is closed in $\mathbb{R} \times \mathbb{R} = \mathbb{R}^2$ but f is discontinuous at $x = 0$.

(b) "$\Longrightarrow$": This comes from part **(a)**.

"$\Longleftarrow$": Let $x_n \longrightarrow x$ in X. Then, because of the compactness of Y, we can find a subsequence $\{x_{n_k}\}_{k\geqslant 1}$ of $\{x_n\}_{n\geqslant 1}$ such that

$$f(x_{n_k}) \longrightarrow y \quad \text{in } Y.$$

Since $\operatorname{Gr} f$ is closed, we have $y = f(x)$. So, every subsequence of $\{f(x_n)\}_{n\geqslant 1}$ has a further subsequence, which converges to $f(x)$. This implies that $f(x_n) \longrightarrow f(x)$ (see Problem 1.3) and proves the continuity of f.

Solution of Problem 1.125

We proceed by contradiction. So, suppose that

$$[0,1] = \bigcup_{n\geqslant 1} C_n,$$

with C_n being nonempty closed sets which are pairwise disjoint. We can find an open set U_2 such that

$$C_2 \subseteq U_2 \quad \text{and} \quad U_2 \cap C_1 = \emptyset$$

(see Problem 1.119). Then some component D_2 of $\overline{U}_2$ (the component is of course a closed interval) satisfies

$$D_2 \cap C_2 \neq \emptyset.$$

If $D_2 \subseteq U_2$, then we can find an open interval (d, d^*) such that

$$D_2 \subseteq (d, d^*) \subseteq U_2$$

(see Problem 1.119 with $K = D_2$ and $C = U_2^c$), which implies that D_2 is not a component of $\overline{D}_2$, a contradiction. Therefore $D_2 \subseteq U_2$ cannot occur and we have $D_2 \setminus U_2 \neq \emptyset$, hence

$$D_2 \setminus C_2 \neq \emptyset.$$

It follows that

$$\emptyset \neq D_2 \setminus C_2 \subseteq \bigcup_{n\geqslant 3} (D_2 \cap C_n)$$

and so for some integer $n \geqslant 3$, we have $D_2 \cap C_n \neq \emptyset$. We repeat this argument to the closed interval D_2 and this way by induction we generate a decreasing sequence of closed intervals $\{D_m\}_{m\geqslant 2}$ such that

$$D_m \neq \emptyset, \quad D_{m+1} \subseteq D_m, \quad \text{and} \quad D_m \cap C_{m-1} = \emptyset \qquad \forall\, m \geqslant 2.$$

Therefore $\bigcap\limits_{m\geqslant 2} D_m \neq \emptyset$ (see Proposition 1.67) and

$$\Big(\bigcap_{m\geqslant 2} D_m\Big) \cap \Big(\bigcup_{n\geqslant 1} C_n\Big) = \bigcup_{n\geqslant 1} \Big(C_n \cap \Big(\bigcap_{m\geqslant 2} D_m\Big)\Big) = \emptyset,$$

a contradiction.

Solution of Problem 1.126

Since $\bigcap\limits_{k=1}^{N} C_k = \emptyset$, we have

$$\bigcup_{k=1}^{N} U_k = X, \quad \text{with } U_k = C_k^c \ (k = 1, \dots, N).$$

So, if $x \in X$, then we can find $k_0 \in \{1, \dots, N\}$ such that $x \in U_{k_0}$. The set U_{k_0} is open, hence there exists $r_x > 0$ such that $B_{2r_x}(x) \subseteq U_{k_0}$. The family $\{B_{r_x}(x)\}_{x \in X}$ is an open cover of X, so there is a finite subcover $\{B_{r_{x_s}}(x_s)\}_{s=1}^{M}$. Let

$$\delta = \min\{r_{x_s} : s = 1, \dots, M\}.$$

Let $E \subseteq X$ be a set which intersects all the sets C_k for $k = 1, \dots, N$. Suppose that $\operatorname{diam} E < \delta$ and let $y \in E$. Then we can find $s_0 \in \{1, \dots, M\}$ and $k_0 \in \{1, \dots, N\}$ such that

$$y \in B_{r_{x_{s_0}}}(x_{s_0}) \subseteq B_{2r_{x_{s_0}}}(x_{s_0}) \subseteq U_{k_0}.$$

For every other point $u \in E$, we have

$$d_X(u, x_{s_0}) \leqslant d_X(u, y) + d_X(y, x_{s_0}) < \delta + r_{x_{s_0}} \leqslant 2r_{x_{s_0}},$$

so

$$E \subseteq B_{2r_{x_{s_0}}}(x_{s_0}) \subseteq U_{k_0}$$

and thus

$$E \cap C_{k_0} = \emptyset,$$

a contradiction. Therefore $\operatorname{diam} E \geqslant \delta$.

Solution of Problem 1.127

First we show that $\sup\limits_{\mathbb{R}} f < +\infty$. If this is not the case, then for every $n \geqslant 1$, we can find $x_n \in X$ such that $f(x_n) > n$. Then $\{f^{-1}([n, +\infty))\}_{n \geqslant 1}$ is a closed subsets of X with the finite intersection property (see Definition 1.66). By Proposition 1.67, it has a nonempty intersection. Let u be an element in that intersection. Then

$$f(u) \geqslant n \qquad \forall\, n \geqslant 1,$$

a contradiction. Hence

$$\sup_{\mathbb{R}} f = M < +\infty.$$

Next, let

$$C_n = \left\{x \in X : f(x) \geqslant M - \tfrac{1}{n}\right\} = f^{-1}\left(\left[M - \tfrac{1}{n}, +\infty\right)\right) \qquad \forall\, n \geqslant 1.$$

By hypothesis, each C_n is a closed set in X and

$$C_n \supseteq C_{n+1} \qquad \forall\, n \geqslant 1.$$

It follows that

$$\bigcap_{n\geqslant 1} C_n \neq \emptyset$$

(see Proposition 1.67). We choose $x_0 \in \bigcap\limits_{n\geqslant 1} C_n$. Then

$$f(x_0) = M = \sup_{\mathbb{R}} f < +\infty.$$

Solution of Problem 1.128

Let $\{x_n\}_{n\geqslant 1}$ be a Cauchy sequence in X. The sequence $\{x_n\}_{n\geqslant 1}$ is bounded, i.e.,

$$\operatorname{diam}\left\{x_n : n \geqslant 1\right\} < +\infty.$$

So, we can find $r > 0$ and $y \in X$ such that $\{x_n\}_{n\geqslant 1} \subseteq \overline{B}_r(y)$. But by hypothesis the latter is compact. Hence, we can find a subsequence $\left\{x_{n_k}\right\}_{k\geqslant 1}$ of $\{x_n\}_{n\geqslant 1}$ such that

$$x_{n_k} \longrightarrow x \in \overline{B}_r(y) \quad \text{as } k \to +\infty.$$

Then $x_n \longrightarrow x$ and this proves the completeness of X. Recall that every compact set is closed and bounded (see Proposition 1.69(a)). On the other hand every closed and bounded set, is a closed subset of some closed ball, hence it is compact (see Proposition 1.69(b)).

Solution of Problem 1.129
By hypothesis, there exist $\beta > 0$ and $M = M(\beta) > 0$ such that

$$f(x) \geqslant \beta\|x\| \qquad \forall\ \|x\| \geqslant M.$$

For a given $\lambda \in \mathbb{R}$. Let $r = \max\left\{M, \frac{\lambda}{\beta}\right\}$. Then

$$L_\lambda = \left\{x \in \mathbb{R}^N : f(x) \leqslant \lambda\right\} \subseteq \overline{B}_r(0),$$

so $\overline{L}_\lambda$ is compact (see Proposition 1.69(b)).

Solution of Problem 1.130
Since $E \subseteq B_r(0)$ for some $r > 0$, it suffices to prove the problem for the case when $E = B_r(0)$. We set

$$\|f\|_\infty = \max_{t\in[0,1]} |f(t)| \qquad \forall\ f \in C([0,1])$$

and

$$\|G\|_\infty = \max\left\{|G(t,s)| : (t,s) \in [0,1]\times[0,1]\right\}.$$

Then for all $t \in [0,1]$, we have

$$|L(f)(t)| \leqslant \int_0^1 |G(t,s)||f(s)|\,ds \leqslant \|G\|_\infty\|f\|_\infty \leqslant r\|G\|_\infty,$$

so

$$L(B_r(0)) \subseteq \overline{B}_{r\|G\|_\infty}(0).$$

Also, if $f \in B_r(0)$ and $t, \tau \in [0,1]$, then

$$\begin{aligned}|L(f)(t) - L(f)(\tau)| &\leqslant \int_0^1 |G(t,s) - G(\tau,s)||f(s)|\,dz \\ &\leqslant r\int_0^1 |G(t,s) - G(\tau,s)|\,ds.\end{aligned}$$

Note that $[0,1]\times[0,1]$ is compact and by hypothesis G is continuous on $[0,1]\times[0,1]$. Hence it is uniformly continuous (see Definition 1.45

and Proposition 1.77). Therefore, for a given $\varepsilon > 0$, we can find $\delta = \delta(\varepsilon) > 0$ such that

$$\left|G(t,s) - G(\tau,s)\right| < \tfrac{\varepsilon}{r} \qquad \forall\, s,t,\tau \in [0,1], \text{ with } |t-\tau| < \delta.$$

Then

$$\left|L(f)(t) - L(f)(\tau)\right| < \varepsilon \qquad \forall\, f \in B_r(0),\ t,\tau \in [0,1], \text{ with } |t-\tau| < \delta,$$

so $L\big(B_r(0)\big)$ is equicontinuous. Invoking the Arzela–Ascoli theorem (see Theorem 1.84), we conclude that

$$\overline{L\big(B_r(0)\big)} \text{ is compact in } C\big([0,1]\big).$$

Solution of Problem 1.131

No. According to the Arzela–Ascoli theorem (see Theorem 1.84), if the set $\{\sin(nx) :\ n \geqslant 1\}$ is relatively compact (for the d^∞ metric), it must be equicontinuous (uniformly since $[-\pi,\pi]$ is compact) and uniformly bounded (see Definition 1.83). Clearly $\{\sin(nx) :\ n \geqslant 1\}$ is uniformly bounded. However, for a given δ, we can find $n_0 \geqslant 1$ large enough such that $\frac{\pi}{n_0} < \delta$. If

$$x = -\tfrac{\pi}{2n_0} \quad \text{and} \quad y = \tfrac{\pi}{2n_0},$$

then

$$|x-y| = \tfrac{\pi}{n_0} < \delta \quad \text{and} \quad \left|\sin(n_0x) - \sin(n_0y)\right| = 2.$$

Therefore the set $\{\sin(nx) :\ n \geqslant 1\}$ is not equicontinuous and so it cannot be relatively compact in $C\big([-\pi,\pi]\big)$.

Solution of Problem 1.132

Let $x \in E$. Since by hypothesis E is locally compact with respect to the metric d_X^1 (see Definition 1.78), we can find an open set U in (E, d_X^1) such that $\overline{U}^E$ is compact in (E, d_X^1) (by $\overline{U}^E$ we denote the closure of U in (E, d_X^1)). Since

$$\overline{U}^E = E \cap \overline{U},$$

it follows that $\overline{U}^E$ is compact in (X, d_X^1). Also since U is open in (E, d_X^1), we can find a set V open in (X, d_X^1) such that

$$U = E \cap V.$$

Since spaces (X, d_X^1) and (X, d_X^2) are topologically equivalent (see Definition 1.52), we have that $\overline{U}^E$ is compact in (X, d_X^2) and V is open in (X, d_X^2). Hence $U = E \cap V$ is open in (E, d_X^2) and $\overline{U}^E$ is compact in (E, d_X^2). Because $x \in E$ was arbitrary, we conclude that (E, d_X^2) is locally compact.

Solution of Problem 1.133

Let $r > 0$ be arbitrary. We will show that the closed ball $\overline{B}_r\big((0,0)\big)$ is not compact. Let $s = \min\big\{\frac{r}{2}, 1\big\}$. Let us fix $x_0 \in (0, s)$ such that $\sin \frac{1}{x_0} = s$. Let

$$x_k = \frac{1}{\frac{1}{x_0} + 2k\pi} \qquad \forall\, k = 1, 2, \ldots .$$

Then

$$\sin \tfrac{1}{x_k} = \sin\big(\tfrac{1}{x_0} + 2k\pi\big) = \sin \tfrac{1}{x_0} = s,$$

so $\lim\limits_{k \to +\infty} \big(x_k, \sin \frac{1}{x_k}\big) = (0, s) \notin X$ and the sequence $\big\{\big(x_k, \sin \frac{1}{x_k}\big)\big\}_{k \geqslant 1}$ has no limit in $\overline{B}_r\big((0,0)\big)$. Thus the closed ball $\overline{B}_r\big((0,0)\big)$ is not compact. This proves that X is not locally compact (see Definition 1.78).

Solution of Problem 1.134

(a) Let $y \in f(X)$. Then $y = f(x)$ for some $x \in X$. Since X is locally compact, we can find $r > 0$ such that $\overline{B}_r(x)$ is compact. Since by hypothesis f is open, the set $f\big(B_r(x)\big)$ is open and $y \in f\big(B_r(x)\big)$. Also $f\big(\overline{B}_r(x)\big)$ is compact (see Proposition 1.74), hence closed in Y. By the continuity of f, we have

$$f\big(\overline{B}_r(x)\big) \subseteq \overline{f\big(B_r(x)\big)}$$

(see Proposition 1.32(d)) and since $f(\overline{B}_r(x))$ is compact, we have

$$f(\overline{B}_r(x)) = \overline{f(B_r(x))}$$

(see Problem 1.120). Therefore, $f(B_r(x))$ is a relatively compact neighbourhood of $y \in f(X)$. Since $y \in f(X)$ was arbitrary, we conclude that $f(X)$ is relatively compact.

(b) No. Let

$$\begin{aligned} X &= \{(-1,0)\} \cup ((0,+\infty) \times \mathbb{R}), \\ Y &= \{(0,0)\} \cup ((0,+\infty) \times \mathbb{R}), \end{aligned}$$

both with the natural metric induced by the Euclidean metric in $\mathbb{R}^2$ and let $f\colon X \longrightarrow Y$ be defined by

$$f(x) = \begin{cases} x & \text{if } x \in (0,+\infty) \times \mathbb{R}, \\ (0,0) & \text{if } x = (-1,0). \end{cases}$$

Then f is a continuous bijection (but not open), X is locally compact and $Y = f(X)$ is not locally compact.

(c) No. Let $X = [0,+\infty)$ (with the natural metric induced by the Euclidean metric in $\mathbb{R}$) and let

$$Y = \{(0,0)\} \cup \{(x, \sin \tfrac{1}{x}) : x > 0\}$$

(with the natural metric induced by the Euclidean metric in $\mathbb{R}^2$). Let $f\colon X \longrightarrow Y$ be the function, defined by

$$f(x) = \begin{cases} (x, \sin \frac{1}{x}) & \text{if } x > 0, \\ (0,0) & \text{if } x = 0. \end{cases}$$

The f is an open bijection (but not continuous), X is locally compact and $Y = f(X)$ is not locally compact (see Problem 1.133).

Solution of Problem 1.135

From Proposition 1.80, we know that X is an open subset of its completion. Invoking the Alexandrov theorem (see Theorem 1.58), we infer that X is topologically complete. In particular, it is homeomorphic (see Definition 1.39) to a complete metric space (see Remark 1.53).

Solution of Problem 1.136
If $\mathbb{Q}$ is homeomorphic (see Definition 1.39) to a complete metric space, then $\mathbb{Q}$ is a Baire metric space(see Problem 1.114). Let $\{q_n\}_{n\geqslant 1}$ be an enumeration of $\mathbb{Q}$. Then for each $n \geqslant 1$, the set $\mathbb{Q} \setminus \{q_n\}$ is open and dense (see Definition 1.20). Hence the set $\bigcap\limits_{n\geqslant 1} (\mathbb{Q} \setminus \{q_n\})$ is dense in $\mathbb{Q}$. But

$$\bigcap_{n\geqslant 1} (\mathbb{Q} \setminus \{q_n\}) = \emptyset,$$

a contradiction.

Solution of Problem 1.137
Let $x \in X$. Then $x \in U$ for some $U \in \mathcal{Y}$. Since U is open, we can find $r_x > 0$ such that $B_{2r_x}(x) \subseteq U$. The balls $\{B_{2r_x}(x)\}_{x\in X}$ form an open cover of X and so we can find a finite subcover $\{B_{2r_{x_k}}(x_k)\}_{k=1}^n$. Let $\delta = \min\{r_{x_k}\}_{k=1}^n$. We will show that this is the desired $\delta > 0$. To this end, let $A \subseteq X$ be a nonempty set with $\operatorname{diam} A < \delta$. Let $u \in A$. We can find $k \in \{1, \ldots, n\}$ such that $u \in B_{r_{x_k}}(x_k)$. Let $y \in A$. Using the triangle inequality, we have

$$d_X(y, x_k) \leqslant d_X(y, u) + d_X(u, x_k) < \delta + r_{x_k} \leqslant 2r_{x_k},$$

so $y \in B_{2r_{x_k}}(x_k) \subseteq U$ and thus $A \subseteq U$.

Solution of Problem 1.138
Let $f\colon \mathbb{R} \longrightarrow \mathbb{R}$ be the function, defined by

$$f(x) = \begin{cases} 0 & \text{if } x \text{ is irrational or } x = 0, \\ \frac{(-1)^q(q-1)}{q} & \text{if } x \in \mathbb{Q} \setminus \{0\} \text{ and } x = \frac{p}{q} \text{ with } (p,q) \\ & \quad \text{irreducible representation of } x \text{ and } q > 0. \end{cases}$$

For every rational $r \neq 0$, we have

$$|f(r)| \leqslant \tfrac{q-1}{q} < 1.$$

So, f is bounded above by 1 and bounded below by -1. Let T be a compact interval with nonempty interior. Let

$$A = \{q > 0 : (p,q) \text{ is the irreducible representation with positive denominator of a rational } x \in T\}.$$

We will show that A is unbounded. Let $M > 0$ be arbitrary and we will find $k > M$ such that $k \in A$. Let $T = [a,b]$ with $a < b$ and let $x = \frac{p}{q} \in (a,b)$ with (p,q) being irreducible representation of x, with $q \geqslant 2$. Note that the function

$$\varphi(t) = \tfrac{tp}{tq+p} \qquad \forall\, t \in \mathbb{R} \setminus \{-\tfrac{p}{q}\}$$

is continuous and

$$\lim_{t \to \pm\infty} \varphi(t) = \tfrac{p}{q}.$$

So, we can choose $r \in \mathbb{N}$ large such that

$$a < \tfrac{q^r p}{q^r q + p} < b \quad \text{and} \quad q^r q + p > M.$$

But note that the fraction

$$\tfrac{q^r p}{q^r q+p} \text{ is irreducible.}$$

So $k = q^r q + p$ satisfies $k \in A$ and $k > M$. Hence $A \subseteq \mathbb{N}$ is unbounded. Since $\frac{n-1}{n} \longrightarrow 1$ as $n \to +\infty$, we infer that

$$-1 \in \inf f(T) \quad \text{and} \quad 1 = \sup f(T).$$

But

$$-1 < f(x) < 1 \qquad \forall\, x \in T.$$

Solution of Problem 1.139

Suppose that (X, d_X) is totally bounded metric space (see Definitions 1.70 and 1.21(d)). Then for every $n \geqslant 1$, we can find a finite set $F_n \subseteq X$ such that

$$X = \bigcup_{x \in F_n} B_{\frac{1}{n}}(x),$$

where $B_{\frac{1}{n}}(x) = \{u \in X : d_X(u,x) < \frac{1}{n}\}$. We set

$$D = \bigcup_{n \geqslant 1} F_n.$$

Then D is countable and dense in X (see Definition 1.20). Therefore X is separable.

Solution of Problem 1.140

"**(a)** $\Longrightarrow$ **(b)**": Since the identity $i\colon (X,d_X) \longrightarrow (X,\hat{d}_X)$ is a homeomorphism (see Remark 1.53 and Definition 1.52), then $(X,\hat{d}_X)$ is compact (see Proposition 1.74), hence complete too (see Theorem 1.71).

"**(b)** $\Longrightarrow$ **(a)**": We argue by contradiction. So, suppose that (X,d_X) is not compact. We may assume that d_X is bounded and more precisely $d_X \leqslant 1$ (if this is not the case, we replace d_X by $\frac{d_X}{1+d_X}$; note that the new metric is topologically equivalent (or even uniformly equivalent) to the previous one (see the solution of Problem 1.70) and the new metric still remains non-compact (see Remark 1.53 and Proposition 1.74)). Since we assume that (X,d_X) is not compact, we can find a sequence $\{x_n\}_{n\geqslant 1} \subseteq X$ with no convergent subsequence.

Let

$$L = \{f\colon X \longrightarrow \mathbb{R} : |f(x) - f(u)| \leqslant d_X(x,u) \text{ for all } x,u \in X\}.$$

Then

$$d_X(x,u) = \sup_{f \in L} |f(x) - f(u)|.$$

Let

$$\hat{L}_n = \{f\colon X \longrightarrow \mathbb{R} : |f(x) - f(u)| \leqslant \tfrac{1}{n} d_X(x,u) \text{ and } f(x_k) = 0 \ \forall k > n\}$$

and let us set

$$\hat{L} = \bigcup_{n \geqslant 1} \hat{L}_n.$$

We define

$$\hat{d}_X(x,u) = \sup_{f \in \hat{L}} |f(x) - f(u)|.$$

Claim 1. $\{x_n\}_{n\geqslant 1}$ is a Cauchy sequence in $(X, \hat{d}_X)$.

Let $\varepsilon > 0$, $N \geqslant \frac{1}{\varepsilon}$, $m, k \geqslant N$ and $f \in \hat{L}$. If $f \in \hat{L}_n$ and $n \geqslant N$, then

$$|f(x_m) - f(x_k)| \leqslant \tfrac{1}{n} d_X(x_m, x_k) \leqslant \tfrac{1}{n} \leqslant \tfrac{1}{N} \leqslant \varepsilon.$$

If $f \in \hat{L}_n$ and $n < N$, then $m, k > n$ and so

$$f(x_m) = f(x_k) = 0.$$

Hence

$$\hat{d}_X(x_m, x_k) \leqslant \varepsilon \qquad \forall\, m, k \geqslant N.$$

So, we conclude that $\{x_n\}_{n\geqslant 1}$ is a Cauchy sequence in $(X, \hat{d}_X)$. This proves Claim 1.

Claim 2. d_X is topologically equivalent to $\hat{d}_X$.

Since $\hat{d}_X \leqslant d_X$, the identity function $i\colon (X, d_X) \longrightarrow (X, \hat{d}_X)$ is continuous. We consider the inverse function

$$i^{-1}\colon (X, \hat{d}_X) \longrightarrow (X, d_X).$$

Let $u \in X$ and $\varepsilon > 0$. By hypothesis u is not an accumulation point of the sequence $\{x_n\}_{n\geqslant 1}$ (see Definition 1.13(b) and Theorem 1.14(d)). So, we can find $n_0 \geqslant 1$ such that

$$d_X(x_n, u) \geqslant \varepsilon \qquad \forall\, n \geqslant n_0.$$

Let

$$f(x) = \tfrac{1}{n_0}\big(\varepsilon - d_X(x, u)\big)^+.$$

Evidently f is $\frac{1}{n_0}$-Lipschitz continuous and so $f \in L_{n_0}$. Let $\delta = \frac{\varepsilon}{n_0} > 0$. Suppose that

$$\hat{d}_X(x, u) < \delta.$$

Then

$$\tfrac{1}{n_0}\left|\varepsilon - \big(\varepsilon - d_X(x, u)\big)^+\right| = |f(x) - f(u)| \leqslant \hat{d}_X(x, u) < \delta,$$

so

$$d_X(x, u) < \varepsilon.$$

This proves the continuity of i^{-1}. Therefore d_X and $\hat{d}_X$ are topologically equivalent. This proves Claim 2.

But the sequence $\{x_n\}_{n\geqslant 1} \subseteq (X, \hat{d}_X)$ has no accumulation point (since from Definition 1.52, the sequence $\{x_n\}_{n\geqslant 1}$ has the same accumulation points in (X, d_X) and in $(X, \hat{d}_X)$) and so by Claim 1, the metric space $(X, \hat{d}_X)$ is not complete, a contradiction.

Solution of Problem 1.141
Arguing by contradiction, suppose that we could find two sequences $\{x_n\}_{n\geqslant 1} \subseteq K$ and $\{u_n\}_{n\geqslant 1} \subseteq X$ such that

$$d_X(x_n, u_n) \longrightarrow 0$$

and $\varepsilon > 0$ such that

$$d_Y\big(f(x_n), f(u_n)\big) > \varepsilon \qquad \forall\, n \geqslant 1.$$

Since K is compact and $\{x_n\}_{n\geqslant 1} \subseteq K$, by passing to a subsequence if necessary, we may assume that $x_n \longrightarrow x \in K$. Then

$$u_n \longrightarrow x \quad \text{in } X.$$

So,

$$f(u_n) \longrightarrow f(x) \quad \text{in } Y \quad \text{and} \quad f(x_n) \longrightarrow f(x) \quad \text{in } Y.$$

Thus, by the triangle inequality, we have

$$d_Y\big(f(x_n), f(u_n)\big) \longrightarrow 0,$$

a contradiction.

Solution of Problem 1.142
(a) Recall that if B is compact, then B is totally bounded (see Theorem 1.71). So, $\alpha(B) = 0$ by Definition 1.70(b).

(b) Let $B \subseteq X$ be a set such that $\alpha(B) = 0$. Let $\varepsilon > 0$. From the definition of α, we can find a finite family of sets $A_1, \ldots, A_n$ such that

$$B \subseteq \bigcup_{k=1}^{n} A_k \quad \text{with } \operatorname{diam} A_k \leqslant \tfrac{\varepsilon}{2} \text{ for all } k \in \{1, \ldots, n\}.$$

Let us fix $x_k \in A_k$ for all $k \in \{1, \ldots, n\}$. Because $\operatorname{diam} A_k \leqslant \frac{\varepsilon}{2}$, we have

$$A_k \subseteq B_\varepsilon(x_k) \qquad \forall\, k \in \{1, \ldots, n\},$$

Thus

$$B \subseteq \bigcup_{k=1}^{n} A_k \subseteq \bigcup_{k=1}^{n} B_\varepsilon(x_k)$$

and so the set B is totally bounded (see Definition 1.70(b)).

(c) If

$$B' \subseteq \bigcup_{k=1}^{n} C_k \quad \text{with } \operatorname{diam} C_k \leqslant d \text{ for all } k \in \{1, \ldots, n\},$$

then

$$B \subseteq \bigcup_{k=1}^{n} C_k$$

and so $\alpha(B) \leqslant \alpha(B')$.

(d) Since $B \subseteq \overline{B}$, from part **(c)**, we have $\alpha(B) \leqslant \alpha(\overline{B})$. On the other hand, if $B \subseteq \bigcup_{k=1}^{n} C_k$, then

$$\overline{B} \subseteq \overline{\bigcup_{k=1}^{n} C_k} = \bigcup_{k=1}^{n} \overline{C}_k$$

and since $\operatorname{diam} C_k = \operatorname{diam} \overline{C}_k$, it follows that $\alpha(\overline{B}) \leqslant \alpha(B)$, hence $\alpha(\overline{B}) = \alpha(B)$.

(e) First note that

$$\alpha\Big(\bigcap_{n \geqslant 1} C_n\Big) \leqslant \alpha(C_k) \qquad \forall\, k \geqslant 1$$

(see part **(c)**) and so

$$\alpha\Big(\bigcap_{n \geqslant 1} C_n\Big) = 0,$$

thus in particular the set $\bigcap_{n \geqslant 1} C_n$ is totally bounded (see **(b)**). Hence $\bigcap_{n \geqslant 1} C_n$ being also closed is complete, thus compact (see Theorem 1.71).

Passing to a subsequence if necessary, we may assume that

$$\alpha(C_k) \leqslant \tfrac{1}{k+1} \qquad \forall\, k \geqslant 1$$

(cf. Problem 1.4). The set C_1 can be covered by a finite family of sets: $A_{1,1}, \ldots, A_{1,s_1}$, all of them of diameter less or equal to 1. We can assume that sets $A_{1,1}, \ldots, A_{1,s_1}$ are closed (as for any set A, we have that $A \subseteq \overline{A}$ and $\operatorname{diam} A = \operatorname{diam} \overline{A}$). Note that there exists $i_1 \in \{1, \ldots, s_1\}$ such that

$$A_{1,i_1} \cap C_k \neq \emptyset \qquad \forall\, k \geqslant 1.$$

Indeed, if this is not the case, then we can find $k_0 \geqslant 1$ such that

$$A_{1,i} \cap C_k = \emptyset \qquad \forall\, k \geqslant k_0,\ i \in \{1, \ldots, s_1\}.$$

But sets $A_{1,1}, \ldots, A_{1,s_1}$ cover C_1, so they also cover C_{k_0} (because $C_{k_0} \subseteq C_1$), a contradiction.

Let us replace sets $C_1, C_2, \ldots$ by sets $A_{1,i_1} \cap C_1, A_{1,i_1} \cap C_2, \ldots$ obtaining a decreasing sequence of nonempty closed sets (still denoted by $\{C_k\}_{k \geqslant 1}$), with $\alpha(C_k) \longrightarrow 0$ and additionally such that

$$\operatorname{diam}(C_k) \leqslant 1 \qquad \forall\, k \geqslant 1.$$

Now, we proceed by induction. So, suppose that we have changed sets $C_1, C_2, \ldots$ is such a way that they are nonempty, closed, $C_{k+1} \subseteq C_k$ for $k \geqslant 1$,

$$\operatorname{diam} C_k \leqslant \tfrac{1}{k} \qquad \forall\, k \in \{1, \ldots, n\},$$

$$\alpha(C_k) \leqslant \tfrac{1}{k+1} \qquad \forall\, k \geqslant 1$$

and $\alpha(C_k) \longrightarrow 0$.

Now the set C_{n+1} can be covered by a finite family of sets: $A_{n+1,1}, \ldots, A_{n+1,s_{n+1}}$, all of them of diameter less or equal to $\frac{1}{n+1}$. We can assume that sets $A_{n+1,1}, \ldots, A_{n+1,s_{n+1}}$ are closed. Note that there exists $i_{n+1} \in \{1, \ldots, s_{n+1}\}$ such that

$$A_{n+1,i_{n+1}} \cap C_k \neq \emptyset \qquad \forall\, k \geqslant 1.$$

Let us replace sets $C_{n+1}, C_{n+2}, \ldots$ by sets

$$A_{n+1,i_{n+1}} \cap C_{n+1}, A_{n+1,i_{n+1}} \cap C_{n+2}, \ldots$$

obtaining a decreasing sequence of nonempty closed sets (still denoted by $\{C_k\}_{k \geqslant 1}$), with $\alpha(C_k) \longrightarrow 0$ and additionally such that

$$\operatorname{diam} C_k \leqslant \tfrac{1}{k} \qquad \forall\, k \in \{1, \ldots, n+1\},$$

$$\alpha(C_k) \leqslant \tfrac{1}{k+1} \qquad \forall\, k \geqslant 1$$

and $\alpha(C_k) \longrightarrow 0$.

So, finally, we obtain a decreasing sequence of nonempty closed sets $\{C_n\}_{n \geqslant 1}$ such that

$$\operatorname{diam} C_n \leqslant \tfrac{1}{n} \qquad \forall\, n \geqslant 1.$$

Let us choose $x_n \in C_n$ for all $n \geqslant 1$. By the above procedure, we know that

$$d_X(x_n, x_m) \leqslant \tfrac{1}{n} \qquad \forall\, m \geqslant n,$$

so $\{x_n\}_{n \geqslant 1} \subseteq X$ is a Cauchy sequence. But by hypothesis, X is complete. So,

$$x_n \longrightarrow x \in X$$

and $x \in C_n$ for all $n \geqslant 1$. Hence

$$x \in \bigcap_{n \geqslant 1} C_n \neq \emptyset.$$

Solution of Problem 1.143

(a) Let $x_n \longrightarrow x$ in X. Then by hypothesis the sets

$$\{f(x)\} \cup \{f(x_n) : n \geqslant k\} \subseteq Y$$

are compact for all $k \geqslant 1$. From Proposition 1.67, we have that

$$\bigcap_{n \geqslant 1} \left(\{f(x_n) : n \geqslant k\} \cup \{f(x)\}\right) \neq \emptyset.$$

Since f is injective,

$$\bigcap_{n \geqslant 1} \{f(x_n) : n \geqslant k\} = \emptyset$$

and so

$$\bigcap_{n\geqslant 1}\big(\{f(x_n):\ n\geqslant k\}\cup\{f(x)\}\big) \;=\; \{f(x)\}.$$

From Problem 1.94, we have

$$\operatorname{diam}\big(\{f(x_n):\ n\geqslant k\}\big) \;\longrightarrow\; \operatorname{diam}\{f(x)\} \;=\; 0 \quad \text{as } k\to+\infty,$$

thus $\{f(x_n)\}_{n\geqslant 1}$ is a Cauchy sequence. As the set

$$\{f(x)\}\cup\{f(x_n):\ n\geqslant 1\}$$

is complete (being compact), we have that $f(x_n)\longrightarrow y$ for some y and of course $y=f(x)$, i.e., f is continuous.

(b) Let $X=[0,1]$ and $Y=\{0,1\}$ and

$$f(x) \;=\; \begin{cases} 0 & \text{if } \ x=0,\\ 1 & \text{if } \ x\in(0,1]. \end{cases}$$

Then f maps compact sets into compact sets, but f is neither injection nor continuous.

Solution of Problem 1.144
(a) For every $k\in\{1,\ldots,m\}$ and $n\geqslant 1$, let

$$U_k^n \;=\; \big\{x\in X:\ \operatorname{dist}(x,C_k)<\tfrac{1}{n}\big\}.$$

These are open sets and $C_k\subseteq U_k^n$. Suppose that for every $n\geqslant 1$, we have $\bigcap\limits_{k=1}^{m}U_k^n\neq\emptyset$. Then we can find $x_n\in\bigcap\limits_{k=1}^{m}U_k^n$ and so

$$\operatorname{dist}(x_n,C_k) \;<\; \tfrac{1}{n} \qquad \forall\, k=1,\ldots,m$$

(see Definition 1.6). Since X is compact and $\{x_n\}_{n\geqslant 1}\subseteq X$, so passing to a subsequence if necessary, we may assume that $x_n\longrightarrow x$. Then $\operatorname{dist}(x,C_k)=0$ for all $k=1,\ldots,m$, so $x\in C_k$ for all $k=1,\ldots,m$ (since each C_k is closed) and thus $x\in\bigcap\limits_{k=1}^{m}C_k$, a contradiction. So, for some $n\geqslant 1$, we must have

$$\bigcap_{k=1}^{m}U_k^n \;=\; \emptyset \qquad \forall\, k=1,\ldots,m$$

and of course $C_k\subseteq U_k$ for all $k\in\{1,\ldots,m\}$.

(b) Yes. In the above proof of part **(a)**, we have used the compactness of X. But the result can be proved in another way (and the compactness is in fact not needed).

For $k \in \{1, \ldots, m\}$, let

$$f_k(x) \; = \; \operatorname{dist}(x, C_k) \qquad \forall\, x \in X.$$

We know that functions f_k are continuous (in fact Lipschitz continuous; see Problem 1.78). Let us define

$$U_k \; = \; \Big\{x \in X : \; f_k(x) < \tfrac{1}{m} \sum_{\substack{i=1 \\ i \neq k}}^{m} f_i(x)\Big\} \qquad \forall\, k \in \{1, \ldots, m\}.$$

The sets U_k are open and $C_k \subseteq U_k$ for all $k \in \{1, \ldots, m\}$. We will show that

$$\bigcap_{k=1}^{m} U_k \; = \; \emptyset.$$

If in contrast, we assume that there exists $x \in \bigcap\limits_{k=1}^{m} U_k$, then

$$f_k(x) \; < \; \tfrac{1}{m} \sum_{\substack{i=1 \\ i \neq k}}^{m} f_i(x) \qquad \forall\, k \in \{1, \ldots, m\}.$$

Adding the above inequalities for $k = 1, \ldots, m$, we get

$$\sum_{k=1}^{m} f_k(x) \; < \; \frac{m-1}{m} \sum_{i=1}^{m} f_i(x),$$

a contradiction. This ends the proof.

Solution of Problem 1.145

Arguing by contradiction, suppose that we cannot find such Lipschitz constant. Then for every $n \geqslant 1$, we can find $u_n, u_n' \in X$ such that

$$\big\|f(u_n) - f(u_n')\big\| \; > \; n\|u_n - u_n'\|. \tag{1.6}$$

Let

$$M = \max_{u \in C} \|f(u)\|$$

(M is finite by the Weierstrass theorem; see Theorem 1.75). From (1.6), we have

$$\|u_n - u'_n\| \leqslant \frac{2M}{n}$$

and so $\|u_n - u'_n\| \longrightarrow 0$. On the other hand, by the compactness of K and by passing to a suitable subsequences if necessary, we may assume that

$$u_n \longrightarrow u \quad \text{in } K \quad \text{and} \quad u'_n \longrightarrow u' \quad \text{in } K.$$

We infer that $u = u'$. Let U_u be an open set containing u and $k_u > 0$ such that

$$\|f(x) - f(x')\| \leqslant k_u \|x - x'\| \qquad \forall\, x, x' \in U_u.$$

We can find $n_0 \geqslant 1$ such that $u_n, u'_n \in U_u$ for all $n \geqslant n_0$ and so

$$\|f(u_n) - f(u'_n)\| \leqslant k_u \|u_n - u'_n\| \qquad \forall\, n \geqslant n_0$$

which contradicts (1.6).

Solution of Problem 1.146

Let $\{x_n\}_{n \geqslant 1} \subseteq \operatorname{proj}_X(C)$ be a sequence and suppose that $x_n \longrightarrow x$. For every $n \geqslant 1$, we can find $y_n \in Y$ such that $(x_n, y_n) \in C$ for all $n \geqslant 1$. The compactness of Y implies that by passing to a subsequence if necessary, we may assume that $y_n \longrightarrow y$ in Y. Then $(x_n, y_n) \longrightarrow (x, y)$ in $X \times Y$ and $(x, y) \in C$ (since C is by hypothesis closed). Hence $x \in \operatorname{proj}_X(C)$ and so we conclude that $\operatorname{proj}_X(C)$ is closed in X.

Solution of Problem 1.147

(a) No. Let

$$D = \left\{\tfrac{1}{n} : n \in \mathbb{N}_+\right\} \quad \text{and} \quad X = D \cup \{0\}$$

(with the natural metric induced from $\mathbb{R}$). Then D has property S and it is dense in X (see Definition 1.20), while X has not property S.

(b) No. Let $X = \mathbb{R}$ and $D = \mathbb{Q}$. Clearly X has property S but D does not.

Solution of Problem 1.148

First note that the set A is closed (see Definition 1.13(b), Theorem 1.14(d) and Problem 1.24(b)).To prove that A is connected, we argue by contradiction. So, suppose that A is not connected, thus $A = C_1 \cup C_2$, with $C_1, C_2 \subseteq X$ being nonempty, closed and disjoint sets (see Definition 1.85 and Remark 1.86). Sets C_1 and C_2 are also closed in X and so they are compact. Thus there exists $\varepsilon > 0$ such that

$$\operatorname{dist}(C_1, C_2) = \inf \{d_X(u, u') : u \in C_1, u' \in C_2\} = 3\varepsilon > 0$$

(see Definition 1.6 and Problem 1.118). We set

$$\begin{aligned} D_1 &= (C_1)_\varepsilon = \{x \in X : \operatorname{dist}(x, C_1) < \varepsilon\} \\ D_2 &= (C_2)_\varepsilon = \{x \in X : \operatorname{dist}(x, C_2) < \varepsilon\}. \end{aligned}$$

Then, the sets D_1 and D_2 are open and $\operatorname{dist}(D_1, D_2) \geqslant \varepsilon > 0$. Inductively, we can generate a subsequence $\{x_{n_k}\}_{k\geqslant 1}$ of $\{x_n\}_{n\geqslant 1}$ such that $\{x_{n_k}\}_{k\geqslant 1} \subseteq X \setminus (D_1 \cup D_2)$. Because $d_X(x_{n+1}, x_n) \longrightarrow 0$, then there exists $\overline{n} \geqslant 1$ such that

$$d_X(x_{n+1}, x_n) < \varepsilon \qquad \forall\, n \geqslant \overline{n}.$$

First let $l, m \geqslant \overline{n}_0$ be such that $x_l \in D_1$ and $x_m \in D_2$. We choose $n_1 \in (l, m)$ such that $x_{n_1} \in X \setminus (D_1 \cup D_2)$ (the existence of such an element x_{n_1} comes from the fact that $\operatorname{dist}(D_1, D_2) \geqslant \varepsilon$). Then, suppose that we have produced $x_{n_1}, x_{n_2}, \ldots, x_{n_k}$. Then, let $l, m \geqslant n_k + 1$ be such that $x_l \in D_1$ and $x_m \in D_2$. We choose $n_{k+1} \in (l, m)$ such that $x_{n_k} \in X \setminus (D_1 \cup D_2)$.

Because the set $X \setminus (D_1 \cup D_2)$ is compact (as a closed subset of a compact metric space; see Proposition 1.69(b)), and $\{x_{n_k}\}_{k\geqslant 1} \subseteq X \setminus (D_1 \cup D_2)$, there exists a convergent subsequence of $\{x_{n_k}\}_{k\geqslant 1}$ with a limit $x \in X \setminus (D_1 \cup D_2)$. Thus x is an accumulation point of $\{x_n\}_{n \geqslant 1}$ and $x \notin A$, a contradiction to the definition of A.

Solution of Problem 1.149

(a) Since f is continuous, the set $f([0,1])$ is a closed interval $[a,b] \subseteq [0,1]$. Let

$$h(x) = x - f(x).$$

Evidently h is continuous and

$$h(a) = a - f(a) \leqslant 0 \leqslant b - f(b) = h(b).$$

Then the intermediate value theorem (Bolzano theorem; see also Theorem 1.90) implies that there exist $x \in [0,1]$ such that $h(x) = 0$, hence $f(x) = x$.

(b) No. Let $f\colon (0,1) \longrightarrow (0,1)$ be the function, defined by

$$f(x) = x^2.$$

Then, f is continuous, but it has no fixed point.

(c) No. Let $X = [-2,-1] \cup [1,2]$ and let $f\colon X \longrightarrow X$ be defined by $f(x) = -x$. Then, f is continuous, but it has no fixed point.

Solution of Problem 1.150

We argue indirectly. So, suppose that A is not a singleton and let $x, u \in A$ with $x \neq u$. We may assume that $x < u$. Let $y \in \mathbb{R} \setminus \mathbb{Q}$ be such that $x < y < u$. Let

$$U = (-\infty, y) \cap A \quad \text{and} \quad V = (y, +\infty) \cap A.$$

Both U and V are nonempty open subsets of A such that $A = U \cup V$ (as $y \notin A$). So, A admits a separation and this contradicts the hypothesis that A is connected (see Definition 1.85). This proves that A is a singleton.

Solution of Problem 1.151

"**(a) $\Longrightarrow$ (b)**": Since X is disconnected, $X = U \cup V$ with U and V being nonempty disjoint open sets (see Definition 1.85).

Let $h\colon X \longrightarrow \{0,1\}$ be defined by

$$h(x) = \begin{cases} 0 & \text{if } x \in U, \\ 1 & \text{if } x \in V. \end{cases}$$

Clearly h is surjective. We claim that h is continuous. The open sets of the discrete space $\{0,1\}$ are: $\emptyset$, $\{0\}$, $\{1\}$ and $\{0,1\}$. We have

$$h^{-1}(\emptyset) = \emptyset, \quad h^{-1}(\{0\}) = U, \quad h^{-1}(\{1\}) = V, \quad h^{-1}(\{0,1\}) = X.$$

So, all these inverse images are open sets in X and this implies that h is continuous (see Proposition 1.32).

"**(b) $\Longrightarrow$ (a)**": Let $U = h^{-1}(\{0\})$ and $V = h^{-1}(\{1\})$. Both are nonempty (since h is surjective) and open (since h is continuous). Moreover $U \cap V = \emptyset$ and $X = U \cup V$. Therefore X is disconnected (see Definition 1.85).

Solution of Problem 1.152

(a) Suppose that $\mathbb{R}^N \setminus \{0\}$ is disconnected. So, we can find nonempty disjoint open sets U and V such that $\mathbb{R}^N \setminus \{0\} = U \cup V$. Let $x \in U$ and $y \in V$. Consider the line interval joining x and y, i.e.,

$$[x,y] = \{z = (1-t)x + ty : 0 \leqslant t \leqslant 1\}.$$

If $0 \notin [x,y]$, then $[x,y] \in \mathbb{R}^N \setminus \{0\}$ and we contradict the fact that $[x,y]$ is connected (as the continuous image of the interval $[0,1] \subseteq \mathbb{R}$ which is connected; see Proposition 1.88 and Theorem 1.90). So, $0 \in [x,y]$. Let $z \in \mathbb{R}^N$ be such that it is not on a straight line through x and y. Then $z \in \mathbb{R}^N \setminus \{0\}$ and so $z \in U$ or $z \in V$. If $z \in U$, then we replace x by z and we consider $[z,y] \subseteq \mathbb{R}^N \setminus \{0\}$, a contradiction to the fact that $[z,y]$ is connected. Similarly, if $z \in V$, then we consider $[x,z] \subseteq \mathbb{R}^N \setminus \{0\}$, again a contradiction. This proves that $\mathbb{R}^N \setminus \{0\}$ is connected (cf. Theorem 1.90).

(b) Suppose that $\mathbb{R}$ and $\mathbb{R}^N \setminus \{0\}$ are homeomorphic (see Definition 1.39). Let h be a homeomorphism such that $h(0) = 0$ (clearly we can always have this). Then $\mathbb{R} \setminus \{0\}$ and $\mathbb{R}^N \setminus \{0\}$ are homeomorphic under h, a contradiction since $\mathbb{R} \setminus \{0\}$ is disconnected, while $\mathbb{R}^N \setminus \{0\}$ is connected.

Solution of Problem 1.153

We argue by contradiction. So, suppose that for some $u \in \mathbb{R}$, the set $\varphi^{-1}(\{u\})$ is bounded. Then we can find $R > 0$ such that $\varphi^{-1}(\{u\}) \subseteq \overline{B}_R(0)$. We have $u \notin \varphi(\mathbb{R}^N \setminus \overline{B}_R(0))$. The set $\mathbb{R}^N \setminus \overline{B}_R(0)$ is connected and so by the continuity of φ, the set $\varphi(\mathbb{R}^N \setminus \overline{B}_R(0))$ is connected (see Theorem 1.90), hence it is an interval $I \subseteq \mathbb{R}$ (see Proposition 1.88). Because $u \notin I$, we have $I \subseteq (-\infty, u)$ or $I \subseteq (u, +\infty)$. To fix things assume that $I \subseteq (u, +\infty)$ (the argument is similar if $I \subseteq (-\infty, u)$). Note that $\varphi(\overline{B}_R(0)) \subseteq \mathbb{R}$ is compact (see Proposition 1.74) and so it is bounded above by some $\lambda \in \mathbb{R}$. Then

$$\varphi(\mathbb{R}^N) = \varphi(\overline{B}_R(0)) \cup \varphi(\mathbb{R}^N \setminus \overline{B}_R(0)) \subseteq (-\infty, \mu],$$

with $\mu = \max\{\lambda, u\}$, which contradicts the hypothesis that φ is surjective.

Solution of Problem 1.154

No. If $h\colon [0,1] \longrightarrow (0,1)$ is a homeomorphism, then $[0,1] \setminus \{1\} = [0,1)$ must be homeomorphic to $(0,1) \setminus \{h(1)\}$. But $[0,1] \setminus \{1\} = [0,1)$ is connected, while $(0,1) \setminus \{h(1)\}$ is not (it is not an interval; see Proposition 1.88 and Theorem 1.90). So $[0,1]$ and $(0,1)$ cannot be homeomorphic.

Solution of Problem 1.155

Yes. Let $u_0 = \left(u_0^k = e\right)_{k=1}^N \in \mathbb{R}^N$. Let $u \in E$ and suppose that $u^i \in \mathbb{R} \setminus \mathbb{Q}$, for some $i \in \{1, \ldots, N\}$. Then we connect u_0 and $\hat{u} \in \mathbb{R}^N$, the vector with all components equal to e except $\hat{u}^i = u^i$, by the line segment starting from u_0 and ending at $\hat{u}$. Evidently this line segment lies in E. Next connect $\hat{u}$ and u again with the line segment determined by these two vectors. This line segment is also in E (since the ith component of all its elements is $u^i \in \mathbb{R}\setminus\mathbb{Q}$). Therefore the path from u_0 to u which results by concatenating the two line segments is in E and so by Proposition 1.98, E is path-connected.

Solution of Problem 1.156

(a) "(i) $\Longrightarrow$ (ii)": Since X is locally connected (see Definition 1.95), for every $x \in X$ there is a connected open set U_x containing $x \in X$ such that

$$U_x \subseteq B_{\frac{\varepsilon}{2}}(x).$$

The family $\{U_x\}_{x\in X}$ is an open cover of X and since X is compact, we can find a finite subcover $\{U_{x_k}\}_{k=1}^m$. So,

$$\begin{aligned} X &= \bigcup_{k=1}^m U_{x_k}, \quad U_{x_k} \text{ is connected and } \quad \operatorname{diam} U_{x_k} \\ &\leqslant \operatorname{diam} B_{\frac{\varepsilon}{2}}(x_k) = \varepsilon. \end{aligned}$$

"(ii) $\Longrightarrow$ (i)": Let $\varepsilon > 0$ and let $\{E_k\}_{k=1}^m$ be a finite cover of X with each E_k connected and

$$\operatorname{diam} E_k \leqslant \varepsilon.$$

Since $\overline{E}_k$ is connected too (see Proposition 1.89) and $\operatorname{diam} E_k = \operatorname{diam} \overline{E}_k$, we may assume that each E_k is closed. Let $u \in X$ and let

$$\begin{aligned} I_u &= \{k \in \{1,\ldots,m\} : u \in E_k\}, \\ \hat{E} &= \bigcup_{k\in I_u} E_k \quad \text{and} \quad U = \bigcap_{k\in I_u^c} E_k^c. \end{aligned}$$

Evidently U is open and $u \in U \subseteq \hat{E}$. Also $\hat{E}$ is connected (see Theorem 1.91). Also, if $x \in \hat{E}$, then $x \in E_k$ for some $k \in I_u$ and so

$$d_X(x,u) \leqslant \operatorname{diam} E_k \leqslant \varepsilon.$$

Hence $\hat{E} \subseteq \overline{B}_\varepsilon(u)$. Therefore u admits a local basis consisting of connected sets. Since u is arbitrary, we conclude that X is locally connected.

(b) The implication "**(ii)** $\Longrightarrow$ **(i)**" remains true, as we do not exploit the compactness of X in the above proof.

The implication "**(i)** $\Longrightarrow$ **(ii)**" does not hold. Let $X = \mathbb{R}$. Then X is locally connected, but X is not a finite union of sets with diameter less or equal to 1.

Solution of Problem 1.157

(a) We argue by induction on $m \geqslant 1$. For $m = 1$ by hypothesis A_1 is connected. Suppose that the result is true for some $m \geqslant 1$. Then we set

$$B_m = \bigcup_{k=1}^{m} A_k,$$

which by hypothesis is connected and by hypothesis $B_m \cap A_{m+1} \neq \emptyset$. So, Theorem 1.91 implies that the set $B_m \cup A_{m+1} = \bigcup_{k=1}^{m+1} A_k$ is connected.

(b) Yes. We proceed by contradiction. So let us suppose that the set

$$A = \bigcup_{k=1}^{\infty} A_k$$

is not connected. Then, there exists a pair of nonempty, disjoint sets $U, V \subseteq A$ both open in A such that $A = U \cup V$, both open (in A). First, note that for all $k \geqslant 1$, we have

$$A_k \subseteq U \quad \text{or} \quad A_k \subseteq V$$

(if this is not the case for some $k \geqslant 1$, then the pair $A_k \cap U$, $A_k \cap V$ is a separation of the set A_k and so A_k is not connected, a contradiction). Let us define

$$k_U = \inf\{k : A_k \subseteq U\} \quad \text{and} \quad k_V = \inf\{k : A_k \subseteq V\}.$$

One of the integers k_U or k_V is equal to 1 and the other is greater than 1. Without loss of generality, we can assume that $k_U > 1$. Then

$A_{k_U} \subseteq U$ and $A_{k_U-1} \subseteq V$. But from the assumption, we have that $A_{k_U} \cap A_{k_U-1} \neq \emptyset$, so also $U \cap V \neq \emptyset$, a contradiction to the fact that the pair U, V is a separation for A.

Solution of Problem 1.158
Note that

$$C_\varepsilon = C \cup \bigcup_{x \in C} B_\varepsilon(x)$$

(see Definition 1.6). By hypothesis C and $B_\varepsilon(x)$ (with $x \in C$) are connected sets and

$$C \cap B_\varepsilon(x) \neq \emptyset \qquad \forall\, x \in C.$$

So, we can apply Theorem 1.91 and conclude that for every $\varepsilon > 0$, the set C_ε is connected.

Solution of Problem 1.159
Let $u_1 = (x_1, \lambda_1)$ and $u_2 = (x_2, \lambda_2)$ be two elements in $\operatorname{epi} f$ (see Definition 1.132(b)). Let M be the maximum of f on the bounded closed interval $[x_1, x_2]$. We connect u_1 and u_2 with the following continuous path. First we connect u_1 and $v_1 = (x_1, M)$ with the vertical line segment $[u_1, v_1]$. Then we connect v_1 and $v_2 = (x_2, M)$ with the horizontal line segment $[v_1, v_2]$. Finally we connect v_2 and u_2 with the vertical line segment $[v_2, u_2]$. The result is a piecewise linear path which is continuous and connects u_1 and u_2. Therefore we conclude that $\operatorname{epi} f$ is path-connected (see Definition 1.97).

Solution of Problem 1.160
(a) Let $x \in X$ and $\varepsilon > 0$ be fixed and define

$$C_\varepsilon = \{u \in X : x \overset{\varepsilon}{\sim} u\}.$$

We show that C_ε is nonempty, open and closed in X. Since $x \in C_\varepsilon$, we have that $C_\varepsilon \neq \emptyset$.

Let $u \in C_\varepsilon$ and let $y \in B_\varepsilon(u)$. Then $(\{u, y\}$ is an ε-chain connecting u and y and so $u \overset{\varepsilon}{\sim} y$, hence by transitivity $x \overset{\varepsilon}{\sim} y$ and so $B_\varepsilon(u) \subseteq C_\varepsilon$ which proves that C_ε is open.

Next, let $u \in \overline{C}_\varepsilon$ and let $y \in \overline{B}_\varepsilon(u) \cap C_\varepsilon$. As before, we have $y \overset{\varepsilon}{\sim} u$, while $u \overset{\varepsilon}{\sim} x$. Hence, by transitivity, $y \overset{\varepsilon}{\sim} x$ and so $y \in C_\varepsilon$, which proves that C_ε is also closed.

Since by hypothesis X is connected, we have $X = C_\varepsilon$ (see Proposition 1.87) and this is true for all $\varepsilon > 0$. Therefore X is well-chained (see Definition 1.108).

(b) No. The space $\mathbb{Q}$ (with the metric induced from $\mathbb{R}$) is well-chained but not connected.

Solution of Problem 1.161

"$\Longrightarrow$": This is Problem 1.160.

"$\Longleftarrow$": Suppose that X is well-chained (see Definition 1.108). Let $\xi\colon X \longrightarrow \{0,1\}$ be any continuous function. Since X is compact, ξ is uniformly continuous (see Definition 1.45 and Proposition 1.77) and so we can find $\varepsilon > 0$ such that

$$d_X(x,u) \leqslant \varepsilon \quad \Longrightarrow \quad |\xi(x) - \xi(u)| < 1.$$

Therefore $\xi(x) = \xi(u)$. For $x, u \in X$, let $(c_0, \ldots, c_n)$ be the ε-chain connecting x and u. Then $d_X(c_k, c_{k+1}) \leqslant \varepsilon$ for $k = 1, \ldots, n-1$, implies that $\xi(c_k) = \xi(c_{k+1})$ and so

$$\xi(x) = \xi(c_1) = \ldots = \xi(c_{n-1}) = \xi(c_n) = \xi(u),$$

hence ξ is constant. This, by Problem 1.151 implies that X is connected.

Solution of Problem 1.162
Let $u = f(x)$. The set $f\big(C(x)\big)$ (see Definition 1.92) is a connected subset of Y containing u (see Theorem 1.90 and Definition 1.39). So, we have $f\big(C(x)\big) \subseteq C(u)$ (see Definition 1.92). Similarly, we show that $f^{-1}\big(C(u)\big) \subseteq C(x)$. Acting with f, we obtain $C(u) \subseteq f\big(C(x)\big)$. Therefore

$$f\big(C(x)\big) \;=\; C(u) \;=\; C\big(f(x)\big) \qquad \forall\, x \in X.$$

Solution of Problem 1.163
Let U be the open set in E, defined by

$$U \;=\; E \cap \big(\mathbb{R} \times \big(-\tfrac{1}{2}, \tfrac{1}{2}\big)\big).$$

One of the connected components of U is the interval $\{0\} \times \big(-\tfrac{1}{2}, \tfrac{1}{2}\big)$, which is not open in E. So, by Proposition 1.96, the set E is not locally connected.

Solution of Problem 1.164
"$\Longrightarrow$": Follows from Theorem 1.90.

"$\Longleftarrow$": Without any loss of generality, we can assume that f is increasing (the proof for f being decreasing is similar). Let $t_0 \in T$. If f is not continuous at t_0, then

$$f(t_0^-) \;=\; \lim_{\substack{t \to t_0 \\ t < t_0}} f(t) \;<\; f(t_0) \quad \text{or} \quad f(t_0) \;<\; f(t_0^+) \;=\; \lim_{\substack{t \to t_0 \\ t > t_0}} f(t).$$

To fix things, we assume that $f(t_0) < f(t_0^+)$ (the reasoning is similar if the other possibility is true). Since the limit $f(t_0^+)$ exists, we can find $t \in T \cap (t_0, +\infty)$ such that

$$f(t_0) \;<\; f(t_0^+) \;\leqslant\; f(t).$$

Let $y \in \big(f(t_0), f(t_0^+)\big) \subseteq \big(f(t_0), f(t)\big)$. Since by hypothesis $f(T)$ is an interval, we have $y \in f(T)$ and so, we can find $s \in T$ such that

$y = f(s)$. If $s \leqslant t_0$, then $y \leqslant f(t_0)$, a contradiction. If $s > t_0$, then $y \geqslant f(t_0^+)$ again a contradiction. This prove the continuity of f on T.

Solution of Problem 1.165
Let

$$C = \{(x,u) \in T \times T : x < u\}$$

and let $g\colon C \longrightarrow \mathbb{R}$ be defined by

$$g(x,u) = \tfrac{f(u)-f(x)}{u-x} \qquad \forall\, (x,u) \in C.$$

Let $(x,u) \in C$. Then by the mean value theorem, we can find $y \in (x,u)$ such that

$$g(x,u) = f'(y),$$

so

$$g(C) \subseteq f'(T).$$

On the other hand, from the definition of the derivative, we have

$$f'(T) \subseteq \overline{g(C)}.$$

So, we have

$$g(C) \subseteq f'(T) \subseteq \overline{g(C)}.$$

But the set C is connected (in fact it is convex in $\mathbb{R}^2$) and g is continuous. So $g(C)$ is connected (see Theorem 1.90). Invoking Proposition 1.89, we infer that $f'(T)$ is connected which implies that f' has the Darboux property.

Solution of Problem 1.166
Let $x_0, x_1 \in X$ with $x_0 \neq x_1$ and let $\vartheta\colon X \longrightarrow \mathbb{R}$ be defined by

$$\vartheta(x) = d_X(x, x_0) \qquad \forall\, x \in X.$$

Then ϑ is continuous (see Proposition 1.36) and so $\vartheta(X)$ is connected (see Theorem 1.90), thus an interval (see Proposition 1.88), which is not a singleton (as $0 \in \vartheta(X)$ and $0 \neq d_X(x_0, x_1) \in \vartheta(X)$). So the set $\vartheta(X)$ is uncountable, and hence X is uncountable.

Solution of Problem 1.167

The nonemptiness of C follows from Proposition 1.67 and the compactness of C follows from the fact that C is a closed subset of the compact set C_1. It remains to show the connectedness of C. Suppose that $C = K_1 \cup K_2$ with K_1, K_2 being closed subsets of C which are disjoint. Evidently K_1 and K_2 are compact and so we can find two disjoint open sets U_1 and U_2 such that $K_1 \subseteq U_1$ and $K_2 \subseteq U_2$. Then we can find $n_0 \geqslant 1$ such that $C_n \subseteq U_1 \cup U_2$ for all $n \geqslant n_0$. The connectedness of C_n implies that $C_n \subseteq U_1$ (or $C_n \subseteq U_2$) for all $n \geqslant n_0$, hence $C \subseteq K_1$ (or $C \subseteq K_2$) and so we conclude that $K_2 = \emptyset$ (or $K_1 = \emptyset$), which proves that C is connected.

Solution of Problem 1.168

Let $V = C_b(X;Y)$ be furnished with the supremum metric

$$d^\infty(f,h) \;=\; \sup_{x\in X} d_Y\big(f(x), h(x)\big).$$

From Theorem 1.131(b), we know that (V, d^∞) is a complete metric space. Let $\xi\colon V \longrightarrow V$ be defined by

$$\xi(h)(x) \;=\; g\big(x, h(x)\big) \qquad \forall\, h \in V,\ x \in X.$$

Let us check that this function is well defined, i.e., for all $h \in V$, $\xi(h) \in V$. First, note that the function $x \longmapsto \xi(h)(x) = g\big(x, h(x)\big)$ is continuous, being the composition of two continuous functions. Also, we will show that $\xi(h)(\cdot)$ is bounded. To this end, fix $\hat{x} \in X$ and $\hat{y} \in Y$. Then for all $u \in V$ and all $x \in X$, we have

$$\begin{aligned} d_Y\left(\xi(u)(x), g(\hat{x},\hat{y})\right) &\leqslant \sup_{x\in X} d_Y\left(\xi(u)(x), g(x,\hat{y})\right) \\ &= \sup_{x\in X} d_Y\left(g\big(x,u(x)\big), g(x,\hat{y})\right) \\ &\leqslant k \sup_{x\in X} d_Y\left(u(x),\hat{y}\right) \;<\; +\infty \end{aligned}$$

(since $u \in V$). So, indeed $\xi \colon V \longrightarrow V$. For all $u, v \in V$, we have

$$\begin{aligned} d^\infty\big(\xi(u), \xi(v)\big) &= \sup_{x \in X} d_Y\big(g(x, u(x)), g(x, v(x))\big) \\ &\leqslant k \sup_{x \in X} d_Y\big(u(x), v(x)\big) = kd^\infty(u, v), \end{aligned}$$

so ξ is a contraction on V.

Because (V, d^∞) is a complete metric space, we can apply the Banach fixed point theorem (see Theorem 1.49) and obtain $f \in V = C_b(X; Y)$ such that

$$\xi(f) = f,$$

so

$$g\big(x, f(x)\big) = f(x) \qquad \forall\, x \in X.$$

Solution of Problem 1.169

The function

$$X \times X \ni (x, u) \longmapsto d_X(x, u) \in \mathbb{R}$$

is continuous (see Proposition 1.36). Also the set $K \times C$ is compact in $X \times X$ (see Proposition 1.130). So, invoking the Weierstrass theorem (see Theorem 1.75), we can find $(a, c) \in K \times C$ such that

$$d_X(a, c) = \inf\big\{d_X(x, u) : x \in K,\ u \in C\big\} = \operatorname{dist}(K, C)$$

(see Definition 1.6).

Also, if $K = C$, then again from the Weierstrass theorem (see Theorem 1.75), we can find $a', c' \in K$ such that

$$d_X(a', c') = \sup\big\{d_X(x, u) : x, u \in K\big\} = \operatorname{diam} K$$

(see Definition 1.6(c)).

Solution of Problem 1.170

Recall that all the metrics d_p, $1 \leqslant p \leqslant +\infty$ are Lipschitz equivalent (see Example 1.3(a) and Remark 1.119). So, without any loss of generality, we may assume that the metric on X is the metric

$$d_1(x, u) = \sum_{k=1}^{m} d_k(x^k, u^k) \qquad \forall\, x, u \in X.$$

The projection function $\operatorname{proj}_k : X \longrightarrow X_k$, defined by $\operatorname{proj}_k(x) = x^k$ for all $x = (x_1, \ldots, x_m) \in X$ is 1-Lipschitz continuous, hence uniformly continuous (see Definition 1.45). So, by Proposition 1.46(b), proj_k maps Cauchy sequences to Cauchy sequences. Therefore, we conclude that $\{x_n^k\}_{n\geqslant 1}$ is a Cauchy sequence in X_k.

Conversely, if each $\{x_n^k\}_{n\geqslant 1}$, for $k = 1, \ldots, m$ is a Cauchy sequence, then from

$$d_1(x_n, x_l) = \sum_{k=1}^{m} d_k(x_n^k, x_l^k)$$

we infer that $\{x_n\}_{n\geqslant 1}$ is a Cauchy sequence in X.

Solution of Problem 1.171

Let $x \in \liminf\limits_{n\to+\infty} E_n$ (see Definition 1.133). Then for every integer $m \geqslant 1$ there exists an index $n_m \geqslant 1$ such that

$$B_{\frac{1}{m}}(x) \cap E_n \neq \emptyset \qquad \forall\, n \geqslant n_m. \tag{1.7}$$

We may assume that $n_m > n_{m-1}$ for $m \geqslant 2$. Let $x_n \in B_{\frac{1}{m}} \cap E_n$ for $n_m \leqslant n < n_{m+1}$. We see that

$$d_X(x_n, x) < \tfrac{1}{m} \qquad \forall\, n \geqslant n_m,$$

hence

$$d_X(x_n, x) \longrightarrow 0 \quad \text{as } n \to +\infty.$$

This proves the expressions for $\liminf\limits_{n\to+\infty} E_n$.

Next let $x \in \limsup\limits_{n\to+\infty} E_n$. Then by definition for every integers $m, k \geqslant 1$, we can find an integer $n_{m,k} \geqslant k$ such that

$$B_{\frac{1}{m}}(x) \cap E_{n_{m,k}} \neq \emptyset. \tag{1.8}$$

Therefore the sequence $\{\operatorname{dist}(x, E_n)\}_{n\geqslant 1}$ of nonnegative numbers has 0 as an accumulation point (see Definitions 1.6, 1.13(b) and Theorem 1.14(d)) and so

$$\liminf_{n\to+\infty} \operatorname{dist}(x, E_n) = 0.$$

So, we can find a subsequence $\{E_{n_k}\}_{k\geqslant 1}$ of $\{E_n\}_{n\geqslant 1}$ such that

$$\lim_{k\to+\infty} \operatorname{dist}(x, E_{n_k}) = \liminf_{n\to+\infty} \operatorname{dist}(x, E_n) = 0.$$

Let $x_{n_k} \in E_{n_k}$ be such that

$$d_X(x, x_{n_k}) \leqslant \operatorname{dist}(x, E_{n_k}) + \tfrac{1}{k} \longrightarrow 0 \quad \text{as } k \to +\infty.$$

Hence

$$x_{n_k} \longrightarrow x \quad \text{in } X,$$

with $x_{n_k} \in E_{n_k}$ for all $k \geqslant 1$. Finally directly from the definition of $\limsup\limits_{n\to+\infty} E_n$, we see that for all $m \geqslant 1$, we have $x \in \limsup\limits_{n\to+\infty} E_n$ if and only if

$$x \in \overline{\bigcup_{n\geqslant m} E_n} \qquad \forall\, m \geqslant 1.$$

This proves the different expressions for the set $\limsup\limits_{n\to+\infty} E_n$.

From the above expressions it is clear that we always have

$$\liminf_{n\to+\infty} E_n \subseteq \limsup_{n\to+\infty} E_n.$$

Moreover, it is clear that $\limsup\limits_{n\to+\infty} E_n$ is closed (being the intersection of the closed sets $C_m = \overline{\bigcup\limits_{n\geqslant 1} E_n}$). Also, note that the condition (1.8) is equivalent to

$$\forall m, k \geqslant 1\ \exists n_{m,k} \geqslant 1 \ \ B_{\frac{1}{m}}(x) \cap \overline{E}_{n_{m,k}} \neq \emptyset.$$

Therefore, we infer that

$$\overline{\limsup_{n\to+\infty} E_n} = \limsup_{n\to+\infty} E_n = \limsup_{n\to+\infty} \overline{E}_n.$$

Next let $x \in \overline{\liminf\limits_{n\to+\infty} E_n}$. Then for every $r > 0$, we have

$$B_r(x) \cap \liminf_{n\to+\infty} E_n \neq \emptyset$$

(see Definition 1.13(b)), hence we can find an integer $n_r \geqslant 1$ such that

$$B_r(x) \cap E_n \neq \emptyset \qquad \forall\, n \geqslant n_r,$$

hence $x \in \liminf_{n\to+\infty} E_n$. In addition, as before, since the equivalence of (1.7) to

$$\forall m \geqslant 1\ \exists n_m \geqslant 1\ \forall n \geqslant m_n \quad B_r(x) \cap \overline{E}_n \neq \emptyset,$$

we conclude that

$$\overline{\liminf_{n\to+\infty} E_n} \;=\; \liminf_{n\to+\infty} E_n \;=\; \liminf_{n\to+\infty} \overline{E}_n.$$

Hence it follows that the two sets $\liminf\limits_{n\to+\infty} E_n$ and $\limsup\limits_{n\to+\infty} E_n$ are closed (possibly empty).

Solution of Problem 1.172

If $\liminf\limits_{n\to+\infty} C_n = \emptyset$, then

$$\operatorname{dist}\big(x, \liminf_{n\to+\infty} C_n\big) \;=\; +\infty$$

(see Definition 1.133, Problem 1.171 and Definition 1.6) and so the inequality is trivially true. Therefore, we may assume that $\liminf\limits_{n\to+\infty} C_n \neq \emptyset$. So, let $u \in \liminf\limits_{n\to+\infty} C_n$. Then, we can find $u_n \in C_n$ such that

$$d_X(u_n, u) \;\longrightarrow\; 0$$

(see Problem 1.171). We have

$$\operatorname{dist}(x, C_n) \;\leqslant\; d_X(x, u_n) \qquad \forall\ n \geqslant 1$$

and so

$$\limsup_{n\to+\infty} \operatorname{dist}(x, C_n) \;\leqslant\; d_X(x, u).$$

Since $u \in \liminf\limits_{n\to+\infty} C_n$ is arbitrary, it follows that

$$\limsup_{n\to+\infty} \operatorname{dist}(x, C_n) \;\leqslant\; \operatorname{dist}\big(x, \liminf_{n\to+\infty} C_n\big).$$

Solution of Problem 1.173
Since the sets C_n are compact for $n \geqslant 1$, we can find $u_n \in C_n$ such that

$$\operatorname{dist}(x, C_n) = d_{\mathbb{R}^N}(x, u_n)$$

(see Definition 1.6). From the definition of lim inf (see Definition 1.133 and Problem 1.171), there exists a subsequence $\{u_{n_k}\}_{k\geqslant 1}$ of $\{u_n\}_{n\geqslant 1}$ such that

$$\liminf_{n\to+\infty} d_{\mathbb{R}^N}(x, u_n) = \lim_{k\to+\infty} d_{\mathbb{R}^N}(x, u_{n_k}).$$

Next, since $\{u_n\}_{n\geqslant 1} \subseteq \overline{B}_R$ and the latter is compact, there exists a further subsequence $\{u_{n_{k_l}}\}_{l\geqslant 1}$ of $\{u_{n_k}\}_{k\geqslant 1}$ such that $u_{n_{k_l}} \longrightarrow u$ in $\mathbb{R}^N$. From the definition of Kuratowski upper limit, we know that $u \in C$. From the continuity of the distance function (see Proposition 1.36), we have

$$\operatorname{dist}(x, C_{n_{k_l}}) = d_{\mathbb{R}^N}(x, u_{n_{k_l}}) \longrightarrow d_{\mathbb{R}^N}(x, u) \geqslant \operatorname{dist}(x, C),$$

so

$$\liminf_{n\to+\infty} \operatorname{dist}(x, C_n) = \lim_{l\to+\infty} d_{\mathbb{R}^N}(x, u_{n_{k_l}}) = d_{\mathbb{R}^N}(x, u) \geqslant \operatorname{dist}(x, C).$$

Because $C = \liminf\limits_{n\to+\infty} C_n$, from Problem 1.172, we also have

$$\limsup_{n\to+\infty} \operatorname{dist}(x, C_n) \leqslant \operatorname{dist}(x, C).$$

Thus, we conclude that $\operatorname{dist}(x, C_n) \longrightarrow \operatorname{dist}(x, C)$.

Solution of Problem 1.174
For every $\varepsilon > 0$, we can find $x \in C$ such that

$$m \leqslant \varphi(x) \leqslant m + \varepsilon.$$

Let $\{x_n\}_{n\geqslant 1} \subseteq X$ be such that $x_n \in C_n$ for all $n \geqslant 1$ and $d_X(x_n, x) \longrightarrow 0$ (Problem 1.171). For every $n \geqslant 1$, we have $m_n \leqslant \varphi(x_n)$. Due to the continuity of φ, we have

$$\varphi(x_n) \longrightarrow \varphi(x),$$

so

$$\limsup_{n \to +\infty} m_n \leqslant m + \varepsilon.$$

Since $\varepsilon > 0$ was arbitrary, we let $\varepsilon \searrow 0$ to conclude that

$$\limsup_{n \to +\infty} m_n \leqslant m.$$

Solution of Problem 1.175

Let us set

$$E_n = \{x_{m,n} : m \geqslant 1\} \qquad \forall n \geqslant 1.$$

Then from Problem 1.171, we have

$$x_m \in \liminf_{n \to +\infty} E_n \qquad \forall m \geqslant 1$$

(see Definition 1.132). Because $\liminf\limits_{n \to +\infty} E_n$ is closed (see Problem 1.171) and $x_m \longrightarrow x$, we have that $x \in \liminf\limits_{n \to +\infty} E_n$. So, $x = \lim\limits_{n \to +\infty} y_n$, with $y_n \in E_n$. But from the definition of the sets E_n, these elements have the form $y_n = x_{m_n,n}$. Therefore

$$x = \lim_{n \to +\infty} x_{m_n,n}.$$

Next let us set

$$G_m = \{x_{m,n} : n \geqslant 1\} \qquad \forall m \geqslant 1.$$

Note that

$$x_m \in \overline{G}_m \qquad \forall m \geqslant 1.$$

So

$$x = \lim_{m \to +\infty} x_m \in \liminf_{m \to +\infty} \overline{G}_m = \liminf_{m \to +\infty} G_m$$

(see Problem 1.171). Therefore $x = \lim\limits_{m \to +\infty} u_m$, with $u_m \in G_m$ for all $m \geqslant 1$. Every u_m has the form $u_m = x_{m,n_m}$. Hence $x = \lim\limits_{m \to +\infty} x_{m,n_m}$.

Solution of Problem 1.176

(a) Of course $h(A,A) = 0$. Let $A,B \in P_f(X)$ be such that $h(A,B)=0$. Let $a \in A$. Then

$$\begin{aligned} 0 &= h(A,B) = \sup_{x\in X} \big|\operatorname{dist}(x,A)-\operatorname{dist}(x,B)\big| \\ &\geqslant \big|\operatorname{dist}(a,A)-\operatorname{dist}(a,B)\big| = \operatorname{dist}(a,B), \end{aligned}$$

thus $a \in B$ (as B is closed). Analogously, we show that for any $b \in B$, we have $b \in A$. Thus $A = B$.

The symmetry

$$h(A,B) = h(B,A) \qquad \forall\, A,B \in P_f(X)$$

is obvious.

To show the triangle inequality, let $A,B,C \in P_f(X)$. Then

$$\begin{aligned} h(A,C) &= \sup_{x\in X} \big|\operatorname{dist}(x,A)-\operatorname{dist}(x,C)\big| \\ &\leqslant \sup_{x\in X} \Big(\big|\operatorname{dist}(x,A)-\operatorname{dist}(x,B)\big| + \big|\operatorname{dist}(x,B)-\operatorname{dist}(x,C)\big|\Big) \\ &\leqslant \sup_{x\in X} \big|\operatorname{dist}(x,A)-\operatorname{dist}(x,B)\big| + \sup_{x\in X} \big|\operatorname{dist}(x,B)-\operatorname{dist}(x,C)\big| \\ &= h(A,B)+h(B,C). \end{aligned}$$

(b) Let $A,B \in P_f(X)$. Note that

$$\begin{aligned} \sup_{a\in A} \operatorname{dist}(a,B) &= \sup_{a\in A} \big|\operatorname{dist}(a,A)-\operatorname{dist}(a,B)\big| \\ &\leqslant \sup_{x\in X} \big|\operatorname{dist}(a,A)-\operatorname{dist}(a,B)\big| = h(A,B). \end{aligned}$$

Analogously, we have

$$\sup_{b\in B} \operatorname{dist}(b,A) \leqslant h(A,B).$$

Thus

$$\overline{h}(A,B) \leqslant h(A,B).$$

On the other hand, let $x \in X$ be fixed. Then

$$\begin{aligned} &\operatorname{dist}(x,A)-\operatorname{dist}(x,B) = \operatorname{dist}(x,A) - \inf_{b\in B} d_X(x,b) \\ &= \sup_{b\in B} \big(\operatorname{dist}(x,A)-d_X(x,b)\big) = \sup_{b\in B} \big(\inf_{a\in A} d_X(x,a)-d_X(x,b)\big) \\ &= \sup_{b\in B}\inf_{a\in A} \big(d_X(x,a)-d_X(x,b)\big) \leqslant \sup_{b\in B}\inf_{a\in A} d_X(a,b) = \sup_{b\in B}\operatorname{dist}(b,A). \end{aligned}$$

Analogously, we show that

$$\operatorname{dist}(x,B)-\operatorname{dist}(x,A) \leqslant \sup_{a\in A}\operatorname{dist}(a,B).$$

Thus

$$\begin{aligned}\left|\operatorname{dist}(x,A)-\operatorname{dist}(x,B)\right| &\leqslant \max\Big\{\sup_{a\in A}\operatorname{dist}(a,B),\ \sup_{b\in B}\operatorname{dist}(b,A)\Big\}\\ &= \overline{h}(A,B).\end{aligned}$$

As the above inequality holds for every $x \in X$, taking the supremum on the left hand side, we obtain

$$h(A,B) \leqslant \overline{h}(A,B).$$

(c) Let $A, B \in P_f(X)$. Let

$$\mathcal{E} = \big\{\varepsilon > 0:\ A \subseteq B_\varepsilon \text{ and } B \subseteq A_\varepsilon\big\}.$$

If $\varepsilon \in \mathcal{E}$, then $A \subseteq B_\varepsilon$ and $B \subseteq A_\varepsilon$. Thus

$$\begin{cases}\operatorname{dist}(a,B) \leqslant \varepsilon & \forall\, a \in A,\\ \operatorname{dist}(b,A) \leqslant \varepsilon & \forall\, b \in B.\end{cases}$$

So

$$\sup_{a\in A}\operatorname{dist}(a,B) \leqslant \varepsilon \quad\text{and}\quad \sup_{b\in B}\operatorname{dist}(b,BA) \leqslant \varepsilon$$

and hence

$$\overline{h}(A,B) \leqslant \varepsilon.$$

As, it holds for any $\varepsilon \in \mathcal{E}$, thus

$$\overline{h}(A,B) \leqslant \overline{\overline{h}}(A,B).$$

On the other hand, from the definition of $\overline{h}$, we have that

$$\operatorname{dist}(a,B) \leqslant \overline{h}(A,B) \qquad \forall\, a \in A,$$

so

$$a \in B_{\overline{h}(A,B)} \qquad \forall\, a \in A$$

and thus

$$A \subseteq B_{\overline{h}(A,B)}.$$

Analogously, we show that

$$B \subseteq A_{\overline{h}(A,B)}.$$

Thus finally

$$\overline{\overline{h}}(A,B) \leqslant \overline{h}(A,B).$$

Solution of Problem 1.177

(a) Let $D_1, D_2, E_1, E_2 \in P_f(X)$ and let us set

$$\begin{aligned} \varepsilon &= \max\{h(D_1,E_1),\ h(D_2,E_2)\} \quad \text{and} \quad D = D_1 \cup D_2, \\ E &= E_1 \cup E_2, \end{aligned}$$

then

$$\begin{aligned} D &\subseteq (E_1)_\varepsilon \cup (E_2)_\varepsilon = (E_1 \cup E_2)_\varepsilon = E_\varepsilon \\ E &\subseteq (D_1)_\varepsilon \cup (D_2)_\varepsilon = (D_1 \cup D_2)_\varepsilon = D_\varepsilon, \end{aligned}$$

where for each $S \in P_f(X)$, $S_\varepsilon = \{x \in X : \ \operatorname{dist}(x,S) \leqslant \varepsilon\}$ (see Definition 1.6 and Problem 1.176(c)). From these inclusions and the definition of the Hausdorff metric (see Problem 1.176(c)), we obtain that

$$h(D_1 \cup D_2,\ E_1 \cup E_2) \leqslant \varepsilon = \max\{h(D_1,E_1),\ h(D_2,E_2)\}.$$

(b) Let $\eta\colon X \longrightarrow X$ be a k-contraction and let $D, E \in P_f(X)$. Let $\varepsilon = h(D,E)$ and let us set

$$\hat{D} = \eta(D), \quad \hat{E} = \eta(E).$$

If $y = \eta(x) \in \hat{D}$, let $u \in E$ be such that $d_X(x,u) \leqslant \varepsilon$. Then, we have

$$\operatorname{dist}(y,\hat{E}) \leqslant d_X(y,\eta(u)) = d_X(\eta(x),\ \eta(u)) \leqslant k d_X(x,u) \leqslant k\varepsilon,$$

so $\hat{D} \subseteq \hat{E}_{k\varepsilon}$ and similarly $\hat{E} \subseteq \hat{D}_{k\varepsilon}$. Thus

$$h(\eta(D),\eta(E)) = h(\hat{D},\hat{E}) \leqslant k\varepsilon = kh(D,E).$$

Solution of Problem 1.178

(a) Let (X, d_X) be a complete bounded metric space (see Definition 1.7). To show that the Hausdorff metric space $\big(P_f(X), h\big)$ (see Definition 1.134 and Problem 1.176) is complete, let us take any Cauchy sequence $\{D_k\}_{k\geqslant 1} \subseteq P_f(X)$ (see Definition 1.7).

Let us start with two observations. First note that

$$\forall i \geqslant 1\ \exists N_i \in \mathbb{N}\ \forall m, n \geqslant N_i :\ h(D_m, D_n)\ \leqslant\ \tfrac{\varepsilon}{2^i} \tag{1.9}$$

(see Definition 1.7). Next, note that

$$\forall i \geqslant 1\ \forall k \geqslant N_i\ \forall x \in D_k\ \forall j \geqslant k\ \exists y \in D_j :\ d_X(x, y)\ <\ \tfrac{\varepsilon}{2^i}. \tag{1.10}$$

To see this, let us fix $i \geqslant 1$, $k \geqslant N_i$, $x \in D_k$ and $j \geqslant k$. From (1.9), we have that

$$h(D_j, D_k)\ <\ \tfrac{\varepsilon}{2^i},$$

so

$$D_k\ \subseteq\ (D_j)_{\frac{\varepsilon}{2^i}}$$

and so, there exists $y \in D_j$ such that $x \in B_{\frac{\varepsilon}{2^i}}(y)$. This proves (1.10).

Let us define

$$\begin{aligned} F\ &=\ \big\{x \in X :\ x \text{ is an accumulation point of some} \\ &\qquad \text{sequence } \{d_k\}_{k\geqslant 1} \subseteq X, \text{ such that } d_k \in D_k \text{ for } k \geqslant 1\big\} \end{aligned}$$

(see Definition 1.13(b) and Theorem 1.14(d)). We will show that $D_k \xrightarrow{h} F$. To this end, let $\varepsilon > 0$. We will show that

$$\forall k \geqslant N_2 : h(F, D_k)\ \leqslant\ \varepsilon$$

(see (1.10)). In other words, we need to show that

$$\forall k \geqslant N_2 : F\ \subseteq (D_k)_\varepsilon \text{ and } D_k\ \subseteq\ F_\varepsilon.$$

First, let us fix $k \geqslant N_2$ and $x \in F$. From the definition of F, there exists a sequence $\{d_n\}_{n\geqslant 1} \subseteq X$ such that $d_n \in D_n$ for $n \geqslant 1$ and x is an accumulation point of $\{d_n\}_{n\geqslant 1}$. From (1.9), for all $m, n \geqslant N_2$, we have

$$h(D_m, D_n)\ <\ \frac{\varepsilon}{2},$$

so for all $n \geqslant N_2$, we have $D_n \subseteq (D_k)_{\frac{\varepsilon}{2}}$ and thus $d_n \in (D_k)_{\frac{\varepsilon}{2}}$ for all $n \geqslant N_2$. Hence all accumulation points of the sequence $\{d_n\}_{n\geqslant 1}$ are in $\overline{(D_k)_{\frac{\varepsilon}{2}}}$, so $x \in (D_k)_\varepsilon$. Thus, we have proved that $F \subseteq (D_k)_\varepsilon$.

Next, let us fix $k \geqslant N_2$ (see (1.9)) and $x \in D_k$. We will show that $x \in F_{2\varepsilon}$. To this end, let i_0 be the biggest $i \geqslant 1$ such that $k \geqslant N_i$, so $k \geqslant N_{i_0}$. We define

$$f_k = x \in D_k.$$

From (1.10), there exists

$$f_{N_{i_0+1}} \in D_{N_{i_0+1}}, \quad \text{with } d_X\big(f_{N_{i_0+1}}, x\big) < \tfrac{\varepsilon}{2^{i_0}}.$$

Then proceeding by induction, for any $i > i_0$ we choose

$$f_{N_i} \in D_{N_i}, \quad \text{with } d_X\big(f_{N_i}, f_{N_{i-1}}\big) < \tfrac{\varepsilon}{2^{i-1}}.$$

For any other $l > n_{i_0}$ (namely for $l \neq N_i$ for all $i \geqslant n_0$), we choose any $f_l \in D_l$. Note that

$$\begin{aligned} d_X(x, f_{N_i}) &\leqslant d_X(x, f_{N_{i_0}}) + d_X(f_{N_{i_0}}, f_{N_{i_0}+1}) + \ldots + d_X(f_{N_i-1}, f_{N_i}) \\ &< \frac{c}{2^{i_0}} + \frac{c}{2^{i_0+1}} + \ldots + \frac{\varepsilon}{2^s} < \tfrac{\varepsilon}{2}, \end{aligned}$$

so $\{f_{N_i}\}_{i\geqslant i_0} \subseteq B_{\frac{\varepsilon}{2}}(x)$ and thus for any accumulation point y of $\{f_{N_i}\}_{i\geqslant i_0}$ (so also for some accumulation point of $\{f_n\}_{n\geqslant 1}$), we have $d_X(x,y) \leqslant \frac{\varepsilon}{2}$. As $y \in F$, so $x \in F_\varepsilon$.

(b) Since X is compact, X is complete (see Theorem 1.71) and so $(P_f(X), h)$ is complete (see part **(a)**). Let us fix $b \in X$ and let $g\colon (P_f(X), h) \longrightarrow (C(X;\mathbb{R}), d^\infty)$ be the function, defined by

$$g(A)(x) = \operatorname{dist}(x, A) - d_X(x, b).$$

First we show that g is an isometry. Note that

$$d^\infty\big(g(A), g(B)\big) = \sup_{x\in X} \big|\operatorname{dist}(x,A) - \operatorname{dist}(x,B)\big| = h(A,B),$$

so g is an isometry.

Next we show that $g(P_f(X)) \subseteq C(X;\mathbb{R})$ is equicontinuous. For all $A \in P_f(X)$, we have

$$\begin{aligned} \big|g(A)(x) - g(A)(u)\big| &= \big|\operatorname{dist}(x,A) - \operatorname{dist}(u,B) - \big(d_X(x,b) - d_X(u,b)\big)\big| \\ &\leqslant 2d_X(x,u), \end{aligned}$$

so $g(P_f(X))$ is (uniformly) equicontinuous in $C(X;\mathbb{R})$.

Because g is isometry, the set $g\big(P_f(X)\big)$ is closed in $C(X,\mathbb{R})$. Also $g\big(P_f(X)\big)$ is equicontinuous and for all $x \in X$, the set $\big\{g(A)(x) : A \in P_f(X)\big\}$ is bounded (since by hypothesis X is bounded). Therefore we can apply the Arzela–Ascoli theorem (see Theorem 1.84) and obtain that $g\big(P_f(X)\big)$ is compact in $C(X;\mathbb{R})$. Because g is an isometry, $\big(P_f(X), h\big)$ is a compact metric space.

Solution of Problem 1.179

Let h be the Hausdorff metric (see Definition 1.134 and Problem 1.176). From Problem 1.178, we know that $\big(P_f(X), h\big)$ is a complete metric space.

Using Problem 1.177(a) and (b), we have

$$\begin{aligned} h\big(\xi(D), \xi(E)\big) &= h\big(f(D) \cup g(D),\ f(E) \cup g(E)\big) \\ &\leqslant \max\big\{h\big(f(D), f(E)\big),\ h\big(g(D), g(E)\big)\big\} \\ &\leqslant kh(D, E). \end{aligned}$$

Since the space $\big(P_f(X), h\big)$ is complete and ξ is a k-contraction with $k \in [0, 1)$, we can apply the Banach fixed point theorem (see Theorem 1.49) and obtain a unique $C_0 \in P_f(X)$ such that $\xi(C_0) = C_0$.

Solution of Problem 1.180

(a) Let $\{C_n\}_{n\geqslant 1} \subseteq P_k(X)$ be a sequence such that

$$C_n \xrightarrow{h} C \in P_f(X).$$

For a given $\varepsilon > 0$, we can find $n_0 = n_0(\varepsilon) \geqslant 1$ such that

$$h(C_n, C) \leqslant \tfrac{\varepsilon}{2} \qquad \forall\, n \geqslant n_0.$$

The set C_{n_0} is compact, hence it is also totally bounded (see Theorem 1.71). Thus we can find a finite set $\{x_k\}_{k=1}^m \subseteq X$ such that

$$C_{n_0} \subseteq \bigcup_{k=1}^m B_{\frac{\varepsilon}{2}}(x_k).$$

It follows that

$$C \subseteq \bigcup_{k=1}^{m} B_\varepsilon(x_k).$$

and so C is totally bounded. Since $C \in P_f(X)$, once again Theorem 1.71 implies that $C \in P_k(X)$. Therefore $P_k(X)$ is a closed subset of $P_f(X)$.

(b) Now, if X is complete, then from Problem 1.178(a), we know that $(P_f(X), h)$ is complete too. Thus $(P_k(X), h)$ is complete (as a closed subset of a complete space).

Solution of Problem 1.181

"$\Longrightarrow$": This is exactly Problem 1.178(b).

"$\Longleftarrow$": Let $\{x_n\}_{n \geqslant 1}$ be a sequence in X and let

$$C_n = \{x_n\} \in P_f(X) \qquad \forall\, n \geqslant 1.$$

The compactness of $(P_f(X), h)$ implies that we can find a subsequence $\{C_{n_k}\}_{k \geqslant 1}$ of $\{C_n\}_{n \geqslant 1}$ such that

$$C_{n_k} \xrightarrow{h} C \in P_f(X)$$

(see Definition 1.134 and Problem 1.176). So, for a given $\varepsilon > 0$, we can find $k_0 = k_0(\varepsilon) \geqslant 1$ such that

$$C \subseteq (C_{n_k})_\varepsilon = \overline{B}_\varepsilon(x_{n_k}) \qquad \forall\, k \geqslant k_0$$

(see Problem 1.176(c)). Hence $\operatorname{diam} C \leqslant 2\varepsilon$. Since $\varepsilon > 0$ was arbitrary, we conclude that $\operatorname{diam} C = 0$, i.e., $C = \{x\}$. Then, we have

$$x_{n_k} \in \overline{B}_\varepsilon(x) \qquad \forall\, k \geqslant k_0$$

and so $x_{n_k} \longrightarrow x$. This proves the compactness of X.

Solution of Problem 1.182

We argue by contradiction. So, suppose that C is not connected. Then we can find two open nonempty sets $U_1, U_2 \subseteq X$ such that

$$C \subseteq U_1 \cup U_2, \quad C \cap U_1 \neq \emptyset, \quad C \cap U_2 \neq \emptyset \quad \text{and} \quad U_1 \cap U_2 = \emptyset.$$

Let $x \in C \cap U_1$. Since by hypothesis $C_n \xrightarrow{h} C$, we have that

$$\text{dist}(x, C_n) \longrightarrow 0$$

(see Problem 1.176(b)). For every $n \geqslant 1$, we can find $x_n \in C_n$ such that

$$d_X(x, x_n) \leqslant \text{dist}(x, C_n) + \frac{1}{n},$$

hence $d_X(x, x_n) \longrightarrow 0$. Since U_1 is open and $x \in U_1$, we can find $n_1 \geqslant 1$ such that

$$x_n \in U_1 \qquad \forall\, n \geqslant n_1.$$

So

$$C_n \cap U_1 \neq \emptyset \qquad \forall\, n \geqslant n_1.$$

Similarly, we show that

$$C_n \cap U_2 \neq \emptyset \qquad \forall\, n \geqslant n_2,$$

for some $n_2 \geqslant 1$. The connectedness of C_n implies that

$$C_n \nsubseteq U_1 \cup U_2 \qquad \forall\, n \geqslant n_0 = \max\{n_1, n_2\}.$$

So, for all $n \geqslant n_0$, we can find $u_n \in C_n$, $u_n \notin U_1 \cup U_2$. Since the set $(U_1 \cup U_2)^c$ is closed and the set C is compact, from Problem 1.118(a), we have

$$0 < \text{dist}\big((U_1 \cup U_2)^c, C\big) \leqslant \text{dist}(u_n, C) \leqslant h(C_n, C) \longrightarrow 0,$$

a contradiction. This proves that the set C is connected too.

Solution of Problem 1.183

From Problem 1.180, we already have that $\big(P_k(X), h\big)$ is complete. So, it remain to show the separability of $\big(P_k(X), h\big)$ (see Definition 1.21). Let $\{x_n\}_{n\geqslant 1}$ be dense in X (it exists since X is separable; Definition 1.20) and let $\mathcal{F}$ be the family of all finite subsets of $\{x_n\}_{n\geqslant 1}$.

Evidently $\mathcal{F}$ is countable and clearly it is dense in $(P_k(X), h)$ (see Definition 1.20).

Solution of Problem 1.184

Let $X = \mathbb{N}$ and consider two metrics

$$\begin{aligned} d_1(n,m) &= \begin{cases} 1 & \text{if} \quad n \neq m, \\ 0 & \text{if} \quad n = m, \end{cases} \qquad \forall\, n, m \geqslant 1 \\ d_2(n,m) &= \left|\tfrac{1}{n} - \tfrac{1}{m}\right| \qquad \forall\, n, m \geqslant 1. \end{aligned}$$

The two metrics are topologically equivalent (they both generate the discrete topology). Let h_1 be the Hausdorff metric for d_1 and let h_2 be the Hausdorff metric for d_2 (see Definition 1.134 and Problem 1.176). Setting $C_n = \{1, \ldots, n\}$, we have

$$\begin{aligned} h_1(C_n, \mathbb{N}) &= 1 \qquad \forall\, n \geqslant 1, \\ h_2(C_n, \mathbb{N}) &= \sup_k \operatorname{dist}_2(k, C_n) = \lim_{k \to +\infty} \left(\tfrac{1}{n} - \tfrac{1}{k}\right) = \tfrac{1}{n}, \end{aligned}$$

hence $h_2(C_n, \mathbb{N}) \longrightarrow 0$. So, the Hausdorff metrics h_1 and h_2 are not topologically equivalent.

Solution of Problem 1.185

Consider $\mathbb{N}$ furnished with the discrete metric. Let $C_n = \{1, \ldots, n\}$. Then these are closed sets and

$$\liminf_{n \to +\infty} C_n = \limsup_{n \to +\infty} C_n = \mathbb{N}$$

(see Definition 1.133 and Problem 1.171). On the other hand,

$$h(C_n, \mathbb{N}) = 1 \qquad \forall\, n \geqslant 1,$$

so

$$C_n \not\longrightarrow \mathbb{N} \quad \text{in the } h\text{-metric.}$$

Solution of Problem 1.186
First note, that we always have

$$\liminf_{n\to+\infty} C_n \subseteq \limsup_{n\to+\infty} C_n$$

(see Definition 1.133 and Problem 1.171). Let $x \in C$. Since $C_n \xrightarrow{h} C$ see Definition 1.134 and Problem 1.176, we have

$$\mathrm{dist}(x, C_n) \longrightarrow \mathrm{dist}(x, C) = 0,$$

hence $x \in \liminf\limits_{n\to+\infty} C_n$ (see Problem 1.171) and thus

$$C \subseteq \liminf_{n\to+\infty} C_n.$$

Next, let $x \in \limsup\limits_{n\to+\infty} C_n$. Then we can find a subsequence $\{n_k\}_{k\geqslant 1}$ of $\{n\}_{n\geqslant 1}$ and $x_{n_k} \in C_{n_k}$ such that $x_{n_k} \longrightarrow x$ in X (see Problem 1.171). We have

$$\mathrm{dist}(x, C_{n_k}) \leqslant d_X(x, x_{n_k}) \longrightarrow 0$$

and since $C_{n_k} \xrightarrow{h} C$, we also have

$$\mathrm{dist}(x, C_{n_k}) \longrightarrow \mathrm{dist}(x, C)$$

(see Problem 1.176). Therefore $\mathrm{dist}(x, C) = 0$, hence $x \in C$ (since $C \in P_f(X)$). This proves that

$$\limsup_{n\to+\infty} C_n \subseteq C \subseteq \liminf_{n\to+\infty} C_n$$

and so finally

$$\liminf_{n\to+\infty} C_n = \liminf_{n\to+\infty} C_n = C.$$

Solution of Problem 1.187
Let

$$A = \bigcup_{n\geqslant 1} C_n$$

It suffices to show that (A, d_X) is totally bounded. For a given $\varepsilon > 0$, we can find $N = N(\varepsilon) \geqslant 1$ such that

$$h(C_n, C_m) < \tfrac{\varepsilon}{2} \qquad \forall\, n, m \geqslant N.$$

The set $\bigcup_{i=1}^{N} C_i$ is compact, hence totally bounded. So, we can find an $\frac{\varepsilon}{2}$-net $\{x_1, \ldots, x_k\} \subseteq \bigcup_{i=1}^{N} C_i$. We will show that $\{x_1, \ldots, x_k\}$ is an ε-net in A. To this end let $x \in A$. If $x \in \bigcup_{i=1}^{N} C_1$, then clearly $d_X(x, x_i)$ for some $i \in \{1, \ldots, N\}$. If $x \in C_i$ with $i > N$, then $h(C_i, C_N) < \frac{\varepsilon}{2}$, hence $\operatorname{dist}(x, C_n) < \frac{\varepsilon}{2}$. Choosing $y \in C_N$ with $d_X(x, y) < \frac{\varepsilon}{2}$ and $j \geqslant 1$ such that $d_X(y, x_j) < \frac{\varepsilon}{2}$, we set $d_X(x, x_j) < \varepsilon$.

Bibliography

[1] Bruckner, A.M., Bruckner, J.B., Thomson, B.S.: Real Analysis. Prentice-Hall, Upper Saddle River, NJ (1997)

[2] Dieudonné, J.: Foundations of Modern Analysis. Pure and Applied Mathematics, vol. 10-I. Academic, New York (1969)

[3] Goebel, K., Kirk, W.A.: A fixed point theorem for asymptotically nonexpansive mappings. Proc. Am. Math. Soc. **35**, 171–174 (1972)

[4] Hoffman, K.: Analysis in Euclidean Spaces. Prentice-Hall, Englewood Cliffs (1975)

[5] Giaquinta, M., Modica, G.: Mathematical Analysis: Linear and Metric Structures and Continuity. Birkhäuser Boston, Boston (2007)

[6] Knapp, A.W.: Basic Real Analysis. Birkhäuser Boston, Boston (2005)

[7] Royden, H.L.: Real Analysis. Macmillan, New York (1988)

[8] Rosenlicht, M.: Introduction to Analysis. Scott, Foresman and Co., Glenview (1968)

[9] Rudin, W.: Principles of Mathematical Analysis. International Series in Pure and Applied Mathematics. McGraw-Hill Book Co., New York (1976)

[10] Shirali, S., Vasudeva, H.L.: Metric Spaces. Springer, London (2006)

[11] Simmons, G.F.: Introduction to Topology and Modern Analysis. McGraw-Hill Book Co., New York (1963)

Chapter 2

Topological Spaces

2.1 Introduction

2.1.1 Basic Definitions and Notation

Definition 2.1
*A **topological space** is a pair (X, τ) where X is a nonempty set and τ is a family of subsets of X called **open sets**, which satisfies the following three requirements:*
(a) $\emptyset, X \in \tau$;
(b) *τ is closed under arbitrary unions, i.e.,*

$$\text{if } \{U_i\}_{i\in I} \subseteq \tau, \text{ then } \bigcup_{i\in I} U_i \in \tau;$$

(c) *τ is closed under finite intersections, i.e.,*

$$\text{if } \{U_i\}_{i\in I} \subseteq \tau \text{ and } I \text{ is finite, then } \bigcap_{i\in I} U_i \in \tau.$$

*The complements of open sets are called **closed sets**,*

Remark 2.2
By the De Morgan law, if $\{C_i\}_{i\in I}$ is any family of closed sets, the set $\bigcap\limits_{i\in I} C_i$ is a closed set. Evidently many different topologies can be defined on the same set X. The smallest of all topologies on X (i.e., the one with the fewest open sets) is the topology $\tau = \{\emptyset, X\}$ and it is called the ***indiscrete topology***. The biggest of all topologies on

L. Gasiński and N.S. Papageorgiou, *Exercises in Analysis: Part 1*,
Problem Books in Mathematics, DOI 10.1007/978-3-319-06176-4_2,

X (i.e., the one with the most open sets) is the topology consisting of all subsets of X and it is called the ***discrete topology***. The space X with the discrete topology is a ***discrete space***. If τ_1, τ_2 are two topologies on X and $\tau_1 \subseteq \tau_2$ then we say that τ_1 is ***weaker*** than τ_2 or that τ_2 is ***stronger*** than τ_1. Finally note that a subset of X may be both open and closed (for example $\emptyset$ and X are open and closed in any topology on X or in a discrete space every set is open and closed). A set which is both open and closed is usually called ***clopen***.

If (X, d_X) is a metric space then the family τ of all open sets (in the sense of Definition 1.8) is a topology. We call it the topology ***induced*** by the metric d_X. So every metric space is in a natural way a topological space.

Finally, if for the topological space (X, τ) there exists a metric d such that τ is induced by the metric d, then we say that the topological space (X, τ) is ***metrizable***.

Definition 2.3
Let (X, τ) be a topological space. A **neighbourhood** *of x is a set $U \in \tau$ such that $x \in U$. The family of neighbourhoods of $x \in X$ is denoted by $\mathcal{N}(x)$. Similarly, a* **neighbourhood** *of a set $C \subseteq X$ is an open set U such that $C \subseteq U$. If $U \in \mathcal{N}(x)$, then $U \setminus \{x\}$ is called the* **deleted neighbourhood** *of x.*

Definition 2.4
Let (X, τ) be a topological space.
*(a) X is a T_0-**space** (or **Kolmogorov space**) if for every pair of distinct points $x, y \in X$, we can find $U \in \tau$ such that $x \in U$, $y \notin U$.*
*(b) X is a T_1-**space** (or **Hausdorff** or **separated space**) if for every pair of distinct points $x, y \in X$, we can find $U, V \in \tau$ such that $x \in U$, $y \notin U$ and $x \notin V$, $y \in V$.*
*(c) X is a T_2-**space** if for every pair of distinct points $x, y \in X$, we can find $U, V \in \tau$ such that $x \in U$, $y \in V$ and $U \cap V = \emptyset$.*
*(d) X is a T_3-**space** (or **regular** or **Vietoris space**) if for every $x \in X$ and every closed set $C \subseteq X$ not containing x, we can find $U, V \in \tau$ such that $x \in U$, $C \subseteq V$ and $U \cap V = \emptyset$.*
*(e) X is a T_4-**space** (or **normal** or **Tietze space**) if for every pair of disjoint closed sets $C, D \subseteq X$, we can find $U, V \in \tau$ such that $C \subseteq U$, $D \subseteq V$ and $U \cap V = \emptyset$.*

Remark 2.5
In general a T_1-space is always a T_0-space, a T_2-space is a T_1-space, but a T_3 space does not need to be a T_2-space. A T_4 space needs to be neither T_3 nor T_2 (consider $X = \{x, y\}$ with topology $\tau = \{\emptyset, \{x\}, X\}$). A discrete topological space (see Remark 2.2) is T_0, T_1, T_2, T_3 and T_4, while the indiscrete topological space X (where X has at least two elements) is T_3 and T_4 but not T_0, T_1 and T_2 (in particular it is not a Hausdorff space).

Remark 2.6
Strictly speaking, for a space X to be non-Hausdorff, seems rather unreasonable. However, it turns out that there exist useful topologies (such as the Zariski topology in algebraic geometry), which are non-Hausdorff. Nevertheless, for the purpose of Analysis, the Hausdorff separation axiom suffices. For this reason in this book

all topological spaces are assumed to be Hausdorff.

Every metric space is Hausdorff and in a Hausdorff space singletons are closed sets.

Definition 2.7
Let (X, τ) be a topological space and let $\{x_n\}_{n\geqslant 1} \subseteq X$ be a sequence. We say that $x \in X$ is the ***limit*** *of the sequence $\{x_n\}_{n\geqslant 1}$, if for every $U \in \mathcal{N}(x)$, we can find $n_0 = n_0(U) \geqslant 1$ such that $x_n \in U$ for all $n \geqslant n_0$.*

Remark 2.8
In a Hausdorff space a sequence can have at most one limit.

Definition 2.9
Let (X, τ) be a topological space and let $A \subseteq X$ be a nonempty set.
(a) *A point $x \in X$ is an* ***interior point*** *of A, if we can find $U \in \mathcal{N}(x)$ such that $U \subseteq A$. The set of interior points of A is called the* ***interior*** *of A and is denoted by $\operatorname{int} A$ (or $\mathring{A}$).*
(b) *A point $x \in X$ is an* ***exterior point*** *of A, if $x \in \operatorname{int} A^c$ (recall that $A^c = X \setminus A$). The set of exterior points of A is called the* ***exterior*** *of A and is denoted by $\operatorname{ext} A$.*
(c) *A point $x \in X$ is a* ***boundary point*** *of A, if for every $U \in \mathcal{N}(x)$, we have $A \cap U \neq \emptyset$ and $A^c \cap U \neq \emptyset$. The set of boundary points of A is called the* ***boundary*** *of A and is denoted by ∂A (or $\operatorname{bd} A$).*

(d) *A point* $x \in X$ *is a* ***limit point*** *(or* ***accumulation point*** *or* ***cluster point****) of* A*, if every* $U \in \mathcal{N}(x)$ *satisfies* $A \cap (U \setminus \{x\}) \neq \emptyset$*. The set of limit points of* A *is called the* ***derived set*** *of* A *and is denoted by* A'*. The set* $A \cup A'$ *is called the* ***closure*** *of* A *and is denoted by* $\overline{A}$ *(or* $\operatorname{cl} A$*). A point* $x \in A$ *is an isolated point of* A *if* $x \in A \setminus A'$ *(i.e., there exists* $U \in \mathcal{N}(x)$ *such that* $A \cap (U \setminus \{x\} = \emptyset)$*).*
(e) *We say that a set* A *is* ***dense*** *in* X *if* $\overline{A} = X$*. We say that a set* A *is* ***nowhere dense*** *in* X *if* $\operatorname{int} \overline{A} = \emptyset$*.*
(f) *We say that the space* X *is* ***separable*** *if it has a countable dense subset.*

Proposition 2.10
If (X, τ) *is a topological space and* $A, B \subseteq X$ *are nonempty sets, then*
(a) *A is open if and only if* $A = \operatorname{int} A$*;*
(b) $\operatorname{int} A$ *is the biggest open set contained in* A*, i.e., if* $\mathcal{U}$ *is the family of all open sets contained in* A*, then* $\operatorname{int} A = \bigcup_{U \in \mathcal{U}} U$*;*
(c) $\operatorname{int}(A \cap B) = \operatorname{int} A \cap \operatorname{int} B$*;*
(d) $\operatorname{int}(A \cup B) \supseteq \operatorname{int} A \cup \operatorname{int} B$*.*

Proposition 2.11
If (X, τ) *is a topological space and* $A, B \subseteq X$ *are nonempty sets, then*
(a) *A is closed if and only if* $A = \overline{A}$*;*
(b) *A is closed if and only if* $A' \subseteq A$*;*
(c) $x \in \overline{A}$ *if and only if for all* $V \in \mathcal{N}(x)$*, we have* $V \cap A \neq \emptyset$*;*
(d) $\overline{A}$ *is the smallest closed set which contains* A*, i.e., if* $\mathcal{F}$ *is the family of all closed set containing* A*, then* $\overline{A} = \bigcap_{C \in \mathcal{F}} C$*;*
(e) $\overline{A \cup B} = \overline{A} \cup \overline{B}$*;*
(f) $\overline{A \cap B} \subseteq \overline{A} \cap \overline{B}$*.*

Proposition 2.12
If (X, τ) *is a topological space and* $\{A_i\}_{i \in I} \subseteq X$ *is a family of sets, then*
(a) $\bigcup_{i \in I} \overline{A_i} \subseteq \overline{\bigcup_{i \in I} A_i}$*;*
(b) $\overline{\bigcap_{i \in I} A_i} \subseteq \bigcap_{i \in I} \overline{A_i}$*..*

Proposition 2.13
Let (X, τ) be a topological space. The following statements are equivalent:
(a) *D is dense in X;*
(b) *if C is a closed subset of X and $D \subseteq C$, then $C = X$;*
(c) $\operatorname{int} D^c = \emptyset$;
(d) *for every $U \in \tau$, we have $U \cap D \neq \emptyset$.*

Definition 2.14
*Let (X, τ) be a topological space and let $C \subseteq X$ be a nonempty set. The **subspace topology** (or **relative topology** or **induced topology**) τ_C on C is defined by $\tau_C = \{C \cap U : U \in \tau\}$. With these open sets C becomes a topological space in its own right and we refer to C as a **subspace** of X.*

Proposition 2.15
Let (X, τ) be a topological space and let $C \subseteq X$ be a nonempty subspace. Every open (respectively, closed) set in C is open (respectively, closed) in X if and only if C is open (respectively, closed) in X.

Proposition 2.16
Suppose that (X, τ) is a topological space, $C \subseteq X$ is a nonempty subspace and D is a nonempty subspace of C. Then the subspace topologies on D induced by X and C coincide.

Proposition 2.17
The Hausdorff property and the regularity property (see Definition 2.4) are hereditary, i.e., if (X, τ) is Hausdorff (respectively, regular) and $C \subseteq X$ is nonempty, then (C, τ_C) is Hausdorff (respectively, regular) too.

Definition 2.18
Let (X, τ) be a topological space.
(a) *A set $E \subseteq X$ is said to be a G_δ-set if $E = \bigcap_{n \geqslant 1} U_n$ with $U_n \subseteq X$ open for all $n \geqslant 1$.*
(b) *A set $D \subseteq X$ is said to be a F_σ-set if $D = \bigcup_{n \geqslant 1} C_n$ with $C_n \subseteq X$ closed for all $n \geqslant 1$.*

2.1.2 Topological Basis and Subbasis

Definition 2.19
Let (X, τ) be a topological space.
(a) *A family $\mathcal{B} \subseteq \tau$ is a* ***basis*** *for τ if every $U \in \tau$ is a union of elements in $\mathcal{B}$.*
(b) *A family $\mathcal{Y} \subseteq \tau$ is a* ***subbasis*** *for τ if the collection of all finite intersections of sets from $\mathcal{Y}$ form a basis for τ.*

Proposition 2.20
If (X, τ) is a topological space and $\mathcal{B} \subseteq \tau$,
then $\mathcal{B}$ is a basis if and only if for every $x \in X$ and every $U \in \mathcal{N}(x)$, we can find $V \in \mathcal{B}$ such that $x \in V \subseteq U$.

Proposition 2.21
If (X, τ) is a topological space and $\mathcal{B}$ is a basis for τ,
then U is open (i.e., $U \in \tau$) if and only if for every $x \in U$, we can find $V \in \mathcal{B}$ such that $x \in V \subseteq U$.

Definition 2.22
Let $\mathcal{A}$ be a family of subsets of a set X. The topology on X ***generated*** *by $\mathcal{A}$ (denoted by $\tau(\mathcal{A})$), is the smallest topology on X containing $\mathcal{A}$. This topology can be obtained from $\mathcal{A}$ by taking finite intersections of elements of $\mathcal{A}$ and creating this way a basis for a topology which is unique. Clearly $\tau(\mathcal{A})$ is the intersection of all topologies containing $\mathcal{A}$.*

Definition 2.23
Let (X, τ) be a topological space and let $x \in X$. We say that $\mathcal{D} \subseteq \mathcal{N}(x)$ is a ***local basis*** *at x (or a* ***neighbourhood basis*** *at x) if for every $U \in \mathcal{N}(x)$, we can find $V \in \mathcal{D}$ such that $x \in V \subseteq U$.*

Definition 2.24
A topological space (X, τ) is said to be:
(a) first countable *if there is a countable local basis at every $x \in X$;*
(b) second countable *if it has a countable basis.*

Remark 2.25
Every metric space is first countable, but need not be second countable. From Proposition 1.24 we know that a metric space is second countable if and only if it is separable. For example the open intervals

with rational endpoints form a countable basis for the separable space $X = \mathbb{R}$ (recall $\overline{\mathbb{Q}} = \mathbb{R}$). Clearly every second countable space is first countable. The converse is not in general true.

Definition 2.26
Let (X, τ) be a topological space.
(a) *Let $\mathcal{Y} \subseteq \tau$. We say that $\mathcal{Y}$ is an* ***open cover*** *of X if $X = \bigcup_{U \in \mathcal{Y}} U$. A subfamily $\mathcal{Y}' \subseteq \mathcal{Y}$ which is also a cover is called a* ***subcover*** *of $\mathcal{Y}$.*
(b) *X is said to be a* ***Lindelöf space*** *if every open cover of X has a countable subcover.*

Theorem 2.27
Every second countable space is Lindelöf and separable.

Remark 2.28
While for metric spaces separability (or the Lindelöf property) is equivalent to second countability (see Theorem 2.27), for general spaces this is not true. Consider $X = \mathbb{R}$ and

$$\mathcal{A} = \{(\lambda, +\infty) : \lambda \in \mathbb{R}\} \cup \{(-\infty, \mu] : \mu \in \mathbb{R}\}.$$

We consider the topology $\tau(\mathcal{A})$ (see Definition 2.22). The topology has a basis consisting of the intervals $(\lambda, \mu]$, $\lambda \leqslant \mu$. Note that $\tau(\mathcal{A})$ is not the usual (Euclidean) topology on $\mathbb{R}$ since the sets $(\lambda, \mu]$ are not open sets in the Euclidean topology. In fact the Euclidean topology is strictly weaker than $\tau(\mathcal{A})$ (usually called the ***upper limit topology***). Then $(\mathbb{R}, \tau(\mathcal{A}))$ is Lindelöf and separable ($\mathbb{Q}$ is a countable dense subset), but it is not second countable.

2.1.3 Nets

Sequences which were very useful in the study of metric spaces are of limited interest in general topological spaces which need not be first countable. Instead we use nets, which are a suitable generalization of sequences. In the indexing of a net the set of positive integers (used in sequences) is replaced by the more general notion of directed set.

Definition 2.29

(a) A relation $\leqslant$ on a set I is a ***partial ordering*** *if it is reflexive (i.e., $x \leqslant x$ for all $x \in I$), antisymmetric (i.e., $x \leqslant y$ and $y \leqslant x$ imply $x = y$ for all $x, y \in I$) and transitive (i.e., $x \leqslant y$ and $y \leqslant z$ imply $x \leqslant z$ for all $x, y, z \in I$).*
(b) A pair $(I, \leqslant)$ consisting of a set I and a partial ordering $\leqslant$ is said to be a ***directed set*** *if for any $x, y \in I$, we can find $z \in I$ such that $x \leqslant z$ and $y \leqslant z$.*

Example 2.30

(a) If $I = \mathbb{N}$ and $\leqslant$ is the usual ordering in $\mathbb{N}$, then $(\mathbb{N}, \leqslant)$ is a directed set.
(b) Let X be a nonempty set and let I be the family of finite subsets of X and let $\leqslant$ be the relation on I defined by $F_1 \leqslant F_2$ (for $F_1, F_2 \in I$) if and only if $F_1 \subseteq F_2$. Then $(I, \leqslant)$ is a directed set.
(c) Let (X, τ) be a topological space, $x \in X$ and let $I = \mathcal{N}(x)$. On I we consider the relation $\leqslant$ defined by $U \leqslant V$ (for $U, V \in \mathcal{N}(x)$) if and only if $U \supseteq V$ (the partial order is the reverse inclusion). Then $(\mathcal{N}(x), \leqslant)$ is a directed set.
(d) Let $(I_1, \leqslant_1)$ and $(I_2, \leqslant_2)$ be two directed sets. If on $I = I_1 \times I_2$ we consider the partial order $\leqslant$ defined by $(x_1, x_2) \leqslant (u_1, u_2)$ if and only if $x_1 \leqslant_1 u_1$ and $x_2 \leqslant_2 u_2$, then $(I, \leqslant)$ is a directed set.

Definition 2.31

Let X be any set. A net $\{x_i\}_{i \in I} \subseteq X$ is any function $x \colon I \longrightarrow X$ with $(I, \leqslant)$ being a directed set. In particular, a sequence is a net defined on the directed set $\mathbb{N}$ with the usual ordering. If (X, τ) is a topological space, then a net $\{x_i\}_{i \in I}$ is said to ***converge*** *to x, written $x_i \longrightarrow x$ if and only if for every $U \in \mathcal{N}(x)$ there is $i_0 \in I$ such that $x_i \in U$ for all $i_0 \leqslant i$. This convergence can also be expressed by saying that $\{x_i\}_{i \in I}$ is* ***eventually*** *in U for every $U \in \mathcal{N}(x)$. We say that the net $\{x_i\}_{i \in I}$ is* ***frequently*** *in a set $C \subseteq X$, if for any $i \in I$, there is a $j \in I$ such that $i \leqslant j$ and $x_j \in C$. Note that the negation of the statement "the net is eventually in C" is the statement "the net is frequently in C^c" Evidently if a net $\{x_i\}_{i \in I}$ is eventually in $C_1 \subseteq X$ and eventually in $C_2 \subseteq X$, then it is eventually in $C_1 \cap C_2$.*

Proposition 2.32
The Hausdorff property of (X, τ) is equivalent to saying that every convergent net has a unique limit.

Proposition 2.33
If (X, τ) is a topological space and $C \subseteq X$ is nonempty, then
(a) *x is a limit point of C if and only if there exists a net $\{x_i\}_{i \in I} \subseteq C \setminus \{x\}$ such that $x_i \longrightarrow x$.*
(b) *C is closed if and only if every convergent net $\{x_i\}_{i \in I} \subseteq C$ has its limit in C.*

In Theorem 1.71 we saw that in metric spaces compactness can be characterized using convergent subsequences. We shall see that the same can be done for general topological spaces provided we replace sequences by nets and subsequences by subnets, which we define next.

Definition 2.34
Let I and J be directed sets and let $\{x_i\}_{i \in I}$ and $\{y_j\}_{j \in J}$ be nets in a set X. We say that $\{y_j\}_{j \in J}$ is a subnet of $\{x_i\}_{i \in I}$, if there is a function $\psi \colon J \longrightarrow I$ such that:
(a) *$y_j = x_{\psi(j)}$ for each $j \in J$;*
(b) *for every $i_0 \in I$, there is a $j_0 \in J$ such that for every $j \in J$, $j_0 \leqslant j$ implies $i_0 \leqslant \psi(j)$.*

Proposition 2.35
If (X, τ) is a topological space and $\{x_i\}_{i \in I}$ is a net in X, then $x \in X$ is a limit point of $\{x_i\}_{i \in I}$ if and only if x is the limit of some subnet of $\{x_i\}_{i \in I}$.

Corollary 2.36
If (X, τ) is a topological space, then a net $\{x_i\}_{i \in I}$ converges to $x \in X$ if and only if every subnet of $\{x_i\}_{i \in I}$ converges to x.

2.1.4 Continuous and Semicontinuous Functions

The notion of "continuous function", which we are about to introduce, is in the core of Topology. We give both the local and global definitions of continuity.

Definition 2.37
Let (X, τ_X) and (Y, τ_Y) be two topological spaces and let $f\colon X \longrightarrow Y$ be a function.
(a) *We say that f is* ***continuous*** *at $x \in X$ if for every $V \in \mathcal{N}(f(x))$, we can find $U \in \mathcal{N}(x)$ such that $f(U) \subseteq V$. If f is not continuous at $x \in X$, then we say that f is* ***discontinuous*** *at x.*
(b) *We say that f is* ***continuous*** *if for every $V \in \tau_Y$ we have $f^{-1}(V) \in \tau_X$.*

Proposition 2.38
If (X, τ_X) and (Y, τ_Y) are two topological spaces and $f\colon X \longrightarrow Y$ is a function,
then f is continuous if and only if f is continuous at every $x \in X$.

Recalling that inverse images preserve all set theoretic operations, we have the following equivalence.

Proposition 2.39
If (X, τ_X) and (Y, τ_Y) are two topological spaces and $f\colon X \longrightarrow Y$ is a function,
then f is continuous if and only if for all $C \subseteq Y$ closed, $f^{-1}(C) \subseteq X$ is closed.

Proposition 2.40
If (X, τ_X) and (Y, τ_Y) are two topological spaces and $f\colon X \longrightarrow Y$ is a function,
then the following statements are equivalent:
(a) *f is continuous;*
(b) *the inverse image of every basis element of τ_Y is in τ_X;*
(c) *the inverse image of every subbasis element of τ_Y is in τ_X;*
(d) *for every $A \subseteq X$, we have $f(\overline{A}) \subseteq \overline{f(A)}$;*
(e) *for every $B \subseteq Y$, we have $\overline{f^{-1}(B)} \subseteq f^{-1}(\overline{B})$;*
(f) *for every net $\{x_i\}_{i \in I}$, if $x_i \longrightarrow x$, then $f(x_i) \longrightarrow f(x)$.*

Proposition 2.41
If X, Y, Z are topological spaces and $f\colon X \longrightarrow Y$ is continuous at x_0 and $g\colon Y \longrightarrow Z$ is continuous at $y_0 = f(x_0)$,
then $g \circ f\colon X \longrightarrow Z$ is continuous at x_0.

If Y is a metric space, the notion of continuity of a function $f\colon X \longrightarrow Y$ can be expressed in terms of oscillation (see Problem 2.23). The following definition extends Definition 1.37 to a more general notion.

Definition 2.42
Suppose that X is a topological space, (Y, d_Y) is a metric space and $f\colon X \longrightarrow Y$ is a function. For $x \in X$, the ***oscillation*** *of f at x, denoted by $\omega_f(x)$, is defined by*

$$\omega_f(x) = \inf_{U \in \mathcal{N}(x)} \operatorname{diam} f(U).$$

Next we will see how we can "glue" together continuous functions (piecewise definition of functions). First a definition.

Definition 2.43
Let X be a topological space and let $\{C_\alpha\}_{\alpha \in A}$ be a family of subsets of X. We say that $\{C_\alpha\}_{\alpha \in A}$ is ***locally finite*** *(or* ***neighbourhood finite****) if for every $x \in X$, we can find $U \in \mathcal{N}(x)$ such that $U \cap C_\alpha \neq \emptyset$ for only finite number of indices $\alpha \in A$.*

Using this notion, we can have the following "gluing theorem".

Theorem 2.44
If X is a topological space, $\{C_\alpha\}_{\alpha \in A}$ is a cover of X, for all $\alpha \in A$, functions $f_\alpha\colon C_\alpha \longrightarrow Y$ are continuous, we have

$$f_\alpha\big|_{C_\alpha \cap C_\beta} = f_\beta\big|_{C_\alpha \cap C_\beta} \qquad \forall\ (\alpha, \beta) \in A \times A$$

and at least one of the following conditions holds:
(i) all sets C_α are open; or
(ii) all sets C_α are closed and the family $\{C_\alpha\}_{\alpha \in A}$ is locally finite,
then there exists a unique continuous function $f\colon X \longrightarrow Y$ such that

$$f\big|_{C_\alpha} = f_\alpha \qquad \forall\ \alpha \in A$$

(i.e., f extends f_α for all $\alpha \in A$).

For real functions (i.e., functions into $\mathbb{R}$ or $\mathbb{R}^* = \mathbb{R} \cup \{\pm\infty\}$), we can exploit the fact that half lines $\{(a, +\infty)\}_{a \in \mathbb{R}} \cup \{(-\infty, a)\}_{a \in \mathbb{R}}$ form a subbasis for the usual topology on $\mathbb{R}$, to have the following result.

Proposition 2.45
If X is a topological space and $f\colon X \longrightarrow \mathbb{R}$ is a function, then f is continuous if and only if for every $\lambda, \mu \in \mathbb{R}$ both sets

$$\{x \in X : \ f(x) < \lambda\} \text{ and } \{x \in X : \ f(x) > \mu\}$$

are open in X.

If we require openness only for one of the two sets in the above proposition, then we have the notion of semicontinuous function (upper or lower).

Definition 2.46
Let X be a topological space and let $f\colon X \longrightarrow \mathbb{R}^* = \mathbb{R} \cup \{\pm\infty\}$ be a function.
(a) *We say that f is **lower semicontinuous** at $x \in X$, if for every $\lambda \in \mathbb{R}$ with $\lambda < f(x)$, we can find $U \in \mathcal{N}(x)$ such that $\lambda < f(y)$ for all $y \in U$. We say that f is **lower semicontinuous**, if f is lower semicontinuous at each $x \in X$.*
(b) *We say that f is **upper semicontinuous** at $x \in X$, if for every $\lambda \in \mathbb{R}$ with $\lambda > f(x)$, we can find $U \in \mathcal{N}(x)$ such that $\lambda > f(y)$ for all $y \in U$. We say that f is **upper semicontinuous**, if f is upper semicontinuous at each $x \in X$.*

Remark 2.47
According to the above definition $f\colon X \longrightarrow \mathbb{R}^*$ is lower semicontinuous if and only if $-f$ is upper semicontinuous.

Also $f\colon X \longrightarrow \mathbb{R}^*$ is lower semicontinuous at $x \in X$ if and only if

$$f(x) \leqslant \sup_{U \in \mathcal{N}(x)} \inf_{y \in U} f(y).$$

Since

$$\inf_{y \in U} f(y) \leqslant f(x) \qquad \forall\, U \in \mathcal{N}(x),$$

we infer that f is lower semicontinuous at $x \in X$ if and only if

$$f(x) = \sup_{U \in \mathcal{N}(x)} \inf_{y \in U} f(y).$$

Similarly $f\colon X \longrightarrow \mathbb{R}^*$ is upper semicontinuous at $x \in X$ if and only if

$$f(x) = \inf_{U \in \mathcal{N}(x)} \sup_{y \in U} f(y).$$

It follows immediately that if f is lower semicontinuous (respectively, upper semicontinuous) at x, then

$$f(x) \leqslant \liminf_{n\to\infty} f(x_n) \quad \left(\text{respectively } \limsup_{n\to\infty} f(x_n) \leqslant f(x)\right)$$

for every sequence $\{x_n\}_{n\geqslant 1}$ converging to x in X. The converse is true if we assume additional structure on the topological space X.

Proposition 2.48
<u>If</u> X is a first countable topological space, $f\colon X \longrightarrow \mathbb{R}^*$ is a function and $x \in X$,
<u>then</u> the following statements are equivalent:
(a) f is lower semicontinuous at $x \in X$ (respectively, upper semicontinuous at $x \in X$);
(b) $f(x) \leqslant \liminf_{n\to+\infty} f(x_n)$ (respectively, $\limsup_{n\to+\infty} f(x_n) \leqslant f(x)$) for every sequence $\{x_n\}_{n\geqslant 1} \subseteq X$ such that $x_n \longrightarrow x \in X$;
(c) $f(x) \leqslant \lim_{n\to+\infty} f(x_n)$ (respectively, $\lim_{n\to+\infty} f(x_n) \leqslant f(x)$) for every sequence $\{x_n\}_{n\geqslant 1} \subseteq X$ such that $x_n \longrightarrow x \in X$ and $\lim_{n\to+\infty} f(x_n) < +\infty$ (respectively, $\lim_{n\to+\infty} f(x_n) > -\infty$).

Definition 2.49
Let X, Y be two topological spaces.
*(a) A function $f\colon X \longrightarrow Y$ is said to be **sequentially continuous** at $x \in X$ if for every sequence $\{x_n\}_{n\geqslant 1} \subseteq X$ such that $x_n \longrightarrow x$ in X, we have that $f(x_n) \longrightarrow f(x)$ in Y. We say that f is **sequentially continuous** if it is sequentially continuous at every $x \in X$;*
*(b) A function $f\colon X \longrightarrow \mathbb{R}^*$ is said to be **sequentially lower semicontinuous** at $x \in X$ (respectively, **sequentially upper semicontinuous** at $x \in X$) if for every sequence $\{x_n\}_{n\geqslant 1} \subseteq X$ such that $x_n \longrightarrow x$ in X, we have that*

$$f(x) \leqslant \liminf_{n\to+\infty} f(x_n) \quad \left(\textit{respectively } \limsup_{n\to+\infty} f(x_n) \leqslant f(x)\right).$$

*We say that f is **sequentially lower semicontinuous** (respectively, **sequentially upper semicontinuous**) if it is sequentially lower semicontinuous at every $x \in X$ (respectively, sequentially upper semicontinuous at every $x \in X$).*

Proposition 2.50
If X, Y are topological spaces, then
(a) *$f\colon X \longrightarrow Y$ is continuous (at $x \in X$) implies that f is sequentially continuous (at $x \in X$) and the converse is true if X is first countable;*
(b) *$f\colon X \longrightarrow \mathbb{R}^*$ is lower semicontinuous (at $x \in X$) (respectively, upper semicontinuous (at $x \in X$)) implies that f is sequentially lower semicontinuous (at $x \in X$) (respectively, upper semicontinuous (at $x \in X$)).*

Definition 2.51
Let (X, τ) be a topological space and let $A \subseteq X$ be a set. We say that A is ***sequentially open*** *in X, if for every $x \in A$ and every sequence $\{x_n\}_{n\geqslant 1} \subseteq X$, such that $x_n \longrightarrow x$ in X, there exists $n_0 \geqslant 1$ such that $x_n \in A$ for all $n \geqslant n_0$. We say that A is* ***sequentially closed*** *in X, if for every sequence $\{x_n\}_{n\geqslant 1} \subseteq X$ such that $x_n \longrightarrow x$ in X, we have $x \in A$. Evidently A is sequentially open if and only if A^c is sequentially closed.*

Remark 2.52
The sequentially open sets form a topology τ_{seq}. This is the strongest topology on X for which the converging sequences are the τ-converging sequences. We have $\tau \subseteq \tau_{seq}$ and the inclusion can be strict (for example an infinite dimensional Banach space endowed with the weak topology, see Chap. 5). We have $\tau = \tau_{seq}$ if and only if X is first countable.

Proposition 2.53
If X is a topological space and $f\colon X \longrightarrow \mathbb{R}^*$ is a function, then the following statements are equivalent:
(a) *f is lower semicontinuous (respectively, sequentially lower semicontinuous);*
(b) *for every $\lambda \in \mathbb{R}$ the set $\{x \in X : f(x) > \lambda\}$ is open (respectively, sequentially open);*
(c) *for every $\lambda \in \mathbb{R}$ the set $\{x \in X : f(x) \leqslant \lambda\}$ is closed (respectively, sequentially closed);*

Proposition 2.54
If X is a topological space and $f\colon X \longrightarrow \mathbb{R}^*$ is a function, then the following statements are equivalent:

(a) *f is upper semicontinuous (respectively, sequentially upper semicontinuous);*
(b) *for every $\lambda \in \mathbb{R}$ the set $\{x \in X : f(x) < \lambda\}$ is open (respectively, sequentially open);*
(c) *for every $\lambda \in \mathbb{R}$ the set $\{x \in X : f(x) \geqslant \lambda\}$ is closed (respectively, sequentially closed);*

Proposition 2.55
If X is a topological space, $f_i \colon X \longrightarrow \mathbb{R}^*$ are functions for $i \in I$, then
(a) *if each f_i is lower semicontinuous (respectively, sequentially lower semicontinuous), then $\sup\limits_{i \in I} f_i(x)$ is lower semicontinuous (respectively, sequentially lower semicontinuous);*
(b) *if I is finite and each f_i is lower semicontinuous (respectively, sequentially lower semicontinuous), then $\inf\limits_{i \in I} f_i(x)$ is lower semicontinuous (respectively, sequentially lower semicontinuous);*
(c) *if each f_i is upper semicontinuous (respectively, sequentially upper semicontinuous), then $\inf\limits_{i \in I} f_i(x)$ is upper semicontinuous (respectively, sequentially upper semicontinuous);*
(d) *if I is finite and each f_i is upper semicontinuous (respectively, sequentially upper semicontinuous), then $\sup\limits_{i \in I} f_i(x)$ is upper semicontinuous (respectively, sequentially upper semicontinuous).*

Corollary 2.56
If X is a topological space and $f_i \colon X \longrightarrow \mathbb{R}^*$ for $i \in I$ are continuous (respectively, sequentially continuous) functions, then $f = \sup\limits_{i \in I} f_i$ is lower semicontinuous (respectively, sequentially lower semicontinuous) and $f = \inf\limits_{i \in I} f_i$ is upper semicontinuous (respectively, sequentially upper semicontinuous).

Definition 2.57
*Suppose that X is a topological space and $f \colon X \longrightarrow \mathbb{R}^*$ is a function. The **relaxed function** of f, denoted by $\overline{f}$, is the biggest lower semicontinuous function majorized by f, i.e., $\overline{f} = \sup\limits_{g \in \mathcal{L}(f)} g$, where*

$$\mathcal{L}(f) = \{g \colon X \longrightarrow \mathbb{R}^* : \ g \text{ is lower semicontinuous and } g \leqslant f\}.$$

Proposition 2.58
Let X be a topological space. The set of all functions $f\colon X \longrightarrow \overline{\mathbb{R}} = \mathbb{R} \cup \{+\infty\}$ which are lower semicontinuous (respectively, sequentially lower semicontinuous) is a cone, i.e., it is closed under addition and scalar multiplication with $\lambda \geqslant 0$. Similarly for the set of functions $f\colon X \longrightarrow \overline{\mathbb{R}}^{*} = \mathbb{R} \cup \{-\infty\}$ which are upper semicontinuous (respectively, sequentially upper semicontinuous).

2.1.5 Open and Closed Maps: Homeomorphisms

Definition 2.59
Let (X, τ_X) and (Y, τ_Y) be two topological spaces and let $f\colon X \longrightarrow Y$ be a function.
(a) *We say that f is an* ***open function*** *if for every $U \in \tau_X$, $f(U) \in \tau_Y$.*
(b) *We say that f is a* ***closed function*** *if for every $C \subseteq X$ closed, $f(C)$ is closed in Y.*

The notion of homeomorphism that we are about to introduce, is of fundamental importance in Topology.

Definition 2.60
Let (X, τ_X) and (Y, τ_Y) be two topological spaces and let $f\colon X \longrightarrow Y$ be a function. We say that f is ***homeomorphism*** *if f is bijective and $U \in \tau_X$ if and only if $f(U) \in \tau_Y$. Evidently $f^{-1}\colon Y \longrightarrow X$ is a homeomorphism too and we say that the spaces X and Y are homeomorphic. We say that $f\colon X \longrightarrow Y$ is a* ***local homeomorphism*** *if each $x \in X$ has a neighbourhood which is mapped homeomorphically by f onto an open subset of Y.*

Remark 2.61
The relation "X homeomorphic to Y" is an equivalence relation in the family of topological spaces. It is the fundamental equivalence relation in Topology. A homeomorphism preserves the topological structure (it is an isomorphism of topological structures). From a purely topological point of view there is no difference between two homeomorphic topological spaces. They can be considered as two representatives of the same geometric entity. When a space X has several different structures (such as algebraic, metric, linear etc.), we say that a property of X is

"topological", if it is preserved by homeomorphisms. Roughly speaking, a property which can be stated in terms of open sets and derived notions such as closed sets, limit points, dense sets, etc., is topological. For example, the Hausdorff property is a topological property, but in a metric space completeness is not a topological property.

2.1.6 Weak (or Initial) and Strong (or Final) Topologies

From the definition of continuity (see Definition 2.37), we see that the continuity of a function $f\colon X \longrightarrow Y$ is preserved if we strengthen the topology on X or weaken the topology on Y. More generally, let X be a set and let $\{(Y_i, \tau_i)\}_{i\in I}$ be a family of topological spaces (finite or infinite). For a given family of functions $\{f_i\colon X \longrightarrow Y_i\}_{i\in I}$, we are looking for those topologies on X which ensure the continuity of all functions f_i, $i \in I$.

Definition 2.62
*Let X be a set and let $\{(Y_i, \tau_i)\}_{i\in I}$ be a family of topological spaces. Also for each $i \in I$, let $f_i\colon X \longrightarrow Y_i$ be a function. The weakest topology on X which makes all the functions f_i, $i \in I$ continuous is called the **weak topology** (or **initial topology**) on X. This is the topology $w(\{f_i\}_{i\in I})$ generated by the subbasis*

$$\mathcal{Y} = \{f_i^{-1}(U) : \ i \in I, \ U \in \tau_i\}.$$

Remark 2.63
If I is a singleton, then the defining subbasis $\mathcal{Y} = \{f^{-1}(U) : \ U \in \tau_Y\}$ is in fact the weak topology $w(f)$.

Since we are interested on Hausdorff topological spaces, we want to know when this topology $w(\{f_i\}_{i\in I})$ is Hausdorff. To check this we need the following definition.

Definition 2.64
*Let X be a set and let $\{f_i\}_{i\in I}$ be a family of functions defined on X. We say that the family $\{f_i\}_{i\in I}$ is **separating** (or **total**) if for each pair of points $x, u \in X$, $x \neq u$ we can find $i \in I$ such that $f_i(x) \neq f_i(u)$.*

Proposition 2.65
<u>If</u> X is a set endowed with the weak topology $w(\{f_i\}_{i\in I})$ determined by a separating family of functions $\{f_i\colon X \longrightarrow Y_i\}_{i\in I}$ (where $\{(Y_i, \tau_i)\}_{i\in I}$ is a family of topological spaces),
<u>then</u> $(X, w(\{f_i\}_{i\in I}))$ is Hausdorff.

Proposition 2.66
If X is a set endowed with the weak topology $w(\{f_i\}_{i\in I})$ determined by a separating family of functions $\{f_i\colon X \longrightarrow Y_i\}_{i\in I}$ (where $\{(Y_i,\tau_i)\}_{i\in I}$ is a family of topological spaces) and $\{x_\alpha\}_{\alpha\in A} \subseteq X$ is a net, then $x_\alpha \longrightarrow x$ in X if and only if $f_i(x_\alpha) \longrightarrow f_i(x)$ for all $i \in I$.

Example 2.67
(a) Let Y be a topological space, $X \subseteq Y$ and consider the identity function $i_X : X \longrightarrow Y$ (i.e., $i_X(x) = x$ for all $x \in X$). Then the weak topology $w(i_X)$ is nothing else but the subspace topology on X (see Definition 2.14). So, we see that the subspace topology is a particular case of a weak (initial) topology.
(b) Let V be a set and let $\mathcal{X}$ be a separating family of functions $f\colon V \longrightarrow \mathbb{R}$. For every $v \in V$, we define $e_V : \mathcal{X} \longrightarrow \mathbb{R}$ by

$$e_V(f) \;=\; f(v) \qquad \forall\, f \in \mathcal{X}.$$

Then the family $\{e_v\}_{v\in V}$ induces a weak topology on $\mathcal{X}$ which is Hausdorff. Moreover $f_\alpha \longrightarrow f$ in $\mathcal{X}$ with this weak topology if and only if $f_\alpha(v) \longrightarrow f(v)$ for all $v \in V$ (i.e., weak convergence coincides with pointwise convergence).

Proposition 2.68
If X is a set, $\{f_i\colon X \longrightarrow \mathbb{R}\}_{i\in I}$ is a separating family of functions and $C \subseteq X$, then $w(\{f_i\}_{i\in I})\Big|_C = w(\{f_i|_C\}_{i\in I})$.

The notion of weak or initial topology leads easily to the notion of product topology.

Definition 2.69
Let $\{(X_i,\tau_i)\}_{i\in I}$ be a family (finite or infinite) of topological spaces and let

$$X \;=\; \prod_{i\in I} X_i \;=\; X^I.$$

We consider the canonical projections $p_i\colon X \longrightarrow X_i$ *for* $i \in I$*, defined by*

$$p_i\big((x_j)_{j\in I}\big) = x_i.$$

Then the weak topology $w\big(\{p_i\}_{i\in I}\big)$ *on* X *is called the* ***product topology*** *on* X*. The space* $\big(X, w\big(\{p_i\}_{i\in I}\big)\big)$ *is called the* ***product space*** *of the family* $\{X_i\}_{i\in I}$*.*

Remark 2.70
Based on Definition 2.62, we see that a basis for the product topology (obtained by taking finite intersections of subbasis elements) is the collection of all sets of the form $\prod\limits_{i\in I} U_i$ which satisfy two conditions:

(a) each U_i is open in X_i;
(b) only for a finite number of indices $i \in I$, we have $U_i \neq X_i$.

When I is infinite, if requirement (b) is not satisfied and we permit in the collection set $\prod\limits_{i\subset I} U_i$, where $U_i \neq X_i$ for an infinite number of indices, we get a stronger topology, known as the ***box topology***. In what follows on product sets $\prod\limits_{i\in I} X_i$ we will always consider the product topology.

Proposition 2.71
(a) *$\prod\limits_{i\in I} X_i$ is Hausdorff (respectively, regular; see Definition 2.4) if and only if each X_i if Hausdorff (respectively, regular).*
(b) *The product $\prod\limits_{i\in I} X_i$ of normal spaces X_i need not be normal.*
(c) *The product $\prod\limits_{n\geqslant 1} X_n$ is second countable (see Definition 2.24) if and only if each X_n is second countable.*
(d) *The product $\prod\limits_{n\geqslant 1} X_n$ is separable if and only if each X_n is separable.*
(e) *The product $\prod\limits_{i\in I} X_n$ of Lindelöf spaces need not be Lindelöf (see Definition 2.26).*

Next we consider the opposite case from the one that led to the weak (initial) topology. So, now we are given a family of functions $\{f_i\colon X_i \longrightarrow Y\}_{i\in I}$, with $\{X_i\}_{i\in I}$ being a family of topological spaces and we are looking for the strongest topology which can be placed on Y for which all functions $\{f_i\}_{i\in I}$ are continuous.

Definition 2.72
Suppose that $\{(X_i, \tau_i)\}_{i\in I}$ *is a family of topological spaces,* Y *is a set and* $\{f_i\colon X_i \longrightarrow Y\}_{i\in I}$ *is a family of functions. The strongest topology on* Y *which makes all functions* $\{f_i\}_{i\in I}$ *continuous, is called the* ***strong topology*** *(or* ***final topology****) on* X*. It is given by*

$$\{U \in Y : \ f_i^{-1}(U) \in \tau_i \text{ for all } i \in I\}$$

and it is denoted by $s(\{f_i\}_{i\in I})$*.*

Proposition 2.73
If Y *is endowed with the strong topology* $s(\{f_i\}_{i\in I})$ *determined by a family of functions* $\{f_i\colon X_i \longrightarrow Y\}_{i\in I}$*,* V *is a topological space and* $g\colon Y \longrightarrow V$*,*
then g *is continuous if and only if for every* $i \in I$*,* $g \circ f_i\colon X_i \longrightarrow V$ *is continuous.*

We concentrate our attention on the case of a single function $f\colon X \longrightarrow Y$. Let Y have the strong topology $s(f)$. Let $Y_0 = Y \setminus f(X)$. If $y \in Y_0$, then $f^{-1}(y) = \emptyset$ and so $\{y\}$ is clopen in Y. Also, $f(X)$ is closed in Y. Therefore Y is the disjoint union of $f(X)$ with a discrete space. Hence we may assume that f is surjective.

Definition 2.74
Let X, Y *be two topological spaces and let* $f\colon X \longrightarrow Y$ *be a function. We say that* f *is an* ***identification function*** *if the following two requirements are satisfied:*
(a) f *is a surjective; and*
(b) Y *is endowed with the strong topology* $s(f)$*.*
Then $s(f)$ *is called the* ***identification topology*** *with respect to* f *and* Y *is an* ***identification space*** *of* f*. A set* $A \subseteq X$ *is said to be* f***-saturated*** *if* $f^{-1}(f(A)) = A$*.*

Proposition 2.75
If $f\colon X \longrightarrow Y$ *is an identification function,*
then

$$\begin{aligned} s(f) &= \{f(V) : \text{ for all } V \subseteq X \text{ open, } f\text{-saturated}\} \\ &= \{f(C) : \text{ for all } C \subseteq X \text{ closed, } f\text{-saturated}\}. \end{aligned}$$

A special case of the identification topology is the quotient topology.

Definition 2.76
Let X, Y be topological spaces and let $\sim$ be an equivalence relation on X. Let $p\colon X \longrightarrow X/_\sim$ be the quotient map. Then

$$s(p) = \{U \subseteq X/_\sim : p^{-1}(U) \text{ is open in } X\}$$

*is called the **quotient topology** on $X/_\sim$ and $(X/_\sim, s(p))$ is called the **quotient space** of X by $\sim$.*

Remark 2.77
Quotients of Hausdorff spaces need not be Hausdorff. However, if $\mathrm{Gr} \sim \subseteq X \times X$ is closed, then $X/_\sim$ is Hausdorff.

2.1.7 Compact Topological Spaces

One of the most important topological properties is compactness.

Definition 2.78
Let (X, τ) be a topological space.
(a) *We say that X is **compact** if every open cover of X admits a finite subcover (see Definition 2.26).*
(b) *A set $C \subseteq X$ is **compact** if every open cover of C in X (i.e., a family $\{U_i\}_{i \in I} \subseteq \tau$ such that $C \subseteq \bigcup_{i \in I} U_i$) admits a finite subcover.*

Proposition 2.79
If X is a topological space and $C \subseteq X$,
then C is a compact subset of X if and only if C with the subspace topology is a compact topological space.

Definition 2.80
*Let X be a set and let $\mathcal{D} \subseteq 2^X \setminus \{\emptyset\}$. We say that $\mathcal{D}$ has the **finite intersection property** if for every nonempty finite subset $\mathcal{F} \subseteq \mathcal{D}$, we have $\bigcap_{A \in \mathcal{F}} A \neq \emptyset$.*

The next theorem provides important alternative characterization of compactness.

Theorem 2.81
Let X be a topological space. The following statements are equivalent:

(a) *X is compact;*
(b) *Every nonempty family of closed subsets of X with the finite intersection property has a nonempty intersection;*
(c) *Every net in X has a convergent subnet.*

Proposition 2.82
If X, Y are two topological spaces, $f\colon X \longrightarrow Y$ is a continuous function and $C \subseteq X$ is a compact set,
then $f(C) \subseteq Y$ is compact.

In the next proposition we have gathered some properties of compact spaces which are useful in Analysis.

Proposition 2.83
(a) *Every closed subset of a compact set is compact.*
(b) *Every compact set in a topological space is closed.*
(c) *If C_1 and C_2 are disjoint compact sets in a topological space X, then there exist disjoint open sets U_1 and U_2 such that $C_1 \subseteq U_1$ and $C_2 \subseteq U_2$.*
(d) *A compact topological space is normal (see (c)).*
(e) *A quotient space of a compact space is compact.*

Theorem 2.84
If X, Y are two topological spaces, X is compact and $f\colon X \longrightarrow Y$ is a continuous bijection,
then f is homeomorphism.

Theorem 2.85 *(Heine–Borel Theorem)*
A set $C \subseteq \mathbb{R}^N$ is compact if and only if C is closed and bounded.

Theorem 2.86
If X is a compact topological space and $f\colon X \longrightarrow \overline{\mathbb{R}} = \mathbb{R} \cup \{+\infty\}$ is lower semicontinuous (respectively, $f\colon X \longrightarrow \overline{\mathbb{R}}^* = \mathbb{R} \cup \{-\infty\}$ is upper semicontinuous),
then f attains its infimum (respectively, supremum).

Corollary 2.87
If X is a compact topological space and $f\colon X \longrightarrow \mathbb{R}$ is continuous,
then f attains both its infimum and supremum.

Definition 2.88
Let X be a topological space and let $C \subseteq X$.
(a) We say that C is ***sequentially compact*** *if every sequence in C has a subsequence which converges to a point in C.*
(b) We say that C is ***countably compact*** *(or* ***Fréchet compact*** *or* ***a set with Bolzano–Weierstrass property****) if every infinite set $D \subseteq C$ has a limit point in C.*

Remark 2.89
From Theorem 1.71, we know that in metrizable spaces the above two notions coincide with compactness. For general topological spaces this is no longer true.

Proposition 2.90

(a) A compact topological space is countably compact, but the converse is not in general true.
(b) A sequentially compact topological space is countably compact, but the converse is not in general true.
(c) A first countable space is sequentially compact if and only if it is countably compact.
(d) A first countable compact space is sequentially compact.
(e) A topological space is countably compact if and only if every countable cover has a finite subcover.
(f) A topological space is countably compact if and only if every sequence has a limit point.

Theorem 2.91 *(Tichonov Theorem)*
If $\{X_i\}_{i \in I}$ is a family of topological spaces,
then $X = \prod_{i \in I} X_i$ with the product topology (see Definition 2.69) is compact if and only if for each $i \in I$, X_i is compact.

Definition 2.92
A topological space X is ***locally compact*** *if for every point $x \in X$ there exists $U \in \mathcal{N}(x)$ such that $\overline{U}$ is compact.*

Remark 2.93
Compact spaces are locally compact, but the converse is not true (consider $X = \mathbb{R}^N$, $N \geqslant 1$). Also discrete spaces (see Remark 2.2) are

locally compact and as will see in Chap. 5, a normed space is locally compact if and only if it is finite dimensional.

Proposition 2.94
If (X, τ) is a locally compact topological space, $C \subseteq X$ is a nonempty compact subset and $U \in \tau$ is such that $C \subseteq U$,
then we can find $V \in \tau$ which is ***relatively compact*** *(i.e., $\overline{V}$ is compact in X) and $C \subseteq V \subseteq \overline{V} \subseteq U$.*

Corollary 2.95
X is a locally compact topological space if and only if every $x \in X$ has a local basis consisting of relatively compact open sets.

Definition 2.96
Let X be a topological space. A compact topological space Y is said to be a ***compactification*** *of X, if X is homeomorphic to a dense subspace of Y.*

Remark 2.97
Such a compactification always exists. The simplest example is the so called Alexandrov one-point compactification. Let (X, τ) be a non-compact topological space and define a new set X^* by $X^* = X \cup \{\infty\}$, where ∞ is not already an element of X. We define

$$\tau^* = \tau \cup \big\{K^c \cup \{\infty\} : K \subseteq X \text{ compact}\big\}.$$

Then τ^* is a topology on X^* and (X^*, τ^*) is the ***Alexandrov one-point compactification*** of X. Since for us only Hausdorff spaces matter, the following theorem is important.

Theorem 2.98
If (X, τ) is a topological space and (X^*, τ^*) is its Alexandrov one-point compactification,
then (X^*, τ^*) is Hausdorff if and only if X is locally compact.

Definition 2.99
A topological space X is said to be ***σ-compact*** *if there is a sequence $\{C_n\}_{n \geqslant 1}$ of compact subsets of X such that*

$$X = \bigcup_{n \geqslant 1} C_n.$$

Proposition 2.100
If X is a locally compact, σ-compact topological space,
then there exists an increasing sequence $\{C_n\}_{n\geqslant 1}$ of compact subsets of X such that

$$X = \bigcup_{n\geqslant 1} C_n \quad \text{and} \quad C_n \subseteq \operatorname{int} C_{n+1} \qquad \forall\, n \geqslant 1.$$

Proposition 2.101
If X is a locally compact topological space,
then X is σ-compact if and only if it is Lindelöf (see Definition 2.26).

Corollary 2.102
A locally compact, σ-compact topological space is normal.

Definition 2.103
Let X be a topological space and let $f\colon X \longrightarrow \mathbb{R}^*$ be a function. We say that f is ***coercive*** *(respectively,* ***sequentially coercive****) on X if for every $\lambda \in \mathbb{R}$, the set $\overline{\{x \in X : f(x) \leqslant \lambda\}}$ is countably compact (respectively, sequentially compact; see Definition 2.88).*

2.1.8 Connectedness

Connectedness is another important topological property of a space.

Definition 2.104
Let X be a topological space. We say that X is ***disconnected*** *if we can find two disjoint, nonempty, open sets U, V such that $X = U \cup V$ (i.e., the pair $\{U, V\}$ forms a partition of X). The pair $\{U, V\}$ is called a* ***disconnection*** *of X. We say that the topological space X is* ***connected*** *if it is not disconnected. A subset C of X is* ***connected*** *if it is connected as a topological space with the subspace topology (see Definition 2.14).*

Theorem 2.105
Let X be a topological space. The following statements are equivalent:
(a) X is connected;
(b) The only clopen subsets of X are $\emptyset$ and X;
(c) There is no continuous surjection $f\colon X \longrightarrow \{0, 1\}$.

Proposition 2.106
<u>If</u> X is a topological space and $C \subseteq X$,
<u>then</u> C is connected if and only if for every U, V open subsets of X such that $C \subseteq U \cup V$ and $U \cap V = \emptyset$, we have $C \cap U = \emptyset$ or $C \cap V = \emptyset$.

Corollary 2.107
<u>If</u> X is a topological space and $C \subseteq X$ is a connected set,
<u>then</u> every D satisfying $C \subseteq D \subseteq \overline{C}$ is connected. In particular the closure of a connected set is connected.

Proposition 2.108
<u>If</u> X, Y are two topological spaces, $f\colon X \longrightarrow Y$ is continuous and $C \subseteq X$ is a connected set,
<u>then</u> $f(C)$ is connected.

Proposition 2.109
<u>If</u> X is a topological space and $\{C_i\}_{i\in I}$ is a family of connected subsets of X such that there exists $i_0 \in I$ with $C_i \cap C_{i_0} \neq \emptyset$ for all $i \in I$ or $C_i \cap C_j \neq \emptyset$ for all $i, j \in I$,
<u>then</u> the set $\bigcup_{i\in I} C_i$ is connected.

Proposition 2.110
<u>If</u> $\{X_i\}_{i\in I}$ is a family of topological spaces,
<u>then</u> $X = \prod_{i\in I} X_i$ with the product topology (see Definition 2.69) is connected if and only if for every $i \in I$ the space X_i is connected.

Every topological space can be decomposed uniquely into "connected components", which are the maximal connected subsets of the space. Of course, if X is connected, then it has just one connected component (X itself).

Definition 2.111
*Let X be a topological space. A **connected component** of X is a connected subset of X which is not properly contained in another connected subset of X.*

Remark 2.112
Here is another way to define the connected components of X. Let $x \in X$ and let $C(x)$ be the union of all connected subsets of X which

contain $x \in X$. By Proposition 2.109, $C(x)$ is connected and it is maximal with respect to inclusion among all connected subsets of X containing $x \in X$. This is the connected component of X containing $x \in X$.

Proposition 2.113
If X is a topological space,
then the connected components of X are closed and form a partition of X.

Remark 2.114
The connected components of a topological space need not be open subsets of X. For example, let $X = \mathbb{Q}$ with the subspace topology as a subset of $\mathbb{R}$. Then the connected components of $X = \mathbb{Q}$ are the singletons. Indeed, suppose that $A \subseteq X$ and $\operatorname{card} A > 1$. Let $x, y \in A$ with $x < y$. We can find $u \in \mathbb{Q}$ such that $x < u < y$. Then $\{U = A \cap (-\infty, u), V = A \cap (u, +\infty)\}$ is a disconnection of $A \subseteq \mathbb{Q}$. So, every subset of $X = \mathbb{Q}$ with more than one point is disconnected. The connected components $\{x\}$, $x \in X = \mathbb{Q}$ are closed but not open. This example leads to the following definition.

Definition 2.115
A topological space X is said to be ***totally disconnected*** *if its connected components are its singletons.*

Sometimes it is more important that the space exhibits connectivity "locally". This leads to the following definition.

Definition 2.116
A topological space X is said to be ***locally connected*** *at $x \in X$ if every $U \in \mathcal{N}(x)$ contains a connected open set belonging in $\mathcal{N}(x)$. We say that X is* ***locally connected*** *if it is locally connected at each point $x \in X$.*

Remark 2.117
Another equivalent way to define local connectedness for a topological space X is to say that X has a basis containing of connected sets.

Proposition 2.118
A topological space X is locally connected if and only if the connected components of open sets are open sets (see Definitions 2.111 and 2.116).

Remark 2.119
A connected space need not be locally connected. To see this, let $Y = \{(x, y) : y = \sin\frac{1}{x}, x \in (0, 1]\}$ and let $X = Y \cup \{(0, 0)\}$. Then X is connected (note that Y is connected being the image of $(0, 1]$ under the action of the continuous function $\psi(x) = \sin\frac{1}{x}$, $x \in (0, 1]$; see Proposition 2.108, note that $X \subseteq \overline{Y}$ and use Corollary 2.107). However X is not locally connected since the connected components of $X \cap \{(x, y) : y < \frac{1}{2}\}$ are not open in X; see Proposition 2.118. A discrete space is a totally disconnected and locally connected topological space.

Let X, Y be two topological spaces and let $f: X \longrightarrow Y$ be a homeomorphism. If C is a connected component and $x \in C$, then $f(C)$ is a connected component of Y containing $f(x)$. Consequently f induces a bijection from the connected components of X to the connected components of Y. Therefore the number of topological components of X is a topological invariant of X. The drawback of this topological invariant is that it fails to distinguish between different connected spaces. For this reason we introduce the following notion.

Definition 2.120
Let X be a connected topological space and let $k \in \mathbb{N} \cup \{\infty\}$. A point $x \in X$ is a ***cut point of order*** *k if $X \setminus \{x\}$ has k connected components (see Definition 2.111). The number of cut points of order k is a topological invariant of X.*

Refining Definition 2.120, we introduce the following notion.

Definition 2.121
Let X be a connected topological space and let $x, y \in X$. We say that $\{x, y\}$ is a ***cut pair of order*** *k if $X \setminus \{x, y\}$ has k connected components (see Definition 2.111).*

In many cases in Analysis, only connected spaces of special type are used. For example, in Global Analysis we deal with spaces which locally look like (i.e., are homeomorphic to) $\mathbb{R}^N$. For this reason we introduce the following notions.

Definition 2.122
Let X be a topological space and $x, u \in X$.
(a) *A* ***path*** *from x to u in X is a continuous function $f: I = [0, 1] \longrightarrow X$ such that $f(0) = x$ and $f(1) = u$. We say that x is the* ***initial point*** *and u is the* ***finale point****.*

*(b) We say that X is **path-connected** (or **arcwise-connected**) if every pair of points $x, u \in X$ can be connected by a path.*

Proposition 2.123
<u>If</u> X is a topological space and $x_0 \in X$,
<u>then</u> X is path-connected if and only if each $x \in X$ can be connected to x_0 be a path (see Definition 2.122).

Proposition 2.124
<u>If</u> X, Y are two topological spaces, X is path-connected (see Definition 2.122) and $f\colon X \longrightarrow Y$ is a continuous function,
<u>then</u> $f(X) \subseteq Y$ is path-connected. In particular path-connectedness is a topological invariant.

Proposition 2.125
A path-connected topological space is connected (see Definitions 2.122 and 2.104). The converse is not in general true.

Example 2.126
Let

$$C = \{(x, u) : u = \sin \tfrac{1}{x}, \ 0 < x \leqslant 1\}.$$

Evidently X is connected being the image of the connected set $(0, 1]$ under the continuous function $\psi(x) = (x, \sin \frac{1}{x})$. Hence $\overline{C}$ is connected (see Corollary 2.107) and

$$\overline{C} = C \cup (\{0\} \times [-1, 1]).$$

However, $\overline{C}$ is not path-connected since there is no path joining $(0, 0)$ to $(\frac{1}{\pi}, 0)$ (see Definition 2.122).

To produce conditions which guarantee that the converse of Proposition 2.125 holds, we need the following analog of Definition 2.116.

Definition 2.127
*A topological space X is said to be **locally path-connected** at $x \in X$, if every $U \in \mathcal{N}(x)$ contains a path-connected open set belonging in $\mathcal{N}(x)$. We say that X is **locally path-connected** if it is locally path-connected at each point $x \in X$.*

Theorem 2.128
A topological space X is path-connected if and only if it is connected and locally path-connected (see Definitions 2.122 and 2.127).

Proposition 2.129
*If X is a topological space and $\{A_i\}_{i\in I}$ is a family of path-connected subspaces of X (see Definition 2.122) such that there exists $i_0 \in I$ with $A_i \cap A_{i_0} \neq \emptyset$ for all $i \in I$ or $A_i \cap A_j \neq \emptyset$ for all $i, j \in I$,
then $\bigcup_{i\in I} A_i$ is path-connected too.*

Because of this proposition, we can define path-connected components of a topological space X, as maximal path-connected subsets of X.

Definition 2.130
Let X be a topological space. A ***path-connected component*** *of X, is a path-connected subset of X which is not properly contained in another path-connected subset of X.*

Remark 2.131
The path-connected components of X form a partition of X (as did the connected components; see Remark 2.112). However, in contrast to the connected components (see Definition 2.111), path-connected components of X need not to be closed subsets of X. For example,

$$C = \left\{(x,u) : u = \sin \tfrac{1}{x},\ 0 < x \leqslant 1\right\}$$

is a path-connected component of $\overline{C}$ (see Example 2.126), but it is not closed.

Proposition 2.132
*If X is a topological space,
then X is locally path-connected if and only if the path-connected components of every open set of X are open (see Definitions 2.122 and 2.130).*

Corollary 2.133
X is locally path-connected (see Definition 2.127) if and only if its connected components and path-connected components coincide (see Definitions 2.111 and 2.130) (and are both open and closed, i.e., clopen).

Corollary 2.134
An open set in $\mathbb{R}^N$ *(or in* $S^N = \{x \in \mathbb{R}^{N+1} : \|x\| = 1\}$*) is connected if and only if it is path-connected.*

Proposition 2.135
If $\{X_i\}_{i \in I}$ *are topological spaces,*
then $X = \prod_{i \in I} X_i$ *with the product topology (see Definition 2.69) is path-connected if and only if for every* $i \in I$, X_i *is path-connected (see Definition 2.122).*

2.1.9 Urysohn and Tietze Theorems

The next two theorems provide alternative characterization of normality (see Definition 2.4).

Theorem 2.136 *(Urysohn Lemma)*
If X *is a topological space,*
then the following two statements are equivalent:
(a) X *is normal (see Definition 2.4);*
(b) *if* C *and* D *are two disjoint closed sets in* X*, then there exists a continuous function* $f \colon X \longrightarrow [0,1]$ *such that* $f|_C = 0$ *and* $f|_D = 1$.

Definition 2.137
We say that a topological space X *is* ***perfectly normal*** *if for every pair of disjoint closed sets* $C, D \subseteq X$*, we can find a continuous function* $f \colon X \longrightarrow [0,1]$ *such that* $C = f^{-1}(\{0\})$ *and* $D = f^{-1}(\{1\})$.

Theorem 2.138 *(Tietze Extension Theorem)*
If X *is a topological space,*
then the following two statements are equivalent:
(a) X *is normal (see Definition 2.4);*
(b) *for every closed set* $C \subseteq X$ *and every continuous function* $f \colon C \longrightarrow \mathbb{R}$*, we can find a continuous extension* $\hat{f} \colon X \longrightarrow \mathbb{R}$ *of* f *(i.e.,* $\hat{f}|_C = f$*); moreover, if* $|f(x)| \leqslant M$ *for all* $x \in C$ *and some* $M > 0$*, then we can choose* $\hat{f}$ *such that* $|\hat{f}(x)| \leqslant M$ *for all* $x \in X$.

One of the main problems in point set topology is the so called "metrization problem". Namely, for a given topological space (X, τ), when is it possible to define a metric d_X on X such that the metric topology induced by d_X coincide with τ? In this direction we have the following important result.

Theorem 2.139 *(Urysohn Metrization Theorem)*
A second countable topological (see Definition 2.24) space is metrizable if and only if it is regular (see Definition 2.4).

In fact a second countable regular space is homeomorphic to a subset of the Hilbert cube $[0,1]^{\mathbb{N}}$.

2.1.10 Paracompact and Baire Spaces

One of the most important generalizations of the notion of compactness is paracompactness. Before introducing it, we need some additional information about coverings.

Definition 2.140
Let $\{U_i\}_{i\in I}$ and $\{V_j\}_{j\in J}$ be two coverings of a set X. We say that $\{U_i\}_{i\in I}$ is a ***refinement*** *of $\{V_j\}_{j\in J}$, if for each $i \in I$ there is a $j \in J$ such that $U_i \subseteq V_j$. We often write $\{U_i\}_{i\in I} \prec \{V_j\}_{j\in J}$ to denote refinement. A refinement of a covering may contain more sets than the given covering. We say that a refinement is* ***precise*** *if $I = J$ and $U_j \subseteq V_j$ for all $j \in J$.*

Proposition 2.141
If the covering $\{V_j\}_{j\in J}$ has a locally finite refinement $\{U_i\}_{i\in I}$ (see Definition 2.43),
then it also has a precise locally finite refinement $\{\hat{U}_j\}_{j\in J}$. Moreover, if for each $j \in J$, the set U_j is open, then each $\hat{U}_j$ can be chosen to be open too.

Now we can introduce the notion of paracompactness.

Definition 2.142
A topological space X is ***paracompact*** *if every open covering of X has a locally finite refinement.*

Proposition 2.143
If X is a regular topological space (see Definition 2.4),
then X is paracompact if and only if every open covering of X has a closed locally finite refinement.

Theorem 2.144
Compact spaces and metrizable spaces are paracompact. Every paracompact space is normal (see Definition 2.4).

Proposition 2.145
(a) *Paracompactness is invariant under continuous closed surjections.*
(b) *Paracompactness is not a hereditary property; however every F_σ-subset of a paracompact space is paracompact.*

One of the main reasons that paracompactness is important is that paracompact spaces admit arbitrarily fine partitions of unity and the latter is a major tool in modern analysis.

Definition 2.146
Let X be a topological space.
(a) *For any function f, the **support** of f is the following closed set*

$$\operatorname{supp} f = \overline{\{x \in X : f(x) \neq 0\}}.$$

(b) *A family $\{\psi_i\}_{i\in I}$ of continuous functions $\psi_i\colon X \longrightarrow [0,1]$, $i \in I$ is a **partition of unity on** X if $\{\operatorname{supp}\psi_i\}_{i\in I}$ form a closed, locally finite covering of X and $\sum\limits_{i\in I}\psi_i(x) = 1$ for all $x \in X$ (in this sum only finitely many terms are nonzero since $\{\operatorname{supp}\psi_i\}_{i\in I}$ is locally finite).*
(c) *If $\{U_i\}_{i\in I}$ is an open covering of X, then we say that the partition of unity $\{\psi_i\}_{i\in I}$ is **subordinated** to $\{U_i\}_{i\in I}$, if $\operatorname{supp}\psi_i \subseteq U_i$ for every $i \in I$.*

Theorem 2.147
If X is a paracompact space and $\{U_i\}_{i\in I}$ is an open covering of X, then there is a partition of unity $\{\psi_i\}_{i\in I}$ subordinated to $\{U_i\}_{i\in I}$.

Definition 2.148
*A topological space X is said to be a **Baire space** if every nontrivial countable intersection of dense open sets in X is dense too.*

From the Baire category theorem (Theorem 1.26), we know that complete metric spaces are Baire spaces. Next we add one more class of spaces to Baire spaces.

Theorem 2.149
Locally compact spaces are Baire spaces.

2.1.11 Polish and Souslin Sets

Next we pass to a generalization of separable metric spaces.

Definition 2.150
*A topological space X is a **Polish space** if it is separable and metrizable by means of a complete metric space.*

Remark 2.151
There are many topological spaces which are Polish, but have no complete metric which is particularly natural or simple. However, many constructions and results in Analysis depend on the existence of a complete metric but not on a specific choice of it. So, working with Polish spaces instead of separable complete metric spaces provides a more general framework of Analysis.

Proposition 2.152
Every closed or open subset of a Polish space is Polish.

Proposition 2.153

(a) *Any finite or countable product of Polish spaces with the product topology (see Definition 2.69) is Polish.*
(b) *Countable intersection of Polish spaces is Polish.*
(c) *A locally compact, σ-compact metrizable space is Polish.*
(d) *The set of irrationals $\mathbb{R} \setminus \mathbb{Q}$ with the subspace topology induced by $\mathbb{R}$ is a Polish space.*

Theorem 2.154

(a) *A subspace of a Polish space X is Polish if and only if it is G_δ in X.*
(b) *A topological space X is Polish if and only if it is homeomorphic to a G_δ-subset of a Hilbert cube $[0,1]^{\mathbb{N}}$.*

Proposition 2.155
Every Polish space X is a continuous image of $\mathbb{N}^\infty$.

Definition 2.156
A topological space (X, τ) is said to be ***Souslin space*** *if there exists a Polish space Y and a continuous surjection from Y to X. Equivalently, we can say that (X, τ) is a Souslin space, if there exists a topology $\hat{\tau}$ on X stronger than τ such that $(X, \hat{\tau})$ is homeomorphic to a quotient of a Polish space.*

Remark 2.157
A Souslin space is separable and Lindelöf (see Definitions 2.156 and 2.26) but need not be metrizable (for example an infinite dimensional separable Banach space with the weak topology; see Chap. 5). A Souslin subset of a Polish space is called in the literature an ***analytic set***.

Proposition 2.158
If X is a Souslin space (see Definition 2.156) and $C \subseteq X$ is nonempty, then C has a countable subset D which is sequentially dense in C, i.e., every $x \in C$ is the limit of a sequence in D.

Proposition 2.159
(a) *Closed and open subsets of a Souslin space are Souslin spaces (see Definition 2.156).*
(b) *Countable products of Souslin spaces with the product topology (see Definition 2.69), are Souslin spaces.*
(c) *In a topological space, countable intersections or unions of Souslin subspaces are Souslin spaces.*

Theorem 2.160
If X is a locally compact topological space,
then the following statements are equivalent:
(a) *X is Polish (see Definition 2.150);*
(b) *X is Souslin (see Definition 2.156);*
(c) *X is second countable (see Definition 2.24);*
(d) *X is separable metrizable (see Definition 2.9);*
(e) *the Alexandrov one-point compactification X^* of X has any of the above properties (see Remark 2.97).*

Definition 2.161

(a) A topological space X is said to be ***dispersed*** *if it is Polish and every point has a local basis consisting of clopen sets.*
(b) A topological space is said to be ***dispersible*** *if it is the continuous image of a dispersed space.*

Proposition 2.162

(a) Countable products of dispersible spaces are dispersible.
(b) Countable intersections of dispersible spaces are dispersible.
(c) A subspace of a dispersible space is dispersible if and only if it is a G_δ-set. Also disjoint unions of dispersible spaces are dispersible.

We introduce the following stronger version of a Lindelöf space (see Definition 2.26).

Definition 2.163

A topological space X is said to be ***strongly Lindelöf*** *if every open cover of any open set of X has a countable subcover.*

Proposition 2.164

(a) Arbitrary intersections and countable unions of strongly Lindelöf spaces are strongly Lindelöf.
(b) A subspace of a strongly Lindelöf space is strongly Lindelöf.
(c) If (X, τ) is strongly Lindelöf, then for any topology $\tau' \subseteq \tau$, we have that (X, τ') is strongly Lindelöf.
(d) A second countable topological space (see Definition 2.24) is strongly Lindelöf.
(e) The continuous image of a strongly Lindelöf space is strongly Lindelöf.

Corollary 2.165

Every Souslin space is strongly Lindelöf.

Remark 2.166

The cartesian product of two strongly Lindelöf spaces need not be strongly Lindelöf. However, if $\{X_n\}_{n\geqslant 1}$ are second countable (see Definition 2.24) or Souslin (see Definition 2.156), then so is $\prod X_n$ and therefore it is strongly Lindelöf (see Proposition 2.164(d) and Corollary 2.165).

Definition 2.167
We say that a topological space (X, τ) *is a* ***Lusin space*** *if there exists a Polish space* Y *and a continuous bijection from* Y *to* X*.*

Remark 2.168
The above definition of a Lusin space (X, τ) is equivalent to saying that there is a topology τ^* on X stronger than τ such that (X, τ^*) is a Polish space.

2.1.12 Michael Selection Theorem

In this section, we introduce some continuity notions from the theory of multifunctions (set valued functions) and state the celebrated selection theorem of Michael.

Definition 2.169
Let X, Y *be two topological spaces and let* $F \colon X \longrightarrow 2^Y$ *be a multifunction.*
(a) *We say that* F *is* ***upper semicontinuous*** *at* $x_0 \in X$ *(****usc*** *at* x_0*, for short), if for any open set* $V \subseteq Y$ *such that* $F(x_0) \subseteq V$*, we can find* $U \in \mathcal{N}(x_0)$ *such that* $F(U) \subseteq V$*. If this is true at every* $x_0 \in X$*, then we say that* F *is* ***upper semicontinuous*** *(****usc****, for short).*
(b) *We say that* F *is* ***lower semicontinuous*** *at* $x_0 \in X$ *(****lsc*** *at* x_0*, for short), if for any open set* $V \subseteq Y$ *such that* $F(x_0) \cap V \neq \emptyset$*, we can find* $U \in \mathcal{N}(x_0)$ *such that* $F(x) \cap V \neq \emptyset$ *for all* $x \in U$*. If this is true at every* $x_0 \in X$*, then we say that* F *is* ***lower semicontinuous*** *(****lsc****, for short). In fact in the above definition* V *can be taken to be a basis open set.*
(c) *We say that* F *is* ***continuous*** *at* $x_0 \in X$ *(or* ***Vietoris continuous*** *at* $x_0 \in X$*), if it is both upper semicontinuous and lower semicontinuous at* x_0*. If this is true at every* $x_0 \in X$*, then we say that* F *is* ***continuous*** *(or* ***Vietoris continuous****).*

Theorem 2.170 *(Michael Selection Theorem)*
If X *is paracompact (see Definition 2.142),* Y *is a Banach space and* $F \colon X \longrightarrow 2^Y$ *is a lower semicontinuous multifunction with values which are nonempty, closed and convex subsets in* Y *(a set* $C \subseteq Y$ *is convex if for every* $u, v \in C$ *and* $t \in [0, 1]$*, we have* $tu + (1 - t)v \in C$*; see Chap. 5),*

then there exist a continuous function $f\colon X \longrightarrow Y$ such that

$$f(x) \in F(x) \quad \forall\, x \in X$$

*(we call f, a **continuous selector** of F).*

Theorem 2.171
If X is a metric space, Y is a Banach space and $F\colon X \longrightarrow 2^Y$ is a lower semicontinuous multifunction with nonempty, closed and convex values,
then there exists a sequence $\{f_n\colon X \longrightarrow Y\}_{n\geqslant 1}$ of continuous selectors of F such that $F(x) = \overline{\{f_n(x)\}_{n\geqslant 1}}$ for all $x \in X$.

Theorem 2.172
If X is a metric space, Y is a Banach space and $F\colon X \longrightarrow 2^Y$ is a lower semicontinuous multifunction with nonempty, convex values such that $\operatorname{int} F(x) \neq \emptyset$ for all $x \in X$,
then F admits a continuous selector.

Definition 2.173
*Let X and Y be two topological spaces and let $F\colon X \longrightarrow 2^Y$ be a multifunction. The **graph** of F is the set $\operatorname{Gr} F \subseteq X \times Y$, defined by*

$$\operatorname{Gr} F \;=\; \{(x, y) \in X \times Y : \; y \in F(x)\}.$$

2.1.13 The Space $C(X;Y)$

Next we will topologize the space of continuous functions between two topological spaces using a topology that corresponds to the concept of uniform convergence on compacta.

Definition 2.174
*Let X and Y be two topological spaces and let $C(X;Y)$ denote the space of all continuous functions from X into Y. The **compact-open topology** on $C(X;Y)$ is the one having as subbasis all sets of the form:*

$$(K, V) \;=\; \{f \in C(X;Y) : \; f(K) \subseteq V\}$$

*with any compact $K \subseteq X$, and open $V \subseteq Y$. In what follows the compact-open topology is abbreviated as the c-**topology** and is denoted by c.*

Proposition 2.175

(a) *$C(X;Y)$ with the c-topology is Hausdorff (in fact $\big(C(X;Y),c\big)$ is Hausdorff if and only if Y is Hausdorff).*
(b) *$C(X;Y)$ with the c-topology is regular (see Definition 2.4) if and only if Y is regular.*

Definition 2.176

Let X,Y,Z be three topological spaces. We introduce the following two functions.
(a) *The* ***composition function*** *$\tau\colon C(X;Y)\times C(Y;Z)\longrightarrow C(X;Z)$, defined by*

$$\tau(f,g) \;=\; g\circ f \qquad \forall\, f\in C(X;Y),\; g\in C(Y;Z).$$

(b) *The* ***evaluation function*** *$e\colon C(X,Y)\times X\longrightarrow Y$, defined by*

$$e(f,x) \;=\; f(x) \qquad \forall\, f\in C(X;Y),\; x\in X.$$

Theorem 2.177

(a) *The composition function $\tau\colon C(X;Y)\times C(Y;Z)\longrightarrow C(X;Z)$ is continuous on each argument separately, when $C(X;Y)$, $C(Y;Z)$ and $C(X;Z)$ are endowed with the respective c-topology.*
(b) *If Y is locally compact, then $\tau(\cdot,\cdot)$ is jointly continuous, when $C(X;Y)$, $C(Y;Z)$ and $C(X;Z)$ are endowed with the respective c-topology.*

Theorem 2.178

(a) *For fixed $x_0\in X$, the function $e_{x_0}\colon C(X;Y)\longrightarrow Y$, defined by*

$$e_{x_0}(f) \;=\; e(f,x_0) \;=\; f(x_0)$$

is continuous, when $C(X;Y)$ is endowed with the c-topology.
(b) *If Y is additionally locally compact, then the evaluation function $e\colon C(X;Y)\times X\longrightarrow Y$ is jointly continuous, when $C(X;Y)$ is endowed with the c-topology.*

Theorem 2.179

(a) *If $f\colon X\times Y\longrightarrow Z$ is continuous and $\hat{f}\colon X\longrightarrow C(Y;Z)$ is defined by*

$$\hat{f}(x)(\cdot) \;=\; f(x,\cdot),$$

then $\hat{f}$ is continuous, when $C(Y;Z)$ is endowed with the c-topology.
(b) *If $\hat{f}\colon X\longrightarrow C(Y;Z)$ is continuous, when $C(Y;Z)$ is endowed with the c-topology and Y is locally compact, then $f\colon X\times Y\longrightarrow Z$ is jointly continuous.*

Assuming that Y is a metric space, we will identify the compact subsets of $\big(C(X;Y),c\big)$ (the Arzela–Ascoli theorem). We start with a definition.

Definition 2.180

Let X be a topological space and let (Y,d) be a metric space. A subset $A\subseteq C(X;Y)$ is said to be ***equicontinuous*** *at $x_0\in X$, if for every $\varepsilon>0$, we can find $U\in\mathcal{N}(x_0)$ such that $f(U)\subseteq B_\varepsilon\big(f(x_0)\big)$ for all $f\in A$. The set $U\in\mathcal{N}(x_0)$ is called a* ***neighbourhood of ε-continuity for*** *A. We say that A is* ***equicontinuous*** *if it is equicontinuous at every $x_0\in X$.*

Theorem 2.181 *(Arzela–Ascoli Theorem)*

If X is a topological space, (Y,d) is a metric space, $A\subseteq C(X;Y)$ and
(i) A is equicontinuous;
(ii) for every $x\in X$, $\overline{e_x(A)}\subseteq Y$ is compact (i.e., $\{f(x):\ f\in A\}$ is relatively compact in Y),
then $\overline{A}^c$ (c denotes the c-topology on $C(X;Y)$) is compact and equicontinuous.

Next we have a closer look at the sequential convergence in the c-topology.

Definition 2.182

Let X be a topological space and let (Y,d_Y) be a metric space. A sequence $\{f_n\}_{n\geqslant 1}\subseteq C(X;Y)$ is said to converge to $f\in C(X;Y)$ ***uniformly on compacta*** *if for every compact set $K\subseteq X$ and every $\varepsilon>0$, we can find $n_0=n_0(K,\varepsilon)\geqslant 1$ such that*

$$d_Y\big(f_n(x),f(x)\big) \;<\; \varepsilon \qquad \forall\, n\geqslant n_0,\ x\in K.$$

Theorem 2.183
If X is a topological space, (Y, d_Y) is a metric space, $\{f_n\}_{n\geqslant 1} \subseteq C(X;Y)$ and $f \in C(X;Y)$,
then

$$\left[f_n \longrightarrow f \text{ uniformly on compacta}\right] \iff \left[f_n \xrightarrow{c} f \text{ in } C(X;Y)\right].$$

Definition 2.184
Let X, Y be two topological space, $\{f_n\}_{n\geqslant 1} \subseteq C(X;Y)$ and $f \in C(X;Y)$. We say that the sequence $\{f_n\}_{n\geqslant 1}$ ***converges continuously*** *to f, if for every $x \in X$ and every sequence $\{x_n\}_{n\geqslant 1} \subseteq X$ such that $x_n \longrightarrow x \in X$, we have $f_n(x_n) \longrightarrow f(x)$ in Y.*

Theorem 2.185
If X is first countable, (Y, d_Y) is a metric space, $\{f_n\}_{n\geqslant 1} \subseteq C(X;Y)$ is a sequence and $f \in C(X;Y)$,
then

$$\left[f_n \longrightarrow f \text{ continuously}\right] \iff \left[f_n \longrightarrow f \text{ uniformly on compacta}\right].$$

2.1.14 Elements of Algebraic Topology I: Homotopy

The main problem of topology is the classification of topological spaces. Given two topological spaces X and Y, are they homeomorphic? In general, this is a very difficult question to answer, without the use of tools from other fields such as algebra. This then leads us to the realm of "algebraic topology". In the rest of this chapter we present some introductory notions and results from this field, which can be useful to the analyst.

Definition 2.186
(a) A ***topological pair*** *(X, C) consists of a topological space X and a subspace $C \subseteq X$. If $C = \emptyset$, then the pair $(X, \emptyset)$ stands for the space X. A* ***subpair*** *$(X', C') \subseteq (X, C)$ is a topological pair such that $X' \subseteq X$ and $C' \subseteq C$. A function $f\colon (X, C) \longrightarrow (Y, D)$ between topological pairs, is a continuous function $f\colon X \longrightarrow Y$ such that $f(C) \subseteq D$. If $C = \{x\}$ with $x \in X$, then the topological pair (X, C) is said to be a* ***pointed topological space*** *and is denoted by (X, x).*

(b) *For a topological pair (X, C) and $I = [0,1]$, we set*

$$I \times (X, C) = \big(I \times X, I \times C\big).$$

Let $V \subseteq X$ and suppose that $f_0, f_1 \colon (X, C) \longrightarrow (Y, D)$ are functions between topological pairs such that $f_0|_V = f_1|_V$. We say that f_0 is ***homotopic to f_1 relative to V*** *(denoted by $f_0 \overset{V}{\simeq} f_1$), if there exists a continuous function $h \colon I \times (X, C) \longrightarrow (Y, D)$ such that*

$$\begin{aligned} h(0,x) &= f_0(x) && \forall\, x \in X, \\ h(1,x) &= f_1(x) && \forall\, x \in X, \\ h(t,x) &= f_0(x) = f_1(x) && \forall\, t \in I,\ x \in V. \end{aligned}$$

The function h is a ***homotopy relative to V*** *from f_0 to f_1. If $V = \emptyset$, then we omit the expression* ***relative to V*** *and write $\simeq$ instead of $\overset{V}{\simeq}$.*
(c) *A continuous function $f \colon X \longrightarrow Y$ is said to be* ***nullhomotopic*** *if it is homotopic to a constant function. In this case we write $f \simeq 0$.*
(d) *Two paths $f, g \colon [0,1] \longrightarrow X$ are said to be* ***path homotopic*** *if $f(0) = g(0) = x_0$, $f(1) = g(1) = x_1$ (i.e., the two paths have the same initial and finale points) and there exists a continuous map $h \colon [0,1] \times [0,1] \longrightarrow X$ such that*

$$h(s,0) = f(s), \quad h(s,1) = g(s) \quad s \in [0,1]$$

$$h(0,t) = x_0, \quad h(1,t) = x_1 \quad t \in [0,1].$$

We call h a ***path homotopy*** *between f and g and we write $f \simeq_p g$.*

Proposition 2.187
If $W \subseteq V \subseteq X$ and $f_0 \overset{V}{\simeq} f_1$,
then $f_0 \overset{W}{\simeq} f_1$.

Proposition 2.188
The homotopy relation is an equivalence relation.

Definition 2.189
For a continuous function $f \colon X \longrightarrow Y$ the equivalence class under $\simeq$ is denoted by $[f]$ and is called the ***homotopy class of f.***

The next proposition shows that the equivalence relation $\simeq$ (see Proposition 2.188) is compatible with composition.

Proposition 2.190
If X, Y, Z *are topological spaces and* $f, \hat{f}: X \longrightarrow Y$, $g, \hat{g}: Y \longrightarrow Z$ *are continuous functions such that* $f \simeq \hat{f}$ *and* $g \simeq \hat{g}$,
then $g \circ f \simeq \hat{g} \circ \hat{f}$.
Therefore, we can define the composition of homotopy classes, by

$$[g] \circ [f] = [g \circ f].$$

Recall that two topological spaces X, Y are said to be homeomorphic, if there exist continuous functions $f: X \longrightarrow Y$ and $g: Y \longrightarrow X$ such that $g \circ f = id_X$ and $f \circ g = id_Y$. We can extend this notion by replacing "=" by "$\simeq$". This leads to the following definition extending the idea of homeomorphic spaces.

Definition 2.191
Two topological pairs (X, C) *and* (Y, D) *are said to be* ***homotopy equivalent*** *if there exist functions* $f: (X, C) \longrightarrow (Y, D)$ *and* $g: (Y, D) \longrightarrow (X, C)$ *such that* $g \circ f \simeq id_{(X,C)}$ *(i.e., the homotopy keeps the image of* C *in* C*) and* $f \circ g \simeq id_{(Y,D)}$ *(i.e., the homotopy keeps the image of* D *in* D*). We write* $(X, C) \simeq (Y, D)$*. If* $C = D = \emptyset$*, then* $g \circ f \simeq id_X$, $f \circ g \simeq id_Y$ *and we write* $X \simeq Y$*. Clearly homotopy equivalence between topological spaces is an equivalence relation. The equivalence classes are called* ***homotopy types****.*

The simplest topological spaces are the singletons. We characterize the homotopy type of a singleton as follows.

Definition 2.192
A topological space X *is said to be* ***contractible*** *if it is homotopy equivalent to a singleton. This is equivalent to saying that the identity function on* X *is nullhomotopic (i.e.,* $id_X \simeq 0$*). Another equivalent definition is that there exists a continuous function* $h: I \times X \longrightarrow X$ *such that*

$$h(0, \cdot) = id_X, \quad \text{and} \quad h(1, x) = u_0 \quad \text{for some } u_0 \in X \text{ and all } x \in X.$$

Remark 2.193
In other words contractible is a topological space which can be continuously shrunk to a point. Some obvious examples of contractible spaces are convex sets in $\mathbb{R}^N$ and more generally, any star-shaped set

$C \subseteq \mathbb{R}^N$. This means that there is a point $u_0 \in C$ such that for every $x \in C$, the line segment

$$[u_0, x] \stackrel{\text{def}}{=} \{u = (1-t)u_0 + tx : 0 \leqslant t \leqslant 1\}$$

is contained in C.

Proposition 2.194
A topological space X is contractible (see Definition 2.192) if and only if for any topological space Y any two continuous functions $f, g \colon Y \longrightarrow X$ are homotopic.

The next theorem establishes an important relation between null-homotopy and extendability of the special class of functions defined on spheres.

Theorem 2.195
If $S^N = \{x \in \mathbb{R}^{N+1} : \|x\| = 1\}$, $u_0 \in S^N$, Y is a topological space and $f \colon S^N \longrightarrow Y$ is a continuous function, then the following statements are equivalent:
(a) $f \simeq 0$;
(b) $f \stackrel{\{u_0\}}{\simeq} 0$;
(c) there exists a continuous function

$$\hat{f} \colon \overline{B}_1^{N+1} = \{x \in \mathbb{R}^{N+1} : \|x\| \leqslant 1\} \longrightarrow X$$

such that $\hat{f}|_{S^N} = f$.

Definition 2.196
Let X be a topological space and let $C \subseteq X$ be a subset.
*(a) We say that C is a **retract** of X if there is a continuous function $r \colon X \longrightarrow C$ such that $r|_C = id_C$. The function r is called a **retraction**.*
*(b) We say that C is a **deformation retract** of X if there exists a retraction $r \colon X \longrightarrow C$ which is homotopic (as a function into X) to the identity function id_X. Therefore C is a deformation retract of X, if there exists a continuous function $h \colon I \times X \longrightarrow X$ such that*

$$h(0, \cdot) = id_X, \quad h(1, x) \in C \quad \forall\, x \in X \quad \text{and} \quad h(1, \cdot)|_C = id_C.$$

*This homotopy $h(t, x)$ is called **deformation retraction**.*

(c) We say that C is a ***strong deformation retract*** *of X if there exists a retraction $r\colon X \longrightarrow C$ which is homotopic (as a function into X) relative to C to the identity function id_X. Therefore C is a strong deformation retract of X, if there exists a continuous function $h\colon I \times X \longrightarrow X$ such that*

$$\begin{aligned} h(0,\cdot) &= id_X, \\ h(1,x) &\in C \qquad \forall\, x \in X, \\ h(t,\cdot)|_C &= id_C \qquad \forall\, t \in [0,1]. \end{aligned}$$

This homotopy $h(t,x)$ is called ***strong deformation retraction.***

Proposition 2.197
A topological space X is contractible if and only if any singleton in X is a deformation retract of X (see Definitions 2.192 and 2.196). Hence a contractible space is path-connected (see Definition 2.122).

Theorem 2.198
If X is a topological space and C is a deformation retract of X (see Definition 2.196),
then C and X are homotopy equivalent (i.e., of the same homotopy type).

Next we will briefly discuss the fundamental group, an algebraic group, which in some sense measures the number of "holes" in a space. Consider for example the plane and the punctured plane, which are both path-connected. Using the standard topological properties, we cannot distinguish between the two and this may suggest that they are homeomorphic. Yet intuition suggests that due to the "hole" in the punctured plane, the two cannot be homeomorphic. To see that this can be detected topologically, observe that every closed curve in $\mathbb{R}^2$ can be continuously shrunk to a point, while by contrast it is clear that this cannot be done in the punctured plane, without leaving it. This observation is formalized by using the fundamental group.

Definition 2.199
Let X be a topological space and let $x_0 \in X$. We say that a continuous function $f\colon I = [0,1] \longrightarrow X$ is a ***loop based at*** *x_0 if $f(0) = f(1) = x_0$. Two loops $f, g\colon I \longrightarrow X$ are said to be* ***homotopic***

if $f \overset{\{0,1\}}{\simeq} g$. *The homotopy relation between loops is an equivalence relation and it partitions the set of all loops based at* x_0 *into equivalence classes. This set of equivalence classes is denoted by* $\pi_1(X, x_0)$ *and if* f *is a loop based at* x_0, *then* $[f]$ *denotes its equivalence class.*

On $\pi_1(X, x_0)$ we can define an operation "$\star$" as follows. For any two loops f, g based at x_0, we set

$$(f \star g)(t) = \begin{cases} f(2t) & \text{if } t \in \left[0, \frac{1}{2}\right], \\ g(2t-1) & \text{if } t \in \left[\frac{1}{2}, 1\right]. \end{cases}$$

Evidently $(f \star g)(\cdot)$ is again a loop based at x_0. We can lift this operation to the set $\pi_1(X, x_0)$ by setting

$$[f] \star [g] = [f \star g].$$

This operation "$\star$" is well defined on $\pi_1(X, x_0)$, namely, if $f \overset{\{0,1\}}{\simeq} g$ and $\hat{f} \overset{\{0,1\}}{\simeq} \hat{g}$, then $f \star \hat{f} \overset{\{0,1\}}{\simeq} g \star \hat{g}$.

Theorem 2.200
$\big(\pi_1(X, x_0), \star\big)$ *is a group (not necessarily abelian).*

Definition 2.201
The group $\pi_1(X, x_0)$ *is called the* ***fundamental group of*** X ***at*** x_0.

Remark 2.202
In general the dependence of the fundamental group on the base point cannot be omitted. However, for an important special class of spaces it can be ignored.

Theorem 2.203
If X *is path-connected (see Definition 2.122) and* $x_0, x_1 \in X$, *then* $\pi_1(X, x_0)$ *and* $\pi_1(X, x_1)$ *are isomorphic.*

Remark 2.204
So, for path-connected topological spaces we can speak of the fundamental group $\pi_1(X)$ without any reference to a base point.

The power of the fundamental group lies on the fact that continuous functions between pointed topological spaces induce homomorphisms between the corresponding algebraic structures.

Theorem 2.205
Every continuous function $h\colon (X, x_0) \longrightarrow (Y, y_0)$ *induces a homomorphism* $h_*\colon \pi_1(X, x_0) \longrightarrow \pi_1(Y, y_0)$*, defined by*

$$h_*([f]) \stackrel{\text{def}}{=} [h \circ f].$$

The function h_* *is called the* ***homomorphism induced by*** h.

The induced homomorphism introduced above has two "functional properties", which are crucial in applications.

Proposition 2.206
(a) *If* $h\colon (X, x_0) \longrightarrow (Y, y_0)$ *and* $g\colon (Y, y_0) \longrightarrow (Z, z_0)$ *are continuous functions,*
then $(g \circ h)_* = g_* \circ h_*$.
(b) *If* $i\colon (X, x_0) \longrightarrow (X, x_0)$ *is the identity function,*
then $i_*\colon \pi_1(X, x_0) \longrightarrow \pi_1(X, x_0)$ *is the identity homomorphism.*

Corollary 2.207
If $h\colon (X, x_0) \longrightarrow (Y, y_0)$ *is a homotopy equivalence,*
then $h_*\colon \pi_1(X, x_0) \longrightarrow \pi_1(Y, y_0)$ *is an isomorphism.*

Definition 2.208
A topological space S *is* ***simply connected*** *if it is path-connected (see Definition 2.122) and* $\pi_1(X)$ *is trivial.*

Theorem 2.209
A contractible topological space is simply connected (recall that a contractible space is automatically path-connected; see Proposition 2.197).

Proposition 2.210
If $h, g\colon (X, x_0) \longrightarrow (Y, y_0)$ *are homotopic,*
then $h_* = g_*$.

Next let us have an alternative description of the fundamental group which motivates the introduction of higher order homotopy groups. Consider the loop $\varepsilon\colon I = [0, 1] \longrightarrow S^1 = \{x \in \mathbb{R}^2 : \|x\| = 1\}$, defined by

$$\varepsilon(t) \stackrel{\text{def}}{=} e^{2\pi i t} \qquad \forall\, t \in I.$$

This function, known as the ***exponential function***, wraps the interval once around the circle counterclockwise and functions 0 and 1 to the base point 1 of S^1. From Problem 2.51, we know that ε is an identification function (see Definition 2.74). If X is a topological space and $f\colon I = [0,1] \longrightarrow X$ is a loop, then there exists a unique continuous function $\widetilde{f}\colon S^1 \longrightarrow X$ such that

$$\widetilde{f} \circ \varepsilon = f,$$

i.e., we have the commutative diagram

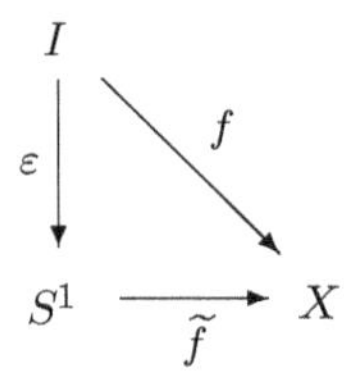

Definition 2.211
*We call $\widetilde{f}$ the **circle representative** of the loop f.*

The next proposition gives a convenient criterion for detecting nullhomotopic loops using their circle representatives (see also Theorem 2.195).

Proposition 2.212
A loop in a topological space X is nullhomotopic if and only if its circle representative $\hat{f}\colon S^1 \longrightarrow X$ extends to a continuous function $\hat{f}_0\colon \overline{B}_1 \longrightarrow X$, where

$$\overline{B}_1 \stackrel{\text{def}}{=} \{x \in \mathbb{R}^2 : \|x\| \leqslant 1\} \quad \textit{and} \quad \hat{f} = \hat{f}_0\big|_{S^1}.$$

Here by identifying loops with their circle representatives, we can view the fundamental group $\pi_1(X,x)$ as the set of equivalence classes of functions from S^1 into X sending 1 to x modulo homotopy relative to the base point 1. Generalizing this, we have the following definition.

Definition 2.213
For every integer $n \geqslant 1$, we define $\pi_n(X,x)$ to be the set of all equivalence classes of continuous functions from $S^n = \{x \in \mathbb{R}^{n+1} : \|x\| = 1\}$

into X sending $(1,0,\ldots,0)$ *to* x *modulo homotopy relative to the base point* $(1,0,\ldots,0)$. *For given such elements* f,g, *we define* $f\#g$, *by*

$$f\#g(x_1,\ldots,x_n) \;=\; \begin{cases} g(x_1,\ldots,x_{n-1},2x_n) & \text{if } x_n \leqslant \frac{1}{2}, \\ f(x_1,\ldots,x_{n-1},2x_n-1) & \text{if } x_n \geqslant \frac{1}{2}. \end{cases}$$

Using $\#$, *we introduce the operation "+" on* $\pi_n(X,x)$, *defined by*

$$[f]+[g] \;=\; [f\#g].$$

We have that $(\pi_n(X,x),+)$ is a group, which is abelian if $n \geqslant 2$. We call $(\pi_n(X,x),+)$ the ***n*th homotopy group** of the pointed space (X,x).

Remark 2.214
For every $n \geqslant 1$, $\pi_n(X,x)$ is a topological invariant. The higher homotopy groups ($n > 1$) are notoriously difficult to compute. If X,Y are pointed topological spaces, then

$$\pi_n(X \times Y) \;=\; \pi_n(X) \times \pi_n(Y) \qquad \forall\, n \geqslant 1.$$

Theorem 2.215
(a) $\pi_1(S^1) \simeq \mathbb{Z}$.
(b) *If* $n \geqslant 2$, *then* $\pi_k(S^n) = 0$ *for all* $k < n$ *and* $\pi_n(S^n) \simeq \mathbb{Z}$. *So, for* $n \geqslant 2$, S^n *is simply connected (see Definition 2.208).*

Remark 2.216
However, if $k > n$, then $\pi_k(S^n)$ can be nontrivial. For example, it can be shown that $\pi_3(S^2) \simeq \mathbb{Z}$.

Definition 2.217
Let $\hat{X}$ *and* X *be two topological spaces and let* $p\colon \hat{X} \longrightarrow X$ *be a continuous function. We say that a set* $U \subseteq X$ *is* ***evenly covered by*** p *if* U *is open, connected and each component of* $p^{-1}(U)$ *is an open set which is mapped homeomorphically onto* U.

The next definition introduces an important tool in computing critical groups.

Definition 2.218
Let $\hat{X}$ and X be two topological spaces with $\hat{X}$ being path-connected and locally path-connected (see Definitions 2.122 and 2.127) and let $p\colon \hat{X} \longrightarrow X$ be a continuous surjection. If every $x \in X$ has a neighbourhood U which is evenly covered by p, then p is called a ***covering function*** *and $\hat{X}$ is a* ***covering space*** *of X.*

Example 2.219
(a) The exponential function $\varepsilon\colon \mathbb{R} \longrightarrow S^1$, defined by $\varepsilon(x) = e^{2\pi i x}$, i.e.,

$$\varepsilon(x) = \big(\cos 2\pi x, \sin 2\pi x\big) \qquad \forall\, x \in \mathbb{R}$$

is a covering function.
(b) The function $p_n\colon S^1 \longrightarrow S^1$, given by

$$p_n(z) = z^n \qquad \forall\, z \in S^1$$

(here we consider S^1 as a subset of $\mathbb{C}$ with $|z| = 1$, $z \in \mathbb{C}$), is a covering function for every $n \geqslant 1$.
(c) Consider the n-torus $T^n = \underbrace{S^1 \times \ldots \times S^1}_{n-\text{times}}$ (if $n = 2$ we have the usual torus $T = S^1 \times S^1$, i.e., the doughnut) and let $\varepsilon_n\colon \mathbb{R}^n \longrightarrow T^n$ be the function, defined by

$$\varepsilon_n(x) = \big(\varepsilon(x_k)\big)_{k=1}^n \qquad \forall\, x = (x_k)_{k=1}^n \in \mathbb{R}^n.$$

(d) Let $\mathbb{P}^n$ be the n-dimensional projective plane (this is the set of one-dimensional linear subspaces (lines through the origin) in $\mathbb{R}^{n+1}$). There is a natural function $\hat{p}\colon \mathbb{R}^{n+1} \setminus \{0\} \longrightarrow \mathbb{P}^n$, defined by sending a point $x \in \mathbb{R}^{n+1} \setminus \{0\}$ to its span. We topologize $\mathbb{P}^n$ be giving it the identification (quotient) topology (see Definition 2.74) with respect to $\hat{p}$ (an alternative way to define $\mathbb{P}^n$ is to say that it is the space obtained from $S^n = \big\{x \in \mathbb{R}^{n+1} : \|x\| = 1\big\}$ by identifying antipodal points, i.e., the points where every line through the origin intersects S^n). Consider the function $p\colon S^n \longrightarrow \mathbb{P}^n$ ($n \geqslant 1$) which sends each $x \in S^n$ to the line through the origin and x, thought as an element of $\mathbb{P}^n$. Then p is a covering function.

Definition 2.220
Let $\hat{X}, X$ and Y be three topological spaces, let $p\colon \hat{X} \longrightarrow X$ be a covering function and let $f\colon Y \longrightarrow X$ be a continuous function. A ***lift***

of f is a continuous function $\hat{f}\colon Y \longrightarrow \hat{X}$ such that $p \circ \hat{f} = f$. So we have the following commutative diagram

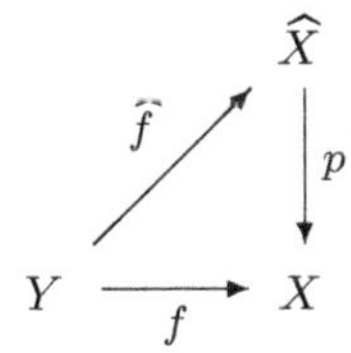

The next theorem establishes a crucial link between covering spaces and the fundamental groups.

Theorem 2.221
<u>If</u> $p\colon \hat{X} \longrightarrow X$ is a covering function, $\gamma_0, \gamma_1\colon I = [0,1] \longrightarrow X$ are path homotopic and $\hat{\gamma}_0, \hat{\gamma}_1\colon I \longrightarrow \hat{X}$ are lifts of γ_0 and γ_1 such that $\hat{\gamma}_0(0) = \hat{\gamma}_1(0)$,
<u>then</u> $\hat{\gamma}_0$ and $\hat{\gamma}_1$ have the same end point and are path homotopic.

The next theorem characterizes the fundamental group homomorphism induced by a covering function.

Theorem 2.222
<u>If</u> $p\colon \hat{X} \longrightarrow X$ is a covering function and $\hat{x} \in \hat{X}$,
<u>then</u> $p_*\colon \pi_1(\hat{X}, \hat{x}) \longrightarrow \pi_1(X, p(\hat{x}))$ is injective.

We still need to determine when a given continuous function $f\colon Y \longrightarrow X$ admits a lift. The theorem that follows transforms this topological problem to an equivalent algebraic one.

Theorem 2.223
<u>If</u> $p\colon \hat{X} \longrightarrow X$ is a covering function, Y is a connected and locally path-connected topological space (see Definitions 2.104 and 2.127) and $f\colon Y \longrightarrow X$ is a continuous function,
<u>then</u> for a given $y_0 \in Y$ and $\hat{x}_0 \in \hat{X}$ such that $p(\hat{x}_0) = f(y_0)$, f has a lift $\hat{f}\colon Y \longrightarrow \hat{X}$ such that $\hat{f}(y_0) = \hat{x}_0$ if and only if the subgroup $f_*\big(\pi_1(Y, y_0)\big)$ of $\pi_1\big(X, f(y_0)\big)$ is contained in $p_*\big(\pi_1(\hat{X}, \hat{x}_0)\big)$.

Two immediate consequences of this theorem are useful in applications.

Corollary 2.224
If $p\colon \hat{X} \longrightarrow X$ *is a covering function and* Y *is a simply connected and locally path-connected topological space (see Definitions 2.104, 2.208 and 2.127),*
then every continuous function $f\colon Y \longrightarrow X$ *admits a lift.*
Moreover, for a given $y_0 \in Y$, *we can choose the lift to map* y_0 *to any point in the fibre over* $f(y_0)$.

Corollary 2.225
If $p\colon \hat{X} \longrightarrow X$ *is a covering function,* $\hat{X}$ *is simply connected,* Y *is connected and locally path-connected (see Definitions 2.104, 2.208 and 2.127) and* $f\colon Y \longrightarrow X$ *is a continuous function,*
then f *admits a lift to* $\hat{X}$ *if and only if* f_* *is the trivial homomorphism for any base point* $y_0 \in Y$ *and the lift can be chosen to map* y_0 *to any point in the fiber above* $f(y_0)$.

Theorem 2.226
If $p\colon (\hat{X}, \hat{x}_0) \longrightarrow (X, x_0)$ *is a covering function,*
then there is a surjection $\xi\colon \pi_1(X, x_0) \longrightarrow p^{-1}(x_0)$.
Moreover, if $\hat{X}$ *is also simply connected, then* ξ *is a bijection.*

Definition 2.227
If $\hat{X}$ *is simply connected covering space of* X *(see Definition 2.208), then we say that* $\hat{X}$ *is a* ***universal covering space*** *of* X.

For easy reference let us recall the fundamental groups of some well known spaces.

Example 2.228
(a) $\pi_1(S^1) = \mathbb{Z}$ and $\pi_k(S^n) = 0$ for all $n \geqslant 2$ and $k < n$, while $\pi_n(S^n) = \mathbb{Z}$ (see Theorem 2.215).
(b) $\pi_1(\mathbb{R}^n \setminus \{0\}) = \pi_1(S^1)$ (here $\mathbb{R}^n \setminus \{0\}$ is simply connected; Definition 2.208).
(c) $\pi_1(T) = \mathbb{Z} \times \mathbb{Z}$ (here we use the fact that $\pi_1(X \times Y,\ (x_0, y_0)) = \pi_1(X, x_0) \times \pi_2(Y, y_0))$.
(d) $\pi_1(\mathbb{P}^2) = \mathbb{Z}_2$.
(e) The "double torus" T_2 is the surface obtained by taking two copies of the torus T deleting a small open disk from each of them and pasting the remaining pieces. The fundamental group of T_2 is not abelian.

Before passing to homology theory, we state an important theorem, which is a useful tool in the computation of fundamental groups. It describes the fundamental group of a space in terms of the fundamental groups of its subspaces, subject to certain conditions.

Theorem 2.229 *(Van Kampen Theorem)*
If $X = U \cup V$ where $U, V \subseteq X$ are open, $U \cap V$ is path-connected (see Definition 2.122) and $x_0 \in U \cap V$,
then every element $\vartheta \in \pi_1(X, x_0)$ can be written as a sum

$$\vartheta = \eta_1 + \ldots + \eta_n,$$

where for each $k \geqslant 1$, we have

$$\eta_k \in \pi_1(U, x_0) \quad or \quad \eta_k \in \pi_1(V, x_0).$$

2.1.15 Elements of Algebraic Topology II: Homology

Algebraic topology came into existence by the efforts to solve topological problems using algebraic tools. Homology theory is the core subject in algebraic topology. As usual, we start with the geometric approach to the subject which is called "simplicial homology theory" and then pass to the abstract approach which has two branches: "singular homology theory" and "Čech homology theory".

Definition 2.230
Let $\{x_k\}_{k=0}^n \subseteq \mathbb{R}^N$ with $n \leqslant N$. We say that these vectors are ***affinely independent*** *if the vectors $\{x_k - x_0\}_{k=1}^n \subseteq \mathbb{R}^N$ are linearly independent. The convex hull of $n+1$ affinely independent vectors in $\mathbb{R}^N$ (with $n \leqslant N$) is called an n-**simplex** and is denoted by σ_n. All these $n+1$ vectors are called **vertices** of σ_n and n is the **dimension** of σ_n.*

Remark 2.231
If $u \in \sigma_n$, then $u = \sum\limits_{k=0}^{n} \lambda_k x_k$ with $\lambda_k \geqslant 0$ for all $k \in \{0, 1, \ldots, n\}$ and $\sum\limits_{k=0}^{n} \lambda_k = 1$. The numbers $\{\lambda_k\}_{k=0}^n$ are called the ***barycentric coordinates*** of u. Every $u \in \sigma_n$ admits a unique set of barycentric coordinates. The boundary of σ_n consists of $(n-1)$-simplices.

Definition 2.232
*An m-**face** of an n-simplex σ_n is the convex hull of any set of m vertices of σ_n, where $m < n$.*

Definition 2.233
K is a ***geometric simplicial complex*** *if it is a finite set of simplexes in some Euclidean space* $\mathbb{R}^N$ *satisfying the following requirements:*
(a) *If* $\sigma_n \in K$ *and* σ_m *is a face of* σ_n*, then* $\sigma_m \in K$.
(b) *If* $\sigma_n, \sigma_m \in K$*, then* $\sigma_n \cap \sigma_m = \emptyset$ *or otherwise* $\sigma_n \cap \sigma_m$ *is a common face of both simplexes and so by* ***(a)*** *belongs in* K.
The ***dimension*** *of* K*, denoted by* $\dim K$*, is the maximum of the dimensions of its simplexes. A* ***subcomplex*** L *of* K *is a subfamily of simplexes of* K *satisfying* ***(a)***.

Definition 2.234
The union of all simplexes of a geometric simplicial complex K *in* $\mathbb{R}^N$*, topologized as a subset of* $\mathbb{R}^N$ *is called the* ***polyhedron*** *of* K *and is denoted by* $|K|$. *If* L *is a subcomplex of* K*, then* $|L|$ *is called a* ***subpolyhedron*** *of* K.

Proposition 2.235
A subset C *of* $|K|$ *is closed if and only if* $C \cap \sigma$ *is closed in* σ *for every simplex* $\sigma \in K$.

Definition 2.236
For given two simplicial complexes K *and* L*, a function* $f\colon |K| \longrightarrow |L|$ *is a* ***simplicial function*** *if it satisfies the following requirements:*
(a) *If* x *is a vertex of a simplex in* K*, then* $f(x)$ *is a vertex of a simplex in* L.
(b) *If* σ *is a simplex of* K *with vertices* $\{x_k\}_{k=0}^n$*, then* $\tau = \operatorname{conv}\{f(x_k)\}_{k=1}^n$ *is a simplex of* L *(possibly of lower dimension).*
(c) *If* $x = \sum_{k=0}^{n} \lambda_k x_k$ *is in a simplex* σ *of* K *with vertices* $\{x_k\}_{k=0}^n$*, then* $f(x) = \sum_{k=0}^{n} \lambda_k f(x_k)$*, i.e.,* f *is "linear on each simplex".*
A ***simplicial function of simplicial pairs*** $f\colon (|K|, |L|) \longrightarrow (|M|, |N|)$ *is a simplicial function* $f\colon |K| \longrightarrow |M|$ *such that* $f(|L|) \subseteq |N|$.

Proposition 2.237
A simplicial function $f\colon |K| \longrightarrow |L|$ *is continuous.*

There is a slight difficulty in the use of polyhedra. Namely, not every topological space which is homeomorphic to a polyhedron is itself a polyhedron. We rectify this problem with the following definition.

Definition 2.238
If X is a topological space, a ***triangulation*** *of X is a pair (K, f), where K is a simplicial complex and $f\colon |K| \longrightarrow X$ is a homeomorphism. A topological space X with a triangulation is called a* ***triangulated space****. Similarly, if (X, A) is a pair of topological spaces, a* ***triangulation*** *consists of a simplicial pair (K, L) and a homeomorphism of pairs $f\colon (|K|, |L|) \longrightarrow (X, A)$. A pair (X, A) with a triangulation is called a* ***triangulated pair****.*

Remark 2.239
In general the particular homeomorphism f involved does not matter and so we often mention only K when referring to a triangulation of X.

Definition 2.240
Let K be a simplicial complex and $x \in |K|$.
(a) The ***simplicial neighbourhood*** *of x, denoted by $N_K(x)$, is the set of all simplexes of K which contains x together with all their faces.*
(b) The ***link*** *of x, denoted by $Lk_K(x)$, is the subset of simplexes of $N_K(x)$, which do not contain x.*
(c) For every simplex $\sigma \in K$, the ***star*** *of σ, denoted by $st_K(\sigma)$, is the union of the interiors of the simplexes of K that have σ as a face.*

Remark 2.241
We usually omit the suffix K, when it is clear from the context which complex is used.

Proposition 2.242
For every simplex $\sigma \in K$, $st(\sigma)$ is an open set and if x is a point in the interior of σ, then $st(\sigma) = |N(x)| - |Lk(x)|$.

Theorem 2.243
If K and L are simplicial complexes and $f\colon |K| \longrightarrow |L|$ is a homeomorphism,
then for every $x \in |K|$, $\big|Lk_K(x)\big|$ and $\big|Lk_L\big(f(x)\big)\big|$ are homeomorphic.

Definition 2.244
Let K and L be simplicial complexes and let $f\colon |K| \longrightarrow |L|$ be a continuous function. A simplicial function $g\colon |K| \longrightarrow |L|$ is a ***simplicial approximation*** *of f, if for every vertex u of K, we have $f\big(st_K(u)\big) \subseteq st_L\big(g(x)\big)$.*

Remark 2.245
A simplicial function is a simplicial approximation of itself.

Theorem 2.246
If K and L are simplicial complexes and $f\colon |K| \longrightarrow |L|$ is a continuous function,
then there exists a simplicial approximation g of f and every such simplicial approximation is homotopic to f.

Orientation of a simplex means an ordering of its vertices.

Definition 2.247
Two orderings of the vertices are said to determine the same ***orientation*** *of the simplex if every even perturbation of the vertices transforms one ordering to the other. If the permutation is odd, then we say that the orientations are opposite. Hence there are basically two orientations of a simplex. An* ***oriented simplex*** *is a simplex together with a choice of orientation. So, for a k-simplex σ ordered by the linear ordering $(u_0, \dots, u_k)$ of its vertices, by $-\sigma$ we will denote the same simplex but with the opposite orientation.*

This definition allows us to produce an abelian group out of a simplicial complex.

Definition 2.248
*Let K be a simplicial complex. A k-**chain** is a function g from the set of all oriented k-simplexes of K to the integers such that:*
(a) *$g(\sigma) = -g(\tau)$, if the k-simplexes σ, τ have opposite orientation;*
(b) *$g(\sigma) = 0$ for all but finitely many oriented k-simplexes σ.*

We add k-chains in the usual way by adding their values. Similarly for the multiplication of k-chains with $c \in \mathbb{Z}$. The resulting group is called the ***group of (oriented)*** *k-**chains** and is denoted by $C_k(K)$. If $k < 0$ or $k > \dim K$, then $C_k(K) = 0$.*

If σ is an oriented simplex, the ***elementary chain*** *g corresponding to σ is the function g defined by $g(\sigma) = 1$, $g(-\sigma) = -1$ and $g(\tau) = 0$ for all other oriented simplexes.*

Proposition 2.249
$C_k(K)$ is a free abelian group. A basis for $C_k(K)$ can be obtained by orienting each k-simplex and the using of corresponding elementary chains as a basis.

Remark 2.250
The group $C_0(K)$ differs from the others, since it has a natural basis (note that the 0-simplex, being a point, has only one orientation). In contrast, the group $C_k(K)$, for $k \geqslant 1$, has no "natural basis". We need to orient the k-simplexes of K in some arbitrary fashion in order to produce a basis.

Proposition 2.251
A function f from the oriented k-simplexes of K to an abelian group G extends uniquely to a homomorphism from $C_k(K)$ to G provided $f(-\sigma) = -f(\sigma)$ for every oriented k-simplex σ.

Definition 2.252
We define a homomorphism $\partial_k \colon C_k(K) \longrightarrow C_{k-1}(K)$ called the ***boundary operator*** *as follows. If $\sigma = [u_0, \ldots, u_k]$ is an oriented k-simplex ($k \geqslant 1$), then*

$$\partial_k \sigma \;=\; \sum_{k=0}^{n} (-1)^n \big[u_0, \ldots, \hat{u}_n, \ldots, u_k\big],$$

where the symbol $\hat{u}_n$ means that the vertex u_n is deleted from the array. Since $C_k(K) = 0$ for $k < 0$, the operator ∂_k is the trivial homomorphism for $k \leqslant 0$.

Proposition 2.253
$\partial_{k-1} \circ \partial_k \;=\; 0.$

This fundamental equation leads to the definition of the (simplicial) homology groups.

Definition 2.254
Consider the boundary operator $\partial_k \colon C_k(K) \longrightarrow C_{k-1}(C)$. The kernel of ∂_k is an abelian subgroup of $C_k(K)$ and is denoted by $Z_k(K)$. It is the group of ***k-cycles****. The image of $\partial_{k+1} \colon C_{k+1}(K) \longrightarrow C_k(K)$ is an*

*abelian subgroup of $C_k(K)$ and is denoted by $B_k(K)$. It is the group of k-**boundaries**. By Proposition 2.253, we have*

$$B_k(K) \subseteq Z_k(K).$$

We can define the quotient (factor) group

$$H_k(K) = Z_k(K)/_{B_k(K)},$$

*which is called k**th (simplicial) homology group** of K.*

Theorem 2.255
$H_0(K) \simeq \mathbb{Z}$ if and only if $|K|$ is connected.

Theorem 2.256
If $|K| = \bigcup_{i=1}^{n} |K_i|$, where $|K_i|$ are pairwise disjoint and connected, then

$$H_k(K) = \sum_{i=1}^{n} H_k(K_i) \qquad \forall\, k \geqslant 0.$$

Sometimes it is more convenient to consider an alternative version of the 0-dimensional homology.

Definition 2.257
Let $\varepsilon\colon C_0(K) \longrightarrow \mathbb{Z}$ be the surjective homomorphism, defined by

$$\varepsilon(u) = 1 \quad \text{for every vertex } u \text{ of } K.$$

*Then, if g is a 0-chain, $\varepsilon(g)$ equals the sum of the values of g on the vertices of K. We call ε the **augmentation function** for $C_0(K)$. Note that $\varepsilon(\partial_1 c) = 0$ if c is a 1-chain. The **reduced homology group** of K in dimension 0, denoted by $\widetilde{H}_0(K)$, is defined by*

$$\widetilde{H}_0(K) = \ker \varepsilon/_{\operatorname{im} \partial_1}.$$

If $k \geqslant 1$, then $\widetilde{H}_k(K) = H_k(K)$.

Proposition 2.258
$\widetilde{H}_0(K)$ is a free abelian group and

$$H_0(K) \simeq \widetilde{H}_0(K) \oplus \mathbb{Z}.$$

Hence, if $|K|$ is connected, then

$$\widetilde{H}_0(K) = 0.$$

Definition 2.259
If L is a subcomplex of K, the quotient group $C_k(K)/_{C_k(L)}$ is called the group of **relative chains of** *$K \bmod L$ and is denoted by $C_k(K, L)$.*

Remark 2.260
The group $C_k(K, L)$ is a free abelian group. It has as basis all cosets of the form $[\sigma_i] = \sigma_i + C_k(L)$, where $\{\sigma_i\}$ is a basis for $C_k(K)$ and $\sigma_i \notin L$.

Proposition 2.261
The boundary operator $\partial_k \colon C_k(L) \longrightarrow C_{k-1}(L)$ is the restriction of the boundary operator $\partial_k \colon C_k(K) \longrightarrow C_{k-1}(K)$ (we use the same symbol for both). Then this homomorphism induces a homomorphism $\partial_k \colon C_k(K, L) \longrightarrow C_{k-1}(K, L)$ (again the same symbol is used). We have $\partial_{k-1} \circ \partial_k = 0$. We set

$$\begin{aligned} Z_k(K, L) &= \ker \partial_k, \\ B_k(K, L) &= \operatorname{im} \partial_{k+1}, \\ H_k(K, L) &= Z_k(K, L)/_{B_k(K, L)}. \end{aligned}$$

We call $Z_k(K, L)$ the **group of relative k-cycles,** *$B_k(K, L)$ the* **group of relative k-boundaries** *and $H_k(K, L)$ the* **kth relative (simplicial) homology group** *of $K \bmod L$.*

Remark 2.262
A relative k-chain is a coset $c + C_k(K)$. This is a relative k-cycle if and only if $\partial_k c$ is carried by L. It is a relative k-boundary if and only if there is a $(k+1)$-chain g_{k+1} of K such that $c - \partial_{k+1} g_{k+1}$ is carried by L. If $L = \{v\}$ (a singleton), the

$$H_0(K, L) \simeq \widetilde{H}_0(K).$$

If K consists of an n-simplex and its faces and L is the set of all proper faces of K, then

$$H_k(K, L) = \delta_{k,n}\mathbb{Z} \qquad \forall\, k \geqslant 0,$$

where n is the dimension of the simplex.

One of the greatest advances in the algebraic topology has been the extension of homology theory to general topological spaces. So far homology groups have been defined for a special kind of spaces, namely compact polyhedra and complexes resulting from them. Singular homology theory extends the notion of homology groups to general topological spaces.

Definition 2.263
Let $e_n = (0, \ldots, 0, 1) \in \mathbb{R}^n$ $(n \geqslant 1)$. By means of the standard identification of $\mathbb{R}^n$ with the subspace $\mathbb{R}^n \times \{0\}$ of $\mathbb{R}^n \times \mathbb{R}^m = \mathbb{R}^{n+m}$, e_n can be viewed as an element of $\mathbb{R}^{n+m}$ for any $m \geqslant 0$. Let $e_0 = (0, \ldots, 0) \in \mathbb{R}^n$ $(n \geqslant 1)$. Clearly $\{e_0, \ldots, e_n\}$ are affinely independent and so they may be taken as vertices of an n-simplex denoted by Δ_n. We call Δ_n the ***standard n-simplex****. If X is a topological space, a* ***singular n-simplex****, is a continuous function $T\colon \Delta_n \longrightarrow X$.*

Definition 2.264
The free abelian group with the singular n-simplexes as generators and coefficients in $\mathbb{Z}$ is called the ***nth singular chain group*** *and is denoted by $S_n(X)$. For $n < 0$, $S_n(X) = 0$. If $c \in S_n(X)$, then c is called a* ***singular n-chain****.*

Definition 2.265
Let X, Y be two topological spaces and let $f\colon X \longrightarrow Y$ be a continuous function. If $T\colon \Delta_n \longrightarrow X$ is a singular n-simplex in X, then the composition $f \circ T\colon \Delta_n \longrightarrow Y$ is a singular n-simplex in Y denoted by fT. If $c = \sum_{i=1}^{n} a_i T_i$, $a_i \in \mathbb{Z}$, belongs in $C_n(X)$ (an n-chain in X), then

$$f_*(c) = \sum_{i=1}^{n} a_i\, fT_i \in C_n(Y)$$

(an n-chain in Y). The homomorphism $f_*\colon C_n(X) \longrightarrow C_n(Y)$ is called the ***homomorphism induced by f****. Clearly*

$$(f_2 \circ f_1)_* = (f_2)_* \circ (f_1)_*.$$

Definition 2.266
For each $i \in \{1, \ldots, n\}$, let $d_i\colon \Delta_{n-1} \longrightarrow \Delta_n$ be the affine function, defined by

$$d_i(x_0, \ldots, x_{n-1}) = (x_0, \ldots, x_{i-1}, 0, x_i, \ldots, x_{n-1}),$$

called the i-***face function****. It maps* Δ_{n-1} *onto* Δ_n^i *(the* i*th face of* Δ_n *opposite to the vertex* e_i*). For a given singular* n*-simplex* $T\colon \Delta_n \longrightarrow X$*, then* $T \circ d_i\colon \Delta_{n-1} \longrightarrow X$ *are* $n+1$ *different singular* $(n-1)$*-simplexes, which can be thought of as the boundary of* T*. So, for each* $n \geqslant 1$*, the* ***boundary operator*** $\partial_n\colon C_n(X) \longrightarrow C_{n-1}(X)$ *is the homomorphism, defined by*

$$\partial_n T = \sum_{i=0}^{n} (-1)^i T \circ d_i.$$

Proposition 2.267
For all $n \geqslant 1$*, we have* $\partial_n \circ \partial_{n+1} = 0$*.*

Definition 2.268
The collection $C_*(X) = \{C_n(X), \partial_n\}_{n \geqslant 0}$ *of abelian groups and boundary operators is called a* ***singular chain complex for*** X*. The remainder of the construction of singular homology copies that of simplicial homology. So, we set*

$$Z_n(X) = \ker \partial_n \qquad \forall\, n \geqslant 1$$

and

$$Z_0(X) = C_0(X).$$

These are abelian subgroups of $C_n(X)$ *and their elements are called* ***singular*** (n)***-cycles****. Also we set*

$$B_n(X) = \operatorname{im} \partial_{n+1} \qquad \forall\, n \geqslant 0.$$

These are abelian subgroups of $C_n(X)$ *too and their elements are called* ***singular*** (n)***-boundaries****. From Proposition 2.267 we see that* $B_n(X) \subseteq Z_n(X)$ *and so we can define the quotient groups*

$$H_n(X) = Z_n(X)/B_n(X) = \begin{cases} \ker \partial_n / \operatorname{im} \partial_{n+1} & \text{if } n \geqslant 1, \\ C_0(X)/\operatorname{im} \partial_1 & \text{if } n = 0. \end{cases}$$

This is the n***th singular homology group*** *of* X*. The* ***singular homology*** *of* X *is the collection*

$$H_*(X) = \{H_n(X)\}_{n \geqslant 0}.$$

Remark 2.269
The elements of $H_n(X)$ are called singular homology classes, the coset $u + B_n(X)$ being the class for the singular n-cycle u. If two singular n-cycles u and u' belong to the same singular homology class, they are said to be homologous. Evidently u and u' are homologous if and only if $u - u' = \partial_{n+1}c$ for some singular $(n+1)$-chain. If $H_n(X)$ is finitely generated, then $\operatorname{rank} H_n(X)$ is the ***nth Betti number*** of X. Note that $Z_n(X)$ and $B_n(X)$ being subgroups of an abelian group are normal. Finally note that in contrast to the simplicial homology, if $X \neq \emptyset$, then

$$C_n(X) \neq 0 \qquad \forall\, n \geqslant 0.$$

Proposition 2.270
(a) *If $X = \{\star\}$ (a singleton), then $H_0(\star) \simeq \mathbb{Z}$ and*

$$H_n(\star) = 0 \qquad \forall\, n \geqslant 1.$$

(b) *If X is path-connected, then $H_0(X) \simeq \mathbb{Z}$ and if X has path-connected components $\{X_i\}_{i\in I}$ (see Definitions 2.122 and 2.130), then*

$$H_n(X) = \sum_{i\in I} H_n(X_i) \qquad \forall\, n \geqslant 0.$$

(b) *If $X \subseteq \mathbb{R}^N$ is convex, then $H_0(X) \simeq \mathbb{Z}$ and*

$$H_n(X) = 0 \qquad \forall\, n \geqslant 1.$$

Remark 2.271
As with simplicial homology (see Theorems 2.255 and 2.256), H_0 counts the number of path-connected components of X, thus giving the same information as the zero homotopy group $\pi_0(X)$ (in fact $H_0(X)$ is isomorphic to the abelianization $\mathbb{Z}\big[\pi_0(X)\big]$ of $\pi_0(X)$). However, unlike higher homotopy groups, the singular homology groups give information about all the path-connected components of X (see Definition 2.130).

Proposition 2.272
If $\overline{B}_1^n = \{x \in \mathbb{R}^n : \|x\| \leqslant 1\}$ and $S^n = \partial B_1^{n+1} = \{x \in \mathbb{R}^{n+1} : \|x\| = 1\}$, then

(a) *we have*

$$H_k(B^n) \simeq \delta_{k,0}\mathbb{Z} \qquad \forall\, k \geqslant 0,$$

where

$$\delta_{k,0} = \begin{cases} 1 & \text{if } k = 0, \\ 0 & \text{if } k \neq 0. \end{cases}$$

(b) *we have*

$$H_k(S^n) \simeq \begin{cases} \mathbb{Z} & \text{if } k = 0 \text{ or } k = n, \\ 0 & \text{otherwise.} \end{cases}$$

We can also define relative singular homology. In this theory chains on a certain subspace $A \subset X$ are identified with 0. That is, for a chain to be a singular cycle of $X \bmod A$, it must have a boundary which is a singular chain on A rather than zero. So, we make the following definition.

Definition 2.273
Let (X, A) be a topological pair. We set

$$C_n(X, A) = C_n(X)/_{C_n(A)} \qquad \forall\, n \geqslant 1.$$

This is the ***relative n-singular chain group of*** *$X \bmod A$. It is a free abelian group with generators those singular n-simplexes $T\colon \Delta_n \longrightarrow X$ whose image is not completely contained in A. The elements of $C_n(X, A)$ are called* ***relative singular n-chains of*** *$X \bmod A$. Since $\partial_n\colon C_n(X) \longrightarrow C_{n-1}(X)$ $(n \geqslant 1)$ is a homomorphism and $\partial\big(C_n(A)\big) \subseteq C_{n-1}(A)$, there exists a unique homomorphism $\partial_n\colon C_n(X, A) \longrightarrow C_{n-1}(X, A)$ (for notational economy it is denoted by the same symbol). This is the* ***boundary operator*** *for the relative singular chain groups. Again*

$$\partial_{n-1} \circ \partial_n = 0 \qquad \forall\, n \geqslant 1.$$

As before (see Definition 2.268), we set

$$Z_n(X, A) = \ker \partial_n \quad \text{and} \quad B_n(X, A) = \operatorname{im} \partial_{n+1} \qquad \forall\, n \geqslant 0.$$

Then $Z_n(X, A)$ is the group of ***relative singular n-cycles of*** *$X \bmod A$ and $B_n(X, A)$ is the group of* ***relative singular n-boundaries of*** *$X \bmod A$. Both are abelian subgroups of $C_n(X, A)$ and*

$$B_n(X, A) \subseteq Z_n(X, A) \qquad \forall\, n \geqslant 0.$$

We set

$$H_n(X,A) = Z_n(X,A)/_{B_n(X,A)} \qquad \forall\, n \geqslant 0.$$

This is the n**th relative singular homology group of** $X \bmod A$ *and it is a free abelian group. If it is finitely generated, then* $\operatorname{rank} H_n(X,A)$ *is the* n**th Betti number** *of the topological pair* (X,A).

Remark 2.274
Note that

$$Z_n(X,A) = \begin{cases} \{c \in C_n(X) : \partial_n c \in C_{n-1}(A)\} & \text{if} \quad n \geqslant 1, \\ C_0(X) & \text{if} \quad n \geqslant 0, \end{cases}$$

and

$$B_n(X,A) = B_n(X) + C_n(A)$$

(the subgroup of $C_n(X)$ generated by $B_n(X)$ and $C_n(A)$). If $A = \emptyset$, the

$$H_n(X,\emptyset) = H_n(X).$$

One of the characteristic features of homology theory is a long exact sequence, which relates the homologies of X, A and $X \bmod A$. To describe it we will need the following proposition.

Proposition 2.275
If $f\colon (X,A) \longrightarrow (Y,B)$ *is a continuous function between topological pairs,*
then f *induces a homomorphism* $f_*\colon H_n(X,A) \longrightarrow H_n(Y,B)$, $n \geqslant 1$ *such that:*
(a) *if* f *is the identity function* $i\colon (X,A) \longrightarrow (X,A)$, *then* f_* *is the identity isomorphism;*
(b) *for another continuous function between topological pairs* $g\colon (Y,B) \longrightarrow (Z,C)$, *we have* $(g \circ f)_* = g_* \circ f_*$.

Let (X,A) be a topological pair and let $i\colon A \longrightarrow X$ and $j\colon (X,\emptyset) \longrightarrow (X,A)$ be the inclusion functions. They induce the homomorphisms $i_*\colon H_n(A) \longrightarrow H_n(X)$ and $j_*\colon H_n(X) \longrightarrow H_n(X,A)$ (see Proposition 2.275). Also, there is a unique homomorphism $\partial_*\colon H_n(X,A) \longrightarrow H_{n-1}(A)$, defined by

$$\partial_*\big([c + B_n(X) + C_n(A)]\big) = [\partial_n c + B_{n-1}(A)]$$

(see Remark 2.274).

Theorem 2.276
If (X, A) *is a topological space,*
then the following infinite homology sequence is exact

$$\begin{aligned} \ldots \longrightarrow H_n(A) \xrightarrow{i_*} H_n(X) \xrightarrow{j_*} H_n(X, A) \xrightarrow{\partial_*} H_{n-1}(A) \longrightarrow \ldots \\ \ldots \longrightarrow H_0(A) \xrightarrow{i_*} H_0(X) \xrightarrow{j_*} H_0(X, A) \xrightarrow{\partial_*} 0. \end{aligned}$$

Moreover, a continuous function $f\colon (X, A) \longrightarrow (Y, D)$ *between topological pairs induces a homomorphism of the infinite exact sequence of* (X, A) *into that of* (Y, D)*, namely in the following diagram every square is commutative.*

$$\begin{array}{ccccccccc} \longrightarrow & H_n(A) & \xrightarrow{i_*} & H_n(X) & \xrightarrow{j_*} & H_n(X,A) & \xrightarrow{\partial_*} & H_n-1(A)) & \longrightarrow \\ & \downarrow{\scriptstyle (f|_A)_*} & & \downarrow{\scriptstyle f_*} & & \downarrow{\scriptstyle f_*} & & \downarrow{\scriptstyle (f|_A)_*} & \\ \longrightarrow & H_n(D) & \xrightarrow{i_*} & H_n(Y) & \xrightarrow{j_*} & H_n(Y,D) & \xrightarrow{\partial_*} & H_n-1(D) & \longrightarrow \end{array}$$

Theorem 2.277
If $A \subseteq D \subseteq X$ *is a triple of topological spaces,*
then there is an exact sequence

$$\begin{aligned} \ldots \longrightarrow H_n(D, A) \longrightarrow H_n(X, A) \longrightarrow H_n(X, D) \xrightarrow{\hat{\partial}_*} H_{n-1}(D, A) \longrightarrow \ldots \\ \ldots \longrightarrow H_0(D, A) \longrightarrow 0, \end{aligned}$$

where the unlabelled homomorphisms result from the corresponding inclusion functions and $\hat{\partial}_* = j_* \circ \partial_*$*, i.e., the composition*

$$H_n(D, A) \xrightarrow{\partial_*} H_n(D) \xrightarrow{j_*} H_{n-1}(D, A).$$

Moreover, a continuous function $f\colon (X, D, A) \longrightarrow (X', D', A')$ *between topological triples induces a homomorphism of the exact sequence of* (X, D, A) *into that of* (X', D', A') *(see Theorem 2.276).*

In the relative singular homology of $X \bmod A$, the singular chains on A are identified with 0. This suggests that removal of a set $E \subseteq A$ should not affect the homology. This is true under the stronger condition $\overline{E} \subseteq \operatorname{int} A$ and is known as "excision property" of singular homology.

Theorem 2.278 *(Excision Property)*
If (X, A) *is a topological pair and* $\overline{E} \subseteq \operatorname{int} A$,
then $H_n(X \setminus E, A \setminus E) \simeq H_n(X, A)$ *for all* $n \geqslant 0$.

This property leads to the following definition.

Definition 2.279
Let X *be a topological space and let* $A, D \subseteq X$ *be two sets such that* $X = A \cup D$. *We say that the pair* (A, D) *is* ***excisive*** *if the inclusion function* $i\colon (A, A \cap D) \longrightarrow (A \cup D, D) = (X, D)$ *induces an isomorphism of the singular homology groups.*

Theorem 2.280
If X *is a topological space and* $A, D \subseteq X$ *are such that* $X = A \cup D = (\operatorname{int} A) \cup (\operatorname{int} D)$,
then the pair (A, D) *is excisive.*

Theorem 2.281
If X *is a topological space and* (A, D) *is an excisive pair,*
then there is a homomorphism $\partial_*^n\colon H_n(X) \longrightarrow H_{n-1}(A \cap D)$ *such that the following infinite homology sequence is exact:*

$$\ldots \longrightarrow H_n(A \cap D) \overset{(j_*^1, -j_*^2)}{\longrightarrow} H_n(A) \oplus H_n(D) \overset{i_*^1 + i_*^2}{\longrightarrow} H_n(X) \overset{\partial_*^n}{\longrightarrow} H_{n-1}(A \cap D) \longrightarrow \ldots$$
$$\ldots \longrightarrow H_1(X) \overset{\partial_*^1}{\longrightarrow} H_0(A \cap D) \overset{(j_*^1, -j_*^2)}{\longrightarrow} H_0(A) \oplus H_0(D) \overset{i_*^1 + i_*^2}{\longrightarrow} H_0(X) \longrightarrow 0,$$

where $i^1\colon A \longrightarrow X$, $i^2\colon D \longrightarrow X$, $j^1\colon A \cap D \longrightarrow A$, $j^2\colon A \cap D \longrightarrow X$ *are the inclusion functions. This exact sequence is known as the* ***Mayer–Vietoris sequence****.*

Theorem 2.282
If (X, A) *is a topological pair and* $\star \in A$,
then the following infinite singular homology sequence is exact:

$$\ldots \longrightarrow H_n(A, \star) \longrightarrow H_n(X, \star) \longrightarrow H_n(X, A) \longrightarrow H_{n-1}(A, \star) \longrightarrow \ldots \longrightarrow 0,$$

where the homomorphisms are induced by inclusions or boundary operators (see Theorem 2.277). This sequence is known as the ***reduced exact homology sequence****.*

Theorem 2.283
Let $f\colon (X, A) \longrightarrow (Y, D)$ be a continuous function between topological pairs. If $f\colon X \longrightarrow Y$ and $f|_A\colon A \longrightarrow D$ are homotopy equivalences (see Definition 2.191),
then f_* is an isomorphism in relative singular homology.

First we defined homology groups for a particular class of spaces, namely polyhedra. Then we passed to general topological spaces, by means of the singular homology theory. In fact there is a plethora of homology theories, a fact that led to an axiomatic unification by Eilenberg–Steenrod.

Definition 2.284
Let $\mathcal{P}$ be a class of topological pairs such that:
(a) If $(X, A) \in \mathcal{P}$, then $(X, X), (X, \emptyset), (A, A), (A, \emptyset) \in \mathcal{P}$.
(b) If $(X, A) \in \mathcal{P}$, then $(I \times X, I \times A) \in \mathcal{P}$ (where $I = [0, 1]$).
(c) There is a singleton $\{\star\}$ such that $(\{\star\}, \emptyset) \in \mathcal{P}$.

*Then we call $\mathcal{P}$ an **admissible class of spaces** for a homology theory*

Using this notion we can produce an axiomatic definition of a homology theory.

Definition 2.285
*Let $\mathcal{P}$ be an admissible class of spaces. A **homology theory** on $\mathcal{P}$ consists of the following three "functions":*
(a) A function H from $\mathbb{Z} \times \mathcal{P}$ into the family of abelian groups with the image of $(n, (X, A))$ denoted by $H_n(X, A)$.
(b) A function which to every $n \in \mathbb{Z}$ and every continuous function $f\colon (X, A) \longrightarrow (Y, D)$ with $(X, A), (Y, D) \in \mathcal{P}$, assigns a homomorphism $(f_*)_n\colon H_n(X, A) \longrightarrow H_n(Y, D)$.
(c) A function which to every $n \in \mathbb{Z}$ and every $(X, A) \in \mathcal{P}$, assigns a homomorphism $(\partial_*)_n\colon H_n(X, A) \longrightarrow H_n(A)$ (where A denotes the element $(A, \emptyset) \in \mathcal{P}$).
These three functions should satisfy the following axioms:
Axiom 1. *If i is the identity, then i_* is the identity homomorphism.*
Axiom 2. *$(g \circ f)_\star = g_* \circ f_*$.*
Axiom 3. *If $f\colon (X, A) \longrightarrow (Y, D)$ is a continuous function between topological pairs, then the following diagram commutes:*

$$\begin{array}{ccc} H_n(X,A) & \xrightarrow{f_*} & H_n(Y,D) \\ \Big\downarrow \partial_* & & \Big\downarrow (\partial_*)_n \\ H_{n-1}(A) & \xrightarrow{(f|_A)_*} & H_{n-1}(D) \end{array}$$

Axiom 4. *If $i\colon A \longrightarrow X$ and $j\colon X = (X,\emptyset) \longrightarrow (X,A)$ are the inclusion functions, then the sequence*

$$\ldots \longrightarrow H_n(A) \xrightarrow{i_*} H_n(X) \xrightarrow{j_*} H_n(X,A) \xrightarrow{\partial_*} H_{n-1}(A) \longrightarrow \ldots$$

is exact (this is known as the ***exactness axiom****).*
Axiom 5. *If f and g are homotopic, then $f_* = g_*$ (this is known as the* ***homotopy axiom****).*
Axiom 6. *Let $(X,A) \in \mathcal{P}$ and let U be an open subset of X such that $\overline{U} \subseteq \operatorname{int} A$. If $(X \setminus U, A \setminus U) \in \mathcal{P}$, then*

$$H_n(X \setminus U, A \setminus U) \simeq H_n(X,A) \qquad \forall\, n \geqslant 0$$

(this is known as the ***excision axiom****).*
Axiom 7. *If $\{\star\}$ is the singleton space in $\mathcal{P}$, then*

$$H_n(\{\star\}) = \delta_{n,0}\mathbb{Z} \qquad \forall\, n \geqslant 0$$

(this is known as the ***dimension axiom****).*

Theorem 2.286
(a) *Simplicial homology theory on the class of triangulated pairs (see Definition 2.238) satisfies the Eilenberg–Steenrod axioms of Definition 2.285.*
(b) *Singular homology theory on the class of topological pairs satisfies the Eilenberg–Steenrod axioms.*

Theorem 2.287
If X is a topological space, K is a simplicial complex and $X, |K|$ are homeomorphic,
then for every integer $H_n(X)$ (singular homology group) is isomorphic to $H_n(K)$ (simplicial homology group).

Cohomology theory can be axiomatized in the same way as homology theory.

Definition 2.288
Let $\mathcal{P}$ be an admissible class of spaces and let G be an abelian group (the group of coefficients). A ***cohomology theory*** *on $\mathcal{P}$ consists of the following three "functions":*
(a) *A function H from $G \times \mathcal{P}$ into the family of abelian groups with the image of $(n, (X, A))$ denoted by $H^n(X, A; G)$.*
(b) *A function which to every $n \in \mathbb{Z}$ and every continuous function $f\colon (X, A) \longrightarrow (Y, D)$ with $(X, A), (Y, D) \in \mathcal{P}$ assigns a homomorphism $(f^*)_n\colon H^n(Y, D; G) \longrightarrow H^n(X, A; G)$.*
(c) *A function which to every $n \in \mathbb{Z}$ and every $(X, A) \in \mathcal{P}$ assigns a homomorphism $(\delta^*)_n\colon H^{n-1}(A; G) \longrightarrow H^n(X, A; G)$.*

These three functions should satisfy the following axioms:
Axiom 1. *If i is the identity, then i^* is the identity homomorphism.*
Axiom 2. *$(g \circ f)^* = f^* \circ g^*$.*
Axiom 3. *If $f\colon (X, A) \longrightarrow (Y, D)$ is a continuous function between topological pairs, then the following diagram commutes:*

$$\begin{array}{ccc} H^n(X, A; G) & \xrightarrow{\ f^*\ } & H^n(Y, D; G) \\ \uparrow{\scriptstyle (\delta^*)_n} & & \uparrow{\scriptstyle (\delta^*)_n} \\ H^{n-1}(A; G) & \xrightarrow{\ (f|_A)^*\ } & H^{n-1}(D; G) \end{array}$$

Axiom 4. *If $i\colon A \longrightarrow X$ and $j\colon X = (X, \emptyset) \longrightarrow (X, A)$ are the inclusion functions, then the sequence*

$$\ldots \longleftarrow H^n(A; G) \xleftarrow{\ i^*\ } H^n(X; G) \xleftarrow{\ j^*\ } H^n(X, A; G) \xleftarrow{\ (\partial^*)_n\ } H^{n-1}(A; G) \longleftarrow \ldots$$

is exact.
Axiom 5. *If f and g are homotopic, then $f^* = g^*$.*
Axiom 6. *Let $(X, A) \in \mathcal{P}$ and let U be an open subset of X such that $\overline{U} \subseteq \operatorname{int} A$. If $(X \setminus U, A \setminus U) \in \mathcal{P}$, then*

$$H^n(X \setminus U, A \setminus U) \ \simeq\ H^n(X, A) \qquad \forall\, n \in \mathbb{Z}.$$

Axiom 7. *If $\{\star\}$ is the singleton space in $\mathcal{P}$, then*

$$H^n(\{\star\}) \ =\ \delta_{n,0}\mathbb{Z} \qquad \forall\, n \geqslant 0.$$

Definition 2.289
The ***singular*** n***-cochain*** *is defined to be the homomorphism* $c\colon C_n(X;G) \longrightarrow G$. *We use the bracket notation* $[\cdot, c]$ *and we have*

$$[T_1 + T_2, c] = [T_1, c] + [T_2, c] \qquad \forall\, T_1, T_2 \in C_n(X;G)$$

and

$$[gT, c] = g[T, c] \qquad \forall\, g \in G,\ T \in C_n(X;G).$$

The set of all singular n*-cochains* $\operatorname{Hom}\big(C_n(X;G), G\big)$ *is denoted by* $C^n(X;G)$. *We have*

$$[T, c_1 + c_2] = [T, c_1] + [T, c_2] \qquad \forall\, T \in C_n(X;G),\ c_1, c_2 \in C^n(X;G)$$

and

$$[T, gc] = g[T, c] \qquad \forall\, g \in G,\ T \in C_n(X;G),\ c \in C^n(X;G).$$

So the ***duality brackets*** $[\cdot,\cdot]$ *is a bilinear form on* $C_n(X;G) \times C^n(X;G)$. *Then the dual operator of the boundary operator* $(\partial_*)_n$ *with respect to* $[\cdot,\cdot]$ *is called the* ***coboundary operator*** *and is denoted by* $(\delta^*)_n$. *We have*

$$[(\partial_*)_n T, c] = [T, (\delta^*)_n c] \qquad \forall\, T \in C_n(X;G),\ c \in C^n(X;G).$$

Hence $(\delta^*)_n\colon C^{n-1}(X;G) \longrightarrow C^n(X;G)$ *is a homomorphism and* $(\partial_*)_{n-1} \circ (\partial_*)_n = 0$ *implies* $(\delta^*)_n \circ (\delta^*)_{n-1} = 0$. *Then singular cohomology is defined as follows. Let* (X, A) *be a topological pair and let*

$$\hat{C}^n(X, A; G) = \operatorname{Hom}\Big(\big(C_n(X;G)\big)/_{C_n(A;G)}, G\Big)$$

and

$$(\hat{\delta}^*)_n\colon \hat{C}^{n-1}(X, A) \longrightarrow \hat{C}^n(X, A)$$

is the dual operator of the boundary operator

$$(\hat{\partial}_*)_n\colon C_n(X, A; G) \longrightarrow C_{n-1}(X, A; G).$$

We define

$$H^n(X, A) = \ker{(\hat{\delta}^*)_n}/_{\operatorname{im}{(\hat{\delta}^*)_{n-1}}}.$$

This is the ***relative*** n***th singular cohomology group*** *of* $X \bmod A$.

Remark 2.290
Note that

$$\hat{C}^n(X,A;G) \simeq \{c \in C^n(X;G):\ [T,c]=0 \text{ for all } T \in C_n(A;G)\}.$$

The isomorphism is realized by the dual homomorphism

$$\xi^*\colon \hat{C}^n(X,A;G) \longrightarrow \hat{C}^n(X;G)$$

of the homomorphism $\xi\colon C_n(X;G) \longrightarrow C_n(X;A;G)$. Therefore

$$\begin{aligned} \ker(\delta^*)_n &= Z^n(X,A;G) \\ &= \{c \in C^n(X,A;G):\ [T,c]=0 \text{ for all } T \in B_n(X,A;G)\} \end{aligned}$$

and

$$\begin{aligned} \operatorname{im}(\delta^*)_{n-1} &= B^n(X,A;G) \\ &= \{c \in C^n(X,A;G):\ [T,c]=0 \text{ for all } T \in Z_n(X,A;G)\}. \end{aligned}$$

In general we have a canonical homomorphism

$$\beta\colon H^n(X,A;G) \longrightarrow H_n(X,A;G)^*.$$

If G is a field, then β is surjective. Note that in this case $H_n(X,A;G)$ is a vector space and $H^n(X,A;G)$ is its dual.

Proposition 2.291
The singular cohomology theory satisfies the Eilenberg–Steenrod axioms.

Definition 2.292
Let $\{C_n(X),(\partial_*)_n,\varepsilon\}_{n\geqslant 1}$ *be an augmented chain complex (see Definition 2.257). Let* $\hat{\varepsilon}\colon \operatorname{Hom}(\mathbb{Z},G) \longrightarrow \hat{C}^0(X;G)$ *be the dual to the augmentation function. We define the* ***reduced singular cohomology groups*** *by setting*

$$\widetilde{\hat{H}}^n(X;G) = \hat{H}^n(X;G) \quad \text{if } n \geqslant 1, \quad \text{and} \quad \widetilde{\hat{H}}^0(X;G) = \ker(\delta^*)_1/_{\operatorname{im}\hat{\varepsilon}}.$$

The next theorem relates homology and cohomology groups.

Theorem 2.293 *(Alexander Duality Theorem)*
<u>If</u> A is a proper nonempty subset of $S^n = \{x \in \mathbb{R}^{n+1}:\ \|x\| = 1\}$ and (S^n, A) is a triangulated pair,
<u>then</u> $\widetilde{\hat{H}}^k(A) \simeq \widetilde{H}_{n-k-1}(S^n \setminus A)$.

We conclude this section with the diagram summarizing the relations between the various types of spaces introduced in the theory and in the problems (arrows stand for inclusions).

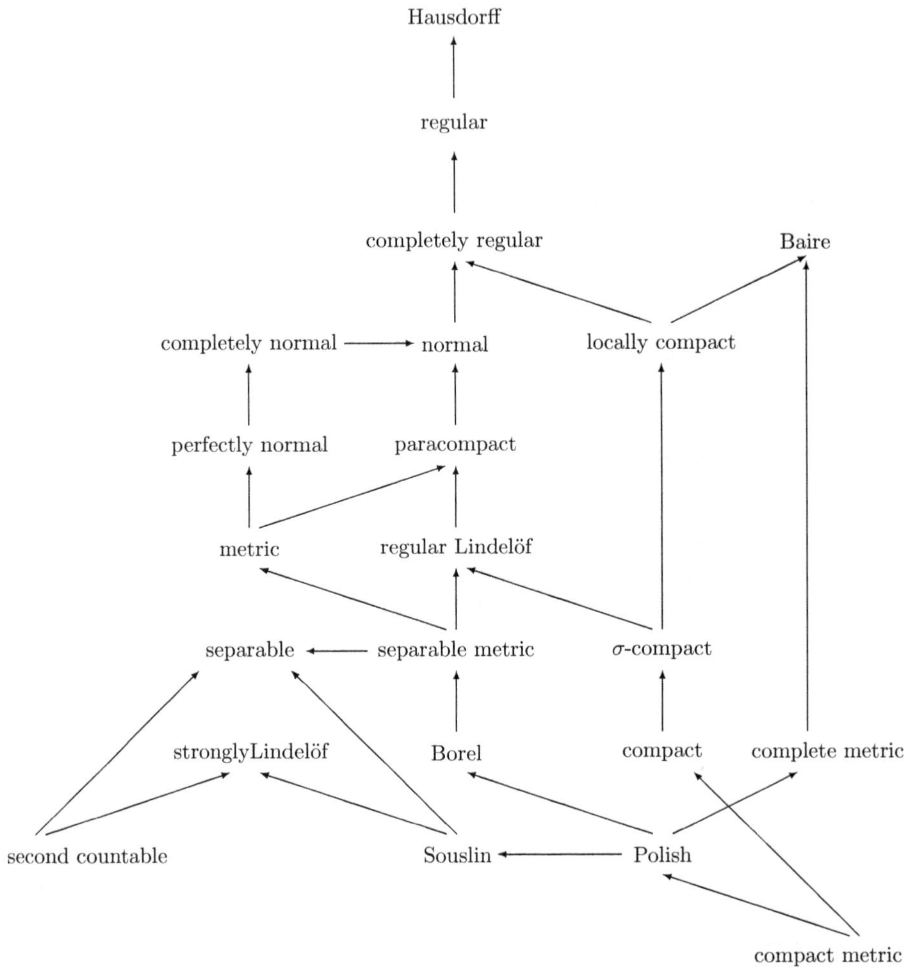

2.2 Problems

Reminder: All topological spaces are taken to be Hausdorff.

Problem 2.1 *
Suppose that X is a topological space and $U \subseteq X$ is an open set. Show that for every set $A \subseteq X$, we have $U \cap \overline{A} \subseteq \overline{U \cap A}$ and the inclusion can be proper.

Problem 2.2 *
Suppose that X is a topological space and $A, C \subseteq X$. Show that:
(a) if $\partial C \subseteq A \subseteq C$, then $\partial C \subseteq \partial A$;
(b) ∂A need not be equal to $\partial \overline{A}$;
(c) $\partial(A \cup C) \subseteq \partial A \cup \partial C$ and the inclusion can be proper;
(d) $\partial(A \cap C) \subseteq \partial A \cup \partial C$ and the inclusion can be proper.

Problem 2.3 **
Let $A \subseteq C[0,1]$ be the subset of Lipschitz functions. Furnish $C[0,1]$ with the supremum metric topology (see Example 1.3(d)). Show that $\operatorname{int} A = \emptyset$.

Problem 2.4 *
Suppose that X is a topological space and $C \subseteq X$ be a closed set. Show that C is nowhere dense in X if and only if $X \setminus C$ is dense in X.

Problem 2.5 *
Suppose that X is a topological space, $A \subseteq X$ is a set furnished with the subspace topology, $C \subseteq A$ and let $\overline{C}^A$ (respectively, $\overline{C}$) denote the closure of C as a subspace of A (respectively, of X). Similarly let $\operatorname{int}_A C$ (respectively, $\operatorname{int} C$) be the interior of C as a subspace of A (respectively, of X). Show that:
(a) $\overline{C}^A = \overline{C} \cap A$.
(b) $\operatorname{int} C \subseteq \operatorname{int}_A C$ and the inclusion can be strict.

Problem 2.6 *
Suppose that X is a topological space and A, C, D are three subsets of X such that $D \subseteq A \cap C$. Show that if D is open (respectively, closed) in both A and C (with their respective subspace topologies), then D is also open (respectively, closed) in $A \cup C$ (with the subspace topology).

Problem 2.7 *
Let X be a topological space and let $C \subseteq X$ be a nonempty subspace. Show that C is closed if and only if every $x \in \overline{C}$ has a neighbourhood U such that $C \cap U$ is closed in U.

Problem 2.8 *
Let X be a topological space and let $C, D \subseteq X$ be two nonempty subsets. Show that if there exists $E \subseteq X$ such that $E \setminus D$ is closed in X and contains $C \setminus D$, then we have

$$\overline{\overline{C} \cap D} = \overline{C \cap D}.$$

Problem 2.9 *
Let X be a topological space with a countable subbasis. Show that X is second countable.

Problem 2.10 *
Show that a closed subspace of a Lindelöf space is also a Lindelöf space.

Problem 2.11 **
Suppose that X is a topological space and $\mathcal{B}$ is a basis for the topology. Show that there exists a dense subset $D \subseteq X$ such that $\#D \leqslant \#\mathcal{B}$ (where $\#$ denotes the cardinality of a set).

Problem 2.12 **
Suppose that X is a topological space and $\{U_i\}_{i \in I}$ is an open cover of X. Show that $C \subseteq X$ is closed if and only if $U_i \cap C$ is closed in U_i (with the subspace topology) for all $i \in I$.

Problem 2.13 *
Can $\mathbb{R}$ be the continuous image of $[0,1)$? Justify the answer.

Problem 2.14 *
Show that the uniform limit of a net of continuous $\mathbb{R}$-valued functions on a topological space X is continuous.

Problem 2.15 **
Suppose that X is a topological space and $f\colon X \longrightarrow \mathbb{R}^* = \mathbb{R} \cup \{\pm\infty\}$. Show that f is lower semicontinuous (respectively, sequentially lower semicontinuous) if and only if $\operatorname{epi} f = \big\{(x, \lambda) \in X \times \mathbb{R} : \ f(x) \leqslant \lambda\big\}$ is closed (respectively, sequentially closed).

Problem 2.16 ***

Suppose that X is a topological space, $f\colon X \longrightarrow \mathbb{R}^* = \mathbb{R} \cup \{\pm\infty\}$ is a function and $\overline{f}\colon X \longrightarrow \mathbb{R}^* = \mathbb{R} \cup \{\pm\infty\}$ is the relaxed function of f. Show that for every $\lambda \in \mathbb{R}$, we have

$$\{x \in X : \overline{f}(x) \leqslant \lambda\} = \bigcap_{\mu > \lambda} \overline{\{x \in X : f(x) \leqslant \mu\}}$$

and

$$\operatorname{epi} \overline{f} = \overline{\operatorname{epi} f} \quad \text{in } X \times \mathbb{R}.$$

Problem 2.17 **

Suppose that X is a topological space and $f\colon X \longrightarrow \mathbb{R}^* = \mathbb{R} \cup \{\pm\infty\}$ is a function. Let $\overline{f}$ be the relaxed function of f (see Definition 2.57) Corollary 2.56 guarantees that $\overline{f}$ is lower semicontinuous. Show that

$$\overline{f}(x) = \sup_{U \in \mathcal{N}(x)} \inf_{y \in U} f(y) \qquad \forall\, x \in X.$$

Problem 2.18 **

Suppose that X is a first countable topological space, $f\colon X \longrightarrow \overline{\mathbb{R}} = \mathbb{R} \cup \{+\infty\}$ is a function and $\overline{f}\colon X \longrightarrow \mathbb{R}^* = \mathbb{R} \cup \{\pm\infty\}$ is the relaxed function of f. Show that for every $x \in X$, $\overline{f}(x)$ is characterized by the following two properties:

(a) For every sequence $x_n \longrightarrow x$, we have $\overline{f}(x) \leqslant \liminf_{n \to +\infty} f(x_n)$.

(b) There exists a sequence $y_n \longrightarrow x$ in X such that $\liminf_{n \to +\infty} f(y_n) \leqslant \overline{f}(x)$.

Problem 2.19 ***

Suppose that (X, d_X) is a metric space, $f\colon X \longrightarrow \overline{\mathbb{R}}_+ = [0, +\infty]$ is a function and $\overline{f}\colon X \longrightarrow \mathbb{R}^* = \mathbb{R} \cup \{\pm\infty\}$ is the relaxed function of f. For every $\lambda > 0$, we define

$$f_\lambda(x) \stackrel{\text{def}}{=} \inf_{y \in X} \big(f(y) + \lambda d_X(x, y)\big).$$

Show that:

(a) f_λ is λ-Lipschitz;

(b) $\lim_{\lambda \to +\infty} f_\lambda(x) = \overline{f}(x)$ for all $x \in X$.

Problem 2.20 ***
Let (X, d_X) be a metric space and let $f\colon X \longrightarrow \mathbb{R}^* = \mathbb{R} \cup \{\pm\infty\}$ be a function. Show that:
(a) f is lower semicontinuous and bounded below if and only if there is a sequence $\{f_n\}_{n\geqslant 1}$ of bounded continuous functions such that $f_n \nearrow f$ and $n \to +\infty$.
(b) f is upper semicontinuous and bounded above if and only if there is a sequence $\{f_n\}_{n\geqslant 1}$ of bounded continuous functions such that $f_n \searrow f$ and $n \to +\infty$.

Problem 2.21 **
Is the pointwise limit of upper semicontinuous functions, an upper semicontinuous function too? How about the uniform limit? Justify your answer.

Problem 2.22 *
Suppose that X and Y are two topological spaces, $D \subseteq X$ is a dense subset and $f, g\colon X \longrightarrow Y$ are two continuous functions such that $f(x) = g(x)$ for all $x \in D$. Show that $f(x) = g(x)$ for all $x \in X$.

Problem 2.23 *
Suppose that X is a topological space, (Y, d_Y) is a metric space, $f\colon X \longrightarrow Y$ is a function and ω_f is the oscillation function of f (see Definition 2.42). Show that f is continuous at x if and only if $\omega_f(x) = 0$ (this problem extends to general topological spaces Problem 1.46).

Problem 2.24 **
Suppose that X is a topological space, (Y, d_Y) is a metric space, $f\colon X \longrightarrow Y$ is a function and $\operatorname{cont} f = \{x \in X : f \text{ is continuous at } x\}$. Show that $\operatorname{cont} f$ is a G_δ-set in X.

Problem 2.25 *
Suppose that X is a topological space and $f\colon X \longrightarrow \mathbb{R}$ is a function. Show that f is continuous if and only if for every $\lambda \in \mathbb{R}$, the sets $\{x \in X : f(x) \geqslant \lambda\}$ and $\{x \in X : f(x) > \lambda\}$ are closed and open respectively.

Problem 2.26 **
Suppose that X is a topological space and $\{f_n\colon X \longrightarrow \mathbb{R}\}_{n\geqslant 1}$ is a sequence of continuous functions such that $f_n(x) \longrightarrow f(x)$ for all $x \in X$. Show that the following two statements are equivalent:

(a) f is continuous on X;
(b) For every $\varepsilon > 0$ and $k \geqslant 1$, we can find $n \geqslant k$ such that the set $\{x \in X : |f(x) - f_n(x)| < \varepsilon\}$ is open in X.

Problem 2.27 **
Let X be a normal space and let $\mathcal{Y}$ be a locally finite open cover of X. Show that X has a locally finite open cover $\mathcal{D}_*$ such that $\{\overline{V} : V \in \mathcal{D}_*\}$ is a refinement of $\mathcal{Y}$.

Problem 2.28 *
Let X be a topological space and let $\mathcal{Y} = \{U_\alpha\}_{\alpha \in J}$ be an open cover of X which is locally finite. Show that for any $K \subseteq J$, the set $\bigcup_{x \in K} \overline{U}_\alpha$ is closed.

Problem 2.29 **
Suppose that X, Y, Z are three topological spaces and $f\colon X \longrightarrow Y$, $g\colon Y \longrightarrow Z$ are two functions. Show that:
(a) If f, g are open (respectively, closed), then $g \circ f$ is open (respectively, closed) too.
(b) If $g \circ f$ is open (respectively, closed) and g is bijective and continuous, then f is open (respectively, closed).
(c) If $g \circ f$ is open (respectively, closed) and f is surjective and continuous, then g is open (respectively, closed).

Problem 2.30 *
Suppose that X is a topological space, $A \subseteq X$ is a set and $i_A\colon A \longrightarrow X$ is the canonical injection function (i.e., $i_A(x) = x$ for all $x \in A$). Show that i_A is open (respectively, closed) if and only if A is open (respectively, closed).

Problem 2.31 *
Suppose that X and Y are two topological spaces and $f\colon X \longrightarrow Y$ is a surjection. Show that f is a homeomorphism if and only if

$$\overline{A} = f^{-1}\big(\overline{f(A)}\big) \qquad \forall\, A \subseteq X.$$

Problem 2.32 **
Suppose that X and Y are two topological spaces, X is normal and $f\colon X \longrightarrow Y$ is a continuous closed surjection. Show that Y is normal too.

Problem 2.33 ***

Suppose that X and Y are two topological spaces, Y is second countable and $E \subseteq X \times Y$ is a nowhere dense set. For every $x \in X$, we set $E_x \stackrel{\text{def}}{=} \{y \in Y : (x, y) \in E\}$. Show that E_x is nowhere dense for all $x \in X$ except on a set of first category.

Remark. The result of this problem is known in the literature as the ***Kuratowski–Ulam theorem*** and can be found in Kuratowski [11, p. 249] and Oxtoby [16, p. 56] The result can be viewed as a topological analogue of the Fubini theorem (see Chap. 3).

Problem 2.34 **

Suppose that X and Y are two topological spaces with Y being second countable. Let $C \subseteq X$ and $D \subseteq Y$ be nonempty sets. Show that $C \times D$ is of first category in $X \times Y$ if and only if at least one of the sets C or D is of first category.

Problem 2.35 *

Suppose that X and Y are two topological spaces and $f: X \longrightarrow Y$ is a continuous surjection. Show that if f is open or closed, then f is an identification function.

Problem 2.36 **

Suppose that X and Y are two topological spaces and $f: X \longrightarrow Y$ is a function. Show that f is continuous and closed if and only if the projections $\text{proj}_X : \text{Gr}\, f \longrightarrow X$ and $\text{proj}_Y : \text{Gr}\, f \longrightarrow Y$ are closed functions (recall that $\text{Gr}\, f = \{(x, y) \in X \times Y : y = f(x)\}$ and on it we consider the subspace product topology).

Problem 2.37 **

Suppose that X and Y are two topological spaces and $A \subseteq X$, $C \subseteq Y$ are two nonempty sets. Show that $A \times C$ is closed (respectively, open, dense) if and only if both sets A and C are closed (respectively, open, dense).

Problem 2.38 ***

Suppose that I is an uncountable set and $\{(X_i, d_{X_i})\}_{i \in I}$ is a family of metric spaces. Assume that for each $i \in I$, the metric space (X_i, d_{X_i}) is nontrivial (i.e., X_i is not a singleton). Consider the set $X \stackrel{\text{def}}{=} \prod_{i \in I} X_i$ with the product topology. Show that X is not metrizable.

Problem 2.39 ⋆
Suppose that (X, τ_X) and (Y, τ_Y) are two topological spaces and $f: X \longrightarrow Y$ is a continuous open function. Show that
(a) if f is a surjection, then $\tau_Y = s(f)$.
(b) if f is an injection, then $\tau_X = w(f)$.

Problem 2.40 ⋆
Suppose that X and Y are two topological spaces and $f: X \longrightarrow Y$ is a bijection. Suppose that either the topology of X is $w(f)$ or the topology of Y is $s(f)$. Show that f is a homeomorphism.

Problem 2.41 ⋆
Suppose that X is a topological space, Δ is an equivalence relation on X and $p: X \longrightarrow X/_\Delta$ is the quotient function. We say that Δ is ***open*** (respectively, ***closed***) if for every open (respectively, closed) set A, its p-saturation $p^{-1}(p(A))$ (see Definition 2.74) is open (respectively, closed). Show that Δ is open (respectively, closed) if and only if the quotient function is open (respectively, closed).

Problem 2.42 ⋆⋆⋆
A topological space X is said to be ***completely regular*** if for every nonempty closed set $C \subseteq X$ such that $x \notin C$, we can find a continuous function $f: X \longrightarrow [0,1]$ such that $f(x) = 0$ and $f|_C = 1$. Let $C(X)$ (respectively, $C_b(X)$) be the space of all continuous (respectively, bounded continuous) functions $f: X \longrightarrow \mathbb{R}$. Show that a topological space (X, τ) is completely regular if and only if $\tau = w\big(C(X)\big) = w\big(C_b(X)\big)$.

Problem 2.43 ⋆⋆
Let K be a nonempty compact set in a completely regular topological space X (see Problem 2.42) and let U be an open set containing K. Show that there exists a continuous function $f: X \longrightarrow [0,1]$ such that $f|_K = 1$ and $f|_{X \setminus U} = 0$.

Problem 2.44 ⋆⋆
Let X be a topological space and let $f: X \longrightarrow \mathbb{R}^* = \mathbb{R} \cup \{\pm\infty\}$ be a function. Suppose that f is coercive and lower semicontinuous (respectively, sequentially coercive and sequentially lower semicontinuous). Show that:
(a) There exists $x_0 \in X$ such that $f(x_0) = \inf\limits_X f$.

(b) If $\{x_n\}_{n\geqslant 1} \subseteq X$ is a minimizing sequence for f (i.e., $f(x_n) \longrightarrow \inf\limits_X f$) and $\hat{x}$ is a limit point of $\{x_n\}_{n\geqslant 1}$ (respectively, $\hat{x}$ is the limit point of a subsequence of $\{x_n\}_{n\geqslant 1}$), then $\hat{x}$ is a minimizer of f.
(c) If f is not identically $+\infty$, then every minimizing sequence for f has a limit point (respectively, convergent subsequence).

Problem 2.45 **
Is a convergent net necessarily compact? Justify the answer.

Problem 2.46 **
Suppose that X is a topological space and assume that $\{f_n \colon X \longrightarrow \mathbb{R}^* = \mathbb{R} \cup \{\pm\infty\}\}_{n\geqslant 1}$ is a sequence of functions. We say that the sequence $\{f_n\}_{n\geqslant 1}$ is ***equicoercive*** if for every $\lambda \in \mathbb{R}$, there exists a closed, countably compact set K_λ such that

$$\bigcup_{n\geqslant 1} \{x \in X : \ f_n(x) \leqslant \lambda\} \ \subseteq \ K_\lambda.$$

Show that the sequence $\{f_n\}_{n\geqslant 1}$ is equicoercive if and only if there exists a lower semicontinuous, coercive (see Definition 2.103) function $\psi \colon X \longrightarrow \mathbb{R}^* = \mathbb{R} \cup \{\pm\infty\}$ such that

$$\psi(x) \ \leqslant \ f_n(x) \qquad \forall\, x \in X,\ n \geqslant 1.$$

Problem 2.47 **
Suppose that X is a topological space, $f \colon X \longrightarrow \overline{\mathbb{R}} = \mathbb{R} \cup \{+\infty\}$ is a coercive function. Show that the relaxed function $\overline{f}$ is coercive too.

Problem 2.48 **
Suppose that X is a topological space, $f \colon X \longrightarrow \mathbb{R}^* = \mathbb{R} \cup \{\pm\infty\}$ is a coercive function and $\overline{f} \colon X \longrightarrow \mathbb{R}^*$ is the relaxed function of f. Show the following statements:
(a) $\overline{f}$ admits a minimizer in X and $\min\limits_X \overline{f} = \inf\limits_X f$.
(b) Every limit point of a minimizing sequence for f is a minimizer for $\overline{f}$.
(c) If X is first countable, then every minimizer of $\overline{f}$ is the limit point of a minimizing sequence for f.

Problem 2.49 *
Suppose that X and Y are two topological spaces with Y being compact and C is a closed subset of $X \times Y$. Show that $\mathrm{proj}_X(C)$ is closed

in X (here $\mathrm{proj}_X : X \times Y \longrightarrow X$ is the natural projection on the first factor X).

Problem 2.50 *
Show that a topological space X is compact if and only if every open cover consisting of basic open sets has a finite subcover.

Problem 2.51 *
Suppose that X and Y are two topological spaces and $f\colon X \longrightarrow Y$ is a continuous surjection. Show that if X is compact, then f is an identification function.

Problem 2.52 **
Suppose that X and Y are two topological spaces and $f\colon X \longrightarrow Y$ is a continuous function. On X we introduce the equivalence relation Δ, defined by

$$x \,\Delta\, u \quad \Longleftrightarrow \quad f(x) = f(u)$$

Let $\hat{X} = X/_\Delta$. Show that if X is compact, then $\hat{X}$ is homeomorphic to $f(X)$.

Problem 2.53 **

(a) Show that two comparable compact Hausdorff topologies on a set X are equal.
(b) Give an example of two topologies on a set which are both compact Hausdorff but not comparable.

Problem 2.54 *
Suppose that X is a topological space, Y is a compact topological space and $f\colon X \longrightarrow Y$ is a function. Show that f is continuous if and only if $\mathrm{Gr}\, f \subseteq X \times Y$ is closed (see Definition 1.132). Is any of the implications false if we do not assume the compactness of Y?

Problem 2.55 **
Let X be a topological space. Show that a locally compact dense subset of X is open.

Problem 2.56 **

(a) Suppose that X is a compact topological space and assume that $\{f_i\colon X \longrightarrow \mathbb{R}\}_{i\in I}$ is a net of continuous functions converging to a continuous function $f\colon X \longrightarrow \mathbb{R}$. Suppose that the net $\{f_i\}_{i\in I}$ is monotone (increasing or decreasing). Show that $f_i \rightrightarrows f$ (this is the topological version of the so called ***Dini theorem***; see Problem 1.104).
(b) Show that the above result fails if we drop the assumption on the compactness of the space X.
(c) Show that the above result fails if we drop the assumption on the continuity of the limit function f.

Problem 2.57 **

Show that, if $I = [0,1]$, then $I/_{(0\sim 1)}$ is homeomorphic to S^1, where $S^1 = \{x \in \mathbb{R}^2 : \ \|x\| = 1\}$.

Problem 2.58 **

Show that the Alexandrov one-point compactification of $\mathbb{R}^N$ is homeomorphic to $S^N = \{x \in \mathbb{R}^{N+1} : \ \|x\| = 1\}$.

Problem 2.59 *

Suppose that X is a topological space and $\{x_n\}_{n\geqslant 1}$ is a sequence in X such that $x_n \longrightarrow x$ in X. Show that the set $K = \{x_n : \ n \geqslant 1\} \cup \{x\}$ is compact in X. (Compare with Problem 2.45.)

Problem 2.60 **

Suppose that X is a topological space and $C, D \subseteq X$ are two nonempty, compact and disjoint sets. Show that we can find two open sets $U, V \subseteq X$ such that $C \subseteq U$, $D \subseteq V$ and $U \cap V = \emptyset$.

Problem 2.61 **

Suppose that X is a topological space, $K \subseteq X$ is a nonempty compact set and $U_1, U_2 \subseteq X$ are two nonempty open sets such that $K \subseteq U_1 \cup U_2$. Show that we can find two compact sets $K_1, K_2 \subseteq X$ such that $K_1 \subseteq U_1$, $K_2 \subseteq U_2$ and $K = K_1 \cup K_2$.

Problem 2.62 *

Suppose that X and Y are two topological spaces and $f\colon X \longrightarrow Y$ is a function with a closed graph $\operatorname{Gr} f$. Show that for every compact set $K \subseteq Y$, the set $f^{-1}(K)$ is closed in X.

Problem 2.63 **
Suppose that X and Y are two topological spaces and $f: X \longrightarrow Y$ is a closed function such that for every $y \in Y$, the set $f^{-1}(y) \subseteq X$ is compact. Show that for every compact set $K \subseteq Y$, the set $f^{-1}(K) \subseteq X$ is compact.

Problem 2.64 **
Suppose that X is a compact topological space, $f: X \longrightarrow \mathbb{R}$ is a function and for every $\lambda \in \mathbb{R}$, $f^{-1}([\lambda, +\infty))$ is closed. Show that we can find $x_0 \in X$ such that $f(x_0) = \sup_X f < +\infty$.

Problem 2.65 **
Suppose that X and Y are two topological spaces, with X being compact and $p_Y: X \times Y \longrightarrow Y$ is the canonical projection on the second factor Y. Show that p_Y is continuous, closed and for all $y \in Y$, the set $p_Y^{-1}(y)$ is compact.

Problem 2.66 **
Suppose that X is a Lindelöf topological space in which every sequence has a limit point. Show that X is compact.

Problem 2.67 **
Suppose that I is an infinite index set and $\{X_i\}_{i \in I}$ is a family of topological spaces. Assume that an infinite number of these spaces are noncompact and let $K \subseteq \prod_{i \in I} X_i$ be a compact set (for the product topology see Definition 2.69). Show that $\operatorname{int} K = \emptyset$ (i.e., K is nowhere dense).

Problem 2.68 *
A topological space X is said to be a ***k-space***, if the following is true: "$A \subseteq X$ is closed if and only if for every compact set $K \subseteq X$, the set $A \cap K$ is closed". Suppose that X and Y are topological spaces, with X being a k-space and $f: X \longrightarrow Y$ is a function such that for every compact set $K \subseteq X$, $f|_K$ is continuous. Show that f is continuous.

Problem 2.69 **
Suppose that X is a locally compact topological space and $A \subseteq X$. Show that if A is open or closed in X, then A with the subspace topology is locally compact.

Problem 2.70 **
Suppose that X is a compact topological space and $x \in X$. Show that $X \setminus \{x\}$ is locally compact.

Problem 2.71 ***
Let K be a compact, convex subset of $\mathbb{R}^N$ with nonempty interior. Show that K is homeomorphic to the closed unit ball $\overline{B}_1^N$ by homeomorphism which sends $S^{N-1} = \partial B_1^N$ to ∂K.

Problem 2.72 **
Suppose that X and Y are two topological spaces with Y being locally compact and $f\colon X \longrightarrow Y$ is a continuous bijection which returns compact sets in Y to compact sets in X (i.e., if $K \subseteq Y$ is compact, then $f^{-1}(K) \subseteq X$ is compact). Show that f is closed.

Problem 2.73 **
Suppose that X is a compact topological space and there is a countable separating family Φ of continuous functions from X into a metric space (Y, d_Y). Show that the topology of X is metrizable.

Problem 2.74 **
Suppose that X is a locally compact topological space and X^* is the Alexander one-point compactification of X. Show that X is σ-compact if and only if the point $\infty \in X^*$ has a countable local basis.

Problem 2.75 **

(a) Suppose that X is a compact topological space and $f\colon X \longrightarrow X$ is a continuous function. Show that there exists a nonempty and closed subset $A \subseteq X$ such that $f(A) = A$.
(b) Show that the result fails if X is not compact.

Problem 2.76 **
Suppose that X is a compact topological space, $f, g \in C(X)$, $f(x) \geqslant 0$ for all $x \in X$ and if $f(x) = 0$, then $g(x) > 0$. Show that there exists $\lambda > 0$ such that $\lambda f(x) + g(x) > 0$ for all $x \in X$.

Problem 2.77 *
Let X be a topological space and let $\{C_i\}_{i\in I}$ be a family of compact subsets of X. Let $U \subseteq X$ be an open set such that $\bigcap\limits_{i\in I} C_i \subseteq U$. Show that there exists a finite set $F \subseteq I$ such that $\bigcap\limits_{i\in F} C_i \subseteq U$.

Problem 2.78 *
Let X be a topological space and let $f\colon X \longrightarrow \mathbb{R}$ be a functional. We say that f is ***locally bounded*** if for each $x \in X$, there exist $U \in \mathcal{N}(u)$ and $M = M(U) > 0$ such that

$$\left|f(x)\right| \leqslant M \qquad \forall\, x \in U.$$

Show that, if X is compact, then every locally bounded function is bounded.

Problem 2.79 *
Let D be a locally compact dense subset of a metrizable space X. Show that D is open in X.

Problem 2.80 **
Show that every locally compact metrizable space X is completely metrizable.

Problem 2.81 **
Is the circle $S^1 = \big\{x \in \mathbb{R}^2 : \; \|x\| = 1\big\}$ homeomorphic to the closed interval $[0,1]$?

Problem 2.82 *
Suppose that X is a topological space and $K \subseteq X$ is a nonempty, compact and disconnected set. Show that we can find two disjoint open sets $U, V \subseteq X$ such that $K \subseteq U \cup V$, $K \cap U \neq \emptyset$ and $K \cap V \neq \emptyset$.

Problem 2.83 **
Suppose that X is a topological space and $\{K_n\}_{n\geqslant 1}$ is a decreasing sequence of compact sets in X. Let $K = \bigcap\limits_{n\geqslant 1} K_n$. Show the following:

(a) $K \subseteq X$ is nonempty and compact.
(b) If U is open and $K \subseteq U$, then we can find $n_0 \geqslant 1$ such that $K_n \subseteq U$ for all $n \geqslant n_0$.
(c) If for every $n \geqslant 1$, the set K_n is also connected, then K is connected too.

Problem 2.84 *
Is it true that, if X is a topological space and $\{C_n\}_{n\geqslant 1}$ is a decreasing sequence of nonempty closed connected sets in X, then $\bigcap\limits_{n\geqslant 1} C_n$ is connected too? Justify your answer.

Problem 2.85 ⋆⋆
Let $f\colon [0,1] \longrightarrow A \times C$ be a homeomorphism. Show that one of A and C is a singleton.

Problem 2.86 ⋆⋆
(a) Let $U \subseteq \mathbb{R}^N$ be an open and connected set. Show that U is path-connected.
(b) Give an example showing that the above is no longer true if we replace $\mathbb{R}^N$ by any topological space.

Problem 2.87 ⋆⋆
Are the intervals (x,u) and $[y,z)$ in $\mathbb{R}$ homeomorphic? Justify jour answer.

Problem 2.88 ⋆⋆⋆
Suppose that X is a topological space, $\{A_i\}_{i\in I}$ is a family of connected subsets of X and for every pair $(i,j) \in I \times I$, there exists a finite family $\{i_k\}_{k=0}^n \subseteq I$ such that $i_0 = i$ and $i_n = j$ and

$$A_{i_{k-1}} \cap A_{i_k} \neq \emptyset \qquad \forall\, k \in \{1,\ldots,n\}.$$

Show that the set $\bigcup_{i\in I} A_i$ is connected.

Problem 2.89 ⋆⋆⋆
Suppose that X and Y are two topological spaces with X being compact and $f\colon X \longrightarrow Y$ is a continuous function. Suppose that for every $y \in Y$, the set $f^{-1}(y) \subseteq X$ is connected. Show that the inverse image of every connected set in Y is connected in X.

Problem 2.90 ⋆
Suppose that X is a topological space, $A \subseteq X$ is a nonempty set and $C \subseteq X$ is a connected set such that $C \cap A \neq \emptyset$ and $C \cap (X \setminus A) \neq \emptyset$. Show that $C \cap \partial A \neq \emptyset$.

Problem 2.91 ⋆⋆
Let X and Y be two topological spaces such that X is path-connected and Y is not. Show that there can be no continuous surjection $f\colon X \longrightarrow Y$.

Problem 2.92 **
Show that for every $N \geqslant 1$, the sphere $S^N \stackrel{\text{def}}{=} \{x \in \mathbb{R}^{N+1} : \|x\| = 1\}$ is path-connected.

Problem 2.93 *
Suppose that X and Y are two topological spaces and $f\colon X \longrightarrow Y$ is an open, continuous surjection. Show that, if X is locally connected (respectively, locally compact), then so is Y.

Problem 2.94 *
Assume that X is a disconnected topological space and assume that $X = C \cup D$, where both C and D are nonempty and connected. Show that C and D are the connected components of X.

Problem 2.95 **
Suppose that X is a topological space and C and D are two nonempty connected subsets of X such that $\overline{C} \cap D \neq \emptyset$. Show that $C \cup D$ is connected.

Problem 2.96 **
Let X be a connected topological space and let $A \subseteq X$ be a nonempty subset. Show that, if ∂A is connected, then $\overline{A}$ is connected too.

Problem 2.97 *
Suppose that X is a topological space, $A \subseteq X$ is a nonempty subset, $x \in A$, $u \in X \setminus A$ and $\gamma\colon [0,1] \longrightarrow X$ is a path such that $\gamma(0) = x$ and $\gamma(1) = u$. Show that $\gamma([0,1]) \cap \partial A \neq \emptyset$.

Problem 2.98 *
Consider the following subsets of $\mathbb{R}^2$:
(a) the set of points with both coordinates rational;
(b) the set of points with at least one coordinate rational;
(c) the set of points with coordinates which are either both rational or both irrational.
Determine which of the above sets is connected. Justify your answer.

Problem 2.99 *
Suppose that X is a connected normal topological space. Show that X is either a singleton or uncountable.

Problem 2.100 **
Suppose that X is a normal topological space and $\mathcal{Y}$ is locally finite open cover of X. Show that for every $U \in \mathcal{Y}$, we can find a continuous function $f_U \colon X \longrightarrow [0,1]$ such that $f|_{X \setminus U} = 0$ and $\sum\limits_{U \in \mathcal{Y}} f_U(x) = 1$ for all $x \in X$.

Problem 2.101 *
Suppose that X is a normal topological space and A, C are two nonempty, disjoint closed subsets in X. Show that there exist open sets U and V such that $A \subseteq U$, $C \subseteq V$ and $\overline{U} \cap \overline{V} = \emptyset$.

Problem 2.102 ***
Suppose that X is a normal topological space and $A \subseteq X$ is a nonempty closed set. Show the following:
(a) There exists a continuous function $f \colon X \longrightarrow [0,1]$ such that

$$A = f^{-1}(\{0\}) \quad \Longleftrightarrow \quad A \text{ is a } G_\delta\text{-subset of } X.$$

(b) If A is also a G_δ-set and $C \subseteq X$ is nonempty, closed and disjoint from A, then we can find a continuous function $f \colon X \longrightarrow [0,1]$ such that $A = f^{-1}(\{0\})$ and $f|_C = 1$.

Problem 2.103 **
Suppose that X is a locally compact topological space, $K \subseteq X$ is a compact set and $U \subseteq X$ is an open set such that $K \subseteq U$. Show that there is a continuous function $f \colon X \longrightarrow [0,1]$ such that $f|_K = 0$ and $f_{X \setminus U} = 1$.

Problem 2.104 **
Suppose that X is a locally compact topological space and $K \subseteq X$ is a nonempty, compact set. Show that K is a G_δ-set if and only if there exists a continuous function $f \colon X \longrightarrow [0,1]$ such that $K = f^{-1}(\{0\})$.

Problem 2.105 **
Show that a normal topological space X is perfectly normal (see Definition 2.137) if and only if every closed subset of X is a G_δ-set.

Problem 2.106 **
Suppose that X is a second countable, regular topological space and $U \subseteq X$ is a nonempty and open set. Show that there exists a continuous function $f \colon X \longrightarrow [0,1]$ such that $f|_U > 0$ and $f_{X \setminus U} = 0$.

Problem 2.107 **

Let X be a locally compact topological space. Show that X is second countable if and only if X is separable and metrizable.

Problem 2.108 **

Let X be a locally compact and paracompact topological space and let $\mathcal{Y}$ be a locally finite open cover of X. Show that X has a locally finite open cover $\mathcal{D}$ such that for every $V \in \mathcal{D}$, the set $\overline{V}$ is compact and $\{\overline{V} : V \in \mathcal{D}\}$ is a refinement of $\mathcal{Y}$.

Problem 2.109 **

Let X be a topological space in which the union of every countable family of closed nowhere dense sets is nowhere dense. Show that X is a Baire space. Is the converse true? Justify your answer.

Problem 2.110 **

Suppose that X is a Baire space and $\{A_n\}_{n \geqslant 1}$ is a sequence of nowhere dense sets. Show that $\bigcup_{n \geqslant 1} A_n \neq X$.

Problem 2.111 *

Suppose that X is a Baire space and $D_1, D_2 \subseteq X$ are two dense subsets. Is $D = D_1 \cap D_2$ necessarily dense in X? Justify your answer.

Problem 2.112 **

Suppose that X is a topological space in which every point has an open neighbourhood, which is a Baire space (with the subspace topology). Show that X is a Baire space.

Problem 2.113 **

Let X be a Baire topological space and let $f\colon X \longrightarrow \mathbb{R}^* = \mathbb{R} \cup \{\pm\infty\}$ be a lower semicontinuous function. Suppose that the set $\{x \in X : f(x) < +\infty\}$ is not of the first category. Show that we can find an open set $U \subseteq X$ and $M > 0$ such that $f(x) \leqslant M$ for all $x \in U$.

Problem 2.114 **

Suppose that X is a Baire topological space and $\{C_n\}_{n \geqslant 1}$ is a sequence of closed subsets of X such that $X = \bigcup_{n \geqslant 1} C_n$. Show that the set $\bigcup_{n \geqslant 1} \operatorname{int} C_n$ is dense in X.

Problem 2.115 **

Suppose that X is a Baire topological space, Y is a separable metric space and $f\colon X \longrightarrow Y$ is a function such that the inverse image of every open set is an F_σ-set. Show that f is continuous at every point of a dense G_δ-set.

Problem 2.116 **

Suppose that X is a Baire topological space in which every open set is F_δ (for example, a complete metric space; see Theorem 1.26 or a perfectly normal Baire space; see Problem 2.105). Show that a lower semicontinuous (or an upper semicontinuous) function $f\colon X \longrightarrow \mathbb{R}$ is continuous at every point of a dense G_δ-set.

Problem 2.117 **

Show that a closed subset C of a paracompact space X is paracompact.

Remark. In fact something more general is true. Every F_σ-subset of a paracompact space is paracompact (see Proposition 2.145(b)).

Problem 2.118 **

Let X be a Baire space. Show the following:

(a) Every open subset of X with the subspace topology is Baire too.
(b) If $\sim$ is an equivalence relation, $X/_\sim$ is the quotient space with the quotient topology and if $\hat{U} \subseteq X/_\sim$ is open, then $\hat{U}$ is Baire too.

Problem 2.119 **

Let X be a separable Baire space and let $U \subseteq X$ be a nonempty open set. Consider a function $f\colon U\times X \longrightarrow \mathbb{R}$ such that for every $u \in X$, the function $x \longmapsto f(x,u)$ is lower semicontinuous and for every $x \in U$, the function $u \longmapsto f(x,u)$ is continuous. Show that there exists a dense subset D of U such that for all $u \in X$, the function $x \longmapsto f(x,u)$ is continuous on D.

Problem 2.120 **

Suppose that X is a paracompact space and $C, D \subseteq X$ are two disjoint closed sets. Assume that the set C is covered by open sets $\{U_\alpha\}_{\alpha\in I}$ with $\overline{U}_\alpha \cap D = \emptyset$. Show that the sets C and D have disjoint neighbourhoods.

Problem 2.121 **
Let X be a normal space and let $\mathcal{Y}$ be a locally finite open cover of X. Show that there exists a partition of unity $\{\psi_i\}_{i \in I}$ subordinate to $\mathcal{Y}$.

Problem 2.122 *
Show that every open or closed subset of a Lusin space is itself Lusin.

Problem 2.123 *
Let $\{X_n\}_{n \geqslant 1}$ be a sequence of Lusin spaces. Show that the product space $X = \prod\limits_{n \geqslant 1} X_n$ is a Lusin space too.

Problem 2.124 **
Show that a Souslin space is dispersible.

Problem 2.125 **
Suppose that X is a strongly Lindelöf topological space and assume that $\{f_i \colon X \longrightarrow \mathbb{R}\}_{i \in I}$ is a family of lower semicontinuous functions. Show that there is a countable subfamily $J \subseteq I$ such that $f = \sup\limits_{i \in I} f_i = \sup\limits_{j \in J} f_j$.

Problem 2.126 **
Let X be a regular strongly Lindelöf topological space. Show that every open set in X is a F_σ-set and every closed set in X is a G_δ-set.

Problem 2.127 **
Suppose that X and Y are two topological spaces and $X \times X$ is strongly Lindelöf. Suppose that $\{f_i \colon X \longrightarrow Y\}_{i \in I}$ is a separating family of continuous functions. Show that there is a countable subset $J \subseteq I$ such that the sequence $\{f_j \colon X \longrightarrow Y\}_{j \in J}$ is separating too.

Problem 2.128 **
Suppose that X is a Souslin space, Y is a topological space and $f \colon X \longrightarrow Y$ is a function such that $\operatorname{Gr} f$ is a Souslin subspace of $X \times Y$. Show that the inverse image of every Souslin subspace of Y is a Souslin subspace of X.

Problem 2.129 **
Suppose that X and Y is two topological spaces with Y being regular and $F \colon X \longrightarrow 2^Y \setminus \{\emptyset\}$ is a multifunction which is upper semicontinuous. Show that $\operatorname{Gr} F$ is closed in $X \times Y$. Is the converse true? Justify your answer.

Problem 2.130 ***
Let X and Y be two topological spaces. Show that a multifunction $F\colon X \longrightarrow 2^Y \setminus \{\emptyset\}$ is compact valued (i.e., $F(x) \subseteq Y$ is a compact set for every $x \in X$) and upper semicontinuous if and only if for every net $\{(x_i, y_i)\}_{i\in I} \subseteq \operatorname{Gr} F$ such that $x_i \longrightarrow x$ in X, $\{y_i\}_{i\in I}$ has a limit point in $F(x)$.

Problem 2.131 **
Suppose that X and Y are two topological spaces and $F\colon X \longrightarrow 2^Y \setminus \{\emptyset\}$ is a compact valued multifunction (i.e., $F(x) \subseteq Y$ is compact for every $x \in X$) which is upper semicontinuous. Show that:
(a) for every compact set $K \subseteq X$, the set $F(K) \subseteq Y$ is compact;
(b) $\operatorname{Gr} F \subseteq X \times Y$ is closed (note that no regularity on Y is assumed; cf. Problem 2.129).

Problem 2.132 ***
Suppose that X is a paracompact topological space and $f\colon X \longrightarrow \mathbb{R}$ is an upper semicontinuous function, $g\colon X \longrightarrow \mathbb{R}$ a lower semicontinuous function such that

$$f(x) < g(x) \qquad \forall\, x \in X.$$

Show that there exists a continuous function $h\colon X \longrightarrow \mathbb{R}$ such that

$$f(x) < h(x) < g(x) \qquad \forall\, x \in X.$$

Problem 2.133 **
Suppose that X is a paracompact space, Y is a Banach space and $F\colon X \longrightarrow 2^Y \setminus \{\emptyset\}$ is a lower semicontinuous function with closed and convex values. Suppose that $(\hat{x}, \hat{y}) \in \operatorname{Gr} F$. Show that we can find a continuous function $g\colon X \longrightarrow Y$ such that $g(x) \in F(x)$ for all $x \in X$ and $g(\hat{x}) = \hat{y}$.

Problem 2.134 **
Suppose that X is a paracompact space, Y is a Banach space, $C \subseteq X$ is a nonempty, closed set and $F\colon X \longrightarrow 2^Y \setminus \{\emptyset\}$ is a lower semicontinuous multifunction with closed, convex values. Show that every continuous function $f\colon C \longrightarrow Y$ such that $f(x) \in F(x)$ for all $x \in C$ can be extended to a continuous function $\hat{f}\colon X \longrightarrow Y$ such that $\hat{f}(x) \in F(x)$ for all $x \in X$.

Problem 2.135 **
Suppose that X is a compact topological space, Y is a topological vector space and $F\colon X \longrightarrow 2^Y \setminus \{\emptyset\}$ is a multifunction with convex values such that for every $y \in Y$, the set

$$F^-(\{y\}) \stackrel{\text{def}}{=} \{x \in X : \ y \in F(x)\}$$

is open. Show that there is a continuous function $f\colon X \longrightarrow Y$ such that $f(x) \in F(x)$ for all $x \in X$.

Problem 2.136 **
Suppose that X and Y are topological spaces and $F\colon X \longrightarrow 2^Y \setminus \{\emptyset\}$ is a multifunction which is lower semicontinuous or upper semicontinuous with connected values. Show that F maps connected sets to connected sets.

Problem 2.137 **
Suppose that X is a topological space and $f\colon X \longrightarrow \mathbb{R}$ is a function. Show the following:
(a) f is lower semicontinuous if and only if the multifunction

$$X \ni x \longmapsto L_f(x) = \{\lambda \in \mathbb{R} : \ f(x) \leqslant \lambda\} \in 2^{\mathbb{R}}$$

is upper semicontinuous;
(b) f is upper semicontinuous if and only if the multifunction

$$X \ni x \longmapsto L_f(x) \in 2^{\mathbb{R}}$$

is lower semicontinuous.

Problem 2.138 **
Suppose that X and Y are topological spaces, $f\colon X \times Y \longrightarrow \mathbb{R}^* = \mathbb{R} \cup \{\pm\infty\}$ is a lower semicontinuous function and $F\colon Y \longrightarrow 2^X \setminus \{\emptyset\}$ is a lower semicontinuous multifunction. Let $m(y) \stackrel{\text{def}}{=} \sup\limits_{x \in F(y)} f(x,y)$. Show that the function $m\colon Y \longrightarrow \mathbb{R} \cup \{+\infty\}$ is lower semicontinuous.

Problem 2.139 **
Suppose that X and Y are two topological spaces, $f\colon X \times Y \longrightarrow \mathbb{R}^* = \mathbb{R} \cup \{\pm\infty\}$ is an upper semicontinuous function and $F\colon Y \longrightarrow 2^X \setminus \{\emptyset\}$ is an upper semicontinuous multifunction with compact values. Let $m(y) \stackrel{\text{def}}{=} \sup\limits_{x \in F(y)} f(x,y)$. Show that the function $m\colon Y \longrightarrow \mathbb{R} \cup \{+\infty\}$ is upper semicontinuous.

Problem 2.140 ***
Suppose that X is a compact topological space and $F\colon X \longrightarrow 2^X \setminus \{\emptyset\}$ is an upper semicontinuous with compact values. Then there exists a compact set $C \subseteq X$ such that $F(C) = C$.

Problem 2.141 ***
Let (X, d_X) be a compact metric space and let (Y, d_Y) be a complete separable metric space (a Polish space). Show that $C(X;Y)$ with the c-topology (see Definition 2.174) is a Polish space.

Problem 2.142 ***
Let X be a topological space and let

$$P_k(X) \stackrel{\text{def}}{=} \big\{C \subseteq X : \ C \text{ is nonempty and compact}\big\}.$$

The topology on $P_k(X)$ generated by the sets of form $\big\{C \in P_k(X) : C \subseteq U\big\}$ and $\big\{C \in P_k(X) : \ C \cap U \neq \emptyset\big\}$, where $U \subseteq X$ is open, is called the ***Vietoris topology*** on $P_k(X)$. Show that
(a) The set of all finite nonempty sets in X is dense in $P_k(X)$ with the Vietoris topology.
(b) If X is separable, then so is $P_k(X)$ with the Vietoris topology.

Problem 2.143 ***
Let (X, d_X) be a bounded metric space and let h be the Hausdorff metric on $P_k(X)$ (see Definition 1.134). Show that the Hausdorff metric h induces the Vietoris topology on $P_k(X)$ (see Problem 2.142).

Problem 2.144 *
Suppose that X is a topological space and A is a retract of X. Show that A is closed. Also, if Y is another topological space and $f\colon A \longrightarrow Y$ is a continuous function, then show that f admits a continuous extension, i.e., there exists a continuous function $\hat{f}\colon X \longrightarrow Y$ such that $\hat{f}|_A = f$.

Problem 2.145 **
Show that a retract of a normal space is itself normal.

Problem 2.146 *
Show that $S^N = \partial B_1^{N+1} = \big\{x \in \mathbb{R}^{N+1} : \ \|x\| = 1\big\}$ is a strong deformation retract of $\mathbb{R}^{N+1} \setminus \{0\}$.

Problem 2.147 **

Show the following:

(a) A retract of a compact (respectively, (path-)connected) topological space is itself compact (respectively, (path-)connected).

(b) A retract of a simply connected topological space is itself simply connected.

(c) A retract of a retract is itself a retract (i.e., if $C \subseteq A \subseteq X$ and A is a retract of X and C is a retract of A, then C is a retract of X).

Problem 2.148 ***

Show that for $N \geqslant 3$, the sets $S^{N-1} = \partial B_1^N = \{x \in \mathbb{R}^N : \|x\| = 1\}$ and $\mathbb{R}^N \setminus \{0\}$ are simply connected.

Problem 2.149 **

By constructing a homotopy equivalence show that $[0,1]$ and $(0,1)$ are homotopy equivalent to the singleton.

Remark. Since every open interval (a,c) is homeomorphic to $(0,1)$, it follows that every open interval (a,c) is homotopy equivalent to $\{0\}$. This also holds for half-lines $(a,+\infty)$ and $(-\infty,a)$.

Problem 2.150 **

Consider the annulus $A \stackrel{\text{def}}{=} \{(x,y) \in \mathbb{R}^2 : 1 \leqslant \sqrt{x^2+y^2} \leqslant 2\}$. Show that A is homotopy equivalent to the unit sphere $S^1 \stackrel{\text{def}}{=} \{(x,y) \in \mathbb{R}^2 : x^2+y^2 = 1\}$.

Problem 2.151 *

Show that the cylinder $I \times S^1$ and the circle S^1 are homotopy equivalent.

Problem 2.152 *

Suppose that X is a topological space, $S^N \stackrel{\text{def}}{=} \{x \in \mathbb{R}^{N+1} : \|x\| = 1\}$ and $f\colon X \longrightarrow S^N$ is a continuous function, which is not surjective. Show that f is nullhomotopic, i.e., $f \simeq 0$.

Problem 2.153 *

Let X and Y be two topological spaces and let $f\colon X \longrightarrow Y$ be a continuous function. Show that $\operatorname{Gr} f$ is a retract of $X \times Y$.

Problem 2.154 **
Compute the fundamental group of the following figure (the union of two tangent circles):

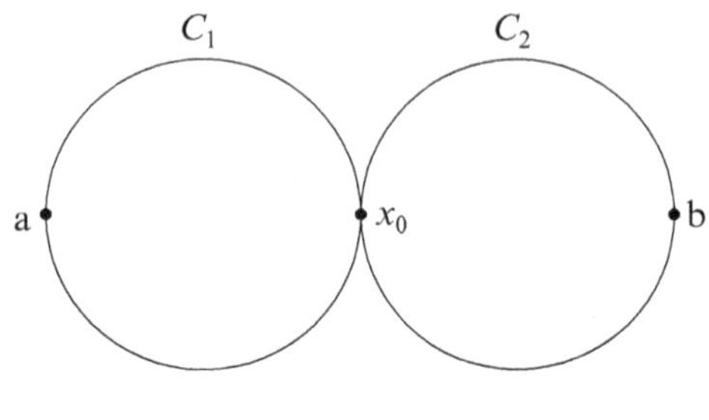

Problem 2.155 ***
Suppose that X is a compact topological space, $A \subseteq X$ is a connected, closed subset of X and $f\colon A \longrightarrow A$ is a continuous function.
(a) Show that the set $C = \bigcap_{n \geqslant 1} f^n(A)$ is connected, where $f^n = \underbrace{f \circ \ldots \circ f}_{n\text{-times}}$.
(b) Let $g\colon S^1 \longrightarrow X \setminus C$ be a nullhomotopic function. Show that there exists an integer $n \geqslant 1$ such that $g(S^1) \subseteq X \setminus f^n(A)$.

Problem 2.156 **
Consider the torus $T = S^1 \times S^1$ and let $A \subseteq T$ be defined by

$$A \stackrel{\text{def}}{=} \left[S^1 \times \{1\}\right] \cup \left[\{1\} \times S^1\right].$$

Is A a retract of T? Justify your answer.
Hint: Use the fact that $\pi_1(A)$ is nonabelian with two generators a and b.

Problem 2.157 ***
Suppose that X and Y are two topological spaces, with X being path-connected, $f\colon X \longrightarrow Y$ is a continuous function and $x_0, x_1 \in X$. Suppose that the induced homomorphism $f_*\colon \pi_1(X, x_0) \longrightarrow \pi_1\big(Y, f(x_0)\big)$ is surjective. Show that $f_*\colon \pi_1(X, x_1) \longrightarrow \pi_1\big(Y, f(x_1)\big)$ is surjective.

Problem 2.158 **
Find the fundamental group of the projective n-space $\mathbb{P}^N$ $(N \geqslant 2)$.

Problem 2.159 *
Find the fundamental group of $\mathbb{P}^2 \times S^2$.

Problem 2.160 **

(a) Suppose that X is a topological space, A is a retract of X, $r\colon X \longrightarrow A$ is a retraction and $i\colon A \longrightarrow X$ is an inclusion function. Show that for all $x \in A$, $(i_A)_*\colon \pi_1(A,x) \longrightarrow \pi_1(X,x)$ is injective and $r_*\colon \pi_1(X,x) \longrightarrow \pi_1(A,x)$ is surjective.
(b) Show that the fundamental groups of $\mathbb{R}^2 \setminus \{0\}$ and $T = S^1 \times S^1$ (the torus) have infinite cyclic subgroups.

Problem 2.161 **

Let X be a path-connected topological space with a compact universal cover and let $x \in X$. Show that $\pi_1(X,x)$ is finite.

Problem 2.162 **

Let X be a locally path-connected and simply connected topological space. Show that every continuous function $f\colon X \longrightarrow S^1$ is nullhomotopic. (Compare with Problem 2.173.)

Problem 2.163 **

Suppose that X is a path-connected and locally path-connected topological space and $\pi_1(X)$ is finite. Show that every continuous function $f\colon X \longrightarrow S^1$ is nullhomotopic.

Problem 2.164 **

Show that homotopy equivalent spaces have the same number of path-connected components.

Problem 2.165 ***

Let $p\colon \hat{X} \longrightarrow X$ be a covering map and suppose that $\hat{x}_0, \hat{x}_1 \in \hat{X}$ belong to the same path-connected component of $\hat{X}$. Let $\widetilde{u} = p(\hat{x}_0) = p(\hat{x}_1)$. Show that $p_*\big(\pi_1(\hat{X},\hat{x}_0)\big)$ and $p_*\big(\pi_1(\hat{X},\hat{x}_1)\big)$ are ***conjugate subgroups*** of $\pi_1(X,\widetilde{u})$, i.e., there exists $v \in \pi_1(X,\widetilde{u})$ such that $p_*\big(\pi_1(\hat{X},\hat{x}_1)\big) = v^{-1}p_*\big(\pi_1(\hat{X},\hat{x}_0)\big)v$.

Problem 2.166 **

Suppose that X is a topological space and A is a retract of X. Show that

$$H_n(X) \simeq H_n(A) \oplus H_n(X,A) \qquad \forall\, n \geqslant 0.$$

Problem 2.167 **

Suppose that X is a topological space and $A \subseteq X$ is a deformation retract. Show that

$$H_n(X, A) = 0 \qquad \forall\, n \geqslant 0.$$

Remark. In particular this implies that $H_n(X, X) = 0$ for all $n \geqslant 0$.

Problem 2.168 **

Suppose that X is a topological space and $\star \in X$. Show that

$$H_n(X) = H_n(X, \star) \oplus H_n(\star) \qquad \forall\, n \geqslant 0.$$

Problem 2.169 **

Suppose that X is a topological space and $\{A_k\}_{k=1}^m$ are disjoint closed subsets of X such that $X \stackrel{\text{def}}{=} \bigcup\limits_{k=1}^m A_k$. Show that $H_n(X) = \bigoplus\limits_{k=1}^m H_n(A_k)$ for all $n \geqslant 0$.

Problem 2.170 **

Show that, if X is a contractible topological space, then

$$H_n(X, \star) = 0 \qquad \forall\, n \geqslant 0,\ \star \in X.$$

Problem 2.171 **

Suppose that X is a topological space and $A \subseteq X$ is a contractible subspace. Show that

$$H_n(X, A) \simeq H_n(X, \star) \qquad \forall\, n \geqslant 1,\ \star \in X.$$

Problem 2.172 ***

Show that

$$H_0(S^1) = H_1(S^1) = \mathbb{Z} \quad \text{and} \quad H_n(S^1) = 0 \qquad \forall\, n \geqslant 2.$$

Then show that

$$H_0(S^2) = H_2(S^2) = \mathbb{Z} \quad \text{and} \quad H_n(S^2) = 0 \qquad \forall\, n \in \mathbb{N} \setminus \{0, 2\}.$$

Problem 2.173 **

(a) Suppose that $N \geqslant 2$. Is there a continuous function $f\colon S^N \longrightarrow S^1$ which is not nullhomotopic? Justify your answer.

(b) Let $T = S^1 \times S^1$ be the torus. Is there a continuous function $f\colon T \longrightarrow S^1$ which is not nullhomotopic? Justify your answer.

Problem 2.174 **
Let X be a topological space. The ***suspension*** ΣX of X is the identification space obtained from $X \times [-1,1]$ by identifying $X \times \{-1\}$ to a point and $X \times \{1\}$ to another point (we call them ***identification points***). Compute the homology groups of ΣX in terms of the homology groups of X.

Problem 2.175 **
Let X be a topological space such that $X = U \cup V$, with $U, V \subseteq X$ being open, path-connected with $U \cap V$ nonempty and path-connected and $H_1(U) = 0$. Let the inclusion $j^2\colon U \cap V \longrightarrow V$ induce a surjective homomorphism $j^2_*\colon H_1(U \cap V) \longrightarrow H_1(V)$. Show that $H_1(X) = 0$. Is the result true if throughout H_1 is replaced by H_2? Justify your answer.

Problem 2.176 ***
Let $\overline{B}_1^N = \{x \in \mathbb{R}^N\colon \|x\| \leqslant 1\}$ with $N \geqslant 2$ and let $f\colon \overline{B}_1^N \longrightarrow \overline{B}_1^N$ be a continuous function such that $f|_{S^{N-1}}$ is a homeomorphism from S^{N-1} to S^{N-1}. Show that f is surjective.

Problem 2.177 ***
Let $X = \{(x,y,z) \in \mathbb{R}^3 : xy = 0\}$. Show that X is not homeomorphic to $\mathbb{R}^2$ but it is of the same homotopy type.

Problem 2.178 ***
Let $X \subseteq \mathbb{R}^3$ be the union of S^2, of $\overline{B}_1^2 = \{u \in \mathbb{R}^2 : \|u\| \leqslant 1\}$ in the xy-plane and of C being the portion of the z-axes which is inside S^2. Compute the fundamental and homology groups of X.

Problem 2.179 *
Find the homology groups of a cylinder $C = S^1 \times \mathbb{R}$.

Problem 2.180 *
Show that S^N is not a retract of $\overline{B}_1^{N+1}$ $(N \geqslant 1)$.

Remark. In contrast, in an infinite dimensional Banach space X, the set $\partial\overline{B}_1$ is a retract of $\overline{B}_1 = \{x \in X : \|x\| \leqslant 1\}$.

Problem 2.181 ***
Let $U \subseteq \mathbb{R}^N$ $(N \geqslant 2)$ be an open set and let $x \in U$. Show that $H^{N-1}(U \setminus \{x\}) \neq 0$.

Problem 2.182 ***
Let $U \subseteq \mathbb{R}^N$, $V \subseteq \mathbb{R}^M$ be two nonempty open sets and $N \neq M$. Show that U and V cannot be homeomorphic.

Problem 2.183 ***
Show that

$$H_k(S^N, \star) = \begin{cases} H_{k-N}(\star) & \text{if } k \geqslant N, \\ 0 & \text{if } k < N, \end{cases}$$

for $\star \in S^N$.

Problem 2.184 ***
Let $X_1 \subseteq X_2 \subseteq X_3 \subseteq X_4$ be topological spaces. Show that

$$\operatorname{rank} H_k(X_3, X_2) - \operatorname{rank} H_k(X_4, X_1) \leqslant \operatorname{rank} H_{k-1}(X_2, X_1) + \operatorname{rank} H_{k+1}(X_4, X_3),$$

for all $k \geqslant 1$.

2.3 Solutions

Solution of Problem 2.1

Let $x \in U \cap \overline{A}$ and let $V \in \mathcal{N}(x)$. Then $V \cap U \in \mathcal{N}(x)$. Because $x \in \overline{A}$, we have that

$$V \cap (U \cap A) = (V \cap U) \cap A \neq \emptyset,$$

which shows that $x \in \overline{U \cap A}$. Therefore we conclude that $U \cap \overline{A} \subseteq \overline{U \cap A}$.

Now, let $X = \mathbb{R}$ with the natural topology. Let $U = (0,2)$ and $A = (1,3)$. Then

$$U \cap \overline{A} = [1,2) \subsetneq [1,2] = \overline{U \cap A}.$$

Solution of Problem 2.2

(a) Let $x \in \partial C$ and $U \in \mathcal{N}(x)$. Then $x \in A \cap U$ and

$$\emptyset \neq (X \setminus C) \cap U \subseteq (X \setminus A) \cap U,$$

which implies that $x \in \partial A$. Hence $\partial C \subseteq \partial A$.

(b) Let $A = \overline{B}_1^N \setminus \{0\}$, where $\overline{B}_1^N = \{x \in \mathbb{R}^N : \|x\| \leqslant 1\}$. Then $0 \in \partial A$ but $0 \notin \partial \overline{B}_1^N$. Therefore $\partial A \neq \partial \overline{A}$.

(c) From Proposition 2.11(e) and (f), we have

$$\begin{aligned}\partial(A \cup C) &= (\overline{A \cup C}) \cap (\overline{X \setminus (A \cup C)}) = (\overline{A} \cup \overline{C}) \cap (\overline{(X \setminus A) \cap (X \setminus C)}) \\ &\subseteq (\overline{A} \cup \overline{C}) \cap (\overline{X \setminus A}) \cap (\overline{X \setminus C}) \\ &\subseteq (\overline{A} \cap (\overline{X \setminus A})) \cup (\overline{C} \cap (\overline{X \setminus C})) = \partial A \cup \partial C.\end{aligned}$$

To see that this inclusion can be proper, let $X = \mathbb{R}$, $A = [0,1)$ and $C = [1,2]$. Then

$$\partial A = \{0,1\}, \quad \partial C = \{1,2\} \quad \text{and} \quad \partial(A \cup C) = \{0,2\}.$$

(d) We have

$$\partial(A \cap C) = \partial(X \setminus (A^c \cup C^c)) = \partial(A^c \cup C^c) \subseteq \partial A^c \cup \partial C^c = \partial A \cup \partial C$$

(see (c) above). Again, if $X = \mathbb{R}$, $A = [0,1)$ and $C = [1,2]$. Then

$$\partial A = \{0,1\}, \quad \partial C = \{1,2\} \quad \text{and} \quad \partial(A \cap C) = \emptyset$$

and so we see that the inclusion can be proper.

Solution of Problem 2.3
Clearly A is a vector subspace of $C[0,1]$. Consider the function $f(x) = \sqrt{x}$. Then $f \in C[0,1] \setminus A$, since

$$\frac{f(x)}{x} \longrightarrow +\infty \quad \text{as } x \to 0^+.$$

Let $g \in A$. For every $\varepsilon > 0$ the ball

$$B_\varepsilon(g) = \{h \in C[0,1] : d^\infty(h,g) < \varepsilon\}$$

contains the function $g + \frac{\varepsilon}{2} f$ which is not Lipschitz continuous. Therefore $\operatorname{int} A = \emptyset$.

Solution of Problem 2.4
We have

$$\operatorname{int} \overline{C} = \emptyset \iff \operatorname{int} C = \emptyset \iff X = \operatorname{int}(X \setminus C) \cup \partial(X \setminus C) = \overline{X \setminus C}$$

(see Definition 2.9(e)) and so we conclude that C is nowhere dense if and only if $X \setminus C$ is dense.

Solution of Problem 2.5
(a) The set $\overline{C} \cap A$ is closed in A (see Definition 2.14) and so $\overline{C}^A \subseteq \overline{C} \cap A$. Let $x \in \overline{C} \cap A$. Let U be a neighbourhood of x in A. Then $U = V \cap A$, with V being a neighbourhood of x in X. Since $x \in \overline{C}$, we have $V \cap C \neq \emptyset$. Since $C \subseteq A$, we have $U \cap C \neq \emptyset$, hence $x \in \overline{C}^A$. Therefore $\overline{C} \cap A \subseteq \overline{C}^A$ and we conclude that $\overline{C}^A = \overline{C} \cap A$.

(b) Note that $\operatorname{int} C$ is an open subset of X contained in A (see Proposition 2.15). Hence from the definition of the subspace topology (see Definition 2.14), we have that $\operatorname{int} C$ is also open in A. Hence $\operatorname{int} C \subseteq \operatorname{int}_A C$.

This inclusion can be strict. To see this let $X = \mathbb{R} \times \mathbb{R} = \mathbb{R}^2$, $A = \mathbb{R} \times \{0\} \subseteq \mathbb{R} \times \mathbb{R}$ and $C = (0,1) \times \{0\}$. Then $\operatorname{int}_A C = C$, while $\operatorname{int} C = \emptyset$.

Solution of Problem 2.6

Since by hypothesis D is open in both A and C, we can find two open subsets U, V of X such that

$$D = U \cap A \quad \text{and} \quad D = V \cap A.$$

Then

$$(U \cap V) \cap (A \cup C) = (U \cap V \cap A) \cup (U \cap V \cap C) = D.$$

Because $U \cap V$ is open in X, we conclude that D is open in $A \cup C$.

Solution of Problem 2.7

"$\Longrightarrow$": Since the set C is closed, from the definition of the subspace topology (see Definition 2.14), the set $C \cap U$ is closed in U (in fact for every subset $U \subseteq X$).

"$\Longleftarrow$": Arguing by contradiction, suppose that the set C is not closed. Then $\overline{C} \neq C$ and so we can find $x \in \overline{C} \setminus C$. For every neighbourhood U of x, every open set $V \subseteq U$ with $x \in V$ satisfies $C \cap V \neq \emptyset$ (recall that $x \in \overline{C}$). So, we can find $u \in C \cap V = C \cap U \cap V$ and this means that $x \in U \setminus C$ belongs in the closure of $C \cap U$, which contradicts our hypothesis.

Solution of Problem 2.8

Evidently, we always have $\overline{\overline{C} \cap D} \supseteq \overline{C \cap D}$. So, we need to show that the opposite inclusion also holds. Let $K = E \setminus D$. We have $C \subseteq (C \cap D) \cup K$, hence

$$\overline{C} \subseteq \overline{C \cap D} \cup \overline{K} = \overline{C \cap D} \cup K,$$

so

$$\overline{C} \cap D = (\overline{C \cap D} \cup K) \cap D = \overline{C \cap D} \cup (K \cap D) = \overline{C \cap D},$$

thus $\overline{\overline{C} \cap D} \subseteq \overline{C \cap D}$.

Solution of Problem 2.9

Let $\mathcal{Y}$ be the countable subbasis (see Definition 2.19) and let $\mathcal{F} \subseteq 2^{\mathcal{Y}}$ be the collection of all finite elements. Then $\mathcal{F}$ is countable. Every basic element in X is of the form $\bigcap \widetilde{F}$ with $\widetilde{F} \in \mathcal{F}$. Hence the basis $\mathcal{B}$ is countable (see Definition 2.24).

Solution of Problem 2.10

Let C be a closed subset of a Lindelöf space X (see Definition 2.26) and let $\mathcal{Y}$ be an open cover of C. Every $V \in \mathcal{Y}$ has the form $V = U \cap C$, where U is an open subset of X. Let $\mathcal{Y}^*$ be the set of all such open sets $U \subseteq X$ together with $X \setminus C$. Evidently this is an open cover of X. Because X is a Lindelöf space, we can find a countable subcover of X. Removing $X \setminus C$ (if it is included in the countable subcover), we see that $\mathcal{Y}$ has been reduced to a countable subcover of C. This shows that C with the subspace topology is a Lindelöf space.

Solution of Problem 2.11

Let $U \in \mathcal{B}$, $U \neq \emptyset$ and choose $x_U \in U$. We claim that

$$D = \{x_U\}_{U \in \mathcal{B} \setminus \{\emptyset\}}$$

is the desired set. First we show that D is dense in X (see Definition 2.9(e)). To this end, let V be an open set in X. If $u \in V$, then we can find $U \in \mathcal{B}$ such that $x_U \in U \subseteq V$ (see Definition 2.19). Hence $x_U \in V$ and so $V \cap D \neq \emptyset$, which proves the density of D in X.

Next we show that $\#D \leqslant \#\mathcal{B}$. For this purpose, we introduce the function $\xi \colon \mathcal{B} \setminus \{\emptyset\} \longrightarrow D$, defined by $\xi(U) = x_U$. This function is surjective. The Axiom of Choice implies that there exists $\mathcal{Y} \subseteq \mathcal{B}$ such that for every $x \in D$, $\mathcal{Y} \cap \xi^{-1}(\{x\})$ is a singleton. Then $\xi \colon \mathcal{Y} \longrightarrow D$ is a bijection and so we conclude that $\#D = \#\mathcal{Y} \leqslant \#\mathcal{B}$.

Solution of Problem 2.12

Let $C \subseteq X$ be closed. Then from the definition of the subspace topology (see Definition 2.14), it follows that $U_i \cap C$ is closed in U_i for all $i \in I$.

Conversely, suppose that for every $i \in I$, the set $U_i \cap C$ is closed in U_i. Let $\{x_j\}_{j \in J} \subseteq C$ be a net such that $x_j \longrightarrow x$ in X (see Definition 2.31). Because $\{U_j\}_{j \in J}$ is an open cover of X, we can find $i_0 \in I$ such that $x \in U_{i_0}$. Since $x_j \longrightarrow x$, we can find $j_0 \in J$ such that $x_j \in U_{i_0}$ for all $j \geqslant j_0$, hence $x_j \in U_{i_0} \cap C$ for all $j \geqslant j_0$. But $U_{i_0} \cap C$ is closed in U_{i_0} and $x_j \longrightarrow x$. Hence $x \in U_{i_0} \cap C$ and so $x \in C$, which proves that C is closed in X.

Solution of Problem 2.13

Yes. Consider the function $f \colon [0,1) \longrightarrow \mathbb{R}$, defined by

$$f(x) = \tfrac{1}{1-x} \sin\left(\tfrac{1}{1-x}\right).$$

Evidently f is continuous and $f\big([0,1)\big) = \mathbb{R}$.

Solution of Problem 2.14

Let $\{f_i \colon X \longrightarrow \mathbb{R}\}_{i \in I}$ be a net of continuous functions and suppose that

$$\sup_{x \in X} \left|f_i(x) - f(x)\right| \longrightarrow 0$$

To show that $f\colon X \longrightarrow \mathbb{R}$ is continuous, it suffices to show that, if $\{x_j\}_{j\in J}$ is a net such that $x_j \longrightarrow x$, then $f(x_j) \longrightarrow f(x)$. To this end, let $\varepsilon > 0$ be given and we pick $i_0 \in I$ such that

$$\left|f_i(u) - f(u)\right| \;<\; \tfrac{\varepsilon}{3} \qquad \forall\, i \geqslant i_0,\; u \in X.$$

Exploiting the continuity of f_{i_0}, we can find $j_0 \in J$ such that

$$\left|f_{i_0}(x_j) - f_{i_0}(x)\right| \;<\; \tfrac{\varepsilon}{3} \qquad \forall\, j \geqslant j_0.$$

Therefore for $j \geqslant j_0$, we have

$$\begin{aligned}\left|f(x_j) - f(x)\right| \;&\leqslant\; \left|f(x_j) - f_{i_0}(x_j)\right| + \left|f_{i_0}(x_j) - f_{i_0}(x)\right| + \left|f_{i_0}(x) - f(x)\right| \\ &<\; \tfrac{\varepsilon}{3} + \tfrac{\varepsilon}{3} + \tfrac{\varepsilon}{3} \;=\; \varepsilon,\end{aligned}$$

so $f(x_j) \longrightarrow f(x)$ and thus f is continuous.

Solution of Problem 2.15

Suppose that f is lower semicontinuous (respectively, sequentially lower semicontinuous; see Definitions 2.46 and 2.49). Let $\lambda \in \mathbb{R}$ and consider the function

$$h(x, \lambda) \;=\; f(x) - \lambda.$$

Evidently h is lower semicontinuous (respectively, sequentially lower semicontinuous) on $X \times \mathbb{R}$. Hence the set

$$\big\{(x, \lambda) \in X\times\mathbb{R} : \; h(x, \lambda) \leqslant 0\big\} \;=\; \big\{(x, \lambda) \in X\times\mathbb{R} : \; f(x) \leqslant \lambda\big\} \;=\; \operatorname{epi} f$$

is closed (respectively, sequentially closed; see Definition 2.51) in $X\times\mathbb{R}$.

Now suppose that $\operatorname{epi} f$ is closed (respectively, sequentially closed) in $X \times \mathbb{R}$. We need to show that for every $\lambda \in \mathbb{R}$, the set

$$L_\lambda \;=\; \big\{x \in X : \; f(x) \leqslant \lambda\big\}$$

is closed (respectively, sequentially closed) in X. So let $\{x_\alpha\}_{\alpha\in J} \subseteq L_\lambda$ (respectively, $\{x_n\}_{n\geqslant 1} \subseteq L_\lambda$) be a net (respectively, sequence) such that $x_\alpha \longrightarrow x$ (respectively, $x_n \longrightarrow x$) in X. Note that $\big\{(x_\alpha, \lambda)\big\}_{\alpha\in J} \subseteq \operatorname{epi} f$ (respectively, $\big\{(x_n, \lambda)\big\}_{n\geqslant 1} \subseteq \operatorname{epi} f$) and $(x_\alpha, \lambda) \longrightarrow (x, \lambda)$ (respectively, $(x_n, \lambda) \longrightarrow (x, \lambda)$) in $X \times \mathbb{R}$. Then $(x, \lambda) \in \operatorname{epi} f$ and so $f(x) \leqslant \lambda$, which proves that L_λ is closed (respectively, sequentially closed), hence f is lower semicontinuous (respectively, sequentially lower semicontinuous).

Solution of Problem 2.16

From the definition of $\overline{f}$ (see Definition 2.57), we see that

$$\overline{f}(x) = \min\left\{\liminf f(x_i) : \{x_i\}_{i\in I} \text{ is a net converging to } x\right\}.$$

Let $x \in \left\{x \in X : \overline{f}(x) \leqslant \lambda\right\}$. For a given $\mu > \lambda$, we can find a net $\{x_i\}_{i\in I} \subseteq X$ such that $x_i \longrightarrow x$ in X and

$$\overline{f}(x) \leqslant \liminf f(x_i) = \sup_{i_0\in I} \inf_{i\geqslant i_0} f(x_i) < \mu.$$

So, we can find a subnet $\{x_j\}_{j\in J}$ of $\{x_i\}_{i\in I}$ such that

$$f(x_j) \longrightarrow \liminf f(x_i) \quad \text{and} \quad f(x_j) \leqslant \mu.$$

Hence

$$x_j \in \left\{x \in X : f(x) \leqslant \mu\right\} \qquad \forall\, j \in J$$

and since $x_j \longrightarrow x$ in X, we have that $x \in \overline{\left\{x \in X : f(x) \leqslant \mu\right\}}$. But $\mu > \lambda$ was arbitrary. So, we infer that

$$\left\{x \in X : \overline{f}(x) \leqslant \lambda\right\} \subseteq \bigcap_{\mu>\lambda} \overline{\left\{x \in X : f(x) \leqslant \mu\right\}}.$$

On the other hand let

$$x \in \bigcap_{\mu>\lambda} \overline{\left\{x \in X : f(x) \leqslant \mu\right\}}.$$

Then

$$x \in \overline{\left\{x \in X : f(x) \leqslant \mu\right\}} \qquad \forall\, \mu > \lambda.$$

So, for a given $\mu > \lambda$, we can find a net $\{x_i\}_{i\in I} \subseteq \left\{x \in X : f(x) \leqslant \mu\right\}$ such that $x_i \longrightarrow x$ in X. Then $f(x_i) \leqslant \mu$ and so

$$\overline{f}(x) \leqslant \liminf \overline{f}(x_i) \leqslant \liminf f(x_i) \leqslant \mu.$$

Because $\mu > \lambda$ was arbitrary, we let $\mu \longrightarrow \lambda^+$ and so we have $\overline{f}(x) \leqslant \lambda$. Hence we have

$$\bigcap_{\mu>\lambda} \overline{\left\{x \in X : f(x) \leqslant \mu\right\}} \subseteq \left\{x \in X : \overline{f}(x) \leqslant \lambda\right\},$$

so

$$\{x \in X : \overline{f}(x) \leqslant \lambda\} = \bigcap_{\mu > \lambda} \overline{\{x \in X : \overline{f}(x) \leqslant \mu\}}.$$

Since $\overline{f} \leqslant f$, we have $\operatorname{epi} f \subseteq \operatorname{epi} \overline{f}$. Because $\overline{f}$ is lower semicontinuous (see Definition 2.46), $\operatorname{epi} \overline{f}$ is closed in $X \times \mathbb{R}$ (see Problem 2.15) and so $\overline{\operatorname{epi} f} \subseteq \operatorname{epi} \overline{f}$. Suppose that the inclusion is strict. So, we can find $(x, \lambda) \in \operatorname{epi} \overline{f}$ such that $(x, \lambda) \notin \overline{\operatorname{epi} f}$. Hence there exist $U \in \mathcal{N}(x)$ and $\varepsilon > 0$ such that

$$\big(U \times (\lambda - \varepsilon, \lambda + \varepsilon)\big) \cap \overline{\operatorname{epi} f} = \emptyset.$$

Let $\mu \in (\lambda, \lambda + \varepsilon)$. We claim that $x \notin \overline{\{y \in X : f(y) \leqslant \mu\}}$. Indeed, if $\{x_i\}_{i \in I}$ is a net in X converging to x, then we can find $i_0 \in I$ such that

$$(x_i, \mu) \in U \times (\lambda - \varepsilon, \lambda + \varepsilon) \qquad \forall i \geqslant i_0,$$

hence $(x_i, \mu) \notin \overline{\operatorname{epi} f}$ and so

$$f(x_i) > \mu \qquad \forall i \geqslant i_0,$$

which proves the claim. But then this contradicts the first part of the problem.

Solution of Problem 2.17

Let

$$g_*(x) = \sup_{U \in \mathcal{N}(x)} \inf_{y \in U} f(y).$$

According to Remark 2.47, g_* is lower semicontinuous (see Definition 2.46) and of course $g_* \leqslant f$. Therefore $g_* \leqslant \overline{f}$. On the other hand, let $g \in \mathcal{L}(f)$ (see Definition 2.57). Then again by Remark 2.47 and since $g \leqslant f$, we have

$$g(x) = \sup_{U \in \mathcal{N}(x)} \inf_{y \in U} g(y) \leqslant \sup_{U \in \mathcal{N}(x)} \inf_{y \in U} f(y) = g_*(x),$$

so $\overline{f} \leqslant g_*$. Hence we conclude that

$$\overline{f}(x) = \sup_{U \in \mathcal{N}(x)} \inf_{y \in U} f(y) \qquad \forall\, x \in X.$$

Solution of Problem 2.18

Since $\overline{f} \leqslant f$ is lower semicontinuous (see Definitions 2.46 and 2.57), for every sequence $x_n \longrightarrow x$ in X, we have

$$\overline{f}(x) \leqslant \liminf_{n \to +\infty} \overline{f}(x_n) \leqslant \liminf_{n \to +\infty} f(x_n)$$

(see Proposition 2.48). This proves property (a). To prove (b), we may assume that $\overline{f}(x) < +\infty$. Let $\{U_n\}_{n \geqslant 1}$ be a local basis at $x \in X$ such that $U_{n+1} \subseteq U_n$ for all $n \geqslant 1$ (recall that by hypothesis X is first countable) and let $\lambda_n \longrightarrow \overline{f}(x)$ in $\mathbb{R}^* = \mathbb{R} \cup \{\pm\infty\}$ with $\lambda_n > \overline{f}(x)$ for all $n \geqslant 1$. From Problem 2.17, we have

$$\inf_{y \in U_n} f(y) < \lambda_n \qquad \forall\, n \geqslant 1$$

and so there exists $y_n \in U_n$ such that $f(y_n) < \lambda_n$. Evidently $y_n \longrightarrow y$ in X and so

$$\limsup_{n \to +\infty} f(y_n) \leqslant \lim_{n \to +\infty} \lambda_n = \overline{f}(x).$$

This proves property (b).

Solution of Problem 2.19

(a) Let $x, u, y \in X$. Then using the triangle inequality, we have

$$f(y) + \lambda d_X(x, y) \leqslant f(y) + \lambda d_X(y, u) + \lambda d_X(u, x),$$

so

$$f_\lambda(x) \leqslant f_\lambda(u) + \lambda d_X(u, x).$$

Reversing the roles of x and u in the above argument, we also have

$$f_\lambda(u) \leqslant f_\lambda(x) + \lambda d_X(u, x)$$

and so we conclude that f_λ is λ-Lipschitz.

(b) Evidently $f_\lambda \leqslant f$ and from part (a), f_λ is λ-Lipschitz. So $f_\lambda \leqslant \overline{f}$. Let $x \in X$ and assume that

$$\sup_{\lambda>0} f_\lambda(x) = M < +\infty.$$

Choose $x_\lambda \in X$ such that

$$f(x_\lambda) + \lambda d_X(x, x_\lambda) \leqslant f_\lambda(x) + \tfrac{1}{\lambda},$$

so

$$d_X(x, x_\lambda) \leqslant \tfrac{1}{\lambda} M + \tfrac{1}{\lambda^2}$$

(since $f \geqslant 0$) and thus

$$d_X(x, x_\lambda) \longrightarrow 0 \quad \text{as } \lambda \to +\infty,$$

i.e., $x_\lambda \longrightarrow x$ in X. Then from the definition of $\overline{f}$ (see Definition 2.57), we have

$$\overline{f}(x) \leqslant \liminf_{\lambda\to+\infty} f(x_\lambda) \leqslant \liminf_{\lambda\to+\infty} \big(f(x_\lambda)+\lambda d_X(x, x_\lambda)\big) \leqslant \liminf_{\lambda\to+\infty} f_\lambda(x).$$

Therefore, we conclude that

$$\lim_{\lambda\to+\infty} f_\lambda(x) = \overline{f}(x) \qquad \forall\, x \in X.$$

Solution of Problem 2.20

(a) "$\Longrightarrow$": We may assume that for some $x_0 \in X$, $f(x_0) < +\infty$. Let

$$\hat{f}_n(x) = \inf_{y\in X} \big(f(y) + n d_X(x, y)\big).$$

For every $x \in X$, we have

$$\xi \leqslant \hat{f}_n(x) \leqslant f(x) \quad \text{and} \quad \xi \leqslant \hat{f}_n(x) \leqslant f(x_0)+n d_X(x, x_0) < +\infty.$$

Therefore, we have

$$\xi \leqslant \hat{f}_1 \leqslant \hat{f}_2 \leqslant \ldots \leqslant \hat{f}_n$$

and each $\hat{f}_n$ is $\mathbb{R}$-valued. From triangle inequality, for every $x, u, y \in X$, we have

$$f(y) + nd_X(x, y) \leqslant f(y) + nd_X(y, u) + nd_X(u, x)$$

so

$$\hat{f}_n(x) \leqslant \hat{f}_n(u) + nd_X(u, x).$$

Reversing the roles of x, u, we also have

$$\hat{f}_n(u) \leqslant \hat{f}_n(x) + nd_X(u, x),$$

so

$$\left|\hat{f}_n(u) - \hat{f}_n(x)\right| \leqslant nd_X(u, x)$$

(i.e., $\hat{f}_n$ is n-Lipschitz). We have

$$\lim_{n\to+\infty} \hat{f}_n(x) \leqslant f(x) \qquad \forall\, x \in X.$$

For a given $\varepsilon > 0$, choose $y_n \in X$ for $n \geqslant 1$ such that

$$f(y_n) + nd_X(x, y_n) \leqslant \hat{f}_n(x) + \varepsilon.$$

As $n \to +\infty$, either $\hat{f}_n(x) \nearrow +\infty$ and so $\hat{f}_n(x) \nearrow f(x) = +\infty$, since $\lim\limits_{n\to+\infty} \hat{f}_n \leqslant f$ or else $d_X(x, y_n) \longrightarrow 0$. Therefore

$$f(x) \leqslant \liminf_{n\to+\infty} f(y_n) \leqslant \lim_{n\to+\infty} \hat{f}_n(x) + \varepsilon.$$

Since $\varepsilon > 0$ was arbitrary, we let $\varepsilon \searrow 0$, to obtain

$$f(x) \leqslant \lim_{n\to+\infty} \hat{f}_n(x),$$

hence $\hat{f}_n \nearrow f$ as $n \to +\infty$. Let us define

$$f_n = \min\{n, \hat{f}_n\}.$$

Then for every $n \geqslant 1$, the function f_n is bounded continuous and

$$f_n \nearrow f.$$

"$\Longleftarrow$": This part is an immediate consequence of Corollary 2.56.

(b) Consider $-f$ which is lower semicontinuous and bounded below and use part (a).

Solution of Problem 2.21

The pointwise limit of upper semicontinuous functions (see Definition 2.46) need not be an upper semicontinuous function. To see this, consider the following sequence $f_n\colon [0,1] \longrightarrow \mathbb{R}$ of continuous functions

$$f_n(x) = \begin{cases} nx & \text{if } x \in [0, \frac{1}{n}], \\ 1 & \text{if } x \in (\frac{1}{n}, 1], \end{cases} \qquad \forall\, n \geqslant 1.$$

For every $x \in [0,1]$, we have $f_n(x) \longrightarrow f(x)$ as $n \to +\infty$, where

$$f_n(x) = \begin{cases} 0 & \text{if } x = 0, \\ 1 & \text{if } x \in (0, 1], \end{cases}$$

which is not upper semicontinuous (if $x_n \longrightarrow 0$, $x_n \neq 0$, then $\lim\limits_{n\to+\infty} f(x_n) = 1 > 0 = f(0)$).

On the other hand, the uniform limit of upper semicontinuous functions is an upper semicontinuous function. Let $f_n\colon \mathbb{R} \longrightarrow \mathbb{R}$, $n \geqslant 1$, be a sequence of upper semicontinuous functions and assume that $f_n \rightrightarrows f$ (see Definition 1.59). Let $\lambda \in \mathbb{R}$ and $x \in \{y \in \mathbb{R} : f(y) > \lambda\}$. Since $f_n \rightrightarrows f$, we can find $\delta > 0$ small enough such that

$$f_n(x) > \lambda + \delta \qquad \forall\, n \geqslant n_1.$$

The upper semicontinuity of each f_n, implies that we can find $\varepsilon_n > 0$ such that

$$f_n(y) > \lambda + \delta \qquad \forall\, y \in (x - \varepsilon_n, x + \varepsilon_n).$$

Because $f_n \rightrightarrows f$, we can find an integer $n_2 \geqslant n_1$ such that

$$\big|f_n(u) - f(u)\big| < \delta \qquad \forall\, u \in \mathbb{R},\ n \geqslant n_2.$$

Hence

$$f_n(u) < f(u) + \delta \qquad \forall\, n \geqslant n_2,\ u \in \mathbb{R}$$

and so

$$\lambda < f(u) \qquad \forall\, u \in (x - \varepsilon_n, x + \varepsilon_n),$$

which shows that the set $\{y \in \mathbb{R} : \lambda < f(y)\}$ is open. This in turn implies the upper semicontinuity of f.

Solution of Problem 2.22

Since $\overline{D} = X$, for a given $x \in X$, we can find a net $\{x_i\}_{i \in I} \subseteq D$ such that $x_i \longrightarrow x$ (see Proposition 2.33(a)). By hypothesis $f(x_i) = g(x_i)$ for all $i \in I$. Also since f and g are continuous, from Proposition 2.40, we have

$$f(x_i) \longrightarrow f(x) \quad \text{and} \quad g(x_i) \longrightarrow g(x) \quad \text{in } Y.$$

Therefore $f(x) = g(x)$ (see Proposition 2.32).

Solution of Problem 2.23

"$\Longrightarrow$": Since f is continuous, for a given $\varepsilon > 0$, we can find $U \in \mathcal{N}(x)$ such that

$$d_Y\big(f(u), f(x)\big) < \tfrac{\varepsilon}{2} \qquad \forall\, u \in U.$$

Hence, if $u, v \in U$, then

$$d_Y\big(f(u), f(v)\big) \leqslant d_Y\big(f(u), f(x)\big) + d_Y\big(f(x), f(v)\big),$$

so

$$0 \leqslant \omega_f(x) \leqslant \operatorname{diam} f(U) \leqslant \varepsilon.$$

Because $\varepsilon > 0$ was arbitrary, we let $\varepsilon \searrow 0$ and obtain $\omega_f(x) = 0$.

"$\Longleftarrow$": Since $\omega_f(x) = 0$, from the definition of the oscillation, we see that for a given $\varepsilon > 0$, we can find $U \in \mathcal{N}(x)$ such that

$$\operatorname{diam} f(U) < \varepsilon,$$

so

$$d_Y\big(f(u), f(x)\big) < \varepsilon \qquad \forall\, u, x \in U$$

and thus f is continuous at $x \in X$.

Solution of Problem 2.24

Consider the oscillation function $\omega_f : X \longrightarrow \mathbb{R}_+ = [0, +\infty)$ (see Definition 2.42). From Problem 2.23, we know that

$$\operatorname{cont} f \;=\; \big\{x \in X : \; \omega_f(x) = 0\big\}.$$

For every $\lambda > 0$, let

$$L_\lambda \;=\; \big\{x \in X : \; \omega_f(x) < \lambda\big\}.$$

Let $x \in L_\lambda$, then we can find $U \in \mathcal{N}(x)$ such that

$$\operatorname{diam} f(U) \;<\; \lambda$$

(from the definition of ω_f). So, if $u \in U$, then $U \in \mathcal{N}(u)$ and we have $\omega_f(u) < \lambda$, which proves that L_λ is an open set in X (hence ω_f is upper semicontinuous; see Definition 2.46). But

$$\operatorname{cont} f \;=\; \bigcap_{n \geqslant 1} \{x \in X : \; \omega_f(x) < \tfrac{1}{n}\}$$

and each set $\{x \in X : \; \omega_f(x) < \frac{1}{n}\}$ is open. Therefore, $\operatorname{cont} f$ is a G_δ-set in X (see Definition 2.18).

Solution of Problem 2.25

We know that open half-lines $(-\infty, \lambda)$ and $(\lambda, +\infty)$ with $\lambda \in \mathbb{R}$ form a subbasis for the usual topology on $\mathbb{R}$. Then by Proposition 2.40, the function f is continuous if and only if for every $\lambda \in \mathbb{R}$, the sets $f^{-1}\big((-\infty, \lambda)\big)$ and $f^{-1}\big((\lambda, +\infty)\big)$ are open in X. Therefore, the function f is continuous if and only if the sets $\big\{x \in X : \; f(x) < \lambda\big\}$ and $\big\{x \in X : \; f(x) > \lambda\big\}$ are open. Hence we conclude that the function f is continuous if and only if the sets $\big\{x \in X : \; f(x) \geqslant \lambda\big\}$ and $\big\{x \in X : \; f(x) > \lambda\big\}$ are closed and open respectively.

Solution of Problem 2.26

"**(a) $\Longrightarrow$ (b)**": For any $n \geqslant 1$, the function $x \longmapsto (f - f_n)(x)$ is continuous and so for every $\varepsilon > 0$, the set $\{x \in X : |f(x) - f_n(x)| < \varepsilon\}$ is open in X.

"**(b) $\Longrightarrow$ (a)**": Let $\varepsilon > 0$ be given. By hypothesis, we can find strictly increasing sequence $\{n_k\}_{k \geqslant 1}$ such that sets

$$C_k = \{x \in X : |f(x) - f_{n_k}(x)| < \tfrac{\varepsilon}{3}\}$$

are all open in X. Since

$$f_n(x) \longrightarrow f(x) \qquad \forall x \in X,$$

we have

$$X = \bigcup_{k \geqslant 1} C_k.$$

Let $x_0 \in X$. Then, we can find $k_0 \geqslant 1$ such that $x_0 \in C_{k_0}$. Let $x \in C_{k_0}$. Then

$$\begin{aligned} |f(x) - f(x_0)| &\leqslant |f(x) - f_{n_{k_0}}(x)| + |f_{n_{k_0}}(x) - f_{n_{k_0}}(x_0)| + |f_{n_{k_0}}(x_0) - f(x_0)| \\ &< \tfrac{2\varepsilon}{3} + |f_{n_{k_0}}(x) - f_{n_{k_0}}(x_0)|. \end{aligned}$$

The continuity of $f_{n_{k_0}}$ implies that there exists $U \in \mathcal{N}(x_0)$ such that $U \subseteq C_{k_0}$ and

$$|f_{n_{k_0}}(x) - f_{n_{k_0}}(x_0)| < \tfrac{\varepsilon}{3} \qquad \forall x \in U,$$

so

$$|f(x) - f(x_0)| < \varepsilon \qquad \forall x \in U,$$

which establishes the continuity of f at $x_0 \in X$. Since $x_0 \in X$ was arbitrary, we conclude that f is continuous.

Solution of Problem 2.27

Let $\mathcal{L}$ be the set of all open covers $\mathcal{D}$ of X such that for every $V \in \mathcal{D}$, we have

$$V \in \mathcal{Y} \quad \text{or} \quad \overline{V} \subseteq U \quad \text{for some } U \in \mathcal{Y}.$$

For each $\mathcal{D} \in \mathcal{L}$, let

$$\Gamma_{\mathcal{D}} = \{V \in \mathcal{D} : \overline{V} \subseteq U \text{ with } U \in \mathcal{Y}\}.$$

We introduce the order by inclusion on $\{\Gamma_{\mathcal{D}} : \mathcal{D} \in \mathcal{L}\}$. Then every chain has an upper bound. Thus by the Kuratowski–Zorn lemma, there exists a maximal element $\Gamma_{\mathcal{D}_*}$ which coincides with $\mathcal{Y}$. Then $\mathcal{D}_*$ is the desired locally finite open cover of X (see Definition 2.43).

Solution of Problem 2.28

Let $x \notin \bigcup\limits_{x\in K} \overline{U}_\alpha$. Let V be an open neighbourhood of u such that V intersects only $U_{\alpha_1}, \ldots, U_{\alpha_n}$ (see Definition 2.43). Since $u \notin \overline{U}_{\alpha_i}$ for $i \in \{1, \ldots, n\}$, we can find $W_i \in \mathcal{N}(u)$ such that

$$W_i \cap \overline{U}_i \;=\; \emptyset \qquad \forall\, i \in \{1, \ldots, n\}.$$

We set

$$W \;=\; V \cap W_1 \cap \ldots \cap W_n.$$

Then $W \in \mathcal{N}(x)$ and

$$W \cap \Big(\bigcup_{\alpha\in K} \overline{U}_\alpha\Big) \;=\; \emptyset,$$

so the set

$$\bigcup_{x\in K} \overline{U}_\alpha$$

is closed.

Solution of Problem 2.29

(a) For every set $D \subseteq X$, we have

$$(g \circ f)(D) \;=\; g(f(D)),$$

from which we immediately infer that, if f and g are open (respectively, closed), then $g \circ f$ is open (respectively, closed) too (see Definition 2.59).

(b) Since g is bijective, then for every $D \subseteq X$, we have

$$f(D) \;=\; \big(g^{-1} \circ (g \circ f)\big)(D).$$

By hypothesis, we know that $g \circ f$ is open (respectively, closed). So, if A is open (respectively, closed), then $(g \circ f)(A) \subseteq Z$ is open (respectively, closed). The continuity of g implies that

$$g^{-1}\big((g \circ f)(A)\big) = f(A)$$

is open (respectively, closed), which proves that f is an open (respectively, closed) function.
(c) Because f is surjective, for every $D \subseteq Y$, we have

$$g(D) = (g \circ f)\big(f^{-1}(D)\big).$$

By hypothesis f is continuous and so, if $D \subseteq Y$ is open (respectively, closed), then $f^{-1}(D) \subseteq X$ is open (respectively, closed). Since $g \circ f$ is assumed to be open (respectively, closed), we have that the set $(g \circ f)\big(f^{-1}(D)\big) = g(D)$ is open (respectively, closed). Therefore g is open (respectively, closed).

Solution of Problem 2.30
First suppose that A is open. Then the open subsets of A (with the subspace topology) are the open subsets of X contained in A. Therefore their images under i are open in X and so i is an open function (see Definition 2.59). Next suppose that $i_A : A \longrightarrow X$ is an open function. Since A is an open subset of itself, $i_A(A) = A \subseteq X$ is open. Similarly for the "closed case".

Solution of Problem 2.31
"$\Longrightarrow$": Since f is bijection, we have

$$f^{-1}\big(\overline{f(A)}\big) \supseteq f^{-1}\big(f(A)\big) = A \qquad \forall\, A \subseteq X.$$

The set $f^{-1}\big(\overline{f(A)}\big)$ is closed and so we have

$$\overline{A} \subseteq f^{-1}\big(\overline{f(A)}\big) \qquad \forall\, A \subseteq X.$$

On the other hand, since f is surjective, we have

$$f\big(f^{-1}\big(\overline{f(A)}\big)\big) = \overline{f(A)} \qquad \forall\, A \subseteq X$$

and because f is closed (see Definition 2.59), we have

$$\overline{f(A)} \subseteq f(\overline{A}) \qquad \forall\, A \subseteq X.$$

Hence

$$f\big(f^{-1}(\overline{f(A)})\big) \subseteq f(\overline{A}) \qquad \forall\, A \subseteq X.$$

Acting with f^{-1} and using the injectivity of f, we obtain that

$$f^{-1}\big(\overline{f(A)}\big) \subseteq \overline{A} \qquad \forall\, A \subseteq X.$$

We conclude that

$$f^{-1}\big(\overline{f(A)}\big) = \overline{A} \qquad \forall\, A \subseteq X.$$

"$\Longleftarrow$": We show that f is also injective. So, suppose that $f(x) = f(y)$. Then

$$\{x\} = \overline{\{x\}} = f^{-1}\big(\overline{\{f(x)\}}\big) = f^{-1}\big(\overline{\{f(y)\}}\big) = \overline{\{y\}} = \{y\}.$$

Now let $A \subseteq X$ be closed. Then, since $f^{-1}\big(\overline{f(A)}\big) = \overline{A} = A$ and f is surjective, we obtain $\overline{f(A)} = f(A)$, which proves that f is continuous and closed, hence a homeomorphism.

Solution of Problem 2.32

Let C and D be disjoint closed subsets of Y. Because f is continuous, we have that $f^{-1}(C)$ and $f^{-1}(D)$ are disjoint closed subsets of X. Exploiting the normality of X (see Definition 2.4), we can find two disjoint open sets $U, V \subseteq X$ such that

$$U \supseteq f^{-1}(C) \quad \text{and} \quad V \supseteq f^{-1}(D).$$

Let us set $E = Y \setminus f(X \setminus U)$. Since by hypothesis f is a closed function (see Definition 2.59), the set $f(X \setminus U) \subseteq Y$ is closed and so E is open in Y. Moreover, using the fact that $f^{-1}(C) \subseteq U$ and the surjectivity of f, we have

$$\begin{aligned} E &= Y \setminus f(X \setminus U) \supseteq Y \setminus f\big(X \setminus f^{-1}(C)\big) \\ &= Y \setminus f\big(f^{-1}(Y \setminus C)\big) = Y \setminus (Y \setminus C) = C. \end{aligned}$$

Similarly, if $G = Y \setminus f(X \setminus V)$, then we show that G is open and $G \supseteq D$. Evidently E and G are disjoint open neighbourhoods of C and D respectively and we have proved that Y is normal.

Solution of Problem 2.33

Note that $\overline{E}$ is nowhere dense too and E_x is nowhere dense whenever $\overline{E}_x$ is of the first category (see Definition 1.25). So, we may assume that E is closed.

Let $\{V_n\}_{n\geqslant 1}$ be a countable basis of Y (see Definition 2.24) and let $D = (X \times Y) \setminus E$. Then D is an open dense subset of $X \times Y$. For each $n \geqslant 1$, let

$$D_n = \operatorname{proj}_X\big((X \times V_n) \cap D\big),$$

where proj_X is the projection from $X \times Y$ onto X. Hence

$$D_n = \big\{x \in X : (x, y) \in D \text{ and } y \in V_n\big\}.$$

Recalling that the projections are open functions (see Definition 2.59), we have that $D_n \subseteq X$ is an open set. For any nonempty and open set $U \subseteq X$, we have

$$D \cap (U \times V_n) \neq \emptyset$$

(since D is dense) and so $D_n \cap U \neq \emptyset$, which proves the density of D_n in X. The set $\bigcap\limits_{n\geqslant 1} D_n$ is the complement of a set of first category in X. Let $x \in \bigcap\limits_{n\geqslant 1} D_n$. Then

$$D_x \cap V_n \neq \emptyset \qquad \forall\, n \geqslant 1.$$

Hence D_x is an open dense subset of V and so $E_x = Y \setminus D_x$ is nowhere dense. So for all $x \in X \setminus C$, with C of first category, E_x is a nowhere dense set.

Solution of Problem 2.34

If S is an open dense subset of X, then $S \times Y$ is open dense in $X \times Y$. So, if $C \subseteq X$ is nowhere dense (see Definition 2.9(e)), then so is $C \times Y$ in $X \times Y$. Since

$$\Big(\bigcup_{k\geqslant 1} D_k\Big) \times Y = \bigcup_{k\geqslant 1} (D_k \times Y),$$

it follows that $C \times D$ is of the first category in $X \times Y$ whenever C is in X. Similarly for D in Y.

Conversely, suppose that $C \times D$ is of first category in $X \times Y$ and C is not of first category in X. As Y is second countable (see Definition 2.24), then by Problem 2.33, we can find $x \in C$ such that the set $(C \times D)_x$ is of first category in Y. But note that $(C \times D)_x = D$. So, D is of first category in Y.

Solution of Problem 2.35

Suppose that f is open (see Definition 2.59). Let τ_X and τ_Y be the topologies of X and Y respectively. We need to show that for $U \subseteq Y$, $f^{-1}(U) \in \tau_X$ if and only if $U \in \tau_Y$ (see Definition 2.74). From the continuity of f, we know that, if $U \in \tau_Y$, then $f^{-1}(U) \in \tau_X$. Conversely, suppose that $f^{-1}(U) \in \tau_X$. Then since by hypothesis f is an open function, we have $f\big(f^{-1}(U)\big) \in \tau_Y$. Because f is surjective, we have $f\big(f^{-1}(U)\big) = U$, hence $U \in \tau_Y$. This proves that f is an identification function.

If f is closed, then the argument is similar by replacing open sets with closed ones and open functions with closed functions.

Solution of Problem 2.36

"$\Longrightarrow$": First note that the continuity of f implies that $\mathrm{proj}_X \colon \mathrm{Gr}\, f \longrightarrow X$ is a homeomorphism. Indeed, proj_X is continuous, injective and $\mathrm{proj}_X^{-1} = \varphi$, where $\varphi \colon X \longrightarrow \mathrm{Gr}\, f$ is defined by

$$\varphi(x) = \big(x, f(x)\big),$$

which is continuous because each component function is continuous. Therefore proj_X is a closed function (see Definition 2.59). Because f is continuous, $\mathrm{Gr}\, f \subseteq X \times Y$ is closed (see Problem 2.54). We endow $\mathrm{Gr}\, f$ with the subspace product topology (see Definition 2.69). Since $\mathrm{Gr}\, f$ is closed in $X \times Y$, every closed subset C of $\mathrm{Gr}\, f$ is also closed in $X \times Y$. Then $\mathrm{proj}_Y(C) = f\big(\mathrm{proj}_X(C)\big)$. But since f and proj_X are both closed, we infer that $\mathrm{proj}_Y(C)$ is closed in Y and so we conclude that proj_Y is a closed function.

"$\Longleftarrow$": Now we assume that $\operatorname{proj}_X \colon \operatorname{Gr} f \longrightarrow X$ and $\operatorname{proj}_Y \colon \operatorname{Gr} f \longrightarrow Y$ are both closed functions. We have

$$\operatorname{proj}_X^{-1}(x) = (x, f(x)) \quad \text{and} \quad \operatorname{proj}_Y\big(\operatorname{proj}_X^{-1}(x)\big) = f(x),$$

i.e., $\operatorname{proj}_Y \circ \operatorname{proj}_X^{-1} = f$. Let $C \subseteq X$ be closed. Then

$$f(C) = \operatorname{proj}_Y\big(\operatorname{proj}_X^{-1}(C)\big).$$

But proj_X is continuous, hence $\operatorname{proj}_X^{-1}(C)$ is closed. By hypothesis proj_Y is a closed function, hence $\operatorname{proj}_Y\big(\operatorname{proj}_X^{-1}(C)\big)$ is closed in Y. This proves that $\operatorname{proj}_Y \circ \operatorname{proj}_X^{-1} = f$ is a closed function. Also, let $D \subseteq Y$ be a closed set. Then $f^{-1}(D) = \operatorname{proj}_X\big(\operatorname{proj}_Y^{-1}(D)\big)$ (since $f = \operatorname{proj}_Y \circ \operatorname{proj}_X^{-1}$). But proj_Y is continuous on $\operatorname{Gr} f$ and so $\operatorname{proj}_Y^{-1}(D) \subseteq \operatorname{Gr} f$ is closed. By hypothesis proj_X is closed, hence $\operatorname{proj}_X\big(\operatorname{proj}_Y^{-1}(D)\big) = f^{-1}(D)$ is closed in X and this proves the continuity of f.

Solution of Problem 2.37

First we show that

$$\overline{A \times C} = \overline{A} \times \overline{C}.$$

Evidently, $A \times C \subseteq \overline{A} \times \overline{C}$ and because $\overline{A} \times \overline{C}$ is closed, we have that

$$\overline{A \times C} \subseteq \overline{A} \times \overline{C}.$$

On the other hand, let $(x, y) \in \overline{A} \times \overline{C}$ and let W be a neighbourhood of (x, y). We know that we can find $U \in \mathcal{N}(x)$ and $V \in \mathcal{N}(y)$ such that $(x, y) \in U \times V \subseteq W$ (see Remark 2.70). Since $x \in \overline{A}$ and $y \in \overline{C}$, we have

$$A \cap U \neq \emptyset \quad \text{and} \quad C \cap V \neq \emptyset,$$

hence

$$(A \times C) \cap (U \times V) \neq \emptyset$$

and so

$$(A \times C) \cap W \neq \emptyset.$$

Since W was an arbitrary neighbourhood of (x, y), we conclude that $(x, y) \in \overline{A \times C}$ and so, finally

$$\overline{A \times C} = \overline{A} \times \overline{C}.$$

From this we infer that $A \times C$ is closed if and only if $A \subseteq X$ and $C \subseteq Y$ are both closed.

Next we show that

$$\operatorname{int}(A \times C) \;=\; \operatorname{int} A \times \operatorname{int} C.$$

Evidently, $\operatorname{int} A \times \operatorname{int} C \subseteq A \times C$ and because $\operatorname{int} A \times \operatorname{int} C$ is open, we have $\operatorname{int} A \times \operatorname{int} C \subseteq \operatorname{int}(A \times C)$. Next, let $(x, y) \in \operatorname{int}(A \times C)$. We can find $U \in \mathcal{N}(x)$ and $V \in \mathcal{N}(y)$ such that $U \times V \subseteq A \times C$. Hence $U \subseteq A$ and $V \subseteq C$ and these two inclusions imply that $x \in \operatorname{int} A$ and $y \in \operatorname{int} C$, hence $(x, y) \in \operatorname{int} A \times \operatorname{int} C$. So, we conclude that $\operatorname{int}(A \times C) = \operatorname{int} A \times \operatorname{int} C$. From this we infer that $A \times C$ is open if and only if $A \subseteq X$ and $C \subseteq Y$ are both open.

Finally, note that

$$\begin{gathered} \overline{A \times C} \;=\; X \times Y \\ \Updownarrow \\ \overline{A} \times \overline{C} \;=\; X \times Y \\ \Updownarrow \\ \overline{A} \;=\; X \quad \text{and} \quad \overline{C} \;=\; Y. \end{gathered}$$

So, we conclude that $A \times C$ is dense in $X \times Y$ if and only if A is dense in X and C is dense in Y.

Solution of Problem 2.38
Let $x = (x_i)_{i \in I} \in X$. For every finite set $F \subseteq I$ and $r > 0$, we set

$$U_{F,r} \;=\; \prod_{i \in I} U_{F,r,i},$$

where

$$U_{F,R,i} \;=\; \begin{cases} B_r^i(x_i) & \text{if} \quad i \in F, \\ X_i & \text{if} \quad i \notin F, \end{cases}$$

with

$$B_r^i(x_i) \;=\; \{u \in X_i \colon d_{X_i}(u, x_i) < r\}.$$

We know that $\{U_{F,r}\}_{F,r}$ form a local basis at the point $x = (x_i)_{i \in I} \in X$ in the product topology (see Definition 2.69 and Remark 2.70). Proceeding by contradiction, suppose that X is metrizable. So, there exists a metric d_X on X generating the product topology. The balls

$$B_{2^{-n}}(x) \;=\; \{u \in X : \; d_X(u, x) < 2^{-n}\}, \quad n \geqslant 1,$$

form a local basis at the point $x \in X$. Therefore, for every $n \in \mathbb{N}$, we can find a finite set $F_n \subseteq I$ and $r_n > 0$ such that $U_{F_n,r_n} \subseteq B_{2^{-n}}(x)$. Let

$$S = \bigcup_{n \geqslant 1} F_n.$$

This set being a countable union of finite sets is countable. Therefore the inclusion $S \subseteq I$ is strict. Let $y = (y_i)_{i \in I} \subseteq X$ be such that $y_i = x_i$ for all $i \in S$. Then

$$y_n \in U_{F_n,r_n} \subseteq B_{2^{-n}}(x) \qquad \forall\, n \geqslant 1.$$

Hence

$$y \in \bigcap_{n \geqslant 1} B_{2^{-n}}(x) = \{x\}$$

and so $y = x$. This implies that $y_i = x_i$ for all $i \in I \setminus S$. So we have shown that for all $i \in I \setminus S$, X_i is a singleton, a contradiction to our hypothesis. Therefore X equipped with the product topology is not metrizable.

Solution of Problem 2.39

(a) By Definition 2.72, $s(f)$ is the strongest topology on Y for which f is continuous. Hence $\tau_Y \subseteq s(f)$. On the other hand, if $U \in s(f)$, then $f^{-1}(U) \in \tau_X$ (from the definition of $s(f)$). The openness of f (see Definition 2.59) implies that $f\big(f^{-1}(U)\big) \in \tau_Y$. But because f is surjective, we have

$$U = f\big(f^{-1}(U)\big) \in \tau_Y$$

and so we conclude that $\tau_Y = s(f)$.

(b) By Definition 2.62, $w(f)$ is the weakest topology on X for which f is continuous. Hence $w(f) \subseteq \tau_X$. On the other hand, let $U \in \tau_X$. Because f is open, we have $f(U) \in \tau_Y$. The injectivity of f implies that $f^{-1}(f(U)) = U$ and from definition of the weak (initial) topology, we have

$$f^{-1}(f(U)) = U \in w(f).$$

Therefore, we conclude that $\tau_X = w(f)$.

Solution of Problem 2.40

From the definitions of $w(f)$ and $s(f)$ (see Definitions 2.62 and 2.72 respectively), we see that in both cases f is continuous. Moreover, if X is furnished with the weak (initial) topology $w(f)$, then given $V \in w(f)$, we know that $V = f^{-1}(U)$ with $U \subseteq Y$ open (see Remark 2.63). Hence due to surjectivity of f, we have $f(V) = f(f^{-1}(U)) = U$ and so we infer that f is an open function (see Definition 2.59). Thus f is a homeomorphism.

On the other hand, if Y is furnished with the strong (final) topology $s(f)$, then for a given $V \subseteq X$ open, we have $f^{-1}(f(V)) = V$ (since f is injective) and so from the definition of $s(f)$ (see Definition 2.72), we have that $f(V) \in s(f)$ and so f is an open function. Thus f is a homeomorphism.

Solution of Problem 2.41

For every open (respectively, closed) set $A \subseteq X$, we have that the set $A_s = p^{-1}(p(A))$ is open (respectively, closed) if and only if $p(A)$ is open (respectively, closed) (see the definition of the quotient topology; Definition 2.76) if and only if the quotient function p is open (respectively, closed).

Solution of Problem 2.42

First note that directly from Definition 2.62, we have that $w(C(X)) \subseteq \tau$. Also, we have

$$w(C_b(X)) \subseteq w(C(X)).$$

On the other hand, let $f \in C(X)$, $x \in X$ and $\varepsilon > 0$ and consider the subbasic element

$$U(x, f, \varepsilon) = \{u \in X : |f(u) - f(x)| < \varepsilon\}.$$

We set

$$h(u) = \min\{f(x) + \varepsilon, \max\{f(x) - \varepsilon, f(u)\}\}.$$

Evidently $h \in C_b(X)$ and

$$U(x, f, \varepsilon) = U(x, h, \varepsilon).$$

It follows that $w(C(X)) \subseteq w(C_b(X))$, hence $w(C(X)) = w(C_b(X))$. So, the two weak topologies $w(C(X))$ and $w(C_b(X))$ are equal for any topological space.

Now assume that (X, τ) is completely regular. Let $U \in \tau$ and $x \in U$. Because of the complete regularity of X, we can find $f \in C_b(X)$ such that

$$f(x) = 0 \quad \text{and} \quad f|_{X \setminus U} = 1.$$

Then

$$V(f, x, 1) = \{u \in X : f(u) < 1\}$$

is a $w(C(X))$-neighbourhood of x and $V(f, x, 1) \subseteq U$. Therefore

$$\tau \subseteq w(C(X))$$

and so from the observation of the beginning of proof, we conclude that

$$\tau = w(C(X)) = w(C_b(X)).$$

Conversely, suppose that $\tau = w(C(X)) = w(C_b(X))$. Let $C \subseteq X$ be a nonempty closed set and let $x \notin C$. Then $X \setminus C = U$ is $w(C(X))$-open, contains x and we can find

$$V = \bigcap_{k=1}^{n} \{u \in X : |f_k(u) - f_k(x)| < 1\},$$

with $f_k \in C(X)$, $k = 1, \dots, n$ such that $x \in V \subseteq U$. For every $k \in \{1, \dots, n\}$, let

$$h_k(u) = \min\{1, |f_k(u) - f_k(x)|\} \qquad \forall\, u \in X$$

and

$$h(u) = \max_k h_k(u) \qquad \forall\, u \in X.$$

Evidently $h\colon X \longrightarrow [0,1]$ is a continuous function satisfying

$$h(x) = 0 \quad \text{and} \quad h|_C = 1.$$

This proves that (X,τ) is completely regular.

Solution of Problem 2.43

From the definition of complete regularity (see Problem 2.42), for every $x \in K$, we can find a continuous function $f_x\colon X \longrightarrow [0,1]$ such that

$$f_x(x) = 1 \quad \text{and} \quad f_x|_{X\setminus U} = 0.$$

Let

$$U_x = \{y \in X : f_x(y) > \tfrac{1}{2}\}.$$

Then U_x is open, $x \in U_x$ and $\{U_x\}_{x\in K}$ is an open cover of K. By the compactness of K, we can find a finite subcover $\{U_{x_k}\}_{k=1}^N$. Let

$$g(x) = \tfrac{1}{N}\sum_{k=1}^{N} f_{x_k}(x) \qquad \forall\, x \in X.$$

Then the function $g\colon X \longrightarrow [0,1]$ is continuous,

$$g|_{X\setminus U} = 0 \quad \text{and} \quad g(x) \geqslant \tfrac{1}{2N} \qquad \forall\, x \in K.$$

Let $\xi\colon [0,1] \longrightarrow [0,1]$ be a continuous function such that

$$\xi(0) = 0 \quad \text{and} \quad \xi|_{[\frac{1}{2N},1]} = 1.$$

Then $f = \xi \circ g$ is the desired continuous function.

Solution of Problem 2.44

Clearly for (a) and (b) also, we may assume that f is not identically $+\infty$.

Let f be coercive (respectively, sequentially coercive; see Definition 2.103) and lower semicontinuous (respectively, sequentially lower semicontinuous; see Definitions 2.46 and 2.49). Let $\{x_n\}_{n\geqslant 1} \subseteq X$ be a minimizing sequence for f. So

$$f(x_n) \longrightarrow m = \inf_{x\in X} f(x) < +\infty.$$

Let us consider the set

$$L = \begin{cases} \{x \in X : f(x) \leqslant m+1\} & \text{if } m \in \mathbb{R}, \\ \{x \in X : f(x) \leqslant 0\} & \text{if } m = -\infty. \end{cases}$$

Because f is coercive (respectively, sequentially coercive), the set L is countably compact (respectively, sequentially compact; see Definition 2.88). Note that we can find an integer $n_0 \geqslant 1$ such that

$$\{x_n\}_{n\geqslant n_0} \subseteq L.$$

So the sequence $\{x_n\}_{n\geqslant 1}$ must have a limit point (respectively, convergent subsequence); see Definition 2.88. This proves statement (c). Due to the lower semicontinuity (respectively, sequential lower semicontinuity; see Definitions 2.46 and 2.49) of f, we have

$$f(x) \leqslant \liminf_{n\to+\infty} f(x_n) = m$$

(see Remark 2.47) and so

$$m = f(x).$$

This proves statements (a) and (b).

Solution of Problem 2.45

No. Let I be the set of all rational numbers in $(0,1)$ equipped with the usual ordering of $\mathbb{R}$. We introduce a net $\{x_i\}_{i\in I} \subseteq [0,1]$, defined by $x_i = i$ for all $i \in I$. Evidently $x_i \longrightarrow 1$ in $[0,1]$ (see Definition 2.31). If $i_0 \in I$, then

$$\{x_i\}_{i\geqslant i_0} \cup \{1\} = \{g \in [i_0, 1] : g \text{ is a rational number}\}$$

and the latter fails to be compact for any $i_0 \in I$.

Solution of Problem 2.46

"$\Longrightarrow$": We assume that the sequence $\{f_n\}_{n\geqslant 1}$ is equicoercive. Then for every $\lambda \in \mathbb{R}$, we can find a closed, countably compact set K_λ (see Definition 2.88) such that

$$\{x \in X : f_n(x) \leqslant \lambda\} \subseteq K_\lambda \qquad \forall\, n \geqslant 1.$$

Let $\psi\colon X \longrightarrow \mathbb{R}^* = \mathbb{R} \cup \{\pm\infty\}$ be defined by

$$\psi(x) = \inf\{\mu \in \mathbb{R} : x \in K_\lambda \text{ for all } \lambda > \mu\},$$

with the usual convention that $\inf \emptyset = +\infty$. If $f_n(x) \leqslant \mu$ for all $n \geqslant 1$, the $x \in K_\lambda$ for all $\lambda > \mu$, hence $\psi(x) \leqslant \mu$. Therefore $\psi \leqslant f_n$ for all $n \geqslant 1$. Note that

$$\{x \in X : \psi(x) \leqslant \mu\} = \bigcap_{\lambda > \mu} K_\lambda,$$

so the set $\{x \in X : \psi(x) \leqslant \mu\}$ is closed and countably compact and thus ψ is lower semicontinuous (see Definition 2.46), coercive (see Definition 2.103).

"$\Longleftarrow$": Suppose that there is a lower semicontinuous, coercive function $\psi\colon X \longrightarrow \mathbb{R}^* = \mathbb{R} \cup \{\pm\infty\}$ such that $\psi \leqslant f_n$ for all $n \geqslant 1$. Then for all $n \geqslant 1$ and all $\lambda \in \mathbb{R}$, we have

$$\{x \in X : f_n(x) \leqslant \lambda\} \subseteq \{x \in X : \psi(x) \leqslant \lambda\} = K_\lambda$$

and the set K_λ is closed and countably compact. This shows that the sequence $\{f_n\}_{n\geqslant 1}$ is equicoercive.

Solution of Problem 2.47

From Problem 2.16, we know that

$$\{x \in X : \overline{f}(x) \leqslant \lambda\} = \bigcap_{\mu > \lambda} \overline{\{x \in X : f(x) \leqslant \mu\}} \qquad \forall\, \lambda \in \mathbb{R}$$

(see Definition 2.57). Since f is coercive (see Definition 2.103), the set $\overline{\{x \in X : f(x) \leqslant \mu\}}$ is countably compact (see Definition 2.88), hence the set $\overline{\{x \in X : \overline{f}(x) \leqslant \lambda\}}$ is countably compact too, which establishes the coercivity of $\overline{f}$.

Solution of Problem 2.48

(a) Note that $\overline{f}$ is coercive (see Definition 2.103 and Problem 2.47) and lower semicontinuous (see Definition 2.46). So, we can find $x_0 \in X$ such that $f(x_0) = \inf_X f$ (see Problem 2.44(a)). Clearly the constant function $\inf_X f$ is lower semicontinuous and so $\inf_X f \leqslant \overline{f}$. Hence

$$\inf_X f \;\leqslant\; \min_X \overline{f}.$$

Because $\overline{f} \leqslant f$, we also have $\min_X \overline{f} \leqslant \inf_X f$ and so we conclude that $\min_X \overline{f} = \inf_X f$.

(b) Let x be a limit point of $\{x_n\}_{n\geqslant 1} \subseteq X$ which is a minimizing sequence for f. Then from Remark 2.47, we have

$$\overline{f}(x) \;\leqslant\; \liminf_{n\to+\infty} \overline{f}(x_n) \;\leqslant\; \liminf_{n\to+\infty} f(x_n) \;=\; \inf_X f \;=\; \min_X \overline{f}$$

(see (a)), so

$$\overline{f}(x) \;=\; \min_X \overline{f}.$$

(c) Suppose that X is first countable and let $x \in X$ be a minimizer of $\overline{f}$ (see (a)). Then by Problem 2.18(b), we can find a sequence $\{y_n\}_{n\geqslant 1}$ such that $y_n \longrightarrow x$ and

$$\limsup_{n\to+\infty} f(y_n) \;\leqslant\; \overline{f}(x) \;=\; \min_X \overline{f} \;=\; \inf_X f.$$

Since $\overline{f} \leqslant f$, we conclude that

$$\lim_{n\to+\infty} f(y_n) \;=\; \min_X \overline{f} \;=\; \inf_X f.$$

Solution of Problem 2.49
Let $\{x_i\}_{i\in I}$ be a net in $\operatorname{proj}_X(C)$ and assume that $x_i \longrightarrow x$ in X. Then we can find $y_i \in Y$ such that $(x_i, y_i) \in C$. Because of the compactness of Y, we can find $\{y_j\}_{j\in J}$ a subnet of $\{y_i\}_{i\in I}$ such that $y_j \longrightarrow y \in Y$ (see Theorem 2.81). We consider the corresponding subnet $\{x_j\}_{j\in J}$ of $\{x_i\}_{i\in I}$ and we have

$$(x_i, y_i) \longrightarrow (x, y) \quad \text{in } X \times Y.$$

Because $\{(x_j, y_j)\}_{j\in J} \subseteq C$ and the latter set is closed, we have $(x, y) \in C$. Moreover, from the continuity of the natural projection proj_X (see Definition 2.69), we have

$$\operatorname{proj}_X(x_j, y_j) \longrightarrow \operatorname{proj}_X(x, y).$$

Hence $x \in \operatorname{proj}_X(C)$ and so we have proved that $\operatorname{proj}_X(C)$ is closed in X.

Solution of Problem 2.50
"$\Longrightarrow$": It follows from the definition of compactness (see Definition 2.78).

"$\Longleftarrow$": Let $\mathcal{Y} = \{U_\alpha\}_{\alpha\in I}$ be an open cover of X. We know that each open set U_α is a union of basic open sets. Let $\mathcal{M}_\alpha$ be the family of basic open sets whose union is U_α. Then the collection of all basic open sets V such that V belongs in some $\mathcal{M}_\alpha$ is an open cover of X. By hypothesis this open cover has a finite subcover $\{V_k\}_{k=1}^n$. For each V_k, let $\alpha_k \in I$ be the index such that $V_k \in \mathcal{M}_{\alpha_k}$. Then $\{U_{\alpha_k}\}_{k=1}^n$ is the desired finite subcover of X and so X is compact.

Solution of Problem 2.51
Let $C \subseteq X$ be a closed set. Then C is compact and so $f(C) \subseteq Y$ is compact too (see Proposition 2.82), in particular then closed (see

Proposition 2.83). This proves that f is a closed function. Then Problem 2.35 implies that f is an identification function (see Definition 2.74).

Solution of Problem 2.52

From Problem 2.51, we know that the function $f\colon X \longrightarrow f(X) \subseteq Y$ is an identification function (see Definition 2.74). The points of $\hat{X}$ are the sets $\{f^{-1}(y)\}$, $y \in Y$. Define $h\colon \hat{X} \longrightarrow Y$, by

$$h(f^{-1}(y)) = y \qquad \forall\, y \in Y.$$

Then h is a bijection and if $p\colon X \longrightarrow \hat{X} = X/_{\vartriangle}$ is the quotient function (see Definition 2.76), then $h \circ p = f$. So, by Proposition 2.73, h is continuous. Because X is compact, $\hat{X}$ is compact (as the image of X under the action of the continuous quotient function p). Therefore we can apply Theorem 2.84 and conclude that $h\colon \hat{X} \longrightarrow f(X) \subseteq Y$.

Solution of Problem 2.53

(a) Let τ and $\hat{\tau}$ be two comparable compact Hausdorff topologies on a set X. Suppose that $\tau \subseteq \hat{\tau}$. Then the identity function $i\colon (X, \hat{\tau}) \longrightarrow (X, \tau)$ is continuous. Then Theorem 2.84 implies that i is a homeomorphism, hence $\hat{\tau} = \tau$.

(b) Let X be any countable set, let $a, b \in X$ with $a \neq b$. Let

$$A = \left\{\tfrac{1}{n} : n \geqslant 1\right\} \cup \{0\}$$

be the topological space with the natural topology (induced from $\mathbb{R}$). Let $f, g\colon A \longrightarrow X$ be two bijections such that $f(a) = 0$ and $g(b) = 0$. On X we consider two topologies $\tau = s(f)$ and $\hat{\tau} = s(g)$ (see Definition 2.72). Then both topologies are Hausdorff compact, but not comparable (as $\{a\} \in \hat{\tau} \setminus \tau$ and $\{b\} \in \tau \setminus \hat{\tau}$).

Solution of Problem 2.54

"$\Longrightarrow$": Let $\{(x_i, y_i)\}_{i\in I} \subseteq \operatorname{Gr} f$ be a net such that $(x_i, y_i) \longrightarrow (x, y)$ in $X \times Y$ (see Definition 2.31). Then $x_i \longrightarrow x$ in X and $y_i \longrightarrow y$ in Y. The continuity of f implies that $f(x_i) \longrightarrow f(x)$ in Y. Hence $y = f(x)$ and so $(x, y) \in \operatorname{Gr} f$, which proves that $\operatorname{Gr} f$ is closed in $X \times Y$ (see Proposition 2.40). For this implication we do not need the compactness of Y (we need only the continuity of f).

"$\Longleftarrow$": We argue by contradiction. So, suppose that f is not continuous. Then we can find a net $\{x_i\}_{i\in I}$ such that $x_i \longrightarrow x$ in X and $f(x_i) \not\longrightarrow f(x)$ (see Proposition 2.40). So, there is $U \in \mathcal{N}(f(x))$ and a subnet of $\{f(x_i)\}_{i\in I}$ (which after renaming, we also denote by $\{f(x_i)\}_{i\in I}$) such that $f(x_i) \notin U$ for all $i \in I$. The compactness of Y guarantees that a subnet of $\{f(x_i)\}_{i\in I}$ (still denoted by $\{f(x_i)\}_{i\in I}$) converges to $y \in Y$. Then

$$(x_i, f(x_i)) \longrightarrow (x, y) \quad \text{in } X \times Y$$

and so $y = f(x)$, which contradicts the fact that $f(x_i) \notin U$ for all $i \in I$. Therefore f is continuous.

Finally we show that the second implication does not hold if we drop the assumption on the compactness of Y. Let $X = Y = \mathbb{R}$. Let

$$f(x) = \begin{cases} \frac{1}{x} & \text{if } x > 0, \\ 0 & \text{if } x \leqslant 0. \end{cases}$$

Then $Grf \subseteq \mathbb{R} \times \mathbb{R}$ is closed, but f is not continuous.

Solution of Problem 2.55

Let D be a locally compact dense subset of X and let $x \in D$. Due to the local compactness of D, we can find U open in D (with the subspace topology) such that $K = \overline{U}^D$ (the closure in D) is compact in D. We know that $U = V \cap D$, with $V \in \mathcal{N}(x)$ and clearly K is also compact in X. A fortiori then K is closed in X and so $X \setminus K$ is open in X. Let $u \in V \setminus K$. We claim that $u \in D$. Indeed, if this is not the case, because D is dense in X, u is a limit point of D and because $V \setminus K \in \mathcal{N}(u)$, we have

$$(V \setminus K) \cap D \neq \emptyset \quad \Longrightarrow \quad U \cap K^c \neq \emptyset,$$

a contradiction. So, we infer that $V \setminus K \subseteq D$ and so $V \subseteq D$, hence $V = U$. This shows that every point $x \in D$ has an X-open neighbourhood which remains in D, hence D is open.

Solution of Problem 2.56

We will do the proof for a decreasing net $\{f_i\}_{i \in I}$ (the proof being similar if the net is increasing). Replacing f_i by $f_i - f$ if necessary, we may assume that $f = 0$.

Let $\varepsilon > 0$ be given. Then for every $x \in X$ we can find $i_x \in I$ such that

$$0 \leqslant f_{i_x}(x) < \varepsilon.$$

The continuity of f_{i_x} implies that we can find $U_x \in \mathcal{N}(x)$ such that

$$0 \leqslant f_{i_x}(u) < \varepsilon \qquad \forall\, u \in U_x.$$

Since

$$f_i \leqslant f_{i_x} \qquad \forall\, i \geqslant i_x,$$

we see that

$$0 \leqslant f_i(u) < \varepsilon \qquad \forall\, u \in U_x,\ i \geqslant i_x.$$

Note that $\{U_x\}_{x \in X}$ is an open cover of X. The compactness of X (see Definition 2.78) implies that we can find a finite subcover $\{U_{x_k}\}_{k=1}^n$ of $\{U_x\}_{x \in X}$. Choose $i^* \in I$ such that $i^* \geqslant i_{x_k}$ for all $k \in \{1, \ldots, n\}$. Then

$$0 \leqslant f_i(u) < \varepsilon \qquad \forall\, u \in X,\ i \geqslant i^*,$$

so

$$f_i \rightrightarrows f$$

(see Definition 1.59).

(b) The functions

$$f_n \colon [0,1) \ni x \longmapsto x^n$$

satisfy $f_n \searrow 0$, but the convergence is not uniform. In this case the space $X = [0,1)$ is not compact.

(c) The functions

$$f_n\colon [0,1] \ni x \longmapsto x^n$$

satisfy $f_n \searrow \chi_{\{1\}}$, but the convergence is not uniform. In this case the limit function $f(x) = \chi_{\{1\}}$ is not continuous.

Solution of Problem 2.57

Let $\sigma\colon [0,1] \longrightarrow S^1$ be the map, defined by

$$\sigma(t) \;=\; \big(\cos 2\pi t,\; \sin 2\pi t\big).$$

Let $P(f)$ be the partition of $[0,1]$ determined by the σ-preimages of the elements of S^1. Then $P(f)$ is the same as the partition induced by the equivalence relation $0 \sim 1$. Finally note that $\sigma\colon I/_{P(f)} \longrightarrow S^1$ is a homeomorphism onto, since $I/_{P(f)} \longrightarrow S^1$ is compact and σ is a bijection (see Proposition 2.83(e) and Theorem 2.84).

Solution of Problem 2.58

Let $e_n = \{\delta_{k,N+1}\}_{k=1}^{N+1}$ be the "north pole" for the sphere S^N and let $s\colon S^N \setminus \{e_n\} \longrightarrow \mathbb{R}^N$ be the stereographic projection, defined by

$$s(u) \;=\; \tfrac{1}{1-u_{N+1}}(u_1,\ldots,u_N) \qquad \forall\, u \in S^N \setminus \{e_n\}.$$

Clearly, if $U \subseteq S^N$ is an open set containing the north pole e_n, then $s(S^N \setminus U)$ is compact and $s|_{S^N\setminus U}$ is homeomorphism. Therefore the Alexandrov one-point compactification of $\mathbb{R}^N$ (see Remark 2.97) is homeomorphism to S^N.

Solution of Problem 2.59

Let $\{U_i\}_{i\in I}$ be a family of open sets in X whose union contains K. Since $x \in K$, we can find an element U_{i^*} of this family such that $x \in U_{i^*}$. Because $x_n \longrightarrow x$, we can find $n_0 \geqslant 1$ such that $x_n \in U_{i^*}$ for all $n \geqslant n_0$. For every $1 \leqslant n \leqslant n_0$, we have $x_n \in K$ and so we can find $i_n \in I$ such that $x_n \in U_{i_n}$. Then the subfamily $\{U_{i_n}\}_{n=1}^{n_0-1} \cup \{U_{i^*}\}$

is a finite subcover of K. This proves that K is compact (see Definition 2.78(b)).

Solution of Problem 2.60

First assume that $D = \{x\}$. Since X is Hausdorff, for every $y \in C$, we can find $U_y \in \mathcal{N}(y)$ and $V_y \in \mathcal{N}(x)$ such that $U_y \cap V_y = \emptyset$. The family $\{U_y\}_{y \in C}$ is an open cover of C and so we can find a finite subfamily $\{U_{y_k}\}_{k=1}^n$ such that $C \subseteq \bigcup\limits_{k=1}^n U_{y_k}$. We set

$$U_x = \bigcup_{k=1}^n U_{y_k} \quad \text{and} \quad V_x = \bigcap_{k=1}^n V_{y_k}.$$

Both are open sets and

$$C \subseteq U_x, \quad x \in V_x \quad \text{and} \quad U_x \cap V_x = \emptyset.$$

So, we have established the separation when D is a singleton.

For the general case, we proceed as follows. From the first part of the proof, we know that for every $x \in D$, we can find an open set U_x with $C \subseteq U_x$ and $V_x \in \mathcal{N}(x)$ such that $U_x \cap V_x = \emptyset$. The family $\{V_x\}_{x \in D}$ is an open cover of D. The compactness of D implies that we can find a finite subcover $\{V_{x_k}\}_{k=1}^n$ of $\{V_x\}_{x \in D}$. Let

$$U = \bigcap_{k=1}^n U_{x_k} \quad \text{and} \quad V = \bigcup_{k=1}^n V_{n_k}.$$

Then U and V are two open sets in X such that $C \subseteq U$, $D \subseteq V$ and $U \cap V = \emptyset$.

Solution of Problem 2.61

Let

$$C_1 = K \setminus U_1 \quad \text{and} \quad C_2 = K \setminus U_2.$$

Both are closed subsets of K, hence they are compact. Moreover, $C_1 \cap C_2 = \emptyset$. By Problem 2.60, we can find two open sets V_1 and V_2 such that

$$C_1 \subseteq V_1 \subseteq U_2, \quad C_2 \subseteq V_2 \subseteq U_1 \quad \text{and} \quad V_1 \cap V_2 = \emptyset.$$

Let

$$K_1 = K \setminus V_1 \quad \text{and} \quad K_2 = K \setminus V_2.$$

Then both are compact, $K_1 \subseteq U_1$, $K_2 \subseteq U_2$ and $K = K_1 \cup K_2$.

Solution of Problem 2.62

Let $\{x_i\}_{i \in I} \subseteq f^{-1}(K)$ be a net such that $x_i \longrightarrow x$. Then

$$f(x_i) = y_i \in K \qquad \forall\, i \in I.$$

The compactness of K (see Definition 2.78) implies that there is a subnet $\{y_j\}_{j \in J}$ of $\{y_i\}_{i \in I}$ such that $y_j \longrightarrow y \in K$. Then $(x_j, y_j) \in \operatorname{Gr} f$ and $(x_j, y_j) \longrightarrow (x, y)$ in $X \times Y$. The closedness of $\operatorname{Gr} f$ implies that $(x, y) \in \operatorname{Gr} f$. Hence

$$x \in f^{-1}(y) \subseteq f^{-1}(K)$$

and this proves that the set $f^{-1}(K) \subseteq X$ is closed.

Solution of Problem 2.63

Let $K \subseteq Y$ be a compact set and consider a net $\{x_i\}_{i \in I} \subseteq f^{-1}(K)$. Then $\{y_i = f(x_i) : i \in I\} \subseteq K$. Because K is compact, we can find a subnet of $\{y_i\}_{i \in I}$ (which for the sake of notational simplicity, we denote by the same index) such that $y_i \longrightarrow y \in K$. For every $i_0 \in I$, we define $A_{i_0} = \{x_i : i \geqslant i_0\}$. Since f is closed (see Definition 2.59), we have

$$\overline{f(A_i)} \subseteq f(\overline{A}_i) \qquad \forall\, i \in I$$

and because $y_i = f(x_i) \longrightarrow y$ in Y, it follows that

$$\{y\} = \bigcap_{i \in I} \overline{f(A_i)} \subseteq \bigcap_{i \in I} f(\overline{A}_i).$$

Let $C = f^{-1}(y)$. Then for every $x \in C$, we have

$$f(x) \in \bigcap_{i \in I} f(\overline{A}_i)$$

and so the closed sets $D_i = C \cap \overline{A}_i \subseteq C$ are nonempty and clearly have the finite intersection property. Since by hypothesis C is compact, it follows that

$$\bigcap_{i\in I} D_i \;=\; C \cap \big(\bigcap_{i\in I} \overline{A}_i\big) \;\neq\; \emptyset.$$

So, we can find a limit point of $\{x_i\}_{i\in I}$ belonging in C. Thus we can find a suitable subnet of $\{x_i\}_{i\in I}$ (still labelled by the same index set I) such that $x_i \longrightarrow x \in C \subseteq f^{-1}(K)$. This proves that $f^{-1}(K)$ is compact.

Solution of Problem 2.64

We claim that f is bounded from above. Indeed, if this is not the case, then for every integer $n \geqslant 1$, we can find $x_n \in X$ such that $f(x_n) \geqslant n$. We consider the decreasing sequence of nonempty closed set

$$\big\{C_n = f^{-1}\big([n, +\infty)\big)\big\}_{n\geqslant 1}.$$

Evidently this has the finite intersection property. Since X is compact, by Theorem 2.81, we have that $\bigcap_{n\geqslant 1} C_n \neq \emptyset$. Let $y \in \bigcap\limits_{n\geqslant 1} C_n$. Then

$$f(y) \;\geqslant\; n \qquad \forall\, n \geqslant 1,$$

hence $f(y) = +\infty$, a contradiction. Therefore $\sup\limits_X f = M < +\infty$. Let

$$E_n \;=\; \big\{x \in X :\; f(x) \geqslant M - \tfrac{1}{n}\big\} \qquad \forall\, n \geqslant 1.$$

By hypothesis each E_n, $n \geqslant 1$ is closed and $\{E_n\}_{n\geqslant 1}$ is decreasing. Once again the compactness of X via Theorem 2.81 implies that

$$\bigcap_{n\geqslant 1} E_n \;\neq\; \emptyset.$$

Then, if $x_0 \in \bigcap\limits_{n\geqslant 1} E_n$, we have

$$f(x_0) \;\geqslant\; M - \tfrac{1}{n} \qquad \forall\, n \geqslant 1,$$

hence

$$f(x_0) \;=\; M \;=\; \sup_X f \;<\; +\infty.$$

Solution of Problem 2.65

The continuity of p_Y follows from the definition of the product topology (see Definition 2.69). Let $C \subseteq X \times Y$ be closed and let $\{y_i\}_{i \in I} \subseteq p_Y(C)$ be a net such that $y_i \longrightarrow y$. Then we can find a net $\{x_i\}_{i \in I} \subseteq X$ such that $(x_i, y_i) \in C$ for all $i \in I$. The compactness of X implies that we can find a subnet of $\{x_i\}_{i \in I}$ (which we denote as the original net) such that $x_i \longrightarrow x \in X$. Then $(x_i, y_i) \longrightarrow (x, y)$ and because by hypothesis C is closed, we have that $(x, y) \in C$ and so $y \in p_Y(C)$, which proves that p_Y is closed (see Definition 2.59). Also, if $y \in Y$, then $p_Y^{-1}(y) \subseteq X \times \{y\}$ and $X \times \{y\} \subseteq X \times Y$ is compact (see Theorem 2.91). Because $p_Y^{-1}(y)$ is closed (due to the continuity of p_Y), we conclude that $p_Y^{-1}(y) \subseteq X \times Y$ is compact.

Solution of Problem 2.66

Let $\{U_i\}_{i \in I}$ be an open cover of X. Because X is Lindelöf (see Definition 2.26), we can find a countable subcover $\{U_n\}_{n \geqslant 1}$. We need to extract from $\{U_n\}_{n \geqslant 1}$, a finite subcover. We proceed by induction. Let $V_1 = U_1$ and for every integer $n \geqslant 2$, let V_n be the first set in the sequence $\{U_n\}_{n \geqslant 1}$ not covered by $U_1 \cup \ldots \cup U_{n-1}$. If such a choice is not possible, then this means that $\{U_k\}_{k=1}^{n-1}$ is a cover of K and so we are done. Otherwise, for every $n \geqslant 1$, choose $x_n \in V_n$ such that $x_n \notin V_k$ for all integers $1 \leqslant k < n$. By hypothesis the sequence $\{x_n\}_{n \geqslant 1}$ has a limit point $x \in X$. Then $x \in V_m$ for some integer $m \geqslant 1$ and since it is a limit point of $\{x_n\}_{n \geqslant 1}$, we can find $n > m$ such that $x_n \in V_m$, a contradiction to the choice of the sequence $\{x_n\}_{n \geqslant 1}$. So the cover $\{U_n\}_{n \geqslant 1}$ has a finite subcover, which proves that X is compact (see Definition 2.78).

Solution of Problem 2.67

Arguing indirectly, suppose that $\operatorname{int} K \neq \emptyset$ and let $x \in \operatorname{int} K$. Then we can find a basic element U such that $x \in U \subseteq K$. We know that

$$U = \bigcap_{i \in F} p_i^{-1}(V_i),$$

with finite set $F \subseteq I$ and open sets $V_i \subseteq X_i$ (see Definition 2.69). If $j \in I \setminus F$, then

$$p_j(K) = X_j$$

and so X_j is compact being the continuous image of a compact space (see Proposition 2.82). So, all but a finite number of the spaces X_i are compact, a contradiction to our hypothesis. This prove that $\operatorname{int} K = \emptyset$ (i.e., K is nowhere dense; see Definition 1.25).

Solution of Problem 2.68

Let $C \subseteq Y$ be a closed set. Since by hypothesis $f|_K$ is continuous, we have that $f^{-1}(C) \cap K$ is closed in K and because K is compact, it is also closed in X (see Proposition 2.83). But then, because X is a k-space, we infer that $f^{-1}(C)$ is closed in X, which in turn implies the continuity of f (see Proposition 2.39).

Remark. It can be shown that locally compact or first countable spaces are k-spaces. In particular metric spaces are k-spaces.

Solution of Problem 2.69

Since X is locally compact (see Definition 2.92), for every $x \in A$, we can find $U \in \mathcal{N}(x)$ such that $\overline{U}$ is compact in X.

If A is open and since $\{x\}$ is compact, invoking Proposition 2.94, we can choose $U \in \mathcal{N}(x)$ such that

$$x \in U \subseteq \overline{U}^A \subseteq \overline{U} \subseteq A$$

and this proves that A with the subspace topology is locally compact.

If A is closed, then

$$U \cap A \in \mathcal{N}_A(x) \quad \text{and} \quad \overline{U \cap A}^A = \overline{U \cap A} \subseteq \overline{U}$$

with the latter being compact. Therefore, A with the subspace topology is again locally compact.

Solution of Problem 2.70

Let $y \in X \setminus \{x\}$. We can find $U \in \mathcal{N}(x)$ and $V \in \mathcal{N}(y)$ such that $U \cap V = \emptyset$. Note that $X \setminus U$ is closed in X, hence compact in X (see Proposition 2.83). Also

$$V \in \mathcal{N}_{X\setminus\{x\}}(y) \quad \text{and} \quad \overline{V}^{X\setminus\{x\}} \subseteq X \setminus U,$$

hence

$$\overline{V}^{X\setminus\{x\}} \text{ is compact in} X \setminus \{x\}$$

(see Proposition 2.79). So, we conclude that $X\setminus\{x\}$ is locally compact (see Definition 2.92).

Solution of Problem 2.71

Let $x \in \operatorname{int} K$. By replacing K with its image under the translation $y \longmapsto y - x$ (which is a homeomorphism of $\mathbb{R}^N$ with itself), we may assume that $x = 0 \in \operatorname{int} K$. So, we can find $\varepsilon > 0$, small such that

$$B_\varepsilon^N = \{y \in \mathbb{R}^N : \|y\| < \varepsilon\} \subseteq K.$$

Then using the dilation $x \longmapsto \frac{1}{\varepsilon}x$ (another homeomorphism on $\mathbb{R}^N$ with itself), we see that without any loss of generality we may assume that $B_1^N \subseteq K$.

Now we show that every ray originating at origin intersects ∂K at exact one point. Because K is compact, the intersection with every closed ray is compact. Denote this compact intersection by D. Since the distance function from the origin is continuous, it attains its supremum on D. Let $x_0 \in D$ be a point where it realizes this supremum. Evidently $x_0 \in \partial K$. Consider the line segment

$$[0, x_0] = \{x \in \mathbb{R}^N : tx_0,\ t \in [0,1]\}.$$

Since $B_1^N \subseteq K$, every line segment from x_0 to $u \in B_1^N$ is contained in K. As y moves over all B_1^N, we sweep a set C (which looks like an ice-cream cone). For every point tx_0, $t \in [0,1)$, we see that

$$B_{1-t}(tx_0) = \{y \in \mathbb{R}^N : \|y - tx_0\| < 1 - t\}$$

is in C, hence in K too. This proves that x_0 is unique and so we have proved that every ray originating at origin intersects ∂K at exact one point.

Now let $f: \partial K \longrightarrow S^{N-1} = \partial \overline{B}_1^N$ be a function, defined by

$$f(x) = \tfrac{x}{\|x\|} \qquad \forall\, x \in \partial K.$$

Evidently $f(x)$ corresponds to the point where the line segment joining the origin and $x \in \partial K$ intersects the unit sphere $S^{N-1} = \overline{B}_1^N$ (recall that $B_1^N \subseteq K$). Clearly f is continuous and a bijection (see the previous discussion). Because ∂K is compact, we can apply Theorem 2.84 and infer that f is a homeomorphism. Next let $h: \overline{B}_1^N \longrightarrow K$ be a function defined by

$$h(x) = \|x\| f^{-1}\big(\tfrac{x}{\|x\|}\big).$$

Clearly h is continuous and takes every line segment joining the origin with $u \in S^{N-1}$ onto the line segment joining the origin with $f^{-1}(u) \in \partial K$. The convexity of K implies that h has its values in K. Evidently h is injective and onto, therefore a homeomorphism (again Theorem 2.84).

Solution of Problem 2.72

Let $C \subseteq X$ be a closed set and let $y \in \partial f(C)$. Since Y is locally compact (see Definition 2.92) we can find $U \in \mathcal{N}(y)$ such that $\overline{U}$ is compact in Y. Because $y \in \partial f(C)$, for every $V \in \mathcal{N}(y)$, we have

$$f(C) \cap V \neq \emptyset \quad \text{and} \quad \big(X \setminus f(C)\big) \cap V \neq \emptyset.$$

Using this we can easily verify that we also have $y \in \partial\big(f(C) \cap \overline{U}\big)$. From the hypothesis on f, we have that $f^{-1}(\overline{U})$ is compact in X, which implies that $C \cap f^{-1}(\overline{U}) \subseteq X$ is compact too. Because f is a continuous bijection we have

$$f\big(C \cap f^{-1}(\overline{U})\big) = f(C) \cap f\big(f^{-1}(\overline{U})\big) = f(C) \cap \overline{U}$$

is compact (see Proposition 2.82). So, $y \in f(C) \cap \overline{U} \subseteq f(C)$, which shows that $\partial f(C) \subseteq f(C)$ and this proves that $f(C)$ is closed. Thus f is closed (see Definition 2.59).

Solution of Problem 2.73

First we note that for every $f \in \Phi$, the weak topology $w(\{f\})$ on X (see Definition 2.62) is semimetrizable (i.e., it is generated by a semimetric; see Remark 1.2). To see this, let

$$d_f(x, x') = d_Y\big(f(x), f(x')\big) \qquad \forall\, x, x' \in X.$$

Then $w(\{f\})$ is generated by the semimetric d_f. To establish this it is enough to show that the two topologies $w(\{f\})$ and τ_{d_f} have the same convergent nets (see Definition 2.31). By Proposition 2.66, $x_i \longrightarrow x$ in $w(\{f\})$ if and only if $d_Y\big(f(x_i), f(x)\big) = d_f(x_1, x) \longrightarrow 0$. Therefore $w(\{f\}) = \tau_{d_f}$ and so $w(\{f\})$ is semimetrizable. Because Φ is a countable family and the supremum of a countable family of semimetrizable topologies is semimetrizable, it follows that

$$\bigvee_{f \in \Phi} w(\{f\}) = w(\Phi)$$

is semimetrizable ($\bigvee_{f \in \Phi} w(\{f\})$ denoting the supremum of the topologies $\big\{w(\{f\})\big\}_{f \in \Phi}$).In fact since by hypothesis Φ is separating (see Definition 2.64), $w(\Phi)$ is metrizable. By the definition of the weak topology, we have that $w(\Phi) \subseteq \tau$, where τ denotes the topology on X. Invoking Theorem 2.84, we conclude that $w(\Phi) = \tau$ and so τ is metrizable (see Remark 2.2).

Solution of Problem 2.74

"$\Longrightarrow$": Since X is locally compact (see Definition 2.92) and σ-compact (see Definition 2.99), by Proposition 2.100, we can find an increasing sequence $\{C_n\}_{n \geqslant 1}$ of compact subsets of X such that

$$X = \bigcup_{n \geqslant 1} C_n \quad \text{and} \quad C_n \subseteq C_{n+1} \qquad \forall\, n \geqslant 1.$$

Then $\big\{X^* \setminus C_n\big\}_{n \geqslant 1}$ is a local basis of ∞ in X^* (see Definition 2.23 and Remark 2.97).

"$\Longleftarrow$": Let $\{U_n^*\}_{n \geqslant 1}$ be a local basis of ∞ in X^*. We have

$$U_n^* = X^* \setminus K_n, \qquad \forall\, n \geqslant 1,$$

with $K_n \subseteq X$ being compact. Then

$$X = \bigcup_{n \geqslant 1} K_n$$

and so X is σ-compact.

Solution of Problem 2.75
(a) Let $C_1 = f(X)$ and inductively define

$$C_{n+1} = f(C_n) \qquad \forall\, n \geqslant 1.$$

This way we produce a decreasing sequence $\{C_n\}_{n \geqslant 1}$ of nonempty and compact sets (hence closed too; see Proposition 2.83). Due to the compactness of X,

$$\text{the set } \bigcap_{n \geqslant 1} C_n = A \text{ is nonempty and compact.}$$

Clearly $f(A) = A$.

(b) Let $X = (0, 1]$ and consider the continuous function

$$f(x) = \tfrac{x}{2} \qquad \forall\, x \in X.$$

Suppose that there is a nonempty and closed subset $A \subseteq (0, 1]$ such that $f(A) = A$. Let $\hat{x} = \sup A \in A$. Because $f(A) = A$, we can find $u \in A$ such that $f(u) = \frac{1}{2}u = \hat{x}$. Therefore we have

$$u = 2\hat{x} \leqslant \hat{x},$$

a contradiction (since $\hat{x} \neq 0$).

Solution of Problem 2.76
Arguing by contradiction, suppose that no such $\lambda > 0$ exists. We define the sets

$$C_n = \big\{x \in X : nf(x) + g(x) \leqslant 0\big\} \qquad \forall\, n \geqslant 1.$$

This is a decreasing sequence of nonempty closed subsets of X. By Theorem 2.81, we have that

$$\bigcap_{n\geqslant 1} C_n \neq \emptyset.$$

Let $u \in \bigcap_{n\geqslant 1} C_n$. Then

$$f(u) \leqslant -\frac{g(u)}{n} \qquad \forall\, n \geqslant 1,$$

so

$$f(u) = 0$$

(recall that $f \geqslant 0$). Thus, by hypothesis $g(u) > 0$ and so $f(u) < 0$, a contradiction.

Solution of Problem 2.77

Let $i_0 \in I$ and consider the set $D = C_{i_0} \setminus U$. Then D is compact and $\{X \setminus C_i\}_{i\in I}$ is an open cover of D. Therefore we can find a finite subcover $\{X \setminus C_i\}_{i\in F}$. Hence

$$\bigcap_{i\in F} C_i \subseteq U.$$

Solution of Problem 2.78

Since f is locally bounded, for every $x \in X$, we can find a neighbourhood U_x of x and $M_x > 0$ such that

$$|f(y)| \leqslant M_x \qquad \forall\, y \in U_x.$$

Then $\{U_x\}_{x\in X}$ is an open cover of X. Since X is compact, we can find a finite subcover $\{U_{x_k}\}_{k=1}^N$ of X (see Definition 2.78). Let

$$M = \max_{1\leqslant k\leqslant N} M_{x_k}.$$

Then

$$\left|f(x)\right| \leqslant M \qquad \forall\, x \in X$$

and so we conclude that f is bounded.

Solution of Problem 2.79

Let $x \in D$. Since by hypothesis D is locally compact, we can find an open set U in D such that $x \in U$ and $\overline{U} \cap D$ is compact, hence closed in X. We have

$$\overline{U} \subseteq \overline{U} \cap D \subseteq D.$$

Let V be an open subset of X such that $U = V \cap D$. From the density of D in X and the fact that V is open in X, we have

$$\overline{V} = \overline{V \cap D},$$

so

$$x \in V \subseteq \overline{V} = \overline{V \cap D} = \overline{U} \subseteq D.$$

So, for every $x \in D$, there exists an open set V in X such that $x \in V \subseteq X$, which means that D is open in X.

Solution of Problem 2.80

Let d_X be a metric on X generating the topology of X and let $\hat{X}$ be the completion of (X, d_X). Then X is a locally compact dense subset of the metric space $\hat{X}$. By Problem 2.79, X is open in $\hat{X}$. Invoking the Alexandrov theorem (see Theorem 1.58), we conclude that X is completely metrizable.

Solution of Problem 2.81

No. To show this we argue indirectly. So, suppose that

$$f \colon S^1 \longrightarrow [0,1]$$

is a homeomorphism. Let $h = f^{-1} \colon [0,1] \longrightarrow S^1$. Then consider

$$X = \left[0, \tfrac{1}{2}\right) \cup \left(\tfrac{1}{2}, 1\right] \quad \text{and} \quad Y = S^1 \setminus h\left(\tfrac{1}{2}\right).$$

Hence $h(X) = Y$ and $f(Y) = X$. Equip X and Y with the corresponding subspace topologies. Then $f: Y \longrightarrow X$ and $h: X \longrightarrow Y$ are continuous bijections and $h = f^{-1}$. Therefore X and Y are homeomorphic. However, note that X is disconnected ($U = [0, \frac{1}{2})$ and $V = (\frac{1}{2}, 1]$ is a disconnection of X; see Definition 2.104) while Y is connected (being homeomorphic to $(0, 1)$ by the stereographic projection; cf. the solution of Problem 2.58). So, X and Y cannot be homeomorphic, a contradiction. This proves that S^1 and $[0, 1]$ cannot be homeomorphic.

Solution of Problem 2.82

Since K is disconnected (see Definition 2.104), we can find two nonempty disjoint closed sets $C, D \subseteq K$ such that $K = C \cup D$. The sets C and D are compact and disjoint. So, we can use Problem 2.60 and obtain two disjoint open sets $U, V \subseteq X$ such that $C \subseteq U$ and $D \subseteq V$. Then

$$K = C \cup D \subseteq U \cup V.$$

Moreover, we have

$$K \cap U = C \neq \emptyset \quad \text{and} \quad K \cap V = D \neq \emptyset.$$

Solution of Problem 2.83

(a) Note that the sets K_n are closed and $K_n \subseteq K_1$ for all $n \geqslant 1$ with K_1 being compact. Moreover, the sequence $\{K_n\}_{n \geqslant 1}$ has finite intersection property. Therefore, by Theorem 2.81, we have that $K = \bigcap_{n \geqslant 1} K_n$ is nonempty and closed, hence compact.

(b) We argue by contradiction. So, suppose that for every $n \geqslant 1$, we have $K_n \cap U^c \neq \emptyset$. Let $x_n \in K_n \cap U^c$ for all $n \geqslant 1$. Then $\{x_n\}_{n \geqslant 1}$ is a sequence in K_1, which is compact and so $\{x_n\}_{n \geqslant 1}$ has a limit point $x \in K_1$. Note that $x \in K_n$ for every $n \geqslant 1$ and so $x \in K$. Also, $x \in U^c$, since the latter is closed. Therefore $K \cap U^c \neq \emptyset$, a contradiction to the fact that $K \subseteq U$. This means that there is an integer $n_0 \geqslant 1$ such that $K_n \subseteq U$ for all $n \geqslant n_0$ (recall that the sequence $\{K_n\}_{n \geqslant 1}$ is decreasing).

(c) Arguing indirectly, suppose that K is disconnected (see Definition 2.104). Then according to Problem 2.82, we can find two open and disjoint sets $U, V \subseteq X$ such that $K \subseteq U \cup V$, $K \cap U \neq \emptyset$, $K \cap V \neq \emptyset$. From part **(b)** above, we know that there exists an integer $n_0 \geqslant 1$ such that $K_n \subseteq U \cup V$ for all $n \geqslant n_0$. Then $K_n \cap U$ and $K_n \cap V$ is a disconnection of K_n for all $n \geqslant 1$, a contradiction to the fact that for every $n \geqslant 1$, the set K_n is connected.

Solution of Problem 2.84

No. Consider the following counterexample:

$$C_n = \{(x_1, x_2) \in \mathbb{R}^2 : |x_1| \geqslant n \text{ or } |x_2| \geqslant 1\} \qquad \forall\, n \geqslant 1.$$

Evidently $\{C_n\}_{n\geqslant 1}$ is a decreasing sequence of nonempty closed sets which are connected (in fact path-connected; see Definitions 2.104 and 2.122). However

$$C = \bigcap_{n\geqslant 1} C_n = \{(x_1, x_2) \in \mathbb{R}^2 : |x_2| \geqslant 1\}$$

which is obviously disconnected.

Solution of Problem 2.85

Because $[0,1]$ is compact and connected and f is a homeomorphism, $A \times C$ is compact and connected too (see Theorem 2.91 and Proposition 2.110). Then from Theorem 2.91 and Proposition 2.110, it follows that both sets A and C are compact and connected. If $x, u \in A$, $x \neq u$ and $y, v \in C$, $y \neq v$, then since f^{-1} is a homeomorphism, we have

$$\begin{aligned} E_x &= f^{-1}(\{x\} \times C), & E_u &= f^{-1}(\{u\} \times C), \\ G_y &= f^{-1}(A \times \{y\}), & G_v &= f^{-1}(A \times \{v\}) \end{aligned}$$

are all compact and connected in $[0,1]$, hence closed subintervals of $[0,1]$. Note that

$$E_x \cap G_y = f^{-1}(\{x,y\}), \quad E_x \cap G_v = f^{-1}(\{x,v\}),$$
$$E_u \cap G_y = f^{-1}(\{u,y\}), \quad E_u \cap G_v = f^{-1}(\{u,v\})$$

and

$$E_x \cap E_u = \emptyset, \quad G_y \cap G_v = \emptyset.$$

Let $a \in E_x \cap G_y$, $b \in E_x \cap G_v$, $c \in E_u \cap G_y$ and $d \in E_u \cap G_v$. Without any loss of generality assume that $a \leqslant b \leqslant c \leqslant d$. Then $[a,c] \subseteq G_y$ and $[b,d] \subseteq G_v$, hence $[b,c] \subseteq G_y \cap G_v \neq \emptyset$, a contradiction. This proves that one of the sets A or C must be a singleton.

Solution of Problem 2.86

Let $x \in U$ and let $V(x)$ be the path-connected component of U containing x (see Definition 2.130). Clearly $\mathbb{R}^N$ is locally path-connected (since every open set contains a convex open set; see Definition 2.127). So, by Proposition 2.132, we have that $V(x)$ is open. We claim that $U \setminus V(x)$ is open too. Let $u \in U \setminus V(x)$. Since U is open, we can find $\varepsilon > 0$ small enough such that

$$B_\varepsilon(u) = \{y \in \mathbb{R}^N : \|y-u\| < \varepsilon\} \subseteq U.$$

All points $y \in B_\varepsilon(u)$ are contained in $X \setminus V(x)$ since they can be connected to u by a straight-line path (a line segment) contained in $B_\varepsilon(u)$ (see Definition 2.122). This proves that $U \setminus V(x)$ is open. But because by hypothesis U is connected (see Definition 2.104), we must have $U \setminus V(x) = \emptyset$, hence

$$V(x) = U$$

and so U is path-connected (see Definition 2.122).

(b) It is enough to take any connected space which is not path-connected (see, e.g., Example 1.100).

Solution of Problem 2.87

No. The reason being that every point in (x,u) is a cut point of order 2 (see Definition 2.120), while in $[y,z)$ the point y is a cut point of order 1. Recall that the number of cut points of order k is a topological invariant. This implies that the two intervals cannot be homeomorphic.

Solution of Problem 2.88

Let

$$x \in \bigcup_{i\in I} A_i$$

and let $i_x \in I$ be such that $x \in A_{i_x}$. For every other $u \in \bigcup\limits_{i\in I} A_i$, we can find $i_u \in I$ such that $u \in A_{i_u}$. Then by hypothesis, we can find a finite set $\{i_k\}_{k=0}^n \subseteq I$ such that $i_0 = i_x$ and $i_n = i_u$ and

$$A_{i_{k-1}} \cap A_{i_k} \neq \emptyset \qquad \forall\, k \in \{1,\ldots,n\}.$$

We claim that for every $m \in \{0,1,\ldots,n\}$ the set $\bigcup\limits_{k=0}^{m} A_{i_k}$ is connected (see Definition 2.104). To see this note that for $m=0$, this is true since $A_{i_0} = A_{i_x}$ is by hypothesis connected. Suppose that the claim is true for $m = r \in \{0,1,\ldots,n-1\}$. We will show that it is also true for $m = r+1$. The sets $\bigcup\limits_{k=0}^{r} A_{i_k}$ and $A_{i_{r+1}}$ are connected with nonempty intersection (since $A_{i_r} \cap A_{i_{r+1}} \neq \emptyset$). Proposition 2.109 implies that the set $\bigcup\limits_{k=0}^{r+1} A_{i_k}$ is connected. This proves the claim and so we have shown that the set $\bigcup\limits_{k=0}^{n} A_{i_k}$ is connected. This means that the points x and u are in the same connected component of $\bigcup\limits_{i\in I} A_i$ (see Definition 2.111). But $u \in \bigcup\limits_{i\in I} A_i$ was arbitrary. Therefore the set $\bigcup\limits_{i\in I} A_i$ has only one connected component, i.e., it is connected.

Solution of Problem 2.89
Let $C \subseteq Y$ be a connected set (see Definition 2.104) and $A = f^{-1}(C)$. We proceed by contradiction. So, suppose that A is not connected. Then we can find two closed sets $D_1, D_2 \subseteq X$ such that

$$\begin{aligned} A &= (D_1 \cap A) \cup (D_2 \cap A), \quad (D_1 \cap A) \cap (D_2 \cap A) = \emptyset, \\ D_1 \cap A &\neq \emptyset \quad \text{and} \quad D_2 \cap A \neq \emptyset. \end{aligned}$$

If there exist $x_1 \in D_1 \cap A$, $x_2 \in D_2 \cap A$ such that $f(x_1) = f(x_2) = y$, then

$$y \in C \quad \text{and} \quad x_1, x_2 \in f^{-1}(y).$$

But by hypothesis the set $E = f^{-1}(y)$ is connected,

$$E = (D_1 \cap E) \cup (D_2 \cap E) \quad \text{and} \quad (D_1 \cap E) \cap (D_2 \cap E) = \emptyset,$$

which contradicts the connectedness of E. Therefore, we must have

$$C = f(D_1 \cap A) \cup f(D_2 \cap A) \quad \text{with} \quad f(D_1 \cap A) \cap f(D_2 \cap A) = \emptyset.$$

Recall that C is connected. Since the sets $f(D_1 \cap A)$ and $f(D_2 \cap A)$ are nonempty, if we show that they are closed in C, we would have a contradiction and so we are done. To this end let $\{y_i\}_{i \in I} \subseteq f(D_1 \cap A)$ be a net and assume that $y_i \longrightarrow y \in C$. Then

$$y_i = f(x_i) \quad \text{with } x_i \in D_1 \cap A \qquad \forall\, i \in I.$$

Because of the compactness of X, we can find a subnet $\{x_j\}_{j \in J}$ of $\{x_i\}_{i \in I}$ such that $x_j \longrightarrow x$ in X. Since D_1 is closed in X and $\{x_j\}_{j \in J} \subseteq D_1$, we infer that $x \in D_1$. Also $y_j \longrightarrow y \in C$ in Y and by the continuity of the function f, we have $f(x_j) \longrightarrow f(x)$. Hence $y = f(x) \in C$ and so $x \in A = f^{-1}(C)$. Therefore $x \in D_1 \cap A$ and so $y = f(x) \in f(D_1 \cap A)$. This proves that $f(D_1 \cap A)$ is closed in C. Similarly, we show that $f(D_2 \cap A)$ is closed in C. So, finally we have a contradiction to the fact that C is connected. This proves that $f^{-1}(C) = A$ is connected.

Solution of Problem 2.90
If $C \cap \partial A = \emptyset$, then

$$C \cap \operatorname{int} A \neq \emptyset \quad \text{and} \quad C \cap \operatorname{ext} A \neq \emptyset$$

(see Definition 2.9(b)). Then $C \cap \operatorname{int} A$ and $C \cap \operatorname{ext} A$ is a disconnection of C, contradicting the hypothesis that C is connected (see Definition 2.104).

Solution of Problem 2.91

Since Y is not path-connected (see Definition 2.122), we can find $y_1, y_2 \in Y$ which cannot be joined by a path in Y. Arguing by contradiction, suppose that we could find a continuous surjection $f\colon X \longrightarrow Y$. Then there are points $x_1, x_2 \in X$ such that

$$f(x_1) = y_1 \quad \text{and} \quad f(x_2) = y_2.$$

Because X is path-connected, we can find a path $\vartheta\colon [0,1] \longrightarrow X$ such that

$$\vartheta(0) = x_1 \quad \text{and} \quad \vartheta(1) = x_2.$$

Then $f \circ \vartheta\colon [0,1] \longrightarrow Y$ is a path in Y such that

$$(f \circ \vartheta)(0) = f(x_1) = y_1 \quad \text{and} \quad (f \circ \vartheta)(u) = f(x_2) = y_2,$$

a contradiction.

Solution of Problem 2.92

For $N = 1$, we already know that S^1 in path-connected (see Definition 2.122), since it is the image of $\mathbb{R}$ under the exponential function

$$e(x) = \big(\cos(2\pi x),\ \sin(2\pi x)\big) \qquad \forall\, x \in \mathbb{R}.$$

For the general case ($N \geqslant 1$), we know that $\mathbb{R}^{N+1} \setminus \{0\}$ is path-connected and $f\colon \mathbb{R}^{N+1} \setminus \{0\} \longrightarrow S^N$, defined by

$$f(x) = \tfrac{x}{\|x\|} \qquad \forall\, x \in \mathbb{R}^{N+1}$$

is a continuous surjection. Problem 2.91 implies that S^N must be path-connected.

Alternative Solution

Let e_n be the "north pole" of S^N (i.e., $e_n = \{\delta_{k,N}\}_{k=1}^N$) and e_s is the "south pole" of S^N (i.e., $e_s = \{-\delta_{k,N}\}_{k=1}^N$). Using the stereographic projection (cf. the solution of Problem 2.58), we have that $S^N \setminus \{e_n\}$ and $S^N \setminus \{e_s\}$ are both homeomorphic to $\mathbb{R}^N$. Since $\mathbb{R}^N$ is path-connected (being convex), it follows that both $S^N \setminus \{e_n\}$ and $S^N \setminus \{e_s\}$ are path-connected (see Definition 2.122). Invoking Proposition 2.129, we conclude that

$$S^N = (S^N \setminus \{e_n\}) \cup (S^N \setminus \{e_s\})$$

is path-connected.

Solution of Problem 2.93

We do the solution for the locally connected case (the locally compact case being similar; Definition 2.92). So, assume that X is locally connected (see Definition 2.116). Let $y \in Y$ and $V \in \mathcal{N}_Y(y)$. Then $y = f(x)$ for some $x \in X$ and $f^{-1}(V) \in \mathcal{N}_X(x)$. Let $U \in \mathcal{N}_X(x)$ be such that $U \subseteq f^{-1}(V)$ and since f is an open continuous surjection (see Definition 2.59), we have

$$y \in f(U) \subseteq f(f^{-1}(V)) = V$$

and $f(U)$ is an open connected neighbourhood of y (see Proposition 2.108). Since $y \in Y$ and $V \in \mathcal{N}_Y(y)$ were arbitrary, we conclude that Y is locally connected.

Solution of Problem 2.94

Let $x \in C$ and let $C(x)$ be the connected component containing x (see Definition 2.111). Then $C \subseteq C(x)$. If $C(x) \cap D \neq \emptyset$, then in a similar fashion $C(x) \supseteq D$ and so $C(x) = X$, a contradiction to the fact that X is disconnected (see Definition 2.104). So $C(x) \cap D = \emptyset$ and it follows that $C(x) = C$. Similarly, $C(u) = D$ for $u \in D$. This proves that C and D are both the connected components of X.

Solution of Problem 2.95

Let $f\colon C \cup D \longrightarrow \{0,1\}$ be a continuous function. Since C and D are connected (see Definition 2.104), the functions $f|_C$ and $f|_D$ are constant functions (see Theorem 2.105). Note that the set $\overline{C} \cap (C \cup D)$ is the closure of C in $C \cup D$ and so by continuity the value of $f|_C$ is also the value of $f|_{\overline{C}\cap(C\cup D)}$. This is also the value of $f|_{\overline{C}\cap D}$ and so we conclude that

$$f|_C = f|_D$$

which by Theorem 2.105 establishes the connectedness of $C \cup D$.

Solution of Problem 2.96

We proceed by contradiction. So, suppose that $\overline{A}$ is disconnected (see Definition 2.104). Then

$$\overline{A} = C_1 \cup C_2,$$

with $C_1, C_2 \subseteq X$ being nonempty, disjoint, closed sets.

Since $C_1 \cup C_2$ is the smallest closed set containing A (see Proposition 2.11(d)), we have that

$$A \cap C_1 \neq \emptyset \quad \text{and} \quad A \cap C_2 \neq \emptyset.$$

We also have

$$A = (A \cap C_1) \cup (A \cap C_2).$$

Note that

$$\partial(A \cap C_1) \neq \emptyset \quad \text{and} \quad \partial(A \cap C_2) \neq \emptyset.$$

Indeed, if this is not the case, say $\partial(A \cap C_1) = 0$, then

$$A \cap C_1 = \overline{A \cap C_1} = \operatorname{int}(A \cap C_1)$$

and so, we have a nontrivial clopen set, contradicting the connectedness of X. Since

$$\partial A = \partial(A \cap C_1) \cup \partial(A \cap C_2)$$

and by hypothesis ∂A is connected, we have a contradiction.

Solution of Problem 2.97
We know that $D = \gamma\big([0,1]\big)$ is a closed, connected set (see Definition 2.104 and Proposition 2.108) and by hypothesis

$$D \cap A \neq \emptyset \quad \text{and} \quad D \cap (X \setminus A) \neq \emptyset.$$

If $D \cap \partial A = \emptyset$, then

$$D = (D \cap \overline{A}) \cup (D \cap \overline{X \setminus A})$$

and the sets $D \cap \overline{A}$ and $D \cap \overline{X \setminus A}$ are closed in D and

$$(D \cap \overline{A}) \cap (D \cap \overline{X \setminus A}) = D \cap (\overline{A} \cap \overline{X \setminus A}) = D \cap \partial A = \emptyset.$$

So, we infer that the set D is disconnected, a contradiction.

Solution of Problem 2.98
(a) This set is disconnected (see Definition 2.104), since its projection on the x-axis is disconnected.

(b) This set is connected, since any two points in it can be connected by a piecewise linear path (with at most two segments).

(c) Let

$$C = \bigcup_{\vartheta \in \mathbb{Q}\setminus\{0\}} \big\{(x,y) \in \mathbb{R}^2 : \; y = \vartheta x\big\}.$$

Clearly the coordinates of an element in C are either both rational or both irrational. The set C is connected (in fact it is path-connected; see Definition 2.122). Note that our set contains C and is contained in $\overline{C} = \mathbb{R}^2$ and so it is connected (see Corollary 2.107).

Solution of Problem 2.99
If X is not a singleton, then we can find $x, u \in X$ with $x \neq u$. Since $\{x\}$ and $\{u\}$ are two disjoint closed sets, invoking the Urysohn lemma (see Theorem 2.136), we can find a continuous function $f: X \longrightarrow [0,1]$ such that $f(x) = 0$ and $f(u) = 1$. Since X is connected, $F(X)$ is an

interval in $[0,1]$, hence $f(X)=[0,1]$. This means that $\operatorname{card} X \geqslant \mathfrak{c}$ (the cardinality of the continuum).

Solution of Problem 2.100

For every $x \in X$ choose $U_x \in \mathcal{Y}$ such that $x \in U_x$. Since X is normal, we can apply the Urysohn lemma (see Theorem 2.136) and produce a continuous function $g_x \colon X \longrightarrow [0,1]$ such that

$$g_x|_{X \setminus U_x} = 0 \quad \text{and} \quad g_x(x) = 1.$$

Let us set

$$V_x = \{u \in X : g_x(u) > 0\}.$$

Evidently $V_x \subseteq U_x$ and so $\{V_x\}_{x \in X}$ is a locally finite open cover of X. Therefore

$$g(x) = \sum_{U \in \mathcal{Y}} g_U(x)$$

is a well defined continuous function on X and

$$g(x) > 0 \qquad \forall\, x \in X.$$

Replacing g_U by $f_U = \frac{g_U}{g}$, we obtain $f_U \colon X \longrightarrow [0,1]$, continuous functions, such that

$$\sum_{U \in \mathcal{Y}} f_U(x) = 1 \qquad \forall\, x \in X.$$

Solution of Problem 2.101

Because of the normality of X (see Definition 2.4), we can find two open sets $U, V_1 \subseteq X$ such that $A \subseteq U$, $C \subseteq V_1$ and $U \cap V_1 = \emptyset$. We claim that $\overline{U} \cap V_1 = \emptyset$. Proceeding by contradiction, suppose that $x \in \overline{U} \cap V_1$. Since $x \in \overline{U}$ and $V_1 \in \mathcal{N}(x)$, we must have $U \cap V_1 \neq \emptyset$, a contradiction. This proves the claim.

We have $\overline{U} \cap C = \emptyset$ and so the normality of X implies that we can find two open sets $U_1, V \subseteq X$ such that

$$\overline{U} \subseteq U_1, \quad C \subseteq V \quad \text{and} \quad U_1 \cap V = \emptyset.$$

As before, we show that $U_1 \cap \overline{V} = \emptyset$. Evidently U and V are the desired open sets.

Alternative Solution

Because X is normal, the Urysohn lemma (see Theorem 2.136) implies that there exists a continuous function $f: X \longrightarrow [0,1]$ such that

$$f|_A = 0 \quad \text{and} \quad f|_C = 1.$$

Let

$$U = f^{-1}([0, \tfrac{1}{4})) \quad \text{and} \quad V = f^{-1}((\tfrac{3}{4}, 1)].$$

Both are open sets (due to the continuity of f) and they satisfy

$$A \subseteq \overline{U}, \quad C \subseteq \overline{V} \quad \text{and} \quad \overline{U} \cap \overline{V} = \emptyset.$$

Solution of Problem 2.102

(a) "$\Longrightarrow$": We have

$$A = f^{-1}(\{0\}) = f^{-1}\Big(\bigcap_{n \geqslant 1} [0, \tfrac{1}{n})\Big) = \bigcap_{n \geqslant 1} f^{-1}\big([0, \tfrac{1}{n})\big)$$

and each set $f^{-1}\big([0, \frac{1}{n})\big)$, for $n \geqslant 1$, is open by the continuity of f and of the fact that the set $[0, \frac{1}{n})$ is open in $[0,1]$. So A is a G_δ-set (see Definition 2.18).

"$\Longleftarrow$": Since A is a G_δ-set in X, $A = \bigcap\limits_{n \geqslant 1} U_n$ with $U_n \subseteq X$ being open for $n \geqslant 1$. Note that A and $X \setminus U_n$ are disjoint closed sets. Since X is normal (see Definition 2.4), by the Urysohn lemma (see Theorem 2.136), for every $n \geqslant 1$, we can find a continuous function $f_n: X \longrightarrow [0,1]$ such that

$$f_n|_A = 0 \quad \text{and} \quad f_n|_{X \setminus U_n} = 1.$$

Let $f: X \longrightarrow [0,1]$ be defined by

$$f(x) = \sum_{n \geqslant 1} \tfrac{1}{2^n} f_n(x) \qquad \forall\, x \in X.$$

Since

$$\tfrac{1}{2^n}\left|f_n(x)\right| \leqslant \tfrac{1}{2^n} \qquad \forall\, n \geqslant 1,\ x \in X,$$

by the Weierstrass M-test we infer that

$$\sum_{n\geqslant 1} \tfrac{1}{2^n} f_n(x) \text{ converges uniformly on } X$$

and so f is continuous. We claim that $f^{-1}(\{0\}) = A$. Clearly $f|_A = 0$. On the other hand, if $f(x) = 0$, then $f_n(x) = 0$ for all $n \geqslant 1$ (recall that $f_n \geqslant 0$) and so $x \in U_n$ for all $n \geqslant 1$ (since $f_n|_{X\setminus U_n} = 1$ for all $n \geqslant 1$). Therefore $x \in \bigcap_{n\geqslant 1} U_n = A$ and we have proved that $A = f^{-1}(\{0\})$.

(b) Since X is normal, we can find two open sets $U, V \subseteq X$ such that

$$A \subseteq U, \quad C \subseteq V \quad \text{and} \quad U \cap V = \emptyset.$$

Since A is a G_δ-set, we have

$$A = \bigcap_{n\geqslant 1} U_n,$$

with U_n being open sets and we may assume that $U_n \subseteq U$ for all $n \geqslant 1$. As in part **(a)** we obtain continuous functions $f_n\colon X \longrightarrow [0,1]$ for $n \geqslant 1$ such that

$$f_n|_A = 0 \quad \text{and} \quad f_n|_{X\setminus U_n} = 1.$$

Evidently $f_n|_C = 1$ and we set

$$f(x) = \sum_{n\geqslant 1} \tfrac{1}{2^n} f_n(x) \qquad \forall\, x \in X,$$

as in part **(a)**, we show that the function $f\colon X \longrightarrow [0,1]$ is continuous, $A = f^{-1}(\{0\})$ and $f|_C = 1$.

Solution of Problem 2.103

Since X is locally compact (see Definition 2.92), for every $x \in K$, we can find $V_x \in \mathcal{N}(x)$ such that $\overline{V}_x$ is compact in X. Then $\{V_x\}_{x\in K}$ is an open cover of K. We can also assume that $V_x \subseteq U$ for all $x \in K$. The compactness of K implies that we can find a finite subcover $\{V_{x_k}\}_{k=1}^n$ of $\{V_x\}_{x\in K}$. We have

$$K \subseteq V = \bigcup_{k=1}^{n} V_{x_k} \quad \text{and} \quad \overline{V} = \bigcup_{k=1}^{n} \overline{V}_{x_k} \text{ is compact in } X.$$

The set $\overline{V} \subseteq X$ furnished with the subspace topology is compact (see Proposition 2.79) and so it is normal (see Definition 2.4 and Proposition 2.83(d)). Then the Urysohn lemma (see Theorem 2.136) implies that there exists a continuous function $g\colon V \longrightarrow [0,1]$ such that

$$g|_K = 0 \quad \text{and} \quad g|_{\partial V} = 1.$$

Let $f\colon X \longrightarrow [0,1]$ be defined by

$$f(x) = \begin{cases} g(x) & \text{for } x \in \overline{V}, \\ 1 & \text{for } x \in X \setminus \overline{V}. \end{cases}$$

Clearly f is continuous and

$$f|_K = 0 \quad \text{and} \quad f|_{X\setminus U} = 1$$

(since $X \setminus U \subseteq X \setminus \overline{V}$).

Solution of Problem 2.104

"$\Longrightarrow$": Since by hypothesis K is a G_δ-set (see Definition 2.18), we can find a sequence $\{U_n\}_{n\geqslant 1}$ of open sets in X such that

$$K = \bigcap_{n\geqslant 1} U_n.$$

By Problem 2.103, for every $n \geqslant 1$, we can find a continuous function $f_n\colon X \longrightarrow [0,1]$ such that

$$f_n|_K = 0 \quad \text{and} \quad f_n|_{X\setminus U_n} = 1.$$

Then, if we set

$$f(x) = \sum_{n\geqslant 1} \tfrac{1}{2^n} f_n(x) \qquad \forall\, x \in X,$$

we have that the function $f\colon X \longrightarrow [0,1]$ is continuous (see the proof of Problem 2.102) and $K = f^{-1}(\{0\})$.

"$\Longleftarrow$": This part is identical as the first part of the solution of Problem 2.102(a).

Solution of Problem 2.105

"$\Longrightarrow$": Let X be a normal topological space Let $C \subseteq X$ be a nonempty, proper, closed set. If $u \in X \setminus C$, then by the perfect normality of X, we can find a continuous function $f\colon X \longrightarrow [0,1]$ such that $C = f^{-1}(\{0\})$, $\{u\} = f^{-1}(\{1\})$. Then, from Problem 2.102(a), we infer that C is a G_δ-set.

"$\Longleftarrow$": Suppose that every closed subset of X is a G_δ-set. Let $C, D \subseteq X$ be two nonempty disjoint closed sets. Then Problem 2.102(b) implies the existence of two continuous functions $g, h\colon X \longrightarrow [0,1]$ such that

$$g^{-1}(\{0\}) = C \quad \text{and} \quad g|_D = 1$$

and

$$h^{-1}(\{0\}) = D \quad \text{and} \quad g|_C = 1.$$

We set

$$f = \tfrac{1}{2}g + \tfrac{1}{2}(1-h).$$

Evidently $f\colon X \longrightarrow [0,1]$ is continuous and

$$f^{-1}(\{0\}) = C \quad \text{and} \quad f^{-1}(\{1\}) = D,$$

which shows that X is perfectly normal.

Solution of Problem 2.106

By the Urysohn metrization theorem (see Theorem 2.139), the space X is metrizable. Then Problems 1.79 and 2.105 imply that X is perfectly normal (see Definition 2.137). Thus, if $C = X \setminus U$, we can find a continuous function $f\colon X \longrightarrow [0,1]$ such that $f^{-1}(\{0\}) = C$. So,

$$f|_U > 0 \quad \text{and} \quad f_{X\setminus U} = 0.$$

Solution of Problem 2.107

"$\Longrightarrow$": Since X is second countable (see Definition 2.24), it is a Lindelöf separable space (see Definition 2.26 and Theorem 2.27). Also, because X is locally compact (see Definition 2.92), it is the union of relatively compact open sets. The Lindelöf property implies that we can find a sequence $\{U_n\}_{n\geqslant 1}$ of relatively compact sets such that

$$X = \bigcup_{n\geqslant 1} U_n.$$

Let $\{V_k\}_{k\geqslant 1}$ be a basis for the topology of X. If by X^* we denote the one-point compactification of X (see Remark 2.97 and Theorem 2.98), then

$$\{V_k\}_{k\geqslant 1} \cup \left\{X^* \setminus \overline{U}_n\right\}_{n\geqslant 1} \text{ is a countable basis for } X^*.$$

So X^* is compact and second countable. But a compact space is normal, hence regular too. Invoking the Urysohn metrization theorem (see Theorem 2.139), we infer that X^* is metrizable, hence X is metrizable too.

"$\Longleftarrow$": Recall that in a metric space, separability is equivalent to second countability (see Proposition 1.24).

Solution of Problem 2.108

Since X is locally compact (see Definition 2.92), for every $x \in U_\alpha \in \mathcal{Y}$, we can find a neighbourhood $V_{\alpha,x}$ of x such that $\overline{V}_{\alpha,x}$ is compact. Then $\mathcal{D}' = \{V_{\alpha,x}\}_{x\in X}$ is an open cover of the space X. Since by hypothesis X is paracompact (see Definition 2.142), it has a locally finite refinement $\mathcal{D}$ (see Definition 2.140). This is the desired cover.

Solution of Problem 2.109
Let $\{D_n\}_{n\geqslant 1}$ be a sequence of open dense subsets of X. Then $X \setminus D_n$ for $n \geqslant 1$ are closed nowhere dense sets. Then by hypothesis, the set $\bigcup\limits_{n\geqslant 1} (X \setminus D_n)$ is a nowhere dense set. Hence the set $\operatorname{int} \overline{\bigcup\limits_{n\geqslant 1} (X \setminus D_n)}$ is empty and so the set

$$X \setminus \bigcup_{n\geqslant 1} (X \setminus D_n) = \bigcap_{n\geqslant 1} D_n$$

is dense. This proves that X is Baire (see Definition 2.148).

The converse is not true. Let $X = \mathbb{R}$. From the Baire category theorem (see Theorem 1.26), we know that X is a Baire space. Let $\{g_n\}_{n\geqslant 1}$ be an enumeration of the rationals and let $D_n = \{g_n\}$ for every $n \geqslant 1$. Then each D_n is closed and nowhere dense, but

$$\overline{\bigcup_{n\geqslant 1} D_n} = X.$$

Solution of Problem 2.110
Suppose that $\bigcup\limits_{n\geqslant 1} A_n = X$. Then $\bigcup\limits_{n\geqslant 1} \overline{A}_n = X$ and so $\bigcap\limits_{n\geqslant 1} (X \setminus \overline{A}_n) = \emptyset$. But for every $n \geqslant 1$, the set $X \setminus \overline{A}_n$ is open and dense (see Definition 1.25 and Problem 1.29). Because X is a Baire space (see Definition 2.148), the set $\bigcap\limits_{n\geqslant 1} (X \setminus \overline{A}_n)$ is dense in X, a contradiction. This proves that

$$\bigcup_{n\geqslant 1} A_n \neq X.$$

Solution of Problem 2.111

No. Consider $X = \mathbb{R}$, which is a Baire space by Theorem 1.26 and let $D_1 = \mathbb{Q}$ and $D_2 = \mathbb{R} \setminus \mathbb{Q}$. Both are dense subsets of $\mathbb{R}$ but $D_1 \cap D_2 = \emptyset$.

Solution of Problem 2.112

Let $\{U_n\}_{n\geqslant 1}$ be a sequence of open and dense subsets of X. We need to show that

$$\bigcap_{n\geqslant 1} U_n \text{ is dense.}$$

Let $x \in X$ and let $V \in \mathcal{N}(x)$ such that V is a Baire space (see Definition 2.148). Then every $W \in \mathcal{N}(x)$, $W \subseteq V$ is open in V (with the subspace topology) and so W is a Baire space itself. Then $U_n \cap W \subseteq W$ is open and dense in W, hence

$$\bigcap_{n\geqslant 1} (U_n \cap W) = \Big(\bigcap_{n\geqslant 1} U_n\Big) \cap W$$

is dense in W and so

$$x \text{ is a limit point of } \bigcap_{n\geqslant 1} U_n,$$

which proves that $\bigcap_{n\geqslant 1} U_n$ is dense and this proves that X is a Baire space.

Solution of Problem 2.113

Let X be a Baire topological space (see Definition 2.148). Since by hypothesis f is lower semicontinuous (see Definition 2.46), for every $n \geqslant 1$ the set

$$C_n = \big(x \in X : f(x) \leqslant n\big) \text{ is closed.}$$

Then

$$\bigcup_{n\geqslant 1} C_n = \big\{x \in X : f(x) < +\infty\big\}$$

and by hypothesis this set is not of first category (i.e., is not included in the union of a sequence of closed nowhere dense sets; see Definition 1.25). So, for some integer $n \geqslant 1$, we have $\operatorname{int} C_n \neq \emptyset$ (because X is a Baire space). This means that we can find $x \in \operatorname{int} C_n$ and $U \in \mathcal{N}(x)$ such that

$$f(x) \leqslant n \qquad \forall\, x \in U.$$

Solution of Problem 2.114

Let $U \subseteq X$ be a nonempty and open set. Then U is a Baire space too. We have

$$U = U \cap X = \bigcup_{n \geqslant 1} (C_n \cap U)$$

and for each $n \geqslant 1$, the set $C_n \cap U$ is closed in U (with the subspace topology). Hence for some $n_0 \geqslant 1$, we have

$$\operatorname{int}_U(C_{n_0} \cap U) \neq \emptyset$$

(here int_U denotes the interior in U with the subspace topology). But because U is open, we have

$$\operatorname{int}_U(C_{n_0} \cap U) = (\operatorname{int} C_{n_0}) \cap U \neq \emptyset.$$

It follows that

$$\Big(\bigcup_{n \geqslant 1} \operatorname{int} C_n\Big) \cap U \neq \emptyset$$

and as U was an arbitrary open set in X, we conclude that

$$\bigcup_{n \geqslant 1} \operatorname{int} C_n \text{ is dense in } X.$$

Solution of Problem 2.115

Since Y is a separable metric space, for every $n \geqslant 1$, we can find a sequence of open sets $\{U_k^n\}_{k \geqslant 1}$ with $\operatorname{diam} U_k^n < \frac{1}{n}$ and which cover Y. Then by hypothesis

$$f^{-1}(U_k^n) = \bigcup_{m \geqslant 1} C_{k,m}^n,$$

where for every $m \geqslant 1$, the set $C^n_{k,m} \subseteq X$ is closed. Clearly

$$X = \bigcup_{k,m \geqslant 1} C^n_{k,m}$$

and because X is a Baire space, we have that

$$V_n = \bigcup_{k,m \geqslant 1} \operatorname{int} C^n_{k,m} \text{ is open, dense in } X$$

(see Problem 2.114). Let ω_f be the oscillation function for f (see Definition 2.42). Then

$$\omega_f(x) < \tfrac{1}{n} \qquad \forall\, x \in V_n.$$

Therefore

$$f \text{ is continuous on } V = \bigcap_{n \geqslant 1} V_n$$

(see Problem 2.23), which is a dense G_δ-set.

Solution of Problem 2.116

Let $f\colon X \longrightarrow \mathbb{R}$ be a lower semicontinuous function (see Definition 2.46). For every $a, c \in \mathbb{R}$, $a < c$, we have

$$f^{-1}\big((a,c)\big) = \bigcup_{n \geqslant 1} \big(f^{-1}(-\infty, c - \tfrac{1}{n}] \cap f^{-1}(a, +\infty)\big).$$

Due to the lower semicontinuity of f, we have that the set $f^{-1}\big((-\infty, c - \frac{1}{n}]\big)$ is closed and the set $f^{-1}\big((a,+\infty)\big)$ is open in X. By hypothesis $f^{-1}(a, +\infty)$ is an F_σ-set (see Definition 2.18). Therefore $f^{-1}\big((a,c)\big)$ is an F_σ-set and since the open intervals form a basis for the usual topology on $\mathbb{R}$ which is a separable metric space, we conclude that for every open set $V \subseteq \mathbb{R}$, the set $f^{-1}(V)$ is an F_σ-set. Because X is also a Baire topological space (see Definition 2.148), we can apply Problem 2.115 and conclude that f is continuous at every point of a dense G_δ-set.

When f is an upper semicontinuous function, we apply the above result to $-f$.

Solution of Problem 2.117

Let C be a closed subset of a paracompact space X (see Definition 2.142) and let $\{V_i\}_{i\in I}$ be an open cover of C. Then for every $i \in I$, there is an open set $U_i \subseteq X$ such that $V_i = U_i \cap C$. Let

$$\mathcal{Y} = \{U_i\}_{i\in I} \cup \{X \setminus C\}.$$

Then $\mathcal{Y}$ is an open cover of X and because X is paracompact, $\mathcal{Y}$ has a locally finite refinement $\mathcal{Y}'$ (see Definition 2.140). The family $\{U' \cap C\}_{U'\in\mathcal{Y}'}$ is a locally finite refinement of $\{V_i\}_{i\in I}$. This proves that C is paracompact.

Solution of Problem 2.118

(a) Let $U \subseteq X$ be a nonempty open set and let $\{U_n\}_{n\geqslant 1}$ be a sequence of open dense subsets of U. Then the set $V_n = U_n \cup (\overline{U})^c$ is open in X and

$$\overline{V_n} \supseteq \overline{U_n} \cup (\overline{U})^c = \overline{U} \cup (\overline{U})^c = X.$$

So the set V_n is also dense in X. Then, since X is Baire (see Definition 2.148), we have

$$\bigcap_{n\geqslant 1} (V_n \cap U) \neq \emptyset$$

and so

$$\bigcap_{n\geqslant 1} (U_n \cap U) \neq \emptyset,$$

hence the set $\bigcap\limits_{n\geqslant 1} U_n$ is dense in U, which proves that U is a Baire space.

(b) Suppose that $p\colon X \longrightarrow Y = X/_\sim$ is the quotient map and $\hat{V} \subseteq Y$ is an open dense set. Then the set $p^{-1}(\hat{V})$ is open (see Definition 2.76). Also it is dense in X. Indeed, let V be any open set in X. Then, since

p is an open map, the set $p(V) \subseteq Y$ is open and so $\hat{V} \cap p(V) \neq \emptyset$ (since $\hat{V}$ is dense). Hence

$$p^{-1}(\hat{V}) \cap V \neq \emptyset,$$

which shows that the set $p^{-1}(\hat{V})$ is open dense in X. Now, let $\{\hat{U}_n\}_{n \geqslant 1}$ be a sequence of open dense sets in X. Then from the previous part of the solution, the sequence $\{p^{-1}(\hat{U}_n)\}_{n \geqslant 1}$ consists of open dense sets in X and by hypothesis X is a Baire space. So, the set $\bigcap\limits_{n \geqslant 1} p^{-1}(\hat{U}_n)$ is dense in X. So, we have

$$p^{-1}\Big(\bigcap_{n \geqslant 1} \hat{U}_n \cap \hat{U}\Big) = \bigcap_{n \geqslant 1} p^{-1}(\hat{U}_n) \cap p^{-1}(\hat{U}) \neq \emptyset,$$

so

$$\bigcap_{n \geqslant 1} \hat{U}_n \cap \hat{U} \neq \emptyset.$$

Thus the set $\bigcap\limits_{n \geqslant 1} \hat{U}_n$ is dense in $\hat{U}$ and so $\hat{U}$ is a Baire space.

Solution of Problem 2.119

From Problem 2.118(a), we know that U is a Baire space too. Let $\{u_n\}_{n \geqslant 1}$ be a countable dense subset of X. By Problem 2.116, for every integer $n \geqslant 1$, there exists a dense G_δ-set D_n of U such that the function $x \longmapsto f(x, u)$ is continuous on D_n. Let

$$D = \bigcap_{n \geqslant 1} D_n.$$

Since U is a Baire space, we have that $D \subseteq U$ is dense (see Definition 2.148) and for every $n \geqslant 1$, the function $x \longmapsto f(x, u_n)$ is continuous on D. The density of $\{u_n\}_{n \geqslant 1}$ in X and the continuity of the function $u \longmapsto f(x, u)$ imply that for every $u \in X$, the function $x \longmapsto f(x, u)$ is continuous on D.

Solution of Problem 2.120

Let

$$\mathcal{Y} = \{X \setminus C\} \cup \{U_\alpha\}_{\alpha \in I}.$$

This is an open cover of X. Because of the paracompactness of X (see Definition 2.142), we can find a locally finite refinement $\{V_\beta\}_{\beta\in J}$ of $\mathcal{Y}$ (see Definition 2.140). Let

$$K = \{\beta \in J : V_\beta \cap C \neq \emptyset\}$$

and then

$$\mathcal{D} = \{V_\beta\}_{\beta\in K}$$

is an open cover of C. Let

$$W = \bigcup_{\beta\in K} V_\beta.$$

Since $\mathcal{D}$ is locally finite, the set $E = \bigcup\limits_{\beta\in\mathcal{D}} \overline{V}_\beta$ is closed. Then W and $X \setminus E$ are the two disjoint neighbourhoods of C and D respectively.

Solution of Problem 2.121

Let $\mathcal{Y} = \{U_i\}_{i\in I}$. By Problem 2.27, we can find a locally finite open cover $\{V_i\}_{i\in I}$ of X such that

$$\overline{V}_i \subseteq U_i \qquad \forall\, i \in I$$

(see Proposition 2.141). Invoking the Urysohn lemma (see Theorem 2.136), we can find a continuous function $\hat{\psi}_i \colon X \longrightarrow [0,1]$ such that

$$\operatorname{supp}\hat{\psi}_i = X \setminus U_i \quad \text{and} \quad \hat{\psi}_i^{-1}(1) \supseteq \overline{V}_i.$$

Let

$$\hat{\psi}(x) = \sum_{i\in I} \hat{\psi}_i(x).$$

This is well defined since $\mathcal{Y}$ is locally finite. Let

$$\psi_i(x) = \tfrac{\hat{\psi}_i(x)}{\hat{\psi}(x)} \qquad \forall\, i \in I.$$

Then clearly $\{\psi_i\}_{i\in I}$ is the desired partition of unity (see Definition 2.146).

Solution of Problem 2.122

Let X be a Lusin space (see Definition 2.167 and Remark 2.168). Then by definition there is a Polish space Y (see Definition 2.150) and a continuous bijection $f\colon Y \longrightarrow X$. Suppose that $D \subseteq X$ is open (or closed). Then $f^{-1}(D)$ is open (or closed) in Y (due to the continuity of f). Hence by Proposition 2.152, the set $f^{-1}(D)$ is a Polish space. The function f restricted on $f^{-1}(D)$ is a continuous bijection from $f^{-1}(D)$ to D. Hence by definition D is a Lusin space.

Solution of Problem 2.123

From Definition 2.167, for every $n \geqslant 1$, we can find a Polish space Y_n (see Definition 2.150) and a continuous bijection

$$f_n\colon Y_n \longrightarrow X_n.$$

Let

$$Y = \prod_{n\geqslant 1} Y_n$$

and let $f\colon Y \longrightarrow X$ be a function, defined by $f = (f_n)_{n\geqslant 1}$. From Proposition 2.153, we know that Y is a Polish space. Also note that f is a continuous bijection. Hence by definition X is a Lusin space.

Solution of Problem 2.124

Let $I = [0,1]$ and let $\mathbb{Q} \subseteq \mathbb{R}$ be a set of rational numbers. Then the set $I \setminus (I \cap \mathbb{Q})$ is a G_δ-subset of I. So, by Theorem 2.154(a), it is also a Polish space. Evidently $I \setminus (I \cap \mathbb{Q})$ is dispersed (see Definition 2.161) and so is $I \cap \mathbb{Q}$. Therefore by Proposition 2.162(c), the set I is dispersible and then so is $I^{\mathbb{N}}$ (see Proposition 2.162(a)). If Y is a Polish space, then Y is homeomorphic to a G_δ-subset of $I^{\mathbb{N}}$ and so Proposition 2.162(c) implies that Y is dispersible. Therefore, if X is a Souslin space (see Definition 2.156), then X is dispersible.

Solution of Problem 2.125

For every $q \in \mathbb{Q}$ and $i \in I$, let

$$U_{i,q} = \{x \in X : f_i(x) > q\}.$$

Then each set $U_{i,q} \subseteq X$ is open (see Definition 2.46). Since X is a strongly Lindelöf topological space (see Definition 2.163), there exist a countable subset J_q of I such that

$$\bigcup_{i \in I} U_{i,q} = \bigcup_{j \in J_q} U_{j,q}.$$

Let us set

$$J = \bigcup_{q \in \mathbb{Q}} J_q.$$

Then J is countable. Also, if

$$f = \sup_{i \subset I} f_i \quad \text{and} \quad f(x) > q,$$

then $x \in \bigcup\limits_{i \in I} U_{i,q}$ and so $x \in \bigcup\limits_{j \in J_q} U_{j,q}$, hence $\sup\limits_{j \in J} f_j(x) > q$, from which we infer that $f = \sup\limits_{j \in J} f_j$.

Solution of Problem 2.126

It is enough to show the result for open sets, since the one for closed sets can be obtained by complementation. Let $U \subseteq X$ be an open set. Due to regularity, U is the union of open sets V such that $V \subseteq \overline{V} \subseteq U$. Then the strong Lindelöf property (see Definition 2.163) implies that there is a countable subfamily of such closed sets $\overline{V}$ whose union is U. Therefore U is a F_σ-set (see Definition 2.18).

Solution of Problem 2.127

Let Δ_X be the diagonal of $X \times X$ and Δ_Y the diagonal of $Y \times Y$. As $\{f_i \colon X \longrightarrow Y\}_{i \in I}$ is a separating family (see Definition 2.64), for every $(x, x') \in (X \times X) \setminus \Delta_X$, we can find $i \in I$ such that $\big(f_i(x), f_i(x')\big) \notin \Delta_Y$.

This means that the open sets $(f_i, f_i)^{-1}\big((Y \times Y) \setminus \Delta_Y\big)$ form an open cover of $(X \times X) \setminus \Delta_X$. Because by hypothesis $X \times X$ is strongly Lindelöf (see Definition 2.163), we can find a countable subset $J \subseteq I$ such that the family $\{(f_j, f_j)^{-1}\big((Y \times Y) \setminus \Delta_Y\big)\}_{j \in J}$ is still an open cover of $(X \times X) \setminus \Delta_X$. Then the countable family $\{f_j\}_{j \in J}$ is separating.

Solution of Problem 2.128

In what follows let proj_X and proj_Y be the canonical projections of $X \times Y$ onto X and Y respectively. Let A be a Souslin subspace of Y (see Definition 2.156). Then $\text{proj}_Y^{-1}(A) = X \times A$ and since by hypothesis X is Souslin, $X \times A$ is a Souslin subspace of $X \times Y$ (see Proposition 2.159(b)). Since $\text{Gr}\, f$ is a Souslin subspace of $X \times Y$, then

$$\text{Gr}\, f \cap \text{proj}_Y^{-1}(A) = \text{Gr}\, f \cap (X \times A) = D$$

is Souslin (see Proposition 2.159(c)). So, there exists a Polish space V and a continuous surjection $h \colon V \longrightarrow D$. Then $\text{proj}_X \circ h$ is a continuous surjection from V onto $f^{-1}(A)$. This proves that $f^{-1}(A) \subseteq X$ is a Souslin set.

Solution of Problem 2.129

Let $\{(x_i, y_i)\}_{i \in I} \subseteq \text{Gr}\, F$ be a net such that

$$(x_i, y_i) \longrightarrow (x, y) \quad \text{in } X \times Y.$$

Suppose that $y \notin F(x)$. By the regularity of Y we can find $U \in \mathcal{N}(y)$ and an open set $V \subseteq Y$ such that $V \supset F(x)$ and $U \cap V = \emptyset$. Because F is upper semicontinuous (see Definition 2.169), we can find $W \in \mathcal{N}(x)$ such that $F(W) \subseteq V$. Since $x_i \longrightarrow x$ in X and $y_i \longrightarrow y$ in Y, we can find $i_0 \in I$ such that

$$x_i \in W \quad \text{and} \quad y_i \in U \qquad \forall\, i \geqslant i_0,$$

so

$$y_i \in F(x_i) \subseteq V \quad \text{and} \quad y_i \in U \qquad \forall\, i \geqslant i_0.$$

But $U \cap V = \emptyset$. So, we reach a contradiction. Therefore $y \in F(x)$ and so we conclude that the set $\text{Gr}\, f \subseteq X \times Y$ is closed.

Note that when the multifunction is actually single-valued (i.e., a usual function), then the notion of upper semicontinuity coincides with that of continuity. For usual functions (i.e., single-valued), closedness of the graph does not imply continuity. For example, consider a function $f\colon \mathbb{R} \longrightarrow \mathbb{R}$, defined by

$$f(x) \;=\; \begin{cases} 0 & \text{if} \quad x = 0, \\ \frac{1}{x} & \text{if} \quad x \neq 0. \end{cases}$$

Then $\operatorname{Gr} f \subseteq \mathbb{R} \times \mathbb{R}$ is closed, but f is not continuous.

Solution of Problem 2.130

"$\Longrightarrow$": We argue by contradiction. So, suppose that $\{(x_i, y_i)\}_{i\in I} \subseteq \operatorname{Gr} F$ is a net such that $x_i \longrightarrow x$ in X and $\{y_i\}_{i\in I}$ does not have a limit point in $F(x)$. So, for every $y \in F(x)$, we can find $i_0(y) \in I$ and $V(y) \in \mathcal{N}(y)$ such that $y_i \notin V(y)$ for all $i \geqslant i_0(y)$. The family $\big\{V(y)\big\}_{y\in F(x)}$ is an open cover of $F(x)$ and because $F(x)$ is compact, we can find a finite subcover $\big\{V(y_k)\big\}_{k=1}^{n}$. Let

$$V \;=\; \bigcup_{k=1}^{n} V(y_k) \;\supset F(x).$$

Evidently we can find $i_0 \in I$ such that $y_i \notin V$ for all $i \geqslant i_1$. On the other hand the upper semicontinuity of F (see Definition 2.169) implies that there is $U \in \mathcal{N}(x)$ such that $F(U) \subseteq V$. Then we can find $i_2 \geqslant i_1$ such that $x_i \in U$ for all $i \geqslant i_2$. So, $y_i \in F(x_i) \subseteq F(U) \subseteq V$ for all $i \geqslant i_2$, a contradiction. Therefore $\{y_i\}_{i\in I}$ has a limit point in $F(x)$.

"$\Longleftarrow$": By hypothesis every net $\{y_i\}_{i\in I} \subseteq F(x)$ has a limit point $y \in F(x)$, hence a subnet converging to y. This means that $F(x) \subseteq Y$ is compact (see Theorem 2.81). Suppose that F is not upper semicontinuous at x. Then we can find a net $\{x_i\}_{i\in I}$ and an open set V such that $x_i \longrightarrow x$ in X, $F(x) \subseteq V$ and $F(x_i) \cap (X \setminus V) \neq \emptyset$ for all $i \in I$. Let $y_i \in F(x_i) \cap (X \setminus V)$. Because $x_i \longrightarrow x$, by hypothesis $\{y_i\}_{i\in I}$ has a limit point $y \in F(x)$. So, we can find a subnet $\{y_j\}_{j\in J}$ of $\{y_i\}_{i\in I}$

such that $y_j \longrightarrow y$ in Y and $y \in F(x) \cap (X \setminus V)$ since the latter set is closed. But recall that $F(x) \subseteq V$, a contradiction.

Solution of Problem 2.131

(a) Let K be a compact set and let $\{y_i\}_{i \in I}$ be a net in $F(K)$. We need to show that it has a convergent subnet to some element in $F(K)$. We have $y_i \in F(x_i)$ with $x_i \in K$ for all $i \in I$. Because K is compact, we can find a subnet $\{x_j\}_{j \in J}$ of $\{x_i\}_{i \in I}$ such that $x_j \longrightarrow x \in K$. Then by Problem 2.130, the net $\{y_j\}_{j \in J}$ has a limit point $y \in F(x)$ and so we can find a subnet $\{y_\lambda\}_{\lambda \in \Lambda}$ of $\{y_j\}_{j \in J}$ such that

$$y_\lambda \longrightarrow y \in F(x) \subseteq F(K).$$

This proves the compactness of $F(K)$ (see Theorem 2.81).

(b) Let $\{(x_i, y_i)_i\}_{i \in I} \subseteq \operatorname{Gr} F$ be a net and assume that

$$(x_i, y_i) \longrightarrow (x, y) \text{ in } X \times Y.$$

Then by Problem 2.130, we have $y \in F(x)$, hence $\operatorname{Gr} F \subseteq X \times Y$ is closed.

Solution of Problem 2.132

Let $F \colon X \longrightarrow 2^{\mathbb{R}} \setminus \{\emptyset\}$ be the multifunction, defined by

$$F(x) = [f(x), g(x)] \qquad \forall\, x \in X.$$

Clearly F has closed, convex values and

$$\operatorname{int} F(x) \neq \emptyset \qquad \forall\, x \in X.$$

We claim that F is a lower semicontinuous multifunction. Let us fix $x \in X$. By Definition 2.169(b), it suffices to show that for every $a, c \in \mathbb{R}$, $a < c$ such that $F(x) \cap (a, c) \neq \emptyset$, we can find $U \in \mathcal{N}(x)$ such that

$$F(u) \cap (a, c) \neq \emptyset \qquad \forall\, u \in U.$$

Since

$$\emptyset \neq F(x) \cap (a, c) = [f(x), g(x)] \cap (a, c),$$

we have

$$f(x) \ < \ c \quad \text{and} \quad g(x) \ > \ a.$$

Because f is an upper semicontinuous function and g is a lower semicontinuous function (see Definition 2.46), we can find $U \in \mathcal{N}(x)$ such that

$$f(u) \ < \ c \quad \text{and} \quad g(u) \ > \ a \qquad \forall\, u \in U,$$

so

$$F(u) \cap (a,c) \ = \ \big[f(u), g(u)\big] \cap (a,c) \ \neq \ \emptyset \qquad \forall\, u \in U$$

and thus F is lower semicontinuous.

Observing that

$$\big[\operatorname{int} F(x) \cap (a,c) \neq \emptyset\big] \quad \Longleftrightarrow \quad \big[F(x) \cap (a,c) = \overline{\operatorname{int} F(x)} \cap (a,c) \neq \emptyset\big]$$

we infer that the function $x \longmapsto \operatorname{int} F(x)$ is lower semicontinuous. So, we can apply Theorem 2.172 and obtain a continuous function $h\colon X \longrightarrow \mathbb{R}$ such that

$$h(x) \ \in \ \operatorname{int} F(x) \ = \ \big(f(x), g(x)\big) \qquad \forall\, x \in X$$

and so

$$f(x) \ < \ h(x) \ < \ g(x) \qquad \forall\, x \in X.$$

Solution of Problem 2.133

Let $G\colon X \longrightarrow 2^Y \setminus \{\emptyset\}$ be the multifunction, defined by

$$G(x) \ = \ \begin{cases} F(x) & \text{if} \ \ x \neq \hat{x}, \\ \{\hat{y}\} & \text{if} \ \ x = \hat{x}. \end{cases}$$

Clearly G is a lower semicontinuous multifunction (see Definition 2.169) with values which are nonempty, closed and convex sets in Y. So, we can apply the Michael selection theorem (see Theorem 2.170) and obtain a continuous function $g\colon X \longrightarrow Y$ such that

$$g(x) \ \in \ G(x) \qquad \forall\, x \in X.$$

Then

$$g(x) \ \in \ F(x) \qquad \forall\, x \in X$$

and $g(\hat{x}) = \hat{y}$.

Solution of Problem 2.134
Consider the multifunction $G\colon X \longrightarrow 2^Y \setminus \{\emptyset\}$, defined by

$$G(x) = \begin{cases} f(x) & \text{if } x \in C, \\ F(x) & \text{if } x \in X \setminus C. \end{cases}$$

Evidently G is lower semicontinuous (see Definition 2.169) and has closed and convex values. So, we can apply the Michael selection theorem (see Theorem 2.170) and obtain a continuous function $\hat{f}\colon X \longrightarrow Y$ such that

$$\hat{f}(x) \in G(x) \qquad \forall\, x \in X.$$

Then

$$\hat{f}(x) \in F(x) \qquad \forall\, x \in X$$

and $\hat{f}|_C = f$.

Solution of Problem 2.135
The family $\{F^-(\{y\})\}_{y \in Y}$ is an open cover of X. Because X is compact, we can find a finite subcover $\{F^-(\{y_k\})\}_{k=1}^n$ of $\{F^-(\{y\})\}_{y \in Y}$. Recalling that a compact space is paracompact (see Definition 2.142 and Theorem 2.144) and invoking Theorem 2.147, we can find a partition of unity $\{p_k\}_{k=1}^n$ subordinated to $\{F^-(\{y_k\})\}_{k=1}^n$. We set

$$f(x) = \sum_{k=1}^n p_k(x) y_k \qquad \forall\, x \in X.$$

Evidently $f\colon X \longrightarrow Y$ is continuous. Moreover, if $p_k(x) \neq 0$, then

$$x \in F^-(\{y_k\})$$

and so

$$y_k \in F(x).$$

Therefore $f(x)$ is a convex combination of elements of $F(x)$ and so the convexity of $F(x)$ implies that

$$f(x) \in F(x) \qquad \forall\, x \in X.$$

Solution of Problem 2.136

First we do the lower semicontinuous case. In what follows, for every $C \subseteq Y$, we set

$$F^{-}(C) = \{x \in X : F(x) \cap C \neq \emptyset\}.$$

Let $A \subseteq X$ be a connected set (see Definition 2.104) and suppose that $V_1, V_2 \subseteq Y$ are open sets such that

$$F(A) \subseteq V_1 \cup V_2, \quad F(A) \cap V_1 \neq \emptyset, \quad F(A) \cap V_2 \neq \emptyset.$$

We need to show that

$$F(A) \cap V_1 \cap V_2 \neq \emptyset.$$

To this end, suppose that

$$A \cap F^{-}(V_1) \cap F^{-}(V_2) = \emptyset.$$

Because F is lower semicontinuous, both sets $F^{-}(V_1)$ and $F^{-}(V_2)$ are open in X and

$$A \subseteq F^{-}(V_1) \cup F^{-}(V_2), \quad A \cap F^{-}(V_1) \neq \emptyset, \quad A \cap F^{-}(V_2) \neq \emptyset.$$

These facts contradict the connectedness of A. Hence

$$A \cap F^{-}(V_1) \cap F^{-}(V_2) \neq \emptyset$$

and let $x \in A \cap F^{-}(V_1) \cap F^{-}(V_2)$. Then

$$F(x) \cap V_1 \neq \emptyset, \quad F(x) \cap V_2 \neq \emptyset \quad \text{and} \quad F(x) \subseteq V_1 \cup V_2.$$

Because the set $F(x)$ is connected in Y, it follows that

$$F(x) \cap V_1 \cap V_2 \neq \emptyset,$$

hence

$$F(A) \cap V_1 \cap V_2 \neq \emptyset$$

and this proves the connectedness of $F(C)$.

For the upper semicontinuous case, the proof is similar, checking this time a disconnection consisting of closed sets. Recall from

Definition 2.169(a) that F is upper semicontinuous if and only if for every closed set $D \subseteq Y$, the set

$$F^{-}(D) = \{x \in X : F(x) \cap D \neq \emptyset\}$$

is closed.

Solution of Problem 2.137

(a) "$\Longrightarrow$": Let $V \subseteq \mathbb{R}$ be an open set. We need to show that the set

$$L_f^{+}(V) = \{x \in X : L_f(x) \subseteq V\}$$

is open (see Definition 2.169). Let $x \in L_f^{+}(V)$. Then $L_f(x) \subseteq V$ and this means that $(\mu, +\infty) \subseteq V$ with some $\mu < f(x)$. Because f is lower semicontinuous (see Definition 2.46), we can find $U \in \mathcal{N}(x)$ such that

$$\mu < f(u) \qquad \forall\, u \in U.$$

Then

$$L_f(u) \subseteq (\mu, +\infty) \subseteq V \qquad \forall\, u \in U$$

and so we have proved that the multifunction $x \longmapsto L_f(x)$ is upper semicontinuous.

"$\Longleftarrow$": Since L_f is upper semicontinuous, for every $\mu \in \mathbb{R}$, the set

$$L_f^{+}\big((\mu, +\infty)\big) = \{x \in X : L_f(x) \subseteq (\mu, +\infty)\} = \{x \in X : \mu < f(x)\}$$

is open, which is equivalent to saying that f is lower semicontinuous (see Proposition 2.53).

(b) "$\Longrightarrow$": Let $V \subseteq \mathbb{R}$ be an open set. We need to show that the set

$$L_f^{-}(V) = \{x \in X : L_f(x) \cap V \neq \emptyset\}$$

is open. Let $x \in L_f^{-}(V)$. Then we can find $\lambda \in V$ such that $f(x) \leqslant \lambda$. Because V is open we can assume that $f(x) < \lambda$. The upper semicontinuity of f implies that there exists $U \in \mathcal{N}(x)$ such that

$$f(u) < \lambda \qquad \forall\, u \in U.$$

Then

$$L_f(u) \cap V \neq \emptyset \qquad \forall\, u \in U$$

and so we have proved that the multifunction $x \longmapsto L_f(x)$ is lower semicontinuous.

"$\Longleftarrow$": Since L_f is lower semicontinuous, for every $\lambda \in \mathbb{R}$, the set

$$L_f^-\big((-\infty, \lambda)\big) \;=\; \big\{x \in X : \; f(x) < \lambda\big\}$$

is open, which is equivalent to saying that f is upper semicontinuous (see Proposition 2.53).

Solution of Problem 2.138

We need to show that for every $\lambda \in \mathbb{R}$, the set

$$L_\lambda \;=\; \big\{y \in Y : \; m(y) \leqslant \lambda\big\}$$

is closed (see Definition 2.46 and Proposition 2.53). To this end, let $\{y_i\}_{i \in I}$ be a net in L_λ and assume that $y_i \longrightarrow y$. Let $x \in F(y)$, take any $V \in \mathcal{N}(x)$ and consider the set $F^-(V) \stackrel{\text{def}}{=} \big\{y' \in Y : \; F(y') \cap V \neq \emptyset\big\}$. We see that $y \in F^-(V)$ and since F is lower semicontinuous (see Definition 2.169), the set $F^-(V)$ is open. Because $y_i \longrightarrow y$ in Y, we can find $i_V \in I$ such that

$$y_i \;\in\; F^-(V) \qquad \forall\, i \geqslant i_V.$$

Hence we can find

$$x_i \;\in\; F(y_i) \cap V \qquad \forall\, i \geqslant i_V,$$

so

$$x_i \;\in\; F(y_i) \quad \text{and} \quad x_i \in V \qquad \forall\, i \geqslant i_V.$$

Because $V \in \mathcal{N}(x)$ was arbitrary, we conclude that $x_i \longrightarrow x$ in X. We have

$$f(x_i, y_i) \;\leqslant\; m(y_i) \;\leqslant\; \lambda \qquad \forall\, i \in I,$$

so $f(x, y) \leqslant \lambda$ (since f is lower semicontinuous on $X \times Y$ and $(x_i, y_i) \longrightarrow (x, y)$ in $X \times Y$).

Because $x \in F(y)$ is arbitrary, we conclude that $m(y) \leqslant \lambda$ and so $y \in L_\lambda$. This proves that L_λ is closed, hence m is lower semicontinuous.

Solution of Problem 2.139

We need to show that for any $\varepsilon > 0$ and $y \in Y$, we can find $U_y \in \mathcal{N}(y)$ such that for all $y' \in U_y$, we have

$$m(y') \leqslant m(y) + \varepsilon$$

(see Definition 2.46). Let $\varepsilon > 0$ and let $y \in Y$. Because f is upper semicontinuous, for any $x \in X$, we can find two sets $V_x \in \mathcal{N}(x)$ and $U'_x(y) \in \mathcal{N}(y)$ such that

$$f(x', y') \leqslant f(x, y) + \varepsilon \qquad \forall\, x' \in V_x,\ y' \in U'_x(y).$$

Note that $\{V_x\}_{x \in F(y)}$ is an open cover of the set $F(y)$ and by hypothesis $F(y)$ is compact. So, we can find a finite subcover $\{V_{x_k}\}_{k=1}^n$ of $\{V_x\}_{x \in F(y)}$. Then

$$F(y) \subseteq \bigcup_{k=1}^n V_{x_k} = V$$

and because F is upper semicontinuous (see Definition 2.169), we can find $W_y \in \mathcal{N}(y)$ such that

$$F(y') \subseteq V \qquad \forall\, y' \in W_y.$$

Let

$$U_y = W_y \cap \Big(\bigcap_{k=1}^n U'_{x_k}(y) \Big) \in \mathcal{N}(y).$$

If $y' \in U_y$ and $x' \in F(y')$, we have $x' \in V$ and so $x' \in V_{x_k}$ for some $k \in \{1, \ldots, n\}$. Since $y' \in U'_{x_k}$, we have

$$f(x', y') \leqslant f(x_k, y) + \varepsilon \leqslant m(y) + \varepsilon,$$

hence

$$m(y') \leqslant m(y) + \varepsilon,$$

which proves the upper semicontinuity of m.

Solution of Problem 2.140
Consider the sequence $\{K_n\}_{n\geqslant 1}$, with

$$K_n = F^{(n)}(X) \qquad \forall\, n \geqslant 1$$

(here $F^{(n)}$ denotes the composition of F with itself n times which is easily seen to be upper semicontinuous; see Definition 2.169). From Problem 2.131, we know that this is a decreasing sequence of compact subsets of X. Let us set

$$C = \bigcap_{n\geqslant 1} K_n.$$

Since the sequence $\{K_n\}_{n\geqslant 1}$ has the finite intersection property (see Definition 2.80), from Theorem 2.81, we infer that

$$C \neq \emptyset \quad \text{and} \quad F(C) \subseteq C.$$

We claim that equality holds. Arguing indirectly, suppose that we can find $x \in C \setminus F(C)$. Recall that a compact space is paracompact and a paracompact space is normal (see Definitions 2.4, 2.142 and Theorem 2.144). So, we can find open sets $U_1 \subseteq \mathcal{N}(x)$ and $U_2 \supseteq F(C)$ such that $U_1 \cap U_2 = \emptyset$. Since F is upper semicontinuous, the set $F^+(U_2) = \{x \in X : F(x) \subseteq U_2\}$ is open and contains C. So, we have

$$C = \bigcap_{n\geqslant 1} K_n \subseteq F^+(U_2),$$

hence

$$X \setminus F^+(U_2) \subseteq X \setminus C = \bigcup_{n\geqslant 1}(X \setminus K_n).$$

Note that $\{X\setminus K_n\}_{n\geqslant 1}$ is an open cover of the set $X\setminus F^+(U_2)$ which is compact. So, we can find a finite subcover $\{X \setminus K_{n_k}\}_{k=1}^N$ of the cover $\{X \setminus K_n\}_{n\geqslant 1}$, i.e.,

$$X \setminus F^+(U_2) \subseteq \bigcup_{k=1}^{N}(X \setminus K_{n_k}).$$

For every $n > \max\{n_1, \ldots, n_N\}$, we have

$$K_n \subseteq \bigcap_{k=1}^{N} K_{n_k} \subseteq F^+(U_2).$$

So, the sets $\{K_n\}_{n\geqslant 1}$ eventually reside in U_2. This implies that $C \subseteq U_2$, hence $x \in U_2$, a contradiction. This proves that $F(C) = C$.

Solution of Problem 2.141

From Theorem 2.183, we know that the c-topology on $C(X;Y)$ is induced by the supremum metric

$$d_\infty(f,g) \;=\; \sup_X d_Y\big(f(x),g(x)\big) \qquad \forall\, f,g \in C(X;Y).$$

Since the uniform limit of continuous functions is a continuous function, it follows that $\big(C(X;Y),d_\infty\big)$ is a compete metric space. So, we only need to check that $\big(C(X;Y),d_\infty\big)$ is separable. Let $k,m,n \geqslant 1$ be integers. The compactness of X implies that it has an $\frac{1}{m}$-net $X_m = \{x_1,\dots,x_l\} \subseteq X$. Since Y is separable, there is a countable open cover $\mathcal{D}_k = \{U_i\}_{i\geqslant 1}$ such that

$$\operatorname{diam} U_i \;<\; \tfrac{1}{k} \qquad \forall\, i \geqslant 1.$$

Let

$$A_{m,n} = \big\{f \in C(X;Y) :\ \forall x,u \in X :\ d_X(x,u) < \tfrac{1}{m} \implies d_Y\big(f(x),f(y)\big) < \tfrac{1}{n}\big\}.$$

For each l-tupe $s = (i_1,\dots,i_l)$, let us choose $f_s \in A_{m,n}$ (whenever possible) such that $f_s(x_j) \in U_j$ for all $1 \leqslant j \leqslant l$. Let $B_{m,n,k}$ be the collection of all these f_s and let

$$B_{m,n} \;=\; \bigcup_{k\geqslant 1} B_{m,n,k}.$$

Claim: For every $f \in A_{m,n}$ and every $\varepsilon > 0$, we can find $g \in B_{m,n}$ such that

$$d_Y\big(f(u),g(u)\big) \;<\; \varepsilon \qquad \forall\, u \in X_m.$$

To this end, take $k > \frac{1}{\varepsilon}$ and choose $i_1,i_2,\dots,i_k$ such that $f(x_j) \in U_{i_j}$ for all $1 \leqslant j \leqslant k$. Hence for $s = (i_1,\dots,i_k)$, f_s exists and we can take $g = f_s$. This proves the Claim.

Let

$$B \;=\; \bigcup_{m,n\geqslant 1} B_{m,n}.$$

Evidently B is countable. We will show that it is dense in $C(X;Y)$. So, let $f \in C(X;Y)$ and let $\varepsilon > 0$. Let us take $n > \frac{3}{\varepsilon}$. The uniform continuity of f (recall that the domain X is compact), implies that $f \in A_{m,n}$ for some m. By the Claim, we can find $g \in B_{m,n}$ such that

$$d_Y\big(f(u), g(u)\big) < \tfrac{\varepsilon}{3} \qquad \forall\, u \in X_m.$$

Since X_m is an $\frac{1}{m}$-net, via the triangle inequality, we have

$$d_Y\big(f(x), g(x)\big) < \varepsilon \qquad \forall\, x \in X,$$

so the set B is dense in $C(X;Y)$ and so $\big(C(X;Y), d_\infty\big)$ is separable, hence Polish.

Solution of Problem 2.142

(a) Evidently the sets of the form

$$[U_0; U_1, \ldots U_n] = \big\{C \in P_k(X) : C \subseteq U_0 \text{ and } C \cap U_i \neq \emptyset \text{ for all } 1 \leqslant i \leqslant n\big\}$$

form a base for the Vietoris topology on $P_k(X)$. So, let $[U_0; U_1, \ldots U_n]$ be such a nonempty basic open set. Then $U_0 \cap U_i \neq \emptyset$ for all $1 \leqslant i \leqslant n$. We choose $x_i \in U_0 \cap U_i$ for $1 \leqslant i \leqslant n$. Then

$$\{x_1, \ldots, x_n\} \in [U_0; U_1, \ldots U_n]$$

and this proves the density of the finite sets in $P_k(X)$ with the Vietoris topology.

(b) Let D be a countable dense set in X and let F be the set of all nonempty finite subsets of D. In the proof of **(a)**, choose x_i to belong in D (since D is dense in X). Then F is dense in $P_k(X)$ and F is countable.

Solution of Problem 2.143

First we show that if a set is open in the metric space $\big(P_k(X), h\big)$, then it is also Vietoris open. Let $K \in P_k(X)$ and let $\varepsilon > 0$. The compactness of K implies that there exists an $\frac{\varepsilon}{2}$-net $\{x_1, \ldots, x_n\} \subseteq K$. Let

$$U_0 = B_\varepsilon(K) = \{x \in X : \operatorname{dist}(x, K) < \varepsilon\}$$

and

$$U_i = B_\varepsilon(x_i) \qquad \forall\, i \in \{1, \ldots, n\}.$$

It suffices to show that

$$K \in [U_0; U_1, \ldots, U_n] = \{C \in P_k(X) : h(K, C) < \varepsilon\}$$

(see the solution of Problem 2.142). We have that $K \subseteq U$ and

$$x_i \in K \cap U_i \quad \forall 1 \leqslant i \leqslant n.$$

Therefore $K \subseteq [U_0; U_1, \ldots, U_n]$. Let $C \in [U_0; U_1, \ldots, U_n]$. We need to show that $h(K, C) < \varepsilon$. Since $C \subseteq U_0 = B_\varepsilon(K)$, from the definition of the Hausdorff metric, it is sufficient to show that

$$K \subseteq B_\varepsilon(C) = \{x \in X : \operatorname{dist}(x, C) < \varepsilon\}.$$

Let $x \in K$ and choose x_i such that $d_X(x, x_i) < \frac{\varepsilon}{2}$. Let $y \in K \cap U_i \neq \emptyset$. We have

$$d_X(x, y) \leqslant d_X(x, x_i) + d_X(x_i, y) < \varepsilon,$$

so $x \in B_\varepsilon(C)$, hence $K \subseteq B_\varepsilon(C)$.

Next we show that every Vietoris open set is also open in the metric space $(P_k(X), h)$. It suffices to show that every subbasic open set is open in $(P_k(X), h)$. So, let U be open in X and let $K \in P_k(X)$ be such that $K \subseteq U$. We define

$$\varepsilon = \min\{\operatorname{dist}(x, K) : x \in X \setminus U\} > 0.$$

For every $C \in P_k(X)$ with $h(C, K) < \varepsilon$, we have $C \subseteq B_\varepsilon(K) \subseteq U$, which shows that the subbasic open set $\{C \in P_k(X) : C \subseteq U\}$ is also open in $(P_k(X), h)$.

Also, let $K \in P_k(X)$ and let $U \subseteq X$ be an open set such that $K \cap U \neq \emptyset$. Let $x \in K \cap U$ and let $\varepsilon > 0$ be such that $B_\varepsilon(x) \subseteq U$. Suppose that $h(K, C) < \varepsilon$. Since $x \in K$, we have $\operatorname{dist}(x, C) < \varepsilon$ and so there exists $y \in C$ such that $y \in B_\varepsilon(x) \subseteq U$. Hence $C \cap U \neq \emptyset$ and this proves that the subbasic open set $\{C \in P_k(X) : C \cap U \neq \emptyset\}$ is open in $(P_k(X), h)$.

Solution of Problem 2.144

Let $\{x_i\}_{i \in I} \subseteq A$ be a net such that $x_i \longrightarrow x$. We need to show that $x \in A$. Let $r\colon X \longrightarrow A$ be the retraction function (see Definition 2.196). Then $r(x_i) = x_i$ for all $i \in I$. The continuity of r implies that $r(x_i) \longrightarrow r(x) \in A$. Hence $r(x) = x \in A$ and so we have proved that A is closed.

Next, if Y is another topological space and $f\colon A \longrightarrow Y$ is a continuous function, let us set $\hat{f} = f \circ r\colon X \longrightarrow Y$. Then $\hat{f}$ is continuous (as a composition of two continuous functions; see Proposition 2.41) and $\hat{f}(x) = f\big(r(x)\big)$ for all $x \in A$, i.e., $\hat{f}|_A = f$.

Solution of Problem 2.145

Let X be a normal space and let $A \subseteq X$ be a retract of it (see Definition 2.196). Let $r\colon X \longrightarrow A$ be the retraction of X onto A. Consider $C, D \subseteq A$ closed (these are closed in X too, since A is closed; see Problem 2.144) such that $C \cap D = \emptyset$. The sets $r^{-1}(C)$ and $r^{-1}(D)$ are disjoint, closed subsets of X. The normality of X implies that we can find two open sets $U, V \subseteq X$ such that

$$r^{-1}(C) \subseteq U, \quad r^{-1}(D) \subseteq V \quad \text{and} \quad U \cap V = \emptyset.$$

Consider the sets $U \cap A$ and $V \cap A$. Both are open in A and of course disjoint. Since $r|_A = id_A$, we have

$$(r|_A)^{-1}(C) = C \subseteq U \cap A \quad \text{and} \quad (r|_A)^{-1}(D) = D \subseteq V \cap A.$$

So, we have two disjoint open sets in A containing C and D respectively. This implies the normality of A.

Solution of Problem 2.146

Let $h\colon [0,1] \times (\mathbb{R}^{N+1} \setminus \{0\}) \longrightarrow \mathbb{R}^{N+1} \setminus \{0\}$ be a function, defined by

$$h(t,x) = (1-t)x + t\tfrac{x}{\|x\|} \qquad \forall\, (t,x) \in [0,1] \times (\mathbb{R}^{N+1} \setminus \{0\}).$$

Then

$$\begin{aligned} h(0,\cdot) &= id_{\mathbb{R}^{N+1}\setminus\{0\}}, \\ h(1,x) &= \tfrac{x}{\|x\|} \in S^N \qquad \forall\, x \in \mathbb{R}^{N+1} \setminus \{0\}, \\ h(t,\cdot)|_{S^N} &= id_{S^N}. \end{aligned}$$

Therefore h is a strong deformation retraction and so we conclude that S^N is a strong deformation retract of $\mathbb{R}^{N+1} \setminus \{0\}$ (see Definition 2.196).

Solution of Problem 2.147

(a) Let X be a compact (respectively, (path-)connected) topological space (see Definitions 2.78, 2.104 and 2.122) and let $A \subseteq X$ be a retract (see Definition 2.196). Then there exists a continuous function $r\colon X \longrightarrow A$ such that $r|_A = id_A$ (a retraction from X to A). Because of the continuity of the retraction r, we have that $r(X) = A$ is compact (respectively, (path-)connected).

(b) Let X be a simply connected topological space (see Definition 2.208) and let $A \subseteq X$ be a retract. Let $r\colon X \longrightarrow A$ be a retraction from X to A and let $i_A\colon A \longrightarrow X$ be the inclusion function (i.e., $i_A(x) = x$ for all $x \in A$). Then

$$r \circ i_A = id_A$$

and so

$$r_* \circ (i_A)_* = (r \circ i_A)_* = (id_A)_*$$

is the identity function on $\pi_1(A)$. Since $(i_A)_*\colon \pi_1(A) \longrightarrow \pi_1(X)$ and $r_*\colon \pi_1(X) \longrightarrow \pi_1(A)$, it follows that $(i_A)_*$ is injective and r_* is surjective. Hence

$$r_*\big(\pi_1(X)\big) = \pi_1(A).$$

But since X is simply connected, we have $\pi_1(X) = 0$. Hence

$$0 = r_*\big(\pi_1(X)\big) = \pi_1(A)$$

and this implies that A is simply connected too.
(c) Since A is a retract of X, there is a retraction $r_A\colon X \longrightarrow A$ from X to A. Also, because C is a retract of A, there is a retraction

$\hat{r}_C : A \longrightarrow C$ from A to C. Let $r_C = \hat{r}_C \circ r_A$. Then $r_C : X \longrightarrow C$ is continuous and it $x \in C$, then

$$r_C(x) = \hat{r}_C(r_A(x)) = x,$$

hence

$$r_C|_C = id_C,$$

which means that r_C is a retraction from X to C. Therefore C is a retract of X.

Solution of Problem 2.148

Let $e_n = (0, \ldots, 0, 1)$ be the "north pole" of S^{N-1} and let $e_s = -e_n$ be the "south pole". Consider the open sets

$$U = S^{N-1} \setminus \{e_n\} \quad \text{and} \quad V = S^{N-1} \setminus \{e_s\}.$$

Both are homeomorphic to $\mathbb{R}^N$ be means of the stereographic projection (cf. the solution of Problem 2.58). Let $f : [0,1] \longrightarrow S^{N-1}$ be a path. By Problem 1.137, we can find an integer $m \geqslant 1$ such that for every $k \in \{0, 1, \ldots, m-1\}$, we have

$$f\left(\left[\tfrac{k}{m}, \tfrac{k+1}{m}\right]\right) \subseteq U \quad \text{or} \quad f\left(\left[\tfrac{k}{m}, \tfrac{k+1}{m}\right]\right) \subseteq V.$$

Note that $V \setminus \{e_n\}$ is homeomorphic to $\mathbb{R}^{N-1} \setminus \{0\}$ which is connected (here the restriction $N \geqslant 3$ plays a crucial role, since $\mathbb{R} \setminus \{0\}$ is disconnected; Definition 2.104). Then for every path in V, there is another one also located in V with the same initial and final points which avoids e_n. Since V is simply connected (being homeomorphic to $\mathbb{R}^{N-1}$; Definition 2.208), the two paths are homotopic in V (see Definition 2.186) and so also in S^{N-1}. Every path located in U already avoids e_n. Therefore we conclude that f is homotopic to a path in

$S^{N-1} \setminus \{e_n\}$ and the latter is homeomorphic to $\mathbb{R}^{N-1}$. Therefore f is nullhomotopic (see Proposition 2.194) and so we conclude that

$$\pi_1(S^{N-1}) = 0,$$

i.e., S^{N-1} is simply connected (see Definition 2.208).

The function $g\colon \mathbb{R}^N \setminus \{0\} \longrightarrow S^{N-1}$ given by

$$g(x) = \frac{x}{\|x\|}$$

is a strong deformation retraction (see Definition 2.196 and Problem 2.146). Hence

$$\pi_1\big(\mathbb{R}^N \setminus \{0\}\big) = \pi_1(S^{N-1}) = 0$$

by the first part of the proof (see Corollary 2.207).

Solution of Problem 2.149

First we consider the closed unit interval $[0,1]$. Let $f\colon [0,1] \longrightarrow \{0\}$ be defined by $f(x) = 0$ for all $x \in [0,1]$ and let $h\colon \{0\} \longrightarrow [0,1]$ be defined by $h(0) = 0$. Then $f \circ h\colon \{0\} \longrightarrow \{0\}$ is the identity function and so it is trivially homotopic to the identity function (see Definition 2.186). Conversely, note that $h \circ f\colon [0,1] \longrightarrow [0,1]$ and

$$(h \circ f)(x) = 0 \qquad \forall\, x \in [0,1].$$

Consider the homotopy $G\colon [0,1] \times [0,1] \longrightarrow [0,1]$, defined by

$$G(t,x) = (1-t)x \qquad \forall\, (t,x) \in [0,1] \times [0,1].$$

This shows that $h \circ f \equiv 0$ and the identity are homotopic. Therefore $[0,1]$ and $\{0\}$ are homotopy equivalent (see Definition 2.191).

For the open unit interval $(0,1)$, the proof is similar. In this case, we need to change the definition of $h\colon \{0\} \longrightarrow (0,1)$, say to $h(0) = \frac{1}{2}$. Then a homotopy $G\colon [0,1] \times (0,1) \longrightarrow (0,1)$ which passes from $h \circ f$ to the identity of $(0,1)$ is given by

$$G(t,x) = \tfrac{1}{2}(1 - t + 2tx) \qquad \forall\, (t,x) \in [0,1] \times (0,1).$$

Evidently

$$G(0,x) = \tfrac{1}{2} = h \circ f \quad \text{and} \quad G(1,x) = x.$$

Therefore $(0,1)$ and $\{0\}$ are homotopy equivalent.

Solution of Problem 2.150

Let $f\colon S^1 \longrightarrow A$ be the canonical inclusion of S^1 into A, i.e.,

$$f(x,y) = (x,y) \qquad \forall\,(x,y) \in A$$

and let $h\colon A \longrightarrow S^1$ be the radial projection inwards (it moves the outer boundary of A onto S^1 which is the inner boundary of A). We have

$$h(x,y) = \tfrac{1}{\sqrt{x^2+y^2}}(x,y) \qquad \forall\,(x,y) \in A.$$

Then $h \circ f\colon S^1 \longrightarrow S^1$ and for all $(x,y) \in S^1$, we have

$$(h \circ f)(x,y) = h(x,y) = (x,y)$$

(since $\sqrt{x^2+y^2} = 1$). Of course $h \circ f = id_{S^1}$ is trivially homotopic to id_{S^1}.

On the other hand $f \circ h\colon A \longrightarrow A$ is defined by

$$(f \circ h)(x,y) = \tfrac{1}{\sqrt{x^2+y^2}}(x,y) \qquad \forall\,(x,y) \in A.$$

Consider the homotopy $G\colon [0,1] \times A \longrightarrow A$, defined by

$$G\big(t,(x,y)\big) = \tfrac{t\sqrt{x^2+y^2}+(1-t)}{\sqrt{x^2+y^2}}(x,y) \qquad \forall\,\big(t,(x,y)\big) \in [0,1] \times A.$$

Clearly G is continuous and

$$G\big(0,(x,y)\big) = (f \circ h)(x,y), \quad G\big(1,(x,y)\big) = (x,y) \qquad \forall\,(x,y) \in A.$$

This proves that A and S^1 are homotopy equivalent, i.e., $A \simeq S^1$ (see Definition 2.191).

Solution of Problem 2.151

The homotopy equivalence $f\colon S^1 \longrightarrow I \times S^1$ is given by

$$f(x) \;=\; (0,x)$$

(see Definition 2.191). The homotopy equivalence $g\colon I \times S^1 \longrightarrow S^1$ is given by $g = g_2 \circ g_1$, where

$$g_1(t,x) \;=\; (0,x) \qquad \forall\ (t,x) \in I \times S^1$$

and $g_2 = f^{-1}$.

Solution of Problem 2.152

Since by hypothesis f is not surjective, we have that

$$f(X) \;\subseteq\; S^N \setminus \{y_0\},$$

for some $y_0 \in S^N$. But via the stereographic projection (cf. the solution of Problem 2.58) the set $S^N \setminus \{y_0\}$ is homeomorphic to $\mathbb{R}^N$. So, without any loss of generality, we may assume that f is a continuous function from X into $\mathbb{R}^N$. Consider the homotopy $G\colon [0,1] \times X \longrightarrow S^N$, defined by

$$G(t,x) \;=\; tf(x) \qquad \forall\ (t,x) \in [0,1] \times X.$$

It is continuous and shows that f is nullhomotopic (see Definition 2.186).

Solution of Problem 2.153

Let $\operatorname{Gr} f$ be a graph of f (see Definition 1.132) and let $h\colon X \times Y \longrightarrow \operatorname{Gr} f$ be defined by

$$h(x,y) \;=\; \big(x, f(x)\big).$$

Evidently h is continuous and

$$h|_{\mathrm{Gr}\, f} = id_{\mathrm{Gr}\, f}.$$

This means that $\mathrm{Gr}\, f$ is a retract of $X \times Y$ (see Definition 2.196).

Solution of Problem 2.154

Let X be the union of two tangent circles C_1 and C_2. Let $C_1 \cap C_2 = x_0$, $a \in C_1 \setminus \{x_0\}$, $b \in C_2 \setminus \{x_0\}$. Let

$$U = X \setminus \{a\} \quad \text{and} \quad V = X \setminus \{b\}.$$

Both sets U and V are open and $X = U \cup V$. Invoking the Van Kampen theorem (see Theorem 2.229), we infer that $\pi_1(X, x_0)$ is a free group with two generators $\{\vartheta, \eta\}$, which are illustrated by the circles C_1 and C_2 in the above figure.

Solution of Problem 2.155

(a) Because A is a closed subset of the compact space X, it is itself compact. Hence due to the continuity of f, the sets $f^n(A)$ are compact and connected (see Propositions 2.82 and 2.108). We have

$$f^{n+1}(A) \subseteq f^n(A) \subseteq A \qquad \forall\, n \geqslant 1$$

and since A is compact, we have that

$$C = \bigcap_{n \geqslant 1} f^n(A) \quad \text{is nonempty and coampct too}$$

(see Theorem 2.81). We will show that the set C is also connected. We argue by contradiction. So, suppose that C is not connected. Then $C = D \cup E$, with D, E being nonempty, disjoint, closed subsets of C, hence of A too. The space A is normal (see Definition 2.4 and

Proposition 2.83(d)). Therefore, we can find U and V disjoint, open sets in A such that $D \subseteq U$ and $E \subseteq V$. Then $C \subseteq U \cup V = W$ and W is open in A.

Now we show that there exists $n_0 \geqslant 1$ such that

$$f^{n_0}(A) \subseteq W.$$

If this is not the case, we can find a sequence $\{x_n\}_{n\geqslant 1} \subseteq A$ such that $x_n \in f^n(A) \cap (A \setminus W)$ for all $n \geqslant 1$. The set $A \setminus W$ is compact and $\{x_n\}_{n\geqslant 1}$ admits a limit point x. Evidently $x \in C \subseteq W$, which is a contradiction. This proves that $f^{n_0}(A) \subseteq W$ for some $n_0 \geqslant 1$.

Thus

$$f^{n_0}(A) = \big(f^{n_0}(A) \cap U\big) \cup \big(f^{n_0}(A) \cap V\big),$$

which contradicts the fact that $f^{n_0}(A)$ is connected.

(b) Let $h\colon [0,1] \times S^1 \longrightarrow X \setminus C$ be a homotopy between f and a constant function (see Definition 2.186(c)). The set $h\big([0,1] \times S^1\big)$ is compact in $X \setminus C$ and the latter is open. Hence $h\big([0,1] \times S^1\big)$ is compact in X. Because X is normal (being compact), we can find two open sets $U, V \subseteq X$ such that

$$h\big([0,1] \times S^1\big) \subseteq U \quad \text{and} \quad C \subseteq V.$$

From **(a)**, we know that $f^{n_0}(A) \subseteq V$ for some $n_0 \geqslant 1$. Therefore

$$h\big([0,1] \times S^1\big) \subseteq U \subseteq X \setminus f^{n_0}(A)$$

and so

$$g(S^1) \subseteq X \setminus f^{n_0}(A).$$

Solution of Problem 2.156

No. Suppose that A is a retract of T (see Definition 2.196) and let $r\colon T \longrightarrow A$ be a retraction (i.e., r is continuous and $r|_A = id_A$). Let $i\colon A \longrightarrow T$ be the inclusion function, then $r \circ i = id_A$. So, by Proposition 2.206, the induced homomorphism

$$(id_A)_* = (r \circ i)_* = r_* \circ i_*\colon \pi_1(A) \longrightarrow \pi_1(A)$$

is the identity homomorphism. We know that $\pi_1(T)$ is abelian, while $\pi_1(A)$ is nonabelian free with two generators a and b. We have $ab \neq ba$ and

$$i_*(ab) = i_*(a)i_*(b) = i_*(b)i_*(a) = i_*(ba).$$

Therefore

$$ab = r_*i_*(ab) = r_*\big(i_*(ba)\big) = r_*i_*(b)r_*i_*(a) = ba,$$

a contradiction. This proves that A is not a retract of T.

Solution of Problem 2.157

Since X is path-connected (see Definition 2.122), we can find a path $\gamma\colon [0,1] \longrightarrow X$ such that $\gamma(0) = x_0$ and $\gamma(1) = x_1$. Then $\hat{\gamma} = f \circ \gamma\colon [0,1] \longrightarrow Y$ is a path joining $f(x_0)$ and $f(x_1)$. Let $[a] \in \pi_1\big(Y, f(x_1)\big)$. Then a is a loop in Y based on $f(x_1)$ and $\hat{\gamma}a\hat{\gamma}^{-1}$ is a loop in Y based on $f(x_0)$, hence

$$[\hat{\gamma}a\hat{\gamma}^{-1}] \in \pi_1\big(Y, f(x_0)\big).$$

By assumption, we can find a loop β in X based on x_0 such that

$$f_*([\beta]) = [\hat{\gamma}a\hat{\gamma}^{-1}].$$

So $f \circ \beta$ and $\hat{\gamma}a\hat{\gamma}^{-1}$ are homotopic. It follows that $f \circ \big(\hat{\gamma}\beta\hat{\gamma}^{-1}\big)$ and a are homotopic and so

$$pf_*([\hat{\gamma}\beta\hat{\gamma}^{-1}]) = [a].$$

Therefore $f_*\colon \pi_1(X, x_1) \longrightarrow \pi_1\big(Y, f(x_1)\big)$ is surjective.

Solution of Problem 2.158

From Example 2.219(d), we know that the projection $p\colon S^N \longrightarrow \mathbb{P}^N$ is a covering function. Since S^N is simply connected (see Definition 2.208 and Theorem 2.215(b)), we can apply Theorem 2.226 and obtain an isomorphism $\eta\colon \pi_1(\mathbb{P}^N, u) \longrightarrow p^{-1}(u)$. The latter is a two element set, hence $\pi_1(\mathbb{P}^N, u)$ is a group of order 2, i.e., $\pi_1(\mathbb{P}^N, u) = \mathbb{Z}_2$.

Solution of Problem 2.159

From Example 2.228(c), we know that

$$\pi_1\big(\mathbb{P}^2 \times S^2\big) \;=\; \pi_1\big(\mathbb{P}^2\big) \times \pi_1(S^2).$$

But from Problem 2.158 (see also Example 2.228(d)) and Theorem 2.215(b), we know that

$$\pi_1(\mathbb{P}^2) \;=\; \mathbb{Z}_2 \quad \text{and} \quad \pi_1(S^2) \;=\; 0.$$

Therefore, we conclude that

$$\pi_1\big(\mathbb{P}^2 \times S^2\big) \;=\; \mathbb{Z}_2.$$

Solution of Problem 2.160

(a) Note that $r \circ i_A = id_A$ and so

$$(r \circ i_A)_* \;=\; r_* \circ (i_A)_*$$

is the identity homomorphism on $\pi_1(A, x)$ (see Proposition 2.206(b)). From this it follows that $(i_A)_*$ is injective and r_* is surjective.

(b) Consider the function $r\colon \mathbb{R}^2 \setminus \{0\} \longrightarrow S^1$, defined by

$$r(x) \;=\; \tfrac{x}{\|x\|}.$$

Clearly this is a retraction (see Definition 2.196), i.e., S^1 is a retract of $\mathbb{R}^2 \setminus \{0\}$. By part **(a)**, the homomorphism $\pi_1(S^2) \longrightarrow \pi_1\big(\mathbb{R}^2 \setminus \{0\}\big)$ induced by the inclusion function is injective. But from Theorem 2.215(a), we know that $\pi_1(S^1) = \mathbb{Z}$. Hence $\pi_1\big(\mathbb{R}^2 \setminus \{0\}\big)$ has an infinite cyclic subgroup.

Also let $A = S^1 \times \{1\}$. This is homeomorphic to S^1. Hence by Corollary 2.207, we have that

$$\pi_1(A) \;=\; \pi_1(S^1) \;=\; \mathbb{Z}.$$

Moreover, the function $r\colon T \longrightarrow A$, defined by

$$r(x,y) \;=\; (x,1)$$

is a retraction, i.e., A is a retract of T. Hence the homomorphism $(i_A)_*\colon \pi_1(A) \longrightarrow \pi_1(T)$ is injective, which by what was observed above, implies that $\pi_1(T)$ has an infinite cyclic subgroup.

Solution of Problem 2.161

Let $p\colon \hat{X} \longrightarrow X$ be the universal covering function (see Definition 2.227). Suppose that $\pi_1(X,x)$ is not finite. Then $p^{-1}(x)$ is an infinite closed subset of $\hat{X}$. Because $\hat{X}$ is compact, the set $p^{-1}(x)$ has at least one limit point $\hat{x}$ such that

$$p(\hat{x}) \;=\; x.$$

Hence p cannot be a local homeomorphism at $\hat{x}$, which contradicts Definition 2.218.

Solution of Problem 2.162

Let $\varepsilon\colon \mathbb{R} \longrightarrow S^1$ be the exponential covering function, defined by

$$\varepsilon(x) \;=\; e^{2\pi i x}$$

(i.e., ε is the universal covering function of S^1; see Definition 2.227). Since X is locally path-connected (see Definition 2.127) and simply connected (see Definition 2.208), Theorem 2.221 implies that f has a lift $\hat{f}\colon X \longrightarrow \mathbb{R}$ (see Definition 2.220) such that

$$\varepsilon \circ \hat{f} \;=\; f.$$

Because $\mathbb{R}$ is simply connected, $\hat{f}$ is nullhomotopic (see Definition 2.186). Let h be the homotopy transforming $\hat{f}$ continuously to a constant function $\hat{c}$. Then $\varepsilon \circ h = b_0$ is a homotopy transforming f continuously to a constant function c. Therefore f is nullhomotopic.

Solution of Problem 2.163

Since $\pi_1(S^1) = \mathbb{Z}$ (see Theorem 2.215(a)), $\pi_1(X)$ is finite and $f_*\colon \pi_1(X) \longrightarrow \pi_1(S^1)$ is a homomorphism, we see that $f_*(\pi_1(X))$ is a finite subgroup of $\pi_1(S^1) = \mathbb{Z}$, hence

$$f_*(\pi_1(X)) = 0.$$

So, the argument in the solution of Problem 2.162 works and we have the desired result.

Solution of Problem 2.164

Let X and Y be two homotopy equivalent spaces (see Definition 2.191) and let $\pi_0(X)$, $\pi_0(Y)$ denote the sets of path-connected components of X and Y respectively (see Definition 2.130). Let $f\colon X \longrightarrow Y$ and $g\colon Y \longrightarrow X$ be two mutually inverse homotopy equivalences. Since f, g are continuous, they map path-connected sets to path-connected ones. So, f, g induce maps $\widetilde{f}\colon \pi_0(X) \longrightarrow \pi_0(Y)$ and $\widetilde{g}\colon \pi_0(Y) \longrightarrow \pi_0(X)$ respectively. Since $g \circ f \simeq id_X$, it follows that each $x \in X$ lies in the same path-connected component as $g(f(x))$. Hence $\widetilde{g} \circ \widetilde{f} = id_{\pi_0(X)}$. Similarly $\widetilde{f} \circ \widetilde{g} = id_{\pi_0(Y)}$. Therefore $\widetilde{g} \circ \widetilde{f}$ and $\widetilde{f} \circ \widetilde{g}$ are mutually inverse homotopy inverses, hence

$$\operatorname{card} \pi_0(X) = \operatorname{card} \pi_0(Y).$$

So, X and Y have the same number of path-connected components.

Solution of Problem 2.165

Let $\hat{\gamma}$ be a path in $\hat{X}$ connecting $\hat{x}_0$ and $\hat{x}_1$ and let $v = [p \circ \hat{\gamma}]$. Let $\tau_{\hat{\gamma}}\colon \pi_1(\hat{X}, \hat{x}_0) \longrightarrow \pi_1(\hat{X}, \hat{x}_1)$ be defined by

$$\tau_{\hat{\gamma}}([\hat{u}]) = [\hat{\gamma}]^{-1} * [\hat{u}] * [\hat{\gamma}]$$

and let $h\colon \pi_1(X, \widetilde{u}) \longrightarrow \pi_1(X, \widetilde{u})$ be defined by

$$h(w) \stackrel{\text{def}}{=} v^{-1} * w * v.$$

We consider the following diagram

$$\begin{array}{ccc}
\pi_1(\widehat{X}, \widehat{x}_0) & \xrightarrow{\tau_{\widehat{\gamma}}} & \pi_1(\widehat{X}, \widehat{x}_1) \\
{\scriptstyle p^*}\big\downarrow & & \big\downarrow{\scriptstyle p_*} \\
\pi_1(X, \widetilde{u}) & \xrightarrow[h]{} & \pi_1(X, \widetilde{u})
\end{array}$$

We claim that the above diagram is commutative. Note that

$$\tau_{\hat{\gamma}}\big([\hat{u}]\big) = \big[\hat{\gamma}^{-1} * \hat{u} * \hat{\gamma}\big],$$

so

$$\begin{aligned}
p_*\big(\tau_{\hat{\gamma}}([\hat{u}])\big) &= \big[p \circ (\hat{\gamma}^{-1} * \hat{u} * \hat{\gamma})\big] = \big[(p \circ \hat{\gamma}^{-1}) * (p \circ \hat{u}) * (p \circ \hat{\gamma})\big] \\
&= v^{-1} * p_*\big([\hat{u}]\big) * v = h\big(p_*([\hat{u}])\big).
\end{aligned}$$

So, the diagram is commutative. This proves that the two subgroups of $\pi_1(X, \hat{u})$ are conjugate as claimed by the problem.

Solution of Problem 2.166

Let $r\colon X \longrightarrow A$ be the retraction (see Definition 2.196) and let $i\colon A \longrightarrow X$ be the inclusion function. Then $r \circ i = id_A$ and so

$$r_* \circ i_* = (r \circ i)_* = (id_A)_*.$$

It follows that the homomorphism $i_*\colon H_n(A) \longrightarrow H_n(X)$ is injective and the homomorphism $r_*\colon H_n(X) \longrightarrow H_n(A)$ is surjective. Consider the short exact sequence

$$0 \longrightarrow H_n(A) \stackrel{i_*}{\longrightarrow} H_n(X) \stackrel{j_*}{\longrightarrow} H_n(X, A) \longrightarrow 0,$$

where $j\colon X = (X, \emptyset) \longrightarrow (X, A)$ is the inclusion function. Since i_* is injective, we see that the above sequence splits as

$$H_n(X) \simeq \operatorname{im} i_* \oplus \ker r_* \simeq H_n(A) \oplus \ker r_*.$$

But we know that

$$\ker r_* = H_n(X)/_{H_n(A)} \simeq H_n(X, A).$$

Therefore, finally

$$H_n(X) \simeq H_n(A) \oplus H_n(X, A).$$

Solution of Problem 2.167

Since A is deformation retract of X (see Definition 2.196), we can find a homotopy $h\colon [0,1] \times X \longrightarrow X$ such that

$$\begin{aligned} h(0,\cdot) &= id_X, \\ h(1,x) &\in A \qquad \forall\, x \in X \\ h(1,\cdot)|_A &= id_A \end{aligned}$$

(see Definition 2.196(b)). Then the function $r\colon X \longrightarrow A$, defined by

$$r(x) = h(1,x) \qquad \forall\, x \in X$$

is a retraction of X onto A. Therefore A is a retract of X and we can apply the result of Problem 2.166 and obtain

$$H_n(X) \simeq H_n(A) \oplus H_n(X, A) \qquad \forall\, n \geqslant 0.$$

But X and A are homotopy equivalent (see Definition 2.191 and Theorem 2.198). Therefore Theorem 2.283 implies that

$$H_n(X) \simeq H_n(A) \qquad \forall\, n \geqslant 0,$$

so

$$H_n(X, A) = 0 \qquad \forall\, n \geqslant 0.$$

Solution of Problem 2.168

Note that $\star$ is a retract of X (indeed consider the retraction $r(x) = \star$ for all $x \in X$; Definition 2.196). Then according to Problem 2.166, we have

$$H_n(X) = H_n(X, \star) \oplus H_n(\star) \qquad \forall\, n \geqslant 0.$$

Solution of Problem 2.169
We do the proof for $m = 2$, the general case following by induction. Then we have

$$X = A_1 \cup A_2 \quad \text{and} \quad A_1 \cap A_2 = \emptyset.$$

Let $e \in A_1$ and let $r\colon X \longrightarrow X$ be defined by

$$r(x) = \begin{cases} x & \text{if } x \in A_1, \\ e & \text{if } x \in A_2. \end{cases}$$

Evidently r is continuous and $r|_{A_1} = id_{A_1}$. Hence A_1 is a retract of X and so by Problem 2.166, we have

$$H_n(X) = H_n(A_1) \oplus H_n(X, A_1) \qquad \forall\, n \geqslant 0.$$

But by the excision property (see Theorem 2.278), we have

$$H_n(X, A_1) \simeq H_n(A_2, \emptyset) = H_n(A_2) \qquad \forall\, n \geqslant 0.$$

Hence

$$H_n(X) = H_n(A_1) \oplus H_n(A_2) \qquad \forall\, n \geqslant 0.$$

By induction we have the general case $m \geqslant 2$.

Solution of Problem 2.170
Let $\star \in X$ and let $f\colon X \longrightarrow \{\star\}$ be the constant function

$$f(x) = \star \in X \qquad \forall\, x \in X$$

(hence f is continuous). Also let $i\colon \{\star\} \longrightarrow X$ be the inclusion function, i.e., $i(\star) = \star \in X$. Then $f \circ i = id_{\{\star\}}$. Also, due to the fact that X is contractible (see Definition 2.192), we have that $i \circ f$ is homotopic to id_X (see Definition 2.186). Therefore X and $\{\star\}$ are homotopy equivalent (see Definition 2.191) and so

$$H_n(X) = H_n(\star) \qquad \forall\, n \geqslant 0.$$

Invoking Problem 2.168, we conclude that

$$H_n(X, \star) = 0 \qquad \forall\, n \geqslant 0.$$

Solution of Problem 2.171

From Theorem 2.282, we have the following exact sequence of homological groups:

$$\ldots\ H_n(A, \star) \xrightarrow{i_*} H_n(X, \star) \xrightarrow{j_*} H_n(X, A) \xrightarrow{\partial_*} H_{n-1}(A, \star)\ \ldots \qquad \forall\, n \geqslant 0,$$

where i_* and j_* are the homomorphisms induced by the corresponding inclusion functions and ∂_* is the boundary homomorphism. Since A is contractible, from Problem 2.170, we have

$$H_n(A, \star) = 0 \qquad \forall\, n \geqslant 0$$

(note that by definition $H_{-1}(A, \star) = 0$). Then from the long exact sequence, we have

$$\{0\} = \operatorname{im} i_* = \ker j_* \quad \text{and} \quad \operatorname{im} j_* = \ker \partial_* = H_n(X, A).$$

Therefore j_* is an isomorphism and so we conclude that

$$H_n(X, A) \simeq H_n(X, \star) \quad \forall n \geqslant 0.$$

Solution of Problem 2.172

Let e_n and e_s denote the "north" and "south" poles respectively in S^1. Let $U = S^1 \setminus \{e_n\}$ and $V = S^1 \setminus \{e_s\}$. Then $S^1 = U \cup V$. Moreover, using the stereographic projection (cf. the solution of Problem 2.58), we see that both U and V are homeomorphic to $\mathbb{R}$ and so

$$H_n(U) = H_n(V) = 0 \qquad \forall\, n \geqslant 1$$

(see Proposition 2.270(c)). Note that $U \cap V$ is the disjoint union of two spaces (two half-circles) each homeomorphic to $\mathbb{R}$. Using Proposition 2.270(b), the Mayer–Vietoris exact sequence in Theorem 2.281 becomes

$$\begin{aligned} \ldots \longrightarrow 0 \longrightarrow H_n(S^1) \longrightarrow \ldots \longrightarrow 0 \longrightarrow \\ \longrightarrow H_1(S^1) \longrightarrow \mathbb{Z} \oplus \mathbb{Z} \longrightarrow \mathbb{Z} \oplus \mathbb{Z} \longrightarrow H_0(S^1) \longrightarrow 0. \end{aligned}$$

Since S_1 is path-connected, we have

$$H_0(S^1) \;=\; \mathbb{Z}$$

(see Proposition 2.270(b)). So, the kernel of the last function $\mathbb{Z}\oplus\mathbb{Z} \longrightarrow H_0(S^1)$ is $\mathbb{Z}$. Similarly, the kernel of the function $\mathbb{Z}\oplus\mathbb{Z} \longrightarrow \mathbb{Z}\oplus\mathbb{Z}$ must be $\mathbb{Z}$. The exactness of the Mayer–Vietoris sequence, implies that

$$H_1(S^1) \;=\; \mathbb{Z}.$$

For $n > 1$, $H_n(S^1)$ is between two trivial groups and so

$$H_n(S^1) \;=\; 0 \qquad \forall\, n \geqslant 2.$$

Similarly, for the 2-sphere S^2. Again we consider

$$U \;=\; S^2 \setminus \{e_n\} \quad \text{and} \quad V = S^2 \setminus \{e_s\}.$$

We have

$$S^2 \;=\; U \cup V.$$

Again using the stereographic projection, we have that U and V are both homeomorphic to $\mathbb{R}^2$ and $U \cap V$ is homeomorphic to $\mathbb{R} \times S^2$ which is homotopy equivalent to S^1 (see Definition 2.191). Therefore from the first part of the proof, we have

$$H_0(U \cap V) \;=\; H_1(U \cap V) \;=\; \mathbb{Z} \quad \text{and} \quad H_n(U \cap V) \;=\; 0 \qquad \forall\, n \geqslant 2.$$

Then the Mayer–Vietoris exact sequence (see Theorem 2.281) becomes

$$\begin{aligned} \ldots \longrightarrow 0 \longrightarrow H_n(S^2) \longrightarrow 0 \ldots \longrightarrow 0 \longrightarrow H_1(S^2) \longrightarrow \mathbb{Z} \longrightarrow \\ \longrightarrow 0 \longrightarrow H_1(S^2) \longrightarrow \mathbb{Z} \longrightarrow \mathbb{Z} \oplus \mathbb{Z} \longrightarrow \mathbb{Z} \longrightarrow 0. \end{aligned}$$

It follows that

$$H_n(S^2) \;=\; 0 \qquad \forall\, n \geqslant 3 \quad \text{and} \quad H_2(S^2) \;=\; \mathbb{Z}.$$

Also as for S^1, we have

$$H_0(S^2) \;=\; \mathbb{Z}.$$

Finally because the function $\mathbb{Z} \longrightarrow \mathbb{Z} \oplus \mathbb{Z}$ has image $\mathbb{Z}$ and is injective, we have

$$H_1(S^2) = 0.$$

Solution of Problem 2.173

(a) No. Let $p\colon \mathbb{R} \longrightarrow S^1$ be the covering function, defined by

$$p(t) = e^{2\pi i t}$$

and let $f\colon S^N \longrightarrow S^1$ be a continuous function. From Theorem 2.215(b), we know that

$$\pi_1(S^N) = 0 \qquad \forall\, N \geqslant 2.$$

Hence by Theorem 2.221, f admits a lift $\widetilde{f}\colon S^N \longrightarrow \mathbb{R}$ (see Definition 2.220) such that $p \circ \widetilde{f} = f$. But $\mathbb{R}$ is contractible (see Definition 2.192 and Remark 2.193). Hence $\widetilde{f}$ is homotopic to a constant function $\widetilde{c}$. This implies that f is homotopic to $p \circ \widetilde{c}$ (see Proposition 2.190) which is a constant function. Therefore, we conclude that there is no continuous function $f\colon S^N \longrightarrow S^1$ $(N \geqslant 2)$ which is not nullhomotopic.

(b) Yes. Note that $(x, y) \in T$ if and only if $x = e^{2\pi i t_1}$, $y = e^{2\pi i t_2}$ with $t_1, t_2 \in [0, 1]$. Let $f\colon T \longrightarrow S^1$ be defined by

$$f\left(e^{2\pi i t_1, 2\pi i t_2}\right) = e^{2\pi i t_1} \in S.$$

Then the induced homomorphism $f_*\colon H_1(T) \longrightarrow H_1(S^1)$ maps one generator of $H_1(T)$ to the generator of $H_1(S^1)$ and the other generator of $H_1(T)$ to zero. So, f_* is not trivial, which implies that f is not nullhomotopic.

Solution of Problem 2.174

Let y_1 and y_2 be the two identification points. We introduce the sets

$$U = \Sigma X \setminus \{y_1\} \quad \text{and} \quad V = \Sigma X \setminus \{y_2\}.$$

Then U and V are open subsets of ΣX and $\Sigma X = U \cup V$. Moreover, U and V are contractible spaces and deformation retracts of $U \cap V$ (see Definitions 2.192 2.196). By the Mayer–Vietoris sequence (see Theorem 2.281), we have

$$H_n(\Sigma X) = \begin{cases} \mathbb{Z} & \text{if } n = 0, \\ H_{n-1}(X, \star) & \text{if } n \geqslant 1, \end{cases}$$

with $\star \in X$.

Solution of Problem 2.175

Let $\star \in U \cap V$ and consider the Mayer–Vietoris sequence for (U, V) (see Theorem 2.281)

$$\ldots \longrightarrow H_2(X, \star) \xrightarrow{\partial_*} H_1(U \cap V, \star) \xrightarrow{(j^1_*, -j^2_*)} H_1(U, \star) \oplus H_1(V, \star) \xrightarrow{(i^1_*, i^2_*)}$$
$$\xrightarrow{(i^1_*, i^2_*)} H_1(X, \star) \xrightarrow{\partial_*} H_0(U \cap V, \star),$$

where $i^1 \colon U \longrightarrow X$, $i^2 \colon V \longrightarrow X$, $j^1 \colon U \cap V \longrightarrow U$, $j^2 \colon U \cap V \longrightarrow V$ are the inclusion functions. Since $U \cap V$ is path-connected, we have $H_0(U \cap V, \star) = 0$. Also the exactness of the Mayer–Vietoris sequence implies that

$$\operatorname{im}(i^1_*, i^2_*) = \ker \partial_* = H_1(X, \star) = H_1(X).$$

But by hypothesis $(j^1_*, -j^2_*)$ is surjective and from the exactness, we have

$$\ker(i^1_*, i^2_*) = \operatorname{im}(j^1_*, -j^2_*) = H_1(U, \star) \oplus H_1(V, \star),$$

so

$$\operatorname{im}(i^1_*, i^2_*) = H_1(X) = 0.$$

The result fails if throughout H_1 is replaced by H_2. To see this let $X = S^2$ and let

$$U = \left\{(x, y, z) \in S^2 : z > -\tfrac{1}{2}\right\} \quad \text{and} \quad V = \left\{(x, y, z) \in S^2 : z < \tfrac{1}{2}\right\}.$$

Then $X = U \cup V$. Clearly U, V and $U \cap V$ are path-connected and $H_2(U) = H_2(V) = 0$. So, the homomorphism $H_2(U \cap V) \longrightarrow H_2(V)$

induced by the corresponding inclusion function is surjective. But $H_2(S^2) = \mathbb{Z}$ (see Proposition 2.272(b)).

Solution of Problem 2.176

We argue by contradiction. So suppose that there is $x \in \overline{B}_1^N \setminus f(\overline{B}_1^N)$. Then $x \in \overline{B}_1^N \setminus S^{N-1}$. Consider the function $r\colon \overline{B}_1^N \setminus \{x\} \longrightarrow S^{N-1}$, defined as follows. Consider $u \in \overline{B}_1^N \setminus \{x\}$ and draw the line between u and x. Extend this line beyond u until it meets S^{N-1} and let $r(u)$ be the point where this happens. Clearly r is continuous and $r|_{S^{N-1}} = id_{S^{N-1}}$. Hence r is a retraction (see Definition 2.196) and so S^{N-1} is a retract of $\overline{B}_1^N \setminus \{x\}$. Then $h = r \circ f\colon \overline{B}_1^N \longrightarrow S^{N-1}$ is a continuous function and $h|_{S^{N-1}} = f|_{S^{N-1}}$ which by hypothesis is a homeomorphism. Hence if $i\colon S^{N-1} \longrightarrow \overline{B}_1^N$ is the inclusion function, then $h \circ i\colon S^{N-1} \longrightarrow S^{N-1}$ is a homeomorphism. This implies that $(h \circ i)_*\colon H_{N-1}(S^{N-1}) \longrightarrow H_{N-1}(S^{N-1})$ is an isomorphism (see Theorem 2.283). But $(h \circ i)_* = h_* \circ i_*$ and $i_*\colon H_{N-1}(S^{N-1}) \longrightarrow H_{N-1}(\overline{B}_1^N)$ is the trivial homomorphism since $H_{N-1}(\overline{B}_1^N) = 0$ (see Proposition 2.272(a)). Therefore $(h \circ i)_*$ is the trivial homomorphism, a contradiction.

Solution of Problem 2.177

Evidently $X = P_x \cup P_y$, where P_x is the yz-plane and P_y is the xz-plane. Let C_x be the unit circle in P_x and let C_y be the unit circle in P_y. Then $C_x \cup C_y$ is a strong deformation retract of $X \setminus \{(0,0,0)\}$ (see Definition 2.196). Let

$$U = C_x \cup C_y \setminus \{(0,0,1)\} \quad \text{and} \quad V = C_x \cup C_y \setminus \{(0,0,-1)\}.$$

These sets are contractible, $U \cup V = C_x \cup C_y$ and $U \cap V$ has four path-connected components which are contractible. The Mayer–Vietoris sequence for the pair (U, V) (see Theorem 2.281), implies that

$$H_1(X \setminus \{(0,0,0)\}) = H_1(U \cup V) = H_1(C_x \cup C_y) = \mathbb{Z} \oplus \mathbb{Z} \oplus \mathbb{Z}.$$

Suppose that X and $\mathbb{R}^2$ are homeomorphic and let $f\colon X \longrightarrow \mathbb{R}^2$ be a homeomorphism. Then $f|_{X\setminus\{(0,0,0)\}}\colon X\setminus\{(0,0,0)\} \longrightarrow \mathbb{R}^2\setminus\{f(0,0,0)\}$ is a homeomorphism too. Then by Theorem 2.283, the groups

$$H_1\big((X \setminus \{(0,0,0)\}\big)\big) \text{ and } H_1\big((\mathbb{R}^2 \setminus \{(0,0)\}\big)\big)$$

are isomorphic. But

$$H_1\big((\mathbb{R}^2 \setminus \{(0,0)\}\big)\big) \;=\; H_1(S^1) \;=\; \mathbb{Z}$$

(see Proposition 2.272(b)), a contradiction. This proves that X and $\mathbb{R}^2$ are not homeomorphic.

Let $h\colon [0,1] \times X \longrightarrow X$ be define dy

$$h\big(t,(x,y,z)\big) \;=\; \begin{cases} (x,y,z) & \text{if } \; t = 0, \\ (tx,y,z) & \text{if } \; t \in (0,1]. \end{cases}$$

Clearly this is a deformation retraction of X onto P_y. This proves that X and $\mathbb{R}^2$ are of the same homotopy type (see Definition 2.191).

Solution of Problem 2.178

Let A_1 and A_2 be two copies of the unit sphere in $\mathbb{R}^2$ together with the portion of the z-axis inside them, which are touching at the origin. The union of A_1 and A_2 is called the ***wedge product*** of A_1 and A_2 and is denoted by $A_1 \vee A_2$. For example, the figure eight in Problem 2.154 is such a wedge product and is a deformation retract of $T = S^1 \times S^2$ (the torus) minus a point by the function

$$h\big(t,(x,y)\big) \;=\; \big((1-t)x + t\tfrac{x}{|x|},\; (1-t)y + t\tfrac{y}{|y|}\big)$$

(see Definition 2.196). Let $U = A_1 \vee A_2 \setminus \{e_n\}$ (e_n being the north pole of S^2) and $V = A_1 \vee A_2 \setminus \{e_s\}$ (e_s being the south pole of S^2). Then $U \cup V = A_1 \vee A_2$. Note that $U \cap V$ is contractible (hence path-connected). So, by the Van Kampen theorem (see Theorem 2.229), we have that $\pi_1(A_1 \vee A_2)$ is the free product of the groups $\pi_1(U)$ and $\pi_1(V)$. Note that

$$\pi_1(U) \;=\; \pi_1(V) \;=\; \pi_1(A_1) \;=\; \pi_1(A_2) \;=\; \pi_1(S^1).$$

Since X is of the same homotopy type as $A_1 \vee A_2$, we have

$$\pi_1(X) = \pi_1(A_1 \vee A_2)$$

(see Corollary 2.207) and so $\pi_1(X)$ is a free product generated by two generators.

Using the Mayer–Vietoris sequence (see Theorem 2.281) for (U, V), we obtain

$$H_k(X) = H_k(U) \oplus H_k(V) = H_k(A_1) \oplus H_k(A_2) \qquad \forall\, k \geqslant 0.$$

Note that

$$H_k(A_1) = H_k(A_2) \qquad \forall\, k \geqslant 0$$

and it is infinite cyclic for $k = 0, 1, 2$. Hence

$$H_k(A) = \begin{cases} \mathbb{Z} & \text{if } k = 0, \\ \mathbb{Z} \times \mathbb{Z} & \text{if } k \in \{1, 2\}, \\ 0 & \text{if } k \geqslant 3. \end{cases}$$

Solution of Problem 2.179

Using the homotopy $h\colon [0,1] \times C \longrightarrow C$, defined by

$$h\big(t, (x, u)\big) = (x, tu) \qquad \forall\, t \in [0, 1],\ (x, u) \in C,$$

we see that C deformation retracts to S^1 (see Definition 2.196). Hence

$$H_k(C) = H_k(S^1) = \begin{cases} \mathbb{Z} & \text{if } k \in \{0, 1\}, \\ 0 & \text{otherwise} \end{cases}$$

(see Proposition 2.272(b)).

Solution of Problem 2.180

We argue by contradiction. So, suppose that S^N is a retract of $\overline{B}_1^{N+1}$ (see Definition 2.196) and let $r\colon \overline{B}_1^{N+1} \longrightarrow S^N$ be a retraction. Also let $i\colon S^N \longrightarrow \overline{B}_1^{N+1}$ be the inclusion function. We have that $r \circ i\colon S^N \longrightarrow S^N$ and $r \circ i = id_{S^N}$. Note that by Proposition 2.272, we have

$$i_*\colon H_N(S^N) = \mathbb{Z} \longrightarrow H_N(\overline{B}^{N+1}) = 0.$$

So $(r \circ i)_* = r_* \circ i_* = 0$. On the other hand, since $r \circ i = id_{S^N}$, we have that $(r \circ i)_* = r_* \circ i_*$ is the identity homomorphism, a contradiction.

Solution of Problem 2.181

Choose $r > 0$ small such that $\overline{B}_r(x) \subseteq U$. Using the inclusion functions, we have the following commutative diagram

$$\begin{array}{ccc} \partial B_r(x) & & \\ {\scriptstyle i}\downarrow & \searrow^{s} & \\ U \setminus \{x\} & \xrightarrow[j]{} & \mathbb{R}^N \setminus \{x\} \end{array}$$

These functions induce homology homomorphisms:

$$\begin{array}{ccc} H_{N-1}(\partial B_r(x)) & & \\ {\scriptstyle i_*}\downarrow & \searrow^{s_*} & \\ H_{N-1}(U \setminus \{x\}) & \xrightarrow[j_*]{} & H_{N-1}(\mathbb{R}^N \setminus \{x\}) \end{array}$$

Note that s is a homotopy equivalence (see Definition 2.191). Hence s_* is an isomorphism (see Corollary 2.207) and so i_* is injective and j_* is surjective. But by Proposition 2.272(b), we know that $H_{N-1}\big(\partial B_r(x)\big) \neq 0$. Hence

$$H_{N-1}\big(U \setminus \{x\}\big) \neq 0.$$

Solution of Problem 2.182

We argue by contradiction. So, suppose that U and V are homeomorphic. Let $h\colon U \longrightarrow V$ be the homeomorphism. Then by Problem 2.181, we have that

$$H_{N-1}\big(V \setminus \{h(x)\}\big) \neq \emptyset \qquad \forall\, x \in U.$$

On the other hand, the set $V \setminus \big\{h(x)\big\}$ is homeomorphic to $\mathbb{R}^M \setminus \{0\}$ which in turn is homotopy equivalent to S^{M-1} (see Definition 2.191; in fact S^{M-1} is a strong deformation retract of $\mathbb{R}^M \setminus \{0\}$; see Definition 2.196). So

$$H_{N-1}\big(V \setminus \{h(x)\}\big) = H_{N-1}(S^{M-1}) = 0$$

(since $N \neq M$; see Proposition 2.272), a contradiction.

Solution of Problem 2.183

We write $S^N = S^N_+ \cup S^N_-$, where S^N_+ is the "northern" hemisphere and S^N_- is the "southern" hemisphere. By the Mayer–Vietoris sequence (see Theorem 2.281), for $\star \in S^N_+ \cap S^N_-$ we have the exact sequence:

$$\begin{aligned} \ldots \longrightarrow H_k(S^N_+,\star) \oplus H_k(S^N_-,\star) \longrightarrow H_k(S^N_+ \cup S^N_-,\star) \longrightarrow \\ \longrightarrow H_{k-1}(S^N_+ \cup S^N_-,\star) \longrightarrow H_{k-1}(S^N_+,\star) \oplus H_{k-1}(S^N_-,\star) \longrightarrow \ldots. \end{aligned}$$

So, because S^N_+ and S^N_- are both contractible (see Definition 2.192), we have

$$H_k(S^N_+,\star) = H_k(S^N_-,\star) = 0$$

(see Problem 2.170). It follows that

$$H_k(S^N,\star) = \begin{cases} H_{k-1}(S^{N-1},\star) = \ldots = H_0(S^{N-k},\star) & \text{if } k < N, \\ H_0(S^0,\star) & \text{if } k = N, \\ H_{k-N}(S^0,\star) & \text{if } k > N. \end{cases}$$

Since

$$H_0(S^{N-k}_+,\star) \oplus H_0(S^{N-k}_-,\star) \stackrel{\xi}{\longrightarrow} H_0(S^{N-k}_+ \cup S^{N-k}_-,\star) \longrightarrow 0,$$

the direct sum is trivial and the chain is exact, we infer that

$$H_0(S^{N-k}_+,\star) = 0 \qquad \forall\, k < N.$$

Also, we have

$$H_{k-N}(S^0,\star) = H_{k-N}\big(\{x\} \cup \{\star\},\star\big) = H_{k-N}(\star) \qquad \forall\, k \geqslant N.$$

Solution of Problem 2.184

We consider the triple (X_1, X_2, X_3) and the corresponding long exact sequence

$$\ldots \longrightarrow H_k(X_3,X_1) \stackrel{i_*}{\longrightarrow} H_k(X_3,X_2) \stackrel{\hat{\partial}_*}{\longrightarrow} H_{k-1}(X_2,X_1) \longrightarrow \ldots$$

(see Theorem 2.277). From the rank theorem, we have

$$\begin{aligned} \operatorname{rank} H_k(X_3,X_2) &= \operatorname{rank}\ker\hat{\partial}_* + \operatorname{rank}\operatorname{im}\hat{\partial}_* = \operatorname{rank}\operatorname{im} i_* + \operatorname{rank}\operatorname{im}\hat{\partial}_* \\ &\leqslant \operatorname{rank} H_k(X_3,X_1) + \operatorname{rank} H_{k-1}(X_2,X_1) \qquad (2.1) \end{aligned}$$

(using the exactness of the sequence). Similarly considering the triple (X_1, X_3, X_4), we obtain

$$\operatorname{rank} H_k(X_3, X_1) \leqslant \operatorname{rank} H_k(X_4, X_1) + \operatorname{rank} H_{k+1}(X_4, X_3). \quad (2.2)$$

Adding (2.1) and (2.2), we obtain

$$\operatorname{rank} H_k(X_3, X_2) - \operatorname{rank} H_k(X_4, X_1) \leqslant \operatorname{rank} H_{k-1}(X_2, X_1) + \operatorname{rank} H_{k+1}(X_4, X_3).$$

Bibliography

[1] Armostrong, M.A.: Basic Topology. Undergraduate Texts in Mathematics. Springer, New York (1983)

[2] Berge, C.: Espaces Topologiques: Fonctions Multivoques. Collection Universitaire de Mathématiques, vol. III. Dunod, Paris (1959)

[3] Buskes, G., van Rooij, A.: Topological Spaces. From Distance to Neighborhood. Undergraduate Texts in Mathematics. Springer, New York (1997)

[4] Choquet, G.: In: Marsden, J., Lance, T., Gelbart, S. (eds.) Lectures on Analysis. Volume I: Integration and Topological Vector Spaces. W.A. Benjamin, Inc., New York (1969)

[5] Choquet, G.: In: Marsden, J., Lance, T., Gelbart, S. (eds.) Lectures on Analysis. Volume II: Representation Theory. W.A. Benjamin, Inc., New York (1969)

[6] Dold, A.: Lectures on Algebraic Topology. Grundlehren der Mathematischen Wissenschaften, vol. 200. Springer-Verlag, Berlin (1980)

[7] Dugundji, J.: Topology. Allyn and Bacon Series in Advanced Mathematics. Allyn and Bacon, Inc., Boston (1978)

[8] Engelking, R.: General Topology. Sigma Series in Pure Mathematics, vol. 6. Heldermann Verlag, Berlin (1989)

[9] Jänisch, K.: Algebraic Topology. Springer, New York (1984)

[10] Kelley, J.L.: General Topology. Graduate Texts in Mathematics, vol. 27. Springer, New York (1975)

[11] Kuratowski, K.: Topology: Volume I. Academic, New York (1966)

[12] Kuratowski, K.: Topology: Volume II. Academic, New York (1968)

[13] Munkres, J.R.: Topology, a First Course. Prentice-Hall, Englewood Cliffs (1975)

[14] Maunder, C.R.F.: Algebraic Topology. Dover, Mineola (1996)

[15] Nagata, J.-I.: Modern General Topology. North-Holland Mathematical Library, vol. 33. North-Holland Publishing Co., Amsterdam (1985)

[16] Oxtoby, J.C.: Measure and Category. A Survey of the Analogies between Topological and Measure Spaces. Graduate Texts in Mathematics, vol. 2. Springer, New York (1971)

[17] Spanier, E.H.: Algebraic Topology. Springer, New York (1981)

[18] Steen, L.A., Seebach, J.A.J.: Counterexamples in Topology. Springer, New York (1978)

Chapter 3

Measure, Integral and Martingales

3.1 Introduction

3.1.1 Basic Definitions and Notation

We start by defining some families of subsets of a set Ω that play a central role in measure theory.

Definition 3.1
Let Ω be a set and let $\mathcal{Y} \subseteq 2^{\Omega}$ be a family of subsets of Ω.
(a) *We say that $\mathcal{Y}$ is a **semiring** if for any $A, C \in \mathcal{Y}$, we have that $A \cap C \in \mathcal{Y}$ and there exists a finite family $\{D_k\}_{k=1}^n \subseteq \mathcal{Y}$ of pairwise disjoint sets such that $A \setminus C = \bigcup_{k=1}^{n} D_k$.*
(b) *We say that $\mathcal{Y}$ is a **ring** if for any $A, C \in \mathcal{Y}$, we have that $A \cup C \in \mathcal{Y}$ and $A \setminus C \in \mathcal{Y}$ (i.e., a ring is closed under finite unions and relative complements).*
(c) *We say that $\mathcal{Y}$ is an **algebra** (or **field**), if it is a ring and $\Omega \in \mathcal{Y}$.*
(d) *We say that $\mathcal{Y}$ is a **σ-ring** if it is a ring which is closed under countable unions, i.e., if $\{A_n\}_{n\geqslant 1} \subseteq \mathcal{Y}$, then $\bigcup_{n\geqslant 1} A_n \in \mathcal{Y}$.*
(e) *We say that $\mathcal{Y}$ is a **σ-algebra** (or **σ-field**), if it is a σ-ring and $\Omega \in \mathcal{Y}$.*

L. Gasiński and N.S. Papageorgiou, *Exercises in Analysis: Part 1*, Problem Books in Mathematics, DOI 10.1007/978-3-319-06176-4_3,

Remark 3.2
It is clear from the above definitions that we have the following relations with these notions (arrows stand for inclusions):

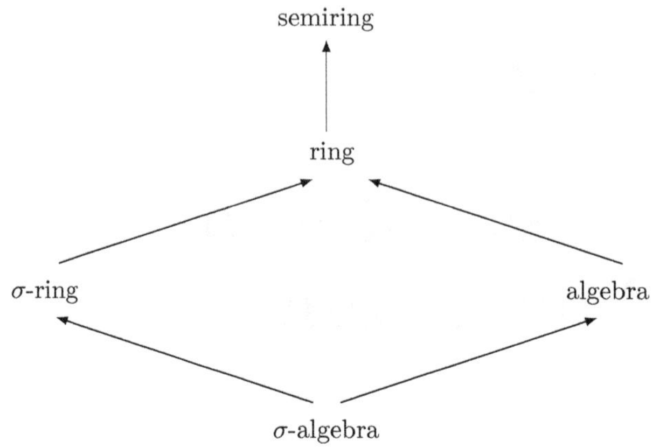

Remark 3.3
Note that if $\mathcal{Y}$ is a ring, then $\emptyset \in \mathcal{Y}$ since $\emptyset = A \setminus A$ for all $A \in \mathcal{Y}$. Moreover, it is easy to see that a ring is closed under symmetric difference (i.e., if $A, C \in \mathcal{Y}$, then $A \triangle C = (A \setminus C) \cup (C \setminus A) \in \mathcal{Y}$) and under finite intersections (i.e., if $\{A_k\}_{k=1}^n \subseteq \mathcal{Y}$, then $\bigcap\limits_{k=1}^{n} A_k \in \mathcal{Y}$). An algebra is closed under complementation, finite unions and finite intersections. Finally a σ-algebra is an algebra which is closed under countable unions and countable intersections.

Example 3.4
Let Ω be an uncountable set.
(a) Let $\mathcal{Y}$ be the family of all singletons of Ω and $\emptyset$. Then $\mathcal{Y}$ is a semiring but not a ring.
(b) Let $\mathcal{Y}$ be the family of all finite subsets of the set Ω. Then $\mathcal{Y}$ is a ring but not an algebra or a σ-ring (as always we consider $\emptyset$ as a finite set).
(c) Let $\mathcal{Y}$ be the family of all subsets of the set Ω, which are finite or cofinite (i.e., have a finite complement). Then $\mathcal{Y}$ is an algebra, but not a σ-algebra or even a σ-ring.
(d) Let $\mathcal{Y}$ be the family of all subsets of the set Ω, which are countable. Then $\mathcal{Y}$ is a σ-ring but not an algebra.

(e) Let $\mathcal{Y}$ be the family of all subsets of the set Ω, which are countable or co-countable (i.e., have a countable complement). Then $\mathcal{Y}$ is a σ-algebra and it is generated by the collection of all singletons (see Definition 3.6 below).

Remark 3.5
It is clear from Definition 3.1 that the intersection of σ-algebras (respectively, σ-rings) is again a σ-algebra (respectively, σ-ring). So, every $\mathcal{Y} \subseteq 2^\Omega$ is included in a smallest σ-algebra (respectively, σ-ring), known as the σ-algebra (respectively, σ-ring) generated by $\mathcal{Y}$.

Definition 3.6
(a) *Let Ω be a set and let $\mathcal{Y} \subseteq 2^\Omega$. The* ***$\sigma$-algebra generated by $\mathcal{Y}$*** *is denoted by $\sigma(\mathcal{Y})$ and is given by*

$$\sigma(\mathcal{Y}) \stackrel{\text{def}}{=} \bigcap \{\mathcal{F} : \mathcal{F} \text{ is a } \sigma\text{-algebra and } \mathcal{Y} \subseteq \mathcal{F}\}.$$

(b) *Let (X, τ) be a topological space. Then the σ-algebra $\sigma(\tau)$ (i.e., σ-algebra generated by the open sets) is called the* ***Borel σ-algebra*** *of X and is denoted by $\mathcal{B}(X)$. The elements of $\mathcal{B}(X)$ are called* ***Borel sets****.*

Definition 3.7
Let Ω be a set and let $\mathcal{Y} \subseteq 2^\Omega$.
(a) *We say that $\mathcal{Y}$ is a* ***π-class*** *if for any $A, C \in \mathcal{Y}$, we have $A \cap C \in \mathcal{Y}$. So, $\mathcal{Y}$ is also closed under finite intersections (the letter π is used because it brings in mind products, whose set theoretic analogs are intersections).*
(b) *We say that $\mathcal{Y}$ is a* ***λ-class*** *(or* ***Dynkin class****) if $\Omega \in \mathcal{Y}$ and for any $A, C \in \mathcal{Y}$ with $A \subseteq C$, we have $C \setminus A \in \mathcal{Y}$ and for any increasing family $\{A_n\}_{n \geqslant 1} \subseteq \mathcal{Y}$, we have $\bigcup\limits_{n \geqslant 1} A_n \in \mathcal{Y}$ (the letter λ is used because it brings in mind limits).*

Remark 3.8
An alternative equivalent definition of a λ-class is the following: $\emptyset \in \mathcal{Y}$, for every $A \in \mathcal{Y}$, we have $A^c \in \mathcal{Y}$ (closed under complementation) and if $\{A_n\}_{n \geqslant 1} \subseteq \mathcal{Y}$ are pairwise disjoint sets, then we have $\bigcup\limits_{n \geqslant 1} A_n \in \mathcal{Y}$.

A σ-algebra is a λ-class but the converse is not in general true. An algebra though need not be a λ-class. The intersection of a family of λ-classes is a λ-class. So, we can speak about the smallest λ-class containing a family $\mathcal{F} \subseteq 2^{\Omega}$, which is the λ-class generated by $\mathcal{F}$ and is denoted by $\lambda(\mathcal{F})$.

Theorem 3.9 *(π-λ Theorem; Dynkin Theorem)*
If $\mathcal{Y}$ is a π-class, then $\sigma(\mathcal{Y}) = \lambda(\mathcal{Y})$.

Definition 3.10
Let Ω be a set and let $\mathcal{Y} \subseteq 2^{\Omega}$. We say that $\mathcal{Y}$ is a ***monotone class*** *if for any sequence of increasing or decreasing sets $\{A_n\}_{n \geqslant 1} \subseteq \mathcal{Y}$, we have $\bigcup\limits_{n \geqslant 1} A_n \in \mathcal{Y}$ and $\bigcap\limits_{n \geqslant 1} A_n \in \mathcal{Y}$.*

Remark 3.11
We have the following relations:

$$\sigma\text{-algebra} \implies \lambda\text{-class} \implies \text{monotone class}.$$

The converse implications are not in general true. Clearly the intersection of a family of monotone classes is again a monotone class. Therefore, any $\mathcal{F} \subseteq 2^{\Omega}$ is included in a smallest monotone class, the ***monotone class generated by*** $\mathcal{F}$ and denoted by $m(\mathcal{F})$.

Theorem 3.12 *(Monotone Class Theorem)*
Let Ω be a set and let $\mathcal{F}$ be an algebra of subsets of Ω.
(a) *$\mathcal{F}$ is a monotone class if and only if $\mathcal{F}$ is a σ-algebra.*
(a) *$\sigma(\mathcal{F}) = m(\mathcal{F})$.*

Definition 3.13
Let Σ be a σ-algebra of subsets of Ω and let $A \subseteq \Omega$. We set

$$\Sigma_A \stackrel{\text{def}}{=} \{A \cap C : C \in \Sigma\}.$$

Then Σ_A is a σ-algebra of subsets of A and it is called the ***relative σ-algebra*** *of Σ on A (or* ***trace*** *of Σ on A).*

Definition 3.14
Let Σ be a σ-algebra of subsets of Ω.
(a) *We say that Σ is* **countably generated** *if there exists a countable subfamily $\mathcal{Y}$ of Σ such that $\Sigma = \sigma(\mathcal{Y})$.*

(b) *We say that Σ is* **separable** *if it is countable generated and it* **separates points** *of Ω, i.e., if $\omega, \omega' \in \Omega$, then we can find $A \in \Sigma$ such that $\chi_A(\omega) \neq \chi_A(\omega')$.*

3.1.2 Measures and Outer Measures

Now we are ready to introduce one of the central notions in measure and integration theory, namely the notion of measure.

Definition 3.15
If Ω is a set and $\Sigma \subseteq 2^{\Omega}$ is a σ-algebra, then the pair (Ω, Σ) is said to be a **measurable space**. *Let $\mu\colon \Sigma \longrightarrow \mathbb{R}^* = \mathbb{R} \cup \{\pm\infty\}$ be a function.*
(a) *We say that μ is a* **signed measure** *if $\mu(\emptyset) = 0$, it takes only one of the values $+\infty$ or $-\infty$ and it is σ-additive, i.e., for any family $\{A_n\}_{n\geqslant 1} \subseteq \Sigma$ of pairwise disjoint sets, we have $\mu(\bigcup\limits_{n\geqslant 1} A_n) = \sum\limits_{n\geqslant 1} \mu(A_n)$.*
(b) *We say that μ is a* **measure** *if it is a signed measure which takes nonnegative values only.*

Definition 3.16
In general, if $\mathcal{Y}$ is a semiring and $\mu\colon \mathcal{Y} \longrightarrow \mathbb{R}^* = \mathbb{R} \cup \{\pm\infty\}$ is a set function, we say that:
(a) *μ is* **monotone** *if for any $A, C \in \mathcal{Y}$ with $A \subseteq C$, we have $\mu(A) \leqslant \mu(C)$;*
(b) *μ is* **additive** *(or* **finitely additive***) if for any finite family $\{A_k\}_{k=1}^n \subseteq \mathcal{Y}$ of pairwise disjoint sets such that $\bigcup\limits_{k=1}^n A_k \in \mathcal{Y}$, we have $\mu(\bigcup\limits_{k=1}^n A_k) = \sum\limits_{k=1}^n \mu(A_k)$;*

(c) *μ is **σ-additive** (or **countably additive**) if for any sequence $\{A_n\}_{n\geqslant 1} \subseteq \mathcal{Y}$ of pairwise disjoint sets such that $\bigcup\limits_{n\geqslant 1} A_n \in \mathcal{Y}$, we have $\mu(\bigcup\limits_{n\geqslant 1} A_n) = \sum\limits_{n\geqslant 1} \mu(A_n)$;*
(d) *μ is **subadditive** if for any finite family $\{A_k\}_{k=1}^n \subseteq \mathcal{Y}$ such that $\bigcup\limits_{k=1}^n A_k \in \mathcal{Y}$, we have $\mu(\bigcup\limits_{k=1}^n A_k) \leqslant \sum\limits_{k=1}^n \mu(A_k)$;*
(e) *μ is **superadditive** if for any finite family $\{A_k\}_{k=1}^n \subseteq \mathcal{Y}$ such that $\bigcup\limits_{k=1}^n A_k \in \mathcal{Y}$, we have $\mu(\bigcup\limits_{k=1}^n A_k) \geqslant \sum\limits_{k=1}^n \mu(A_k)$;*
(f) *μ is **σ-subadditive** if for any sequence $\{A_n\}_{n\geqslant 1} \subseteq \mathcal{Y}$ such that $\bigcup\limits_{n\geqslant 1} A_n \in \mathcal{Y}$, we have $\mu(\bigcup\limits_{n\geqslant 1} A_n) \leqslant \sum\limits_{n\geqslant 1} \mu(A_n)$.*

Definition 3.17
*The triple (Ω, Σ, μ) where Ω is a set, Σ is a σ-algebra of subsets of Ω and $\mu\colon \Sigma \longrightarrow \overline{\mathbb{R}}_+ = [0, +\infty]$ is a measure is said to be a **measure space**. We say that a measure spaces is:*
(a) ***finite*** *if $\mu(\Omega) < +\infty$ (which implies that $\mu(A) < +\infty$ for all $A \in \Sigma$);*
(b) *a **probability space** if $\mu(\Omega) = 1$;*
(c) ***σ-finite*** *if*

$$\Omega = \bigcup_{n\geqslant 1} A_n, \quad A_n \in \Sigma \quad \text{and} \quad \mu(A_n) < +\infty \qquad \forall\, n \geqslant 1;$$

(d) ***semifinite*** *if for every $A \in \Sigma$ with $\mu(A) = \infty$, we can find $C \in \Sigma$ with $C \subseteq A$ and $0 < \mu(C) < +\infty$.*

Remark 3.18
In a σ-finite measure space (Ω, Σ, μ), we can always find a pairwise disjoint sequence $\{C_n\}_{n\geqslant 1} \subseteq \Sigma$ such that

$$\Omega = \bigcup_{n\geqslant 1} C_n \quad \text{and} \quad \mu(C_n) < +\infty \qquad \forall\, n \geqslant 1.$$

Indeed, according to Definition 3.17(c), we can find a sequence $\{A_n\}_{n\geqslant 1} \subseteq \Sigma$ such that

$$\Omega = \bigcup_{n\geqslant 1} A_n \quad \text{and} \quad \mu(A_n) < +\infty \qquad \forall\, n \geqslant 1.$$

Let

$$C_1 \stackrel{\text{def}}{=} A_1 \quad \text{and} \quad C_n \stackrel{\text{def}}{=} A_n \setminus \bigcup_{k=1}^{n-1} A_k \qquad \forall\, n \geqslant 2.$$

Every σ-finite measure space is semifinite, but the converse is not in general true. We often refer also to finite, σ-finite and semifinite measures. Most measures that arise in practice are σ-finite, which is fortunate since non-σ-finite measures tend to exhibit pathological behaviour.

The basic properties of measures are summarized in the next theorem.

Theorem 3.19
If (Ω, Σ, μ) is a measure space, then
(a) *μ is monotone and σ-subadditive;*
(b) *if $\{A_n\}_{n\geqslant 1} \subseteq \Sigma$ is increasing, then $\mu\big(\bigcup_{n\geqslant 1} A_n\big) = \lim_{n\to+\infty} \mu(A_n)$ (continuity from below).*
(c) *if $\{A_n\}_{n\geqslant 1} \subseteq \Sigma$ is decreasing and $\mu(A_1) < +\infty$, then $\mu\big(\bigcap_{n\geqslant 1} A_n\big) = \lim_{n\to+\infty} \mu(A_n)$ (continuity from above).*

Proposition 3.20
If (Ω, Σ) is a measurable space and $\mu\colon \Sigma \longrightarrow \overline{\mathbb{R}}_+ = [0, +\infty]$ is an additive set function which satisfies one of the following two properties:

(i) μ is continuous from below (see Theorem 3.19(b)); or
(ii) μ is continuous at 0, i.e., if $\{A_n\}_{n\geqslant 1} \subseteq \Sigma$ is decreasing and $\bigcap_{n\geqslant 1} A_n = \emptyset$, then $\lim_{n\to+\infty} \mu(A_n) = 0$,

then μ is a measure.

Definition 3.21
*Let (Ω, Σ, μ) be a measure space. A set $A \in \Sigma$ such that $\mu(A) = 0$ is said to be a **μ-null set**.*

Remark 3.22
By subadditivity, any countable union of μ-null sets is a μ-null set. If $\mu(A) = 0$ and $C \subseteq A$, $C \in \Sigma$, then by monotonicity, we have that $\mu(C) = 0$, i.e., C is μ-null too.

Definition 3.23
A measure space (Ω, Σ, μ) *where* Σ *contains all subsets of* μ*-null sets is said to be a* **complete measure space**. *For any measure space* (Ω, Σ, μ)*, the* **completion** *of* Σ *under* μ *is the collection*

$$\Sigma_\mu = \big\{A \subseteq \Omega : \text{ there exist } C, \widehat{C} \in \Sigma, \text{ such that } C \subseteq A \subseteq \widehat{C} \text{ and } \mu(\widehat{C} \setminus C) = 0\big\}.$$

If $C, \widehat{C} \in \Sigma$ *are such that* $\mu(\widehat{C} \setminus C) = 0$*, then evidently* $\mu(\widehat{C}) = \mu(C)$ *and we define* $\overline{\mu} \colon \Sigma_\mu \longrightarrow \overline{\mathbb{R}}_+ = [0, +\infty]$*, by*

$$\overline{\mu}(A) \stackrel{\text{def}}{=} \mu(\widehat{C}) = \mu(C).$$

Then $\overline{\mu}$ *is called the* **completion** *of* μ.

Proposition 3.24
If (Ω, Σ, μ) *is any measure space,*
then Σ_μ *is a* σ*-algebra,* $\Sigma \subseteq \Sigma_\mu$*,* $(\Omega, \Sigma_\mu, \overline{\mu})$ *is a complete measure space and* $\overline{\mu}|_\Sigma = \mu$.

Remark 3.25
Note that

$$\Sigma_\mu = \big\{A \cup D : A \in \Sigma, \ D \subseteq N \in \mathcal{N}\big\},$$

where $\mathcal{N} = \big\{N \in \Sigma : \mu(N) = 0\big\}$ and $\overline{\mu}(A \cup D) = \mu(A)$.

Proposition 3.26
If (Ω, Σ) *is any measurable space,* $\mathcal{Y} \subseteq \Sigma$ *is a* π*-class such that* $\Sigma = \sigma(\mathcal{Y})$*,* $\Omega = \bigcup\limits_{n \geqslant 1} A_n$ *with* $\{A_n\}_{n \geqslant 1} \subseteq \mathcal{Y}$ *being an increasing sequence of sets and* μ_1, μ_2 *are two measures on* Σ *such that*

$$\mu_1|_\mathcal{Y} = \mu_2|_\mathcal{Y} \quad \text{and} \quad \mu_1(A_n) = \mu_2(A_n) < +\infty \qquad \forall\, n \geqslant 1,$$

then $\mu_1 = \mu_2$.

Now we introduce the tools that will help us construct measures. We start with a notion due to C. Carathéodory, which is an abstract generalization of the notion of outer area.

Definition 3.27
Let Ω *be a set and let* $\mu^\star \colon 2^\Omega \longrightarrow \overline{\mathbb{R}}_+ = [0, +\infty]$ *be a set function. We say that* μ *is an* **outer measure** *if it satisfies:*

(a) $\mu^\star(\emptyset) = 0$;
(b) $\mu^\star$ *is monotone, i.e.,* $\mu^\star(C) \leqslant \mu^\star(A)$ *when* $C \subseteq A$; *and*
(c) $\mu^\star$ *is* σ*-subadditive, i.e.,* $\mu^\star\big(\bigcup\limits_{n\geqslant 1} A_n\big) \leqslant \sum\limits_{n\geqslant 1} \mu^\star(A_n)$, *for any sequence* $\{A_n\}_{n\geqslant 1} \subseteq \Omega$.

A common way to produce outer measures is to start with a small semiring at sets where the notion of measure is defined (such as half-open intervals in $\mathbb{R}$) and then approximate arbitrary sets "from the outside" by countable unions of elements in the semiring.

Proposition 3.28
If Ω *is a set,* $\mathcal{Y} \subseteq 2^\Omega$ *is a semiring,* $\mu\colon \mathcal{Y} \longrightarrow \overline{\mathbb{R}}_+ = [0, +\infty]$ *is a* σ*-additive set function and* $\mu^\star\colon 2^\Omega \longrightarrow \overline{\mathbb{R}}_+ = [0, +\infty]$ *is defined by*

$$\mu^\star(A) \stackrel{\text{def}}{=} \inf\Big\{\sum_{n\geqslant 1} \mu(A_n) : \{A_n\}_{n\geqslant 1} \subseteq \mathcal{Y},\ A \subseteq \bigcup_{n\geqslant 1} A_n\Big\} \qquad \forall\, A \in 2^\Omega,$$

with the usual convention that $\inf \emptyset = +\infty$,
then $\mu^\star$ *is an outer measure.*

The basic step that leads to the construction of a measure is the following notion due to C. Carathéodory.

Definition 3.29
Let Ω *be a set and let* $\mu^\star$ *be an outer measure on* 2^Ω. *A set* $A \subseteq \Omega$ *is said to be* $\mu^\star$***-measurable****, if*

$$\mu^\star(D) \;=\; \mu^\star(D \cap A) + \mu^\star(D \cap A^c) \qquad \forall\, D \subseteq \Omega,$$

i.e., A *splits all sets in* Ω *additively for* $\mu^\star$. *The family of all* $\mu^\star$*-measurable sets is denoted by* $\Sigma_{\mu^\star}$.

The next theorem produces a complete measure out of an outer measure.

Theorem 3.30 *(Carathéodory Theorem)*
If Ω *is a set,* $\mu^\star$ *is an outer measure on* 2^Ω *(see Definition 3.27) and* $\Sigma_{\mu^\star}$ *is the family of all* $\mu^\star$*-measurable sets (see Definition 3.29),*
then $\Sigma_{\mu^\star}$ *is a* σ*-algebra and* $(\Omega, \Sigma_{\mu^\star}, \mu^\star)$ *is a complete measure space.*

3.1.3 The Lebesgue Measure

Let us now apply the Carathéodory theorem (see Theorem 3.30) to establish the existence of the N-dimensional Lebesgue measure λ^N, $N \geqslant 1$. So, let $\mathcal{Y}^N$ be the family of half-open rectangles

$$R = \prod_{k=1}^{N} [a_k, b_k)$$

and let

$$\lambda^N\Big(\prod_{k=1}^{N} [a_k, b_k)\Big) = \prod_{k=1}^{N} (b_k - a_k) = \operatorname{vol} R$$

(the usual volume of R) and for any $A \subseteq \mathbb{R}^N$, we set

$$(\lambda^N)^\star(A) = \inf\Big\{\sum_{k\geqslant 1} \operatorname{vol} R_k : R_k \in \mathcal{Y}^N,\ A \subseteq \bigcup_{k\geqslant 1} R_k\Big\}.$$

We can check that $\mathcal{Y}^N$ is a semiring and then from Proposition 3.28, we have that $(\lambda^N)^\star$ is an outer measure on $2^{\mathbb{R}^N}$. Invoking Theorem 3.30, we have the following theorem.

Theorem 3.31
*There exists a unique extension of λ^N to the σ-algebra $\Sigma_{(\lambda^N)^\star}$ of $(\lambda^N)^\star$-measurable sets (see Definition 3.29). This measure, still denoted by λ^N is called the N-**dimensional Lebesgue measure** and the σ-algebra $\Sigma_{(\lambda^N)^\star}$ denoted for simplicity by $\mathcal{L}(\mathbb{R}^N)$ is the σ-**algebra of Lebesgue measurable sets** and $\mathcal{B}(\mathbb{R}^N) \subseteq \mathcal{L}(\mathbb{R}^N)$ (where $\mathcal{B}(\mathbb{R}^N)$ is the Borel σ-algebra of $\mathbb{R}^N$; see Definition 3.6(b)). The Lebesgue measure is translation invariant, i.e., if $A \in \mathcal{L}(\mathbb{R}^N)$ and $x \in \mathbb{R}^N$, then $\lambda^N(A) = \lambda^N(A + x)$ and $A \subseteq \mathbb{R}^N$ belongs in $\mathcal{L}(\mathbb{R}^N)$ if and only if $A + x$ belongs in $\mathcal{L}(\mathbb{R}^N)$ for some (respectively, all) $x \in \mathbb{R}^N$.*

Remark 3.32
The uniqueness of the extension follows from Proposition 3.26. When $N = 1$, we write $\lambda^1 = \lambda$.

Proposition 3.33
<u>If</u> μ is a nontrivial measure on $\big(\mathbb{R}^N, \mathcal{B}(\mathbb{R}^N)\big)$ which is translation invariant (i.e., $\mu(A + x) = \mu(A)$ for all $A \in \mathcal{B}(\mathbb{R}^N)$ and all $x \in \mathbb{R}^N$) and finite on bounded Borel sets in $\mathbb{R}^N$,
<u>then</u> there is a $\xi > 0$ such that $\mu = \xi\lambda^N$.

Remark 3.34
Next we will discuss the celebrated Cantor set, which turns out to be a rich source of examples. The ***Cantor set*** is defined by

$$C \stackrel{\text{def}}{=} \Big\{\sum_{n\geqslant 1}\frac{x_n}{3^n} : x_n \in \{0,2\} \text{ for all } n \geqslant 1\Big\},$$

i.e., it is the set of all $x \in [0,1]$, which have a base-3 expansion $x = \sum_{n\geqslant 1} \frac{x_n}{3^n}$ with coefficients x_n that are all different from 1. The set C is obtained from $[0,1]$ by removing the open middle third $\left(\frac{1}{3},\frac{2}{3}\right)$, then for the remaining two closed intervals $\left[0,\frac{1}{3}\right]$ and $\left[\frac{2}{3},1\right]$, we remove their respective middle thirds $\left(\frac{1}{9},\frac{2}{9}\right)$ and $\left(\frac{7}{9},\frac{8}{9}\right)$ and we continue in this fashion. The resulting set C is the Cantor set.

Proposition 3.35
The Cantor set $C \subseteq [0,1]$ has the following properties:
(a) *C is compact, nowhere dense, totally disconnected and has no isolated points (hence C is a perfect set).*
(b) $\operatorname{card} C = \mathfrak{c}$ *($\mathfrak{c}$ being the cardinality of the continuum).*
(c) $\lambda(C) = 0$ *(λ being the Lebesgue measure on $\mathbb{R}$).*

Remark 3.36
With the help of the Cantor set we can compare $\mathcal{L}(\mathbb{R})$ and $\mathcal{B}(\mathbb{R})$; see Problem 3.61. A natural question that arises is whether $\mathcal{L}(\mathbb{R}^N) = 2^{\mathbb{R}}$. A cardinality argument cannot help because of Problem 3.61.

Theorem 3.37
Assuming the axiom of choice (which is always the case in this volume), there exists a set $E \subseteq \mathbb{R}$ which does not belong in $\mathcal{L}(\mathbb{R})$. In fact, there is a set $E \subseteq [0,1]$ such that $\lambda^*(E) = \lambda^*\big([0,1] \setminus E\big) = 1$.

Remark 3.38
So, the Lebesgue measure is not defined on all subsets of $\mathbb{R}$. This is also true for $\mathbb{R}^N$, $N \geqslant 2$. In fact there is the famous "Banach–Tarski paradox", which says that for $N \geqslant 3$, any two subsets $A, C \subseteq \mathbb{R}^N$ with nonempty interior are congruent by finite decomposition, i.e.,

$$A = \bigcup_{k=1}^{m} D_k \quad \text{and} \quad C = \bigcup_{k=1}^{m} E_k,$$

with $\{D_k\}_{k=1}^m$ and $\{E_k\}_{k=1}^m$ pairwise disjoint families and for every $k \in \{1, \ldots, m\}$, the sets D_k and E_k are congruent, i.e., there exists a bijective isometry $f_k \colon D_k \longrightarrow E_k$. So, for example, any two balls are geometrically congruent by finite decomposition, regardless of the difference in their radii. Hence in these decompositions not all sets are Lebesgue measurable or otherwise the balls would have had the same Lebesgue measure (the same volume), which of course is absurd if the two balls have different radii. For dimensions $N = 1$ and $N = 2$, then A and C are congruent by countable decomposition, i.e., the decompositions of A and C into mutually congruent pieces can be accomplished by using countably many pieces. The Banach–Tarski paradox reveals that for $N \geqslant 3$, there cannot be even an additive set function defined on all $2^{\mathbb{R}^N}$ which is invariant under Euclidean motions. For $N = 1$ and $N = 2$ there is no measure (σ-additive function), defined on all $2^{\mathbb{R}}$ and $2^{\mathbb{R}^2}$. In any case, we infer that in all dimensions, nonmeasurable sets must exist. One then may ask whether λ^N can be extended as a measure to all of $2^{\mathbb{R}^N}$. Again the answer is no, at least if we assume that the continuum hypothesis. Finally every Lebesgue measurable set in $\mathbb{R}^N$ with positive measure contains a nonmeasurable subset.

Theorem 3.39
Assuming the continuum hypothesis, there is no measure μ defined on all subsets of $[0,1]$ such that

$$\mu([0,1]) = 1 \quad \text{and} \quad \mu(\{x\}) = 0 \qquad \forall\, x \in [0,1].$$

3.1.4 Atoms and Nonatomic Measures

Definition 3.40
*Let (Ω, Σ, μ) be a measure space. We say that $A \in \Sigma$ is an **atom** of μ if $0 < \mu(A) < +\infty$ and for every $C \subseteq A$ with $C \in \Sigma$, we have*

$$\mu(C) = 0 \quad \text{or} \quad \mu(C) = \mu(A).$$

*If μ has no atoms, then we say that the measure μ is **nonatomic**. We say that μ is **purely atomic** if whenever $A \in \Sigma$ and A is not μ-null, there is an atom of μ included in A.*

The following result gives a fundamental property of nonatomic measures.

Theorem 3.41
If (Ω, Σ, μ) *is a nonatomic measure space,* $A \in \Sigma$ *and* $0 \leqslant \eta \leqslant \mu(A) < +\infty$,
then there exists a set $C \in \Sigma$ *with* $C \subseteq A$ *such that* $\mu(C) = \eta$.

The next definition extends the notion of σ-finiteness (see Definition 3.17).

Definition 3.42
A measure space (Ω, Σ, μ) *is said to be* ***localizable*** *if there is a collection* $\mathcal{Y}$ *of disjoint* Σ*-sets of finite measure such that for every* $A \subseteq \Omega$, *we have*

$$A \in \Sigma \quad \Longleftrightarrow \quad \left[\mu(A) = \sum_{C \in \mathcal{Y}} \mu(A \cap C) \qquad \forall\, C \in \mathcal{Y},\ A \cap C \in \Sigma\right].$$

Proposition 3.43
If (Ω, Σ, μ) *is a nonatomic localizable measure space,*
then for every $\varepsilon > 0$, *we can find a family* $\{A_i\}_{i \in I} \subseteq \Sigma$ *of pairwise disjoint sets such that*

$$\Omega = \bigcup_{i \in I} A_i \quad \text{and} \quad \mu(A_i) \leqslant \varepsilon \qquad \forall\, i \in I.$$

If μ *is* σ*-finite, then* $I = \mathbb{N}$ *(i.e., the partition of* Ω *is countable).*
If μ *is finite, then* $I = \{1, \ldots, N\}$ *for some* $N \geqslant 1$ *(i.e., the partition of* Ω *is finite).*

3.1.5 Product Measures

Let (Ω, Σ, μ) and (Ω', Σ', ν) be two measure spaces. Let

$$\mathcal{R} \overset{\text{def}}{=} \{A \times C :\ A \in \Sigma,\ C \in \Sigma'\}$$

(the collection of all "rectangles" in $\Omega \times \Omega'$). This is a semiring of sets in $\Omega \times \Omega'$. For the elements of this semiring, we let

$$\varrho(A \times C) \overset{\text{def}}{=} \mu(A)\nu(C).$$

It is easily seen that ϱ is σ-additive on $\mathcal{R}$.

Definition 3.44
*The **product σ-algebra** $\Sigma \otimes \Sigma'$ is defined by*

$$\Sigma \otimes \Sigma' = \sigma(\mathcal{R}).$$

Let $\mathcal{A}$ be the ring generated by $\mathcal{R}$. Then $\mathcal{A}$ is the collection of finite disjoint unions of rectangles. Since $\Omega \times \Omega' \in \mathcal{R}$, it follows that $\mathcal{A}$ is an algebra (see Definition 3.1(c)). Then ϱ is well defined on $\mathcal{A}$ and extends to a measure on $\sigma(\mathcal{A}) = \Sigma \otimes \Sigma'$ (see Definition 3.44), called the ***product measure*** of μ and ν and denoted by $\mu \times \nu$. In general, this extension is not unique. However, we have the following result.

Proposition 3.45
If (Ω, Σ, μ) and (Ω', Σ', ν) are σ-finite measure spaces,
*then ϱ extends uniquely to a measure $\mu \times \nu$ (called the **product measure**) on $\Sigma \otimes \Sigma'$.*

Definition 3.46
*Let (Ω, Σ, μ) and (Ω', Σ', ν) be two measure spaces and $D \in \Sigma \otimes \Sigma'$. Then for $x \in \Omega$ and $y \in \Omega'$, we define the **x-section** D_x of D and the **y-section** D_y of D by*

$$D_x \stackrel{\text{def}}{=} \{y \in \Omega' : (x, y) \in D\} \quad \text{and} \quad D_y \stackrel{\text{def}}{=} \{x \in \Omega : (x, y) \in D\}.$$

Proposition 3.47
If (Ω, Σ, μ) and (Ω', Σ', ν) be two measure spaces and $D \in \Sigma \otimes \Sigma'$,
then for every $x \in \Omega$ and every $y \in \Omega'$, we have $D_x \in \Sigma'$ and $D_y \in \Sigma$.

3.1.6 Lebesgue–Stieltjes Measures

Consider the following two semirings:

$$\mathcal{Y}_1 \stackrel{\text{def}}{=} \{(a, b] : a, b \in \mathbb{R}\} \quad \text{and} \quad \mathcal{Y}_2 \stackrel{\text{def}}{=} \{[a, b) : a, b \in \mathbb{R}\}.$$

Let $f \colon \mathbb{R} \longrightarrow \mathbb{R}$ be a nondecreasing, right continuous function (i.e., $\lim\limits_{x \to u^+} f(x) = f(u)$). We set

$$\mu_f\big((a, b]\big) \stackrel{\text{def}}{=} f(b) - f(a) \qquad \forall\, a, b \in \mathbb{R},\ a \leqslant b.$$

Proposition 3.48
If $f\colon \mathbb{R} \longrightarrow \mathbb{R}$ *is nondecreasing, right continuous and* μ_f *is the set function defined on* $\mathcal{Y}_1$ *as above,*
then μ_f *is* σ*-additive on* $\mathcal{Y}_1$ *and by the Carathéodory procedure described earlier extends uniquely to a measure on* $\sigma(\mathcal{Y}_1) = \mathcal{B}(\mathbb{R})$.

The same can be done starting with a nondecreasing, left continuous function $f\colon \mathbb{R} \longrightarrow \mathbb{R}$ (i.e., $\lim\limits_{x \to u^-} f(x) = f(u)$). In this case, we set

$$\nu_f\big([a,b)\big) \stackrel{\text{def}}{=} f(b) - f(a) \qquad \forall\, a, b \in \mathbb{R},\ a \leqslant b.$$

Proposition 3.49
If $f\colon \mathbb{R} \longrightarrow \mathbb{R}$ *is nondecreasing, left continuous and* ν_f *is the set function defined on* $\mathcal{Y}_2$ *as above,*
then ν_f *is* σ*-additive on* $\mathcal{Y}_2$ *and by the Carathéodory procedure described earlier extends uniquely to a measure on* $\sigma(\mathcal{Y}_2) = \mathcal{B}(\mathbb{R})$.

More generally, if $f\colon \mathbb{R} \longrightarrow \mathbb{R}$ is a nondecreasing function and we introduce the set functions $\mu_f\colon \mathcal{Y}_1 \longrightarrow \mathbb{R}$, $\nu_f\colon \mathcal{Y}_2 \longrightarrow \mathbb{R}$, defined by

$$\begin{aligned} \mu_f\big((a,b]\big) &= f(b^+) - f(a^+) \qquad \forall\, a, b \in \mathbb{R},\ a \leqslant b, \\ \nu_f\big([a,b)\big) &= f(b^-) - f(a^-) \qquad \forall\, a, b \in \mathbb{R},\ a \leqslant b, \end{aligned}$$

where recall that

$$f(u^+) = \lim_{x \to u^+} f(x) \quad \text{and} \quad f(u^-) = \lim_{x \to u^-} f(x),$$

then μ_f and ν_f extend uniquely to identical measures on $\mathcal{B}(\mathbb{R})$. So, we have seen that every nondecreasing right or left continuous function $f\colon \mathbb{R} \longrightarrow \mathbb{R}$ defines a unique measure on $\mathbb{R}$. The converse is also true.

Definition 3.50
If X *is a topological space and* $\mathcal{B}(X)$ *denotes the corresponding Borel* σ*-algebra, then a measure* μ *on* $\mathcal{B}(X)$ *is said to be a* ***Borel measure*** *on* X.

Proposition 3.51

(a) *If* μ *is a Borel measure which is finite on bounded subintervals of* $\mathbb{R}$ *and*

$$f(x) = \begin{cases} -\mu\big((x,0]\big) & \text{if } x < 0, \\ 0 & \text{if } x = 0, \\ \mu\big((0,x]\big) & \text{if } x > 0, \end{cases}$$

then f *is nondecreasing, right continuous and* $\mu = \mu_f$.

(b) *<u>If</u> μ is a Borel measure which is finite on bounded subintervals of $\mathbb{R}$ and*

$$f(x) = \begin{cases} -\mu([x,0)) & \text{if } x < 0, \\ 0 & \text{if } x = 0, \\ \mu([0,x)) & \text{if } x > 0, \end{cases}$$

<u>then</u> f is nondecreasing, left continuous and $\mu = \mu_f$.

Remark 3.52
In fact the Carathéodory procedure gives us not only Borel measure μ_f but also a complete measure $\overline{\mu}_f$ whose domain includes $\mathcal{B}(\mathbb{R})$. One can show that the domain of $\overline{\mu}_f$ is always strictly bigger than $\mathcal{B}(\mathbb{R})$. We usually denote this complete measure also by μ_f and is called the ***Lebesgue–Stieltjes measure*** associated with f. When $f(x) = x$, then the resulting measure μ_f is the classical Lebesgue measure on $\mathbb{R}$.

3.1.7 Measurable Functions

Definition 3.53
*Let (Ω, Σ) and (Ω', Σ') be two measurable spaces. A function $f\colon \Omega \longrightarrow \Omega'$ is said to be (Σ, Σ')-**measurable** (or just **measurable** when Σ and Σ' are unambiguously understood), if $f^{-1}(A) \in \Sigma$ for all $A \in \Sigma'$.*

Proposition 3.54
<u>If</u> (Ω, Σ) and (Ω', Σ') are two measurable space, $\Sigma' \in \sigma(\mathcal{A})$ and $f\colon \Omega \longrightarrow \Omega'$,
<u>then</u> f is measurable if and only if $f^{-1}(A) \in \Sigma$ for all $A \in \mathcal{A}$.

Corollary 3.55
<u>If</u> X and Y are two topological spaces and $f\colon X \longrightarrow Y$ is a continuous function,
<u>then</u> f is $\big(\mathcal{B}(X), \mathcal{B}(Y)\big)$-measurable.

Proposition 3.56
<u>If</u> (Ω_k, Σ_k) for $k = 1, 2, 3$ are measurable spaces and $f\colon \Omega_1 \longrightarrow \Omega_2$ is (Σ_1, Σ_2)-measurable and $f\colon \Omega_2 \longrightarrow \Omega_3$ is (Σ_2, Σ_3)-measurable,
<u>then</u> $g \circ f\colon \Omega_1 \longrightarrow \Omega_3$ is (Σ_1, Σ_3)-measurable.

Often we deal with functions $f\colon \Omega \longrightarrow Y$, where Y is equipped with a natural σ-algebra $\mathcal{Y}$ (for example, if Y is a topological spaces, then $\mathcal{Y} = \mathcal{B}(Y)$), but no σ-algebra is specified on Ω. Then it is natural to ask the following question: "Is there a smallest σ-algebra on Ω that makes f measurable?" Such a σ-algebra exists as the following proposition says.

Proposition 3.57
If Ω is a set, $(Y_i, \mathcal{Y}_i)$ for $i \in I$, are measurable spaces and $f_i \colon \Omega \longrightarrow Y_i$ are functions,
then there is a smallest σ-algebra on Ω that makes all $\{f_i\}_{i\in I}$ measurable and it is given by

$$\sigma\big(\{f_i\}_{i\in I}\big) = \sigma\Big(\bigcup_{i\in I} f_i^{-1}(\mathcal{Y}_i)\Big).$$

Definition 3.58
The σ-algebra $\sigma\big(\{f_i\}_{i\in I}\big)$ provided by the previous proposition is called the ***σ-algebra generated by*** *$\{f_i\}_{i\in I}$.*

Remark 3.59
For every $i \in I$, $f_i^{-1}(\mathcal{Y}_i)$ is a σ-algebra. But the union of σ-algebras is not in general a σ-algebra.

Theorem 3.60
If (Ω, Σ) and (Ω', Σ') are two measurable spaces, $f \colon \Omega \longrightarrow \Omega'$ is a (Σ, Σ')-measurable function and μ is a measure on (Ω, Σ),
then the set function

$$\nu(C) \stackrel{\text{def}}{=} \mu\big(f^{-1}(C)\big) \qquad \forall\, C \in \mathcal{Y}$$

is a measure on (Ω', Σ').

Remark 3.61
The measure ν on (Ω', Σ') obtained in the previous theorem is called the ***image measure*** of μ under f (or the ***measure induced by*** f) and it is denoted by μf^{-1} or $\mu \circ f^{-1}$.

Remark 3.62
If $\mu(\Omega) = 1$ (i.e., it is a probability measure), then $\mu \circ f^{-1}(\Omega') = 1$. In fact, if (Ω, Σ, μ) is a probability space, $\Omega' = \mathbb{R}$ and $\Sigma' = \mathcal{B}(\mathbb{R})$, then a measurable function $f \colon \Omega \longrightarrow \mathbb{R}$ is said to be a ***random variable*** and $\mu \circ f^{-1}$ is the ***probability distribution*** (or the ***law***) of f.

Next we focus on $\mathbb{R}$ or $\mathbb{R}^* = \mathbb{R} \cup \{\pm\infty\}$-valued measurable functions. On $\mathbb{R}$ we consider the Borel σ-algebra $\mathcal{B}(\mathbb{R})$. So, let (Ω, Σ) be a measurable space and let $f\colon \Omega \longrightarrow \mathbb{R}$ be a $\big(\Sigma, \mathcal{B}(\mathbb{R})\big)$-measurable function. Recall that $\mathcal{B}(\mathbb{R})$ is generated by all sets of the form $[a, +\infty)$ or $(a, +\infty)$ or $(-\infty, b]$ or $(-\infty, b)$ ($a, b \in \mathbb{R}$ or even $a, b \in \mathbb{Q}$). This fact leads to the following equivalent characterization of measurability of $f\colon \Omega \longrightarrow \mathbb{R}$.

Proposition 3.63
If (Ω, Σ) *is a measurable space and* $f\colon \Omega \longrightarrow \mathbb{R}$ *is a function,*
then f *is* $\big(\Sigma, \mathcal{B}(\mathbb{R})\big)$*-measurable if and only if one (hence all) of the following conditions holds:*
(a) $\big\{\omega \in \Omega : f(\omega) \geqslant \lambda\big\} \in \Sigma$ *for all* $\lambda \in \mathbb{R}$ *(or all* $\lambda \in \mathbb{Q}$*);*
(b) $\big\{\omega \in \Omega : f(\omega) > \lambda\big\} \in \Sigma$ *for all* $\lambda \in \mathbb{R}$ *(or all* $\lambda \in \mathbb{Q}$*);*
(c) $\big\{\omega \in \Omega : f(\omega) \leqslant \lambda\big\} \in \Sigma$ *for all* $\lambda \in \mathbb{R}$ *(or all* $\lambda \in \mathbb{Q}$*);*
(d) $\big\{\omega \in \Omega : f(\omega) < \lambda\big\} \in \Sigma$ *for all* $\lambda \in \mathbb{R}$ *(or all* $\lambda \in \mathbb{Q}$*).*

Remark 3.64
If X is a topological space with the Borel σ-algebra $\mathcal{B}(X)$ and $f\colon X \longrightarrow \mathbb{R}$ is a $\big(\mathcal{B}(X), \mathcal{B}(\mathbb{R})\big)$-measurable function, then we say that f is ***Borel measurable***. We emphasize that on $\mathbb{R}$, we consider always the σ-algebra $\mathcal{B}(\mathbb{R})$.

Sometimes it is convenient to allow f to take values in the extended real line $\mathbb{R}^* = \mathbb{R} \cup \{\pm\infty\}$. Of course $\mathbb{R}^*$ is no longer a field and expressions like $\infty - \infty$ or $\frac{\pm\infty}{\pm\infty}$ make no sense and so they must be avoided.

Proposition 3.65
The collection

$$\mathcal{B}(\mathbb{R}^*) = \big\{A \cup D : A \in \mathcal{B}(\mathbb{R}) \text{ and } D \in \big\{\emptyset, \{+\infty\}, \{-\infty\}, \{-\infty, +\infty\}\big\}\big\}$$

is a σ*-algebra and it is the Borel* σ*-algebra of* $\mathbb{R}^* = \mathbb{R} \cup \{\pm\infty\}$*. Evidently*

$$\mathcal{B}(\mathbb{R}) = \mathcal{B}(\mathbb{R}^*) \cap \mathbb{R}$$

(the trace on $\mathbb{R}$ *of* $\mathcal{B}(\mathbb{R}^*)$*). Moreover,* $\mathcal{B}(\mathbb{R}^*)$ *is generated by all sets of the form*

$$[a, +\infty] \text{ or } (a, +\infty) \text{ or } [-\infty, b] \text{ or } [-\infty, b) \qquad \forall\, a, b \in \mathbb{R} \quad (\text{or } a, b \in \mathbb{Q}).$$

Next we introduce the measurable functions which are the building blocks of integration theory. So, let (Ω, Σ) be a measurable space and $A \subseteq X$. We consider the function

$$\chi_A(\omega) \stackrel{\text{def}}{=} \begin{cases} 1 & \text{if } \omega \in A, \\ 0 & \text{if } \omega \notin A. \end{cases}$$

Definition 3.66
The function χ_A is called the ***characteristic function*** *(or* ***indicator function****) of A.*

Definition 3.67
Let (Ω, Σ) be a measurable space. A function $s \colon \Omega \longrightarrow \mathbb{R}$ is a ***simple function*** *if it is a finite linear combination, with real coefficients, of characteristic functions of sets in Σ. This is equivalent to saying that $s \colon \Omega \longrightarrow \mathbb{R}$ is simple if and only if s is $(\Sigma, \mathcal{B}(\mathbb{R}))$-measurable and the range of s is a finite set on $\mathbb{R}$. Then we have*

$$s = \sum_{k=1}^{n} a_k \chi_{A_k},$$

with $A_k = f^{-1}(\{a_k\})$, $k \in \{1, \ldots, n\}$. This representation is called the ***standard representation*** *of s and it exhibits s as a linear combination with different coefficients of characteristic functions of disjoint Σ-sets whose union is Ω.*

One reason that simple functions are important is because they approximate in a nice way every measurable function.

Theorem 3.68
Let (Ω, Σ) be a measurable space.
(a) *If $f \colon \Omega \longrightarrow \overline{\mathbb{R}} = \mathbb{R} \cup \{+\infty\}$ is a measurable function, then there exists a sequence $\{s_n\}_{n \geqslant 1}$ of simple functions such that $0 \leqslant s_1 \leqslant s_2 \leqslant \ldots \leqslant f$ and $s_n(\omega) \longrightarrow f(\omega)$ as $n \to +\infty$ for all $\omega \in \Omega$.*
(b) *If $f \colon \Omega \longrightarrow \mathbb{R}^* = \mathbb{R} \cup \{\pm\infty\}$ is a measurable function, then there exists a sequence $\{s_n\}_{n \geqslant 1}$ of simple functions such that $0 \leqslant |s_1| \leqslant |s_2| \leqslant \ldots \leqslant |f|$ and $s_n(\omega) \longrightarrow f(\omega)$ as $n \to +\infty$ for all $\omega \in \Omega$ and the convergence is uniform on any subset of Ω where f is bounded.*

Corollary 3.69
If (Ω, Σ) *is a measurable space and* $f_n \colon \Omega \longrightarrow \mathbb{R}^* = \mathbb{R} \cup \{\pm\infty\}$ *for* $n \in \mathbb{N}$ *are measurable functions,*
then so are

$$\sup_{n\geqslant 1} f_n, \quad \inf_{n\geqslant 1} f_n, \quad \limsup_{n\to+\infty} f_n, \quad \liminf_{n\to+\infty} f_n, \quad \text{and} \quad \lim_{n\to+\infty} f_n$$

when it exists.

Proposition 3.70
If (Ω, Σ) *is a measurable space,* (Y, d) *is a metric space,* $\{f_n \colon \Omega \longrightarrow Y\}_{n\geqslant 1}$ *is a sequence of* $(\Sigma, \mathcal{B}(Y))$*-measurable functions and for all* $\omega \in \Omega$*, we have that*

$$f_n(\omega) \longrightarrow f(\omega) \quad \text{in } Y \quad \text{as } n \to +\infty,$$

then f *is* $(\Sigma, \mathcal{B}(Y))$*-measurable.*

Another straightforward consequence of Theorem 3.68 is the following corollary.

Corollary 3.71
If (Ω, Σ) *is a measurable space and* $f, g \colon \Omega \longrightarrow \mathbb{R}^* = \mathbb{R} \cup \{\pm\infty\}$ *are measurable functions,*
then so are $f+g$*,* fg*,* $\max\{f,g\}$ *and* $\min\{f,g\}$*.*

Remark 3.72
In particular then the measurability of $f \colon \Omega \longrightarrow \mathbb{R}^* = \mathbb{R} \cup \{\pm\infty\}$ implies the measurability of its positive and negative parts, defined, respectively, by

$$f^+ = \max\{f, 0\} \quad \text{and} \quad f^- = \max\{-f, 0\} = -\min\{f, 0\}.$$

Note that $f = f^+ - f^-$ and $|f| = f^+ + f^-$. Hence $|f|$ is measurable too. Moreover, if $f, g \colon \Omega \longrightarrow \mathbb{R}^* = \mathbb{R} \cup \{\pm\infty\}$ are both measurable functions, then

$$\{f < g\}, \ \{f \leqslant g\}, \ \{f = g\}, \ \{f \neq g\} \in \Sigma.$$

Definition 3.73
Let (Ω, Σ, μ) *be a measure space. If a property is true for all* $\omega \in \Omega \setminus N$*, where* N *is a subset of a* μ*-null set (see Definition 3.21), then we say that the property holds* ***almost everywhere*** *(or* μ***-almost everywhere*** *or for short* ***a.e.*** *or* μ***-a.e.****).*

Proposition 3.74
If (Ω, Σ, μ) *is a complete measure space (see Definition 3.23), then*
(a) *if* $f: \Omega \longrightarrow \mathbb{R}^* = \mathbb{R} \cup \{\pm\infty\}$ *is measurable and* $f = g$ μ*-almost everywhere, then* g *is measurable too;*
(b) *if* $\{f_n: \Omega \longrightarrow \mathbb{R}^* = \mathbb{R} \cup \{\pm\infty\}\}_{n \geqslant 1}$ *is a sequence of measurable functions and* $f_n \longrightarrow f$ μ*-almost everywhere, then* f *is measurable.*

Proposition 3.75
If (Ω, Σ, μ) *is a measure space,* $(\Omega, \Sigma_\mu, \overline{\mu})$ *is its completion (see Definition 3.23) and* $f: \Omega \longrightarrow \mathbb{R}^* = \mathbb{R} \cup \{\pm\infty\}$ *is a* $(\Sigma_\mu, \mathcal{B}(\mathbb{R}^*))$*-measurable function,*
then there is a $(\Sigma, \mathcal{B}(\mathbb{R}^*))$*-measurable function* $g: \Omega \longrightarrow \mathbb{R}^* = \mathbb{R} \cup \{\pm\infty\}$ *such that*

$$f(\omega) = g(\omega) \quad \overline{\mu}\text{-almost everywhere on } \Omega.$$

The next two theorems are important tools in the study of measurable functions. Both guarantee that outside certain "small" sets, we have nice structural properties.

Theorem 3.76 *(Egorov Theorem)*
If (Ω, Σ, μ) *is a finite measure space,* (Y, d_Y) *is a metric space,* $\{f_n: \Omega \longrightarrow Y\}_{n \geqslant 1}$ *is a sequence of measurable functions,* $f: \Omega \longrightarrow Y$ *is a measurable function and*

$$d_Y\big(f_n(\omega), f(\omega)\big) \longrightarrow 0 \quad \mu\text{-almost everywhere on } \Omega,$$

then for every $\varepsilon > 0$*, there is a set* $A \in \Sigma$ *with* $\mu(\Omega \setminus A) < \varepsilon$ *such that* $f_n \rightrightarrows f$ *on* A *(*$\rightrightarrows$ *denoting uniform convergence), i.e.,* $\lim_{n \to +\infty} \sup\big\{d_Y\big(f_n(\omega), f(\omega)\big) : \omega \in A\big\} = 0$.

Theorem 3.77 *(Lusin Theorem)*
If X *is a metric space,* $\mu: \mathcal{B}(X) \longrightarrow \mathbb{R}$ *is a finite measure and* $f: X \longrightarrow \mathbb{R}$ *is a Borel measurable function,*
then for every $\varepsilon > 0$*, there is a closed set* $C \subseteq X$ *such that* $\mu(X \setminus C) < \varepsilon$ *and* $f|_C$ *is continuous.*

Proposition 3.78
If (Ω, Σ) *and* (Ω', Σ') *are two measurable spaces,* $\{A_n\}_{n\geqslant 1} \subseteq \Sigma$ *are pairwise disjoint sets,* $\Omega = \bigcup\limits_{n\geqslant 1} A_n$*, for every* $n \geqslant 1$*,* Σ_{A_n} *is the relative* σ*-algebra of* Σ *on* A_n *(see Definition 3.13),* $f_n \colon A_n \longrightarrow Y$ *is a* (Σ_{A_n}, Σ')*-measurable function and* $f \colon \Omega \longrightarrow \Omega'$ *is defined by*

$$f(\omega) = f_n(\omega) \qquad \forall\, \omega \in A_n,\ n \geqslant 1,$$

then f *is* (Σ, Σ')*-measurable.*

Remark 3.79
Evidently, if (Ω, Σ) is a measurable space and $f \colon \Omega \longrightarrow \mathbb{R}$ is a measurable function, then for any $A \subseteq \Omega$, the restriction $f|_A$ is Σ_A-measurable. This reminds us of the analogous topological result which says that a continuous function on a topological space, when restricted to a subset with the subspace topology, is continuous. But, recall that for continuous functions the converse is not in general true, namely a continuous function from a set with the subspace topology in general does not extend continuously (additional conditions are needed; see Theorem 2.138, the Tietze extension theorem). In contrast, this extension is always possible for any set $A \subseteq \Omega$ (measurable or nonmeasurable). If $A \in \Sigma$, then a trivial extension of f, by setting it equal to some fixed value on $\Omega \setminus A$ is measurable. What is not immediate clear is that such a measurable extension is also true if $A \notin \Sigma$.

Theorem 3.80
If (Ω, Σ) *is a measurable space,* $A \subseteq \Omega$ *(not necessarily in* Σ*) and* $f \colon A \longrightarrow \mathbb{R}$ *is* $(\Sigma_A, \mathcal{B}(\mathbb{R}))$*-measurable function,*
then there exists a $(\Sigma, \mathcal{B}(\mathbb{R}))$*-measurable function* $\widehat{f} \colon \Omega \longrightarrow \mathbb{R}$ *such that* $\widehat{f}|_A = f$*.*

Remark 3.81
In fact the above theorem remains true if $\mathbb{R}$ is replaced by Polish space Y (see Definition 2.150).

As we already mentioned (see Remark 3.64), for $\mathbb{R}$-valued (or $\mathbb{R}^*$-valued) functions on the range space we always consider the Borel σ-algebra $\mathcal{B}(\mathbb{R})$ (or $\mathcal{B}(\mathbb{R}^*)$). The following proposition shows why the Lebesgue σ-algebra $\mathcal{L}(\mathbb{R})$ may be too large.

Proposition 3.82
There exists a continuous nondecreasing function $f\colon [0,1] \longrightarrow [0,1]$ and a Lebesgue measurable set $D \subseteq [0,1]$ such that $f^{-1}(D)$ is not Lebesgue measurable.

Remark 3.83
The function f is given by $f = h^{-1}$, where $h\colon [0,1] \longrightarrow [0,1]$ is a continuous, strictly increasing surjection (hence a homeomorphism), defined by

$$h(x) \stackrel{\text{def}}{=} \tfrac{1}{2}\big(g(x)+x\big),$$

with $g\colon [0,1] \longrightarrow [0,1]$ being the Cantor function. The ***Cantor function*** is defined as follows. First we define g on the Cantor set C (see Proposition 3.35). For $x \in C$, we define $g(x)$ by means of the binary expansion

$$g(x) \stackrel{\text{def}}{=} \frac{1}{2}\sum_{n\geqslant 1} \frac{a_n}{2^n},$$

where a_n are the ternary "components" of $x \in C$. So, $g|_C$ is monotone increasing. On each of the missing open middle thirds, we define g to be locally constant. In fact, the missing open middle third (a_k, b_k) is defined by the requirement that the kth digit in the ternary expansion of a_k is not 1, so $g(a_k) = g(b_k)$, which is the constant value chosen for g on $[a_k, b_k]$ (for example, $g(x) = \frac{1}{2}$ on $\big(\frac{1}{3}, \frac{2}{3}\big)$, $g(x) = \frac{1}{4}$ on $\big(\frac{1}{9}, \frac{2}{9}\big)$, $g(x) = \frac{3}{4}$ on $\big(\frac{7}{9}, \frac{8}{9}\big)$ and so on). Then g is nondecreasing and continuous. The function $\varphi\colon [0,1] \longrightarrow [0,2]$, defined by

$$\varphi(x) = g(x) + x$$

is a homeomorphism,

$$\lambda\big(h([0,1]) \setminus C\big) = 1 \quad \text{and} \quad \lambda\big(h(C)\big) = 1$$

(λ being the Lebesgue measure on $\mathbb{R}$) and if E is a nonmeasurable subset of $\varphi(C)$ (assuming as always the axiom of choice), then $\varphi^{-1}(E)$ is Lebesgue measurable but not a Borel set.

3.1.8 The Lebesgue Integral

Now we are ready to introduce the Lebesgue integral and study its properties. We start with nonnegative functions.

Definition 3.84
Let (Ω, Σ, μ) be a measure space and let

$$\mathcal{M}_+ \stackrel{\text{def}}{=} \{f : \Omega \longrightarrow [0, +\infty] : \ f \text{ is measurable}\}.$$

(a) *If $s \in \mathcal{M}_+$ is a simple function with standard representation $s = \sum\limits_{k=1}^{n} a_k \chi_{A_k}$, then the **integral** of s with respect to μ is defined by*

$$\int\limits_{\Omega} s \, d\mu \stackrel{\text{def}}{=} \sum_{k=1}^{n} a_k \mu(A_k)$$

(as always we use the convention $0 \cdot \infty = 0$).
(b) *If $f \in \mathcal{M}_+$ is a general $\overline{\mathbb{R}}_+ = [0, +\infty]$-valued measurable function, then*

$$\int\limits_{\Omega} f \, d\mu \stackrel{\text{def}}{=} \sup\Big\{ \int\limits_{\Omega} s \, d\mu : \ 0 \leqslant s \leqslant f, \ s \text{ is simple} \Big\}.$$

Remark 3.85
One can show that Definition 3.84(a) is independent of the representation of s (i.e., $\int\limits_{\Omega} s \, d\mu$ is well defined). Thanks to Theorem 3.68(a), the integral of a nonnegative measurable function can be expressed as the limit of a sequence, rather than a more general supremum. So, by Theorem 3.68(a), we can find a sequence of simple function $\{s_n\}_{n \geqslant 1}$ such that $s_n \nearrow f$. Then for any such sequence, we have

$$\int\limits_{\Omega} f \, d\mu = \lim_{n \to +\infty} \int\limits_{\Omega} s_n \, d\mu.$$

Finally, if $A \in \Sigma$, then

$$\int\limits_{A} f \, d\mu = \int\limits_{\Omega} \chi_A f \, d\mu.$$

In the next proposition we summarize the most important properties of the integral on $\mathcal{M}^+$.

Proposition 3.86

(a) *For all $c \geqslant 0$ and all $f, g \in \mathcal{M}_+$, we have*

$$\int_\Omega (cf + g)\, d\mu \;=\; c \int_\Omega f\, d\mu + \int_\Omega g\, d\mu.$$

(b) *If $f, g \in \mathcal{M}_+$ and $f \leqslant g$ μ-almost everywhere, then*

$$\int_\Omega f\, d\mu \;\leqslant\; \int_\Omega g\, d\mu.$$

(c) *The set function $m \colon \Sigma \longrightarrow [0, +\infty]$, defined by*

$$m(A) \;=\; \int_A d\mu \;=\; \int_\Omega \chi_A\, d\mu$$

is a measure.

(d) *If $\{f_n\}_{n \geqslant 1} \subseteq \mathcal{M}_+$ is a sequence such that $f_n \leqslant f_{n+1}$ for all $n \geqslant 1$ and $f = \lim\limits_{n \to +\infty} f_n = \sup\limits_{n \geqslant 1} f_n$, then*

$$\int_\Omega f\, d\mu \;=\; \lim_{n \to +\infty} \int_\Omega f_n\, d\mu$$

(monotone convergence theorem).

(e) *If $\{f_n\}_{n \geqslant 1} \subseteq \mathcal{M}_+$ is a sequence and $f = \sum\limits_{n \geqslant 1} f_n$, then*

$$\int_\Omega f\, d\mu \;=\; \sum_{n \geqslant 1} \int_\Omega f_n\, d\mu$$

(f) *If $f \in \mathcal{M}_+$, then*

$$\int_\Omega f\, d\mu \;=\; 0 \quad \Longleftrightarrow \quad f \;=\; 0 \ \mu\text{-almost everywhere on } \Omega.$$

(g) *If $\{f_n\}_{n \geqslant 1} \subseteq \mathcal{M}_+$ is a sequence, then*

$$\int_\Omega \liminf_{n \to +\infty} f_n\, d\mu \;\leqslant\; \liminf_{n \to +\infty} \int_\Omega f_n\, d\mu$$

(Fatou Lemma).

Next we extend the integral to measurable functions which are not necessarily positive. So, as before let (Ω, Σ, μ) be a measure space and set

$$\mathcal{M} \stackrel{\text{def}}{=} \big\{f\colon \Omega \longrightarrow \mathbb{R}^* = \mathbb{R} \cup \{\pm\infty\} : \ f \text{ is measurable}\big\}.$$

Recall that, if $f \in \mathcal{M}$, then

$$f^+ = \max\{f, 0\}, \quad f^- = \max\{-f, 0\}, \quad f = f^+ - f^- \quad \text{and} \quad f^+, f^- \in \mathcal{M}_+.$$

Definition 3.87
*A function $f\colon \Omega \longrightarrow \mathbb{R}^* = \mathbb{R}\cup\{\pm\infty\}$ is said to be **μ-quasi integrable** if $f \in \mathcal{M}$ and $\int_\Omega f^+ \, d\mu$, $\int_\Omega f^- \, d\mu$ are not both equal to $+\infty$. In this case the **integral** of f is defined by*

$$\int_\Omega f \, d\mu = \int_\Omega f^+ \, d\mu - \int_\Omega f^- \, d\mu.$$

*We say that $f\colon \Omega \longrightarrow \mathbb{R}^* = \mathbb{R}\cup\{\pm\infty\}$ is **μ-integrable** (or **integrable**) if $f \in \mathcal{M}$ and both $\int_\Omega f^+ \, d\mu$ and $\int_\Omega f^- \, d\mu$ are finite. By $\mathcal{L}^1(\mu)$ we denote the set of all $\mathbb{R}^*$-valued, μ-integrable functions.*

The next proposition summarizes some basic integrability criteria.

Proposition 3.88
Let $f \in \mathcal{M}_+$. The following statements are equivalent:
(a) *$f \in \mathcal{L}^1(\mu)$;*
(b) *$f^+, f^- \in \mathcal{L}^1(\mu)$;*
(c) *$|f| \in \mathcal{L}^1(\mu)$;*
(d) *there exists $h \in \mathcal{L}^1(\mu)$, $h \geqslant 0$ such that $|f| \leqslant h$.*

In what follows, we focus on the space $\mathcal{L}^1(\mu)$.

Definition 3.89
For $f \in \mathcal{L}^1(\mu)$ and $A \in \Sigma$, we set

$$\int_A f \, d\mu = \int_\Omega \chi_A f \, d\mu.$$

The next proposition summarizes the basic properties of the integral.

Proposition 3.90
(a) *For any* $f, g \in \mathcal{L}^1(\mu)$ *and* $c \in \mathbb{R}$, *we have*

$$\int_\Omega (cf + g)\, d\mu = c \int_\Omega f\, d\mu + \int_\Omega g\, d\mu$$

(i.e., the integral is a linear operator).
(b) *If* $f, g \in \mathcal{L}^1(\mu)$, *then* $\min\{f, g\}$, $\max\{f, g\} \in \mathcal{L}^1(\mu)$ *(lattice property).*
(c) *If* $f, g \in \mathcal{L}^1(\mu)$ *and* $f \leqslant g$ μ*-almost everywhere, then*

$$\int_\Omega f\, d\mu \leqslant \int_\Omega g\, d\mu$$

(monotonicity).
(d) *If* $f \in \mathcal{L}^1(\mu)$, *then*

$$\Big|\int_\Omega f\, d\mu\Big| \leqslant \int_\Omega |f|\, d\mu$$

(triangle inequality).
(e) *If* $f \in \mathcal{L}^1(\mu)$ *and* $f \geqslant 0$ μ*-almost everywhere, then*

$$\text{the function } A \longmapsto \int_A f\, d\mu \text{ is a finite measure on } \Sigma.$$

(f) *If* $f \in \mathcal{L}^1(\mu)$, $A \in \Sigma$, $c > 0$ *and* $0 < p < +\infty$, *then*

$$\mu(\{|f|^p \geqslant c\} \cap A) \leqslant \frac{1}{c} \int_A |f|^p\, d\mu$$

(Markov inequality).
(g) *If* $f, g \in \mathcal{L}^1(\mu)$ *and* $f = g$ μ*-almost everywhere, then*

$$\int_A f\, d\mu = \int_A g\, d\mu \qquad \forall\, A \in \Sigma.$$

Remark 3.91
Usually for $p \neq 1$, the inequality in ***(f)*** is called ***Chebyshev inequality***.

3.1.9 Convergence Theorems

Next we present the convergence theorems for the (Lebesgue) integral. These theorems are the strong feature of Lebesgue integration compared to the Riemann integral.

Theorem 3.92 *(Lebesgue Monotone Convergence Theorem)*
If (Ω, Σ, μ) is a measure space, $\{f_n\}_{n\geqslant 1} \subseteq \mathcal{M}$ is an increasing sequence such that $f_n \nearrow f$ (i.e., $f = \sup\limits_{n\geqslant 1} f_n$) and $\int\limits_\Omega f_1\, d\mu > -\infty$ (i.e., the functions f_n are quasi-integrable),
then

$$\int\limits_\Omega f_n\, d\mu \nearrow \int\limits_\Omega f\, d\mu$$

$$\textit{i.e.,}\ \sup_{n\geqslant 1} \int\limits_\Omega f_n\, d\mu = \int\limits_\Omega \sup_{n\geqslant 1} f_n\, d\mu = \int\limits_\Omega f\, d\mu.$$

Remark 3.93
There is a symmetric version of this result, with $\{f_n\}_{n\geqslant 1} \subseteq \mathcal{M}$ decreasing, $f_n \searrow f$ and $\int\limits_\Omega f_1\, d\mu < +\infty$, then

$$\int\limits_\Omega f_n\, d\mu \searrow \int\limits_\Omega f\, d\mu,$$

$$\text{i.e.,}\ \inf_{n\geqslant 1} \int\limits_\Omega f_n\, d\mu = \int\limits_\Omega \inf_{n\geqslant 1} f_n\, d\mu = \int\limits_\Omega f\, d\mu.$$

Moreover, if $\{f_n\}_{n\geqslant 1} \subseteq \mathcal{L}^1(\mu)$, then $f \in \mathcal{L}^1(\mu)$ if and only if

$$\sup_{n\geqslant 1} \int\limits_\Omega f_n\, d\mu < +\infty \quad \text{in the increasing case}$$

and $f \in \mathcal{L}^1(\mu)$ if and only if

$$\inf_{n\geqslant 1} \int\limits_\Omega f_n\, d\mu > -\infty \quad \text{in the decreasing case.}$$

Theorem 3.94 *(Lebesgue Dominated Convergence Theorem)*
If (Ω, Σ, μ) *is a measure space and* $\{f_n\}_{n\geqslant 1} \subseteq \mathcal{M}$ *is a sequence satisfying:*
(a) $f_n(x) \longrightarrow f(x)$ *μ-almost everywhere on* Ω *as* $n \to +\infty$*;*
(b) *there exists* $h \in \mathcal{L}^1(\mu)$ *such that* $|f_n(x)| \leqslant h(x)$ *μ-almost everywhere on* Ω *for all* $n \geqslant 1$*,*
then

$$\lim_{n\to+\infty} \int_\Omega |f_n - f| \, d\mu \, d\mu \;=\; 0$$

and

$$\lim_{n\to+\infty} \int_\Omega f_n \, d\mu \;=\; \int_\Omega \lim_{n\to+\infty} f_n \, d\mu \;=\; \int_\Omega f \, d\mu.$$

Theorem 3.95 *(Fatou Lemma)*
(a) *If* (Ω, Σ, μ) *is a measure space,* $\{f_n\}_{n\geqslant 1} \subseteq \mathcal{M}$ *and* $f_n \geqslant h$ *μ almost everywhere for all* $n \geqslant 1$ *and some* $h \in \mathcal{M}$ *such that* $\int_\Omega h \, d\mu > -\infty$*,*
then

$$\int_\Omega \liminf_{n\to+\infty} f_n \, d\mu \;\leqslant\; \liminf_{n\to+\infty} \int_\Omega f_n \, d\mu.$$

(b) *If* (Ω, Σ, μ) *is a measure space,* $\{f_n\}_{n\geqslant 1} \subseteq \mathcal{M}$ *and* $f_n \leqslant h$ *μ-almost everywhere for all* $n \geqslant 1$ *and some* $h \in \mathcal{M}$ *such that* $\int_\Omega h \, d\mu < +\infty$*,*
then

$$\limsup_{n\to+\infty} \int_\Omega f_n \, d\mu \;\leqslant\; \int_\Omega \limsup_{n\to+\infty} f_n \, d\mu.$$

3.1.10 L^p-Spaces

From the viewpoint of integration, there is no loss of generality in assuming that a function $f \in \mathcal{L}^1(\mu)$ is $\mathbb{R}$-valued. Indeed, note that if $f \in \mathcal{L}^1(\mu)$, then $\chi_{\{|f|<+\infty\}} f \in \mathcal{L}^1(\mu)$ and we have

$$\int_\Omega |f - \chi_{\{|f|<+\infty\}} f| \, d\mu \;=\; 0.$$

So, the integrals of f and $\chi_{\{|f|<+\infty\}} f$ are equal. So, from now on, when dealing with functions in $\mathcal{L}^1(\mu)$, we will assume that they are $\mathbb{R}$-valued. This way we can take linear combinations of such functions, without having to worry about undefined expressions like $\infty - \infty$. Next note that if $\int_\Omega |f|\, d\mu = 0$, then f need not be identically 0, as a function. We can only conclude that $f(\omega) = 0$ μ-almost everywhere on Ω. For this reason we make the following definition.

Definition 3.96
Let

$$N(\mu) \stackrel{\text{def}}{=} \{f \in \mathcal{L}^1(\mu):\ \mu\{f \neq 0\} = 0\}$$

and introduce the following equivalence relation on $\mathcal{L}^1(\mu)$*:*

$$f \sim g \quad \Longleftrightarrow \quad f - g \in N(\mu).$$

We set

$$L^1(\mu) \;=\; L^1(\Omega) \stackrel{\text{def}}{=} \mathcal{L}^1(\mu)/_{N(\mu)}.$$

So, the elements of $L^1(\Omega)$ *are in fact equivalence classes, but we still denote them as those of* $\mathcal{L}^1(\mu)$ *(i.e., using a representative). Similarly, if* $1 < p < +\infty$ *and*

$$\mathcal{L}^p(\mu) \stackrel{\text{def}}{=} \{f\colon \Omega \longrightarrow \mathbb{R}:\ f \in \mathcal{M} \text{ and } \int_\Omega |f|^p\, d\mu < +\infty\},$$

then

$$L^p(\mu) \;=\; L^p(\Omega) \stackrel{\text{def}}{=} \mathcal{L}^p(\mu)/_{N(\mu)}.$$

For $1 \leqslant p < +\infty$ *and* $f \in L^p(\Omega)$*, we define*

$$\|f\|_p \stackrel{\text{def}}{=} \Big(\int_\Omega |f|^p\, d\mu\Big)^{\frac{1}{p}}$$

and this now is a bona fide norm. To complete the picture, we also introduce a space corresponding to the limit case $p = +\infty$*. So, we define*

$$\mathcal{L}^\infty(\mu) \stackrel{\text{def}}{=} \{f\colon \Omega \longrightarrow \mathbb{R}^*:\ f \in \mathcal{M} \text{ and } \inf\{\lambda \geqslant 0:\ \mu\{|f| > \lambda\} = 0\} < +\infty\}.$$

We set

$$L^\infty(\mu) \;=\; L^\infty(\Omega) \;\stackrel{\text{def}}{=}\; \mathcal{L}^\infty(\mu)/_{N(\mu)}$$

and

$$\|f\|_\infty \;\stackrel{\text{def}}{=}\; \inf\left\{\lambda \geqslant 0:\; \mu\{|f| \geqslant \lambda\} = 0\right\} \;=\; \operatorname{ess\,sup}|f|.$$

This is a norm too.

Finally, we say that $f \in L^1_{\mathrm{loc}}(\Omega)$ *if for all* $C \in \Sigma$, *with* $\mu(C) < +\infty$, *we have* $\int\limits_C |f|\,d\mu < +\infty$.

Remark 3.97
Note that the norms $\|\cdot\|_p$, $1 \leqslant p < +\infty$ are defined by integrals, while $\|\cdot\|_\infty$ is defined pointwise. This difference is reflected in the structure of the resulting normed spaces $L^p(\Omega)$, $1 \leqslant p < +\infty$ and $L^\infty(\Omega)$. Also note that $L^p(\Omega)$, $1 \leqslant p \leqslant +\infty$ are Banach lattices (i.e., $|f| \leqslant |h|$ implies that $\|f\|_p \leqslant \|h\|_p$).

The following version of the Fatou lemma (see Theorem 3.95) can be useful in many circumstances.

Theorem 3.98 *(Brezis–Lieb Lemma)*
If (Ω, Σ, μ) *is a measure space,* $\{f_n\}_{n\geqslant 1} \subseteq L^1(\Omega)$, $f \in L^1(\Omega)$ *and* $f_n \longrightarrow f$ μ*-almost everywhere on* Ω,
then

$$\lim_{n\to+\infty} \big|\|f_n\|_1 - \|f\|_1 - \|f_n - f\|_1\big| \;=\; \lim_{n\to+\infty} \int\limits_\Omega \big||f_n| - |f| - |f_n - f|\big|\,d\mu \;=\; 0.$$

In particular, if $\|f_n\|_1 \longrightarrow \|f\|_1$ *and* $f_n \longrightarrow f$ μ*-almost everywhere on* Ω, *then* $\|f_n - f\|_1 \longrightarrow 0$.

Recall that a function $\varphi\colon \mathbb{R} \longrightarrow \mathbb{R}$ is said to be ***convex*** if for all $x, y \in \mathbb{R}$ and all $\lambda \in [0,1]$, we have

$$\varphi\big(\lambda x + (1-\lambda)y\big) \;\leqslant\; \lambda\varphi(x) + (1-\lambda)\varphi(y).$$

If this inequality is strict when $x \neq y$ and $\lambda \in (0,1)$, then φ is said to be ***strictly convex***. A convex function is continuous. The defining inequality takes a continuous form using integrals as follows.

Theorem 3.99 *(Jensen Inequality)*
If (Ω, Σ, μ) *is a finite measure space,* $f \in L^1(\Omega)$*,* $\varphi\colon \mathbb{R} \longrightarrow \mathbb{R}$ *is convex and* $(\varphi \circ f)^- \in L^1(\Omega)$*, then*

$$\varphi\Big(\frac{1}{\mu(\Omega)} \int_\Omega f \, d\mu\Big) \leqslant \frac{1}{\mu(\Omega)} \int_\Omega (\varphi \circ f) \, d\mu.$$

If φ *is strictly convex, then we have equality if and only if* f *is a constant function.*

Corollary 3.100
If (Ω, Σ, μ) *is a measure space,* $f \in L^1(\Omega)$*,* $f \geqslant 0$ *and* $\varphi\colon \mathbb{R} \longrightarrow \mathbb{R}$ *is convex, then*

$$\varphi\Big(\frac{\int_\Omega fg \, d\mu}{\int_\Omega f \, d\mu}\Big) \leqslant \frac{\int_\Omega \varphi(g) f \, d\mu}{\int_\Omega f \, d\mu} \qquad \forall \, g \in \mathcal{M}_+.$$

There are a few more important inequalities related to the space $L^p(\Omega)$ $(1 \leqslant p \leqslant +\infty)$.

Theorem 3.101 *(Minkowski Inequality)*
If (Ω, Σ, μ) *is a measure space,* $1 \leqslant p \leqslant +\infty$*,* $f, g \in L^p(\Omega)$*, then* $\|f + g\|_p \leqslant \|f\|_p + \|g\|_p$.

In fact some more is true.

Theorem 3.102
If (Ω, Σ, μ) *is a measure space,* $1 \leqslant p \leqslant +\infty$*, then* $L^p(\Omega)$ *is a Banach space (i.e., a normed space which is complete metric space for the metric introduced by the norm,* $d(f, g) = \|f - g\|_p$*).*

Theorem 3.103 *(Hölder Inequality)*
If (Ω, Σ, μ) *is a measure space,* $1 \leqslant p, p' \leqslant +\infty$ *satisfy* $\frac{1}{p} + \frac{1}{p'} = 1$ *(if* $p = 1$*, then* $p' = \infty$ *and if* $p = \infty$ *then* $p' = 1$*), then for every* $f \in L^p(\Omega)$ *and every* $g \in L^{p'}(\Omega)$*, we have* $fg \in L^1(\Omega)$ *and* $\|fg\|_1 \leqslant \|f\|_p \|g\|_{p'}$ *and equality holds if and only if* $\alpha|f|^p = \beta|g|^{p'}$ μ*-almost everywhere, with* $\alpha\beta \neq 0$.

Remark 3.104
If $p = 2$, then this inequality is known as ***Cauchy–Schwarz–Bunyakowski inequality***.

The next theorem gives a generalization of the Hölder inequality.

Theorem 3.105
If (Ω, Σ, μ) *is a measure space,* $\{p_k\}_{k=1}^m \subseteq [1, +\infty]$ *with* $\frac{1}{p} = \sum_{k=1}^m \frac{1}{p_k} \leqslant 1$ *and* $f_k \in L^{p_k}(\Omega)$ *for all* $k \in \{1, \ldots, m\}$,
then $f_1, \ldots, f_m \in L^p(\Omega)$ *and*

$$\|f_1 \cdots f_m\|_p \leqslant \|f_1\|_{p_1} \cdots \|f_m\|_{p_m}.$$

The next theorem gives another related inequality.

Theorem 3.106 *(Interpolation Inequality)*
If (Ω, Σ, μ) *is a measure space,* $1 \leqslant p \leqslant q \leqslant +\infty$ *and* $f \in L^p(\Omega) \cap L^q(\Omega)$,
then $f \in L^r(\Omega)$ *for every* $r \in [p, q]$ *and we have*

$$\|f\|_r \leqslant \|f\|_p^{1-t} \|f\|_q^t,$$

where $t \in [0, 1]$ *and* $\frac{1}{r} = \frac{1-t}{p} + \frac{t}{q}$.

The next inequality is rather obvious, but nevertheless is very useful to warrant special mention (see also Proposition 3.90(f) and Remark 3.91).

Theorem 3.107 *(Chebyshev Inequality)*
If (Ω, Σ, μ) *is a measure space,* $1 \leqslant p < +\infty$ *and* $f \in L^p(\Omega)$,
then for all $\lambda > 0$, *we have*

$$\mu\{|f| > \lambda\} \leqslant \left(\frac{\|f\|_p}{\lambda}\right)^p.$$

The above theorem in case $p = 1$ is also known as the ***Markov Inequality***.

The next theorem has important implications on the geometry of the Banach spaces $L^p(\Omega)$ $(1 < p < +\infty)$, namely it implies that they are uniformly convex (see Chap. 5).

Theorem 3.108 *(Clarkson Inequalities)*
Suppose that (Ω, Σ, μ) *is a measure space,* $1 < p < +\infty$ *and* $f, g \in L^p(\Omega)$. *We have:*
(a) If $1 < p < 2$, *then*

$$\left\|\frac{f+g}{2}\right\|_p^{p'} + \left\|\frac{f-g}{2}\right\|_p^{p'} \leqslant \left(\frac{1}{2}\|f\|_p^p + \frac{1}{2}\|g\|_p^p\right)^{\frac{1}{p-1}}$$

(recall that $p' = \frac{p}{p-1}$*).*

(b) If $p \geqslant 2$, then

$$\left\| \frac{f+g}{2} \right\|_p^p + \left\| \frac{f-g}{2} \right\|_p^p \leqslant \frac{1}{2}\|f\|_p^p + \frac{1}{2}\|g\|_p^p.$$

Remark 3.109
There are two other closely related inequalities, known as ***Hanner inequalities***, which also lead to the uniform convexity of the Banach spaces $L^p(\Omega)$, $1 < p < +\infty$. They read as follows:
(a) For $1 < p < 2$, we have

$$\left(\|f\|_p + \|g\|_p\right)^p + \left|\|f\|_p - \|g\|_p\right|^p \leqslant \|f+g\|_p^p + \|f-g\|_p^p$$

and

$$\left(\|f+g\|_p + \|f-g\|_p\right)^p + \left|\|f+g\|_p - \|f-g\|_p\right|^p \leqslant 2^p\left(\|f\|_p^p + \|g\|_p^p\right).$$

(b) For $2 \leqslant p$, the above inequalities are reversed.

We have already seen in Theorem 3.102 that $L^p(\Omega)$, $1 \leqslant p \leqslant +\infty$, are Banach spaces. Let us identify some useful dense subsets of these spaces.

Proposition 3.110
If (Ω, Σ, μ) is a measure spaces and $1 \leqslant p \leqslant +\infty$,
then the simple functions are dense in $L^p(\Omega)$.

Proposition 3.111
If X is a locally compact topological space which is second countable (see Definitions 2.92 and 2.24), μ is a Borel measure on X which is finite on compact sets and $1 \leqslant p < +\infty$,
then the space $C_c(X)$ of continuous functions on X with compact supports is dense in $L^p(X, \mu)$.

Proposition 3.112
If $\Omega \subseteq \mathbb{R}^N$ is an open set and $1 \leqslant p < +\infty$,
then the space $C_c^\infty(\Omega)$ of C^∞-functions on Ω with compact supports is dense in $L^p(\Omega)$.

3.1.11 Multiple Integrals: Change of Variables

Let (Ω, Σ), (Ω', Σ') be measurable spaces and let $f\colon \Omega \times \Omega' \longrightarrow \mathbb{R}$ be a function. In analogy to Definition 3.46, we define the x-section f_x and the y-section f_y of f, by

$$f_x(y) \;=\; f_y(x) \;=\; f(x,y) \qquad \forall\, x \in \Omega,\; y \in \Omega'.$$

Proposition 3.113
If $f\colon \Omega \times \Omega' \longrightarrow \mathbb{R}$ *is* $\Sigma \otimes \Sigma'$*-measurable,*
then for every $x \in \Omega$*,* f_x *is* Σ'*-measurable and for every* $y \in \Omega'$*,* f_y *is* Σ*-measurable.*

The next two theorems relate integrals on $\Omega \times \Omega'$ to integrals on Ω and on Ω'.

Theorem 3.114
If (Ω, Σ, μ) *and* (Ω', Σ', μ') *are two* σ *finite measure spaces and* $D \in \Sigma \otimes \Sigma'$*,*
then the functions

$$x \longmapsto \mu'(D_x) \quad \text{and} \quad y \longmapsto \mu(D_y)$$

are integrable on Ω *and* Ω'*, respectively, and*

$$(\mu \times \mu')(D) \;=\; \int_{\Omega} \mu'(D_x)\, d\mu \;=\; \int_{\Omega'} \mu(D_y)\, d\mu'.$$

Theorem 3.115 *(Fubini–Tonelli Theorem)*
If (Ω, Σ, μ) *and* (Ω', Σ', μ') *are two* σ*-finite measure spaces, then*
(a) *for* $f \in \mathcal{M}_+(\Omega \times \Omega')$*, we have*

$$g(x) \;=\; \int_{\Omega'} f_x\, d\mu' \;\in\; \mathcal{M}_+(\Omega), \quad h(y) \;=\; \int_{\Omega} f_y\, d\mu \;\in\; \mathcal{M}_+(\Omega')$$

and

$$\int_{\Omega\times\Omega'} f\, d(\mu \times \mu') \;=\; \int_{\Omega} \Big(\int_{\Omega'} f(x,y)\, d\mu' \Big)\, d\mu \;=\; \int_{\Omega'} \Big(\int_{\Omega} f(x,y)\, d\mu \Big)\, d\mu'$$

(b) for $f \in L^1(\Omega \times \Omega', \mu \times \mu')$*, we have*

$$f_x \in L^1(\Omega', \mu'), \quad f_y \in L^1(\Omega, \mu), \quad g \in L^1(\Omega, \mu), \quad h \in L^1(\Omega', \mu')$$

*and the same equations as in **(a)** hold.*

As for the product topologies (see Definition 2.69), using coordinate projections, we can have an alternative way to introduce the product σ-algebra $\Sigma_1 \otimes \Sigma_2$. Recall that the projection maps $p_1 \colon \Omega_1 \times \Omega_2 \longrightarrow \Omega_1$ and $p_2 \colon \Omega_1 \times \Omega_2 \longrightarrow \Omega_2$ are defined by

$$p_1(x,y) \stackrel{\text{def}}{=} x \quad \text{and} \quad p_1(x,y) \stackrel{\text{def}}{=} y \qquad \forall\, (x,y) \in \Omega \times \Omega'.$$

Let $\sigma(p_1, p_2)$ be the σ-algebra generated by the maps $\{p_1, p_2\}$. This is the smallest σ-algebra on $\Omega_1 \times \Omega_2$ for which both maps p_1 and p_2 are measurable (see Definition 3.58).

Theorem 3.116
If (Ω_1, Σ_1), (Ω_2, Σ_2) *and* $(Y, \mathcal{Y})$ *are measurable spaces, then*
(a) $\Sigma_1 \otimes \Sigma_2 = \sigma(p_1, p_2)$*;*
(b) *the function* $\xi \colon Y \longrightarrow \Omega_1 \times \Omega_2$ *is* $(\mathcal{Y}, \Sigma_1 \otimes \Sigma_2)$*-measurable if and only if* $p_1 \circ \xi$ *is* (Y, Σ_1)*-measurable and* $p_2 \circ \xi$ *is* (Y, Σ_2)*-measurable.*
(c) *the function* $\eta \colon \Omega_1 \times \Omega_2 \longrightarrow Y$ *is* $(\Sigma_1 \otimes \Sigma_2, \mathcal{Y})$*-measurable if and only if for every* $x_1 \in \Omega_1$*, the function* $\eta_{x_1}(\cdot) = \eta(x_1, \cdot)$ *if* $(\Sigma_2, \mathcal{Y})$*-measurable and for every* $x_2 \in \Omega_2$*, the function* $\eta_{x_2}(\cdot) = \eta(\cdot, x_2)$ *if* $(\Sigma_1, \mathcal{Y})$*-measurable.*

The next result can be viewed as a continuous version of the Minkowski inequality (see Theorem 3.101). More precisely, sums are replaced by integrals.

Theorem 3.117
Let (Ω, Σ, μ), $(Y, \mathcal{Y}, \nu)$ *be* σ*-finite measure spaces and let* $f \colon \Omega \times Y \longrightarrow \mathbb{R}$ *be a* $\Sigma \otimes \mathcal{Y}$*-measurable function.*
(a) *If* $f \geqslant 0$ *and* $1 \leqslant p < +\infty$*, then*

$$\Big(\int_\Omega \Big(\int_Y f(x,y)\, d\nu\Big)^p d\mu\Big)^{\frac{1}{p}} \leqslant \int_Y \Big(\int_\Omega f(x,y)^p\, d\mu\Big)^{\frac{1}{p}} d\nu.$$

(b) *If* $1 \leqslant p \leqslant +\infty$, $f(\cdot, y) \in L^p(\Omega)$ *for* ν*-almost every* $y \in Y$ *and the function* $y \longmapsto \|f(\cdot, y)\|_p$ *is in* $L^1(Y)$*,* *then* $f(x,\cdot) \in L^1(Y)$ *for* μ*-almost every* $x \in \Omega$*, the function* $x \longmapsto \int\limits_Y f(x,y)\,d\nu$ *belongs in* $L^1(\Omega)$ *and*

$$\Big\| \int\limits_Y f(\cdot,y)\,d\nu \Big\|_1 \leqslant \int\limits_Y \big\| f(\cdot,y) \big\|_p \,d\nu.$$

In Theorem 3.60 and Remark 3.61 we introduced the notion of ***image measure***. So, we saw that if (Ω,Σ,μ) is a measure space, $(Y,\mathcal{Y})$ is a measurable space and $f\colon \Omega \longrightarrow Y$ is a $(\Sigma,\mathcal{Y})$-measurable function, then we can use f to transport the measure μ defined on (Ω,Σ) to a measure $\nu = \mu f^{-1}$, defined by

$$\nu(A) = \mu\big(f^{-1}(A)\big) \qquad \forall\, A \in \mathcal{Y}.$$

Theorem 3.118 *(Change of Variable Formula)*
If (Ω,Σ,μ) *is a measure space,* $(Y,\mathcal{Y})$ *is a measurable space and* $f\colon \Omega \longrightarrow Y$ *is* $(\Sigma,\mathcal{Y})$*-measurable,*
then for every $u \in L^1(Y,\mu f^{-1})$*, we have that*

$$u \circ f \in L^1(\Omega,\mu) \quad \text{and} \quad \int\limits_Y u\,d(\mu f^{-1}) = \int\limits_\Omega (u \circ f)\,d\mu.$$

If $u\colon Y \longrightarrow [0,+\infty]$ *is* $\mathcal{Y}$*-measurable, then the last equality of integrals holds without assuming the* $\mu \circ f^{-1}$*-integrability of* u*.*

Definition 3.119
Let $D \subseteq \mathbb{R}^N$*,* $\varphi\colon D \longrightarrow \mathbb{R}^M$ *and* $\alpha \in (0,1]$*. We say that* φ *is an* ***α-Hölder continuous function*** *if*

$$\big\|\varphi(x) - \varphi(y)\big\| \leqslant \xi \|x-y\|^\alpha \qquad \forall\, x,y \in D,$$

for some $\xi > 0$*. If* $\alpha = 1$*, then* φ *is Lipschitz continuous with Lipschitz constant* $\xi > 0$*.*

Theorem 3.120
If $D \subseteq \mathbb{R}^N$ *is an* F_σ*-set and* $\varphi\colon D \longrightarrow \mathbb{R}^M$ *is an* α*-Hölder continuous function,*
then for every F_σ*-set* $E \subseteq \mathbb{R}^N$*, the set* $\varphi(D \cap E) \subseteq \mathbb{R}^M$ *is* F_σ *(hence Borel) and, if* $M \geqslant \alpha N$*, we also have*

$$\lambda^M\big(\varphi(D \cap E)\big) \leqslant \xi^M \lambda^N(E),$$

where λ^M (respectively, λ^N) denotes the Lebesgue measure on $\mathbb{R}^M$ (respectively, on $\mathbb{R}^N$).

Using the Lebesgue σ-algebras $\mathcal{L}(D)$ and $\mathcal{L}(\mathbb{R}^M)$ (there are the completions of $\mathcal{B}(D)$ and $\mathcal{B}(\mathbb{R}^M)$ with respect to λ^N and λ^M, respectively; see Definition 3.23), we can strengthen the previous theorem as follows.

Theorem 3.121
If $D \subseteq \mathbb{R}^N$ is an F_σ-set, $\varphi\colon D \longrightarrow \mathbb{R}^M$ is an α-Hölder continuous function and $M \geqslant \alpha N$, then

$$\varphi(\mathcal{L}(D)) = \varphi(\mathcal{L}(\mathbb{R}^N) \cap D) \subseteq \mathcal{L}(\mathbb{R}^M)$$

and

$$\lambda^M(\varphi(D \cap E)) \leqslant \xi^M \lambda^N(E) \qquad \forall\, E \in \mathcal{L}(D) = \mathcal{L}(\mathbb{R}^N) \cap D.$$

Theorem 3.122 *(Jacobi Transformation Theorem; Change of Variable Formula)*
If $\varphi\colon \mathbb{R}^N \longrightarrow \mathbb{R}^N$ is a C^1-diffeomorphism (i.e., a C^1-function for which φ^{-1} exists and is C^1 too),
then for every set $A \in \mathcal{B}(\mathbb{R}^N)$, we have

$$\lambda^N(\varphi(A)) = \int_A |\det \varphi'(x)|\, d\lambda^N.$$

Theorem 3.123
If $U, V \subseteq \mathbb{R}^N$ are two open sets and $\varphi\colon U \longrightarrow V$ is a C^1-diffeomorphism,
then $u\colon V \longrightarrow \mathbb{R}$ belongs in $L^1(V, \lambda^N)$ if and only if $u \circ \varphi |\det \varphi'(\cdot)| \in L^1(U, \lambda^N)$.
Moreover, in this case, we have

$$\int_V u(y)\, d\lambda^N = \int_U u(\varphi(x))|\det \varphi'(x)|\, d\lambda^N.$$

3.1.12 Uniform Integrability: Modes of Convergence

The Lebesgue dominated convergence theorem (see Theorem 3.94) describes one of the most important features of Lebesgue integration, since it provides sufficient conditions which allow to interchange limits and integrals. The crucial assumption is the existence of a dominant function $h \in L^1(\Omega)$ (i.e., $|f_n(x)| \leqslant h(x)$ μ-almost everywhere on Ω, for all $n \geqslant 1$) which controls things. However, this condition is not necessary. A slight weaker one is necessary and sufficient in order to be able to interchange limits and integrals. This notion is more interesting in the context of finite measure spaces and the idea there is to control the size of the set $\{|f_n| \geqslant c\}$, $c > 0$.

Definition 3.124
Let (Ω, Σ, μ) be a measure space and let $\mathcal{F} \subseteq L^1(\Omega)$ be a family of functions. We say that the family $\mathcal{F}$ is ***uniformly integrable*** *(or* ***equiintegrable****) if the following two conditions are satisfied:*
(a) *for every $\varepsilon > 0$, we can find a set $C_\varepsilon \in \Sigma$ such that*

$$\mu(C_\varepsilon) < +\infty \quad \text{and} \quad \sup_{f \in \mathcal{F}} \int_{C_\varepsilon^c} |f| \, d\mu \; < \; \varepsilon;$$

(b) *we have*

$$\lim_{c \to +\infty} \sup_{f \in \mathcal{F}} \int_{\{|f|>c\}} |f| \, d\mu \; = \; 0.$$

The next theorem provides alternative equivalent definitions of uniform integrability.

Theorem 3.125
If (Ω, Σ, μ) is a measure space and $\mathcal{F} \subseteq L^1(\Omega)$,
then the following statements are equivalent:
(a) *$\mathcal{F}$ is uniformly integrable.*
(b) • *$\mathcal{F}$ is L^1-bounded, i.e.,*

$$\sup_{f \in \mathcal{F}} \int_\Omega |f| \, d\mu \; < \; \infty;$$

• *for some $\varepsilon > 0$, we can find a set $C_\varepsilon \in \Sigma$ such that*

$$\mu(C_\varepsilon) \; < \; +\infty \quad \text{and} \quad \sup_{f \in \mathcal{F}} \int_{C_\varepsilon^c} |f| \, d\mu \; < \; \varepsilon,$$

• *for every* $\varepsilon > 0$*, there exists* $\delta > 0$ *such that for all* $A \in \Sigma$ *with* $\mu(A) < \delta$*, we have*

$$\sup_{f \in \mathcal{F}} \int_A |f| \, d\mu \; < \; \varepsilon.$$

If (Ω, Σ, μ) *is* σ*-finite, then conditions* ***(a)*** *and* ***(b)*** *are also equivalent to:*

(c) • $\mathcal{F}$ *is* L^1*-bounded, i.e.,*

$$\sup_{f \in \mathcal{F}} \int_\Omega |f| \, d\mu \; < \; \infty;$$

• *for every decreasing sequence* $\{A_n\}_{n \geqslant 1} \subseteq \mathcal{F}$ *such that* $A_n \searrow \emptyset$*, we have*

$$\lim_{n \to +\infty} \sup_{f \in \mathcal{F}} \int_{A_n} |f| \, d\mu \; = \; 0.$$

If (Ω, Σ, μ) *is finite, then conditions* ***(a)****,* ***(b)*** *and* ***(c)*** *are also equivalent to:*

(d) *we have*

$$\lim_{c \to +\infty} \sup_{f \in \mathcal{F}} \int_{\{|f| > c\}} |f| \, d\mu \; = \; 0.$$

(e) *(de la Vallée Poussin condition) for some increasing convex function* $\varphi \colon \mathbb{R}_+ \longrightarrow \mathbb{R}_+$ *such that* $\lim\limits_{r \to +\infty} \frac{\varphi(r)}{r} = +\infty$*, we have*

$$\sup_{f \in \mathcal{F}} \int_\Omega \varphi(|f|) \, d\mu \; < \; \infty.$$

The other key assumption in the Lebesgue dominated convergence theorem (see Theorem 3.94) is that $f_n \longrightarrow f$ μ-almost everywhere on Ω. This too can be weakened by using the following mode of convergence.

Definition 3.126

Let (Ω, Σ, μ) *be a measure space and let* $\{f_n\}_{n \geqslant 1} \subseteq \mathcal{M}$ *(i.e., a sequence of* $\mathbb{R}^*$*-valued* Σ*-measurable functions). We say that:*

(a) $\{f_n\}_{n \geqslant 1}$ *is* ***Cauchy in*** ***(***μ***-)measure*** *if for every* $\varepsilon > 0$*, we have*

$$\mu(\{|f_n - f_m| \geqslant \varepsilon\}) \;\longrightarrow\; 0 \quad \text{as } n, m \to +\infty.$$

(b) $\{f_n\}_{n\geqslant 1}$ ***converges in (μ-)measure*** *to f, if for every $\varepsilon > 0$, we have*

$$\mu(\{|f_n - f| \geqslant \varepsilon\}) \longrightarrow 0 \quad \textit{as } n \to +\infty.$$

If μ is a probability measure, then we say that we have ***convergence in probability****. We denote the convergence in measure by $f_n \xrightarrow{\mu} f$.*

Remark 3.127
Convergence in measure is a mode of convergence strictly weaker than μ-almost everywhere convergence. To see this, consider the measure space $(\Omega, \Sigma, \mu) = \big([0,1], \mathcal{B}([0,1]), \lambda\big)$ and the sequence $\{f_n\}_{n\geqslant 1}$, defined by

$$f_n = \chi_{\left[\frac{i}{2^k}, \frac{i+1}{2^k}\right]} \qquad \text{where } n = 2^k + i,\ 0 \leqslant i < 2^k.$$

Note that for all $\varepsilon \in (0,1)$, we have

$$\lambda(\{|f| \geqslant \varepsilon\}) = \frac{1}{2^k} \longrightarrow 0 \quad \text{as } n - n(k) \to +\infty$$

and so

$$f_n \xrightarrow{\lambda} 0.$$

However, the pointwise limit does not exist for any $x \in [0,1]$. On the measure space $(\Omega, \Sigma, \mu) = \big(\mathbb{R}, \mathcal{B}(\mathbb{R}), \lambda\big)$, the sequence

$$f_n = \chi_{(n,n+1)} \qquad \forall\, n \geqslant 1$$

is not Cauchy in measure.

Now we are ready for the generalized version of the Lebesgue dominated convergence theorem.

Theorem 3.128 *(Vitali Theorem)*
If (Ω, Σ, μ) is a σ-finite measure space, $1 \leqslant p < +\infty$ and $\{f_n\}_{n\geqslant 1} \subseteq L^p(\Omega)$ is a sequence such that $f_n \xrightarrow{\mu} f$,
then the following assertions are equivalent:
(a) $f_n \longrightarrow f$ *in* $L^p(\Omega)$, *i.e.,* $\|f_n - f\|_p \longrightarrow 0$.

(b) The sequence $\{|f_n|^p\}_{n\geqslant 1} \subseteq L^1(\Omega)$ *is uniformly integrable.*
(c) $\|f_n\|_p \longrightarrow \|f\|_p$.

Remark 3.129
The result is also true for measure spaces which are not σ-finite. However, in this case we cannot identify f with the L^p-limit. So, **(a)** reads as "$\{f_n\}_{n\geqslant 1}$ converges in $L^p(\Omega)$" and **(c)** reads as "the sequence $\{\|f_n\|_p\}_{n\geqslant 1}$ converges in $\mathbb{R}$".

So far we have introduced and used the following four modes for a sequence $\{f_n\}_{n\geqslant 1}$ of measurable functions: almost everywhere convergence, convergence in measure, L^p-convergence and almost uniform convergence. The last mode comes from the Egorov theorem (see Theorem 3.76). Next we examine how these notions are related.

Proposition 3.130
If (Ω, Σ, μ) *is a finite measure space,*
then almost everywhere convergence implies convergence in measure.

In Remark 3.127, we have seen that convergence in measure is a notion strictly weaker than almost everywhere convergence. Nevertheless, we have the following result.

Proposition 3.131
If (Ω, Σ, μ) *is any measure space,*
then every sequence which converges in measure has a subsequence which converges almost everywhere.

The next result is a straightforward consequence of the Chebyshev inequality (see Theorem 3.107).

Proposition 3.132
If (Ω, Σ, μ) *is any measure space,*
then norm convergence in $L^p(\Omega)$ $(1 \leqslant p < +\infty)$ *implies convergence in measure.*

Let (Ω, Σ, μ) be a measure space and let $L^0(\Omega)$ be the vector space of all equivalence classes of measurable functions.

Proposition 3.133
<u>If</u> (Ω, Σ, μ) is a finite measure space,
<u>then</u> convergence in measure in the vector space $L^0(\Omega)$ is equivalent to convergence with respect to the translation invariant measure

$$d_0(f,g) = \int_\Omega \tfrac{|f-g|}{1+|f-g|}\,d\mu \qquad \forall\ f,g \in L^0(\Omega),$$

i.e., $f_n \xrightarrow{\mu} f$ if and only if $d_0(f_n,f) \longrightarrow 0$.

Remark 3.134
In all above propositions, the limit function of the sequence belongs in $L^0(\Omega)$, since the latter is the vector space of equivalence classes. Another equivalent metric for the convergence in measure is the following one due to Fréchet. For $f,g \in L^0(\Omega)$, we set

$$K(f,g) = \big\{(\varepsilon,\xi) \in \mathbb{R}_+ \times \mathbb{R}_+ :\ \mu\big(\{|f-g| > \xi\}\big) < \varepsilon\big\}.$$

We set

$$e(f,g) = \inf\big\{\varepsilon + \xi :\ (\varepsilon,\xi) \in K(f,g)\big\}.$$

If $K(f,g) = \emptyset$, then $e(f,g) = +\infty$. We can also set

$$\widehat{e}(f,g) = 2\inf\big\{\lambda :\ (\lambda,\lambda) \in K(f,g)\big\}.$$

Note that

$$e(f,g) \leqslant \widehat{e}(f,g) \leqslant 2e(f,g).$$

We set

$$d_1(f,g) = \tfrac{e(f,g)}{1+e(f,g)}.$$

Then $f_n \xrightarrow{\mu} f$ if and only if $d_1(f_n,f) \longrightarrow 0$. Moreover, the metric spaces $\big(L^0(\Omega), d_0\big)$ and $\big(L^0(\Omega), d_1\big)$ are complete.

Proposition 3.135
<u>If</u> (Ω, Σ, μ) is a finite measure space and $\{f_n\}_{n\geqslant 1} \subseteq \mathcal{M}$,
<u>then</u> $f_n \longrightarrow f$ almost everywhere if and only if for every $\varepsilon > 0$, we have

$$\lim_{n\to+\infty} \mu\Big(\bigcup_{k\geqslant n} \{|f_k - f| \geqslant \varepsilon\}\Big) = 0.$$

Proposition 3.136
If (Ω, Σ, μ) is any measure space, $\{f_n\}_{n\geqslant 1}, \{g_n\}_{n\geqslant 1} \subseteq \mathcal{M}$ are two sequences and $f, g \in \mathcal{M}$, then
(a) *if $f_n \xrightarrow{\mu} f$ and $g_n \xrightarrow{\mu} g$, then $\lambda f_n + \eta g_n \xrightarrow{\mu} \lambda f + \eta g$ for all $(\lambda, \eta) \in \mathbb{R} \times \mathbb{R}$;*
(b) *if $f_n \xrightarrow{\mu} f$ and $g_n \xrightarrow{\mu} g$, then $\max\{f_n, g_n\} \xrightarrow{\mu} \max\{f, g\}$ and $\min\{f_n, g_n\} \xrightarrow{\mu} \min\{f, g\}$;*
(c) *if $f_n \xrightarrow{\mu} f$, then $f_n^+ \xrightarrow{\mu} f^+$, $f_n^- \xrightarrow{\mu} f^-$, $|f_n| \xrightarrow{\mu} |f|$.*

Definition 3.137
Let (Ω, Σ, μ) be a complete finite measure space. We say that sets $A, C \in \Sigma$ are μ-equivalent if and only if $\mu(A \triangle C) = 0$. This defines an equivalence relation on Σ and we denote the set of equivalence classes by $\overset{\circ}{\Sigma}$ (i.e., we identify μ-equivalent sets). Then consider the map

$$\overset{\circ}{\Sigma} \ni A \longmapsto \chi_A \in L^1(\Omega).$$

This defines a natural embedding of $\overset{\circ}{\Sigma}$ into $L^1(\Omega)$. Identifying $\overset{\circ}{\Sigma}$ with its image in $L^1(\Omega)$, we can think of $\overset{\circ}{\Sigma}$ as a subset of $L^1(\Omega)$ and as such is a metric space with the induced metric

$$\widetilde{d}(A, C) = \|\chi_A - \chi_C\|_1 = \int_\Omega |\chi_A - \chi_C| \, d\mu = \mu(A \triangle C).$$

Proposition 3.138
The set $\overset{\circ}{\Sigma}$ is closed in $L^1(\Omega)$ and so $\left(\overset{\circ}{\Sigma}, \widetilde{d}\right)$ is a compete metric space.

Proposition 3.139
The metric space $\left(\overset{\circ}{\Sigma}, \widetilde{d}\right)$ is separable if and only if the Banach space $L^1(\Omega)$ is separable (see Definition 2.9).

Remark 3.140
In this case we say that the measure μ is ***separable***.

3.1.13 Signed Measures

So far we have considered measures with values in $\overline{\mathbb{R}}_+ = [0, +\infty]$. However, for the purpose of differentiating a measure with respect to another measure on the same σ-algebra, it is useful to generalize the notion of measure and allow also negative values.

Definition 3.141
Let (Ω, Σ) be a measurable space. A set function $\mu\colon \Sigma \longrightarrow \mathbb{R}^* = \mathbb{R} \cup \{\pm\infty\}$ is said to be a ***signed measure*** *if*
(a) $\mu(\emptyset) = 0$;
(b) *μ takes on at most one of the two values $-\infty$ and $+\infty$;*
(c) *μ is σ-additive, i.e., for any disjoint family $\{A_n\}_{n\geqslant 1} \subseteq \Sigma$, we have $\mu(\bigcup_{n\geqslant 1} A_n) = \sum_{n\geqslant 1} \mu(A_n)$. In practice signed measures are $\mathbb{R}$-valued.*

Proposition 3.142
If $\mu\colon (\Omega, \Sigma) \longrightarrow \mathbb{R}^* = \mathbb{R} \cup \{\pm\infty\}$ is a signed measure, then
(a) *for an increasing sequence $\{A_n\}_{n\geqslant 1} \subseteq \Sigma$, we have $\lim_{n\to+\infty} \mu(A_n) = \mu(\bigcup_{n\geqslant 1} A_n)$;*
(b) *for a decreasing sequence $\{A_n\}_{n\geqslant 1} \subseteq \Sigma$ with $\mu(A_1) < +\infty$, we have $\lim_{n\to+\infty} \mu(A_n) = \mu(\bigcap_{n\geqslant 1} A_n)$.*

Proposition 3.143
If $\mu\colon (\Omega, \Sigma) \longrightarrow \mathbb{R}^* = \mathbb{R} \cup \{\pm\infty\}$ is a signed measure, then there exist $A^*, A_* \in \Sigma$ such that

$$\mu(A^*) = \sup_{A\in\Sigma} \mu(A) \quad \text{and} \quad \mu(A_*) = \inf_{A\in\Sigma} \mu(A).$$

Definition 3.144
Let $\mu\colon (\Omega, \Sigma) \longrightarrow \mathbb{R}^* = \mathbb{R} \cup \{\pm\infty\}$ be a signed measure. We say that $A \subseteq X$ is a ***positive set*** *for μ, if $A \in \Sigma$ and for each $C \subseteq A$, $C \in \Sigma$, we have $\mu(C) \geqslant 0$. We say that $A \subseteq X$ is a* ***negative set*** *for μ, if $A \in \Sigma$ and for each $C \subseteq A$, $C \in \Sigma$, we have $\mu(C) \leqslant 0$.*

The next two theorems give the standard decompositions for signed measures.

Theorem 3.145 *(Hahn Decomposition Theorem)*
If $\mu\colon (\Omega, \Sigma) \longrightarrow \mathbb{R}^* = \mathbb{R} \cup \{\pm\infty\}$ is a signed measure, then there exist a positive set P and a negative set N such that

$$P \cup N = \Omega \quad \text{and} \quad P \cap N = \emptyset.$$

This decomposition known as the ***Hahn decomposition*** *for μ is essentially unique in the sense that if (P', N') is another such decomposition, then*

$$\mu(P \bigtriangleup P') = \mu(N \bigtriangleup N') = 0.$$

Theorem 3.146 *(Jordan Decomposition Theorem)*
If $\mu\colon (\Omega, \Sigma) \longrightarrow \mathbb{R}^* = \mathbb{R} \cup \{\pm\infty\}$ *is a signed measure,*
then $\mu = \mu^+ - \mu^-$ *with* μ^+ *and* μ^- *being two measures at least one of which is finite. We have*

$$\begin{aligned} \mu^+(A) &= \sup\big\{\mu(C) : \ C \in \Sigma, \ C \subseteq A\big\}, \\ \mu^-(A) &= \inf\big\{\mu(C) : \ C \in \Sigma, \ C \subseteq A\big\}. \end{aligned}$$

This decomposition of μ *is known as the* ***Jordan decomposition.***

Remark 3.147
If (P, N) is a Hahn decomposition for μ, then

$$\mu^+(A) \ = \ \mu(A \cap P) \quad \text{and} \quad \mu^-(A) \ = \ -\mu(A \cap N).$$

Definition 3.148
In the Jordan decomposition of a signed measure, μ^+ *is the* ***positive part*** *of* μ*,* μ^- *is the* ***negative part*** *of* μ *and* $|\mu| = \mu^+ + \mu^-$ *(a measure on* (Ω, Σ)*) is the* ***total variation*** *of* μ*. The* ***total variation norm*** *of the signed measure* μ *is defined by*

$$\|\mu\| \ = \ |\mu|(\Omega).$$

We say that μ *is* ***finite*** *(respectively,* σ***-finite****) if* $|\mu|$ *is finite (respectively,* σ*-finite).*

Proposition 3.149
If (Ω, Σ) *is a measurable space and* $M(\Sigma)$ *is the space of all finite signed measures furnished with the total variation norm,*
then $M(\Sigma)$ *is complete, i.e., it is a Banach space.*
For any $A \in \Sigma$*, we have*

$$\sup\big\{|\mu(C)| : \ C \subseteq A, \ C \in \Sigma\big\} \ \leqslant \ |\mu|(A) \ \leqslant \ 4\sup\big\{|\mu(C)| : \ C \subseteq A, \ C \in \Sigma\big\}$$

and

$$|\mu|(A) \ = \ \sup \sum_{k=1}^{n} |\mu(A_k)|,$$

where the supremum is taken over all mutually disjoint subsets $\{A_k\}_{k=1}^n \subseteq \Sigma$ *of* A*. Of course* $|\mu|(\Omega) = \|\mu\|$*.*

3.1.14 Radon–Nikodym Theorem

Definition 3.150
Suppose that (Ω, Σ) be a measurable space, μ is a measure on Σ and ν is a signed measure on Σ. We say that ν is ***absolutely continuous*** *with respect to μ, denoted by $\nu \ll \mu$, if $\mu(A) = 0$ implies $\nu(A) = 0$.*

Theorem 3.151 *(Vitali–Hahn–Saks Theorem)*
If $\{\mu_n\}_{n \geqslant 1} \subseteq M(\Sigma)$ is a sequence of signed measures, $m \in M(\Sigma)$, $m \geqslant 0$, $\mu_n \ll m$ for all $n \geqslant 1$ and $\mu_n(A) \longrightarrow \mu(A)$ for all $A \in \Sigma$, then $\mu \in M(\Sigma)$ and $\mu \ll m$.

The next theorem is one of the most important results in measure theory.

Theorem 3.152 *(Radon–Nikodym Theorem)*
If (Ω, Σ) is a measurable space, μ is a σ-finite measure on Σ, ν is a σ-finite signed measure on Σ and $\nu \ll \mu$,
then there exists a unique function $f \in L^1(\Omega, \mu)$ such that

$$\nu(A) = \int_A f \, d\mu \qquad \forall\, A \in \Sigma.$$

This function f is known as the ***Radon–Nikodym derivative*** *of ν with respect to μ and it is denoted by $\frac{d\nu}{d\mu} = f$.*

Next we introduce a notion which in a sense is the opposite of absolute continuity (see Definition 3.150).

Definition 3.153
Let (Ω, Σ) be a measurable space and let μ_1, μ_2 be two measures on Σ. We say that μ_1 is ***singular*** *with respect to μ_2, denoted by $\mu_1 \perp \mu_2$, if there exists a set $A \in \Sigma$ such that*

$$\mu_1(A) = 0 \quad \text{and} \quad \mu_2(\Omega \setminus A) = 0.$$

Evidently $\mu_1 \perp \mu_2$ if and only if $\mu_2 \perp \mu_1$ and so we often say that μ_1 and μ_2 are ***mutually singular****. If ν_1, ν_2 are two signed measures on Σ, then we say that ν_1, ν_2 are* ***mutually singular*** *if $|\nu_1| \perp |\nu_2|$.*

The two notions of absolute continuity (see Definition 3.150) and of singularity (see Definition 3.153) are adequate to describe the relation between a measure μ and a signed measure ν, both σ-finite.

Theorem 3.154 *(Lebesgue Decomposition Theorem)*
<u>If</u> (Ω, Σ) is a measurable space, μ is a σ-finite measure on Σ and ν is a σ-finite signed measure on Σ,
<u>then</u> $\nu = \nu_{ac} + \nu_s$, with ν_{ac} and ν_s both signed measures, $\nu_{ac} \ll \mu$, $\nu_s \perp \mu$ and this decomposition (known as the Lebesgue decomposition of ν with respect to μ) is unique.

Definition 3.155
*Let (Ω, Σ) be a measurable space and let μ_1, μ_2 be two measures on Σ. We say that μ_1 and μ_2 are **equivalent**, denoted by $\mu_1 \equiv \mu_2$, if both $\mu_1 \ll \mu_2$ and $\mu_2 \ll \mu_1$ hold.*

Proposition 3.156
<u>If</u> (Ω, Σ) is a measurable space, μ_1, μ_2 are two equivalent σ-finite measures on Σ and $f_1 = \frac{d\mu_1}{d\mu_2}$, $f_2 = \frac{d\mu_2}{d\mu_1}$ (the Radon–Nikodym derivatives; see Theorem 3.152),
<u>then</u> $f_1 = \frac{1}{f_2}$ μ_2-almost everywhere and $f_2 = \frac{1}{f_1}$ μ_1-almost everywhere on Ω.
Moreover, $L^1(\Omega, \mu_1) = L^1(\Omega, \mu_2)$.

3.1.15 Maximal Function and Lyapunov Convexity Theorem

Definition 3.157
*A measurable function $f\colon \mathbb{R}^N \longrightarrow \mathbb{R}$ is said to be **locally integrable** (with respect to the Lebesgue measure λ^N on $\mathbb{R}^N$), if for every compact set $K \subseteq \mathbb{R}^N$, we have*

$$\int_K |f(x)|\, dx \;<\; +\infty.$$

By $L^1_{\text{loc}}(\mathbb{R}^N)$ we denote the linear space of locally integrable functions on $\mathbb{R}^N$. If $f \in L^1_{\text{loc}}(\mathbb{R}^N)$, $x \in \mathbb{R}^N$ and $r > 0$, we set

$$f^*(x) \stackrel{\text{def}}{=} \sup_{r>0} \frac{1}{\lambda^N(\overline{B}_r(x))} \int_{\overline{B}_r(x)} |f(y)|\, dy.$$

*The function f^* is known as the **Hardy–Littlewood function** for f.*

Remark 3.158
It can be shown that f^* is Borel measurable (see Problem 3.180).

Theorem 3.159 *(Hardy–Littlewood–Wiener Maximal Theorem)*
<u>If</u> $f \in L^p(\mathbb{R}^N)$, $1 \leqslant p < +\infty$,
<u>then</u> there exists a constant $\widehat{c}(p, N) > 0$ such that

$$\lambda^N\big(\{f^* \geqslant \eta\}\big) \leqslant \tfrac{\widehat{c}}{\eta}\|f\|_1 \qquad \forall\, \eta > 0,$$

when $p = 1$ and

$$\|f^*\|_p \leqslant \tfrac{p\widehat{c}}{p-1}\|f\|_p,$$

when $1 < p < +\infty$.

The next theorem is a significant extension of Theorem 3.41 and has important applications in control theory, optimization, mathematical economics and game theory.

Theorem 3.160 *(Lyapunov Convexity Theorem)*
<u>If</u> (Ω, Σ) is a measurable space and $\mu_k \colon \Sigma \longrightarrow \mathbb{R}_+ = [0, +\infty)$, for $k = 1, \ldots, N$, are finite nonatomic measures (see Definition 3.40),
<u>then</u> the set $C \overset{\text{def}}{=} \big\{\big(\mu_1(A), \ldots, \mu_N(A)\big) \in \mathbb{R}^N : A \in \Sigma\big\}$ is compact and convex.

3.1.16 Conditional Expectation and Martingales

In the last part of this chapter, we will present some basic facts about martingales.

Definition 3.161
Let (Ω, Σ, μ) be a finite measure space (usually a probability space), let Σ_0 be a sub-σ-algebra of Σ and let $f \colon \Omega \longrightarrow \mathbb{R}$ be a Σ-measurable function. Then the ***conditional expectation*** *of f with respect to Σ_0, if it exists, it is a Σ_0-measurable function $g \colon \Omega \longrightarrow \mathbb{R}$ such that*

$$\int_A f\, d\mu = \int_A g\, d\mu \qquad \forall\, A \in \Sigma_0.$$

Then we write $g = E^{\Sigma_0} f$.

Remark 3.162
If $\Sigma_0 = \Sigma$ or more generally, if f is Σ_0-measurable, then $f = E^{\Sigma_0} f$.

Proposition 3.163
If $f \in L^1(\Omega, \Sigma)$ *and* Σ_0 *is a sub-*σ*-algebra of* Σ,
then $E^{\Sigma_0} f$ *exists, belongs in* $L^1(\Omega, \Sigma_0)$ *and is unique up to a* Σ_0*-set of* μ*-measure zero.*

Remark 3.164
So, $E^{\Sigma_0} \colon L^1(\Omega, \Sigma) \longrightarrow L^1(\Omega, \Sigma_0)$ is a well-defined linear operator. If Σ_1 is a sub-σ-algebra of Σ_0, then clearly from Definition 3.161, we have

$$E^{\Sigma_1} f = E^{\Sigma_1}(E^{\Sigma_0} f) \qquad \forall\, f \in L^1(\Omega, \Sigma).$$

If $\Sigma_0 = \{\emptyset, \Omega\}$ (the trivial sub-σ-algebra), then

$$E^{\Sigma_0} f = \int_\Omega f \, d\mu \qquad \forall\, f \in L^1(\Omega, \Sigma).$$

Finally, if $f, g \in L^1(\Omega, \Sigma)$ and $f \leqslant g$, then

$$E^{\Sigma_0} f \leqslant E^{\Sigma_0} g$$

(i.e., E^{Σ_0} is a positive operator).

The main convergence theorems for integrals extend also to the conditional expectations.

Theorem 3.165
If (Ω, Σ, μ) *is a finite measure space and* Σ_0 *is a sub-*σ*-algebra of* Σ,
then
(a) *(Monotone Convergence)*
for every sequence $\{f_n\}_{n \geqslant 1}$ *of* Σ*-measurable functions such that* $f_n \geqslant 0$ *and* $f_n \searrow f$ μ*-almost everywhere on* Ω*, we have* $\lim\limits_{n \to +\infty} E^{\Sigma_0} f_n = E^{\Sigma_0} f$.
(b) *(Fatou Lemma)*
for every sequence $\{f_n\}_{n \geqslant 1}$ *of* Σ*-measurable functions such that* $f_n \geqslant 0$*, we have* $E^{\Sigma_0}\big(\liminf\limits_{n \to +\infty} f_n\big) \leqslant \liminf\limits_{n \to +\infty} E^{\Sigma_0} f_n$.
(c) *(Dominated Convergence)*
for every sequence $\{f_n\}_{n \geqslant 1}$ *of* Σ*-measurable functions such that* $f_n \longrightarrow f$ μ*-almost everywhere and* $|f_n| \leqslant h$ μ*-almost everywhere on* Ω *for all* $n \geqslant 1$ *with* $h \in L^1(\Omega, \Sigma)$*, we have* $E^{\Sigma_0} f = \lim\limits_{n \to +\infty} E^{\Sigma_0} f_n$.

The Jensen inequality (see Theorem 3.99) remains true for conditional expectations.

Theorem 3.166 *(Jensen Inequality)*
If (Ω, Σ, μ) *is a finite measure space,* $\varphi\colon \mathbb{R} \longrightarrow \mathbb{R}$ *is convex,* f *and* $\varphi(f)$ *both belong to* $L^1(\Omega, \Sigma)$ *and* Σ_0 *is a sub-*σ*-algebra of* Σ
then $\varphi(E^{\Sigma_0} f) \leqslant E^{\Sigma_0}(\varphi(f))$.

Martingales are a key tool in modern probability theory and are also useful in Banach space theory as we will see in Chap. 5.

Definition 3.167
Let (Ω, Σ, μ) *be a finite measure space, let* $\{\Sigma_n\}_{n\geqslant 0}$ *be an increasing sequence of sub-*σ*-algebras of* Σ *and let* $f_n \in L^1(\Omega, \Sigma_n)$ *for* $n \geqslant 0$*. We say that*
(a) $\{f_n, \Sigma_n\}_{n\geqslant 0}$ *is a* ***martingale*** *if* $E^{\Sigma_n} f_{n+1} = f_n$ *for all* $n \geqslant 0$.
(b) $\{f_n, \Sigma_n\}_{n\geqslant 0}$ *is a* ***submartingale*** *if* $E^{\Sigma_n} f_{n+1} \geqslant f_n$ *for all* $n \geqslant 0$.
(c) $\{f_n, \Sigma_n\}_{n\geqslant 0}$ *is a* ***supermartingale*** *if* $E^{\Sigma_n} f_{n+1} \leqslant f_n$ *for all* $n \geqslant 0$.

Remark 3.168
Using the language of gambling, a martingale is a "fair" game, a submartingale is a "favourable" game and a supermartingale is an "unfavourable" game. If $\{f_n, \Sigma_n\}_{n\geqslant 0}$ is a martingale (respectively, submartingale, supermartingale), then $\left\{\int_\Omega f_n \, d\mu\right\}_{n\geqslant 0}$ is a constant sequence (respectively, increasing sequence, a decreasing sequence). It is clear from Definition 3.167 that $\{f_n, \Sigma_n\}_{n\geqslant 0}$ is a martingale if and only if it is both a submartingale and supermartingale. Also, $\{f_n, \Sigma_n\}_{n\geqslant 0}$ is a submartingale if and only if $\{-f_n, \Sigma_n\}_{n\geqslant 0}$ is a supermartingale. Finally, if $\{f_n, \Sigma_n\}_{n\geqslant 0}$ and $\{g_n, \Sigma_n\}_{n\geqslant 0}$ are martingales (respectively, submartingales) and $\eta, \xi \geqslant 0$, then $\{\eta f_n + \xi g_n, \Sigma_n\}_{n\geqslant 0}$ is a martingale (respectively, submartingale) too.

When dealing with martingales, it is often necessary to consider the index k (discrete time instant), after which the process stays above or below a certain threshold value. So, we need to consider indices which depend on $x \in \Omega$ (random indices) and this leads to the notion of stopping time.

Definition 3.169
Let (Ω, Σ, μ) be a finite measure space and let $\{\Sigma_n\}_{n\geqslant 1}$ be an increasing sequence of sub-σ-algebras of Σ.

*A **stopping time** is a map $\sigma\colon \Omega \longrightarrow \mathbb{N}_0 \cup \{\infty\}$ which satisfies*

$$\{\sigma \leqslant k\} \in \Sigma_k \qquad \forall\, k \in \mathbb{N}_0.$$

*A set $A \in \Sigma$ is said to be **prior to** σ, if*

$$A \cap \{\sigma \leqslant k\} \in \Sigma_k \qquad \forall\, k \in \mathbb{N}_0,$$

or equivalently

$$A \cap \{\sigma = k\} \in \Sigma_k \qquad \forall\, k \in \mathbb{N}_0.$$

The collection of all sets prior to σ is denoted by Σ_σ and it readily follows that Σ_σ is a σ-algebra.

Theorem 3.170
If (Ω, Σ, μ) is a finite measure space, $\{\Sigma_n\}_{n\geqslant 0}$ is an increasing sequence of sub-σ-algebras of Σ and $f_n \in L^1(\Omega, \Sigma_n)$ for $n \geqslant 1$, then the following statements are equivalent:
(a) *$\{f_n, \Sigma_n\}_{n\geqslant 0}$ is a martingale.*
(b) *for all bounded stopping times $\sigma \leqslant \tau$, we have*

$$\int_\Omega f_\sigma \, d\mu \leqslant \int_\Omega f_\tau \, d\mu.$$

(c) *for all bounded stopping times $\sigma \leqslant \tau$ and all $A \in \Sigma_\sigma$, we have*

$$\int_A f_\sigma \, d\mu \leqslant \int_A f_\tau \, d\mu.$$

The next theorem establishes a connection between submartingales and martingales.

Theorem 3.171 *(Doob Decomposition Theorem)*
If $\{f_n, \Sigma_n\}_{n\geqslant 0}$ is a submartingale, then there exist a martingale $\{g_n, \Sigma_n\}_{n\geqslant 0}$ and a discrete-time process $\{h_n\}_{n\geqslant 0}$ such that h_{n+1} is Σ_n-measurable, $h_n \leqslant h_{n+1}$ for all $n \geqslant 1$ and

$$f_n = f_0 + g_n + h_n \quad \text{with } g_0 = h_0 = 0.$$

Moreover, this decomposition is unique.

Of course there is a corresponding connection between supermartingales and martingales.

Theorem 3.172
If $\{f_n, \Sigma_n\}_{n \geqslant 0}$ *is a supermartingale,*
then there exist a martingale $\{g_n, \Sigma_n\}_{n \geqslant 0}$ *and a discrete-time process* $\{h_n\}_{n \geqslant 0}$ *such that* h_{n+1} *is* Σ_n*-measurable,* $h_n \leqslant h_{n+1}$ *for all* $n \geqslant 1$ *and*

$$f_n = f_0 + g_n - h_n \quad \text{with } g_0 = h_0 = 0.$$

Moreover, this decomposition is unique.

If the index $n \in \mathbb{N}_0$ in a martingale (respectively, submartingale, supermartingale) is replaced by two stopping times, does the main defining the process equality (respectively, inequality) remain true? The answer is "yes", provided that two stopping times are bounded.

Theorem 3.173 *(Stopping Time Theorem)*
If $\{f_n, \Sigma_n\}_{n \geqslant 0}$ *is a martingale (respectively, submartingale, supermartingale) and* σ, τ *are two bounded stopping times,*
then $E^{\Sigma_\sigma} f_\tau = f_\sigma$ *(respectively* $E^{\Sigma_\sigma} f_\tau \geqslant f_\sigma$, $E^{\Sigma_\sigma} f_\tau \leqslant f_\sigma$*).*
Hence

$$\int_\Omega f_\sigma \, d\mu = \int_\Omega f_0 \, d\mu \quad \Big(\text{respectively } \int_\Omega f_\sigma \, d\mu \geqslant \int_\Omega f_0 \, d\mu, \quad \int_\Omega f_\sigma \, d\mu \leqslant \int_\Omega f_0 \, d\mu\Big)$$

for every bounded stopping time σ.

Remark 3.174
In this theorem, the assumption of the boundedness of the stopping times is essential and cannot be removed without additional hypotheses.

One of the main reasons for the importance of martingales in probability theory is the fact that their structure leads to some useful inequalities, known in the literature as ***Doob inequalities*** or ***maximal inequalities***. In the following theorems we present these inequalities.

As before (Ω, Σ, μ) is a finite measure space, $\{\Sigma_n\}_{n\geqslant 0}$ is an increasing sequence of sub-σ-algebra of Σ. Consider $f_n \in L^1(\Omega, \Sigma)$, $n \geqslant 0$ and set

$$f_n^* \stackrel{\text{def}}{=} \sup_{k\leqslant n} |f_k|.$$

Then $f_n^* \in L^1(\Omega, \Sigma_n)$, $f_n^* \leqslant f_{n+1}^*$ for all $n \geqslant 0$ and $\{f_n^*, \Sigma_n\}_{n\geqslant 0}$ is easily seen to be a submartingale. From the Markov inequality (see Proposition 3.90(f)), we have

$$\mu(\{f_n^* \geqslant c\}) \leqslant \frac{\int_\Omega f_n^* \, d\mu}{c} \qquad \forall\, c > 0.$$

In the case of a martingale, in the right-hand side, we can replace f_n^* be $|f_n|$ only.

Theorem 3.175 *(Doob First Martingale Inequality)*
If $\{f_n, \Sigma_n\}_{n\geqslant 0}$ *is a martingale or a positive submartingale, then*

$$\mu(\{f_n^* \geqslant c\}) \leqslant \frac{\|f_n\|_1}{c} \qquad \forall\, c > 0,\ n \geqslant 0.$$

If the martingale belongs in L^p, then the above theorem takes the following form.

Theorem 3.176 *(Doob L^p Martingale Inequality)*
If $\{f_n, \Sigma_n\}_{n\geqslant 0}$ *is a martingale or a positive submartingale such that* $f_n \in L^p(\Omega, \Sigma_n)$ *for all* $n \geqslant 0$ *with* $1 < p < +\infty$, *then*

$$\|f_n^*\|_p \leqslant \frac{p}{p-1}\|f_n\|_p = \frac{1}{p'}\|f_n\|_p \qquad \forall\, n \geqslant 0$$

(recall $\frac{1}{p} + \frac{1}{p'} = 1$*).*

Undoubtedly the most important results in martingale theory are the convergence theorems. The main tool in this direction is the so-called Doob upcrossing inequality. To understand better this inequality, consider the following simple situation with real numbers $\{\xi_n\}_{n\geqslant 0}$. If $\eta = \lim_{n\to+\infty} \xi_n$ exists and we know that $\eta \in (a, b)$, then only a finite number of the ξ_n's can be outside of (a, b). So, if infinitely many of the ξ_n's are bigger than b and infinitely many of the ξ_n's are smaller than a, then the sequence $\{\xi_n\}_{n\geqslant 0}$ has no limit. An occurrence of

$$\xi_n \leqslant a \quad \text{and} \quad \xi_{n+m} \geqslant b \quad \text{for some } m \in \mathbb{N}$$

is called an ***upcrossing*** of $[a, b]$. From the previous discussion, we know that

$$\text{there is infinitely many upcrossing on } [a,b] \implies \{\xi_n\}_{n\geqslant 0} \text{ has no limit.}$$

For submartingales we can estimate the average number of upcrossings over any interval. We can express this using stopping times. So, let $\{f_n, \Sigma_n\}_{n\geqslant 0}$ be the submartingale and let us set $\sigma_0 = 0$ and inductively for $k \geqslant 0$, let

$$\vartheta_{k+1} \stackrel{\text{def}}{=} \min\{i > \sigma_k : f_i \leqslant a\}, \quad \sigma_{k+1} \stackrel{\text{def}}{=} \min\{i > \vartheta_{k+1} : f_k \geqslant b\},$$

with the usual convention that the minimum of an empty set in $\mathbb{N}_0$ is $+\infty$. Using the dual convention that the maximum of an empty set is 0, we can define

$$U_n \stackrel{\text{def}}{=} \max\{i : \sigma_i \leqslant n\} \qquad \forall\, n \geqslant 0.$$

Then U_n represents the number of upcrossings of $[a, b]$ before time n.

Theorem 3.177 *(Doob Upcrossing Inequality)*
__If__ $\{f_n, \Sigma_n\}_{n\geqslant 0}$ is a submartingale, $a < b$ and U_n is the number of upcrossings of $[a, b]$ before time $n \geqslant 0$ as defined above, __then__

$$\int_\Omega U_n\, d\mu \leqslant \tfrac{1}{b-a} \int_\Omega \left|(f_n - a)^+\right| d\mu.$$

The above upcrossing inequality is the basis for all martingale convergence theorems.

Theorem 3.178 *(Submartingale Convergence Theorem)*
__If__ $\{f_n, \Sigma_n\}_{n\geqslant 0}$ is a submartingale such that $\sup\limits_{n\geqslant 1} \|f_n^+\|_1 < +\infty$, __then__ there exists $f \in L^1(\Omega)$ such that $f_n \longrightarrow f$ μ-almost everywhere on Ω.

Corollary 3.179
__If__ $\{f_n, \Sigma_n\}_{n\geqslant 0}$ is
(i) a supermartingale such that $\sup\limits_{n\geqslant 1} \|f_n^-\|_1 < +\infty$; or

(ii) a positive supermartingale; or
(iii) a martingale with $\sup_{n \geqslant 1} \|f_n\|_1 < +\infty$,
then there exists $f \in L^1(\Omega)$ *such that* $f_n \longrightarrow f$ *μ-almost everywhere on* Ω.

Remark 3.180
In both the above two convergence results, we can only conclude pointwise convergence μ-almost everywhere. In general, it is not true that we have L^1-convergence. Below we will see when we can conclude this. The limit function f is measurable with respect to $\Sigma_\infty = \sigma\big(\bigcup_{n \geqslant 0} \Sigma_n\big)$.

We can also consider discrete time processes adopted to a decreasing sequence $\{\Sigma_n\}_{n \geqslant 0}$ of sub-σ-algebras of Σ (i.e., $\Sigma_{n+1} \subseteq \Sigma_n$ for all $n \geqslant 0$). We can think of such processes as being indexed by $-\mathbb{N}_0$.

Definition 3.181
Let (Ω, Σ, μ) *be a finite measure space, let* $\{\Sigma_n\}_{n \geqslant 0}$ *be a decreasing sequence of sub-σ-algebras of* Σ *and let* $f_n \in L^1(\Omega, \Sigma_n)$ *for all* $n \geqslant 0$. *We say that*
(a) $\{f_n, \Sigma_n\}_{n \geqslant 0}$ *is a* ***reverse martingale*** *if* $E^{\Sigma_{n+1}} f_n = f_{n+1}$ *for all* $n \geqslant 0$;
(b) $\{f_n, \Sigma_n\}_{n \geqslant 0}$ *is a* ***reverse submartingale*** *if* $E^{\Sigma_{n+1}} f_n \geqslant f_{n+1}$ *for all* $n \geqslant 0$;
(c) $\{f_n, \Sigma_n\}_{n \geqslant 0}$ *is a* ***reverse supermartingale*** *if* $E^{\Sigma_{n+1}} f_n \leqslant f_{n+1}$ *for all* $n \geqslant 0$.

Theorem 3.182
If $\{f_n, \Sigma_n\}_{n \geqslant 0}$ *is a reverse submartingale such that* $\sup_{n \geqslant 1} \|f_n^-\|_1 < +\infty$ *and* $\Sigma_{-\infty} = \bigcap_{n \geqslant 0} \Sigma_n$,
then there exists $f \in L^1(\Omega, \Sigma_{-\infty})$ *such that* $f_n \longrightarrow f$ *μ-almost everywhere on* Ω *and* $f_n \longrightarrow f$ *in* $L^1(\Omega)$.

Remark 3.183
The same is true for reverse supermartingales satisfying $\sup_{n \geqslant 1} \|f_n^+\|_1 < +\infty$.

In the next theorem we see when for a direct martingale we can guarantee L^1-convergence.

Theorem 3.184
If $\{f_n, \Sigma_n\}_{n \geqslant 0}$ *is a martingale,*
then the following statements are equivalent:
(a) *There exists* $f \in L^1(\Omega, \Sigma_\infty)$ *(where* $\Sigma_\infty = \sigma\big(\bigcup_{n \geqslant 0} \Sigma_n\big)$*) such that* $f_n \longrightarrow f$ *in* $L^1(\Omega)$*.*
(b) $\sup_{n \geqslant 0} \|f_n\|_1 < +\infty$*,* $f_n \longrightarrow f$ *μ-almost everywhere on Ω with* $f \in L^1(\Omega, \Sigma_\infty)$ *and* $f_n = E^{\Sigma_n} f$ *for all* $n \geqslant 0$*.*
(c) *There exists* $f \in L^1(\Omega, \Sigma_\infty)$ *such that* $f_n = E^{\Sigma_n} f$ *for all* $n \geqslant 0$*.*
(d) *The sequence* $\{f_n\}_{n \geqslant 1} \subseteq L^1(\Omega)$ *is uniformly integrable.*

Definition 3.185
The martingale $\{f_n, \Sigma_n\}_{n \geqslant 0}$ *is said to be* ***regular*** *if there is* $f \in L^1(\Omega)$ *such that* $f_n = E^{\Sigma_n} f$ *for all* $n \geqslant 0$*.*

Remark 3.186
Theorem 3.184 implies that a martingale is regular if and only if it is L^1-convergent.

For square integrable martingales, we have the following result.

Theorem 3.187
If $\{f_n, \Sigma_n\}_{n \geqslant 0}$ *is a martingale such that* $\sup_{n \geqslant 0} \|f_n\|_2 < +\infty$*,*
then $\{f_n\}_{n \geqslant 0}$ *converges μ-almost everywhere on Ω and in* $L^2(\Omega)$*.*

We conclude with a theorem that is useful in the study of the L^1 space.

Theorem 3.188
If (Ω, Σ, μ) *is a finite measure space and* $\{f_n\}_{n \geqslant 1} \subseteq L^1(\Omega)$ *satisfies the property that for every* $A \in \Sigma$*, the limit* $\lim_{n \to +\infty} \int_A f\, d\mu$ *exists and is finite,*
then the sequence $\{f_n\}_{n \geqslant 1}$ *is uniformly integrable.*

3.2 Problems

Problem 3.1 *

Suppose that X and Y are two sets, $f\colon X \longrightarrow Y$ is a function and $\mathcal{Y} \subseteq 2^Y$. Show that

$$m(f^{-1}(\mathcal{Y})) = f^{-1}(m(\mathcal{Y})),$$

where recall that $m(\mathcal{Y})$ denotes the monotone class generated by $\mathcal{Y}$.

Problem 3.2 *

Suppose that X is an uncountable set and

$$\mathcal{X} \stackrel{\text{def}}{=} \{A \subseteq X : A \text{ or } X \setminus A \text{ is finite or countable}\}.$$

Show that $\mathcal{X}$ is the σ-algebra generated by the singletons of X.

Problem 3.3 **

Show that there is no countable σ-algebra.

Problem 3.4 *

Let X be a topological space. Determine the σ-algebra generated by all nowhere dense subsets of X.

Problem 3.5 *

Suppose that X is a topological space, $\mathcal{B}(X)$ is its Borel σ-algebra and $Y \subset X$ is a subset furnished with the subspace topology. Show that

$$\mathcal{B}(Y) = Y \cap \mathcal{B}(X) = \{Y \cap A : A \in \mathcal{B}(X)\}.$$

Problem 3.6 **

Suppose that $\mathcal{D}$ is a semiring and $\{A_k\}_{k=1}^m \subseteq \mathcal{D}$. Show that there exists a family $\{C_i\}_{i=1}^n \subseteq \mathcal{D}$ of mutually disjoint sets such that each A_k can be written as the union of some sets from $\{C_i\}_{i=1}^n$.

Problem 3.7 **

Suppose that X is a set and $\mathcal{X} \subseteq 2^X$ is a semiring. Suppose that $A_1, \ldots, A_n, A \in \mathcal{X}$. Show that the set $A \setminus \bigcup\limits_{k=1}^{n} A_k$ can be written as a union of finite family of mutually disjoint sets of $\mathcal{X}$.

Problem 3.8 *
Suppose that X is a set and $\mathcal{X} \subseteq 2^X$ is a semiring. Show that every countable union of elements of $\mathcal{X}$ can be written as a countable union of mutually disjoint elements of $\mathcal{X}$.

Problem 3.9 *
Determine the metric spaces in which the open sets (i.e., the metric topology) form a σ-algebra.

Problem 3.10 *
Suppose that X is a set and $\mathcal{X} \subseteq 2^X$ is a π-class (i.e., it is closed under finite intersections; see Definition 3.7(a)). Let $\mathcal{R}$ be the ring generated by $\mathcal{X}$. Moreover, let $\mathcal{F}$ be the smallest family in 2^X such that
(a) $\mathcal{X} \subseteq \mathcal{F}$;
(b) it is closed under finite unions; and
(c) it is closed under ***proper differences*** (i.e., if $A, C \in \mathcal{F}$ and $C \subseteq A$ then $A \setminus C \in \mathcal{F}$).
Show that $\mathcal{F} = \mathcal{R}$.

Problem 3.11 *
Suppose that (Ω, Σ) is a measurable space, $A \subseteq \Omega$ and $\Sigma_A = A \cap \Sigma$ is the trace of the σ-algebra Σ on A (see Definition 3.13). Show that if $\Sigma = \sigma(\mathcal{Y})$, then $\Sigma_A = \sigma(\mathcal{Y}_A)$, where $\mathcal{Y}_A = A \cap \mathcal{Y}$.

Problem 3.12 **
Suppose that A and Y are two metric spaces and $f \colon X \longrightarrow Y$ is a function. Let

$$C_f \stackrel{\text{def}}{=} \big\{x \in X : \ f \text{ is continuous at } x\big\}.$$

Show that $C_f \in \mathcal{B}(X)$.

Problem 3.13 **
Suppose that Ω is a set, $\mathcal{X}$ is a ring of subsets of Ω and $\mu \colon \mathcal{X} \longrightarrow [0, +\infty]$ is a set function not identically $+\infty$. Show that the following conditions are equivalent:
(a) μ is additive;
(b) $\mu(\emptyset) = 0$ and for all $A, C \in \mathcal{X}$, we have $\mu(A \cup C) + \mu(A \cap C)) = \mu(A) + \mu(C)$.

Problem 3.14 **
Suppose that (Ω, Σ, μ) is a finite measure space, $\{A_n\}_{n\geqslant 1} \subseteq \Sigma$ and $\sum_{n\geqslant 1} \mu(A_n) < +\infty$. Show that $\mu\big(\limsup_{n\to+\infty} A_n\big) = 0$.
Remark. This result is known as the ***Borel–Cantelli Lemma***.

Problem 3.15 *
Suppose that X is a set and $\mu^\star$ is an outer measure on 2^X. Let $A \subseteq X$ be a set such that $\mu^\star(A) < +\infty$ and suppose that $C \in \Sigma_{\mu^\star}$ satisfies $A \subseteq C$ and $\mu^\star(C) = \mu^\star(A)$. Show that

$$\mu^\star(A \cap E) \;=\; \mu^\star(C \cap E) \qquad \forall\, E \in \Sigma_{\mu^\star}.$$

Problem 3.16 *
Suppose that X is a set, $\mu^\star$ is an outer measure on 2^X and $A \subseteq X$. Show that

$$A \in \Sigma_{\mu^\star} \quad \Longleftrightarrow \quad \big[\mu^\star(D \cup E) = \mu^\star(D) + \mu^\star(E) \quad D \subseteq A,\ E \subseteq X \setminus A\big].$$

Problem 3.17 ***
Suppose that (Ω, Σ) is a measurable space and $\mu\colon \Sigma \longrightarrow \mathbb{R}$ is an additive set function. For every $A \in \Sigma$, we set

$$m(A) \;\stackrel{\text{def}}{=}\; \inf\big\{\lim_{n\to+\infty} \mu(C_n)\big\},$$

where the infimum is taken over all increasing sequences $\{C_n\}_{n\geqslant 1} \subseteq \Sigma$ such that $A = \bigcup_{n\geqslant 1} C_n$. Show that $m\colon \Sigma \longrightarrow \mathbb{R}_+$ is a measure.

Problem 3.18 **
Suppose that X is an uncountable set and

$$\Sigma \;\stackrel{\text{def}}{=}\; \big\{A \subseteq X : \ A \text{ or } A^c \text{ is finite or countable}\big\}.$$

From Problem 3.2, we know that Σ is a σ-algebra. Let $m\colon \Sigma \longrightarrow \mathbb{R}_+$ be defined by

$$m(A) \;\stackrel{\text{def}}{=}\; \begin{cases} 0 & \text{if } A \text{ is finite or countable}, \\ 1 & \text{if } A^c \text{ is finite or countable}. \end{cases}$$

Show that (X, Σ, m) is a measure space.

Problem 3.19 **

Suppose that $\mathcal{Y}$ is a semiring and $\mu\colon \mathcal{Y} \longrightarrow \overline{\mathbb{R}}_+ = [0, +\infty]$ is an additive and σ-subadditive set function. Show that μ is σ-additive on $\mathcal{Y}$.

Problem 3.20 *

Suppose that $\mathcal{Y}$ is a semiring and $\mu\colon \mathcal{Y} \longrightarrow \overline{\mathbb{R}}_+ = [0, +\infty]$ is an additive set function. Show that μ is monotone.

Problem 3.21 **

Suppose that $\mathcal{Y}$ is a semiring and $\mu_n\colon \mathcal{Y} \longrightarrow \overline{\mathbb{R}}_+ = [0, +\infty]$ is a sequence of σ-additive set functions such that

$$\mu_n(A) \leqslant \mu_{n+1}(A) \qquad \forall\, n \geqslant 1,\ A \in \mathcal{Y}.$$

We set

$$\mu(A) \stackrel{\text{def}}{=} \sup_{n \geqslant 1} \mu_n(A) \qquad \forall\, A \in \mathcal{Y}.$$

Show that $\mu\colon \mathcal{Y} \longrightarrow \overline{\mathbb{R}}_+ = [0, +\infty]$ is σ-additive too and $\mu(\emptyset) = 0$.

Problem 3.22 ***

Suppose that X is a metric space, $\mathcal{B}(X)$ is its Borel σ-algebra and μ, ϑ are two finite measures on $\mathcal{B}(X)$. Show that

(a) If μ and ϑ are equal on open or closed sets, then $\mu = \vartheta$.

(b) If X is σ-compact and μ and ϑ are equal on compact sets, then $\mu = \vartheta$.

(c) Statements **(a)** and **(b)** remain true if μ and ϑ are σ-finite.

Problem 3.23 *

Suppose that $\mu^\star$ is an outer measure on a set X and $A \subseteq X$ is a $\mu^\star$-null set. Show that for every $C \subseteq X$, we have $\mu^\star(C) = \mu^\star(A \cup C) = \mu^\star(C \setminus A)$.

Problem 3.24 *

Suppose that $\mu^\star$ is an outer measure and let $\{A_k\}_{k=1}^n \subseteq \Sigma_{\mu^\star}$ be mutually disjoint sets. Show that

$$\mu^\star\Big(D \cap \Big(\bigcup_{k=1}^n A_k\Big)\Big) = \sum_{k=1}^n \mu^\star(D \cap A_k) \qquad \forall\, D \subseteq X.$$

Problem 3.25 **
Let $\mu^\star$ be an outer measure on a set X. Show that
(a) If $A \subseteq X$ and $\{C_n\}_{n\geqslant 1} \subseteq \Sigma_{\mu^\star}$ is a sequence of mutually disjoint sets, then

$$\mu^\star\Big(\bigcup_{n\geqslant 1}(A \cap C_n)\Big) \;=\; \sum_{n\geqslant 1}\mu(A \cap C_n).$$

(b) If $\{A_n\}_{n\geqslant 1}$ is a sequence of subsets of X and there is a sequence $\{C_n\}_{n\geqslant 1} \subseteq \Sigma_{\mu^\star}$ of mutually disjoint sets such that $A_n \subseteq C_n$ for all $n \geqslant 1$, then

$$\mu^\star\Big(\bigcup_{n\geqslant 1} A_n\Big) \;=\; \sum_{n\geqslant 1}\mu^\star(A_n).$$

Problem 3.26 **
Suppose that (Ω, Σ) is a measurable space, $\{\mu_n\}_{n\geqslant 1}$ is a sequence of finite measures on Σ and $\{\vartheta_n\}_{n\geqslant 1} \subseteq (0, +\infty)$ is such that

$$\mu(\Omega) \;=\; \sum_{n\geqslant 1}\vartheta_n\mu_n(\Omega) \;<\; +\infty.$$

Show that

$$\Sigma \ni A \longmapsto \mu(A) = \sum_{n\geqslant 1}\vartheta_n\mu_n(A) \in \mathbb{R}_+ = [0, +\infty)$$

is a measure on Σ.

Problem 3.27 **
Let $\mu^\star$ be an outer measure on a set X. Show that
(a) $A \in \Sigma_{\mu^\star}$ if and only if for every $\varepsilon > 0$, we can find $C \in \Sigma_{\mu^\star}$ such that $C \subseteq A$ and $\mu^\star(A \setminus C) \leqslant \varepsilon$.
(b) If for a set $A \subseteq X$, we have that given $\varepsilon > 0$, we can find $C \in \Sigma_{\mu^\star}$ such that $\mu^\star(A \triangle C) < \varepsilon$, then $A \in \Sigma_{\mu^\star}$.

Problem 3.28 **
Suppose that (Ω, Σ, μ) is a measure space, $\mu^\star$ is the corresponding outer measure (see Proposition 3.28) and $\{A_n\}_{n\geqslant 1}$ is an increasing sequence of subsets of X. Let us set $A \overset{\text{def}}{=} \bigcup\limits_{n\geqslant 1} A_n$. Show that $\mu^\star(A_n) \nearrow \mu^\star(A)$ as $n \to +\infty$.

Problem 3.29 **
Let (Ω, Σ, μ) be a measure space such that "for all $A \in \Sigma$ for which we have $\mu(A) = +\infty$, there exists $C \in \Sigma$ such that $C \subseteq A$ and $0 < \mu(C) < +\infty$". Let $A \in \Sigma$ with $\mu(A) = +\infty$. Show that we can find $E \in \Sigma$, $E \subseteq A$ which is σ-finite and $\mu(E) = +\infty$.

Problem 3.30 **
Suppose that (Ω, Σ, μ) is a probability space and $A \subseteq \Omega$ is such that for every $C \in \Sigma$ with $A \subseteq C$, we have that $\mu(C) = 1$. Let $\Sigma_A = \Sigma \cap A$ and let $\mu_A \colon \Sigma_A \longrightarrow [0,1]$ be defined by

$$\mu_A(A \cap C) \stackrel{\text{def}}{=} \mu(C) \qquad \forall\, C \in \Sigma.$$

Show that (A, Σ_A, μ_A) is a probability space.

Problem 3.31 **
Let $\mu^\star$ be an outer measure on a set X. Show that:
(a) For every $A \subseteq X$ and $C \in \Sigma_{\mu^\star}$, we have

$$\mu^\star(A \cap C) + \mu^\star(A \cup C) \;=\; \mu^\star(A) + \mu^\star(C).$$

(b) If $A \notin \Sigma_{\mu^\star}$ and $C \in \Sigma_{\mu^\star}$ with $A \subseteq C$, then $\mu^\star(C \setminus A) > 0$.

Problem 3.32 **
Let $(\Omega, \Sigma_1, \mu_1)$ and $(\Omega, \Sigma_2, \mu_2)$ be two measure spaces on the same set Ω. Let $\mu_1^\star$ and $\mu_2^\star$ be the two outer measures on 2^Ω such that $\mu_1 = \mu_2^\star|_{\Sigma_1}$ and $\mu_2 = \mu_1^\star|_{\Sigma_2}$. Show that $\mu_1^\star = \mu_2^\star$.

Problem 3.33 **
Let (Ω, Σ, μ) be a measure space and let $\{A_n\}_{n \geqslant 1} \subseteq \Sigma$ be a sequence. Show that

$$\mu\big(\liminf_{n \to +\infty} A_n\big) \;\leqslant\; \liminf_{n \to +\infty} \mu(A_n)$$

and when μ is finite, we also have

$$\limsup_{n \to +\infty} \mu(A_n) \;\leqslant\; \mu\big(\limsup_{n \to +\infty} A_n\big).$$

Recall that

$$\liminf_{n \to +\infty} A_n \;=\; \bigcup_{k \geqslant 1} \bigcap_{n \geqslant k} A_n \;\in\; \Sigma \quad \text{and} \quad \limsup_{n \to +\infty} A_n \;=\; \bigcap_{k \geqslant 1} \bigcup_{n \geqslant k} A_n \;\in\; \Sigma.$$

Problem 3.34 **
Let μ be a Borel measure on $\mathbb{R}^N$ which is finite on compact sets. For every $x \in \mathbb{R}^N$ let $f(x) = \mu\big(B_1(x)\big)$. Show that f attains its infimum on every compact set.

Problem 3.35 *
Let (Ω, Σ) be a measurable space and let $\mathcal{A}$ be an algebra such that $\sigma(\mathcal{A}) = \Sigma$. Suppose that μ_1 and μ_2 are two finite measures on (Ω, Σ) such that $\mu_1(A) = \mu_2(A)$ for all $A \in \mathcal{A}$. Show that $\mu_1 = \mu_2$.

Problem 3.36 **
Suppose that (Ω, Σ) is a measurable space, $\mathcal{Y} \subseteq \Sigma$ is a π-class (see Definition 3.7) such that $\Sigma = \sigma(\mathcal{Y})$, μ_1, μ_2 are two measures on Σ such that $\mu_1|_{\mathcal{Y}} = \mu_2|_{\mathcal{Y}}$ and suppose that there is an increasing sequence $\{C_n\}_{n \geqslant 1} \subseteq \mathcal{Y}$ such that $\Omega = \bigcup\limits_{n \geqslant 1} C_n$ and $\mu_1(C_n) = \mu_2(C_n) < +\infty$ for all $n \geqslant 1$. Show that $\mu_1 = \mu_2$.

Problem 3.37 ***
Show that a sub-σ-algebra $\mathcal{Y}$ of a countably generated σ-algebra Σ need not be countably generated.

Problem 3.38 ***
Let (Ω, Σ) be a measurable space. Show that Σ is countably generated (see Definition 3.14) if and only if there exists a Σ-measurable function $f \colon \Omega \longrightarrow [0,1]$ such that $\Sigma = \big\{f^{-1}(A) : \ A \in \mathcal{B}\big([0,1]\big)\big\}$.

Problem 3.39 **
Suppose that Ω is a set, $\mathcal{Y} \subseteq 2^{\Omega}$ and $A \in \sigma(\mathcal{Y})$. Show that there is a countable subfamily $\mathcal{D}$ of $\mathcal{Y}$ such that $A \in \sigma(\mathcal{D})$.

Problem 3.40 **
Suppose that (Ω, Σ) is a measurable space, μ and ν are two probability measures on (Ω, Σ). Suppose that $\mathcal{Y} \subseteq \Sigma$ is a π-class (i.e., it is a family closed under finite intersections). Show that $\mu = \nu$ on $\sigma(\mathcal{Y})$.

Problem 3.41 *
For a given $\varepsilon > 0$, find an open dense set $U \subseteq \mathbb{R}$ such that $\lambda(U) < \varepsilon$ (λ being the Lebesgue measure on $\mathbb{R}$).

Problem 3.42 **
Suppose that λ^* is the Lebesgue outer measure on $\mathbb{R}$ and $A, C \subseteq \mathbb{R}$ are nonempty sets such that $\operatorname{dist}(A, C) = \inf\big\{|x-u| : x \in A,\ u \in C\big\} > 0$. Show that $\lambda^*(A \cup C) = \lambda^*(A) + \lambda^*(C)$.

Problem 3.43 **
Suppose that (Ω, Σ, μ) is a measure space and $\{A_n\}_{n\geqslant 1} \subseteq \Sigma$ is a sequence of Σ-sets such that $\mu\big(\bigcup\limits_{n\geqslant 1} A_n\big) < +\infty$. Show that $\mu\big(\limsup\limits_{n\to+\infty} A_n\big) \geqslant \liminf\limits_{n\to+\infty} \mu(A_n) = \eta$.
Can we drop the hypothesis that $\mu\big(\bigcup\limits_{n\geqslant 1} A_n\big) < +\infty$?

Problem 3.44 **
Suppose that $A \subseteq \mathbb{R}$ is a Lebesgue measurable set and $\lambda(A) < +\infty$ (λ being the Lebesgue measure on $\mathbb{R}$). Show that the function $\varphi\colon \mathbb{R} \longrightarrow [0, +\infty)$, defined by $\varphi(x) \stackrel{\text{def}}{=} \lambda\big(A \cap (-\infty, x]\big)$ is continuous.

Problem 3.45 ***
Suppose that $f\colon [0,1] \longrightarrow \mathbb{R}$ is a function and $A \subseteq \big\{x \in [0,1] : f'(x) \text{ exists}\big\}$. Suppose that $\lambda(A) = 0$ (λ is the Lebesgue measure on $\mathbb{R}$). Show that $\lambda\big(f(A)\big) = 0$.

Problem 3.46 ***
Suppose that $\Omega \subseteq \mathbb{R}^N$ is an open set, $f\colon \Omega \longrightarrow \mathbb{R}^N$ is a function and $D \subseteq \Omega$ is a Lebesgue-null set such that f is differentiable on D. Show that $\lambda^N\big(f(D)\big) = 0$
(λ^N being the Lebesgue measure on $\mathbb{R}^N$).

Problem 3.47 ***
Suppose that $u\colon [a,b] \longrightarrow \mathbb{R}$ is a function and there exists a set $A \subseteq [a,b]$ (not necessarily measurable) and $M > 0$ such that u is differentiable at every $x \in A$ and

$$\big|u'(x)\big| \leqslant M \qquad \forall\, x \in A.$$

Show that $\lambda^*\big(u(A)\big) \leqslant M\lambda^*(A)$ (λ^* being the Lebesgue outer measure on $\mathbb{R}$).

Problem 3.48 ***
Suppose that $u\colon [a,b] \longrightarrow \mathbb{R}$ is a measurable function, $A \subseteq [a,b]$ is a Lebesgue measurable set and u is differentiable at every point of A. Show that

$$\lambda^*\big(u(A)\big) \leqslant \int_A \big|u'(t)\big|\,dt,$$

(λ^* being the Lebesgue outer measure on $\mathbb{R}$).

Problem 3.49 **
Let $f\colon [a,b] \longrightarrow \mathbb{R}$ be a measurable function and let $A \stackrel{\text{def}}{=} \big\{x \in [a,b] : f'(x) = 0\big\}$. Show that the set $f(A)$ is Lebesgue-null.

Problem 3.50 ***
Assume that T is an interval, $u\colon T \longrightarrow \mathbb{R}$ is a function such that u has derivative (finite or infinite) on a set $A \subseteq T$ (possibly not Lebesgue measurable) and assume that $\lambda\big(u(A)\big) = 0$ (λ being the Lebesgue measure on $\mathbb{R}$). Show that $u'(x) = 0$ for almost all $x \in A$. (Compare with Problem 3.45.)

Problem 3.51 **
Choose $0 < \vartheta < 1$ and consider a Cantor-like set C_ϑ as follows: From the unit interval $[0,1]$ we remove an open interval of length ϑ. We are left with two disjoint closed intervals $I_{1,1}$ and $I_{1,2}$ each of length less than $\frac{1}{2}$. From each of those intervals we remove an open interval of length $\vartheta\lambda(I_{1,2})$, $k = 1,2$ (as before λ denotes the Lebesgue measure on $\mathbb{R}$). We keep doing this to infinity. At the end we are left with a closed set C_ϑ. Show that $\lambda(C_\vartheta) = 0$. (Compare with Remark 3.34.)

Problem 3.52 ***
Suppose that μ is a Borel measure on $\mathbb{R}$ such that
(a) $\mu\big([0,1]\big) = 1$; and
(b) $\mu(A) = \mu(A + x)$ for every $A \in \mathcal{B}(\mathbb{R})$ and every $x \in \mathbb{R}$.
Show that $\mu = \lambda$ (λ being the Lebesgue measure on $\mathbb{R}$).

Problem 3.53 ***
A function $f\colon \mathbb{R} \longrightarrow \mathbb{R}$ is said to be an ***N-function***, if it maps Lebesgue-null sets to Lebesgue-null sets. Show that a continuous function f is an N-function if and only if it maps Lebesgue measurable sets to Lebesgue measurable sets.

Problem 3.54 *
Show that every countable subset of $\mathbb{R}$ has zero Lebesgue measure.

Problem 3.55 **
Suppose that $A \subseteq \mathbb{R}$ is a Lebesgue measurable set and $\lambda(A) = 1$ (λ being the Lebesgue measure on $\mathbb{R}$). Without using the Lyapunov convexity theorem (see Theorem 3.160), show that we can find a set $C \in \Sigma$, $C \subseteq A$ such that $\lambda(C) = \frac{1}{2}$.

Problem 3.56 **
Suppose that $A \subseteq \mathbb{R}$ is a Lebesgue measurable set with $\lambda(A) > 0$. Show that for any given $\varepsilon > 0$, we can find a bounded interval $I_\varepsilon = [a, b]$ (with $a < b$) such that

$$\lambda(A \cap I_\varepsilon) \geqslant (1 - \varepsilon)\lambda(I_\varepsilon).$$

Problem 3.57 **
Suppose that $A \subseteq \mathbb{R}$ is a Lebesgue measurable set with $\lambda(A) > 0$. Show that for some $\varepsilon > 0$, we have

$$A - A \supseteq [-\varepsilon, \varepsilon],$$

where $A - A = \{x - u : x, u \in A\}$

Problem 3.58 ***
Suppose that μ is a Borel measure on $\mathbb{R}^N$, finite on compact sets and $f : \mathbb{R}_+ \longrightarrow \mathbb{R}_+$ is defined by

$$f(r) \stackrel{\text{def}}{=} \sup\{\mu(B_r(x)) : x \in \mathbb{R}^N\}.$$

Suppose that f is $\mathbb{R}$-valued and assume that $\liminf\limits_{n\to+\infty} \frac{f(r)}{r^N} = 0$. Show that $\mu \equiv 0$.

Problem 3.59 *
Suppose that $A \subseteq [0, 1]$ is a Lebesgue measurable set and $\lambda(A) = 1$. Show that A is dense in $[0, 1]$.

Problem 3.60 *
Suppose that $A \subseteq \mathbb{R}^N$ is a Lebesgue measurable set and $\lambda^N(A) = 0$. Show that $\operatorname{int} A = \emptyset$.

Problem 3.61 ***
Show that there are $2^{\mathfrak{c}}$ Lebesgue measurable sets in $\mathbb{R}$ ($\mathfrak{c}$ is the cardinality of $[0,1]$ or $\mathbb{R}$).

Problem 3.62 ***
Find a Borel set $C \subseteq \mathbb{R}$ such that for every nonempty interval T, the sets $C \cap T$ and $C^c \cap T = (\mathbb{R} \setminus C) \cap T$ both have positive Lebesgue measure.

Problem 3.63 **
Suppose that (Ω, Σ, μ) is a nonatomic σ-finite measure space and ν is a measure on Σ such that for every $\varepsilon > 0$, we can find $\delta > 0$ for which we have

$$\text{if } A \in \Sigma \text{ and } \mu(A) \leqslant \delta, \quad \text{then } \nu(A) \leqslant \varepsilon.$$

Show that ν is σ-finite too.

Problem 3.64 **
Let $(\Omega_1, \Sigma_1, \mu_1)$ and $(\Omega_2, \Sigma_2, \mu_2)$ be two complete σ-finite measure spaces. Is the product measure $\mu = \mu_1 \times \mu_2$ complete on the measurable space $(\Omega_1 \times \Omega_2, \Sigma_1 \otimes \Sigma_2)$ (see Proposition 3.45)? Justify your answer.

Problem 3.65 **
Suppose that (Ω, Σ, μ) is a σ-finite measure space, Σ_μ is the μ-completion of Σ and $f\colon \Omega \longrightarrow \mathbb{R}$ is a Σ_μ-measurable function. Show that there exists a Σ-measurable function $h\colon \Omega \longrightarrow \mathbb{R}$ such that

$$|h(\omega)| \leqslant |f(\omega)| \quad \text{and} \quad h(\omega) = f(\omega) \quad \mu\text{-almost everywhere on } \Omega.$$

Problem 3.66 ***
Produce a nonmeasurable function $f\colon [0,1] \longrightarrow \mathbb{R}$ such that $|f|$ is measurable and for every $c \in \mathbb{R}$, the set $f^{-1}(\{c\})$ is Lebesgue measurable.

Problem 3.67 **
Suppose that (Ω, Σ) is a measurable space and $\{f_n\colon \Omega \longrightarrow \mathbb{R}\}_{n \geqslant 1}$ is a sequence of Σ-measurable functions. Let

$$A \overset{\text{def}}{=} \{x \in \Omega : \lim_{n \to +\infty} f_n(x) \text{ exists}\}.$$

Show that $A \in \Sigma$.

Problem 3.68 **

Suppose that (Ω, Σ) is a measurable space and $\{f_n\colon \Omega \longrightarrow \mathbb{R}\}_{n\geqslant 1}$ is a sequence of Σ-measurable functions. Let us set

$$A^+ \stackrel{\text{def}}{=} \{x \in \Omega : \lim_{n\to+\infty} f_n(x) = +\infty\},$$

$$A^- \stackrel{\text{def}}{=} \{x \in \Omega : \lim_{n\to+\infty} f_n(x) = -\infty\}.$$

Show that $A^+, A^- \in \Sigma$.

Problem 3.69 **

Suppose that (Ω, Σ) is a measurable space and (Y, d_Y) is a metric space. Show that $f\colon \Omega \longrightarrow Y$ is measurable if and only if for every continuous function $\varphi\colon Y \longrightarrow \mathbb{R}$, the function $\varphi \circ f\colon \Omega \longrightarrow \mathbb{R}$ is Σ-measurable.

Problem 3.70 **

Suppose that (Ω, Σ, μ) is a measure space and $f\colon \Omega \longrightarrow [0,1]$ is a Σ-measurable function. Show that one of the following is true:

(a) $f = \chi_A$ for some $A \in \Sigma$; or

(b) there exists $a \in \left(0, \frac{1}{2}\right)$ such that $\mu\big(\{x \in \Omega : 0 < f(x) < 1 - a\}\big) > 0$.

Problem 3.71 ***

Let $\{f_n\colon [0,1] \longrightarrow \mathbb{R}\}_{n\geqslant 1}$ be a sequence of measurable functions. Show that the following statements are equivalent:

(a) Sequence $\{f_n\}_{n\geqslant 1}$ has a subsequence which converges to zero almost everywhere.

(b) There exists a sequence $\{\beta_n\}_{n\geqslant 1} \subseteq \mathbb{R}$ such that $\limsup_{n\to+\infty} |\beta_n| > 0$ and $\sum_{n\geqslant 1} \beta_n f_n(x)$ converges for almost all $x \in [0,1]$.

(c) There exists a sequence $\{\beta\}_{n\geqslant 1} \subseteq \mathbb{R}$ such that $\sum_{n\geqslant 1} |\beta_n| = +\infty$ and $\sum_{n\geqslant 1} \beta_n f_n(x)$ is absolutely convergent for almost all $x \in [0,1]$.

Problem 3.72 **

Suppose that X is a metric space, (X, Σ, μ) is a complete measure space with $\mathcal{B}(X) \subseteq \Sigma$ and $f\colon X \longrightarrow \mathbb{R}$ is a function such that there exists a sequence $\{A_n\}_{n\geqslant 0} \subseteq \Sigma$ of mutually disjoint sets with A_0 being μ-null, $X = \bigcup_{n\geqslant 0} A_n$ and for every $n \geqslant 1$, the function $f|_{A_n}$ is lower semicontinuous (respectively, upper semicontinuous). Show that f is Σ-measurable.

Problem 3.73 **

Show that a monotone function $f\colon \mathbb{R} \longrightarrow \mathbb{R}$ is Borel measurable.

Problem 3.74 ***

Show that the supremum of an uncountable family of measurable $\mathbb{R}$-valued functions need not be measurable. (Compare with Corollary 3.69.)

Problem 3.75 **

Suppose that (Ω, Σ) is a measurable space and $f\colon \Omega \longrightarrow \mathbb{R}$ is a Σ-measurable function. Show that $\operatorname{Gr} f \in \Sigma \otimes \mathcal{B}(\mathbb{R})$.

Problem 3.76 **

Let (Ω, Σ) be a measurable space and let X be a separable metric space. Suppose that $f\colon \Omega \longrightarrow X$ is a measurable function. Show that $\operatorname{Gr} f \in \Sigma \times \mathcal{B}(X)$.

Problem 3.77 **

Suppose that (Ω, Σ, μ) is a measure space and $\{f_n\}_{n\geqslant 1} \subseteq L^2(\Omega)$ is a bounded sequence. Show that $\frac{1}{n} f_n \longrightarrow 0$ μ-almost everywhere on Ω.

Problem 3.78 *

Let (Ω, Σ, μ) be a measure space and let $\{C_n\}_{n\geqslant 1} \subseteq \Sigma$ be a sequence such that $\sum\limits_{n\geqslant 1} \mu(C_n) < +\infty$. For every $k \geqslant 1$, let D_k be the set of all elements of Ω, which belong to at least k of the sets C_n. Show that for every integer $k \geqslant 1$, we have $k\mu(D_k) \leqslant \sum\limits_{n\geqslant 1} \mu(C_n)$.

Problem 3.79 **

Suppose that (Ω, Σ, μ) is a semifinite measure space and assume that $\{f_n\colon \Omega \longrightarrow \mathbb{R}_+\}_{n\geqslant 1}$ is a sequence of Σ-measurable functions such that $f_n(x) \longrightarrow f(x)$ μ-almost everywhere on Ω and

$$\lim_{n\to+\infty} \int\limits_{\Omega} f_n(x)\,dx = 0.$$

Show that $f(x) = 0$ μ-almost everywhere on Ω.

Problem 3.80 **

Suppose that (Ω, Σ, μ) is a σ-finite measure space and assume that $f, g \colon \Omega \longrightarrow \mathbb{R}^* = \mathbb{R} \cup \{\pm\infty\}$ are Σ-measurable functions such that $\int_\Omega f \, d\mu$, $\int_\Omega g \, d\mu$ exist and

$$\int_A f \, d\mu \leqslant \int_A g \, d\mu \qquad \forall\, A \in \Sigma.$$

Show that $f(x) \leqslant g(x)$ μ-almost everywhere on Ω. (Compare with Proposition 3.90(c).)

Problem 3.81 **

Let $f, h \colon \mathbb{R} \longrightarrow \mathbb{R}$ be two Lebesgue integrable functions with compact supports. Is it true that $f \circ h \colon \mathbb{R} \longrightarrow \mathbb{R}$ is Lebesgue integrable too? Justify your answer.

Problem 3.82 **

Find a σ-finite measure μ on $\mathcal{B}(\mathbb{R}^N)$ $(N \geqslant 1)$ such that for every $f \in C(\mathbb{R}^N) \setminus \{0\}$, we have

$$\int_{\mathbb{R}^N} |f| \, d\mu = +\infty.$$

Problem 3.83 ***

Assuming the continuum hypothesis (i.e., that every subset of $\mathbb{R}$ is finite, countable or is equipotent to $\mathbb{R}$), show that there is a set $A \subseteq [0,1] \times [0,1]$ such that for every $x \in [0,1]$, the set $A_x = \{y \in [0,1] : (x,y) \in A\}$ is countable, for every $y \in [0,1]$, the set $A_y = \{x \in [0,1] : (x,y) \in A\}$ is co-countable (i.e., its complement is countable) and A is not Lebesgue measurable.

Problem 3.84 **

Suppose that (Ω, Σ, μ) is a finite measure space and $f, h \colon \Omega \longrightarrow \mathbb{R}$ are two Σ-measurable functions such that $\int_\Omega f \, d\mu = \int_\Omega h \, d\mu$. Show that the following alternative holds:

(a) $f = h$ μ-almost everywhere on Ω; or

(b) there exists $A \in \Sigma$ such that

$$\int_A h \, d\mu < \int_A f \, d\mu.$$

Problem 3.85 *
If $C \subseteq [0,1]$ is the Cantor set, show that χ_C is Riemann integrable and

$$\int_0^1 \chi_C \, dx = 0.$$

Problem 3.86 **
Suppose that $f\colon [0,1] \longrightarrow (0,+\infty)$ is a Lebesgue measurable functions and $\theta \in (0,1]$. Show that

$$\inf \Big\{ \int_A f \, dx : \ A \subseteq [0,1], \ A \text{ is Legesgue measurable with } \lambda(A) \geqslant \theta \Big\} > 0$$

(as always λ denotes the Lebesgue measure on $\mathbb{R}$).

Problem 3.87 **
Let $f \in C^\infty(\mathbb{R}^N)$ and $g \in C_c(\mathbb{R}^N)$ (i.e., g has compact support). Show that the function

$$(f * g)(x) = \int_{\mathbb{R}^N} f(x-y) g(y) \, dy$$

belongs to $C^\infty(\mathbb{R}^N)$.

Problem 3.88 **
Let $f \in C_c(\mathbb{R}^N)$. Show that there exists a compact set $C \subseteq \mathbb{R}^N$ such that $\operatorname{supp} f \subseteq C$ and a sequence $\{f_n\}_{n \geqslant 1} \subseteq C_c^\infty(\mathbb{R}^N)$ with $\operatorname{supp} f_n \subseteq C$ for all $n \geqslant 1$ such that $f_n \longrightarrow f$ uniformly on C.

Problem 3.89 **
Let (Ω, Σ, μ) and $(Y, \mathcal{Y}, m)$ be two σ-finite measure spaces and let $f\colon \Omega \times Y \longrightarrow \mathbb{R}$ be a $\mu \times m$-measurable function. Suppose that for μ-almost all $\omega \in \Omega$, the function $y \longmapsto f(\omega, y)$ is m-integrable. Show that the function

$$\Omega \ni \omega \longrightarrow \psi(\omega) = \int_Y f(\omega, y) \, dm(y)$$

is μ-measurable.

Problem 3.90 ***
Show that the Lebesgue dominated convergence theorem (see Theorem 3.94) does not hold for nets of functions.

Problem 3.91 **
Find a Borel measure μ on $\mathbb{R}$ such that

$$\int_{\mathbb{R}} f\, d\mu = f(0) \qquad \forall\, f \in C_c(\mathbb{R}).$$

Is there a Borel measure m on $\mathbb{R}$ such that

$$\int_{\mathbb{R}} f\, dm = f'(0) \qquad \forall\, f \in C_c^1(\mathbb{R})?$$

Justify your answer.

Problem 3.92 **
Let $f \in C(\mathbb{R}_+; \mathbb{R})$ and assume that $\lim\limits_{x \to +\infty} f(x) = \theta \in \mathbb{R}$. Show that

$$\lim_{n \to +\infty} \int_0^a f(nx)\, dx = a\theta \qquad \forall\, a > 0.$$

Problem 3.93 ***
(a) Show that if $f \in L^1(\mathbb{R}^N)$ and $K \subseteq \mathbb{R}^N$ is compact, then

$$\lim_{\|x\|_{\mathbb{R}^N} \to +\infty} \int_{x+K} |f(z)|\, dz = 0.$$

(b) Show that if $f \colon \mathbb{R}^N \longrightarrow \mathbb{R}$ is uniformly continuous and $f \in L^p(\mathbb{R}^N)$ for some $p \geqslant 1$, then

$$\lim_{\|x\|_{\mathbb{R}^N} \to +\infty} f(x) = 0.$$

Problem 3.94 **

Suppose that $\xi \in C(\mathbb{R})$ is such that $\xi(0) = 0$ and $\xi(x) > 0$ for all $x \neq 0$ and $f\colon \mathbb{R}^N \longrightarrow \mathbb{R}$ is a uniformly continuous and bounded function. Assume that

$$\int\limits_{\mathbb{R}^N} \xi(f(x))\, dx \;<\; +\infty.$$

Show that $f(x) \longrightarrow 0$ as $\|x\| \to +\infty$. (Compare with Problem 3.93(b).)

Problem 3.95 ***

Is it true or false: "Every $f \in L^1(\mathbb{R})$ with $f \geqslant 0$ satisfies $f(x) \longrightarrow 0$ as $|x| \to +\infty$". (Compare with Problem 3.93.)

Problem 3.96 **

Suppose that (Ω, Σ, μ) is a measure space and $f \in L^1(\Omega)$. Show that:

(a) If $\int_A f\, dx = 0$ for all $A \in \Sigma$, then $f(x) = 0$ μ-almost everywhere on Ω.

(b) The set $D = \{x \in \Omega :\ f(x) \neq 0\} \in \Sigma$ is σ-finite.

Problem 3.97 **

Let $f \in L^1(\mathbb{R})$ and set

$$F(x) \;=\; \int\limits_{-\infty}^{x} f(t)\, dt.$$

Show that $F\colon \mathbb{R} \longrightarrow \mathbb{R}$ is continuous.

Problem 3.98 **

Suppose that (Ω, Σ, μ) is a measure space and $\{f_n\}_{n\geqslant 1}$ is a sequence of nonnegative Σ-measurable functions such that $f_n \longrightarrow f$ μ-almost everywhere on Ω and

$$\int\limits_{\Omega} f_n\, d\mu \;\longrightarrow\; \int\limits_{\Omega} f\, d\mu,$$

with $\int\limits_{\Omega} f\, d\mu < +\infty$. Show that $f_n \longrightarrow f$ in $L^1(\Omega)$.

Problem 3.99 **
Let $f \in L^1(\mathbb{R}^N)$. Show that $\lim\limits_{\lambda^N(C)\to 0} \int_C |f|\, d\lambda^N = 0$.

Problem 3.100 **
Let (Ω, Σ, μ) be a σ-finite measure space, $1 \leqslant p_1 < p_2 < +\infty$ and $f \in L^{p_1}(\Omega) \cap L^{p_2}(\Omega)$. Show that $f \in L^p(\Omega)$ for all $p \in [p_1, p_2]$ and the function $[p_1, p_2] \ni p \longmapsto \|f\|_p$ is continuous.

Problem 3.101 *
Suppose that (Ω, Σ, μ) is a measure space, $f \in L^1(\Omega)$, $f \geqslant 0$ and $g\colon \Omega \longrightarrow \mathbb{R}$ is a Σ-measurable function. Suppose that there exist $a, b \in \mathbb{R}$ such that $a \leqslant g(x) \leqslant b$ μ-almost everywhere on Ω. Show that there exists $c \in [a, b]$ such that

$$\int\limits_\Omega fg\, d\mu \;=\; c \int\limits_\Omega f\, d\mu.$$

Problem 3.102 **
Find a function $f \in L^1([0,1])$ such that $f \notin L^p([0,1])$ for all $p > 1$.

Problem 3.103 **
Find a function $f \in \bigcap\limits_{p\geqslant 1} L^p([0,1]) \setminus L^\infty([0,1])$.

Problem 3.104 ***
Suppose that (Ω, Σ, μ) is a σ-finite measure space and $p_0 \geqslant 1$ is such that $f \in L^p(\Omega)$ for all $p \in [p_0, +\infty)$. Show that $\|f\|_\infty = \liminf\limits_{p\to+\infty} \|f\|_p$. (Compare with Problem 3.103.)

Problem 3.105 ***
Let $\{f_n\}_{n\geqslant 1} \subseteq L^1(\mathbb{R})$ be a sequence such that $f_n \longrightarrow f$ almost everywhere on $\mathbb{R}$.
(a) Show that $f \in L^1(\mathbb{R})$.
(b) Show that if

$$\int\limits_{\mathbb{R}} f_n\, dx \longrightarrow \int\limits_{\mathbb{R}} f\, dx,$$

then for every $\varepsilon > 0$, we can find a Lebesgue measurable set $C \subseteq \mathbb{R}$ with finite Lebesgue measure, $h \in L^1(\mathbb{R})$, $h \geqslant 0$ and $n_0 \geqslant 1$ such that for all $n \geqslant n_0$, we have

$$\Big|\int_{C^c} f_n\,dx\Big| \leqslant \varepsilon \quad \text{and} \quad \big|f_n(x)\big| \leqslant h(x) \qquad \forall\, x \in C.$$

(c) Is the converse in **(b)** true? Justify your answer.

Problem 3.106 ***

Let $\{f_n\}_{n\geqslant 1} \subseteq L^1(\mathbb{R})$ be a sequence such that $f_n \longrightarrow f$ almost everywhere on $\mathbb{R}$ and assume that for every $\varepsilon > 0$, there exist a Lebesgue measurable set $A \subseteq \mathbb{R}$, a function $h \in L^1(\mathbb{R})$, $h \geqslant 0$ and an integer $n_0 \geqslant 1$ such that

$$\int_{A^c} |f_n|\,dx \leqslant \varepsilon \qquad \forall\, n \geqslant n_0$$

and

$$\big|f_n(x)\big| \leqslant h(x) \qquad \forall\, x \in A,\ n \geqslant n_0.$$

Show that $f \in L^1(\mathbb{R})$ and

$$f_n \longrightarrow f \text{ in } L^1(\mathbb{R}).$$

Is the last condition necessary in order to have f in $L^1(\mathbb{R})$ and $f_n \longrightarrow f$ in $L^1(\mathbb{R})$? Justify your answer.

Problem 3.107 *

Let (Ω, Σ, μ) be a measure space and let $\{f_n\}_{n\geqslant 1} \subseteq L^1(\Omega)$ be a sequence such that $0 \leqslant f_{n+1} \leqslant f_n$ μ-almost everywhere for all $n \geqslant 1$. Show that

$$f_n \searrow 0 \ \mu\text{-almost everywhere} \qquad \Longleftrightarrow \qquad \int_\Omega f_n\,d\mu \searrow 0.$$

Problem 3.108 *

Suppose that (Ω, Σ, μ) is a σ-finite measure space, $g\colon \Omega \longrightarrow \mathbb{R}$ is a Σ-measurable function and $1 \leqslant p < +\infty$. Assume that for every $f \in L^p(\Omega)$, we have $fg \in L^1(\Omega)$. Show that:

(a) If $\int\limits_{\Omega} fg\,d\mu = 0$ for all $f \in L^p(\Omega)$, then $g(x) = 0$ μ-almost everywhere on Ω.
(b) If $f_n \longrightarrow f$ μ-almost everywhere on Ω and $|f_n(x)| \leqslant h(x)$ μ-almost everywhere on Ω with $h \in L^p(\Omega)$, then $f_n g \longrightarrow fg$ in $L^1(\Omega)$ as $n \to +\infty$.

Problem 3.109 ★★
Suppose that (Ω, Σ, μ) is a finite measure space and $f\colon \Omega \longrightarrow [0, +\infty)$ is a Σ-measurable function. Show that

$$f \in L^1(\Omega) \qquad \Longleftrightarrow \qquad \sum_{n \geqslant 0} \mu(\{f \geqslant n\}) \text{ converges.}$$

Problem 3.110 ★
Let $f \in L^1(\mathbb{R})$ and for every finite interval I, let us set

$$m(I) \stackrel{\text{def}}{=} \frac{1}{\lambda(I)} \int\limits_I f\,dx$$

(λ being the Lebesgue measure on $\mathbb{R}$) and $A(I) = \{f > m(I)\} \cap I$. Show that

$$\int\limits_I |f - m(I)|\,dx = 2 \int\limits_{A(I)} (f - m(I))\,dx.$$

Problem 3.111 ★★
Let $f \in L^1(\mathbb{R}_+)$ and assume that

$$\int\limits_0^y f(x)\,dx = 0 \qquad \forall\, y \geqslant 0.$$

Show that $f(x) = 0$ almost everywhere on $\mathbb{R}_+$.

Problem 3.112 ★★
Suppose that $A \subseteq \mathbb{R}_+$ is Lebesgue measurable with $\lambda^N(A) = 1$ (λ^N being the Lebesgue measure on $\mathbb{R}^N$) and $f, g\colon \mathbb{R}^N \longrightarrow \mathbb{R}$ are two Lebesgue measurable functions, $f, g \geqslant 0$. Show that, if $fg \geqslant 1$ almost everywhere on A, then

$$\Big(\int\limits_A f\,dx\Big)\Big(\int\limits_A g\,dx\Big) \geqslant 1.$$

Problem 3.113 **
Suppose that (Ω, Σ, μ) is a finite measure space and $1 \leqslant p < q \leqslant +\infty$. Show that $L^q(\Omega) \subseteq L^p(\Omega)$. Is the result true if $\mu(\Omega)$ is not finite? Justify your answer.

Problem 3.114 **
Let $g\colon [0,1] \longrightarrow \mathbb{R}$ is a continuous and nondecreasing function. Is it true that

$$\int_0^1 dg = \int_0^1 g'\,dx?$$

Justify your answer.

Problem 3.115 **
Let (Ω, Σ, μ) be a measure space and let $C \subseteq \mathbb{R}^N$ be a closed set. Assume that $f \in L^1(\Omega; \mathbb{R}^N)$ and

$$\frac{1}{\mu(A)} \int_A f\,d\mu \in C \qquad \forall\, A \in \Sigma, \text{ with } \mu(A) > 0.$$

Show that $f(x) \in C$ μ-almost everywhere on Ω.

Problem 3.116 **
Let (Ω, Σ, μ) be a measure space and let $\mathcal{Y} \subseteq \Sigma$ be such that for a given $\varepsilon > 0$, for every $A \in \Sigma$ with $\mu(A) < +\infty$, we can find $C \in \mathcal{Y}$ for which we have $\mu(A \triangle C) \leqslant \varepsilon$. Show that, if $\mathcal{Y}$ is countable, then $L^p(\Omega)$ is separable for every $p \in [1, +\infty)$.

Problem 3.117 **
Suppose that (Ω, Σ, μ) is a measure space and $\{f_n\colon \Omega \longrightarrow \mathbb{R}\}_{n \geqslant 1}$ is a sequence of Σ-measurable functions such that $\sum\limits_{n\geqslant 1} \|f_n\|_1 < +\infty$. Show that the series $\sum\limits_{n \geqslant 1} f_n$ converges μ-almost everywhere to a finite valued function and

$$\int_\Omega \Big(\sum_{n\geqslant 1} f_n\Big)\,d\mu = \sum_{n\geqslant 1} \int_\Omega f_n\,d\mu.$$

Problem 3.118 **
Suppose that $f \in L^2([0,1]) \setminus \{0\}$ and let us set

$$F(x) = \int_0^x f(s)\,ds.$$

Show that $\|F\|_2 < \|f\|_2$.

Problem 3.119 **
Let $\{f_n\}_{n\geqslant 1} \subseteq L^1(\mathbb{R})$ be a sequence, $f \in L^1(\mathbb{R})$ and assume that

$$\int_{\mathbb{R}} |f_n(t) - f(t)|\, dt \leqslant \frac{1}{n^2} \qquad \forall\, n \geqslant 1.$$

Show that $f_n \longrightarrow f$ almost everywhere on $\mathbb{R}$.

Problem 3.120 **
Suppose that $f \in L^2([0,1])$ with $\|f\|_2 = 1$ and $\int_0^1 f\, dx \geqslant \theta > 0$. For every $\eta \in \mathbb{R}$, let

$$A_\eta \stackrel{\text{def}}{=} \{x \in [0,1] : f(x) \geqslant \eta\}.$$

Show that, if $\eta \in (0, \theta)$, then $(\theta - \eta)^2 \leqslant \lambda(A_\eta)$ (λ being the Lebesgue measure on $\mathbb{R}$).

Problem 3.121 **
Let (Ω, Σ, μ) be a finite measure space and let $\{f_n\}_{n\geqslant 1} \subseteq L^2(\Omega)$ be a sequence, such that $|f_n(x)| \leqslant M$ μ-almost everywhere on Ω for all $n \geqslant 1$ and $\|f_n\|_2 = 1$ for all $n \geqslant 1$. Suppose that $\{\beta_n\}_{n\geqslant 1} \subseteq \mathbb{R}$ is a sequence such that $\sum\limits_{n\geqslant 1} \beta_n f_n(x)$ converges for μ-almost all $x \in \Omega$. Show that $\beta_n \longrightarrow 0$ as $n \to +\infty$.

Problem 3.122 **
Let (Ω, Σ, μ) be a measure space and let $f \in L^1(\Omega) \cap L^2(\Omega)$. Show the following:
(a) $f \in L^p(\Omega)$ for all $p \in [1,2]$;
(b) $\lim\limits_{p \to 1^+} \|f\|_p = \|f\|_1$.
(Compare with Problem 3.100.)

Problem 3.123 ***
Let (Ω, Σ, μ) be a probability space and let $h\colon \Omega \longrightarrow [0, +\infty]$ be a Σ-measurable function. Let us set $A = \int_\Omega h\, d\mu$. Show that

$$\sqrt{1 + A^2} \leqslant \int_\Omega \sqrt{1 + h^2}\, d\mu \leqslant 1 + A.$$

If $\Omega = [0,1]$, Σ is the σ-algebra of Lebesgue measurable sets, $\mu = \lambda$ is the Lebesgue measure on $[0,1]$ and $h = f'$ with $f \in C^1([0,1])$, then the result has a simple geometric interpretation. From this geometric interpretation deduce (for general Ω) under what conditions on h equalities hold in the above inequalities.

Problem 3.124 **
Suppose that (Ω, Σ, μ) is a measure space, $p \in (0,1)$ and $f, g \in L^1(\Omega)$, $f, g \geqslant 0$. Show that

$$\int_\Omega (f^p + g^p)^{\frac{1}{p}}\, d\mu \leqslant \Big(\int_\Omega f\, d\mu\Big)^p + \Big(\int_\Omega g\, d\mu\Big)^p.$$

Problem 3.125 **
(a) Let $\{f_n\}_{n\geqslant 1} \subseteq L^p(\mathbb{R})$ $(1 \leqslant p < +\infty)$ be a bounded sequence such that $f_n \longrightarrow f$ almost everywhere on $\mathbb{R}$. Is it true that $f_n \longrightarrow f$ in $L^p(\mathbb{R})$? Justify your answer.
(b) Let $\{f_n\}_{n\geqslant 1} \subseteq L^p(\mathbb{R})$ $(1 \leqslant p < +\infty)$ be a sequence such that $f_n \longrightarrow f$ almost everywhere on $\mathbb{R}$ and $\|f_n\|_p \longrightarrow M < \infty$. Show that $\|f\|_p \leqslant M$.

Problem 3.126 **
Suppose that (Ω, Σ, μ) is a measure space and $f \in L^1(\Omega)$. Find

$$\lim_{n\to+\infty} \int_\Omega n \ln\big(1 + \tfrac{|f|^2}{n^2}\big)\, dx$$

Problem 3.127 ***
Let (Ω, Σ, μ) be a measure space and let $p \in (0,1)$. Show the following:
(a) $\|f\|_p = \Big(\int_\Omega |f|^p\, d\mu\Big)^{\frac{1}{p}}$ is not a norm on $L^p(\Omega)$.
(b) If for $f, h \in L^p(\Omega)$, we set

$$d(f,h) = \int_\Omega |f-h|^p\, d\mu = \|f-h\|_p^p,$$

then d is a metric on $L^p(\Omega)$ and $\big(L^p(\Omega), d\big)$ is a complete metric space.

Problem 3.128 ***
Find a function $f \in L^1(\mathbb{R})$ such that for any $a, b \in \mathbb{R}$, $a < b$ and any $M > 0$, we have that the set $(a,b) \cap \big\{x \in \mathbb{R} : f(x) \geqslant M\big\}$ has a positive measure.

Problem 3.129 ***

Find a function $f \in L^1(\mathbb{R})$ such that $f \notin L^2\big((a,b)\big)$ for any $a, b \in \mathbb{R}$, $a < b$.

Problem 3.130 **

Suppose that (Ω, Σ, μ) is a measure space, $\{A_n\}_{n\geqslant 1} \subseteq \Sigma$ is a sequence such that $\mu(A_n) \longrightarrow 0$ as $n \to +\infty$, let $1 < p < +\infty$ and let

$$f_n \overset{\text{def}}{=} \frac{1}{\mu(A_n)^{\frac{1}{p'}}} \chi_{A_n} \qquad \forall\, n \geqslant 1,$$

where $\frac{1}{p} + \frac{1}{p'} = 1$. Show that for every $h \in L^p(\Omega)$, we have

$$\int_\Omega h f_n \, d\mu \;\longrightarrow\; 0 \quad \text{as } n \to +\infty.$$

Problem 3.131 **

Suppose that $\{f_n\}_{n\geqslant 1} \subseteq L^1(\mathbb{R})$ and $f_n \rightrightarrows f$ ($\rightrightarrows$ denotes uniform convergence). Is it true that $f \in L^1(\mathbb{R})$? Justify your answer.

Problem 3.132 **

Suppose that I is a nontrivial interval in $\mathbb{R}$, $1 \leqslant p < +\infty$, $h \in L^p(I)$ and

$$S \overset{\text{def}}{=} \big\{f \in L^p(I) : \ f \leqslant h \text{ almost everywhere on } I\big\}.$$

Show that S is a closed nowhere dense set in $L^p(I)$.

Problem 3.133 *

Let $f \in L^1(a,b)$, $f \geqslant 0$ and $h \in C\big([a,b]\big)$. Show that there exists $s_0 \in [a,b]$ such that

$$\int_a^b f(t)h(t)\,dt \;=\; h(s_0)\int_a^b f(t)\,dt.$$

Problem 3.134 **

Suppose that $f \in L^p(\mathbb{R}^N)$ with $1 \leqslant p < +\infty$ and $T_f \colon \mathbb{R}^N \longrightarrow L^p(\mathbb{R}^N)$ is defined by

$$T_f(z)(x) \overset{\text{def}}{=} f(x+z).$$

Show that T_f is continuous and bounded.

Problem 3.135 **
Let (Ω, Σ, μ) be a finite measure space and let $\mathcal{F} \subseteq L^1(\Omega)_+$. Suppose that there exists $h \in L^1(\Omega)_+$ such that for all $\lambda > 0$, we have

$$\int\limits_{\{f>\lambda\}} f \, d\mu \leqslant \int\limits_{\{f>\lambda\}} h \, d\mu \qquad \forall\, f \in \mathcal{F}.$$

Show that $\mathcal{F} \subseteq L^1(\Omega)$ is uniformly integrable.

Problem 3.136 **
Let (Ω, Σ, μ) be a measure space with the following property: there is a Σ-partition $\{\Omega_n\}_{n\geqslant 1}$ of Ω such that $\mu(\Omega_n) \neq 0$ for all $n \geqslant 1$. Show that the space $L^\infty(\Omega)$ is not separable.

Problem 3.137 **
Suppose that (Ω, Σ, μ) is a measure space, $f \in L^1(\Omega)$ and $\varepsilon > 0$. Show that there exist $g, h \in L^1(\Omega)$, $g, h \geqslant 0$ such that

$$f = g - h \quad \text{and} \quad \int\limits_{\Omega} h \, d\mu \leqslant \varepsilon.$$

Problem 3.138 ***
Let (Ω, Σ, μ) be a measurable space and let $\{u_n\}_{n\geqslant 1} \subseteq L^p(\Omega)$ (with $1 \leqslant p < +\infty$). Suppose that:
(a) $\sup\limits_{n\geqslant 1} \|u_n\|_p < +\infty$;
(b) $u_n(\omega) \longrightarrow u(\omega)$ μ-almost everywhere in Ω.
Show that $u \in L^p(\Omega)$ and $\lim\limits_{n\to+\infty} \left(\|u_n\|_p^p - \|u_n - u\|_p^p\right) = \|u\|_p^p$ (see Theorem 3.98).

Problem 3.139 *
Let (Ω, Σ, μ) be a measure space, let $u \in L^p(\Omega)$ and let $\{u_n\}_{n\geqslant 1} \subseteq L^p(\Omega)$ with $1 \leqslant p < +\infty$. Suppose that:
(a) $\|u_n\|_p \longrightarrow \|u\|_p$;
(b) $u_n(\omega) \longrightarrow u(\omega)$ μ-almost everywhere in Ω.
Show that $u_n \longrightarrow u$ in $L^p(\Omega)$.

Problem 3.140 *
Let (Ω, Σ, μ) be a measure space. Show the following:
(a) If $f, g \in L^p(\Omega)$, $1 \leqslant p \leqslant +\infty$, then $\max\{f, g\} \in L^p(\Omega)$.
(b) If $\{f_n\}_{n\geqslant 1}, \{g_n\}_{n\geqslant 1} \subseteq L^p(\Omega)$ and $f_n \longrightarrow f$, $g_n \longrightarrow g$ in $L^p(\Omega)$, then $\max\{f_n, g_n\} \longrightarrow \max\{f, g\}$ in $L^p(\Omega)$.

Problem 3.141 **
Suppose that (Ω, Σ, μ) is a measure space, $\{f_n\}_{n\geqslant 1} \subseteq L^p(\Omega)$ (with $1 \leqslant p \leqslant +\infty$) and $\{g_n\}_{n\geqslant 1} \subseteq L^\infty(\Omega)$ is a bounded sequence. Assume that $f_n \longrightarrow f$ in $L^p(\Omega)$ and $g_n(\omega) \longrightarrow g(\omega)$ μ-almost everywhere. Show that $f_n g_n \longrightarrow fg$ in $L^p(\Omega)$.

Problem 3.142 **
Let (Ω, Σ, μ) be a finite measure space and let $\{f_n\}_{n\geqslant 1}$, $\{h_n\}_{n\geqslant 1}, \{g_n\}_{n\geqslant 1} \subseteq L^1(\Omega)$ be three sequences such that $f_n(\omega) \leqslant h_n(\omega) \leqslant g_n(\omega)$ μ-almost everywhere on Ω and

$$f_n \longrightarrow f, \quad h_n \longrightarrow h, \quad g_n \longrightarrow g \quad \mu\text{-almost everywhere on } \Omega.$$

Suppose that $f, g \in L^1(\Omega)$ and

$$\int_\Omega f_n \, d\mu \longrightarrow \int_\Omega f \, d\mu \quad \text{and} \quad \int_\Omega g_n \, d\mu \longrightarrow \int_\Omega g \, d\mu.$$

Show that $h \in L^1(\Omega)$ and

$$\int_\Omega h_n \, d\mu \longrightarrow \int_\Omega h \, d\mu$$

Problem 3.143 **
Let (Ω, Σ, μ) be a finite measure space and let $\{f_n\}_{n\geqslant 1} \subseteq L^1(\Omega)$ be a sequence such that $f_n \geqslant 0$ for $n \geqslant 1$, $f_n \longrightarrow f$ μ-almost everywhere on Ω, with $f \in L^1(\Omega)$ and

$$\int_\Omega f_n \, d\mu \longrightarrow \int_\Omega f \, d\mu.$$

Show that $\|f_n - f\|_1 \longrightarrow 0$.

Problem 3.144 **

Let (Ω, Σ, μ) be a probability space and let $f \in L^p(\Omega)$ with $1 < p < +\infty$. Show that

$$\mu\big(\{\omega \in \Omega : |f(\omega)| \geqslant \lambda \|f\|_1\}\big) \geqslant (1-\lambda)^{p'} \frac{\|f\|_1^{p'}}{\|f\|_p^{p'}} \qquad \forall\, \lambda \in [0,1],$$

where $p' = \frac{p}{p-1}$.

Problem 3.145 ***

Let (X, d_X) be a metric space and let μ be a finite measure on $\big(X, \mathcal{B}(X)\big)$. Let $C_b(X)$ be the space of all continuous bounded functions on X. Show that for any $p \in [1, +\infty)$, the space $C_b(X)$ is dense in $L^p\big(X, \mathcal{B}(X); \mu\big)$. (Compare with Proposition 3.111.)

Problem 3.146 ***

Let (Ω, Σ, μ) be a measure space and let $\{u_n\}_{n\geqslant 1} \subseteq L^1(\Omega)$ be a sequence such that $u_n \geqslant 0$ for all $n \geqslant 1$ and $u(\omega) \leqslant \liminf\limits_{n\to+\infty} u_n(\omega)$ μ-almost everywhere on Ω. Suppose that

$$\int_\Omega u_n \, d\mu = \int_\Omega u \, d\mu = 1 \qquad \forall\, n \geqslant 1.$$

Show that $u_n \longrightarrow u$ in $L^1(\Omega)$.

Problem 3.147 **

Let (Ω, Σ, μ) be a measure space and let $\{f_n\}_{n\geqslant 1}, \{g_n\}_{n\geqslant 1} \subseteq L^1(\Omega)$ be two sequences such that $g_n \longrightarrow g$ in $L^1(\Omega)$, $g_n \geqslant 0$ and $-g_n \leqslant f_n$ for all $n \geqslant 1$. Show that

$$\int_\Omega \liminf_{n\to+\infty} f_n \, d\mu \leqslant \liminf_{n\to+\infty} \int_\Omega f_n \, d\mu.$$

Problem 3.148 ***

Let (Ω, Σ, μ) be a measure space and let $\{f_n\}_{n\geqslant 1} \subseteq L^1(\Omega)$ be a sequence such that $f_n \geqslant 0$ for all $n \geqslant 1$, $f(\omega) \leqslant \liminf\limits_{n\to+\infty} f_n(\omega)$ μ-almost everywhere on Ω and

$$\limsup_{n\to+\infty} \int_\Omega f_n \, d\mu \leqslant \int_\Omega f \, d\mu < +\infty.$$

Show that $f_n \longrightarrow f$ in $L^1(\Omega)$.

Problem 3.149 ***
Let $f\colon \mathbb{R}^N \longrightarrow \mathbb{R}_+ = [0,+\infty)$ be a function and let

$$S(f) \stackrel{\text{def}}{=} \{(x,\eta) \in \mathbb{R}^N \times \mathbb{R} : 0 \leqslant \eta \leqslant f(x)\}.$$

(a) Show that, f is a Borel measurable function if and only if $S(f) \subseteq \mathbb{R}^N \times \mathbb{R}$ is a Borel set.
(b) Show that, if f is Borel measurable and $r > 0$, then

$$\int\limits_{\mathbb{R}^N} f(x)^r \, dx = r \int\limits_0^\infty \eta^{r-1} \lambda^N(\{f > \eta\})\, d\eta.$$

Recall that λ^N denotes the Lebesgue measure on $\mathbb{R}^N$.
(c) Show that, if f is Borel measurable, then $\operatorname{Gr} f$ has Lebesgue measure zero.

Problem 3.150 **
Suppose that (Ω, Σ, μ) is a σ-finite measure space, $f \in L^1(\Omega)$, $f \geqslant 0$ and λ is the Lebesgue measure on $\mathbb{R}$. Show that

$$\int\limits_\Omega f \, d\mu = (\mu \otimes \lambda)(\{\omega, \eta) : 0 \leqslant \eta \leqslant f(\omega)).$$

Problem 3.151 ***
Suppose that $C \subseteq \mathbb{R}^N$ is a set with finite Lebesgue measure and $f, h\colon C \longrightarrow \mathbb{R}$ are two Lebesgue measurable functions, $f, h \geqslant 0$. Assume that for every $\eta > 0$, we have

$$\lambda^N(\{h > \eta\}) \leqslant \tfrac{1}{\eta} \int\limits_{\{h > \eta\}} f \, dx$$

(λ^N being the Lebesgue measure on $\mathbb{R}^N$). Show that for $1 < p < +\infty$, we have

$$\Big(\int\limits_C h^p \, dx\Big)^{\frac{1}{p}} \leqslant \tfrac{p}{p-1} \Big(\int\limits_C f^p \, dx\Big)^{\frac{1}{p}}.$$

Problem 3.152 ***
Suppose that (Ω, Σ) is a measurable space, $\{\mu_n\}_{n \geqslant 1}$ is a sequence of measures on Σ and μ is a measure on Σ. Show that the following statements are equivalent:

(a) $\int_\Omega f\,d\mu \leqslant \liminf_{n\to+\infty} \int_\Omega f_n\,d\mu_n$ for any nonnegative Σ-measurable functions f_n such that $f = \liminf_{n\to+\infty} f_n$;
(b) $\int_\Omega f\,d\mu \leqslant \liminf_{n\to+\infty} \int_\Omega f\,d\mu_n$ for any nonnegative Σ-measurable functions f;
(c) $\mu(A) \leqslant \liminf_{n\to+\infty} \mu_n(A)$ for all $A \in \Sigma$.

Problem 3.153 ***
Let $f \in L^1(\mathbb{R})$ and for every $y > 0$, we define

$$h_y(x) = \frac{1}{y}\int_0^y \big|f(x+s) - f(x)\big|\,ds.$$

Show that:
(a) h_y is Lebesgue measurable and

$$\int_{\mathbb{R}} h_y(x)\,dx \leqslant 2\int_{\mathbb{R}} \big|f(x)\big|\,dx.$$

(b) $h_y \longrightarrow 0$ in $L^1(\mathbb{R})$ as $y \to 0^+$.

Problem 3.154 ***
Assume that $\Omega = [0,1]$, Σ is the Lebesgue σ-algebra on $[0,1]$, λ is the Lebesgue measure on $[0,1]$, $Y = [0,1]$, $\mathcal{Y} = 2^Y$, ν is the counting measure on $[0,1]$. Let us set

$$\Delta = \big\{(x,x) : x \in [0,1]\big\} \subseteq \Omega \times Y$$

(the "diagonal" in $\Omega \times Y$). Show the following properties:
(a) Δ is $\Sigma \otimes \mathcal{Y}$-measurable in $\Omega \times Y$;
(b) the iterated integrals

$$\int_Y \int_\Omega \chi_\Delta(x,y)\,d\lambda\,d\nu \quad \text{and} \quad \int_\Omega \int_Y \chi_\Delta(x,y)\,d\nu\,d\lambda$$

exist;
(c) χ_Δ is not $\lambda \times \nu$-integrable.

Problem 3.155 **

Suppose that (Ω, Σ, μ) is a probability space and $f\colon \Omega \longrightarrow [1, +\infty)$ is a Σ-measurable function. Is it true that

$$\Big(\int_\Omega f\, d\mu\Big)\Big(\int_\Omega \ln f\, d\mu\Big) \leqslant \int_\Omega f(\ln f)\, d\mu?$$

Justify your answer.

Problem 3.156 **

Suppose that (Ω, Σ, μ) is a finite measure space, $f \in L^1(\Omega)$ and for $\lambda > 0$ let us define

$$\xi(\lambda) \stackrel{\text{def}}{=} \mu\big(\{f > \lambda\}\big) \quad \text{and} \quad \eta(\lambda) \stackrel{\text{def}}{=} \mu\big(\{f < -\lambda\}\big).$$

Show that the functions ξ and η are Borel measurable and

$$\|f\|_1 = \int_0^\infty \big(\xi(\lambda) + \eta(\lambda)\big)\, d\lambda.$$

Problem 3.157 **

Suppose that $f, h \in L^1\big([0,1]\big)$ and

$$f(x)h(y) = f(y)h(x) \qquad \forall\, x, y \in [0,1].$$

Show that

$$\int_0^1\int_0^1 f(x)h(y)\, dx\, dy = 2\int_E f(x)h(y)\, d\lambda^2(x,y),$$

where λ^2 is the Lebesgue measure on $\mathbb{R}^2$ and

$$E \stackrel{\text{def}}{=} \big\{(x,y) \in \mathbb{R}^2 :\ 0 \leqslant x < y \leqslant 1\big\}.$$

Problem 3.158 **

Let

$$f(x,y) \stackrel{\text{def}}{=} \begin{cases} \frac{x^2-y^2}{(x^2+y^2)^2} & \text{if } x, y \in [0,1] \setminus \{0\}, \\ 0 & \text{if } x = y = 0. \end{cases}$$

Show that

$$\int_0^1\int_0^1 \big|f(x,y)\big|\, dy\, dx = \int_0^1\int_0^1 \big|f(x,y)\big|\, dx\, dy < +\infty.$$

Is it true that $f \in L^1\big([0,1] \times [0,1]\big)$? Justify your answer.

Problem 3.159 **

Let $f\colon \mathbb{R}^N \longrightarrow \mathbb{R}$ be a function such that for every $\varepsilon > 0$, we can find an open set $U \subseteq \mathbb{R}^N$ with $\lambda^N(U) < \varepsilon$ and $f|_{\mathbb{R}^N \setminus U}$ is continuous. Show that f is measurable. (Compare with Theorem 3.77.)

Problem 3.160 **

Suppose that (Ω, Σ, μ) is a finite measure space, and $\{f_n\}_{n\geqslant 1}$, $\{g_n\}_{n\geqslant 1}$ are two sequences of Σ-measurable functions such that $f_n \xrightarrow{\mu} f$ and $g_n \xrightarrow{\mu} g$. Show that $f_n g_n \xrightarrow{\mu} fg$.
Is the result true if we drop the hypothesis that μ is finite? Justify your answer.

Problem 3.161 ***

Let (Ω, Σ, μ) be a measure space, let $\{f_n\colon \Omega \longrightarrow \mathbb{R}_+\}_{n\geqslant 1}$ be a sequence of Σ-measurable functions and let $f\colon \Omega \longrightarrow \mathbb{R}$ be a Σ-measurable function. Assume that $f_n \xrightarrow{\mu} f$. Show that $f_n^\theta \xrightarrow{\mu} f^\theta$ for all $\theta > 0$.

Problem 3.162 ***

Suppose that (Ω, Σ, μ) is a measure space and $\{f_n\colon \Omega \longrightarrow \mathbb{R}\}_{n\geqslant 1}$ is a sequence of Σ-measurable functions. Show that the sequence $\{f_n\}_{n\geqslant 1}$ is Cauchy in μ-measure if and only if there exists a Σ-measurable function $f\colon \Omega \longrightarrow \mathbb{R}$ such that $f_n \xrightarrow{\mu} f$ as $n \to +\infty$.

Problem 3.163 **

Suppose that (Ω, Σ, μ) is a σ-finite measure space, $\{f_n\}_{n\geqslant 1} \subseteq L^1(\Omega)$ is a sequence such that $f_n \geqslant 0$ for $n \geqslant 1$, $f_n \longrightarrow f$ μ-almost everywhere on Ω and

$$\int_\Omega f_n \, d\mu \longrightarrow \int_\Omega f \, d\mu,$$

with $f \in L^1(\Omega)$, $f \geqslant 0$. Show that

$$\int_A f_n \, d\mu \longrightarrow \int_A f \, d\mu \qquad \forall\, A \in \Sigma.$$

Problem 3.164 **

Let Ω be any set and consider the σ-algebra 2^Ω. For a given $x \in \Omega$, let δ_x be the Dirac measure concentrated at x, i.e.,

$$\delta_x(C) \stackrel{\text{def}}{=} \begin{cases} 1 & \text{if } x \in C, \\ 0 & \text{if } x \notin C. \end{cases}$$

Also let $x_0 \in \Omega$ and let $\{f_n = n\chi_{\{x_0\}}\}_{n\geqslant 1}$ be a sequence of functions. Determine whether the sequence $\{f_n\}_{n\geqslant 1}$ is δ_{x_0}-uniformly integrable.

Problem 3.165 **

Let $\{f_n\}_{n\geqslant 1} \subseteq L^p(\mathbb{R})$ $(1 < p < +\infty)$ be a sequence such that $f_n \geqslant 0$ for all $n \geqslant 1$. Show that

$$f_n \longrightarrow f \quad \text{in } L^p(\mathbb{R}) \qquad \iff \qquad f_n^p \longrightarrow f^p \quad \text{in } L^1(\mathbb{R}).$$

Problem 3.166 **

Suppose that (Ω, Σ, μ) is a measure space, $1 \leqslant p < +\infty$ and $0 < r < p$. Show the following:

(a) The function $K\colon L^p(\Omega) \longrightarrow L^{\frac{p}{r}}(\Omega)$, defined by $K(f) = |f|^r$ is continuous.

(b) If $f_n \longrightarrow f$ and $h_n \longrightarrow h$ in $L^p(\Omega)$, then

$$\int_\Omega |f_n|^{p-r}|h_n|^r \, d\mu \longrightarrow \int_\Omega |f|^{p-r}|h|^r \, d\mu.$$

Problem 3.167 **

Suppose that (Ω, Σ, μ) is a finite measure space, $f\colon \Omega \longrightarrow \mathbb{R}$ and $f_n\colon \Omega \longrightarrow \mathbb{R}$ for $n \geqslant 1$ are Σ-measurable functions. We say that f_n converges to f ***almost uniformly*** (denoted by $f_n \xrightarrow{au} f$), if for every $\varepsilon > 0$, we can find $A \in \Sigma$ with $\mu(A) \leqslant \varepsilon$ and $f_n \rightrightarrows f$ on $\Omega \setminus A$. Show that, if $f_n \xrightarrow{au} f$, then $f_n \xrightarrow{\mu} f$.

Problem 3.168 ***

Let (Ω, Σ, μ) be a finite measure space and let $V \subseteq L^1(\Omega)$ be a closed vector subspace. Suppose that $V \subseteq \bigcup\limits_{1<r\leqslant\infty} L^r(\Omega)$. Show that there exists $q > 1$ such that $V \subseteq L^q(\Omega)$.

Problem 3.169 ***

Suppose that (Ω, Σ) is a measurable space and $\{\mu_n\}_{n\geqslant 1}$ are signed measures on Σ. Suppose that $\mu_n(A) \longrightarrow \mu(A) \in \mathbb{R}$ for all $A \in \Sigma$. Show that μ is a signed measure too.

Problem 3.170 **

Suppose that (Ω, Σ) is a measurable space, μ is a measure on Σ and ν is a finite measure on Σ. Show that $\nu \ll \mu$ if and only if for every $\varepsilon > 0$, we can find $\delta > 0$ such that for every $A \in \Sigma$, with $\mu(A) < \delta$, we have $\nu(A) < \varepsilon$.

Problem 3.171 **

Suppose that (Ω, Σ) is a measurable space and μ, ν are two measures on Σ. Consider the following two statements:

(a) $\nu \ll \mu$;

(b) for every $\varepsilon > 0$ there exists $\delta > 0$ such that

$$\text{if } A \in \Sigma \text{ and } \mu(A) \leqslant \delta, \quad \text{then } \nu(A) \leqslant \varepsilon.$$

Show that "**(b)** $\Longrightarrow$ **(a)**", but in general the opposite implication is not true. (Compare with Problem 3.63.)

Problem 3.172 **

Show that in the Egorov theorem (see Theorem 3.76), the hypothesis $\mu(\Omega) < +\infty$ can be replaced by the hypothesis that

$$|f_n(x)| \leqslant g(x) \qquad \forall\, x \in \Omega,\ n \geqslant 1,$$

with $g \in L^1(\Omega, \mu)$.

Problem 3.173 **

Produce an example to show that in the Radon–Nikodym theorem (see Theorem 3.152), the assumption on the σ-finiteness of the measure cannot be dropped.

Problem 3.174 **

Suppose that (Ω, Σ, μ) is a measure space, $f \in L^1(\Omega) \setminus \{0\}$, $f \geqslant 0$. $\nu(A) = \int\limits_A f\, d\mu$ for all $A \in \Sigma$, M_μ is the essential supremum of f with respect to μ and M_ν is the essential supremum of f with respect to ν. Then $M_\mu, M_\nu \in [0, +\infty]$. Show that $M_\mu = M_\nu$.

Problem 3.175 **

(a) Suppose that $f \in L^p(\mathbb{R})$ $(1 \leqslant p < +\infty)$ and let us set

$$F(x) = \int_0^x f(s)\,ds.$$

Show that

$$\frac{F(x+h)-F(x)}{|h|^{\frac{1}{p'}}} \longrightarrow 0 \quad \text{as } h \to 0 \quad \text{uniformly in } x \in \mathbb{R}$$

(recall that $\frac{1}{p} + \frac{1}{p'} = 1$).

(b) Let $f \in L^1(\mathbb{R}) \cap C^1(\mathbb{R})$ and assume that $f' \in L^p(\mathbb{R})$ for some $p \geqslant 1$. Show that $\lim\limits_{|x| \to +\infty} f(x) = 0$.

Problem 3.176 **

Suppose that (Ω, Σ, μ) is a σ finite measure space and m and ν are two probability measures on Σ such that $m \ll \mu$ and $\nu \ll \mu$. Let

$$f \overset{\text{def}}{=} \frac{dm}{d\mu} \in L^1(\Omega, \mu) \quad \text{and} \quad h \overset{\text{def}}{=} \frac{d\nu}{d\mu} \in L^1(\Omega, \mu)$$

(see Theorem 3.152). Show that

$$2\|m - \nu\|_* = \int_\Omega |f - h|\,d\mu,$$

where for any signed measure ξ on (Ω, Σ),

$$\|\xi\|_* = 2 \sup_{A \in \Sigma} |\xi(A)|$$

(this is equivalent to the total variation norm; see Definition 3.148)

Problem 3.177 **

Let $\{\mu_n\}_{n \geqslant 1}$ be a sequence of probability measures on the measurable space (Ω, Σ). Show that there exist a probability measure ν on (Ω, Σ) and a sequence $\{f_n\}_{n \geqslant 1} \subseteq L^1(\Omega, \nu)_+$ such that

$$\mu_n(A) = \int_A f_n\,d\nu \qquad \forall\, n \geqslant 1.$$

Problem 3.178 **
Is the Radon–Nikodym theorem (see Theorem 3.152) true, if μ is not σ-finite? Justify your answer.

Problem 3.179 *
Let (Ω, Σ) be a measurable space and let ξ, μ, ν be three measures on Σ. Suppose that $\xi \leqslant \mu + \nu$ and $\xi \perp \nu$. Show that $\xi \leqslant \mu$.

Problem 3.180 **
Show that the Hardy–Littlewood maximal function f^* (see Definition 3.157) is measurable.

Problem 3.181 **
Let $f \in L^1_{\text{loc}}(\mathbb{R})$ and suppose that the sequence $\left\{ \int_0^n f\, dx \right\}_{n \geqslant 1}$ does not have a limit in $\mathbb{R}$. Show that

$$\int_0^\infty |f|\, dx = +\infty.$$

Is the converse true? Justify your answer.

Problem 3.182 **
Let (Ω, Σ, μ) be a complete measure space and $f \in L^1_{\text{loc}}(\Omega)$ (see Definition 3.96). Show that for every $\varepsilon > 0$, we can find $\delta > 0$ such that for all $C \in \Sigma$, with $\mu(C) < \delta$, we have $\int_C |f|\, d\mu < \varepsilon$.

Problem 3.183 **
Suppose that (Ω, Σ, μ) is a finite measure space, $K \subseteq L^1(\Omega)$ is a uniformly integrable set and K^* is the sequential closure with respect to the μ-almost everywhere convergence in K. Show that K^* is uniformly integrable too.

Problem 3.184 ***
Suppose that (Ω, Σ, μ) is a probability space and let $\{\Sigma_n\}_{n \geqslant 1}$ be an increasing sequence of sub-σ-algebras of Σ and τ, σ are stopping times for $\{\Sigma_n\}_{n \geqslant 1}$ and Σ_τ (respectively, Σ_σ) is the σ-algebra of events prior to τ (respectively, σ). Show the following:
(a) If $\tau \equiv k \in \mathbb{N}_0$, then $\Sigma_\tau = \Sigma_k$.
(b) $\min\{\tau, \sigma\}$, $\max\{\tau, \sigma\}$ and $\tau + \sigma$ are also stopping times for $\{\Sigma_n\}_{n \geqslant 1}$;

(c) If $\tau \leqslant \sigma$, then $\Sigma_\tau \subseteq \Sigma_\sigma$;
(d) $\Sigma_{\min\{\tau,\sigma\}} = \Sigma_\tau \cap \Sigma_\sigma$;
(e) $\{\tau < \sigma\} \in \Sigma_\tau \cap \Sigma_\sigma$ and $\{\tau = \sigma\} \in \Sigma_\tau \cap \Sigma_\sigma$.

Problem 3.185 **
Suppose that $\{f_n, \Sigma_n\}_{n \geqslant 0}$ is a martingale (respectively, submartingale, supermartingale) and τ is a stopping time for $\{\Sigma_n\}_{n \geqslant 1}$. Show that $\{f_n^\tau = f_{\min\{\tau,n\}}, \Sigma_n\}_{n \geqslant 0}$ is a martingale (respectively, submartingale, supermartingale).

Problem 3.186 **
Suppose that $\{f_n, \Sigma_n\}_{n \geqslant 0}$ is a submartingale and τ is a stopping time for $\{\Sigma_n\}_{n \geqslant 1}$ such that $\tau \leqslant k$ almost everywhere. Show that $Ef_0 \leqslant Ef_\tau \leqslant Ef_k$.

Problem 3.187 ***
Suppose that (Ω, Σ, μ) is a probability space and $\{\Sigma_n\}_{n \geqslant 0}$ is an increasing sequence of sub-σ-algebras of Σ. Show the following:
(a) If $\{f_n, \Sigma_n\}_{n \geqslant 0}$ is a supermartingale and

$$Ef_n = \int_\Omega f_n \, d\mu = c \in \mathbb{R} \qquad \forall\, n \geqslant 0,$$

then $\{f_n, \Sigma_n\}_{n \geqslant 0}$ is a martingale.
(b) If $f_n \in L^1(\Omega, \Sigma_n)$, then $\{f_n, \Sigma_n\}_{n \geqslant 0}$ is a martingale if and only if $Ef_\tau = Ef_0$ for every bounded stopping time τ for $\{\Sigma_n\}_{n \geqslant 0}$

Problem 3.188 **
Let (Ω, Σ, μ) be a probability space and let $\{\Sigma_n\}_{n \geqslant 0}$ be an increasing sequence of sub-σ-algebras of Σ. Suppose that $\{f_n, \Sigma_n\}_{n \geqslant 0}$ and $\{g_n, \Sigma_n\}_{n \geqslant 0}$ are two supermartingales (respectively, martingales) and τ is a stopping time for $\{\Sigma_n\}_{n \geqslant 0}$ such that $f_\tau \leqslant g_\tau$ (respectively, $f_\tau = g_\tau$) μ-almost everywhere on $\{\tau < +\infty\}$. Let

$$h_n \overset{\text{def}}{=} \begin{cases} g_n & \text{on} \quad \{n < \tau\}, \\ f_n & \text{on} \quad \{\tau \leqslant n\}. \end{cases}$$

Show that $\{h_n, \Sigma_n\}_{n \geqslant 0}$ is a supermartingale (respectively, martingale).

Problem 3.189 **
Suppose that (Ω, Σ, μ) is a probability space, $\{\Sigma_n\}_{n\geqslant 0}$ is an increasing sequence of sub-σ-algebras of Σ and $\{f_n, \Sigma_n\}_{n\geqslant 0}$ is a martingale such that $|f_n| \leqslant M$ μ-almost everywhere on Ω for all $n \geqslant 0$. We define $h_n \stackrel{\text{def}}{=} \sum\limits_{k=1}^{n} \frac{1}{k}(f_k - f_{k-1})$ for all $n \geqslant 1$. Show that $\{h_n, \Sigma_n\}_{n\geqslant 0}$ is a martingale which converges μ-almost everywhere and in $L^2(\Omega)$.

Problem 3.190 **
Suppose that (Ω, Σ, μ) is a probability space, $\{\Sigma_n\}_{n\geqslant 0}$ is an increasing sequence of sub-σ-algebras of Σ, $\Sigma_\infty = \sigma\big(\bigcup\limits_{n\geqslant 0} \Sigma_n\big)$ and m is a finite measure on Σ. Suppose that $m_n = m|_{\Sigma_n} \ll \mu_n = \mu|_{\Sigma_n}$ for all $n \geqslant 0$ and let

$$f_n \stackrel{\text{def}}{=} \frac{dm_n}{d\mu_n} \in L^1(\Omega, \Sigma_n).$$

Show the following:
(a) $\{f_n, \Sigma_n\}_{n\geqslant 0}$ is a martingale.
(b) There exists $f \in L^1(\Omega, \Sigma_\infty)$ such that $f_n \longrightarrow f$ μ-almost everywhere on Ω.
(c) $m \ll \mu_\infty = \mu|_{\Sigma_\infty}$ if and only if $\{f_n, \Sigma_n\}_{n\geqslant 0}$ is a regular martingale and in this case $f = \frac{dm}{d\mu_\infty} \in L^1(\Omega, \Sigma_\infty)$ and $f_n \longrightarrow f$ in $L^1(\Omega)$.

Problem 3.191 **
Suppose that (Ω, Σ, μ) is a probability space, $\{\Sigma_n\}_{n\geqslant 0}$ is an increasing sequence of sub-σ-algebras of Σ, $\{f_n, \Sigma_n\}_{n\geqslant 0}$ is a martingale and τ is a stopping time for $\{\Sigma_n\}_{n\geqslant 0}$ such that

$$\mu(\{\tau < +\infty\}) = 1, \quad \int_\Omega |f_\tau|\, d\mu < +\infty \quad \text{and} \quad \int_{\{\tau > n\}} |f_n|\, d\mu \longrightarrow 0.$$

Show the following:
(a) $\int\limits_{\{\tau>n\}} |f_\tau|\, d\mu \longrightarrow 0.$
(b) $\int\limits_\Omega \big|f_{\min\{\tau,n\}} - f_\tau\big|\, d\mu \longrightarrow 0.$
(c) $\int\limits_\Omega f_\tau\, d\mu = \int\limits_\Omega f_0\, d\mu.$

3.3 Solutions

Solution of Problem 3.1

Since the inverse function f^{-1} preserves all set theoretic operations, we have that $f^{-1}(m(\mathcal{Y}))$ is a monotone class (see Definition 3.10) which contains $f^{-1}(\mathcal{Y})$. Therefore, we have

$$m(f^{-1}(\mathcal{Y})) \subseteq f^{-1}(m(\mathcal{Y}))$$

(see Remark 3.11). We need to show the opposite inclusion. To this end let

$$\mathcal{F} = \{A \subseteq Y : A \in m(\mathcal{Y}) \text{ and } f^{-1}(A) \in m(f^{-1}(\mathcal{Y}))\}.$$

Evidently $\mathcal{F}$ is a monotone class and $m(\mathcal{Y}) \subseteq \mathcal{F}$. Hence

$$f^{-1}(m(\mathcal{Y})) \subseteq m(f^{-1}(\mathcal{Y}))$$

and so finally we have equality.

Solution of Problem 3.2

Let $\mathcal{F}$ be the σ-algebra generated by the singletons of X (see Definitions 3.1 and 3.6). Evidently $\mathcal{X} \subseteq \mathcal{F}$ and also $\mathcal{X}$ contains all singletons of X. So, it is enough to show that $\mathcal{X}$ is a σ-algebra. Note that $\emptyset$ and X both belong in $\mathcal{X}$ and $\mathcal{X}$ is closed under complementations. Let $\{A_n\}_{n\geqslant 1} \subseteq \mathcal{X}$. Suppose that every A_n is finite or countable. Then $\bigcup\limits_{n\geqslant 1} A_n$ is finite or countable and so $\bigcup\limits_{n\geqslant 1} A_n \in \mathcal{X}$. Next suppose that some A_{n_0} is uncountable for some $n_0 \geqslant 1$. Then from the definition of $\mathcal{X}$, we have that $X \setminus A_{n_0}$ is finite or countable. The inclusion

$$X \setminus \bigcup_{n\geqslant 1} A_n \subseteq X \setminus A_{n_0},$$

implies that

$$\bigcup_{n\geqslant 1} A_n \in \mathcal{X}$$

and so $\mathcal{X}$ is a σ-algebra, hence $\mathcal{X} = \mathcal{F}$.

Solution of Problem 3.3

Let X be a set and let $\mathcal{F}$ be a countable family of subsets of X which form a σ-algebra (see Definition 3.1). Let $x \in X$ and define

$$\mathcal{F}_x = \{A \in \mathcal{F} : x \in A\}.$$

Evidently $\mathcal{F}_x$ is finite or countable and so

$$A_x = \bigcap_{A \in \mathcal{F}_x} A \in \mathcal{F}.$$

Clearly A_x is the smallest set in $\mathcal{F}$ which contains $x \in X$. For every $A \in \mathcal{F}$, either $A_x \cap A = \emptyset$ or $A_x \subseteq A$. In particular, for $x, y \in X$, either $A_x \cap A_y = \emptyset$ or $A_x = A_y$.

Since by hypothesis $\mathcal{F}$ is countable, we can find a countable index set I such that

$$\{A_x\}_{x \in X} = \{A_k\}_{k \in I} \quad \text{and} \quad A_k \cap A_i = \emptyset \quad \text{for } k \neq i.$$

For every subset $D \subseteq I$, we have

$$A_D = \bigcup_{k \in D} A_k \in \mathcal{F}$$

and

$$A_D \neq A_E \text{ for any } D, E \subseteq I \text{ with } D \neq E.$$

Therefore the function $2^I \ni D \longmapsto A_D \in \mathcal{F}$ is a bijection. But 2^I is uncountable. Therefore $\mathcal{F}$ is uncountable too, a contradiction.

Solution of Problem 3.4

Let

$$\mathcal{X} = \{A \subseteq X : A \text{ or } X \setminus A \text{ is of first category}\}$$

(recall that a set is of first category if it is a countable union of nowhere dense sets; Definition 1.25). We will show that $\mathcal{X}$ is the σ-algebra generated by the nowhere dense sets (see Definitions 3.1 and 3.6). Evidently every nowhere dense set belongs in $\mathcal{X}$ and $\mathcal{X}$ is included in the σ-algebra generated by the nowhere dense sets. So, it is enough to show that $\mathcal{X}$ is a σ-algebra.

Note that $\emptyset, X \in \mathcal{X}$ and $\mathcal{X}$ is closed under complementation. Consider $\{A_n\}_{n\geqslant 1} \subseteq \mathcal{X}$. If each A_n is of first category, then

$$\bigcup_{n\geqslant 1} A_n \in \mathcal{X}.$$

Otherwise, for some $n_0 \geqslant 1$, the set $X \setminus A_{n_0}$ is of first category and then the inclusion

$$X \setminus \bigcup_{n\geqslant 1} A_n \subseteq X \setminus A_{n_0}$$

implies that

$$\bigcup_{n\geqslant 1} A_n \in \mathcal{X}.$$

Therefore $\mathcal{X}$ is a σ-algebra and so $\mathcal{X}$ is the σ-algebra generated by all nowhere dense subsets of X.

Solution of Problem 3.5

Let

$$\mathcal{F} = Y \cap \mathcal{B}(X) = \{Y \cap A : A \in \mathcal{B}(X)\}$$

(see Definition 3.6). Evidently $\mathcal{F}$ is a σ-algebra of subsets of Y and $\mathcal{F}$ contains all open sets of Y (recall that V is open in Y for the subspace topology if and only if $V = Y \cap U$ with $U \subseteq X$ open; see Definition 2.14). So, $\mathcal{B}(Y) \subseteq \mathcal{F}$.

Next consider the family

$$\mathcal{X} = \{A \in \mathcal{B}(X) : Y \cap A \in \mathcal{B}(Y)\}.$$

Clearly $\mathcal{X}$ is a σ-algebra of subsets of X and includes all open subsets of X. Therefore $\mathcal{X} = \mathcal{B}(X)$ and this implies that $\mathcal{F} = \mathcal{B}(Y)$. So, we conclude that $\mathcal{F} = \mathcal{B}(Y)$.

Solution of Problem 3.6

We use induction. For $m = 1$ the result is obvious. Now suppose that the result is true for $m > 1$ and let $\{C_i\}_{i=1}^n \subseteq \mathcal{D}$ be mutually disjoint sets such that each A_k (for $k = 1, \ldots, m$) can be written as a union of some sets from $\{C_i\}_{i=1}^m$. Hence

$$\bigcup_{k=1}^m A_k \subseteq \bigcup_{i=1}^n C_i.$$

The family $\{C_i \cap A_{m+1}\}_{i=1}^n$ consists of mutually disjoint elements of the semiring $\mathcal{D}$. On the other hand from the definition of a semiring (see Definition 3.1(a)), for every $i \in \{1, \ldots, n\}$, we can find a mutually disjoint family $\mathcal{Y}_i \subseteq \mathcal{D}$ such that

$$C_i \setminus A_{m+1} = \bigcup_{C \in \mathcal{Y}_i} C.$$

Then the family

$$\mathcal{Y} = \Big(\bigcup_{i=1}^n \mathcal{Y}_i\Big) \cup \{C_i \cap A_{m+1}\}_{i=1}^n \subseteq \mathcal{D}$$

is finite and mutually disjoint. Moreover, every A_k (for $k = 1, \ldots, m$) can be written as a union of elements of $\mathcal{Y}$. Observe that

$$A_{m+1} = \Big(A_{m+1} \setminus \bigcup_{i=1}^n C_i\Big) \cup (C_1 \cap A_{m+1}) \cup \ldots \cup (C_n \cap A_{m+1}).$$

We can find $\{E_j\}_{j=1}^l \subseteq \mathcal{D}$ mutually disjoint sets such that

$$A_{m+1} \setminus \bigcup_{i=1}^n C_i = \bigcup_{j=1}^l E_j$$

(see Definition 3.1(a)).

The family $\mathcal{Y} \cup \{E_j\}_{j=1}^l \subseteq \mathcal{D}$ is finite and mutually disjoint. Moreover, every set A_k (for $k \in \{1, \ldots, m+1\}$) can be written as a union of some of these sets. Hence, by induction we have proved the result.

Solution of Problem 3.7

The solution proceeds by induction. For $n = 1$, this follows from the definition of a semiring (see Definition 3.1(a)). We assume that the result is true for some $n > 1$ and let $A_1, \ldots, A_n, A_{n+1}, A \in \mathcal{X}$. By the induction hypothesis, we can find mutually disjoint sets $\{C_k\}_{k=1}^m \subseteq \mathcal{X}$ such that

$$A \setminus \bigcup_{i=1}^{n} A_i = \bigcup_{k=1}^{m} C_k.$$

Then we have

$$A \setminus \bigcup_{i=1}^{n+1} A_i = (A \setminus \bigcup_{i=1}^{n} A_i) \setminus A_{n+1} = \bigcup_{k=1}^{m} C_k \setminus A_{n+1} = \bigcup_{k=1}^{m} (C_k \setminus A_{n+1}).$$

Using once again the definition of a semiring for each $k \in \{1, \ldots, m\}$, we can find $\{E_{k,j}\}_{j=1}^{l_k} \subseteq \mathcal{X}$ mutually disjoint such that

$$\bigcup_{k=1}^{m} (C_k \setminus A_{n+1}) = \bigcup_{k=1}^{m} \bigcup_{j=1}^{l_k} E_{k,j},$$

so

$$A \setminus \bigcup_{i=1}^{n+1} A_i = \bigcup_{k=1}^{m} \bigcup_{j=1}^{l_k} E_{k,j}.$$

Solution of Problem 3.8

Let $\{A_n\}_{n \geqslant 1} \subseteq \mathcal{X}$ and let us set

$$A = \bigcup_{n \geqslant 1} A_n.$$

Let

$$C_1 = A_1 \quad \text{and} \quad C_{n_1} = A_{n+1} \setminus \bigcup_{k=1}^{n} A_k \qquad \forall\, n \geqslant 1.$$

Then $\{C_n\}_{n \geqslant 1}$ are mutually disjoint sets (i.e., $C_n \cap C_m = \emptyset$ if $n \neq m$) and

$$A = \bigcup_{n \geqslant 1} C_n.$$

By Problem 3.7, each C_n can be written as a union of a finite family of mutually disjoint $\mathcal{X}$-sets. Therefore, A can be written as a union of a sequence of mutually disjoint $\mathcal{X}$-sets.

Solution of Problem 3.9

Let τ be the collection of all open sets of the metric space X. If

$$\tau = 2^X$$

(i.e., X is a discrete space; see Example 1.3), then clearly τ is a σ-algebra (see Definition 3.1). Conversely, if τ is a σ-algebra and $x \in X$, then

$$\{x\} = \bigcap_{n \geqslant 1} B_{\frac{1}{n}}(x) \in \tau,$$

i.e., the singletons are open sets and so $\tau = 2^X$. Therefore we conclude that the metric spaces in which the open sets form a σ-algebra are the discrete metric spaces.

Solution of Problem 3.10

First note that a family in 2^X which is closed under finite intersections, proper differences and disjoint unions is a ring (see Definition 3.1). Indeed note that for all A, C belonging in the family, we have

$$\begin{aligned} A \setminus C &= (A \cup C) \setminus C, \\ A \cup C &= (A \setminus (A \cap C)) \cup (C \setminus (A \cap C)) \cup (A \cap C). \end{aligned}$$

Now, let $A_1, A_2 \in \mathcal{F}$. Then

$$\begin{aligned} A_1 \vartriangle A_2 &= (A_1 \setminus A_2) \cup (A_2 \setminus A_1) \\ &= \big((A_1 \cup A_2) \setminus A_2\big) \cup \big((A_1 \cup A_2) \setminus A_1\big) \in \mathcal{F}. \end{aligned}$$

Then since $A_1 \cup A_2 \in \mathcal{F}$ (see property **(b)** of $\mathcal{F}$), we infer that

$$A_1 \cap A_2 = (A_1 \cup A_2) \setminus (A_1 \vartriangle A_2) \in \mathcal{F}$$

(see property **(c)** of $\mathcal{F}$).

Therefore $\mathcal{F}$ is closed under finite intersections. Since $\mathcal{F}$ is closed under finite intersections, finite unions (see **(b)**) and under proper differences (see **(c)**), it is a ring. But $\mathcal{X} \subseteq \mathcal{F}$. Thus we conclude that

$$\mathcal{F} = \mathcal{R}.$$

Solution of Problem 3.11
We have $\mathcal{Y}_A \subseteq \Sigma_A$ and so $\sigma(\mathcal{Y}_A) \subseteq \Sigma_A$. Let

$$\mathcal{Y}_0 = \{C \in \Sigma : A \cap C \in \sigma(\mathcal{Y}_A)\}.$$

Then $\emptyset \in \mathcal{Y}_0$. Also, if $C \in \mathcal{Y}_0$, then

$$A \cap (\Omega \setminus C) = A \setminus (A \cap C) \in \sigma(\mathcal{Y}_A),$$

hence $\mathcal{Y}_0$ is closed under complementation. Also, if $\{C_n\}_{n\geqslant 1} \subseteq \mathcal{Y}_0$, then

$$A \cap \Big(\bigcup_{n\geqslant 1} C_n\Big) = \bigcup_{n\geqslant 1} (A \cap C_n) \in \Sigma_A$$

and so we have shown that $\mathcal{Y}_0$ is closed under countable unions. Therefore $\mathcal{Y}_0$ is a σ-algebra (see Definition 3.1). Note that $\mathcal{Y} \subseteq \mathcal{Y}_0$. Therefore $\sigma(\mathcal{Y}) = \Sigma$ and so it follows that $\sigma(\mathcal{Y}_A) = \Sigma_A$ (see Definition 3.6).

Solution of Problem 3.12
From Problem 1.62, we know that C_f is a G_δ-set in X. Hence $C_f \in \mathcal{B}(X)$ (see Definition 3.6).

Solution of Problem 3.13
"**(a)** $\Longrightarrow$ **(b)**": For $A, C \in \mathcal{X}$, the sets $A \cap C$, $A \setminus C$ and $C \setminus A$ are all in $\mathcal{X}$ (recall that $\mathcal{X}$ is a ring; see Definition 3.1) and are mutually disjoint. Also, we have

$$A = (A \setminus C) \cup (A \cap C), \quad C = (C \setminus A) \cup (A \cap C)$$

and

$$A \cup C = (A \setminus C) \cup (C \setminus A) \cup (A \cap C).$$

Then from the additivity of μ (see Definition 3.16), we have

$$\mu(A)+\mu(C) = \mu(A\setminus C)+2\mu(A\cap C)+\mu(C\setminus A) = \mu(A\cup C)+\mu(A\cap C).$$

Moreover, choosing $A \in \mathcal{X}$ such that $\mu(A) < +\infty$ and $C = \emptyset$, we have

$$\mu(A) = \mu(A \cup \emptyset) = \mu(A) + \mu(\emptyset),$$

so $\mu(\emptyset) = 0$.

"**(b)** $\Longrightarrow$ **(a)**": This is obvious.

Solution of Problem 3.14

From the monotonicity and σ-subadditivity of μ (see Theorem 3.19), we have

$$\mu\big(\limsup_{n\to+\infty} A_n\big) \leqslant \mu\Big(\bigcup_{k\geqslant n} A_k\Big) \leqslant \sum_{k\geqslant n} \mu(A_k) \qquad \forall\, n \geqslant 1.$$

But since by hypothesis

$$\sum_{n\geqslant 1} \mu(A_n) < +\infty,$$

we have

$$\sum_{k\geqslant n} \mu(A_k) \longrightarrow 0,$$

hence

$$\mu\big(\limsup_{n\to+\infty} A_n\big) = 0.$$

Solution of Problem 3.15

Since $A \subseteq C$, from the monotonicity of $\mu^\star$ (see Definition 3.27), we have

$$\mu^\star(A \cap E) \leqslant \mu^\star(C \cap E)$$

and

$$\mu^\star(A \cap E^c) \leqslant \mu^\star(C \cap E^c)$$

(here $E^c = X \setminus E$). Also, since $E \in \Sigma_{\mu^\star}$, we have

$$\mu^\star(A) = \mu^\star(A\cap E)+\mu^\star(A\cap E^c) \leqslant \mu^\star(C\cap E)+\mu^\star(C\cap E^c) = \mu^\star(C)$$

(see Definition 3.29). Because by hypothesis $\mu^\star(A) = \mu^\star(C)$, we conclude that

$$\mu^\star(A \cap E) = \mu^\star(C \cap E).$$

Solution of Problem 3.16
"$\Longrightarrow$": Let $D \subseteq A$, $E \subseteq A^c = X \setminus A$ and $A \in \Sigma_{\mu^\star}$. Then, we have

$$\mu^\star(D) = \mu^\star(D \cap A), \quad \mu^\star(E) = \mu^\star(E \cap A^c)$$

and

$$\begin{aligned} \mu^\star(D \cup E) &= \mu^\star\big((D \cup E) \cap A\big) + \mu^\star\big((D \cup E) \cap A^c\big) \\ &= \mu^\star(D \cap A) + \mu^\star(E \cap A^c) = \mu^\star(D) + \mu^\star(E) \end{aligned}$$

(see Definition 3.29).

"$\Longleftarrow$": Let $S \in 2^X$. Then, by hypothesis, we have

$$\mu^\star(S) = \mu^\star\big((S \cap A) \cup (S \cap A^c)\big) = \mu^\star(S \cap A) + \mu^\star(S \cap A^c).$$

So, $A \in \Sigma_{\mu^\star}$ (see Definition 3.29).

Solution of Problem 3.17
Step 1. We show that m is monotone (see Definition 3.16). To this end let $A, D \in \Sigma$, with $A \subseteq D$. We set

$$M_D = \Big\{ \lim_{n\to+\infty} \mu(E_n) : \{E_n\}_{n\geqslant 1} \subseteq \Sigma \text{ is increasing and } D = \bigcup_{n\geqslant 1} E_n \Big\}.$$

We have that

$$m(D) = \inf\big\{\eta : \eta \in M_D\big\}.$$

For $\eta \in M_D$, let $\{E_n\}_{n\geqslant 1} \subseteq \Sigma$ be an increasing sequence such that

$$D = \bigcup_{n\geqslant 1} E_n \quad \text{and} \quad \lim_{n\to+\infty} \mu(E_n) = \eta.$$

We have that $\{A \cap E_n = C_n\}_{n\geqslant 1} \subseteq \Sigma$ defines a corresponding sequence for M_A. So,

$$m(A) \leqslant \lim_{n\to+\infty} \mu(A \cap E_n) \leqslant \lim_{n\to+\infty} \mu(E_n) = \eta \in M_D,$$

thus $m(A) \leqslant m(D)$ (since $\eta \in M_D$ was arbitrary). This proves the monotonicity of m.

Step 2. We show that m is σ-subadditive (see Definition 3.16). So, let $\{A_n\}_{n\geqslant 1} \subseteq \Sigma$ be a sequence of sets such that

$$A = \bigcup_{n\geqslant 1} A_n.$$

For every $n \geqslant 1$, let $\{C_{nk}\}_{k\geqslant 1} \subseteq \Sigma$ be an increasing sequence of sets as in the definition of M_{A_n} and set

$$C_k = \bigcup_{n=1}^{k} C_{nk}.$$

Then $\{C_k\}_{k\geqslant 1} \subseteq \Sigma$ is an increasing sequence such that

$$A = \bigcup_{k\geqslant 1} C_k.$$

We have

$$\begin{aligned} m(A) &\leqslant \lim_{k\to+\infty} \sum_{n=1}^{k} \mu(C_{nk}) \leqslant \lim_{k\to+\infty} \sum_{n=1}^{k} \lim_{i\to+\infty} \mu(C_{ni}) \\ &\leqslant \sum_{n\geqslant 1} \lim_{i\to+\infty} \mu(C_{ni}). \end{aligned}$$

If

$$\sum_{n\geqslant 1} m(A_n) = +\infty,$$

then there is nothing to prove. So, we may assume that

$$\sum_{n\geqslant 1} m(A_n) < +\infty.$$

Then, for a given $\varepsilon > 0$, we can find an increasing sequence of sets $\{C_{ni}\}_{i\geqslant 1} \subseteq \Sigma$ as in the definition of M_{A_n} such that

$$\lim_{i\to+\infty} \mu(C_{ni}) \leqslant m(A_n) + \tfrac{\varepsilon}{2^n},$$

so

$$m(A) \leqslant \sum_{n\geqslant 1} \lim_{i\to+\infty} \mu(C_{ni}) \leqslant \sum_{n\geqslant 1} \left(m(A_n) + \tfrac{\varepsilon}{2^n}\right) = \sum_{n\geqslant 1} m(A_n) + \varepsilon.$$

Since $\varepsilon > 0$ was arbitrary, we let $\varepsilon \searrow 0$, to conclude that m is σ-subadditive.

Step 3. We show that m is superadditive. To this end let $A_1, A_2 \in \Sigma$ be such that $A_1 \cap A_2 = \emptyset$. For a given $\varepsilon > 0$, we can find an increasing sequence of sets $\{C_n\}_{n\geqslant 1} \subseteq \Sigma$ such that

$$A_1 \cup A_2 = \bigcup_{n\geqslant 1} C_n$$

and

$$\lim_{n\to+\infty} \mu(C_n) \leqslant m(A_1 \cup C_2) + \varepsilon.$$

Note that

$$\mu(C_n) = \mu(C_n \cap A_1) + \mu(C_n \cap A_2) \qquad \forall\, n \geqslant 1$$

(by the additivity of μ) and the sequences

$$\{C_n \cap A_1\}_{n\geqslant 1} \subseteq \Sigma \quad \text{and} \quad \{C_n \cap A_2\}_{n\geqslant 1} \subseteq \Sigma$$

are as in the definitions of M_{A_1} and M_{A_2}, respectively. Therefore

$$m(A_1) + m(A_2) \leqslant m(A_1 \cup A_2) + \varepsilon.$$

Since $\varepsilon > 0$ is arbitrary, we let $\varepsilon \searrow 0$ and obtain the superadditivity of m. Using the mathematical induction we can show the above for any finite family $A_1, \ldots, A_n \in \Sigma$ of mutually disjoint sets.

Step 4. We show that m is σ-additive. So, let $\{A_n\}_{n\geqslant 1} \subseteq \Sigma$ be a sequence of mutually disjoint sets such that

$$A = \bigcup_{n\geqslant 1} A_n \in \Sigma.$$

From the monotonicity of m (see (a)), we have

$$m\Big(\bigcup_{k=1}^{n} A_k\Big) \leqslant m(A) \qquad \forall\, n \geqslant 1,$$

so

$$\lim_{n\to+\infty} m\Big(\bigcup_{k=1}^{n} A_k\Big) \leqslant m(A).$$

From the superadditivity of m (see Step 3), we have

$$\sum_{k=1}^{n} m(A_k) \leqslant m\Big(\bigcup_{k=1}^{n} A_k\Big),$$

so

$$\sum_{k\geqslant 1} m(A_k) = \lim_{n\to+\infty} \sum_{k=1}^{n} m(A_k) \leqslant \lim_{n\to+\infty} m\Big(\bigcup_{k=1}^{n} A_k\Big) \leqslant m(A).$$

This combined with the σ-subadditivity of m (see *Step 2*), we conclude that m is σ-additive.

From Step 4, we conclude that $m\colon \Sigma \longrightarrow \mathbb{R}_+$ is a measure.

Solution of Problem 3.18

Clearly $m(\emptyset) = 0$. Let $\{A_n\}_{n\geqslant 1} \subseteq \Sigma$ be a sequence of mutually disjoint sets and let

$$A = \bigcup_{n\geqslant 1} A_n \in \Sigma.$$

If each A_n is finite or countable, then $m(A_n) = 0$ for all $n \geqslant 1$ and so

$$m(A) = \sum_{n\geqslant 1} m(A_n) = 0.$$

If for some $n_0 \geqslant 1$, the set $A_{n_0}^c$ is finite or countable, then by the disjointness of the sets $\{A_n\}_{n\geqslant 1}$, we have

$$A_n \subseteq A_{n_0}^c \qquad \forall\, n \neq n_0$$

and so the sets A_n are finite or countable for all $n \neq n_0$, which implies that

$$m(A_n) = 0 \qquad \forall\, n \neq n_0.$$

Therefore

$$m(A) = m(A_{n_0}) = 1 = \sum_{n\geqslant 1} m(A_n).$$

This proves that m is a measure (see Definition 3.15).

Solution of Problem 3.19

Let $\{A_n\}_{n\geqslant 1} \subseteq \mathcal{Y}$ be a sequence of mutually disjoint sets and assume that

$$A = \bigcup_{n\geqslant 1} A_n \in \mathcal{Y}.$$

Since by hypothesis μ is σ-subadditive (see Definition 3.16), we have

$$\mu(A) \leqslant \sum_{n\geqslant 1} \mu(A_n).$$

On the other hand, for $n \geqslant 1$ fixed, by Problem 3.7, we can find $\{C_i\}_{i=1}^m \subseteq \mathcal{Y}$ mutually disjoint such that $A \setminus \bigcup_{k=1}^n A_k = \bigcup_{i=1}^m C_i$. Then

$$A = \Big(\bigcup_{k=1}^n A_k\Big) \cup \Big(\bigcup_{i=1}^m C_i\Big)$$

and so A is a finite union of mutually disjoint $\mathcal{Y}$-sets. Hence the additivity of μ implies that

$$\sum_{k=1}^n \mu(A_k) + \sum_{i=1}^m \mu(C_i) = \mu(A) \geqslant \sum_{k=1}^n \mu(A_k).$$

Since $n \geqslant 1$ was arbitrary, we infer that $\sum\limits_{k\geqslant 1} \mu(A_k) \leqslant \mu(A)$ and so μ is σ-additive.

Solution of Problem 3.20

Let $A, C \in \mathcal{Y}$ with $A \subseteq C$. Since $\mathcal{Y}$ is a semiring (see Definition 3.1), we can find $\{E_k\}_{k=1}^m \subseteq \mathcal{Y}$ mutually disjoint such that $C \setminus A = \bigcup\limits_{k=1}^m E_k$.

Then $C = A \cup \big(\bigcup_{k=1}^{m} E_k\big)$ and the right-hand side is a disjoint union of $\mathcal{Y}$-sets. Then from the additivity of μ, we have

$$\mu(A) \leqslant \mu(A) + \sum_{k=1}^{m} \mu(E_k) = \mu(C),$$

which proves that μ is monotone (see Definition 3.16).

Solution of Problem 3.21

Clearly $\mu(\emptyset) = 0$. Let $\{A_k\}_{k\geqslant 1} \subseteq \mathcal{Y}$ be mutually disjoint sets such that

$$\bigcup_{k\geqslant 1} A_k = A \in \mathcal{Y}.$$

Since each μ_n is σ-additive (see Definition 3.16), we have

$$\mu_n(A) = \sum_{k\geqslant 1} \mu_n(A_k) \leqslant \sum_{k\geqslant 1} \mu(A_k),$$

so

$$\mu(A) \leqslant \sum_{k\geqslant 1} \mu(A_k).$$

On the other hand, for every $m \geqslant 1$, we have

$$\sum_{k=1}^{m} \mu(A_k) = \lim_{n\to+\infty} \sum_{k=1}^{m} \mu_n(A_k) = \lim_{n\to+\infty} \mu_n\Big(\bigcup_{k=1}^{m} A_k\Big) \leqslant \mu(A)$$

(see Problem 3.20), so $\sum\limits_{k\geqslant 1} \mu(A_k) \leqslant \mu(A)$. Therefore we conclude that μ is σ-additive.

Solution of Problem 3.22

(a) Let

$$\mathcal{Y} = \big\{A \in \mathcal{B}(X) : \ \mu(A) = \vartheta(A)\big\}.$$

We claim that $\mathcal{Y}$ is a monotone class (see Definition 3.10). To this end suppose that $\{A_n\}_{n\geqslant 1} \subseteq \mathcal{Y}$ is an increasing sequence. Then

$$\mu(A_n) = \vartheta(A_n) \qquad \forall\, n \geqslant 1$$

and the continuity of μ and ϑ from below (see Theorem 3.19(b)) implies that

$$\mu\Big(\bigcup_{n\geqslant 1} A_n\Big) = \lim_{n\to+\infty} \mu(A_n) \quad \text{and} \quad \vartheta\Big(\bigcup_{n\geqslant 1} A_n\Big) = \lim_{n\to+\infty} \vartheta(A_n),$$

so

$$\mu\Big(\bigcup_{n\geqslant 1} A_n\Big) = \vartheta\Big(\bigcup_{n\geqslant 1} A_n\Big),$$

thus

$$\bigcup_{n\geqslant 1} A_n \in \mathcal{Y}.$$

Next suppose that $\{C_n\}_{n\geqslant 1} \subseteq \mathcal{Y}$ is a decreasing sequence. Then

$$\mu(A_n) = \vartheta(A_n) \qquad \forall\, n \geqslant 1.$$

Because μ and ϑ are finite, from their continuity from above (see Theorem 3.19(c)), we have

$$\mu\Big(\bigcap_{n\geqslant 1} C_n\Big) = \lim_{n\to+\infty} \mu(C_n) \quad \text{and} \quad \vartheta\Big(\bigcap_{n\geqslant 1} C_n\Big) - \lim_{n\to+\infty} \vartheta(C_n),$$

so

$$\mu\Big(\bigcap_{n\geqslant 1} C_n\Big) = \vartheta\Big(\bigcap_{n\geqslant 1} C_n\Big)$$

and thus

$$\bigcap_{n\geqslant 1} C_n \in \mathcal{Y}.$$

This proves that $\mathcal{Y}$ is a monotone class. Since by hypothesis the open or closed sets are included in $\mathcal{Y}$, from Theorem 3.12(b) (the monotone class theorem), we have

$$\mathcal{B}(X) = \mathcal{Y}$$

and so

$$\mu = \vartheta.$$

(b) If X is σ-compact (see Definition 2.99), then every closed set $C \subseteq X$ can be written as

$$C = \bigcup_{n\geqslant 1} K_n,$$

with an increasing sequence $\{K_n\}_{n\geqslant 1}$ of compact sets. Therefore $\mathcal{B}(X)$ is the monotone class generated by the compact sets and since on compact sets μ and ϑ coincide, we have $\mu = \vartheta$.

(c) If μ and ϑ are σ-finite, then we can find two open covers $\{U_n\}_{n\geqslant 1}$ and $\{V_m\}_{m\geqslant 1}$ of X such that

$$\mu(U_n) \ < \ +\infty, \quad \vartheta(V_m) \ < \ +\infty \qquad \forall\, n, m \geqslant 1.$$

So, we can find an open cover of X, namely

$$\{W_k\}_{k\geqslant 1} \ = \ \{U_n \cap V_m\}_{(n,m)\in\mathbb{N}\times\mathbb{N}}$$

such that

$$\mu(W_k) \ < \ +\infty \quad \text{and} \quad \vartheta(W_k) \ < \ +\infty \qquad \forall\, k \geqslant 1.$$

We may always assume that $\{W_k\}_{k\geqslant 1}$ is an increasing family. From **(a)**, we know that

$$\mu\big|_{\mathcal{B}(W_k)} \ = \ \vartheta\big|_{\mathcal{B}(W_k)} \qquad \forall\, k \geqslant 1.$$

But note that

$$\mathcal{B}(W_k) \ = \ W_k \cap \mathcal{B}(X)$$

(see Problem 3.5). So, for all $A \in \mathcal{B}(X)$, we have

$$\mu(W_k \cap A) \ = \ \vartheta(W_k \cap A) \qquad \forall\, k \geqslant 1,$$

so, using the continuity of μ and ϑ from below (see Theorem 3.19(b)), we have

$$\mu(A) \ = \ \vartheta(A)$$

and thus $\mu = \vartheta$.

Similarly, if X is σ-compact and μ, ϑ coincide on compact sets, using this time **(b)**.

Solution of Problem 3.23

Note that $A \cup C = (C \setminus A) \cup A$. Then from the monotonicity and subadditivity of $\mu^\star$ (see Definition 3.27), we have

$$\begin{aligned} \mu^\star(C) \ &\leqslant \ \mu^\star(A \cup C) \ = \ \mu^\star\big((C \setminus A) \cup A\big) \ \leqslant \ \mu^\star(C \setminus A) + \mu^\star(A) \\ &= \ \mu^\star(C \setminus A) \ \leqslant \ \mu^\star(C), \end{aligned}$$

so

$$\mu^\star(C) = \mu^\star(A \cup C) = \mu^\star(C \setminus A).$$

Solution of Problem 3.24
Let $A_1, A_2 \subseteq \Sigma_{\mu^\star}$ be such that $A_1 \cap A_2 = \emptyset$. Then for any $D \subseteq X$, from the definition of $\Sigma_{\mu^\star}$ (see Definition 3.29), we have

$$\begin{aligned}\mu^\star(D \cap (A_1 \cup A_2)) &= \mu^\star(D \cap (A_1 \cup A_2) \cap A_1) + \mu^\star(D \cap (A_1 \cup A_2) \cap A_1^c) \\ &= \mu^\star(D \cap A_1) + \mu^\star(D \cap A_2).\end{aligned}$$

The general case follows by induction.

Solution of Problem 3.25
(a) From the σ-subadditivity (see Definition 3.16) of the outer measure $\mu^\star$ (see Definition 3.27), we have

$$\mu^\star\Big(A \cap \big(\bigcup_{n\geqslant 1} C_n\big)\Big) = \mu^\star\Big(\bigcup_{n\geqslant 1}(A \cap C_n)\Big) \leqslant \sum_{n\geqslant 1} \mu^\star(A \cap C_n).$$

Also, from Problem 3.24 and the monotonicity of $\mu^\star$, we have

$$\sum_{n=1}^{m} \mu^\star(A\cap C_n) = \mu^\star\Big(A\cap\big(\bigcup_{n=1}^{m} C_n\big)\Big) \leqslant \mu^\star\Big(A\cap\big(\bigcup_{n\geqslant 1} C_n\big)\Big) \qquad \forall\, m \geqslant 1,$$

so

$$\sum_{n\geqslant 1} \mu^\star(A \cap C_n) \leqslant \mu^\star\Big(A \cap \big(\bigcup_{n\geqslant 1} C_n\big)\Big)$$

and thus

$$\mu^\star\Big(\bigcup_{n\geqslant 1}(A \cap C_n)\Big) = \sum_{n\geqslant 1} \mu^\star(A \cap C_n).$$

(b) Let

$$A = \bigcup_{n\geqslant 1} A_n.$$

Note that from properties of the family $\{C_n\}_{n\geqslant 1}$, we have

$$A \cap C_n = A_n \qquad \forall\, n \geqslant 1.$$

Then from part **(a)**, we have

$$\mu^\star\Big(\bigcup_{n\geqslant 1} A_n\Big) \;=\; \mu^\star\Big(\bigcup_{n\geqslant 1}(A\cap C_n)\Big) \;=\; \sum_{n\geqslant 1}\mu^\star(A\cap C_n) \;=\; \sum_{n\geqslant 1}\mu^\star(A_n).$$

Solution of Problem 3.26

Clearly $\mu(\emptyset)=0$ and for all $A\in\Sigma$, we have

$$\mu(A) \;\leqslant\; \mu(\Omega) \;<\; +\infty.$$

Also, let $\{A_k\}_{k\geqslant 1}\subseteq\Sigma$ be mutually disjoint sets with

$$A \;=\; \bigcup_{k\geqslant 1} A_k.$$

Then for every $n\geqslant 1$, we have

$$\mu_n(A) \;=\; \sum_{k\geqslant 1}\mu_n(A_k),$$

so

$$\sum_{n=1}^{m}\vartheta_n\mu_n(A) \;=\; \sum_{n=1}^{m}\vartheta_n\sum_{k\geqslant 0}\mu_n(A_k) \qquad \forall\, m\geqslant 1,$$

thus

$$\sum_{n\geqslant 1}\vartheta_n\mu_n(A) \;=\; \sum_{n\geqslant 1}\sum_{k\geqslant 1}\vartheta_n\mu_n(A_k) \;=\; \sum_{k\geqslant 1}\sum_{n\geqslant 1}\vartheta_n\mu_n(A_k) \;=\; \sum_{k\geqslant 1}\mu(A_k)$$

and hence μ is σ-additive (see Definition 3.15). Thus μ is a measure on Σ.

Solution of Problem 3.27

(a) "$\Longrightarrow$": Immediate, as we can take $C=A$.

"$\Longleftarrow$": By hypothesis, for every $n\geqslant 1$, we can find $C_n\in\Sigma_{\mu^\star}$ such that

$$\mu^\star(A\setminus C_n) \;\leqslant\; \tfrac{1}{n}.$$

Let us set

$$C = \bigcup_{n \geqslant 1} C_n \in \Sigma_{\mu^\star},$$

so $C \subseteq A$. From the monotonicity of $\mu^\star$ (see Definition 3.27), we have

$$\mu^\star(A \setminus C) \leqslant \mu^\star(A \setminus C_n) \leqslant \tfrac{1}{n} \qquad \forall\, n \geqslant 1,$$

hence

$$\mu^\star(A \setminus C) = 0.$$

Since $\Sigma_{\mu^\star}$ is $\mu^\star$-complete, it follows that $A \in \Sigma_{\mu^\star}$.

(b) By part **(a)**, it suffices to show that for every $\varepsilon > 0$, we can find $E \in \Sigma_{\mu^\star}$ such that

$$\mu^\star(A \setminus E) \leqslant \varepsilon.$$

For every $n \geqslant 1$, we choose $C_n \subset \Sigma_{\mu^\star}$ such that

$$\mu^\star(A \vartriangle C_n) \leqslant \tfrac{\varepsilon}{2^n}.$$

Let us set

$$C = \bigcap_{n \geqslant 1} C_n \in \Sigma_{\mu^\star}.$$

Then

$$C \setminus A \subseteq C_n \setminus A$$

and so from the monotonicity of $\mu^\star$, we have

$$\mu^\star(C \setminus A) \leqslant \mu^\star(C_n \setminus A) \leqslant \mu^\star(A \vartriangle C_n) \leqslant \tfrac{\varepsilon}{2^n} \qquad \forall\, n \geqslant 1,$$

so

$$\mu^\star(C \setminus A) = 0$$

and thus $C \setminus A \in \Sigma_{\mu^\star}$ (recall that $\Sigma_{\mu^\star}$ is $\mu^\star$-complete). Hence

$$A \cap C = C \setminus (C \setminus A) \in \Sigma_{\mu^\star}.$$

Note that $A \cap C \subseteq A$ and from the σ-subadditivity of $\mu^\star$, we have

$$\mu^\star(A\setminus(A\cap C)) = \mu^\star(A\setminus C) = \mu^\star\Big(\bigcup_{n\geqslant 1}(A\setminus C_n)\Big) \leqslant \sum_{n\geqslant 1}\mu^\star(A\setminus C_n) \leqslant \varepsilon,$$

so $A \in \Sigma_{\mu^\star}$ (see part **(a)**).

Solution of Problem 3.28

From the solution of Problem 3.27(a), we know that we can find $C \in \Sigma$ such that

$$A \subseteq C \quad \text{and} \quad \mu^\star(A) = \mu(C).$$

Similarly, for every $n \geqslant 1$, we can find $C_n \in \Sigma$ such that

$$A_n \subseteq C_n \subseteq C \quad \text{and} \quad \mu^\star(A_n) = \mu(C_n).$$

Let

$$E_n = \bigcap_{k\geqslant n} C_k \quad \text{and} \quad E = \bigcup_{n\geqslant 1} E_n$$

(i.e., $E = \liminf\limits_{n\to+\infty} C_n$). Since

$$A_n \subseteq E_n \subseteq C_n \quad \text{and} \quad \mu^\star(A_n) = \mu(C_n),$$

we have

$$\mu^\star(A_n) = \mu(E_n) \qquad \forall\, n \geqslant 1.$$

Moreover, since

$$A_n \subseteq E_n \subseteq C_n \subseteq C \qquad \forall\, n \geqslant 1,$$

we have

$$A \subseteq E \subseteq C$$

and since $\mu^\star(A) = \mu(C)$, we infer that

$$\mu^\star(A) = \mu(E).$$

From Theorem 3.19(b), we have

$$\mu(E_n) \nearrow \mu(E) \quad \text{as } n \to +\infty.$$

Therefore

$$\mu^\star(A_n) \nearrow \mu^\star(A) \quad \text{as } n \to +\infty.$$

Solution of Problem 3.29

Let

$$\eta = \sup\{\mu(C) : C \in \Sigma, \ C \subseteq A, \ \mu(C) < +\infty\}.$$

By the hypothesis of the problem, the supremum is taken over a nonempty set and so $0 < \eta \leqslant +\infty$. Let $\{C_n\}_{n\geqslant 1} \subseteq \Sigma$, $C_n \subseteq A$, $\mu(C_n) < +\infty$ for all $n \geqslant 1$ such that

$$\mu(C_n) \nearrow \eta \quad \text{as } n \to +\infty.$$

From the definition of η, we have

$$\mu\Big(\bigcup_{k=1}^{n} C_k\Big) \leqslant \eta.$$

Let us set

$$C = \bigcup_{n\geqslant 1} C_n.$$

We have

$$\eta = \lim_{n\to+\infty} \mu(C_n) \leqslant \mu(C) = \lim_{n\to+\infty} \mu\Big(\bigcup_{k=1}^{n} C_k\Big) \leqslant \eta$$

(see Theorem 3.19), so $\mu(C) = \eta$.

If $\eta = +\infty$, then we are done. If $\eta < +\infty$, then for all $E \subseteq A \setminus C$, with $\mu(E) < +\infty$, we have

$$\eta + \mu(E) = \mu(C) + \mu(E) = \mu(C \cup E) \leqslant \eta$$

(recall the definition of η), so $\mu(E) = 0$ and hence by hypothesis, we have

$$\mu(A \setminus C) = 0.$$

But then

$$\mu(A) = \mu(A \setminus C) + \mu(C) \;=\; \mu(C) \;<\; +\infty,$$

a contradiction. So, $\eta < +\infty$ cannot occur and we have proved the result.

Solution of Problem 3.30

First let us show that the set function μ_A is well defined. To this end, we need to show that, if $C, E \in \Sigma$ and $A \cap C = A \cap E$, then $\mu(C) = \mu(E)$. Note that

$$A \subseteq C^c \cup (C \cap E) \quad \text{and} \quad A \subseteq E^c \cup (C \cap E).$$

Therefore, by hypothesis, we have

$$1 - \mu(C) + \mu(C \cap E) \;=\; 1 \quad \text{and} \quad 1 - \mu(E) + \mu(C \cap E) \;=\; 1,$$

so

$$\mu(C) \;=\; \mu(E)$$

and this proves that μ_A is well defined.

We have

$$\mu_A(\emptyset) \;=\; \mu(A \cap \emptyset) \;=\; \mu(\emptyset) \;=\; 0$$

and

$$\mu_A(A) \;=\; \mu_A(A \cap \Omega) \;=\; \mu(\Omega) \;=\; 1.$$

Also, suppose that $C, E \in \Sigma$ are such that $A \cap (C \cap E) = \emptyset$. Then

$$0 \;=\; \mu_A\big(A \cap (C \cap E)\big) \;=\; \mu(C \cap E).$$

Since $\mu(C) + \mu(E) = \mu(C \cup E) + \mu(C \cap E)$ and $\mu(C \cap E) = 0$, we have

$$\begin{aligned} \mu_A\big((A \cap C) \cup (A \cap E)\big) &= \mu_A\big(A \cap (C \cup E)\big) \\ &= \mu(C \cup E) \;=\; \mu(C) + \mu(E) \\ &= \mu_A(A \cap C) + \mu_A(A \cap E), \end{aligned}$$

so μ_A is additive.

Finally, if $\{C_n\}_{n \geqslant 1} \subseteq \Sigma$ is such that the sequence $\{A \cap C_n\}_{n \geqslant 1}$ is increasing. Let us set

$$\widehat{C}_n \;=\; \bigcup_{k=1}^{n} C_k \qquad \forall\, n \geqslant 1.$$

Then

$$A \cap \widehat{C}_n = A \cap \big(\bigcup_{k=1}^{n} C_k\big) = \bigcup_{k=1}^{n} A \cap C_k = A \cap C_n$$

and the sequence $\{\widehat{C}_n\}_{n \geqslant 1} \subseteq \Sigma$ is increasing. Let us set

$$\widehat{C} = \bigcup_{n \geqslant 1} \widehat{C}_n \in \Sigma.$$

Then

$$\begin{aligned} \lim_{n \to +\infty} \mu_A(A \cap \widehat{C}_n) &= \lim_{n \to +\infty} \mu(\widehat{C}_n) = \mu(\widehat{C}) \\ &= \mu_A(A \cap \widehat{C}) = \mu_A\big(A \cap \big(\bigcup_{n \geqslant 1} \widehat{C}_n\big)\big), \end{aligned}$$

so μ_A is continuous from below and thus μ_A is a probability measure (see Proposition 3.20).

Solution of Problem 3.31

(a) Since $C \in \Sigma_{\mu^\star}$, we have

$$\begin{aligned} \mu^\star(A \cup C) &= \mu^\star\big((A \cup C) \cap C\big) + \mu^\star\big((A \cup C) \cap C^c\big) \\ &= \mu^\star(C) + \mu^\star(A \cap C^c), \end{aligned}$$

so

$$\begin{aligned} \mu^\star(A \cup C) + \mu^\star(A \cap C) &= \mu^\star(C) + \mu^\star(A \cap C^c) + \mu^\star(A \cap C) \\ &= \mu^\star(C) + \mu^\star(A) \end{aligned}$$

(see Definition 3.29).

(b) If $\mu^\star(C \setminus A) = 0$, then $C \setminus A \in \Sigma_{\mu^\star}$ (since the latter is $\mu^\star$-complete; see Definition 3.30). Hence

$$A = C \setminus (C \setminus A) \in \Sigma_{\mu^\star},$$

a contradiction. This proves that $\mu^\star(C \setminus A) > 0$.

Solution of Problem 3.32

Let $A \subseteq \Omega$. If $\mu_2^\star(A) = +\infty$, then clearly

$$\mu_1^\star(A) \leqslant \mu_2^\star(A).$$

So, suppose that $\mu_2^\star(A) < +\infty$. Then by Proposition 3.28, for a given $\varepsilon > 0$, we can find a sequence $\{A_n\}_{n\geqslant 1} \subseteq \Sigma_2$ such that

$$A \subseteq \bigcup_{n\geqslant 1} A_n \quad \text{and} \quad \sum_{n\geqslant 1} \mu_2(A_n) \leqslant \mu_2^\star(A) + \varepsilon.$$

Hence from the σ-subadditivity of $\mu_1^\star$ (see Definition 3.27) and the hypothesis that $\mu_2 = \mu_1^\star|_{\Sigma_1}$, we have

$$\mu_1^\star(A) \leqslant \sum_{n\geqslant 1} \mu_1^\star(A_n) = \sum_{n\geqslant 1} \mu_2(A_n) \leqslant \mu_2^\star(A) + \varepsilon.$$

Because $\varepsilon > 0$ was arbitrary, we conclude that

$$\mu_1^\star(A) \leqslant \mu_2^\star(A).$$

By reversing the roles of $\mu_1^\star$ and $\mu_2^\star$ in the above argument, we also have that

$$\mu_2^\star(A) \leqslant \mu_1^\star(A) \qquad \forall\, A \subseteq \Omega.$$

Therefore we conclude that $\mu_1^\star = \mu_2^\star$.

Solution of Problem 3.33

Let

$$C_k = \bigcap_{n\geqslant k} A_n \qquad \forall\, k \geqslant 1.$$

Then $\{C_k\}_{k\geqslant 1} \subseteq \Sigma$ is an increasing sequence and

$$C_k \nearrow \liminf_{n\to+\infty} A_n \in \Sigma \quad \text{as } k \to +\infty.$$

Theorem 3.19(b) implies that

$$\lim_{k\to+\infty} \mu(C_k) = \mu\big(\liminf_{n\to+\infty} A_n\big).$$

But note that

$$\mu(C_k) \leqslant \mu(A_k) \qquad \forall\, k \geqslant 1.$$

Therefore we conclude that

$$\mu\big(\liminf_{n\to+\infty} A_n\big) \leqslant \liminf_{n\to+\infty} \mu(A_n).$$

Next suppose that μ is a finite measure and let

$$E_k = \bigcup_{n\geqslant k} A_n.$$

Then $\{E_k\}_{k\geqslant 1} \subseteq \Sigma$ is a decreasing sequence and

$$E_k \searrow \limsup_{n\to+\infty} A_n \in \Sigma \quad \text{as } k\to+\infty.$$

Invoking Theorem 3.19(c) (here we need the finiteness of μ), we have

$$\lim_{k\to+\infty} \mu(E_k) = \mu\big(\limsup_{n\to+\infty} A_n\big).$$

But in this case

$$\mu(E_k) \geqslant \mu(A_k) \qquad \forall\, k\geqslant 1.$$

Therefore we conclude that

$$\limsup_{n\to+\infty} \mu(A_n) \leqslant \mu\big(\limsup_{n\to+\infty} A_n\big).$$

Alternative Solution
Note that

$$\liminf_{n\to+\infty} \chi_{A_n} = \chi_{\liminf\limits_{n\to+\infty} A_n} \quad \text{and} \quad \limsup_{n\to+\infty} \chi_{A_n} = \chi_{\limsup\limits_{n\to+\infty} A_n}.$$

Invoking the Fatou lemma (see Theorem 3.95), we have

$$\begin{aligned}\mu\big(\liminf_{n\to+\infty} A_n\big) &= \int_\Omega \chi_{\liminf\limits_{n\to+\infty} A_n}\, d\mu = \int_\Omega \liminf_{n\to+\infty} \chi_{A_n}\, d\mu \\ &\leqslant \liminf_{n\to+\infty} \int_\Omega \chi_{A_n}\, d\mu = \liminf_{n\to+\infty} \mu(A_n)\end{aligned}$$

and

$$\begin{aligned}\mu\big(\limsup_{n\to+\infty} A_n\big) &= \int_\Omega \chi_{\limsup\limits_{n\to+\infty} A_n}\, d\mu = \int_\Omega \limsup_{n\to+\infty} \chi_{A_n}\, d\mu \\ &\geqslant \limsup_{n\to+\infty} \int_\Omega \chi_{A_n}\, d\mu = \limsup_{n\to+\infty} \mu(A_n).\end{aligned}$$

Note that in the lim sup case the application of the Fatou lemma is possible since $0 \leqslant \chi_{A_n} \leqslant 1$ and $1 \in L^1(\Omega)$ due to the fact that μ is finite in this case.

Solution of Problem 3.34

Let $K \subseteq \mathbb{R}^N$ be a compact set and let

$$m = \inf_{x \in K} f(x).$$

Choose a minimizing sequence $\{x_n\}_{n \geqslant 1} \subseteq K$, i.e.,

$$f(x_n) \searrow m.$$

Since K is compact, by passing to a suitable subsequence if necessary, we may assume that

$$x_n \longrightarrow x \quad \text{in } K.$$

We have

$$B_1(x) \subseteq \liminf_{n \to +\infty} B_1(x_n),$$

so

$$\mu\big(B_1(x)\big) \leqslant \mu\big(\liminf_{n \to +\infty} B_1(x_n)\big) \leqslant \liminf_{n \to +\infty} \mu\big(B_1(x_n)\big) = m$$

(see Problem 3.33) and thus

$$f(x) = \mu\big(B_1(x)\big) = m.$$

Solution of Problem 3.35

Let

$$\mathcal{Y} = \big\{A \in \Sigma : \mu_1(A) = \mu_2(A)\big\}.$$

Evidently $\mathcal{A} \subseteq \mathcal{Y}$. Theorem 3.19 implies that $\mathcal{Y}$ is a monotone class. Then from the monotone class theorem (see Theorem 3.12), it follows that

$$\Sigma = \sigma(\mathcal{A}) = \mathcal{Y}$$

and so we conclude that $\mu_1 = \mu_2$.

Solution of Problem 3.36
For each integer $n \geqslant 1$, we introduce the measures μ_1^n and μ_2^n on Σ, defined by

$$\mu_1^n(A) = \mu_1(A \cap C_n) \qquad \forall A \in \Sigma.$$

and

$$\mu_2^n(A) = \mu_2(A \cap C_n) \qquad \forall A \in \Sigma.$$

These are finite measures on Σ and so we can apply Proposition 3.26 and infer that

$$\mu_1^n = \mu_2^n \qquad \forall n \geqslant 1.$$

Since

$$\mu_1(A) = \lim_{n \to +\infty} \mu_1^n(A) = \lim_{n \to +\infty} \mu_2^n(A) = \mu_2(A) \qquad \forall A \in \Sigma$$

(see Theorem 3.19(b)), we conclude that $\mu_1 = \mu_2$.

Solution of Problem 3.37
Let $\Omega = [0,1]$, $\Sigma = \mathcal{B}([0,1])$ (the Borel σ-algebra of $[0,1]$; see Definition 3.6). Let $\mathcal{Y}$ be the sub-σ-algebra of Σ formed by the countable and co-countable subsets of $[0,1]$ (co-countable set means that its complement is countable). Note that Σ is countable generated (see Definition 3.14) since

$$\Sigma = \sigma\big(\{(a,b) : \ a,b \in \mathbb{Q}\}\big).$$

Suppose that $\mathcal{Y}$ is countably generated too, i.e., $\mathcal{Y} = \sigma\big(\{A_n\}_{n\geqslant 1}\big)$ with $A_n \subseteq [0,1]$, $n \geqslant 1$. From the definition of $\mathcal{Y}$, we see that without any loss of generality, we may assume that each A_n is countable. Let

$$C = \bigcup_{n\geqslant 1} A_n.$$

Then C is countable and so we can find $x \in [0,1]$, $x \notin C$. Let

$$\mathcal{F} = \{A : \ A \subseteq C\} \cup \{A^c : \ A \subseteq C\}.$$

Then $\mathcal{F}$ is a sub-σ-algebra of $\mathcal{Y}$ and we have

$$\sigma\big(\{A_n\}_{n\geqslant 1}\big) \subseteq \mathcal{F} \subseteq \mathcal{Y},$$

hence $\mathcal{F} = \mathcal{Y}$. But note that $\{x\} \in \mathcal{Y}$ while $\{x\} \notin \mathcal{F}$, hence $\mathcal{F} \neq \mathcal{Y}$, a contradiction. This proves that $\mathcal{Y}$ cannot be countably generated.

Solution of Problem 3.38

"$\Longrightarrow$": Suppose that $\Sigma = \sigma(\{A_n\}_{n\geqslant 1})$ and let

$$f = \sum_{n\geqslant 1} \tfrac{1}{3^n}\chi_{A_n}.$$

Evidently f is Σ-measurable. Let $\mathcal{Y} = \{f^{-1}(A) : A \in \mathcal{B}(\mathbb{R})\}$. This is a σ-algebra. Note that

$$A_1 = f^{-1}([\tfrac{1}{3}, \tfrac{2}{3}]), \quad A_2 = f^{-1}([\tfrac{1}{9}, \tfrac{2}{9}] \cup [\tfrac{1}{3} + \tfrac{1}{9}, \tfrac{1}{3} + \tfrac{2}{9}]), \quad \text{etc.}$$

Thus $A_n \in \mathcal{Y}$ and so $\mathcal{Y} = \Sigma$.

"$\Longleftarrow$": Let $\{q_n\}_{n\geqslant 1} \subseteq [0,1]$ be the rational numbers in the unit interval. Let $A_n = f^{-1}([0, r_n])$. Then $\Sigma = \sigma(\{A_n\}_{n\geqslant 1})$ and so Σ is countably generated.

Solution of Problem 3.39

Let

$$\mathcal{F} = \{A \in \sigma(\mathcal{Y}) : A \in \sigma(\mathcal{D}) \text{ for some } \mathcal{D} \subseteq \mathcal{Y} \text{ countable}\}$$

(see Definition 3.6). If $C \in \mathcal{Y}$, then $C \in \sigma(C)$ and so we have

$$\mathcal{Y} \subseteq \mathcal{F} \subseteq \sigma(\mathcal{Y}).$$

Note that $\emptyset \in \mathcal{F}$ and $\mathcal{F}$ is closed under complementation. Moreover, if $\{C_n\}_{n\geqslant 1} \subseteq \mathcal{F}$, then

$$\bigcup_{n\geqslant 1} C_n \in \mathcal{F},$$

since countable union of countable sets is countable. Therefore $\mathcal{F}$ is a σ-algebra and so $\mathcal{F} = \sigma(\mathcal{Y})$.

Solution of Problem 3.40
Let

$$\mathcal{D} \;=\; \{A \in \Sigma :\ \mu(A) = \nu(A)\}.$$

Since $\mu(\Omega) = \nu(\Omega)$, we have $\Omega \in \mathcal{D}$. If $A, C \in \mathcal{D}$ and $A \subseteq C$, then

$$\mu(C \setminus A) \;=\; \nu(C \setminus A)$$

and so $C \setminus A \in \mathcal{D}$. Finally, if $\{A_n\}_{n\geqslant 1} \subseteq \mathcal{D}$ is an increasing sequence, then

$$\mu\Big(\bigcup_{n\geqslant 1} A_n\Big) \;=\; \lim_{n\to+\infty} \mu(A_n) \;=\; \lim_{n\to+\infty} \nu(A_n) \;=\; \nu\Big(\bigcup_{n\geqslant 1} A_n\Big)$$

(see Theorem 3.19) and so $\bigcup_{n\geqslant 1} A_n \in \mathcal{D}$. Therefore $\mathcal{D}$ is a λ-class (see Definition 3.7(b)). By Theorem 3.9, we have $\mu = \nu$ on $\sigma(\mathcal{Y})$.

Solution of Problem 3.41
Let $\{q_n\}_{n\geqslant 1}$ be an enumeration of the rationals of $\mathbb{R}$ and for every $n \geqslant 1$ consider the open interval

$$T_n \;=\; \big(q_n - \tfrac{\varepsilon}{2^{n+1}}, q_n + \tfrac{\varepsilon}{2^{n+1}}\big).$$

Let us set

$$U \;=\; \bigcup_{n\geqslant 1} T_n.$$

Clearly U is open and because $\mathbb{Q} \subseteq U$, we see that U is dense in $\mathbb{R}$. Moreover

$$\lambda(U) \;\leqslant\; \sum_{n\geqslant 1} \lambda(T_n) \;=\; \varepsilon \sum_{n\geqslant 1} \tfrac{1}{2^n} \;=\; \varepsilon.$$

Solution of Problem 3.42
Let $r = \operatorname{dist}(A, C) > 0$ and consider the following sets

$$U \;=\; \{x \in \mathbb{R} :\ \operatorname{dist}(x, A) < \tfrac{1}{r}\},$$
$$V \;=\; \{x \in \mathbb{R} :\ \operatorname{dist}(x, C) < \tfrac{1}{r}\}.$$

Then U and V are disjoint open sets and $A \subseteq U$, $C \subseteq V$. Using the monotonicity of λ and the fact that

$$\inf(S_1 + S_2) = \inf S_1 + \inf S_2 \qquad \forall\, S_1, S_2 \subseteq \mathbb{R}$$

we have

$$\begin{aligned}
\lambda^*(A \cup C) &= \inf\big\{\lambda(D) : \ A \cup C \subseteq D,\ D \text{ is open}\big\} \\
&= \inf\big\{\lambda(D) : \ A \cup C \subseteq D \subseteq U \cup V,\ D \text{ is open}\big\} \\
&= \inf\big\{\lambda(D_1) + \lambda(D_2) : \ A \subseteq D_1 \subseteq U,\ C \subseteq D_2 \subseteq V, \\
&\qquad\qquad D_1, D_2 \text{ are both open}\big\} \\
&= \inf\big\{\lambda(D_1) : \ A \subseteq D_1 \subseteq U,\ D_1 \text{ is open}\big\} \\
&\quad + \inf\big\{\lambda(D_2) : \ C \subseteq D_2 \subseteq V,\ D_2 \text{ is open}\big\} \\
&= \lambda^*(A) + \lambda^*(C).
\end{aligned}$$

Solution of Problem 3.43

We know that

$$\limsup_{n \to +\infty} A_n = \bigcap_{n \geqslant 1} \bigcup_{k \geqslant n} A_n$$

(see Problem 3.14). We have

$$\mu\big(\bigcup_{k \geqslant n} A_k\big) \geqslant \eta \qquad \forall\, n \geqslant 1.$$

The family $\big\{\bigcup_{k \geqslant n} A_k\big\}_{n \geqslant 1} \subseteq \Sigma$ is decreasing and by hypothesis they have finite measure. So, by Theorem 3.19(c), we have

$$\mu\big(\limsup_{n \to +\infty} A_n\big) = \lim_{n \to +\infty} \mu\big(\bigcup_{k \geqslant n} A_k\big) \geqslant \liminf_{n \to +\infty} \mu(A_n).$$

We cannot drop the hypothesis

$$\mu\big(\bigcup_{n \geqslant 1} A_n\big) < +\infty.$$

To see this, let $\Omega = \mathbb{R}$, let Σ be the Lebesgue σ-algebra and let $\mu = \lambda$ be the Lebesgue measure on $\mathbb{R}$ (see Theorem 3.31). If we consider the sequence $\big\{A_n = [n, n+1]\big\}_{n \geqslant 1} \subseteq \Sigma$ then

$$\limsup_{n \to +\infty} A_n = \emptyset, \qquad \text{but} \quad \liminf_{n \to +\infty} \mu(A_n) = 1.$$

Note that in this case

$$\mu\Big(\bigcup_{n\geqslant 1} A_n\Big) \;=\; +\infty.$$

Solution of Problem 3.44
Suppose that $x_n \searrow x$ in $\mathbb{R}$. Then

$$A\cap(-\infty,x] \;=\; \bigcap_{n\geqslant 1}\big(A\cap(\infty,x_n]\big),$$

so

$$\varphi(x) \;=\; \lambda\big(A\cap(-\infty,x]\big) \;=\; \lim_{n\to+\infty}\lambda\big(A\cap(-\infty,x_n]\big) \;=\; \lim_{n\to+\infty}\varphi(x_n)$$

(because $\lambda(A) < +\infty$ by Theorem 3.19(c)).

On the other hand, if $x_n \nearrow x$, then

$$A\cap(-\infty,x) \;=\; \bigcup_{n\geqslant 1}\big(A\cap(-\infty,x_n]\big).$$

Because $\lambda\big(\{x\}\big)=0$, we have

$$\lambda\big(A\cap(\infty,x]\big) \;=\; \lambda\big(A\cap(\infty,x)\big) \;=\; \lim_{n\to+\infty}\lambda\big(A\cap(-\infty.x_n]\big)$$

(see Theorem 3.19(b)), so

$$\varphi(x) \;=\; \lim_{n\to+\infty}\varphi(x_n).$$

Therefore φ is both right and left continuous, hence it is continuous.

Solution of Problem 3.45
Let $\eta>0$ and consider the set

$$C_\eta \;=\; \big\{x\in A,\ \big|f'(x)\big|<\eta\big\}.$$

Evidently, it suffices to show that $\lambda\big(f(C_\eta)\big) = 0$. Let us set

$$E_n = \big\{x \in C_\eta,\ \big|f(u) - f(x)\big| \leqslant \eta|u-x| \text{ if } |u-x| \leqslant \tfrac{1}{n}\big\} \qquad \forall\, n \geqslant 1.$$

Note that $\{E_n\}_{n\geqslant 1}$ is increasing and

$$\lambda(E_n) = 0 \qquad \forall\, n \geqslant 1.$$

Also we have

$$f(C_\eta) \subseteq \bigcup_{n\geqslant 1} f(E_n) = \lim_{n\to+\infty} f(E_n).$$

For every $n \geqslant 1$ and for a given $\varepsilon > 0$, we can find a sequence $\{I_{n,k}\}_{k\geqslant 1}$ of open intervals such that

$$E_n \subseteq \bigcup_{k\geqslant 1} I_{n,k}, \quad \lambda(I_{n,k}) < \tfrac{1}{n} \quad \text{and} \quad \sum_{k\geqslant 1} \lambda(I_{n,k}) < \varepsilon.$$

If $x, u \in E_n \cap I_{n,k}$, then

$$\big|f(u) - f(x)\big| \leqslant \eta\lambda(I_{n,k}).$$

Hence, we have

$$\begin{aligned}\lambda^*\big(f(E_n)\big) &= \lambda^*\Big(f\Big(E_n \cap \Big(\bigcup_{k\geqslant 1} I_{n,k}\Big)\Big)\Big) \leqslant \sum_{k\geqslant 1} \lambda^*\big(f(E_n \cap I_{n,k})\big) \\ &\leqslant \eta \sum_{k\geqslant 1} \lambda(I_{n,k}) \leqslant \eta\varepsilon.\end{aligned}$$

Because $\varepsilon > 0$ was arbitrary, we let $\varepsilon \searrow 0$ to conclude that

$$\lambda^*\big(f(E_n)\big) = 0,$$

hence

$$\lambda^*\big(f(C_\eta)\big) = 0.$$

As we already indicated, this implies that

$$\lambda\big(f(A)\big) = 0.$$

Solution of Problem 3.46
For every integers $n, k \geqslant 1$, we define

$$\begin{aligned} D_{n,k} &= \big\{x \in \Omega : \ \operatorname{dist}(x, \partial\Omega) > \tfrac{1}{n} \text{ and} \\ &\quad \big\|f(x) - f(y)\big\| \leqslant k\|x - y\| \text{ for all } y \in \Omega \\ &\quad \text{such that } \|x - y\| < \tfrac{1}{n}\big\}. \end{aligned}$$

Because f is differentiable on D, we have $D \subseteq \bigcup\limits_{n,k \geqslant 1} D_{n,k}$. Therefore, it suffices to show that

$$\lambda^N\big(f(D_{n,k})\big) = 0.$$

To this end we fix $n, k \geqslant 1$ and $\varepsilon > 0$. Because $\lambda^N(D_{n,k}) = 0$, we can find a sequence of cubes $\Big\{C_{r_m}(x_m) = x_m + \big(-\frac{r_m}{2}, \frac{r_m}{2}\big)^N\Big\}_{m \geqslant 1}$ with centre x_m and side length $r_m \leqslant \frac{1}{2n\sqrt{N}}$ such that

$$D_{n,k} \subseteq \bigcup_{m \geqslant 1} C_{r_m}(x_m) \quad \text{and} \quad \sum_{m \geqslant 1} r_m^N \leqslant \varepsilon.$$

If $x \in D_{n,k} \cap C_{r_m}(x_m)$, then

$$\|x - x_m\| < r_m\sqrt{N} \leqslant \tfrac{1}{2n} \quad \text{and} \quad x_n \in \Omega.$$

Then we have

$$\big\|f(x) - f(x_m)\big\| \leqslant k\|x - x_m\| < kr_m\sqrt{N} \qquad \forall\, x \in D_{n,k} \cap C_{r_m}(x_m).$$

It follows that the set $f\big(D_{n,k} \cap C_{r_m}(x_m)\big)$ is contained in a cube which has centre $f(x_m)$ and side length $kr_m\sqrt{N}$. Hence, if $(\lambda^N)^*$ denotes the Lebesgue outer measure on $\mathbb{R}^N$, then

$$(\lambda^N)^*\big(f\big(D_{n,k} \cap C_{r_m}(x_m)\big)\big) \leqslant k^N r_m^N N^{\frac{N}{2}}.$$

Summing over $m \geqslant 1$, we obtain

$$\begin{aligned} (\lambda^N)^*\big(f(D_{n,k})\big) &\leqslant \sum_{m \geqslant 1} (\lambda^N)^*\big(f\big(D_{n,k} \cap C_{r_m}(x_m)\big)\big) \\ &\leqslant k^N N^{\frac{N}{2}} \sum_{m \geqslant 1} r_m^N \leqslant k^N N^{\frac{N}{2}} \varepsilon. \end{aligned}$$

Because $\varepsilon > 0$ is arbitrary, we conclude that $\lambda^N\big(f(D_{n,k})\big) = 0$, hence $\lambda^N\big(f(D)\big) = 0$.

Solution of Problem 3.47

Without any loss of generality, we may assume that $A \subseteq (a, b)$. For a given $\varepsilon > 0$, for every integer $n \geqslant 1$, let

$$A_n = \big\{x \in A : \lambda^*\big(u(I)\big) \leqslant (M+\varepsilon)\lambda^*(I) \text{ for all intervals } I \text{ such that } x \in I \text{ and } 0 < \lambda(I) < \tfrac{1}{n}\big\}.$$

Note that the sequence $\{A_n\}_{n\geqslant 1}$ is increasing.

Claim 1. $A = \bigcup\limits_{n\geqslant 1} A_n$.

Evidently

$$\bigcup_{n\geqslant 1} A_n \subseteq A.$$

Let $x \in A$. We have $\big|u'(x)\big| \leqslant M$. So, there exists $\delta > 0$ such that

$$\big|u(y) - u(x)\big| \leqslant (M+\varepsilon)|y - x| \qquad \forall\, y \in [a, b],\ |y - x| < \delta.$$

So, if $y, y' \in [a, b]$ with $y < x < y'$ and $|y - y'| < \delta$, then

$$\begin{aligned} \big|u(y) - u(y')\big| &\leqslant \big|u(y) - u(x)\big| + \big|u(x) - u(y')\big| \\ &\leqslant (M+\varepsilon)(x - y) + (M+\varepsilon)(y' - x) \\ &= (M+\varepsilon)(y' - y), \end{aligned}$$

so $x \in A_n$ for every integer $n \geqslant \frac{1}{\delta}$ and so we infer that $x \in \bigcup\limits_{n=1}^{\infty} A_n$. This proves Claim 1.

Claim 2. $\lambda^*\big(u(A_n)\big) \leqslant (M+\varepsilon)\big(\lambda^*(A_n) + \varepsilon\big)$ for all $n \geqslant 1$.

We can find open sets $U_n \supseteq A_n$ such that

$$\lambda(U_n) \leqslant \lambda^*(A_n) + \varepsilon \qquad \forall\, n \geqslant 1.$$

We may assume that $U_n \subseteq (a, b)$ (otherwise we replace U_n by $U_n \cap (a, b)$). Let

$$U_n = \bigcup_{k\geqslant 1} I_{n,k} \cup \big\{y_0, \ldots, y_{s_n}\big\},$$

where $\{I_{n,k}\}_{k\geqslant 1}$ is a family of pairwise disjoint open intervals with

$$0 < \lambda(I_{n,k}) < \tfrac{1}{n} \qquad \forall\, k \geqslant 1$$

and $y_0, \ldots, y_{s_n} \in (a,b)$. Let

$$S = \big\{k \geqslant 1 :\ I_{n,k} \cap A_n \neq \emptyset\big\}.$$

If $k \in S$, then from the definition of the set A_n, we have

$$\lambda^*\big(u(I_{n,k})\big) \leqslant (M+\varepsilon)\lambda(I_{n,k}),$$

so

$$\begin{aligned}\lambda^*\big(u(A_n)\big) &\leqslant \lambda^*\Big(\bigcup_{k\in S} u(I_{n,k})\Big) \leqslant \sum_{k\in S}\lambda^*\big(u(I_{n,k})\big) \leqslant (M+\varepsilon)\sum_{k\in S}\lambda(I_{n,k})\\ &\leqslant (M+\varepsilon)\lambda\Big(\bigcup_{k\geqslant 1} I_{n,k}\Big) = (M+\varepsilon)\lambda(U_n) \leqslant (M+\varepsilon)\big(\lambda^*(A_n)+\varepsilon\big).\end{aligned}$$

This proves Claim 2.

Using Claims 1 and 2 and the fact that the sequence $\{A_n\}_{n\geqslant 1}$ is increasing, passing to the limit as $n \to +\infty$, we obtain

$$\lambda^*\big(u(A)\big) \leqslant (M+\varepsilon)\big(\lambda^*(A)+\varepsilon\big).$$

Let $\varepsilon \searrow 0$, to conclude that

$$\lambda^*\big(u(A)\big) \leqslant M\lambda^*(A).$$

Solution of Problem 3.48
For a given $\varepsilon > 0$, we set

$$A_n = \big\{x \in A :\ (n-1)\varepsilon \leqslant |f'(x)| < n\varepsilon\big\} \qquad \forall\, n \geqslant 1.$$

Evidently A_n is Lebesgue measurable and by Problem 3.47, we have

$$\begin{aligned}\lambda^*\big(u(A)\big) &\leqslant \sum_{n\geqslant 1}\lambda^*\big(u(A_n)\big) \leqslant \sum_{n\geqslant 1} n\varepsilon\lambda(A_n)\\ &= \sum_{n\geqslant 1}(n-1)\varepsilon\lambda(A_n) + \sum_{n\geqslant 1}\varepsilon\lambda(A_n) \leqslant \int_A |u'(t)|\,dt + \varepsilon\lambda(A).\end{aligned}$$

Since $\varepsilon > 0$ is arbitrary, we let $\varepsilon \searrow 0$, to conclude that

$$\lambda^*\big(u(A)\big) \leqslant \int_A \big|u'(t)\big|\, dt.$$

Solution of Problem 3.49

First note that $A \subseteq [a, b]$ is Lebesgue measurable. From Problem 3.48, we have

$$\lambda^*\big(f(A)\big) \leqslant \int_A \big|f'(t)\big|\, dt = 0,$$

so $\lambda\big(f(A)\big) = 0$.

Solution of Problem 3.50

Let

$$P = \big\{x \in A : \big|u'(x)\big| > 0\big\}$$

and for every integer $k \geqslant 1$, let

$$P_k = \big\{x \in P : \big|u(x) - u(y)\big| \geqslant \tfrac{|x-y|}{k} \text{ for all } y \in \big(x - \tfrac{1}{k}, x + \tfrac{1}{k}\big) \cap T\big\}.$$

We have

$$P = \bigcup_{k \geqslant 1} P_k.$$

For fixed $k \geqslant 1$, let

$$C = I \cap P_k,$$

where I is an interval of length less than $\frac{1}{k}$. Since $C \subseteq A$ and $\lambda\big(u(A)\big) = 0$, we have $\lambda\big(u(C)\big) = 0$ and so for a given $\varepsilon > 0$, we can find a sequence of intervals $\{I_n\}_{n \geqslant 1}$ such that

$$u(C) \subseteq \bigcup_{n \geqslant 1} I_n \quad \text{and} \quad \sum_{n \geqslant 1} \lambda(I_n) < \varepsilon.$$

Let

$$S_n = u^{-1}(I_n) \cap A \qquad \forall\, n \geqslant 1.$$

Then

$$C \subseteq \bigcup_{n \geqslant 1} S_n$$

and so with λ^* being the Lebesgue outer measure on $\mathbb{R}$, we have

$$\lambda^*(C) \leqslant \sum_{n\geqslant 1} \lambda^*(S_n) \leqslant \sum_{n\geqslant 1} \operatorname{diam} S_n \leqslant \sum_{n\geqslant 1} k \sup_{x,y\in S_n} \left|u(x)-u(y)\right| = \xi$$

(since $S_n \subseteq I \cap P_k$). Since $u(S_n) \subseteq I_n$, we have

$$\sup_{x,y\in S_n} \left|u(x) - u(y)\right| \leqslant \lambda(I_n),$$

so

$$\xi \leqslant k \sum_{n\geqslant 1} \lambda(I_n) < k\varepsilon.$$

Letting $\varepsilon \searrow 0$, we conclude that $\lambda^*(C) = 0$, hence $\lambda^*(P) = 0$.

Solution of Problem 3.51

First we show that at the nth step in the construction of C_ϑ, the total length of the intervals removed from $[0,1]$ is $\vartheta(1-\vartheta)^{n-1}$. For $n = 1$, we have that the interval removed has length ϑ which is true. Suppose that the formula holds for n. Then the total length of the intervals removed at the $n+1$ step is

$$\vartheta\Big(1 - \sum_{k=1}^{n} \vartheta(1-\vartheta)^{k-1}\Big) = \vartheta(1-\vartheta)^n.$$

So, by induction the expression is true for every integer $n \geqslant 1$. Then the Lebesgue measure of C_ϑ is given by

$$1 - \sum_{n\geqslant 1} \vartheta(1-\vartheta)^{n-1} = 0.$$

Solution of Problem 3.52

Let

$$h(x) = \begin{cases} \mu\big((0,x]\big) & \text{if } x \geqslant 0, \\ -\mu\big((x,0]\big) & \text{if } x < 0, \end{cases} \qquad \forall\, x \in \mathbb{R}.$$

The function $h\colon \mathbb{R} \longrightarrow \mathbb{R}$ is increasing and right continuous (see Theorem 3.19(c)). Also, for every $x, u \in \mathbb{R}$ such that $x < u$, we have

$$h(u) - h(x) = \mu\big((x, u]\big).$$

By hypothesis for every $x, u \in \mathbb{R}$, we have

$$\begin{aligned} \mu\big((x, x+u]\big) &= \mu\big((0, u]\big) \quad \text{if } u \geqslant 0, \\ \mu\big((x+u, x]\big) &= \mu\big((u, 0]\big) \quad \text{if } u < 0. \end{aligned}$$

So, we have

$$h(x+u) = h(x) + h(u)$$

(i.e., h is additive). From the additivity, we have that

$$h(rx) = rh(x) \qquad \forall\, r \in \mathbb{Q},\ x \in \mathbb{R}.$$

Then the right continuity of h implies that

$$h(x) = xh(1) \qquad \forall\, x \in \mathbb{R}.$$

Note that

$$\begin{aligned} h(1) &= \mu\big((0,1]\big) = \mu\big([0,1]\big) - \mu\big(\{0\}\big) \\ &= 1 - \lim_{n \to +\infty} \mu\big(\big(-\tfrac{1}{n}, 0\big]\big) \\ &= 1 - \lim_{n \to +\infty} \big(h(0) - h\big(-\tfrac{1}{n}\big)\big) = 1, \end{aligned}$$

so $h(x) = x$, hence $\mu = \lambda$.

Solution of Problem 3.53

"$\Longrightarrow$": Let $A \subseteq \mathbb{R}$ be a Lebesgue measurable set. Then we can find a sequence $\{K_n\}_{n\geqslant 1}$ of mutually disjoint compact sets and a Lebesgue-null set D such that

$$A = \Big(\bigcup_{n\geqslant 1} K_n\Big) \cup D.$$

Then

$$f(A) = f\Big(\bigcup_{n\geqslant 1} K_n\Big) \cup f(D) = \Big(\bigcup_{n\geqslant 1} f(K_n)\Big) \cup f(D).$$

By the continuity of f, the set $f(K_n)$ is compact for every $n \geqslant 1$ and so the set $\bigcup_{n \geqslant 1} f(K_n)$ is Borel (see Definition 3.6). Also, the set $f(D)$ is Lebesgue-null since by hypothesis f is N-function. Therefore

$$f(A) = \Big(\bigcup_{n \geqslant 1} f(K_n)\Big) \cup f(D)$$

is Lebesgue measurable (see Remark 3.25).

"$\Longleftarrow$": We proceed by contradiction. So, suppose that f is not an N-function. Then we can find a Lebesgue-null set $D \subseteq \mathbb{R}$ such that $f(D)$ is not a Lebesgue-null set. For every $C \subseteq f(D)$, we have

$$f\big(f^{-1}(C) \cap D\big) = C$$

and since the set $f^{-1}(C) \cap D$ is Lebesgue measurable (being Lebesgue-null), we have that C is Lebesgue measurable. Since $C \subseteq f(D)$ was arbitrary, we have a contradiction (see Remark 3.38).

Solution of Problem 3.54

For every $x \in \mathbb{R}$, we have

$$\{x\} \subseteq (x - \varepsilon, x + \varepsilon) \qquad \forall\, \varepsilon > 0$$

and so

$$\lambda(\{x\}) \leqslant 2\varepsilon \qquad \forall\, \varepsilon > 0,$$

hence

$$\lambda(\{x\}) = 0$$

(see Theorem 3.19). If $A = \{x_n\}_{n \geqslant 1}$, then

$$\lambda(A) = \sum_{n \geqslant 1} \lambda(\{x_n\}) = 0.$$

Solution of Problem 3.55

Let $f\colon \mathbb{R} \longrightarrow [0,1]$ be defined by

$$f(x) = \lambda\big(A \cap (-\infty, x]\big) \qquad \forall\, x \in \mathbb{R}.$$

From Problem 3.44, we know that f is continuous. Note that

$$\lim_{x\to-\infty} f(x) = 0 \quad \text{and} \quad \lim_{x\to+\infty} f(x) = 1.$$

Therefore, by Bolzano theorem, we can find $\widehat{x} \in \mathbb{R}$ such that $f(\widehat{x}) = \frac{1}{2}$. Let

$$C = A \cap (-\infty, \widehat{x}].$$

Then C is Lebesgue measurable (see Theorem 3.31), $C \subseteq A$ and $\lambda(C) = \frac{1}{2}$.

Solution of Problem 3.56

Clearly we may assume that $\lambda(A) < +\infty$ and $\varepsilon \in (0,1)$. Consider intervals $T_n = (a_n, b_n]$ such that

$$A \subseteq \bigcup_{n\geqslant 1} T_n \quad \text{and} \quad \sum_{n\geqslant 1} \lambda(T_n) \leqslant \tfrac{1}{1-\varepsilon}\lambda(A)$$

(see Proposition 3.28). Also, we have

$$\lambda(A) \leqslant \sum_{n\geqslant 1} \lambda(A \cap T_n).$$

Therefore for some $n_0 \geqslant 1$, we have

$$(1-\varepsilon)\lambda(T_{n_0}) \leqslant \lambda(A \cap T_{n_0}).$$

So, setting $I_\varepsilon = T_{n_0}$, we have the result.

Solution of Problem 3.57

By Problem 3.56, we can find a nontrivial bounded interval I such that

$$\lambda(A \cap I) \geqslant \tfrac{3}{4}\lambda(I).$$

Let $\varepsilon = \frac{1}{2}\lambda(I)$. If $|x| \leqslant \varepsilon$, then

$$(A\cap I)\cap\big(x+(A\cap I)\big) \subseteq I\cap(x+I) \quad \text{and} \quad \lambda\big(I\cup(x+I)\big) \leqslant \tfrac{3}{2}\lambda(I).$$

But from the translation invariance of the Lebesgue measure (see Theorem 3.31), we have

$$\lambda(A \cap I) = \lambda\big(x + (A \cap I)\big),$$

so

$$(A \cap I) \cap \big(x + (A \cap I)\big) \neq \emptyset,$$

thus

$$x \in (A \cap I) - (A \cap I) \subseteq A - A$$

and so finally

$$[-\varepsilon, \varepsilon] \subseteq A - A.$$

Solution of Problem 3.58

For $x = (x_1, \ldots, x_N) \in \mathbb{R}^N$, we define

$$E_r(x) = \big\{y = (y_1, \ldots, y_N) \in \mathbb{R}^N : |x_k - y_k| < r \text{ for all } k \in \{1, \ldots, N\}\big\}$$

and

$$h(r) = \sup\big\{\mu\big(E_r(x)\big) : x \in \mathbb{R}^N\big\}.$$

Note that

$$E_r(x) \subseteq B_{r\sqrt{N+1}}(x) \quad \text{and} \quad \liminf_{r \to +\infty} \tfrac{h(r)}{r^N} = 0.$$

Let $K \subseteq \mathbb{R}^N$ be a compact set. We can find $\varrho > 0$ such that $K \subseteq E_\varrho(0)$. For a given $\varepsilon > 0$, we can find small $r > 0$ such that

$$h(r) < (2r)^N \varepsilon.$$

Due to the compactness of K, we can find $x^1, \ldots, x^l \in \mathbb{R}^N$ such that $\big\{E_r(x^i)\big\}_{i=1}^l$ are mutually disjoint and

$$K \subseteq \bigcup_{i=1}^{l} \overline{E_r(x^i)} \subseteq E_{2\varrho}(0).$$

If by λ^N we denote the Lebesgue measure on $\mathbb{R}^N$, then

$$\begin{aligned}
\mu(K) &\leqslant \sum_{i=1}^{l} \mu\big(E_r(x^i)\big) \leqslant \sum_{i=1}^{l} (2r)^N \varepsilon \\
&= \sum_{i=1}^{l} \lambda^N\big(E_r(x^i)\big)\varepsilon \leqslant \lambda^N\big(E_{2\varrho}(0)\big)\varepsilon = (4\varrho)^N \cdot \varepsilon.
\end{aligned}$$

Letting $\varepsilon \searrow 0$, we obtain that $\mu(K) = 0$. Since $\mathbb{R}^N$ is σ-compact (see Definition 2.99) and the compact sets form a π-class generating the Borel σ-algebra of $\mathbb{R}^N$ (see Definitions 3.6 and 3.7), we conclude that $\mu \equiv 0$ (see Proposition 3.26).

Solution of Problem 3.59

Let $U \subseteq [0,1]$ be a nonempty open set such that $U \cap A = \emptyset$. Then

$$\lambda(A) + \lambda(U) \;=\; \lambda(A \cup U) \;\leqslant\; \lambda(I) \;=\; 1,$$

so

$$\lambda(U) \;=\; 0,$$

a contradiction, since U contains an open interval.

Solution of Problem 3.60

Evidently $\lambda^N(\operatorname{int} A) = 0$. On the other hand for every nonempty and open set $U \subseteq \mathbb{R}^N$, we have that U contains a ball and so $\lambda^N(U) > 0$. Therefore the open set $\operatorname{int} A$ must be empty.

Solution of Problem 3.61

Every subset of the Cantor set C is Lebesgue measurable and $\operatorname{card} C = \mathfrak{c}$ (see Proposition 3.35). Hence, if $\mathcal{L}(\mathbb{R})$ denotes the σ-algebra of Lebesgue measurable sets in $\mathbb{R}$, then

$$2^{\mathfrak{c}} \;\leqslant\; \operatorname{card} \mathcal{L}(\mathbb{R}) \;\leqslant\; 2^{\mathfrak{c}},$$

so

$$\operatorname{card} \mathcal{L}(\mathbb{R}) \;=\; 2^{\mathfrak{c}}.$$

Solution of Problem 3.62

Let $\{I_n\}_{n \geqslant 1}$ be the sequence of all nontrivial intervals in $[0,1]$ with rational endpoints. Let $K_1 \subseteq I_1$ be a nowhere dense compact set of positive Lebesgue measure. Then $I_1 \setminus K_1$ contains an interval and so we can find a nowhere dense compact set $L_1 \subseteq I_1 \setminus K_1$ of positive Lebesgue measure. Similarly, there exist nowhere dense compact sets $K_2 \subseteq I_2 \setminus (K_1 \cup L_1)$ and $L_2 \subseteq I_2 \setminus (K_1 \cup L_1 \cup K_2)$ both with positive Lebesgue measure. Suppose we have constructed sets $\{K_i\}_{i=1}^n$ and $\{L_i\}_{i=1}^n$ with these properties. So, K_n and L_n are nowhere dense compact sets of positive Lebesgue measure such that $L_n \subseteq I_n \setminus K_n$. The set $I_{n+1} \setminus \bigcup\limits_{i=1}^{n} (K_i \cup L_i)$ contains an interval since the set $\bigcup\limits_{i=1}^{n} (K_i \cup L_i)$ is nowhere dense and compact. In this interval, we can find disjoint nowhere dense compact sets K_{n+1} and L_{n+1} of positive Lebesgue measure and by induction, we have two sequences $\{K_n\}_{n \geqslant 1}$ and $\{L_n\}_{n \geqslant 1}$ with the above properties. Let

$$E = \bigcup_{n \geqslant 1} L_n.$$

If J is an interval in $[0,1]$, then $I_m \subseteq J$ for some $m \geqslant 1$. By the construction described above, we have

$$K_{m+1} \subseteq I_m \quad \text{and} \quad L_{m+1} \subseteq I_m.$$

So, both sets $E \cap I_m$ and $\big([0,1] \setminus E\big) \cap I_m$ have positive Lebesgue measure. We set

$$C = \bigcup_{r \in \mathbb{Z}} (E + r).$$

This is the desired Borel set in $\mathbb{R}$.

Solution of Problem 3.63

Since μ is σ-finite (see Definition 3.17), we can find a sequence $\{A_n\}_{n \geqslant 1} \subseteq \Sigma$ such that

$$\Omega = \bigcup_{n \geqslant 1} A_n \quad \text{and} \quad \mu(A_n) < +\infty \qquad \forall\, n \geqslant 1.$$

By the hypothesis, we can find $\delta > 0$ such that

$$\text{if } A \in \Sigma \text{ and } \mu(A) \leqslant \delta, \quad \text{then } \nu(A) \leqslant 1.$$

Because μ is nonatomic, invoking Proposition 3.43, for every $n \geqslant 1$, we can find pairwise disjoint sets $\{C_{n,k}\}_{k=1}^{m_n} \subseteq \Sigma$ such that

$$A_n = \bigcup_{k=1}^{m_n} C_{n,k} \quad \text{and} \quad \mu(C_{n,k}) \leqslant \delta.$$

Then, we have

$$\nu(C_{n,k}) \leqslant 1$$

and so

$$\nu(A_n) = \sum_{k=1}^{m_n} \nu(C_{n,k}) \leqslant m_n < +\infty,$$

thus ν is σ-finite too.

Solution of Problem 3.64

No. Let $D_1 \subseteq \Omega_1$ be such that $D_1 \notin \Sigma_1$ and let $C_2 \subseteq \Omega_2$ be a μ_2-null set. Then we claim that $D_1 \times C_2 \notin \Sigma_1 \otimes \Sigma_2$ (see Definition 3.23). Indeed, if $\omega \in C_2$, then

$$(D_1 \times C_2)_\omega = \{z \in \Omega_1 : (z, \omega) \in D_1 \times C_2\} = D_1 \notin \Sigma_1.$$

Hence, by Proposition 3.47, we have

$$D_1 \times C_2 \notin \Sigma_1 \otimes \Sigma_2.$$

Since $\mu_2(C_2) = 0$, we have

$$\mu(\Omega_1 \times C_2) = 0 \quad \text{and} \quad D_1 \times C_2 \subseteq \Omega_1 \times C_2.$$

This shows that μ is not complete on $(\Omega_1 \times \Omega_2, \Sigma_1 \otimes \Sigma_2)$ (see Definition 3.23).

Solution of Problem 3.65

First assume that f is a positive simple function. Then

$$f(\omega) = \sum_{k=1}^{m} a_k \chi_{C_k}(\omega),$$

with $a_k \geqslant 0$, $C_k \in \Sigma_\mu$. By Definition 3.23, we can find $A_k \in \Sigma$, $A_k \subseteq C_k$ such that

$$\mu(C_k \setminus A_k) = 0 \qquad \forall\, k \in \{1, \ldots, m\}.$$

We set

$$h(\omega) = \sum_{k=1}^{m} a_k \chi_{A_k}(\omega).$$

Evidently $h \geqslant 0$, h is Σ-measurable and

$$h(\omega) = f(\omega) \quad \mu\text{-almost everywhere on } \Omega.$$

Next suppose that f is an arbitrary positive Σ_μ-measurable function. Then by Theorem 3.68(a), we can find $s_n \geqslant 0$, $n \geqslant 1$, a sequence of Σ_μ-simple functions such that

$$s_n(\omega) \nearrow f(\omega) \qquad \forall\, \omega \in \Omega.$$

From the first part of the proof, we know that we can replace $\{s_n\}_{n\geqslant 1}$ by a sequence $\{r_n\}_{n\geqslant 1}$ of positive, Σ-simple functions such that

$$r_n(\omega) \nearrow f(\omega) \quad \mu\text{-almost everywhere on } \Omega.$$

Let

$$h(\omega) = \lim_{n\to+\infty} r_n(\omega) = \sup_{n\geqslant 1} r_n(\omega) \qquad \forall\, \omega \in \Omega.$$

Then $h \geqslant 0$, is Σ-measurable and

$$h(\omega) = f(\omega) \quad \mu\text{-almost everywhere on } \Omega.$$

Finally for any arbitrary Σ_μ-measurable function f, we write $f = f^+ - f^-$. Both f^+ and f^- are Σ_μ-measurable, positive functions. We apply the previous part of the solution and obtain two positive, Σ-measurable functions h^+ and h^- such that

$$h^+(\omega) = f^+(\omega) \quad \text{and} \quad h^-(\omega) = f^-(\omega) \quad \mu\text{-almost everywhere on } \Omega.$$

We set $h = h^+ - h^-$ and then this is a Σ-measurable function such that

$$|h(\omega)| \leqslant |f(\omega)| \quad \text{and} \quad h(\omega) = f(\omega) \quad \mu\text{-almost everywhere on } \Omega.$$

Solution of Problem 3.66

Let $E \subseteq [0,1]$ be a nonmeasurable set (it exists if we assume the axiom of choice; see Theorem 3.37). Let $f\colon [0,1] \longrightarrow \mathbb{R}$ be defined by

$$f(x) = \begin{cases} x & \text{if } x \in E, \\ -x & \text{if } x \in [0,1] \setminus E. \end{cases}$$

Then $f^{-1}(\mathbb{R}_+) = E$ and so f is nonmeasurable (see Definition 3.53). On the other hand

$$|f|(x) = x$$

and so $|f|$ is measurable and for every $c \in \mathbb{R}$, the set $f^{-1}(\{c\})$ is a singleton or empty, hence a measurable set (in fact closed).

Solution of Problem 3.67

Note that A is the set of points $x \in \Omega$ at which $\{f_n(x)\}_{n\geqslant 1} \subseteq \mathbb{R}$ is a Cauchy sequence (see Definition 1.7). So, we can write

$$A = \left\{x \in \Omega : \forall n \geqslant 1 \ \exists N \geqslant 1 \ \forall m,k \geqslant N : \ \left|f_m(x) - f_k(x)\right| \leqslant \tfrac{1}{n}\right\}.$$

Let us set

$$A(n,N,m,k) = \left\{x \in \Omega : \ \left|f_m(x) - f_k(x)\right| \leqslant \tfrac{1}{n}\right\}.$$

Since $f_m - f_k\colon \Omega \longrightarrow \mathbb{R}$ is Σ-measurable, it follows that $A(n,N,m,k) \in \Sigma$ (see Definition 3.53) and

$$A = \bigcap_{n\geqslant 1} \bigcup_{N\geqslant 1} \bigcap_{m,k\geqslant N} A(n,N,m,k)$$

and so $A \in \Sigma$ (see Remark 3.3).

Solution of Problem 3.68

For every integers $m,k \geqslant 1$, we set

$$A^+_{m,k} = \left\{x \in \Omega : \ f_n(x) \geqslant k \text{ for all } n \geqslant m\right\}.$$

Note that

$$A^+_{m,k} = \bigcap_{n\geqslant m} \left\{x \in \Omega : \ f_n(x) \geqslant k\right\},$$

so $A^+_{m,k} \in \Sigma$ (since each f_n is Σ-measurable; see Definition 3.53). Also, we have

$$A^+ = \bigcap_{k\geqslant 1} \bigcup_{m\geqslant 1} A^+_{m,k}$$

and so we conclude that $A^+ \in \Sigma$ (see Remark 3.3).

Similarly, we set

$$A^-_{m,k} = \{x \in \Omega : f_n(x) \leqslant -k \text{ for all } n \geqslant m\}$$

and we have $A^-_{m,k} \in \Sigma$ and

$$A^- = \bigcap_{k\geqslant 1} \bigcup_{m\geqslant 1} A^-_{m,k},$$

so $A^- \in \Sigma$.

Solution of Problem 3.69

"$\Longrightarrow$": This follows from Proposition 3.56, since φ is Borel measurable.

"$\Longleftarrow$": Let $U \subseteq Y$ be an open set and let $h_U \in C(Y)$ be defined by

$$h_U(y) = \begin{cases} \operatorname{dist}(y, U^c) & \text{if } U \neq Y, \\ 1 & \text{if } U = Y \end{cases}$$

(see Definition 1.6) Note that

$$U = \{y \in Y : h_U(y) > 0\}.$$

Then, by hypothesis, we have

$$f^{-1}(U) = \{x \in \Omega : (h_U \circ f)(x) > 0\} \in \Sigma$$

Since open sets generate $\mathcal{B}(Y)$ (see Definition 3.6), from Proposition 3.54, we conclude that f is Σ-measurable.

Solution of Problem 3.70

Let

$$C_n = \{x \in \Omega : \tfrac{1}{3n} < f(x) < 1 - \tfrac{1}{3n}\} \qquad \forall\, n \geqslant 1.$$

If $\mu(C_n) > 0$ for some $n \geqslant 1$, then $a = \frac{1}{3n}$ satisfies

$$\mu\big(\{x \in \Omega : 0 < f(x) < 1 - a\}\big) > 0.$$

Otherwise

$$\mu(C_n) = 0 \qquad \forall\, n \geqslant 1.$$

Note that

$$C_n \nearrow \{x \in \Omega : 0 < f(x) < 1\},$$

so

$$\mu\big(\{x \in \Omega : 0 < f(x) < 1\}\big) = 0$$

(continuity from below; see Theorem 3.19(b)). Therefore $f = \chi_A$, where $A = f^{-1}(\{1\}) \in \Sigma$ (see Definition 3.53).

Solution of Problem 3.71

"**(a)** $\Longrightarrow$ **(b)**, **(c)**": For notational simplicity we assume that

$$f_n(x) \longrightarrow 0 \quad \text{almost everywhere on } [0,1].$$

Invoking the Egorov theorem (see Theorem 3.76), we can find an increasing sequence of sets $\{C_k\}_{k\geqslant 1} \subseteq \mathcal{B}(I)$, where $I = [0,1]$ such that

$$\lambda(I \setminus C_k) < \tfrac{1}{k}$$

and

$$f_k \rightrightarrows 0 \quad \text{on } C_k.$$

So, we can find an increasing sequence $\{n_k\}_{k\geqslant 1}$ such that

$$\big|f_n(x)\big| \leqslant \tfrac{1}{2^k} \qquad \forall\, n \geqslant n_k.$$

Let

$$C = \bigcup_{k\geqslant 1} C_k \in \Sigma.$$

Then $\lambda(I \setminus C) = 0$. We set

$$\beta_n = \begin{cases} 0 & \text{if} \quad n \neq n_k \text{ for all } k \geqslant 1, \\ 1 & \text{if} \quad n = n_k \text{ for some } k \geqslant 1. \end{cases}$$

We have

$$\sum_{n \geqslant 1} \beta_n f_n(x) = \sum_{k \geqslant 1} f_{n_k}(x) \qquad \forall\, x \in I = [0, 1.]$$

Note that for all $x \in C$, we have

$$\left|f_{n_k}(x)\right| < \tfrac{1}{2^k} \qquad \forall\, k \geqslant k_0,$$

with k_0 such that $x \in C_{k_0}$. Evidently then **(b)** and **(c)** hold.

"**(b)** $\Longrightarrow$ **(a)**": Since

$$\limsup_{n \to +\infty} |\beta_n| > 0,$$

we can find $\xi > 0$ and a subsequence $\left\{\beta_{n_k}\right\}_{k \geqslant 1} \subseteq \{\beta_n\}_{n \geqslant 1}$ such that

$$|\beta_{n_k}| \geqslant \xi \qquad \forall\, k \geqslant 1.$$

By hypothesis

$$\sum_{n \geqslant 1} \beta_n f_n(x) \text{ converges for almost all } x \in I.$$

Hence

$$\beta_{n_k} f_{n_k}(x) \longrightarrow 0 \quad \text{for almost all } x \in I,$$

which necessarily implies that

$$f_{n_k}(x) \longrightarrow 0 \quad \text{for almost all } x \in I.$$

"**(c)** $\Longrightarrow$ **(a)**": Let

$$h(x) = \sum_{n \geqslant 1} \left|\beta_n f_n(x)\right|.$$

By hypothesis

$$h(x) < +\infty \quad \text{for almost all } z \in I.$$

Let

$$C_k = \{x \in I : h(x) \leqslant k\} \in \Sigma$$

and

$$C = \bigcup_{n \geqslant 1} C_k \in \Sigma.$$

Note that $\lambda(I \setminus C) = 0$. We have

$$\int_{C_k} h\,dx = \sum_{n \geqslant 1} |\beta_k| \int_{C_k} |f_k|\,dx \leqslant k.$$

Since by hypothesis

$$\sum_{k \geqslant 1} |\beta_k| = +\infty,$$

we must have

$$\liminf_{k \to +\infty} \int_{C_k} |f_k|\,dx = 0.$$

So, we can find a subsequence $\{f_{n_k}\}_{k \geqslant 1} \subseteq \{f_n\}_{n \geqslant 1}$ such that

$$\int_{C_k} |f_{n_k}|\,dx \leqslant \tfrac{1}{2^k} \qquad \forall\, k \geqslant 1.$$

For every integer $m \geqslant 1$, we have

$$\sum_{k \geqslant 1} \int_{C_m} |f_{n_k}|\,dx < +\infty$$

since $C_m \subseteq C_k$ for all $k \geqslant m$. So, for almost all $x \in C_m$, we have

$$\sum_{k \geqslant 1} |f_n(x)| < +\infty$$

and so

$$f_{n_k}(x) \longrightarrow 0 \quad \text{for almost all } x \in C_m.$$

Since $m \geqslant 1$ was arbitrary and $\lambda(I \setminus C) = 0$, we conclude that

$$f_{n_k}(x) \longrightarrow 0 \quad \text{for almsot all } x \in I.$$

Solution of Problem 3.72
We do the proof for the lower semicontinuous case, the proof being similar for the upper semicontinuous case (see Definition 2.46). For every $\eta \in \mathbb{R}$ and every $n \geqslant 1$, we have

$$(f|_{A_n})^{-1}((\eta, +\infty)) \quad \text{is open in } A_n,$$

so

$$(f|_{A_n})^{-1}((\eta, +\infty)) = A_n \cap U_n \quad \text{with } U_n \subseteq \mathbb{R} \text{ being open}$$

and thus

$$(f|_{A_n})^{-1}((\eta, +\infty)) \in \Sigma \qquad \forall\, n \geqslant 1.$$

Also

$$(f|_{A_0})^{-1}((\eta, +\infty)) \subseteq A_0$$

and due to the μ-completeness of Σ (see Definition 3.23), we have

$$(f|_{A_0})^{-1}((\eta, +\infty)) \in \Sigma \quad \text{and is } \mu\text{-null}.$$

Therefore, since

$$X = A_0 \cup \Big(\bigcup_{n\geqslant 1} A_n\Big),$$

we have

$$f^{-1}((\eta, +\infty)) = \bigcup_{n\geqslant 0} (f|_{A_n})^{-1}((\eta, +\infty)) \in \Sigma,$$

and so f is Σ-measurable (see Proposition 3.63).

Solution of Problem 3.73
We know that f has at most a countable number of jump discontinuities (see Theorem 4.98). Let $\{c_n\}_{n\geqslant 1}$ be the increasing sequence of discontinuity points. For every $n \geqslant 1$, $f|_{(c_n, c_{n+1})}$ is continuous, hence Borel measurable. Similarly for $f|_{\mathbb{R}\setminus \bigcup\limits_{n\geqslant 1}[c_n, c_{n+1}]}$. Finally for every $n \geqslant 1$, we have

$$\{x \in \mathbb{R} : f(x) = f(c_n)\} = \begin{cases} [u, c_n] & \text{for some } u < c_n, \\ \{c_n\}, & \\ [c_n, y] & \text{for some } y > c_n, \end{cases}$$

so

$$\{x \in \mathbb{R} : f(x) = f(c_n)\} \in \mathcal{B}(\mathbb{R})$$

and thus

$$f|_{\{c_n\}_{n\geqslant 1}} \text{ is Borel measurable.}$$

Since

$$\mathbb{R} = \{c_n\}_{n\geqslant 1} \cup \Big(\bigcup_{n\geqslant 1}(c_n, c_{n+1})\Big) \cup \Big(\mathbb{R} \setminus \bigcup_{n\geqslant 1}[c_n, c_{n+1}]\Big),$$

from Proposition 3.78, we conclude that f is Borel measurable.

Solution of Problem 3.74

Let $E \subseteq \mathbb{R}$ be a nonmeasurable set and let $\mathcal{Y}$ be the family of all finite subsets of E. Then $\mathcal{Y}$ is uncountable. We consider the family $\{\chi_A\}_{A\in\mathcal{Y}}$. We have

$$\sup_{\mathcal{Y}} \chi_A = \chi_E,$$

which is not measurable.

Solution of Problem 3.75

Let $r \in \mathbb{Q}$ and consider the open interval

$$I_{r,n} = \big(r - \tfrac{1}{2n}, \, r + \tfrac{1}{2n}\big).$$

Then for every $x \in \Omega$ and every $n \geqslant 1$, we can find $r \in \mathbb{Q}$ such that $f(x) \in I_{r,n}$. Hence

$$\big(x, f(x)\big) \in f^{-1}(I_{r,n}) \times I_{r,n}.$$

Conversely, if for every $n \geqslant 1$, we can find $r \in \mathbb{Q}$ such that

$$(x, y) \in f^{-1}(I_{r,n}) \times I_{r,n},$$

then $f(x), y \in I_{r,n}$ and so

$$|f(x) - y| \leqslant \tfrac{1}{n}$$

and thus $y = f(x)$.

Therefore we conclude that $\operatorname{Gr} f \in \Sigma \otimes \mathcal{B}(\mathbb{R})$ (see Definition 3.44).

Solution of Problem 3.76

We know that X is second countable (see Proposition 1.24). Let $\{U_n\}_{n \geqslant 1}$ be a base for X. We have

$$x \neq f(\omega) \quad \Longleftrightarrow \quad \exists n \geqslant 1 : \ f(\omega) \in U_n \text{ and } x \notin U_n.$$

Therefore,

$$\operatorname{Gr} f = \Big(\bigcup_{n \geqslant 1} \big(f^{-1}(U_n) \times U_n^c\big)\Big)^c \in \Sigma \times \mathcal{B}(X).$$

Solution of Problem 3.77

We have

$$\sum_{n \geqslant 1} \int_\Omega \big(\tfrac{1}{n} f_n\big)^2 \, dx \leqslant M \sum_{n \geqslant 1} \tfrac{1}{n^2} < +\infty,$$

for some $M > 0$, so

$$f = \sum_{n \geqslant 1} \big(\tfrac{1}{n} f_n\big)^2 \in L^1(\Omega).$$

Hence f is μ-almost everywhere finite and so we conclude that

$$\tfrac{1}{n} f_n \longrightarrow 0 \quad \mu\text{-almost everywhere on } \Omega.$$

Solution of Problem 3.78

Let $f_n = \chi_{C_n}$ for $n \geqslant 1$ and let

$$f = \sum_{n \geqslant 1} f_n.$$

Then

$$\sum_{n\geqslant 1}\mu(C_n) \;=\; \int_{\Omega} f\,d\mu \;\geqslant\; \int_{D_k} f\,d\mu \;\geqslant\; k\mu(D_k)$$

(see Definition 3.84 and Proposition 3.86)

Solution of Problem 3.79

We know that $f \geqslant 0$ is Σ-measurable (see Corollary 3.69). From the Fatou lemma (see Theorem 3.95), we have

$$\int_{\Omega} f(x)\,dx \;=\; \int_{\Omega} \liminf_{n\to+\infty} f(x)\,dx \;\leqslant\; \liminf_{n\to+\infty} \int_{\Omega} f_n(x)\,dx \;=\; 0.$$

But $\int_{\Omega} f(x)\,dx \geqslant 0$. Hence

$$\int_{\Omega} f(x)\,dx \;=\; 0.$$

Let

$$A \;=\; \big\{x\in\Omega:\; f(x)>0\big\}$$

and

$$A_n \;=\; \big\{x\in\Omega:\; f(x)\geqslant \tfrac{1}{n}\big\} \qquad \forall\, n\geqslant 1.$$

We see that $A_n \nearrow A$ and

$$\tfrac{1}{n}\mu(A_n) \;\leqslant\; \int_{A_n} f(x)\,dx \;\leqslant\; \int_{\Omega} f(x)\,dx \;=\; 0 \qquad \forall\, n\geqslant 1,$$

so

$$\mu(A_n) \;=\; 0 \qquad \forall\, n\geqslant 1$$

and thus

$$\lim_{n\to+\infty}\mu(A_n) \;=\; \mu(A) \;=\; 0,$$

i.e.,

$$f(x) \;=\; 0 \quad \mu\text{-almost everywhere on } \Omega.$$

Alternative Solution

Arguing by contradiction, suppose that we can find $A \in \Sigma$, with $\mu(A) > 0$ such that $f(x) > 0$ for all $x \in A$. Then we can find $\delta > 0$ and a set $C \in \Sigma$, with $0 < \mu(C) < +\infty$ (since μ is semifinite; see Definition 3.17(c)) such that $f(x) \geqslant \delta$ for all $x \in C$. By the Egorov theorem (see Theorem 3.76), we can find $E \in \Sigma$, $E \subseteq C$ such that

$$\mu(E) \geqslant \tfrac{\delta}{2} \quad \text{and} \quad f_n \rightrightarrows f \quad \text{on } E.$$

But then

$$\lim_{n \to +\infty} \int_{\Omega} f_n(x)\, dx \geqslant \lim_{n \to +\infty} \int_{E} f_n(x)\, dx \neq 0,$$

a contradiction to the hypothesis.

Solution of Problem 3.80

First assume that μ is finite. Let

$$A_n = \left\{ x \in \Omega : \ f(x) \geqslant g(x) + \tfrac{1}{n},\ |g(x)| \leqslant n \right\} \qquad \forall\, n \geqslant 1.$$

Then $A_n \in \Sigma$ (see Definition 3.53) and we have

$$\int_{A_n} g\, d\mu \geqslant \int_{A_n} f\, d\mu \geqslant \int_{A_n} g\, d\mu + \tfrac{1}{n}\mu(A_n)$$

(see Proposition 3.90(c)). Note that

$$\Big| \int_{A_n} g\, d\mu \Big| \leqslant \int_{A_n} |g|\, d\mu \leqslant n\mu(A_n) < +\infty$$

(see Proposition 3.90(d)) and so from the above inequality, we obtain

$$\tfrac{1}{n}\mu(A_n) \leqslant 0 \qquad \forall\, n \geqslant 1,$$

i.e.,

$$\mu(A_n) = 0 \qquad \forall\, n \geqslant 1,$$

so

$$\mu\Big(\bigcup_{n \geqslant 1} A_n\Big) = 0$$

and thus

$$\mu\big(\{x \in \Omega : \ f(x) > g(x) \text{ and } g(x) \in \mathbb{R}\}\big) \ = \ 0.$$

So, we have proved that

$$f \ \leqslant \ g \quad \text{almost everywhere on } \big\{x \in \Omega : \ g(x) \in \mathbb{R}\big\}.$$

Clearly $f \leqslant g$ also on $\big\{x \in \Omega : \ g(x) = +\infty\big\}$. Finally, let

$$C_n \ = \ \big\{x \in \Omega : \ g(x) = -\infty, \ f(x) \geqslant -n\big\} \ \in \ \Sigma \qquad \forall \, n \geqslant 1.$$

Then

$$-\infty \mu(C_n) \ = \ \int_{C_n} g \, d\mu \ \geqslant \ \int_{C_n} f \, d\mu \ \geqslant \ -n\mu(C_n),$$

hence

$$\mu(C_n) \ = \ 0$$

and so

$$\mu\Big(\bigcup_{n \geqslant 1} C_n\Big) \ = \ 0.$$

Thus

$$\mu\big(\{x \in \Omega : \ f(x) > g(x), \ g(x) = -\infty\}\big) \ = \ 0,$$

i.e.,

$$f \ \leqslant \ g \quad \mu\text{-almost everywhere on } \big\{x \in \Omega : \ g(x) = -\infty\big\}.$$

Next, let μ be σ-finite. Then

$$\Omega \ = \ \bigcup_{n \geqslant 1} \Omega_n,$$

with

$$\mu(\Omega_n) \ < \ +\infty \qquad \forall \, n \geqslant 1.$$

Then by the first part of the solution, we have

$$f(x) \ \leqslant \ g(x) \quad \mu\text{-almost everywhere on } \Omega_n \text{ for all } n \geqslant 1,$$

hence

$$f(x) \leqslant g(x) \quad \mu\text{-almost everywhere on } \Omega.$$

Solution of Problem 3.81

No. To see this let

$$f(x) = \chi_{\{0,1\}}(x) \quad \text{and} \quad h(x) = \chi_{\{0\}} \qquad \forall\, x \in \mathbb{R}.$$

Evidently $f, h\colon \mathbb{R} \longrightarrow \mathbb{R}$ are two Lebesgue integrable functions with compact supports (see Definition 2.146), but $f \circ h \equiv 1 \notin L^1(\mathbb{R})$.

Solution of Problem 3.82

Let $\mathbb{Q}^N \subseteq \mathbb{R}^N$ be the set of all vectors in $\mathbb{R}^N$ with rational coordinates. This set is countable and dense in $\mathbb{R}^N$. We introduce the set function $\mu\colon \mathcal{B}(\mathbb{R}^N) \longrightarrow \overline{R}_+ = [0, +\infty]$, defined by

$$\mu(A) = \sum_{x \in \mathbb{Q}^N} \chi_A(x) \qquad \forall\, A \in \mathcal{B}(\mathbb{R}^N).$$

Evidently μ being the countable sum of measures is itself a measure. Note that any Borel set with a nonempty interior contains infinitely many points in $\mathbb{Q}^N$. So, its μ-measure is $+\infty$. Therefore, if $f \in C(\mathbb{R}^N) \setminus \{0\}$, then f is bounded away from zero on an open set and so

$$\int_{\mathbb{R}^N} |f|\, d\mu = +\infty.$$

Solution of Problem 3.83

Let Ω be the set of all ordinals less than the first uncountable ordinal. The continuum hypothesis implies that Ω is equipotent to $[0, 1]$ (i.e., it has the cardinality of the continuum). Let $h\colon [0, 1] \longrightarrow \Omega$ be a bijection. Using h we define a set $A \subseteq [0, 1] \times [0, 1]$ as follows

$$(x, y) \in A \quad \Longleftrightarrow \quad h(y) < h(x).$$

For $x \in [0,1]$, the set A_x is countable since it is the set of all $y \in [0,1]$ such that $h(y) < h(x)$ and by the continuum hypothesis there are finitely or countable many such y's. Similarly, we infer that for every $y \in [0,1]$, the set A_y is co-countable.

Let $\xi\colon [0,1]\times[0,1] \longrightarrow [0,1]$ be defined by $\xi = \chi_A$. Then the functions $x \longmapsto \xi(x,y)$ and $y \longmapsto \xi(x,y)$ are both Lebesgue measurable, since both are almost everywhere constant. But

$$\int_0^1 \xi(x,y)\,dx \;=\; 1 \quad \text{and} \quad \int_0^1 \xi(x,y)\,dy \;=\; 0.$$

So, the Fubini–Tonelli theorem (see Theorem 3.115) does not hold and so A cannot be Lebesgue measurable.

Solution of Problem 3.84

Suppose that **(b)** does not hold. Then

$$\int_A f\,d\mu \;\leqslant\; \int_A h\,d\mu \qquad \forall\, A \in \Sigma.$$

Since, by hypothesis $\int_\Omega f\,d\mu = \int_\Omega h\,d\mu$, it follows that

$$\int_A f\,d\mu \;=\; \int_A h\,d\mu \qquad \forall\, A \in \Sigma.$$

For a given $\varepsilon > 0$, let

$$\begin{aligned} A_1 &= \big(x \in \Omega :\ h(x) + \varepsilon \leqslant f(x)\big), \\ A_2 &= \big(x \in \Omega :\ f(x) \leqslant h(x) + \varepsilon\big). \end{aligned}$$

Evidently $A_1, A_2 \in \Sigma$ and so we have

$$\int_{A_1} f\,d\mu \;=\; \int_{A_1} h\,d\mu \quad \text{and} \quad \int_{A_2} f\,d\mu \;=\; \int_{A_2} h\,d\mu.$$

It follows that $\mu(A_1) = \mu(A_2) = 0$. Since $\varepsilon > 0$ was arbitrary, we infer that

$$\mu(\{f \neq h\}) = \mu\Big(\bigcup_{n\geqslant 1}\{|f-h| \geqslant \tfrac{1}{n}\}\Big) = 0$$

and so **(a)** holds.

Solution of Problem 3.85

From Proposition 3.35, we know that C is compact and nowhere dense. Hence χ_C is continuous on $[0,1]\setminus C$ and discontinuous on C. Because C is Lebesgue-null (see Proposition 3.35(c)), from elementary calculus, we infer that χ_C is Riemann integrable on $[0,1]$. Since $\chi_C = 0$ almost everywhere on $[0,1]$, it follows that

$$\int_0^1 \chi_C \, dx = 0.$$

Solution of Problem 3.86

Note that

$$[0,1] = \Big(\bigcup_{n\geqslant 1}\{\tfrac{1}{n+1} \leqslant f < \tfrac{1}{n}\}\Big) \cup \{1 \leqslant f\},$$

so

$$\sum_{n\geqslant 1}\lambda(\{\tfrac{1}{n+1} \leqslant f < \tfrac{1}{n}\}) + \lambda(\{1 \leqslant f\}) = 1.$$

So, we can find $m \geqslant 1$ such that

$$\sum_{n\geqslant m}\lambda(\{\tfrac{1}{n+1} \leqslant f < \tfrac{1}{n}\}) \leqslant \tfrac{\theta}{2}.$$

Consider a Lebesgue measurable set $A \subseteq [0,1]$ with $\lambda(A) \geqslant \theta$. Let

$$A_1 = A \cap \{f \geqslant \tfrac{1}{m}\} \quad \text{and} \quad A_2 = A \setminus A_1.$$

Then $\lambda(A_2) \leqslant \frac{\theta}{2}$ and so $\lambda(A_1) \geqslant \frac{\theta}{2}$. Hence

$$\int_A f\,dx \ \geqslant\ \int_{A_1} f\,dx \ \geqslant\ \tfrac{1}{m}\tfrac{\theta}{2} \ >\ 0.$$

Solution of Problem 3.87

First we show that $f*g \in C(\mathbb{R}^N)$. Let $h \in \mathbb{R}^N$. Then for $C = \operatorname{supp} g$, we have

$$\begin{aligned}\left|(f*g)(x+h)-(f*g)(x)\right| &\leqslant \int_{\mathbb{R}^N} \left|f(x+h-y)-f(x-y)\right|\left|g(y)\right| dy \\ &= \int_C \left|f(x+h-y)-f(x-y)\right|\left|g(y)\right| dy.\end{aligned}$$

For fixed $x \in \mathbb{R}^N$, the set $x-C = \{x-y:\ y \in C\}$ is compact and f is uniformly continuous over it. Therefore, for a given $\varepsilon > 0$, we can find $\delta > 0$ such that

$$\left|f(x+h-y)-f(x-y)\right| \ \leqslant\ \varepsilon \qquad \forall\ \|h\| \leqslant \delta.$$

So, we have

$$\left|(f*g)(x+h)-(f*g)(x)\right| \ \leqslant\ \int_C \varepsilon\left|g(y)\right| dy \ \leqslant\ \varepsilon\|g\|_1,$$

so $f*g$ is continuous.

Let $e_k = \{\delta_{i,k}\}_{i=1}^N$, where $\delta_{i,k}$ being the Kronecker symbol defined by

$$\delta_{i,k} \ =\ \begin{cases} 1 & \text{if} \quad i=k \\ 0 & \text{if} \quad i \neq k \end{cases}.$$

Then

$$\begin{aligned}\tfrac{\partial}{\partial x_k}(f*g)(x) &= \lim_{\lambda\to 0} \tfrac{(f*g)(x+\lambda e_k)-(f*g)(x)}{\lambda} \\ &= \lim_{\lambda\to 0} \tfrac{1}{\lambda} \int_C \big(f(x+\lambda e_k-y)-f(x-y)\big)g(y)\,dy\end{aligned}$$

$$= \lim_{\lambda \to 0} \frac{1}{\lambda} \int\limits_C \frac{\partial f}{\partial x_k}\big((x-y) + \mu\lambda e_k\big) g(y)\, dy$$

$$= \int\limits_{\mathbb{R}^N} \frac{\partial f}{\partial x_k}(x-y) g(y)\, dy \;=\; \Big(\frac{\partial f}{\partial x_k} * g\Big)(x)$$

with $0 < \mu < 1$ (by the Lebesgue dominated convergence theorem; see Theorem 3.94). Therefore $f * g \in C^\infty(\mathbb{R}^N)$.

Solution of Problem 3.88

Let $C = \operatorname{supp} f + \overline{B}_1$ (where $\overline{B}_r = \big\{x \in \mathbb{R}^N : \; \|x\| \leqslant r\big\}$ for all $r > 0$). Let $u \in C_c^\infty\big(\mathbb{R}^N\big)$, $u \geqslant 0$ such that

$$\int\limits_{\mathbb{R}^N} u(x)\, dx \;=\; 1,$$

for example, we can take

$$u(x) \;=\; \begin{cases} \exp\big(\frac{1}{\|x\|^2 - 1}\big) & \text{if} \quad \|x\| < 1, \\ 0 & \text{if} \quad \|x\| \geqslant 1. \end{cases}$$

Let $u_n = n^N u(nx)$ and define $f_n = u_n * f$. From Problem 3.87, we know that $f_n \in C^\infty(\mathbb{R}^N)$. Moreover,

$$\operatorname{supp} f_n \;\subseteq\; \operatorname{supp} f + \overline{B}_{\frac{1}{n}} \;\subseteq\; \operatorname{supp} f + \overline{B}_1 \;=\; C \;\subseteq\; \mathbb{R}^N \qquad \forall\, n \geqslant 1.$$

Since f is uniformly continuous on C, for a given $\varepsilon > 0$, we can find $\delta = \delta(\varepsilon) > 0$ such that

$$\big|f(x-y) - f(x)\big| \;\leqslant\; \varepsilon \qquad \forall\, x \in C, \; \|y\| \leqslant \delta.$$

Let us choose $n_0 \geqslant 1$ such that $\frac{1}{n} \leqslant \varepsilon$ for all $n \geqslant n_0$. Then

$$\begin{aligned} \big|u_n(x) - f(x)\big| \;&\leqslant\; \int\limits_{\overline{B}_{\frac{1}{n}}} \big|f(x-y) - f(x)\big| u_n(y)\, dy \\ &\leqslant\; \varepsilon \int\limits_{\overline{B}_{\frac{1}{n}}} u_n(y)\, dy \;=\; \varepsilon \qquad \forall\, n \geqslant n_0, \end{aligned}$$

so $u_n \rightrightarrows f$ on C.

Solution of Problem 3.89

First we assume that the measures μ and m are finite. Let

$$f_n(\omega,y) = \begin{cases} f(\omega,y) & \text{if } |f(\omega,y)| \leqslant n, \\ n & \text{if } f(\omega,y) \geqslant n, \\ -n & \text{if } f(\omega,y) \leqslant -n. \end{cases}$$

Then for every $n \geqslant 1$, the function f_n is $\mu \times m$-measurable and bounded, hence it is integrable. By the Fubini theorem, for every $n \geqslant 1$, the function

$$\Omega \ni \omega \longrightarrow \psi_n(\omega) = \int\limits_Y f_n(\omega,y)\,dm(y)$$

is μ-measurable. Since $f_n \longrightarrow f$ pointwise and

$$|f_n| \leqslant |f| \qquad \forall\, n \geqslant 1,$$

from the Lebesgue dominated convergence theorem (see Theorem 3.94), we have

$$\psi_n(\omega) \longrightarrow \psi(\omega) \quad \text{for } \mu\text{-almost all } \omega \in \Omega.$$

Hence ψ is μ-measurable.

In the general case (i.e., when μ and m are σ-finite) we can find a sequence $\{\Omega_n \times Y_n\}_{n\geqslant 1}$ of subsets of $\Omega \times Y$ with finite measure and $\mu \times m$ is concentrated on their union. Then from the first part of the solution, we have that functions $\Omega \ni \omega \longrightarrow \psi_n(\omega) = \int\limits_Y f_n(\omega,y)\,dm(y)$ are $\mu|_{\Omega_n}$-measurable. We extend by zero on all of Ω. Note that $\psi_n(\omega) \longrightarrow \psi(\omega)$ for μ-almost all $\omega \in \Omega$ (by the Lebesgue dominated convergence theorem). Hence ψ is μ-measurable.

Solution of Problem 3.90

Let I be the family of all finite subsets of $[0,1]$ ordered by inclusion. Clearly I is a directed set. We consider the net $\{\chi_A\}_{A\in I}$. Evidently

$$\chi_A \longrightarrow 1.$$

However,

$$\int_\Omega \chi_A(x)\,dx \;=\; 0 \qquad \forall\, A \in I.$$

Solution of Problem 3.91

Let $\mu(A) = \chi_A(0)$ for all $A \in \mathcal{B}(\mathbb{R})$ (see Definition 3.6). Then for every $f \in C_c(\mathbb{R})$, we have $\int\limits_{\mathbb{R}} f\,d\mu = f(0)$.

For the second part of the problem the answer is no. Arguing indirectly, suppose that we could find such a Borel measure m. Let $f \in C_c^1(\mathbb{R})$, $f \geqslant 0$ with $f|_{[-1,1]} = 1$. We have

$$\int_{\mathbb{R}} f(x)e^{\frac{x}{n}}\,dm \;=\; \left(\tfrac{d}{dx}f(x)e^{\frac{x}{n}}\right)\Big|_{x=0} \;=\; \frac{e^{\frac{x}{n}}}{n}\Big|_{x=0} \;=\; \frac{1}{n},$$

so

$$\lim_{n\to+\infty}\int_{\mathbb{R}} f(x)e^{\frac{x}{n}}\,dm \;=\; 0.$$

On the other hand from the Lebesgue dominated convergence theorem (see Theorem 3.94), we have

$$\lim_{n\to+\infty}\int_{\mathbb{R}} f(x)e^{\frac{x}{n}}\,dm \;=\; \int_{\mathbb{R}} f(x)\,dm \;>\; 0,$$

a contradiction.

Solution of Problem 3.92

Let

$$h_n(x) \;=\; f(nx) \qquad \forall\, n \geqslant 1,\; x \in \mathbb{R}_+.$$

Then for all $x \in (0,a]$, we have

$$h_n(x) \;\longrightarrow\; \theta \quad \text{as } n \to +\infty.$$

Since by hypothesis $\lim\limits_{x\to+\infty} f(x) = \theta$, we can find $M > 0$ such that

$$|f(x)| \;\leqslant\; |\theta| + 1 \qquad \forall\, x \geqslant M.$$

On the other hand, due to the continuity of f, there is $\xi > 0$ such that

$$|f(x)| \leqslant \xi \qquad \forall\, x \in [0, M].$$

Let $\xi^* = \max\{\xi, |\theta| + 1\}$. Then we have

$$|f(x)| \leqslant \xi^* \qquad \forall\, x \in \mathbb{R}_+.$$

Hence

$$|h_n(x)| = |f(nx)| \leqslant \xi^* \qquad \forall\, x \in [0, a].$$

Invoking the Lebesgue dominated convergence theorem (see Theorem 3.94), we have

$$\lim_{n\to+\infty} \int_0^a f(nx)\, dx = \lim_{n\to+\infty} \int_0^a h_n(x)\, dx = \int_0^a \theta\, dx = a\theta.$$

Solution of Problem 3.93

(a) Let $\eta > 0$ and let us set $f_\eta = \chi_{\{\|x\| \geqslant \eta\}} f$. Note that

$$|f_\eta| \leqslant f \quad \text{and} \quad |f_\eta| \longrightarrow 0 \quad \text{almost everywhere on } \mathbb{R}^N, \text{ as } n \to +\infty.$$

So, by the Lebesgue dominated convergence theorem (see Theorem 3.94), we have

$$\lim_{\eta\to+\infty} \int_{\mathbb{R}^N} |f_\eta|\, dz = \lim_{\eta\to+\infty} \int_{\{\|z\| \geqslant \eta\}} |f|\, dz = 0.$$

Let

$$\xi = \sup\{\|u\| : u \in K\} < +\infty$$

(since $K \subseteq \mathbb{R}^N$ is compact). We have

$$x + K = \{x + u : u \in K\} \subseteq \{z \in \mathbb{R}^N : \|z\| \geqslant \|x\| - \xi\}$$

(triangle inequality), so, by the first part, we have

$$\int_{x+K} |f(z)|\, dz \leqslant \int_{\{\|z\| \geqslant \|x\| - \xi\}} |f(z)|\, dz \longrightarrow 0 \quad \text{as } \|x\| \to +\infty.$$

(b) We argue by contradiction. So, suppose that $f(x) \not\to 0$ as $\|x\| \to +\infty$. Then, we can find a sequence $\{x_n\}_{n\geqslant 1} \subseteq \mathbb{R}^N$ and $\varepsilon > 0$ such that

$$\big|f(x_n)\big| \geqslant \varepsilon \qquad \forall\, n \geqslant 1 \quad \text{and} \quad \|x_n\| \longrightarrow +\infty \quad \text{as } n \to +\infty.$$

By the uniform continuity of f, we can find $\delta = \delta(\varepsilon) > 0$ such that

$$\big|f(z) - f(x)\big| \leqslant \tfrac{\varepsilon}{2} \qquad \forall\, z \in x + \overline{B}_\delta(0)$$

(recall that $\overline{B}_\delta = \{u \in \mathbb{R}^N : \|u\| \leqslant \delta\}$). Hence

$$\big|f(z)\big| \geqslant \tfrac{\varepsilon}{2} \qquad \forall\, z \in x_n + \overline{B}_\delta(0) \qquad \forall\, n \geqslant 1.$$

It follows that

$$\int\limits_{x_n+\overline{B}_\delta(0)} \big|f(z)\big|^p\,dz \geqslant \big(\tfrac{\varepsilon}{2}\big)^p \lambda^N\big(\overline{B}_\delta(0)\big) > 0$$

(see Proposition 3.90), so

$$\lim_{n\to+\infty} \int\limits_{x_n+\overline{B}_\delta(0)} \big|f(z)\big|^p\,dz > 0,$$

which contradicts part **(a)** of the problem.

Solution of Problem 3.94

We proceed by contradiction. So, suppose that the conclusion of the problem is not true. Then we can find a sequence $\{x_n\}_{n\geqslant 1} \subseteq \mathbb{R}^N$ and $\varepsilon > 0$ such that

$$\|x_n\| \longrightarrow +\infty \quad \text{and} \quad \big|f(x_n)\big| \geqslant \varepsilon.$$

By the uniform continuity of f, we can find $\delta = \delta(\varepsilon) > 0$ such that

$$\big|f(x) - f(u)\big| \leqslant \tfrac{\varepsilon}{2} \qquad \forall\, u \in x + \overline{B}_\delta(0),$$

so

$$\tfrac{\varepsilon}{2} \leqslant \big|f(u)\big| \leqslant \|f\|_\infty \qquad \forall\, u \in x_n + \overline{B}_\delta(0).$$

From the hypothesis on ξ, we have

$$\min\left\{\xi(y) : \tfrac{\varepsilon}{2} \leqslant |y| \leqslant \|f\|_\infty\right\} = m > 0,$$

so

$$\int\limits_{x_n+\overline{B}_\delta(0)} \xi(f(x))\, dx \geqslant m\lambda^N(\overline{B}_\delta(0)).$$

Since $\xi(f(\cdot)) \in L^1(\mathbb{R}^N)$, this last inequality contradicts Problem 3.93(a).

Solution of Problem 3.95

False. Consider the function $f\colon \mathbb{R} \longrightarrow \mathbb{R}_+$, defined by

$$f(x) = \begin{cases} n & \text{if } x \in I_n = \left[n, n+\frac{1}{n^3}\right],\ n \geqslant 2, \\ 0 & \text{if } x \notin \bigcup\limits_{n\geqslant 1} I_n. \end{cases}$$

Then we have

$$\int\limits_{\mathbb{R}} f(x)\, dx = \sum_{n\geqslant 2} n\tfrac{1}{n^3} = \sum_{n\geqslant 2} \tfrac{1}{n^2} < +\infty,$$

so $f \in L^1(\mathbb{R})$ (see Definition 3.96), but clearly $f(x) \not\longrightarrow 0$ as $|x| \longrightarrow +\infty$.

Solution of Problem 3.96

(a) Let

$$A^+ = \left\{x \in \Omega : f(x) > 0\right\} \in \Sigma.$$

Then by hypothesis

$$\int\limits_{A^+} f(x)\, d\mu = 0.$$

Since $f|_{A^+} > 0$, as in the solution of Problem 3.79, we infer that $\mu(A^+) = 0$. Using the same argument on $-f$, we show that $\mu(A^-) = 0$, where

$$A^- = \left\{x \in \Omega : f(x) < 0\right\}.$$

Therefore $f = 0$ μ-almost everywhere on Ω.

(b) Let $\{S_n\}_{n\geqslant 1}$ be a sequence of simple functions such that

$$S_n(x) \longrightarrow f(x) \quad \mu\text{-almost everywhere on } \Omega$$

and

$$|S_n| \leqslant |f| \quad \mu\text{-almost everywhere on } \Omega \text{ and for all } n \geqslant 1.$$

Let

$$D_n = \{x \in \Omega : S_n(x) \neq 0\}.$$

Then for every $n \geqslant 1$, the set $D_n \in \Sigma$ has a finite μ-measure. Let

$$A = D \setminus \bigcup_{n\geqslant 1} D_n \in \Sigma$$

and let $C \subseteq A$, $C \in \Sigma$. We have

$$0 = \lim_{n\to+\infty} \int_C S_n \, d\mu = \int_C f \, d\mu$$

(see Remark 3.85), so

$$f = 0 \quad \text{almost everywhere on } A$$

(see part **(a)**). By the definition of D, we see that $\mu(A) = 0$. Then

$$D \subseteq \Big(\bigcup_{n\geqslant 1} D_n\Big) \cup A$$

and so D is σ-finite.

Alternative Solution

(b) For every $\eta > 0$, we have

$$\eta\mu\big(\{x \in \Omega : |f(x)| \geqslant \eta\}\big) \leqslant \int_{\{|f|\geqslant\eta\}} |f| \, d\mu \leqslant \int_\Omega |f| \, d\mu < +\infty.$$

Then

$$\{x \in \Omega : f(x) \neq 0\} = \bigcup_{n\geqslant 1} \{x \in \Omega : |f(x)| \geqslant \tfrac{1}{n}\},$$

so the set D is σ-finite (see Definition 3.17).

Solution of Problem 3.97

Let $x_n \longrightarrow x$ and set

$$f_n = \chi_{(-\infty,x_n]} f.$$

Then

$$|f_n| \leqslant |f| \qquad \forall\, n \geqslant 1$$

and

$$f_n(t) \longrightarrow \chi_{(-\infty,x]}(t) f(t) \qquad \forall\, t \in \mathbb{R}.$$

Invoking the Lebesgue dominated convergence theorem (see Theorem 3.94), we have

$$F(x_n) = \int_{-\infty}^{x_n} f(t)\,dt = \int_{\mathbb{R}} f_n(t)\,dt \longrightarrow \int_{\mathbb{R}} \chi_{(-\infty,x]} f(t)\,dt = \int_{-\infty}^{x} f(t)\,dt = F(x),$$

so F is continuous.

Solution of Problem 3.98

Let

$$h_n = f - f_n \qquad \forall\, n \geqslant 1.$$

Since by hypothesis

$$h_n \longrightarrow 0 \quad \mu\text{-almost everywhere on } \Omega,$$

we have that

$$h_n^+,\ h_n^- \longrightarrow 0 \quad \mu\text{-almost everywhere on } \Omega.$$

Because

$$0 \leqslant h_n^+ \leqslant f \qquad \forall\, n \geqslant 1$$

and $\int_\Omega f\,d\mu < +\infty$, by the dominated convergence theorem (see Theorem 3.94), we have

$$\int_\Omega h_n^+\,d\mu \longrightarrow 0.$$

By hypothesis $\int_\Omega h_n\,d\mu \longrightarrow 0$. Therefore

$$\int_\Omega h_n^-\,d\mu = \int_\Omega (h_n - h_n^+)\,d\mu = \int_\Omega h_n\,d\mu - \int_\Omega h_n^+\,d\mu \longrightarrow 0 \quad \text{as } n \to +\infty$$

and so finally

$$\int_{\Omega} |f - f_n| \, d\mu \;=\; \int_{\Omega} |h_n| \, d\mu \;=\; \int_{\Omega} h_n^+ \, d\mu + \int_{\Omega} h_n^- \, d\mu \;\longrightarrow\; 0 \quad \text{as } n \to +\infty.$$

Solution of Problem 3.99

Note that only $|f|$ appears in the statement of the problem. Therefore without any loss of generality, we may assume that $f \geqslant 0$. We define

$$\begin{aligned} A_0 &= \{x \in \mathbb{R}^N : \; f(x) = 0\}, \\ A_n &= \{x \in \mathbb{R}^N : \; \tfrac{1}{n} \leqslant f(x) \leqslant n\}, \\ A_\infty &= \{x \in \mathbb{R}^N : \; f(x) = +\infty\}. \end{aligned}$$

Evidently A_0, A_n and A_∞ are Lebesgue measurable sets, the sequence $\{A_n\}_{n \geqslant 1}$ is increasing and

$$\Big(\bigcup_{n \geqslant 1} A_n\Big)^c \;=\; \bigcap_{n \geqslant 1} A_n^c \;=\; A_0 \cup A_\infty.$$

Because $f \in L^1(\mathbb{R}^N)$, $\lambda^N(A_\infty) = 0$ and by the Lebesgue monotone convergence theorem (see Theorem 3.92), we have

$$\lim_{n \to +\infty} \int_{A_n^c} f \, d\lambda^N \;=\; \int_{A_0} f \, d\lambda^N + \int_{A_\infty} f \, d\lambda^N \;=\; 0.$$

Hence, for a given $\varepsilon > 0$, we can find an integer $n_0 \geqslant 1$ such that

$$\int_{A_{n_0}^c} f \, d\lambda^N \;\leqslant\; \varepsilon, \quad f|_{A_{n_0}} \text{ is bounded and } \quad \lambda^N(A_{n_0}) \;<\; +\infty,$$

since

$$\tfrac{1}{n} \lambda^N(A_{n_0}) \;\leqslant\; \int_{A_{n_0}} f \, d\lambda^N \;<\; +\infty.$$

Therefore the above argument shows that for a given $\varepsilon > 0$, we can find a Lebesgue measurable set $A \subseteq \mathbb{R}^N$, with finite measure such that $f|_A$ is bounded and

$$\int_{A^c} |f| \, d\lambda^N \;\leqslant\; \varepsilon.$$

Now let $A \subseteq \mathbb{R}^N$ be such a set corresponding to $\frac{\varepsilon}{2}$ and set $M = \sup\limits_A f$. If $C \subseteq \mathbb{R}^N$ is a Lebesgue measurable set with $\lambda^N(C) \leqslant \frac{\varepsilon}{2M}$, we have

$$\int\limits_C f\,d\lambda^N \leqslant \int\limits_{A^c} f\,d\lambda^N + \int\limits_{C\cap A} f\,d\lambda^N \leqslant \frac{\varepsilon}{2} + \frac{\varepsilon}{2M}M = \varepsilon,$$

so

$$\lim_{\lambda^N(C)\to+\infty} \int\limits_C |f|\,d\lambda^N = 0.$$

Solution of Problem 3.100

Note that for $p \in [p_1, p_2]$, we have

$$\begin{aligned} |f|^p &\leqslant |f|^{p_2} && \text{if } |f| > 1,\\ |f|^p &\leqslant |f|^{p_1} && \text{if } |f| < 1. \end{aligned}$$

Therefore, it follows that

$$|f|^p \leqslant |f|^{p_1} + |f|^{p_2}.$$

But by hypothesis $|f|^{p_1}, |f|^{p_2} \in L^1(\Omega)$. Hence

$$f \in L^p(\Omega) \qquad \forall\, p \in [p_1, p_2].$$

Suppose that $\{p_n\}_{n\geqslant 1} \subseteq [p_1, p_2]$ and $p_n \longrightarrow p \in [p_1, p_2]$. Then

$$|f|^{p_n} \longrightarrow |f|^p$$

and because

$$|f|^{p_n} \leqslant |f|^{p_1} + |f|^{p_2} \qquad \forall\, n \geqslant 1,$$

form the Lebesgue dominated convergence theorem (see Theorem 3.94), we conclude that $\|f\|_{p_n} \longrightarrow \|f\|_p$, which proves the continuity of the function $p \longmapsto \|f\|_p$ on $[p_1, p_2]$.

Solution of Problem 3.101

Note that

$$\big|(fg)(x)\big| \leqslant \max\big\{|a|, |b|\big\}\big|f(x)\big| \quad \mu\text{-almost everywhere on } \Omega.$$

Hence $fg \in L^1(\Omega)$. Moreover, we have

$$a \int_\Omega f \, d\mu \leqslant \int_\Omega fg \, d\mu \leqslant b \int_\Omega f \, d\mu.$$

Therefore, we can find $c \in [0, b]$ such that

$$\int_\Omega fg \, d\mu = c \int_\Omega f \, d\mu.$$

Solution of Problem 3.102

Let

$$f(x) = \frac{1}{x(1-\ln x)^2} \qquad \forall \, x \in [0,1].$$

First we show that $f \in L^1([0,1])$. Indeed, for every $n \geqslant 1$, we have

$$\int_{\frac{1}{n}}^{1} \frac{1}{x(1-\ln x)^2} \, dx = 1 - \frac{1}{1+\ln n} \longrightarrow 1 \quad \text{as } n \to +\infty,$$

so

$$\frac{1}{x(1-\ln x)^2} \in L^1([0,1]).$$

On the other hand, if $1 < p < +\infty$, then we can find $\xi > 0$ such that

$$\frac{1}{x^p(1-\ln x)^{2p}} \geqslant \frac{\xi}{x} \qquad \forall \, x \in (0,1]$$

and

$$\int_{\frac{1}{n}}^{1} \frac{1}{x} \, dx \longrightarrow +\infty \quad \text{as } n \to +\infty,$$

hence $f \notin L^p([0,1])$ for all $1 < p < +\infty$.

Finally, since

$$\frac{1}{x(1-\ln x)^2} \geqslant \frac{1}{x} \qquad \forall \, x \in (0,1],$$

we conclude that $f \notin L^\infty([0,1])$.

Solution of Problem 3.103
For $f(x) = \ln x$, we have

$$\int_0^1 |\ln x|^n \, dx \;=\; n! \qquad \forall\, n \geqslant 1$$

and of course $f \notin L^\infty([0,1])$.

Solution of Problem 3.104
First we show that

$$\|f\|_\infty \;\leqslant\; \lim_{p\to+\infty} \|f\|_p.$$

We may assume that $\|f\|_\infty > 0$ (otherwise the inequality is clearly true). Then for every $\xi \in \big(0, \|f\|_\infty\big)$, the set

$$A_\xi \;=\; \big\{x \in \Omega : \big|f(x)\big| > \xi\big\}$$

has positive μ-measure and for all $p \geqslant p_0$, we have

$$\xi \mu(A_\xi)^{\frac{1}{p}} \;\leqslant\; \|f\|_p.$$

Since

$$\mu(A_\xi)^{\frac{1}{p}} \;\longrightarrow\; 1 \quad \text{as } p \to +\infty,$$

we obtain

$$\xi \;\leqslant\; \lim_{p\to+\infty} \|f\|_p.$$

Because $\xi \in \big(0, \|f\|_\infty\big)$ was arbitrary, we conclude that

$$\|f\|_\infty \;\leqslant\; \liminf_{p\to+\infty} \|f\|_p.$$

Next assuming that $f \in L^\infty(\Omega)$, we will show that

$$\limsup_{p\to+\infty} \|f\|_p \;\leqslant\; \|f\|_\infty.$$

We may assume that $\|f\|_\infty > 0$ or otherwise for all $p \geqslant 1$, we have $\|f\|_p = 0$ and so we are done. Invoking the interpolation inequality (see Theorem 3.106), we have

$$\|f\|_p \;\leqslant\; \|f\|_{p_0}^{\frac{p_0}{p}} \|f\|_\infty^{1-\frac{p_0}{p}},$$

so

$$\limsup_{p \to +\infty} \|f\|_p \leqslant \|f\|_\infty$$

(since $\|f\|_{p_0}^{\frac{p_0}{p}} \longrightarrow 1$ as $p \to +\infty$).

Finally, if $f \in L^\infty(\Omega)$, then from the above two cases, we infer that

$$\|f\|_\infty = \lim_{p \to +\infty} \|f\|_p.$$

If $f \notin L^\infty(\Omega)$, then by the first inequality, we have

$$\limsup_{p \to +\infty} \|f\|_p = +\infty$$

and all the sequences extracted from $\{\|f\|_p\}_{p \in [p_0, +\infty)}$ also converge to $+\infty$. Indeed, let $p_n \longrightarrow +\infty$ with $p_n \in [p_0, +\infty)$ and suppose that

$$\|f\|_{p_n} \leqslant M \qquad \forall\, n \geqslant 1,$$

for some $M > 0$. Then from the interpolation inequality (see Theorem 3.106) for $p \in [p_n, p_{n+1}]$, we have

$$\|f\|_p \leqslant M^{1-t} M^t = M,$$

where $t \in [0,1]$ satisfies $\frac{1}{p} = \frac{1-t}{p_n} + \frac{t}{p_{n+1}}$ and so

$$\|f\|_p \leqslant M \qquad \forall\, p \geqslant p_0,$$

a contradiction.

Solution of Problem 3.105

(a) and **(b)** From the solution of Problem 3.99, we know that for a given $\varepsilon > 0$, we can find a Lebesgue measurable set $A \subseteq \mathbb{R}$ of finite Lebesgue measure such that

$$|f(x)| \leqslant M \qquad \forall\, x \in A \quad \text{and} \quad \int_{A^c} |f|\, dx \leqslant \tfrac{\varepsilon}{4}.$$

Hence $f \in L^1(\mathbb{R})$. By the Egorov theorem (see Theorem 3.76), we can find a Lebesgue measurable set $C \subseteq A$ such that

$$\lambda(A \setminus C) \leqslant \tfrac{\varepsilon}{4M}$$

(recall that λ denotes the Lebesgue measure on $\mathbb{R}$) and

$$f_n \rightrightarrows f \quad \text{on } C.$$

So, we can find an integer $n_0 \geqslant 1$ such that for all $n \geqslant n_0$, we have

$$\big|f_n(x) - f(x)\big| \leqslant \tfrac{\varepsilon}{4\lambda(C)} \qquad \forall\, x \in C$$

and

$$\Big|\int_{\mathbb{R}} (f_n - f)(x)\, dx\Big| \leqslant \tfrac{\varepsilon}{4}.$$

Then for $n \geqslant n_0$, we have

$$\begin{aligned} \Big|\int_{C^c} f_n\, dx\Big| &\leqslant \Big|\int_{A\setminus C} f\, dx\Big| + \Big|\int_{A^c} f\, dx\Big| + \Big|\int_{\mathbb{R}} (f_n - f)\, dx\Big| + \Big|\int_{C} (f_n - f)\, dx\Big| \\ &\leqslant \tfrac{\varepsilon}{4M} M + \tfrac{\varepsilon}{4} + \tfrac{\varepsilon}{4} + \lambda(C)\tfrac{\varepsilon}{4\lambda(C)} = \varepsilon. \end{aligned}$$

Moreover, if

$$h(x) = \begin{cases} \big|f(x)\big| + \frac{\varepsilon}{\lambda(C)} & \text{for } x \in C, \\ 0 & \text{for } x \in C^c, \end{cases}$$

Then

$$\big|f_n(x)\big| \leqslant h(x) \qquad \forall\, x \in C,\ n \geqslant n_0.$$

(c) The converse of **(b)** is not true. Indeed, for all $n \geqslant 2$, let

$$f_n(x) = \begin{cases} -\frac{n-2}{2} & \text{if } x \in \big[0, \frac{1}{n}\big], \\ 1 & \text{if } x \in \big(\frac{1}{n}, 1\big], \\ 0 & \text{if } x \notin \big[0, 1\big]. \end{cases}$$

Then

$$f_n \in L^1(\mathbb{R}) \qquad \forall\, n \geqslant 2$$

and

$$f_n \longrightarrow \chi_{[0,1]} \quad \text{almost everywhere on } \mathbb{R}.$$

If $C = \big[\frac{1}{2}, 1\big]$, then

$$\big|f_n(x)\big| \leqslant 1 \quad \forall x \in C,\ n \geqslant 2,$$

so

$$\int_{C^c} f_n\,dx = 0 \quad \text{and} \quad \int_{\mathbb{R}} f_n\,dx = \tfrac{1}{2} \qquad \forall\, n \geqslant 2.$$

Solution of Problem 3.106

Using the Fatou lemma (see Theorem 3.95(a)), we have

$$\int_{A^c} |f|\,dx \leqslant \varepsilon.$$

On the other hand, by the Lebesgue dominated convergence theorem (see Theorem 3.94), we have

$$\int_A |f_n - f|\,dx \longrightarrow 0 \quad \text{as } n \to +\infty.$$

Then, from the inequality

$$\int_{\mathbb{R}} |f_n - f|\,dx \leqslant \int_A |f_n - f|\,dx + \int_{A^c} |f|\,dx + \int_{A^c} |f_n|\,dx,$$

we infer that

$$\limsup_{n\to+\infty} \int_{\mathbb{R}} |f_n - f|\,dx \leqslant 2\varepsilon$$

and since $\varepsilon > 0$ was arbitrary, we conclude that

$$f_n \longrightarrow f \quad \text{in } L^1(\mathbb{R}).$$

The condition is not necessary. Indeed, for $n \geqslant 1$, let

$$f_n(x) = \begin{cases} \frac{1}{x}\sin\frac{1}{x} & \text{if } x \in \left[\frac{1}{2n}, \frac{1}{n}\right], \\ 0 & \text{otherwise.} \end{cases}$$

Evidently

$$f_n \longrightarrow 0 \quad \text{and} \quad \int_{\mathbb{R}} f_n\,dx \longrightarrow 0.$$

Suppose that the condition is satisfied by the sequence $\{f_n\}_{n\geqslant 1}$. Then it is also satisfied by the sequence $\{|f_n|\}_{n\geqslant 1}$ and so

$$\lim_{n\to+\infty}\int_n^{2n}\frac{|\sin x|}{x}\,dx \;=\; \lim_{n\to+\infty}\int_{\mathbb{R}}|f_n|\,dx \;=\; 0$$

(see Problem 3.105). But

$$\lim_{n\to+\infty}\int_n^{2n}\frac{|\sin x|}{x}\,dx \;=\; \frac{2\ln 2}{\pi},$$

a contradiction.

Solution of Problem 3.107

"$\Longrightarrow$": This implication is the Lebesgue monotone convergence theorem (see Theorem 3.92 and Remark 3.93).

"$\Longleftarrow$": Let $f_n \searrow f$ μ-almost everywhere. Then $f \geqslant 0$ and $f \in L^1(\Omega)$. Hence

$$\int_\Omega f\,d\mu \;=\; 0$$

(see Theorem 3.92). Since $f \geqslant 0$, we conclude that $f = 0$ μ-almost everywhere. Hence $f_n \searrow 0$ μ-almost everywhere on Ω.

Solution of Problem 3.108

(a) We argue by contradiction. So, suppose that $g \neq 0$. Then since the measure space is σ-finite, we can find $\varepsilon > 0$ and $C \in \Sigma$ with $0 < \mu(C) < +\infty$ such that

$$|g|\Big|_C \;\geqslant\; \varepsilon$$

and

$$\int_\Omega g(\operatorname{sgn} g)\chi_C\,d\mu \;=\; \int_\Omega |g|\chi_C\,d\mu \;\geqslant\; \varepsilon\mu(C) \;>\; 0,$$

a contradiction to the hypothesis.

(b) Note that $f_n g \longrightarrow fg$ μ-almost everywhere on Ω and

$$|f_n g| \leqslant h|g| \in L^1(\Omega)$$

(by hypothesis). Applying the Lebesgue dominated convergence theorem (see Theorem 3.94), we conclude that

$$f_n g \longrightarrow fg \quad \text{in } L^1(\Omega) \text{ as } n \to +\infty.$$

Solution of Problem 3.109

"$\Longrightarrow$": We have

$$\begin{aligned}
\sum_{n\geqslant 0} \mu(\{f \geqslant n\}) &= \sum_{n\geqslant 0}\sum_{m\geqslant n} \mu(\{m \leqslant f < m+1\}) \\
&= \sum_{m\geqslant 0}\sum_{n=0}^{m} \mu(\{m \leqslant f < m+1\}) \\
&= \sum_{m\geqslant 0} (m+1)\mu(\{m \leqslant f < m+1\}) \\
&- \sum_{m\geqslant 0} m\mu(\{m \leqslant f < m+1\}) \\
&\quad + \sum_{m\geqslant 0} \mu(\{m \leqslant f < m+1\}) \\
&\leqslant \sum_{m\geqslant 0} \int\limits_{\{m\leqslant f<m+1\}} f\,dx + \mu(\Omega) \\
&= \int\limits_{\Omega} (f+1)\,d\mu \; < \; +\infty.
\end{aligned}$$

"$\Longleftarrow$": We have

$$\begin{aligned}
\int\limits_{\Omega} f\,d\mu &= \sum_{m\geqslant 0} \int\limits_{\{m\leqslant f<m+1\}} f\,d\mu \leqslant \sum_{m\geqslant 0} (m+1)\mu(\{m \leqslant f < m+1\}) \\
&= \sum_{n\geqslant 0} \mu(\{f \geqslant n\}) \; < \; +\infty,
\end{aligned}$$

so $f \in L^1(\Omega)$.

Solution of Problem 3.110
We have

$$\int_{A(I)} (f - m(I))\, dx + \int_{I\setminus A(I)} (f - m(I))\, dx = \int_I (f - m(I))\, dx = 0,$$

so

$$\int_{A(I)} (f - m(I))\, dx = \int_{I\setminus A(I)} (m(I) - f)\, dx.$$

Then, we have

$$\begin{aligned} \int_I \left|f - m(I)\right| dx &= \int_{A(I)} \left|f - m(I)\right| dx + \int_{I\setminus A(I)} \left|f - m(I)\right| dx \\ &= \int_{A(I)} (f - m(I))\, dx + \int_{I\setminus A(I)} (m(I) - f)\, dx \\ &= 2\int_{A(I)} (f - m(I))\, dx. \end{aligned}$$

Solution of Problem 3.111
For every $a, b \geqslant 0$, we have

$$\int_a^b f(x)\, dx = \int_0^b f(x)\, dx - \int_0^a f(x)\, dx = 0.$$

Since every open set is the countable union of disjoint open intervals, it follows that

$$\int_U f(x)\, dx = 0 \qquad \forall\, U \subseteq \mathbb{R}_+\ U \text{ being open.}$$

Then from Problem 3.22, we conclude that

$$\int_A f(x)\, dx = 0 \qquad \forall\, A \subseteq \mathbb{R}_+\ A \text{ being Lebesgue measurable.}$$

Invoking Problem 3.96(a), we conclude that $f(x) = 0$ almost everywhere on $\mathbb{R}_+$.

Solution of Problem 3.112

Since $fg \geqslant 1$ almost everywhere on A, it follows that $f > 0$, $g > 0$ almost everywhere on A and so

$$\int_A f\,dx > 0 \quad \text{and} \quad \int_A g\,dx > 0.$$

We may assume that $\int_A f\,dx < +\infty$ (or otherwise the inequality is clearly true). Let

$$\varphi(x) = \tfrac{1}{x} \qquad \forall\, x > 0.$$

Clearly φ is a convex function. Applying the Jensen inequality (see Theorem 3.99), we have

$$\frac{1}{\int_A f\,dx} \leqslant \int_A \tfrac{1}{f}\,dx \leqslant \int_A g\,dx$$

(since $fg \geqslant 1$ almost everywhere on A), so

$$1 \leqslant \Big(\int_A f\,dx\Big)\Big(\int_A g\,dx\Big).$$

Solution of Problem 3.113

Let $f = \chi_\Omega$. Then since $\mu(\Omega) < +\infty$, $f \in L^r(\Omega)$ for all $1 \leqslant r \leqslant +\infty$. There exists $r \geqslant 1$ such that $\frac{1}{p} = \frac{1}{q} + \frac{1}{r}$. Invoking the generalized Hölder inequality (see Theorem 3.105), for every $h \in L^q(\Omega)$, we have

$$\|h\|_p = \|hf\|_p \leqslant \|h\|_q \|f\|_r \leqslant \|h\|_q \mu(\Omega)^{\frac{1}{p}-\frac{1}{q}} < +\infty,$$

hence $h \in L^p(\Omega)$ and so $L^q(\Omega) \subseteq L^p(\Omega)$.

If $\mu(\Omega)$ is not finite, the result fails. For example, $h = \chi_\Omega \in L^\infty(\Omega)$, but $h \notin L^p(\Omega)$ for all $1 \leqslant p < +\infty$. Also, if $f(x) = x^{-\frac{2}{3}}$, then $f \in L^2(\mathbb{R})$ abut $f \notin L^1(\mathbb{R})$.

Solution of Problem 3.114

The answer is no. Consider the Cantor function $g\colon [0,1] \longrightarrow [0,1]$ introduced in Remark 3.83. We know that g is continuous, nondecreasing, $g(0) = 0$, $g(1) = 1$ and $g'(x) = 0$ for almost all $x \in [0,1]$. Therefore

$$\int_0^1 g'\,dx = 0 < 1 = \int_0^1 dg.$$

Remark. As we will see in Chap. 4, the reason that the equality fails for g is that the Cantor function g is not absolutely continuous (see Definition 4.120).

Solution of Problem 3.115

Let

$$D = \{x \in \Omega :\ f(x) \notin C\} = \{x \in \Omega :\ \operatorname{dist}(f(x), C) > 0\}.$$

We need to show that $\mu(D) = 0$. So suppose that $\mu(D) > 0$. Of course, the set D is open. Since $\mathbb{R}^N$ is a separable, locally compact metric space (see Definitions 2.9 and 2.92), we can find a sequence of closed balls

$$\{\overline{B}_{r_n}(u_n) = \{u \in \mathbb{R}^N :\ \|u - u_n\| \leqslant r_n\}\}_{n \geqslant 1}$$

such that

$$\mathbb{R}^N \setminus C = \bigcup_{n \geqslant 1} \overline{B}_{r_n}(u_n).$$

Let $D_n = \{x \in \Omega :\ \|f(x) - u_n\| \leqslant r_n\}$. Then $D \subseteq \bigcup\limits_{n \geqslant 1} D_n$ and so as $\mu(D) > 0$, from the countable subadditivity of μ, we can find an integer $n \geqslant 1$ such that $\mu(D_n) > 0$. Let $A = D_n$. Then

$$\frac{1}{\mu(A)} \int_A f \, d\mu \in \overline{B}_{r_n}(u_n) \subseteq \mathbb{R}^N \setminus C,$$

a contradiction.

Solution of Problem 3.116

Let $S(\Omega)$ be the linear space of Σ-simple functions. From Proposition 3.110, we know that $S(\Omega) \cap L^p(\Omega)$ is dense in $L^p(\Omega)$. Let $S(\Omega, \mathcal{Y})$ denote the linear space generated by indicator functions of elements of $\mathcal{Y}$. It suffices to show that $S(\Omega, \mathcal{Y}) \cap L^p(\Omega)$ is dense in $S(\Omega) \cap L^p(\Omega)$ for the $\|\cdot\|_p$-norm. To this end, let $f \in S(\Omega) \cap L^p(\Omega)$. Then

$$f = \sum_{k=1}^N a_k \chi_{A_k},$$

with $a_k \in \mathbb{R}$, $A_k \in \Sigma$ disjoint (standard representation). Evidently $\mu(A_k) < +\infty$ for all $k \in \{1, \ldots, N\}$ (since $f \in S(\Omega) \cap L^p(\Omega)$). For a given $\varepsilon > 0$, we can find $C_k \in \mathcal{Y}$ such that $\mu(A_k \vartriangle C_k) \leqslant \varepsilon$. Let us set

$$h = \sum_{k=1}^N a_k \chi_{C_k} \in S(\Omega, \mathcal{Y}) \cap L^p(\Omega).$$

We have

$$|f(x) - h(x)| \leqslant \sum_{k=1}^N a_k \chi_{A_k \vartriangle C_k}(x) \qquad \forall\, x \in \Omega,$$

so

$$\|f - h\|_p \leqslant \varepsilon^{\frac{1}{p}} \sum_{k=1}^N |a_k|,$$

with $1 \leqslant p < +\infty$. This proves the density of $S(\Omega, \mathcal{Y}) \cap L^p(\Omega)$ in $S(\Omega) \cap L^p(\Omega)$ for the $\|\cdot\|_p$-norm. Hence $S(\Omega, \mathcal{Y}) \cap L^p(\Omega)$ is dense in $L^p(\Omega)$ too. But $S(\Omega, \mathcal{Y})$ has a countable dense subset (take $a_k \in \mathbb{Q}$). Therefore $L^p(\Omega)$ $(1 \leqslant p < +\infty)$ is separable for all $1 \leqslant p < +\infty$ (see Definition 2.9).

Solution of Problem 3.117

Using the Lebesgue monotone convergence theorem (see Theorem 3.92), we show that

$$\int_\Omega \sum_{n\geqslant 1} |f_n|\, d\mu \;=\; \sum_{n\geqslant 1} \int_\Omega |f_n|\, d\mu \;<\; +\infty,$$

so $\sum\limits_{n\geqslant 1} f_n \in L^1(\Omega)$. Therefore $\sum\limits_{n\geqslant 1} f_n$ is μ-almost everywhere finite and so it converges μ-almost everywhere to a Σ-measurable function $h\colon \Omega \longrightarrow \mathbb{R}$. Let us set

$$h_n \;=\; \sum_{k=1}^{n} f_k \qquad \forall\, n \geqslant 1.$$

Then

$$|h_n| \;\leqslant\; \sum_{k=1}^{n} |f_k| \;\in\; L^1(\Omega) \qquad \forall\, n \geqslant 1$$

and $h_n \longrightarrow h$ as $n \to +\infty$. So, by the dominated convergence theorem (see Theorem 3.94), we have

$$\int_\Omega h_n\, d\mu \;\longrightarrow\; \int_\Omega h\, d\mu, \quad \text{so} \quad \sum_{n\geqslant 1} \int_\Omega f_n\, d\mu \;=\; \int_\Omega \Big(\sum_{n\geqslant 1} f_n\Big)\, d\mu.$$

Solution of Problem 3.118

Using the Cauchy–Schwarz–Bunyakowski inequality (see Theorem 3.103 and Remark 3.104), we have

$$\begin{aligned} \|F\|_2^2 \;&=\; \int_0^1 \Big|\int_0^x f(s)\, ds\Big|^2 dx \;\leqslant\; \int_0^1 \Big(\int_0^x 1^2\, ds\Big)\Big(\int_0^x |f(s)|^2\, ds\Big)\, dx \\ &\leqslant\; \int_0^1 x\|f\|_2^2\, dx \;=\; \tfrac{1}{2}\|f\|_2^2 \;<\; \|f\|_2^2, \end{aligned}$$

So $\|F\|_2 < \|f\|_2$.

Solution of Problem 3.119
We have

$$\sup_{n\geqslant 1} \int_{\mathbb{R}} \sum_{k=1}^{n} |f_{k+1} - f_k|\, dt \;\leqslant\; \sum_{k\geqslant 1} \left(\tfrac{1}{(k+1)^2} + \tfrac{1}{k^2}\right) \;<\; +\infty,$$

so

$$\sum_{k\geqslant 1} |f_{k+1} - f_k| \;<\; +\infty \quad \text{almost everywhere on } \mathbb{R}.$$

Therefore for all $x \in \mathbb{R} \setminus N$ with $\lambda(N) = 0$, we have

$$f_n(x) \;=\; f_1(x) + \sum_{k=2}^{n} (f_k - f_{k-1})(x)$$

and so the sequence $\{f_n(x)\}_{n\geqslant 1}$ converges on $\mathbb{R} \setminus N$ to $f(x)$ (recall that $f_n \longrightarrow f$ in $L^1(\mathbb{R})$).

Solution of Problem 3.120
Note that

$$f - \eta \;\leqslant\; (f - \eta)\chi_{A_\eta} \;\leqslant\; f\chi_{A_\eta},$$

so, using Cauchy–Schwarz–Bunyakowski inequality (see Theorem 3.103 and Remark 3.104), we have

$$\begin{aligned} 0 \;<\; \theta - \eta \;\leqslant\; \int_0^1 f\, dx - \eta \;&=\; \int_0^1 \big(f(x) - \eta\big)\, dx \\ \leqslant\; \int_0^1 f\chi_{A_\eta}\, dx \;&\leqslant\; \|f\|_2 \lambda(A_\eta)^{\frac{1}{2}} \;=\; \lambda(A_\eta)^{\frac{1}{2}}. \end{aligned}$$

Solution of Problem 3.121
Since $\sum\limits_{n\geqslant 1} \beta_n f_n(x)$ converges for μ-almost all $x \in \Omega$, we have

$$\beta_n f_n(x) \;\longrightarrow\; 0 \quad \text{for } \mu\text{-almost all } x \in \Omega.$$

Choose $\delta > 0$ such that $M^2\delta < 1$. By the Egorov theorem (see Theorem 3.76), we can find $A \in \Sigma$ such that

$$\mu(\Omega \setminus A) \leqslant \delta$$

and

$$\xi_n = \sup_{x\in A} \left|\beta_n f_n(x)\right| \longrightarrow 0 \quad \text{as } n \to +\infty.$$

Hence we have

$$\beta_n^2 = \int_\Omega \left(\beta_n f_n\right)^2 dx = \int_A \left(\beta_n f_n\right)^2 dx + \int_{\Omega\setminus A} \left(\beta_n f_n\right)^2 dx \leqslant \xi_n^2 \mu(\Omega) + \beta_n^2 M^2 \delta,$$

so

$$\beta_n^2 \leqslant \tfrac{\xi_n^2}{1-M^2\delta}\mu(\Omega) \longrightarrow 0 \quad \text{as } n \to +\infty.$$

Solution of Problem 3.122

(a) Since $f \in L^1(\Omega) \cap L^2(\Omega)$, we may assume that $f(x) \in \mathbb{R}$ for all $x \in \mathbb{R}$. Let $A = \left\{|f| \geqslant 1\right\}$ and consider the function

$$h(x) = \begin{cases} f(x)^2 & \text{if } x \in A, \\ \left|f(x)\right| & \text{if } x \in \Omega \setminus A, \end{cases}$$

i.e., $h = \chi_A f^2 + \chi_{\Omega\setminus A}|f|$. It follows that $h \in L^1(\Omega)$. For $p \in (1,2)$, we have

$$\left|f(x)\right|^p \leqslant h(x),$$

so $f \in L^p(\Omega)$.

(b) Let $\{p_n\}_{n\geqslant 1} \subseteq [0,1]$ be a sequence such that $p_n \longrightarrow 1^+$. Then

$$|f|^{p_n} \longrightarrow |f| \quad \text{on } \Omega$$

and from part **(a)**, we have

$$\left|f(x)\right|^{p_n} \leqslant h(x) \qquad \forall\, x \in \Omega.$$

So, we can apply the Lebesgue dominated convergence theorem (see Theorem 3.94) and obtain $\|f\|_{p_n} \longrightarrow \|f\|_1$. Therefore $\lim\limits_{p\to 1^+} \|f\|_p = \|f\|_1$.

Solution of Problem 3.123

Let $\varphi(x) = \sqrt{1+x^2}$, $x \in \mathbb{R}$. Note that

$$\varphi''(x) = \frac{1}{(1+x^2)^{\frac{3}{2}}} \geqslant 0 \qquad \forall\, x \in \mathbb{R},$$

hence φ is convex. If, h is integrable, then by the Jensen inequality (see Theorem 3.99), we have

$$\sqrt{1+A^2} \leqslant \int\limits_A \sqrt{1+h^2}\, d\mu.$$

If $A = +\infty$, then since $h < \sqrt{1+h^2}$, equality holds.

Finally, note that

$$\sqrt{1+h^2} \leqslant 1+h$$

and so we obtain

$$\int\limits_\Omega \sqrt{1+h^2}\, d\mu \leqslant 1+A.$$

Next, suppose that $\Omega = [0,1]$, Σ is the σ-algebra of Lebesgue measurable sets, $\mu = \lambda$ and $h = f'$ with $f \in C^1([0,1])$. Consider the figure

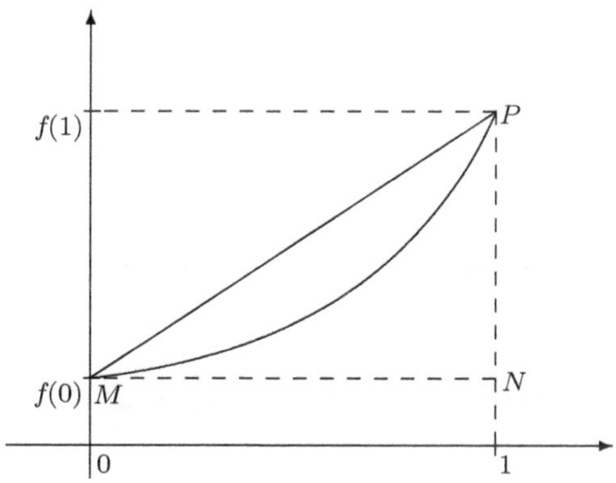

Then $A = f(1) - f(0)$ and $\int_0^1 \sqrt{1+f^2}\, dx$ is the length of the arc $\widehat{\mathrm{MP}}$. From the inequalities provided above, we have

$$\overline{MP} \leqslant \widehat{\mathrm{MP}} \leqslant \overline{MN} + \overline{NP}.$$

Therefore, the first inequality is an equality provided $f(x) = ax+b$ (with $b, a+b \in [0,1]$), while the second inequality is in fact an equality if f is a constant. So, we conjecture that the first inequality is an equality if $h = constant$ and the second inequality is an equality if $h = 0$. Let us prove these. So, first we assume that

$$\sqrt{1+A^2} = \int_\Omega \sqrt{1+h^2}\, d\mu.$$

Note that

$$(1+Ah)^2 \leqslant (1+h^2)(1+A^2),$$

so

$$\sqrt{1+A^2} + \tfrac{A(h-A)}{\sqrt{1+A^2}} \leqslant \sqrt{1+h^2}.$$

But the integrals over Ω of the two sides of this inequality are equal. Therefore

$$\sqrt{1+A^2} + \tfrac{A(h-A)}{\sqrt{1+A^2}} = \sqrt{1+h^2},$$

so

$$2Ah = A^2 + h^2,$$

i.e., $h = A$ and so h is a constant function.

Next assume that

$$\int_\Omega \sqrt{1+h^2}\, d\mu = 1 + A,$$

so

$$\int_\Omega \sqrt{1+h^2}\, d\mu = \int_\Omega (1+h)\, d\mu.$$

But $\sqrt{1+h^2} \leqslant 1+h$ and so it follows that $\sqrt{1+h^2} = 1+h$, hence $h = 0$.

Solution of Problem 3.124

Let $q = \frac{1}{p}$ and let $h = f^p$, $u = g^p$. Since $f, g \in L^1(\Omega)$, it follows that $h, u \in L^q(\Omega)$, $1 < q$. Using the Minkowski inequality (see Theorem 3.101), we have

$$\int_\Omega (f^p+g^p)^{\frac{1}{p}}\,d\mu \;=\; \int_\Omega (h+u)^q\,d\mu \;=\; \|h+u\|_q^q \;\leqslant\; \big(\|h\|_q+\|u\|_q\big)^q$$
$$=\; \Big(\int_\Omega f\,d\mu\Big)^p + \Big(\int_\Omega g\,d\mu\Big)^p.$$

Solution of Problem 3.125

(a) The answer is negative. To see this, let $f_n = n^{\frac{1}{p}}\chi_{(0,\frac{1}{n})}$ for $n \geqslant 1$. Then

$$f_n(x) \longrightarrow 0 \qquad \forall\, x \in [0,1].$$

On the other hand

$$\|f_n - f\|_p \;=\; \|f_n\|_p \;=\; 1 \qquad \forall\, n \geqslant 1.$$

(b) By the Fatou lemma (see Theorem 3.95), we have

$$\|f\|_p \;=\; \Big(\int_{\mathbb{R}} |f|^p\,dx\Big)^{\frac{1}{p}} \;\leqslant\; \liminf_{n\to+\infty} \Big(\int_{\mathbb{R}} |f_n|^p\,dx\Big)^{\frac{1}{p}} \;\leqslant\; M.$$

Solution of Problem 3.126

Note that for every $x \geqslant 0$:

$$\text{if } 1+x^2 \leqslant e^x, \quad \text{then } \ln(1+x^2) \leqslant x.$$

Hence

$$n\ln\Big(1+\tfrac{|f(x)|^2}{n^2}\Big) \;\leqslant\; \big|f(x)\big| \qquad \forall\, x \in \mathbb{R}.$$

Also recall that

$$n\ln\Big(1+\tfrac{|f(x)|^2}{n^2}\Big) \;\leqslant\; n\tfrac{|f(x)|^2}{n^2} \;=\; \tfrac{|f(x)|^2}{n} \;\longrightarrow\; 0 \quad \text{for almost all } x \in \Omega.$$

Therefore by the Lebesgue dominated convergence theorem (see Theorem 3.94), we have

$$\lim_{n\to+\infty} \int_{\mathbb{R}} n\ln\Big(1+\tfrac{|f(x)|^2}{n^2}\Big) \;=\; 0.$$

Solution of Problem 3.127

(a) Let $\Omega = [0,1]$, let Σ be the Lebesgue σ-algebra of $[0,1]$ and let $\mu = \lambda$ be the Lebesgue measure on $[0,1]$. Let

$$f = \chi_{[0,\frac{1}{2})} \quad \text{and} \quad h = \chi_{(\frac{1}{2},1]}.$$

Then $f, g \in L^p([0,1])$, $p \in (0,1)$. We have

$$\|f\|_p + \|h\|_p = \left(\tfrac{1}{2}\right)^{\frac{1}{p}} + \left(\tfrac{1}{2}\right)^{\frac{1}{p}} = 2^{1-\frac{1}{p}} < 1 = \|f+h\|_p$$

So, $\|\cdot\|_p$ is not a norm since the triangle inequality fails.

(b) Clearly $d(f,h) > 0$, if $f \neq h$ and $d(f,f) = 0$. Also

$$d(f,h) = d(h,f) \qquad \forall\, f,h \in L^p(\Omega).$$

Finally, if $f,h \in L^p(\Omega)$, then we have

$$\begin{aligned}\int_\Omega |f-h|^p\,d\mu &= \int_\Omega |f-g+g-h|^p\,d\mu \leqslant \int_\Omega \big(|f-g|+|g-h|\big)^p\,d\mu \\ &\leqslant \int_\Omega \big(|f-g|^p+|g-h|^p\big)\,d\mu = d(f,g)+d(g,h).\end{aligned}$$

This proves that d is a metric on $L^p(\Omega)$.

Next we check the completeness of $\big(L^p(\Omega),d\big)$. To this end, let $\{f_n\}_{n\geqslant 1}$ be a d-Cauchy sequence. By passing to a suitable subsequence if necessary, we may assume that

$$d(f_{n+1},f_n) = \int_\Omega |f_{n+1}-f_n|^p\,d\mu < \tfrac{1}{2^n} \qquad \forall\, n \geqslant 1.$$

Let $g_1 = 0$ and

$$g_n = |f_1| + |f_2-f_1| + \ldots + |f_n-f_{n-1}| \qquad \forall\, n \geqslant 2.$$

Then $g_n \geqslant 0$ and $\{g_n\}_{n\geqslant 1}$ is an increasing sequence. We have

$$\int_\Omega g_n^p\,d\mu \leqslant \int_\Omega |f_1|^p\,d\mu + \sum_{k=2}^n \int_\Omega |f_k-f_{k-1}|^p\,d\mu \leqslant \int_\Omega |f_1|^p\,d\mu + 1 < +\infty.$$

So, we can find $g \in L^p(\Omega)$ such that $g_n \nearrow g$. Note that

$$|f_{n+k} - f_n| = \Big|\sum_{k=n+1}^{n+k} (f_k - f_{k-1})\Big| \leqslant \sum_{k=n+1}^{n+k} |f_k - f_{k-1}| = g_{n+k} - g_n,$$

so

$$f_n \longrightarrow f \quad \mu\text{-almost everywhere on } \Omega,$$

with $f\colon \Omega \longrightarrow \mathbb{R}$ Σ-measurable.

Observe that

$$|f_n| = \Big|f_1 + \sum_{k=2}^{n}(f_k - f_{k-1})\Big| \leqslant g_n \leqslant g \quad \mu\text{-almost everywhere on } \Omega,$$

so

$$|f| \leqslant g \quad \mu\text{-almost everywhere on } \Omega$$

and thus $f \in L^p(\Omega)$.

Since

$$|f_n - f| \leqslant 2g \quad \mu\text{-almost everywhere on } \Omega$$

and

$$|f_n - f|^p \longrightarrow 0 \quad \mu\text{-almost everywhere on } \Omega,$$

by the Lebesgue dominated convergence theorem (see Theorem 3.94), we have

$$d(f_n, f) = \int_\Omega |f_n - f|^p \, d\mu \longrightarrow 0 \quad \text{as } n \to +\infty,$$

so $\big(L^p(\Omega), d\big)$ is a complete metric space.

Solution of Problem 3.128

Let $\{q_n\}_{n\geqslant 1}$ be an enumeration of the rationals and for every $n \geqslant \mathbb{N}$, we set

$$f_n(x) = \begin{cases} \dfrac{1}{\sqrt{x - q_n}} & \text{if } x \in (q_n, q_n + 1), \\ 0 & \text{otherwise.} \end{cases}$$

Then we define

$$f(x) = \sum_{n\geqslant 1} \tfrac{1}{2^n} f_n(x).$$

We have

$$\int_{\mathbb{R}} f(x)\,dx \;=\; \sum_{n\geqslant 1} \tfrac{1}{2^n} \int_{\mathbb{R}} f_n(x)\,dx \;=\; \sum_{n\geqslant 1} \tfrac{2}{2^n} \;=\; 2,$$

so $f \in L^1(\mathbb{R})$.

If $a, b \in \mathbb{R}$, $a < b$ and $M > 0$, then we can find a rational $q \in (a, b)$ and an open interval $(q, r) \subseteq (a, b)$ such that

$$f(x) \;\geqslant\; M \qquad \forall\, x \in (q, r).$$

Therefore, the set

$$(a, b) \cap \big\{x \in \mathbb{R} :\; f(x) \geqslant M\big\}$$

has positive measure.

Solution of Problem 3.129

Let $h\colon \mathbb{R} \longrightarrow \mathbb{R}$ be defined by

$$h(x) \;=\; \chi_{(0,1)}(x)\tfrac{1}{\sqrt{x}}.$$

Then

$$h \;\in\; L^1(\mathbb{R}) \quad \text{and} \quad \int_{\mathbb{R}} h\,dx \;=\; 2.$$

Also, let $\{q_n\}_{n\geqslant 1}$ be an enumeration of the rationals and let $\{\beta_n\}_{n\geqslant 1}$ be a sequence of positive real numbers such that

$$\sum_{n\geqslant 1} \beta_n \;<\; +\infty.$$

We introduce $f\colon \mathbb{R} \longrightarrow \mathbb{R}$, defined by

$$f(x) \;=\; \sum_{n\geqslant 1} \beta_n h(x - q_n).$$

Clearly $f \in L^1(\mathbb{R})$. Let $a, b \in \mathbb{R}$, $a < b$ and consider the open interval (a, b). We can find a rational $q_k \in (a, b)$. Then

$$\int_a^b f^2\,dx \;>\; \int_{q_k}^b f^2\,dx \;>\; \beta_k^2 \int_{q_k}^b h(x - q_k)^2\,dx \;=\; \beta_k^2 \int_0^{b-q_k} \tfrac{1}{x}\,dx \;=\; +\infty.$$

Remark. The function f produced in the above solution is not bounded over any neighbourhood of any point.

Solution of Problem 3.130

Evidently $f_n \in L^{p'}(\Omega)$. So, using the Hölder inequality (see Theorem 3.103), we have

$$\begin{aligned}\Big|\int_{\Omega} h f_n\,d\mu\Big| &= \Big|\int_{\Omega} (h\chi_{A_n}) f_n\,d\mu\Big| \leqslant \Big(\int_{\Omega} (h\chi_{A_n})^p\,d\mu\Big)^{\frac{1}{p}}\Big(\int_{\Omega} f_n^{p'}\,d\mu\Big)^{\frac{1}{p'}}\\ &= \Big(\int_{A_n} h^p\,d\mu\Big)^{\frac{1}{p}} \longrightarrow 0,\end{aligned}$$

since $\mu(A_n) \longrightarrow 0$ as $r \to +\infty$.

Solution of Problem 3.131

The answer is negative. Let $f \in L^\infty(\mathbb{R})$ be such that $f \notin L^1(\mathbb{R})$ and

$$\lim_{|x|\to+\infty} f(x) = 0$$

(for example, consider the function $f(x) = \frac{1}{1+|x|}$, for $x \in \mathbb{R}$). Let $f_n = \chi_{[-n,n]} f$. Then $\{f_n\}_{n\geqslant 1} \subseteq L^1(\mathbb{R})$ and

$$f_n \rightrightarrows f.$$

Solution of Problem 3.132

First we show that S is closed. So, let $\{f_n\}_{n\geqslant 1} \subseteq S$ be a sequence such that $f_n \longrightarrow f$ in $L^p(I)$. Then by passing to a suitable subsequence, we may assume that

$$f_n \longrightarrow f \quad \text{almost everywhere on } I$$

(see Theorem 3.98). Since $f_n \leqslant h$ almost everywhere on I, we infer that $f \leqslant h$ almost everywhere on I and so $f \in S$. This proves that

$S \subseteq L^p(\Omega)$ is closed. Next we show that $\operatorname{int} S = \emptyset$ proving that S is nowhere dense in $L^p(I)$ (see Definition 1.25). So, let $f \in S$, $\varepsilon > 0$ and choose a subinterval $T \subseteq I$ with $\lambda(I \setminus T) > 0$ such that

$$\int_T |h + 1 - f|^p \, dx \leqslant \varepsilon^p.$$

Let us set

$$\widehat{f}(x) = \begin{cases} h(x) + 1 & \text{if } x \in T, \\ f(x) & \text{if } x \in I \setminus T. \end{cases}$$

Then $\widehat{f} \in L^p(I)$ and $\widehat{f} \notin S$. Moreover,

$$\|\widehat{f} - f\|_p = \Big(\int_T |h + 1 - f|^p \, dx\Big)^{\frac{1}{p}} \leqslant \varepsilon.$$

Since $\varepsilon > 0$ was arbitrary, we conclude that $\operatorname{int} S = \emptyset$, hence S is nowhere dense in $L^p(I)$.

Solution of Problem 3.133

Let $\eta = \int_a^b f(t)\, dt$. Then

$$\eta \min_{[a,b]} h \leqslant \int_a^b f(t)h(t)\, dt \leqslant \eta \max_{[a,b]} h.$$

Since the map $s \longmapsto \eta h(s)$ is continuous, by Bolzano theorem, we can find $s_0 \in [a, b]$ such that

$$\eta h(s_0) = \int_a^b f(t)h(t)\, dt.$$

Solution of Problem 3.134

If $f \in C_c^1(\mathbb{R}^N)$ and $z_n \longrightarrow z$, then

$$\big\|T_f(z_n) - T_f(z)\big\|_p^p = \int_{\mathbb{R}^N} \big|f(x + z_n) - f(x + z)\big|^p \, dx \longrightarrow 0.$$

For the general $f \in L^p(\mathbb{R}^N)$, we can find a sequence $\{f_n\}_{n\geqslant 1} \subseteq C_c^1(\mathbb{R}^N)$ such that $f_n \longrightarrow f$ in $L^p(\mathbb{R}^N)$ (see Proposition 3.112). For every $z \in \mathbb{R}^N$, we have

$$\|T_{f_n}(z) - T_f(z)\|_p^p = \int\limits_{\mathbb{R}^N} |f_n(x+z) - f(x+z)|^p \, dx = \int\limits_{\mathbb{R}^N} |f_n(x) - f(x)|^p \, dx = \|f_n - f\|_p^p,$$

so

$$T_{f_n} \rightrightarrows T_f.$$

From the first part of the proof, we know that T_{f_n} is continuous. Therefore, T_f is continuous too (see Proposition 1.62).

Solution of Problem 3.135

Let $\varepsilon > 0$ be given. We can find $\delta > 0$ such that, if $A \in \Sigma$ with $\mu(A) \leqslant \delta$, then $\int_A h \, d\mu \leqslant \varepsilon$. Let $\lambda = \frac{1}{\delta}\|h\|_1$. Then using the Markov inequality (see Proposition 3.90(f)), we have

$$\mu(\{f > \lambda\}) \leqslant \frac{1}{\lambda} \int\limits_{\{f>\lambda\}} f \, d\mu \leqslant \frac{1}{\lambda} \int\limits_{\{f>\lambda\}} h \, d\mu \leqslant \frac{1}{\lambda}\|h\|_1 = \delta,$$

so

$$\int\limits_{\{f>\lambda\}} f \, d\mu \leqslant \int\limits_{\{f>\lambda\}} h \, d\mu \leqslant \varepsilon$$

and thus $\mathcal{F}$ is uniformly integrable (see Definition 3.124).

Solution of Problem 3.136

Let

$$\mathcal{Y} = \{f \colon \Omega \longrightarrow \{0,1\} : f|_{\Omega_n} \text{ is constant for every } n \geqslant 1\}.$$

Then $\operatorname{card} \mathcal{Y} = \operatorname{card} \{0,1\}^{\mathbb{N}}$ and so $\mathcal{Y}$ is not countable. If $u, v \in \mathcal{Y}$, $u \neq v$, then $\|u - v\|_\infty = 1$. Therefore $\mathcal{Y}$ is not separable (see Definition 2.9). But $\mathcal{Y} \subseteq L^\infty(\Omega)$. Hence $L^\infty(\Omega)$ cannot be separable.

Solution of Problem 3.137
We know that

$$f = \widehat{g} - \widehat{h} \quad \text{with } \widehat{g}, \widehat{h} \in L^1(\Omega),\ \widehat{g}, \widehat{h} \geqslant 0$$

(see Remark 3.72). We know that we can find an increasing sequence $\{s_n\}_{n \geqslant 1}$ of positive simple functions such that

$$\int_\Omega s_n \, d\mu \longrightarrow \int_\Omega \widehat{h} \, d\mu.$$

Hence we can find $n_0 \geqslant 1$ such that

$$\int_\Omega (\widehat{h} - s_{n_0}) \, d\mu \leqslant \varepsilon.$$

Then we choose $g = \widehat{g} - s_{n_0}$ and $h = \widehat{h} - s_{n_0} \geqslant 0$. Evidently $g - h = \widehat{g} - \widehat{h} = f$.

Solution of Problem 3.138
By the Fatou lemma (see Theorem 3.95), we have

$$\|u\|_p \leqslant \liminf_{n \to +\infty} \|u_n\|_p < +\infty,$$

so $u \in L^p(\Omega)$. Note that

$$\lim_{|t| \to +\infty} \frac{|t+1|^p - |t|^p - 1}{|t|^p} = 0.$$

From this, it follows that for a given $\varepsilon > 0$, we can find $c(\varepsilon) > 0$ such that

$$\big||a+b|^p - |a|^p - |b|^p\big| \leqslant \varepsilon |a|^p + c(\varepsilon)|b|^p \qquad \forall\, a, b \in \mathbb{R}.$$

Using this inequality with $a = u_n(\omega) - u(\omega)$ and $b = u(\omega)$, we have

$$\big||u_n(\omega)|^p - |u_n(\omega) - u(\omega)|^p - |u(\omega)|^p\big| \leqslant \varepsilon \big|u_n(\omega) - u(\omega)\big|^p + c(\varepsilon)|u(\omega)|^p.$$

Setting

$$v_n(\omega) = \big|u_n(\omega)\big|^p - \big|u_n(\omega) - u(\omega)\big|^p - \big|u(\omega)\big|^p,$$

we have

$$\left|v_n(\omega)\right| - \varepsilon\left|u_n(\omega) - u(\omega)\right|^p \leqslant c(\varepsilon)\left|u(\omega)\right|^p.$$

By the Lebesgue dominated convergence theorem (see Theorem 3.94), we have

$$\lim_{n\to+\infty} \int\limits_\Omega \left(|v_n| - \varepsilon|u_n - u|^p\right) d\mu = 0.$$

But

$$\int\limits_\Omega |v_n|\, d\mu = \int\limits_\Omega \left(|v_n| - \varepsilon|u_n - u|^p\right) d\mu + \varepsilon\|u_n - u\|_p^p,$$

so

$$\int\limits_\Omega |v_n|\, d\mu \leqslant \int\limits_\Omega \left(|v_n| - \varepsilon|u_n - u|^p\right) d\mu + \varepsilon M \qquad \forall\, n \geqslant 1,$$

for some $M > 0$ and thus

$$\limsup_{n\to+\infty} \int\limits_\Omega |v_n|\, d\mu \leqslant \varepsilon M.$$

Since $\varepsilon > 0$ was arbitrary, we conclude that

$$\|u_n\|_p^p - \|u_n - u\|_p^p \longrightarrow \|u\|_p^p.$$

Solution of Problem 3.139

From hypothesis **(a)**, we have that

$$\sup_{n\geqslant 1} \|u_n\|_p < +\infty.$$

Using Problem 3.138, we have

$$\|u_n\|_p^p - \|u_n - u\|_p^p \longrightarrow \|u\|_p^p,$$

so, using also hypothesis **(a)**, we have

$$\|u_n - u\|_p \longrightarrow 0.$$

Solution of Problem 3.140

(a) Let $h = \max\{f, g\}$ and $1 \leqslant p < +\infty$. Then

$$\begin{aligned}\int_\Omega |h|^p\, d\mu &= \int_{\{f\leqslant g\}} |h|^p\, d\mu + \int_{\{f>g\}} |h|^p\, d\mu \\ &= \int_{\{f\leqslant g\}} |g|^p\, d\mu + \int_{\{f>g\}} |f|^p\, d\mu \leqslant \|g\|_p^p + \|f\|_p^p < +\infty,\end{aligned}$$

so $h \in L^p(\Omega)$.

When $p = +\infty$ the result is clear.

(b) Let $h_n = \max\{f_n, g_n\}$ and let $h = \max\{f, g\}$. From part **(a)**, we know that $h \in L^p(\Omega)$ and $\{h_n\}_{n\geqslant 1} \subseteq L^p(\Omega)$ (for $1 \leqslant p \leqslant +\infty$). Note that

$$h_n = \tfrac{1}{2}\big(|f_n - g_n| + f_n + g_n\big) \quad \text{and} \quad h = \tfrac{1}{2}\big(|f - g| + f + g\big).$$

Therefore $h_n \longrightarrow h$ in $L^p(\Omega)$ (with $1 \leqslant p \leqslant +\infty$).

Solution of Problem 3.141

Let $M = \sup\limits_{n\geqslant 1} \|g_n\|_\infty < +\infty$. Evidently $\|g\|_\infty \leqslant M$. Note that $fg \in L^p(\Omega)$ and $\{f_n g_n\}_{n\geqslant 1} \subseteq L^p(\Omega)$. We have

$$\begin{aligned}\int_\Omega |f_n g_n - fg|^p\, d\mu &\leqslant 2^{p-1}\int_\Omega |f_n g_n - f g_n|^p\, d\mu + 2^{p-1}\int_\Omega |f g_n - fg|^p\, d\mu \\ &\leqslant 2^{p-1}M^p \int_\Omega |f_n - f|^p\, d\mu + 2^{p-1}\int_\Omega |f|^p |g_n - g|^p\, d\mu.\end{aligned}$$

By hypothesis, we have $2^{p-1}M^p \int\limits_\Omega |f_n - f|^p\, d\mu \longrightarrow 0$. Also, we have

$$|f|^p |g_n - g|^p \longrightarrow 0 \quad \text{almost everywhere in } \Omega$$

and $|f|^p|g_n - g|^p \leqslant 2M^p|f|^p$, with $2M\big|f(\cdot)\big|^p \in L^1(\Omega)$.

So, by the Lebesgue dominated convergence theorem (see Theorem 3.94), we have

$$2^{p-1}\int_\Omega |f|^p |g_n - g|^p\, d\mu \longrightarrow 0.$$

Therefore, finally

$$\int_\Omega |f_n g_n - fg|^p \, d\mu \longrightarrow 0,$$

so

$$f_n g_n \longrightarrow fg \quad \text{in } L^p(\Omega).$$

Solution of Problem 3.142
Clearly

$$f(\omega) \leqslant h(\omega) \leqslant g(\omega) \quad \mu\text{-almost everywhere on } \Omega$$

and so $h \in L^1(\Omega)$. We also have

$$|h(\omega)| \leqslant |f(\omega)| + |g(\omega)| \quad \mu\text{-almost everywhere on } \Omega$$

and so by the Fatou lemma (see Theorem 3.95), we have

$$\begin{aligned}\int_\Omega h \, d\mu - \int_\Omega f \, d\mu &= \int_\Omega \lim_{n \to +\infty} (h_n - f_n) \, d\mu \\ &\leqslant \liminf_{n \to +\infty} \int_\Omega (h_n - f_n) \, d\mu = \liminf_{n \to +\infty} \int_\Omega h_n \, d\mu - \int_\Omega f \, d\mu,\end{aligned}$$

so

$$\int_\Omega h \, d\mu \leqslant \liminf \int_\Omega h_n \, d\mu.$$

Similarly, using g_n, we have

$$\limsup_{n \to +\infty} \int_\Omega h_n \, d\mu \leqslant \int_\Omega h \, d\mu.$$

Therefore, finally we have

$$\int_\Omega h_n \, d\mu \longrightarrow \int_\Omega h \, d\mu.$$

Solution of Problem 3.143
Note that

$$0 \leqslant |f_n - f| \leqslant |f_n| + |f|$$

and apply Problem 3.142, to conclude that

$$\|f_n - f\|_1 = \int_\Omega |f_n - f| \, d\mu \longrightarrow 0.$$

Solution of Problem 3.144
Let

$$C = \{\omega \in \Omega : |f(\omega)| \geqslant \lambda \|f\|_1\}$$

and set $h = \chi_C |f|$. Then

$$\Big(\int_\Omega h \, d\mu\Big)^p \leqslant \mu(C)^{p-1} \int_\Omega h^p \, d\mu \leqslant \mu(C)^{p-1} \int_\Omega |f|^p \, d\mu.$$

Since $(1-\lambda)\|f\|_1 \leqslant \|h\|_1$, we have

$$(1-\lambda)^p \|f\|_1^p \leqslant \mu(C)^{p-1} \|f\|_p^p,$$

so

$$(1-\lambda)^{p'} \frac{\|f\|_1^{p'}}{\|f\|_p^{p'}} \leqslant \mu(C),$$

which is what we wanted.

Solution of Problem 3.145
By Proposition 3.110, it suffices to show that for any $A \in \mathcal{B}(X)$, we can find a sequence $\{u_n\}_{n \geqslant 1} \subseteq C_b(X)$ such that

$$u_n \longrightarrow \chi_A \quad \text{in } L^p(X).$$

First we assume that A is closed. We set

$$u_n(x) = \begin{cases} 1 - n \operatorname{dist}(x, A) & \text{if } \operatorname{dist}(x, A) \leqslant \frac{1}{n}, \\ 0 & \text{if } \operatorname{dist}(x, A) > \frac{1}{n}. \end{cases}$$

Evidently $u_n \in C_b(X)$, $0 \leqslant u_n \leqslant 1$ for all $n \geqslant 1$ and

$$u_n(x) \longrightarrow \chi_A(x) \qquad \forall\, x \in X.$$

The Lebesgue dominated convergence theorem (see Theorem 3.94) implies that

$$u_n \longrightarrow \chi_A \quad \text{in } L^p(X).$$

Next, let $\widehat{C}$ be the L^p-closure of $C_b(X)$ and let us set

$$\mathcal{D} = \{A \in \mathcal{B}(X) : \chi_A \in \widehat{C}\}.$$

Then, we can easily check that $\mathcal{D}$ is a Dynkin class with generating π-class the closed sets. Hence by the Dynkin theorem (see Theorem 3.9), we conclude that $\mathcal{D} = \mathcal{B}(X)$.

Solution of Problem 3.146

Note that

$$\int_\Omega (u - u_n)^+\, d\mu = \int_\Omega (u - u_n)^-\, d\mu \qquad \forall\, n \geqslant 1.$$

So, we have

$$\int_\Omega |u - u_n|\, d\mu = 2\int_\Omega (u - u_n)^+\, d\mu.$$

Since by hypothesis

$$\liminf_{n\to+\infty} u_n(\omega) \geqslant u(\omega) \quad \mu\text{-almost everywhere on } \Omega,$$

we infer that

$$(u - u_n)^+(\omega) \longrightarrow 0 \quad \mu\text{-almost everywhere in } \Omega.$$

Moreover

$$(u - u_n)^+(\omega) \leqslant u^+(\omega) \quad \mu\text{-almost everywhere in } \Omega.$$

So, by the Lebesgue dominated convergence theorem (see Theorem 3.94), we obtain

$$\int_\Omega (u - u_n)^+\, d\mu \longrightarrow 0,$$

so

$$u_n \longrightarrow u \quad \text{in } L^1(\Omega).$$

Solution of Problem 3.147

First, suppose that

$$g_n(\omega) \longrightarrow g(\omega) \quad \mu\text{-almost everywhere on } \Omega.$$

From the Fatou lemma (see Theorem 3.95), we have

$$\liminf_{n\to+\infty} \int_\Omega (f_n + g_n)\, d\mu \geqslant \int_\Omega \big(\liminf_{n\to+\infty} f_n + g\big)\, d\mu,$$

so

$$\liminf_{n\to+\infty} \int_\Omega f_n\, d\mu + \limsup_{n\to+\infty} \int_\Omega g_n\, d\mu \geqslant \int_\Omega \liminf_{n\to+\infty} f_n\, d\mu + \int_\Omega g\, d\mu$$

and thus

$$\liminf_{n\to+\infty} \int_\Omega f_n\, d\mu \geqslant \int_\Omega \liminf_{n\to+\infty} f_n\, d\mu$$

(since $g_n \longrightarrow g$ in $L^1(\Omega)$).

Now the general case. Let $\{f_{n_k}\}_{k\geqslant 1}$ be a subsequence of the sequence $\{f_n\}_{n\geqslant 1}$ such that

$$\lim_{k\to+\infty} \int_\Omega f_{n_k}\, d\mu = \liminf_{n\to+\infty} \int_\Omega f_n\, d\mu.$$

Then, let $\{f_{n_{k_i}}\}_{i\geqslant 1}$ be a subsequence of the sequence $\{f_{n_k}\}_{k\geqslant 1}$ such that

$$f_{n_{k_i}}(\omega) \longrightarrow \liminf_{n\to+\infty} f_n(\omega) \quad \mu\text{-almost everywhere on } \Omega.$$

Then

$$\liminf_{n\to+\infty} \int_\Omega f_n\, d\mu = \lim_{i\to+\infty} \int_\Omega f_{n_{k_i}}\, d\mu \geqslant \int_\Omega \lim_{i\to+\infty} f_{n_{k_i}}\, d\mu = \int_\Omega \liminf_{n\to+\infty} f_n\, d\mu.$$

Solution of Problem 3.148

By the Fatou lemma (see Theorem 3.95), we have

$$\int_\Omega f\,d\mu \leqslant \int_\Omega \liminf_{n\to+\infty} f_n\,d\mu \leqslant \liminf_{n\to+\infty}\int_\Omega f_n\,d\mu,$$

so

$$\vartheta_n = \int_\Omega f_n\,d\mu \longrightarrow \vartheta = \int_\Omega f\,d\mu.$$

Let

$$g = \frac{f}{\vartheta} \quad \text{and} \quad g_n = \frac{f_n}{\vartheta_n} \qquad \forall\, n \geqslant 1.$$

We have

$$\int_\Omega g_n\,d\mu = \int_\Omega g\,d\mu = 1 \qquad \forall\, n \geqslant 1.$$

So, we can use Problem 3.146 and infer that $g_n \longrightarrow g$ in $L^1(\Omega)$. Since $\vartheta_n \longrightarrow \vartheta$, we conclude that $f_n \longrightarrow f$ in $L^1(\Omega)$.

Solution of Problem 3.149

(a) "$\Longrightarrow$": Let $h\colon \mathbb{R}^N \times \mathbb{R}_+ \longrightarrow \mathbb{R}$ be the function defined by

$$h(x,\eta) = f(x) - \eta.$$

The Borel measurability of f implies that h is Borel measurable. Hence the set

$$S(f) = \big\{(x,\eta) \in \mathbb{R}^N \times \mathbb{R} :\ h(x,\eta) \geqslant 0\big\}$$

is Borel.

"$\Longleftarrow$": For every $x \in \mathbb{R}^N$, we have

$$S(f)_x = \big\{\eta \in \mathbb{R}_+ :\ (x,\eta) \in S(f)\big\} \in \mathcal{B}(\mathbb{R}_+)$$

(see Proposition 3.47) and

$$\lambda\big(S(f)_x\big) = f(x)$$

with λ being the Lebesgue measure on $\mathbb{R}$. Therefore f is Borel measurable.

(b) Using the Fubini–Tonelli theorem (see Theorem 3.115), we have

$$\int_{\mathbb{R}^N} f(x)^r \, dx = r \int_{\mathbb{R}^N} \int_0^{f(x)} \eta^{r-1} \, d\eta \, dx = r \int_0^{\infty} \eta^{r-1} \int_{\{f>\eta\}} dx \, d\eta$$
$$= r \int_0^{\infty} \eta^{r-1} \lambda^N\big(\{f > \eta\}\big) \, d\eta.$$

(c) Reasoning as in part **(a)**, we can show that

$$U(f) = \big\{(x,\eta) \in \mathbb{R}^N \times \mathbb{R} : 0 \leqslant \eta < f(x)\big\}$$

is Borel.

Moreover, using part **(b)** with $r = 1$, we obtain

$$\int_{\mathbb{R}^N} f(x) \, dx = \int_{\mathbb{R}^N} \lambda^N\big(S(f)_x\big) \, dx = \lambda^{N+1}\big(S(f)\big) = \lambda^{N+1}\big(U(f)\big),$$

where by λ^{N+1} we denote the Lebesgue measure on $\mathbb{R}^{N+1} = \mathbb{R}^N \times \mathbb{R}$. Therefore

$$\operatorname{Gr} f = S(f) \setminus U(f) \quad \text{is Lebesgue-null.}$$

Solution of Problem 3.150

Note that

$$S(f) = \big\{(\omega,\eta) \in \Omega \times \mathbb{R} : 0 \leqslant \eta \leqslant f(\omega)\big\}$$

is $\mu \otimes \lambda$-measurable. Also, we have

$$\lambda\big(S(f)_\omega\big) = f(\omega)$$

(see also the solution of Problem 3.149). Therefore, by Theorem 3.114, we have

$$\int_{\Omega} f \, d\mu = (\mu \otimes \lambda)\big(\{\omega,\eta) : 0 \leqslant \eta \leqslant f(\omega)\big).$$

Solution of Problem 3.151

We may assume that $f \in L^p(C)$ (otherwise the inequality is clearly true). Let

$$h_n = \min\{h, n\} \qquad \forall\, n \geqslant 1.$$

Then

$$\{h_n > \eta\} \subseteq \{h > \eta\} \qquad \forall\, n \geqslant 1$$

and

$$\{h_n > \eta\} = \emptyset \qquad \forall\, \eta \geqslant n.$$

Using Problem 3.149(b) (with $\mathbb{R}^N$ replaced by C) and our hypothesis, we have

$$\begin{aligned} \int_C h_n^p\, dx &= p \int_0^n \eta^{p-1} \lambda^N(\{h_n > \eta\})\, d\eta \leqslant p \int_0^n \eta^{p-2} \int_{\{h>\eta\}} f\, dx\, d\eta \\ &\leqslant p \int_C f \int_0^{\min\{h(x),n\}} \eta^{p-2}\, d\eta\, dx = \tfrac{p}{p-1} \int_C f h_n^{p-1}\, dx. \end{aligned}$$

If $1 < p' < +\infty$, $\frac{1}{p} + \frac{1}{p'} = 1$, then by the Hölder inequality (see Theorem 3.103), we have

$$\int_C f h_n^{p-1}\, dx \leqslant \Big(\int_C f^p\, dx\Big)^{\frac{1}{p}} \Big(\int_C h_n^p\, dx\Big)^{\frac{1}{p'}},$$

so

$$\Big(\int_C h_n^p\, dx\Big)^{\frac{1}{p}} \leqslant \tfrac{p}{p-1} \Big(\int_C f^p\, dx\Big)^{\frac{1}{p}}.$$

Passing to the limit as $n \to +\infty$ and using the Lebesgue monotone convergence theorem (see Theorem 3.92), we obtain

$$\Big(\int_C h^p\, dx\Big)^{\frac{1}{p}} \leqslant \tfrac{p}{p-1} \Big(\int_C f^p\, dx\Big)^{\frac{1}{p}}.$$

Solution of Problem 3.152
"**(a)** $\Longrightarrow$ **(b)**": Clear.

"**(b)** $\Longrightarrow$ **(c)**": Use $f = \chi_A$.

"**(c)** $\Longrightarrow$ **(a)**": Fix $\eta > 0$. Then, if $f(x) > \eta$, we have

$$f_n(x) > \eta \qquad \forall\, n \geqslant 1 \text{ large.}$$

Hence

$$\{f > \eta\} \subseteq \liminf_{n \to +\infty} \{f_n > \eta\} = \bigcup_{k \geqslant 1} \bigcap_{n \geqslant k} \{f_n > \eta\}.$$

Let us set

$$C_k = \bigcap_{n \geqslant k} \{f_n > \eta\} \in \Sigma.$$

Evidently $\{C_k\}_{k \geqslant 1} \subseteq \Sigma$ is an increasing sequence of sets and for every $k \geqslant 1$, we have

$$C_k \subseteq \{f_n > \eta\} \qquad \forall\, n \geqslant k.$$

Therefore, since **(c)** holds, we have

$$\begin{aligned}\mu(\{f > \eta\}) &\leqslant \lim_{k \to +\infty} \mu(C_k) \leqslant \lim_{k \to +\infty} \liminf_{n \to +\infty} \mu_n(C_k) \\ &\leqslant \liminf_{n \to +\infty} \mu_n(\{f_n > \eta\}).\end{aligned}$$

Performing a change of variable and using Problem 3.149(b) (with $\Omega = \mathbb{R}^N$), we have

$$\int_\Omega f\, d\mu = \int_{\mathbb{R}_+} \eta\, d(\mu \circ f^{-1}) = \int_{\mathbb{R}_+} \mu(\{f > \eta\})\, d\eta$$

(see Theorem 3.118). Then using the Fatou lemma (see Theorem 3.95) and once again Problem 3.149(b) (with $\Omega = \mathbb{R}^N$), we obtain

$$\begin{aligned}\int_\Omega f\, d\mu &\leqslant \int_\Omega \liminf_{n \to +\infty} \mu_n(\{f_n > \eta\})\, d\eta \leqslant \liminf_{n \to +\infty} \int_\Omega \mu_n(\{f_n > \eta\})\, d\eta \\ &= \liminf_{n \to +\infty} \int_\Omega f_n\, d\mu_n.\end{aligned}$$

Solution of Problem 3.153

(a) Let

$$\xi_y(x,s) \;=\; \tfrac{1}{y}\big|f(x+s)-f(x)\big| \qquad \forall\,(x,s)\in\mathbb{R}\times[0,y].$$

Clearly ξ_y is measurable. Then by the Fubini–Tonelli theorem (see Theorem 3.115), we have that h_y is Lebesgue measurable. Also note that by the translation invariance of the Lebesgue measure (see Theorem 3.31), we have

$$\int\limits_{\mathbb{R}} \big|f(x+s)-f(x)\big|\,dx \;\leqslant\; \int\limits_{\mathbb{R}} \big|f(x+s)\big|\,dx + \int\limits_{\mathbb{R}} \big|f(x)\big|\,dx \;=\; 2\int\limits_{\mathbb{R}} \big|f(x)\big|\,dx,$$

so

$$\int\limits_{\mathbb{R}} h_y(x)\,dx \;=\; \int\limits_{\mathbb{R}\times[0,y]} \xi_y(x,s)\,d(\lambda\otimes\lambda)(x,s) \;\leqslant\; 2\|f\|_1.$$

(b) First we show that the function $\mathbb{R}\ni s\longmapsto f_s\in L^1(\mathbb{R})$, where $f_s(x)=f(x+s)$ is continuous. Note that for every Lebesgue set $A\subseteq\mathbb{R}$ with finite Lebesgue measure and $s,t\in\mathbb{R}$, we have

$$\chi_A(x+s) \;=\; \chi_{A-s}(x)$$

and

$$\big|\chi_A(x+s)-\chi_A(x+t)\big| \;=\; \chi_{(A-s)\triangle(A-t)}(x).$$

Then

$$\big|\chi_A(x+s)-\chi_A(x+t)\big| \;=\; \chi_{A-s}(x)+\chi_{A-t}(x)-2\chi_{A-s}(x)\chi_{A-t}(x),$$

so, using the translation invariance of λ, we have

$$\int\limits_{\mathbb{R}} \big|\chi_A(x+s)-\chi_A(x+t)\big|\,dx = 2\lambda(A)-2\lambda\big((A-s)\cap(A-t)\big).$$

Similarly, as in Problem 3.44, we establish the continuity of the function $y\longmapsto\lambda\big(A\cap(A+y)\big)$. So, if $y=s-t$, then

$$\int\limits_{\mathbb{R}} \big|\chi_A(x+s)-\chi_A(x+t)\big|\,dx \;\longrightarrow\; 0 \quad \text{as } s-t\to 0,$$

and so the function $\mathbb{R}\ni y\longmapsto\big(\chi_A\big)_y\in L^1(\mathbb{R})$ is continuous.

This result is then true for all simple functions in $L^1(\mathbb{R})$ and by their density in $L^1(\mathbb{R})$ (see Proposition 3.110), we infer that the function $\mathbb{R} \ni s \longmapsto f_s \in L^1(\mathbb{R})$, where $f_s(x) = f(x+s)$ is continuous. Therefore

$$\int_{\mathbb{R}} |f(x+s) - f(x)|\, dx \longrightarrow 0 \quad \text{as } s \to 0^+.$$

Hence, for a given $\varepsilon > 0$, we can find $\delta > 0$ such that

$$\int_{\mathbb{R}} |f(x+s) - f(x)|\, dx \leqslant \varepsilon \qquad \forall\, s \in [0, \delta],$$

so

$$\int_{\mathbb{R}} h_y(x)\, dx = \int_0^y \tfrac{1}{y} \int_{\mathbb{R}} |f(x+s) - f(x)|\, dx\, ds \leqslant \varepsilon \qquad \forall\, y \in [0, \delta]$$

and thus

$$h_y \longrightarrow 0 \quad \text{in } L^1(\Omega) \quad \text{as } y \to 0^+.$$

Solution of Problem 3.154

(a) Let

$$A = \{(x, y) \in \Omega \times Y : x > y\} \quad \text{and} \quad C = \{(x, y) \in \Omega \times Y : x < y\}.$$

Note that

$$A = \bigcup \{[a, b] \times [c, d] : a, b, c, d \in \mathbb{Q} \cap [0, 1],\ a > d\}.$$

this union is countable and each rectangle belongs in $\Sigma \otimes \mathcal{Y}$ (see Definition 3.44). Hence $A \in \Sigma \otimes \mathcal{Y}$. Similarly, we show that

$$C \in \Sigma \otimes \mathcal{Y}$$

(in this case $a < d$). Since

$$\Delta = (\Omega \times Y) \setminus (A \cup C),$$

we infer that $\Delta \in \Sigma \otimes \mathcal{Y}$.

(b) We have

$$\begin{aligned}\int_Y \int_\Omega \chi_\Delta(x,y)\,d\lambda\,d\nu &= \int_0^1 \Big(\int_0^1 \chi_\Delta(x,y)\,d\lambda(x)\Big)\,d\nu(y)\\ &= \int_0^1 \Big(\int_0^1 \chi_{\{y\}}(x)\,d\lambda(x)\Big)\,d\nu(y) = \int_0^1 0\,d\nu(y) = 0\end{aligned}$$

and

$$\begin{aligned}\int_\Omega \int_Y \chi_\Delta(x,y)\,d\nu\,d\lambda &= \int_0^1 \Big(\int_0^1 \chi_\Delta(x,y)\,d\nu(y)\Big)\,d\lambda(x)\\ &= \int_0^1 \Big(\int_0^1 \chi_{\{x\}}(y)\,d\nu(y)\Big)\,d\lambda(x) = \int_0^1 1\lambda(x) = 1\end{aligned}$$

(c) If χ_Δ was $\lambda \times \nu$-integrable, then by the Fubini–Tonelli theorem (see Theorem 3.115), the two iterated integrals in **(b)** would have been equal, a contradiction.

Solution of Problem 3.155

The inequality is valid. To show this, we may assume without any loss of generality that $f \ln f \in L^1(\Omega)$ (otherwise the inequality is obvious). Recall that

$$\ln x \leqslant x \ln x \qquad \forall\, x > 0.$$

So, we have

$$0 \leqslant \ln f \leqslant f \ln f$$

and also

$$f(x) \leqslant \begin{cases} e & \text{if } f(x) \leqslant e,\\ f(x)\ln f(x) & \text{if } f(x) > e.\end{cases}$$

So, we infer that $f, \ln f \in L^1(\Omega)$, $f \geqslant 0$, $\ln f \geqslant 0$ (since f has values in $[1,+\infty)$). We have

$$\begin{aligned}
&\Big(\int_\Omega f\,d\mu\Big)\Big(\int_\Omega \ln f\,d\mu\Big) - \int_\Omega f(\ln f)\,d\mu \\
&= \tfrac{1}{2}\int_{\Omega\times\Omega} \big(f(x)(\ln f(y)) + f(y)(\ln f(x))\big)\,d\mu(x)\,d\mu(y) \\
&\quad -\tfrac{1}{2}\int_{\Omega\times\Omega} \big(f(x)(\ln f(x)) + f(y)(\ln f(y))\big)\,d\mu(x)\,d\mu(y) \\
&= \tfrac{1}{2}\int_{\Omega\times\Omega} f(x)\Big(\ln\tfrac{f(y)}{f(x)} - \tfrac{f(y)}{f(x)}\ln\tfrac{f(y)}{f(x)}\Big)\,d\mu(x)\,d\mu(y) \;\leqslant\; 0.
\end{aligned}$$

Solution of Problem 3.156

Note that ξ and η are monotone functions. So, their Borel measurability follows from Problem 3.73. The function

$$(x,\lambda) \longmapsto \chi_{(0,+\infty)}\big(|f|(x)-\lambda\big)$$

is $\Sigma\otimes\mathcal{B}(\mathbb{R})$-measurable (see Definition 3.44) and

$$\begin{aligned}
&\int_0^\infty\Big(\int_\Omega \chi_{(0,+\infty)}\big(|f|(x)-\lambda\big)\,d\mu\Big)\,d\lambda \;=\; \int_0^\infty\Big(\int_\Omega \chi_{\{|f|>\lambda\}}(x)\,d\mu\Big)\,d\lambda \\
&= \int_0^\infty \mu\big(\{|f|>\lambda\}\big)\,d\lambda \;=\; \int_0^\infty \big(\xi(\lambda)+\eta(\lambda)\big)\,d\lambda.
\end{aligned}$$

On the other hand

$$\int_0^\infty\Big(\int_\Omega \chi_{(0,+\infty)}\big(|f|(x)-\lambda\big)\,d\mu\Big)\,d\lambda \;=\; \int_\Omega\int_0^{|f|(x)} d\lambda\,d\mu \;=\; \int_\Omega |f|(x)\,d\mu \;=\; \|f\|_1.$$

So, finally the Fubini–Tonelli theorem (see Theorem 3.115) implies that

$$\|f\|_1 \;=\; \int_0^\infty \big(\xi(\lambda)+\eta(\lambda)\big)\,d\lambda.$$

Solution of Problem 3.157

By hypothesis and the Fubini–Tonelli theorem (see Theorem 3.115), we have

$$\begin{aligned}\int_0^1\int_0^1 f(x)h(y)\,dx\,dy &= \int_{[0,1]\times[0,1]} f(x)h(y)\,d\lambda^2(x,y)\\ &= \int_E f(x)h(y)\,d\lambda^2(x,y) + \int_{\{(x,y):0\leqslant x=y\leqslant 1\}} f(x)h(y)\,d\lambda^2(x,y)\\ &\quad + \int_{\{(x,y):0\leqslant y\leqslant x\leqslant 1\}} f(x)h(y)\,d\lambda^2(x,y)\\ &= 2\int_E f(x)h(y)\,d\lambda^2(x,y).\end{aligned}$$

Solution of Problem 3.158

We have

$$\begin{aligned}\int_0^1\int_0^1 \tfrac{x^2-y^2}{(x^2+y^2)^2}\,dy\,dx &= \int_0^1 \tfrac{y}{x^2+y^2}\Big|_{y=0}^{y=1}\,dx\\ &= \int_0^1 \tfrac{1}{1+x^2}\,dx = \arctan x\Big|_0^1 = \tfrac{\pi}{4}.\end{aligned}$$

Similarly

$$\int_0^1\int_0^1 \tfrac{x^2-y^2}{(x^2+y^2)^2}\,dx\,dy = \int_0^1 \tfrac{-x}{x^2+y^2}\Big|_{x=0}^{x=1}\,dy = -\tfrac{\pi}{4}.$$

Next we show that $f \notin L^1\big([0,1]\times[0,1]\big)$. To this end, we have

$$\begin{aligned}\int_{[0,1]\times[0,1]} \big|f(x,y)\big|\,dx\,dy &\geqslant \int_{\{(x,y):x,y\geqslant 0,x^2+y^2\leqslant 1\}} \big|f(x,y)\big|\,dx\,dy\\ &= \int_0^1\int_0^{\frac{\pi}{2}} \tfrac{|\cos 2\theta|}{r^4} r\,dr\,d\vartheta \geqslant \int_0^1\int_0^{\frac{\pi}{4}} \tfrac{\cos 2\theta}{r^3}\,dr\,d\vartheta = \tfrac{1}{2}\int_0^1 \tfrac{dr}{r^3}. = +\infty,\end{aligned}$$

Solution of Problem 3.159

By hypothesis, for every $n \geqslant 1$ we can find an open set $U_n \subseteq \mathbb{R}^N$ such that $\lambda^N(U_n) < \frac{1}{n}$ and $f|_{\mathbb{R}^N \setminus U_n}$ is continuous. Let

$$f_n = \chi_{\mathbb{R}^N \setminus U_n} f.$$

Then f_n is measurable and for any $\varepsilon > 0$, we have

$$\lambda^N(\{x \in \mathbb{R}^N : |f_n(x) - f(x)| \geqslant \varepsilon\}) \leqslant \lambda^N(U_n) < \tfrac{1}{n},$$

so

$$f_n \xrightarrow{\lambda^N} f$$

(see Definition 3.126) and thus f is measurable (see Proposition 3.70).

Solution of Problem 3.160

By Proposition 3.131, the only possible limit (for the convergence in μ-measure; see Definition 3.126) is fg. Then arguing indirectly, suppose that

$$f_n g_n \not\longrightarrow fg \quad \text{in } \mu\text{-measure}$$

(see Definition 3.126). Then we can find $\varepsilon, \xi > 0$ and a subsequence $\{f_{n_k} g_{n_k}\}_{k \geqslant 1} \subseteq \{f_n g_n\}_{n \geqslant 1}$ such that

$$\mu(\{x \in \Omega : |f_{n_k}(x) g_{n_k}(x) - f(x)g(x)| \geqslant \varepsilon\}) \geqslant \xi \qquad \forall\, k \geqslant 1.$$

But by Proposition 3.131, we can find a further subsequence $\{f_{n_{k_i}} g_{n_{k_i}}\}_{i \geqslant 1} \subseteq \{f_{n_k} g_{n_k}\}_{k \geqslant 1}$ such that

$$f_{n_{k_i}}(x) g_{n_{k_i}}(x) \longrightarrow f(x)g(x) \quad \mu\text{-almost everywhere on } \Omega.$$

Then Proposition 3.130 implies that

$$f_{n_{k_i}} g_{n_{k_i}} \xrightarrow{\mu} fg \quad \text{as } i \to +\infty$$

(see Definition 3.126), a contradiction. This proves that

$$f_n g_n \xrightarrow{\mu} fg \quad \text{as } n \to +\infty.$$

The result fails if μ is not finite. Indeed, let $\Omega = \mathbb{R}_+$, $\Sigma = \mathcal{B}(\Omega)$ and $\mu = \lambda$. Consider the sequence

$$f_n(x) \;=\; \sqrt{x^4 + \tfrac{x}{n}} \qquad \forall\, n \geqslant 1.$$

Then

$$f_n \xrightarrow{\;\lambda\;} f,$$

where $f(x) = x^2$. But

$$f_n^2 \not\longrightarrow f^2 \quad \text{in } \lambda\text{-measure}.$$

Solution of Problem 3.161

First suppose that $\theta \in (0,1]$. Then we have

$$\big|f_n(x)^\theta - f(x)^\theta\big| \;\leqslant\; \big|(f_n - f)(x)\big|^\theta \qquad \forall\, x \in \Omega.$$

Then for every $\varepsilon > 0$, we have

$$\mu(\{f_n^\theta - f^\theta\} \geqslant \varepsilon) \;\leqslant\; \mu(\{|f_n - f|^\theta \geqslant \varepsilon\}) \;\longrightarrow\; 0 \quad \text{as } n \to +\infty,$$

so

$$f_n^\theta \xrightarrow{\;\mu\;} f^\theta \quad \text{as } n \to +\infty$$

(see Definition 3.126), where $\theta \in (0,1]$.

Next assume that $\theta > 1$. From the mean value theorem, we have

$$|f_n^\theta - f^\theta| \;\leqslant\; \theta |f_n - f| (f_n \vee f)^{\theta - 1},$$

where $f_n \vee f = \max\{f_n, f\}$. Let $\varepsilon > 0$ and $M > 0$. We have

$$\mu(\{|f_n^\theta - f^\theta| \geqslant \varepsilon\}) \;\leqslant\; \mu(\{|f_n - f| \geqslant \tfrac{\varepsilon}{\theta M^{\theta-1}}\}) + \mu(\{f_n \vee f \geqslant M\}).$$

If $f_n \vee f \geqslant M$, then $f \geqslant \frac{M}{2}$ or $|f_n - f| \geqslant \frac{M}{2}$ or otherwise

$$f_n \;=\; f + (f_n - f) \;<\; \tfrac{M}{2} + \tfrac{M}{2} \;=\; M.$$

So, we have

$$\begin{aligned} &\mu(\{|f_n^\theta - f^\theta| \geqslant \varepsilon\}) \\ \leqslant\;& \mu(\{|f_n - f| \geqslant \tfrac{\varepsilon}{\theta M^{\theta-1}}\}) + \mu(\{|f_n - f| \geqslant \tfrac{M}{2}\}) + \mu(\{f \geqslant \tfrac{M}{2}\}) \\ &\forall\, n \geqslant 1, \end{aligned}$$

so

$$\limsup_{n\to+\infty} \mu(\{|f_n^\theta - f^\theta| \geqslant \varepsilon\}) \leqslant \mu(\{f \geqslant \tfrac{M}{2}\})$$

(since $f_n \xrightarrow{\mu} f$ as $n \to +\infty$). Recall that $M > 0$ was arbitrary and

$$\lim_{M\to+\infty} \mu(\{f \geqslant \tfrac{M}{2}\}) = 0.$$

Therefore

$$\lim_{n\to+\infty} \mu(\{|f_n^\theta - f^\theta| \geqslant \varepsilon\}) = 0,$$

i.e., $f_n^\theta \xrightarrow{\mu} f^\theta$ as $n \to +\infty$.

Solution of Problem 3.162

"$\Longrightarrow$": We assume that the sequence $\{f_n\}_{n\geqslant1}$ is Cauchy in μ-measure (see Definition 3.126(a)). By Proposition 3.131, it suffices to show that we can find a subsequence of $\{f_n\}_{n\geqslant1}$ which converges in μ-measure. So, we choose a subsequence $\{h_n\}_{n\geqslant1}$ of $\{f_n\}_{n\geqslant1}$ such that

$$\mu(\{|h_n - h_m| \geqslant \tfrac{1}{2^n}\}) < \tfrac{1}{2^n} \qquad \forall\, m \geqslant n$$

(this can be done since the sequence $\{f_n\}_{n\geqslant1}$ is by hypothesis Cauchy in μ-measure).

Let

$$A_n = \{|h_n - h_{n+1}| \geqslant \tfrac{1}{2^n}\}, \qquad C_n = \bigcup_{k\geqslant n} A_k \qquad \forall\, n \geqslant 1$$

and

$$C = \limsup_{n\to+\infty} A_n = \bigcap_{n\geqslant1} \bigcup_{k\geqslant n} A_k.$$

Note that

$$\mu(C_n) \leqslant \sum_{k\geqslant n} \mu(A_k) \leqslant \tfrac{1}{2^{n-1}} \longrightarrow 0 \quad \text{as } n \to +\infty,$$

so $\mu(C) = 0$.

If $x \in C^c = \Omega \setminus C$, then we can find an integer $k_x \geqslant 1$ such that $x \notin C_n$ for all $n \geqslant k_x$. Therefore for $n \geqslant k_x$, we have

$$|h_{n+m}(x) - h_n(x)| \leqslant \sum_{i=1}^{m} |h_{n+i}(x) - h_n(x)| \leqslant \tfrac{1}{2^{n-1}},$$

so for all $x \in C^c = \Omega \setminus C$, the sequence $\{h_n(x)\}_{n \geqslant 1} \subseteq \mathbb{R}$ is Cauchy for all $x \in C^c$, with $\mu(C) = 0$.

So, we can find a Σ-measurable function $f\colon \Omega \longrightarrow \mathbb{R}$ such that

$$h_n(x) \longrightarrow f(x) \qquad \forall\, x \in C^c.$$

If $n \geqslant k$ and $x \notin C_n$, then

$$\big|h_{n+m}(x) - h_{n+1}(x)\big| \leqslant \sum_{i=2}^{m} \big|h_{n+i}(x) - h_{n+1}(x)\big| \leqslant \tfrac{1}{2^n},$$

so

$$\big|h_{n+m}(x) - h_{n+1}(x)\big| \leqslant \tfrac{1}{2^n} \leqslant \tfrac{1}{2^k} \qquad \forall\, n \geqslant k$$

and thus

$$\big\{|f - h_{n+1}| \geqslant \tfrac{1}{2^k}\big\} \subseteq C_n \qquad \forall\, k \geqslant n.$$

Finally note that for $n \geqslant k$, we have

$$\begin{aligned} \big\{|f - h_n| \geqslant \tfrac{1}{2^{k-1}}\big\} &\subseteq \big\{|h_{n+1} - h_n| \geqslant \tfrac{1}{2^k}\big\} \cup \big\{|f - h_{n+1}| \geqslant \tfrac{1}{2^k}\big\} \\ &\subseteq A_n \cup C_n = C_n, \end{aligned}$$

so

$$h_n \xrightarrow{\mu} f \quad \text{as } n \to +\infty$$

(see Definition 3.126).

"$\Longleftarrow$": Assume that $f_n \xrightarrow{\mu} f$. Then for every $\varepsilon > 0$ and all $m, n \geqslant 1$, we have

$$\big\{|f_m - f_n| \geqslant 2\varepsilon\big\} \subseteq \big\{|f_m - f| \geqslant \varepsilon\big\} \cup \big\{|f_n - f| \geqslant \varepsilon\big\},$$

so the sequence $\{f_n\}_{n \geqslant 1}$ is Cauchy in μ-measure.

Solution of Problem 3.163

From Problem 3.98, we know that $f_n \longrightarrow f$ in $L^1(\Omega)$. Then from the Vitali theorem (see Theorem 3.128), we have that the sequence $\{|f_n|\}_{n \geqslant 1} \subseteq L^1(\Omega)$ is uniformly integrable. Let $A \in \Sigma$. Note that

$$\chi_A |f_n - f| \leqslant |f_n| + |f| \qquad \forall\, n \geqslant 1$$

and

$$\chi_A f_n \longrightarrow \chi_A f \quad \mu\text{-almost everywhere on } \Omega.$$

Therefore invoking once again the Vitali theorem (see Theorem 3.128), we conclude that

$$\int_A f_n \, d\mu \longrightarrow \int_A f \, d\mu.$$

Solution of Problem 3.164

For every $\beta > 0$ and every $n > \beta$, we have

$$A(\beta, n) = \{x \in \Omega : |f_n(x)| \geqslant \beta\} = \{x_0\}.$$

Hence

$$\int_{A(\beta,n)} f_n \, d\delta_{x_0} = n \int_{\{x_0\}} d\delta_{x_0} = n,$$

so

$$\sup_{n \geqslant 1} \int_{A(\beta,n)} f_n \, d\delta_{x_0} = +\infty,$$

which means that the sequence $\{f_n\}_{n\geqslant 1}$ is not δ_{x_0}-uniformly integrable (see Definition 3.124).

Solution of Problem 3.165

"$\Longrightarrow$": From the Markov inequality (see Proposition 3.90(f)), we have that

$$|f_n - f|^p \xrightarrow{\mu} 0.$$

Invoking Problem 3.161, we infer that $f_n \xrightarrow{\mu} f$ and then $f_n^p \xrightarrow{\mu} f^p$. Note that

$$f_n^p \leqslant 2^{p-1}|f_n - f|^p + 2^{p-1} f^p$$

and from the Vitali theorem (see Theorem 3.128), we have that the sequence $\{|f_n - f|^p\}_{n\geq 1}$ is uniformly integrable. Therefore a new application of the Vitali theorem (see Theorem 3.128) implies that

$$\|f_n^p - f^p\|_1 \longrightarrow 0.$$

"$\Longleftarrow$": Since $f_n^p \longrightarrow f^p$ in $L^1(\mathbb{R})$, it follows that

$$\|f_n\|_p \longrightarrow \|f\|_p$$

and

$$f_n^p \xrightarrow{\mu} f^p,$$

hence

$$f_n \xrightarrow{\mu} f$$

(see Problem 3.161). Invoking the Vitali theorem (see Theorem 3.128), we conclude that

$$f_n \longrightarrow f \quad \text{in } L^p(\mathbb{R}).$$

Solution of Problem 3.166

(a) Suppose that $f_n \longrightarrow f$ in $L^p(\Omega)$. Then from the Markov inequality (see Theorem 3.107), Proposition 3.136 and by passing to a suitable subsequence if necessary, we may assume that

$$f_n(x) \longrightarrow f(x) \quad \mu\text{-almost everywhere on } \Omega.$$

Hence

$$|f_n(x)|^r \longrightarrow |f(x)|^r \quad \mu\text{-almost everywhere on } \Omega.$$

Note that

$$\big||f_n(x)|^r - |f(x)|^r\big|^{\frac{p}{r}} \leqslant 2^{\frac{p}{r}-1}\big(|f_n(x)|^p + |f(x)|^p\big).$$

But from the Vitali theorem (see Theorem 3.128), we know that the sequence $\{|f_n|^p\}_{n\geqslant 1} \subseteq L^1(\Omega)$ is uniformly integrable. So, a new application of the Vitali theorem implies that

$$\int\limits_\Omega \big||f_n|^r - |f|^r\big|^{\frac{p}{r}}\, d\mu \longrightarrow 0 \quad \text{as } n \to +\infty,$$

so

$$K(f_n) \longrightarrow K(f) \quad \text{in } L^{\frac{p}{r}}(\Omega) \quad \text{as } n \to +\infty$$

and thus K is continuous.

(b) As above, we have (at least for a subsequence), that

$$|f_n|^{p-r} \longrightarrow |f|^{p-r} \quad \text{and} \quad |h_n|^r \longrightarrow |h|^r \quad \mu\text{-almost everywhere on } \Omega,$$

so

$$|f_n|^{p-r}|h_n|^r \longrightarrow |f|^{p-r}|h|^r \quad \mu\text{-almost everywhere on } \Omega.$$

Since $L^p(\Omega)$ is a Banach lattice (see Remark 3.97), we have that

$$g_n = \max\{|f_n|, |h_n|\} \longrightarrow g = \max\{|f|, |h|\} \quad \text{in } L^p(\Omega),$$

so the sequence $\{g_n^p\}_{n \geqslant 1} \subseteq L^1(\Omega)$ is uniformly integrable (see the Vitali theorem; Theorem 3.128).

Note that

$$|f_n|^{p-r}|h_n|^r \leqslant g_n^p \qquad \forall\, n \geqslant 1.$$

So, an application of the Vitali theorem (see Theorem 3.128) implies that

$$\int_\Omega |f_n|^{p-r}|h_n|^r \, d\mu \longrightarrow \int_\Omega |f|^{p-r}|h|^r \, d\mu.$$

Remark. Using the notion of the dual of a Banach space and the Riesz representation theorem (see Chap. 5), we have that $L^{\frac{p}{p-r}}(\Omega)$ is the dual of $L^{\frac{p}{r}}(\Omega)$ and using this fact, the proof of **(b)** above is simplified considerably.

Solution of Problem 3.167

We show that $f_n \longrightarrow f$ μ-almost everywhere and then appeal to Proposition 3.130. For every $k \geqslant 1$, we can find $A_k \in \Sigma$ with $\mu(A_k^c) \leqslant \frac{1}{k}$ such that $f_n \rightrightarrows f$ on A_k. Let

$$C = \bigcup_{k \geqslant 1} A_k \in \Sigma.$$

Then for $\omega \in C$, we have $f_n(\omega) \longrightarrow f(\omega)$. Also,

$$C^c = \bigcap_{k \geqslant 1} A_k^c$$

and so

$$\mu(C^c) \leqslant \mu(A_k^c) \leqslant \tfrac{1}{k} \qquad \forall\, k \geqslant 1.$$

Hence $\mu(C^c) = 0$. Therefore $f_n \longrightarrow f$ μ-almost everywhere and so by Proposition 3.130, we have that $f_n \xrightarrow{\mu} f$ (see Definition 3.126).

Solution of Problem 3.168

Since V is closed, $(V, \|\cdot\|_1)$ is a Banach space. For every $n \geqslant 1$, we introduce the set

$$V_n = \{u \in V \cap L^{1+\frac{1}{n}}(\Omega) : \|u\|_{1+\frac{1}{n}} \leqslant n\}.$$

We claim that $V_n \subseteq V$ is closed. So, let $\{u_k\}_{k\geqslant 1} \subseteq V_n$ be a sequence such that $u_k \longrightarrow u$ in $L^1(\Omega)$. Then by passing to a suitable subsequence if necessary, we may assume that $u_n(\omega) \longrightarrow u(\omega)$ μ-almost everywhere in Ω (see Propositions 3.132 and 3.131). By the Fatou lemma (see Theorem 3.95), we have

$$\|u\|_{1+\frac{1}{n}} \leqslant \liminf_{n\to+\infty} \|u_k\|_{1+\frac{1}{n}} \leqslant n,$$

so $u \in V_n$, which proves the closedness of V_n in V.

Next, we show that

$$V = \bigcup_{n\geqslant 1} V_n.$$

To see this, let $u \in V$. By hypothesis, there exists $q > 1$ such that $u \in L^q(\Omega)$. Then Theorem 3.106 implies that for $n \geqslant 1$ large so that $1 + \frac{1}{n} \leqslant q$, we have $u \in L^{1+\frac{1}{n}}(\Omega)$ and

$$\|u\|_{1+\frac{1}{n}} \leqslant \|u\|_q^{1-t_n} \|u\|_1^{t_n},$$

where $\frac{1}{1+\frac{1}{n}} = \frac{1-t_n}{q} + t_n$. So, by choosing $n \geqslant 1$ even bigger if necessary, we will have $\|u\|_{1+\frac{1}{n}} \leqslant n$, hence $v \in V_n$.

Since V is a Banach space, from the Baire category theorem (see Theorem 1.26), we can find $n_0 \geqslant 1$ such that $\operatorname{int} V_{n_0} \neq \emptyset$, therefore $V \subseteq L^{1+\frac{1}{n_0}}(\Omega)$.

Solution of Problem 3.169

For every $A \in \Sigma$, we define

$$m(A) = \sum_{n\geqslant 1} \frac{1}{n^2} \frac{|\mu_n|(A)}{1+|\mu_n|(\Omega)}.$$

Clearly m is a measure on Σ and $\mu_n \ll m$ for all $n \geqslant 1$ (see Definition 3.150). If $\{C_k\}_{k\geq 1} \subseteq \Sigma$ are mutually disjoint, then by the Vitali–Hahn–Saks theorem (see Theorem 3.151), we have

$$\lim_{m\to+\infty} \sup_{n\geqslant 1} \Big|\mu_n\big(\bigcup_{k\geqslant m} C_k\big)\Big| = 0.$$

Then

$$\begin{aligned}
\mu\big(\bigcup_{k\geqslant 1} C_k\big) &= \lim_{n\to+\infty} \mu_n\big(\bigcup_{k\geqslant 1} C_k\big) = \lim_{n\to+\infty}\lim_{m\to+\infty} \Big[\mu_n\big(\bigcup_{k=1}^{m} C_k\big) + \mu_n\big(\bigcup_{k\geqslant m+1} C_k\big)\Big] \\
&= \lim_{m\to+\infty} \sum_{k=1}^{m} \mu(C_k) + \lim_{m\to+\infty}\lim_{n\to+\infty} \mu_n\big(\bigcup_{k\geqslant m+1} C_k\big) \\
&= \sum_{k\geqslant 1} \mu(C_k),
\end{aligned}$$

hence μ is σ-additive and so μ is a signed measure.

Solution of Problem 3.170

"$\Longrightarrow$": Suppose that the implication is not true. Then we can find $\varepsilon > 0$ such that for every integer $k \geqslant 1$, there is $A_k \in \Sigma$ satisfying

$$\mu(A_k) < \tfrac{1}{2^k} \quad \text{and} \quad \nu(A_k) \geqslant \varepsilon.$$

We have

$$\mu\big(\bigcup_{k\geqslant n} A_k\big) \leqslant \sum_{k\geqslant n} \mu(A_k) < \tfrac{1}{2^{n-1}} \qquad \forall\, n \geqslant 1$$

and

$$\nu\big(\bigcup_{k\geqslant n} A_k\big) \geqslant \nu(A_n) \geqslant \varepsilon \qquad \forall\, n \geqslant 1.$$

Therefore, if

$$A = \limsup_{n\to+\infty} A_n = \bigcap_{n\geqslant 1}\bigcup_{k\geqslant n} A_k,$$

then

$$\mu(A) = 0 \quad \text{and} \quad \nu(A) \geqslant \varepsilon,$$

which contradicts the hypothesis that $\nu \ll \mu$ (see Definition 3.150).

"$\Longleftarrow$": Let $A \in \Sigma$ be such that $\mu(A) = 0$. Then $\mu(A) < \delta$ for every $\delta > 0$ and so $\nu(A) < \varepsilon$ for every $\varepsilon > 0$, hence $\nu(A) = 0$ and this proves that $\nu \ll \mu$.

Solution of Problem 3.171

From Problem 3.170, we know that if ν is finite, then "**(b)** $\Longleftrightarrow$ **(a)**". In fact from the solution of Problem 3.170, we see that in proving "**(b)** $\Longrightarrow$ **(a)**" we did not use the finiteness of ν. So, the implication is true in general.

To show that the converse implication does not hold, we will produce a counterexample. So, let $\Omega = \mathbb{R}$ and $\Sigma = \mathcal{L}(\mathbb{R})$ (the Lebesgue σ-algebra on $\mathbb{R}$; see Theorem 3.31). Consider $\mu = \lambda$ (the Lebesgue measure) and let

$$\nu(A) = \sum_{n \geqslant 1} n\lambda\big(A \cap [n, n+1]\big) \qquad \forall\, A \in \mathcal{L}(\mathbb{R}).$$

Clearly $\nu \ll \lambda$ (see Definition 3.150) and so **(a)** is satisfied. Moreover, ν is σ-finite but not finite and for every $k \geqslant 1$, we have

$$\lambda\big([2^k, 2^k + \tfrac{1}{2^k}]\big) = \tfrac{1}{2^k}$$

and

$$\nu\big([2^k, 2^k + \tfrac{1}{2^k}]\big) = 2^k \tfrac{1}{2^k} = 1,$$

which implies that **(b)** is not satisfied.

Solution of Problem 3.172

Let

$$m(A) = \int_A g\, d\mu \qquad \forall\, A \in \Sigma.$$

Then m is a finite measure on (Ω, Σ) and $m \ll \mu$ (see Definition 3.150). So,

$$f_n(x) \longrightarrow f(x) \quad m\text{-almost everywhere on } \Omega.$$

Invoking the Egorov theorem (see Theorem 3.76), for the finite measure space (Ω, Σ, m), for a given $\varepsilon > 0$, we can find a set $A \in \Sigma$, with

$m(A^c) < \varepsilon$ such that $f_n \rightrightarrows f$ on A. By Problem 3.170, we can find $\delta = \delta(\varepsilon) > 0$ such that $\mu(A^c) < \delta$.

Solution of Problem 3.173

Let $\Omega = [0,1]$, $\Sigma = \mathcal{B}([0,1])$ (see Definition 3.6) and $\mu = \lambda$ (the Lebesgue measure on $[0,1]$). Also, let ν be the counting measure on $[0,1]$. Evidently $\mu = \lambda \ll \nu$. Assume that there is a Borel function $f\colon [0,1] \longrightarrow \mathbb{R}$ such that

$$\nu(A) = \int_A f\,d\lambda \qquad \forall\, A \in \mathcal{B}([0,1]).$$

Since

$$\lambda(\{x\}) = 0 \quad \text{and} \quad \nu(\{x\}) = 1 \qquad \forall\, x \in [0,1],$$

we infer that

$$f(x) = 0 \qquad \forall\, x \in [0,1].$$

This implies that

$$\nu(A) = 0 \qquad \forall\, A \in \mathcal{B}([0,1]),$$

a contradiction. So, the Radon–Nikodym theorem without the assumption on the σ-finiteness of the measure does not hold.

Solution of Problem 3.174

We have $\nu \ll \mu$. The set $\{f > M_\mu\}$ is μ-null, hence ν-null too. Therefore $M_\nu \leqslant M_\mu$. Suppose that the inequality is strict. Let

$$A = \left\{f > \tfrac{M_\nu + M_\mu}{2}\right\}.$$

Then A is ν-null and so

$$\int_A f\,d\mu = \nu(A) = 0.$$

Since $f \geqslant 0$, $f \neq 0$, it follows that $\mu(A) = 0$ and so $\frac{M_\nu + M_\mu}{2} \geqslant M_\mu$, hence $M_\nu \geqslant M_\mu$, a contradiction.

Solution of Problem 3.175

(a) By Problem 3.170, for a given $\varepsilon > 0$, we can find $\delta > 0$ such that

$$\text{if } \lambda(A) \leqslant \delta, \quad \text{then } \int_A |f|^p \, dx \leqslant \varepsilon^p$$

(λ being the Lebesgue measure on $\mathbb{R}$).

Let $h \in (0, \delta]$ and assume that $1 < p < +\infty$. Then using the Hölder inequality (see Theorem 3.103), we have

$$\left|F(x+h) - F(x)\right| = \Big|\int_x^{x+h} f \, ds\Big| \leqslant h^{\frac{1}{p'}} \Big(\int_x^{x+h} |f|^p \, ds\Big)^{\frac{1}{p}} \leqslant \varepsilon h^{\frac{1}{p'}}.$$

If $p = 1$, then $p' = \infty$ and we have

$$\left|F(x+h) - F(x)\right| \leqslant \int_x^{x+h} |f| \, ds \leqslant \varepsilon.$$

(b) Since

$$f(x) - f(0) = \int_0^x f' \, ds,$$

from **(a)** above, it follows that f is uniformly continuous. Hence Problem 3.93(b) implies that $f(x) \longrightarrow 0$ as $|x| \to +\infty$.

Solution of Problem 3.176

Because m and ν are probability measures, we have

$$\int_\Omega f \, d\mu = \int_\Omega h \, d\mu = 1.$$

Then for any $A \in \Sigma$, we have

$$0 = \int_\Omega (f - h) \, d\mu = \int_A (f - h) \, d\mu + \int_{A^c} (f - h) \, d\mu,$$

so

$$\int_A (f - h) \, d\mu = -\int_{A^c} (f - h) \, d\mu,$$

thus

$$2\Big|\int_A (f-h)\,d\mu\Big| \;=\; \Big|\int_A (f-h)\,d\mu\Big| + \Big|\int_{A^c} (f-h)\,d\mu\Big| \;\leqslant\; \int_\Omega |f-h|\,d\mu,$$

so finally

$$2\|m-\nu\|_* \;\leqslant\; \int_\Omega |f-h|\,d\mu \;=\; \|f-h\|_1.$$

On the other hand

$$\int_\Omega |f-h|\,d\mu \;\leqslant\; \int_{\{f-h>0\}} (f-h)\,d\mu - \int_{\{f-h<0\}} (f-h)\,d\mu \;\leqslant\; 2\|m-\nu\|_*$$

(see Theorem 3.152). Therefore finally we have

$$2\|m-\nu\|_* \;=\; \int_\Omega |f-h|\,d\mu.$$

Solution of Problem 3.177
Let

$$\nu \;=\; \sum_{n\geqslant 1} \tfrac{1}{2^n}\mu_n.$$

Evidently ν is a probability measure on (Ω, Σ) and $\nu \ll \mu_n$ for all $n \geqslant 1$ (see Definition 3.150). So, applying the Radon–Nikodym theorem (see Theorem 3.152), we can find a sequence $\{f_n\}_{n\geqslant 1} \subseteq L^1(\Omega)$ such that

$$\mu_n(A) \;=\; \int_A f_n\,d\nu \qquad \forall\; A \in \Sigma,\; n \geqslant 1.$$

Solution of Problem 3.178
No. Suppose that $\Omega = [0,1]$, $\Sigma = \mathcal{B}(\Omega)$, μ is the counting measure and $\nu = \lambda$ is the Lebesgue measure. Then $\nu \ll \mu$ (since the only μ-null set

is the empty set). However there is no function $f\colon [0,1] \longrightarrow [0,+\infty]$, satisfying

$$\nu(A) = \int_A f\, d\mu \qquad \forall\, A \in \mathcal{B}(\Omega).$$

Indeed, f is μ-integrable and so it can be nonzero only on a set which is at most countable.

Solution of Problem 3.179

Since by hypothesis $\xi \perp \nu$, we can find a ν-null set $A \in \Sigma$ on which ξ is concentrated (i.e., $\xi(\Omega \setminus A) = 0$). Then, for all $C \in \Sigma$, we have

$$\xi(C) = \xi(C \cap A) \leqslant \mu(C \cap A) + \nu(C \cap A) = \mu(C \cap A) \leqslant \mu(C),$$

so $\xi \leqslant \mu$.

Solution of Problem 3.180

We have $\lambda^N(\overline{B}_r(x)) = \xi r^N$, where $\xi = \lambda^N(\overline{B}_1(0))$. Also, if $r \to r_0$ and $x_n \to x_0$, then

$$\chi_{\overline{B}_r(x)} \longrightarrow \chi_{\overline{B}_{r_0}(x)} \quad \text{on } \mathbb{R}^N \setminus \partial B_{r_0}(x_0).$$

Since $\lambda^N(\partial B_{r_0}(x_0)) = 0$, we have $\chi_{\overline{B}_r(x)} \longrightarrow \chi_{\overline{B}_{r_0}(x)}$ almost everywhere on $\mathbb{R}^N$ and

$$\left|\chi_{\overline{B}_r(x)}\right| \leqslant \chi_{\overline{B}_{r_0+1}(x_0)} \qquad \forall\, r < r_0 + \tfrac{1}{2},\ \|x - x_0\| < \tfrac{1}{2}.$$

So, by the Lebesgue dominated convergence theorem (see Theorem 3.94), we have that the function

$$u(r,x) = \int_{\overline{B}_r(x)} \left|f(y)\right| dy \quad \text{is continuous}$$

and so the function $(r,u) \longmapsto \frac{1}{(\xi r)^N} u(r,x)$ is continuous. Note that

$$(f^*)^{-1}(a,+\infty) = \bigcup_{r>0} u(r,\cdot)^{-1}(a,+\infty),$$

which is open for all $a \in \mathbb{R}$ and this proves the measurability of the Hardy–Littlewood maximal function f^*.

Solution of Problem 3.181

Let $f \in L^1_{\text{loc}}(\mathbb{R})$ (see Definition 3.157). Suppose that

$$\int_0^\infty |f|\,dx \;<\; +\infty.$$

Then f is integrable (see Proposition 3.88). Also, if $f_n = \chi_{[0,n]} f$, then $|f_n| \leqslant |f|$ and $f_n \longrightarrow \chi_{\mathbb{R}_+} f$ almost everywhere on $\mathbb{R}$. So, by the Lebesgue dominated convergence theorem (see Theorem 3.94), we have

$$\int_{\mathbb{R}} f_n\,dx \;=\; \int_0^n f\,dx \;\longrightarrow\; \int_0^\infty f\,dx,$$

a contradiction to the assumption that the sequence $\left\{\int_0^n f\,dx\right\}_{n\geqslant 1}$ does not have a limit in $\mathbb{R}$. So, we have

$$\int_0^\infty |f|\,dx \;=\; +\infty.$$

The converse is not true, namely it can happen that

$$\int_0^\infty |f|\,dx \;=\; +\infty \quad \text{and} \quad \lim_{n\to+\infty}\int_0^n f\,dx \text{ exists.}$$

Consider the function $f(x) = \sum\limits_{k\geqslant 1} \frac{(-1)^k}{k}\chi_{[k,k+1]}(x)$.

Solution of Problem 3.182

We argue by contradiction. So, suppose we can find $\varepsilon > 0$ such that for every $n \geqslant 1$, there exists $C_n \in \Sigma$, with $\mu(C_n) < \frac{1}{2^n}$ such that

$$\int_{C_n} |f|\,d\mu \;>\; \varepsilon.$$

Let

$$E_n = \bigcup_{k \geqslant n} C_k \qquad \forall\, n \geqslant 1.$$

Then

$$\mu(E_n) \leqslant \frac{1}{2^{n-1}}$$

and so by hypothesis $\int\limits_{E_n} |f|\, d\mu < +\infty$.

Also

$$\varepsilon < \int\limits_{E_n} |f|\, d\mu \qquad \forall\, n \geqslant 1.$$

The sequence $\{E_n\}_{n \geqslant 1} \subseteq \Sigma$ is decreasing and so, by the Lebesgue monotone convergence theorem (see Theorem 3.92), we have

$$0 = \int\limits_{\bigcap\limits_{n \geqslant 1} E_n} |f|\, d\mu = \lim_{n \to +\infty} \int\limits_{E_n} |f|\, d\mu \geqslant \varepsilon$$

(since $\mu\big(\bigcap\limits_{n \geqslant 1} E_n\big) = 0$), a contradiction.

Solution of Problem 3.183

Let $\varepsilon > 0$ and let us set

$$\xi(\varepsilon) = \sup\Big\{ \int\limits_A |f|\, d\mu : \ A \in \Sigma,\ \mu(A) < \varepsilon,\ f \in K \Big\}.$$

Let $\{f_n\}_{n \geqslant 1} \subseteq K$ be a sequence and suppose that $f_n \longrightarrow f$ μ-almost everywhere on Ω. By the Fatou lemma (see Theorem 3.95), for all $A \in \Sigma$ with $\mu(A) < \varepsilon$, we have

$$\xi(\varepsilon) \geqslant \liminf_{n \to +\infty} \int\limits_A |f_n|\, d\mu \geqslant \int\limits_A |f|\, d\mu,$$

so

$$\xi(\varepsilon) \geqslant \sup\Big\{ \int\limits_A |h|\, d\mu : \ A \in \Sigma,\ \mu(A) < \varepsilon \text{ and } h \in K^* \Big\}$$

and thus K^* is uniformly integrable (see Theorem 3.184).

Solution of Problem 3.184

(a) We have

$$\{\tau \leqslant n\} = \emptyset \qquad \forall\, k > n$$

and

$$\{\tau \leqslant n\} = \Omega \qquad \forall\, k \leqslant n.$$

Hence, if

$$\Sigma_\infty = \sigma\Big(\bigcup_{n\geqslant 1} \Sigma_n\Big),$$

then

$$\Sigma_\tau = \big\{A \in \Sigma_\infty :\ A \in \Sigma_n \text{ for every } n \geqslant k\big\} = \bigcap_{n\geqslant k} \Sigma_n = \Sigma_k.$$

(b) Let $n \in \mathbb{N}_0$. We have

$$\big\{\min\{\tau,\sigma\} \leqslant n\big\} = \{\tau \leqslant n\} \cup \{\sigma \leqslant n\} \in \Sigma_n$$

(since τ, σ are stopping times). Similarly, we have

$$\big\{\max\{\tau,\sigma\} \leqslant n\big\} = \{\tau \leqslant n\} \cap \{\sigma \leqslant n\} \in \Sigma_n.$$

Finally, note that

$$\{\tau + \sigma \leqslant n\} = \bigcup_{m=0}^{n} \{\tau \leqslant m\} \cap \{\sigma \leqslant n-m\} \in \Sigma_n.$$

From the above we conclude that $\min\{\tau,\sigma\}$, $\max\{\tau,\sigma\}$ and $\tau+\sigma$ are stopping times for $\{\Sigma_n\}_{n\geqslant 1}$ (see Definition 3.169).

(c) Let $A \in \Sigma_\tau$ and $n \geqslant 0$. We have $A \cap \{\tau \leqslant n\} \in \Sigma_n$. Also because $\tau \leqslant \sigma$, we have $\{\sigma \leqslant n\} \subseteq \{\tau \leqslant n\}$. Therefore

$$A \cap \{\sigma \leqslant n\} = A \cap \{\tau \leqslant n\} \cap \{\sigma \leqslant n\} \in \Sigma_n,$$

since $\{\sigma \leqslant n\} \in \Sigma_n$. Hence $A \in \Sigma_\sigma$ and we have proved that $\Sigma_\tau \subseteq \Sigma_\sigma$.

(d) From **(b)** and **(c)**, we have

$$\Sigma_{\min\{\tau,\sigma\}} \subseteq \Sigma_\tau \quad \text{and} \quad \Sigma_{\min\{\tau,\sigma\}} \subseteq \Sigma_\sigma,$$

so

$$\Sigma_{\min\{\tau,\sigma\}} \subseteq \Sigma_\tau \cap \Sigma_\sigma.$$

Let $A \in \Sigma_\tau \cap \Sigma_\sigma \subseteq \Sigma_\infty$. Recall that

$$\{\min\{\tau,\sigma\} \leqslant n\} = \{\tau \leqslant n\} \cup \{\sigma \leqslant n\}.$$

Hence

$$\begin{aligned} A \cap \{\min\{\tau,\sigma\} \leqslant n\} &= A \cap \{\{\tau \leqslant n\} \cup \{\sigma \leqslant n\}\} \\ &= (A \cap \{\tau \leqslant n\}) \cup (A \cap \{\sigma \leqslant n\}) \in \Sigma_n \end{aligned}$$

(since $A \in \Sigma_\tau \cap \Sigma_\sigma$), hence $A \in \Sigma_{\min\{\tau,\sigma\}}$. So, we conclude that $\Sigma_{\min\{\tau,\sigma\}} = \Sigma_\tau \cap \Sigma_\sigma$.

(e) Let $n \in \mathbb{N}_0$. We have

$$\{\tau < \sigma\} \cap \{\tau \leqslant n\} = \bigcup_{m=0}^{n} \{\tau = m\} \cap \{\sigma > m\}.$$

But for every $m \in \{0, 1, \dots, n\}$, we have

$$\{\tau = m\} = \{\tau \leqslant m\} \cap \{\tau \leqslant m-1\}^c \in \Sigma_m \subseteq \Sigma_n$$

and

$$\{\sigma > m\} = \{\sigma \leqslant m\}^c \in \Sigma_m \subseteq \Sigma_n.$$

Therefore, it follows that

$$\{\tau < \sigma\} \cap \{\tau \leqslant n\} \in \Sigma_n,$$

hence $\{\tau < \sigma\} \in \Sigma_\tau$. Similarly, since

$$\{\tau < \sigma\} \cap \{\sigma \leqslant n\} = \bigcup_{m=0}^{n} \{\sigma = m\} \cap \{\tau < m\}$$

and for $m \in \{0, 1, \dots, n\}$, we have

$$\{\sigma = m\} = \{\sigma \leqslant m\} \cap \{\sigma \leqslant m-1\}^c \in \Sigma_m \subseteq \Sigma_n$$

and

$$\{\tau < m\} = \{\tau \leqslant m-1\} \in \Sigma_{m-1} \subseteq \Sigma_n,$$

we infer that

$$\{\tau < \sigma\} \cap \{\sigma \leqslant n\} \in \Sigma_n,$$

hence $\{\tau < \sigma\} \in \Sigma_\tau$ and so $\{\tau < \sigma\} \in \Sigma_\tau \cap \Sigma_\sigma$. Finally note that

$$\{\tau = \sigma\} = \{\tau < \sigma\}^c \cap \{\sigma < \tau\}^c \in \Sigma_\tau \cap \Sigma_\sigma$$

(from what was proved earlier).

Solution of Problem 3.185

Since τ is a stopping time (see Definition 3.169), we have

$$\{\tau \geqslant n+1\} = \{\tau \leqslant n\}^c \in \Sigma_n.$$

Therefore

$$\begin{aligned} E^{\Sigma_n}\left(f_{n+1}^{\tau} - f_n^{\tau}\right) &= E^{\Sigma_n}\left((f_{n+1} - f_n)\chi_{\{\tau \geqslant n+1\}}\right) \\ &= \chi_{\{\tau \geqslant n+1\}} E^{\Sigma_n}(f_{n+1} - f_n) = 0. \end{aligned}$$

Similarly for submartingale (with $\geqslant$ at the end) and supermartingale (with $\leqslant$ at the end).

Solution of Problem 3.186

From Problem 3.185, we know that $\left\{f_{\min\{\tau,n\}}, \Sigma_n\right\}_{n \geqslant 0}$ is a submartingale. Hence, we have

$$Ef_0 = Ef_{\min\{\tau,0\}} \leqslant Ef_{\min\{\tau,k\}} = Ef_{\tau}.$$

For every $m \in \{0, 1, \ldots, k\}$, we have

$$E\left(f_k \chi_{\{\tau=m\}}\right) = E^{\Sigma_m}\left(f_k \chi_{\{\tau=m\}}\right) \geqslant E\left(f_m \chi_{\{\tau=m\}}\right) = E\left(f_{\tau} \chi_{\{\tau=m\}}\right),$$

so, summing in m, we get

$$Ef_0 \leqslant Ef_{\tau} \leqslant Ef_k.$$

Solution of Problem 3.187

(a) Since $\{f_n, \Sigma_n\}_{n \geqslant 0}$ is a supermartingale, we have

$$E^{\Sigma_n} f_{n+1} \leqslant f_n \qquad \forall\, n \geqslant 0$$

(see Definition 3.167), so

$$h_n = f_n - E^{\Sigma_n} f_{n+1} \geqslant 0 \qquad \forall\, n \geqslant 0.$$

Note that

$$\begin{aligned}\int_\Omega h_n\,d\mu &= \int_\Omega f_n\,d\mu - \int_\Omega E^{\Sigma_n} f_{n+1}\,d\mu \\ &= \int_\Omega f_n\,d\mu - \int_\Omega f_{n+1}\,d\mu = c - c = 0 \qquad \forall\, n \geqslant 0.\end{aligned}$$

Therefore

$$h_n = 0 \quad \mu\text{-almost everywhere on } \Omega$$

and so

$$E^{\Sigma_n} f_{n+1} = f_n \qquad \forall\, n \geqslant 0,$$

hence $\{f_n, \Sigma_n\}_{n\geqslant 0}$ is a martingale.
(b) "$\Longrightarrow$": This follows from the stopping time theorem (see Theorem 3.173).

"$\Longleftarrow$": Let $A \in \Sigma_n$ and define $\tau\colon \Omega \longrightarrow \mathbb{N}_0$ by

$$\tau(\omega) = \begin{cases} n & \text{if } \omega \in A, \\ n+1 & \text{if } \omega \notin A \end{cases}$$

and $\sigma \equiv n+1$. We claim that τ is a stopping time for $\{\Sigma_n\}_{n\geqslant 0}$. To see this let $m \in \mathbb{N}_0$. Then

$$\{\tau \leqslant m\} = \begin{cases} \emptyset & \text{if } m \leqslant n-1, \\ A & \text{if } m = n, \\ \Omega & \text{if } m \geqslant n+1, \end{cases}$$

so $\{\tau \leqslant m\} \in \Sigma_m$ and thus τ is a stopping time for $\{\Sigma_n\}_{n\geqslant 0}$ as claimed. Then

$$f_\tau = f_n \chi_A + f_{n+1} \chi_{A^c}$$

and by hypothesis, we have

$$\int_\Omega f_{n+1}\,d\mu = \int_\Omega f_0\,d\mu.$$

Hence

$$\int_A f_n \, d\mu + \int_{A^c} f_{n+1} \, d\mu = \int_\Omega f_\tau \, d\mu = \int_\Omega f_0 \, d\mu = \int_\Omega f_{n+1} \, d\mu$$

$$= \int_A f_{n+1} \, d\mu + \int_{A^c} f_{n+1} \, d\mu,$$

so

$$\int_A f_n \, d\mu = \int_A f_{n+1} \, d\mu \qquad \forall\, A \in \Sigma_n,$$

thus

$$f_n = E^{\Sigma_n} f_{n+1}$$

(see Definition 3.161) and so finally we get that $\{f_n, \Sigma_n\}_{n \geqslant 0}$ is a martingale.

Solution of Problem 3.188

Note that

$$|h_n| \leqslant |f_n| + |g_n| \qquad \forall\, n \geqslant 1$$

and so

$$h_n \in L^1(\Omega, \Sigma_n) \qquad \forall\, n \geqslant 1.$$

We have

$$g_{n+1} = g_\tau \geqslant f_\tau = f_{n+1} \quad \mu\text{-almost everywhere on } \{\tau = n+1\}.$$

Hence

$$\begin{aligned} h_{n+1} &= g_{n+1}\chi_{\{n+1<\tau\}} + f_{n+1}\chi_{\{\tau \leqslant n+1\}} \\ &\leqslant g_{n+1}\chi_{\{n+1<\tau\}} + g_{n+1}\chi_{\{n+1=\tau\}} + f_{n+1}\chi_{\{\tau \leqslant n\}} \\ &= g_{n+1}\chi_{\{n<\tau\}} + f_{n+1}\chi_{\{\tau \leqslant n\}}. \end{aligned}$$

Note that

$$\{n < \tau\} = \{\tau \leqslant n\}^c \in \Sigma_n \quad \text{and} \quad \{\tau \leqslant n\} \in \Sigma_n.$$

Hence

$$\begin{aligned} E^{\Sigma_n} h_{n+1} &\leqslant \chi_{\{n<\tau\}} E^{\Sigma_n} g_{n+1} + \chi_{\{\tau \leqslant n\}} E^{\Sigma_n} f_{n+1} \\ &\leqslant g_n \chi_{\{n<\tau\}} + f_n \chi_{\{\tau \leqslant n\}} = h_n \quad \mu\text{-almost everywhere on } \Omega, \end{aligned}$$

so $\{h_n, \Sigma_n\}_{n \geqslant 0}$ is a supermartingale.

The same argument applies in the case of martingales, with $\leqslant$ replaced by $=$.

Solution of Problem 3.189

First let us show that $\{h_n, \Sigma_n\}_{n \geqslant 0}$ is a martingale. We have

$$E^{\Sigma_n} h_{n+1} = h_n + \tfrac{1}{n+1} E^{\Sigma_n}(f_{n+1} - f_n) = h_n,$$

so $\{h_n, \Sigma_n\}_{n \geqslant 0}$ is a martingale. Next, we have

$$h_n^2 = \sum_{k=1}^{n} \tfrac{1}{k^2}(f_k - f_{k-1})^2 + 2 \sum_{1 \leqslant i < j \leqslant n} \tfrac{1}{ij}(f_j - f_{j-1})(f_i - f_{i-1}).$$

But

$$\begin{aligned} \int_\Omega (f_j - f_{j-1})(f_i - f_{i-1})\, d\mu &= \int_\Omega E^{\Sigma_{j-1}}(f_j - f_{j-1})(f_i - f_{i-1})\, d\mu \\ &= \int_\Omega (f_i - f_{i-1}) E^{\Sigma_{j-1}}(f_j - f_{j-1})\, d\mu = 0. \end{aligned}$$

Therefore

$$\int_\Omega h_n^2\, d\mu = \sum_{k=1}^{n} \tfrac{1}{k^2} E(f_k - f_{k-1})^2 \leqslant 4M^2 \sum_{k \geqslant 1} \tfrac{1}{k^2} < +\infty,$$

so the sequence $\{h_n\}_{n \geqslant 1} \subseteq L^2(\Omega)$ is bounded. Invoking Theorem 3.187, we conclude that

$$h_n \longrightarrow h \quad \mu\text{-almost everywhere on } \Omega \text{ and in } L^2(\Omega),$$

for some $h \in L^2(\Omega)$.

Solution of Problem 3.190

(a) Let $A \in \Sigma_n \subseteq \Sigma_{n+1}$. We have

$$m(A) = \int_A f_n \, d\mu = \int_A f_{n+1} \, d\mu,$$

so

$$E^{\Sigma_n} f_{n+1} = f_n$$

and thus $\{f_n, \Sigma_n\}_{n \geqslant 0}$ is a martingale.

(b) Since $\{f_n, \Sigma_n\}_{n \geqslant 0}$ is a positive martingale, by Corollary 3.179, there exists $f \in L^1(\Omega, \Sigma_\infty)$ such that $f_n \longrightarrow f$ μ-almost everywhere on Ω.

(c) "$\Longrightarrow$": Since $m \ll \mu_\infty = \mu|_{\Sigma_\infty}$ (see Definition 3.150), by the Radon–Nikodym theorem (see Theorem 3.152), there exists $f \in L^1(\Omega, \Sigma_\infty)$, $f \geqslant 0$ such that

$$m(A) = \int_A f \, d\mu \qquad \forall\, A \in \Sigma_\infty.$$

If $A \in \Sigma_n$, then

$$m(A) = \int_A f_n \, d\mu = \int_A f \, d\mu,$$

so

$$E^{\Sigma_n} f = f_n \qquad \forall\, n \geqslant 1$$

and thus $\{f_n, \Sigma_n\}_{n \geqslant 0}$ is a regular martingale (see Definition 3.185). Invoking Theorem 3.184, we have

$$f_n \longrightarrow f \quad \text{in } L^1(\Omega) \quad \text{and} \quad \mu\text{-almost everywhere on } \Omega$$

(see part **(a)**). We need to show that $f = \frac{dm}{d\mu_\infty}$. For every $A \in \Sigma_n$, $n \geqslant 1$, we have

$$\int_A f \, d\mu = \int_A E^{\Sigma_n} f \, d\mu = \int_A f_n \, d\mu \longrightarrow \int_A f \, d\mu.$$

Since $m(A) = \int_A f_n \, d\mu$, it follows that

$$m(A) = \int_A f \, d\mu \qquad \forall\, A \in \Sigma_n,\ n \geqslant 0$$

and so

$$m(A) = \int_A f \, d\mu \qquad \forall\, A \in \bigcup_{n \geqslant 0} \Sigma_n.$$

By the monotone class theorem (see Theorem 3.12), it follows that

$$m(A) = \int_A f \, d\mu \qquad \forall\, A \in \Sigma_\infty.$$

Since f is Σ_∞-measurable, we conclude that $f = \frac{dm}{d\mu_\infty}$.

Solution of Problem 3.191

(a) Because $\tau < +\infty$ and $|f_\tau| < +\infty$ μ-almost everywhere on Ω, we have

$$|f_\tau|\chi_{\{\tau > n\}} \longrightarrow 0 \quad \mu\text{-almost everywhere on } \Omega.$$

Also

$$|f_\tau|\chi_{\{\tau > n\}} \leqslant |f_\tau| \quad \text{and} \quad f_\tau \in L^1(\Omega)$$

(by hypothesis). Therefore, we can apply the Lebesgue dominated convergence theorem (see Theorem 3.94) and conclude that

$$\int_\Omega |f_\tau|\chi_{\{\tau > n\}} \, d\mu = \int_{\{\tau > n\}} |f_\tau| \, d\mu \longrightarrow 0.$$

(b) By hypotheses and part **(a)**, we have

$$\begin{aligned} \int_\Omega \left| f_{\min\{\tau, n\}} - f_\tau \right| d\mu &= \int_\Omega |f_n - f_\tau|\chi_{\{\tau > n\}} \, d\mu \\ &\leqslant \int_\Omega |f_n|\chi_{\{\tau > n\}} \, d\mu + \int_\Omega |f_\tau|\chi_{\{\tau > n\}} \, d\mu \longrightarrow 0. \end{aligned}$$

(c) Since $\min\{\tau, n\}$ is a bounded stopping time (see Problem 3.184(b)), from the stopping time theorem (see Theorem 3.173), we have

$$\int\limits_{\Omega} f_{\min\{\tau,n\}}\, d\mu = \int\limits_{\Omega} f_0\, d\mu.$$

Then

$$\begin{aligned} \Big|\int\limits_{\Omega} f_\tau\, d\mu - \int\limits_{\Omega} f_0\, d\mu\Big| &= \Big|\int\limits_{\Omega} f_\tau\, d\mu - \int\limits_{\Omega} f_{\min\{\tau,n\}}\, d\mu\Big| \\ &\leqslant \int\limits_{\Omega} \big|f_\tau - f_{\min\{\tau,n\}}\big|\, d\mu \longrightarrow 0, \end{aligned}$$

so

$$\int\limits_{\Omega} f_\tau\, d\mu = \int\limits_{\Omega} f_0\, d\mu.$$

Bibliography

[1] Di Benedetto, E.: Real Analysis. Birkhäuser Advanced Texts: Basler Lehrbücher. Birkhäuser Boston, Boston (2002)

[2] Benedetto, J.J., Czaja, W.: Integration and Modern Analysis. Birkhäuser Advanced Texts: Basler Lehrbücher. Birkhäuser Boston, Boston (2009)

[3] Denkowski, Z., Migórski, S., Papageorgiou, N.S.: An Introduction to Nonlinear Analysis: Theory. Kluwer/Plenum, New York (2003)

[4] Doob, J.L.: Measure Theory. Graduate Texts in Mathematics, vol. 143. Springer, New York (1994)

[5] Dudley, R.M.: Real Analysis and Probability. Cambridge Studies in Advanced Mathematics, vol. 74. Cambridge University Press, Cambridge (2004)

[6] Fremlin, D.H.: Measure Theory: Volume I: The Irreducible Minimum. Torres Fremlin, Colchester (2004)

[7] Fremlin, D.H.: Measure Theory: Volume II: Broad Foundations. Torres Fremlin, Colchester (2003)

[8] Halmos, P.R.: Measure Theory. Graduate Texts in Mathematics, vol. 18. Springer, New York (1974)

[9] Hewitt, E., Stromberg, K.: Real and Abstract Analysis. Amodern Treatment of the Theory of Functions of a Real Variable. Graduate Texts in Mathematics, vol. 25. Springer, New York (1975)

[10] Lieb, E.H., Loss, M.: Analysis. Graduate Studies in Mathematics, vol. 14. American Mathematical Society, Providence (2001)

[11] Rudin, W.: Real and Complex Analysis. McGraw-Hill Book Co., New York (1987)

[12] Schilling, R.L.: Measures, Integrals and Martingales. Cambridge University Press, New York (2005)

[13] Stein, E.M., Shakarchi, R.: Real Analysis: Measure Theory, Integration and Hilbert Spaces. Princeton Lectures in Analysis, vol. III. Princeton University Press, Princeton (2005)

[14] Widom, H.: Lectures on Measure and Integration. Van Nostrand Mathematical Studies, vol. 17. Van Nostrand Reinhold Co., New York (1969)

Chapter 4

Measures and Topology

4.1 Introduction

4.1.1 Borel and Baire σ-Algebras

In the previous chapter, we considered measures defined on abstract σ-algebras of sets. However, in most cases the underlying measure space has a natural topological structure. When we combine the measure theoretic and topological structures, we get a richer theory which we outline in this chapter.

Definition 4.1
*Let (X, τ) be a topological space. The **Borel σ-algebra** of X, denoted by $\mathcal{B}(X)$, is the σ-algebra generated by τ (i.e., by the open sets, $\mathcal{B}(X) = \sigma(\tau)$).*

Definition 4.2
Let (X, τ) be a topological space and let

$$C_c(X) \stackrel{\text{def}}{=} \{f \colon X \longrightarrow \mathbb{R} : f \textit{ is } \tau\textit{-continuous and has compact support}\}.$$

*The **Baire σ-algebra** of X, denoted by $\mathcal{B}a(X)$, is the smallest σ-algebra of subsets of X, which makes each function in $C_c(X)$ measurable.*

Remark 4.3
Therefore $\mathcal{B}a(X)$ is the σ-algebra generated by the sets $\{x \in X : f(x) \geqslant \lambda\}$ with $f \in C_c(X)$, $\lambda \in \mathbb{R}$. Evidently $\{f \geqslant \lambda\}$ can be replaced by $\{f \leqslant \lambda\}$ or $\{f > \lambda\}$ or $\{f < \lambda\}$.

L. Gasiński and N.S. Papageorgiou, *Exercises in Analysis: Part 1*, Problem Books in Mathematics, DOI 10.1007/978-3-319-06176-4_4,

The Baire σ-algebra is most interesting when X is locally compact.

Theorem 4.4
If X is a locally compact topological space (see Definition 2.92), then $\mathcal{B}a(X)$ is the σ-algebra generated by the compact G_δ-sets.

The next theorem compares the two σ-algebras $\mathcal{B}(X)$ and $\mathcal{B}a(X)$.

Theorem 4.5
If X is a topological space, then
(a) *$\mathcal{B}a(X) \subseteq \mathcal{B}(X)$.*
(b) *When X is locally compact, C is compact, U is open and $C \subseteq U$, then there exist $V, K \in \mathcal{B}a(X)$ such that V is σ-compact open and K is compact G_δ such that $C \subseteq V \subseteq K \subseteq U$.*
(c) *When X is locally compact separable and metrizable, then $\mathcal{B}(X) = \mathcal{B}a(X)$.*

We consider the Borel and Baire σ-algebras for products of topological spaces.

Theorem 4.6
If X and Y are topological spaces, then
(a) *$\mathcal{B}(X) \otimes \mathcal{B}(Y) \subseteq \mathcal{B}(X \times Y)$.*
(b) *When X and Y are both second countable (see Definition 2.24), then $\mathcal{B}(X \times Y) = \mathcal{B}(X) \otimes \mathcal{B}(Y)$.*
(c) *When X and Y are both second countable and locally compact, then $\mathcal{B}a(X \times Y) = \mathcal{B}a(X) \otimes \mathcal{B}a(Y)$.*

Definition 4.7
Let X be a topological space and let $f\colon X \longrightarrow \mathbb{R}$ be a function.
(a) *We say that f is **Borel measurable** (or a **Borel function**) if it is measurable when X is endowed with the Borel σ-algebra $\mathcal{B}(X)$ (on $\mathbb{R}$ as always we consider the Borel σ-algebra).*
(b) *We say that f is **Baire measurable** (or a **Baire function**) if it is measurable when X is endowed with the Baire σ-algebra $\mathcal{B}a(X)$.*

Remark 4.8
Evidently the elements of $C_c(X)$ are Baire functions and every Baire function is Borel.

4.1.2 Regular and Radon Measures

Definition 4.9
Let X be a topological space, let Σ be a σ-algebra and let μ be a measure defined on Σ.
(a) We say that μ is ***outer regular*** *if for all $A \in \Sigma$, we have*

$$\mu(A) \; = \; \inf\left\{\mu(U): \; U \in \Sigma, \; U \textit{ is open and } A \subseteq U\right\}.$$

(b) We say that μ is ***inner regular*** *if for all $A \in \Sigma$, we have*

$$\mu(A) \; = \; \sup\left\{\mu(C): \; C \in \Sigma, \; C \textit{ is closed and } C \subseteq A\right\}.$$

(c) We say that μ is ***regular*** *if it is both outer and inner regular.*
(d) We say that μ is ***inner regular with respect to compact sets*** *if for all $A \in \Sigma$, we have*

$$\mu(A) \; = \; \sup\left\{\mu(K): \; K \in \Sigma, \; K \textit{ is compact and } K \subseteq A\right\}.$$

(e) We say that μ is ***Radon*** *(or* ***tight****), if $\mu(K) < +\infty$ for all compact sets $K \subseteq X$, μ is regular and inner regular with respect to compact sets.*

Remark 4.10
If Σ contains $\mathcal{B}(X)$, then in the above definitions the requirements that $U \in \Sigma$, $C \in \Sigma$ and $K \in \Sigma$ are of course redundant. The notions of regularity and Radoness will be applied usually on σ-algebras Σ that contain at least $\mathcal{B}a(X)$. Finally, if μ is a signed measure, then μ is regular (respectively, Radon) if and only if $|\mu|$ is regular (respectively, Radon).

Theorem 4.11
<u>If</u> X is a metrizable space,
<u>then</u> every finite Borel measure (i.e., a measure defined on $\mathcal{B}(X)$) is regular.

In Polish spaces, we can improve this theorem.

Theorem 4.12
<u>If</u> X is a Polish space (see Definition 2.150),
<u>then</u> every finite Borel measure is Radon.

Definition 4.13
Let X be a topological space and let $\mu\colon \mathcal{B}(X) \longrightarrow \mathbb{R}^* = \mathbb{R} \cup \{\pm\infty\}$ be a signed measure. The ***support of*** μ*, denoted by* $\operatorname{supp}\mu$*, is a closed set in X such that*
(a) $|\mu|\big((\operatorname{supp}\mu)^c\big) = 0$*; and*
(b) if U is an open set such that $U \cap \operatorname{supp}\mu \neq 0$, then

$$|\mu|(U \cap \operatorname{supp}\mu) > 0.$$

Remark 4.14
The support of μ need not exist. If it exists, it is unique. According to the above definition, the support of μ is the smallest closed set whose complement has $|\mu|$-measure zero.

Theorem 4.15
If X is a topological space and $\mu\colon \mathcal{B}(X) \longrightarrow [0, +\infty]$ is a measure, then if either X is second countable (see Definition 2.24) or μ is Radon, then $\operatorname{supp}\mu$ *exists.*

Definition 4.16
For any set X and $x \in X$, δ_x denotes the ***Dirac measure*** *at x, which is the probability measure on 2^X having all its mass at x, that is*

$$\delta_x(A) \stackrel{\text{def}}{=} \begin{cases} 1 & \text{if } x \in A, \\ 0 & \text{if } x \notin A. \end{cases}$$

Proposition 4.17
If X is a topological space, then δ_x is a Radon measure.

Concerning image measures, we have the following result.

Theorem 4.18
If X and Y are topological spaces, X is compact, $f\colon X \longrightarrow Y$ is a continuous function, μ is a finite tight Borel measure on X and μf^{-1} is its image measure by f on Y (i.e., $(\mu f^{-1})(A) = \mu\big(f^{-1}(A)\big)$ for all $A \in \mathcal{B}(Y)$),
then μf^{-1} is a finite Radon measure on $\mathcal{B}(Y)$.

4.1.3 Riesz Representation Theorem for Continuous Functions

Definition 4.19
Let X be a locally compact topological space (see Definition 2.92).

We introduce the following vector spaces of continuous functions $f: X \longrightarrow \mathbb{R}$*:*

- $C_c(X)$ *is the space of continuous functions* $f: X \longrightarrow \mathbb{R}$ *which have compact supports.*
- $C_0(X)$ *is the space of continuous functions* $f: X \longrightarrow \mathbb{R}$ *which vanish at infinity, i.e., for every* $\varepsilon > 0$*, there exists a compact set* $K \subseteq X$ *such that* $|f(x)| < \varepsilon$ *for* $x \notin K$*.*
- $C_b(X)$*is the space of bounded continuous functions* $f: X \longrightarrow \mathbb{R}$*.*

Clearly we have

$$C_c(X) \subseteq C_0(X) \subseteq C_b(X).$$

Moreover, if X *is compact, then*

$$C_c(X) = C_0(X) = C_b(X).$$

If X *is not compact, then each of the inclusions is strict. On* $C_b(X)$ *we introduce the supremum norm* $\|\cdot\|_{C_b}$*, defined by*

$$\|f\|_{C_b} \stackrel{\text{def}}{=} \sup_{x\in X} |f(x)|.$$

This norm is restricted on $C_0(X)$ *and* $C_c(X)$*. Note that on them the supremum in the above definition is a maximum. We denote these restrictions by* $\|\cdot\|_{C_0}$ *and* $\|\cdot\|_{C_c}$*, respectively.*

Proposition 4.20
If X *is a locally compact topological space,*
then $(C_b(X), \|\cdot\|_{C_b})$ *is complete and* $C_0(X)$ *is a closed subset of it. Finally* $C_c(X)$ *is dense in* $C_0(X)$*.*

For locally compact spaces X, we can introduce their Alexandrov one-point compactification X^* (see Remark 2.97 and Theorem 2.98). We have the following result.

Proposition 4.21
If X *is a locally compact topological space and* X^* *is its Alexandrov one-point compactification (see Remark 2.97),* $Y = \{f \in C(X^*) : f(\infty) = 0\}$ *and for every* $f \in Y$, $\widehat{f} = f|_X$*,*
then $f \longmapsto \widehat{f}$ *is a linear isometry of* Y *onto* $C_0(X)$*.*

Definition 4.22
Let X be a locally compact topological space. By $M_b(X)$ we denote the space of signed Radon (tight) Borel measures μ (i.e., $|\mu|$ is Radon; see Remark 4.10) and bounded (i.e., $|\mu|(X) < +\infty$). We equip $M_b(X)$ with the norm $\|\mu\| = |\mu|(X)$ and then $\big(M_b(X), \|\cdot\|\big)$ becomes a Banach space. The elements of $M_b(X)$ are also known as ***bounded signed Radon measures****.*

The following theorem relates the Banach space $M_b(X)$ with the dual of $C_0(X)$ (i.e., the space of continuous linear functionals on $C_0(X)$, denoted by $C_0(X)^*$; see Chap. 5).

Theorem 4.23 *(Riesz Representation Theorem)*
For every continuous linear functional l on $C_0(X)$, there exists a unique $\mu \in M_b(X)$ such that

$$l(f) = \int_X f \, d\mu \qquad \forall\, f \in C_0(X).$$

Moreover, we have $\|l\|_* = \sup\big\{|l(f)| : \|f\|_\infty \leqslant 1\big\} = \|\mu\| = |\mu|(X)$.

Remark 4.24
Using the terminology of Chap. 5, the above theorem says that $C_0(X)^* = M_b(X)$ (i.e., the two Banach spaces $C_0(X)^*$ and $M_b(X)$ are isometrically isomorphic). If l is positive (i.e., $l(f) \geqslant 0$ for all $f \geqslant 0$), then $\mu \geqslant 0$.

4.1.4 Space of Probability Measures: Prohorov Theorem

Definition 4.25
Let X be a metric space and let $M_1^+(X)$ consist of all probability measures on $\mathcal{B}(X)$. For every $f \in C_b(X)$ consider the functional $\theta_f \colon M_1^+(X) \longrightarrow \mathbb{R}$, defined by

$$\theta_f(\mu) = \int_X f(x) \, d\mu.$$

Then the ***weak topology*** *on $M_1^+(X)$ is defined to be the topology $w\big(\{\theta_f\}_{f \in C_b(X)}\big)$ (see Definition 2.62).*

In the next theorem, we state several equivalent useful definitions of this topology.

Theorem 4.26 *(Portmanteau Theorem)*
Let X be a metric space and let $\{\mu_\alpha\}_{\alpha\in J} \subseteq M_+^1(X)$ be a net. The following statements are equivalent:
(a) $\mu_\alpha \xrightarrow{w} \mu$;
(b) *for every $f \in UC_b(X)$ where*

$$UC_b(X) \stackrel{\text{def}}{=} \{f\colon X \longrightarrow \mathbb{R} : f \text{ is bounded and uniformly continuous}\},$$

we have

$$\int_\Omega f\, d\mu_\alpha \longrightarrow \int_\Omega f\, d\mu;$$

(c) *for every closed set $C \subseteq X$, we have $\limsup_{\alpha\in J} \mu_\alpha(C) \leqslant \mu(C)$;*
(d) *for every open set $U \subseteq X$, we have $\mu(U) \leqslant \liminf_{\alpha\in J} \mu_\alpha(U)$;*
(e) *for every $A \in \mathcal{B}(X)$ such that $\mu(\partial A) = 0$, we have $\lim_{\alpha\in J} \mu_\alpha(A) = \mu(A)$.*

It is interesting to know when this weak topology on $M_1^+(X)$ is metrizable, in which case we can use sequences. The next theorem provides conditions for the metrizability of the weak topology.

Theorem 4.27
The space $\big(M_1^+(X), w\big)$ is separable metrizable if and only if X is a separable metric space.

In the next theorem, we indicate a concrete countable dense subsets of $\big(M_1^+(X), w\big)$.

Theorem 4.28
If X is a separable metric space,
then the set of all convex combinations of Dirac measures (see Definition 4.16) is dense in $\big(M_1^+(X), w\big)$.

Theorem 4.29
The space $\big(M_1^+(X), w\big)$ is compact metrizable if and only if X is a compact metric space.

Theorem 4.30
The space $(M_1^+(X), w)$ is Polish if and only if X is a Polish space.

In the case of Polish spaces, we can have a nice description of the compact subsets of $(M_1^+(X), w)$.

Theorem 4.31 *(Prohorov Theorem)*
If X is a Polish space and $C \subseteq M_1^+(X)$,
then $\overline{C}^w$ is w-compact if and only if C is uniformly tight (i.e., for every $\varepsilon > 0$, we can find a compact set $K_\varepsilon \subseteq X$ such that $\mu(K_\varepsilon) \geqslant 1 - \varepsilon$ for all $\mu \in C$).

Theorem 4.32
If X is a separable metric space and $\{\mu_n\}_{n\geqslant 1} \subseteq M_1^+(X)$ is a sequence,
then $\mu_n \xrightarrow{w} \mu$ if and only if

$$\lim_{n\to+\infty} \sup_{f\in C} \Big| \int_X f\, d\mu_n - \int_\Omega f\, d\mu \Big| = 0$$

for every equicontinuous and uniformly bounded family $C \subseteq C_b(X)$ (see Definition 1.83).

Proposition 4.33
If X and Y are locally compact, second countable topological spaces (see Definitions 2.92 and 2.24) and μ and ν are Radon measures on X and Y, respectively,
then $\mu \times \nu$ is a Radon measure on $X \times Y$.

4.1.5 Polish, Souslin and Borel Spaces

In Definitions 2.150 and 2.156 we introduced the notions of Polish and Souslin spaces, respectively. In the sequel, we will relate these notions to the Borel sets. We start with an interesting property of Souslin spaces known as the "separation property".

Theorem 4.34 *(Separation Property)*
If X is a topological space and $\{A_n\}_{n\geqslant 1}$ is a sequence of pairwise disjoint Souslin subspaces of X,
then there exists a sequence $\{C_n\}_{n\geqslant 1}$ of pairwise disjoint Borel subsets of X such that $A_n \subseteq C_n$ for all $n \geqslant 1$.

Corollary 4.35
If X *is a topological space,* $X = \bigcup_{n \geqslant 1} A_n$*, with* $\{A_n\}_{n \geqslant 1}$ *being pairwise disjoint Souslin subsets of* X*,*
then $A_n \in \mathcal{B}(X)$ *for every* $n \geqslant 1$*.*
In particular, if two complementary subsets of X *are Souslin, then both are Borel sets.*

Corollary 4.36
If τ_1 *and* τ_2 *are two comparable Souslin topologies on a set* X*,*
then $\mathcal{B}(X_{\tau_1}) = \mathcal{B}(X_{\tau_2})$*.*

Proposition 4.37
If X *and* Y *are two topological spaces,* $f\colon X \longrightarrow Y$ *is a function and* $\operatorname{Gr} f = \{(x, y) \in X \times Y : y = f(x)\}$ *is a Souslin subspace of* $X \times Y$*,*
then f *is a Borel function (i.e., the inverse image of every Borel set of* Y *is a Borel set of* X*).*

Corollary 4.38
If X *and* Y *are Souslin spaces,*
then $f\colon X \longrightarrow Y$ *is a Borel function if and only if* $\operatorname{Gr} f \subseteq X \times Y$ *is Souslin if and only if* $\operatorname{Gr} f \subseteq X \times Y$ *is Borel. Also direct and inverse images by* f *of Souslin sets are Souslin sets.*

Definition 4.39
Let (X, Σ) *and* $(Y, \mathcal{Y})$ *be measurable spaces. A bijection function* $f\colon X \longrightarrow Y$ *is an* ***isomorphism*** *if* f *is* $(\Sigma, \mathcal{Y})$*-measurable and* f^{-1} *is* $(\mathcal{Y}, \Sigma)$*-measurable. Then we say that spaces* X *and* Y *are* ***isomorphic****. If* $E \subseteq X$ *and* $F \subseteq Y$*, then we say that* E *and* F *are* ***isomorphic*** *if* $(E, \mathcal{B}(X) \cap E)$ *and* $(F, \mathcal{B}(Y) \cap F)$ *are isomorphic. If* X *and* Y *are separable metrizable and* $\Sigma = \mathcal{B}(X)$*,* $\mathcal{Y} = \mathcal{B}(Y)$ *(the Borel* σ*-algebras), then we use the term* ***Borel isomorphic****.*

Definition 4.40
A topological space X *is said to be a* ***Borel space*** *if there exists a Polish space* Y *and* $A \in \mathcal{B}(Y)$ *such that* X *is homeomorphic to* A*. The empty set is by definition a Borel space.*

Proposition 4.41
Every Borel space is metrizable and separable.

Theorem 4.42
Let X and Y be Borel spaces. Then X and Y are Borel isomorphic if and only if they have the same cardinality.

Corollary 4.43
Every uncountable Borel space is Borel isomorphic to every other uncountable Borel space. In particular, every uncountable Borel space is isomorphic to $[0,1]$ and has the cardinality of the continuum.

Theorem 4.44 *(Kuratowski Theorem)*
If X is a Borel space, Y is a separable metrizable space and $f\colon X \longrightarrow Y$ is an injective Borel function, then $f(X) \in \mathcal{B}(Y)$ and f^{-1} is a Borel set. In particular, if Y is a Borel space, then X and $f(X)$ are Borel isomorphic (see Definition 4.39).

Definition 4.45
Let (X, Σ) be a measurable space, let μ be a probability measure on X and let Σ_μ be the μ-completion of Σ (see Definition 3.23). Then

$$\widehat{\Sigma} \stackrel{\text{def}}{=} \bigcap \{\Sigma_\mu : \mu \text{ is a probability measure on } \Sigma\}$$

*is called the **universal completion** of Σ.*

Proposition 4.46
(a) *Every probability measure μ on Σ can be extended uniquely to a probability measure $\widehat{\mu}$ on $\widehat{\Sigma}$ and the function $\mu \longmapsto \widehat{\mu}$ is a bijection from the set of probability measures on Σ onto the set of probability measures on $\widehat{\Sigma}$.*
(b) *If (X, Σ) and $(Y, \mathcal{Y})$ are measurable spaces and $f\colon X \longrightarrow Y$ is a $(\Sigma, \mathcal{Y})$-measurable function, then f is also $\big(\widehat{\Sigma}, \widehat{\mathcal{Y}}\big)$-measurable.*

Definition 4.47
*Let X be a topological space and let $\mathcal{B}(X)$ be its Borel σ-algebra. The universal completion of $\mathcal{B}(X)$ is denoted by $\mathcal{B}_u(X)$ and it is called the **universal σ-algebra of** X. If X and Y are two topological spaces and $f\colon X \longrightarrow Y$ is a function, then we say that f is **universally measurable** if and only if f is $\big(\mathcal{B}_u(X), \mathcal{B}_u(Y)\big)$-measurable.*

Proposition 4.48
If X is a Borel space, then every Souslin subset of X (see Definition 2.156) is universally measurable.

4.1.6 Measurable Multifunctions: Selection Theorems

Next we will present a few basic things about measurable multifunctions, culminating to the main selection theorem. So, let (Ω, Σ) be a measurable space and let (X, d_X) be a separable metric space. Additional hypotheses will be introduced as needed.

Definition 4.49
Let $F\colon \Omega \longrightarrow 2^X$ be a multifunction.
(a) *We say that F is* ***measurable*** *if for every open set $U \subseteq X$, we have $F^-(U) = \{\omega \in \Omega: F(\omega) \cap U \neq \emptyset\} \in \Sigma$.*
(b) *We say that F is* ***graph measurable*** *if $\operatorname{Gr} f = \{(\omega, x) \in \Omega \times X : x \in F(\omega)\} \in \Sigma \otimes \mathcal{B}(X)$.*

Remark 4.50
It is customary to say that the ***domain of*** F, denoted by $\operatorname{dom} F$, is the set

$$\operatorname{dom} F \stackrel{\text{def}}{=} \{\omega \in \Omega: F(\omega) \neq \emptyset\}.$$

It is clear from Definition 4.49(a), that $\operatorname{dom} F \in \Sigma$ and so, when dealing with measurable multifunctions, it is not a loss generality to assume that their domain is all of Ω.

In what follows for a topological space X, we use the following notation:

$$\begin{aligned} P_f(X) &\stackrel{\text{def}}{=} \{A \subseteq X: A \text{ is nonempty and closed}\} \\ P_k(X) &\stackrel{\text{def}}{=} \{A \subseteq X: A \text{ is nonempty and compact}\} \\ \widehat{P}_f(X) &\stackrel{\text{def}}{=} P_f(X) \cup \{\emptyset\}. \end{aligned}$$

Moreover, if X is a normed space, then

$$\begin{aligned} P_{fc}(X) &\stackrel{\text{def}}{=} \{A \in P_f(X): A \text{ is convex}\} \\ P_{kc}(X) &\stackrel{\text{def}}{=} \{A \in P_k(X): A \text{ is convex}\} \\ P_{wkc}(X) &\stackrel{\text{def}}{=} \{A \subseteq X: A \text{ is nonempty, } w\text{-compact and convex}\}. \end{aligned}$$

The next theorem summarizes some important properties and equivalent definitions of measurable multifunctions.

Theorem 4.51
Let (Ω, Σ) be a measurable space and let (X, d_X) be a separable metric space. Consider a multifunction $F\colon \Omega \longrightarrow \widehat{P}_f(X)$ and the following properties:
(1) *For every $A \in \mathcal{B}(X)$, we have*

$$F^-(A) = \{\omega \in \Omega : \ F(\omega) \cap A \neq \emptyset\} \in \Sigma.$$

(2) *For every closed set $C \subseteq X$, we have*

$$F^-(C) = \{\omega \in \Omega : \ F(\omega) \cap C \neq \emptyset\} \in \Sigma.$$

(3) *F is measurable (see Definition 4.49(a)).*
(4) *For every $x \in X$, the $\overline{\mathbb{R}}_+$-valued function*

$$\omega \longmapsto d_X(x, F(\omega)) = \inf \{d_X(x, u) : \ u \in F(\omega)\}$$

is Σ-measurable (where $\overline{\mathbb{R}}_+ = [0, +\infty]$).
(5) *F is graph measurable (see Definition 4.49(b)).*

Then we have the following relations between these properties:
(a) $(1) \Longrightarrow (2) \Longrightarrow (3) \Longleftrightarrow (4) \Longrightarrow (5)$.
(b) *If X is σ-compact, then* $(2) \Longleftrightarrow (3)$.
(c) *If $\Sigma = \widehat{\Sigma}$ (see Definition 4.45) and X is complete (i.e., a Polish space), then all five properties are equivalent.*

Remark 4.52
If (Ω, Σ, μ) is a complete σ-finite measure space, then $\widehat{\Sigma} = \Sigma$.

In Theorem 2.170, we stated an important selection theorem, the celebrated Michael selection theorem. There the emphasis was topological and so the selector produced was continuous. Here the emphasis is measure theoretic and so we are looking for measurable selectors, i.e., a Σ-measurable single valued function $f\colon \Omega \longrightarrow X$ such that $f(\omega) \in F(\omega)$ for all $\omega \in \Omega$.

Theorem 4.53 *(Kuratowski–Ryll Nardzewski Selection Theorem)*
<u>If</u> (Ω, Σ) is measurable space, X is a Polish space and $F\colon \Omega \longrightarrow P_f(X)$ is a measurable multifunction,
<u>then</u> there exists a Σ-measurable function $f\colon \Omega \longrightarrow X$ such that $f(\omega) \in F(\omega)$ for all $\omega \in \Omega$.

Remark 4.54
In fact the above selection theorem remains true, if we drop the completeness hypothesis on X (i.e., X is only separable metrizable) and instead assume that for every $\omega \in \Omega$, the set $F(\omega)$ is a complete subset of X.

Theorem 4.53 can be strengthened as follows.

Theorem 4.55
If (Ω, Σ) *is a measurable space,* X *is a Polish space and* $F\colon \Omega \longrightarrow P_f(X)$ *is a multifunction,*
then the following two statements are equivalent:
(a) F *is measurable (see Definition 4.49(a)).*
(b) There exists a sequence $\{f_n\}_{n \geqslant 1}$ *of* Σ*-measurable functions* $f_n\colon \Omega \longrightarrow X$, $n \geqslant 1$ *such that for all* $\omega \in \Omega$, *we have*

$$f_n(\omega) \in F(\omega) \qquad \forall\, \omega \in \Omega,\ n \geqslant 1$$

and

$$F(\omega) = \overline{\{f_n(\omega)\}_{n \geqslant 1}} \qquad \forall\, \omega \in \Omega.$$

This theorem leads to the following stronger version of Theorem 4.51.

Theorem 4.56
Let (Ω, Σ) *be a measurable space, let* X *be a separable metric space and let* $F\colon \Omega \longrightarrow P_f(X)$ *be a multifunction. We consider the following properties:*
(1) For every set $A \in \mathcal{B}(X)$, *we have* $F^-(A) \in \Sigma$.
(2) For every closed set $C \subseteq X$, *we have* $F^-(C) \in \Sigma$.
(3) F *is measurable (see Definition 4.49(a)).*
(4) For every $x \in X$, *the function* $\omega \longmapsto \operatorname{dist}(x, F(\omega))$ *is* Σ*-measurable.*
(5) There exists a sequence $\{f_n\colon \Omega \longrightarrow X\}_{n \geqslant 1}$ *of* Σ*-measurable functions such that*

$$f_n(\omega) \in F(\omega) \qquad \forall\, \omega \in \Omega,\ n \geqslant 1$$

and

$$F(\omega) = \overline{\{f_n(\omega)\}_{n \geqslant 1}} \qquad \forall\, \omega \in \Omega.$$

(6) $\operatorname{Gr} F \in \Sigma \otimes \mathcal{B}(X)$ *(i.e.,* F *is graph measurable; see Definition 4.49(b)).*

Then we have the following relations between these properties:
(a) (1) $\Longrightarrow$ (2) $\Longrightarrow$ (3) $\Longleftrightarrow$ (4) $\Longrightarrow$ (6).
(b) *If X is complete (i.e., a Polish space), then* (4) $\Longleftrightarrow$ (5).
(c) *If X is σ-compact, then* (2) $\Longleftrightarrow$ (3).
(d) *If $\Sigma = \widehat{\Sigma}$ (see Definition 4.45) and X is complete (i.e., a Polish space), then all six properties are equivalent.*

In the second measurable selection theorem the hypotheses on the measurable space are stronger, but the conditions on the range space are weakened considerably.

Theorem 4.57 *(Yankov–von Neumann–Aumann Selection Theorem)*
If (Ω, Σ) is a complete measurable space (i.e., $\Sigma = \widehat{\Sigma}$), X is a Souslin space and $F\colon \Omega \longrightarrow 2^X \setminus \{\emptyset\}$ is a multifunction such that $\operatorname{Gr} F \in \Sigma \otimes \mathcal{B}(X)$,
then there exists a Σ-measurable function $f\colon \Omega \longrightarrow X$ such that $f(\omega) \in F(\omega)$ for all $\omega \in \Omega$.

Remark 4.58
If instead we assume that (Ω, Σ, μ) is a σ-finite measurable space and the other hypotheses remain unchanged, then we can have a Σ-measurable function $f\colon \Omega \longrightarrow X$ such that $f(\omega) \in F(\omega)$ μ-almost everywhere on Ω.

Again we can improve Theorem 4.57 and have a whole sequence of selectors which are dense in F.

Theorem 4.59
If (Ω, Σ) is a complete measurable space (i.e., $\Sigma = \widehat{\Sigma}$), X is a Souslin space and $F\colon \Omega \longrightarrow 2^X \setminus \{\emptyset\}$ is a multifunction such that $\operatorname{Gr} F \in \Sigma \otimes \mathcal{B}(X)$,
then there exists a sequence of Σ-measurable functions $f_n\colon \Omega \longrightarrow X$, $n \geqslant 1$ such that

$$f_n(\omega) \in F(\omega) \qquad \forall\, \omega \in \Omega,\ n \geqslant 1$$

and

$$F(\omega) \subseteq \overline{\{f_n(\omega)\}_{n \geqslant 1}} \qquad \forall\, \omega \in \Omega.$$

Remark 4.60
Again, if instead we assume that (Ω, Σ, μ) is a σ-finite measure space and the rest are the same, then we can find a sequence of Σ-measurable functions $f_n \colon \Omega \longrightarrow X$, $n \geqslant 1$ such that

$$f_n(\omega) \in F(\omega) \quad \mu\text{-almost everywhere on } \Omega, \text{ for all } n \geqslant 1$$

and

$$F(\omega) \subseteq \overline{\{f_n(\omega)\}_{n \geqslant 1}} \quad \mu\text{-almost everywhere on } \Omega.$$

4.1.7 Projection Theorems

The projection of a Borel set in $\mathbb{R}^2$ on a coordinate axis need not be Borel. In fact this observation was the starting point for Souslin to develop his theory of Souslin (or analytic) sets. This raises two important questions:
(a) When can we guarantee that the projection of a Borel set is still Borel?
(b) More generally, can we characterize the projection of a measurable set?

Proposition 4.61
If X and Y are Polish spaces, $C \subseteq \mathcal{B}(X \times Y) = \mathcal{B}(X) \otimes \mathcal{B}(Y)$ and for every $x \in X$, we have $C(x) = \{y \in Y : (x, y) \in C\}$ is σ-compact, then $\operatorname{proj}_X C \in \mathcal{B}(X)$.

There is another such projection result, for which we need to introduce the following class of spaces.

Definition 4.62
*A topological space X is said to be of **class** σMK if $X = \bigcup_{n \geqslant 1} K_n$, where K_n is metrizable and compact for all $n \geqslant 1$.*

Remark 4.63
Evidently a separable metrizable locally compact space is of class σMK. But a σMK space need not be metrizable. Anticipating some material from the theory of Banach spaces (see Chap. 5), we can see that, if X^* is the topological dual of a separable Banach space, then X^* is a σMK-space, since $X^* = \bigcup_{n \geqslant 1} n\overline{B}_1^*$, with $\overline{B}_1^* \stackrel{\text{def}}{=} \{x^* \in X^* : \|x^*\|_* \leqslant 1\}$.

Proposition 4.64
If X is a Borel space, Y is a space of class σMK, $C \in \mathcal{B}(X \times Y)$ and for all $x \in X$, the set $C(x) = \{y \in Y : (x,y) \in C\}$ is closed, then $\operatorname{proj}_X C \in \mathcal{B}(X)$.

Now we are ready to answer the second question and characterize the projection of a measurable sets.

Theorem 4.65 *(Yankov–von Neumann–Aumann Projection Theorem)*
If (Ω, Σ) is a measurable space, X is a Souslin space and $C \in \Sigma \otimes \mathcal{B}(X)$, then $\operatorname{proj}_\Omega C \in \widehat{\Sigma}$.

Finally, concerning the Borel σ-algebra of a Souslin space X, we have the following interesting property.

Proposition 4.66
If X is a Souslin space, then $\mathcal{B}(X)$ is separable (see Definition 3.14(b)).

4.1.8 Dual of $L^p(\Omega)$ for $1 \leqslant p \leqslant \infty$

In Chap. 3 we introduced the Banach spaces $L^p(\Omega)$, $1 \leqslant p \leqslant +\infty$. By $L^p(\Omega)^*$ we denote the dual of the Banach space $L^p(\Omega)$, namely the linear space of all continuous, linear functionals $\xi \colon L^p(\Omega) \longrightarrow \mathbb{R}$ (see also Chap. 5). Endowed with the norm

$$\|\xi\|_* = \sup\{|\xi(u)| : \|u\|_p \leqslant 1\},$$

the space $L^p(\Omega)^*$ becomes a Banach space.

In the next theorem, we give a very convenient characterization of the dual space $L^p(\Omega)^*$ for $p \in [1, +\infty)$.

Theorem 4.67 *(Riesz Representation Theorem)*
If (Ω, Σ, μ) is a measure space and $1 \leqslant p < +\infty$, then $\xi \in L^p(\Omega)^*$ if and only if there exists a unique $v \in L^{p'}(\Omega)$ (where $\frac{1}{p} + \frac{1}{p'} = 1$) such that

$$\xi(u) = \int_\Omega uv \, d\mu \qquad \forall\, u \in L^p(\Omega) \text{ and } \|\xi\|_* = \|v\|_{p'}.$$

Therefore, the Banach space $v \in L^{p'}(\Omega)$ (where $\frac{1}{p} + \frac{1}{p'} = 1$) and $L^p(\Omega)^*$ are isometrically isomorphic.

Remark 4.68
The function $L^p(\Omega)^* \ni \xi \longmapsto v \in L^{p'}(\Omega)$ introduced above is a linear isometry which is surjective and permits the identification of $L^p(\Omega)^*$ with $L^{p'}(\Omega)$ (where $\frac{1}{p} + \frac{1}{p'} = 1$). In the sequel we will always use this identification, i.e., that

$$L^p(\Omega)^* \;=\; L^{p'}(\Omega) \qquad \forall\, p \in [1, +\infty),$$

with $\frac{1}{p} + \frac{1}{p'} = 1$. We also say that p and p' are ***conjugate exponents***. Note that, if $p = 1$, then $p' = +\infty$ and so $L^1(\Omega)^* = L^\infty(\Omega)$. Then the dual of $L^\infty(\Omega)$ contains $L^1(\Omega)$ and the inclusion is strict.

To describe the elements of $L^\infty(\Omega)^*$, we need the following definition.

Definition 4.69
Let (Ω, Σ, μ) be a σ-finite measure space.
(a) A function $\xi \in L^\infty(\Omega)^*$ is said to be ***absolutely continuous*** *if there exists $u \in L^1(\Omega)$ such that*

$$\xi(v) \;=\; \int_\Omega vu \, d\mu \qquad \forall\, v \in L^\infty(\Omega).$$

(b) A function $\xi \in L^\infty(\Omega)^*$ is said to be ***singular*** *if there exists a decreasing sequence $\{C_n\}_{n \geqslant 1} \subseteq \Sigma$ such that $\mu(C_n) \searrow 0$ and*

$$\xi(v) \;=\; \xi(\chi_{C_n} v) \qquad \forall\, n \geqslant 1,\; v \in L^\infty(\Omega).$$

The next theorem characterizes the dual of $L^\infty(\Omega)$.

Theorem 4.70 *(Yosida–Hewitt Theorem)*
Every $\xi \in L^\infty(\Omega)^*$ admits a unique decomposition $\xi = \xi_a + \xi_s$, with ξ_a absolutely continuous and ξ_s singular. Moreover, $\|\xi\|_* = \|\xi_a\|_* + \|\xi_s\|_*$.

4.1.9 Sequences of Measures: Weak Convergence in $L^p(\Omega)$

Next we turn our attention to sequences of measures. We start by generalizing the notion of uniform integrability of a sequence of functions (see Definition 3.124), to sequences of signed measures.

Definition 4.71
Let (Ω, Σ, ν) be a measure space.
(a) *If $\mathcal{Y} \subseteq 2^\Omega$ and μ is a set function on $\mathcal{Y}$, we say that μ is* ***Vitali continuous*** *if for every decreasing sequence $\{C_n\}_{n\geqslant 1} \subseteq \mathcal{Y}$ such that $\bigcap\limits_{n\geqslant 1} C_n = \emptyset$, we have $\mu(C_n) \longrightarrow 0$.*
(b) *A sequence $\{\mu_k\}_{k\geqslant 1}$ of Vitali continuous set functions μ_k on $\mathcal{Y}$ is said to be* ***Vitali equicontinuous*** *if for every decreasing sequence $\{C_n\}_{n\geqslant 1} \subseteq \mathcal{Y}$ such that $\bigcap\limits_{n\geqslant 1} C_n = \emptyset$, we have that for every $\varepsilon > 0$, we can find $n_0 = n_0(\varepsilon) \geqslant 1$ such that*

$$\big|\mu_k(C_n)\big| \leqslant \varepsilon \qquad \forall\, k \geqslant 1,\ n \geqslant n_0.$$

(c) *If $\mathcal{Y} = \Sigma$, then a sequence $\{\mu_k\}_{k\geqslant 1}$ of set functions on Σ is* ***uniformly ν-absolutely continuous*** *provided that for a given $\varepsilon > 0$, we can find $\delta = \delta(\varepsilon) > 0$ such that for all $C \in \Sigma$ with $\nu(C) < \delta$, we have*

$$\big|\mu_k(C)\big| < \varepsilon \qquad \forall\, k \geqslant 1.$$

Now, for a given measurable space (Ω, Σ), by $M(\Sigma)$ we denote the linear space of bounded signed measures on Σ. For $\mu \in M(\Sigma)$, we set

$$\|\mu\| = \text{variation of } \mu = |\mu|(\Omega)$$

(the total variation norm on $M(\Omega)$; see Definition 3.148). Then $\big(M(\Sigma), \|\cdot\|\big)$ is a Banach space (see Proposition 3.149). Also let

$$\|\mu\|_\infty = \sup\big\{\big|\mu(C)\big| : C \in \Sigma\big\}.$$

This is another norm on $M(\Sigma)$ equivalent to $\|\cdot\|$, since

$$\|\mu\|_\infty \leqslant \|\mu\| \leqslant 4\|\mu\|_\infty \qquad \forall\, \mu \in M(\Sigma).$$

Recall that, if $\mu \in M(\Sigma)$, then $|\mu| \in M(\Sigma)$.

Theorem 4.72 *(Nikodym Theorem)*
If (Ω, Σ) is a measurable space, $\{\mu_n\}_{n\geqslant 1} \subseteq M(\Sigma)$ and $\lim\limits_{n\to+\infty} \mu_n(C) = \mu(C)$ for $C \in \Sigma$,
then $\mu \in M(\Sigma)$ and $\{\mu_n\}_{n\geqslant 1} \subseteq M(\Sigma)$ is Vitali equicontinuous (see Definition 4.71(b)).

The next theorem is a slight improvement of the Vitali–Hahn–Saks theorem (see Theorem 3.151) and can be proved using the Nikodym theorem (see Theorem 4.72).

Theorem 4.73
If (Ω, Σ) *is a measurable space,* ν *is a nonnegative element in* $M(\Sigma)$ *and* $\{\mu_n\}_{n\geqslant 1} \subseteq M(\Sigma)$ *is a sequence such that*
(a) *for all* $A \in \Sigma$, $\mu_n(A) \longrightarrow \mu(A)$; *and*
(b) $\mu_n \ll \nu$ *for all* $n \geqslant 1$,
then the sequence $\{\mu_n\}_{n\geqslant 1}$ *is uniformly* ν*-absolutely continuous,* $\mu \in M(\Sigma)$ *and* $\mu \ll \nu$.

From Theorem 4.67, we know that $L^1(\Omega)^* = L^\infty(\Omega)$. So, we can speak about weak convergence of sequences in $L^\infty(\Omega)$.

Definition 4.74
Let (Ω, Σ, μ) *be a measure space. We say that the sequence* $\{u_n\}_{n\geqslant 1} \subseteq L^1(\Omega)$ *is* ***weakly convergent*** *to* $u \in L^1(\Omega)$ *(denoted by* $u_n \xrightarrow{w} u$ *in* $L^1(\Omega)$*), if for every* $h \in L^\infty(\Omega)$*, we have*

$$\langle u_n, h\rangle = \int_\Omega u_n h \, d\mu \longrightarrow \int_\Omega uh \, d\mu = \langle u, h\rangle .$$

Similarly, we say that the sequence $\{u_n\}_{n\geqslant 1} \subseteq L^p(\Omega)$ *is weakly convergent to* $u \in L^p(\Omega)$*, with* $1 < p < +\infty$ *(denoted by* $u_n \xrightarrow{w} u$ *in* $L^p(\Omega)$*), if for every* $h \in L^{p'}(\Omega)$ *(where* $\frac{1}{p} + \frac{1}{p'} = 1$*), we have*

$$\int_\Omega u_n h \, d\mu \longrightarrow \int_\Omega uh \, d\mu.$$

The next theorem characterizes weak convergent sequences in $L^1(\Omega)$.

Theorem 4.75 *(Dunford–Pettis Theorem)*
Let (Ω, Σ, μ) *be a finite measure space. A sequence* $\{u_n\}_{n\geqslant 1} \subseteq L^1(\Omega)$ *admits a weakly convergent subsequence (converging to some* $u \in L^1(\Omega)$*) if and only if the sequence* $\{u_n\}_{n\geqslant 1} \subseteq L^1(\Omega)$ *is uniformly integrable (see Definition 3.124).*

By introducing topological structure on the space Ω, we can have the following alternative version of Dunford–Pettis Theorem (see also Theorem 3.125).

Theorem 4.76
If Ω *is a locally compact topological space,* (Ω, Σ, μ) *is a measure space with a tight measure* μ *(see Definition 4.9(e)) and* $\{u_n\}_{n\geqslant 1} \subseteq L^1(\Omega)$ *is a sequence,*
then $u_n \xrightarrow{w} u \in L^1(\Omega)$ *if and only if*
(a) *for every* $\varepsilon > 0$ *there exists* $\delta > 0$ *such that*

$$\mu(A) < \delta \quad \Longrightarrow \quad \int_A |u_n| \, d\mu < \varepsilon \qquad \forall\, n \geqslant 1;$$

(b) *for every* $\varepsilon > 0$*, there exists a compact set* $K \subseteq X$ *such that*

$$\int_{K^c} |u_n| \, d\mu < \varepsilon \qquad \forall\, n \geqslant 1;$$

(c) *the sequence* $\{u_n\}_{n\geqslant 1}$ *is* L^1*-bounded, i.e., there exists* $M > 0$ *such that* $\|u_n\|_1 \leqslant M$ *for all* $n \geqslant 1$.

4.1.10 Covering Theorems

Now we turn out attention to the relation between Lebesgue integration and differentiation of functions, leading to generalizations of the fundamental theorem of calculus. A first result in this direction was the Radon–Nikodym theorem (see Theorem 3.152).

We start with some covering results.

Definition 4.77
(a) *A collection* $\mathcal{F}$ *of closed balls in* $\mathbb{R}^N$ *is a* **cover** *of* $A \subseteq \mathbb{R}^N$ *if* $A \subseteq \bigcup\limits_{B\in\mathcal{F}} B$.
(b) *A cover* $\mathcal{F}$ *is a* **Vitali** *or* **fine cover** *of* $A \subseteq \mathbb{R}^N$ *if*

$$\inf\big\{\operatorname{diam} B : \ x \in B, \ B \in \mathcal{F}\big\} = 0 \qquad \forall\, x \in A.$$

Remark 4.78
We could define Vitali covers using cubes instead of balls. The condition for a cover to be a Vitali cover is equivalent to saying that "for every $\varepsilon > 0$ and every $x \in A$ there exists $B \in \mathcal{F}$ such that $x \in B$ and $\operatorname{diam} B < \varepsilon$".

Theorem 4.79 *(Vitali Covering Theorem)*
If $A \subseteq \mathbb{R}^N$ is a Lebesgue measurable set with finite Lebesgue measure and $\mathcal{F}$ is a Vitali cover of A,
then for every $\varepsilon > 0$, we can find a sequence $\{B_n\}_{n\geqslant 1} \subseteq \mathcal{F}$ of pairwise disjoint balls such that

$$\lambda^N\Big(A \setminus \bigcup_{n\geqslant 1} B_n\Big) = 0 \quad \text{and} \quad \lambda^N\Big(\bigcup_{n\geqslant 1} B_n\Big) \leqslant \lambda^N(A) + \varepsilon$$

(here by λ^N we denote the N-dimensional Lebesgue measure).

Remark 4.80
The above theorem does not claim that $\bigcup_{n\geqslant 1} B_n$ covers A. The covering is only in a measure theoretic sense. However, as a by-product of the proof, we obtain that

$$A \subseteq \bigcup_{B\in\mathcal{F}} B \subseteq \bigcup_{n\geqslant 1} \widehat{B}_n,$$

where $\widehat{B}_n$ denotes the concentric closed ball with radius 5 times the radius of B_n. Theorem 4.79 remains valid if balls are replaced by cubes. Its proof relies on the structure of the Lebesgue measures λ^N and so it cannot be extended if λ^N is replaced by a general tight measure on $\mathbb{R}^N$ (a Radon measure on $\mathbb{R}^N$). The next covering theorem is more suitable for arbitrary Radon measures on $\mathbb{R}^N$, since it does not require enlargement of the balls (to pass from B to $\widehat{B}$).

Theorem 4.81 *(Besicovitch Covering Theorem)*
If $\mathcal{F}$ is a collection of closed balls in $\mathbb{R}^N$ such that $\sup\big\{\operatorname{diam} B : B \in \mathcal{F}\big\} < +\infty$, and A is the set of centres of balls in $\mathcal{F}$,
then there exist $\mathcal{F}_1, \ldots, \mathcal{F}_{m(N)} \subseteq \mathcal{F}$ such that each $\mathcal{F}_k$ ($k = 1, \ldots, m(N)$) is a countable family of disjoint balls in $\mathcal{F}$, $A \subseteq \bigcup_{k=1}^{m(N)} \bigcup_{B\in\mathcal{F}_k} B$ and the number $m(N)$ of these subcollections depends only on the dimension N of the space.

Remark 4.82
Again the above theorem remains true if the balls are replaced by cubes.

The next result is a consequence of Theorem 4.81 and can be viewed as a measure theoretic reformulation of it. It says that for any given open set, we can fill it up with a countable collection of disjoint balls in such a way that the remainder set has μ-measure zero.

Proposition 4.83
If μ *is a Borel measure on* $\mathbb{R}^N$, $\mathcal{F}$ *is a collection of closed balls in* $\mathbb{R}^N$, A *is the set of centres of these balls,* $\mu(A) < +\infty$ *and for each* $a \in A$, *we have* $\inf\{r : \overline{B}_r(a) \in \mathcal{F}\} = 0$ *(where* $\overline{B}_r(a) \stackrel{\text{def}}{=} \{x \in \mathbb{R}^N : \|x - a\| \leqslant r\}$*),*
then for every open set $U \subseteq \mathbb{R}^N$, *there exists a countable subcollection* $\mathcal{F}_0 \subseteq \mathcal{F}$ *of disjoint balls such that*

$$\bigcup_{B \in \mathcal{F}_0} \subseteq U \quad \text{and} \quad \mu\Big((A \cap U) \setminus \big(\bigcup_{B \in \mathcal{F}_0} B\big)\Big) = 0.$$

4.1.11 Lebesgue Differentiation Theorem

Using the covering theorems presented above, we can study the differentiation of Radon measures on $\mathbb{R}^N$ (recall that a Borel measure μ on $\mathbb{R}^N$ is Radon if it is finite on compact sets, outer regular and for all $A \in \mathcal{B}(\mathbb{R}^N)$, we have $\mu(A) = \sup\{\mu(K) : K \subseteq A, F \text{ is compact}\}$).

Definition 4.84
Let μ, ν *be two Radon measures on* $\mathbb{R}^N$. *For every* $x \in \mathbb{R}^N$, *we define*

$$\overline{D}_\mu \nu(x) \stackrel{\text{def}}{=} \begin{cases} \limsup\limits_{n \to +\infty} \dfrac{\nu(\overline{B}_r(x))}{\mu(\overline{B}_r(x))} & \text{if } \mu(\overline{B}_r(x)) > 0 \text{ for all } r > 0, \\ +\infty & \text{if } \mu(\overline{B}_r(x)) = 0 \text{ for some } r > 0, \end{cases}$$

$$\underline{D}_\mu \nu(x) \stackrel{\text{def}}{=} \begin{cases} \liminf\limits_{n \to +\infty} \dfrac{\nu(\overline{B}_r(x))}{\mu(\overline{B}_r(x))} & \text{if } \mu(\overline{B}_r(x)) > 0 \text{ for all } r > 0, \\ +\infty & \text{if } \mu(\overline{B}_r(x)) = 0 \text{ for some } r > 0, \end{cases}$$

If $\overline{D}_\mu \nu(x) = \underline{D}_\mu \nu(x) < +\infty$, *then we say that* ν ***is differentiable with respect to*** μ ***at*** x *and we write*

$$D_\mu \nu(x) = \overline{D}_\mu \nu(x) = \underline{D}_\mu \nu(x).$$

We call $D_\mu \nu(x)$ *the* ***derivative of*** ν ***with respect to*** μ ***at*** x.

Theorem 4.85
If μ *and* ν *are Radon measures on* $\mathbb{R}^N$,
then $D_\mu \nu$ *exists and is finite* μ*-almost everywhere on* $\mathbb{R}^N$*, it is* μ*-measurable and for every Borel set* $A \subseteq \mathbb{R}^N$*, we have*

$$\nu(A) = \nu_s(A) + \int_A D_\mu \nu \, d\mu,$$

with $\nu_s \perp \mu$ *and* $D_\mu \nu_s(x) = 0$ μ*-almost everywhere on* $\mathbb{R}^N$*. If* $\nu \ll \mu$*, then* $\nu_s = 0$ *and*

$$\nu(A) = \int_A D_\mu \nu \, d\mu.$$

As a consequence of Theorem 4.85, we have the following theorem.

Theorem 4.86 *(Lebesgue Differentiation Theorem on* $\mathbb{R}^N$*)*
If μ *is a Radon measure and* $u \in L^1(\mathbb{R}^N; \mu)$*, then*

$$\lim_{r \to 0^+} \frac{1}{\mu(\overline{B}_r(x))} \int_{\overline{B}_r(x)} u \, d\mu = u(x) \quad \text{for } \mu\text{-almost all } x \in \mathbb{R}^N.$$

Corollary 4.87
If μ *is a Radon measure,* $1 \leqslant p < +\infty$ *and* $u \in L^p_{\mathrm{loc}}(\mathbb{R}^N; \mu)$*, then*

$$\lim_{r \to 0^+} \frac{1}{\mu(\overline{B}_r(x))} \int_{\overline{B}_r(x)} |u - u(x)|^p \, d\mu = 0 \quad \text{for } \mu\text{-almost all } x \in \mathbb{R}^N.$$

In fact, if $\mu = \lambda^N$ (the Lebesgue measure on $\mathbb{R}^N$), then we have a stronger version of the above corollary.

Proposition 4.88
If $1 \leqslant p < +\infty$ *and* $u \in L^p_{\mathrm{loc}}(\mathbb{R}^N)$*, then*

$$\lim_{B \searrow \{x\}} \frac{1}{\lambda^N(B)} \int_{\overline{B}_r(x)} |u - u(x)|^p \, d\lambda^N = 0 \quad \text{for } \lambda^N\text{-almost all } x \in \mathbb{R}^N$$

(here the limit is taken over all closed balls B *such that* $x \in B$ *and* $\operatorname{diam} B \longrightarrow 0$*; we emphasize that balls need not be centred at* x*).*

Definition 4.89
Let μ be a Radon measure on $\mathbb{R}^N$ and let $u\colon \mathbb{R}^N \longrightarrow \mathbb{R}$ be a μ-measurable function. The set of points $x_0 \in \mathbb{R}^N$ for which

$$\lim_{r\to 0} \frac{1}{\mu(\overline{B}_r(x))} \int\limits_{\overline{B}_r(x_0)} \big|u - u(x_0)\big|\, d\mu \;=\; 0,$$

*is the **Lebesgue set of** u. The elements of this set are called the **Lebesgue points** of u.*

Also the set of points $x_0 \in \mathbb{R}^N$ for which

$$\lim_{r\to 0} \frac{1}{\mu(\overline{B}_r(x))} \int\limits_{\overline{B}_r(x_0)} u\, d\mu \;=\; u(x_0),$$

*is the **differentiability set** of u.*

Remark 4.90
According to Corollary 4.87, if $u \in L^p_{\mathrm{loc}}(\mathbb{R}^N;\mu)$ $(1 \leqslant p < +\infty)$, then μ-almost all points $x \in \mathbb{R}^N$ belong to the Lebesgue set of u. Moreover, the Lebesgue set of u is a subset of the differentiability set of u and the inclusion can be strict.

Another interesting consequence of Corollary 4.87 is the following result.

Proposition 4.91
<u>If</u> $A \subseteq \mathbb{R}^N$ is a Lebesgue measurable set, <u>then</u>

$$\begin{aligned} \lim_{r\to 0^+} \frac{\lambda^N(\overline{B}_r(x)\cap A)}{\lambda^N(\overline{B}_r(x))} &= 1 \quad \text{for } \lambda^N\text{-almost all } x \in A \text{ and} \\ \lim_{r\to 0^+} \frac{\lambda^N(\overline{B}_r(x)\cap A)}{\lambda^N(\overline{B}_r(x))} &= 0 \quad \text{for } \lambda^N\text{-almost all } x \in \mathbb{R}^N \setminus A. \end{aligned}$$

Definition 4.92
*Let $A \subseteq \mathbb{R}^N$. A point $x \in \mathbb{R}^N$ is a **point of density for** A if*

$$\lim_{r\to 0^+} \frac{\lambda^N(\overline{B}_r(x)\cap A)}{\lambda^N(\overline{B}_r(x))} \;=\; 1.$$

Remark 4.93
According to Proposition 4.91, λ^N-almost all points of A are density points for A. We can think of the set of points of density for A as a kind of a measure theoretic interior for the set A.

We can generalize the Lebesgue differentiation theorem (see Theorem 4.86), by using the notion of a regular family.

Definition 4.94
Let μ be a Radon measure on $\mathbb{R}^N$. For a given $x \in \mathbb{R}^N$, a family $\mathcal{F}_x$ of μ-measurable subsets of $\mathbb{R}^N$ is said to be **regular at** *x if the following holds:*
(a) *for every $\varepsilon > 0$, there exists $C \in \mathcal{F}_x$ such that $\operatorname{diam} C \leqslant \varepsilon$;*
(b) *there exists a constant $c \geqslant 1$ such that for every $C \in \mathcal{F}_x$, we have $\mu(\overline{B}_x) \leqslant c\mu(C)$, where $\overline{B}_x$ is the smallest closed ball with centre at x, which contains C.*

Remark 4.95
In the above definition property **(a)** says that the sets in $\mathcal{F}_x$ shrink to x, without requiring that x belongs in any of the sets of $\mathcal{F}_x$. Property **(b)** says that each set in $\mathcal{F}_x$ is comparable to a ball centred at x. If $\mu = \lambda^N$, then regular families at the origin (and by translation to any other point in $\mathbb{R}^N$) are the balls, cubes, ellipsoids or regular polygons centred at the origin. On the other hand, as we already mentioned, the sets of the regular family need not contain the origin, for example consider the collection of annuli

$$S_\varrho = \left\{x \in \mathbb{R}^N : \tfrac{1}{2}\varrho < \|x\| < \varrho\right\},$$

with $\varrho > 0$.

Proposition 4.96
If μ is a Radon measure on $\mathbb{R}^N$, $u \in L^1_{\text{loc}}(\mathbb{R}^N;\mu)$, $x \in \mathbb{R}^N$ is a Lebesgue point of u (see Definition 4.89) and $\mathcal{F}_x$ is a regular family at x, then

$$\lim_{\substack{\operatorname{diam} C \to 0 \\ C \in \mathcal{F}_x}} \frac{1}{\mu(C)} \int_C \left|f - f(x)\right| d\mu = 0.$$

In particular, we have

$$\lim_{\substack{\operatorname{diam} C \to 0 \\ C \in \mathcal{F}_x}} \frac{1}{\mu(C)} \int_C f \, d\mu = f(x).$$

Remark 4.97
In the particular case $N = 1$ (functions of one variable), Proposition 4.96 implies that for every $u \in L^1_{\text{loc}}(\mathbb{R})$, we have

$$\lim_{h \to 0} \frac{1}{h} \int_x^{x+h} f \, d\lambda = f(x) \quad \text{for } \lambda\text{-almost all } x \in \mathbb{R},$$

with λ being the Lebesgue measure on $\mathbb{R}$.

4.1.12 Bounded Variation and Absolutely Continuous Functions

Next we turn our attention to functions of bounded variation of one variable. Since, as we will see, every function of bounded variation can be written as the difference of two increasing functions, we start with a few basic facts about monotone functions. By monotone (respectively, strictly monotone), we mean increasing or decreasing (respectively, strictly increasing or strictly decreasing). A monotone function need not be continuous. Nevertheless, the discontinuity set of such a function has a precise description.

Theorem 4.98
If $T \subseteq \mathbb{R}$ *is an interval and* $u\colon T \longrightarrow \mathbb{R}$ *is a monotone function,* *then* u *has at most countably many jump discontinuity points.* *Conversely, for a given countable set* D*, we can find a monotone function* $u\colon \mathbb{R} \longrightarrow \mathbb{R}$ *with discontinuity set* D*.*

Remark 4.99
If $D = \mathbb{Q}$, then we see that there is a monotone function $u\colon \mathbb{R} \longrightarrow \mathbb{R}$ which is continuous on the irrationals and discontinuous on the rationals.

Concerning the differentiability properties of a monotone function, we have the following theorem.

Theorem 4.100 *(Lebesgue Theorem on Monotone Functions)*
If $T \subseteq \mathbb{R}$ *is an interval and* $u\colon T \longrightarrow \mathbb{R}$ *is a monotone function,* *then* u *is differentiable almost everywhere on* T*.*

This theorem is sharp.

Theorem 4.101
<u>If</u> D *is a Lebesgue-null set,*
<u>then</u> *there exists a continuous monotone function which is not differentiable on* D.

The next theorem gives some consequences of the Lebesgue theorem on monotone functions (see Theorem 4.100).

Theorem 4.102
<u>If</u> $T \subseteq \mathbb{R}$ *is an interval and* $u: T \longrightarrow \mathbb{R}$ *is a monotone function,*
<u>then</u> u' *is Lebesgue measurable and for every* $[0, b] \subseteq T$, *we have*

$$\int_a^b |u'(t)|\, dt \leqslant |u(b) - u(a)|,$$

i.e., $u \in L^1_{\text{loc}}(T)$. *Moreover, if* u *is bounded the* $u' \in L^1(T)$ *and for* $m_u = \sup T$, $m_l = \inf T$, *we have*

$$\int_T |u'(t)|\, dt \leqslant \Big| \lim_{t \to m_u^+} u(t) - \lim_{t \to m_l^-} u(t) \Big| \leqslant \sup_T u - \inf_T u.$$

Remark 4.103
The Cantor function (see Remark 3.83) is a continuous increasing function such that $u' = 0$ λ-almost everywhere on $[0, 1]$ (λ being the Lebesgue measure on $\mathbb{R}$). Actually, we can find a continuous strictly increasing function $u: [0, 1] \longrightarrow \mathbb{R}$ such that $u' = 0$ λ-almost everywhere on $[0, 1]$.

For the differentiability of series of monotone functions, we have the following theorem.

Theorem 4.104 *(Fubini Theorem on Series of Monotone Functions)*
<u>If</u> $T \subseteq \mathbb{R}$ *is an interval and* $\{u_n: T \longrightarrow \mathbb{R}\}_{n \geqslant 1}$ *is a sequence of increasing functions such that* $\sum\limits_{n \geqslant 1} u_n(t)$ *converges pointwise to* $u(t)$ *for all* $t \in T$,
<u>then</u> *the convergence is uniform on compact subsets of* T, u *is* λ*-almost everywhere differentiable and* $u'(t) = \sum\limits_{n \geqslant 1} u_n'(t)$ *for almost all* $t \in T$.

The set of monotone functions is not a vector space (clearly the difference of two monotone functions need not be monotone). It is only a cone. In order to characterize the smallest vector space of functions $u\colon T \longrightarrow \mathbb{R}$ which contains the cone of monotone functions, we are led to the notion of functions of bounded variation. In what follows, for a given interval $T \subseteq \mathbb{R}$, by a ***partition*** of T, we mean a finite set $P = \{t_k\}_{k=0}^n \subseteq T$ such that $t_0 < t_1 < \ldots < t_n$.

Definition 4.105
*Let $T \subseteq \mathbb{R}$ be an interval and let $u\colon T \longrightarrow \mathbb{R}$ be a function. The **total variation** of u on T, denoted by $\operatorname{Var} u$, is defined by*

$$\operatorname{Var} u \stackrel{\text{def}}{=} \sup_P \Big\{ \sum_{k=1}^n \big|u(t_k) - u(t_{k-1})\big| \,:\, P = \{u_k\}_{k=0}^n \Big\},$$

*with $P = \{u_k\}_{k=0}^n$ being a partition of T. We say that u is of **bounded variation** if $\operatorname{Var} u < +\infty$. We denote the space of all functions $u\colon T \longrightarrow \mathbb{R}$ of bounded variation, by $BV(T)$.*

Remark 4.106
It is easy to see that if T contains one of its endpoints $a = \inf T$ or $b = \sup T$, then we can always consider partitions with $t_0 = a$ (if $a \in T$) and $t_n = b$ (if $b \in T$). We will say that $u\colon T \longrightarrow \mathbb{R}$ is ***locally of bounded variation*** if $u \in BV[a,b]$ for all intervals $[a,b] \subseteq T$. The space of all functions $u\colon T \longrightarrow \mathbb{R}$ locally of bounded variation is denoted by $BV_{\text{loc}}(T)$. In some occasions, in order to emphasize the dependence on the interval T, we write $\operatorname{Var}_T u$. Moreover, if $U \subseteq \mathbb{R}$ is open, then we know that $U = \bigcup\limits_{n\geqslant 1} T_n$, where $\{T_n\}_{n\geqslant 1}$ are pairwise disjoint open intervals. Then for a function $u\colon U \longrightarrow \mathbb{R}$, we define

$$\operatorname{Var}_U v \stackrel{\text{def}}{=} \sum_{n\geqslant 1} \operatorname{Var}_{T_n} u$$

and we say that u is of ***bounded variation on*** U if $\operatorname{Var}_U u < +\infty$. The corresponding space is denoted by $BV(U)$. Finally, if we have functions $u\colon T \longrightarrow \mathbb{R}^N$, we can still consider the total variation of u and consequently speak of functions of bounded variation, if in Definition 4.105, we replace $|\cdot|$ by $\|\cdot\|$ (the norm of $\mathbb{R}^N$). Then the space are denoted by $BV(T;\mathbb{R}^N)$ and $BV_{\text{loc}}(T;\mathbb{R}^N)$.

A first natural question is whether a monotone function is of bounded variation.

Proposition 4.107
<u>If</u> $T \subseteq \mathbb{R}^N$ is an interval and $u\colon T \longrightarrow \mathbb{R}$ is a monotone function, <u>then</u> for every interval $I \subseteq T$, we have

$$\operatorname{Var}_I u = \sup_I u - \inf_I u.$$

Consequently $u \in BV_{\mathrm{loc}}(T)$. Moreover, $u \in BV(T)$ if and only if u is bounded.

In the study of functions of bounded variation, the following function is important.

Definition 4.108
Let $T \subseteq \mathbb{R}$ be an interval, $t_0 \in T$ and $u \in BV_{\mathrm{loc}}(T)$. The ***indefinite variation*** *of u is the function*

$$V(t) \stackrel{\mathrm{def}}{=} \begin{cases} \operatorname{Var}_{[t_0,t]} u & \text{if } t_0 \leqslant t, \\ -\operatorname{Var}_{[t,t_0]} u & \text{if } t < t_0. \end{cases}$$

Sometimes, in order to emphasize the dependence on t_0 (on (t_0, u)), we write V_{t_0} (or $V_{t_0,u}$). If $\inf T \in T$, then we define

$$V_\infty(t) \stackrel{\mathrm{def}}{=} \operatorname{Var}_{T \cap (-\infty,t]} u \qquad \forall\, t \in T.$$

Proposition 4.109
*<u>If</u> $T \subseteq \mathbb{R}$ is an interval and $u \in BV_{\mathrm{loc}}(T)$,
<u>then</u> for all $t, s \in T$ with $s < t$, we have*

$$\left|u(t) - u(s)\right| \leqslant V(t) - V(s) = \operatorname{Var}_{[s,t]} u$$

and so the functions V and $V + u$ are increasing.

As a consequence of this proposition, we have the following theorem.

Theorem 4.110
*<u>If</u> $T \subseteq \mathbb{R}$ is an interval,
<u>then</u> the smallest vector space containing all monotone functions (respectively, all bounded monotone functions) is the space $BV_{\mathrm{loc}}(T)$ (respectively, the space $BV(T)$).*

Remark 4.111
So, according to this theorem every $u \in BV_{\mathrm{loc}}(T)$ (respectively, every $u \in BV(T)$) can be written as the difference of two increasing functions (respectively, of two bounded increasing functions).

Proposition 4.112
If $T \subseteq \mathbb{R}$ is an interval and $u \in BV_{\mathrm{loc}}(T)$,
then u has at most countably many jump discontinuities, u is differentiable λ-almost everywhere and for every $[a,b] \subseteq T$, we have

$$\int_a^b |u'(t)|\, dt \leqslant \operatorname{Var}_{[a,b]} u.$$

Moreover, if $u \in BV(T)$, then u is bounded, $u' \in L^1(T)$ and

$$\int_T |u'|\, dt \leqslant \int_T |V'|\, dt \leqslant \sup_T V - \inf_T V = \operatorname{Var} u.$$

Now, let us look closer the spaces $BV_{\mathrm{loc}}(T)$ and $BV(T)$. First we examine compositions with functions of bounded variation.

Theorem 4.113
If $T \subseteq \mathbb{R}$ is an interval and $f \colon \mathbb{R} \longrightarrow \mathbb{R}$,
then $f \circ u \in BV_{\mathrm{loc}}(T)$ (respectively, $f \circ u \in BV(T)$) for all $u \in BV_{\mathrm{loc}}(T)$ (respectively, all $u \in BV(T)$) if and only if f is locally Lipschitz.

Remark 4.114
The result is also true for the spaces $BV_{\mathrm{loc}}(T;\mathbb{R}^N)$, $BV(T;\mathbb{R}^N)$ and $f \colon \mathbb{R}^N \longrightarrow \mathbb{R}$.

So far we know that $BV(T)$ is a vector space. Is it possible to equip $BV(T)$ with a norm and make it a normed space or even better a Banach space? Note that

$$\operatorname{Var}(\lambda u) = |\lambda| \operatorname{Var} u \qquad \forall\, \lambda \in \mathbb{R}$$

and

$$\operatorname{Var}(u+v) \leqslant \operatorname{Var} u + \operatorname{Var} v \qquad \forall\, u, v \in BV(T).$$

However, $\operatorname{Var} u = 0$ does not imply that $u = 0$, only that u is constant. So, a possible norm on $BV(T)$ should involve more than $\operatorname{Var} u$.

Proposition 4.115
<u>If</u> $T \subseteq \mathbb{R}$ is an interval,
<u>then</u> for any fixed $t_0 \in T$, $u \longmapsto \|u\| = |u(t_0)| + \operatorname{Var} u$ is a norm on $BV(T)$. Moreover, $(BV(T), \|\cdot\|)$ is a Banach space.

The completeness part in the above proposition is based on the following theorem.

Theorem 4.116 *(Helly Selection Theorem; Helly First Theorem)*
<u>If</u> $T \subseteq \mathbb{R}$ is an interval and $C \subseteq BV(T)$ is an infinite set such that there exist $t_0 \in T$ and $M > 0$ for which, we have $|u(t_0)| \leqslant M$ and $\operatorname{Var} u \leqslant M$ for all $u \in C$,
<u>then</u> there exists a sequence $\{u_n\}_{n \geqslant 1} \subseteq C$ and $u \in BV(T)$ such that $u_n(t) \longrightarrow u(t)$ for all $t \in T$.

Remark 4.117
The Banach space $(BV(T), \|\cdot\|)$ is not separable.

Before moving to absolutely continuous functions, we will characterize those continuous functions which are of bounded variation. First a definition.

Definition 4.118
Let X and Y be two nonempty sets, let $u\colon X \longrightarrow Y$ be a function and let $A \subseteq X$. For every $y \in Y$, let

$$S_u(y; A) \stackrel{\text{def}}{=} \{x \in A : \ u(x) = y\}.$$

The function $N_u(\cdot; A)\colon Y \longrightarrow \mathbb{N}_0 \cup \{+\infty\}$, defined by

$$N_u(y; A) \stackrel{\text{def}}{=} \begin{cases} \operatorname{card} S_u(y; A) & \text{if } S_u(y; A) \text{ is finite,} \\ +\infty & \text{if } S_u(y; A) \text{ is infinite} \end{cases}$$

*is called the **Banach indicatrix** of u on A. If $A = X$, then we drop the dependence on A and we write $N_u(\cdot)$.*

Theorem 4.119
<u>If</u> $T \subseteq \mathbb{R}$ is an interval and $u\colon T \longrightarrow \mathbb{R}$ is continuous,
<u>then</u> N_u is Borel measurable and

$$\int_{-\infty}^{+\infty} N_u(y)\, dy = \operatorname{Var} u.$$

Therefore $u \in BV(T)$ if and only if $N_u \in L^1(\mathbb{R})$.

The problem of reconstructing a function from its derivative (i.e., a fundamental theorem of calculus for Lebesgue integral) leads to the notion of absolutely continuous functions. Continuous monotone functions (which are λ-almost everywhere differentiable; see Theorem 4.100) in general cannot be recovered from the integral of its derivative. The Cantor function $u\colon [0,1] \longrightarrow \mathbb{R}$ is continuous increasing and $u'(t) = 0$ for λ-almost all $t \in [0,1]$ (see Remark 3.83). However, we have

$$\int_0^1 u'(t)\,dt \;=\; 0 \;<\; u(1) - u(0) \;=\; 1.$$

So, we need to restrict ourselves to a subclass of $BV(T)$.

Definition 4.120
Let $T \subseteq \mathbb{R}$ be an interval. A function $u\colon T \longrightarrow \mathbb{R}$ is said to be ***absolutely continuous*** *on T, if for every $\varepsilon > 0$ there exists a $\delta > 0$ such that for every finite family $\{(s_k, t_k)\}_{k=1}^n$ of disjoint subintervals of T, we have*

$$\sum_{k=1}^n (t_k - s_k) \;\leqslant\; \delta \quad\Longrightarrow\quad \sum_{k=1}^n \big|u(t_k) - u(s_k)\big| \;\leqslant\; \varepsilon.$$

The space of all absolutely continuous functions $u\colon T \longrightarrow \mathbb{R}$ is denoted by $AC(T)$. A function $u\colon T \longrightarrow \mathbb{R}$ is ***locally absolutely continuous*** *if it is absolutely continuous on every interval $[a,b] \subseteq T$. The space of all locally absolutely continuous functions $u\colon T \longrightarrow \mathbb{R}$ is denoted by $AC_{\mathrm{loc}}(T)$. If $U \subseteq \mathbb{R}$ is an open set, we can still define the notion of absolute continuity, provided that $[s_k, t_k] \subseteq U$ for all $k = 1, \ldots, n$. We denote the space of such functions by $AC(U)$. Finally, we can define the notion of absolute continuity for functions $u\colon T \longrightarrow \mathbb{R}^N$, if in the above definition we replace $|\cdot|$ by $\|\cdot\|$ the norm of $\mathbb{R}^N$. Then for the corresponding spaces, we use the notions $AC(T;\mathbb{R}^N)$ and $AC_{\mathrm{loc}}(T;\mathbb{R}^N)$.*

Remark 4.121
It is clear from the above that an absolutely continuous function $u\colon T \longrightarrow \mathbb{R}$ is uniformly continuous (just take $n = 1$ in the definition). The converse is not truc.

The next result gives a relation between spaces $BV_{\mathrm{loc}}(T)$ and $AC_{\mathrm{loc}}(T)$.

Proposition 4.122
<u>If</u> $T \subseteq \mathbb{R}$ is an interval,
<u>then</u> $AC_{\text{loc}}(T) \subseteq BV_{\text{loc}}(T)$ and $AC(T) \subseteq BV(T)$.
In particular, if $u \in AC_{\text{loc}}(T)$ (respectively, $u \in AC(T)$), then u' exists λ-almost everywhere on T and $u' \in L^1_{\text{loc}}(T)$ (respectively, $u \in L^1(T)$).

Even if $u \in BV(T) \cap C(T)$, u need not be absolutely continuous (the Cantor function is such an example). What is missing is the so called ***Lusin N-property***.

Definition 4.123
*Let $T \subseteq \mathbb{R}$ be an interval and let $u: T \longrightarrow \mathbb{R}$ be a function. We say that u satisfies the **Lusin N-property** if u maps Lebesgue-null subsets of T onto Lebesgue-null subsets of $\mathbb{R}$.*

Theorem 4.124 *(Banach–Zaretsky Theorem)*
<u>If</u> $T \subseteq \mathbb{R}$ is an interval, <u>then</u> $u \in AC_{\text{loc}}(T)$ if and only if
(a) *u is continuous; and*
(b) *u is differentiable λ-almost everywhere on T with $u' \in L^1_{\text{loc}}(T)$; and*
(c) *u has the Lusin N-property.*

Corollary 4.125
<u>If</u> $T \subseteq \mathbb{R}$ is an interval and $u: T \longrightarrow \mathbb{R}$ is a continuous function such that
(i) *u' exists λ-almost everywhere and $u' \in L^p(T)$ for some $1 \leqslant p < +\infty$; and*
(ii) *u satisfies the Lusin N-property,*
<u>then</u> $u \in AC(T)$.

Corollary 4.126
<u>If</u> $T \subseteq \mathbb{R}$ is an interval, <u>then</u> $u \in AC_{\text{loc}}(T)$ if and only if
(a) *u is continuous; and*
(b) *$u \in BV_{\text{loc}}(T)$; and*
(c) *u has the Lusin N-property.*

Now we can say that the functions in $AC_{\text{loc}}(T)$ are precisely the functions which can be reproduced by their derivatives, i.e., the fundamental theorem of the Lebesgue calculus holds.

Theorem 4.127
If $T \subseteq \mathbb{R}$ *is an interval, then* $u \in AC_{\mathrm{loc}}(T)$ *if and only if*
(a) u *is continuous; and*
(b) u *is* λ*-almost everywhere differentiable on* T *with* $u' \in L^1_{\mathrm{loc}}(T)$*; and*
(c) *for all* $t, t_0 \in T$*, we have* $u(t) = u(t_0) + \int_{t_0}^{t} u'(s)\,ds$*.*

Corollary 4.128
If $T \subseteq \mathbb{R}$ *is an interval and* $u\colon T \longrightarrow \mathbb{R}$ *is everywhere differentiable with* $u \in L^1_{\mathrm{loc}}(T)$*,*
then for all $t, t_0 \in T$*, we have* $u(t) = u(t_0) + \int_{t_0}^{t} u'(s)\,ds$*.*

Theorem 4.129
If $T \subseteq \mathbb{R}$ *is an interval,* $u \in BV_{\mathrm{loc}}(T)$ *(respectively,* $u \in BV(T)$*) and* $[a,b] \subseteq T$*,*
then $u \in AC\big([a,b]\big)$ *(respectively,* $u \in AC(T)$*) if and only if*

$$\int_a^b |u'|\,dt \;=\; \operatorname{Var}_{[a,b]} u \quad \textit{(respectively, } \int_T |u'|\,dt \;=\; \operatorname{Var} u\textit{).}$$

Next we examine the validity of the chain rule and the change of variable formula for absolutely continuous functions.

Theorem 4.130 *(Chain Rule)*
If $T \subseteq \mathbb{R}$ *is an interval,* $u \in BV_{\mathrm{loc}}(T)$ *and* $f\colon \mathbb{R} \longrightarrow \mathbb{R}$ *is locally Lipschitz,*
then $f \circ u\colon T \longrightarrow \mathbb{R}$ *is differentiable* λ*-almost everywhere on* T *and*

$$(f \circ u)'(t) \;=\; f^*\big(u(t)\big)u'(t) \quad \textit{for } \lambda\textit{-almost all } t \in T,$$

where $f^*\colon \mathbb{R} \longrightarrow \mathbb{R}$ *is any Borel function such that* $f^* = f'$ λ*-almost everywhere on* $\mathbb{R}$ *(recall that* f *being locally Lipschitz is differentiable* λ*-almost everywhere on* $\mathbb{R}$*).*

Remark 4.131
If $f\colon \mathbb{R}^N \longrightarrow \mathbb{R}$ is locally Lipschitz ($N \geqslant 2$) and $u \in AC_{\mathrm{loc}}(T;\mathbb{R}^N)$, then $f \circ u \in AC_{\mathrm{loc}}(T)$, but the chain rule of Theorem 4.130 may fail. For the chain rule which is formulated in Theorem 4.130, we need additional conditions on the set

$$S_f \overset{\mathrm{def}}{=} \big\{x \in \mathbb{R}^N : \ f \text{ is not differentiable at } x\big\}.$$

In Theorem 4.160 below we will give these conditions.

Theorem 4.132 *(Change of Variables)*
If $u\colon [a,b] \longrightarrow [c,d]$ *is absolutely continuous and* $f\colon [c,d] \longrightarrow \mathbb{R}$ *is bounded measurable, then*

$$\int_{u(a)}^{u(b)} f(y)\,dy = \int_a^b f\big(u(t)\big)u'(t)\,dt.$$

Theorem 4.133 *(Area Formula)*
If $T \subseteq \mathbb{R}$ *is an interval,* $f\colon T \longrightarrow [0,+\infty]$ *is a Borel measurable function and* $u\colon T \longrightarrow \mathbb{R}$ *is differentiable* λ*-almost everywhere on* T *and has the Lusin* N*-property, then*

$$\int_{\mathbb{R}} \sum_{x\in u^{-1}(y)} f(x)\,dy = \int_T f(t)\big|u'(t)\big|\,dt.$$

Remark 4.134
By definition

$$\sum_{x\in u^{-1}(u)} f(x) = \sup\Big\{\sum_{x\in F} f(x) : F \subseteq u^{-1}(y) \text{ is finite}\Big\}.$$

Theorem 4.135 *(Integration by Parts Formula)*
If $T \subseteq \mathbb{R}$ *is an interval and* $u, v \in AC_{\mathrm{loc}}(T)$*,*
then for all $t, t_0 \in T$*, we have*

$$u(t)v(y) - u(t_0)v(t_0) = \int_{t_0}^{t} uv'\,ds + \int_{t_0}^{t} u'v\,ds.$$

Definition 4.136
(a) *Let* $T \subseteq \mathbb{R}$ *be an interval. A nonconstant function* $u\colon T \longrightarrow \mathbb{R}$ *is said to be* ***singular*** *if it is differentiable* λ*-almost everywhere and* $u'(t) = 0$ *for* λ*-almost all* $t \in T$*.*
(b) *Let* $T \subseteq \mathbb{R}$ *be an interval. An increasing function* $u\colon T \longrightarrow \mathbb{R}$ *is a* ***saltus*** *function, if* $u(t) = \sum_{n\in A} u_n(t)$*, where* $A \subseteq \mathbb{N}$ *and*

$$u_n(t) = \begin{cases} 0 & \text{if } t < r_n, \\ s_n & \text{if } t = r_n, \\ s_n + t_n & \text{if } r_n < t, \end{cases}$$

for some sets $\{r_n\}_{n\in A} \subseteq T$ *and* $\{s_n\}_{n\in A}, \{t_n\}_{n\in A} \subseteq [0,+\infty]$ *with* $s_n + t_n > 0$ *for all* $n \in A$.

Remark 4.137

The Cantor function is singular. Also, if $u\colon [a,b] \longrightarrow \mathbb{R}$ is increasing and for every $t \in (a,b)$, we have

$$u_+(t) \;=\; \lim_{s\to t^+} u(s) \quad \text{and} \quad u_-(t) \;=\; \lim_{s\to t^-} u(s),$$

then

$$u_s(t) \;=\; \sum_{\substack{s\,\in\,[a,b]\\ s\,<\,t}} \big[u_+(t) - u_-(t)\big] + u(t) - u_-(t)$$

is a saltus function.

Theorem 4.138

If $T \subseteq \mathbb{R}$ *is an interval and* $u \in BV_{\mathrm{loc}}(T)$,
then $u = u_{ac} + u_{cs} + u_s$, *where* $u_{ac} \in AC_{\mathrm{loc}}(T)$, u_{cs} *is continuous singular and* u_s *is a saltus function. Moreover*

$$\operatorname{Var} u \;=\; \operatorname{Var} u_{ac} + \operatorname{Var} u_{cs} + \operatorname{Var} u_s.$$

Next we extend the notion of bounded variation to functions of several variables.

Definition 4.139

Let $\Omega \subseteq \mathbb{R}^N$ *be an open set and let* $u \in L^1_{\mathrm{loc}}(\Omega)$. *We say that* u *is of* ***bounded variation*** *if*

$$\sup\Big\{\int_\Omega u \operatorname{div} h\, dz:\ h \in C^1_c(\Omega;\mathbb{R}^N),\ \big\|h(z)\big\| \leqslant 1 \textit{for all} z \in \Omega\Big\} < +\infty.$$

We denote the space of functions of bounded variation by $BV(\Omega)$. *We say that* $u \in L^1_{\mathrm{loc}}(\Omega)$ *is of* ***local bounded variation*** *if for all open* $U \subseteq \Omega$ *with* $\overline{U}$ *compact and* $\overline{U} \subseteq \Omega$, *we have* $u \in BV(U)$. *We denote the space of functions which are locally of bounded variation by* $BV_{\mathrm{loc}}(\Omega)$. *A Lebesgue measurable set* $A \subseteq \mathbb{R}^N$ *has a* ***finite perimeter*** *in* Ω, *if* $\chi_A \in BV(\Omega)$. *We say that the Lebesgue set* $A \subseteq \mathbb{R}^N$ *has* ***locally finite perimeter*** *in* Ω, *if* $\chi_A \in BV_{\mathrm{loc}}(\Omega)$.

Remark 4.140
So, according to the above definition $u \in L^1_{\text{loc}}(\Omega)$ is of bounded variation, if the distributional partial derivatives $\frac{\partial u}{\partial z_i}$, $i \in \{1, \ldots, N\}$ are bounded signed Radon measures (see Definition 4.22), i.e., for every $i \in \{1, \ldots, N\}$, there exists a bounded signed measure $\mu_i \colon \mathcal{B}(\Omega) \longrightarrow \mathbb{R}$ ($\mathcal{B}(\Omega)$ being the Borel σ-algebra of Ω) such that

$$\int_\Omega u \frac{\partial h}{\partial z_i}\, dz = -\int_\Omega h\, d\mu_i \qquad \forall\, i \in \{1, \ldots, N\}\,, h \in C_c^\infty(\Omega).$$

The measures μ_i are the distributional (or weak) derivatives of u with respect to z_i, $i \in \{1, \ldots, N\}$ and we denote them by $D_i u$. Therefore, if $u \in BV(\Omega)$, then $D_i u \in M_b(\Omega)$ for all $i \in \{1, \ldots, N\}$ and $Du = \big(D_1 u, \ldots, D_N u\big) \in M_b\big(\Omega; \mathbb{R}^N\big)$ (Du being the gradient of u). So, we can define the total variation $|Du|$ of the vector measure Du by

$$|Du|(A) \stackrel{\text{def}}{=} \sup \Big\{\sum_{k-1}^{n} \big\|Du(C_k)\big\| : \{C_k\}_{k=1}^n \text{ disjoint Borel partition of } A\Big\}.$$

We know that $|Du|(\cdot)$ is a finite Radon measure. By Theorem 4.23, we have

$$\begin{aligned} |Du|(\Omega) &= \|Du\|_{M_b(\Omega;\mathbb{R}^N)} \\ &= \sup\Big\{\sum_{i=1}^{n} \int_\Omega h_i d(D_i u) : h \in C_0\big(\Omega; \mathbb{R}^N\big),\ \|h\|_\infty \leqslant 1\Big\} \\ &< +\infty. \end{aligned}$$

This leads to the following definition.

Definition 4.141
*Let $\Omega \subseteq \mathbb{R}^N$ be an open set and let $u \in L^1_{\text{loc}}(\Omega)$. The **variation** of u in Ω is defined by*

$$V(u;\Omega) \stackrel{\text{def}}{=} \sup\Big\{\sum_{i=1}^{N} \int_\Omega \frac{\partial h_i}{\partial z_i} u\, dz : h \in C_c^\infty(\Omega; \mathbb{R}^N),\ \|h\|_\infty \leqslant 1\Big\}.$$

Proposition 4.142
If $\Omega \subseteq \mathbb{R}^N$ is an open set,
then $BV(\Omega)$ equipped with the norm $\|u\| = \|u\|_1 + |Du|(\Omega)$ is a Banach space.

Based on Remark 4.140, we have the following structure theorem for $BV_{\mathrm{loc}}(\Omega)$.

Theorem 4.143
If $\Omega \subseteq \mathbb{R}^N$ *is an open set and* $u \in BV_{\mathrm{loc}}(\Omega)$,
then there exists $\mu \in M_b(\Omega)$ *and a* μ*-measurable function* $\xi\colon \Omega \longrightarrow \mathbb{R}^N$ *such that* $\|\xi(z)\| \leqslant 1$ *for* μ*-almost all* $z \in \Omega$ *and*

$$\int_\Omega u \operatorname{div} h\, dz = -\int_\Omega (h, \xi)_{\mathbb{R}^N}\, d\mu \qquad \forall\, h \in C_c^\infty(\Omega; \mathbb{R}^N),$$

the measure μ *is denoted by* $|Du|$.

Remark 4.144
If $A \subseteq \mathbb{R}^N$ is of locally finite perimeter in Ω and $u = \chi_A$, then $u \in BV_{\mathrm{loc}}(\Omega)$ (see Definition 4.139) and then by Theorem 4.143, we have

$$\int_A \operatorname{div} h\, dz = -\int_\Omega (h, \xi)_{\mathbb{R}^N}\, d|D\chi_a| \qquad \forall\, h \in C_c^\infty(\Omega; \mathbb{R}^N)$$

(see also Remark 4.148).

From Remark 4.140, we have that $|Du|(\cdot)$ is the variational measure of u and $|\partial A|(\cdot)$ is the perimeter measure of A, with $|\partial A|(\Omega)$ being the perimeter of A in Ω.

Theorem 4.145
If $\Omega \subseteq \mathbb{R}^N$ *is an open set and* $u \in BV(\Omega)$,
then there exists a sequence of functions $\{u_n\}_{n\geqslant 1} \subseteq BV(\Omega) \cap C^\infty(\Omega)$ *such that* $u_n \longrightarrow u$ *in* $L^1(\Omega)$ *and* $|Du_n|(\Omega) \longrightarrow |Du|(\Omega)$.

Remark 4.146
Note that for $u \in BV(\Omega) \cap C^\infty(\Omega)$, we have

$$|Du|(\Omega) = \int_\Omega \|Du\|\, dz.$$

Definition 4.147
Let $A \subseteq \mathbb{R}^N$ *be a set with locally finite perimeter in* $\mathbb{R}^N$. *The* ***reduced boundary*** *of* A, *denoted by* $\partial^* A \subseteq \mathbb{R}^N$, *is the set of all* $z \in \mathbb{R}^N$ *such that:*

(a) $|D\chi_A|(B_r(z)) > 0$; *and*

(b) the limit $\nu_A(z) = -\lim_{r\to 0+} \frac{D\chi_A(B_r(z))}{|D\chi_A|(B_r(z))}$ *exists in* $\mathbb{R}^N$ *and*

$$\|\nu_A(z)\| = 1.$$

Remark 4.148

The minus sign in the definition of ν_A indicates that the normal vector is directed outward from the set A, i.e., in the direction opposite to the gradient of χ_A. Note that for all $x \in \mathbb{R}^N$

$$n(x) = \lim_{r\to 0+} \frac{D\chi_A(B_r(x))}{|D\chi_a|(B_r(x))},$$

when this limit exists, is the Radon–Nikodym derivative of $D\chi_A$ with respect to $|D\chi_A|$ and so for all $C \in \mathcal{B}(\mathbb{R}^N)$, we have

$$D\chi_A(C) = \int_C n\, d|D\chi_A|,$$

hence

$$\int_A \operatorname{div} h\, dz = \int_{\mathbb{R}^N} (h, n)_{\mathbb{R}^N}\, d|D\chi_A| \qquad \forall\, h \in C_c^\infty(\mathbb{R}^N;\mathbb{R}^N)$$

and thus $\|n(z)\| = 1$ $|D\chi_A|$-almost everywhere and $n = -\xi$ in Theorem 4.143.

Because $|D\chi_A|(\mathbb{R}^N \setminus \partial^* A) = 0$, we see that $\nu_A(\cdot)$ is a multidimensional analogue of the Radon–Nikodym derivative of $D\chi_A(\cdot)$ with respect to $|D\chi_A|(\cdot)$.

Next we generate the notion of boundary for a Lebesgue measurable set.

Definition 4.149

(a) Let $A \subseteq \mathbb{R}^N$ *be a Lebesgue measurable set. A unit vector* n *is a* ***measure-theoretic outer normal*** *of* A *at* z*, if the following two conditions hold:*

$$\lim_{r\to 0+} \tfrac{1}{r^N}\lambda^N\big(B_r(z) \cap \{x \in \mathbb{R}^N :\ (x-z, n)_{\mathbb{R}^N} < 0,\ x \notin A\}\big) = 0$$

and

$$\lim_{r\to 0+} \tfrac{1}{r^N}\lambda^N\big(B_r(z) \cap \{x \in \mathbb{R}^N :\ (x-z, n)_{\mathbb{R}^N} > 0,\ x \in A\}\big) = 0.$$

We denote by $n(x,A)$, the measure-theoretic outer normal of A at $x \in \mathbb{R}^N$.
(b) Let $A \subseteq \mathbb{R}^N$ be a Lebesgue measurable set. The ***measure-theoretic boundary*** *of A is the set*

$$\partial_* A \stackrel{\text{def}}{=} \{x \in \mathbb{R}^N : \ n(x,A) \text{ exists}\}.$$

Remark 4.150
If $A \subseteq \mathbb{R}^N$ is a set with locally finite perimeter, then $\partial^* A \subseteq \partial_* A$. Note that, if for a unit vector $n \in \mathbb{R}^N$, we introduce the hyperplane

$$H(z) \stackrel{\text{def}}{=} \{x \in \mathbb{R}^N : \ (x-z,n)_{\mathbb{R}^N} = 0\}$$

and the corresponding two half-spaces

$$\begin{aligned} H_+(z) &\stackrel{\text{def}}{=} \{x \in \mathbb{R}^N : \ (x-z,n)_{\mathbb{R}^N} > 0\}, \\ H_-(z) &\stackrel{\text{def}}{=} \{x \in \mathbb{R}^N : \ (x-z,n)_{\mathbb{R}^N} < 0\}, \end{aligned}$$

then n is a measure theoretic normal to A at z, if the following two conditions hold:

$$\lim_{r \to 0^+} \frac{\lambda^N(B_r(z) \cap A \cap H_+(z))}{r^N} = 0 \quad \text{and} \quad \lim_{r \to 0^+} \frac{\lambda^N((B_r(z) \setminus A) \cap H_-(z))}{r^N} = 0.$$

4.1.13 Hausdorff Measures: Change of Variables

We will use the notions introduced in Definitions 4.147 and 4.149, to produce multidimensional generalizations of the fundamental theorem of calculus. But first we need to say a few basic things about Hausdorff measures.

For $s \in [0,+\infty)$, let

$$\omega_s \stackrel{\text{def}}{=} \frac{\pi^{\frac{s}{2}}}{\Gamma(\frac{s}{2}+1)}, \quad \text{where} \quad \Gamma(t) = \int_0^\infty s^{t-1} e^{-s}\, ds$$

is the Euler Γ-function. This constant coincides with the Lebesgue measure of the unit ball of $\mathbb{R}^N$ if $s = N$.

Definition 4.151
Let $s \in [0, +\infty)$ and $A \subseteq \mathbb{R}^N$. The ***s-dimensional Hausdorff measure*** *of A is defined by*

$$H^s(A) \stackrel{\text{def}}{=} \lim_{\delta \to 0} H^s_\delta(A),$$

where H^s_δ is defined by

$$H^s_\delta(A) \stackrel{\text{def}}{=} \frac{\omega_s}{2^s} \inf\Big\{\sum_{k\geqslant 1} (\operatorname{diam} C_k)^s : \ A \subseteq \bigcup_{k\geqslant 1} C_k, \ \operatorname{diam} C_k < \delta\Big\}.$$

Remark 4.152
In the above definition the covering sets are arbitrary. However, without any loss of generality, we may assume that they are open convex or closed convex. Also, we can replace the condition $\operatorname{diam} C_k < \delta$, by the condition $\operatorname{diam} C_k \leqslant \delta$ without changing the value of $H^s_\delta(A)$. If $s = 0$, then H^0 is the counting measure.

Theorem 4.153
For $s \in [0, +\infty)$, H^s is a Borel measure.

Remark 4.154
However, H^s is not a Radon measure, since for $s \in (0, N)$, $\mathbb{R}^N$ is not σ-finite with respect to H^s.

The next theorem summarizes some basic properties of the Hausdorff measure.

Theorem 4.155
(a) *$H^N = \lambda^N$.*
(b) *$H^s = 0$ on $\mathbb{R}^N$ for all $s > N$.*
(c) *For all $s > 0$, all $A \subseteq \mathbb{R}^N$ and all $\lambda > 0$, we have $H^s(\lambda A) = \lambda^s H^s(A)$.*
(d) *For all $s \geqslant 0$, all $A \subseteq \mathbb{R}^N$ and every affine isometry $L\colon \mathbb{R}^N \longrightarrow \mathbb{R}^N$, we have $H^s\big(L(A)\big) = H^s(A)$.*
(e) *If $s \geqslant 0$, $A \subseteq \mathbb{R}^N$ and $\xi\colon A \longrightarrow \mathbb{R}^k$ is Lipschitz continuous with Lipschitz constant $\eta > 0$, then $H^s\big(\xi(A)\big) \leqslant \eta^s H^s(A)$.*
(f) *If $s > s' > 0$ and $A \subseteq \mathbb{R}^N$, then*

$$\big[H^s(A) > 0 \quad \Longrightarrow \quad H^{s'}(A) = +\infty\big]$$

and also

$$\left[H^{s'}(A) < +\infty \quad \Longrightarrow \quad H^s(A) = 0\right].$$

Definition 4.156
The ***Hausdorff dimension*** *of a set* $A \subseteq \mathbb{R}^N$ *is defined by*

$$H\text{-}\dim(A) \stackrel{\text{def}}{=} \inf\left\{s \geqslant 0: \; H^s(A) = 0\right\} \;=\; \sup\left\{s \geqslant 0: \; H^s(A) > 0\right\}.$$

Remark 4.157
Note that H- dim may be any number in $[0, +\infty]$, not necessary an integer. Moreover, if $d = \text{H-}\dim(A)$, then $H^s(A) = 0$ for all $s > d$ and $H^s(A) = +\infty$ for all $s \in (0, d)$.

Proposition 4.158
If $A, C \subseteq \mathbb{R}^N$ *and* $\operatorname{dist}(A, C) = \inf\left(\|a - c\| : \; a \in A, \; c \in C\right) > 0$, *then* $H^s(A \cup C) = H^s(A) + H^s(A)$ *for all* $s \geqslant 0$.

Definition 4.159
(a) *A set* $A \subseteq \mathbb{R}^N$ *is said to be* H^1***-rectifiable*** *if there exists a sequence of Lipschitz continuous functions* $u_n\colon \mathbb{R} \longrightarrow \mathbb{R}^N$ *such that* $H^1\big(A \setminus \bigcup\limits_{n \geqslant 1} u_n(\mathbb{R})\big) = 0$.
(b) *A set* $A \subseteq \mathbb{R}^N$ *is said to be* ***purely*** H^1***-rectifiable*** *if*

$$H^1\big(A \cap u(\mathbb{R})\big) = 0,$$

for every Lipschitz continuous $u\colon \mathbb{R} \longrightarrow \mathbb{R}^N$.

Using these notions we can have the multidimensional version of the chain rule (see Theorem 4.130).

Theorem 4.160
If $f\colon \mathbb{R}^N \longrightarrow \mathbb{R}$ *is a locally Lipschitz function and* $T \subseteq \mathbb{R}$ *is an interval,*
then $f \circ u \in AC_{\text{loc}}(T)$ *for every* $u \in AC_{\text{loc}}(T; \mathbb{R}^N)$.
Moreover, if the set $S_f = \left\{x \in \mathbb{R}^N : \; f \text{ is not differentiable at } x\right\}$ *is purely* H^1*-rectifiable, then*

$$(f \circ u)'(x) \;=\; \sum_{i=1}^N \tfrac{\partial f}{\partial z_i}\big(u(x)\big)u_i'(x) \qquad \text{for almost all } x \in T,$$

with $\frac{\partial f}{\partial z_i}\left(u(x)\right) u_i'(x)$ *being zero whenever* $u_i'(x) = 0$.

Using Hausdorff measures, we can extend Theorem 4.133, to functions of several variables.

Theorem 4.161 *(Area Formula)*
If $N \leqslant k$ *and* $f\colon \mathbb{R}^N \longrightarrow \mathbb{R}^k$ *is a Lipschitz continuous function,* *then for every Lebesgue measurable set* $A \subseteq \mathbb{R}^N$*, we have*

$$\int_A Jf(z)\,dz = \int_{\mathbb{R}^k} H^0\big(A \cap f^{-1}(y)\big)\,dH^N(y),$$

with $Jf(z) = [[\det Df(z)]]$*, where for linear* $L\colon \mathbb{R}^N \longrightarrow \mathbb{R}^k$ *(with* $N \leqslant k$*), we write* $L = \mathcal{O} \circ S$*, with* $S\colon \mathbb{R}^k \longrightarrow \mathbb{R}^k$ *symmetric and* $\mathcal{O}\colon \mathbb{R}^N \longrightarrow \mathbb{R}^k$ *orthogonal and* $[[L]] = |\det S|$*.*

Theorem 4.162 *(Change of Variable Formula)*
If $N \leqslant k$ *and* $f\colon \mathbb{R}^N \longrightarrow \mathbb{R}^k$ *is a Lipschitz continuous function and* $u \in L^1(\mathbb{R}^N)$*,*
then

$$\int_{\mathbb{R}^N} u(z)Jf(z)\,dz = \int_{\mathbb{R}^k} \Big(\sum_{z \in u^{-1}(y)} u(z)\Big)\,dH^n(y).$$

Theorem 4.163 *(Isodiametric Inequality)*
For all sets $A \subseteq \mathbb{R}^N$*, we have* $\lambda^N(A) \leqslant \omega_N\big(\frac{\operatorname{diam} A}{2}\big)^N$*.*

Finally, the area formula (see Theorem 4.161) has a dual version.

Theorem 4.164 *(Coarea Formula)*
If $N \geqslant k$ *and* $f\colon \mathbb{R}^N \longrightarrow \mathbb{R}^k$ *is a Lipschitz continuous function,* *then for every Lebesgue measurable set* $A \subseteq \mathbb{R}^N$*, we have*

$$\int_A Jf(z)\,dz = \int_{\mathbb{R}^k} H^{N-k}\big(A \cap f^{-1}(y)\big)\,dy.$$

4.1.14 Carathéodory Functions

We conclude this chapter with three theorems that are very useful both in theory and applications. We start with a definition.

Definition 4.165
Let (Ω, Σ) *be a measurable space and let* X *and* Y *be two topological spaces. We say that* $f\colon \Omega \times X \longrightarrow Y$ *is a* ***Carathéodory function*** *if*

(a) *for every $x \in X$, the function $\omega \longmapsto f(\omega, x)$ is $(\Sigma, \mathcal{B}(Y))$-measurable;*
(b) *for every $\omega \in \Omega$, the function $x \longmapsto f(\omega, x)$ is continuous.*

Theorem 4.166
If (Ω, Σ) is a measurable space, X is a separable metric space, Y is a metric space and $f \colon \Omega \times X \longrightarrow Y$ is a Carathéodory function,
then f is jointly measurable.

The next theorem is a parametric version of the Lusin theorem (see Theorem 3.77).

Theorem 4.167 *(Scorza–Dragoni Theorem)*
If T and X are two Polish spaces, Y is a separable metric space, μ is a finite tight Borel measure on T and $f \colon T \times X \longrightarrow Y$ is a Carathéodory function,
then for every $\varepsilon > 0$ we can find a compact subset $T_\varepsilon \subseteq T$ such that $\mu(T \setminus T_\varepsilon) < \varepsilon$ and $f|_{T_\varepsilon \times X}$ is continuous.

If $Y = \mathbb{R} \cup \{+\infty\}$, then we can extend Theorem 4.167 as follows.

Theorem 4.168
If T and X are Polish spaces, μ is a finite tight Borel measure on T and $f \colon T \times X \longrightarrow \mathbb{R} \cup \{+\infty\}$ is a Borel function such that for all $f \in T$, $f(t, \cdot)$ is lower semicontinuous,
then for every $\varepsilon > 0$, we can find a compact set $T_\varepsilon \subseteq T$ such that $\mu(T \setminus T_\varepsilon) < \varepsilon$ and $f|_{T_\varepsilon \times X}$ is lower semicontinuous.

4.2 Problems

Problem 4.1 **
Suppose that $A \subseteq [0,1]$ is a Lebesgue measurable set and $\lambda(A) > 0$ (λ being the Lebesgue measure on $\mathbb{R}$). Show that we can find $x, u \in A$, $x \neq u$ such that $x - u \in \mathbb{Q}$.

Problem 4.2 **
Show that the Borel σ-algebra of a topological space X is the smallest family of subsets of X which contains all open sets, all closed sets and which is closed under countable intersections and countable disjoint unions.

Problem 4.3 *
Let X be a metrizable set. Show that the Borel σ-algebra $\mathcal{B}(X)$ is the smallest family of subsets of X which includes the open sets and which is closed under countable intersections and under countable disjoint unions.

Problem 4.4 **
Suppose that X is a locally compact topological space and let $\mathcal{X}$ is a subbasis for the topology of X (see Definition 2.19(a)). Show that $\mathcal{B}a(X) \subseteq \sigma(\mathcal{X}) \subseteq \mathcal{B}(X)$.

Problem 4.5 *
Suppose that X is a topological space, $\left(X, \widehat{\mathcal{B}(X)}, \mu\right)$ is a complete measure space ($\widehat{\mathcal{B}(X)}$ is the μ-completion of the Borel σ-algebra with respect to μ; see Definition 3.23) and $u\colon X \longrightarrow \mathbb{R}$ is a function such that: there exists a partition $\{C_n\}_{n \geqslant 1} \subseteq \widehat{\mathcal{B}(X)}$ of X for which C_0 is μ-null and for all $n \geqslant 1$, $u|_{C_n}$ is lower semicontinuous (respectively, upper semicontinuous, continuous). Show that u is $\widehat{\mathcal{B}(X)}$-measurable.

Problem 4.6 **
Suppose that X is a locally compact topological space and $U \subseteq X$ is a σ-compact open set. Show that $U \in \mathcal{B}a(X)$.

Problem 4.7 **
Suppose that X is a locally compact, σ-compact topological space and X^* is the Aleksandrov one-point compactification of X (see Remark 2.97). Show that $\mathcal{B}(X) = X \cap \mathcal{B}(X^*)$.

Problem 4.8 **
Suppose that (Ω, Σ) is a measurable space, X is a topological space and $G \in \Sigma \otimes \mathcal{B}(X)$. Show that there exists $\Sigma_0 \subseteq \Sigma$, a countably generated sub-σ-algebra of Σ (see Definition 3.14(a)) such that $G \in \Sigma_0 \otimes \mathcal{B}(X)$.

Problem 4.9 *
Let X be a locally compact topological space. Show that the open Baire sets form a basis for the topology of X.

Problem 4.10 ***
Let X be a topological space and let $C \subseteq X$ be a nonempty closed set endowed with the subspace topology. Show that in general it is not true that $\mathcal{B}(C) \subseteq \mathcal{B}(X)$.

Problem 4.11 **
Let $C \in \mathcal{B}([0,1])$ with $\lambda(C) > 0$ (λ being the Lebesgue measure). Show that C has the cardinality of the continuum (do not use the continuum hypothesis).

Problem 4.12 **
Suppose that X is a separable metric space, μ is a Borel measure on X and $u\colon X \longrightarrow \overline{\mathbb{R}}_+ = [0, +\infty]$ is a Borel function. Show that

$$\int_X u \, d\mu = \int_0^{+\infty} \mu(\{x \in X : u(x) \geqslant \eta\}) \, d\eta.$$

Problem 4.13 ***
Let λ be the Lebesgue measure on $\mathbb{R}$. Suppose that $A \subseteq \mathbb{R}$ is a Lebesgue measurable set and assume that $\lambda\big(A \cap (a,b)\big) \leqslant \frac{1}{2}(b-a)$ for all $a, b \in \mathbb{R}$ such that $a < b$. Show that $\lambda(A) = 0$.

Problem 4.14 *
Suppose that X is a metrizable space, $\mathcal{B}(X)$ is its Borel σ-algebra and μ, ν are two finite measures on $\mathcal{B}(X)$. Show the following:
(a) If μ and ν coincides on the open sets of X, then $\mu = \nu$.
(b) If X is σ-compact and μ, ν coincide on compact sets, then $\mu = \nu$.

Problem 4.15 ***

Suppose that X is a metrizable space and $\mu\colon \mathcal{B}(X) \longrightarrow \overline{\mathbb{R}}_+ = [0, +\infty]$ is a σ-finite measure. Show that every set $A \in \mathcal{B}(X)$ with finite μ-measure is ***inner regular***, i.e., for every $\varepsilon > 0$ there exists a closed set $C_\varepsilon \subseteq A$ such that $\mu(A \setminus C_\varepsilon) < \varepsilon$.

Problem 4.16 **

Suppose that X is a topological space and $\mu\colon \mathcal{B}(X) \longrightarrow \overline{\mathbb{R}}_+ = [0, +\infty]$ is a measure which is outer regular (see Definition 4.9), finite on compact sets and for every open set $U \subseteq X$, we have $\mu(U) = \sup\{\mu(K) : K \subseteq U,\ K \text{ is compact}\}$. Show that for every $A \in \mathcal{B}(X)$ which is σ-finite for μ, we have $\mu(A) = \sup\{\mu(D) :\ D \subseteq A,\ D \text{ is compact}\}$.

Problem 4.17 **

Suppose that X is a locally compact topological space and let μ, ν be two Radon measures on $\mathcal{B}(X)$ (see Definition 4.9(e)). Show that $\nu \leqslant \mu$ (i.e., for all $A \in \mathcal{B}(X)$, $\nu(A) \leqslant \mu(A)$) if and only if

$$\int_X u\, d\nu \leqslant \int_X u\, d\mu \qquad \forall\, u \in C_c(X),\ u \geqslant 0.$$

Problem 4.18 **

Suppose that X is a topological space, $\mathcal{B}(X)$ is its Borel σ-algebra and $\mu\colon \mathcal{B}(X) \longrightarrow \overline{R}_+ = [0, +\infty]$ is σ-bounded measure, i.e., there exists a countable open cover $\{U_n\}_{n \geqslant 1}$ of X with $\mu(U_n) < +\infty$ for all $n \geqslant 1$. Show the following:

(a) Every compact set in X has finite μ-measure.

(b) Suppose that X is a σ-compact metric space. Then every set $A \in \mathcal{B}(X)$ with $\mu(A) < +\infty$ is ***inner regular with respect to compact sets***, i.e., $\mu(A) = \sup\{\mu(K) :\ K \subseteq A,\ K \text{ is compact}\}$.

Problem 4.19 **

Suppose that X is a locally compact topological space, $\mathcal{B}(X)$ is its Borel σ-algebra and $\mu\colon \mathcal{B}(X) \longrightarrow \overline{R}_+ = [0, +\infty]$ is a σ-finite measure which is outer regular, finite on all compact sets and inner regular with respect to compact sets on all open sets (i.e., for any open set $U \subseteq X$, we have $\mu(U) = \sup\{\mu(K) :\ K \subseteq U,\ K \text{ is compact}\}$). For every $A \in \mathcal{B}(X)$, show that the following statements hold:

(a) For a given $\varepsilon > 0$, we can find an open set U and a closed set C such that $C \subseteq A \subseteq U$ and $\mu(U \setminus C) < \varepsilon$.

(b) There exist an G_δ-set D and an F_σ-set E such that $E \subseteq A \subseteq D$ and $\mu(D \setminus E) = 0$.

Problem 4.20 *
Suppose that X is a topological space, $\mathcal{B}(X)$ is its Borel σ-algebra and $\mu\colon \mathcal{B}(X) \longrightarrow \mathbb{R}_+ = [0, +\infty)$ is an additive set function. Show that
(a) μ is outer regular if and only if it is inner regular (see Definition 4.9);
(b) if $\mu(A) = \sup\{\mu(K) : K \subseteq A, K \text{ is compact}\}$ for all $A \in \mathcal{B}(X)$, then μ is regular.

Problem 4.21 **
Suppose that X is a locally compact topological space, $\mathcal{B}(X)$ is its Borel σ-algebra and $\mu\colon \mathcal{B}(X) \longrightarrow \overline{R}_+ = [0, +\infty]$ is a measure. Suppose that $u \in C_c(X)$, $u \geqslant 0$ and assume that $\int\limits_X u \, d\mu = 0$. Show that $u|_{\operatorname{supp}\mu} = 0$.

Problem 4.22 **
Suppose that X is a locally compact topological space, $\mathcal{B}(X)$ is its Borel σ-algebra and μ is a signed Radon measure on $\mathcal{B}(X)$. Show that

$$\int\limits_X u \, d\mu = 0 \qquad \forall\, u \in C_c(X), u|_{\operatorname{supp}\mu} = 0.$$

Problem 4.23 **
Suppose that X is a topological space, $\mathcal{B}(X)$ is its Borel σ-algebra and $\mu\colon \mathcal{B}(X) \longrightarrow \mathbb{R}_+ = [0, +\infty)]$ is an additive set function, which is outer regular and inner regular with respect to the compact sets (i.e., $\mu(A) = \sup\{\mu(K) : K \subseteq A, K \text{ is compact}\}$ for all $A \in \mathcal{B}(X)$). Show that μ is a measure.

Problem 4.24 **
Let X be a metrizable space. Show that a finite Borel measure μ is Radon if and only if for each $\varepsilon > 0$, there exists a compact set K such that $\mu(X) - \varepsilon < \mu(K)$.

Problem 4.25 **
Suppose that X is a topological space and μ is a Borel measure on X. Suppose that X is second countable or alternatively that μ is

outer regular and also inner regular with respect to compact sets (see Definition 4.9). Show that μ has a support (necessarily unique; see Remark 4.14).

Problem 4.26 *
Suppose that (X, Σ) is a measurable space and μ, ν are two finite measures on Σ. Is it true that the smallest measure on Σ not less than μ or ν is $\max\{\mu(A), \nu(A)\}$, $A \in \Sigma$? Justify your answer.

Problem 4.27 ***
Suppose that X is a locally compact topological space, $\mathcal{B}(X)$ is its Borel σ-algebra and μ and m are two Radon measures on $\mathcal{B}(X)$ with μ being σ-finite. Show that $\sigma = \min\{\mu, m\}$ (the largest measure less than or equal to μ and m) is also a Radon measure on $\mathcal{B}(X)$.

Problem 4.28 **
Show that the N-Lebesgue measure λ^N is Radon.

Problem 4.29 **
Suppose that X is a locally compact topological space, $\mathcal{B}(X)$ is the Borel σ-algebra and μ is a Radon measure on $\mathcal{B}(X)$. Show that $C_c(X)$ is dense in $L^p(X, \mu)$ $(1 \leqslant p < +\infty)$.

Problem 4.30 *
Find a bounded measurable function $f: \mathbb{R} \longrightarrow \mathbb{R}$ for which there is no sequence $\{f_n\}_{n \geqslant 1} \subseteq C(\mathbb{R})$ such that $\|f_n - f\|_\infty \longrightarrow 0$.

Problem 4.31 *
On $\mathcal{B}(\mathbb{R})$ we introduce the set function μ_0, defined by

$$\mu_0(A) \stackrel{\text{def}}{=} \operatorname{card}\{x : x \in \mathbb{Q} \cap A\}.$$

Show that μ_0 is a measure which is σ-finite but it is not regular.

Problem 4.32 **
Suppose that X is a locally compact topological space, $\mathcal{B}(X)$ is its Borel σ-algebra and $\mu: \mathcal{B}(X) \longrightarrow [0, +\infty)$ is a finite additive set function with the following property

$$\mu(A) = \sup\{\mu(K) : K \subseteq A,\ K \text{ is compact}\} \qquad \forall\, A \in \mathcal{B}(X).$$

Show that μ is a measure (i.e., μ is countably additive).

Problem 4.33 *
Consider the space

$$V \stackrel{\text{def}}{=} \{u \in C_c^\infty(\mathbb{R}) : \int_{\mathbb{R}} u\, dx = 0\}.$$

Is V dense in $L^1(\mathbb{R})$? Justify your answer.

Problem 4.34 **
Let (X, d_X) be a metric space. We say that two sets $A, C \subseteq X$ are ***metrically separated*** if $\inf\{d_X(a,c) : a \in A,\ c \in C\} > 0$, i.e., the two sets are a positive distance apart. We say that an outer measure μ is ***additive on metrically separated sets***, if $\mu(A \cup C) = \mu(A) + \mu(C)$, whenever A and C are metrically separated. Show that an outer measure which is additive on metrically separated sets is in fact a Borel measure, i.e., every Borel set of X is μ-measurable.

Problem 4.35 *
Suppose that X is a locally compact metric space and $\{\mu_n\}_{n\geqslant 1}$ is a sequence of Radon measures such that

$$\int_X f\, d\mu_n \longrightarrow \int_X f\, d\mu \qquad \forall\, f \in C_0(X)$$

for some Radon measure μ. Show that:
(a) for all compact sets $K \subseteq X$, we have $\limsup\limits_{n\to+\infty} \mu_n(K) \leqslant \mu(K)$;
(b) for all open sets $U \subseteq X$, we have $\mu(U) \leqslant \liminf\limits_{n\to+\infty} \mu_n(U)$.

Problem 4.36 **
Suppose that X is a metric space and μ is a finite Borel measure on X. We say that μ is ***uniformly distributed*** if

$$0 < \mu(B_r(x)) = \mu(B_r(y)) \qquad \forall\, x, y \in X,\ r > 0.$$

Suppose that μ and ν are two uniformly distributed finite Borel measures on X. Show that $\mu = c\nu$ for some $c > 0$.

Problem 4.37 **
Suppose that $\{f_n\}_{n\geqslant 1} \subseteq L^1(0,1)$ is a sequence such that
(a) $|f_n(x)| \leqslant h(x)$ almost everywhere on $[0,1]$, with $h \in L^1(0,1)$; and

(b) we have

$$\int_0^1 f_n g\, dx \longrightarrow 0 \qquad \forall\, g \in C\big([0,1]\big).$$

Show that for every Borel set $A \subseteq [0,1]$, we have

$$\int_A f_n\, dx \longrightarrow 0.$$

Problem 4.38 **
Suppose that X is a locally compact topological space and μ and ν are two tight measures on $\mathcal{B}(X)$. Suppose that

$$\int_X f\, d\mu = \int_X f\, d\nu \qquad \forall\, f \in C_c(X).$$

Show that $\mu = \nu$.

Problem 4.39 **
Suppose that (X, Σ) is a measurable space, Y is a metric space and $f\colon X \longrightarrow Y$ is a function. Show that the following two statements are equivalent:
(a) f is Σ-measurable;
(b) for every continuous function $\varphi\colon Y \longrightarrow \mathbb{R}$, the function $\varphi \circ f$ is Σ-measurable.

Problem 4.40 *
Let (Ω, Σ) be a measurable space, let X be a topological space and let $f\colon \Omega \longrightarrow X$ be a function. Show that the following statements are equivalent:
(a) f is $\big(\Sigma, \mathcal{B}a(X)\big)$-measurable (see Definition 4.2).
(b) For every continuous function $\varphi\colon X \longrightarrow \mathbb{R}$, $\varphi \circ f$ is Σ-measurable.

Problem 4.41 ***

Let $A \subseteq \mathbb{R}$ be a Lebesgue measurable set with $\lambda(A) < +\infty$ (λ being the Lebesgue measure on $\mathbb{R}$) and consider the function $f\colon \mathbb{R} \longrightarrow [0, +\infty)$, defined by

$$f(x) \stackrel{\text{def}}{=} \lambda\big((A + x) \cap A\big).$$

Show that:

(a) f is continuous; and

(b) $\lim\limits_{x \to +\infty} f(x) = 0$. (Compare with Problem 3.44.)

Problem 4.42 ***

Suppose that $A \subseteq \mathbb{R}$ is a Lebesgue measurable set with positive Lebesgue measure and let us set

$$\varphi(\varrho) \stackrel{\text{def}}{=} \int\limits_{\mathbb{R}} \chi_A(t\varrho)\chi_A(t)\, dt.$$

Show that φ is continuous at $\varrho = 1$.

Problem 4.43 **

Find a decreasing sequence of measures on $\mathcal{B}(\mathbb{R}^N)$ ($N \geqslant 1$) such that the limit set function is not a measure.

Problem 4.44 **

Suppose that X is a compact metric space and μ is a finite Borel measure on X such that for every $x \in X$, we have $\mu(\{x\}) = 0$. Show that for a given $\varepsilon > 0$, we can find $\delta = \delta(\varepsilon) > 0$ such that for all $A \in \mathcal{B}(X)$ with $\operatorname{diam} A < \delta$ and we have $\mu(A) < \varepsilon$.

Problem 4.45 **

Suppose that X is a compact topological space and $\mu\colon \mathcal{B}(X) \longrightarrow [0, +\infty)$ is an additive set function such that

$$\mu(A) = \inf_{\substack{U \text{ is open} \\ A \subseteq U}} \mu(U) = \sup_{\substack{K \text{ is compact} \\ K \subseteq A}} \mu(K) \qquad \forall\, A \in \mathcal{B}(X).$$

Show that μ is σ-additive.

Problem 4.46 ***

Suppose that X is a compact metric space and μ is a finite nonatomic Borel measure on X. Show that there is a basis $\{U_n\}_{n \geqslant 1}$ for the metric topology of X such that $\mu(\partial U_n) = 0$ for all $n \geqslant 1$.

Problem 4.47 **
Suppose that X is a metric space and μ is a Borel measure on X. We say that μ is ***doubling*** if μ is finite on bounded sets and there exists $c > 0$ such that

$$\mu\big(B_{2r}(x)\big) \leqslant c\mu\big(B_r(x)\big) \qquad \forall\, x \in X,\ r > 0.$$

Suppose that μ is a doubling Borel measure on X. Show that there exist $t, \widehat{c} > 0$ such that

$$\frac{\mu(B_r(x))}{\mu(B_R(y))} \geqslant \widehat{c}\big(\tfrac{r}{R}\big)^t \qquad \forall\, x, y \in X,\ R \geqslant r > 0,\ x \in B_R(y).$$

Problem 4.48 *
Suppose that X is a metric space, μ is a Borel measure on X, $A \in \mathcal{B}(X)$ and $\mu(A) < +\infty$. Show that for a given $\varepsilon > 0$, we can find a closed set $C \subseteq A$ such that $\mu(A \setminus C) \leqslant \varepsilon$.

Problem 4.49 **
Let $A \subseteq \mathbb{R}^N$ be a Lebesgue measurable set such that $\lambda^N(A) > 0$ (λ^N being the Lebesgue measure on $\mathbb{R}^N$). Show that card $A = \mathfrak{c}$. (Compare Problem 4.11).

Problem 4.50 *
Suppose that $\{f_n\}_{n\geqslant 1} \subseteq C\big([0,1]\big)$ is a sequence of functions such that $f_n(x) \longrightarrow f(x)$ for almost all $x \in [0,1]$ and let $\vartheta \in (0,1)$. Show that there exists a compact set $K \subseteq [0,1]$ such that $\lambda(K) > \vartheta$ and $f|_K$ is continuous (λ being the Lebesgue measure on $\mathbb{R}$).

Problem 4.51 *
Show that in $\mathbb{R}$, a compact set with positive Lebesgue measure may have empty interior.

Problem 4.52 ***
Let $f\colon [0,1] \longrightarrow \mathbb{R}$ be a Lebesgue measurable function. Show that we can find a sequence of continuous functions $h_n\colon [0,1] \longrightarrow \mathbb{R}$ such that $h_n \xrightarrow{\lambda} f$ (λ being the Lebesgue measure on $\mathbb{R}$).

Problem 4.53 *
Let X and Y be two topological spaces such that

$$\Delta_Y \stackrel{\text{def}}{=} \{(y,y) : y \in Y\} \in \mathcal{B}(Y) \otimes \mathcal{B}(Y)$$

and let $f \colon X \longrightarrow Y$ be a Borel function. Show that

$$\operatorname{Gr} f \in \mathcal{B}(X) \otimes \mathcal{B}(Y).$$

Problem 4.54 *
Let (Ω, Σ) and $(Y, \mathcal{Y})$ be two measurable spaces and assume that $(Y, \mathcal{Y})$ is ***countably separated*** (i.e., there exists a sequence $\{C_n\}_{n \geqslant 1} \subseteq \mathcal{Y}$ such that for any $u, v \in Y$, we can find $n \geqslant 1$ such that $\chi_{C_n}(u) \neq \chi_{C_n}(v)$). Suppose that $f \colon \Omega \longrightarrow Y$ is measurable. Show that $\operatorname{Gr} f \in \Sigma \otimes \mathcal{Y}$.

Problem 4.55 *
Let (Ω, Σ, μ) be a σ-finite measure space, let $f \in L^1(\Omega)$ and suppose that

$$\int\limits_A f(\omega)\, d\mu \leqslant \lambda \qquad \forall\, A \in \Sigma,\ \mu(A) < +\infty$$

for some $\lambda \in \mathbb{R}$. Show that

$$\int\limits_\Omega f(\omega)\, d\mu \leqslant \lambda.$$

Problem 4.56 *
Let $C \subseteq L^p(0,1)$ (with $1 < p < +\infty$) be a bounded set. Show that C is uniformly integrable.

Problem 4.57 *
Suppose that (Ω, Σ) is a measurable space, X is a perfectly normal space and $f \colon \Omega \longrightarrow X$ is a function. Show that f is $\big(\Sigma, \mathcal{B}(X)\big)$-measurable if and only if for every continuous function $\varphi \colon X \longrightarrow \mathbb{R}$, the function $\varphi \circ f$ is Σ-measurable.

Problem 4.58 **
Let (Ω, Σ) be a measurable space and let X be a separable metric space. Suppose that $f \colon \Omega \longrightarrow X$ is Σ-measurable. Show that there exists a sequence of measurable functions $f_n \colon \Omega \longrightarrow X$ with at most countable values such that $f_n \rightrightarrows f$ in X.

Problem 4.59 **
Let $f, g\colon (0,1) \longrightarrow [0,+\infty)$ be two measurable functions. We say that f and g are ***equimeasurable*** if for all $\vartheta > 0$, we have $\lambda(\{f > \vartheta\}) = \lambda(\{g > \vartheta\})$ (λ being the Lebesgue measure on $\mathbb{R}$). Show the following:
(a) If f and g are equimeasurable, then

$$\int_0^1 f\,dx = \int_0^1 g\,dx.$$

(b) If $\xi\colon [0,+\infty) \longrightarrow [0,+\infty)$ is a Borel function and f, g are equimeasurable, then $\xi \circ f$ and $\xi \circ g$ are equimeasurable too.

Problem 4.60 ***
Let $f\colon (0,1) \longrightarrow [0,+\infty)$ be a measurable function. Show the following:
(a) There exists a unique function $f^*\colon (0,1) \longrightarrow [0,+\infty)$ which is decreasing, right continuous and the functions f and f^* are equimeasurable (see Problem 4.59).
(b) For every Lebesgue measurable set $A \subseteq (0,1)$, we have

$$\int_A f\,dx \leqslant \int_0^{\lambda(A)} f^*\,dx$$

and if $\xi\colon (0,1) \longrightarrow [0,+\infty)$ is decreasing, then

$$\int_0^1 f\xi\,dx \leqslant \int_0^1 f^*\xi\,dx.$$

Problem 4.61 ***
Suppose that X is a locally compact topological space, $\mathcal{B}(X)$ is its Borel σ-algebra, μ is a Radon measure on $\mathcal{B}(X)$ and $u \in L^1(X,\mu)$. We set

$$\nu(A) = \int_A u\,d\mu \qquad \forall\, A \in \mathcal{B}(X).$$

Show that $\nu \in M_b(X)$ (see Definition 4.22).

Problem 4.62 *

Let X be a topological space and let μ be a Radon measure on X. Show that, if ν is a measure on X and $\nu \ll \mu$, then ν is Radon too.

Problem 4.63 *

Suppose that X is a locally compact topological space, $\mu \in M_b(X)$ and $u \in C_c(X)$ are such that $|u(x)| \leqslant \xi$ for all $x \in \operatorname{supp} \mu$. Show that

$$\Big| \int_\Omega u \, d\mu \Big| \leqslant \xi \|\mu\|.$$

Problem 4.64 **

Suppose that X is a locally compact topological space, $\mathcal{B}(X)$ is its Borel σ-algebra and $\mu \colon \mathcal{B}(X) \longrightarrow [-\infty, +\infty]$ is a signed measure with compact support. Show that μ is finite.

Problem 4.65 ***

Let X be a locally compact topological space in which every open set is σ-compact (for example when X is strongly Lindelöff, in particular when X is second countable; see Proposition 2.164). Show that every Borel measure which is finite on compact sets is Radon.

Problem 4.66 **

Suppose that X is a locally compact topological space and $\{\mu_n\}_{n \geqslant 1}$ is a sequence of Radon measures defined on the Borel σ-algebra $\mathcal{B}(X)$. Suppose that

$$\int_X f \, d\mu_n \longrightarrow \int_X f \, d\mu \qquad \forall \, f \in C_c(X),$$

for some Radon measure μ defined on the Borel σ-algebra $\mathcal{B}(X)$. Show that $\mu(X) \leqslant \liminf_{n \to +\infty} \mu_n(X)$.

Problem 4.67 **

Suppose that X is a locally compact topological space and $\{\mu_n\}_{n \geqslant 1}$ is a sequence of Borel measures on X. Suppose that $\xi = \sup_{n \geqslant 1} \mu_n(X) < +\infty$ and

$$\int_X f \, d\mu_n \longrightarrow \int_X f \, d\mu \qquad \forall \, f \in C_c(X).$$

Show that $\mu(X) < +\infty$ and

$$\int_X h\, d\mu_n \longrightarrow \int_X h\, d\mu \qquad \forall\, h \in C_0(X).$$

Problem 4.68 **

Let X be a compact topological space. Show that a continuous linear functional φ on $C(X)$ is positive if and only if $\|\varphi\|_* = \varphi(1)$ (1 is the constant function on X identically equal to 1).

Problem 4.69 **

Suppose that X is an uncountable locally compact topological space. Show that $C_0(X)^* = M_b(X)$ is not separable.

Problem 4.70 ***

Suppose that X and Y are two compact metric spaces and $f\colon X \longrightarrow Y$ is a continuous surjection. Show that for every Radon measure ϑ on Y, we can find a Radon measure μ on X such that $\mu f^{-1} = \vartheta$ (see Theorem 4.18).

Problem 4.71 **

Suppose that X is a locally compact topological space and $\varphi \in C_0(X)^*$, $\varphi \geqslant 0$. Show that the following two statements are equivalent:
(a) $\varphi\big(\max\{f, h\}\big) = \max\big\{\varphi(f), \varphi(h)\big\}$ for all $f, h \in C_0(X)$;
(b) $\varphi = c\delta_x$ for some $c \geqslant 0$ and some $x \in X$.

Problem 4.72 ***

Suppose that X is a locally compact topological space, μ is a Radon measure on X and $\mathcal{Y}$ is a family of lower semicontinuous functions from X into $\overline{\mathbb{R}}_+ = [0, +\infty]$ which is upward directed (i.e., if $h_1, h_2 \in \mathcal{Y}$, then there exists $h \in \mathcal{Y}$ such that $h_1 \leqslant h$ and $h_2 \leqslant h$). Let $f = \sup\big\{h : h \in \mathcal{Y}\big\}$. Show that

$$\int_X f\, d\mu = \sup_{h \in \mathcal{Y}} \int_X h\, d\mu.$$

Problem 4.73 **

Suppose that X is a locally compact topological space, μ is a Radon measure on X and $f\colon X \longrightarrow \overline{\mathbb{R}}_+ = [0, +\infty]$ is a Borel function. Show that:

(a) we have

$$\int_X f\,d\mu = \inf\Big\{\int_X h\,d\mu : h \text{ is lower semicontinuous on } X,\ h \geqslant f\Big\}.$$

(b) we have

$$\int_X f\,d\mu = \sup\Big\{\int_X \eta\,d\mu : \eta \text{ is upper semicontinuous on } X,\ 0 \leqslant \eta \leqslant f\Big\}$$

(recall that η is upper semicontinuous if and only if $-\eta$ is lower semicontinuous).

Problem 4.74 *

Suppose that X is a locally compact topological space and μ is a signed Borel measure on X satisfying that for every $A \in \mathcal{B}(X)$ and every $\varepsilon > 0$, we can find a compact set $K \subseteq X$ and an open set $U \subseteq X$ such that $K \subseteq A \subseteq U$ and

$$|\mu(B)| < \varepsilon \qquad \forall\, B \in \mathcal{B}(X),\ B \subseteq U \setminus K.$$

Show that $\mu \in M_b(X)$ (see Definition 4.22). (Compare with Problem 4.19.)

Problem 4.75 ***

Suppose that X and Y are two separable metric spaces and $\xi\colon M_1^+(X) \otimes M_1^+(Y) \longrightarrow M_1^+(X \times Y)$ is the function defined by

$$\xi(m,\mu) \stackrel{\text{def}}{=} m \times \mu,$$

where $m \times \mu$ is the product of the two measures m and μ. Show that ξ is continuous when $M_1^+(X)$, $M_1^+(Y)$ and $M_1^+(X \times Y)$ are furnished with their respective weak topologies (see Definition 4.25).

Problem 4.76 **

Suppose that X is a metric space and $f\colon X \longrightarrow \mathbb{R}^* = \mathbb{R} \cup \{\pm\infty\}$ is a lower semicontinuous, bounded below (respectively, upper semicontinuous, bounded above) function. Show that the function $\xi\colon \big(M_1^+(X), w\big) \longrightarrow \mathbb{R}^*$ (see Definition 4.25), defined by

$$\xi(\mu) = \int_X f\,d\mu$$

is lower semicontinuous (respectively, upper semicontinuous).

Problem 4.77 **

Suppose that X is a metric space, $U \subseteq X$ is an open set, $C \subseteq X$ is a closed set and $\eta \in \mathbb{R}$. Show that:

(a) The set $D = \{\mu \in M_1^+(X) : \mu(U) > \eta\}$ is weakly open.

(b) The set $E = \{\mu \in M_1^+(X) : \mu(C) \geqslant \eta\}$ is weakly closed.

Problem 4.78 **

Suppose that X is a metric space, $\mathcal{B}(X)$ is its Borel σ-algebra and for each $A \in \mathcal{B}(X)$ we consider the function $\xi_A \colon M_1^+(X) \longrightarrow [0,1]$, defined by

$$\xi_A(\mu) \stackrel{\text{def}}{=} \mu(A).$$

We furnish $M_1^+(X)$ with the weak topology (see Definition 4.25). Show that ξ_A is Borel measurable.

Problem 4.79 *

Let X be a metric space and consider the space $C_b(X)$ furnished with the sup-norm and the space $M_1^+(X)$ furnished with the weak topology (see Definition 4.25). Show that the function $\gamma \colon C_b(X) \times M_1^+(X) \longrightarrow \mathbb{R}$, defined by

$$\gamma(u,\mu) \stackrel{\text{def}}{=} \int_X u \, d\mu$$

is jointly continuous.

Problem 4.80 ***

Let X be a metrizable space. Show that X is homeomorphic to a sequentially closed subspace of $M_1^+(X)$ with the weak topology (see Definition 4.25).

Problem 4.81 ***

Suppose that X and Y are two metrizable spaces and $f \colon X \longrightarrow Y$ is a homeomorphism into Y. Let $\vartheta \colon M_1^+(X) \longrightarrow M_1^+(Y)$ be defined by

$$\vartheta(\mu)(A) \stackrel{\text{def}}{=} \mu(f^{-1}(A)) \qquad \forall\, \mu \in M_1^+(X),\ A \in \mathcal{B}(Y).$$

Show that ϑ is a homeomorphism too.

Problem 4.82 ***

Let X be a locally compact metric space. A set $A \subseteq X$ is said to be ***bounded*** if $A \subseteq K$ for some compact set $K \subseteq X$. Let $\mathcal{B}(X)$ be the

Borel σ-algebra of X, let $\{\mu_n\}_{n\geqslant 1}$ be a sequence of Radon measures of X. We say that the sequence $\{\mu_n\}_{n\geqslant 1}$ ***converges vaguely*** to a Radon measure μ (denoted by $\mu_n \xrightarrow{v} \mu$), if

$$\int_X f\,d\mu_n \longrightarrow \int_X f\,d\mu \qquad \forall\, f \in C_c(X).$$

Show that the following two statements are equivalent:
(a) $\mu_n \xrightarrow{v} \mu$;
(b) $\mu_n(A) \longrightarrow \mu(A)$ for all bounded set $A \in \mathcal{B}(X)$ with $\mu(\partial A) = 0$.

Problem 4.83 ***

Let X be a Polish space and let $C \subseteq M_1^+(X)$. Show that C is uniformly tight if and only if there exists a function $\varphi\colon X \longrightarrow [0, +\infty]$ such that for every $\lambda > 0$, the set $\{\varphi \leqslant \lambda\}$ is compact in X (i.e., φ is ***inf-compact***) and

$$\sup_{\mu \in C} \int_X \varphi(x)\,d\mu \;<\; +\infty.$$

Problem 4.84 **

Suppose that $\{\mu_n\}_{n\geqslant 1}$ is a sequence of probability measures defined on $\mathcal{B}(\mathbb{R}^N)$ and

$$\int_{\mathbb{R}^N} f(x)\,d\mu_n \longrightarrow \int_{\mathbb{R}^N} f(x)\,d\mu \qquad \forall\, f \in C_c^\infty(\mathbb{R}^N)$$

for some probability measure μ defined on $\mathcal{B}(\mathbb{R}^N)$.

Show that the set $\big\{\mu_n : n \geqslant 1\big\} \cup \{\mu\}$ is uniformly tight and $\mu_n \xrightarrow{w} \mu$ in $M_1^+(\mathbb{R}^N)$.

Problem 4.85 ***

Suppose that X and Y are two separable metric spaces and $\big\{\vartheta_n\colon X \longrightarrow Y\big\}_{n\geqslant 1}$ is a sequence of Borel functions such that $\vartheta_n \longrightarrow \vartheta$ uniformly on compact sets and ϑ is continuous. Let the sequence $\{\mu_n\}_{n\geqslant 1} \subseteq M_1^+(X)$ be uniformly tight and $\mu_n \xrightarrow{w} \mu$. Show that $\mu_n\vartheta_n^{-1} \xrightarrow{w} \mu\vartheta^{-1}$ in $M_1^+(Y)$.

Problem 4.86 ⋆
Suppose that X_1, X_2 and Y are three separable metric spaces, $\vartheta_1 \colon Y \longrightarrow X_1$ and $\vartheta_2 \colon Y \longrightarrow X_2$ are two continuous functions such that $\vartheta = (\vartheta_1, \vartheta_2) \colon Y \longrightarrow X_1 \times X_2$ is proper (i.e., the inverse image of a compact set is compact) and $E \subseteq M_1^+(Y)$ is such that the sets $E_1 = E\vartheta_1^{-1} \subseteq M_1^+(X_1)$ and $E_2 = E\vartheta_2^{-1} \subseteq M_1^+(X_2)$ are uniformly tight. Show that E is uniformly tight.

Problem 4.87 ⋆⋆
Suppose that X is a metric space, $\mu, \nu \in M_1^+(X)$ and

$$\int\limits_X f\, d\mu = \int\limits_X f\, d\nu \qquad \forall\, f \in UC_b(X).$$

Show that $\mu = \nu$ (see also Problem 4.38).

Problem 4.88 ⋆⋆
Suppose that X and Y are two separable metric spaces, $m \colon X \times \mathcal{B}(Y) \longrightarrow [0,1]$ is a function such that the function $x \longmapsto m(x, \cdot)$ is continuous from X into $M_1^+(Y)$ furnished with the weak topology (see Definition 4.25) and $f \in C_b(X \times Y)$. Show that the function

$$x \longmapsto h(x) = \int\limits_Y f(x, y)\, m(x, dy)$$

is continuous.

Problem 4.89 ⋆
Suppose that X is a separable metric space and $C \subseteq X$ is a nonempty closed set. Let

$$M_1^+(C) \stackrel{\text{def}}{=} \{\mu \in M_1^+(X) : \operatorname{supp} \mu \subseteq C\}.$$

Show that the set $M_1^+(C)$ is closed in $M_1^+(X)$ when the latter is furnished with the weak topology (see Definition 4.25).

Problem 4.90 ⋆⋆
Let X be a separable metric group. For any two measures $\mu, \nu \in M_1^+(X)$, the convolution $\mu \star \nu$ is defined to be the set function

$$(\mu \star \nu)(A) \stackrel{\text{def}}{=} \int\limits_X \mu(Ax^{-1})\, d\nu(x) \qquad \forall\, A \in \mathcal{B}(X).$$

Show that $\star\colon M_1^+(X) \times M_1^+(X) \longrightarrow M_1^+(X)$ and it is continuous ($M_1^+(X)$ is furnished with the weak topology; see Definition 4.25).

Problem 4.91 ***

Suppose that X is a separable metric space and $\mathcal{Y} \subseteq 2^X$ is such that $\sigma(\mathcal{Y}) = \mathcal{B}(X)$ and $\mathcal{Y}$ is closed under finite intersections (for example $\mathcal{Y}$ can be the family of closed subsets of X). For every $A \in \mathcal{B}(X)$, let $\eta_A\colon M_1^+(X) \longrightarrow [0,1]$ be the function, defined by

$$\eta_A(\mu) \stackrel{\text{def}}{=} \mu(A) \qquad \forall\, \mu \in M_1^+(X).$$

Show that

$$\mathcal{B}\big(M_1^+(X)\big) = \sigma\Big(\bigcup_{A\in\mathcal{Y}} \eta_A^{-1}\big(\mathcal{B}(\mathbb{R})\big)\Big)$$

($M_1^+(X)$ is furnished with the weak topology; see Definition 4.25).

Problem 4.92 **

Suppose that (Ω, Σ) is a measurable space, X is a Polish space and $F\colon \Omega \longrightarrow 2^X \setminus \{\emptyset\}$ is a graph measurable multifunction. Let

$$S(\omega) \stackrel{\text{def}}{=} \big\{\mu \in M_1^+(X) :\ \mu\big(F(\omega)\big) = 1\big\} \qquad \forall\, \omega \in \Omega.$$

Show that $S\colon \Omega \longrightarrow 2^{M_1^+(X)} \setminus \{\emptyset\}$ is graph measurable ($M_1^+(X)$ is furnished with the weak topology; see Definition 4.25).

Problem 4.93 **

Suppose that X and Y are separable metric spaces, $\mathcal{Y} \subseteq \mathcal{B}(Y)$ is a family such that $\mathcal{Y}$ is closed under finite intersections and $\sigma(\mathcal{Y}) = \mathcal{B}(X)$ and $\xi\colon X \longrightarrow M_1^+(X)$ is a function (spaces $M_1^+(X)$ and $M_1^+(Y)$ are furnished with the weak topologies; see Definition 4.25). Show that ξ is Borel measurable if and only if for every $A \in \mathcal{Y}$, the function $\vartheta_A\colon X \longrightarrow [0,1]$, defined by

$$\vartheta_A(x) \stackrel{\text{def}}{=} \xi(x)(A)$$

is Borel measurable.

Problem 4.94 **

Suppose that X and Y are separable metric spaces, $\xi\colon X \longrightarrow M_1^+(Y)$ is a continuous function ($M_1^+(Y)$ is furnished with the weak topology; see Definition 4.25). Show that for every function $f\colon X \times Y \longrightarrow \mathbb{R}^* =$

$\mathbb{R} \cup \{+\infty\}$, which is lower semicontinuous and bounded below, the function

$$X \ni x \longmapsto \int_Y f(x,y)\xi(x,dy)$$

is lower semicontinuous and bounded below.

Problem 4.95 ***

Let (X, Σ) be a separable measurable space. Show that there exists a subspace A of $\{0,1\}^{\mathbb{N}}$ such that (X, Σ) and $\big(A, \mathcal{B}(A)\big)$ are isomorphic.

Problem 4.96 **

Let X and Y be two topological spaces and let $f\colon X \longrightarrow Y$ be a function such that $\operatorname{Gr} f$ is a Souslin subset of $X \times Y$. Show that f is Borel function.

Problem 4.97 **

Let X be a Borel space (see Definition 4.40). Show that $M_1^+(X)$ equipped with the weak topology (see Definition 4.25) is a Borel space too.

Problem 4.98 **

Suppose that X is a Polish space and $\{\mu_n\}_{n\geqslant 1} \subseteq M_1^+(X)$ is a sequence such that $\mu_n \xrightarrow{w} \mu$ for some $\mu \in M_1^+(X)$. Suppose that $f\colon X \longrightarrow \mathbb{R}$ is a continuous function, $g\colon X \longrightarrow (-\infty, +\infty]$ is a lower semicontinuous function and

$$\lim_{k\to+\infty} \sup_{n\geqslant 1} \int_{\{|f|\geqslant k\}} |f|\, d\mu_n = 0 \quad \text{and} \quad \lim_{k\to+\infty} \sup_{n\geqslant 1} \int_{\{g^-\geqslant k\}} g^-\, d\mu_n = 0.$$

Show that

$$\lim_{n\to+\infty} \int_X f\, d\mu_n = \int_X f\, d\mu \quad \text{and} \quad \liminf_{n\to+\infty} \int_X g\, d\mu_n \geqslant \int_X g\, d\mu > -\infty.$$

Problem 4.99 **

Suppose that X is a Polish space and $\{\mu_n\}_{n\geqslant 1} \subseteq M_1^+(X)$ is a sequence such that $\mu_n \xrightarrow{w} \mu$ for some $\mu \in M_1^+(X)$. Suppose that $f \in C(X) \cap L^1(X, \mu_n)$ for all $n \geqslant 1$, $f \geqslant 0$ and

$$\limsup_{n\to+\infty} \int_X f\, d\mu_n \leqslant \int_X f\, d\mu < +\infty.$$

Show that

$$\lim_{k\to+\infty} \sup_{n\geqslant 1} \int\limits_{\{f\geqslant k\}} f\, d\mu_n \;=\; 0.$$

Problem 4.100 *
Suppose that X and Y are two Polish spaces, $A \in \mathcal{B}(X)$, $f\colon X \longrightarrow Y$ is a Borel measurable, injective function, $C = f(A) \subseteq Y$ and $C \in \mathcal{B}(Y)$. Show that $f^{-1}\colon C \longrightarrow X$ is Borel measurable.

Problem 4.101 **
Let (X,Σ) be a measurable space such that there exists a Polish space Z and a Souslin (analytic) set $A \subseteq Z$ such that the measurable spaces (X,Σ) and $\big(A,\mathcal{B}(A)\big)$ are isomorphic (see Definition 4.39). Let Y be a Polish space and let $f\colon X \longrightarrow Y$ be a Σ-measurable function. Show that for every $C \in \Sigma$, the set $f(C)$ is Souslin (analytic; see Remark 2.157).

Problem 4.102 **
Suppose that X is a Borel space (see Definition 4.40) and $f\colon X \longrightarrow \mathbb{R}^* = \mathbb{R} \cup \{\pm\infty\}$ is a Borel measurable function. We defined the function $\gamma_f\colon M_1^+(X) \longrightarrow \mathbb{R}^*$, by

$$\gamma_f(\mu) \;\stackrel{\mathrm{def}}{=}\; \int\limits_X f\, d\mu.$$

Show that this function is Borel measurable when $M_1^+(X)$ is furnished with the weak topology (see Definition 4.25).

Problem 4.103 **
Let $f\colon \mathbb{R}^N \longrightarrow \mathbb{R}^M$ be a Borel measurable map and let $D \in \mathcal{B}(\mathbb{R}^N)$. Show that $f(D)$ is a Souslin set.

Problem 4.104 **
Let X be a Souslin space which is completely regular (see Problem 2.42). Show that X is perfectly normal.

Problem 4.105 **
Let X be a topological space and $A \in \mathcal{B}a(X)$. Show that there exists a sequence $\{f_k\}_{k\geqslant 1} \subseteq C(X)$ and a set $B \in \mathcal{B}(\mathbb{R}^\infty)$ such that

$$(\star) \qquad A \;=\; \big\{x \in X:\; \{f_k(x):\; k \geqslant 1\} \in B\big\}.$$

In addition every set of the form $(\star)$ is Baire and we can take $\{f_k\}_{k\geqslant 1} \subseteq C_b(X)$.

Problem 4.106 *
Let X be a topological space and let μ be a Baire measure on X. Show that for every $B \in \mathcal{B}a(X)$ and every $\varepsilon > 0$, there exists a continuous function $\xi\colon X \longrightarrow [0,1]$ such that

$$\Big|\int_X \xi\, d\mu - \mu(B)\Big| < \varepsilon.$$

Problem 4.107 **
Let (Ω, Σ) be a measurable space and let (X, d_X) and (Y, d_Y) be two metric spaces with X being separable. Suppose that D is a countable, dense subset of X and let $f\colon \Omega \times X \longrightarrow Y$ be a map such that:
(i) for every $x \in D$, the function $\omega \longmapsto f(\omega, x)$ is measurable;
(ii) for every $\omega \in \Omega$, the function $x \longmapsto f(\omega, x)$ is continuous.
Show that f is jointly measurable.

Problem 4.108 *
Suppose that Ω is a set, $\mathcal{A}$ is a nonempty subset of 2^Ω, X is a metric space and $G\colon \Omega \longrightarrow 2^X \setminus \{\emptyset\}$ is a multifunction such that

$$G^-(C) \stackrel{\text{def}}{=} \{\omega \in \Omega:\ G(\omega) \cap C \neq \emptyset\} \in \mathcal{A} \qquad \forall\, C \subseteq X,\ C \text{ closed}.$$

Suppose that $\mathcal{A}$ is closed under countable unions. Show that for every open set $U \subseteq X$, we have $G^-(U) \in \mathcal{A}$.

Problem 4.109 **
Suppose that (Ω, Σ) is a measurable space, (X, d_X) is a metric space and $\{f_n\colon \Omega \longrightarrow X\}_{n\geqslant 1}$ is a sequence of $(\Sigma, \mathcal{B}(X))$-measurable functions such that $f_n(\omega) \longrightarrow f(\omega)$ for all $\omega \in \Omega$. Show that f is $(\Sigma, \mathcal{B}(X))$-measurable.

Problem 4.110 *
Suppose that (X, Σ, μ) is a σ-finite measure space, X is a separable Banach space and let $F\colon \Omega \longrightarrow 2^X \setminus \{\emptyset\}$ is a graph measurable multifunction. For $1 \leqslant p \leqslant +\infty$, let

$$S_F^p \stackrel{\text{def}}{=} \{f \in L^p(\Omega; X):\ f(\omega) \in F(\omega)\ \mu\text{-almost everywhere on } \Omega\}.$$

Show that $S_F^p \neq \emptyset$ if and only if $\inf_{x \in F(\omega)} \|x\| = m(\omega) \leqslant h(\omega)$ μ-almost everywhere, with some $h \in L^p(\Omega)$.

Problem 4.111 **
Suppose that (Ω, Σ) is a complete measurable space (i.e., $\Sigma = \widehat{\Sigma}$; see Definition 4.45), X is a Souslin space, $\xi\colon \Omega \times X \longrightarrow \mathbb{R}^* = \mathbb{R} \cup \{\pm\infty\}$ is a jointly measurable function and $F\colon \Omega \longrightarrow 2^X \setminus \{\emptyset\}$ is a graph measurable multifunction. Let us set

$$m(\omega) \stackrel{\text{def}}{=} \inf_{x \in F(\omega)} \xi(\omega, x) \qquad \forall\, \omega \in \Omega.$$

Show that the function $m\colon \Omega \longrightarrow \overline{\mathbb{R}} = \mathbb{R} \cup \{+\infty\}$ is Σ-measurable.

Problem 4.112 *
Let X and Y be two Souslin spaces and let $f\colon X \times Y \longrightarrow \mathbb{R}$ be a Borel measurable and bounded below function (respectively, bounded above). We set

$$m(x) \stackrel{\text{def}}{=} \inf_{y \in Y} f(x, y) \quad \text{(respectively, } M(x) \;=\; \sup_{y \in Y} f(x, y)\text{)}.$$

Show that m (respectively, M) is measurable with respect to every Borel measure on X.

Problem 4.113 *
Suppose that (Ω, Σ) is a measurable space, X is a Polish space, $\xi\colon \Omega \times X \longrightarrow \mathbb{R}^* = \mathbb{R} \cup \{\pm\infty\}$ is a jointly measurable function such that for all $\omega \in \Omega$, the function $x \longmapsto \xi(\omega, x)$ is upper semicontinuous and $F\colon \Omega \longrightarrow P_f(X)$ is a measurable multifunction. Let us set

$$m(\omega) \stackrel{\text{def}}{=} \inf_{x \in F(\omega)} \xi(\omega, x) \qquad \forall\, \omega \in \Omega,$$

Show that the function $m\colon \Omega \longrightarrow \overline{\mathbb{R}} = \mathbb{R} \cup \{+\infty\}$ is Σ-measurable.

Problem 4.114 ***
Suppose that (Ω, Σ, μ) is a σ-finite measure space, X is a separable Banach space, $F\colon \Omega \longrightarrow 2^X \setminus \{\emptyset\}$ is a graph measurable multifunction, $\xi\colon \Omega \times X \longrightarrow \mathbb{R}^* = \mathbb{R} \cup \{\pm\infty\}$ is a jointly measurable function,

$$I_\xi(u) \stackrel{\text{def}}{=} \int_\Omega \xi(\omega, u(\omega))\, d\mu$$

is well defined (maybe $+\infty$ or $-\infty$) for all $u \in S_F^p$ $(1 \leqslant p \leqslant +\infty$; see Problem 4.110) and there exists $u_0 \in S_F^p$ such that $I_\xi(u_0) > -\infty$. Show that

$$\sup_{u \in S_F^p} I_\xi(u) = \int_\Omega \sup_{x \in F(\omega)} \xi(\omega, x)\, d\mu.$$

Problem 4.115 *
Suppose that (Ω, Σ, μ) is a σ-finite measure space, X is a separable Banach space and $F\colon \Omega \longrightarrow 2^X \setminus \{\emptyset\}$ is a graph measurable multifunction (see Definition 4.49). Show that the set $S_F^p \subseteq L^p(\Omega; X)$ (see Problem 4.110) is bounded if and only if the function $\omega \longmapsto |F(\omega)|$ belongs to $L^p(\Omega)$ $(1 \leqslant p \leqslant +\infty)$.

Problem 4.116 *
Suppose that (Ω, Σ) is a measurable space, X is a σ-compact metric space and $F\colon \Omega \longrightarrow P_f(X)$ is a multifunction such that for every compact set $K \subseteq X$, we have

$$F^-(K) \stackrel{\text{def}}{=} \{\omega \in \Omega : F(\omega) \cap K \neq \emptyset\} \in \Sigma.$$

Show that F is measurable (see Definition 4.49(a)).

Problem 4.117 ***
Suppose that (Ω, Σ) is a complete measurable space, X is a Polish space, Y is a σ-compact Polish space and $F\colon \Omega \longrightarrow P_f(X \times Y)$ is a multifunction. Consider the multifunction $G\colon \Omega \times X \longrightarrow P_f(Y)$, defined by

$$G(\omega, x) \stackrel{\text{def}}{=} \{y \in Y : (x, y) \in F(\omega)\}.$$

Show that F is measurable if and only if G is measurable.

Problem 4.118 **
Suppose that (Ω, Σ) is a measurable space, X is a Polish space and $u\colon X \longrightarrow \Omega$ is a function such that
(a) for every $\omega \in \Omega$, we have $u^{-1}(\omega) \in P_f(X)$; and
(b) for every open set $V \subseteq X$, we have $u(V) \in \Sigma$.

Show that there exists a Σ-measurable function $f\colon \Omega \longrightarrow X$ such that $u(f(\omega)) = \omega$ for all $\omega \in \Omega$.

Problem 4.119 **
Suppose that (Ω, Σ) is a measurable space, X is a Polish space, Y is a metric space, $g\colon \Omega \times X \longrightarrow Y$ is a Carathéodory function, $U\colon \Omega \longrightarrow P_f(X)$ is a measurable multifunction, we set

$$G(\omega) \stackrel{\text{def}}{=} g\big(\omega, U(\omega)\big) \qquad \forall\, \omega \in \Omega.$$

Show that for all open sets $V \subseteq Y$, we have $G^-(V) \in \Sigma$.

Problem 4.120 *
Suppose (Ω, Σ) is a complete measurable space, T is a Souslin space, X is a metric space, $g\colon \Omega \times T \longrightarrow X$ is a jointly measurable function, $U\colon \Omega \longrightarrow 2^T \setminus \{\emptyset\}$ is a multifunction such that $\operatorname{Gr} U \in \Sigma \otimes \mathcal{B}(T)$ and $h\colon \Omega \longrightarrow X$ is a Σ-measurable function such that $h(\omega) \in g\big(\omega, U(\omega)\big)$ for all $\omega \in \Omega$. Show that there exists a $\big(\Sigma, \mathcal{B}(T)\big)$-measurable function $u\colon \Omega \longrightarrow T$ such that $u(\omega) \in U(\omega)$ and $h(\omega) = g\big(\omega, u(\omega)\big)$ for all $\omega \in \Omega$.

Problem 4.121 **
Suppose that (Ω, Σ) is a measurable space, X is a Polish space and $F\colon \Omega \longrightarrow 2^X \setminus \{\emptyset\}$ is a multifunction such that for every $x \in X$ and every $r > 0$, we have $F^-\big(\overline{B}_r(x)\big) \in \Sigma$ (here $\overline{B}_r(x) = \big\{x' \in X :\ d(x', x) \leqslant r\big\}$ with d being the metric of X). Show that F is measurable.

Problem 4.122 **
Suppose that (Ω, Σ) is a measurable space, X is a separable metric space, Y is a metric space, $f\colon \Omega \times X \longrightarrow Y$ is a Carathéodory function and $U \subseteq Y$ is an open set and for every $\omega \in \Omega$ there exists $x \in X$ such that $f(\omega, x) \in U$. Show that the multifunction $F\colon \Omega \longrightarrow 2^X \setminus \{\emptyset\}$, defined by $F(\omega) \stackrel{\text{def}}{=} \big\{x \in X :\ f(\omega, x) \in U\big\}$ is graph measurable.

Problem 4.123 **
Suppose that (Ω, Σ) is a measurable space, X is a separable metric space, Y is a topological space and $f\colon \Omega \times X \longrightarrow Y$ is a Carathéodory function. Let $U \subseteq Y$ be an open set and let us set

$$F(\omega) \stackrel{\text{def}}{=} \big\{x \in X :\ f(\omega, u) \in U\big\}.$$

Show that for all closed sets $C \subseteq X$, we have that

$$F^-(C) \stackrel{\text{def}}{=} \big\{\omega \in \Omega :\ F(\omega) \cap C \neq \emptyset\big\} \in \Sigma.$$

Problem 4.124 **
Let (Ω, Σ) be a measurable space and let X be a Polish space. Suppose that $f\colon \Omega \times X \longrightarrow \mathbb{R}$ is a Carathéodory function. Let

$$E(\omega) \stackrel{\text{def}}{=} \operatorname{epi} f(\omega, \cdot) = \big\{(x, \lambda) \in X \times \mathbb{R} :\ f(\omega, x) \leqslant \lambda\big\}$$

for all $\omega \in \Omega$. Show that for every open set $U \subseteq X \times \mathbb{R}$, we have

$$E^{-}(U) \stackrel{\text{def}}{=} \big\{\omega \in \Omega :\ E(\omega) \cap U \neq \emptyset\big\} \in \Sigma.$$

Problem 4.125 ***
Let (Ω, Σ, μ) be a σ-finite measure space and let X be a separable Banach space. Suppose that $F\colon \Omega \longrightarrow 2^X \setminus \{\emptyset\}$ is a graph measurable multifunction with open values. Suppose that $S_F^1 \neq \emptyset$ (see Problem 4.110) and let

$$\int\limits_{\Omega} F(\omega)\, d\mu \;=\; \Big\{ \int\limits_{\Omega} f(\omega)\, d\mu :\ f \in S_F^1 \Big\}.$$

Show that the set $\int\limits_{\Omega} F(\omega)\, d\mu$ is open in X.

Problem 4.126 **
Let (Ω, Σ, μ) be a finite measure space. Suppose that $\{f_n\}_{n \geqslant 1} \subseteq L^p(\Omega)$ (with $1 < p < +\infty$), $\|f_n\|_p \leqslant M$ for all $n \geqslant 1$ with some $M > 0$. Assume also that for all $A \in \Sigma$, we have

$$\int\limits_{A} f_n\, d\mu \;\longrightarrow\; \int\limits_{A} f\, d\mu,$$

for some $f \in L^p(\Omega)$ such that $\|f\|_p \leqslant M$. Show that $f_n \xrightarrow{w} f$ in $L^p(\Omega)$ (see Definition 4.74).

Problem 4.127 ***
Let (Ω, Σ, μ) be a σ-finite measure space and let $f\colon \Omega \longrightarrow \mathbb{R}$ be a Σ-measurable function such that $fg \in L^1(\Omega)$ for all $g \in L^p(\Omega)$ (with $1 \leqslant p < +\infty$). Show that $f \in L^{p'}(\Omega)$ (where $\frac{1}{p} + \frac{1}{p'} = 1$).

Problem 4.128 **
Suppose that (X, Σ, μ) is a measure space and $\{f_n\}_{n \geqslant 1} \subseteq L^1(X)$ is a sequence such that the limit

$$\lim_{n \to +\infty} \int\limits_{A} f_n\, d\mu \quad \text{exists and is finite for all } A \in \Sigma.$$

Show that for every $\varepsilon > 0$, we can find $\delta = \delta(\varepsilon) > 0$ such that

$$\text{if } \mu(A) < \delta, \quad \text{then } \int_A |f_n| \, d\mu < \varepsilon \text{ for all } n \geqslant 1.$$

Problem 4.129 ***
Suppose that (X, Σ, μ) is a finite measure space and $\{f_n\}_{n \geqslant 1} \subseteq L^1(X)$ is a sequence such that

(i) for all $A \in \Sigma$, the limit $\lim\limits_{n \to +\infty} \int_A f_n \, d\mu$ exists and is finite; and

(ii) $\sup\limits_{n \geqslant 1} \|f_n\|_1 = M < +\infty$.

Show that there exists $f \in L^1(X)$ such that $f_n \xrightarrow{w} f$ in $L^1(X)$.

Problem 4.130 **
Let (X, Σ, μ) be a measure space and let $\{\mu_n\}_{n \geqslant 1} \subseteq M(\Sigma)$ be a Vitali equicontinuous sequence (see Definition 4.71) such that $\mu_n \ll \mu$ for all $n \geqslant 1$. Show that the sequence $\{\mu_n\}_{n \geqslant 1}$ is uniformly μ-absolutely continuous.

Problem 4.131 **
Suppose that (X, Σ, μ) is a finite measure space and $\{\mu_n\}_{n \geqslant 1} \subseteq M(\Sigma)$ is a uniformly μ-absolutely continuous sequence (see Definition 4.71). Show that the sequence $\{\mu_n\}_{n \geqslant 1}$ is Vitali equicontinuous.

Problem 4.132 **
Find a sequence of functions $\{f_n\}_{n \geqslant 1} \subseteq L^p(a, b)$ (with $1 \leqslant p \leqslant +\infty$) such that $f_n \xrightarrow{w} 0$ but $f_n \not\longrightarrow 0$ in $L^p(a, b)$ (see Definition 4.74).

Problem 4.133 *
Let (Ω, Σ, μ) be a finite measure space and let $\{u_n\}_{n \geqslant 1} \subseteq L^1(\Omega)$ be a uniformly integrable sequence such that $u_n(\omega) \longrightarrow u(\omega)$ μ-almost everywhere in Ω. Show that $u_n \xrightarrow{w} u$ in $L^1(\Omega)$.

Problem 4.134 **
Let f be a real valued function defined in a neighbourhood of a point $x_0 \in \mathbb{R}$. If there exists a set $D \subseteq \mathbb{R}$ such that

$$\operatorname{dist}(x_0, D) = 0 \quad \text{and} \quad \lim_{\substack{x \to x_0 \\ x \in D}} f(x) = f(x_0),$$

then we say that f is an ***approximately continuous function at*** x_0. If f is approximately continuous at all points of its domain, we say that f is an ***approximately continuous function***. Show that a measurable, almost everywhere finite function is approximately continuous at almost every point.

Problem 4.135 **
Suppose that $f \in L^1(0,1)$ and for every $x \in (0,1)$ and every $\varepsilon > 0$, we can find an open interval $I_x \subseteq (0,1)$ such that

$$x \in I_x, \quad \lambda(I_x) < \varepsilon, \quad \text{and} \quad \int\limits_{I_x} f\, d\lambda = 0.$$

Show that for every open interval $I \subseteq (0,1)$, we have $\int_I f\, d\lambda = 0$.

Problem 4.136 *
Let λ be the Lebesgue measure on $\mathbb{R}$. Does there exist a Lebesgue measurable set $A \subseteq \mathbb{R}$ such that $\lambda(A \cap I) = \frac{1}{2}\lambda(I)$ for every bounded interval I? Justify your answer.

Problem 4.137 **
Let $f \in L^1(0,1)$ and let L_f be the Lebesgue set of f (see Definition 4.89). Show that $L_f = S$, where

$$S \stackrel{\text{def}}{=} \Big\{x \in [0,1] : \tfrac{d}{dx} \int\limits_0^x \big|f(t) - r\big|\, dt = \big|f(x) - r\big| \text{ for some } r \in \mathbb{Q}\Big\}.$$

Problem 4.138 ***
For any set $B \subseteq \mathbb{R}$, any $x \in \mathbb{R}$ (not necessarily in B), we introduce the following two quantities:

$$\limsup_{\lambda(I)\to 0} \frac{\lambda(B \cap I)}{\lambda(I)} \quad \text{is the upper density of } B \text{ at } x$$

and

$$\liminf_{\lambda(I)\to 0} \frac{\lambda(B \cap I)}{\lambda(I)} \quad \text{is the lower density of } B \text{ at } x,$$

where lim sup and lim inf are taken over all intervals I such that $x \in I$ (here λ denotes the Lebesgue measure on $\mathbb{R}$). If these two quantities are equal, then their common value is the ***density of*** B ***at the point*** x (cf. Definition 4.92).

Let $A, C \subseteq \mathbb{R}$ be two nonempty sets which are metrically separated (see Problem 4.34). Show that the density of each of sets A and C at almost all points of the other set is zero.

Problem 4.139 **
Show that the set of points at which the density of a Lebesgue measurable set $A \subseteq \mathbb{R}$ exists but it is not equal to 0 or 1, is of measure zero and of first category.
(see Problem 4.138).

Problem 4.140 *
Suppose that $A \subseteq [0,1]$ is a Lebesgue measurable set and that there exists $\vartheta > 0$ such that $\lambda\big(A \cap [a,b]\big) \geqslant \vartheta(b-a)$ for all $0 \leqslant a \leqslant b \leqslant 1$ (λ being the Lebesgue measure on $\mathbb{R}$). Show that $\lambda(A) = 1$. (Compare with Problem 4.13.)

Problem 4.141 *
Let μ be a finite Borel measure on $[0,1]$ which is absolutely continuous with respect to the Lebesgue measure λ on $\mathbb{R}$. Show that

$$\lim_{h \to 0} \frac{\mu([0,1] \cap (x-h, x+h))}{\lambda([0,1] \cap (x-h, x+h))} \quad \text{exists for } \lambda\text{-almost all } x \in [0,1].$$

Problem 4.142 *
Does there exist a Lebesgue measurable set $A \subseteq [0,1]$ such that

$$\lambda\big(A \cap [a,b]\big) = \frac{b-a}{2} \quad \forall\, 0 \leqslant a < b \leqslant 1$$

(λ being the Lebesgue measure on $\mathbb{R}$).

Problem 4.143 *
Let $f\colon \mathbb{R} \longrightarrow \mathbb{R}$ be a Lipschitz continuous function and let $g\colon [a,b] \longrightarrow \mathbb{R}$ be an absolutely continuous function. Show that $f \circ g\colon [a,b] \longrightarrow \mathbb{R}$ is absolutely continuous.

Problem 4.144 *
Let $A \subseteq [0,1]$ be a Lebesgue measurable set and consider the following sequence

$$f_n(x) = n \int_0^{\frac{1}{n}} \chi_A(x+t)\, dt \qquad \forall\, x \in \mathbb{R},\ n \geqslant 1.$$

Show that:
(a) $0 \leqslant f_n \leqslant 1$ for all $n \geqslant 1$;
(b) $f_n \in AC([0,1])$ for all $n \geqslant 1$;
(c) $f_n \longrightarrow \chi_A$ almost everywhere on $[0,1]$;
(d) $f_n \longrightarrow \chi_A$ in $L^1(\mathbb{R})$.

Problem 4.145 **
Suppose that $A \subseteq \mathbb{R}$ is a Lebesgue measurable set and $qA = A$ for all $q \in \mathbb{Q} \setminus \{0\}$ (here $qA = \{qx : x \in A\}$). Show that A or $\mathbb{R} \setminus A$ is Lebesgue-null.

Problem 4.146 **
Let $D \subseteq [0,1]$ be a Lebesgue-null set. Find an increasing and absolutely continuous function $f: [0,1] \longrightarrow \mathbb{R}$ such that $f'(x) = +\infty$ for all $x \in D$.

Problem 4.147 **
Show that the function $f: [0,1] \longrightarrow \mathbb{R}$, defined by

$$f(x) \stackrel{\text{def}}{=} \begin{cases} 0 & \text{if } x = 0, \\ x^2 \cos \frac{1}{x^2} & \text{if } 0 < x \leqslant 1 \end{cases}$$

is everywhere differentiable but it is not of bounded variation.

Problem 4.148 *
Let $f: [a,b] \longrightarrow \mathbb{R}$ be an absolutely continuous function. Show that f is constant if and only if $f'(x) = 0$ for almost all $x \in [a,b]$.

Problem 4.149 **
Let $f: \mathbb{R} \longrightarrow \mathbb{R}$ be a Lipschitz continuous function and let $D \subseteq \mathbb{R}$ be a Lebesgue-null set. Show that $f'(x) = 0$ for almost all $x \in f^{-1}(D)$.

Problem 4.150 **
Assume that $\{u_n: [a,b] \longrightarrow \mathbb{R}\}_{n \geqslant 1}$ is a sequence of functions such that $u_n(x) \longrightarrow u(x)$ for all $x \in [a,b]$. Show that $\operatorname{Var} u \leqslant \liminf_{n \to +\infty} \operatorname{Var} u_n$.

Problem 4.151 **
Let $f: [a,b] \longrightarrow \mathbb{R}$ be a differentiable function. Show that f is absolutely continuous if and only if f is of bounded variation.

Problem 4.152 *
Suppose that $u\colon [a,b] \longrightarrow \mathbb{R}$ is a continuous function and $\vartheta < \operatorname{Var} u$ ($\operatorname{Var} u$ can be 0 or $+\infty$). Show that there exists $\delta > 0$ such that

$$\sum_{k=1}^{n} \left|u(x_k) - u(x_{k-1})\right| > \vartheta,$$

for every partition $a = x_0 < x_1 < \ldots < x_n = b$ such that $\max_{1\leqslant k\leqslant n} |x_k - x_{k-1}| < \delta$.

Problem 4.153 *
Let $u \in BV([a,b]) \cap C([a,b])$. Show that the set

$$A \stackrel{\text{def}}{=} \{y \in \mathbb{R} : \operatorname{card} u^{-1}(\{y\} = +\infty)\}$$

is Lebesgue-null.

Problem 4.154 **
Let $\{u_n\colon [0,1] \longrightarrow \mathbb{R}\}_{n\geqslant 1}$ be a sequence of absolutely continuous functions such that $u_n(t) \longrightarrow u(t)$ for all $t \in [0,1]$ and $\{u_n'\}_{n\geqslant 1} \subseteq L^1(0,1)$ is uniformly integrable. Show that $u \in AC([0,1])$.

Problem 4.155 *
Suppose that $f\colon [a,b] \longrightarrow \mathbb{R}$ is absolutely continuous and $p \geqslant 1$. Show that $|f|^p\colon [a,b] \longrightarrow \mathbb{R}$ is absolutely continuous too.

Problem 4.156 **
Let $A, C \subseteq \mathbb{R}^N$ be two Borel sets. Show that:
(a) If $A \subseteq C$, then $\text{H-}\dim A \leqslant \text{H-}\dim C$.
(b) $\text{H-}\dim(A \cup C) = \max\{\text{H-}\dim A,\ \text{H-}\dim C\}$.

Problem 4.157 **
Let $\gamma\colon [a,b] \longrightarrow \mathbb{R}^N$ be a continuous curve. Show that

$$\left\|\gamma(b) - \gamma(a)\right\| \leqslant H^1(\gamma[a,b]).$$

Problem 4.158 **
Let $\gamma\colon [a,b] \longrightarrow \mathbb{R}^N$ be a Lipschitz injective curve (a ***simple curve***). Show that $H^1(\gamma([a,b])) = \operatorname{Var}\gamma < +\infty$.

Problem 4.159 **
Suppose that $C \subseteq \mathbb{R}^N$ is a connected compact set and $0 < r < \frac{1}{2}\operatorname{diam} C$. Show that $H^1\big(C \cap \overline{B}_r(x)\big) \geqslant r$ for all $x \in C$.

Problem 4.160 **
Let $(\lambda^N)^*$ denote the Lebesgue outer measure on $\mathbb{R}^N$. Show that there exists $\widehat{c} > 0$ such that $\widehat{c}H^N(A) \leqslant (\lambda^N)^*(A) \leqslant H^N(A)$ for all $A \subseteq \mathbb{R}^N$.

Problem 4.161 **
Let $A \subseteq \mathbb{R}^N$ be a set of positive outer measure and let $f: \mathbb{R}^N \longrightarrow \mathbb{R}^M$ be a function. Let $\operatorname{Gr} f|_A \stackrel{\text{def}}{=} \big\{\big(x, f(x)\big) : x \in A\big\}$. Show that $\text{H-}\dim \operatorname{Gr} f|_A \geqslant N$ and if f is Lipschitz, then $\text{H-}\dim \operatorname{Gr} f|_A = N$.

Problem 4.162 ***
Show that for every connected set $A \subseteq \mathbb{R}^N$, we have $H^1(A) \geqslant \operatorname{diam} A$.

Problem 4.163 **
Show that, if $A \subseteq \mathbb{R}^N$ has $\text{H-}\dim A < 1$, then A is totally disconnected (i.e., all connected components are singletons).

Problem 4.164 **
Suppose that (Ω, Σ) is a measurable space, X is a compact metric space, Y is a separable metric space and $f: \Omega \times X \longrightarrow Y$ is a Carathéodory function. Show that the function $\widehat{f}: \Omega \longrightarrow C(X;Y)$, defined by

$$\widehat{f}(\omega)(\cdot) \stackrel{\text{def}}{=} f(\omega, \cdot) \qquad \forall\, \omega \in \Omega$$

is measurable when $C(X;Y)$ is furnished with the uniform convergence topology (i.e., the metric topology induced by the metric $\widehat{d}(h,g) = \max_{x \in X} d_Y\big(h(x), g(x)\big)$ for all $h, g \in C(X;Y)$ and with d_Y being the metric on Y; this topology is the c-topology; see Definition 2.174 and Theorem 2.183).

Problem 4.165 *
Suppose that (Ω, Σ), X, Y and $C(X;Y)$ are as Problem 4.164 and that $\widehat{f}: \Omega \longrightarrow C(X;Y)$ is a Σ-measurable function. Show that the function

$$\Omega \times X \ni (\omega, x) \longmapsto f(\omega, x) = \widehat{f}(\omega)(x) \in Y$$

is Carathéodory.

Problem 4.166 *
Let X be a locally compact topological space. For every $f \in C_c(X)$, we consider the linear functional $\varphi_f \colon M_b(X) \longrightarrow \mathbb{R}$, defined by

$$\varphi_f(\mu) \stackrel{\text{def}}{=} \mu(f) = \int_X f \, d\mu \qquad \forall \, \mu \in M_b(X).$$

Then we consider the weak (initial) topology on $M_b(X)$ introduced by this family of functions (see Definition 2.62). This topology is known among probabilities and measures theorists as ***vague topology*** (see also Problem 4.82) and among functional analysts as w^****-topology*** (see Definition 5.63). We will denote it by w^*. Suppose that $E \subseteq M_b(X)$. Show that the following statements are equivalent:
(a) $\overline{E}^{w^*}$ is w^*-compact;
(b) for every compact set $K \subseteq X$, there exists $\widehat{\eta}_K > 0$ such that $|\mu|(K) \leqslant \widehat{\eta}_K$ for all $\mu \in E$.

Problem 4.167 **
Suppose that X is a locally compact topological space, $\vartheta > 0$ and consider the set $E \stackrel{\text{def}}{=} \{\mu \in M_b(X) : \|\mu\| \leqslant \vartheta\}$. Show that E is w^*-compact.
Is the set $D \stackrel{\text{def}}{=} \{\mu \in M_b(X) : \|\mu\| = \vartheta\}$ w^*-closed? Justify your answer.
Finally show that if X is compact, then

$$D_+ \stackrel{\text{def}}{=} \{\mu \in M_b(X) : \mu \geqslant 0, \ \|\mu\| = \vartheta\}$$

is w^*-compact (for the definition of the w^*-topology see Problem 4.166).

Problem 4.168 ***
Let X be a locally compact topological space. Show that the function $\sigma \colon X \longrightarrow M_b^+(X)$, defined by $\sigma(x) = \delta_x$ is a homeomorphism of X into $M_b^+(X)$ ($M_b^+(X)$ is endowed with the relative w^*-topology).
Moreover, show that if X is not compact, then for every $\varepsilon > 0$ and $f \in C_c(X)$, there is a compact set $K \subseteq X$ such that $|\delta_x(f)| < \varepsilon$ for all $x \notin K$.

Problem 4.169 **
Let X be a locally compact topological space. Show that $M_b^+(X)$ with the relative w^*-topology (see Problem 4.166) is separable metrizable if and only if X is second countable.

Problem 4.170 **

Suppose that X is a separable metric space and $\{\mu_n\}_{n\geqslant 1} \subseteq M_1^+(X)$ is such that $\mu_n \xrightarrow{w^*} \mu$ (see Problem 4.166). Show that

$$\operatorname{supp}\mu \subseteq \liminf_{n\to+\infty} \operatorname{supp}\mu_n$$

(see Problem 1.171).

Problem 4.171 **

Suppose that X is a locally compact separable metric space and $\{\mu_n\}_{n\geqslant 1} \subseteq M_b(X)$ is a sequence such that $\mu_n \xrightarrow{w^*} \mu$ in $M_b(X)$ for some $\mu \in M_b(X)$ (see Problem 4.166). Show that for every lower semicontinuous function $h\colon X \longrightarrow [0,+\infty)$, we have

$$\int_X h\,d\mu \leqslant \liminf_{n\to+\infty} \int_X h\,d\mu_n.$$

Problem 4.172 **

Suppose that $\{\mu_n\}_{n\geqslant 1} \subseteq M_b(\mathbb{R})$ is a sequence of measures, $\mu \in M_b(\mathbb{R})$, $\varphi_n(x) = \mu_n\big((-\infty, x]\big)$ for $n \geqslant 1$ and $\varphi(x) = \mu\big((-\infty, x]\big)$. Suppose that $\|\mu_n\| \leqslant M$ for all $n \geqslant 1$ and for all x, continuity points of φ, we have $\varphi_n(x) \longrightarrow \varphi(x)$. Show that $\mu_n \xrightarrow{w^*} \mu$ (see Problem 4.166).

4.3 Solutions

Solution of Problem 4.1

Let $\{q_n\}_{n\geqslant 1}$ be the rational points of $[-1,1]$. Let

$$A_n \;=\; A+q_n \;=\; \{x+q_n:\; x\in A\}.$$

From the translation invariance of the Lebesgue measure, we have

$$\lambda(A_n) \;=\; \lambda(A) \;>\; 0$$

and clearly $A_n \subseteq [-1,2]$ for all $n\geqslant 1$. Then $\bigcup_{n\geqslant 1} A_n \subseteq [-1,2]$. Suppose that the sets A_n are pairwise disjoint, i.e., if $n\neq m$, then $A_n\cap A_m=\emptyset$. Then, by σ-additivity, we have

$$\sum_{n\geqslant 1}\lambda(A_n) \;=\; \lambda\Big(\bigcup_{n\geqslant 1} A_n\Big) \;\leqslant\; \lambda([-1,2]) \;=\; 3,$$

but

$$\lambda\Big(\bigcup_{n\geqslant 1} A_n\Big) \;=\; \sum_{n\geqslant 1}\lambda(A_n) \;=\; \lambda(A)\sum_{n\geqslant 1} 1 \;=\; +\infty$$

(since $\lambda(A)>0$), a contradiction. This means that we can find $m,n\geqslant 1$ such that $A_m\cap A_n\neq\emptyset$. Let $a\in A_m\cap A_n$. Then

$$a \;=\; x+q_m \;=\; u+q_n \quad\text{with}\quad x,u\in A,\; x\neq u,$$

so $x-u=q_n-q_m\in\mathbb{Q}$.

Solution of Problem 4.2

Let $\mathcal{B}(X)$ be the Borel σ-algebra of X and let $\mathcal{Y}$ be the family of subsets of X which contains all open sets, all closed sets and which is closed under countable intersections and countable disjoint unions. Evidently $\mathcal{Y}\subseteq\mathcal{B}(X)$. Let

$$\mathcal{F} \;=\; \{A\in\mathcal{Y}:\; A^c\in\mathcal{Y}\}.$$

Then $\mathcal{F}\subseteq\mathcal{Y}\subseteq\mathcal{B}(X)$ and $\mathcal{F}$ contains all open and all closed sets in X. We will show that $\mathcal{F}$ is a σ-algebra and then we will have $\mathcal{F}=\mathcal{Y}=\mathcal{B}(X)$. We do this in steps.

(a) If $A, C \in \mathcal{F}$, then $A \setminus C \in \mathcal{F}$.

By definition $A^c, C^c \in \mathcal{F}$. Since $\mathcal{Y}$ is closed under finite intersections, we have that $A \setminus C = A \cap C^c \in \mathcal{Y}$. Also $\mathcal{Y}$ is closed under countable disjoint unions. Hence

$$(A \setminus C)^c = (A \cap C^c)^c = A^c \cup C = A^c \cup (A \cap C) \in \mathcal{Y}.$$

Therefore $A \setminus C \in \mathcal{F}$.

(b) $\mathcal{F}$ is closed under finite unions.

Let $A, C \in \mathcal{F}$. Then $A, C, A^c, C^c \in \mathcal{F}$ and so $(A \cup C)^c = A^c \cap C^c \in \mathcal{Y}$. From **(a)**, we have that

$$A \setminus C \in \mathcal{F}, \quad C \setminus A \in \mathcal{F} \quad \text{and} \quad A \cap C \in \mathcal{F}.$$

Hence the disjoint union

$$(A \setminus C) \cup (C \setminus A) \cup (A \cap C) = A \cup C \in \mathcal{F},$$

which proves that $A \cup C \in \mathcal{F}$ and by induction, we prove that $\mathcal{F}$ is closed under finite unions.

(c) $\mathcal{F}$ is an algebra of sets.

By part **(b)** and since $X, \emptyset \in \mathcal{F}$, it remains to show that $\mathcal{F}$ is closed under intersections. Let $A, C \in \mathcal{F}$. Then $A \cap C \in \mathcal{Y}$ and $(A \cap C)^c = A^c \cup C^c \in \mathcal{F}$ (see **(b)** and recall that $A^c, C^c \in \mathcal{F}$). Hence $A \cap C \in \mathcal{F}$ and by induction we show that $\mathcal{F}$ is closed under finite intersections, hence $\mathcal{F}$ is an algebra.

(d) $\mathcal{F}$ is a σ-algebra of sets.

Let $\{A_n\}_{n \geqslant 1} \subseteq \mathcal{F}$. Inductively we define

$$C_1 = A_1, \quad C_n = A_n \setminus \Big(\bigcup_{k=1}^{n-1} A_k\Big) \qquad \forall\, n > 1.$$

Then from **(c)**, we have

$$\bigcup_{k=1}^{n-1} A_k \in \mathcal{F} \qquad \forall\, n > 1$$

and so from **(a)**, it follows that

$$C_n = A_n \setminus \Big(\bigcup_{k=1}^{n-1} A_k\Big) \in \mathcal{F}.$$

Therefore $\{C_n\}_{n\geqslant 1}$ is a sequence of disjoint $\mathcal{F}$-sets and so

$$\bigcup_{n\geqslant 1} A_n = \bigcup_{n\geqslant 1} C_n \in \mathcal{F}.$$

On the other hand

$$\Big(\bigcup_{n\geqslant 1} A_n\Big)^c = \bigcap_{n\geqslant 1} A_n^c \in \mathcal{F}$$

and so it follows that $\bigcup_{n\geqslant 1} A_n \in \mathcal{F}$ and this proves that $\mathcal{F}$ is a σ-algebra.

Therefore we conclude that $\mathcal{F} = \mathcal{Y} = \mathcal{B}(X)$.

Solution of Problem 4.3

From Problem 1.79, we know that every closed set is G_δ (i.e., countable intersection of open sets). Hence every family of subsets of X that includes the open sets and is closed under countable intersections must contain also closed sets. Then use Problem 4.2 to conclude that the family of sets coincides with the Borel σ-algebra of X.

Solution of Problem 4.4

Recall that the finite intersections of the elements of $\mathcal{X}$ form a basis of the topology of X. So, $\sigma(\mathcal{X})$ contains a basis for the topology of X. Let $K \subseteq X$ be a compact G_δ-set. Then $K = \bigcap\limits_{n\geqslant 1} U_n$, with U_n being open sets. Since X is locally compact (see Definition 2.92), for every $n \geqslant 1$, we can find an open set V_n such that

$$K \subseteq V_n \subseteq U_n \quad \text{and} \quad V_n \text{ is a finite union of basic open sets.}$$

So $V_n \in \sigma(\mathcal{X})$ and thus

$$K = \bigcap_{n \geqslant 1} V_n \in \sigma(\mathcal{X}).$$

Invoking Theorem 4.4, we conclude that $\mathcal{B}a(X) \subseteq \sigma(X) \subseteq \mathcal{B}(X)$.

Solution of Problem 4.5

We will do the solution for the lower semicontinuous case, the other two cases can be treated similarly. For every $\lambda \in \mathbb{R}$ and every $n \geqslant 1$, we have

$$f|_{C_n}^{-1}((\lambda, +\infty)) \quad \text{is open in } C_n,$$

hence it has the form $C_n \cap U_n$ with $U_n \subseteq X$ being an open set. This means that

$$f|_{C_n}^{-1}((\lambda, +\infty)) \in \widehat{\mathcal{B}(X)}.$$

In addition, since μ is complete, $\mu(C_0) = 0$ and $f|_{C_0}^{-1}((\lambda, +\infty)) \subseteq C_0$, we infer that

$$f|_{C_0}^{-1}((\lambda, +\infty)) \in \widehat{\mathcal{B}(X)}$$

and it is μ-null. Finally, we have

$$f^{-1}((\lambda, +\infty)) = \bigcup_{n \geqslant 0} f|_{C_n}^{-1}((\lambda, +\infty)) \in \widehat{\mathcal{B}(X)},$$

so f is $\widehat{\mathcal{B}(X)}$-measurable.

Solution of Problem 4.6

We have

$$U = \bigcup_{n \geqslant 1} K_n,$$

with $K_n \subseteq U$ being compact for all $n \geqslant 1$. Since X is locally compact, by Proposition 2.94, for all $n \geqslant 1$, we can find an open set $V_n \subseteq X$ such that

$$\overline{V}_n \text{ is compact and } K_n \subseteq V_n \subseteq \overline{V}_n \subseteq U.$$

Invoking Problem 2.103, for all $n \geqslant 1$, we can find a continuous function $g_n : X \longrightarrow [0,1]$ such that

$$g_n|_{K_n} = 1 \quad \text{and} \quad g_n|_{V_n^c} = 0$$

(hence $g_n \in C_c(X)$ for all $n \geqslant 1$). Let

$$\widehat{K}_n = \{x \in X : g_n(x) \geqslant \tfrac{1}{2}\} \qquad \forall\, n \geqslant 1.$$

Evidently the sets $\widehat{K}_n$ are compact and

$$\widehat{K}_n = \bigcap_{m \geqslant 3} \{x \in X : g_n(x) > \tfrac{1}{2} - \tfrac{1}{m}\} \qquad \forall\, n \geqslant 1,$$

hence they are G_δ-sets and $K_n \subseteq \widehat{K}_n \subseteq U$ for all $n \geqslant 1$. So, $U = \bigcup_{n \geqslant 1} \widehat{K}_n$, which expresses U as a countable union of compact G_δ-sets and this by Theorem 4.4 implies that $U \in \mathcal{B}a(X)$.

Solution of Problem 4.7

Note that every compact set in X is also compact in X^*. Because in this case the compact sets generate the Borel σ-algebra of X, we see that

$$\mathcal{B}(X) \subseteq X \cap \mathcal{B}(X^*).$$

To show that the opposite inclusion also holds, it suffices to show that for every open $U \subseteq X^*$, the set $U \setminus \{\infty\} \in \mathcal{B}(X)$. Since by hypothesis X is σ-compact, by Proposition 2.100, we can find an increasing sequence $\{K_n\}_{n \geqslant 1}$ of compact sets on X such that

$$X = \bigcup_{n \geqslant 1} K_n \quad \text{and} \quad K_n \subseteq \operatorname{int} K_{n+1} \qquad \forall\, n \geqslant 1.$$

Then the set $U \cap K_{n+1}$ is open in X^* and relatively compact in X, hence it belongs in $\mathcal{B}(X)$. Note that $U \setminus \{\infty\} = \bigcup_{n \geqslant 1} (U \cap \operatorname{int} K_{n+1})$ and so $U \setminus \{\infty\} \in \mathcal{B}(X)$. This proves that $\mathcal{B}(X) = X \cap \mathcal{B}(X^*)$.

Solution of Problem 4.8

Let $\mathcal{Y}$ be the family of such sets $G \in \Sigma \otimes \mathcal{B}(X)$ for which there exists $\Sigma_0 \subseteq \Sigma$, a countably generated sub-σ-algebra of Σ such that $G \in \Sigma_0 \otimes \mathcal{B}(X)$. Evidently $\mathcal{Y} \neq \emptyset$ and it is closed under complementation. Also, if $\{G_n\}_{n\geqslant 1} \subseteq \mathcal{Y}$, then $G_n \in \Sigma_{0n} \otimes \mathcal{B}(X)$ with Σ_{0n} being a countably generated sub-σ-algebra of Σ (see Definition 3.14). Hence

$$\bigcup_{n\geqslant 1} G_n \in \sigma\Big(\bigcup_{n\geqslant 1} (\Sigma_{0n} \otimes \mathcal{B}(X))\Big) \subseteq \sigma\Big(\bigcup_{n\geqslant 1} \Sigma_{0n}\Big) \otimes \mathcal{B}(X)$$

and

$$\widehat{\Sigma}_0 = \sigma\Big(\bigcup_{n\geqslant 1} \Sigma_{0n}\Big)$$

is a countable generated sub-σ-algebra of Σ. Therefore $\mathcal{Y}$ is a σ-algebra containing all cylinders, therefore $\mathcal{Y} = \Sigma \otimes \mathcal{B}(X)$.

Solution of Problem 4.9

Let $x \in X$ and $U \in \mathcal{N}(x)$. Then from the local compactness of X and Problem 2.103, we know that we can find $f \in C_c(X)$ such that

$$f(x) = 1 \quad \text{and} \quad f(y) = 0 \qquad \forall\, y \in U^c.$$

Let us set

$$V = \{f > \tfrac{1}{2}\}.$$

This is an open Baire set and $x \in V \subseteq U$. Therefore the open Baire sets form a basis for the topology of X (see Definition 4.2).

Solution of Problem 4.10

Let $X = \mathbb{R}^2$ furnished with the topology which has as basis all sets of the form $[a, b) \times [c, d)$ with $a, b, c, d \in \mathbb{R}$. Clearly X endowed with this topology is a topological space (i.e., it is Hausdorff). Also $\mathcal{B}(X) = \mathcal{B}(\mathbb{R}^2)$ where $\mathbb{R}^2$ is the plane equipped with the usual Euclidean topology. Let C be the straight line on the plane passing from the origin and with slope -1, i.e.,

$$C = \{(x, y) \in \mathbb{R}^2 : x = -y\}.$$

Then $C \subseteq X$ is closed. For any $x \in \mathbb{R}$, the set $[x, x+1) \times [-x, -x+1)$ is open and intersects C at precisely one point $(x, -x) \in C$. Hence every point of C is open and so $\mathcal{B}(C) = 2^C$, hence $\mathcal{B}(C)$ is not a subset of $\mathcal{B}(X)$.

Solution of Problem 4.11

We know that λ is regular (see Theorem 4.11). So, without any loss of generality, we may assume that C is closed. Let $A = (0,1) \setminus C$ and let us set

$$\vartheta(t) = \lambda\big([0,t] \cap C\big) \qquad \forall\, t \in [0,1].$$

Then $\vartheta \colon [0,1] \longrightarrow \big[0, \lambda(C)\big]$ is a continuous surjection and is constant in any connected component of A, hence $\vartheta(A)$ is at most countable. Since $\vartheta(C)$ contains $\big[0, \lambda(C)\big] \setminus \vartheta(A)$, it follows that C has cardinality at least that of the continuum. Therefore the cardinality of C equals that of the continuum.

Solution of Problem 4.12

Let

$$G = \big\{(x, \eta) \in X \times \mathbb{R}_+ :\ u(x) \geqslant \eta\big\}.$$

Then, we have

$$\begin{aligned}
\int_0^{+\infty} \mu\big(\{x \in X :\ u(x) \geqslant \eta\}\big)\, d\eta &= \int_0^{+\infty} \mu\big(\{x \in X :\ (x,\eta) \in G\}\big)\, d\eta \\
&= \int_X \lambda\big(\{\eta \geqslant 0 :\ (x,\eta) \in G\}\big)\, d\mu(x) \\
&= \int_X \lambda\big([0, u(x)]\big)\, d\mu(x) = \int_X u\, d\mu
\end{aligned}$$

(λ being the Lebesgue measure).

Solution of Problem 4.13

Suppose that $\lambda(A) > 0$. Then we can find an integer n such that

$$\lambda\big(A \cap (n, n+1)\big) > 0.$$

By Theorem 4.11, for a given $\varepsilon \in \big(0, \lambda\big(A \cap (n, n+1)\big)\big)$, we can find an open set $U \subseteq (n, n+1)$ such that

$$A \cap (n, n+1) \subseteq U \subseteq (n, n+1) \quad \text{and} \quad \lambda(U) < \lambda\big(A \cap (n, n+1)\big) + \varepsilon.$$

We know that

$$U = \bigcup_{k \geqslant 1} (a_k, b_k),$$

where $\big\{(a_k, b_k)\big\}_{k \geqslant 1}$ is a sequence of disjoint intervals. Then we have

$$A \cap (n, n+1) = \bigcup_{k \geqslant 1} \big(A \cap (a_k, b_k)\big),$$

so, using the σ-additivity of λ and the assumption, we have

$$\begin{aligned} \lambda\big(A \cap (n, n+1)\big) &= \sum_{k \geqslant 1} \lambda\big(A \cap (a_k, b_k)\big) \leqslant \sum_{k \geqslant 1} \tfrac{1}{2}(b_k - a_k) \\ &= \tfrac{1}{2}\lambda(U) < \tfrac{1}{2}\big[\lambda\big(A \cap (n, n+1)\big) + \varepsilon\big] \end{aligned}$$

and thus

$$\lambda\big(A \cap (n, n+1)\big) < \varepsilon < \lambda\big(A \cap (n, n+1)\big),$$

a contradiction.

Solution of Problem 4.14

(a) From Theorem 4.11, we know that μ and ν are regular (see Definition 4.9). So, using regularity of μ and ν, for every $A \in \mathcal{B}(X)$, we have

$$\begin{aligned} \mu(A) &= \inf\big\{\mu(U) : U \subseteq X \text{ is open}, A \subseteq U\big\} \\ &= \inf\big\{\nu(U) : U \subseteq X \text{ is open}, A \subseteq U\big\} \\ &= \nu(A) \end{aligned}$$

(b) Since X is σ-compact (see Definition 2.99), every closed set is the union of an increasing sequence of compact sets. Then exploiting the regularity of μ and ν, for every $A \subseteq \mathcal{B}(X)$, we have

$$\begin{aligned} \mu(A) &= \sup\{\mu(K) : K \subseteq X \text{ is compact}, \ K \subseteq A\} \\ &= \sup\{\nu(K) : K \subseteq X \text{ is compact}, \ K \subseteq A\} \\ &= \nu(A). \end{aligned}$$

Solution of Problem 4.15

Since by hypothesis μ is σ-finite, we can find an increasing sequence $\{E_n\}_{n\geqslant 1}$ of Borel sets with

$$\mu(E_n) < +\infty \qquad \forall\, n \geqslant 1 \quad \text{and} \quad X = \bigcup_{n\geqslant 1} E_n.$$

Let $\mu_n \colon \mathcal{B}(X) \longrightarrow \mathbb{R}_+ = [0, +\infty)$ be the finite measure defined by

$$\mu_n(A) = \mu(A \cap E_n) \qquad \forall\, A \in \mathcal{B}(X).$$

Theorem 4.11 implies that μ_n is regular (see Definition 4.9) and so we can find a closed set $C_n \subseteq A$ such that

$$\mu_n(A \setminus C_n) < \tfrac{1}{n}.$$

Let us set

$$\widehat{C}_n = \bigcup_{k=1}^{n} C_k.$$

Evidently $\{\widehat{C}_n\}_{n\geqslant 1}$ is an increasing sequence of closed sets such that

$$\mu_n(A \setminus \widehat{C}_n) \leqslant \mu_n(A \setminus C_n) < \tfrac{1}{n}.$$

Moreover, for every $m \leqslant n$, we have $\mu_m \ll \mu_n$ (recall that the sequence $\{E_n\}_{n\geqslant 1}$ is increasing) and so

$$\mu_m(A \setminus \widehat{C}_n) < \tfrac{1}{n} \qquad \forall\, m \leqslant n.$$

Hence

$$\mu_m(A \setminus \bigcup_{n\geqslant 1} \widehat{C}_n) = 0 \qquad \forall\, m \geqslant 1$$

and so finally

$$\mu(A \setminus \bigcup_{n \geqslant 1} \widehat{C}_n) \;=\; \lim_{m \to +\infty} \mu_m(A \setminus \bigcup_{n \geqslant 1} \widehat{C}_n) \;=\; 0.$$

If $\mu(A) < +\infty$, then $\mu(\bigcup_{n \geqslant 1} \widehat{C}_n) < +\infty$ and so, since $\mu(\widehat{C}_n) \nearrow \mu(A)$ as $n \to +\infty$, we have

$$\mu(A \setminus \widehat{C}_n) \;=\; \mu(A) - \mu(\widehat{C}_n) \;\longrightarrow\; 0,$$

which proves the inner regularity of the set A.

Solution of Problem 4.16

First assume that $\mu(A) < +\infty$. Due to the outer regularity of μ (see Definition 4.9), for a given $\varepsilon > 0$, we can find an open set U such that $A \subseteq U$ and $\mu(U) < \mu(A) + \varepsilon$. Then by hypothesis there exists a compact set $K \subseteq U$ such that $\mu(U) < \mu(K) + \varepsilon$. Note that $\mu(U \setminus A) < \varepsilon$ and pick an open set V such that $U \setminus A \subseteq V$ and $\mu(V) < \varepsilon$ (outer regularity of μ). Let $C = K \setminus V$. Then C is compact, $C \subseteq A$ and

$$\mu(C) \;=\; \mu(K) - \mu(K \cap V) \;>\; \mu(U) - \varepsilon - \mu(V) \;\geqslant\; \mu(A) - 2\varepsilon,$$

so

$$\mu(A) \;=\; \sup\big\{\mu(D) : \; D \subseteq A, \; D \text{ is compact}\big\}.$$

Now assume that $\mu(A) = +\infty$. Then $A = \bigcup_{n \geqslant 1} A_n$, with an increasing sequence $\{A_n\}_{n \geqslant 1} \subseteq \mathcal{B}(X)$ such that $\mu(A_n) < +\infty$ for all $n \geqslant 1$. We have $\mu(A_n) \longrightarrow +\infty$ and so for every $k \in \mathbb{N}$, we can find $n_0 \geqslant 1$ such that $\mu(A_{n_0}) > k$. Then from the first part of the solution, we can find a compact set $K \subseteq A_{n_0}$ such that $\mu(K) > k$. Therefore $+\infty = \mu(A) = \sup\big\{\mu(D) : \; D \subseteq A, \; D \text{ is compact}\big\}$.

Solution of Problem 4.17

"$\Longrightarrow$": Let

$$s \;=\; \sum_{k=1}^{n} a_k \chi_{A_k} \quad \text{with} \quad a_k \geqslant 0,\ A_k \in \mathcal{B}(X)$$

be a positive simple function. Then

$$\int_X s\, d\nu \;\leqslant\; \int_X s\, d\mu.$$

For a given $u \in C_c(X)$, we can find a sequence of simple functions $\{s_n\}_{n\geqslant 1}$ such that $0 \leqslant s_n \leqslant u$ and

$$s_n(x) \nearrow u(x) \qquad \forall\, x \in X.$$

Then by the Lebesgue monotone convergence theorem (see Theorem 3.92), we have

$$\int_X u\, d\nu \;=\; \lim_{n\to+\infty} \int_X s_n\, d\nu \;\leqslant\; \lim_{n\to+\infty} \int_X s_n\, d\mu \;=\; \int_X u\, d\mu.$$

"$\Longleftarrow$": Let $K \subseteq X$ be a compact set. For a given $\varepsilon > 0$, we can find an open set U such that $K \subseteq U$ and

$$\mu(U) \;<\; \mu(K) + \varepsilon$$

(due to the regularity of μ; see Definition 4.9). Since X is locally compact (see Definition 2.92), invoking Proposition 2.94, we can find an open set $V \subseteq X$ such that $\overline{V}$ is compact and $K \subseteq V \subseteq \overline{V} \subseteq U$. Hence, applying Problem 2.103 for the pair $\{K, V\}$, we can find a continuous function $u\colon X \longrightarrow [0,1]$ such that

$$u|_K \;=\; 1 \quad \text{and} \quad \operatorname{supp} u \;\subseteq\; V.$$

Then $u \in C_c(X)$ and $\chi_K \leqslant u \leqslant \chi_V$. Hence

$$\begin{aligned} \nu(K) \;&=\; \int_X \chi_K\, d\nu \;\leqslant\; \int_X u\, d\nu \;\leqslant\; \int_X u\, d\mu \;\leqslant\; \int_X \chi_V\, d\mu \\ &=\; \mu(V) \;\leqslant\; \mu(U) \;<\; \mu(K) + \varepsilon. \end{aligned}$$

Since $\varepsilon > 0$ was arbitrary, we let $\varepsilon \searrow 0$, to conclude that

$$\nu(K) \;\leqslant\; \mu(K) \qquad \forall\, K \subseteq X,\ K \text{ is compact}.$$

Since μ, ν are Radon measures, from the above inequality, we conclude that $\nu \leqslant \mu$.

Solution of Problem 4.18

(a) Let $K \subseteq X$ be a compact set. Then because μ is σ-bounded, we can find an open cover $\{V_n\}_{n\geqslant 1}$ of K such that $\mu(V_n) < +\infty$ for all $n \geqslant 1$. Due to the compactness of K, we can find $N \geqslant 1$ such that

$$K \subseteq \bigcup_{n=1}^{N} V_n.$$

Hence

$$\mu(K) \leqslant \mu\Big(\bigcup_{n=1}^{N} V_n\Big) \leqslant \sum_{n=1}^{N} \mu(V_n) < +\infty.$$

(b) By Problem 4.15, for a given $\varepsilon > 0$, we can find a closed set $C \subseteq A$, with $\mu(A \setminus C) < \frac{\varepsilon}{2}$. Because X is σ-compact (see Definition 2.99), we can find an increasing sequence of compact sets $\{K_n\}_{n\geqslant 1}$ such that

$$X = \bigcup_{n\geqslant 1} K_n.$$

Then

$$C = \bigcup_{n\geqslant 1} (C \cap K_n),$$

with $C \cap K_n$ being compact and $\mu(C \cap K_n) \nearrow \mu(C)$ as $n \to +\infty$. So, we can find $n_0 \geqslant 1$ such that

$$\mu\big(C \setminus (C \cap K_{n_0})\big) < \tfrac{\varepsilon}{2}.$$

Let $K_\varepsilon = C \cap K_{n_0} \subseteq X$ be a compact set. Then

$$\mu(A \setminus K_\varepsilon) = \mu(A \setminus C) + \mu(C \setminus K_\varepsilon) < \tfrac{\varepsilon}{2} + \tfrac{\varepsilon}{2} = \varepsilon.$$

Thus, A is inner regular with respect to compact sets.

Solution of Problem 4.19

(a) Since μ is σ-finite, we can write

$$A = \bigcup_{n\geqslant 1} A_n,$$

with $\mu(A_n) < +\infty$ for all $n \geqslant 1$. Due to the outer regularity of μ (see Definition 4.9), we can find open sets U_n such that

$$A_n \subseteq U_n \quad \text{and} \quad \mu(U_n) < \mu(A_n) + \tfrac{\varepsilon}{2^{n+1}}.$$

Let

$$U = \bigcup_{n\geqslant 1} U_n.$$

Then the set U is open, $A \subseteq U$ and

$$\mu(U \setminus A) \leqslant \sum_{n\geqslant 1} \mu(U_n \setminus A_n) < \tfrac{\varepsilon}{2}.$$

Similarly, working this time with $A^c \in \mathcal{B}(X)$, we produce an open set V such that

$$A^c \subseteq V \quad \text{and} \quad \mu(V \setminus A^c) < \tfrac{\varepsilon}{2}.$$

Let $C = V^c$. Then C is closed, $C \subseteq A \subseteq U$ and

$$\mu(U\setminus C) = \mu(U\setminus A)+\mu(A\setminus C) = \mu(U\setminus A)+\mu(V\setminus A^c) < \tfrac{\varepsilon}{2}+\tfrac{\varepsilon}{2} = \varepsilon.$$

(b) From part **(a)**, for every $n \geqslant 1$, we can find a closed set C_n and an open set U_n such that

$$C_n \subseteq A \subseteq U_n \quad \text{and} \quad \mu(U_n \setminus C_n) < \tfrac{1}{n} \qquad \forall\, n \geqslant 1.$$

Then

$$E = \bigcup_{n\geqslant 1} C_n \quad \text{is an } F_\delta\text{-set}$$

and

$$D = \bigcap_{n\geqslant 1} U_n \quad \text{is a } G_\sigma\text{-set}$$

and

$$E \subseteq A \subseteq D \quad \text{and} \quad \mu(D \setminus E) < \tfrac{1}{n} \qquad \forall\, n \geqslant 1,$$

hence $\mu(D \setminus E) = 0$.

Solution of Problem 4.20

(a) "$\Longrightarrow$": Suppose that μ is an outer regular measure. Then for all $A \in \mathcal{B}(X)$, we have

$$\mu(A) = \inf\{\mu(U) : A \subseteq U, U \text{ is open}\}.$$

Since μ is finite, we have

$$\begin{aligned}\mu(X) - \mu(A) &= \mu(A^c) = \inf\{\mu(U) : A^c \subseteq U, U \text{ is open}\} \\ &= \mu(X) - \sup\{\mu(C) : C \subseteq A, C \text{ is closed}\}.\end{aligned}$$

So,

$$\mu(A) = \sup\{\mu(C) : C \subseteq A, C \text{ is closed}\}$$

and thus μ is inner regular.

"$\Longleftarrow$": The proof of this implication is similar.

(b) Let $A \in \mathcal{B}(X)$. Since every compact set is closed, we have

$$\begin{aligned}\mu(A) &= \sup\{\mu(K) : K \subseteq A, K \text{ is compact}\} \\ &\leqslant \sup\{\mu(C) : C \subseteq A, C \text{ is closed}\} \\ &\leqslant \mu(A),\end{aligned}$$

so μ is inner regular. Hence by part **(a)**, μ is outer regular too, hence regular.

Solution of Problem 4.21

Let $x \in X$ be such that $u(x) > 0$. Then we can find $U \in \mathcal{N}(x)$ with $\overline{U}$ being compact (since X is locally compact) such that

$$u(z) \geqslant \xi > 0 \qquad \forall z \in \overline{U}.$$

Let $h \in C_c(X)$, $h \geqslant 0$ be such that $\operatorname{supp} h \subseteq \overline{U}$. Then

$$h(y) \leqslant \tfrac{\|h\|_\infty}{\xi} u(y) \qquad \forall y \in X,$$

so

$$\int_X h\,d\mu \leqslant \frac{\|h\|_\infty}{\xi} \int_X u\,d\mu = 0$$

and thus

$$\int_X h\,d\mu = 0.$$

This means that $x \notin \operatorname{supp}\mu$ (see Definition 4.13) and so $u|_{\operatorname{supp}\mu} = 0$.

Solution of Problem 4.22

Suppose that $u \in C_c(X)$, $K = \operatorname{supp} u$, $E = \operatorname{supp}\mu$ and $u|_E = 0$ (see Definition 4.13). For a given $\varepsilon > 0$, let

$$U_\varepsilon = \{x \in X : |u(x)| < \varepsilon\}.$$

Then $U_\varepsilon \subseteq X$ is an open set and $E \subseteq U_\varepsilon$. Then E^c is an open set containing the compact set U_ε^c. Invoking Problem 2.103, we can find a continuous function $g\colon X \longrightarrow [0,1]$ such that

$$g|_{U_\varepsilon^c} = 1 \quad \text{and} \quad \operatorname{supp} g \subseteq E^c.$$

Then $\operatorname{supp}(ug) \cap E = \emptyset$ and so by hypothesis

$$\int_X ug\,d\mu = 0.$$

Also,

$$u = ug \quad \text{on } K \cap U_\varepsilon^c \quad \text{and} \quad |ug| \leqslant |u| \quad \text{in } X.$$

Therefore, from the definition of U_ε, we have

$$|u(x) - (ug)(x)| \leqslant 2\varepsilon \qquad \forall\, x \in X.$$

Since $\operatorname{supp}(u - ug) \subseteq K$, we have

$$\Big|\int_X (u - ug)\,d\mu\Big| \leqslant 2M\varepsilon,$$

for some $M = M(K) > 0$ and so

$$\Big|\int_X u\,d\mu\Big| \leqslant 2M\varepsilon.$$

Let $\varepsilon \searrow 0$, to conclude that

$$\int_X u \, d\mu = 0.$$

Solution of Problem 4.23

By Proposition 3.20, it suffices to show that, if $\{A_n\}_{n\geqslant 1} \subseteq \mathcal{B}(X)$ is a decreasing sequence such that

$$\bigcap_{n\geqslant 1} A_n = \emptyset,$$

then $\mu(A_n) \searrow 0$. We argue indirectly. So, suppose that

$$0 < \beta = \lim_{n\to+\infty} \mu(A_n).$$

Let $\varepsilon \in \left(0, \frac{1}{2}\beta\right)$. Then for each $n \geqslant 1$, we can find a compact set $K_n \subseteq A_n$ such that

$$\mu(A_n \setminus K_n) < \tfrac{\varepsilon}{2^n}.$$

We claim that the sequence $\{K_n\}_{n\geqslant 1}$ has the finite intersection property. Indeed, note that

$$\mu\big(A_m \setminus \bigcap_{n=1}^m K_n\big) < \sum_{n=1}^m \tfrac{\varepsilon}{2^n} = \varepsilon,$$

so, from the choice of $\varepsilon > 0$, we have

$$\mu\big(\bigcap_{n=1}^m K_n\big) > 0,$$

thus

$$\bigcap_{n=1}^{m} K_n \neq \emptyset$$

and finally, from Theorem 2.81, we have

$$\bigcap_{n\geqslant 1} K_n \neq \emptyset.$$

But then

$$\bigcap_{n\geqslant 1} A_n \neq \emptyset$$

(as $K_n \subseteq A_n$ for all $n \geqslant 1$), a contradiction.

Solution of Problem 4.24

"$\Longrightarrow$": This is immediate from Definition 4.9(a).

"$\Longleftarrow$": From Theorem 4.11, we know that μ is regular (see Definition 4.9). Therefore to show that μ is Radon, it suffices to show that, for every closed set $C \subseteq X$, we have

$$\mu(C) = \sup\big\{\mu(K) : K \subseteq C, \ K \text{ is compact}\big\}.$$

Arguing by contradiction, suppose that $C \subseteq X$ is a closed set such that there exists $\varepsilon \in \big(0, \mu(X)\big)$ for which, we have

$$\sup\big\{\mu(K) : K \subseteq C, \ K \text{ is compact}\big\} \leqslant \mu(C) - \varepsilon.$$

If $D \subseteq X$ is compact, then $D \cap C \subseteq C$ is also compact and so

$$\mu(D \cap C) \leqslant \mu(C) - \varepsilon.$$

Then

$$\mu(D) = \mu(D\cap C)+\mu(D\cap C^c) \leqslant \mu(C)-\varepsilon+\mu(C^c) = \mu(X)-\varepsilon.$$

Since $D \subseteq X$ was arbitrary compact set, we have

$$\sup\left\{\mu(D) : D \subseteq X, \ D \text{ is compact}\right\} \leqslant \mu(X) - \varepsilon,$$

a contradiction to our hypothesis.

Solution of Problem 4.25

First assume that X is second countable (see Definition 2.24). Let

$$D = \bigcup_{\substack{U \text{ is open} \\ \mu(U)=0}} U \quad \text{and} \quad C = D^c.$$

From Proposition 2.164(a), we know that X is strongly Lindelöff. Hence because D is open, we can find a sequence of open sets $\{U_n\}_{n\geqslant 1}$ with

$$\mu(U_n) = 0 \quad \forall\, n \geqslant 1$$

such that

$$D = \bigcup_{n\geqslant 1} U_n.$$

It follows that $\mu(D) = 0$. Let $V \subseteq X$ be an open set such that $V \cap C \neq \emptyset$. We claim that $\mu(V \cap C) > 0$. Indeed, if $\mu(V \cap C) = 0$, then

$$\mu(V) = \mu(V \cap C) + \mu(V \cap D) = 0$$

and so $V \subseteq D$, a contradiction. Therefore according to Definition 4.13, we have that $C = \operatorname{supp} \mu$.

Next assume that μ is outer regular and also inner regular with respect to compact sets (see Definition 4.9). Let D be the open set introduced in the previous part of the solution. Let $K \subseteq D$ be a compact set. Then we can find open sets $V_1, \ldots, V_n$, with $\mu(V_k) = 0$ for all $k \in \{1, \ldots, n\}$ such that

$$K \subseteq \bigcup_{k=1}^{n} V_k.$$

Hence $\mu(K) = 0$. Exploiting our hypothesis on μ, we have

$$\mu(D) = \sup\{\mu(K) : K \subseteq D, \ K \text{ is compact}\} = 0$$

and so, if $C = D^c$, then as above, we conclude that $C = \operatorname{supp}\mu$.

Solution of Problem 4.26

No. Consider $X = \{0,1\}$ and $\Sigma = 2^X$. Let $\mu = \delta_0$ and $\nu = \delta_1$. Then $\vartheta = \max\{\mu, \nu\}$ satisfies

$$\vartheta(A) = \begin{cases} 1 & \text{if } A \in \Sigma, \ A \neq \emptyset, \\ 0 & \text{if } A = \emptyset \end{cases}$$

and in fact this is not a measure. In his case the smallest measure not less than μ or ν is $\mu + \nu$ and

$$(\mu + \nu)(X) = 2.$$

Solution of Problem 4.27

We know that σ is a measure on $\mathcal{B}(X)$ and in view of the fact that $\sigma \leqslant \mu$, we have that σ is σ-finite. Also $\sigma \ll \mu$ (see Definition 3.150).

Let $A \in \mathcal{B}(X)$ be such that $\mu(A) < +\infty$. We consider the Borel σ-algebra $\mathcal{B}(A)$ of A. We know that

$$\mathcal{B}(A) = A \cap \mathcal{B}(X).$$

The Radon–Nikodym theorem (see Theorem 3.152) implies that there exists a unique function $u \in L^1(A, \mathcal{B}(A), \mu)$, $u \geqslant 0$ such that

$$\sigma(A \cap C) = \int\limits_{A \cap C} u \, d\mu \qquad \forall\, C \in \mathcal{B}(X).$$

Because μ is a Radon measure, for a given $\varepsilon > 0$, we can find a compact set $K \subseteq A$ such that

$$0 \leqslant \sigma(A) - \sigma(K) = \int\limits_A u \, d\mu - \int\limits_K u \, d\mu = \int\limits_{A \setminus K} u \, d\mu \leqslant \varepsilon,$$

so

$$\sigma(A) = \sup\{\sigma(K) : K \subseteq A,\ K \text{ is compact}\}.$$

Since σ is clearly finite on compact sets, to show that it is Radon, it remains to show that it is outer regular (see Definition 4.9). So, let $D \in \mathcal{B}(X)$. We know that

$$\sigma(D) = \inf\{\mu(E) + m(D \setminus E) : E \in \mathcal{B}(X),\ E \subseteq D\}.$$

Let

$$\eta = \inf\{\sigma(U) : U \supseteq D,\ U \text{ is open}\}$$

and $\varepsilon > 0$. For a given $E \in \mathcal{B}(X)$, $E \subseteq D$, we choose open sets U and V such that

$$E \subseteq U, \quad \mu(U) \leqslant \mu(E)+\varepsilon, \quad D \setminus E \subseteq V \quad \text{and} \quad m(V) \leqslant m(D \setminus E)+\varepsilon.$$

Then, we have

$$\begin{aligned} \sigma(D) \leqslant \eta \leqslant \sigma(U \cup V) &\leqslant \sigma(U) + \sigma(V) \leqslant \mu(U) + m(V) \\ &\leqslant \mu(E) + m(D \setminus E) + 2\varepsilon, \end{aligned}$$

so $\sigma(D) \leqslant \eta \leqslant \sigma(D) + 2\varepsilon$. Let $\varepsilon \searrow 0$, to conclude that $\sigma(D) = \eta$ and so σ is Radon.

Solution of Problem 4.28

Let $A \in \mathcal{B}(\mathbb{R}^N)$. Let

$$\overline{B}_n(0) = \{x \in \mathbb{R}^N : \|x\| \leqslant n\}.$$

Then $\overline{B}_n(0)$ is a compact metric space and so by Theorem 4.12, $\lambda^N|_{\overline{B}_n(0)}$ is Radon. Therefore

$$\lambda^N(A \cap B_n(0)) = \sup\{\lambda^N(K) : K \subseteq A \cap B_n(0),\ K \text{ is compact}\}.$$

Since $A \cap B_n(0) \nearrow A$, we infer that $\lambda^N(A \cap B_n(0)) \nearrow \lambda^N(A)$ and so

$$\lambda^N(A) = \sup\{\lambda^N(K) : K \subseteq A,\ K \text{ is compact}\}.$$

If $\lambda^N(A) = +\infty$, then obviously, we have

$$\lambda^N(A) = \inf\{\lambda(U) : A \subseteq U,\ U \text{ is open}\}.$$

Hence, we assume that $\lambda^N(A) < +\infty$. For a given $\varepsilon > 0$ and since $\lambda^N|_{\overline{B}_n(0)}$ is Radon and

$$B_n(0) = \{x \in \mathbb{R}^N : \|x\| < n\}$$

is open, we have

$$\begin{aligned}\lambda^N(A \cap B_n(0)) &= \inf\{\lambda^N(U \cap \overline{B}_n(0)) : A \cap B_n(0) \subseteq U \cap \overline{B}_n(0), \\ &\qquad U \text{ is open}\} \\ &= \inf\{\lambda^N(U \cap B_n(0)) : A \cap B_n(0) \subseteq U \cap B_n(0), \\ &\qquad U \text{ is open}\} \\ &= \inf\{\lambda^N(U) : A \cap B_n(0) \subseteq U,\ U \text{ is open}\}.\end{aligned}$$

So, for every $n \geqslant 1$, we can find open set $U_n \subseteq \mathbb{R}^N$ such that

$$\lambda^N(U_n \setminus (A \cap B_n(0))) \leqslant \tfrac{\varepsilon}{2^n}.$$

Let

$$U = \bigcup_{n \geqslant 1} U_n \subseteq \mathbb{R}^N.$$

Then U is open, $A \subseteq U$ and

$$U \setminus A = \bigcup_{n \geqslant 1} U_n \setminus \bigcup_{n \geqslant 1} (A \cap B_n(0)) \subseteq \bigcup_{n \geqslant 1} (U_n \setminus (A \cap B_n(0))),$$

so

$$0 \leqslant \lambda^N(U) - \lambda^N(A) = \lambda^N(U \setminus A) \leqslant \sum_{n \geqslant 1} \lambda^N(U_n \setminus (A \cap B_n(0))) \leqslant \varepsilon,$$

thus $\lambda^N(A) = \inf\{\lambda^N(U) : A \subseteq U,\ U \text{ is open}\}$ and so λ^N is Radon.

Solution of Problem 4.29

Recall that simple functions are dense in $L^p(X, \mu)$ (see Proposition 3.110). So, it suffices to show that for any $A \in \mathcal{B}(X)$ with $\mu(A) < +\infty$, the characteristic function χ_A can be approximated in the L^p-norm by functions in $C_c(X)$. Because μ is Radon, for a given

$\varepsilon > 0$, we can find a compact set $K \subseteq A$ and an open set $U \supseteq A$ such that

$$\mu(U \setminus K) \leqslant \varepsilon$$

(see Problem 4.19). As before, because X is locally compact (see Definition 2.92), we can find a function $h \in C_c(X)$ such that

$$h\big|_K = 1, \quad 0 \leqslant h \leqslant 1 \quad \text{and} \quad \operatorname{supp} h \subseteq U$$

(see Problem 2.103). Then

$$\|\chi_A - h\|_p^p = \int_X |\chi_A - h|^p \, d\mu \leqslant \int_X \chi_{U\setminus K}^p \, d\mu = \mu(U \setminus K) \leqslant \varepsilon,$$

so $C_c(X)$ is dense in $L^p(X, \mu)$ for all $1 \leqslant p < +\infty$.

Solution of Problem 4.30

Let $a, b \in \mathbb{R}$ with $a < b$ and let

$$f(x) = \chi_{(a,b)}(x) \qquad \forall\, x \in \mathbb{R}.$$

Suppose that f can be approximated in the L^∞-norm by continuous functions. Let $\varepsilon \in \left(0, \frac{1}{2}\right)$ and let $g \in C(\mathbb{R})$ be such that

$$\|f - g\|_\infty < \varepsilon.$$

Then

$$\big|\chi_{(a,b)}(x) - g(x)\big| < \varepsilon \quad \text{almost everywhere on } \mathbb{R}.$$

For every $\delta > 0$, there are points $y \in (a, a+\delta)$ and $z \in (a-\delta, a)$ such that

$$\big|1 - g(y)\big| < \varepsilon \quad \text{and} \quad \big|g(z)\big| < \varepsilon.$$

Hence

$$\limsup_{y \to a^+} g(y) \geqslant 1 - \varepsilon \quad \text{and} \quad \liminf_{y \to a^-} g(y) \leqslant \varepsilon,$$

a contradiction to the continuity of g.

Solution of Problem 4.31

Clearly μ_0 is a measure. Also, since

$$\mathbb{R} = \Big(\bigcup_{q\in\mathbb{Q}}\{q\}\Big) \cup (\mathbb{R}\setminus\mathbb{Q}),$$

we see that μ_0 is σ-finite. For every nonempty open set $U \subseteq \mathbb{R}$, $\operatorname{card}(U \cap \mathbb{Q}) = +\infty$ and for every $x \in \mathbb{R}$, we have $\mu_0(\{x\}) < +\infty$. However

$$\inf\big\{\mu_0(U) : U \text{ is open and } x \in U\big\} = +\infty$$

and so μ_0 is not regular.

Solution of Problem 4.32

Suppose that μ is not a measure. Then by Theorem 3.19 and Proposition 3.20, we can find a decreasing sequence $\{A_n\}_{n\geqslant 1} \subseteq \mathcal{B}(X)$ such that

$$\bigcap_{n\geqslant 1} A_n = \emptyset \quad \text{and} \quad \lim_{n\to+\infty}\mu(A_n) = \inf_{n\geqslant 1}\mu(A_n) > 0.$$

By hypothesis, for every $n \geqslant 1$, we can find a compact set $K_n \subseteq A_n$ such that

$$\mu(A_n) \leqslant \mu(K_n) + \tfrac{1}{2^{n+1}}\inf_{m\geqslant 1}\mu(A_m).$$

So, we have

$$\mu\big(A_n \setminus \bigcap_{m=1}^{n} K_n\big) \leqslant \sum_{m=1}^{n}\mu(A_m\setminus K_m) \leqslant \tfrac{1}{2}\inf_{n\geqslant 1}\mu(A_m),$$

thus

$$\mu\big(\bigcap_{m=1}^{n} K_n\big) \neq 0, \quad \text{hence} \quad \bigcap_{m=1}^{n} K_m \neq \emptyset.$$

Let

$$\widehat{K}_n = \bigcap_{m=1}^{n} K_n.$$

Note that $\{\widehat{K}_n\}_{n\geqslant 1}$ is a decreasing sequence of nonempty, compact sets contained in the compact set K_1. Therefore

$$\bigcap_{m\geqslant 1} K_m \neq \emptyset$$

(see Theorem 2.81), which contradicts the fact that

$$\bigcap_{n\geqslant 1} A_n = \emptyset.$$

Solution of Problem 4.33
No. Consider the function

$$f = \chi_{[0,1]} \in L^1(\mathbb{R}).$$

Suppose that $g \in L^1(\mathbb{R})$ satisfies

$$\|f-g\|_1 < \tfrac{1}{2}.$$

We have

$$1 - \int_{\mathbb{R}} g\,dx = \int_{\mathbb{R}} (f-g)\,dx \leqslant \int_{\mathbb{R}} |f-g|\,dx < \tfrac{1}{2},$$

so

$$\int_{\mathbb{R}} g\,dx > \tfrac{1}{2}$$

and so $g \notin V$.

Solution of Problem 4.34
Evidently, it suffices to show that closed sets are μ-measurable. So, let $C \subseteq X$ be a closed set and let

$$\begin{aligned} E_0 &= \{x \in X :\ \operatorname{dist}(x, C) \geqslant 1\}, \\ E_n &= \{x \in X :\ \tfrac{1}{2^n} \leqslant \operatorname{dist}(x, C) < \tfrac{1}{2^{n-1}}\} \qquad \forall\, n \geqslant 1. \end{aligned}$$

Due to the subadditivity of μ, it suffices to show that

$$\mu(D) \geqslant \mu(D \cap C) + \mu(D \cap C^c) \qquad \forall\, D \subseteq X.$$

Clearly we may assume that $\mu(D) < +\infty$. Note that, if $|n-m| \geqslant 2$, then E_n and E_m are metrically separated and so, we have

$$\mu(D) \geqslant \mu\big(D \cap \big(\bigcup_{m=0}^{n} E_{2m}\big)\big) = \sum_{m=0}^{n} \mu(D \cap E_{2m}) \qquad \forall\, n \geqslant 1.$$

The same inequality is also true with $2m$ replaced by $2m+1$. So, it follows that

$$\sum_{m \geqslant 0} \mu(D \cap E_m) < +\infty.$$

Note that the sets C and $\bigcup\limits_{m=0}^{n} E_m$ $(n \geqslant 0)$ are metrically separated. Hence

$$\begin{aligned} \mu(D) &\geqslant \mu\big((D \cap C) \cup \big(D \cap \big(\bigcup_{m=0}^{n} E_n\big)\big)\big) \\ &= \mu(D \cap C) + \mu\big(D \cap \big(\bigcup_{m=0}^{n} E_n\big)\big) \\ &\geqslant \mu(D \cap C) + \mu(D \cap C^c) - \sum_{k>n} \mu(D \cap E_k) \end{aligned}$$

(from the subadditivity of μ). Passing to the limit as $n \to +\infty$, we conclude that

$$\mu(D) \geqslant \mu(D \cap C) + \mu(D \cap C^c),$$

i.e., C is μ-measurable.

Solution of Problem 4.35

(a) Let $K \subseteq X$ be a compact set. By the regularity of μ and the local compactness of the space X, for a given $\varepsilon > 0$, we can find an open set $V \subseteq X$ such that

$$K \subseteq V, \quad \overline{V} \text{ is compact and } \quad \mu(V) \leqslant \mu(K) + \varepsilon.$$

By Problem 2.103, we can find $f \in C_c(X)$ such that

$$0 \leqslant f \leqslant 1, \quad f|_K = 1 \quad \text{and} \quad f|_{V^c} = 0.$$

Then

$$\begin{aligned} \mu(K) &\geqslant \mu(V) - \varepsilon \geqslant \int_X f \, d\mu - \varepsilon = \lim_{n \to +\infty} \int_X f \, d\mu_n - \varepsilon \\ &\geqslant \limsup_{n \to +\infty} \mu_n(K) - \varepsilon. \end{aligned}$$

Since $\varepsilon > 0$ was arbitrary, we let $\varepsilon \searrow 0$ to conclude that

$$\limsup_{n \to +\infty} \mu_n(K) \leqslant \mu(K).$$

(b) The proof of this part is similar to that of part **(a)** using this time the approximation of $\mu(u)$ by the μ-measure of compact subsets of U (see Definition 4.9).

Solution of Problem 4.36
Let

$$g(r) = \mu\big(B_r(x)\big) \quad \text{and} \quad h(r) = \nu\big(B_r(x)\big) \qquad \forall\, x \in X,\ r > 0.$$

Also let $U \subseteq X$ be a nonempty open set. Then

$$\lim_{r \searrow 0} \tfrac{\nu(U \cap B_r(x))}{h(r)} = 1 \qquad \forall\, x \in U$$

and using the Fatou lemma (see Theorem 3.95) and the Fubini theorem (see Theorem 3.115), we have

$$\begin{aligned} \mu(U) &= \int_U \lim_{r \searrow 0} \tfrac{\nu(U \cap B_r(x))}{h(r)} \, d\mu \leqslant \liminf_{r \searrow 0} \tfrac{1}{h(r)} \int_U \nu\big(U \cap B_r(x)\big) \, d\mu \\ &= \liminf_{r \searrow 0} \tfrac{1}{h(r)} \int_U \mu\big(B_r(x)\big) \, d\nu = \big(\liminf_{r \searrow 0} \tfrac{g(r)}{h(r)}\big) \nu(U). \end{aligned}$$

If in the above argument we interchange μ and ν, we obtain

$$\nu(U) \leqslant \big(\liminf_{r \searrow 0} \tfrac{h(r)}{g(r)}\big)\mu(U).$$

It follows that the limit $c = \lim\limits_{r \searrow} \frac{g(r)}{h(r)}$ exists and so $\mu(U) = c\nu(U)$ for all open sets $U \subseteq X$. Then Theorem 4.11 implies that $\mu = c\nu$.

Solution of Problem 4.37

For a given $\varepsilon > 0$, we can find $\delta = \delta(\varepsilon) > 0$ such that

$$\int\limits_D h\, dx \;<\; \varepsilon \qquad \forall\, D \in \mathcal{B}([0,1]),\ \lambda(D) < \delta$$

(λ being the Lebesgue measure on $\mathbb{R}$). Theorem 4.12 and Problem 4.19 imply that we can find a compact set $K \subseteq A$ and an open set U with $A \subseteq U$ such that $\lambda(U \setminus K) < \delta$. Also, we can find a continuous function $g\colon [0,1] \longrightarrow \mathbb{R}$ such that

$$0 \leqslant g \leqslant 1, \quad g|_K \;=\; 1, \quad g|_{U^c} \;=\; 0.$$

Then, using also hypotheses **(a)** and **(b)** and recalling that $\lambda(U \setminus K) < \delta$, we have

$$\begin{aligned}
\limsup_{n \to +\infty} \Big|\int\limits_A f_n\, dx\Big| \;&=\; \limsup_{n \to +\infty} \Big|\int\limits_0^1 f_n \chi_A\, dz\Big| \\
&\leqslant\; \limsup_{n \to +\infty} \Big|\int\limits_0^1 f_n g\, dz\Big| + \limsup_{n \to +\infty} \Big|\int\limits_0^1 f_n(\chi_A - g)\, dx\Big| \\
&\leqslant\; \limsup_{n \to +\infty} \Big|\int\limits_0^1 f_n g\, dx\Big| + \int\limits_0^1 f\chi_{U \setminus K}\, dx \;\leqslant\; \varepsilon.
\end{aligned}$$

Since $\varepsilon > 0$ was arbitrary, we let $\varepsilon \searrow 0$, to conclude that

$$\lim_{n\to+\infty} \int_A f_n \, dx = 0.$$

Solution of Problem 4.38

Let $K \subseteq X$ be a compact set. The tightness of μ and ν (see Definition 4.9) implies that we can find two sequences of relatively compact open sets $\{U_n\}_{n\geqslant 1}$ and $\{V_n\}_{n\geqslant 1}$ such that

$$\mu(U_n),\ \nu(V_n) < +\infty, \quad K \subseteq U_n, \quad K \subseteq V_n \qquad \forall\, n \geqslant 1$$

and

$$\mu(K) = \lim_{n\to+\infty} \mu(U_n) \quad \text{and} \quad \nu(K) = \lim_{n\to+\infty} \nu(V_n).$$

Let us set

$$W_n = \bigcap_{k=1}^{n} (U_k \cap V_k) \qquad \forall\, n \geqslant 1.$$

Then $\{W_n\}_{n\geqslant 1}$ is a decreasing sequence of open sets in X with

$$\mu(K) = \lim_{n\to+\infty} \mu(W_n), \quad \nu(K) = \lim_{n\to+\infty} \nu(W_n).$$

By Problem 2.103, we can find $f_n \in C_c(X)$ such that

$$f_n|_K = 1, \quad f_n|_{W_n^c} = 0 \quad \text{and} \quad 0 \leqslant f_n \leqslant 1 \qquad \forall\, n \geqslant 1.$$

Let $h_n = \min\{f_1, \ldots, f_n\}$. We have that $\{h_n\}_{n\geqslant 1}$ is a decreasing sequence in $C_c(X)$ and $h_n \searrow \chi_K$. Then using the Lebesgue monotone convergence theorem (see Theorem 3.92) and the hypothesis of the problem, we have

$$\mu(K) = \int_X \chi_K \, d\mu = \lim_{n\to+\infty} \int_X h_n \, d\mu = \lim_{n\to+\infty} \int_X h_n \, d\nu = \int_X \chi_K \, d\nu = \nu(K).$$

So, we have proved that for every compact set $K \subseteq X$, we have $\mu(K) = \nu(K)$. Since μ and ν are tight, we conclude that $\mu = \nu$ (see Definition 4.9(e)).

Solution of Problem 4.39

"(a) $\Longrightarrow$ (b)": This is immediate since φ is Borel measurable.

"(b) $\Longrightarrow$ (a)": Let U be an open subset of Y. Let $h_U \in C(Y)$ be defined by

$$h_U(y) = \operatorname{dist}(y, U^c) \qquad \forall\, y \in Y,$$

if $U \neq Y$ and

$$h_U(y) = 1 \qquad \forall\, y \in Y,$$

if $U = Y$. We have

$$U = \{y \in Y : h_U(y) > 0\}.$$

Note that $f^{-1}(U) = \{x \in X : (h_U \circ f)(x) > 0\} \in \Sigma$. Since U is an arbitrary open subset of Y and open sets generate the Borel σ-algebra of Y, we conclude that f is Σ-measurable.

Solution of Problem 4.40

"(a) $\Longrightarrow$ (b)": Obvious (since the composition of measurable maps is measurable).

"(b) $\Longrightarrow$ (a)": Let $\mathcal{X} = \{A \in \mathcal{B}a(X) : f^{-1}(A) \in \Sigma\}$. Evidently $\mathcal{X}$ is a σ-algebra and since $\varphi \circ f$ is Σ-measurable, it contains all sets of the form $\varphi^{-1}([\lambda, +\infty))$. Therefore, by Remark 4.3, $\mathcal{X} = \mathcal{B}a(X)$ and so f is $(\Sigma, \mathcal{B}a(X))$-measurable.

Solution of Problem 4.41
First we show the following Claim.

Claim. $\lim\limits_{x\to 0}\lambda\big((A+x)\cap A\big)=\lambda(A)$.

If

$$A = \bigcup_{k=1}^{n}(a_k,b_k),$$

where $\{(a_k,b_k)\}_{k=1}^{n}$ are pairwise disjoint open bounded intervals, then

$$\begin{aligned}\lambda(A) &= \sum_{k=1}^{n}(b_k-a_k)=\lim_{x\to 0}\sum_{k=1}^{n}\big(\min\{b_k,b_k+x\}-\max\{a_k,a_k+x\}\big)\\ &= \lim_{x\to 0}\sum_{k=1}^{n}\lambda\big((a_k+x,b_k+x)\cap(a_k,b_k)\big)\\ &= \lim_{x\to 0}\lambda\Big(\bigcup_{k=1}^{n}(a_k+x,b_k+x)\cap(a_k,b_k)\Big)\leqslant\liminf_{x\to 0}\lambda\big((A+x)\cap A\big)\\ &\leqslant \limsup_{x\to 0}\lambda\big((A+x)\cap A\big)\leqslant\lambda(A),\end{aligned}$$

so

$$\lim_{x\to 0}\lambda\big((A+x)\cap A\big) = \lambda(A).$$

Now suppose that the set A is compact. For a given $\varepsilon>0$, we can find an open set $U\subseteq\mathbb{R}$ such that

$$A\subseteq U,\quad \lambda(U\setminus A)<\varepsilon \quad\text{and}\quad U=\bigcup_{k=1}^{n}(a_k,b_k),$$

with $\{(a_k,b_k)\}_{k=1}^{n}$ mutually disjoint. Let $V=U\setminus A$. This is open too. Then

$$\begin{aligned}&\lambda(A)\leqslant\lambda(U)=\lim_{x\to 0}\lambda\big((U+x)\cap U\big)\\ &= \lim_{x\to 0}\lambda\big(((A+x)\cap A)\cup((A+x)\cap V)\cup((V+x)\cap A)\cup((V+x)\cap V)\big)\\ &\leqslant \liminf_{x\to 0}\lambda\big((A+x)\cap A\big)+3\lambda(V)\leqslant\liminf_{x\to 0}\lambda\big((A+x)\cap A\big)+3\varepsilon\\ &\leqslant \limsup_{x\to 0}\lambda\big((A+x)\cap A\big)+3\varepsilon\leqslant\lambda(A)+3\varepsilon.\end{aligned}$$

Since $\varepsilon > 0$ is arbitrary, we let $\varepsilon \searrow 0$, to conclude that

$$\lambda(A) = \lim_{x \to 0} \lambda\big((A+x) \cap A\big),$$

where A is compact.

Finally for the general case, we can find an increasing sequence $\{A_n\}_{n \geqslant 1}$ such that

$$A_n \text{ is compact}, \quad A_n \subseteq A \text{ for all } n \geqslant 1, \quad \lim_{n \to +\infty} \lambda(A_n) = \lambda(A)$$

(since λ is a Radon measure on $\mathbb{R}$; see Problem 4.28). Then

$$\begin{aligned} \lambda(A) &= \lim_{n \to +\infty} \lambda(A_n) = \lim_{n \to +\infty} \lim_{x \to 0} \lambda\big((A_n + x) \cap A_n\big) \\ &\leqslant \liminf_{x \to 0} \lambda\big((A+x) \cap A\big) \leqslant \limsup_{x \to 0} \lambda\big((A+x) \cap A\big) \leqslant \lambda(A), \end{aligned}$$

so

$$\lambda(A) = \lim_{x \to 0} \lambda\big((A+x) \cap A\big).$$

This proves the Claim.

Using this Claim, we can prove the two statements of the problem.

(a) For all $y, x \in \mathbb{R}$, we have

$$\big|f(y) - f(x)\big| = \big|\lambda\big((A+y) \cap A\big) - \lambda\big((A+x) \cap A\big)\big|.$$

Note that

$$\lambda\big((A+y) \cap A\big) = \lambda\big(((A+y) \setminus (A+x)) \cap A\big) + \lambda\big((A+y) \cap (A+x) \cap A\big)$$

and

$$\lambda\big((A+x) \cap A\big) = \lambda\big(((A+x) \setminus (A+y)) \cap A\big) + \lambda\big((A+y) \cap (A+x) \cap A\big).$$

So, subtracting the above two quantities, we obtain

$$\begin{aligned} \big|f(y) - f(x)\big| &= \big|\lambda\big(((A+y) \setminus (A+x)) \cap A\big) \\ &\quad -\lambda\big(((A+x) \setminus (A+y)) \cap A\big)\big| \\ &\leqslant \lambda\big((A+y) \setminus (A+x)\big) + \lambda\big((A+x) \setminus (A+y)\big) \\ &= \lambda\big((A+(y-x)) \setminus A\big) + \lambda\big((A+(x-y)) \setminus A\big) \\ &= \lambda(A) - \lambda\big((A+(y-x)) \cap A\big) + \lambda(A) - \lambda\big((A+(x-y)) \cap A\big). \end{aligned}$$

Using the Claim, we get that

$$\big|f(y) - f(x)\big| \longrightarrow 0 \quad \text{as } y \to x.$$

(b) If A is a compact set, then we can find $R_0 > 0$ such that for any $x \geqslant R_0$, we have $(A+x) \cap A = \emptyset$ and so

$$f(x) = 0 \qquad \forall\, x \geqslant R_0.$$

For the general case, as before exploiting the fact that λ is a Radon measure, we can find an increasing sequence $\{A_n\}_{n\geqslant 1}$ such that

$$A_n \text{ is compact}, \quad A_n \subseteq A \text{ for all } n \geqslant 1, \quad \lim_{n\to+\infty} \lambda(A_n) = \lambda(A)$$

Then we have

$$\begin{aligned}
&\lim_{x\to+\infty} \lambda\big((A+x)\cap A\big)\\
=\ &\lim_{x\to+\infty} \big[\lambda\big((A+x)\cap A\big)-\lambda\big((A_n+x)\cap A_n\big)+\lambda\big((A_n+x)\cap A_n\big)\big]\\
=\ &\lim_{n\to+\infty}\lim_{x\to+\infty} \big[\lambda\big(((A+x)\cap A)\setminus((A_n+x)\cap A_n)\big)+\lambda\big((A_n+x)\cap A_n\big)\big]\\
\leqslant\ &\lim_{n\to+\infty} \big[\lambda\big((A+x)\setminus(A_n+x)\big)+\lambda(A\setminus A_n)\big] = 0.
\end{aligned}$$

Solution of Problem 4.42

Let $B \subseteq \mathbb{R}$ be a bounded open set of the form

$$B = \bigcup_{k=1}^{m} (a_k, b_k)$$

for some $m \geqslant 1$ and the intervals (a_k, b_k) being pairwise disjoint. Then

$$\bigcup_{k=1}^{m} \big((\varrho a_k, \varrho b_k) \cap (a_k, b_k)\big) \subseteq \varrho B \cap B \qquad \forall\, \varrho > 0.$$

If λ denotes the Lebesgue measure on $\mathbb{R}$, then

$$\begin{aligned}
\lambda(B) &= \sum_{k=1}^{m} (b_k - a_k)\\
&= \lim_{\varrho\to 1} \sum_{k=1}^{m} \lambda\big((\varrho a_k, \varrho b_k)\cap(a_k,b_k)\big)\\
&\leqslant \liminf_{\varrho\to 1} \lambda(\varrho B \cap B)\\
&\leqslant \lambda(B).
\end{aligned}$$

Let $K \subseteq \mathbb{R}$ be a compact set. Then we can find a decreasing sequence $\{B_n\}_{n\geqslant 1}$ of bounded open sets of the above form such that

$$K = \bigcap_{n\geqslant 1} B_n.$$

Then from the subadditivity of λ, we have

$$\lambda(\varrho B_n \cap B_n) \leqslant \lambda\big(\varrho(B_n \setminus K)\big) + \lambda(B_n \setminus K) + \lambda(\varrho K \cap K)$$

and so

$$\begin{aligned}
\lambda(K) &= \lim_{n\to+\infty} \lambda(B_n) = \lim_{n\to+\infty} \lim_{\varrho\to 1} \lambda(\varrho B_n \cap B_n) \\
&\leqslant \liminf_{n\to+\infty} \liminf_{\varrho\to 1} \big(\lambda\big(\varrho(B_n \setminus K)\big) + \lambda(B_n \setminus K) + \lambda(\varrho K \cap K)\big) \\
&= \liminf_{n\to+\infty} \big(\liminf_{\varrho\to 1} \lambda(\varrho K \cap K) + 2\lambda(B_n \setminus K)\big) \\
&= \liminf_{\varrho\to 1} \lambda(\varrho K \cap K) \leqslant \limsup_{\varrho\to 1} \lambda(\varrho K \cap K) \leqslant \lambda(K).
\end{aligned}$$

Since λ is Radon (see Problem 4.28), we can find an increasing sequence $\{K_n\}_{n\geqslant 1}$ of compact sets such that

$$\bigcup_{n\geqslant 1} K_n \subseteq A \quad \text{and} \quad \lambda(K_n) \nearrow \lambda(A).$$

So, we have

$$\begin{aligned}
\lambda(A) &= \lim_{n\to+\infty} \lambda(K_n) = \lim_{n\to+\infty} \lim_{\varrho\to 1} \lambda(\varrho K_n \cap K_n) \\
&\leqslant \liminf_{\varrho\to 1} \lambda(\varrho A \cap A) \leqslant \limsup_{\varrho\to 1} \lambda(\varrho A \cap A) \leqslant \lambda(A).
\end{aligned}$$

We have

$$\begin{aligned}
\big|\varphi(\varrho) - \varphi(1)\big| &= \Big| \int_{-\infty}^{+\infty} \big|\chi_{\frac{1}{\varrho}A}(t)\chi_A(t) - \chi_A(t)\big| \, dt \Big| \\
&= \lambda(A) - \lambda(\tfrac{1}{n}A \cap A),
\end{aligned}$$

so

$$\varphi(\varrho) \longrightarrow \varphi(1) \quad \text{as } \varrho \to 1.$$

This proves the continuity of φ at $\varrho = 1$.

Solution of Problem 4.43

Let $N = 1$ and let λ be the Lebesgue measure on $\mathbb{R}$. For each $n \geqslant 1$ and each $A \in \mathcal{B}(\mathbb{R})$, let

$$\lambda_n(A) = \lambda\big(A \cap (n, +\infty)\big).$$

Evidently $\{\lambda_n\}_{n \geqslant 1}$ is a decreasing family of Borel measures. Let

$$\mu(A) = \lim_{n \to +\infty} \lambda_n(A) \qquad \forall\, A \in \mathcal{B}(\mathbb{R}).$$

The set function μ is not a measure, since

$$\mu\big([0, +\infty)\big) = \lim_{n \to +\infty} \lambda_n\big([0, +\infty)\big) = +\infty$$

and it is not equal to

$$\sum_{k \geqslant 1} \mu\big([k-1, k)\big) = \sum_{k \geqslant 1} \lim_{n \to +\infty} \lambda_n\big([k-1, k)\big) = 0.$$

hence μ is not countably additive, thus μ is not measure.

Solution of Problem 4.44

Note that X is a Polish space and so Theorem 4.12 implies that μ is a Radon measure. Since by hypothesis $\mu\big(\{x\}\big) = 0$ so, for each $x \in X$, we can find $\delta(x) > 0$ such that

$$\mu\big(B_{\delta(x)}(x)\big) < \varepsilon.$$

Due to the compactness of X, we can find a finite set $\{x_k\}_{k=1}^n \subseteq X$ such that

$$X = \bigcup_{k=1}^{n} B_{\delta(x_k)}(x_k).$$

Let us set $\delta = \frac{1}{2} \lim_{1 \leqslant k \leqslant n} \delta(x_k) > 0$. Let $A \in \mathcal{B}(X)$ be such that $\operatorname{diam} A < \delta$. We can find $k \in \{1, \ldots, n\}$ such that

$$A \cap B_{\frac{1}{2}\delta(x_k)} \neq \emptyset,$$

so $A \subseteq B_{\delta(x_k)}(x_k)$ and thus $\mu(A) < \varepsilon$.

Solution of Problem 4.45

Let $\{A_k\}_{k \geqslant 1} \subseteq \mathcal{B}(X)$ be a sequence of disjoint Borel sets and let us set

$$A = \bigcup_{k \geqslant 1} A_k \in \mathcal{B}(X).$$

Then for a given $\varepsilon > 0$, we can find a compact set $K_\varepsilon \subseteq A$ such that

$$\mu(A) \leqslant \mu(K_\varepsilon) + \varepsilon.$$

Also, for every $k \geqslant 1$, we can find an open set $U_k \subseteq X$ such that

$$A_k \subseteq U_k \quad \text{and} \quad \mu(U_k) \leqslant \mu(A_k) + \tfrac{\varepsilon}{2^k} \qquad \forall\, k \geqslant 1.$$

For every integer $n \geqslant 1$, we have

$$\sum_{k=1}^{n} \mu(A_k) = \mu\Big(\bigcup_{k=1}^{n} A_k\Big) \leqslant \mu(A)$$

(due to the additivity of μ), so passing to the limit as $n \to +\infty$, we have

$$\sum_{k \geqslant 1} \mu(A_k) \leqslant \mu(A).$$

Since K_ε is compact and $K_\varepsilon \subseteq A$, we can find $n_1 \geqslant 1$ such that

$$K_\varepsilon \subseteq \bigcup_{k=1}^{n_1} U_k$$

and so

$$\mu(A) \leqslant \mu(K_\varepsilon)+\varepsilon \leqslant \sum_{k=1}^{n_1} \mu(U_k)+\varepsilon \leqslant \sum_{k \geqslant 1} \mu(U_k)+\varepsilon \leqslant \sum_{k \geqslant 1} \mu(A_k)+2\varepsilon.$$

Because $\varepsilon > 0$ is arbitrary, we let $\varepsilon \searrow 0$, to obtain

$$\mu(A) \leqslant \sum_{k\geqslant 1} \mu(A_k),$$

so finally

$$\mu(A) = \sum_{k\geqslant 1} \mu(A_k),$$

i.e., μ is σ-additive.

Solution of Problem 4.46
If $x \in X$ and $r_1 \neq r_2$, then

$$\partial B_{r_1}(x) \cap \partial B_{r_2}(x) = \emptyset.$$

Therefore at most countably many of the sets $\{\partial B_r(x)\}_{r>0}$ have positive measure. So, for every $x \in X$, we can find a sequence $\{r_n(x)\}_{n\geqslant 1}$ such that

$$0 < r_n(x) < \tfrac{1}{n} \quad \text{and} \quad \mu\big(\partial B_{r_n(x)}(x)\big) = 0 \qquad \forall\, n \geqslant 1.$$

Since X is compact, for every $n \geqslant 1$, we can find a finite set $\{x_k\}_{k=1}^{N_n} \subseteq X$ such that

$$X = \bigcup_{k=1}^{N_n} B_{r_n(x_k)}(x_k).$$

If $U \subseteq X$ is an open set and $x \in U$, then we can find $r > 0$ such that $B_r(x) \subseteq U$. Let $n \geqslant 1$ be such that $r > \frac{2}{n}$. We have $x \in B_{r_n(x_k)}(x_k)$ for some $k \in \{1, \ldots, N_n\}$. If $y \in B_{r_n(x_k)}(x_k)$, then

$$d(x,y) \leqslant \operatorname{diam} B_{r_n(x_k)}(x_k) < \tfrac{2}{n} < r$$

and so we have

$$x \in B_{r_n(x_k)}(x_k) \subseteq B_r(x) \subseteq U,$$

thus

$$\mathcal{B} = \{B_{r_n(x_k)}(x_k)\}_{1 \leqslant k \leqslant N_n, n \geqslant 1} = \{U_n\}_{n \geqslant 1}$$

is a basis for the topology and $\mu(\partial U_n) = 0$ for all $n \geqslant 1$.

Solution of Problem 4.47

Let $x, y \in X$, $R \geqslant r > 0$ and $x \in B_R(y)$. For every integer $k \geqslant 0$, we set

$$R_k = 2^k r \quad \text{and} \quad m = \min\{k \geqslant 0 : B_{R_k}(x) \supseteq B_R(y)\}.$$

Then from the doubling property of μ, after m-iterations, we have

$$\mu(B_R(y)) \leqslant \mu(B_{R_k}(x)) \leqslant c^m \mu(B_r(x)),$$

so

$$\frac{\mu(B_r(x))}{\mu(B_R(y))} \geqslant \frac{1}{c^m}.$$

But note that $B_R(y) \nsubseteq B_{R_{m-1}}(x)$, hence $R_{m-1} \leqslant 2R$ (recall that $x \in B_R(y)$) and $m \leqslant \log_2\left(\frac{4R}{r}\right)$. So, setting $\widehat{c} = \frac{1}{c^2}$ and $t = -\log_2 c$, we conclude that

$$\frac{\mu(B_r(x))}{\mu(B_R(y))} \geqslant \widehat{c}\left(\frac{r}{R}\right)^t.$$

Solution of Problem 4.48

Let $\mu_A \colon \mathcal{B}(X) \longrightarrow [0, +\infty)$ be the finite Borel measure, defined by

$$\mu_A(C) = \mu(A \cap C) \qquad \forall\, C \in \mathcal{B}(X).$$

Invoking Theorem 4.11, we have that μ_A is regular and so we can find a closed set $C \subseteq A$ such that

$$\mu_A(A) \leqslant \mu_A(C) + \varepsilon.$$

Then

$$\mu(A) \leqslant \mu(A \cap C) + \varepsilon$$

and so

$$\mu(A) - \mu(A \cap C) \;=\; \mu(A \setminus C) \;\leqslant\; \varepsilon.$$

Solution of Problem 4.49

Since λ^N is a Radon measure (see Problem 4.28), we can find a compact set $K \subseteq A$ such that $\lambda^N(K) > 0$. According to the Cantor–Bendixson theorem (see Problem 1.16 and Remark 2.28), we have that $K = P \cup N$ with $P \subseteq \mathbb{R}^N$ being a perfect set (see Definition 1.13) and $N \subseteq \mathbb{R}^N$ being countable set. Moreover, $P \neq \emptyset$ or otherwise $K = N$ and so $\lambda^(K) = 0$, a contradiction. Now recall that for a perfect set P, $\operatorname{card} P = \mathfrak{c}$ (see Problem 1.31). Hence

$$\mathfrak{c} \;\geqslant\; \operatorname{card} A \;\geqslant\; \operatorname{card} P \;=\; \mathfrak{c}$$

so $\operatorname{card} A = \mathfrak{c}$.

Solution of Problem 4.50

By the Egorov theorem (see Theorem 3.76), for a given $\varepsilon \in (0, 1-\vartheta)$, we can find a Lebesgue measurable set $A \subseteq [0,1]$ such that

$$\lambda(A) \;>\; 1-\varepsilon \quad \text{and} \quad f_n \rightrightarrows f \quad \text{on } A.$$

Hence the function $f|_A$ is continuous and because λ is regular, we can find a compact set $K \subseteq A$ such that

$$\lambda(K) \;>\; 1-\varepsilon \;>\; \vartheta.$$

Solution of Problem 4.51

Let $\varepsilon > 0$ and let $\{q_n\}_{n\geqslant 1}$ be an enumeration of the rational numbers in $[0,1]$. Let us set

$$K \;=\; [0,1] \setminus \bigcup_{n\geqslant 1} \big(q_n - \tfrac{\varepsilon}{2\cdot 2^n},\; q_n + \tfrac{\varepsilon}{2\cdot 2^n}\big).$$

Then K is compact and $\operatorname{int} K = \emptyset$. Moreover, if λ denotes the Lebesgue measure on $\mathbb{R}$, then

$$\lambda(K) \geqslant 1 - \sum_{n \geqslant 1} \tfrac{\varepsilon}{2 \cdot 2^n} \geqslant 1 - \varepsilon$$

since $\lambda\big([0,1] \setminus K\big) \leqslant 1 - (1 - \varepsilon) = \varepsilon$.

Solution of Problem 4.52

Let

$$h_n = \begin{cases} f(t) & \text{if } |f(t)| \leqslant n, \\ n \cdot \operatorname{sgn} f(t) & \text{if } |f(t)| > n, \end{cases}$$

where

$$\operatorname{sgn} f(t) = \begin{cases} 1 & \text{if } f(t) > 0, \\ -1 & \text{if } f(t) < 0 \end{cases} \qquad \forall\, n \geqslant 1.$$

Evidently

$$h_n(t) \longrightarrow f(t) \qquad \forall\, t \in [0,1]$$

and so $h_n \xrightarrow{\lambda} f$ (see Proposition 3.130). According to Proposition 3.110, we can find simple functions converging uniformly to h_n for all $n \geqslant 1$. Therefore, if suffices to consider a simple function

$$f(t) = \sum_{k=1}^{m} a_k \chi_{A_k}(t),$$

with A_k disjoint measurable subsets of $[0,1]$. Using Problem 4.28, we may assume that each set A_k is compact. For every $i \geqslant 1$, we can find disjoint sets U_k such that $A_k \subseteq U_k$ and

$$\lambda\Big(\bigcup_{k=1}^{m} (U_k \setminus A_k)\Big) < \tfrac{1}{i}.$$

Let $\xi = \max\limits_{1 \leqslant k \leqslant n} |a_k|$. By Problem 2.103 (see also the Urysohn lemma; Theorem 2.136), we can find a continuous function $f_i \colon [0,1] \longrightarrow [-c,c]$ such that

$$f_i = f \quad \text{on} \ \bigcup_{k=1}^{m} A_k \quad \text{and} \quad f_i = 0 \quad \text{on} \ [0,1] \setminus \bigcup_{k=1}^{m} U_k.$$

So, $\lambda(\{t \in [0,1] : f_i(t) \neq f(t)\}) < \frac{1}{i}$, which means that $f_i \xrightarrow{\lambda} f$.

Solution of Problem 4.53
Let $h \colon X \times Y \longrightarrow Y \times Y$ be the function, defined by

$$h(x,y) = (f(x), y).$$

Evidently h is $(\mathcal{B}(X) \otimes \mathcal{B}(Y),\ \mathcal{B}(Y) \otimes \mathcal{B}(Y))$-measurable. So

$$\operatorname{Gr} f = \{(x,y) \in X \times Y : f(x) = y\} = h^{-1}(\Delta_Y) \in \mathcal{B}(X) \otimes \mathcal{B}(Y).$$

Solution of Problem 4.54
Let $\{C_n\}_{n \geqslant 1} \subseteq \mathcal{Y}$ be a separating sequence for Y. Let

$$\Delta = \{(y,y) : y \in Y\}$$

be the diagonal of $Y \times Y$. Then

$$(Y \times Y) \setminus \Delta = \Big(\bigcup_{n \geqslant 1} C_n \times (Y \setminus C_n)\Big) \cup \Big(\bigcup_{n \geqslant 1} (Y \setminus C_n) \times C_n\Big),$$

so $\Delta \in \Sigma \otimes \mathcal{Y}$.

The function $\xi\colon \Omega \times Y \longrightarrow Y \times Y$, defined by

$$\xi(\omega, u) = \big(f(\omega), u\big)$$

is measurable. Finally, note that $\operatorname{Gr} f = \xi^{-1}(\Delta)$.

Solution of Problem 4.55

By recalling that $f = f^+ - f^-$, we see that we may assume without any loss of generality, that $f \geqslant 0$. Let $\{\Omega_n\}_{n\geqslant 1} \subseteq \Sigma$ be such that $\Omega_n \nearrow \Omega$ and $\mu(\Omega_n) < +\infty$ for all $n \geqslant 1$. Then, let us set $f_n = \chi_{\Omega_n} f$. We see that $f_n \nearrow f$ and so by the Lebesgue monotone convergence theorem (see Theorem 3.92), we have

$$\int_\Omega f_n \, d\mu \nearrow \int_\Omega f \, d\mu.$$

But

$$\int_\Omega f_n \, d\mu = \int_{\Omega_n} f \, d\mu \leqslant \lambda,$$

hence

$$\int_\Omega f(\omega) \, d\mu \leqslant \lambda.$$

Solution of Problem 4.56

Let $M > 0$ be such that

$$\|u\|_p \leqslant M \qquad \forall\, u \in C.$$

For every Lebesgue measurable set $A \subseteq [0,1]$, using the Hölder inequality (see Theorem 3.103), we have

$$\int_A |u| \, d\mu \leqslant \|u\|_p \|\chi_A\|_{p'} \leqslant M\lambda(A)^{\frac{1}{p'}} \qquad \forall\, u \in C,$$

where λ denotes the Lebesgue measure on $\mathbb{R}$. For a given $\varepsilon > 0$, let $\delta = \left(\frac{\varepsilon}{M}\right)^{p'} > 0$. Hence, if $\lambda(A) \leqslant \delta$, then

$$\int_A |u|\, d\mu \leqslant \varepsilon \qquad \forall\, u \in C,$$

so the set C is uniformly integrable.

Solution of Problem 4.57

"$\Longrightarrow$": For every open set $U \subseteq \mathbb{R}^N$, the set $\varphi^{-1}(U) \subseteq X$ is open (since φ is continuous). Hence $(\varphi \circ f)^{-1}(U) \in \Sigma$ and so $\varphi \circ f$ is Σ-measurable.

"$\Longleftarrow$": Let $C \subseteq X$ be a closed set. Since X is perfectly normal, for some continuous function $\varphi\colon X \longrightarrow \mathbb{R}$, we have $C = \varphi^{-1}(0)$. Then

$$f^{-1}(C) = (\varphi \circ f)^{-1}(0) \in \Sigma$$

and so f is $\big(\Sigma, \mathcal{B}(X)\big)$-measurable.

Solution of Problem 4.58

For every $n \geqslant 1$, we cover X by a finite or countable family of balls $\{B_k^n\}_{k \geqslant 1}$ of diameter less than or equal to $\frac{1}{n}$. From this cover, we generate a cover of X consisting of disjoint Borel sets $C_{n,k}$, $k \geqslant 1$ of diameter less than or equal to $\frac{1}{n}$, namely

$$C_{n,k} = B_k^n \setminus \bigcup_{i=1}^{n-1} B_i^n \qquad \forall\, k \geqslant 1.$$

From every $C_{n,k}$ we choose ξ_k and set

$$f_n(x) = \xi_k \qquad \forall\, x \in f^{-1}(C_{n,k}).$$

Then

$$d_X\big(f_n(x), f(x)\big) \leqslant \tfrac{1}{n} \qquad \forall\, x \in X,$$

hence

$$f_n \rightrightarrows f \quad \text{in } X.$$

Solution of Problem 4.59

(a) From Problem 3.149, we know that

$$\int_0^1 f\,dx = \int_0^{+\infty} \lambda(\{f > \vartheta\})\,d\vartheta \quad \text{and} \quad \int_0^1 g\,dx = \int_0^{+\infty} \lambda(\{g > \vartheta\})\,d\vartheta.$$

So, if f and g are equimeasurable, then we conclude that

$$\int_0^1 f\,dx = \int_0^1 g\,dx.$$

(b) Let

$$\begin{aligned}\mathcal{Y} = \big\{&A \subseteq [0,+\infty) :\ f^{-1}(A), g^{-1}(A) \subseteq (0,1) \text{ are Lebesgue} \\ &\text{measurable and } \lambda\big(f^{-1}(A)\big) = \lambda\big(g^{-1}(A)\big)\big\}.\end{aligned}$$

It is easy to see that $\mathcal{Y}$ is a σ-algebra. If f and g are equimeasurable, then $\mathcal{Y}$ contains all half-lines of the form $(\vartheta,+\infty)$, $\vartheta > 0$ and so it contains all Borel sets of $[0,+\infty)$. Then

$$\lambda\big(f^{-1}(\xi^{-1}(\vartheta,+\infty))\big) = \lambda\big(f^{-1}(\xi^{-1}(\vartheta,+\infty))\big) \qquad \forall\, \vartheta > 0,$$

so $\xi \circ f$ and $\xi \circ g$ are equimeasurable too.

Solution of Problem 4.60

(a) We start by observing that if $\eta\colon (0,+\infty) \longrightarrow [0,+\infty)$ is decreasing, right continuous and

$$\lim_{\vartheta \to +\infty} \eta(\vartheta) = 0,$$

then for every $r > 0$, the set $\{\eta > r\}$ is a bounded open interval of the form $\big(0, \eta^*(r)\big)$, with $\eta^*\colon (0,+\infty) \longrightarrow [0,+\infty)$ also decreasing, right continuous and vanishing at infinity (it is the inverse of η and it is defined by

$$\eta^*(r) = \inf\{\vartheta \in [0,+\infty) : \eta(\vartheta) \leqslant r\},$$

hence $(\eta^*)^* = \eta$).

Let

$$\eta(\vartheta) = \lambda(\{f > \vartheta\}) \qquad \forall\, \vartheta > 0$$

(the ***distribution function*** of f). Then η is decreasing, right continuous and

$$\eta(\vartheta) \longrightarrow 0 \quad \text{as } \vartheta \to +\infty.$$

Let $h\colon (0,1) \longrightarrow [0,+\infty)$ be a decreasing, right continuous function. Let $\widehat{h}\colon (0,+\infty) \longrightarrow [0,+\infty)$ be the extension by zero of h. If f and h are equimeasurable, then

$$\eta(\vartheta) = \lambda(\{f > \vartheta\}) = \lambda(\{h > \vartheta\}) = \lambda(\{\widehat{h} > \vartheta\}) = \widehat{h}^*(\vartheta),$$

so $\widehat{h} = \eta^*$.

(b) We have

$$\int_A f\,dx = \int_0^1 \chi_A f\,dx = \int_0^1 (\chi_A f)^*\,dr$$

(see Problem 4.59(a)). For every $\vartheta > 0$, we have

$$\lambda(\{\chi_A f > \vartheta\}) \leqslant \lambda(A)$$

and so

$$(\chi_A f)^*(\vartheta) = 0 \qquad \forall\, \vartheta \geqslant \lambda(A),$$

hence

$$(\chi_A f)^* \leqslant f^*.$$

It follows that

$$\int_A f\,dx = \int_0^{\lambda(A)} (\chi_A f)^*\,dr \leqslant \int_0^{\lambda(A)} f^*\,dr.$$

In particular, if $t \in (0,1)$, the

$$\int_0^t f\,dx \leqslant \int_0^t f^*\,dx$$

and so, we conclude that

$$\int_0^1 f\xi\,dx \;\leqslant\; \int_0^1 f^*\xi\,dx.$$

Solution of Problem 4.61

Since $u = u^+ - u^-$, without any loss of generality, we may assume that $u \geqslant 0$. Consider the set

$$E \;=\; \{x \in X:\; u(x) > 0\}.$$

For every $n \geqslant 1$, let

$$E_n \;=\; \{x \in X:\; u(x) \geqslant \tfrac{1}{n}\}.$$

Then E_n is μ-measurable and since

$$\tfrac{\mu(E_n)}{n} \;\leqslant\; \int_{E_n} u\,d\mu \;\leqslant\; \int_X u\,d\mu \;<\; +\infty,$$

we have

$$\mu(E_n) \;<\; +\infty \qquad \forall\, n \geqslant 1.$$

Since

$$E \;=\; \bigcup_{n\geqslant 1} E_n,$$

we see that E is σ-finite for μ. For every $D \in \mathcal{B}(X)$, we see that

$$\nu(D) \;=\; \nu(D \cap E).$$

So, for a given $\varepsilon > 0$, we can find $n \geqslant 1$ such that

$$\nu(D) - \nu(D \cap E_n) \;<\; \tfrac{\varepsilon}{2}.$$

Since $\nu \ll \mu$ (see Definition 3.150) and μ is a Radon measure, we can find a compact set $K \subseteq D \cap E_n$ such that

$$\nu(D \cap E_n) - \nu(K) \;=\; \int_{D\cap E_n} u\,d\mu - \int_K u\,d\mu \;<\; \tfrac{\varepsilon}{2},$$

so

$$\nu(D)-\nu(K) = \big(\nu(D)-\nu(D\cap E_n)\big)+\big(\nu(D\cap E_n)-\nu(K)\big) < \tfrac{\varepsilon}{2}+\tfrac{\varepsilon}{2} = \varepsilon$$

and thus ν is inner regular with respect to compact sets.

Also, since $\nu \ll \mu$ and μ is a Radon measure (hence outer regular; see Definition 4.9), for every $n \geqslant 1$, we can find an open set U_n such that

$$D \cap E_n \subseteq U_n \quad \text{and} \quad \nu(U_n) - \nu(D \cap E_n) < \tfrac{\varepsilon}{2^n}.$$

Then

$$U = \bigcup_{n\geqslant 1} U_n \quad \text{is an open set}$$

and $D \subseteq U$. Since

$$U \setminus D \subseteq \bigcup_{n\geqslant 1} \big(U_n \setminus (D \cap E_n)\big),$$

we have

$$0 \leqslant \nu(U)-\nu(D) = \nu(U\setminus D) \leqslant \sum_{n\geqslant 1} \nu\big(U_n\setminus(D\cap E_n)\big) \leqslant \sum_{n\geqslant 1} \tfrac{\varepsilon}{2^n} = \varepsilon$$

and so ν is outer regular.

Finally note that ν is finite (since $u \in L^1(X,\mu)$). Hence, we conclude that $\nu \in M_b(X)$.

Solution of Problem 4.62

By the Radon–Nikodym theorem (see Definition 3.150 and Theorem 3.152), we can find $f \in L^1(X,\mu)_+$ such that

$$\nu(A) = \int_A f(x)\,d\mu(x) \qquad \forall\, A \in \Sigma.$$

Recall that the Lebesgue integral is absolutely continuous. So, for a given $\varepsilon > 0$, we can find $\delta = \delta(\varepsilon) > 0$ such that

$$\mu(C) \leqslant \delta \quad \Longrightarrow \quad \nu(C) = \int_C f\,d\mu \leqslant \varepsilon.$$

From this and the fact that μ is a Radon measure, we see that ν is regular and

$$\nu(A) = \sup\{\nu(K) : K \subseteq A, K \text{ is compact}\}$$

(inner regular with respect to compact sets). Therefore ν is a Radon measure too.

Solution of Problem 4.63
Let $g: X \longrightarrow [0,1]$ be a continuous function such that

$$g|_{\operatorname{supp} u} = 1 \quad \text{and} \quad \operatorname{supp} g \text{ is compact}$$

(see Problem 2.103 and recall that X is locally compact). We have

$$|u(x)| \leqslant \xi g(x) \qquad \forall x \in \operatorname{supp} \mu$$

(see Definition 4.13), so

$$\int_X |u| \, d|\mu| \leqslant \xi \int_X g \, d|\mu| \leqslant \xi \|\mu\|$$

(since $\operatorname{supp} \mu = \operatorname{supp} |\mu|$) and thus

$$\Big| \int_X u \, d\mu \Big| \leqslant \int_X |u| \, d|\mu| \leqslant \xi \|\mu\|.$$

Solution of Problem 4.64
Since $|\mu|$ is a measure of the same support as μ (see Definition 4.13), without any loss of generality we may assume that $\mu \geqslant 0$. As in the solution of Problem 4.21, exploiting the local compactness of X, we can find $h \in C_c(X)$, $0 \leqslant h \leqslant 1$ with $h|_{\operatorname{supp} \mu} = 1$ (see Definition 4.13). Then, for all $u \in C_c(X)$, we have

$$|u(x)| \leqslant \|u\|_\infty h(x) \qquad \forall x \in \operatorname{supp} \mu,$$

so

$$\int_X |u|\,d\mu \leqslant \|u\|_\infty \int_X h\,d\mu$$

and thus μ is finite (since $u \in C_c(X)$ was arbitrary; see Theorem 4.23).

Solution of Problem 4.65

Let μ be a Borel measure on X which is finite on compact sets. Then $C_c(X) \subseteq L^1(X,\mu)$ and $l\colon C_c(X) \longrightarrow \mathbb{R}$, defined by

$$l(u) = \int_X u\,d\mu \qquad \forall\, u \in C_c(X)$$

is a positive linear functional, so by Theorem 4.23, we can find a Radon measure m corresponding to l. If $U \subseteq X$ is an open set, then by hypothesis

$$U = \bigcup_{n\geqslant 1} K_n$$

with $K_n \subseteq X$ being compact for all $n \geqslant 1$. Exploiting the local compactness of X and using Problem 2.103, we can find a function $h_1 \in C_c(X)$ such that

$$h_1\big|_{K_1} = 1, \quad 0 \leqslant h_1 \leqslant 1 \quad \text{and} \quad \operatorname{supp} h_1 \subseteq U.$$

Then by induction, for every $n \geqslant 2$, we can find $h_n \in C_c(X)$ such that

$$h_n\Big|_{\bigcup\limits_{i=1}^{n} K_i} = h_n\Big|_{\bigcup\limits_{i=1}^{n} \operatorname{supp} h_i} = 1, \quad 0 \leqslant h_n \leqslant 1 \quad \text{and} \quad \operatorname{supp} h_n \subseteq U.$$

Evidently $h_n \nearrow \chi_U$ and so by the Lebesgue monotone convergence theorem (see Theorem 3.92), we have

$$\mu(U) = \lim_{n\to+\infty} \int_X h_n\,d\mu = \lim_{n\to+\infty} \int_X h_n\,dm = m(U).$$

By Problem 4.19(a), if $A \in \mathcal{B}(X)$ and $\varepsilon > 0$, we can find an open set $W \supseteq A$ and a closed set $D \subseteq A$ such that $m(W \setminus D) < \varepsilon$. But the set $W \setminus D$ is open. So, from the previous part of the proof, we have

$$\mu(W \setminus D) \;=\; m(W \setminus D) \;<\; \varepsilon,$$

so $\mu(W) \leqslant \mu(A) + \varepsilon$ and thus μ is outer regular (see Definition 4.9). Also, we have $\mu(A) \leqslant \mu(D) + \varepsilon$ and D is σ-compact (since X is σ-compact; see Definition 2.99). So, we can find a compact set $K \subseteq A$ such that

$$\mu(A) \;\leqslant\; \mu(K) + 2\varepsilon$$

and so μ is inner regular with respect to compact sets, i.e., μ is Radon.

Solution of Problem 4.66

For every $f \in C_c(X)$ with $0 \leqslant f \leqslant 1$, we have

$$\int_X f \, d\mu_n \;\leqslant\; \mu_n(X) \qquad \forall\, n \geqslant 1.$$

Hence

$$\int_X f \, d\mu \;\leqslant\; \liminf_{n \to +\infty} \mu_n(X)$$

and so

$$\mu(X) \;=\; \sup\Big\{ \int_X f \, d\mu : \; f \in C_c(X), \; 0 \leqslant f \leqslant 1 \Big\} \;\leqslant\; \liminf_{n \to +\infty} \mu_n(X)$$

(see Theorem 4.23).

Solution of Problem 4.67

By Problem 4.66, we have $\mu(X) \leqslant \liminf\limits_{n \to +\infty} \mu_n(X)$ and so $\mu(X) < +\infty$. Recall that the embedding $C_c(X) \subseteq C_0(X)$ is dense. So, for a given $\varepsilon > 0$ and $h \in C_0(X)$, we can find $f_\varepsilon \in C_c(X)$ such that

$$\|f_\varepsilon - h\|_\infty \;\leqslant\; \varepsilon.$$

We have

$$\Big|\int_X f_\varepsilon\,d\mu_n - \int_X h\,d\mu_n\Big| \leqslant \|f_\varepsilon - h\|_\infty \mu_n(X) \leqslant \varepsilon\xi \qquad \forall\, n \geqslant 1$$

and

$$\Big|\int_X f_\varepsilon\,d\mu - \int_X h\,d\mu\Big| \leqslant \varepsilon\xi.$$

So, by the triangle inequality, we have

$$\Big|\int_X h\,d\mu_n - \int_X h\,d\mu\Big| \leqslant 2\varepsilon\xi + \Big|\int_X f_\varepsilon\,d\mu_n - \int_X f_\varepsilon\,d\mu\Big| \qquad \forall\, n \geqslant 1.$$

But, by hypothesis

$$\int_X f_\varepsilon\,d\mu_n \longrightarrow \int_X f_\varepsilon\,d\mu.$$

So, passing to the limit as $n \to +\infty$, we have

$$\limsup_{n\to+\infty}\Big|\int_X h\,d\mu_n - \int_X h\,d\mu\Big| \leqslant 2\varepsilon\xi.$$

Since $\varepsilon > 0$ was arbitrary, we let $\varepsilon \searrow 0$, to conclude that

$$\int_X h\,d\mu_n \longrightarrow \int_X h\,d\mu \qquad \forall\, h \in C_0(X).$$

Solution of Problem 4.68

"$\Longrightarrow$": Suppose that φ is a continuous linear functional on $C(X)$. Note that

$$\big\{u \in C(X):\ \|u\|_\infty \leqslant 1\big\} = \big\{u \in C(X):\ \big|u(x)\big| \leqslant 1 \text{ for all } x \in X\big\}.$$

Therefore, if φ is positive, then by the Riesz representation theorem (see Theorem 4.23; note that in our case $C(X) = C_0(X)$), we have

$$\begin{aligned}\|\varphi\|_* &= \sup\big\{\varphi(u) : u \in C(X), \|u\|_\infty \leqslant 1\big\}\\ &= \sup\big\{\varphi(u) :\ u \in C(X), \big|u(x)\big| \leqslant 1 \text{for all } x \in X\big\} = \varphi(1).\end{aligned}$$

"$\Longleftarrow$": Suppose that $\|\varphi\|_* = \varphi(1)$. Let $u \in C(X)$, $u \geqslant 0$, $u \neq 0$ and let us set $v = \frac{u}{\|u\|_\infty}$. Then $\|1 - v\|_\infty \leqslant 1$ and so $\varphi(1) - \varphi(v) = \varphi(1 - v) \leqslant \|\varphi\|_* = \varphi(1)$, hence $\varphi(v) \geqslant 0$ and thus

$$\varphi(u) = \|u\|_\infty \varphi(v) \geqslant 0 \qquad \forall\, u \in C(X),\ u \geqslant 0,$$

hence φ is positive.

Solution of Problem 4.69

For each $x \in X$, we define the positive continuous linear functional $\varphi_x \colon C(X) \longrightarrow \mathbb{R}$, by

$$\varphi_x(u) = u(x) \qquad \forall\, u \in C(X).$$

Then, for all $x, y \in X$, $x \neq y$, we have $\|\varphi_x - \varphi_y\|_* = 2$. Since by hypothesis X is uncountable, the set $\{\varphi_x\}_{x \in X} \subseteq C_0(X)^*$ is uncountable too. Then $\{B_1(\varphi_x)\}_{x \in X}$ is an uncountable family of mutually disjoint open balls. This implies that no countable subset of $C_0(X)^*$ can be dense in it. Therefore $C_0(X)^* = M_b(X)$ is not separable.

Solution of Problem 4.70

We consider the functional $\xi \colon C_0(X) \longrightarrow \mathbb{R}$, defined by

$$\xi(u) = \vartheta(Y) \max\{u(x) : x \in X\} \qquad \forall\, u \in C_0(X).$$

It is easy to see that $\xi(u + v) \leqslant \xi(u) + \xi(v)$ and

$$\xi(\lambda u) = \lambda \xi(u) \qquad \forall\, u, v \in C_0(X),\ \lambda \geqslant 0.$$

The surjectivity of f allows us to introduce the following linear subspace of $C_0(X)$,

$$V = \{v \circ f : v \in C_0(Y)\}.$$

We consider the linear functional $l \colon V \longrightarrow \mathbb{R}$, defined by

$$l(v \circ f) = \int_Y v \, d\vartheta.$$

Then

$$\begin{aligned} l(u\circ f) &\leqslant \vartheta(Y)\max\{v(y):\ y\in Y\} \\ &= \vartheta(Y)\max\{(v\circ f)(x):\ x\in X\} \ = \ \xi(v\circ f) \end{aligned}$$

(since f is surjective). Then the Hahn–Banach theorem (see Theorem 5.24) implies that we can find a linear functional $\widehat{l}\colon C_0(X)\longrightarrow\mathbb{R}$ such that $\widehat{l}|_V = l$ and $\widehat{l}(u)\leqslant\xi(u)$ for all $u\in C_0(X)$. Note that $\widehat{l}(u)\leqslant 0$ if $u\leqslant 0$ and so

$$\widehat{l}(u) \ = \ -\widehat{l}(-u) \qquad \forall\ u\in C_0(X),\ u\geqslant 0.$$

Clearly $\widehat{l}$ is continuous. Therefore $\widehat{l}$ is a positive, continuous, linear functional on $C_0(X)$ and so by Theorem 4.23, we can fine $\mu\in M_b(X)$, $\mu\geqslant 0$ which is necessarily Radon, since X is compact (see Theorem 4.12) such that

$$\widehat{l}(u) \ = \ \int_X u\,d\mu \qquad \forall\ u\in C_0(X).$$

Since $\widehat{l}$ extends l, we have

$$\int_Y v\,d\vartheta \ = \ \int_X (v\circ f)\,d\mu \ = \ \int_Y v\,d(\mu\circ f^{-1}) \qquad \forall\ v\in C_0(Y)$$

and so $\vartheta=\mu f^{-1}$ by the uniqueness of the measure in Theorem 4.23.

Solution of Problem 4.71

"(b) $\Longrightarrow$ (a)": Clearly, if $\varphi = c\delta_x$ for some $c\geqslant 0$ and $x\in X$, the desired lattice property holds.

"(a) $\Longrightarrow$ (b)": Suppose that $\varphi\in C_0(X)^*$, $\varphi\geqslant 0$ has the described lattice property. By Theorem 4.23, we can find $\mu\in M_b(X)$, $\mu\geqslant 0$ such that

$$\varphi(f) \ = \ \int_X f\,d\mu \qquad \forall\ f\in C_0(X).$$

If $x, y \in \operatorname{supp}\mu$ (see Definition 4.13) and $x \neq y$, then we can find $f, h \in C_c(X)$ such that

$$f \geqslant 0, \quad h \geqslant 0, \quad \min\{f, h\} = 0 \quad \text{and} \quad f(x) = h(y) = 1.$$

So, we have

$$\varphi(\max\{f, h\}) = \varphi(f + h) = \varphi(f) + \varphi(h) > \max\{\varphi(f), \varphi(h)\},$$

a contradiction. Therefore $\operatorname{supp}\mu$ is a singleton, i.e., $\operatorname{supp}\mu = \{x\}$. Let $c = \mu(\{x\}) > 0$. Then, for every $f \in C_0(X)$, we have

$$\varphi(f) = \int_X f\,d\mu = f(x)\mu(\{x\}) = cf(x),$$

so $\varphi = c\delta_x$.

Solution of Problem 4.72

Let $\xi\colon X \longrightarrow \overline{\mathbb{R}}_+ = [0, +\infty]$ be a lower semicontinuous function, let $x \in X$ be such that $\xi(x) > 0$ and choose $\vartheta \in (0, \xi(x))$. Then the set $\{\xi > \vartheta\}$ is nonempty open and so by Problem 2.103, we can find $g \in C_c(X)$ such that

$$g(x) = \vartheta \quad \text{and} \quad 0 \leqslant g \leqslant \vartheta\chi_U \leqslant \xi.$$

Hence, if $\xi(x) > 0$, then

$$\xi(x) = \sup\{g(x) : g \in C_c(X),\ 0 \leqslant g \leqslant \xi\}.$$

Clearly the same is true if $\xi(x) = 0$. Therefore

$$\xi = \sup\{g : g \in C_c(X),\ 0 \leqslant g \leqslant \xi\}.$$

From Proposition 2.55(a) we know that f is lower semicontinuous. Also let X^* be the Alexandrov one-point compactification of X (see Remark 2.97). Since $C(X^*)$ is separable, $\mathcal{Y}$ is directed and using Proposition 4.21 and the last equation, we can find an increasing sequence $\{g_n\}_{n\geqslant 1} \subseteq C_c(X)$ such that

$$0 \leqslant g_n \leqslant h_n \qquad \forall\, n \geqslant 1$$

for some $h_n \in \mathcal{Y}$ and $g_n \nearrow f$. Then from the Lebesgue monotone convergence theorem (see Theorem 3.92), we have

$$\int_X f\,d\mu \;=\; \lim_{n\to+\infty} \int_X g_n\,d\mu \;=\; \sup_{n\geqslant 1} \int_X g_n\,d\mu.$$

But note that

$$\sup_{n\geqslant 1} \int_X g_n\,d\mu \;\leqslant\; \sup_{h\in\mathcal{Y}} \int_X h\,d\mu \;\leqslant\; \int_X f\,d\mu.$$

Therefore, we conclude that

$$\int_X f\,d\mu \;=\; \sup_{h\in\mathcal{Y}} \int_X h\,d\mu.$$

Remark. Therefore, if X is a locally compact topological space, μ is a Radon measure on X and $f\colon X \longrightarrow \overline{\mathbb{R}}_+ = [0,+\infty]$ is lower semicontinuous, then

$$\int_X f\,d\mu \;=\; \sup\Big\{\int_X g\,d\mu:\; g\in C_c(X),\; 0\leqslant g\leqslant f\Big\}.$$

Solution of Problem 4.73

Let $\{s_n\}_{n\geqslant 1}$ be a sequence of nonnegative simple functions such that

$$s_n(x) \nearrow f(x) \qquad \forall\, x\in X.$$

We have

$$f \;=\; s_1 + \sum_{n\geqslant 2} (s_n - s_{n-1})$$

and in the series each term is a nonnegative simple function. So, we write that

$$f \;=\; \sum_{k\geqslant 1} \vartheta_k \chi_{A_k},$$

with $\vartheta_k > 0$ and $A_k \in \mathcal{B}(X)$ for all $k \geqslant 1$.

(a) Since μ is a Radon measure, for a given $\varepsilon > 0$, we can find an open set $U_k \supseteq A_k$ such that

$$\mu(U_k) \leqslant \mu(A_k) + \tfrac{\varepsilon}{2^k \vartheta_k}.$$

Let us set

$$h = \sum_{k \geqslant 1} \vartheta_k \chi_{U_k}.$$

Recall that the characteristic function of an open set is lower semicontinuous. So, h is lower semicontinuous and clearly

$$f \leqslant h \quad \text{and} \quad \int_X h \, d\mu \leqslant \int_X f \, d\mu + \varepsilon.$$

Since $\varepsilon > 0$ was arbitrary, we conclude that

$$\int_X f \, d\mu = \inf \Big\{ \int_X h \, d\mu : \ h \text{ is lower semicontinuous on } X, \ h \geqslant f \Big\}.$$

(b) Let

$$\sigma < \int_X f \, d\mu.$$

Then for $N \geqslant 1$ large enough, we will have

$$\sum_{k=1}^{N} \vartheta_k \mu(A_k) > \sigma.$$

Since μ is a Radon measure, for all $k \geqslant 0$, we can find a compact set $K_k \subseteq A_k$ such that

$$\sum_{k=1}^{N} \vartheta_k \mu(A_k) > \sigma.$$

Recall that the characteristic function of a closed set is upper semicontinuous. Therefore

$$\eta = \sum_{k=1}^{N} \vartheta_k \chi_{K_k} \geqslant 0$$

is upper semicontinuous,

$$0 \leqslant \eta \leqslant f \quad \text{and} \quad \int_X \eta \, d\mu > \sigma.$$

Since $\sigma < \int_X f\,d\mu$ was arbitrary, we conclude that

$$\int_X f\,d\mu = \sup\Big\{\int_X \eta\,d\mu :\ \eta \text{ is upper semicontinuous on } X,\ 0 \leqslant \eta \leqslant f\Big\}.$$

Solution of Problem 4.74

Let $A \in \mathcal{B}(X)$ and let $\varepsilon > 0$ be given. We can find a compact set $K \subseteq X$ and an open set $U \subseteq X$ such that

$$\big|\mu(B)\big| < \varepsilon \qquad \forall\, B \in \mathcal{B}(X),\ B \subseteq U \setminus K.$$

Then we have

$$0 \leqslant \mu^+(A) - \mu^+(K) = \mu^+(A \setminus K) = \sup\big\{\mu(B) :\ B \in \mathcal{B}(X),\ B \subseteq A \setminus K\big\}$$

and

$$0 \leqslant \mu^+(U) - \mu^+(A) = \mu^+(U \setminus A) = \sup\big\{\mu(B) :\ B \in \mathcal{B}(X),\ B \subseteq U \setminus A\big\}$$

(see Theorem 3.146). Therefore

$$\mu^+(A) - \mu^+(K) < \varepsilon \quad \text{and} \quad \mu^+(U) - \mu^+(A) < \varepsilon.$$

This implies that μ^+ is a finite Radon measure on X. Similarly for μ^-. So, we conclude that $\mu = \mu^+ - \mu^- \in M_b(X)$.

Solution of Problem 4.75

Due to the separability of X and Y, we can treat X and Y as subsets of the Hilbert cube $\mathcal{H} = [0,1]^{\mathbb{N}}$ (see Theorem 2.154(b)). Let d be a metric on $\mathcal{H} \times \mathcal{H}$ consistent with its topology and let $g \in UC_b(X \times Y)$ (here $UC_b(X \times Y)$ denotes the space of bounded d-uniformly continuous functions on $X \times Y$ with values in $\mathbb{R}$). Then g admits a uniformly continuous extension $\widehat{g} \in UC(\mathcal{H} \times \mathcal{H}) = C(\mathcal{H} \times \mathcal{H})$. From the Stone–Weierstrass theorem, we know that for a given $\varepsilon > 0$, we can find functions $\widehat{f}_k, \widehat{h}_k \in C(\mathcal{H}) = UC(\mathcal{H})$, for $k \geqslant 1$ such that

$$\sup_{x,y \in \mathcal{H}} \Big|\sum_{k=1}^{m} \widehat{f}_k(x)\widehat{h}_k(y) - \widehat{g}(x,y)\Big| \leqslant \varepsilon.$$

If $m_n \xrightarrow{w} m$ in $M_1^+(X)$ and $\mu_n \xrightarrow{w} \mu$ in $M_1^+(Y)$ and $f_k = \widehat{f}|_X$, $h_k = \widehat{h}|_Y$, then

$$\begin{aligned}
&\limsup_{n \to +\infty} \Big| \int\limits_{X \times Y} g \, d(m_n \times \mu_n) - \int g \, d(m \times \mu) \Big| \\
\leqslant\ & \limsup_{n \to +\infty} \Big| \int\limits_{X \times Y} \big(g - \sum_{k=1}^{m} f_k h_k\big) \, d(m_n \times \mu_n) \Big| \\
& + \sum_{k=1}^{m} \lim_{n \to +\infty} \Big| \int\limits_X f_k \, dm_n \int\limits_Y h_k \, d\mu_n - \int\limits_X f_k \, dm \int\limits_Y h_k \, d\mu \Big| \\
& + \lim_{n \to +\infty} \Big| \int\limits_{X \times Y} \big(\sum_{k=1}^{m} f_k h_k - g\big) \, d(m \times \mu) \Big| \\
\leqslant\ & 2\varepsilon.
\end{aligned}$$

Let $\varepsilon \searrow 0$ and use the Portmanteau theorem (see Theorem 4.26), to conclude that

$$\xi(m_n, \mu_n) \xrightarrow{w} \xi(m, \mu),$$

so ξ is continuous.

Solution of Problem 4.76

We do the solution for f being lower semicontinuous and bounded below. The other case can be done similarly. To show the weak lower semicontinuity of ξ, we need to show that for every $\eta \in \mathbb{R}$, the set

$$I_\eta = \{\mu \in M_1^+(X) : \xi(\mu) \leqslant \eta\}$$

is weakly closed. So, let $\{\mu_\alpha\}_{\alpha \in I} \subseteq I_\eta$ be a net and assume that

$$\mu_\alpha \xrightarrow{w} \mu \quad \text{in } M_1^+(X).$$

From Problem 2.20, we know that there exists a sequence $\{f_n\}_{n \geqslant 1} \subseteq C_b(X)$ such that

$$f_n(x) \nearrow f(x) \qquad \forall\, x \in X.$$

We have

$$\int\limits_X f_n \, d\mu_\alpha \leqslant \int\limits_X f \, d\mu_\alpha = \xi(\mu_\alpha) \leqslant \eta \qquad \forall\, \alpha \in I,\ n \geqslant 1.$$

From the Portmanteau theorem (see Theorem 4.26), we have

$$\lim_{\alpha \in I} \int_X f_n \, d\mu_\alpha = \int_X f_n \, d\mu \leqslant \eta \qquad \forall \, n \geqslant 1$$

and by the Lebesgue monotone convergence theorem (see Theorem 3.92), we have

$$\int_X f_n \, d\mu \nearrow \int_X f \, d\mu = \xi(\mu) \leqslant \eta,$$

thus $\mu \in I_\eta$ and so I_η is weakly closed, i.e., ξ is lower semicontinuous.

Solution of Problem 4.77

(a) Let $f = \chi_U : X \longrightarrow \mathbb{R}$ be the characteristic function of U. Since U is open, χ_U is lower semicontinuous. So, by Problem 4.76 above, the function

$$\xi(\mu) = \int_X \chi_U \, d\mu$$

is weakly lower semicontinuous. So, the set

$$D = \{\mu \in M_1^+(X) : \mu(U) > \eta\} = \{\mu \in M_1^+(X) : \xi(\mu) > \eta\}$$

is open.

(b) In this case χ_C is upper semicontinuous and so the function

$$\xi(\mu) = \int_X \chi_C \, d\mu$$

is upper semicontinuous (see Problem 4.76). Therefore, the set

$$E = \{\mu \in M_1^+(X) : \mu(C) \geqslant \eta\} = \{\mu \in M_1^+(X) : \xi(\mu) \geqslant \eta\}$$

is closed.

Solution of Problem 4.78

Let τ denote the topology of X and let

$$\mathcal{X} = \{A \in \mathcal{B}(X) : \xi_A \text{ is Borel}\}.$$

From Problem 4.77, we see that $\mathcal{X} \supseteq \tau$. In particular $X \in \mathcal{X}$. Also, if $A, C \in \mathcal{X}$ and $A \subseteq C$, then $\xi_{C \setminus A} = \xi_C - \xi_A$, which shows that $\xi_{C\setminus A}$ is Borel and so $C \setminus A \in \mathcal{X}$. Finally, if $\{A_n\}_{n \geqslant 1} \subseteq \mathcal{X}$ is a sequence such that $A_n \nearrow A$, then $\xi_{A_n} \nearrow \xi_A$ and so ξ_A is Borel, hence $A \in \mathcal{X}$. So, we infer that $\mathcal{X}$ is a λ-class (Dynkin class; see Definition 3.7(a)). Because $\tau \subseteq \mathcal{X}$ and τ is closed under finite intersections, we apply Theorem 3.9 and obtain $\mathcal{X} = \sigma(\tau) = \mathcal{B}(X)$.

Solution of Problem 4.79

Let $\{u_\alpha\}_{\alpha \in I} \subseteq C_b(X)$ and $\{\mu_\alpha\}_{\alpha \in I} \subseteq M_1^+(X)$ be nets such that

$$u_\alpha \xrightarrow{\|\cdot\|_\infty} u \quad \text{and} \quad \mu_\alpha \xrightarrow{w} \mu.$$

Then, we have

$$\begin{aligned} \big|\gamma(u_\alpha, \mu_\alpha) - \gamma(u, \mu)\big| &= \Big|\int_X u_\alpha \, d\mu_\alpha - \int_X u \, d\mu\Big| \\ &\leqslant \Big|\int_X (u_\alpha - u) \, d\mu_\alpha\Big| + \Big|\int_X u \, d\mu_\alpha - \int_X u \, d\mu\Big| \\ &\leqslant \|u_\alpha - u\|_\infty + \Big|\int_X u \, d\mu_\alpha - \int_A u \, d\mu\Big| \longrightarrow 0 \end{aligned}$$

(by the Portmanteau theorem; see Theorem 4.26). Thus γ is jointly measurable.

Solution of Problem 4.80

Let $\xi \colon X \longrightarrow M_1^+(X)$ be defined by $\xi(x) = \delta_x$. Evidently ξ is injective. Also, if $\{x_\alpha\}_{\alpha \in I}$ is a net such that $x_\alpha \longrightarrow x$ in X, then for all $u \in C_b(X)$, we have

$$\int_X u \, d\delta_{x_\alpha} = u(x_\alpha) \longrightarrow u(x) = \int_X u \, d\delta_x,$$

so, using the Portmanteau theorem (see Theorem 4.26), we have

$$\xi(x_\alpha) = \delta_{x_\alpha} \xrightarrow{w} \delta_x = \xi(x)$$

and thus ξ is continuous.

Conversely, let

$$\xi(x_\alpha) = \delta_{x_\alpha} \xrightarrow{w} \delta_x = \xi(x) \quad \text{in } M_1^+(X).$$

Suppose that

$$\xi^{-1}(\xi(x_0)) = x_\alpha \not\longrightarrow x = \xi^{-1}(\xi(x)) \quad \text{in } X.$$

Then we can find $U \in \mathcal{N}(x)$ and a subnet $\{x_\beta\}_\beta$ of $\{x_\alpha\}_{\alpha \in I}$ such that

$$x_\beta \in X \setminus U \qquad \forall \beta.$$

Recall that a metric space is completely regular (see Problem 2.42 and Theorem 1.43). So, we can find a continuous function $h \colon X \longrightarrow [0,1]$ such that

$$h(x) = 0 \quad \text{and} \quad h|_{X \setminus U} = 1,$$

so

$$\int\limits_X h \, d\delta_{x_\beta} = h(x_\beta) = 1 \not\longrightarrow 0 = h(x) = \int\limits_X h \, d\delta_x,$$

a contradiction. This proves that ξ^{-1} is continuous too and so ξ is a homeomorphism into $M_1^+(X)$.

Finally, we show that $\xi(X)$ is sequentially closed in $M_1^+(X)$. So, let $\{x_n\}_{n\geqslant 1} \subseteq X$ be a sequence and assume that

$$\delta_{x_n} \xrightarrow{w} \mu \quad \text{in } M_1^+(X).$$

If $\{x_n\}_{n\geqslant 1}$ does not have a convergent subsequence, then the set $D = \{x_n : n \geqslant 1\}$ is a closed subset of X and so is every $C \subseteq D$. By the Portmanteau theorem (see Theorem 4.26), we have

$$\limsup_{n \to +\infty} \delta_{x_n}(C) \leqslant \mu(C).$$

Hence, for every infinite set $A \subseteq D$, we have $\mu(A) = 1$, a contradiction to the fact that μ is a measure. So, we can find a subsequence $\{x_{n_k}\}_{k\geqslant 1}$ of a sequence $\{x_n\}_{n\geqslant 1}$ such that

$$x_{n_k} \longrightarrow x \quad \text{in } X,$$

so

$$\xi(x_{n_k}) = \delta_{x_{n_k}} \xrightarrow{w} \delta_x = \xi(x) \quad \text{in } M_1^+(X).$$

Thus $\mu = \delta_x$ and $\xi(X)$ is sequentially closed in $M_1^+(X)$.

Solution of Problem 4.81

Without any loss of generality, we assume that $f(X) = Y$. We start by showing that ϑ is injective. So, let $\mu, m \in M_1^+(X)$, $\mu \neq m$. By Problem 4.14(a), we can find an open set $U \subseteq X$ such that $\mu(U) \neq m(U)$. Since f is a homeomorphism, we have that the set $V = f(U) \subseteq Y$ is open and $U = f^{-1}(V)$. Then

$$\vartheta(\mu)(V) = \mu(f^{-1}(V)) = \mu(U) \neq m(U) = m(f^{-1}(V)) = \vartheta(m)(V)$$

and so ϑ is injective.

Next we show the continuity of ϑ. So, assume that

$$\mu_\alpha \xrightarrow{w} \mu \quad \text{in } M_1^+(X)$$

and let $V \subseteq Y$ be an open set. Then the set $f^{-1}(V) = U \subseteq X$ is open and so by the Portmanteau theorem (see Theorem 4.26), we have

$$\liminf_\alpha \vartheta(\mu_\alpha)(V) = \liminf_\alpha \mu_\alpha(U) \geqslant \mu(U) = \vartheta(\mu)(V).$$

Since $V \subseteq Y$ was an arbitrary open set, Theorem 4.26 implies that

$$\vartheta(\mu_\alpha) \xrightarrow{w} \vartheta(\mu) \quad \text{in } M_1^+(Y)$$

and this proves the continuity of ϑ.

Finally we show that ϑ^{-1} is continuous. To this end, suppose that

$$\vartheta(\mu_\alpha) \xrightarrow{w} \vartheta(\mu) \quad \text{in } M_1^+(Y).$$

Let $U \subseteq X$ be an open set. Then $f(U) = V \subseteq Y$ is an open set too. We have

$$\begin{aligned} \liminf_\alpha \mu_\alpha(U) &= \liminf_\alpha \mu_\alpha(f^{-1}(V)) = \liminf_\alpha \vartheta(\mu_\alpha)(V) \\ &\geqslant \vartheta(\mu)(V) = \mu(f^{-1}(V)) = \mu(U) \end{aligned}$$

(see Theorem 4.26). Thus

$$\mu_\alpha \xrightarrow{w} \mu \quad \text{in } M_1^+(X)$$

(see Theorem 4.26). Therefore ϑ^{-1} is continuous and so ϑ is a homeomorphism.

Solution of Problem 4.82

"(a) $\Longrightarrow$ (b)": Let $A \in \mathcal{B}(X)$ be a bounded set with $\mu(\partial A) = 0$ and let $U \subseteq X$ be a bounded and open set such that $\overline{A} \subseteq U$. For a given integer $k \geqslant 1$, we define

$$D_k = \{x \in U : \operatorname{dist}(x, \overline{A}) < \tfrac{1}{k}\}.$$

By Problem 2.103, we can find a continuous function $f_k : X \longrightarrow [0,1]$ such that

$$f_k\big|_{\overline{A}} = 1 \quad \text{and} \quad f_k\big|_{D_k^c} = 0.$$

Since the set U is bounded, $f_k \in C_c(X)$. We have

$$\begin{aligned}\big|\mu_n(A) - \mu(A)\big| \leqslant\ & \Big|\int_X (\chi_A - f_k)\, d\mu_n\Big| + \Big|\int_X f_k\, d\mu_n - \int_X f_k\, d\mu\Big| \\ & + \Big|\int_X (f_k - \chi_A)\, d\mu\Big|.\end{aligned}$$

Note that

$$(f_k - \chi_A)\big|_A = 0 \quad \text{and} \quad (f_k - \chi_A)\big|_{D_k^c} = 0.$$

Hence, we have

$$\Big|\int_X (f_k - \chi_A)\, d\mu\Big| \leqslant \mu(D_k \setminus A) \leqslant \mu(\overline{D}_k \setminus \operatorname{int} A).$$

Similarly, we obtain that

$$\Big|\int_X (\chi_A - f_k)\, d\mu_n\Big| \leqslant \mu_n(\overline{D}_k \setminus \operatorname{int} A).$$

Moreover, since by hypothesis (statement **(a)**), $\mu \xrightarrow{v} \mu$ and $f \in C_c(X)$, we have

$$\Big|\int_X f_k\, d\mu_n - \int_X f\, d\mu\Big| \longrightarrow 0.$$

The set $\overline{D}_k \setminus \operatorname{int} A$ is closed and so the function $\chi_{\overline{D}_k \setminus \operatorname{int} A}$ is upper semicontinuous. Then according to Problem 2.20(c), we can find a sequence $\{h_{km}\}_{m \geqslant 1} \subseteq C(X)$ with $\operatorname{supp} h_{km} \subseteq \overline{U}$ such that $h_{km} \searrow \chi_{\overline{D}_k \setminus \operatorname{int} A}$ as $m \to +\infty$. Since $h_{km} \in C_c(X)$, we have

$$\begin{aligned} \mu_n(\overline{D}_k \setminus \operatorname{int} A) &= \int_X \chi_{\overline{D}_k \setminus \operatorname{int} A}(x)\, d\mu_n \\ &\leqslant \int_X h_{km}(x)\, d\mu_n \longrightarrow \int_X h_{km}\, d\mu \quad \text{as } n \to +\infty, \end{aligned}$$

so

$$\limsup_{n \to +\infty} \mu_n(\overline{D}_k \setminus \operatorname{int} A) \leqslant \int_X h_{km}(x)\, d\mu.$$

Because $\operatorname{supp} h_{km} \subseteq \overline{U}$ and $\mu(\overline{U}) < +\infty$ (recall that μ is a Radon measure and the set $\overline{U}$ is compact; see Definition 4.9(e)), we can let $m \to +\infty$ and then by the Lebesgue monotone convergence theorem (see Theorem 3.92), we have

$$\limsup_{n \to +\infty} \mu_n(\overline{D}_k \setminus \operatorname{int} A) \leqslant \mu(\overline{D}_k \setminus \operatorname{int} A).$$

Therefore, finally we have

$$\limsup_{n \to +\infty} \big|\mu_n(A) - \mu(A)\big| \leqslant 2\mu(\overline{D}_k \setminus \operatorname{int} A).$$

Sending $k \to +\infty$, we have

$$\overline{D}_k \setminus \operatorname{int} A \searrow \overline{A} \setminus \operatorname{int} A = \partial A.$$

But by hypothesis $\mu(\partial A) = 0$. Therefore, we conclude that $\mu_n(A) \longrightarrow \mu(A)$.

"(b) $\Longrightarrow$ (a)": Let $f \in C_c(X)$. By Proposition 2.94, we can find two bounded, open sets V and W in X such that

$$K \subseteq V \subseteq \overline{V} \subseteq W,$$

where K is compact and $\operatorname{supp} f \subseteq K$. We have

$$\overline{V} = \bigcap_{\delta > 0} D_\delta,$$

where

$$D_\delta = \{x \in W : \operatorname{dist}(x, \overline{V}) < \delta\}.$$

Evidently each D_δ is open and

$$\partial D_\delta = \{x \in W : \operatorname{dist}(x, \overline{V}) = \delta\}.$$

Hence, the sets $\{D_\delta\}_{\delta>0}$ have disjoint boundaries and because $\mu(\overline{W}) < +\infty$, we have $\mu(D_\delta) = 0$ for some $\delta > 0$. Therefore, without any loss of generality, we can have $\mu(\partial V) = 0$.

If A is a Borel subset of V, the closure of A relative to V is given by $\overline{A}^V = \overline{A} \cap V$. Hence, the boundary of A relative to V is $\partial_V A = (\partial A) \cap V$. Suppose that $\mu(\partial_V A) = 0$. Then $\mu\big((\partial A) \cap V\big) = 0$. Also $\partial A \cap V^c \subseteq \overline{A} \cap V^c \subseteq \overline{V} \setminus V = \partial V$. Hence $\mu(\partial A \cap V^c) = 0$. Therefore $\mu(\partial A) = 0$. If $\mu_n' = \mu_n|_V$ and $\mu' = \mu|_V$, then

$$\mu_n' \xrightarrow{w} \mu' \quad \text{in } M_1^+(V)$$

(see the Portmanteau theorem; Theorem 4.26). Hence

$$\int_V f\, d\mu_n \longrightarrow \int_V f\, d\mu,$$

so

$$\int_X f\, d\mu_n \longrightarrow \int_X f\, d\mu$$

(recall that $\operatorname{supp} f \subseteq K \subseteq V$) and thus $\mu_n \xrightarrow{v} \mu$ (since $f \in C_c(X)$ was arbitrary).

Solution of Problem 4.83

"$\Longrightarrow$": Suppose that $C \subseteq M_1^+(X)$ is uniformly tight (see Theorem 4.31). Let $\{\varepsilon_n\}_{n\geqslant 0} \subseteq \mathbb{R}_+$ be a sequence such that

$$\sum_{n\geqslant 0} \varepsilon_n < +\infty.$$

Let $K_n = K_{\varepsilon_n}$ be an increasing sequence of compact subsets of X such that

$$\mu(K_n) \geqslant 1 - \tfrac{1}{n} \qquad \forall\, n \geqslant 1,\ \mu \in C.$$

Let

$$\varphi(x) = \inf\{n \geqslant 0 : x \in K_n\} = \sum_{n \geqslant 0} \chi_{X \setminus K_n}(x).$$

Then clearly

$$\sup_{\mu \in C} \int_X \varphi(x)\, d\mu < +\infty$$

and φ has compact sublevel sets.

"$\Longleftarrow$": If

$$\sup_{\mu \in C} \int_X \varphi(x)\, d\mu < +\infty,$$

then by Markov inequality (see Proposition 3.90), the definition of uniform tightness is satisfied by the compact sublevel sets of φ.

Solution of Problem 4.84

Let $g \in C_c^\infty(\mathbb{R}^N)$ be such that

$$0 \leqslant g \leqslant 1, \quad g(x) = \begin{cases} 1 & \text{if } \|x\| \leqslant \frac{1}{2}, \\ 0 & \text{if } \|x\| \geqslant 1. \end{cases}$$

Let

$$g_k(x) = g\left(\tfrac{1}{k}x\right) \qquad \forall\, k \geqslant 1.$$

Then by hypothesis, we have

$$\liminf_{n \to +\infty} \mu_n\left(\overline{B}_k(0)\right) \geqslant \lim_{n \to +\infty} \int_{\mathbb{R}^N} g_k(x)\, d\mu_n = \int_{\mathbb{R}^N} g_k(x)\, d\mu.$$

Also, from the Lebesgue dominated convergence theorem (see Theorem 3.94), we have

$$\lim_{k \to +\infty} \int_{\mathbb{R}^N} g_k(x)\, d\mu = 1.$$

So, for a given $\varepsilon > 0$, we can find $n_0 \geqslant 1$ and $k \geqslant 1$ large enough such that

$$\mu_n\left(\overline{B}_k(0)\right) \geqslant 1 - \varepsilon \qquad \forall\, n \geqslant n_0.$$

Then from this and Theorem 4.12, we conclude that the set $\{\mu_n : n \geqslant 1\} \cup \{\mu\}$ is uniformly tight. Then the Prohorov theorem (see Theorem 4.31) implies that we can find a subsequence $\{\mu_{n_k}\}_{k\geqslant 1}$ of $\{\mu_n\}_{n\geqslant 1}$, such that

$$\mu_{n_k} \xrightarrow{w} \widehat{\mu}.$$

Hence

$$\int_X f\, d\mu = \int_X f\, d\widehat{\mu} \qquad \forall\, f \in C_c^\infty(\mathbb{R}^N),$$

so $\mu = \widehat{\mu}$.

But from Theorem 4.30, we know that $\big(M_1^+(\mathbb{R}^N), w\big)$ is a Polish space. So, by the Urysohn criterion (see Problem 1.3), we conclude that the original sequence $\{\mu_n\}_{n\geqslant 1}$ converges weakly to μ.

Solution of Problem 4.85

Let $f \in C_b(Y)$. By adding a constant if necessary, we may assume that $f \geqslant 0$. If $K \subseteq X$ is compact, then

$$f \circ \vartheta_n \rightrightarrows f \circ \vartheta \quad \text{on } K$$

and

$$\lim_{n\to+\infty} \int_K \big((f\circ\vartheta_n) - (f\circ\vartheta)\big)\, d\mu_n = 0$$

(since K is compact and the sequence $\{\mu_n(K)\}_{n\geqslant 1}$ is bounded), so

$$\begin{aligned}
\liminf_{n\to+\infty} \int_X (f\circ\vartheta_n)\, d\mu_n &\geqslant \liminf_{n\to+\infty} \int_K (f\circ\vartheta_n)\, d\mu_n \\
&= \lim_{n\to+\infty} \int_K \big((f\circ\vartheta_n) - (f\circ\vartheta)\big)\, d\mu_n \\
&\quad + \liminf_{n\to+\infty} \int_K (f\circ\vartheta)\, d\mu_n \\
&= \liminf_{n\to+\infty} \int_K (f\circ\vartheta)\, d\mu_n
\end{aligned}$$

$$\geqslant \big(-\sup f\big)\sup_{n\geqslant 1}\mu_n(X\setminus K) + \liminf_{n\to+\infty}\int_X (f\circ\vartheta)\,d\mu_n$$
$$= \big(-\sup f\big)\sup_{n\geqslant 1}\mu_n(X\setminus K) + \int_X (f\circ\vartheta)\,d\mu$$

(since $f\circ\vartheta\in C(K)$). Since $\{\mu_n\}_{n\geq 1}$ is uniformly tight (see Theorem 4.31), we can find an increasing sequence $\{K_m\}_{m\geq 1}$ of compact sets in X such that

$$\lim_{m\to+\infty}\sup_{n\geqslant 1}\mu_n(X\setminus K_m) = 0.$$

So, using in the above inequality $K=K_m$ and letting $m\to+\infty$, we conclude that

$$\liminf_{n\to+\infty}\int_Y f\,d(\mu_n\vartheta_n^{-1}) \geqslant \int_Y f\,d(\mu\vartheta^{-1}).$$

replacing f by $-f$, we also have

$$\limsup_{n\to+\infty}\int_Y f\,d(\mu_n\vartheta_n^{-1}) \leqslant \int_Y f\,d(\mu\vartheta^{-1}),$$

so

$$\lim_{n\to+\infty}\int_Y f\,d(\mu_n\vartheta_n^{-1}) = \int_Y f\,d(\mu\vartheta^{-1})$$

and thus $\mu_n\vartheta_n^{-1}\xrightarrow{w}\mu\vartheta^{-1}$ in $M_1^+(Y)$.

Solution of Problem 4.86

Let $\mu\in M_1^+(Y)$ (see Definition 4.25) and let

$$\mu_1 = \mu\vartheta_1^{-1} \quad\text{and}\quad \mu_2 = \mu\vartheta_2^{-1}.$$

By hypothesis, for every $\varepsilon>0$, we can find compact sets $K_1\subseteq X_1$ and $K_2\subseteq X_2$ such that

$$\mu_1(X_1\setminus K_1) \leqslant \tfrac{\varepsilon}{2} \quad\text{and}\quad \mu_2(X_2\setminus K_2) \leqslant \tfrac{\varepsilon}{2} \qquad \forall\,\mu\in E.$$

It follows that

$$\mu(Y \setminus \vartheta_1^{-1}(K_1)) \leqslant \tfrac{\varepsilon}{2} \quad \text{and} \quad \mu(Y \setminus \vartheta_2^{-1}(K_2)) \leqslant \tfrac{\varepsilon}{2}$$

and so

$$\mu\big(Y \setminus (\vartheta_1^{-1}(K_1) \cap \vartheta_2^{-1}(K_2))\big) \leqslant \mu(Y \setminus \vartheta_1^{-1}(K_1)) + \mu(Y \setminus \vartheta_2^{-1}(K_2)) \leqslant \varepsilon$$
$$\forall\, \mu \in E.$$

But note that

$$\vartheta_1^{-1}(K_1) \cap \vartheta_2^{-1}(K_2) \;=\; \vartheta^{-1}(K_1 \times K_2)$$

is compact, since by hypothesis ϑ is proper and $K_1 \times K_2$ is compact. Therefore, we conclude that $E \subseteq M_1^+(Y)$ is uniformly tight (see Theorem 4.31).

Solution of Problem 4.87

Let $C \subseteq X$ be a nonempty closed set and let

$$U_n \;=\; \big\{x \in X :\ \operatorname{dist}(x, C) < \tfrac{1}{n}\big\} \qquad \forall\, n \geqslant 1.$$

Then U_n is open for every $n \geqslant 1$ and

$$C \;=\; \bigcap_{n \geqslant 1} U_n.$$

Then sets C and U_n^c are disjoint closed and

$$\inf\big\{d(x, y) :\ x \in C,\ y \in U_n^c\big\} \;\geqslant\; \tfrac{1}{n} \;>\; 0.$$

So, if we set

$$f_n(x) \;=\; \frac{\operatorname{dist}(x, U_n^c)}{\operatorname{dist}(x, U_n^c) + \operatorname{dist}(x, C)} \qquad \forall\, x \in X,$$

then

$$f_n \in UC_b(X), \quad 0 \leqslant f_n \leqslant 1, \quad f_n|_{U_n^c} \;=\; 0, \quad f_n|_C \;=\; 1.$$

We have

$$\mu(C) \;\leqslant\; \int_X f_n \, d\mu \;=\; \int_X f_n \; d\nu \;=\; \int_{U_n} f_n \, d\nu \;\leqslant\; \nu(U_n).$$

Letting $n \to +\infty$, we have $\mu(C) \leqslant \nu(C)$. Reversing the roles of μ and ν in the above argument, we infer that

$$\mu(C) = \nu(C) \qquad \forall\, C \subseteq X,\ C \text{ is closed}.$$

But μ and ν are regular (see Theorem 4.11). So, we conclude that $\mu = \nu$.

Solution of Problem 4.88

Let $u\colon X \longrightarrow M_1^+(X \times Y)$ be defined by

$$u(x) = \delta_x \times m(x, \cdot).$$

Since both functions $X \ni x \longmapsto \delta_x \in M_1^+(X)$ and $X \ni x \longmapsto m(x, \cdot) \in M_1^+(Y)$ are continuous, we see that u is continuous. Also, let $\eta_f\colon M_1^+(X \times Y) \longrightarrow \mathbb{R}$ be defined by

$$\eta_f(\nu) = \int\limits_{X \times Y} f\, d\nu \qquad \forall\, \nu \in M_1^+(X \times Y).$$

From the definition of the weak topology this function is continuous too. Finally note that $h = \eta_f \circ u$, so h is continuous.

Solution of Problem 4.89

Because of Theorem 4.27, we can work with sequences. So, let $\{\mu_n\}_{n \geqslant 1} \subseteq M_1^+(C)$ be a sequence such that

$$\mu_n \xrightarrow{w} \mu \quad \text{in } M_1^+(X).$$

Let $U = C^c$. Then $U \subseteq X$ is an open set and by the Portmanteau theorem (see Theorem 4.26), we have

$$0 = \liminf_{n \to +\infty} \mu_n(U) \geqslant \mu(U),$$

i.e., $\mu(U) = 0$. Thus $\operatorname{supp}\mu \subseteq C$ (see Definition 4.13) and so $\mu \in M_1^+(C)$.

Solution of Problem 4.90

That $\mu \star \nu \in M_1^+(X)$ follows at once from the Fubini theorem (see Theorem 3.115) and we also have

$$(\mu \star \nu)(A) = \int\limits_X \nu(x^{-1}A)\, d\mu(x).$$

Because of Theorem 4.27, it suffices to show the sequential continuity. Let $\{\mu_n\}_{n\geqslant 1}, \{\nu_n\}_{n\geqslant 1} \subseteq M_1^+(X)$ be two sequences and assume that

$$\mu_n \xrightarrow{w} \mu, \quad \nu_n \xrightarrow{w} \nu \quad \text{in } M_1^+(X).$$

Then for every $f \in C_b(X)$, we have

$$\int\limits_X f\, d(\mu \star \nu) = \int\limits_X f(xy)\, d\mu(x)\, d\nu(y) = \int\limits_X f(xy)\, d(\mu \times \nu)(x, y).$$

Since $f \in C(X)$, we have that the function $(x, y) \longmapsto f(xy)$ belongs in $C(X \times X)$. From Problem 4.75, we have

$$\int\limits_X f(xy)\, d(\mu_n \times \nu_n) \longrightarrow \int\limits_X f(xy)\, d(\mu \times \nu),$$

so

$$\int\limits_X f\, d(\mu_n \star \nu_n) \longrightarrow \int\limits_X f\, d(\mu \star \nu)$$

and thus

$$\mu_n \star \nu_n \xrightarrow{w} \mu \star \nu \quad \text{in } M_1^+(X),$$

i.e., $\star$ is continuous.

Solution of Problem 4.91

Let $\mathcal{F}$ be the smallest σ-algebra on $M_1^+(X)$ that makes all functions $\{\eta_A\}_{A\in\mathcal{F}}$ measurable. Let

$$\mathcal{D} = \big\{A \in \mathcal{B}(X) : \eta_A \text{ is Borel measurable}\big\}.$$

It is easy to see that $\mathcal{D}$ is a λ-class (a Dynkin class; see Definition 3.7(b)) and contains all closed (or open) sets in X (see the Portmanteau theorem; Theorem 4.26). By the Dynkin theorem (see Theorem 3.9), we have that $\mathcal{D} = \mathcal{B}(X)$. Therefore $\mathcal{F} \subseteq \mathcal{B}(M_1^+(X))$.

The function

$$M_1^+(X) \ni \mu \longmapsto \vartheta_f(\mu) = \int_X f \, d\mu \in \mathbb{R}$$

is $\mathcal{F}$-measurable, when $f = \chi_A$ for all $A \in \mathcal{B}(X)$. It follows that ϑ_f is $\mathcal{F}$-measurable, when f is Borel simple. Let $f \in C_b(X)$. We can find an increasing sequence of simple functions $\{s_n\}_{n \geqslant 1}$ such that

$$-M \leqslant s_n \leqslant f \qquad \forall \, n \geqslant 1$$

for some $M > 0$ and $s_n \nearrow f$. By the Lebesgue monotone convergence theorem (see Theorem 3.92), we have $\vartheta_{s_n} \nearrow \vartheta_f$, hence ϑ_f is $\mathcal{F}$-measurable. Hence for every $\varepsilon > 0$, $\mu \in M_1^+(X)$ and $f \in C_b(X)$, the subbasic open set

$$\mathcal{U}_\varepsilon(\mu; f) = \{\nu \in M_1^+(X) : \Big| \int_X f \, d\nu - \int_X f \, d\mu \Big| < \varepsilon\}$$

is $\mathcal{F}$-measurable. It follows that $\mathcal{F} = \mathcal{B}(M_1^+(X))$.

Solution of Problem 4.92

Let $C \in \Sigma \otimes \mathcal{B}(X)$ and let $\eta_C \colon \Omega \times M_1^+(X) \longrightarrow [0, 1]$ be the function, defined by

$$\eta_C(\omega, \mu) = (\delta_\omega \times \mu)(C) \qquad \forall \, (\omega, \mu) \in \Omega \times M_1^+(X).$$

Let

$$\mathcal{Y} = \{C \in \Sigma \otimes \mathcal{B}(X) : \ \eta_C \text{ is } \Sigma \otimes \mathcal{B}(X)\text{-measurable}\}.$$

If $C = A \times B$ with $A \in \Sigma$, $B \in \mathcal{B}(X)$, then

$$\eta_C(\omega, x) = \chi_A(\omega)\mu(B)$$

and so by Problem 4.91, we have that η_C is $\Sigma \otimes \mathcal{B}(X)$-measurable, hence

$$\{A \times B : A \in \Sigma, B \in \mathcal{B}(X)\} \subseteq \mathcal{Y}.$$

Note that $\mathcal{Y}$ is an algebra and a monotone class (see Definition 3.10). Therefore invoking Theorem 3.12, we conclude that $\mathcal{Y} = \Sigma \otimes \mathcal{B}(X)$.

Now note that

$$\operatorname{Gr} S = \{(\omega, \mu) \in \Omega \times M_1^+(X) : \mu(F(\omega)) = 1\} = \eta_{\operatorname{Gr} F}^{-1}(\{1\}) \in \Sigma \otimes \mathcal{B}(X).$$

Solution of Problem 4.93

"$\Longrightarrow$": Let $A \in \mathcal{Y}$ and let $\eta_A : M_1^+(Y) \longrightarrow [0,1]$ be defined by

$$\eta_A(\mu) = \mu(A) \qquad \forall \mu \in M_1^+(Y).$$

We see that $\vartheta_A = \eta_A \circ \xi$. By hypothesis we know ξ is Borel measurable, while from Problem 4.91, we have that η_A is Borel measurable too. Therefore, it follows that $\vartheta_A = \eta_A \circ \xi$ is Borel measurable.

"$\Longleftarrow$": By Problem 4.91, we have

$$\begin{aligned} \xi^{-1}(\mathcal{B}(M_1^+(Y))) &= \xi^{-1}(\sigma(\bigcup_{A \in \mathcal{Y}} \eta_A^{-1}(\mathcal{B}(\mathbb{R})))) \\ &= \sigma(\bigcup_{A \in \mathcal{Y}} \xi^{-1}(\eta_A^{-1}(\mathcal{B}(\mathbb{R})))) = \sigma(\bigcup_{A \in \mathcal{Y}} \vartheta_A^{-1}(\mathcal{B}(\mathbb{R}))) \subseteq \mathcal{B}(X) \end{aligned}$$

(recall that $\vartheta_A = \eta_A \circ \xi$).

Solution of Problem 4.94

According to Problem 2.20(a), there exists a sequence $\{f_n\}_{n \geqslant 1} \subseteq C_b(X \times Y)$ such that $f_n \nearrow f$. Let

$$\vartheta_n(x) = \int_Y f_n(x,y)\xi(x,dy) \qquad \forall x \in X.$$

From Problem 4.88, we know that $\vartheta_n \in C_b(X)$ and from the Lebesgue monotone convergence theorem (see Theorem 3.92), we have $\vartheta_n \nearrow \vartheta$ and this by Problem 2.20(a) implies that ϑ is lower semicontinuous and of course bounded below.

Solution of Problem 4.95

Since (X, Σ) is separable, Σ is countably generated (see Definition 3.14). Let $\{C_n\}_{n \geqslant 1}$ be a sequence of generators of Σ. We claim that the sequence $\{C_n\}_{n \geqslant 1}$ separates points of X. Suppose that for some $x, x' \in X$, $x \neq x'$, we have

$$\chi_{C_n}(x) = \chi_{C_n}(x') \qquad \forall\, n \geqslant 1.$$

We define $\Sigma_0 = \{C \in 2^X : \chi_C(x) = \chi_C(x')\}$. Evidently Σ_0 is a σ-algebra and by hypothesis $\{C_n\}_{n \geqslant 1} \subseteq \Sigma_0$ and so $\Sigma \subseteq \Sigma_0$, which contradicts the separability of (X, Σ) (see Definition 3.14(b)). Let $f\colon X \longrightarrow \{0,1\}^{\mathbb{N}}$ be defined by $f(x) = \{\chi_{C_n}(x)\}_{n \geqslant 1}$. From the separating property of $\{C_n\}_{n \geqslant 1}$ we see that f is an injection and of course f is measurable. Let $A = f(X)$. We need to show that $f^{-1}\colon A \longrightarrow X$ is measurable. To this end, we need to show that if $C \in \Sigma$, then $f(C) \in \mathcal{B}(A)$. Let

$$\Sigma_1 = \{C \in 2^X : f(C) \in \mathcal{B}(X)\}.$$

Clearly Σ_1 is a σ-algebra and $\{C_n\}_{n \geqslant 1} \subseteq \Sigma_1$, because

$$f(C_n) = \{\{\lambda_k\}_{k \geqslant 1} \in \{0,1\}^{\mathbb{N}} : \lambda_n = 1\} \cap A.$$

Therefore $\Sigma \subseteq \Sigma_1$ and so f^{-1} is measurable, hence f is an isomorphism (see Definition 4.39).

Solution of Problem 4.96

Let $A \in \mathcal{B}(Y)$ and let proj_X (respectively, proj_Y) be the projection on X (respectively, on Y). We have

$$f^{-1}(A) = \mathrm{proj}_X\big(\mathrm{proj}_Y^{-1}(A) \cap \mathrm{Gr}\, f\big).$$

Note that the set $\mathrm{proj}_Y^{-1}(A) \cap \mathrm{Gr}\, f$ is a Borel subset of $\mathrm{Gr}\, f$ and so it is a Souslin subset of $\mathrm{Gr}\, f$ (see Definition 2.156). Therefore $f^{-1}(A) \subseteq X$ is a Souslin set. Similarly, we show that the set $f^{-1}(A^c) \subseteq X$ is a Souslin set. Since $f^{-1}(A)$ and $f^{-1}(A^c)$ form a Souslin partition of X, from Corollary 4.35, we infer that they are Borel sets. Hence f is a Borel function.

Solution of Problem 4.97

Since X is a Borel space, we can find a Polish space Y (see Definition 2.150) and a homeomorphism $f: X \longrightarrow Y$ such that $Z = f(X) \in \mathcal{B}(Y)$. By Problem 4.81, the function $\vartheta: M_1^+(X) \longrightarrow M_1^+(Y)$, defined by

$$\vartheta(\mu)(A) = \mu(f^{-1}(A)) \qquad \forall\, A \in \mathcal{B}(X)$$

is a homeomorphism too. Then

$$\vartheta\big(M_1^+(X)\big) = \big\{m \in M_1^+(Y) : m(Z) = 1\big\} = \xi_Z^{-1}(\{1\}),$$

where $\xi_Z: M_1^+(Y) \longrightarrow [0,1]$ is the function, defined by

$$\xi_Z(m) = m(Z) \qquad \forall\, m \in M_1^+(Y).$$

From Problem 4.78, we know that ξ_Z is Borel. Hence

$$\xi_Z^{-1}(\{1\}) = \vartheta\big(M_1^+(X)\big) \subseteq M_1^+(Y) \quad \text{is Borel.}$$

Since ϑ is a homeomorphism, we conclude that $M_1^+(X)$ is a Borel space too.

Solution of Problem 4.98

Let $g_m = \max\{g, -m\}$ for $m \geqslant 1$. Then g_m is still lower semicontinuous, it is bounded below and $g_m \geqslant g$ for all $m \geqslant 1$. Since $\mu_n \xrightarrow{w} \mu$, by Problem 4.76, we have

$$\liminf_{n \to +\infty} \int_X g_m \, d\mu_n \geqslant \int_X g_m \, d\mu \qquad \forall\, m \geqslant 1.$$

Note that

$$\sup_{n\geqslant 1}\Big(\int\limits_X g_m\,d\mu_n - \int\limits_X g\,d\mu_n\Big) \leqslant \sup_{n\geqslant 1}\int\limits_{\{g^-\geqslant m\}} g^-\,d\mu_n \longrightarrow 0 \quad \text{as } m\to+\infty.$$

Recalling that $g_m \geqslant g$ for all $m \geqslant 1$, we conclude that

$$\int\limits_X g\,d\mu \leqslant \liminf_{n\to+\infty}\int\limits_X g\,d\mu_n.$$

Applying this result to $g = f$ and $g = -f$, we obtain that

$$\lim_{n\to+\infty}\int\limits_X f\,d\mu_n = \int\limits_X f\,d\mu.$$

Solution of Problem 4.99

For every $k \geqslant 1$, let

$$f_k = \min\{f, k\} \quad \text{and} \quad C_k = \{f \geqslant k\}.$$

Clearly f_k is continuous and bounded and C_k is closed. We have

$$\begin{aligned}\limsup_{n\to+\infty}\int\limits_{\{f\geqslant k\}} f\,d\mu_n &= \limsup_{n\to+\infty}\Big(\int\limits_X (f-f_k)\,d\mu_n + k\mu_n(C_k)\Big)\\ &\leqslant \int\limits_X (f-f_k)\,d\mu + k\mu(C_k) = \int\limits_{C_k} f\,d\mu\end{aligned}$$

[see Problem 4.98 and the Portmanteau theorem (Theorem 4.26)]. For any $\varepsilon > 0$, we can find $k \geqslant 1$ large enough such that

$$\int\limits_{C_k} f\,d\mu < \varepsilon.$$

Solution of Problem 4.100

Note that $\operatorname{Gr} f^{-1} = u(\operatorname{Gr} f)$, where $u\colon X \times Y \longrightarrow Y \times X$ is the homeomorphism, defined by

$$u(x,y) = (y,x).$$

Hence $\operatorname{Gr} f^{-1} \in \mathcal{B}(Y \times X) = \mathcal{B}(Y) \otimes \mathcal{B}(X)$ if and only if $\operatorname{Gr} f \in \mathcal{B}(X \times Y) = \mathcal{B}(X) \otimes \mathcal{B}(Y)$ (see Theorem 4.6). Then the result follows from Corollary 4.38.

Solution of Problem 4.101

Let $h\colon A \longrightarrow X$ be the isomorphism between the measurable spaces $(A, \mathcal{B}(A))$ and (X, Σ). Let $D \in \Sigma$. Then $h^{-1}(D) \in \mathcal{B}(A)$ and so there exists a set $E \in \mathcal{B}(Z)$ such that

$$h^{-1}(D) = E \cap A,$$

which is a Souslin (analytic) set in Z (see Definition 2.156). Then

$$f(D) = (f \circ h)(E \cap A)$$

and the latter is a Souslin set (see Corollary 4.38).

Solution of Problem 4.102

First suppose that f is a characteristic functions, i.e., $f = \chi_A$ for some $A \in \mathcal{B}(X)$. Then

$$\gamma_f(\mu) = \gamma_{\chi_A}(\mu) = \int_X \chi_A \, d\mu = \mu(A),$$

so the function $\mu \longmapsto \gamma_f(\mu)$ is Borel measurable by Problem 4.91.

The linearity of the integral implies the measurability of γ_f, when f is a simple function. Finally let f be a general Borel measurable function. Then according to Theorem 3.68, we can find a sequence $\{s_n\}_{n\geqslant 1}$ of simple functions such that

$$0 \leqslant |s_1| \leqslant \ldots \leqslant |s_n| \leqslant \ldots \leqslant |f|$$

and

$$s_n(x) \longrightarrow f(x) \quad \mu\text{-almost everywhere on } X.$$

Then by the Lebesgue monotone convergence theorem (see Theorem 3.92), we have

$$\gamma_{s_n}(\mu) = \int_X s_n \, d\mu \longrightarrow \int_X f \, d\mu = \gamma_f(\mu),$$

so the function $\mu \longmapsto \gamma_f(\mu)$ is Borel measurable.

Solution of Problem 4.103
From Corollary 4.38, we know that

$$\operatorname{Gr} f \in \mathcal{B}(\mathbb{R}^N \times \mathbb{R}^M) = \mathcal{B}(\mathbb{R}^N) \otimes \mathcal{B}(\mathbb{R}^M).$$

Note that $f(D) = \operatorname{proj}_{\mathbb{R}^M} \operatorname{Gr} f$. The projection on $\mathbb{R}^M$ is a continuous map and the continuous image of Souslin sets in $\mathbb{R}^N \times \mathbb{R}^M$ (in particular of Borel sets in $\mathbb{R}^N \times \mathbb{R}^M$) are still Souslin. Therefore $f(D) \subseteq \mathbb{R}^M$ is a Souslin set (see Definition 2.156).

Solution of Problem 4.104
Let $U \subseteq X$ be an nonempty open set. The complete regularity of X implies that for every $x \in U$, we can find a continuous function $f_x \colon X \longrightarrow [0, 1]$ such that

$$f_x(x) = 1 \quad \text{and} \quad f_x|_{U^c} = 0.$$

Let

$$U_x = \{u \in X : f_x(u) > 0\}.$$

Then $\{U_x\}_{x \in U}$ is an open cover of X. Recall that a Souslin space is strongly Lindelöff (see Definitions 2.156 and 2.163 and Corollary 2.165). So, we can find a countable subcover $\{U_{x_n}\}_{n \geqslant 1}$ of U. Let

$$f = \sum_{n \geqslant 1} \tfrac{1}{2^n} f_{x_n}.$$

Evidently f is continuous and $f|_{U^c} = 0$. For every $u \in U$, we have $u \in U_{x_n}$ for some $n \geqslant 1$ and so $f_{x_n}(u) > 0$. Hence

$$U = \{u \in X : f(u) > 0\}$$

and this proves that X is perfectly normal (see Definition 2.137).

Solution of Problem 4.105

First we show that every set of the form

$$(\star) \qquad \{x \in X : \{f_k(x) : k \geqslant 1\} \in B\}.$$

is Baire. Note that this is true if B is closed. Indeed, since $\mathbb{R}^\infty$ is a Polish space, B has the form $B = \eta^{-1}(0)$ for some continuous function $\eta\colon \mathbb{R}^\infty \longrightarrow \mathbb{R}$ and the function $x \longmapsto \eta(\{f_n(x) : n \geqslant 1\})$ is continuous. Let us fix a sequence $\{f_n\}_{n\geqslant 1}$ and let

$$\mathcal{D}_0 = \{B \in \mathcal{B}(\mathbb{R}^\infty) : \{x \in X : \{f_n(x) : n \geqslant 1\} \subset B\} \in \mathcal{B}a(X)\}.$$

It is easy to check that $\mathcal{D}_0$ is a σ-algebra. Hence it contains $\mathcal{B}(\mathbb{R}^\infty)$ and so $\mathcal{D}_0 = \mathcal{B}(\mathbb{R}^\infty)$. On the other hand, the class $\mathcal{X}$ of all sets A representable in the form $(\star)$ with $f_k \in C_b(X)$ contains all sets of the form $\{f > 0\}$ with $f \in C(X)$. In addition, it is easy to see that $\mathcal{X}$ is a σ-algebra. Hence $\mathcal{X} = \mathcal{B}a(X)$.

Solution of Problem 4.106

By Problem 4.105, we may assume that $X = \mathbb{R}^\infty$. Since $\mathbb{R}^\infty$ is a Polish space, we can find a closed set $C \subseteq B$ and an open set $U \supseteq B$ such that $|\mu|(U \setminus C) < \frac{\varepsilon}{2}$. Let $\xi\colon X \longrightarrow [0,1]$ be a continuous function such that

$$\xi|_C = 1 \quad \text{and} \quad \xi|_{U^c} = 0.$$

Clearly ξ is the required function.

Solution of Problem 4.107

Let $C \subseteq X$ be a closed set. Then for $(\omega, x) \in \Omega \times X$, we have that $f(\omega, x) \in C$ if and only if for every $n \geqslant 1$, there exists $x' \in D$ such that

$$d_X(x, x') \leqslant \tfrac{1}{n} \quad \text{and} \quad d_Y\big(f(\omega, x), f(\omega, x')\big) \leqslant \tfrac{1}{n}.$$

Therefore, we have

$$\begin{aligned} f^{-1}(C) &= \bigcap_{n\geqslant 1} \bigcup_{x'\in D} \{\omega \in \Omega : \operatorname{dist}_Y(f(\omega, x'), C) \leqslant \tfrac{1}{n}\} \\ &\quad \times \big\{x \in X : d_X(x, x') \leqslant \tfrac{1}{n}\big\}, \end{aligned}$$

so $f^{-1}(C) \in \Sigma \times \mathcal{B}(X)$ and thus f is jointly measurable.

Solution of Problem 4.108

Let U be an open subset of X. Since X is a metric space, we have that U is an F_σ-set (see Problem 1.79). Hence

$$U = \bigcup_{n\geqslant 1} C_n,$$

with C_n closed in X. Then

$$G^-(U) = \bigcup_{n\geqslant 1} G^-(C_n).$$

But by hypothesis

$$G^-(C_n) \in \mathcal{A} \qquad \forall\, n \geqslant 1$$

and $\mathcal{A}$ is closed under countable unions. Therefore, we conclude that $G^-(U) \in \mathcal{A}$.

Solution of Problem 4.109

For any open set $U \subseteq X$ and any integer $k \geqslant 1$, we set

$$U_k = \big\{x \in U : \operatorname{dist}_X(x, U^c) > \tfrac{1}{k}\big\}.$$

Since U is open, we have $U = \bigcup_{k\geqslant 1} U_k = \bigcup_{k\geqslant 1} \overline{U}_k$. Then, for every $\omega \in \Omega$, we have the following implications:

$$\begin{aligned} f(\omega) \in U &\implies \exists k \geqslant 1: \lim_{n\to+\infty} f_n(\omega) \in U_k \\ &\implies \exists k \geqslant 1\ \exists N \geqslant 1\ \forall n \geqslant N: f_n(\omega) \in U_k \\ &\implies \exists k \geqslant 1: f(\omega) \in \overline{U}_k \\ &\implies f(\omega) \in U. \end{aligned}$$

Therefore, it follows that $f^{-1}(U) = \bigcup_{k\geqslant 1} \liminf_{n\to+\infty} f_n^{-1}(U_k)$, so $f^{-1}(U) \in \Sigma$ and thus f is Σ-measurable.

Solution of Problem 4.110

"$\Longrightarrow$": Immediate from the definition of S_F^p.

"$\Longleftarrow$": According to Theorem 4.59, we can find $\{f_n\colon \Omega \longrightarrow X\}_{n\geqslant 1}$, a sequence of Σ-measurable selections of F such that

$$F(\omega) \subseteq \overline{\{f_n(\omega)\}_{n\geqslant 1}} \quad \mu\text{-almost everywhere on } \Omega.$$

Then

$$m(\omega) = \inf_{n\geqslant 1} \|f_n(\omega)\| \quad \text{for } \mu\text{-almost all } \omega \in \Omega,$$

so $m \in L^p(\Omega)_+$.
Let $\varepsilon\colon \Omega \longrightarrow \mathbb{R}_+ \setminus \{0\}$ be a Σ-measurable function such that $\varepsilon \in L^p(\Omega)$. We introduce the multifunction

$$S_\varepsilon(\omega) = \{x \in F(\omega): \|x\| \leqslant m(\omega) + \varepsilon(\omega)\}.$$

Evidently $\operatorname{Gr} S_\varepsilon \in \Sigma \otimes \mathcal{B}(X)$ and so we can apply the Yankov–von Neumann–Aumann selection theorem (see Theorem 4.57; see also Remark 4.58) and obtain a Σ-measurable function $f\colon \Omega \longrightarrow X$ such that $f(\omega) \in S_\varepsilon(\omega)$ μ-almost everywhere on *Omega*. Then $f \in S_F^p$ and so $S_F^p \neq \emptyset$.

Solution of Problem 4.111

It suffices to show that for every $\vartheta \in \mathbb{R}$, the set $\{\omega \in \Omega : \ m(\omega) < \vartheta\}$ belongs in the σ-algebra Σ. Note that

$$\{\omega \in \Omega : \ m(\omega) < \vartheta\} \ = \ \mathrm{proj}_\Omega \{(\omega, x) \in \mathrm{Gr}\, F : \ \xi(\omega, x) < \vartheta\}.$$

But the joint measurability of ξ and the graph measurability of F imply that

$$\{(\omega, x) \in \mathrm{Gr}\, F : \ \xi(\omega, x) < \vartheta\} \ \in \ \Sigma \otimes \mathcal{B}(X).$$

Then invoking the Yankov–von Neumann–Aumann projection theorem (see Theorem 4.65), we infer that

$$\mathrm{proj}_\Omega \{(\omega, x) \in \mathrm{Gr}\, F : \ \xi(\omega, x) < \vartheta\} \ \in \ \Sigma$$

and so for every $\vartheta \in \mathbb{R}$, we have

$$\{\omega \in \Omega : \ m(\omega) < \vartheta\} \ \in \ \Sigma,$$

which proves that m is Σ-measurable.

Solution of Problem 4.112

Let $\lambda \in \mathbb{R}$ and let

$$L_\lambda \ = \ \{x \in X : \ m(x) < \lambda\}.$$

Then

$$L_\lambda \ = \ \mathrm{proj}_X \{(x, y) \in X \times Y : \ f(x, y) < \lambda\}.$$

Invoking the Yankov–von Neumann–Aumann projection theorem (see Theorem 4.65), we conclude that m is measurable with respect to every Borel measure on X. Similarly for M.

Solution of Problem 4.113

From Theorem 4.55, we know that there is a sequence $\{f_n\}_{n \geqslant 1}$ of Σ-measurable selections of F, i.e., for all $n \geqslant 1$, $f_n : \Omega \longrightarrow X$ is Σ-measurable and $f_n(\omega) \in F(\omega)$ for all $\omega \in \Omega$) such that

$$F(\omega) \ = \ \overline{\{f_n(\omega)\}_{n \geqslant 1}} \qquad \forall\, \omega \in \Omega.$$

The upper semicontinuity of $\xi(\omega, \cdot)$ for all $\omega \in \Omega$ implies that

$$m(\omega) = \inf_{n \geqslant 1} \xi\big(\omega, f_n(\omega)\big).$$

But since ξ is jointly measurable, for every $n \geqslant 1$, the function $\omega \longmapsto \xi\big(\omega, f_n(\omega)\big)$ is Σ-measurable. Hence the function m is Σ-measurable too.

Solution of Problem 4.114

From Problem 4.111, we know that the function

$$\omega \longmapsto \eta(\omega) = \sup_{x \in F(\omega)} \xi(\omega, x)$$

is Σ-measurable and so the integral $\int_\Omega \eta(\omega)\, d\mu$ is well defined (possibly $+\infty$).

For every $u \in S_F^p$, we have

$$\xi\big(\omega, u(\omega)\big) \leqslant \eta(\omega) \quad \mu\text{-almost everywhere}$$

and so

$$\sup_{u \in S_F^p} I_\xi(u) \leqslant \int_\Omega \eta\, d\mu.$$

In particular, we have

$$-\infty < I_\xi(u_0) \leqslant \int_\Omega \eta\, d\mu.$$

If $I_\xi(u_0) = +\infty$, then there is nothing to prove. So, we may assume that $I_\xi(u_0) < +\infty$. This means that $\xi\big(\cdot, u_0(\cdot)\big) \in L^1(\Omega)$. Let

$$\beta < \int_\Omega \eta\, d\mu.$$

If we can find $u \in S_F^p$ such that

$$I_\xi(u) > \beta,$$

then we have the desired equality of the problem. To this end let $\{\Omega_n\}_{n\geqslant 1} \subseteq \Sigma$ be an increasing sequence such that

$$\Omega = \bigcup_{n\geqslant 1} \Omega_n \quad \text{and} \quad \mu(\Omega_n) < +\infty \qquad \forall\, n \geqslant 1.$$

Also, let $\delta \in L^1(\Omega)$ be such that

$$\delta(\omega) > 0 \qquad \forall\, \omega \in \Omega.$$

We set

$$A_n = \Omega_n \cap \{\omega \in \Omega : \xi(\omega, u_0(\omega)) \leqslant n\} \in \Sigma$$

and

$$\eta_n(\omega) = \begin{cases} \eta(\omega) - \frac{\delta(\omega)}{n} & \text{if } \omega \in A_n \text{ and } \eta(\omega) \leqslant n, \\ n - \frac{\delta(\omega)}{n} & \text{if } \omega \in A_n \text{ and } \eta(\omega) > n, \\ \xi(\omega, u_0(\omega)) & \text{if } \omega \in A_n^c. \end{cases}$$

Evidently $\eta_n \in L^1(\Omega)$ and $\eta_n \nearrow \eta$ in μ-measure. Hence by passing to a suitable subsequence if necessary, we may assume that

$$\eta_n(\omega) \longrightarrow \eta(\omega) \quad \mu\text{-almost everywhere on } \Omega$$

(see Proposition 3.131). Invoking the Lebesgue monotone convergence theorem (see Theorem 3.92), we can find $n_0 \geqslant 1$ such that

$$\beta < \int_\Omega \eta_{n_0}\, d\mu.$$

Let $G\colon \Omega \longrightarrow 2^X$ be defined by

$$G(\omega) = F(\omega) \cap \{x \in X : \eta_{n_0}(\omega) \leqslant \xi(\omega, x)\}.$$

By modifying G on a μ-null set, we may assume that

$$G(\omega) \neq 0 \qquad \forall\, \omega \in \Omega.$$

The joint measurability of ξ and the graph measurability of F imply that $\operatorname{Gr} G \in \Sigma_\mu \otimes \mathcal{B}(X)$. Invoking the Yankov–von Neumann–Aumann selection theorem (see Theorem 4.57), we obtain a Σ_μ-measurable selection $g\colon \Omega \longrightarrow X$ of G. Let

$$C_n = \Omega_n \cap \{\omega \in \Omega : \|g(\omega)\| \leqslant n\} \in \Sigma_\mu \qquad \forall\, n \geqslant 1$$

and

$$f_n = \chi_{C_n} g + \chi_{C_n^c} u_0 \qquad \forall\, n \geqslant 1.$$

Evidently $f_n \in S_F^p$ for $n \geqslant 1$ and

$$\begin{aligned} I_\xi(f_n) &= \int\limits_{C_n} \xi(\omega, g(\omega))\, d\mu + \int\limits_{C_n^c} \xi(\omega, u_0(\omega))\, d\mu \\ &\geqslant \int\limits_{\Omega} \eta_{n_0}(\omega)\, d\mu + \int\limits_{C_n^c} \big(\xi(\omega, u_0(\omega)) - \eta_{n_0}(\omega)\big)\, d\mu. \end{aligned}$$

Note that

$$\mu(C_n^c) \searrow 0$$

and recall that

$$\int\limits_{\Omega} \eta_{n_0}(\omega)\, d\mu > \beta.$$

So, for some $n_1 \geqslant 1$, we will have $\beta < I_\xi(f_{n_1})$ as desired.

Solution of Problem 4.115

By Problem 4.113, the function $\omega \longmapsto |F(\omega)|$ is Σ-measurable. Moreover, Problem 4.114 implies that

$$\int\limits_{\Omega} |F(\omega)|^p\, d\mu = \int\limits_{\Omega} \sup_{x \in F(\omega)} \|x\|^p\, d\mu = \sup_{f \in S_F^p} \|f\|_p^p.$$

Therefore, the set $S_F^p \subseteq L^p(\Omega; X)$ is bounded if and only if $|F(\cdot)| \in L^p(\Omega)_+$.

Solution of Problem 4.116

By Theorem 4.51(b), it suffices to show that for every closed set $C \subseteq X$, we have $F^-(C) \in \Sigma$. Since X is σ-compact (see Definition 2.99), we can find a sequence $\{K_n\}_{n \geqslant 1}$ of compact subsets of X such that

$$X = \bigcup_{n \geqslant 1} K_n.$$

We have

$$F^{-}(C) = F^{-}\Big(\bigcup_{n\geqslant 1}(C\cap K_n)\Big) = \bigcup_{n\geqslant 1} F^{-}(C\cap K_n) \in \Sigma.$$

Solution of Problem 4.117

"$\Longrightarrow$": By Problem 4.116, it suffices to show that for every $K \in P_k(Y)$, we have that $G^{-}(K) \in \Sigma \otimes \mathcal{B}(X)$. By definition

$$G^{-}(K) = \{(\omega, x) \in \Omega \times X : \exists y \in Y \text{ such that } (x,y) \in F(\omega)\cap \\ \times(X\times K)\}.$$

Let

$$A_K = \{\omega\in\Omega : F(\omega)\cap(X\times K)\neq\emptyset\}.$$

Then Theorem 4.51(d) implies that $A_K \in \Sigma$.

Let $S\colon A_K \longrightarrow P_f(X\times Y)$ be defined by

$$S(\omega) = F(\omega)\cap(X\times K).$$

Then

$$\operatorname{Gr} S = \operatorname{Gr} F\cap(\Omega\times X\times K) \in \Sigma\otimes\mathcal{B}(X)\otimes\mathcal{B}(Y)$$

and so again by Theorem 4.51(d), we infer that S is a measurable multifunction. Hence Theorem 4.55(b) implies that we can find two sequences of Σ-measurable functions $\{f_n\colon A_K \longrightarrow X\}_{n\geqslant 1}$ and $\{g_n\colon A_K\longrightarrow Y\}_{n\geqslant 1}$ such that

$$S(\omega) = \overline{\{(f_n(\omega),\, g_n(\omega))\}_{n\geqslant 1}} \qquad \forall\,\omega\in\Omega.$$

Let

$$D_K = \{(\omega,x)\in A_K\times X : x\in\overline{\{f_n(\omega)\}_{n\geqslant 1}}\}.$$

We claim that $G^{-}(K) = D_K$. Clearly $G^{-}(K)\subseteq D_K$. We will show that the opposite inclusion also holds. To this end, let $(\omega,x)\in D_K$. Then $x = \lim\limits_{k\to+\infty} f_{n_k}(\omega)$ for some subsequence $\{n_k\}_{k\geqslant 1}$ of $\{n\}_{n\geqslant 1}$. Consider the corresponding subsequence of $\{g_n(\omega)\}_{n\geqslant 1}$. We have

$\{g_{n_k}(\omega)\}_{k\geqslant 1} \subseteq K$ and so by passing to a further subsequence if necessary, we may assume that

$$g_{n_k}(\omega) \longrightarrow y \in K.$$

So $(x,y) \in S(\omega)$, which means that $(\omega,x) \in G^-(K)$ and so $G^-(K) = D_K$. Now note that

$$D_K = \{(\omega,x) \in A_K \times X : \inf_{n\geqslant 1} d(x, f_n(\omega)) = 0\} \in \Sigma \otimes \mathcal{B}(X).$$

Therefore, $G^-(K) \in \Sigma \otimes \mathcal{B}(X)$ which proves the measurability of G.

"$\Longleftarrow$": Note that $\operatorname{Gr} G = \operatorname{Gr} F$ and by hypothesis

$$\operatorname{Gr} G \in \Sigma \otimes \mathcal{B}(X) \otimes \mathcal{B}(Y) = \Sigma \otimes \mathcal{B}(X \times Y)$$

(see Theorem 4.6(b)). Invoking Theorem 4.51(c), we conclude that F is measurable.

Solution of Problem 4.118
Let

$$F(\omega) = u^{-1}(\omega) \qquad \forall\, \omega \in \Omega.$$

By hypothesis **(a)**, the values of F are in $P_f(X)$. Moreover, hypothesis **(b)** implies that for every open set $V \subseteq X$, we have $F^-(V) = u(V) \in \Sigma$. Hence F is measurable (see Definition 4.49(a)). Therefore, we can apply the Kuratowski–Ryll Nardzewski selection theorem (see Theorem 4.53) and find a Σ-measurable function $f\colon \Omega \longrightarrow X$ such that

$$f(\omega) \in F(\omega) \qquad \forall\, \omega \in \Omega.$$

Then

$$u(f(\omega)) = \omega \qquad \forall\, \omega \in \Omega.$$

Solution of Problem 4.119

Since U is a measurable multifunction, by Theorem 4.55, we can find a sequence $\{u_n\colon \Omega \longrightarrow X\}_{n\geqslant 1}$ of Σ-measurable selections of U such that

$$U(\omega) = \overline{\{u_n(\omega)\}_{n\geqslant 1}} \qquad \forall\, \omega \in \Omega.$$

Let $V \subseteq Y$ be a nonempty open set. Because of the continuity of $g(\omega,\cdot)$ for all $\omega \in \Omega$, we have

$$\begin{aligned} G^-(V) &= \{\omega \in \Omega:\ g(\omega, U(\omega)) \cap V \neq \emptyset\} \\ &= \bigcup_{n\geqslant 1} \{\omega \in \Omega:\ g(\omega, u_n(\omega)) \in V\}. \end{aligned}$$

But from Theorem 4.166 for every $n \geqslant 1$, the function $\omega \longmapsto g(\omega, u_n(\omega))$ is $(\Sigma, \mathcal{B}(Y))$-measurable, hence

$$\{\omega \in \Omega:\ g(\omega, u_n(\omega))\} \in \Sigma \qquad \forall\, n \geqslant 1.$$

Therefore $G^-(V) \in \Sigma$.

Solution of Problem 4.120

Let $F\colon \Omega \longrightarrow 2^T \setminus \{\emptyset\}$ be the multifunction, defined by

$$F(\omega) = \{t \in U(\omega):\ h(\omega) = g(\omega, t)\}.$$

We have

$$\operatorname{Gr} F = \operatorname{Gr} U \cap \{(\omega, t) \in \Omega \times T:\ h(\omega) = g(\omega, t)\} \in \Sigma \otimes \mathcal{B}(T).$$

Invoking the Yankov–von Neumann–Aumann selection theorem (see Theorem 4.57), we can find a $(\Sigma, \mathcal{B}(T))$-measurable function $u\colon \Omega \longrightarrow T$ such that

$$u(\omega) \in F(\omega) \qquad \forall\, \omega \in \Omega.$$

Then

$$u(\omega) \in U(\omega) \quad \text{and} \quad h(\omega) = g\big(\omega, u(\omega)\big) \qquad \forall\, \omega \in \Omega.$$

Solution of Problem 4.121

Let $U \subseteq X$ be a nonempty open set. We know that U is the countable union of open balls. Moreover, every open ball is a countable union of closed balls. So, U is a countable union of closed balls, i.e., $U = \bigcup\limits_{n \geqslant 1} \overline{B}(n)$ with each $\overline{B}(n)$ being a closed ball in X. Then

$$F^{-}(U) = \bigcup_{n \geqslant 1} F^{-}\big(\overline{B}(n)\big) \in \Sigma,$$

hence F is measurable.

Solution of Problem 4.122

The set U^c is closed. Let

$$\xi(\omega, x) = \operatorname{dist}\big(f(\omega, x),\ U^c\big) \qquad \forall\, (\omega, x) \in \Omega \times X.$$

Evidently ξ is a Carathéodory function, thus measurable (see Theorem 4.166). We have

$$\operatorname{Gr} F = \big\{(\omega, x) \in \Omega \times X :\ \xi(\omega, x) > 0\big\}$$

and so $\operatorname{Gr} F \in \Sigma \otimes \mathcal{B}(X)$, i.e., F is graph measurable.

Solution of Problem 4.123

Let $\{x_n\}_{n \geqslant 1}$ be dense in X. We have

$$\begin{aligned} F^{-}(C) &= \big\{\omega \in \Omega :\ F(\omega) \cap C \neq \emptyset\big\} \\ &= \big\{\omega \in \Omega :\ f(\omega, x) \in U \text{ for some } x \in C\big\} \\ &= \big\{\omega \in \Omega :\ f(\omega, x_n) \in U \text{ for some } n \geqslant 1\big\} \\ &= \bigcup_{n \geqslant 1} \big\{\omega \in \Omega :\ f(\omega, x_n) \in U\big\} \end{aligned}$$

(since U is open and $f(\omega,\cdot)$ is continuous). Since for every $n \geqslant 1$, the function $f(\cdot,x_n)$ is Σ-measurable, we have that $\{\omega \in \Omega: f(\omega,x_n) \in U\} \in \Sigma$. Hence

$$F^-(C) = \bigcup_{n\geqslant 1} \{\omega \in \Omega: f(\omega,x_n) \in U\} \in \Sigma.$$

Solution of Problem 4.124

Let D be a countable dense subset of X. For any open set $V \subseteq X$ and each open interval $I=(a,b) \subseteq \mathbb{R}$, we have

$$\begin{aligned} E^-(V\times I) &= \{\omega \in \Omega: E(\omega) \cap (V\times I) \neq \emptyset\} \\ &= \{\omega \in \Omega: \operatorname{epi} f(\omega,\cdot) \cap (V\times I) \neq \emptyset\} \\ &= \{\omega \in \Omega: f(\omega,x) < b \text{ for some } x \in D\cap V\} \\ &= \bigcup_{u\in D\cap V} \{\omega \in \Omega: f(\omega,u) < b\} \in \Sigma \end{aligned}$$

(due to the continuity of $f(\omega,\cdot)$ and since $f(\cdot,x)$ is measurable).

Now recall that each open set $U \subseteq X\times\mathbb{R}$ can be written as countable union of open sets of the form $V\times I$, since $\mathbb{R}$ has countable basis consisting of open intervals.

Solution of Problem 4.125

Let $f \in S_F^1$. By considering $\omega \longmapsto F(\omega)-f(\omega)$ if necessary, we may assume that for all $\omega \in \Omega$, we have $0 \in F(\omega)$. Let $\vartheta\colon \Omega \longrightarrow \mathbb{R}_+$ be the function, defined by

$$\vartheta(\omega) = \operatorname{dist}(0, F(\omega)^c)$$

(recall that $F(\omega)^c = X\setminus F(\omega)$). First we show that ϑ is Σ_μ-measurable. To this end, let $\lambda>0$ and let us set

$$\begin{aligned} L_\lambda &= \{\omega \in \Omega: \vartheta(\omega) < \lambda\} \\ G &= \{(\omega,x) \in \Omega\times X: x \in B_\lambda \cap F(\omega)^c\}, \end{aligned}$$

where $B_\lambda = \{x \in X: \|x\| < \lambda\}$. Then $G = (\Omega\times B_\lambda) \cap \operatorname{Gr} F^c \in \Sigma\otimes\mathcal{B}(X)$ (since F is graph measurable). Note that $L_\lambda = \operatorname{proj}_\Omega G$.

Hence by the Yankov–von Neumann–Aumann projection theorem (see Theorem 4.65), we have $L_\lambda \in \Sigma_\mu$ and so ϑ is Σ_μ-measurable. Since F has open values, $\vartheta(\omega) > 0$ for all $\omega \in \Omega$. The Σ_μ-measurability of ϑ implies that we can find $\varepsilon > 0$ and $A \in \Sigma$ with $\mu(A) > 0$ such that $\vartheta(\omega) \geqslant \varepsilon$ for all $\omega \in A$. Then $B_\varepsilon \subseteq F(\omega)$ for all $\omega \in A$ and so

$$\mu(A)B_\varepsilon \subseteq \int_A F\,d\mu \subseteq \int_\Omega F\,d\mu.$$

Solution of Problem 4.126

From the Riesz representation theorem (see Theorem 4.67), we have $L^p(\Omega)^* = L^{p'}(\Omega)$ (where $\frac{1}{p}+\frac{1}{p'}=1$). Let $h \in L^{p'}(\Omega)$. For a given $\varepsilon > 0$, we can find a simple function $s_0 \in L^{p'}(\Omega)$ such that

$$\|h - s_0\|_{p'} \leqslant \tfrac{\varepsilon}{4M}.$$

Since

$$\int_A f_n\,d\mu \longrightarrow \int_A f\,d\mu \qquad \forall\, A \in \Sigma,$$

from the linearity of the integral, it follows that

$$\int_\Omega f_n s_0\,d\mu \longrightarrow \int_\Omega f s_0\,d\mu.$$

Then we can find $n_0 = n_0(\varepsilon) \geqslant 1$ such that

$$\begin{aligned} \left|h(f_n - f)\right| &\leqslant \left|(h-s_0)(f_n-f)\right| + \left|s_0(f_n-f)\right| \\ &\leqslant \|h-s_0\|_{p'}\|f_n - f\|_p + \tfrac{\varepsilon}{2} \\ &\leqslant 2M\tfrac{\varepsilon}{4M} + \tfrac{\varepsilon}{2} = \varepsilon \qquad \forall\, n \geqslant n_0, \end{aligned}$$

so $\int_\Omega (f_n - f)h\,d\mu \longrightarrow 0$ for all $h \in L^{p'}(\Omega)$ and thus $f_n \xrightarrow{w} f$ in $L^p(\Omega)$.

Solution of Problem 4.127

Let $\{\Omega_n\}_{n\geqslant 1} \subseteq \Sigma$ be a sequence such that $\Omega_n \nearrow \Omega$ and $\mu(\Omega_n) < +\infty$ for every $n \geqslant 1$. Let

$$C_n \;=\; \{\omega \in \Omega_n :\ |f(z)| \leqslant n\} \;\in\; \Sigma \qquad \forall\, n \geqslant 1.$$

We set $f_n = \chi_{C_n} f$ and consider the linear functionals on $L^p(\Omega)$, defined by

$$L_n(g) \;=\; \int\limits_\Omega f_n g\, d\mu \quad\text{and}\quad L(g) \;=\; \int\limits_\Omega f g\, d\mu \qquad \forall\, g \in L^p(\Omega).$$

From the Riesz representation theorem (see Theorem 4.67), we have that L_n is continuous. Note that

$$|f_n(\omega)g(\omega)| \;\leqslant\; |f(\omega)g(\omega)| \quad \mu\text{-almost everywhere in } \Omega$$

with $fg \in L^1(\Omega)$ and $f_n g \longrightarrow fg$ μ-almost everywhere in Ω. So, by the Lebesgue dominated convergence theorem (see Theorem 3.94), we have

$$\int\limits_\Omega f_n g\, d\mu \;\longrightarrow\; \int\limits_\Omega f g\, d\mu,$$

hence $L_n(g) \longrightarrow L(g)$. Thus L is continuous either by invoking Corollary 5.42 or directly as follows. Let $g_k \longrightarrow g$ in $L^p(\Omega)$. Then we have

$$L_n(g_k) \;\longrightarrow\; L_n(g) \quad \text{as } k \to +\infty$$

(continuity of L_n) and

$$L_n(g) \;\longrightarrow\; L(g) \quad \text{as } n \to +\infty.$$

Invoking Problem 1.175, we can find a sequence $k \longmapsto n(k)$ such that

$$L_{n(k)}(g_k) \;\longrightarrow L(g) \quad \text{as } k \to +\infty.$$

Then

$$\begin{aligned} |L(g_k) - L(g)| \;&\leqslant\; |L(g_k) - L_{n(k)}(g_k)| \\ &\quad + |L_{n(k)}(g_k) - L_{n(k)}(g)| + |L_{n(k)}(g) - L(g)|. \end{aligned}$$

We have

$$|L(g_k) - L_{n(k)}(g_k)| = \Big|\int\limits_\Omega (f - f_{n_k}) g_k\, d\mu\Big| \leqslant \int\limits_\Omega (1 - \chi_{C_{n(k)}})|f g_k|\, d\mu \longrightarrow 0$$

(by Theorem 3.94) and

$$\left|L_{n(k)}(g_k) - L_{n(k)}(g)\right| = \Big|\int_\Omega f_{n_k}(g_k - g)\,d\mu\Big| \leqslant \|f\|_{p'}\|g_k - g\|_p \longrightarrow 0$$

and

$$\left|L_{n(k)} - L(g)\right| \longrightarrow 0 \quad \text{as } k \to +\infty.$$

Therefore, finally we have

$$\left|L(g_k) - L(g)\right| \longrightarrow 0 \quad \text{as } k \to +\infty,$$

so L is continuous and so $f \in L^{p'}(\Omega)$ (by Theorem 4.67).

Solution of Problem 4.128

For every $n \geqslant 1$, we set

$$\mu_n(A) = \int_A f_n\,d\mu \qquad \forall\, A \in \Sigma.$$

Then $\{\mu_n\}_{n\geqslant 1} \subseteq M(\Sigma)$ and $\mu_n \ll \mu$ for all $n \geqslant 1$ (see Definition 3.150). By hypothesis we know that for every $A \in \Sigma$, the limit $\lim\limits_{n\to+\infty} \mu_n(A)$ exists and is finite. By Theorem 4.73, we have that the sequence $\{\mu_n\}_{n\geqslant 1}$ is uniformly μ-absolutely continuous (see Definition 4.71). Hence, for a given $\varepsilon > 0$, we can find $\delta = \delta(\varepsilon) > 0$ such that for all $A \in \Sigma$, we have

$$\text{if } |\mu|(A) < \delta, \quad \text{then } \left|\mu_n(A)\right| = \Big|\int_A f_n\,d\mu\Big| < \tfrac{\varepsilon}{2} \text{ for all } n \geqslant 1.$$

For each such A, we set

$$A_n^+ = \left\{a \in A : f_n(x) \geqslant 0\right\} \quad \text{and} \quad A_n^- = \left\{a \in A : f_n(x) < 0\right\}.$$

Evidently

$$A_n^+ \in \Sigma, \quad A_n^- \in \Sigma, \quad \mu(A_n^+) \leqslant \mu(A) < \delta, \quad \mu(A_n^-) \leqslant \mu(A) < \delta.$$

So, we have

$$\int_{A_n^+} |f_n|\,d\mu = \Big|\int_{A_n^+} f_n\,d\mu\Big| < \tfrac{\varepsilon}{2} \quad \text{and} \quad \int_{A_n^-} |f_n|\,d\mu = \Big|\int_{A_n^-} f_n\,d\mu\Big| < \tfrac{\varepsilon}{2},$$

hence

$$\int_A |f_n|\,d\mu \;=\; \int_{A_n^+} |f_n|\,d\mu + \int_{A_n^-} |f_n|\,d\mu \;<\; \tfrac{\varepsilon}{2} + \tfrac{\varepsilon}{2} \;=\; \varepsilon.$$

Solution of Problem 4.129

Let $h \in L^\infty(X) = L^1(X)^*$ (see Theorem 4.67). We claim that

$$\Big\{ \langle f_n, h\rangle = \int_X f_n h\,d\mu \Big\}_{n\geqslant 1} \quad \text{is a Cauchy sequence.}$$

To this end let $\varepsilon > 0$ and let s be a simple function such that

$$\|h - s\|_\infty \;<\; \tfrac{\varepsilon}{3M}$$

(see Proposition 3.110). We have

$$\Big|\int_X f_n h\,d\mu - \int_X f_n s\,d\mu\Big| \;\leqslant\; \|h - s\|_\infty \|f_n\|_1 \;<\; \tfrac{\varepsilon M}{3} \qquad \forall\, n \geqslant 1.$$

Also, by hypothesis the limit

$$\lim_{n\to+\infty} \int_X f_n s\,d\mu$$

exists and is finite. So, we can find $n_0 \geqslant 1$ such that

$$\Big|\int_X f_n s\,d\mu - \int_X f_m s\,d\mu\Big| \;<\; \tfrac{\varepsilon}{3} \qquad \forall\, n, m \geqslant n_0,$$

so

$$\Big|\int_X f_n h\,d\mu - \int_X f_m h\,d\mu\Big| \;<\; \varepsilon \qquad \forall\, n, m \geqslant n_0$$

and thus

$$\Big\{ \langle f_n, h\rangle = \int_X f_n h\,d\mu \Big\}_{n\geqslant 1} \quad \text{is a sequence Cauchy.}$$

So, for every $h \in L^\infty(X)$, the limit

$$\lim_{n\to+\infty} \int_X f_n h \, d\mu$$

exists and is finite. Let us set

$$\xi(h) = \lim_{n\to+\infty} \int_X f_n h \, d\mu \qquad \forall\, h \in L^\infty(X).$$

Hence $\xi \in \big(L^\infty(X)\big)^*$ (see hypothesis (ii)). Moreover, if

$$\nu_n(A) = \int_A f_n \, d\mu,$$

then $\{\nu_n\}_{n\geqslant 1} \subseteq M(\Sigma)$ and $\nu_n(A) \longrightarrow \nu(A) = \xi(\chi_A)$ and so $\nu \in M(\Sigma)$ (see the Nikodym theorem; Theorem 4.72) and $\nu \ll \mu$ (see Definition 3.150). Then by the Radon–Nikodym theorem (see Theorem 3.152), we can find $f \in L^1(X)$ such that

$$\nu(A) = \int_A f \, d\mu \qquad \forall\, A \in \Sigma,$$

so

$$\xi(s) = \int_X f s \, d\mu \quad \text{for every simple function } s.$$

As simple functions are dense in $L^\infty(X)$ (see Proposition 3.110), we have

$$\xi(h) = \int_X f h \, d\mu,$$

so

$$\langle f_n, h \rangle \longrightarrow \langle f, h \rangle \qquad \forall\, h \in L^\infty(X)$$

and thus

$$f_n \xrightarrow{w} f \quad \text{in } L^1(X).$$

Solution of Problem 4.130

Arguing by contradiction, suppose that the sequence $\{\mu_n\}_{n\geqslant 1}$ is not uniformly μ-absolutely continuous. So, we can find a sequence $\{A_k\}_{k\geqslant 1} \subseteq \Sigma$ and a subsequence $\{n_k\}_{k\geqslant 1}$ of $\{n\}_{n\geqslant 1}$ such that

$$\mu(A_k) < \tfrac{1}{2^{k+1}} \quad \text{and} \quad \mu_{n_k}(A_k) \geqslant \varepsilon \qquad \forall\, k \geqslant 1$$

(see Definition 4.71(c)). Let

$$C_k = \bigcup_{i\geqslant k} A_i \qquad \forall\, k \geqslant 1.$$

Then $\{C_k\}_{k\geqslant 1}$ is a decreasing Σ-sequence and

$$\mu(C_k) \leqslant \sum_{i\geqslant k} \mu(A_i) < \sum_{i\geqslant k} \tfrac{1}{2^{i+1}} = \tfrac{1}{2^k}.$$

Let

$$C_0 = \bigcap_{k\geqslant 1} C_k$$

and let us set

$$E_k = C_k \setminus C_0 \qquad \forall\, k \geqslant 1.$$

Then $\{E_k\}_{k\geqslant 1}$ is a decreasing Σ-sequence and

$$\bigcap_{k\geqslant 1} E_k = \emptyset.$$

Since by hypothesis the sequence $\{\mu_n\}_{n\geqslant 1}$ is Vitali equicontinuous, we can find $k_0 \geqslant 1$ such that

$$\left|\mu_{n_k}(E_k)\right| < \varepsilon \qquad \forall\, k \geqslant k_0.$$

Because

$$\mu_{n_k} \ll \mu \quad \text{and} \quad \mu(C_k) < \tfrac{1}{2^k},$$

we have $\mu(C_0) = 0$. Therefore

$$\left|\mu_{n_k}(C_k)\right| < \varepsilon \qquad \forall\, k \geqslant k_0.$$

But we know that

$$\mu_{n_k}(C_k) \geqslant \mu_{n_k}(A_k) \geqslant \varepsilon \qquad \forall\, k \geqslant 1.$$

Hence, we have reached a contradiction. This proves that the sequence $\{\mu_n\}_{n\geqslant 1}$ is uniformly μ-absolutely continuous.

Solution of Problem 4.131

Let $\varepsilon > 0$ and consider a decreasing sequence $\{A_n\}_{n\geqslant 1} \subseteq \Sigma$ such that

$$\bigcap_{n\geqslant 1} A_n = \emptyset.$$

By hypothesis, we can find $\delta = \delta(\varepsilon) > 0$ such that

$$\text{if } A \in \Sigma,\ \mu(A) < \delta, \quad \text{then } \left|\mu_k(A)\right| < \varepsilon \ \ \forall k \geqslant 1.$$

Since $\mu(X) < +\infty$, we have $\mu(A_n) \longrightarrow 0$ (see Theorem 3.19(c)). So, we can find $n_0 \geqslant 1$ such that

$$\mu(A_n) < \delta \qquad \forall\, n \geqslant n_0.$$

Hence

$$\left|\mu_k(A_n)\right| < \varepsilon \qquad \forall\, n \geqslant n_0,\ k \geqslant 1$$

and this proves that the sequence $\{\mu_n\}_{n\geqslant 1}$ is Vitali equicontinuous.

Solution of Problem 4.132

Let $[a,b] = [0,2\pi]$, $p = 2$ and consider the sequence

$$f_n(t) = \cos nt \qquad \forall\, n \geqslant 1.$$

Then

$$\int_0^{2\pi} \cos^2 nt\, dt = \pi \qquad \forall\, n \geqslant 1.$$

So, $f_n \not\longrightarrow 0$ in $L^p(a,b)$. On the other hand, let $h = \chi_{[c,d]}$, with $[c,d] \subseteq [0,2\pi]$. Then

$$\int_0^{2\pi} \chi_{[c,d]} \cos nt\, dt = \tfrac{1}{n}\big(\sin(nd) - \sin(nc)\big) \longrightarrow 0.$$

From the linearity of the integral, it follows that

$$\int_0^{2\pi} s(t) f_n(t)\, dt \longrightarrow 0$$

for every step function

$$s(t) = \sum_{k=1}^{n} \beta_k \chi_{[c_k, d_k]}.$$

Finally, recall that step functions are dense in $L^2(0, 2\pi)$, to conclude that

$$\int_0^{2\pi} h f_n\, dt \longrightarrow 0 \qquad \forall\, h \in L^2(0, 2\pi),$$

so

$$f_n \xrightarrow{w} 0 \quad \text{in } L^2(0, 2\pi).$$

Solution of Problem 4.133

By the Dunford–Pettis theorem (see Theorem 4.75), we can find a subsequence $\{u_{n_k}\}_{k\geqslant 1}$ of $\{u_n\}_{n\geqslant 1}$ and a function $h \in L^1(\Omega)$ such that

$$u_{n_k} \xrightarrow{w} h \quad \text{in } L^1(\Omega),$$

so

$$\int_A u_{n_k}\, d\mu \longrightarrow \int_A h\, d\mu \qquad \forall\, A \in \Sigma.$$

On the other hand, by the Vitali theorem (see Theorem 3.128), we have

$$\int_A u_n\, d\mu \longrightarrow \int_A u\, d\mu \qquad \forall\, A \in \Sigma.$$

Therefore

$$\int_A h\,d\mu = \int_A u\,d\mu \qquad \forall\, A \in \Sigma,$$

hence $h = u$. Hence by the Urysohn criterion (see Problem 1.3), for the original sequence, we have $u_n \xrightarrow{w} u = h$ in $L^1(\Omega)$.

Solution of Problem 4.134

By the Lusin theorem (see Theorem 3.77), for a given $\varepsilon > 0$, we can find a continuous function g such that

$$\lambda(\{f \neq g\}) < \varepsilon$$

(λ being the Lebesgue measure on $\mathbb{R}$). Let $C = \{f = g\}$. Then C is Lebesgue measurable and according to Proposition 4.91, almost every point of C is a density point. Let $x \in C$ be a density point of C. Then

$$\lim_{\substack{x' \to x \\ x' \in C}} f(x') = \lim_{x' \to x} g(x') = g(x) = f(x),$$

so f is approximately continuous on C and $\lambda(C^c) < \varepsilon$. Since $\varepsilon > 0$ was arbitrary, we conclude that f is continuous at almost every point.

Solution of Problem 4.135

By Remark 4.97, we can find a Lebesgue-null set $D \subseteq (0,1)$ such that every $x \in (0,1) \setminus D$ is a Lebesgue point of f (see Definition 4.89). Hence

$$\lim_{\substack{x \in [a,b] \\ b-a \searrow 0}} \frac{1}{b-a} \int_a^b f(x)\,d\lambda = f(x) \qquad \forall\, x \in (0,1) \setminus D.$$

By hypothesis, for every $x \in (0,1) \setminus D$ and every integer $n \geqslant 1$, we can find an open interval $I_n \subseteq (0,1)$ such that

$$x \in I_n, \quad \lambda(I_n) < \tfrac{1}{n}, \quad \text{and} \quad \int_{I_n} f\,d\lambda = 0.$$

Hence, it follows that $f(x) = 0$ and so

$$f(y) \;=\; 0 \quad \text{for almost all } y \in (0,1).$$

Therefore

$$\int_I f\,d\lambda \;=\; 0 \quad \text{for all open interval } I \subseteq (0,1).$$

Solution of Problem 4.136

No. By Proposition 4.91, for any Lebesgue measurable set $A \subseteq \mathbb{R}$, we have

$$\tfrac{\lambda(A\cap(x-h,x+h))}{2h} \longrightarrow 1 \quad \text{as } h \searrow 0, \text{ for almost all } x \in A$$

and

$$\tfrac{\lambda(A\cap(x-h,x+h))}{2h} \longrightarrow 0 \quad \text{as } h \searrow 0, \text{ for almost all } x \notin A.$$

Solution of Problem 4.137

According to Theorem 4.86, the set $[0,1] \setminus S$ is Lebesgue-null. Let $\varepsilon > 0$ and let $x \in S$. We choose a rational r such that $|r - f(x)| < \varepsilon$. If $h > 0$, then

$$\begin{aligned}
\tfrac{1}{h}\int_x^{x+h} \big|f(t) - f(x)\big|\,dt \;&\leqslant\; \tfrac{1}{h}\int_x^{x+h} \big|f(t) - r\big|\,dt + \tfrac{1}{h}\int_x^{x+h} \big|r - f(x)\big|\,dt \\
&\leqslant\; \tfrac{1}{h}\int_x^{x+h} \big|f(t) - r\big|\,dt + \varepsilon.
\end{aligned}$$

Let $h \searrow 0$. Then

$$\tfrac{1}{h}\int_x^{x+h} \big|f(t) - r\big|\,dt \;\longrightarrow\; \big|f(x) - r\big| \;<\; \varepsilon,$$

since $x \in S$. Since $\varepsilon > 0$ is arbitrary, we conclude that

$$\lim_{h \to 0^+} \frac{1}{h} \int_x^{x+h} \big|f(t) - f(x)\big| \, dt \;=\; 0.$$

Similarly, we also show that

$$\lim_{h \to 0^+} \frac{1}{h} \int_{x-h}^{x} \big|f(t) - f(x)\big| \, dt \;=\; 0,$$

so, from Theorem 4.86, we get that $x \in L_f$. Thus $S \subseteq L_f$.

Solution of Problem 4.138

Let $\xi_n \searrow 0$ and let $A_n \subseteq A$ be the set of points of A at which the upper density of C is greater than ξ_n. Let $\varepsilon > 0$. Since by hypothesis A and C are metrically separated, we can find two open sets $U, V \subseteq \mathbb{R}$ such that

$$A \subseteq U, \quad C \subseteq V, \quad \lambda(U \cap V) < \varepsilon.$$

For each $x \in A_n$ there is a sequence of closed intervals $\{I_k\}_{k \geqslant 1}$ such that

$$x \in I_k, \quad \lambda(I_k) \searrow 0 \quad \text{and} \quad \tfrac{\lambda(C \cap I_k)}{\lambda(I_k)} > \xi_n \qquad \forall\, k \geqslant 1.$$

The collection of such intervals for all $x \in A_n$ is a Vitali cover of A_n (see Definition 4.77(b)). So, by the Vitali covering theorem (see Theorem 4.79), we can find pairwise disjoint intervals $\{I_i\}_{i=1}^m$ such that

$$\lambda(A_n) - \varepsilon \;\leqslant\; \sum_{i=1}^m \lambda(A_n \cap I_i) \;\leqslant\; \sum_{i=1}^m \lambda(I_i) \;\leqslant\; \lambda(A_n) + \varepsilon.$$

Since $I_i \subseteq U$ and $C \subseteq V$, we have

$$\xi_n\big(\lambda(A_n) - \varepsilon\big) \;\leqslant\; \xi_n \sum_{i=1}^m \lambda(I_i) \;<\; \sum_{i=1}^m \lambda(C \cap I_i) \;\leqslant\; \lambda(U \cap V) \;<\; \varepsilon.$$

Since $\varepsilon > 0$ is arbitrary, we let $\varepsilon \searrow 0$ to conclude that $\lambda(A_n) = 0$.

Note that $\bigcup\limits_{n\geqslant 1} A_n$ is the subset of A at which the upper density of C is nonzero. Evidently $\bigcup\limits_{n\geqslant 1} A_n$ is a Lebesgue-null set. Since the lower density is always less than or equal to the upper density, we conclude that at almost all points of A the density of C is zero. Similarly, if we interchange the roles of A and C.

Solution of Problem 4.139

Let D be the set of points at which the density of A exists. Let $S \subseteq D$ be those points at which the density is equal to 0 or 1. From Proposition 4.91, we know that

$$\lambda(D \setminus S) \;=\; 0$$

(λ being the Lebesgue measure on $\mathbb{R}$). For every positive integer $n \geqslant 1$, let

$$f_n(x) \;=\; \tfrac{n}{2}\lambda\big(A \cap \big(x - \tfrac{1}{n}, x + \tfrac{1}{n}\big)\big) \qquad \forall\, x \in \mathbb{R}.$$

Clearly f_n is continuous and we have

$$f(x) \;=\; \lim_{n\to+\infty} f_n(x) \qquad \forall\, x \in D$$

and $f(x)$ is the density of A at x.

Hence by Problem 1.66, the discontinuity set of $f|_D$ is of first category relative to D (see Definition 1.25). So, every interval I containing a point of $D \setminus S$ also contains a point of S and at that point f is 0 or 1. Since $f|_{D\setminus S}$ is never 0 or 1, it follows that $D \setminus S \subseteq \operatorname{Disc}(f)$ (the discontinuity set of f). Therefore we conclude that $D \setminus S$ is of first category.

Solution of Problem 4.140

By hypothesis, we have

$$\tfrac{1}{b-a}\int\limits_a^b \chi_A\, dx \;\geqslant\; \vartheta.$$

Invoking Proposition 4.96 (see also Remark 4.97), we infer that

$$\chi_A(x) \geqslant \vartheta \quad \text{for almost all } x \in [0,1],$$

so

$$\chi_A(x) = 1 \quad \text{for almost all } x \in [0,1]$$

and thus $\lambda(A) = 1$.

Solution of Problem 4.141

By the Radon–Nikodym theorem (see Theorem 3.152), we have $f = \frac{d\mu}{d\lambda} \in L^1([0,1],\lambda)$. Let $x \in (0,1)$ and $h > 0$ be such that $0 < x-h < x+h < 1$ and let

$$F_h(x) = \frac{\mu([0,1] \cap (x-h,x+h))}{\lambda([0,1] \cap (x-h,x+h))}$$

We have

$$F_h(x) = \frac{1}{2h} \int\limits_{x-h}^{x+h} f(t)\,dt.$$

Then Proposition 4.96 (see also Remark 4.97) implies that

$$\lim_{h \to 0} F_h(x) = f(x) \quad \text{for almost all } x \in [0,1].$$

Solution of Problem 4.142

No. Because, if such set exists, then at every $x \in A$, the density of A would be $\frac{1}{2}$ (see Problem 4.138), which contradicts Proposition 4.91.

Alternative Solution

Another explanation follows immediately from Problem 4.13.

Solution of Problem 4.143

Suppose that

$$\big|f(x)-f(u)\big| \leqslant k|x-u| \qquad \forall\, x,u\in\mathbb{R},$$

for some $k>0$. Since g is absolutely continuous, for a given $\varepsilon>0$, we can find $\delta>0$ such that

$$\sum_{i=1}^{n}(b_i-a_i) < \delta \quad\Longrightarrow\quad \sum_{i=1}^{n}\big|g(b_i)-g(a_i)\big| < \tfrac{\varepsilon}{k},$$

where $\big\{(a_i,b_i)\big\}_{i=1}^{n}$ are disjoint subintervals of $[a,b]$. We have

$$\begin{aligned}\sum_{i=1}^{n}\big|(f\circ g)(b_i)-(f\circ g)(a_i)\big| &= \sum_{i=1}^{n}\big|f(g(b_i))-f(g(a_i))\big|\\ &\leqslant k\sum_{i=1}^{n}\big|g(b_i)-g(a_i)\big| < \varepsilon,\end{aligned}$$

so $f\circ g$ is absolutely continuous.

Solution of Problem 4.144

(a) For every $n\geqslant 1$ and $x\in[0,1]$, we have

$$0 \leqslant f_n(x) = n\int_x^{x+\frac{1}{n}}\chi_A(t)\,dt \leqslant n\tfrac{1}{n} = 1.$$

(b) For $a,b\in[0,1]$, $a<b$ and every $n\geqslant 1$, we have

$$\begin{aligned}\big|f_n(b)-f_n(a)\big| &= n\Big|\int_{a+\frac{1}{n}}^{b+\frac{1}{n}}\chi_A(t)\,dt-\int_a^b\chi_A(t)\,dt\Big|\\ &\leqslant n\Big(\int_{a+\frac{1}{n}}^{b+\frac{1}{n}}\chi_A(t)\,dt+\int_a^b\chi_A(t)\,dt\Big) \leqslant 2n(b-a).\end{aligned}$$

So, if $a_1 < b_1 \leqslant a_2 < b_2 \leqslant \ldots \leqslant a_k < b_k \leqslant a_{k+1} < \ldots \leqslant a_m < b_m$, $\varepsilon > 0$ and

$$\sum_{k=1}^{m} (b_k - a_k) < \frac{\varepsilon}{2n},$$

then

$$\sum_{k=1}^{m} \left| f_n(b_k) - f_n(a_k) \right| < \varepsilon,$$

which proves that for every $n \geqslant 1$, the function f_n is absolutely continuous, i.e., $f_n \in AC([0,1])$.

(c) By Proposition 4.96 (see also Remark 4.97), we have

$$f_n(x) \longrightarrow \chi_A(x) \quad \text{for almost all } x \in \mathbb{R}.$$

(d) Using **(c)** and the Lebesgue dominated convergence theorem (see Theorem 3.94), we have

$$\left\| f_n - \chi_A \right\|_1 = \int_{\mathbb{R}} \left| f_n(x) - \chi_A(x) \right| dx \longrightarrow 0$$

and so we conclude that

$$f_n \longrightarrow \chi_A \quad \text{in } L^1(\mathbb{R}).$$

Solution of Problem 4.145

Let

$$\mathbb{R}_0 = \mathbb{R} \setminus \{0\} \quad \text{and} \quad C = A \setminus \{0\}.$$

Then

$$qC = C \quad \text{and} \quad q(\mathbb{R}_0 \setminus C) = \mathbb{R}_0 \setminus C \qquad \forall\, q \in \mathbb{Q} \setminus \{0\}.$$

Suppose that

$$\lambda(\mathbb{R}_0 \setminus C) = \lambda(\mathbb{R} \setminus A) > 0$$

(λ being the Lebesgue measure on $\mathbb{R}$). Then since λ is a Radon measure (see Problem 4.28), we can find a compact set $K \subseteq \mathbb{R}_0 \setminus C$ with positive

Lebesgue measure. For any compact subset E of C, we consider the function $f\colon \mathbb{R}_0 \longrightarrow \mathbb{R}$, defined by

$$f(x) = \int_{\mathbb{R}_0} \chi_K(y)\chi_{E^{-1}}\left(\tfrac{x}{y}\right)\tfrac{1}{y}\,dy \qquad \forall\, x \in \mathbb{R}_0,$$

where $E^{-1} = \left\{\frac{1}{u} : u \in E\right\}$. Then f is continuous (as E^{-1} is compact) and for any $q \in \mathbb{Q} \setminus \{0\}$, we have

$$\chi_{E^{-1}}\left(\tfrac{q}{y}\right) = \chi_E\left(\tfrac{y}{q}\right) = \chi_{qE}(y),$$

so $f(q) = 0$ (as $K \cap qE \subseteq K \cap C = \emptyset$), thus $f \equiv 0$. Since

$$\begin{aligned} 0 &= \int_{\mathbb{R}_0} f(x)\,dx = \int_{\mathbb{R}_0}\int_{\mathbb{R}_0} \chi_K(y)\chi_{E^{-1}}\left(\tfrac{x}{y}\right)\tfrac{1}{y}\,dy\,dx \\ &= \int_{\mathbb{R}_0}\int_{\mathbb{R}_0} \chi_K(y)\chi_{E^{-1}}(t)\,dt\,dx = \int_{\mathbb{R}_0} \chi_K(y)\,dy \int_{\mathbb{R}_0} \chi_{E^{-1}}(t)\,dt, \end{aligned}$$

we conclude that $\lambda(E^{-1}) = 0$. As the function $u \longmapsto \frac{1}{u}$ is locally Lipschitz, thus Lipschitz on E, from the N-property (see Definition 4.123), we infer that $\lambda(E) = 0$, from which we conclude that $\lambda(C) = \lambda(A) = 0$.

Solution of Problem 4.146

Let $\{U_n\}_{n\geqslant 1}$ be a sequence of open sets in $\mathbb{R}$ such that

$$D \subseteq U_n \quad \text{and} \quad \lambda(U_n) < \tfrac{1}{2^n} \qquad \forall\, n \geqslant 1$$

(λ being the Lebesgue measure on $\mathbb{R}$). We set

$$h = \sum_{n\geqslant 1} \chi_{U_n} \geqslant 0.$$

Evidently h is integrable. We set

$$f(x) = \int_0^x h(s)\,ds \qquad \forall\, x \in [0,1].$$

Clearly f is increasing and absolutely continuous (see Theorem 4.127). Let $x \in D$. For every integer $n \geqslant 1$, there exists $\varrho > 0$ such that

$$\text{if } |y - x| < \varrho, \quad \text{then } y \in U_k \text{ for all } k \in \{1, \ldots, n\}.$$

Then, we have

$$\frac{f(y) - f(x)}{y - x} \geqslant n$$

and so $f'(x) = +\infty$.

Solution of Problem 4.147

Clearly we need to check only the differentiability of f at $x = 0$. We have

$$\left|\tfrac{f(x)-f(0)}{x-0}\right| = \left|x \cos\left(\tfrac{1}{x^2}\right)\right| \leqslant x \qquad \forall\, x \in (0, 1]$$

so $f'(0) = 0$.

Consider the partition

$$P_n = \left\{0, \left(\tfrac{2}{2n\pi}\right)^{\frac{1}{2}}, \left(\tfrac{2}{(2n-1)\pi}\right)^{\frac{1}{2}}, \ldots, \left(\tfrac{2}{3\pi}\right)^{\frac{1}{2}}, \left(\tfrac{2}{2\pi}\right)^{\frac{1}{2}}, 1\right\}.$$

Then the variation of f with respect to this partition satisfies

$$\cos 1 + \tfrac{2}{\pi} \sum_{k=2}^{n} \tfrac{1}{k} \leqslant \operatorname{Var}(f) \qquad \forall\, n \geqslant 1,$$

hence $\operatorname{Var}(f) = +\infty$.

Solution of Problem 4.148

"$\Longleftarrow$": By Theorem 4.127, we have

$$f(x) - f(a) = \int_a^x f'(t)\, dt = 0 \qquad \forall\, x \in [a, b],$$

so

$$f(x) = f(a) \qquad \forall\, x \in [a, b],$$

i.e., f is a constant function.

"$\Longrightarrow$": The function f being absolutely continuous, it is differentiable almost everywhere on $[a,b]$. Since f is constant, we conclude that

$$f'(x) = 0 \quad \text{almost everywhere on } [a,b].$$

Solution of Problem 4.149

Let $g = \chi_D$. Then $g \circ f = \chi_{f^{-1}(D)}$. From the change of variables formula, we have

$$\int_a^b \chi_{f^{-1}(D)} f'\, dx = 0 \quad \text{for all intervals } (a,b).$$

Let

$$\xi_+ = \left(\chi_{f^{-1}(D)} f'\right)^+ \quad \text{and} \quad \xi_- = \left(\chi_{f^{-1}(D)} f'\right)^-.$$

Then

$$\int_a^b \xi_+\, dx = \int_a^b \xi_-\, dx \quad \text{for all intervals } (a,b),$$

so

$$\xi_+ = \xi_- \quad \text{for almost all } x \in \mathbb{R},$$

thus

$$f'(x) = 0 \quad \text{for almost all } x \in f^{-1}(D).$$

Solution of Problem 4.150

Without any loss of generality, we may assume that $\operatorname{Var} u > 0$ (otherwise the conclusion is obvious). Let $\vartheta \in (0, \operatorname{Var} u)$ and let $P = \{x_k\}_{k=0}^m$ be a partition of $[a,b]$ such that

$$\vartheta < \sum_{k=1}^m \left|u(x_k) - u(x_{k-1})\right|$$

(see Definition 4.9) Since $u_n(x_k) \longrightarrow u(x_k)$ for all $k \in \{0, \ldots, m\}$, for a given $\varepsilon > 0$, we can find $n_0 \geqslant 1$ such that

$$\left|u_n(x_k) - u(x_k)\right| \leqslant \tfrac{\varepsilon}{2m} \qquad \forall\, k \in \{0, \ldots, m\},\ n \geqslant n_0.$$

Therefore for $n \geqslant n_0$, we have

$$\begin{aligned}
\vartheta &< \sum_{k=1}^{m} \left|u(x_k) - u(x_{k-1})\right| \\
&\leqslant \sum_{k=1}^{m} \left|u(x_k) - u_n(x_k)\right| + \sum_{k=1}^{m} \left|u_n(x_k) - u_n(x_{k-1})\right| \\
&\quad + \sum_{k=1}^{m} \left|u_n(x_{k-1}) - u(x_{k-1})\right| \\
&\leqslant \sum_{k=1}^{m} \left(\left|u_n(x_k) - u_n(x_{k-1})\right| + \tfrac{\varepsilon}{m}\right) \\
&= \sum_{k=1}^{m} \left|u_n(x_k) - u_n(x_{k-1})\right| + \varepsilon \leqslant \operatorname{Var} u_n + \varepsilon,
\end{aligned}$$

so

$$\vartheta \leqslant \liminf_{n \to +\infty} \operatorname{Var} u_n + \varepsilon.$$

Let $\varepsilon \searrow 0$ and $\vartheta \nearrow \operatorname{Var} u$, to conclude that

$$\operatorname{Var} u \leqslant \liminf_{n \to +\infty} \operatorname{Var} u_n.$$

Solution of Problem 4.151

"$\Longrightarrow$": Follows from Proposition 4.153.

"$\Longleftarrow$": Clearly u is continuous and $u' \in L^1(a,b)$ (see Proposition 4.112). Let $D \subseteq [a,b]$ be a Lebesgue-null set. Then by Problem 3.48, we have

$$\lambda^*\big(u(D)\big) \leqslant \int_D \left|u'(t)\right| dt = 0,$$

so the set $u(D)$ is Lebesgue-null. Therefore u satisfies the Lusin N-property. Invoking Corollary 4.126, we conclude that $u \in AC\big([a,b]\big)$.

Solution of Problem 4.152

Since $\vartheta < \operatorname{Var} u$, we can find a partition

$$P = \{a = t_0 < t_1 < \ldots < t_m = b\}$$

such that

$$\xi \;=\; \sum_{i=1}^{m} \big|u(t_i) - u(t_{i-1})\big| \;>\; \vartheta.$$

Note that u is uniformly continuous. So, we can find $\delta > 0$ such that

$$\text{if } |x-y| < \delta, \quad \text{then } \big|u(x) - u(y)\big| \;<\; \tfrac{\xi-\vartheta}{2m}.$$

Consider a partition $\widehat{P} = \{a = x_0 < x_1 < \ldots < x_n = b\}$ such that $\max\limits_{1 \leqslant k \leqslant n} |x_k - x_{k-1}| < \delta$. Then the contribution in the sum

$$\sum_{k=1}^{n} \big|u(x_k) - u(x_{k-1})\big|$$

of the terms corresponding to k's such that $t_{i-1} \leqslant x_{k-1} < x_k \leqslant t_i$ is greater than

$$\big|u(t_i) - u(t_{i-1})\big| - \tfrac{\xi-\vartheta}{m}.$$

Hence, it follows that

$$\sum_{k=1}^{n} \big|u(x_k) - u(x_{k-1})\big| \;>\; \xi - (\xi - \vartheta) \;=\; \vartheta.$$

Solution of Problem 4.153

By Theorem 4.119, the Banach indicatrix function $N_u \in L^1(\mathbb{R})$. Here $N_u(y) \in \mathbb{R}$ for almost all $y \in \mathbb{R}$. This implies that $\lambda(A) = 0$ (λ being the Lebesgue measure on $\mathbb{R}$).

Solution of Problem 4.154

Since $\{u_n'\}_{n\geqslant 1} \subseteq L^1(0,1)$ is uniformly integrable, by the Dunford–Pettis theorem (see Theorem 4.75) and by passing to a suitable subsequence if necessary, we may assume that $u_n' \xrightarrow{w} v$ in $L^1(0,1)$. From Theorem 4.127, we know that

$$u_n(t) - u_n(s) = \int_s^t u_n'(\tau)\,d\tau \qquad \forall\, n \geqslant 1,\ t, s \in [0,1], s \leqslant t,$$

so

$$u(t) - u(s) = \int_s^t v(\tau)\,d\tau$$

(since $u_n' \longrightarrow v$ in $L^1(0,1)$) and thus $v = u'$ and $u \in AC\big([0,1]\big)$.

Solution of Problem 4.155

The result follows from the following fact:

$$\text{if } x, y \in \mathbb{R} \text{ and } |x|, |y| \leqslant M, \quad \text{then } \big||x|^p - |y|^p\big| \leqslant pM^{p-1}|x-y|$$

(see Definition 4.120).

Solution of Problem 4.156

(a) Suppose that $A \subseteq C$. If $s > \text{H-}\dim C$, then Theorem 4.155(f) implies that

$$H^s(A) \leqslant H^s(C) = 0$$

and so $\text{H-}\dim A \leqslant s$ (see Definition 4.156). Since $s > \text{H-}\dim C$ is arbitrary, we let $s \searrow \text{H-}\dim C$, to conclude that

$$\text{H-}\dim A \leqslant \text{H-}\dim C.$$

(b) Let $s > \max\{\text{H-}\dim A,\ \text{H-}\dim C\}$. Then $s > \text{H-}\dim A$, $s > \text{H-}\dim C$, hence

$$H^s(A) \;=\; H^s(C) \;=\; 0$$

(see Theorem 4.155(f)). We have

$$H^s(A \cup C) \;\leqslant\; H^s(A) + H^s(C) \;=\; 0$$

and so $H^s(A \cup C) = 0$, which implies that

$$\text{H-}\dim(A \cup C) \leqslant s$$

(see Definition 4.156). Let $s \searrow \max\{\text{H-}\dim A,\ \text{H-}\dim C\}$, to obtain

$$\text{H-}\dim(A \cup C) \;\leqslant\; \max\{\text{H-}\dim,\ \text{H-}\dim C\}.$$

On the other hand, from **(a)**, we have

$$\text{H-}\dim A \;\leqslant\; \text{H-}\dim(A \cup C) \quad \text{and} \quad \text{H-}\dim C \;\leqslant\; \text{H-}\dim(A \cup C),$$

so finally

$$\text{H-}\dim(A \cup C) \;=\; \max\{\text{H-}\dim A,\ \text{H-}\dim C\}.$$

Solution of Problem 4.157

Let

$$\xi(x) \;=\; \|x - \gamma(a)\| \qquad \forall\, x \in \mathbb{R}^N.$$

Then $\xi\colon \mathbb{R}^N \longrightarrow \mathbb{R}_+$ is nonexpansive (i.e., Lipschitz continuous with Lipschitz constant 1) and so by Theorem 4.155(e), we have

$$H^1\big(\xi\big(\gamma([a,b])\big)\big) \leqslant H^1\big(\gamma([a,b])\big).$$

But $H^1 = \lambda$ (λ being the Lebesgue measure on $\mathbb{R}$; see Theorem 4.155(a)) and the set $\gamma([a,b]) \subseteq \mathbb{R}^N$ is connected (see Definition 2.104 and Theorem 1.90), therefore

$$\xi\big(\gamma([a,b])\big) \;=\; [0,\vartheta] \;\subseteq\; \mathbb{R}.$$

So, we have

$$H^1\big(\xi\big(\gamma([a,b])\big)\big) = \sup_{t \in [a,b]} \xi\big(\gamma(t)\big) = \sup_{t \in [a,b]} \|\gamma(t) - \gamma(a)\| \geqslant \|\gamma(b) - \gamma(a)\|,$$

thus

$$\|\gamma(b) - \gamma(a)\| \leqslant H^1(\gamma([a,b])).$$

Solution of Problem 4.158

It is well known that such curve can be reparametrized by arc-length and so without any loss of generality we may assume that

$$\gamma'(t) \neq 0 \quad \text{almost everywhere on } [a,b]$$

with $a = 0$ and $b = \operatorname{Var}\gamma$. Moreover, γ is of bounded variation, since it is absolutely continuous, being Lipschitz continuous (hence γ is a continuous, rectifiable simple curve).

For a given $\delta > 0$, choose an integer $n \geqslant 1$ such that

$$\tfrac{1}{n}\operatorname{Var}\gamma < \delta$$

and let

$$h = \tfrac{1}{n}\operatorname{Var}\gamma \quad \text{and} \quad T_k = \big[kh,\ (k+1)h\big] \ \text{ for } k \in \{0,\ldots,n-1\}.$$

Since γ is 1-Lipschitz function (because we have reparametrized γ by arc-length), from Theorem 4.155(e), we have

$$H^1_\delta\big(\gamma([a,b])\big) \leqslant \sum_{k=0}^{n-1} \operatorname{diam} T_k = \operatorname{Var}\gamma.$$

Letting $\delta \searrow 0$, we obtain

$$H^1\big(\gamma([a,b])\big) \leqslant \operatorname{Var}\gamma.$$

On the other hand, let $P = \{a = t_0 < t_1 < \ldots < t_n = b\}$ be a partition of $[a,b]$. From Problem 4.157, we have

$$\sum_{k=1}^{n} \|\gamma(t_k) - \gamma(t_{k-1})\| \leqslant \sum_{k=1}^{n} H^1\big(\gamma([t_{k-1},t_k])\big) \leqslant H^1\big(\gamma([a,b])\big)$$

(the last inequality being a consequence of the injectivity of γ and the additivity of the Hausdorff measure). Since the partition P was arbitrary, we obtain

$$\operatorname{Var}\gamma \leqslant H^1\big(\gamma([a,b])\big),$$

so

$$H^1(\gamma([a,b])) = \operatorname{Var}\gamma.$$

Solution of Problem 4.159

Let $x \in C$ and consider the function $\xi\colon \mathbb{R}^N \longrightarrow \mathbb{R}_+$, defined by

$$\xi(y) = \|y-x\|.$$

Then ξ is nonexpansive and so by Theorem 4.155(e), we have

$$H^1(\xi(C\cap \overline{B}_r(x))) \leqslant H^1(C\cap \overline{B}_r(x)).$$

moreover, we know that on $\mathbb{R}$, we have $H^1 = \lambda$ (λ being the Lebesgue measure on $\mathbb{R}$; see Theorem 4.155(a)) and $\xi(C\cap \overline{B}_r(x)) = [a,b]$ (being connected and compact in $\mathbb{R}$). Hence

$$H^1(\xi(C\cap\overline{B}_r(x))) = \sup_{y,z\in C\cap\overline{B}_r(x)} (\|y-x\|-\|z-x\|) = \sup_{y\in C\cap\overline{B}_r(x)} \|y-x\| \geqslant r.$$

Solution of Problem 4.160

Let $\{U_k\}_{k\geqslant 1}$ be a sequence of open convex subsets of $\mathbb{R}^N$ such that

$$\operatorname{diam} U_k < \varepsilon \qquad \forall\, k \geqslant 1, \quad \text{and} \quad A \subseteq \bigcup_{k\geqslant 1} U_k.$$

Then

$$(\lambda^N)^*(A) \leqslant \sum_{k\geqslant 1} \lambda^N(U_k) \leqslant \sum_{k\geqslant 1} (\operatorname{diam} U_k)^N,$$

so

$$(\lambda^N)^*(A) \leqslant H^N(A)$$

(see Definition 4.151 and Remark 4.152).

If $(\lambda^N)^*(A) = +\infty$, then the conclusion of the problem is clearly true.

So, suppose that $(\lambda^N)^*(A) < +\infty$. Then for a given $\delta > 0$, we can find an open set $U \supseteq A$ such that

$$\lambda^N(U) - \tfrac{\delta}{2} \leqslant (\lambda^N)^*(A).$$

We can find a sequence $\{C_k\}_{k\geqslant 1}$ of half-open pairwise disjoint N-cubes such that

$$U = \bigcup_{k\geqslant 1} C_k.$$

We have

$$\lambda^N(U) = \sum_{k\geqslant 1} \lambda^N(C_k).$$

We can also find a sequence $\{D_k\}_{k\geqslant 1}$ of open cubes such that

$$C_k \subseteq D_k \quad \text{and} \quad \lambda^N(D_k) \leqslant \lambda^N(C_k) + \tfrac{\delta}{2^{n+1}} \qquad \forall\, k \geqslant 1.$$

Therefore

$$\sum_{k\geqslant 1} \lambda^N(D_k) - \tfrac{\delta}{2} \leqslant \lambda^N(U).$$

Note that

$$\lambda^N(D_k) = \tfrac{1}{(\sqrt{N})^N} (\operatorname{diam} D_k)^N.$$

Hence

$$\begin{aligned}
(\lambda^N)^*(A) &\geqslant \lambda^N(U) - \tfrac{\delta}{2} \geqslant \sum_{k\geqslant 1} \lambda^N(D_k) - \delta \\
&= \tfrac{1}{\sqrt{N}^N} \sum_{k\geqslant 1} (\operatorname{diam} D_k)^N - \delta \geqslant \widehat{c} H_\varepsilon^N(A) - \delta,
\end{aligned}$$

with $\widehat{c} = \frac{2^N}{\sqrt{N}^N \omega_N}$. Since $\delta, \varepsilon > 0$ are arbitrary, we let $\delta, \varepsilon \searrow 0$, to conclude that

$$(\lambda^N)^*(A) \geqslant \widehat{c} H^N(A).$$

Solution of Problem 4.161

Let proj_N be the projection of $\mathbb{R}^N \times \mathbb{R}^M$ onto $\mathbb{R}^N$. Note that

$$\operatorname{proj}_N (\operatorname{Gr} f|_A) = A.$$

Also, by Theorem 4.155(e), the H-dimension does not increase under the action of proj_N. So, $\text{H-}\dim \operatorname{Gr} f|_A \geqslant \text{H-}\dim A = N$ (recall that A has a positive outer measure and see Problem 4.160).

If f is Lipschitz, then $\operatorname{Gr} f|_A = \xi(A)$, where $\xi\colon A \longrightarrow \mathbb{R}^N \times \mathbb{R}^M$ is the Lipschitz function, defined by

$$\xi(x) \;=\; \big(x, f(x)\big) \qquad \forall\, x \in A.$$

So, by the fact that $H^s(\mathbb{R}^N) = 0$ for $s > N$ and Theorem 4.155(e), we conclude that

$$\text{H-}\dim \operatorname{Gr} f|_A \;=\; N.$$

Solution of Problem 4.162

Let $\{C_k\}_{k\geqslant 1}$ be an open cover of A. First we show that for every $x, u \in A$, there exists a finite subfamily $\{C_{k_i}\}_{i=1}^n$ such that $x \in C_{k_1}$, $u \in C_{k_n}$ and $C_{k_i} \cap C_{k_{i+1}} \neq \emptyset$ for all $i \in \{1, \ldots, n-1\}$. To see this, let us fix $x \in A$ and consider the set D of all points $u \in A$ for which such a finite sequence exists. Then for every set C_k, we have $C_k \subseteq D$ or $C_k \subseteq A \setminus D$. So, D and $A \setminus D$ are both open sets and since A is connected (see Definition 2.104), we conclude that $D = A$.

Now, let $x, u \in A$ and let $\{C_{k_i}\}_{i=1}^n$ be the finite sequence described above. For every $i \in \{1, \ldots, n-1\}$, pick a point $x_i \in C_{k_i} \cap C_{k_{i+1}}$ and also let $x_0 = x$, $x_n = u$. We have

$$\|x_{i-1} - x_i\| \;\leqslant \operatorname{diam} C_{k_i} \qquad \forall\, i \in \{1, \ldots, n\}$$

(note that $x_{i-1}, x_i \in C_{k_i}$), so

$$\sum_{k\geqslant 1} \operatorname{diam} C_k \;\geqslant\; \sum_{i=1}^n \operatorname{diam} C_{k_i} \;\geqslant\; \sum_{i=1}^n \|x_{i-1} - x_i\| \;\geqslant\; \|x_0 - x_n\| \;=\; \|x - u\|$$

and thus

$$H^1(A) \;\geqslant\; \|x - u\|$$

(see Definition 4.151 and recall that $\frac{\omega_1}{2} = 1$). Since $x, u \in A$ were arbitrary, we conclude that

$$H^1(A) \;\geqslant\; \operatorname{diam} A.$$

Solution of Problem 4.163

Let $x, u \in A$, $x \neq u$ and let $d_x \colon \mathbb{R}^N \longrightarrow \mathbb{R}_+$ be defined by

$$d_x(y) \;=\; \|y - x\|.$$

Since d_x is nonexpansive, from Theorem 4.155(e), we have

$$\text{H-}\dim d_x(A) \;\leqslant\; \text{H-}\dim A \;<\; 1,$$

so

$$d_x(A) \;\subseteq\; \mathbb{R} \text{ is a Lebesgue-null set.}$$

So, there exists $\varrho > 0$ such that

$$\varrho \;<\; d_x(u) \quad \text{and} \quad \varrho \;\notin\; d_x(A).$$

Then

$$A \;=\; \{y \in A :\; d_x(y) < \varrho\} \cup \{y \in A :\; d_x(y) > \varrho\}$$

and x is in one set and u in the other. Therefore x and u are in different connected component (see Definition 2.111), hence A is totally discontinuous.

Solution of Problem 4.164

The space $C(X;Y)$ topologized as above is a separable metric space. So, it has a countable basis consisting of balls centred at a countable dense subset of $C(X;Y)$ with radius in $\mathbb{Q}$. For a fix $g \in C(X;Y)$ and $\varepsilon > 0$ let us set

$$B_\varepsilon(g) \;=\; \{h \in C(X;Y) :\; \widehat{d}(h,g) < \varepsilon\}.$$

It suffices to show that $\widehat{f}^{-1}(B_\varepsilon(g)) \in \Sigma$. But X being a compact metric space, it is separable. Hence

$$B_\varepsilon(g) = \bigcap_{n\geqslant 1} \{h \in C(X;Y) : d_Y(h(x_n), g(x_n)) < \varepsilon\},$$

with $\{x_n\}_{n\geqslant 1}$ being dense in X. Therefore, if we fix $n \geqslant 1$ and set

$$B_n = \{h \in C(X;Y) : d_Y(h(x_n), g(x_n)) < \varepsilon\},$$

it suffices to show that $\widehat{f}^{-1}(B_n) \in \Sigma$. But note that

$$\widehat{f}^{-1}(B_n) = \{\omega \in \Omega : d_Y(f(\omega, x_n), g(x_n)) < \varepsilon\} \in \Sigma,$$

so $\widehat{f}$ is measurable as claimed.

Solution of Problem 4.165

Clearly for all $\omega \in \Omega$, the function $f(\omega, \cdot)$ is continuous on X. Let $x \in X$. We need to show that the function $f(\cdot, x)$ is Σ-measurable. Let $U \subseteq Y$ be an open set and let

$$\widehat{U} = \{h \in C(X;Y) : h(x) \in U\}.$$

Then the set $\widehat{U}$ is open in $C(X;Y)$ (see Definition 2.174). Note that

$$\{\omega \in \Omega : f(\omega, x) \in U\} = (\omega \in \Omega : \widehat{f}(\omega) \in \widehat{U}) \in \Sigma$$

(since $\widehat{f}$ is Σ-measurable), so for all $x \in X$, the function $\omega \longmapsto f(\omega, x)$ is Σ-measurable and thus the function $(\omega, x) \longmapsto f(\omega, x)$ is Carathéodory.

Solution of Problem 4.166

"(a) $\Longrightarrow$ (b)": We know that E is norm bounded in $M_b(X)$ (see Corollary 5.45). So, for every compact set $K \subseteq X$, we can find $\eta_K > 0$ such that

$$|\mu(f)| \leqslant \eta_K \|f\|_\infty \qquad \forall\, f \in C_c(X),\ \operatorname{supp} f \subseteq K.$$

Now let $\widehat{K}$ be a compact superset of K (i.e., $K \subseteq \widehat{K}$) and $f_0 \in C_c(\widehat{K})$ with $\operatorname{supp} f_0 \subseteq \widehat{K}$, $f_0 \geqslant \chi_K$ (see Problem 2.103). Since $\|\mu\| = \||\mu|\| \leqslant \eta_{\widehat{K}}$, we have

$$|\mu|(K) = |\mu|(\chi_K) \leqslant |\mu|(f_0) \leqslant \eta_{\widehat{K}} \|f_0\|_\infty = \widehat{\eta}_K \qquad \forall\, \mu \in E$$

"(b) $\Longrightarrow$ (a)": For every compact set $K \subseteq X$ and every $f \in C_c(X)$ with $\operatorname{supp} f \subseteq K$, we have

$$|\mu(f)| \leqslant |\mu|(f) \leqslant |\mu|\big(\|f\|_\infty \chi_K\big) = \eta_K \|f\|_\infty \qquad \forall\, \mu \in E,$$

with $\eta_K = |\mu|(K)$. So E is norm bounded, hence w^*-bounded too. Thus $\overline{E}^{w^*}$ is w^*-compact (the Alaoglu theorem; see Theorem 5.66).

Solution of Problem 4.167

By Problem 4.166, it suffices to show that E is w^*-closed. So, let $\{\mu_\alpha\}_{\alpha \in J}$ be a net in E and assume that $\mu_\alpha \xrightarrow{w^*} \mu$. If $f \in C_c(X)$ with $0 \leqslant f \leqslant 1$, then $|\mu_\alpha(f)| \leqslant \vartheta$ and so $|\mu(f)| \leqslant \vartheta$, from which it follows that $\|\mu\| \leqslant \vartheta$ (see Theorem 4.23) and so we conclude that E is w^*-closed, hence w^*-compact too.

In general, the set D is not w^*-closed (it is however relatively w^*-compact). To see this, let $X = [-1, 1]$ and let

$$\mu_n = \delta_{\frac{1}{n}} - \delta_{-\frac{1}{n}} \qquad \forall\, n.$$

then $\mu_n \xrightarrow{w^*} 0$ but $\|\mu_n\| = 2$ for all $n \geqslant 1$.

Finally suppose that X is compact and let

$$D_+ = \{\mu \in M_b(X) :\ \mu \geqslant 0,\ \|\mu\| = \vartheta\}.$$

Let

$$M_b^+(X) = \{\mu \in M_b(X) :\ \mu \geqslant 0\}$$

be furnished with the relative w^*-topology and consider the continuous function $\xi\colon M_b^+(X) \longrightarrow \mathbb{R}_+$, defined by

$$\xi(u) = \mu(1) \qquad \forall\, \mu \in M_b^+(X).$$

We see that $D_+ = \xi^{-1}(\vartheta)$ and so D_+ is indeed w^*-compact.

Solution of Problem 4.168

If $x_\alpha \longrightarrow x$ in X, then for all $f \in C_c(X)$, we have

$$\delta_{x_\alpha}(f) = f(x_\alpha) \longrightarrow f(x) = \delta_x(f)$$

and so we see that σ is continuous. Also, since X is locally compact (see Definition 2.92), it is completely regular (see Problem 2.42 and the diagram at the end of Chap. 2). So, it follows that σ is an injection. Suppose that $x_\alpha \not\longrightarrow x$ in X. Then we can find $U \in \mathcal{N}(x)$ and a subnet $\{x_\beta\}_{\beta\in I}$ of $\{x_\alpha\}_{\alpha\in I}$ such that

$$x_\beta \notin U \qquad \forall\, \beta \in I.$$

Due to the local compactness of X, we may assume that $\overline{U}$ is compact. Then the complete regularity of X implies that we can find $f \in C(X)$ such that

$$f(x) = 1 \quad \text{and} \quad f|_{U^c} = 0,$$

so

$$f \in C_c(X) \quad \text{and} \quad \delta_x(f) = f(x) = 1,$$

while

$$\delta_{x_\beta}(f) = f(x_\beta) = 0.$$

Thus $\delta_{x_\beta} \not\longrightarrow \delta_x$ and so

$$\delta_{x_\alpha} \not\longrightarrow \delta_x \quad \text{in the } w^*\text{-topology}.$$

This proves that σ is a homeomorphism of X into $M_b^+(X)$.

If X is not compact, then we consider the Alexandrov one-point compactification $X^* = X \cup \{\infty\}$ of X (see Remark 2.97 and Theorem 2.98). Then clearly from the definition of the compact topology on X^*, we have that $x_\alpha \longrightarrow \infty$ implies $\delta_{x_\alpha} \xrightarrow{w^*} 0$ and so for every $\varepsilon > 0$ and $f \in C_c(X)$, there is a compact set $K \subseteq X$ such that

$$|\delta_x(f)| < \varepsilon \qquad \forall\, x \notin K.$$

Solution of Problem 4.169

"$\Longrightarrow$": Since X is homeomorphic to a subset of $M_b^+(X)$ (see Problem 4.168) and the latter is by hypothesis separable metrizable (hence second countable; see Definition 2.24), we infer that X is second countable.

"$\Longleftarrow$": Since X is by hypothesis second countable and locally compact, it is σ-compact and we have

$$X = \bigcup_{n\geqslant 1} K_n$$

with $\{K_n\}_{n\geqslant 1}$ being an increasing sequence of compact subsets such that

$$K_n \subseteq K_{n+1} \qquad \forall\, n \geqslant 1$$

(see Propositions 2.101 and 2.100). For such compact set K_n, let $\{f_{n,m}\}_{m\geqslant 1} \subseteq C(K_n)_+$ be dense (recall that $C(K_n)$ is separable since K_n is compact and

$$C(K_n)_+ = \{f \in C(K_n) : f \geqslant 0\}).$$

The functionals $\varphi_{n,m} \colon M_b^+(X) \longrightarrow \mathbb{R}$, defined by

$$\varphi_{n,m}(\mu) = \mu(f_{n,m})$$

are w^*-continuous and so, $M_b^+(X)$ is homeomorphic to a subset of $\mathbb{R}^{\mathbb{N}\times\mathbb{N}}$ which is separable metrizable, therefore so is $M_b^+(X)$.

Solution of Problem 4.170

Let $x \in \operatorname{supp}\mu$ (see Definition 4.13) and consider the open ball

$$B_{\frac{1}{m}}(x) = \{y \in X : d(y,x) < \tfrac{1}{m}\}.$$

Since $\mu_n \xrightarrow{w^*} \mu$, we have

$$0 \leqslant \mu\big(B_{\frac{1}{m}}(x)\big) \leqslant \liminf_{n\to+\infty} \mu_n\big(B_{\frac{1}{m}}(x)\big) \qquad \forall\, m \geqslant 1.$$

We define $i_0 = 0$ and

$$i_m = \min\big\{n \geqslant 1 : n > i_{m-1},\ \operatorname{supp}\mu_j \cap B_{\frac{1}{n}} \neq \emptyset \text{ for all } j \geqslant n\big\}.$$

This is a well-defined, strictly increasing sequence. For $n \in [i_k, i_{k+1})$, let

$$x_n \in \operatorname{supp}\mu_n \cap B_{\frac{1}{k}}(x).$$

Then $x_n \longrightarrow x$ and so we conclude that

$$\operatorname{supp}\mu \subseteq \liminf_{n\to+\infty} \operatorname{supp}\mu_n.$$

Solution of Problem 4.171

From Problem 2.20(a), we know that we can find a sequence $\{f_m\colon X \longrightarrow [0,+\infty)\}_{n\geqslant 1}$ of continuous function such that $f_m \nearrow h$. Let $\xi \in C_c(X)$ be such that $0 \leqslant \xi \leqslant 1$. Then from the definition of the w^*-topology (see Problem 4.166), we have

$$\lim_{n\to+\infty} \int_X \xi f_m \, d\mu_n = \int_X \xi f_m \, d\mu.$$

Taking supremum over the ξ's as above and using the fact that $f_m \leqslant h$, we have

$$\int_X \xi f_m \, d\mu_n \leqslant \int_X f_m \, d\mu_n \leqslant \int_X h \, d\mu_n,$$

so

$$\int_X \xi f_m \, d\mu \leqslant \liminf_{n\to+\infty} \int_X h \, d\mu_n.$$

From the Lebesgue monotone convergence theorem (see Theorem 3.92), we have

$$\int_X \xi h \, d\mu \leqslant \liminf_{n \to +\infty} \int_X h \, d\mu_n$$

and taking supremum over the ξ's as above, we have

$$\int_X h \, d\mu \leqslant \liminf_{n \to +\infty} \int_X h \, d\mu_n.$$

Solution of Problem 4.172

Since φ is continuous except possibly on a countable set, we have $\varphi_n \longrightarrow \varphi$ almost everywhere on $\mathbb{R}$ (for the Lebesgue measure). Also

$$\|\varphi_n\|_\infty \leqslant \|\mu_n\| \leqslant M \qquad \forall \, n \geqslant 1.$$

If $f \in C_c^1(\mathbb{R})$, then integration by parts and the Lebesgue dominated convergence theorem (see Theorem 3.94) imply that

$$\int_X f \, d\mu_n = \int_X f' \varphi_n \, dx \longrightarrow \int_X f' \varphi \, dx = \int_X f \, d\mu.$$

But the embedding $C_c^1(\mathbb{R}) \subseteq C_0(X)$ is dense. Therefore, we conclude that

$$\mu_n \xrightarrow{w^*} \mu \quad \text{in } M_b(\mathbb{R}).$$

Bibliography

[1] Bourbaki, N.: Integration: Chapitres 1 à 4. Springer, Berlin (2007)

[2] Cohn, D.L.: Measure Theory. Birkhäuser Boston, Boston (1993)

[3] Dellacherie, C., Meyer, P.A.: Probability and Potential. North-Holland Mathematics Studies, vol. 29. North-Holland Publishing Co., Amsterdam (1978)

[4] Folland, G.B.: Real Analysis. Modern Techniques and Their Applications. Pure and Applied Mathematics. Wiley, New York (1999)

[5] König, H.: Measure and Integration. An Advanced Course in Basic Procedures and Applications. Springer, Berlin (2009)

[6] Malliavin, P.G.: Integration and Probability. Graduate Texts in Mathematics, vol. 157. Springer, New York (1995)

[7] Natanson, I.P.: Theory of Functions of a Real Variable. Frederick Ungar Publishing Co., New York (1961)

[8] Oxtoby, J.C.: Measure and Category. A Survey of the Analogies between Topological and Measure Spaces. Graduate Texts in Mathematics, vol. 2. Springer, New York (1971)

[9] Parthasarathy, K.R.: Probability Measures on Metric Spaces. Probability and Mathematical Statistics, vol. 3. Academic, New York (1967)

[10] Schwartz, L.: Radon Measures on Arbitrary Topological Spaces and Cylindrical Measures. Tata Institute of Fundamental Research Studies in Mathematics, vol. 6. Oxford University Press, London (1973)

[11] Stroock, D.W.: A Concise Introduction to the Theory of Integration. Birkhäuser Boston, Boston (1999)

[12] Topsøe, F.: Topology and Measure. Lecture Notes in Mathematics, vol. 133. Springer, Berlin (1970)

Chapter 5

Functional Analysis

5.1 Introduction

5.1.1 Locally Convex, Normed and Banach Spaces

Banach spaces with their reach structure provide the basic framework for the development of modern nonlinear analysis. The aim of this chapter is to give a compact introduction to the core of Banach space theory (linear functional analysis) with emphasis on the theory concerning abstract spaces.

To fix things, unless stated otherwise, all vector spaces are over the real scalar field.

Definition 5.1
*A **topological vector space** is a vector space X with a Hausdorff topology τ such that the two operations of vector addition*

$$X \times X \ni (x,u) \longmapsto x+u \in X$$

and scalar multiplication

$$\mathbb{R} \times X \ni (\lambda, x) \longmapsto \lambda x \in X$$

are both continuous for the τ-topology.

Remark 5.2
In fact, for every $x_0 \in X$, the ***translation operator***:

$$T_{x_0}: X \ni u \longmapsto x_0 + u \in X$$

L. Gasiński and N.S. Papageorgiou, *Exercises in Analysis: Part 1*, Problem Books in Mathematics, DOI 10.1007/978-3-319-06176-4_5,

and for every $\lambda \in \mathbb{R} \setminus \{0\}$, the multiplication operator

$$V_\lambda \colon X \ni u \longmapsto \lambda u \in X$$

are both homeomorphism from X into X.

Definition 5.3
A topological space X is said to be a ***locally convex space*** *if the linear topology τ has a local basis at the origin consisting of convex sets.*

Remark 5.4
By translation we see that at every $x \in X$, the linear topology has a local basis at x consisting of convex sets.

Definition 5.5
(a) *A locally convex space X is said to be a* ***Fréchet space*** *if its linear topology is induced by a complete invariant metric.*
(b) *Let X be a vector space and let $\|\cdot\|_X \colon X \longrightarrow \mathbb{R}$ be a function. We say that $\|\cdot\|_X$ is a* ***norm*** *of X, if:*
(1) $\|x\|_X \geqslant 0$ for all $x \in X$ and $\|x\|_X = 0$ if and only if $x = 0$;
(2) $\|\lambda x\|_X = |\lambda| \|x\|_X$ for all $\lambda \in \mathbb{R}$ and all $x \in X$ (positive homogenity);
(3) $\|x + u\|_X \leqslant \|x\|_X + \|u\|_X$ for all $x, u \in X$ (subadditivity).
We say that $(X, \|\cdot\|_X)$ is a ***normed space****.*

(c) *A normed space $(X, \|\cdot\|_X)$ is a* ***Banach space*** *if it is complete as a metric space with the metric*

$$d(x,u) = \|x - u\|_X \qquad \forall\, x, u \in X.$$

Remark 5.6
Evidently a normed space is locally convex. A locally convex space is said to be ***normable*** if its linear topology is induced by a norm $\|\cdot\|_X$ on X. Moreover, a Banach space is a Fréchet space.

Definition 5.7
(a) *If X is a vector space and $C \subseteq X$ is a set, then we say that the set C is* ***balanced*** *if $\lambda C \subseteq C$ for all $|\lambda| \leqslant 1$ and* ***absorbing*** *if for every $x \in X$, there is λ_x^* such that $x \in \lambda C$ for all $\lambda > \lambda_x^*$.*

(b) *If X is a topological vector space and $C \subseteq X$ is a set, then we say that the set C is* ***bounded*** *if for every $U \in \mathcal{N}(0)$, we can find $\lambda > 0$ such that $\lambda C \subseteq U$.*
(c) *If X is a topological vector space and $C \subseteq X$, then we say that the set C is* ***totally bounded*** *if for every $U \in \mathcal{N}(0)$, there is a finite set $F \subseteq X$ such that $C \subseteq F + U$.*

Proposition 5.8
A locally convex space X has a local basis at 0 consisting of convex and balanced sets.

Proposition 5.9
(a) *If a topological vector space X has a totally bounded neighbourhood of 0,*
then X is finite dimensional (in particular, all locally compact topological vector spaces are finite dimensional). Moreover, every finite dimensional subspace of a topological vector space is closed.
(b) *If X is a locally convex space with a countable local basis at 0, then X is metrizable (i.e., the linear topology of X is induced by a translation invariant metric).*
(c) *A locally convex space is normable (see Remark 5.6) if and only if it has a bounded neighbourhood of the origin.*

Definition 5.10
(a) *Let X be a vector space. A function $p\colon X \longrightarrow \mathbb{R}$ is a* ***seminorm*** *if:*
(1) $p(\lambda x) = |\lambda| p(x)$ for all $\lambda \in \mathbb{R}$ and all $x \in X$;
(2) $p(x + u) \leqslant p(x) + p(u)$ for all $x, u \in X$.

(b) *Let X be a vector space and let $C \subseteq X$ be an absorbing set. The* ***Minkowski functional*** *(or* ***gauge functional****) of C is defined by*

$$p_C(x) \stackrel{\text{def}}{=} \inf\{\lambda > 0 : x \in \lambda C\}.$$

Remark 5.11
If X is a normed space with norm $\|\cdot\|_X$ and $C \stackrel{\text{def}}{=} \overline{B}_1 = \{x \in X : \|x\|_X \leqslant 1\}$, then $p_C = \|\cdot\|_X$. The Minkowski functional p_C is nonnegative, positively homogeneous and $C \subseteq \{x \in X : p_C(x) \leqslant 1\}$. If $C \subseteq X$ is convex, then p_C is sublinear (i.e., positively homogeneous and subadditive) and $\{x \in X : p_C(x) < 1\} \subseteq C$. Finally, if C is both convex and balanced, then p_C is a seminorm.

5.1.2 Linear Operators: Quotient Spaces—Riesz Lemma

Next we turn our attention to linear operators between normed spaces.

Proposition 5.12
If $(X, \|\cdot\|_X)$ *and* $(Y, \|\cdot\|_Y)$ *are two normed spaces and* $A\colon X \longrightarrow Y$ *is a linear operator,*
then the following statements are equivalent:
(a) A is continuous at the origin;
(b) A is linear bounded, i.e., there exists $M > 0$ *such that* $\|A(x)\|_Y \leqslant M\|x\|_X$ *for all* $x \in X$*;*
(c) the set $A(B_1^X)$ *is bounded in* Y*, where* $B_1^X = \{x \in X : \|x\|_X < 1\}$.

Remark 5.13
In the sequel by $\mathcal{L}(X;Y)$ we denote the vector space of all continuous linear operators from X into Y. If $X = Y$, then we write $\mathcal{L}(X)$. Note that, if $A \in \mathcal{L}(X;Y)$, then the kernel or nullspace of A, defined by

$$\ker A \;=\; N(A) \;\stackrel{\text{def}}{=}\; \{x \in X:\; A(x) = 0\}$$

is a closed vector subspace of X.

Proposition 5.14
If X *and* Y *are two normed spaces, then* $\mathcal{L}(X;Y)$ *furnished with the norm*

$$\|A\|_{\mathcal{L}} \;\stackrel{\text{def}}{=}\; \sup\left\{\|A(x)\|_Y:\; \|x\|_X \leqslant 1\right\} \qquad \forall\, A \in \mathcal{L}(X;Y)$$

is a normed vector space. If Y *is a Banach space, then so is* $\left(\mathcal{L}(X;Y), \|\cdot\|_{\mathcal{L}}\right)$.
If $Y = \mathbb{R}$*, then* $\mathcal{L}(X;\mathbb{R}) = X^*$ *is the dual of* X.

Remark 5.15
Evidently X^* is always a Banach space, even if X is only a normed space. Note that

$$\|A\|_{\mathcal{L}} \;=\; \inf\left\{M > 0:\; \|A(x)\|_Y \leqslant M\|x\|_X \text{ for all } x \in X\right\}$$

(see Proposition 5.12(b)).

Definition 5.16
Let X and Y be two normed spaces.
(a) *$A \in \mathcal{L}(X;Y)$ is an* ***isomorphism*** *if A is a bijection and $A^{-1} \in \mathcal{L}(Y;X)$. Then X and Y are said to be isomorphic.*
(b) *$A \in \mathcal{L}(X;Y)$ is an* ***isometry*** *if A is an isomorphism and $\|A(x)\|_Y = \|x\|_X$ for all $x \in X$. Then X and Y are said to be isometric.*
(c) *Two norms $\|\cdot\|_X$ and $|\cdot|_X$ are said to be equivalent, if the identity map from $(X, \|\cdot\|_X)$ into $(X, |\cdot|_X)$ is an isomorphism.*

Proposition 5.17
On a vector space X two norms $\|\cdot\|_X$ and $|\cdot|_X$ are equivalent if and only if there exist constants $0 < c_0 < c_1$ such that

$$c_0|x|_X \leqslant \|x\|_X \leqslant c_1|x|_X \qquad \forall\, x \in X.$$

Definition 5.18
Let X be a vector space and let V be its vector subspace. We define an equivalence relation $\sim$ on X by setting

$$x \sim u \iff x - u \in V.$$

For $x \in X$, let $[x]$ be the equivalence class of x (i.e., $[x]=x+V$). Then

$$X/_{\sim} = \{[x] : x \in X\}$$

is denoted by $X/_V$ and is a vector space with operations

$$\lambda[x] + [u] = [\lambda x + u] \qquad \forall\, x, u \in X,\ \lambda \in \mathbb{R}.$$

Clearly $[x] = 0$ if and only if $x \in V$. If X is a normed space and V is a closed subspace of X, then we can define a norm on $X/_V$ by setting

$$\big|[x]\big| \stackrel{\text{def}}{=} \inf\{\|u\|_X : u \sim x\} = \inf\{\|x + v\|_X : v \in V\}.$$

We call $|\cdot|$, the ***quotient norm*** *of X and $X/_V$ is the* ***quotient normed space*** *of X by V.*

Proposition 5.19
<u>If</u> X is a Banach space and V is a closed subspace of X, <u>then</u> $X/_V$ with the quotient norm is a Banach space.

The next result establishes the existence of an almost orthogonal element in any normed space.

Theorem 5.20 *(Riesz Lemma)*
If X is a normed space and V is a closed proper subspace of X, then for every $\varepsilon > 0$ there exists $x \in X$, $\|x\|_X = 1$ such that $\operatorname{dist}(x, V) \geqslant 1 - \varepsilon$.

Remark 5.21
If V is finite dimensional (or more generally if V is reflexive), then in Theorem 5.20 above, we can take $\varepsilon = 0$. However, in general this is not true.

Using Theorem 5.20 and an easy contradiction argument, we have the following characterization of finite dimensional normed space.

Theorem 5.22
A normed space X is finite dimensional if and only if $\overline{B}_1 = \{x \in X : \|x\|_X \leqslant 1\}$ is compact.

Remark 5.23
All finite dimensional normed spaces are Banach and any two norms on a finite dimensional Banach space are equivalent.

5.1.3 The Hahn–Banach Theorem

Let X and Y be two normed spaces and let V be a subspace of X. It is important to know whether a bounded linear operator from V into X can be extended to a bounded linear operator from X into Y and whether this can be achieved without increasing the norm. If V is dense in X, then such an extension is possible. For functionals (i.e., when $Y = \mathbb{R}$), we can have such an extension for any subspace V, not necessarily dense. This is the celebrated Hahn–Banach theorem. There are analytic and geometric forms of this theorem. We start with the analytic form.

Theorem 5.24 *(Hahn–Banach Theorem)*
*If X is a vector space, V is a vector subspace of X, $f\colon V \longrightarrow \mathbb{R}$ is a linear functional and $p\colon X \longrightarrow \mathbb{R}$ is **sublinear functional**, i.e.,*

$$p(\lambda x) = \lambda p(x) \qquad \forall\, x \in X,\ \lambda > 0$$

and

$$p(x+u) \leqslant p(x) + p(u) \qquad \forall\, x, u \in X$$

then there exists a linear functional $\widehat{f}\colon X \longrightarrow \mathbb{R}$ *such that* $\widehat{f}|_V = f$ *and* $\widehat{f}(x) \leqslant p(x)$ *for all* $x \in X$.

Recall that, if X is a normed space, then X^* is its dual which is always a Banach space with norm

$$\|x^*\|_* \stackrel{\text{def}}{=} \sup\big\{|\langle x^*, x\rangle| : \|x\|_X \leqslant 1\big\} = \sup\big\{|\langle x^*, x\rangle| : \|x\|_X = 1\big\} \tag{5.1}$$

(see Proposition 5.14 and Remark 5.15). In the sequel, by $\langle\cdot,\cdot\rangle$ we denote the duality brackets of the pair (X^*, X), i.e.,

$$\langle x^*, x\rangle = x^*(x) = x(x^*) \qquad \forall\, x \in X,\ x^* \in X^*$$

(in the last equality, we view x as an element of $(X^*)^* = X^{**}$ by the canonical embedding; see Definition 5.68). Evidently

$$|\langle x^*, x\rangle| \leqslant \|x^*\|_*\|x\|_X \qquad \forall\, x \in X,\ x^* \in X^*.$$

Corollary 5.25
If X *is a normed space,* V *is a vector subspace of* X *and* $u^* \in V^*$, *then there exists* $\widehat{u}^* \in X^*$ *such that* $\widehat{u}^*|_V = u^*$ *and* $\|\widehat{u}^*\|_{X^*} = \|u^*\|_{V^*}$.

Corollary 5.26
If X *is a normed space,*
then for every $x_0 \in X$, *we can find* $x_0^* \in X^*$ *such that* $\|x_0\|_X = \|x_0^*\|_*$, $\langle x_0^*, x_0\rangle = \|x_0\|_X^2$ *and*

$$\|x_0\|_X = \sup\big\{|\langle x^*, x_0\rangle| : x^* \in X^*,\ \|x^*\|_* \leqslant 1\big\}. \tag{5.2}$$

Remark 5.27
The map $\mathcal{F}\colon X \longrightarrow 2^{X^*} \setminus \{\emptyset\}$ (which is in general multivalued), defined by

$$\mathcal{F}(x) \stackrel{\text{def}}{=} \{x^* \in X^* : \|x^*\|_* = \|x\|_X,\ \langle x^*, x\rangle = \|x\|_X^2\}$$

is known as the ***duality map*** from X into X^*. Also, note that (5.1) is a definition, while (5.2) is a result. In (5.1) the supremum need not be achieved, while in (5.2) it is always achieved.

Next we pass to the so-called geometric forms of the Hahn–Banach theorem. These are the "separation theorems" for convex sets.

Theorem 5.28 *(Weak Separation Theorem)*
If X is a locally convex vector space, A and C are two nonempty convex subsets of X, $\operatorname{int} A \neq \emptyset$ and $\operatorname{int} A \cap C = \emptyset$,
then there exists $x^* \in X^*$ such that $x^*(a) \leqslant x^*(c)$ for all $a \in A$, $c \in C$.

Theorem 5.29 *(Strong Separation Theorem)*
If X is a locally convex vector space, A and C are two nonempty convex subsets of X, A is compact, C is closed and $A \cap C = \emptyset$,
then there exists $x^* \in X^* \setminus \{0\}$ and $\varepsilon > 0$ such that $x^*(a) \leqslant x^*(c) - \varepsilon$ for all $a \in A$, $c \in C$.

Corollary 5.30
If X is a locally convex vector space and C is a nonempty convex subset of X,
then C is the intersection of all closed half spaces which contain it.

Corollary 5.31
If X is a locally convex vector and V is a vector subspace of X such that $\overline{V} \neq X$,
then there exists $x^* \in X^* \setminus \{0\}$ such that $x^*(v) = 0$ for all $v \in V$.

Remark 5.32
Corollary 5.31 suggests a way to check whether a vector subspace V of X is dense. If the only dual element $x^* \in X^*$ orthogonal to V (i.e., $x^*(v) = 0$ for all $v \in V$) is the zero vector, then V is dense in X.

5.1.4 Adjoint Operators and Annihilators

Definition 5.33
*Let X and Y be two normed spaces and let $A \in \mathcal{L}(X;Y)$. By $\langle \cdot, \cdot \rangle_X$ we denote the duality brackets for the pair (X^*, X) and by $\langle \cdot, \cdot \rangle_Y$ the duality brackets for the pair (Y^*, Y). The **adjoint operator** $A^* \in \mathcal{L}(Y^*; X^*)$ is defined by*

$$\langle y^*, A(x) \rangle_Y = \langle A^*(y^*), x \rangle_X \qquad \forall\, x \in X,\ y^* \in Y^*.$$

Proposition 5.34
If X and Y are two normed spaces and $A \in \mathcal{L}(X;Y)$,
then $\|A\|_{\mathcal{L}} = \|A^*\|_{\mathcal{L}}$.

Remark 5.35
If X and Y are finite dimensional, then every $A \in \mathcal{L}(X, Y)$ can be represented by a matrix $[A]$. In this case $[A^*]$ is the transpose of $[A]$, provided that the various vector space bases are properly chosen.

Definition 5.36
Let X be a normed space and $A \subseteq X$, $C \subseteq X^*$ are nonempty sets.
*(a) The **annihilator** of A in X^* is the set $A^\perp$, defined by*

$$A^\perp = \{x^* \in X^* : \langle x^*, x\rangle = 0 \text{ for every } x \in A\}.$$

*(b) The **annihilator** of C in X is the set ${}^\perp C$, defined by*

$$^\perp C = \{x \in X : \langle x^*, x\rangle = 0 \text{ for every } x^* \in C\}.$$

Proposition 5.37
<u>If</u> X is a normed space and $A \subseteq X$, $C \subseteq X^*$ are nonempty sets, <u>then</u>
(a) $A^\perp$ is closed in X^* and ${}^\perp C$ is closed in X;
(b) ${}^\perp(A^\perp) = \overline{\operatorname{span}}\, A$;
(c) If A is a vector subspace of X, then ${}^\perp(A^\perp) = \overline{A}$.

Proposition 5.38
<u>If</u> X is a Banach space and V is a nonempty closed vector subspace of X,
<u>then</u> $(X/_V)^* = V^\perp$ and $V^* = X^*/_{V^\perp}$.

5.1.5 The Three Basic Theorems of Linear Functional Analysis

Next we shall present the three basic results of linear functional analysis, which pave the way for a deeper investigation of Banach spaces: Banach–Steinhaus theorem (or uniform boundedness principle), open mapping theorem and closed graph theorem.

Theorem 5.39 *(Banach–Steinhaus Theorem; Uniform Boundedness Principle)*
<u>If</u> X is a Banach space, Y is a normed space, I is an arbitrary index set and $\{A_i\}_{i\in I} \subseteq \mathcal{L}(X;Y)$,
<u>then</u> either there exists $M > 0$ such that $\|A_i\|_{\mathcal{L}} \leqslant M$ for all $i \in I$ or $\sup\limits_{i\in I} \|A_i(x)\| = +\infty$ for all x belonging to some dense G_δ subset of X.

Corollary 5.40
If X is a Banach space, Y is a normed space, I is an arbitrary index set, $\{A_i\}_{i\in I} \subseteq \mathcal{L}(X;Y)$ and $\sup\limits_{i\in I} \|A_i(x)\| < +\infty$ for all $x \in X$,
then there exists $M > 0$ such that $\|A_i\|_{\mathcal{L}} \leqslant M$ for all $i \in I$.

Remark 5.41
So, according to this corollary, if the family $\{A_i\}_{i\in I}$ is pointwise bounded, then it is uniformly bounded (see Definition 1.83). This justifies the name "uniform boundedness principle". In this theorem, the completeness of X is crucial (see Problem 5.118).

Corollary 5.42
If X is a Banach space, Y is a normed space and $\{A_n\}_{n\geqslant 1} \subseteq \mathcal{L}(X;Y)$ is a sequence such that for every $x \in X$, the sequence $\{A_n(x)\}_{n\geqslant 1}$ is convergent in Y,
then setting

$$A(x) \stackrel{\text{def}}{=} \lim_{n\to+\infty} A_n(x) \qquad \forall\, x \in X,$$

we have $A \in \mathcal{L}(X;Y)$ and $\|A\|_{\mathcal{L}} \leqslant \liminf\limits_{n\to+\infty} \|A_n\|_{\mathcal{L}}$.

Corollary 5.43
If X is a Banach space, $C \subseteq X$ is a nonempty set and for every $x^* \in X^*$, the set $x^*(C) \subseteq \mathbb{R}$ is bounded,
then the set C is bounded.

Remark 5.44
In the language of weak topology (to be introduced in the sequel; see Definition 5.54), the above corollary says the set $C \subseteq X$ is bounded if and only if it is weakly bounded (i.e., boundedness is duality invariant).

There is also a result dual to Corollary 5.43.

Corollary 5.45
If X is a Banach space, the set $C^* \subseteq X^*$ is nonempty and for every $x \in X$, the set $\{\langle x^*, x\rangle : x^* \in C^*\}$ is bounded in $\mathbb{R}$,
then the set C^* is bounded.

Remark 5.46
Again in the language of the weak*-topology (to be introduced in the sequel; see Definition 5.63), the above corollary says that the set $C^* \subseteq X^*$ is bounded if and only if it is weakly* bounded.

Theorem 5.47 *(Open Mapping Theorem)*
If X and Y are two Banach spaces and $A \in \mathcal{L}(X;Y)$ is surjective, then A is an open map (i.e., maps open sets to open sets).

An interesting consequence of this theorem is the following result.

Theorem 5.48 *(Banach Theorem)*
If X and Y are two Banach spaces and $A \in \mathcal{L}(X;Y)$ is a bijection, then A is an isomorphism.

Corollary 5.49
If X is a vector space which becomes a Banach space for the norms $\|\cdot\|$ and $|\cdot|$ and there exists $c > 0$ such that $\|x\| \leqslant c|x|$ for all $x \in X$, then the two norms are equivalent.

Finally we state the third fundamental theorem of introductory linear functional analysis. Recall that if X and Y are two topological spaces and $f\colon X \longrightarrow Y$ is a continuous map, then $\operatorname{Gr} f = \{(x,y) \in X \times Y : f(x) = y\}$ is closed. The converse is not in general true. However, if X and Y are two Banach spaces and $f = A \in \mathcal{L}(X;Y)$, then the converse holds.

Theorem 5.50 *(Closed Graph Theorem)*
If X and Y are two Banach spaces, $A\colon X \longrightarrow Y$ is a linear map and $\operatorname{Gr} A \subseteq X \times Y$ is closed,
then $A \in \mathcal{L}(X;Y)$.

Remark 5.51
We point out that in the Banach–Steinhauss theorem (see Theorem 5.39), it was sufficient that the domain X is a Banach space. The range space Y can be any normed space (not necessarily complete). In contrast, in the other two theorems, the open mapping theorem and the closed graph theorem, it is essential that both spaces X and Y are Banach spaces.

Definition 5.52
Let X be a Banach space and let V be a closed vector subspace. A vector subspace Y of X is said to be a ***topological complement*** *of V if Y is closed in X, $V \cap Y = \{0\}$ and $X = V + Y$. Then we say that V and Y are* ***complementary subspaces*** *of X and we write $X = V \oplus Y$.*

Remark 5.53
If $X = V \oplus Y$, then every $x \in X$ can be uniquely written as $x = v + y$ with $v \in V$ and $y \in Y$. The maps

$$p_V(x) \stackrel{\text{def}}{=} v \quad \text{and} \quad p_Y(x) \stackrel{\text{def}}{=} y$$

are well defined and $p_V \in \mathcal{L}(X;Y)$, $p_Y \in \mathcal{L}(X;Y)$. Every finite dimensional subspace of X admits a topological complement. Also, if V is a vector subspace of X of finite codimension, then V admits a topological complement.

Finally we mention that in a Banach space, if $V \subseteq X$ is a closed vector subspace and $F \subseteq X$ is finite dimensional vector subspace (hence automatically closed), we have that $V + F$ is closed. We stress though that in general the sum of two closed vector subspaces need not be closed.

5.1.6 The Weak Topology

From Proposition 5.9(a), we see that, if X is a Banach space and $C \subseteq X$ is (norm) compact, then $\operatorname{int} C = \emptyset$. So, the norm (strong) topology on X is too rich to have compactness for some interesting sets. For this reason, we consider topologies on X and X^* which are weaker than the norm (strong) topologies.

Let X be a Banach space and let X^* be its dual. For every $x^* \in X^*$, let $f_{x^*} : X \longrightarrow \mathbb{R}$ be the linear functional, defined by

$$f_{x^*}(x) \stackrel{\text{def}}{=} \langle x^*, x \rangle \qquad \forall\, x \in X.$$

Definition 5.54
*The **weak topology** $w = w(X, X^*)$ on X is the weakest (coarsest) topology on X for which all the functionals $\{f_{x^*}\}_{x^* \in X^*}$ are continuous, i.e., $w = w\big(\{f_{x^*}\}_{x^* \in X^*}\big)$ (see Definition 2.62).*

Remark 5.55
Consider all topologies τ on X such that $(X_\tau)^* = X^*$. Evidently the norm topology is such a topology τ. The weak topology on X is the weakest such topology τ. A ***local basis*** at x_0 is given by

$$\mathcal{U} \stackrel{\text{def}}{=} \big\{x \in X : \big|\langle x_i^*, x - x_0 \rangle\big| < \varepsilon \text{ for all } i \in I\big\}, \tag{5.3}$$

where $\varepsilon > 0$, I is a finite index set and $x_i^* \in X^*$ for all $i \in I$.

Using the weak separation theorem (see Theorem 5.28) and Definition 5.54, we can easily obtain the following result.

Proposition 5.56
The weak topology on X is Hausdorff and X_w is a regular locally convex vector space. In a finite dimensional normed space, the norm and the weak topologies coincide. Finally, if $\{x_\alpha\}_{\alpha\in J}$ is a net in X, then
(a) $x_\alpha \xrightarrow{w} x$ *in* $X \Longleftrightarrow$ *for all* $x^* \in X^*$, *we have* $\langle x^*, x_\alpha\rangle \longrightarrow \langle x^*, x\rangle$.
(b) $x_\alpha \longrightarrow x$ *in* $X \Longrightarrow x_\alpha \xrightarrow{w} x$ *in* X.
(c) If $x_\alpha \xrightarrow{w} x$ *in* X, *then the set* $\{\|x_\alpha\|\}_{\alpha\in J} \subseteq \mathbb{R}$ *is bounded and* $\|x\| \leqslant \liminf_{\alpha\in J} \|x_\alpha\|$.
(d) If $x_\alpha \xrightarrow{w} x$ *in* X *and* $x^*_\alpha \longrightarrow x^*$ *in* X^*, *then* $\langle x^*, x_\alpha\rangle \longrightarrow \langle x^*, x\rangle$.

Remark 5.57
Note that w-open (respectively, w-closed) sets are always (norm) open (respectively, closed). In an infinite dimensional Banach space the weak topology is strictly coarser than the norm (strong) topology. In the sequel, when mentioning topological notions with respect to the norm (strong) topology, we will drop the word norm (strong) (so, we will say, for example, open instead of norm (strong) open). It is clear from (5.3) that every weak neighbourhood U of x_0 contains an affine space passing through x_0. So, U is unbounded. Hence $B_1 = \{x \in X : \|x\| < 1\}$ is open but not weakly open. However, there are infinite dimensional Banach spaces with the property that every weakly convergent sequence is strongly convergent (Schur property). This is the case of the space

$$l^1 \stackrel{\text{def}}{=} \Big\{\{x_n\}_{n\geqslant 1} : x_n \in \mathbb{R},\ \|x\| = \sum_{n\geqslant 1} |x_n| < +\infty\Big\}.$$

In an infinite dimensional Banach space, the weak topology is never first countable, in particular then it is never metrizable.

According to the above remarks, in an infinite dimensional Banach space, there are closed sets which are not weakly closed. However, for convex sets, the situation is different.

Theorem 5.58 *(Mazur Theorem)*
In a Banach space X, every closed convex set is weakly closed (the converse is always true, even without the convexity property). So, for convex sets in X the norm and weak closures coincide.

Corollary 5.59
If X is a Banach space and $x_n \xrightarrow{w} x$ in X,
then there exists a sequence $\{u_n\}_{n \geqslant 1} \subseteq X$ consisting of convex combinations of the x_n's such that $u_n \longrightarrow x$ in X.

Corollary 5.60
If X is a Banach space and $\varphi \colon X \longrightarrow \mathbb{R}^* = \mathbb{R} \cup \{\pm\infty\}$ is convex,
then φ is lower semicontinuous if and only if φ is weakly lower semicontinuous.

The next result shows that as far as continuity of linear maps is concerned, the topology really does not matter.

Proposition 5.61
If X and Y are two Banach spaces,
then $A \in \mathcal{L}(X;Y)$ if and only if $A \in \mathcal{L}(X_w;Y_w)$.

Remark 5.62
When $A \in \mathcal{L}(X_w;Y_w)$, we say that A is ***weakly continuous***.

5.1.7 The Weak* Topology

Consider X^*. This is a Banach space (even if X is only a normed space). On X^* we have the weak topology, which is the weakest of all topologies τ on X^* such that $(X_\tau)^* = X^{**}$. We can have an even weaker topology on X^*. So, for every $x \in X$, let $f_x \colon X^* \longrightarrow \mathbb{R}$ be the linear functional, defined by

$$f_x(x^*) \stackrel{\text{def}}{=} \langle x^*, x\rangle \qquad \forall\, x^* \in X^*.$$

Definition 5.63
The weak topology $w^* = w(X^*,X)$ on X^* is the weakest (coarsest) topology on X^* for which all the functionals $\{f_x\}_{x \in X}$ are continuous, i.e., $w^* = w(\{f_x\}_{x\in X})$ (see Definition 2.62).*

Remark 5.64
Consider all topologies τ on X^* such that $(X^*_\tau)^* = X$. The weak*-topology is the weakest such topology τ. Since $X \subseteq X^{**}$, the w^*-topology on X is in general coarser than the w-topology. A local basis at $x_0^* \in X^*$ is given by

$$\mathcal{U}^* \stackrel{\text{def}}{=} \big\{x^* \in X^* : \big|\langle x^* - x_0^*, x_i\rangle\big| < \varepsilon \text{ for all } i \in I\big\},$$

where $\varepsilon > 0$, I is a finite index set and $x_i \in X$ for all $i \in I$.

Proposition 5.65
The weak topology on X^* is Hausdorff and $X^*_{w^*}$ is a regular locally convex vector space. In a finite dimensional space X^*, the three topologies (norm, weak and weak*) coincide. Finally, if $\{x^*_\alpha\}_{\alpha \in J}$ is a net in X^*, then*
(a) $x^*_\alpha \xrightarrow{w^*} x^*$ *in* $X^* \iff$ *for all* $x \in X$ *we have* $\langle x^*_\alpha, x\rangle \longrightarrow \langle x^*, x\rangle$.
(b) If $x^*_\alpha \longrightarrow x^*$ *or* $x^*_\alpha \xrightarrow{w} x^*$ *in* X^*, *then* $x^*_\alpha \xrightarrow{w^*} x^*$ *in* X^*.
(c) If $x^*_\alpha \xrightarrow{w^*} x^*$ *in* X^*, *then the set* $\{\|x^*_\alpha\|_*\}_{\alpha \in J}$ *is bounded and* $\|x^*\|_* \leqslant \liminf_{\alpha \in J} \|x^*_\alpha\|_*$.
(d) If $x^*_\alpha \xrightarrow{w^*} x^*$ *in* X^* *and* $x_\alpha \longrightarrow x$ *in* X, *then* $\langle x^*_\alpha, x_\alpha\rangle \longrightarrow \langle x^*, x\rangle$.

The next theorem justifies the introduction of the w^*-topology.

Theorem 5.66 *(Alaoglu Theorem)*
<u>If</u> X is a Banach space and $\overline{B}^*_1 \stackrel{\text{def}}{=} \{x^* \in X^* : \|x^*\|_* \leqslant 1\}$,
<u>then</u> $\overline{B}^*_1$ *is w^*-compact.*

Remark 5.67
Hence every set $C^* \in X^*$ which is bounded and w^*-closed, it is also w^*-compact.

Definition 5.68
Let X be a Banach space. The mapping $j\colon X \longrightarrow X^{**}$, defined by

$$j(x)(x^*) \stackrel{\text{def}}{=} \langle x^*, x\rangle \qquad \forall\, x^* \in X^*$$

is an isometry known as the ***canonical embedding*** *of X into X^{**}.*

Remark 5.69
In the sequel, for economy in the notation, for all $x \in X$, we write $x \in X^{**}$ instead of $j(x) \in X^{**}$ (i.e., in what follows we drop the canonical embedding when viewing X embedded in X^{**}).

The next theorem identifies precisely X as a vector subspace of X^{**}.

Theorem 5.70 *(Goldstine Theorem)*
<u>If</u> X is a Banach space, $\overline{B}_1 \stackrel{\text{def}}{=} \{x \in X : \|x\| \leqslant 1\}$ *and* $\overline{B}^{**}_1 \stackrel{\text{def}}{=} \{x^{**} \in X^{**} : \|x^{**}\| \leqslant 1\}$,
<u>then</u> $\overline{X}^{w^*} = X^{**}$ *and* $\overline{\overline{B}_1}^{w^*} = \overline{B}^{**}_1$.

5.1.8 Reflexive Banach Spaces

In general the canonical embedding $j(\cdot)$ is not surjective. Those Banach spaces for which $j(\cdot)$ is surjective form a special class.

Definition 5.71
*A Banach space X is said to be **reflexive** if the canonical embedding $j(\cdot)$ (see Definition 5.68), is surjective.*

Remark 5.72
In the above definition, it is essential that we use the canonical embedding $j(\cdot)$. There is a celebrated example of James [9] of a nonreflexive Banach space X with the property that there exists a surjective isometry from X onto X^{**}. Consistent with the convention introduced in Remark 5.69, a reflexive Banach space X is identified with X^{**} (i.e., $X = X^{**}$ for X reflexive). Note that finite dimensional Banach spaces are reflexive (since $\dim X = \dim X^* = \dim X^{**}$). Also, Hilbert spaces (to be discussed later in this chapter) are reflexive and so are the L^p-spaces for $p \in (1, +\infty)$. However, L^1, L^∞ and $C(K)$ (with K being a compact metric space) are nonreflexive Banach spaces (with the standard norms).

The next result gives a convenient characterization of reflexive Banach spaces.

Theorem 5.73
A Banach space X is reflexive if and only if $\overline{B}_1 = \{x \in X : \|x\| \leqslant 1\}$ is w-compact.

The next theorem provides a useful description of w-compact sets in a Banach space.

Theorem 5.74 *(James Theorem)*
<u>If</u> X is a Banach space and C is a nonempty, w-closed subset of X, <u>then</u> C is w-compact if and only if every $x^* \in X^*$ attains its supremum on C.

Corollary 5.75
A Banach space X is reflexive if and only if every $x^* \in X^*$ attains its supremum on $\overline{B}_1 = \{x \in X : \|x\| \leqslant 1\}$.

In the next theorem, we summarize the properties of reflexive Banach spaces.

Theorem 5.76
Let X be a Banach space. The following statements are equivalent:
(a) *X is reflexive.*
(b) *X^* is reflexive.*
(c) *Every bounded sequence has a weakly convergent subsequence.*
(d) *If $\{C_n\}_{n\geqslant 1}$ is a decreasing sequence of nonempty bounded closed convex subsets of X, then $\bigcap_{n\geqslant 1} C_n \neq \emptyset$.*
(e) *Every separable closed vector subspace of X is reflexive.*
(f) *X is isomorphic to a reflexive Banach space.*
(g) *Every $x^* \in X^*$ is norm attaining on $\overline{B}_1$ (see Corollary 5.75).*
(h) *The following does not hold: For every $\lambda \in (0,1)$, there is a sequence $\{x_n\}_{n\geqslant 1} \subseteq \partial B_1$ such that $d\big(\operatorname{conv}\{x_k\}_{k=1}^n, \operatorname{conv}\{x_k\}_{k\geqslant n+1}\big) \geqslant \lambda$.*

Proposition 5.77
If X is a reflexive Banach space and V is a closed vector subspace of X,
then V is reflexive.

In Theorem 5.76 the equivalence of **(a)** and **(c)** is a special case of a deeper result given in the following theorem.

Theorem 5.78 *(Eberlein–Smulian Theorem)*
If X is a Banach space and $C \subseteq X$ is a nonempty subset,
then C is relatively w-compact if and only if X is relatively weakly sequentially compact (i.e., every sequence $\{x_n\}_{n\geqslant 1} \subseteq C$ admits a weakly convergent subsequence).

Remark 5.79
The Eberlein–Smulian theorem fails for the w^*-topology on X^* (see Problem 5.98).

5.1.9 Separable Banach Spaces

Next we focus on separable Banach spaces. First let us recall the definition.

Definition 5.80
A Banach space X is said to be ***separable*** *if there exists a countable set D such that $\overline{D} = X$ (i.e., X has a countable* ***dense set****).*

Remark 5.81
Every finite dimensional normed space is separable.

Theorem 5.82
If X is a Banach space and X^* is separable, then X is separable.

Remark 5.83
The converse is not true. Standard example is $X = L^1(0,1)$ with $X^* = L^\infty(0,1)$.

Corollary 5.84
Let X be a Banach space. Then X is separable reflexive if and only if X^* is separable reflexive.

Separability is closely related to the metrizability of the relative weak topology on certain subspaces. The next theorem summarizes these important results.

Theorem 5.85
Let X be a Banach space. Then
(a) *The weak* topology on $\overline{B}_1^* \stackrel{\text{def}}{=} \{x^* \in X^* : \|x^*\|_* \leqslant 1\}$ is metrizable if and only if X is separable.*
(b) *The weak topology on $\overline{B}_1 \stackrel{\text{def}}{=} \{x \in X : \|x\| \leqslant 1\}$ is metrizable if and only if X^* is separable.*
(c) *If X is separable, then the weak topology on a weakly compact subset of X is metrizable.*

Theorem 5.86
If X is a Banach space and $C \subseteq X$ is a compact subset (respectively, w-compact subset),
then $\overline{\operatorname{conv}}\, C$ is compact (respectively, w-compact).

5.1.10 Uniformly Convex Spaces

Remark 5.87
The unit ball of a Banach space is convex (local convexity). However, the boundary of this ball depends on the norm. For example, if $X = \mathbb{R}^2$ equipped with the Euclidean norm

$$\|\widehat{x}\|_2 \stackrel{\text{def}}{=} \left(x_1^2 + x_2^2\right)^{\frac{1}{2}} \qquad \forall\, \widehat{x} = (x_1, x_2) \in \mathbb{R}^2,$$

then the unit sphere has no flat parts. In contrast if we consider the norm

$$\|\widehat{x}\|_1 \stackrel{\text{def}}{=} |x_1| + |x_2| \quad \text{or} \quad \|\widehat{x}\|_\infty \;=\; \max\left\{|x_1|, |x_2|\right\},$$

then $B_1 = \left\{\widehat{x} \in \mathbb{R} : \ \|\widehat{x}\|_p < 1\right\}$ $(p = 1, \infty)$ is a rhombus (if $p = 1$) or the unit square (if $p = +\infty$) and so the unit sphere has flat parts. The notion of uniform convexity makes this situation precise and describes those Banach spaces whose unit sphere "bulges uniformly" in all directions.

Definition 5.88
A Banach space X is said to be ***uniformly convex*** *if for every $\varepsilon > 0$ there exists $\delta = \delta(\varepsilon) > 0$ such that if $x, u \in X$, $\|x\|, \|u\| \leqslant 1$ and $\|x - u\| > \varepsilon$, then $\left\|\frac{x+u}{2}\right\| < 1 - \delta$.*

Theorem 5.89 *(Milman–Pettis Theorem)*
Every uniformly convex Banach space is reflexive.

Remark 5.90
Uniform convexity is a geometric property of the norm. An equivalent norm need not be uniformly convex. On the other hand, reflexivity is a topological property. A reflexive Banach space remains reflexive for an equivalent norm. So, in Theorem 5.89 as well as in Theorem 5.58 (Mazur theorem), we have the striking feature that a geometric property implies a topological one.

Proposition 5.91
If X is a uniformly convex Banach space,
then for every sequence $\{x_n\}_{n\geqslant 1} \subseteq X$ such that $x_n \xrightarrow{w} x$ in X and $\limsup\limits_{n\to+\infty} \|x_n\| \leqslant \|x\|$, we have $x_n \longrightarrow x$ in X (this property of a Banach space is known as the ***Kadec–Klee property****).*

Remark 5.92
Recall from Proposition 5.56(c) that if $x_n \xrightarrow{w} x$ in X, then $\|x\| \leqslant \liminf_{n\to+\infty} \|x_n\|$ (w-lower semicontinuity of the norm functional in any Banach space). So, in the Kadec–Klee property, we actually require that $\|x_n\| \longrightarrow \|x\|$ (see Problem 5.174). If (Ω, Σ, μ) is a σ-finite measure space and $1 < p < +\infty$, then $L^p(\Omega)$ is uniformly convex (this is a consequence of the so-called ***Clarkson inequalities***).

5.1.11 Hilbert Spaces

Next, we turn our attention to Hilbert spaces, which are the infinite dimensional analogs of Euclidean spaces.

Definition 5.93
*Let H be a vector space. An **inner** (or **scalar**) product $(\cdot,\cdot)_H$ on H is a bilinear form from $H \times H$ into $\mathbb{R}$ such that*
- *for all $u \in H$, $(u,u)_H \geqslant 0$ and $(u,u)_H \neq 0$ if $u \neq 0$ (positive definite);*
- *for all $u, y \in H$, $(u,y)_H = (y,u)_H$ (symmetric).*

Remark 5.94
We have the so-called ***Cauchy–Schwarz inequality***, which says that

$$\big|(u,y)_H\big| \leqslant (u,u)_H^{\frac{1}{2}} (y,y)_H^{\frac{1}{2}} \qquad \forall\, u,y \in H$$

(we recall that for the proof of the above Cauchy–Schwarz inequality we do not need the ***definiteness property***, namely that $(u,u)_H \neq 0$ if $u \neq 0$), in which case $(\cdot,\cdot)_H$ is said to be a ***semiinner product***. Using the above Cauchy–Schwarz inequality, we see that $\|u\| = (u,u)_H^{\frac{1}{2}}$ is a norm on H. This norm satisfies the so-called ***parallelogram law***:

$$\left\|\tfrac{u+y}{2}\right\|_H^2 + \left\|\tfrac{u-y}{2}\right\|_H^2 = \tfrac{1}{2}\big(\|u\|^2 + \|y\|^2\big). \qquad \forall\, u,y \in H.$$

There is a theorem due to Fréchet–Jordan–von Neumann which says that if the norm of a Banach space X satisfies the parallelogram law, then X is a Hilbert space.

We also have the ***polarization identity***, which says:

$$(x,u)_H = \tfrac{1}{4}\big(\|x+u\|^2 - \|x-u\|^2\big) \qquad \forall\, x,u \in H.$$

Definition 5.95
An inner product space $\left(H, (\cdot,\cdot)_H\right)$ *which is complete for the norm defined above is called a* ***Hilbert space****.*

Using the parallelogram law (see Remark 5.94), we have the following important fact about Hilbert spaces.

Proposition 5.96
Every Hilbert space is uniformly convex, hence reflexive (see Theorem 5.89).

Theorem 5.97
If H *is a Hilbert space and* C *is a nonempty closed convex subset of* H*,*
then for every $y \in H$*, there exists a unique* $u = \operatorname{proj}_C(y) \in C$ *such that* $\|y-u\| = \inf\limits_{x \in C} \|y - x\|$ *and* $u = \operatorname{proj}_C(y) \in C$ *is characterized by*

$$u \in C \quad \textit{and} \quad (y - u, x - u)_H \leqslant 0 \qquad \forall\, x \in C.$$

Remark 5.98
The map $H \ni y \longmapsto \operatorname{proj}_C(y)$ is called the ***metric projection on*** C. In fact the existence of a unique best approximation $u = \operatorname{proj}_C(y) \in C$ is valid in any uniformly convex Banach space, not necessarily a Hilbert space.

Proposition 5.99
Let H *be a Hilbert space.*
(a) *If* $C \subseteq H$ *is nonempty, closed, convex, then the metric projection* $\operatorname{proj}_C : H \longrightarrow C$ *defined above is nonexpansive, i.e.,*

$$\|\operatorname{proj}_C(y) - \operatorname{proj}_C(v)\| \leqslant \|y - v\| \quad \forall\, y, v \in H.$$

(b) *If* $V \subseteq H$ *is a closed vector subspace, then* $u = \operatorname{proj}_V(y)$*,* $y \in H$ *is characterized by*

$$u \in V \quad \textit{and} \quad (y-u, x)_H = 0 \qquad \forall\, x \in V$$

and $\operatorname{proj}_V \in \mathcal{L}(H)$*, called the* ***orthogonal projection*** *onto* V*.*

The next theorem is very basic in the theory of Hilbert spaces. It says that every continuous linear functional on H can be identified with an element of H (i.e., $H^* = H$).

Theorem 5.100 *(Riesz–Fréchet Representation Theorem)*
<u>If</u> $x^* \in H^*$,
<u>then</u> there exists a unique $x \in H$ such that

$$\langle x^*, u\rangle = (x, u)_H \qquad \forall\, u \in H \quad \textit{and} \quad \|x^*\|_* = \|x\|.$$

Remark 5.101
Theorem 5.100 implies that there is a canonical isometry from H onto H^* and so, we can identify H with H^*. However, we should be careful with this identification. We cannot do it in every case. To see this, consider the following situation (typical in many boundary value problems). Let V and H be two Hilbert spaces and assume that V is embedded densely into H. So, we can find $c > 0$ such that

$$\|u\|_H \leqslant c\|u\|_V \qquad \forall\, u \in V. \tag{5.4}$$

Using Theorem 5.100, we identify H with its dual H^*. Let $y \in H$. Then the linear functional $V \ni u \longmapsto (y, u)_H$ is continuous, since

$$\big|(y, u)_H\big| \leqslant \|y\|_H\|u\|_H \leqslant c\|y\|_H\|u\|_V \qquad \forall\, u \in V. \tag{5.5}$$

We denote this continuous linear functional by $A(y)$. Clearly $A \in \mathcal{L}(H; V^*)$ and

$$\|A(y)\|_{\mathcal{L}} \leqslant c\|y\|_H$$

(see (5.5)). If $\big\|A(y)\big\|_{V^*} = 0$, then

$$(y, u)_H = 0 \qquad \forall\, u \in V$$

and since V is dense in H, we infer that $y = 0$. So, A is injective. Let $h \in V^{**}$ and suppose that

$$\big\langle h, A(y)\big\rangle_{V^*} = 0 \qquad \forall\, y \in H.$$

By reflexivity $V^{**} = V$ and so

$$\big\langle A(y), h\big\rangle_V = 0 \qquad \forall\, y \in H,$$

hence

$$(y, h)_H = 0 \qquad \forall\, y \in H.$$

Then $h = 0$ and so $A(H)$ is dense in V^*(see Remark 5.32). Thus we have the following setting

$$V \subseteq H = H^* \subseteq V^* \tag{5.6}$$

with both embeddings being dense. From (5.6), it is clear that we cannot identify V with V^* although it is a Hilbert space. Therefore, we cannot simultaneously identify H and V with their respective duals. In this case H is called ***pivot space***. The other spaces although Hilbert cannot be identified with their respective duals. As we already mentioned, this situation arises in many applications on boundary value problems.

Moreover, if H is a pivot Hilbert space (i.e., $H = H^*$) and V is a vector subspace of H, the annihilator $V^\perp$ given in Definition 5.36(a) is now a subspace of H, namely

$$V^\perp = \{u \in H : (u, v)_H = 0 \text{ for all } v \in V\}.$$

Then

$$V \cap V^\perp = \{0\} \quad \text{and} \quad H = V + V^\perp.$$

Indeed, every $u \in H$ can be written as

$$u = \operatorname{proj}_V(u) + \big(u - \operatorname{proj}_V(u)\big)$$

and so

$$id - \operatorname{proj}_V = \operatorname{proj}_{V^\perp}.$$

So, in a Hilbert space, every closed subspace is complemented (see Definition 5.52) and $H = V \oplus V^\perp$. In general, if V, Y are vector subspaces of H, then we say that V, Y are ***orthogonal*** if

$$(u, y)_H = 0 \qquad \forall\, v \in V,\ y \in Y$$

and we write $V \perp Y$. Similarly, if $u, y \in H$, then we say that they are orthogonal, if $(u, y)_H = 0$ and we write $u \perp y$. Orthogonality is a very important notion special to Hilbert spaces.

Definition 5.102

*Let H be a Hilbert space and $\mathcal{Y} = \{u_i \in H : i \in I\}$. We say that $\mathcal{Y}$ is an **orthonormal set** if $\|u_i\| = 1$ for all $i \in I$ and*

$$(u_i, u_j)_H = 0 \qquad \forall\, i, j \in I,\ i \neq j.$$

*A maximal orthonormal set (in the sense of inclusion) in H is called an **orthonormal basis** of H.*

Remark 5.103
A compact way to express the orthogonality condition in an orthonormal set $\mathcal{Y}$ is to write

$$(u_i, u_j)_H = \delta_{ij} \qquad \forall\, i, j \in I,$$

with δ_{ij} being the Kronecker function, i.e.,

$$\delta_{ij} \stackrel{\text{def}}{=} \begin{cases} 0 & \text{if } i \neq j \\ 1 & \text{if } i = j \end{cases} \qquad \forall\, i, j \in I.$$

It is easy to see that an orthonormal set $\mathcal{Y} = \{u_i\}_{i \in I}$ is linearly independent.

Theorem 5.104
Every Hilbert space has an orthonormal basis.

Theorem 5.105
If H is a Hilbert space, $\mathcal{Y} = \{e_i\}_{i \in I}$ is an orthonormal set and $x \in H$, then
(a) $\sum_{i \in I} |(x, e_i)_H|^2 \leqslant \|x\|^2$ *(**Basel inequality**).*
(b) If $\mathcal{Y}$ is an orthonormal basis of H, then $\|x\|^2 = \sum_{i \in I} |(x, e_i)_H|^2$ *(**Parseval equality**).*
(c) If the Parseval equality holds for every $x \in H$, then $\mathcal{Y}$ is an orthonormal basis of H.
(d) If $\overline{\operatorname{span}}\, \mathcal{Y} = H$, then $\mathcal{Y}$ is an orthonormal basis of H.

Theorem 5.106
A Hilbert space has a countable orthonormal basis if and only if it is separable.

Remark 5.107
Let X be a Banach space and let $\{e_n\}_{n \geqslant 1}$ be a sequence of vectors in X such that for every $u \in X$, we have

$$u = \sum_{n \geqslant 1} \lambda_n(u) e_n = \lim_{N \to +\infty} \sum_{n=1}^{N} \lambda_n(u) e_n.$$

Then we say that $\{e_n\}_{n \geqslant 1}$ is a ***Schauder basis*** for X. According to Theorem 5.106, every separable Hilbert space has a Schauder basis which is equal to the countable orthonormal basis. More generally, l^p (with $1 \leqslant p < +\infty$) has a Schauder basis.

Definition 5.108
Let H be a Hilbert space and let $A \in \mathcal{L}(H)$.
*(a) The **adjoint operator** $A^* \in \mathcal{L}(H)$ of A is defined by*

$$(A(u), y)_H = \langle u, A^*(y) \rangle_H \qquad \forall\, u, y \in H.$$

*(b) We say that A is **self-adjoint** if $A = A^*$.*
*(c) We say that A is **normal** if $AA^* = A^*A$.*
*(d) We say that A is **unitary** if $AA^* = A^*A = id$.*

Remark 5.109
Since H and H^* can be identified (see Theorem 5.100), Definition 5.108(a) coincides with Definition 5.33. Any orthogonal projection proj_V (with $V \subseteq H$ being a closed vector subspace) is self-adjoint.

Proposition 5.110
If H is a Hilbert space, $A_1, A_2, A \in \mathcal{L}(H)$ and $\lambda \in \mathbb{R}$, then
(a) $\|A\|_{\mathcal{L}} = \|A^*\|_{\mathcal{L}}$.
(b) $\|A^*A\|_{\mathcal{L}} = \|A\|_{\mathcal{L}}^2$.
(c) $A^{**} = A$.
(d) $(A_1 + A_2)^* = A_1^* + A_2^*$ and $(\lambda A)^* = \lambda A^*$.
(e) $(A_1 A_2)^* = A_2^* A_1^*$.

Definition 5.111
Let H be a Hilbert space and let $a\colon H \times H \longrightarrow \mathbb{R}$ be a bilinear form:
*(a) We say that a is **continuous** if there exists $c > 0$ such that $|a(u, y)| \leqslant c\|u\|\,\|y\|$ for all $u, y \in H$.*
*(b) We say that a is **symmetric** if $a(u, y) = a(y, u)$ for all $u, y \in H$.*
*(c) We say that a is **coercive** (or H-**elliptic**) if there exists $\widehat{c} > 0$ such that $a(u, y) \geqslant \widehat{c}\|u\|^2$ for all $u \in H$.*

Remark 5.112
The inner product of a Hilbert space is a continuous, symmetric and bilinear form. Conversely, if a is a continuous, symmetric and coercive bilinear form, then

$$(u, y)_a \stackrel{\text{def}}{=} a(u, y) \qquad \forall\, u, y \in H$$

defines a new inner product on H. The norm corresponding to this new inner product on H is given by

$$\|u\|_a \stackrel{\text{def}}{=} a(u,u)^{\frac{1}{2}} \qquad \forall\, u \in H.$$

From the continuity and the coercivity of a, we have

$$\widehat{c}^{\frac{1}{2}}\|u\| \leqslant \|u\|_a \leqslant c^{\frac{1}{2}}\|u\| \qquad \forall\, u \in H.$$

So, $\|\cdot\|$ and $\|\cdot\|_a$ are equivalent norms on H.

The next theorem is very useful in the study of variational inequalities.

Theorem 5.113 *(Stampacchia Theorem)*
If H is a Hilbert space, $a\colon H \times H \longrightarrow \mathbb{R}$ is a continuous, coercive bilinear form, $C \subseteq H$ is a nonempty, closed and convex set and $u^* \in H^*$, then there exists a unique $u \in C$ such that $a(u, y-u) \geqslant \langle u^*, y-u\rangle$ for all $y \in C$.
Moreover, if a is also symmetric, then u is characterized by

$$u \in C \quad \text{and} \quad \tfrac{1}{2}a(u,u) - \langle u^*, u\rangle = \inf_{y\in H}\left(\tfrac{1}{2}a(y,y) - \langle u^*, y\rangle\right).$$

A useful consequence of this theorem is the following result.

Theorem 5.114 *(Lax–Milgram Theorem)*
If H is a Hilbert space, $a\colon H \times H \longrightarrow \mathbb{R}$ is a continuous, coercive bilinear form and $u^* \in H^*$,
then there exists a unique $u \in H$ such that

$$a(u,y) = \langle u^*, y\rangle \qquad \forall\, y \in H.$$

Moreover, if a is also symmetric, then u is characterized by

$$u \in H \quad \text{and} \quad \tfrac{1}{2}a(u,u) - \langle u^*, u\rangle = \inf_{y\in H}\left(\tfrac{1}{2}a(y,y) - \langle u^*, y\rangle\right).$$

5.1.12 Unbounded Linear Operators

Thus far we have considered linear operators which are everywhere defined and bounded (hence continuous; see Proposition 5.12). Next, we look at linear operators which are not necessarily defined on the whole space and do not map bounded sets to bounded ones (i.e., they are not bounded).

Definition 5.115
Let X and Y be two Banach spaces. An ***unbounded linear operator*** *form X into Y is a linear map $A\colon X \supseteq D(A) \longrightarrow Y$ defined on a linear subspace $D(A)$ of X, called the* ***domain*** *of A, into Y. The vector space $R(A) = A(D(A)) \subseteq Y$ is called the* ***range*** *of A. We say that A is* ***closed*** *if $\operatorname{Gr} A$ is closed, where*

$$\operatorname{Gr} A = \{(x, A(x)) \in X \times Y : x \in D(A)\}$$

is the ***graph*** *of A. We say that A is* ***densely defined*** *if $\overline{D(A)} = X$.*

Remark 5.116
Recall that the operator A is bounded if $D(A) = X$ and there exists $M > 0$ such that

$$\|A(x)\|_Y \leqslant M\|x\|_X \qquad \forall\, x \in X.$$

In this case

$$\|A\|_{\mathcal{L}} = \sup_{x \neq 0} \frac{\|A(x)\|_Y}{\|x\|_X}$$

is the norm of A and $(\mathcal{L}(X,Y), \|\cdot\|_{\mathcal{L}})$ is a Banach space (see Proposition 5.14). Note that A is closed if and only if for every sequence $\{u_n\}_{n \geqslant 1} \subseteq D(A)$ such that $u_n \longrightarrow$ in X and $A(u_n) \longrightarrow y$ in Y, we have $u \in D(A)$ and $y = A(u)$. If A is closed, then $N(A)$, the kernel of A defined by

$$N(A) \stackrel{\text{def}}{=} \{u \in D(A) : A(u) = 0\}$$

is closed, but not $R(A)$.

For densely defined unbounded linear operators, we can define an adjoint map (see Definition 5.33 for bounded linear operators).

Definition 5.117
Let X and Y be two Banach spaces and let $A\colon X \supseteq D(A) \longrightarrow Y$ be an unbounded, densely defined linear operator. Then the ***adjoint*** *of A is the unbounded linear operator $A^*\colon Y^* \supseteq D(A^*) \longrightarrow X^*$, defined by*

$$\langle y^*, A(u)\rangle_Y = \langle A^*(y^*), u\rangle_X \qquad \forall\, u \in D(A), y^* \in D(A^*),$$

where

$$D(A^*) = \{y^* \in Y^* : \text{ there exists } M > 0 \text{ such that } |\langle y^*, A(u)\rangle| \leqslant M\|u\| \text{ for all } u \in D(A)\}.$$

Remark 5.118
The adjoint operator A^* defined above need not be densely defined. If A is closed, then $\overline{D(A^*)}^{w^*} = Y^*$. Hence, if Y is reflexive, then A^* is densely defined.

Proposition 5.119
If X and Y are two Banach spaces and $A\colon X \supseteq D(A) \longrightarrow Y$ is an unbounded, densely defined linear operator,
then A^* is closed.

The graphs of A and A^* are related by a simple relation. So, let X and Y be Banach spaces and let $S\colon Y^* \times X^* \longrightarrow X^* \times Y^*$ be defined by

$$S(y^*, x^*) \stackrel{\text{def}}{=} (-x^*, y^*). \tag{5.7}$$

Proposition 5.120
If X and Y are two Banach spaces and $A\colon X \supseteq D(A) \longrightarrow Y$ is a densely defined linear operator,
then $S(\operatorname{Gr} A^*) = (\operatorname{Gr} A)^{\perp}$.

Corollary 5.121
If X and Y are two Banach spaces and $A\colon X \supseteq D(A) \longrightarrow Y$ is a densely defined, closed linear operator, then
(a) $N(A) = R(A^*)^{\perp}$;
(b) $N(A^*) = R(A)^{\perp}$;
(c) $\overline{R(A^*)} \subseteq N(A)^{\perp}$;
(d) $N(A^*)^{\perp} = \overline{R(A)}$.

Remark 5.122
Even for bounded linear operators, the inclusion in **(c)** may be strict.

Theorem 5.123
If X and Y are two Banach spaces and $A\colon X \supseteq D(A) \longrightarrow Y$ is a densely defined, closed linear operator,
then the following properties are equivalent:
(a) $R(A)$ is closed;
(b) $R(A^*)$ is closed;
(c) $R(A) = N(A^*)^{\perp}$
(d) $R(A^*) = N(A)^{\perp}$.

In the next theorem, we have a useful characterization of surjective operators.

Theorem 5.124
If X and Y are two Banach spaces and $A\colon X \supseteq D(A) \longrightarrow Y$ is a densely defined, closed linear operator,
then the following statements are equivalent:
(a) *A is surjective, i.e., $R(A) = Y$;*
(b) *there exists $c > 0$ such that $\|y^*\|_{Y^*} \leqslant c\|A^*(y^*)\|_{X^*}$ for all $y^* \in D(A^*)$;*
(c) *$N(A^*) = \{0\}$ and $R(A^*) \subseteq X^*$ is closed.*

An analogous result concerning the surjectivity of A^* is also true.

Theorem 5.125
If X and Y are two Banach spaces and $A\colon X \supseteq D(A) \longrightarrow Y$ is a densely defined, closed linear operator,
then the following statements are equivalent:
(a) *A^* is surjective, i.e., $R(A^*) = X^*$;*
(b) *there exists $c > 0$ such that $\|u\|_X \leqslant c\big\|A(u)\big\|_Y$ for all $u \in D(A)$;*
(c) *$N(A) = \{0\}$ and $R(A)$ is closed.*

Remark 5.126
If $\dim X < +\infty$ or $\dim Y < +\infty$, then we have
(a) A is surjective if and only if A^* is injective;
(b) A^* is surjective if and only if A is injective.
In the general case, we have
(c) if A is surjective, then A^* is injective;
(d) if A^* is surjective, then A is injective.

5.1.13 Extremal Structure of Sets

Definition 5.127
*Let X be a vector space and let $C \subseteq X$ be a nonempty convex subset. A point $e \in C$ is said to be an **extreme point** of C, if there are no $x_1, x_2 \in C$ such that $x_1 \neq x_2$ and $e = \lambda x_1 + (1-\lambda)x_2$ for some $\lambda \in (0,1)$. The set of all extreme points of C is denoted by $\operatorname{ext} C$.*

Theorem 5.128 *(Krein–Milman Theorem)*
If X is a locally convex space and $C \subseteq X$ is a nonempty, compact, convex set,
then $\operatorname{ext} C \neq \emptyset$ and $C = \overline{\operatorname{conv}} \operatorname{ext} C$.

Remark 5.129
If C is only compact (not necessarily convex), then $C \subseteq \overline{\text{conv}}\,\text{ext}\,C$.

Corollary 5.130
<u>If</u> X is a normed space and $\overline{B}_1^* = \{x^* \in X^* : \|x^*\|_* \leqslant 1\}$,
<u>then</u> $\overline{B}_1^* = \overline{\text{conv}}^{w^*}\,\text{ext}\,B_1^*$.

Proposition 5.131
<u>If</u> X is a locally convex space, $A \subseteq X$ is a nonempty, compact set and $\overline{\text{conv}}\,A$ is compact too,
<u>then</u> $\text{ext}\,\overline{\text{conv}}\,A \subseteq A$.

Remark 5.132
According to Theorem 5.86, the above proposition is valid if X is a Banach space furnished with the norm or weak topology.

Proposition 5.133
<u>If</u> X is a locally convex set, $C \subseteq X$ is a nonempty, compact, convex set and $A \subseteq C$ is such that $\overline{\text{conv}}\,A = C$,
<u>then</u> $\text{ext}\,C \subseteq \overline{A}$.

More generally, we can define.

Definition 5.134
*Let X be a vector space and let $E \subseteq A \subseteq X$ be nonempty sets. We say that E is an **extreme set** if for any $x, u \in A$ for which $\lambda x + (1-\lambda)u \in E$ for some $\lambda \in (0,1)$, we have $x, u \in E$ (i.e., no point of E is an interior point of a line segment whose end points are in A but not in E).*

Remark 5.135
If E is a singleton, then we recover the definition of extreme point (see Definition 5.127). Note that, if E is an extreme set of a convex set A, then $X \setminus E$ is convex, but the converse is not in general true.

Definition 5.136
*Let X be a Banach space and let $A \subseteq X$ be a nonempty set. By a **slice** of A we mean a set of the form*

$$S_A(x^*, \eta) \stackrel{\text{def}}{=} \{x \in A : \langle x^*, x \rangle > \sigma(x^*, A) - \eta\},$$

where $x^* \in X^*$, $\eta > 0$ *and*

$$\sigma(x^*, A) \stackrel{\text{def}}{=} \sup_{u \in A} \langle x^*, u \rangle$$

(the support function of A*). So, a slice of* A *is the intersection of* A *with an open half-space. If* $A \subseteq X^*$*, then we can also define* w^**-slice of* A *by taking the defining functional in* X *and not in* X^{**}*.*

Theorem 5.137
If X *is a Banach space,* $C \subseteq X$ *is a* w*-compact, convex set (respectively,* $C \subseteq X^*$ *is a* w^**-compact, convex set) and* $e \in \operatorname{ext} C$*,*
then the slices (respectively, w^**-slices) of* C *form a local basis of* e *for the relative weak topology (respectively, weak* topology) of* C*.*

5.1.14 Compact Operators

Next we turn our attention to a special class of linear operators between Banach spaces, which generalize linear maps between finite dimensional Banach spaces. In particular, we have a rich spectral theory for such operators.

Definition 5.138
Let X *and* Y *be two Banach spaces and let* $A \in \mathcal{L}(X;Y)$*.*
(a) *We say that* A *is* **compact** *if* $\overline{A(\overline{B}_1^X)}$ *is compact (here* $\overline{B}_1^X = \{x \in X : \|x\|_X \leqslant 1\}$*). The space of compact operators from* X *into* Y *is denoted by* $\mathcal{L}_c(X;Y)$ *(if* $X = Y$*, then we write* $\mathcal{L}_c(X)$*).*
(b) *We say that* A *is* **finite rank** *if* $\dim A(X) < +\infty$*. The space of finite rank operators from* X *into* Y *is denoted by* $\mathcal{L}_f(X;Y)$ *(if* $X = Y$*, then we write* $\mathcal{L}_f(X)$*).*
(c) *We say that* A *is* **completely continuous** *if* $x_n \xrightarrow{w} x$ *in* X *implies that* $A(x_n) \longrightarrow A(x)$ *in* Y*.*

Remark 5.139
Since in a finite dimensional Banach space, bounded sets are relatively compact, we have $\mathcal{L}_f(X;Y) \subseteq \mathcal{L}_c(X;Y)$. In particular then, if $\dim X < +\infty$ or $\dim Y < +\infty$, then $\mathcal{L}(X;Y) \subseteq \mathcal{L}_c(X;Y)$. If X is an infinite dimensional Banach space and $id : X \longrightarrow X$ is the identity operator, then $id \notin \mathcal{L}_c(X)$ (recall that $\overline{B}_1 = \{x \in X : \|x\| \leqslant 1\}$ is norm compact if and only if $\dim X < +\infty$). It is easy to see that, if

X, Y, Z are three Banach spaces, $A \in \mathcal{L}(X;Y)$, $C \in \mathcal{L}(Y;Z)$ and one of them is compact, then $C \circ A \in \mathcal{L}_c(X;Z)$. These last two observations imply that compact operators in an infinite dimensional Banach space are not invertible.

Proposition 5.140
If X and Y are Banach spaces,
then $(\mathcal{L}_c(X;Y), \|\cdot\|_{\mathcal{L}})$ is a Banach space.

Proposition 5.141
If X and Y are Banach spaces and $\{A_n\}_{n\geqslant 1} \subseteq \mathcal{L}_f(X;Y)$ is a sequence such that $\|A_n - A\|_{\mathcal{L}} \longrightarrow 0$,
then $A \in \mathcal{L}_c(X;Y)$.

Remark 5.142
The converse of the above proposition is the celebrated "approximation problem", which asks the question whether it is true that $\overline{\mathcal{L}_f(X;Y)}^{\|\cdot\|_{\mathcal{L}}} = \mathcal{L}_c(X;Y)$. This question remained open for a long time and finally in 1972 Enflo [7] produced a counterexample. However, for some particular spaces this is true (see Problems 5.155 5.157 and 5.160).

Theorem 5.143
If X and Y are Banach spaces and $A \in \mathcal{L}(X;Y)$,
then $A \in \mathcal{L}_c(X;Y)$ if and only if $A^* \in \mathcal{L}_c(Y^*;X^*)$.

Theorem 5.144
If X is a Banach space, $A \in \mathcal{L}_c(X)$ and $\lambda \in \mathbb{R} \setminus \{0\}$,
then $N(\lambda id - A)$ is finite dimensional and $R(\lambda id - A)$ is closed and finite codimensional (i.e., has a finite dimensional topological complement).

These properties lead to the introduction of the following important class of linear operators.

Definition 5.145
*Let X and Y be two Banach spaces and let $A \in \mathcal{L}(X;Y)$. We say that A is a **Fredholm operator** if $N(A)$ is finite dimensional and $R(A)$ is finite codimensional. The number $i(A) = \dim N(A) - \operatorname{codim} R(A)$ is called the **index** of A.*

Remark 5.146
From Theorem 5.144, we see that if $A \in \mathcal{L}_c(X)$, then for every $\lambda \neq 0$, $\lambda id - A$ is a Fredholm operator.

The next theorem is useful in the study of boundary value problems.

Theorem 5.147 *(Fredholm Alternative Theorem)*
If X is a Banach space, $A \in \mathcal{L}_c(X)$ and $\lambda \neq 0$, then

$$N(\lambda id - A) = \{0\} \quad \Longleftrightarrow \quad R(\lambda id - A) = X.$$

Remark 5.148
The Fredholm alternative deals with the solvability of the operator equation

$$\lambda u - A(u) = h$$

and it says: "Either the equation has a unique solution for every $h \in Y$ or the homogeneous equation $\lambda u - A(u) = 0$ has n-linearly independent solutions ($n = \dim N(\lambda id - A)$)".

5.1.15 Spectral Theory

As we already mentioned, compact operators have interesting spectral properties. So, next we concentrate on the spectral properties of compact operators. In order to have the full strength of the theory, we will assume (and this is the only place in this chapter that we do this) that all Banach spaces are over $\mathbb{C}$.

Definition 5.149
Let X be a Banach space and let $A \in \mathcal{L}(X)$. The ***spectrum*** *$\sigma(A)$ of A is defined by*

$$\sigma(A) \stackrel{\text{def}}{=} \{\lambda \in C : \ A - \lambda id \ \textit{is not invertible}\}.$$

The ***resolvent*** *$\varrho(A)$ of A is defined by $\varrho(A) = \mathbb{C} \setminus \sigma(A)$.*

Remark 5.150
If X is infinite dimensional and $A \in \mathcal{L}_c(X)$, then from Remark 5.139, we have $0 \in \sigma(A)$. Also, for any $A \in \mathcal{L}(X)$, we have

$$\sigma(A) \subseteq \{\lambda \in \mathbb{C} : \ |\lambda| \leqslant r(A) = \lim_{n \to +\infty} \|A^n\|^{\frac{1}{n}}\}.$$

Proposition 5.151
If X is a Banach space and $A \in \mathcal{L}(X)$,
then $\sigma(A) \neq \emptyset$ and it is compact in the disc $\{\lambda \in \mathbb{C} : |\lambda| \leqslant \|A\|_{\mathcal{L}}\}$.

Remark 5.152
The nonemptiness of $\sigma(A)$ may fail if we consider real Banach spaces.

Definition 5.153
*Let X be a Banach space and let $A \in \mathcal{L}(X)$. We say that $\lambda \in \sigma(A)$ is an **eigenvalue** if there exists $u \in X$, $u \neq 0$ such that $A(u) = \lambda u$. Such element u is called an **eigenfunction** corresponding to the eigenvalue λ. The space $N(A - \lambda id)$ is the **eigenspace** corresponding to the eigenvalue $\lambda \in \mathbb{C}$. We denote the set of eigenvalues of A by $\sigma_p(A)$ (**point spectrum** of A).*

Remark 5.154
If $\dim X < +\infty$, then the operator $A - \lambda id$ is invertible if and only if it is injective. Therefore $\sigma(A) = \sigma_p(A)$.

Theorem 5.155
If X is an infinite dimensional Banach space and $A \in \mathcal{L}_c(X)$, then
(a) *$0 \in \sigma(A)$;*
(b) *$\sigma(A) \setminus \{0\} = \sigma_p(A) \setminus \{0\}$;*
(c) *one of the following cases holds:*
- *(i) $\sigma(A) = \{0\}$; or*
- *(ii) $\sigma(A) \setminus \{0\}$ is a finite set; or*
- *(iii) $\sigma(A) \setminus \{0\}$ is a sequence converging to 0.*

Remark 5.156
In fact, for a given sequence $\{\lambda_n\}_{n\geqslant 1} \subseteq \mathbb{C}$ such that $\lambda_n \longrightarrow 0$, we can find $A \in \mathcal{L}(X)$ such that $\sigma(A) = \{\lambda_n\}_{n\geqslant 1} \cup \{0\}$.

Proposition 5.157
If H is a Hilbert space, $A \in \mathcal{L}(H)$ is self-adjoint and

$$m \stackrel{\text{def}}{=} \inf\{(A(u), u)_H : u \in H, \ \|u\| = 1\},$$
$$M \stackrel{\text{def}}{=} \sup\{(A(u), u)_H : u \in H, \ \|u\| = 1\},$$

then $\sigma(A) \subseteq [m, M]$, $m \in \sigma(A)$, $M \in \sigma(A)$ and $\|T\|_{\mathcal{L}} = \max\{|m|, |M|\}$.

Remark 5.158
From the above proposition, it follows that

$$\begin{aligned}\|A\|_{\mathcal{L}} &= \sup\big\{\big|\big(A(u),u\big)_H\big| : u\in H,\ \|u\|=1\big\}\\ &= \sup\big\{\big|\big(A(u),u\big)_H\big| : u\in H,\ \|u\|\leqslant 1\big\}.\end{aligned}$$

Moreover, it is clear that for a self-adjoint operator A, we have $\sigma_p(A)\subseteq\mathbb{R}$ and eigenvector corresponding to distinct eigenvalues are orthogonal.

Proposition 5.159
If H is a Hilbert space, $A\in\mathcal{L}(H)$ is self-adjoint and $\lambda\in\mathbb{C}$,
then $\lambda\in\sigma(A)$ if and only if $\inf\big\{\big\|(A-\lambda id)(u)\big\| : u\in H,\ \|u\|=1\big\}=0$.

The main spectral theorem for compact self-adjoint operators on a Hilbert space is the following.

Theorem 5.160
If H is an infinite dimensional separable Hilbert space and $A\in\mathcal{L}_c(H)$ is self-adjoint,
then there is an orthonormal basis $\{e_k\}_{k\geqslant 1}$ of H formed by cigcnvcctors of A such that

$$A(u) = \sum_{k\geqslant 1}\lambda_k\,(u,e_k)_H\,e_k \qquad \forall\ u\in H,$$

with λ_k being eigenvalues corresponding to e_k.

Proposition 5.161
If H is a Hilbert space and $P\in\mathcal{L}(H)$,
then the following statements are equivalent:
(a) *P is an orthogonal projection (i.e., $N(P)^{\perp}=R(P)$, $H=N(P)\oplus R(P)$);*
(b) *P is self-adjoint;*
(c) *P is a normal operator (see Definition 5.108(c));*
(d) *$\big(u-P(u),P(u)\big)_H=0$ for all $u\in H$;*
(e) *$\|P\|_{\mathcal{L}}=1$.*

We conclude this look at the spectral properties of bounded linear operators on Banach spaces and Hilbert spaces, with the following theorem, which has useful applications in the study of spectral properties of second order elliptic equations.

Theorem 5.162 *(Krein-Rutman Theorem)*
If X is a Banach space, $C \subseteq X$ is a closed, convex cone with vertex 0 (i.e., $\lambda u + \eta y \in C$ for all $\lambda, \eta \geqslant 0$ and all $u, y \in X$), $\operatorname{int} C \neq \emptyset$, $C \neq X$, $A \in \mathcal{L}_c(X)$ and $A(C \setminus \{0\}) \subseteq \operatorname{int} C$,
then there exist $u_0 \in \operatorname{int} C$ and $\lambda_0 > 0$ such that $A(u_0) = \lambda_0 u_0$ and λ_0 is the unique eigenvalue corresponding to an eigenvector of A in C, i.e., $A(u) = \lambda u$ with $u \in P \setminus \{0\}$, imply $\lambda = \lambda_0$ and $u \in \mathbb{R}_+ u_0$. Finally $\lambda_0 = \max\{|\lambda| : \lambda \in \sigma(A)\}$ and the multiplicity of λ_0 (i.e., the $\dim N(A - \lambda_0 id)$) equals 1 (i.e., λ_0 is simple).

5.1.16 Differentiability and the Geometry of Banach Spaces

In this section, we investigate the relation between separability, reflexivity, differentiability of the norm and rotundity of the unit ball. A first notion in this direction was introduced in Definition 5.88, where we defined ***uniform convex spaces***. Now, we go behind this notion, introducing more general ones and establishing a duality with the differentiability properties of the norm. We start our discussion by introducing the main differentiability notions for $\mathbb{R}$-valued functions defined on a Banach spaces.

Definition 5.163
Let X be a Banach space and let $f: X \longrightarrow \mathbb{R}$ be a function.
(a) *We say that f is **Gâteaux differentiable** at $x \in X$ if there exists $f'_G(x) \in X^*$ such that*

$$\lim_{\lambda \to 0} \frac{f(x+\lambda h) - f(x)}{\lambda} = \langle f'_G(x), h \rangle \qquad \forall\, h \in X. \tag{5.8}$$

*We say that f is **Gâteaux differentiable**, if it is Gâteaux differentiable at every $x \in X$.*
(b) *We say that f is **Fréchet differentiable** at $x \in X$ if there exists $f'(x) \in X^*$ such that*

$$\lim_{h \to 0} \frac{f(x+h) - f(x) - \langle f'(x), h \rangle}{\|h\|} = 0.$$

*We say that f is **Fréchet differentiable** if it is Fréchet differentiable at every $x \in X$.*

Remark 5.164
If $f\colon X \longrightarrow \mathbb{R}$ is a continuous function which is Fréchet differentiable at $x \in X$, then it is also Gâteaux differentiable at $x \in X$ and $f'(x) = f'_G(x)$. Moreover, f is Fréchet differentiable at $x \in X$ if it is Gâteaux differentiable at x and the limit in (5.8) is uniform for $\|x\| \leqslant 1$ as $\lambda \to 0$. Finally, in finite dimensional Banach spaces, for continuous convex functions, Gâteaux and Fréchet differentiability coincide.

Definition 5.165
Let X be a Banach space with a norm $\|\cdot\|$. We say that $\|\cdot\|$ is ***Fréchet*** *(respectively,* ***Gâteaux****)* ***differentiable*** *if the (convex) function $x \longmapsto \|x\|$ is Fréchet (respectively, Gâteaux) differentiable at every $x \in X \setminus \{0\}$.*

Remark 5.166
Due to homogeneity, a norm is differentiable at x if it is differentiable at λx for some $\lambda \in \mathbb{R}$. Therefore, it is enough to check the differentiability at points of $\partial B_1 = \{x \in X : \|x\| = 1\}$.

Proposition 5.167
If X is a Banach space and the dual norm of X^* is Fréchet differentiable,
then X is reflexive.

Definition 5.168
Let X be a Banach space.
(a) *A norm $\|\cdot\|$ of X is said to be* ***strictly convex*** *(or* ***rotund****) if $\operatorname{ext} \overline{B}_1 = \partial B_1$ (recall that $\overline{B}_1 = \{x \in X : \|x\| \leqslant 1\}$, $\partial B_1 = \{x \in X : \|x\| = 1\}$).*
(b) *A norm $\|\cdot\|$ of X is said to be* ***locally uniformly convex*** *if $\|x_n - x\| \longrightarrow 0$ whenever $x_n, x \in \overline{B}_1$ and $\|x_n + x\| \longrightarrow 2$.*

Remark 5.169
Evidently $\|\cdot\|$ is strictly convex if $x = u$ whenever $\|x + u\| = 2$ and $\|x\| = \|u\| = 1$. In the sequel we will say that X is ***strictly convex*** (respectively, ***locally uniformly convex***) if it has a norm $\|\cdot\|$ with these properties. A locally uniformly convex space is strictly convex.

Proposition 5.170
Let $(X, \|\cdot\|)$ be a Banach space.
(a) *If the dual norm is strictly convex, then $\|\cdot\|$ is Gâteaux differentiable.*
(b) *If the dual norm is locally uniformly convex, then $\|\cdot\|$ is Fréchet differentiable.*

Theorem 5.171
(a) *If X is a separable Banach space, then X admits an equivalent locally uniformly convex norm.*
(b) *If X^* is separable, then X admits an equivalent norm whose dual is locally uniformly convex (in particular then this equivalent norm is Fréchet differentiable).*

From Proposition 5.170 and Theorem 5.171, we obtain the following Corollary.

Corollary 5.172
(a) *If X is a separable Banach space, then X admits an equivalent norm which is both locally uniformly convex and Gâteaux differentiable.*
(b) *If X^* is separable, then X admits an equivalent norm which is both locally uniformly convex and Fréchet differentiable.*

Definition 5.173
A Banach space X is said to be an ***Asplund space*** *(respectively,* ***weak Asplund space****) if every continuous convex function defined on a nonempty open convex set $U \subseteq X$ is Fréchet (respectively, Gâteaux) differentiable on a dense G_δ-subset of U.*

Theorem 5.174
(a) *A separable Banach space is a weak Asplund space.*
(b) *If X is a Banach space with separable dual, then X is an Asplund space.*

The "Asplund property" is hereditary on closed subspaces.

Proposition 5.175
<u>If</u> X is an Asplund space and V is a closed vector subspace of X, <u>then</u> V is an Asplund space.

We have a converse of this proposition.

Proposition 5.176
If X is a Banach space and every separable, closed vector subspace of X is an Asplund space,
then X is an Asplund space.

Next we introduce a notion weaker than separability which has implications for both Asplund and weak Asplund spaces.

Definition 5.177
A Banach space X is said to be ***weakly compactly generated*** *(**WCG** for short), if there exists a weakly compact set $C \subseteq X$ such that $\overline{\operatorname{span}} C = X$. Since $\overline{\operatorname{span}} C$ is w-compact (see Theorem 5.86), we can always assume that the generating set C is convex.*

Remark 5.178
Clearly a reflexive Banach space is WCG (see Theorem 5.73). Similarly a separable Banach space X is WCG. Indeed, let $\{x_n\}_{n\geqslant 1}$ be a dense sequence in X and let $C = \left\{\frac{1}{n}\frac{x_n}{\|x_n\|}\right\}_{n\geqslant 1} \cup \{0\}$. Note that in this case C is in fact compact. However, a WCG Banach space need not be reflexive nor separable. For example, if (Ω, Σ, μ) is a measure space with μ finite and separable (see Remark 3.140), then $L^1(\Omega)$ is nonreflexive, separable Banach space which is WCG. Consider $C = \{\chi_A : A \in \Sigma\}$ which is relatively w-compact by the Dunford–Pettis theorem (see Theorem 4.75). Finally for an uncountable set Γ, the Banach space $X = c_0(\Gamma)$ is a nonreflexive, nonseparable Banach space, which is WCG. Consider $C = \{e_\gamma\}_{\gamma\in\Gamma} \cup \{0\}$, where e_γ is the usual unit basis vector. Note, though that $l^1(\Gamma)$ is WCG if and only if Γ is countable.

Theorem 5.179
If X is a Banach space such that X^* is WCG,
then X is an Asplund space.

Remark 5.180
The converse of this theorem is not true. Let $X = c_0(\Gamma)$ with Γ being uncountable. Then $X^* = l^1(\Gamma)$ (see Remark 5.178).

Theorem 5.181
If the Banach space X is a subspace of a WCG space,
then X is a weak Asplund space.

5.1.17 Best Approximation: Various Theorems for Banach Spaces

We conclude this chapter with a closer look at the "best approximation" problem. So, let $\big(X, \|\cdot\|\big)$ be a Banach space, $C \subseteq X$ and $y \in X \setminus C$. We consider the following minimization problem:

$$\operatorname{dist}(y, C) = \inf_{u \in C} \|y - u\|. \tag{5.9}$$

Definition 5.182
(a) *A point $u_0 \in C$ such that $\operatorname{dist}(y, C) = \|y - u_0\|$ (see (5.9)) is called a point of* ***best approximation*** *to y in C.*
(b) *The set $C \subseteq X$ is said to be a* ***Chebyshev set*** *if for every $y \in X$, there is a unique point of best approximation to $y \in X$ in C.*
(c) *For a Chebyshev set $C \subseteq X$, the map $y \longmapsto \operatorname{proj}_C(y)$ (where $\operatorname{proj}_C(y)$ is the unique best approximation to y) is called the* ***metric projection*** *of X into C.*
(d) *A closed set $C \subseteq X$ is said to be* ***almost convex*** *if for any closed ball B which does not intersect C, there exists a closed ball $\widehat{B} \supseteq B$ of arbitrary large radius which does not intersect C.*

Theorem 5.183
If X is a Banach space and $C \subseteq X$ is a Chebyshev set with continuous metric projection,
then C is almost convex.

Remark 5.184
Almost convexity is very close to convexity. In fact in certain spaces they coincide.

Theorem 5.185
If X is a Banach space with strict convex dual,
then every Chebyshev set with continuous metric projection is convex.

Corollary 5.186
In a Hilbert space, every Chebyshev set with continuous metric projection is convex.

Corollary 5.187
In a Hilbert space, every weakly closed Chebyshev set is convex.

Remark 5.188
Remains an open question (even in Hilbert spaces), whether every Chebyshev set is convex.

We conclude with three important results about Banach spaces.

Theorem 5.189 *(Banach–Dieudnné Theorem)*
If X is a Banach space and $C \subseteq X^*$ is a nonempty convex set such that $C \cap n\overline{B}_1^*$ is w^*-closed for all $n \geqslant 1$ (here $\overline{B}_1^* \stackrel{\text{def}}{=} \{x^* \in X^* : \|x^*\|_* \leqslant 1\}$),
then the set C is w^*-closed.

Theorem 5.190
If X is a Banach space, $C \subseteq X$ is a nonempty set and $x \in \overline{C}^w$,
then there is a countable set $D \subseteq C$ such that $x \in \overline{D}^w$.

Remark 5.191
According to Theorem 5.190, the weak closure of a set in a Banach space is countably determined. Note that this does not mean that X_w is first countable, which is not true if X is infinite dimensional.

The third theorem is a very helpful tool in theory of maximal monotone maps.

Theorem 5.192 *(Troyanski Renorming Theorem)*
Every reflexive Banach space can be given an equivalent norm so that X and X^* are both locally uniformly convex and have Fréchet differentiable norms.

5.2 Problems

Problem 5.1 **

Let X be a normed space with $\dim X \geqslant 2$ and let $\partial B_1 = \{x \in X : \|x\| = 1\}$. Show that ∂B_1 is path-connected.

Problem 5.2 **

Let X be a topological vector space and let $C \subseteq X$ be a convex set such that $\operatorname{int} C \neq \emptyset$ (such a set is usually called ***convex body***). Show that $\overline{C} = \overline{\operatorname{int} C}$ and $\operatorname{int} \overline{C} = \operatorname{int} C$.

Problem 5.3 ***

Let X be a Banach space and let $V \subseteq X$ be a vector subspace, which is a G_δ-set in X. Show that V is closed (hence a Banach space too). Then show that a normed space X, which is topologically complete (see Definition 1.56), is a Banach space.

Problem 5.4 *

Let X be a normed space and let $V \subseteq X$ be a vector subspace such that $\operatorname{int} V \neq \emptyset$. Show that $V = X$.

Problem 5.5 ***

Let X be a normed space and let $\{x_k\}_{k=1}^m \subseteq X$ be linearly independent vectors. Show that there exists $\varepsilon > 0$ such that, if $\{y_k\}_{k=1}^m \subseteq X$ satisfy $\|y_k\| < \varepsilon$ for all $k \in \{1, \ldots, m\}$, then $\{x_k + y_k\}_{k=1}^m$ are linearly independent.

Problem 5.6 *

Let X be a topological vector space and let $\subseteq X$ be a nonempty open set. Show that $\operatorname{conv} U$ is open too.

Problem 5.7 *

Suppose that X is a Banach space, V is a vector subspace of X and $\{V_n\}_{n \geqslant 1}$ is the sequence of closed vector subspaces of X such that $V = \bigcup\limits_{n \geqslant 1} V_n$. Show that for some $n_0 \geqslant 1$, $V = V_{n_0}$.

Problem 5.8 **

Let X be a Banach space and let $C \subseteq X$ be a nonempty set. Show that the set C is relatively compact if and only if for every $\varepsilon > 0$ there exists a relatively compact set $C_\varepsilon \subseteq X$ such that $C \subseteq \varepsilon \overline{B}_1 + C_\varepsilon$.

Problem 5.9 *
Let X be an infinite dimensional Banach space and let $C \subseteq X$ be a nonempty, compact set. Show that $\operatorname{int} C = \emptyset$.

Problem 5.10 *
Let X be a Banach space and let $V \subseteq X$ be a finite dimensional vector subspace. Show that for every $x \in X$, we can find $u \in V$ such that $\|x - u\| = \operatorname{dist}(x, V)$.

Problem 5.11 *
Let X be a Banach space and let $\{x_n\}_{n\geqslant 1} \subseteq X$ be a sequence such that $\sum\limits_{n\geqslant 1} \|x_n\| < +\infty$. Show that there exists $x \in X$ such that $x = \sum\limits_{n\geqslant 1} x_n$.

Problem 5.12 **
Let X be a normed space and let $\{x_n\}_{n\geqslant 1} \subseteq X$ be a sequence. We say that the (formal) series $\sum\limits_{n\geqslant 1} x_n$ is ***convergent*** if the sequence of partial sums $\{s_m\}_{m\geqslant 1}$ (where $s_m = \sum\limits_{n=1}^{m} x_n$) converges to some $x \in X$ in the norm of X. We say that the series $\sum\limits_{n\geqslant 1} x_n$ is ***absolutely convergent*** if the real series $\sum\limits_{n\geqslant 1} \|x_n\|$ converges. From Problem 5.11, we know that in a Banach space, every absolute convergent series is also convergent. Show that if in a normed space X every absolutely convergent series is also convergent, then X is a Banach space.

Problem 5.13 **
Let X be a normed space and let V be a closed vector subspace of X. Assume that both V and $X/_V$ (with the quotient norm; see Definition 5.18) are Banach spaces. Show that X is a Banach space.

Problem 5.14 **
Let X and Y be two Banach spaces and let $A\colon X \longrightarrow Y$ be a linear operator such that for every sequence $\{x_n\}_{n\geqslant 1} \subseteq X$ with $x_n \longrightarrow 0$, we have that the sequence $\{A(x_n)\}_{n\geqslant 1}$ is bounded. Show that $A \in \mathcal{L}(X; Y)$.

Problem 5.15 *
Suppose that X and Y are two normed spaces, $\dim X < +\infty$ and $A\colon X \longrightarrow Y$ is a linear operator. Show that $A \in \mathcal{L}(X; Y)$.

Problem 5.16 *
Suppose that X is a normed space, Y is a Banach space and $A \in \mathcal{L}(X;Y)$ is a surjective isometry. Show that X is a Banach space.

Problem 5.17 *
Let X and Y be two normed spaces and let $A\colon X \longrightarrow Y$ be an injective linear operator. Show that the following three statements are equivalent:
(a) A is a surjective isometry;
(b) $A(\overline{B}_1^X) = \overline{B}_1^Y$;
(c) $A(\partial B_1^X) = \partial B_1^Y$. Recall that $\overline{B}_1^X \stackrel{\text{def}}{=} \{x \in X : \|x\|_X \leqslant 1\}$, $\overline{B}_1^Y \stackrel{\text{def}}{=} \{y \in Y : \|y\|_Y \leqslant 1\}$.

Problem 5.18 **
Suppose that X and Y are two normed spaces, $\{A_n\colon X \longrightarrow Y\}_{n\geqslant 1}$ is a sequence of linear operators and $A\colon X \longrightarrow Y$ is a linear operator. Let

$$C \stackrel{\text{def}}{=} \{x \in X : \{A_n(x)\}_{n\geqslant 1} \text{ does not converge to } A(x) \text{ in } Y\}.$$

Show that C is either empty or dense in X.

Problem 5.19 **
Let X be a Banach space, let $x^* \in \partial B_1^* = \{x^* \in X^* : \|x^*\|_* = 1\}$ and let $x \in X$. Show that $\operatorname{dist}(x, \ker x^*) = |\langle x^*, x\rangle|$.

Problem 5.20 **
Let $X = C([0,1])$ be furnished with the norm

$$\|u\|_\infty = \max_{0\leqslant t\leqslant 1} |u(t)| \qquad \forall\, u \in X$$

(hence $(X, \|\cdot\|_\infty)$ is a Banach space) and let $\xi\colon [0,1] \longrightarrow [0,1]$ be a continuous function. Let $A\colon X \longrightarrow X$ be the linear operator, defined by $A(u) = u \circ \xi$.
(a) Show that A is an isometry if and only if ξ is surjective;
(b) Produce an example of an isometry $A \in \mathcal{L}(X)$ which is not surjective.

Problem 5.21 **

Let $C^1([0,1])$ be the space of continuous functions on $[0,1]$, which are continuously differentiable on $(0,1)$ and whose derivative can be extended continuously on $[0,1]$. Evidently $C^1([0,1]) \subseteq C([0,1])$. We endow both spaces with the supremum norm

$$\|u\|_\infty = \max_{0 \leqslant t \leqslant 1} |u(t)|$$

and consider the linear operator $A\colon C^1([0,1]) \longrightarrow C([0,1])$, defined by $A(u) \stackrel{\text{def}}{=} u'$. Is A continuous? Justify your answer.

Problem 5.22 *

Let X be a Banach space.
Suppose that $V \subseteq X$ is a vector subspace and there exists a bounded projection of X onto V (i.e., an operator $P \in \mathcal{L}(X;V)$ which is surjective and $P|_V = id_V$). Show that V is closed.

Problem 5.23 **

Let $X \stackrel{\text{def}}{=} l^1 = \{\widehat{x} = \{x_n\}_{n\geqslant 1} : \|\widehat{x}\|_{l^1} = \sum_{n\geqslant 1} |x_n| < +\infty\}$ and let $\{e_k\}_{k\geqslant 1} \subseteq X$ be a sequence, defined by $e_k = \{\delta_{kn}\}_{n\geqslant 1}$ for $k \geqslant 1$ (where δ_{kn} is the Kronecker symbol). Let $C = \overline{\text{conv}}\,\{e_k\}_{k\geqslant 1}$. Is it true that $0 \in C$? Justify your answer.

Problem 5.24 ***

Let X and Y be Banach spaces and let $a\colon X \times Y \longrightarrow \mathbb{R}$ be a functional which is continuous and linear in each variable. Show that a is continuous.

Problem 5.25 ***

Let X and Y be two normed spaces. We know that if Y is a Banach space, then so is $(\mathcal{L}(X;Y), \|\cdot\|_{\mathcal{L}})$ (see Proposition 5.14). Show the converse, namely that if $(\mathcal{L}(X;Y), \|\cdot\|_{\mathcal{L}})$ is a Banach space, then so is Y.

Problem 5.26 **

Let X be a Banach space and let $A \in \mathcal{L}(X)$, $A \neq 0$. Is it true that: if $\|x_n\| \longrightarrow +\infty$, then $\|A(x_n)\| \longrightarrow +\infty$? Justify your answer.

Problem 5.27 *
Let X be a Banach space, let $A \in \mathcal{L}(X)$ and assume that for every $x \in X$, we can find an integer $n \geqslant 1$ such that $x \in N(A^n)$. Show that there exists an integer $n_0 \geqslant 1$ such that $A^{n_0} = 0$ (we say that A is ***nilpotent***).

Problem 5.28 **
Suppose that X and Y are two Banach spaces, $A \in \mathcal{L}(X;Y)$ and $\{A_n\}_{n\geqslant 1} \subseteq \mathcal{L}(X;Y)$ is a $\|\cdot\|_{\mathcal{L}}$-Cauchy sequence such that

$$A_n(x) \longrightarrow A(x) \quad \forall\, x \in X.$$

Show that $\|A_n - A\|_{\mathcal{L}} \longrightarrow 0$.

Problem 5.29 **
Let X and Y be two Banach spaces and let $A \in \mathcal{L}(X;Y)$. Suppose that $R(A)$ is a dense proper vector subspace of Y. Show that there exists $y \in Y$ such that

$$\text{if } \lim_{n\to+\infty} A(x_n) = y, \quad \text{then } \|x_n\|_X \to +\infty.$$

Problem 5.30 *
Suppose that X is a Banach space, Y is a normed space and $A \in \mathcal{L}(X;Y)$ is injective with A^{-1} being continuous on $A(X)$. Show that, if $A(X)$ is dense in Y, then A is surjection.

Problem 5.31 ***
Let X and Y be two normed space with $\dim Y < +\infty$ and let $A\colon X \longrightarrow Y$ be a linear operator. Show that A is continuous (i.e., $A \in \mathcal{L}(X;Y)$) if and only if $\ker A$ is closed.

Problem 5.32 ***
Suppose that X is a normed space, V is a vector subspace of X and $x \in X \setminus V$. Show that there exists $x^* \in X^*$ such that $\|x^*\|_* = 1$, $\langle x^*, x\rangle = \operatorname{dist}(x,V)$ and $x^*|_V = 0$, where $\operatorname{dist}(x,V) = \inf\limits_{u\in V} \|x-u\|$.

Problem 5.33 ***
Let X be a Banach space and let $h\colon X \longrightarrow \mathbb{R}$ be a linear functional. Show that either $N(h) = \ker h$ is closed or $N(h)$ is a dense proper vector subspace of X.

Problem 5.34 **
Suppose that X is a Banach space, $V \subseteq X$ is a closed vector subspace and $x_0 \notin V$. Show that the set span $\{V, x_0\}$ is a closed vector subspace of X.

Problem 5.35 **
Suppose that X is a normed space, $V \subseteq X$ is a vector subspace and $h\colon V \longrightarrow \mathbb{R}$ is a continuous functional. Show that the set of all Hahn–Banach extensions of h is convex.

Problem 5.36 **
Suppose that X is a normed space, $\varepsilon \in (0, \frac{1}{2}]$, $x^*, u^* \in X^*$, $\|x^*\|_* = \|u^*\|_* = 1$ and assume that

$$\big| \langle u^*, x \rangle \big| \leqslant \varepsilon \qquad \forall\, x \in \ker x^*,\ \|x\| \leqslant 1.$$

Show that $\|x^* - u^*\|_* \leqslant 2\varepsilon$ or $\|x^* + u^*\|_* \leqslant 2\varepsilon$.

Problem 5.37 **
Let X be a Banach space and let V and Y be two vector subspaces of X such that $X = V \oplus Y$. Show that $X^* = V^\perp \oplus Y^\perp$ (see Definition 5.36).

Problem 5.38 **
Let $X = C\big([0,1]\big)$. On X we consider the usual L^∞-norm, i.e., $\|u\|_\infty = \max\limits_{0 \leqslant t \leqslant 1} \big|u(t)\big|$ and another norm $\|\cdot\|$ with the property that "if $u_n \xrightarrow{\|\cdot\|} u$, then $u_n(t) \longrightarrow u(t)$ for all $t \in [0,1]$". Suppose that $\big(X, \|\cdot\|\big)$ is a Banach space. Show that the two norms $\|\cdot\|_\infty$ and $\|\cdot\|$ are equivalent.

Problem 5.39 **
Assume that X, V, Y are three Banach spaces, $A \in \mathcal{L}(X;Y)$, $B \in \mathcal{L}(V, Y)$ and for every $x \in X$, there exists a unique $v \in V$ such that $A(x) = B(v)$. Let $S\colon X \longrightarrow V$ be defined by $S(x) = v$. Show that $S \in \mathcal{L}(X;V)$.

Problem 5.40 *
Assume that X and Y are two Banach spaces, $\{A_n\}_{n \geqslant 1} \subseteq \mathcal{L}(X;Y)$ is a sequence such that $A_n(x) \longrightarrow A(x)$ in Y for all $x \in X$. Show that, if $x_n \longrightarrow x$ in X, then $A_n(x_n) \longrightarrow A(x)$ in Y.

Problem 5.41 **
Let X be a Banach space and let $A\colon X \longrightarrow X^*$ be a linear operator such that $\langle A(x), x\rangle \geqslant 0$ for all $x \in X$. Show that $A \in \mathcal{L}(X; X^*)$.

Problem 5.42 **
Let X be a Banach space and let $A\colon X \longrightarrow X^*$ be a linear operator such that $\langle A(x), u\rangle = \langle A(u), x\rangle$ for all $x, u \in X$. Show that $A \in \mathcal{L}(X; X^*)$.

Problem 5.43 **
Let X and Y be two Banach spaces and let $A \in \mathcal{L}(X; Y)$ be surjective.
(a) Show that if $C \subseteq X$ is a nonempty set, then $A(C)$ is closed in Y if and only if $N(A) + C$ is closed in X.
(b) Show that if V is a closed vector subspace of X and $\dim N(A) < +\infty$, then $A(V)$ is closed in Y.

Problem 5.44 **
Let X and Y be two Banach spaces and let $A \in \mathcal{L}(X; Y)$ be a surjection. Show that there exists $M > 0$ such that for a given $y \in Y$, we can find $x \in X$ with $\|x\|_X \leqslant M\|y\|_Y$ for which we have $A(x) = y$.

Problem 5.45 **
Let X and Y be two Banach spaces and let $\mathcal{Y} \subseteq \mathcal{L}(X; Y)$. Show that $\mathcal{Y}$ is equicontinuous if and only if it is equibounded (i.e., there exists $M > 0$ such that $\|A\|_{\mathcal{L}} \leqslant M$ for all $A \in \mathcal{Y}$).

Problem 5.46 **
Let X be a Banach space and let $A \in \mathcal{L}(X)$, with $\|A\|_{\mathcal{L}} < 1$. Show that $id - A$ is an isomorphism and $(id - A)^{-1} = \sum\limits_{n\geqslant 0} A^n$.

Problem 5.47 **
Let X be a Banach space and let $\{A_n\}_{n\geqslant 1} \subseteq \mathcal{L}(X)$ be a sequence of isomorphisms such that $\|A_n^{-1}\|_{\mathcal{L}} < 1$ for all $n \geqslant 1$ and $A_n \longrightarrow A$ in $\mathcal{L}(X)$. Show that A is an isomorphism too.

Problem 5.48 ***
Suppose that X is a Banach space, Y is a normed space and $\{A_n\}_{n\geqslant 1} \subseteq \mathcal{L}(X; Y)$ is a sequence such that $x_n \longrightarrow 0$ in X implies that $A_n(x_n) \longrightarrow 0$ in Y. Show that $\sup\limits_{n\geqslant 1} \|A_n\|_{\mathcal{L}} < +\infty$.

Problem 5.49 ***
Let X be a Banach space and let $V, Y \subseteq X$ be two closed vector subspaces such that $V \cap Y = \{0\}$. Let $\partial B_1^V = \{v \in V : \|v\| = 1\}$ and $\partial B_1^Y = \{y \in Y : \|y\| = 1\}$. We set

$$\eta \stackrel{\text{def}}{=} \operatorname{dist}(\partial B_1^V, \partial B_1^Y) = \inf \{\|v - y\| : v \in \partial B_1^V,\ y \in \partial B_1^Y\}.$$

Show that $V + Y$ is closed if and only if $\eta > 0$.

Problem 5.50 **
Let X be a Banach space, let $A, S \in \mathcal{L}(X)$ and suppose that $AS = SA$. Show that AS is invertible if and only if A and S are invertible.

Problem 5.51 **
Suppose that X and Y are two Banach spaces, $A \in \mathcal{L}(X;Y)$ and $\{A_n\}_{n\geqslant 1} \subseteq \mathcal{L}(X;Y)$ is a sequence such that $A_n(x) \longrightarrow A(x)$ in Y for all $x \in X$. Suppose that $C \subseteq X$ is a compact set. Show that $A_n(x) \longrightarrow A(x)$ in Y uniformly for all $x \in C$.

Problem 5.52 **
Let X and Y be two Banach spaces and let $A \in \mathcal{L}(X;Y)$ be a surjection. Show that the following two statements are equivalent:
(a) There exists an operator $S \in \mathcal{L}(Y;X)$ such that $AS = id_Y$ (i.e., A has a right inverse).
(b) $N(A)$ has a topological complement (see Definition 5.52).

Problem 5.53 *
Let X, Y, Z be three Banach spaces and let $A \in \mathcal{L}(X;Y)$, $S \in \mathcal{L}(Y;Z)$. Show that:
(a) $(S \circ A)^* = A^* \circ S^*$.
(b) If $A \in \mathcal{L}(X;Y)$ is a bijection, then so is $A^* \in \mathcal{L}(Y^*;X^*)$ and $(A^*)^{-1} = (A^{-1})^*$.

Problem 5.54 **
Let X and Y be two Banach spaces and let $A\colon X \longrightarrow Y$ be a linear map. Show that A is continuous (i.e., $A \in \mathcal{L}(X;Y)$) if and only if $D(A^*) = Y^*$.

Problem 5.55 ***
Let X be an infinite dimensional Banach space and let $(\overline{B}_1, w)$ denote the closed unit ball $\overline{B}_1 = \{x \in X : \|x\| \leqslant 1\}$ furnished with the relative w-topology. Show that $\partial B_1 = \{x \in X : \|x\| = 1\}$ is a dense G_δ-subset of $(\overline{B}_1, w)$.

Problem 5.56 **
From Problem 5.55, we know that, if X is an infinite dimensional Banach space, then $\overline{B}_1 = \overline{\partial B_1}^w$. Use this fact to show that X furnished with the weak topology (denoted by X_w) is not metrizable.

Problem 5.57 **
Suppose that X is a normed space, $\dim X = m \geqslant 2$ and $\{x_k\}_{k=1}^{m-1} \subseteq X$ are linearly independent vectors. Show that there exists a vector $x_m \in X$ such that $\|x_m\| = 1$ and $\|x_m - x_k\| > 1$ for all $k \in \{1, \ldots, m-1\}$.

Problem 5.58 *
Suppose that X is a normed space, $\{x_n\}_{n \geqslant 1} \subseteq X$ is a norm Cauchy sequence such that $x_n \xrightarrow{w} 0$. Show that $x_n \longrightarrow 0$ in X.

Problem 5.59 *
Show that an infinite dimensional Banach space X with the weak topology is of first Baire category.

Problem 5.60 *
Let X be a Banach space and let $C \subseteq X$ be a nonempty, norm compact set. Suppose that $\{x_n\}_{n \geqslant 1} \subseteq C$ is a sequence such that $x_n \xrightarrow{w} x$ in X. Show that $x_n \longrightarrow x$ in X

Problem 5.61 **
Let X and Y be two Banach spaces, let $A\colon X \longrightarrow Y$ be a linear operator and suppose that for every $y^* \in Y^*$, we have $y^* \circ A \in X^*$. Show that $A \in \mathcal{L}(X; Y)$.

Problem 5.62 **
Let X and Y be two Banach spaces and let $A\colon X \longrightarrow Y$ be a linear operator. Show that $A \in \mathcal{L}(X; Y)$ if and only if $A \in \mathcal{L}(X; Y_w)$ (here Y_w denotes the Banach space Y furnished with the weak topologies).

Problem 5.63 **
Let X be a normed space and let $V \subseteq X$ be a vector subspace. Show that $w(V, V^*) = w(X, X^*)|_V$, i.e., that the weak topology on V (induced by its dual V^*) is the restriction on V of the weak topology of X.

Problem 5.64 *
Suppose that X is a Banach space, $C, E \subseteq X$ are nonempty sets such that C is weakly closed and E is weakly compact. Show that $C + E$ is weakly closed.

Problem 5.65 **
Let X be a Banach space and let $C^* \subseteq X^*$ be a nonempty set. From Corollary 5.45 and Remark 5.46, we know that the set C^* is bounded if and only if it is w^*-bounded. Show that this is no longer true if X is only a normed space (i.e., completeness of X is necessary).

Problem 5.66 ***
Let $(X, \|\cdot\|)$ be a normed space and let $|\cdot|_*$ be a norm on X^*. We say that $|\cdot|_*$ is an ***equivalent dual norm*** if there exists an equivalent norm $|\cdot|$ on X such that $|\cdot|_*$ is the corresponding dual norm of $|\cdot|$. Show that $|\cdot|_*$ is an equivalent dual norm on X^* if and only if $|\cdot|_*$ is w^*-lower semicontinuous.

Problem 5.67 **
Let X and Y be two Banach spaces and let $A \in \mathcal{L}(X; Y)$. Show that:
(a) If $C \subseteq X$ is a nonempty and w-compact set, then $\operatorname{ext} A(C) \subseteq A(\operatorname{ext} C)$.
(b) If A is an isomorphism and the set C is convex, then $\operatorname{ext} A(C) = A(\operatorname{ext} C)$.

Problem 5.68 **
Let X be a Banach space. Show that X is finite dimensional if and only if X^* is finite dimensional. Moreover, $\dim X = \dim X^*$.

Problem 5.69 **
Let X be a Banach space and let $D \subseteq X^*$ be a set. Show that D ***separates points*** in X (i.e., for every $x \in X \setminus \{0\}$, there exists $x^* \in D$ such that $\langle x^*, x\rangle \neq 0$) if and only if $X^* = \overline{\operatorname{span}}^{w^*} D$.

Problem 5.70 **
Let X and Y be two Banach spaces and let $A \in \mathcal{L}(X;Y)$. Show that A is injective if and only if $A^*(Y^*)$ is w^*-dense in X^*.

Problem 5.71 **
Show that the Mazur theorem (see Theorem 5.58) does not hold for the w^*-topology.

Problem 5.72 **
Show that every normed space is a dense vector subspace of a Banach space.

Problem 5.73 **
Show that every normed space X is isometrically isomorphic to a subspace of $C(K)$ (the Banach space of continuous functions on some compact topological space K).

Problem 5.74 **
Let X and Y be two Banach spaces and let $A \in \mathcal{L}(X;Y)$ be such that $\overline{A(\overline{B}_1^X)} \subseteq Y$ is weakly compact (such an operator is called ***weakly compact***). Show that $A^{**}(X^{**}) \subseteq Y$.

Problem 5.75 **
Suppose that X is a Banach space, $V \subseteq X$ is a vector subspace and $u^* \in X^*$. Show that there exists $\widehat{v}^* \in V^\perp$ such that $\|u^* - \widehat{v}^*\|_* = \inf\limits_{v^* \in V^\perp} \|u^* - v^*\|_* = m^\perp$.

Problem 5.76 **
Let X be a Banach space and let $C \subseteq X$ be a nonempty, w-closed set. Suppose that for every $\varepsilon > 0$, there exists a w-compact set $C_\varepsilon \subseteq X$ such that $C \subseteq C_\varepsilon + \varepsilon \overline{B}_1$. Show that the set C is w-compact.

Problem 5.77 **
Show that no infinite dimensional vector subspace of

$$l^1 \overset{\text{def}}{=} \{\widehat{x} = \{x_n\}_{n\geqslant 1} : \ \|x\|_1 = \sum_{n\geqslant 1} |x_n| < +\infty\}$$

is reflexive.

Problem 5.78 *
Let X be a reflexive Banach space and let $V \subseteq X$ be a closed vector subspace. Show that $X/_V$ is reflexive.

Problem 5.79 *
Let X be a normed space and let $C \subseteq X^*$ be a nonempty and w^*-closed set. Show that for a given $x^* \in X^*$, we can find $u^* \in C$ such that $\|x^* - u^*\|_* = \operatorname{dist}(x^*, C)$ (a set in X^* (or in X) with the "best approximation" property for every element in the space is called ***proximinal***).

Problem 5.80 **
Show that a Banach space X is reflexive if and only if every closed, convex set is proximinal (see Problem 5.79).

Problem 5.81 **
Let X be a reflexive Banach space and let $\varphi \colon X \longrightarrow \overline{\mathbb{R}} = \mathbb{R} \cup \{+\infty\}$ be a convex, lower semicontinuous function not identically $+\infty$. Suppose that φ is weakly coercive, i.e., $\lim\limits_{\|x\| \to +\infty} \varphi(x) = +\infty$. Show that there exists $x_0 \in X$ such that $\varphi(x_0) = \inf\limits_{x \in X} \varphi(x)$.

Problem 5.82 *
Let X and Y be two Banach spaces with X being reflexive and let $A \in \mathcal{L}(X;Y)$. Show that $A(\overline{B}_1^X) \subseteq Y$ is closed (where $\overline{B}_1^X = \{x \in X : \|x\|_X \leqslant 1\}$).

Problem 5.83 **
Let X and Y be two Banach spaces with X being reflexive and let $A \in \mathcal{L}(X;Y)$ be a surjection. Show that Y is reflexive too.

Problem 5.84 **
Let X be a Banach space. Show that the following two statements are equivalent:
(a) X is reflexive;
(b) for every decreasing sequence $\{C_n\}_{n \geqslant 1}$ of nonempty, closed, convex and bounded subsets of X, we have $\bigcap\limits_{n \geqslant 1} C_n \neq \emptyset$.

Problem 5.85 *
Let X and Y be two Banach spaces with X nonreflexive and Y reflexive. Suppose that $A \in \mathcal{L}(X;Y)$ is injective. Show that $A(X)$ is not closed in Y.

Problem 5.86 *

Let

$$l^2 \stackrel{\text{def}}{=} \Big\{\widehat{x} = \{x_n\}_{n\geqslant 1} : \ \|\widehat{x}\|_{l^2} = \big(\sum_{n\geqslant 1} x_n^2\big)^{\frac{1}{2}} < +\infty\Big\}$$

and

$$l^1 \stackrel{\text{def}}{=} \Big\{\widehat{x} = \{x_n\}_{n\geqslant 1} : \ \|x\|_{l^1} = \sum_{n\geqslant 1} |x_n| < +\infty\Big\}.$$

Suppose that $A \in \mathcal{L}(l^1; l^2)$ is injective. Is $A(l^1)$ a closed vector subspace of l^2? Justify your answer.

Problem 5.87 *

Let X be a Banach space. We say that $\{x_n\}_{n\geqslant 1} \subseteq X$ is a ***weak Cauchy*** sequence if for every $x^* \in X^*$, the sequence $\big\{ \langle x^*, x_n\rangle \big\}_{n\geqslant 1} \subseteq \mathbb{R}$ is Cauchy. We say that X is ***weakly sequentially complete*** if every weak Cauchy sequence has a weak limit. Show that a reflexive Banach space is weakly sequentially complete.

Problem 5.88 ***

Suppose that H is a Hilbert space, $C \subseteq H$ is a nonempty, closed, convex, bounded set and $f\colon C \longrightarrow C$ is a nonexpansive function (see Definition 1.48). Show that f has a fixed point (i.e., there exists $x \in C$ such that $f(x) = x$).

Problem 5.89 ***

Let H and V be two Hilbert spaces and let $A \in \mathcal{L}(H; V)$. Assume that there is no $x \in \partial B_1^H = \big\{x \in H : \ \|x\|_H = 1\big\}$ such that $\big\|A(x)\big\|_V = \|A\|_{\mathcal{L}}$. Show that there exists a sequence $\{x_n\}_{n\geqslant 1} \subseteq \partial B_1^H$ such that $x_n \xrightarrow{w} 0$ in H and $\big\|A(x_n)\big\| \longrightarrow \|A\|_{\mathcal{L}}$.

Problem 5.90 **

Let X be a reflexive Banach space with X^* being strictly convex and let $\mathcal{F}\colon X \longrightarrow 2^{X^*} \setminus \{\emptyset\}$ be the duality map (see Remark 5.27). Show that $\mathcal{F}$ is single-valued, sequentially continuous from X into X^*_w (X^*_w denoting the duality space X^* furnished with the weak topology) and $\big\langle \mathcal{F}(x) - \mathcal{F}(u), x - u\big\rangle \geqslant 0$ for all $x, u \in X$ (i.e., $\mathcal{F}$ is monotone).

Problem 5.91 **

Let X be a normed space and let $C \subseteq X$ be a nonempty compact set.

(a) Is $\operatorname{conv} C$ necessarily closed?

(b) Is $\overline{\operatorname{conv}}\, C$ necessarily compact? Justify your answer. (Compare with Theorem 5.86.)

Problem 5.92 ***

Suppose that X is a finite dimensional vector space, $\dim X = n$ and $C \subseteq X$ is a nonempty set. Show that every $x \in \operatorname{conv} C$ can be written as a convex combination of at most $n+1$ vectors of C.

Problem 5.93 **

Let X be a finite dimensional vector space and let $C \subseteq X$ be a nonempty, compact set. Show that $\operatorname{conv} C$ is compact too. (Compare with Theorem 5.86.)

Problem 5.94 **

Let X be a locally convex space and let $C \subseteq X$ be a nonempty and totally bounded set. Show that $\operatorname{conv} C$ is totally bounded too.

Problem 5.95 **

Let X be a Fréchet space and let $C \subseteq X$ be a nonempty, compact set. Show that the set $\overline{\operatorname{conv}}\, C$ is compact. (Compare with Theorem 5.86.)

Problem 5.96 ***

Let X be a Banach space with a separable dual. Let $\mathcal{B}(X^*)$ be the Borel σ-algebra of X^* with the norm topology and let $\mathcal{B}(X^*_{w^*})$ be the Borel σ-algebra of X^* with the w^*-topology. Show that $\mathcal{B}(X^*) = \mathcal{B}(X^*_{w^*})$.

Problem 5.97 ***

Show that the Banach space $C_b\big((0,1)\big)$ of bounded continuous functions on $(0,1)$ with the supremum norm

$$\|u\|_\infty \stackrel{\text{def}}{=} \sup_{0<t<1} \big|u(t)\big| \qquad \forall\, u \in C_b\big((0,1)\big)$$

is not separable.

Problem 5.98 **
Give an example to show that the Eberlein–Smulian theorem is not true for the w^*-topology (see Theorem 5.78 and Remark 5.79).

Problem 5.99 **
Show that every separable Banach space is isometric to a subspace of

$$l^\infty \stackrel{\text{def}}{=} \big\{\widehat{x} = \{x_n\}_{n\geqslant 1} : \ \|\widehat{x}\|_{l^\infty} = \sup_{n\geqslant 1} |x_n| < +\infty\big\}.$$

Problem 5.100 **
Let X be a Banach space, let $C \subseteq X$ and assume that for every separable, closed vector subspace $V \subseteq X$, the set $C \cap V$ is w-compact. Show that C is w-compact.

Problem 5.101 **
Suppose that X is a Banach space which is w-separable (i.e., separable with respect to the weak topology in X). Show that X is separable.

Problem 5.102 **
Let X be a separable Banach space. Show that there exists a sequence $\{x_n^*\}_{n\geqslant 1} \subseteq \partial B_1^* = \big\{x^* \in X^* : \ \|x^*\|_* = 1\big\}$ which ***separates points*** in X, i.e., if $x \in X \setminus \{0\}$, then there exists $n_0 \geqslant 1$ such that $\langle x_{n_0}^*, x\rangle \neq 0$.

Problem 5.103 **
Let X be a closed, infinite dimensional vector subspace of

$$l^1 \stackrel{\text{def}}{=} \big\{\widehat{x} = \{x_n\}_{n\geqslant 1} : \ \|\widehat{x}\|_{l^1} = \sum_{n\geqslant 1} |x_n| < +\infty\big\}.$$

Show that X^* is nonseparable. (Compare with Problem 1.18.)

Problem 5.104 **
Let X be a separable Banach space. Show that $X^*_{w^*}$ (i.e., X^* furnished with the weak* topology) is a Souslin space.

Problem 5.105 ***
Show that every separable Banach space is isomorphic to some quotient space of

$$l^1 \stackrel{\text{def}}{=} \big\{\widehat{\lambda} = \{\lambda_n\}_{n\geqslant 1} : \ \|\widehat{\lambda}\|_{l^1} = \sum_{n\geqslant 1} |\lambda_n| < +\infty\big\}.$$

Problem 5.106 ***

Let X be a separable Banach space and let $C \subseteq X$ be a nonempty subset. Show that C is relatively compact if and only if for every sequence $\{x_n^*\}_{n\geqslant 1} \subseteq X^*$ such that $x_n^* \xrightarrow{w} 0$, we have $\sup\limits_{x\in C} \big| \langle x_n^*, x\rangle \big| \longrightarrow 0$.

Problem 5.107 **

Let X be a separable Banach space and let $C \subseteq X^*$ be a nonempty, weakly compact set. Show that the set C is norm separable.

Problem 5.108 **

Let X be an infinite dimensional Banach space and assume that X^* is separable or X is reflexive. Show that we can find a sequence $\{x_n\}_{n\geqslant 1} \subseteq \partial B_1$ (where $\partial B_1 = \big\{x \in X : \|x\| = 1\big\}$) such that $x_n \xrightarrow{w} 0$ in X.

Problem 5.109 **

Let H be a separable Hilbert space and let $V \subseteq H$ be a vector subspace of H which is dense in H. Show that V contains an orthonormal basis of H.

Problem 5.110 **

Let X be a uniformly convex Banach space and let $V \subseteq X$ be a closed subspace. Show that $X/_V$ is uniformly convex.

Problem 5.111 *

Show that the Banach space $C([-1,1])$ with the supremum norm

$$\|u\|_\infty \stackrel{\text{def}}{=} \max_{-1\leqslant t\leqslant 1} \big|u(t)\big| \qquad \forall\, u \in C([-1,1])$$

is not a Hilbert space.

Problem 5.112 **

Let $\{u_n\}_{n\geqslant 1}$ be an orthonormal sequence of the Hilbert space $L^2\big([0,1]\big)$. Show that, if there exists $M > 0$ such that $\big|u_n(t)\big| \leqslant M$ for almost all $t \in T = [0,1]$ and $\sum_{n\geqslant 1} \lambda_n u_n(t)$ converges for almost all $t \in T = [0,1]$, then $\lambda_n \longrightarrow 0$.

Problem 5.113 **
Let H be a Hilbert space and suppose that $x_n \xrightarrow{w} x$ in X. Show that there exists a subsequence $\{x_{n_k}\}_{k \geqslant 1}$ of $\{x_n\}_{n \geqslant 1}$ such that the sequence of "arithmetic means" $\{\frac{1}{k} \sum_{i=1}^{k} x_{n_i}\}_{k \geqslant 1}$ converges in norm to x.

Problem 5.114 **
Let H be a Hilbert space and let $A \in \mathcal{L}(H)$. We say that A is ***positive*** and write $A \geqslant 0$ if and only if A is self-adjoint and $(A(x), x)_H \geqslant 0$ for all $x \in H$. Show that, if $A \in \mathcal{L}(H)$ is positive, then

$$\left|(A(x), u)_H\right|^2 \leqslant (A(x), x)_H (A(u), u)_H \qquad \forall\, x, u \in H$$

and

$$\left\|A(x)\right\|^2 \leqslant \|A\|_{\mathcal{L}} (A(x), x)_H$$

Problem 5.115 *
Let H be a Hilbert space. For two given self-adjoint operators $A, S \in \mathcal{L}(H)$, we write $A \leqslant S$ if $S - A \geqslant 0$ (see Problem 5.114). Let $\{A_n\}_{n \geqslant 1} \subseteq \mathcal{L}(H)$ be a ***monotone sequence*** of self-adjoint operators (i.e., $A_n \leqslant A_{n+1}$ for all $n \geqslant 1$ or $A_n \geqslant A_{n+1}$ for all $n \geqslant 1$) such that $\|A_n\|_{\mathcal{L}} \leqslant M$ for all $n \geqslant 1$ and some $M > 0$. Show that there exists a self-adjoint operator $A \in \mathcal{L}(H)$ such that $A_n(x) \longrightarrow A(x)$ in H for all $x \in H$.

Problem 5.116 *
Let H be a Hilbert space and assume that $A \in \mathcal{L}(H)$ is self-adjoint.
(a) Show that for every $n \geqslant 1$, we have $A^{2n} \geqslant 0$ (see Problem 5.114).
(b) Show that, if $A \geqslant 0$, then for all $n \geqslant 1$, we have $A^{2n+1} \geqslant 0$.

Problem 5.117 ***
Show that, if V is a closed subspace of $L^2([0,1])$ and $V \subseteq C([0,1])$, then $\dim V < +\infty$.

Problem 5.118 ***
Show that the Banach–Steinhaus theorem (see Corollary 5.40) fails if the space X is not complete.

Problem 5.119 ⋆
Let H be a Hilbert space and let $\{A_n\}_{n\geqslant 1} \subseteq \mathcal{L}(H)$ be a sequence such that $\sup\limits_{n\geqslant 1} \big|\big(A_n(x), y\big)_H\big| < +\infty$ for all $x, y \in H$. Show that $\sup\limits_{n\geqslant 1} \|A_n\|_{\mathcal{L}} < +\infty$.

Problem 5.120 ⋆
Let H be a Hilbert space and let $\{e_n\}_{n\geqslant 1}$ be an orthonormal basis of H. Show that $e_n \xrightarrow{w} 0$ in H.

Problem 5.121 ⋆⋆
Let H be a Hilbert space and let $C \subseteq H$ be a nonempty, closed, convex set. Let $x \notin C$ and let $\operatorname{proj}_C(x)$ be its metric projection on C (see Theorem 5.97 and Remark 5.98). Show that $\operatorname{proj}_C(x) \in \partial C$ (see Definition 2.9(e)) and $\operatorname{dist}(x, C) = \operatorname{dist}(x, \partial C)$.

Problem 5.122 ⋆⋆
Let H be a Hilbert space and let $\{C_n\}_{n\geqslant 1} \subseteq H$ be a decreasing sequence of nonempty, closed convex sets such that $C \stackrel{\text{def}}{=} \bigcap\limits_{n\geqslant 1} C_n \neq \emptyset$. Show that $\operatorname{proj}_{C_n} \longrightarrow \operatorname{proj}_C$ (proj_S being the metric projection on the corresponding set $S \subseteq X$).

Problem 5.123 ⋆
Let H be a Hilbert space and let $A \in \mathcal{L}(H)$. Show that $\|A^*A\|_{\mathcal{L}} = \|A\|_{\mathcal{L}}^2$.

Problem 5.124 ⋆⋆
Let H be a Hilbert space and suppose that $A \in \mathcal{L}(H)$ is normal. Show that $\|A(x)\|^2 \leqslant \|A^2(x)\| \, \|x\|$ for all $x \in H$.

Problem 5.125 ⋆
Let H be a Hilbert space and let $A \in \mathcal{L}(H)$ be normal. Show that, if A^2 is compact, then A is compact too.

Problem 5.126 ⋆
Let H be a Hilbert space and let $A \in \mathcal{L}(H)$ be a surjection. Show that the following three statements are equivalent:
(a) A is unitary (see Definition 5.108).
(b) A is an isometry (see Definition 5.16).

(c) A preserves the inner product (i.e., $\big(A(x), A(u)\big)_H = (x, u)_H$ for all $x, u \in H$).

Problem 5.127 *
Let H be a Hilbert space and let $A \in \mathcal{L}(H)$ be a self-adjoint operator. Show that eigenvectors corresponding to different eigenvalues are orthogonal.

Problem 5.128 **
Let H be a Hilbert space and let $C \subseteq H$ be a nonempty, closed, convex cone (i.e., $\lambda C \subseteq C$ for all $\lambda > 0$). Show that for $x, u \in H$, the following conditions are equivalent:
(a) $u = \mathrm{proj}_C(x)$;
(b) $u \in C$, $x - u \perp u$ and $(x - u, c)_H \leqslant 0$ for all $c \in C$.

Problem 5.129 *
Suppose that H is a Hilbert space, $V \subseteq H$ is a nonempty vector subspace, X is a Banach space and $S \in \mathcal{L}(V; X)$. Show that there exists $A \in \mathcal{L}(H; X)$ such that $A|_V = S$ and $\|A\|_{\mathcal{L}} = \|S\|_{\mathcal{L}}$.

Problem 5.130 ***
Give an example of a Hilbert space and a set in it, which is closed and sequentially weakly closed but not weakly closed.

Problem 5.131 *
Let X be a Banach space and let $P \in \mathcal{L}(X)$ be a projection (i.e., $P^2 = P$). Show that P^* is a projection in X^*.

Problem 5.132 **
Let H be a Hilbert space. Show that every orthogonal projection $P \in \mathcal{L}(H)$ satisfies $0 \leqslant P \leqslant id$ (see Problems 5.114 and 5.115).

Problem 5.133 **
Let X be a normed space and let $x, u \in X$ be two linearly independent vectors such that $\|x\| = \|u\| = 1$ and $\|x + u\| = \|x\| + \|u\|$. Show that the unit sphere of X contains a line segment.

Problem 5.134 **

Let X be a separable Banach space with a Schauder basis $\{e_n\}_{n\geqslant 1}$. Show that there exist $c \geqslant 1$ such that for any $x = \sum\limits_{n\geqslant 1} \lambda_n e_n$, we have $\big\| \sum\limits_{n=1}^{m} \lambda_n e_n \big\| \leqslant c \big\| \sum\limits_{n\geqslant 1} \lambda_n e_n \big\|$ for all $m \geqslant 1$.

Problem 5.135 **

Let H be a Hilbert space and let $P \in \mathcal{L}(H)$. Show that P is an orthogonal projection if and only if $P^2 = P$ and P is self-adjoint.

Problem 5.136 **

Let X and Y be two Banach spaces and let $a\colon X \times Y \longrightarrow \mathbb{R}$ be a bilinear form such that:

(i) a is continuous, i.e., there exists $M > 0$ such that

$$\big|a(x,y)\big| \leqslant M\|x\|_X \|y\|_Y \qquad \forall\, x \in X,\ y \in Y;$$

(ii) there exists $c_1 > 0$ such that

$$\sup_{x\in X\setminus\{0\}} \frac{|a(x,y)|}{\|x\|_X} \geqslant c_1\|y\|_Y \qquad \forall\, y \in Y;$$

(iii) there exists $c_2 > 0$ such that

$$\sup_{y\in Y\setminus\{0\}} \frac{|a(x,y)|}{\|y\|_Y} \geqslant c_2\|x\|_X \qquad \forall\, x \in X.$$

Show that for given $x^* \in X^*$ and $y^* \in Y^*$, there exist unique elements $x_0 \in X$ and $y_0 \in Y$ such that

$$a(x_0, y) = \langle y^*, y\rangle_Y \qquad \forall\, y \in Y$$

and

$$a(x, y_0) = \langle x^*, x\rangle_X \qquad \forall\, x \in X.$$

Problem 5.137 **

Let H be a Hilbert space, let $A \in \mathcal{L}(H)$ and assume that there exists $c > 0$ such that $\big\langle A(x), x\big\rangle \geqslant c\|x\|^2$ for all $x \in H$. Show that A is an isomorphism.

Problem 5.138 **

Let X and Y be two Banach spaces and let $A \in \mathcal{L}(X;Y)$. Show that the following statements are equivalent:
(a) A is an isomorphism;
(b) $R(A)$ is dense in Y and there exists $c > 0$ such that $\|A(x)\|_Y \geqslant c\|x\|_X$ for all $x \in X$.

Problem 5.139 **

Can we find a surjective operator $A \in \mathcal{L}(l^2; l^1)$? Justify your answer.

Problem 5.140 **

Let X and Y be two Banach spaces and let $\mathcal{D}(X;Y)$ be the subset of $\mathcal{L}(X;Y)$ which consists of all $A \in \mathcal{L}(X;Y)$ which are injective and $R(A)$ is closed. Show that $\mathcal{D}(X;Y)$ is an open subset of $\mathcal{L}(X;Y)$.

Problem 5.141 **

Let X and Y be two Banach spaces and let $\mathcal{Y}(X;Y)$ be the subset of $\mathcal{L}(X;Y)$ which consists of all $A \in \mathcal{L}(X;Y)$ which are injective. Show that $\mathcal{Y}(X;Y)$ is an open subset of $\mathcal{L}(X;Y)$.

Problem 5.142 **

Let H be a Hilbert space and let $A \in \mathcal{L}(H)$ be such that $\big(A(x), x\big)_H \geqslant (x,x)_H$ for all $x \in H$. Show that A is an isomorphism.

Problem 5.143 **

Let X be a normed space and let $C \subseteq X$ be a nonempty set. Show that

$$\overline{\operatorname{conv}} C = \big\{x \in X : \langle x^*, x\rangle \leqslant \sigma(x^*, C) \text{ for all } x^* \in X^*\big\}.$$

Problem 5.144 **

Let X be a Banach space and let $\overline{B}_1^* = \big\{x^* \in X^* : \|x^*\|_* \leqslant 1\big\}$. Show that $\operatorname{ext} \overline{B}_1^*$ separates the points of X (i.e., for every $x, u \in X$, $x \neq u$, we can find $e^* \in \operatorname{ext} \overline{B}_1^*$ such that $\langle e^*, x\rangle \neq \langle e^*, u\rangle$).

Problem 5.145 **

Let X be an infinite dimensional Banach space and let $A \in \mathcal{L}_c(X)$. Show that there exists $h \in X$ for which the equation $A(x) = h$ has no solution $x \in X$.

Problem 5.146 **

Let X be a reflexive Banach space and let $A \in \mathcal{L}(X; l^1)$. Show that $A \in \mathcal{L}_c(X; l^1)$.

Problem 5.147 *

Let X and Y be two Banach spaces and let $A \in \mathcal{L}_c(X; Y)$. Show that $A(X)$ is a separable vector subspace of Y.

Problem 5.148 **

Let X and Y be two Banach spaces and let $A \in \mathcal{L}_c(X; Y)$. Show that A is completely continuous (see Definition 5.138(c)).

Problem 5.149 **

Let X be a reflexive Banach space, let Y be a Banach space and let $A \in \mathcal{L}(X; Y)$. Show that A is compact if and only if A is completely continuous.

Problem 5.150 **

Let X and Y be two Banach spaces, with $Y \neq 0$. Show that X is reflexive if and only if for every $A \in \mathcal{L}_c(X; Y)$ we can find $x_0 \in X$ with $\|x_0\| \leqslant 1$ such that $\|A\|_{\mathcal{L}} = \|A(x_0)\|_Y$.

Problem 5.151 *

Let $C \stackrel{\text{def}}{=} \{u \in L^1(0,1) : u(t) \geqslant 1 \text{ for almost all } t \in [0,1]\}$. Is this set w-closed in the Banach space $L^1(0,1)$? Justify your answer.

Problem 5.152 **

Let $(X, \|\cdot\|_X)$ and $(Y, \|\cdot\|_Y)$ be two Banach spaces with $(X, \|\cdot\|_X)$ being reflexive and let $A \in \mathcal{L}_c(X; Y)$. Let $|\cdot|_X$ be another norm on X which generates a weaker metric topology than $\|\cdot\|_X$. Show that for every $\varepsilon > 0$, we can find $c_\varepsilon > 0$ such that

$$\|A(x)\|_Y \leqslant \varepsilon\|x\|_X + c_\varepsilon|x|_X \qquad \forall\, x \in X.$$

Problem 5.153 **

Let X be a Banach space and let $A \in \mathcal{L}_c(X)$ with $R(A)$ being infinite dimensional (i.e., $A \in \mathcal{L}_c(X) \setminus \mathcal{L}_f(X)$). Show that $0 \in \overline{A(\partial B_1)}$ (where $\partial B_1 = \{x \in X : \|x\| = 1\}$).

Problem 5.154 **

Let X and Y be two Banach spaces and let $A \in \mathcal{L}_c(X; Y)$. Show that $R(A)$ is separable.

Problem 5.155 ***
Suppose that X is a Banach space, H is a Hilbert space and $A \in \mathcal{L}_c(X;H)$. Show that A is the limit of a sequence of finite rank operators.

Problem 5.156 ***
Let X and H be two Hilbert spaces such that X is dense in H and X is embedded compactly into H (i.e., the identity operator $id\colon X \longrightarrow H$ is compact). Let $a\colon X \times X \longrightarrow \mathbb{R}$ be a continuous bilinear form such that $a(x,x) = 0$ if and only if $x = 0$. Suppose that there exist constants $c_1, c_2 > 0$ such that $a(x,x) \geqslant c_1\|x\|_X^2 - c_2\|x\|_H^2$ for all $x \in X$. For a given $u_0 \in H$, show that there exists unique $x_0 \in X$ such that $a(x_0, x) = (u_0, x)_H$ for all $x \in X$.

Problem 5.157 **
Let X be a Banach space with a Schauder basis (see Remark 5.107). Show that $\overline{\mathcal{L}_f(X)}^{\|\cdot\|_{\mathcal{L}}} = \mathcal{L}_c(X)$.

Problem 5.158 **
Let X be an infinite dimensional Banach space. Show that $\operatorname{dist}\big(id, \mathcal{L}_f(X)\big) = 1$.

Problem 5.159 ***
Suppose that H is an infinite dimensional separable Hilbert space, $\{e_n\}_{n\geqslant 1}$ is an orthonormal basis for H and $A \in \mathcal{L}(H)$ is defined by

$$A(e_n) \stackrel{\text{def}}{=} e_{n+1} \qquad \forall\, n \geqslant 1.$$

Is A compact? Justify your answer.

Problem 5.160 **
Let $X = C\big([0,1]\big)$ with the supremum norm

$$\|u\|_\infty \stackrel{\text{def}}{=} \max_{0\leqslant t\leqslant 1} |u(t)| \qquad \forall\, u \in C\big([0,1]\big).$$

Show that $\mathcal{L}_c(X) = \overline{\mathcal{L}_f(X)}^{\|\cdot\|_{\mathcal{L}}}$.

Problem 5.161 **
Show that if $1 \leqslant p < q < +\infty$, then the embedding $l^p \subseteq l^q$ is not compact (note that $l^p \subseteq l^q$ and $(l^r)^* = l^{r'}$ with $\frac{1}{r} + \frac{1}{r'} = 1$ for all $1 \leqslant r < +\infty$).

Problem 5.162 **
Let A and Y be two Banach spaces and let $A \in \mathcal{L}(X;Y)$. Show that $A \in \mathcal{L}_c(X;Y)$ if and only if the operator $\widehat{A}\colon X/_{N(A)} \longrightarrow Y$, defined by

$$\widehat{A}([x]) \stackrel{\text{def}}{=} A(x)$$

is compact.

Problem 5.163 **
Consider the Hilbert space $H = l^2$ and $A \in \mathcal{L}(l^2;l^2)$, defined by

$$A\big(\{x_n\}_{n\geqslant 1}\big) \stackrel{\text{def}}{=} \{\lambda_n x_n\}_{n\geqslant 1},$$

with $\lambda_n \longrightarrow 0$. Show that $A \in \mathcal{L}_c(l^2;l^2)$.

Problem 5.164 **
Let X be a Banach space and let $A \in \mathcal{L}_c(X)$. Show that, if $id - A$ is injective, then it has a continuous inverse on $(id - A)(X)$

Problem 5.165 ***
Recall that for $p \in [1,+\infty)$,

$$l^p \stackrel{\text{def}}{=} \Big\{\widehat{x} = \{x_n\}_{n\geqslant 1} : \ \|\widehat{x}\|_{l^p} = \Big(\sum_{n\geqslant 1} |x_n|^p\Big)^{\frac{1}{p}} < +\infty\Big\}$$

and

$$l^\infty \stackrel{\text{def}}{=} \Big\{\widehat{x} = \{x_n\}_{n\geqslant 1} : \ \|\widehat{x}\|_{l^\infty} = \sup_{n\geqslant 1} |x_n| < +\infty\Big\}.$$

These are Banach spaces for $1 \leqslant p \leqslant +\infty$. Let $A\colon l^p \longrightarrow l^p$ be the linear operator, defined by

$$A(\widehat{x}) \stackrel{\text{def}}{=} \{x_n\}_{n\geqslant 2} \qquad \forall\, \widehat{x} = \{x_n\}_{n\geqslant 1} \in l^p$$

(the so-called ***backward shift operator***).
(a) Show that $A \in \mathcal{L}(l^p)$ and find $\|A\|_{\mathcal{L}}$.
(b) Determine $\sigma(A)$ and $\sigma_p(A)$ (in this case the spaces are over $\mathbb{C}$).

Problem 5.166 **
Let H be a (complex) Hilbert space and let $A \in \mathcal{L}(H)$ be a self-adjoint operator. Show that $\sigma(A) \subseteq \mathbb{R}$.

Problem 5.167 **
Let X be a Banach space and let $x_0 \in X$, $x_0^* \in X^*$ be such that $\langle x_0^*, x_0 \rangle \neq 0$. Let $A \in \mathcal{L}(X)$ be defined by

$$A(x) \stackrel{\text{def}}{=} \langle x_0^*, x \rangle \, x_0 \qquad \forall\, x \in X.$$

Show that A is a projection (i.e., $A^2 = A$) if and only if $\langle x_0^*, x_0 \rangle = 1$ and find $\sigma(A)$.

Problem 5.168 **
Let H be a Hilbert space and let $A \in \mathcal{L}(H)$ be a self-adjoint operator. Show that $\|A^n\|_{\mathcal{L}} = \|A\|_{\mathcal{L}}^n$ for all $n \geqslant 1$.

Problem 5.169 ***
Let $X = C\big([0,1]\big)$ be furnished with the supremum norm

$$\|u\|_\infty \stackrel{\text{def}}{=} \max_{0 \leqslant t \leqslant 1} \big|u(t)\big| \qquad \forall\, u \in C\big([0,1]\big),$$

let $k \in C\big([0,1] \times [0,1]\big)$, $M = \sup\limits_{0 \leqslant t,s \leqslant 1} \big|k(t,s)\big|$ and let $A \colon X \longrightarrow X$ be defined by

$$A(u)(t) \stackrel{\text{def}}{=} \int_0^t k(t,\tau) u(\tau)\, d\tau.$$

Show that $A \in \mathcal{L}(X)$ and find $\sigma(A)$.

Problem 5.170 *
Suppose that X is a Banach space, $A \in \mathcal{L}(X)$, $\lambda \in \mathbb{C}$ and assume that there exists a sequence $\{x_n\}_{n \geqslant 1} \subseteq X$ such that $\|x_n\| = 1$ for all $n \geqslant 1$ and $A(x_n) - \lambda x_n \longrightarrow 0$. Show that $\lambda \in \sigma(A)$.

Problem 5.171 *
Let H be a Hilbert space and let $A \in \mathcal{L}(H)$ be a self-adjoint operator. We set

$$\|A\|_{\mathcal{L}} \stackrel{\text{def}}{=} \sup\big\{\big|(A(x), x)\big| : \|x\| = 1\big\}.$$

Show that at least one of $\|A\|_{\mathcal{L}}$ or $-\|A\|_{\mathcal{L}}$ is an element of $\sigma(A)$.

Problem 5.172 **
Let $(X, \|\cdot\|)$ be a Banach space. Show that the following three statements are equivalent:
(a) $\|\cdot\|$ is strictly convex (see Definition 5.168(a)).
(b) If $u, x \in X\setminus\{0\}$ and $\|x+u\| = \|x\|+\|u\|$, then $x = \lambda u$ for some λ.
(c) If $x, u \in X$ and $2\|x\|^2 + 2\|u\|^2 = \|x + u\|^2$, then $x = u$.

Problem 5.173 ***
Consider the Banach space

$$l^1 \stackrel{\text{def}}{=} \Big\{\widehat{x} = \{x_n\}_{n\geqslant 1} : \ \|\widehat{x}\|_{l^1} = \sum_{n\geqslant 1} |x_n| < +\infty\Big\}.$$

Show that the norm $\|\cdot\|_{l^1}$ is Gâteaux differentiable at all $\widehat{x} = \{x_n\}_{n\geqslant 1} \in l^1$ such that $x_n \neq 0$ for all $n \geqslant 1$ and it is not Fréchet differentiable at any $\widehat{x} = \{x_n\}_{n\geqslant 1} \in l^1$.

Problem 5.174 **
Show that a locally uniformly convex Banach space X has the Kadec–Klee property (i.e., if $x_n \xrightarrow{w} x$ in X and $\|x_n\| \longrightarrow \|x\|$, then $x_n \longrightarrow x$ in X; see Proposition 5.91).

Problem 5.175 **
Let X be a uniformly convex Banach space. Show that every $x^* \in X^*$ attains its norm at a unique $u \in \partial B_1 = \big\{x \in X : \ \|x\| = 1\big\}$.

Problem 5.176 **
Let X be a Banach space and let $\varphi\colon X \longrightarrow \overline{\mathbb{R}} = \mathbb{R}\cup\{+\infty\}$ be a convex function which is continuous at $x_0 \in X$. Show that the following two statements are equivalent:
(a) φ is Fréchet differentiable at $x_0 \in X$.
(b) $\lim\limits_{\lambda\to 0} \frac{\varphi(x_0+\lambda h)+\varphi(x_0-\lambda h)-2\varphi(x_0)}{\lambda} = 0$ uniformly for all $h \in \partial B_1$.

Problem 5.177 ***
Let X be a Banach space and let $x \in \partial B_1 = \big\{x \in X : \ \|x\| = 1\big\}$. Show that the following two statements are equivalent:
(a) $\|\cdot\|$ is Fréchet differentiable at $x \in X$.
(b) For every two sequences $\{x_n^*\}_{n\geqslant 1}, \{u_n^*\}_{n\geqslant 1} \subseteq \partial B_1^* = \big\{x^* \in X^* : \|x^*\|_* = 1\big\}$ such that $\lim\limits_{n\to+\infty} \langle x_n^*, x\rangle = \lim\limits_{n\to+\infty} \langle u_n^*, x\rangle = 1$, we have $\|x_n^* - u_n^*\|_* \longrightarrow 0$.

Problem 5.178 **

Let $(X, \|\cdot\|)$ be a Banach space and let $\|\cdot\|_*$ be the dual norm of X^*, which we assume that it is locally uniformly convex. Show that $\|\cdot\|$ is Fréchet differentiable.

Problem 5.179 **

Let X be a Banach space and let $C \subseteq X$ be a nonempty, w-compact set. Show that for every $E \subseteq C$ and $x \in \overline{E}^w$, we can find a sequence $\{x_n\}_{n \geqslant 1} \subseteq E$ such that $x_n \xrightarrow{w} x$ in X.

Problem 5.180 **

Let X be a separable Banach space and let $C^* \subseteq X^*$ be a nonempty, convex and sequentially w^*-closed set. Show that C^* is w^*-closed.

5.3 Solutions

Solution of Problem 5.1

Let $x, u \in \partial B_1$. First suppose that $u \neq -x$ (i.e., x and u are not antipodal). Then, for all $\lambda \in [0,1]$, we have $(1-\lambda)x + \lambda u \neq 0$ and so we can define the continuous path $\gamma_0 \colon [0,1] \longrightarrow \partial B_1$, by

$$\gamma_0(\lambda) = \frac{(1-\lambda)x+\lambda u}{\|(1-\lambda)x+\lambda u\|}$$

(see Definition 2.122). Note that

$$\gamma_0(0) = \frac{x}{\|x\|} = x \quad \text{and} \quad \gamma_0(1) = \frac{u}{\|u\|} = u$$

(since $x, u \in \partial B_1$).

Next suppose that $u = -x$ (i.e., x and u are antipodal). Since by hypothesis $\dim X \geqslant 2$, we can find $y \in X$ such that $\{x, y\}$ is linearly independent. Then $x \neq \frac{y}{\|y\|}$ and $-x \neq \frac{y}{\|y\|}$. From the first part of the solution, we can find a continuous path $\gamma_1 \colon [0,1] \longrightarrow \partial B_1$ such that

$$\gamma_1(0) = x \quad \text{and} \quad \gamma_1(1) = \frac{y}{\|y\|}$$

and a continuous path $\gamma_2 \colon [0,1] \longrightarrow \partial B_1$ such that

$$\gamma_2(0) = -x \quad \text{and} \quad \gamma_2(1) = \frac{y}{\|y\|}.$$

We concatenate the two paths γ_1, γ_2 and produce a new continuous path $\gamma_3 \colon [0,2] \longrightarrow \partial B_1$, defined by

$$\gamma_3(t) = \begin{cases} \gamma_1(t) & \text{if } \ t \in [0,1], \\ \gamma_2(2-t) & \text{if } \ t \in [1,2]. \end{cases}$$

Evidently

$$\gamma_3(0) = \gamma_1(0) = x \quad \text{and} \quad \gamma_3(2) = \gamma_2(0) = -x.$$

Solution of Problem 5.2

We show that for every $\lambda \in [0,1)$, we have

$$\lambda\overline{C} + (1-\lambda)\mathrm{int}\, C \subseteq \mathrm{int}\, C. \tag{5.10}$$

Note that

$$\lambda\overline{C} + (1-\lambda)\operatorname{int} C \;=\; \bigcup_{u\in\overline{C}} (\lambda u + (1-\lambda)\operatorname{int} C),$$

hence the left-hand side in (5.10) is an open set. So, in order to show (5.10), it suffices to show that

$$\lambda\overline{C} + (1-\lambda)\operatorname{int} C \;\subseteq\; C.$$

To this end, let $y \in \operatorname{int} C$. Then

$$(1-\lambda)(\operatorname{int} C - y) \;\in\; \mathcal{N}(0),$$

(where $\mathcal{N}(0)$ is the filter of neighbourhoods of 0) and so

$$\begin{aligned} \lambda\overline{C} \;&=\; \overline{\lambda C} \;\subseteq\; \lambda C + (1-\lambda)(\operatorname{int} C - y) \\ &=\; \lambda C + (1-\lambda)\operatorname{int} C - (1-\lambda)y \;\subseteq\; C - (1-\lambda)y, \end{aligned}$$

so

$$\lambda\overline{C} + (1-\lambda)y \;\subseteq\; C,$$

with $y \in \operatorname{int} C$, which proves (5.10).

Note that $\overline{\operatorname{int} C} \subseteq \overline{C}$. Let $x \in \overline{C}$ and $y \in \operatorname{int} C$. From (5.10), we have

$$\lambda x + (1-\lambda)y \;\in\; \operatorname{int} C \qquad \forall\, \lambda \in [0,1)$$

and so $x \in \overline{\operatorname{int} C}$, which proves that $\overline{\operatorname{int} C} = \overline{C}$.

Next, note that $\operatorname{int} C \subseteq \operatorname{int} \overline{C}$. Let $x \in \operatorname{int} \overline{C}$ and $y \in \operatorname{int} C$. Then

$$x \;=\; \lambda y + (1-\lambda)u$$

for some $\lambda \in (0,1)$ and some $u \in \overline{C}$. Using once again (5.10), we infer that $x \in \operatorname{int} C$, which proves that $\operatorname{int} \overline{C} = \operatorname{int} C$.

Solution of Problem 5.3

Let $D = \overline{V} \setminus V$. Since by hypothesis V is a G_δ-set in X, it is also G_δ in $\overline{V}$ and so $V = \bigcap_{n\geqslant 1} U_n$ with U_n being open subsets of $\overline{V}$. Then $D = \bigcup_{n\geqslant 1} (\overline{V} \setminus U_n)$. Note that for every $n \geqslant 1$, the set $\overline{V} \setminus U_n$ is nowhere dense in $\overline{V}$ (note that $V \subseteq U_n \subseteq \overline{V}$ for all $n \geqslant 1$). Therefore

D is of first category (see Definition 1.25). We claim that $D = \emptyset$. To this end, let $u_0 \in D$ and consider $V_0 = u_0 + V$. We have $V_0 \subseteq D$. To see this, note that $V_0 \subseteq \overline{V}$ and if for some $u \in V$, $u_0 + u \in V$, then $u_0 \in V - u = V$, a contradiction. Therefore, V_0 is of first category in $\overline{V}$ and then so is $V = V_0 - u_0$. Since $\overline{V} = V \cup D$, we infer that $\overline{V}$ is of first category in itself, which is a contradiction since $\overline{V}$ is a Banach space.

If the normed space X is topologically complete, then by the Alexandrov theorem (see Theorem 1.58), X is a G_δ-subset of its completion Y. But then from the first part of the problem, X is a closed subspace of Y, hence a Banach space too.

Solution of Problem 5.4

Let $u \in \operatorname{int} V$ and let $r > 0$ be such that $B_r(u) = \{x \in X : \|x - u\| < r\} \subseteq V$. Then $B_r(0) = B_r(u) - u \subseteq V$. For a given $x \in X \setminus \{0\}$, we have

$$\tfrac{r}{2}\tfrac{x}{\|x\|} \in B_r(0) \subseteq V$$

and so $x \in V$, i.e., $V = X$.

Solution of Problem 5.5

We argue by contradiction. So, suppose that for every $\varepsilon > 0$, we can find $y_k \in \overline{B}_\varepsilon(x_k)$, $k \in \{1, \ldots, m\}$ such that $\{y_k\}_{k=1}^m \subseteq X$ are linearly dependent. Then we can find $\{\lambda_k\}_{k=1}^m \subseteq \mathbb{R}$ such that

$$\sum_{k=1}^m |\lambda_k| > 0 \quad \text{and} \quad \sum_{k=1}^m \lambda_k y_k = 0.$$

We have

$$\sum_{k=1}^m \lambda_k (x_k - y_k) = \sum_{k=1}^m \lambda_k x_k,$$

so

$$\Big\|\sum_{k=1}^m \lambda_k x_k\Big\| \leqslant \varepsilon \sum_{k=1}^m |\lambda_k|$$

and thus

$$\Big\|\sum_{k=1}^{m}\widetilde{\lambda}_k x_k\Big\| \leqslant \varepsilon \quad \text{with } \widetilde{\lambda}_k = \frac{\lambda_k}{\sum\limits_{k=1}^{m}|\lambda_k|} \qquad \forall\, k \in \{1,\ldots,m\}. \quad (5.11)$$

Note that

$$\sum_{k=1}^{m}|\widetilde{\lambda}_k| = 1.$$

Let

$$C = \Big\{\widehat{\lambda} = \{\lambda_k\}_{k=1}^{m} \in \mathbb{R}^m : \sum_{k=1}^{m}|\lambda_k| = 1\Big\}$$

and let $\varphi\colon C \longrightarrow \mathbb{R}$ be defined by

$$\varphi(\widehat{\lambda}) = \Big\|\sum_{k=1}^{m}\lambda_k x_k\Big\|.$$

Evidently C is compact (being closed and bounded) and φ is continuous. By (5.11), for every $\varepsilon > 0$, we can find $\widehat{\lambda}_\varepsilon \in C$ such that $\varphi(\widehat{\lambda}_\varepsilon) < \varepsilon$. Hence $\inf \varphi(C) = 0$ and by the Weierstrass theorem, we can find $\widehat{\lambda}^* \in C$ such that

$$\varphi(\widehat{\lambda}^*) = 0.$$

Then

$$\Big\|\sum_{k=1}^{m}\lambda_k^* x_k\Big\| = 0$$

and this due to the linear independence of the elements of $\{x_k\}_{k=1}^{m}$ implies that

$$\lambda_k^* = 0 \qquad \forall\, k \in \{1,\ldots,m\},$$

a contradiction to the fact that $\sum\limits_{k=1}^{m}|\lambda_k^*| = 1$.

Solution of Problem 5.6

We will show that every element of $\operatorname{conv} U$ is an interior point. Evidently, it suffices to show this for every $x \in \operatorname{conv} U \setminus U$. We have

$$x = \sum_{k=1}^{n}\lambda_k x_k,$$

with $x_k \in U$, $\lambda_k \geqslant 0$ and $\sum\limits_{k=1}^{n} \lambda_k = 1$. By rearranging things if necessary, we may assume that $\lambda_1 > 0$. Let $V \in \mathcal{N}(x_1)$ (with $\mathcal{N}(x_1)$ being the filter of neighbourhoods of x_1) be such that $V \subseteq U$. Then

$$\lambda_1 V + \sum_{k=2}^{n} \lambda_k x_k \ \in\ \mathcal{N}(x)$$

and it is contained in $\operatorname{conv} U$. Hence x is an interior point of $\operatorname{conv} U$ and so we conclude that $\operatorname{conv} U$ is an open set.

Solution of Problem 5.7

By the Baire category theorem (see Theorem 1.26), we have that, for some $n_0 \geqslant 1$,

$$\operatorname{int} V_{n_0} \ \neq\ \emptyset.$$

Then Problem 5.4 implies that $V = V_{n_0}$.

Solution of Problem 5.8

"$\Longrightarrow$": Since the set C is relatively compact, it is totally bounded (see Definition 5.7 and Theorem 1.71). So, for a given $\varepsilon > 0$, we can find a finite set $C_\varepsilon = \{x_1, \ldots, x_{M_\varepsilon}\}$ such that

$$C \ \subseteq\ \bigcup_{k=1}^{M_\varepsilon} B_\varepsilon(x_k) \ \subseteq\ \varepsilon \overline{B}_1 + C_\varepsilon.$$

"$\Longleftarrow$": Let $\varepsilon > 0$. Since by hypothesis, the set $C_{\frac{\varepsilon}{2}}$ is relatively compact, we can find a set $\{x_1^\varepsilon, \ldots, x_{M_\varepsilon}^\varepsilon\}$ such that

$$C_{\frac{\varepsilon}{2}} \ \subseteq\ \bigcup_{k=1}^{M_\varepsilon} B_{\frac{\varepsilon}{2}}(x_k^\varepsilon).$$

We claim that

$$C \ \subseteq\ \bigcup_{k=1}^{M_\varepsilon} B_\varepsilon(x_k^\varepsilon).$$

To see this, let $x \in C$. By hypothesis, we can find $x_{\frac{\varepsilon}{2}} \in C_{\frac{\varepsilon}{2}}$ such that

$$\|x - x_{\frac{\varepsilon}{2}}\| \leqslant \tfrac{\varepsilon}{2}.$$

On the other hand, $x_{\frac{\varepsilon}{2}} \in B_{\frac{\varepsilon}{2}}(x^{\varepsilon}_{k_0})$ for some $k_0 \in \{1, \ldots, M_\varepsilon\}$, hence

$$\|x_{\frac{\varepsilon}{2}} - x^{\varepsilon}_{k_0}\| < \tfrac{\varepsilon}{2}.$$

So, finally

$$\|x - x^{\varepsilon}_{k_0}\| \leqslant \|x - x_{\frac{\varepsilon}{2}}\| + \|x_{\frac{\varepsilon}{2}} + x^{\varepsilon}_{k_0}\| < \varepsilon.$$

Solution of Problem 5.9

Suppose that $\operatorname{int} C \neq \emptyset$ and let $x \in \operatorname{int} C$. Then we can find $r > 0$ such that $\overline{B}_r(x) \subseteq C$, where

$$\overline{B}_r(x) = \{u \in X : \|u - x\| \leqslant r\} = x + r\overline{B}_1.$$

Hence $\overline{B}_1$ is compact and so X is finite dimensional (see Proposition 5.9(a)), a contradiction. Therefore, $\operatorname{int} C = \emptyset$.

Solution of Problem 5.10

We know that V is closed in X (see Proposition 5.9). Let $\{u_n\}_{n \geqslant 1} \subseteq V$ be a sequence such that $\|x - u_n\| \searrow \operatorname{dist}(x, V)$. Clearly the sequence $\{u_n\}_{n \geqslant 1} \subseteq V$ is bounded and since V is closed, by passing to a suitable subsequence, we may assume that $u_n \longrightarrow u \in V$ in X. Then

$$\|x - u_n\| \longrightarrow \|x - u\| = \operatorname{dist}(x, V) \quad \text{and} \quad u \in V.$$

Solution of Problem 5.11

We set

$$s_n = \sum_{k=1}^{n} x_n \qquad \forall\, n \geqslant 1.$$

Then for $n > m \geqslant 1$, we have

$$\|s_n - s_m\| = \Big\| \sum_{k=m+1}^{n} x_k \Big\| \leqslant \sum_{k=m+1}^{n} \|x_k\|,$$

so $\lim_{m \to +\infty} \|s_n - s_m\| = 0$ and thus $\{s_n\}_{n \geqslant 1}$ is a Cauchy sequence in X. So, there exists $x \in X$ such that $x = \lim_{n \to +\infty} s_n$, hence $x = \sum_{n \geqslant 1} x_n$.

Solution of Problem 5.12

Let $\{x_n\}_{n \geqslant 1} \subseteq X$ be a Cauchy sequence. By induction, we can generate a strict increasing subsequence $\{n_k\}_{k \geqslant 1}$ of $\{n\}_{n \geqslant 1}$ such that

$$\|x_n - x_m\| \leqslant \tfrac{1}{2^k} \qquad \forall\, n, m \geqslant n_k,\ k \geqslant 1.$$

We define $x_{n_0} = 0$ and $y_k = x_{n_k} - x_{n_{k-1}}$ for all $k \geqslant 1$. Then, we have

$$\sum_{k \geqslant 1} \|y_k\| \leqslant \|x_{n_1}\| + \sum_{k \geqslant 1} \tfrac{1}{2^k} < +\infty,$$

so

$$\sum_{k \geqslant 1} y_k \text{ is absolutely convergent.}$$

By hypothesis it is convergent. But the partial sums of this series equal x_{n_i}, $i \geqslant 1$ and so the subsequence $\{x_{n_i}\}_{i \geqslant 1}$ of $\{x_n\}_{n \geqslant 1}$ converges in X. But, if a Cauchy sequence has a convergent subsequence, then the sequence itself converges to the same limit.

Solution of Problem 5.13

Let $\{x_n\}_{n \geqslant 1} \subseteq X$ be a Cauchy sequence. By definition

$$\big|[x_n - x_m]\big| \leqslant \|x_n - x_m\| \qquad \forall\, m, n \geqslant 1,$$

hence $\big\{[x_n]\big\}_{n \geqslant 1} \subseteq X/_V$ is a Cauchy sequence. Since by hypothesis $X/_V$ is a Banach space, we have that

$$[x_n] \longrightarrow [y] \in X/_V$$

for some $y \in X$. Then we can find $y_n \in X$ such that

$$[y_n] = [x_n - y] \quad \text{and} \quad \|y_n\| \leqslant \big|[x_n - y]\big| + \tfrac{1}{2^n} \qquad \forall\, n \geqslant 1$$

(see Definition 5.18).

It follows that $y_n \longrightarrow 0$ in X, hence $\{x_n - y - y_n\}_{n\geqslant 1} \subseteq V$ is a Cauchy sequence. Let $u \in V$ be such that $x_n - y - y_n \longrightarrow u$. Then

$$x_n = (x_n - y - y_n) + y_n - y \longrightarrow u + y$$

and so we conclude that X is a Banach space.

Solution of Problem 5.14

We argue by contradiction. So, suppose that A is not continuous. Then according to Proposition 5.12, we can find a sequence $\{x_n\}_{n\geqslant 1} \subseteq X$ such that $x_n \longrightarrow 0$ in X and

$$\big\|A(x_n)\big\|_Y \geqslant \varepsilon > 0 \qquad \forall\, n \geqslant 1.$$

Let

$$u_n = \frac{x_n}{\|x_n\|_X^{1/2}} \qquad \forall\, n \geqslant 1.$$

Then $u_n \longrightarrow 0$ in X and so by hypothesis, we can find $M > 0$ such that

$$M \geqslant \big\|A(u_n)\big\|_Y = \frac{1}{\|x_n\|_X^{1/2}} \big\|A(x_n)\big\|_Y \geqslant \frac{\varepsilon}{\|x_n\|_X^{1/2}} \longrightarrow +\infty,$$

a contradiction. This proves that $A \in \mathcal{L}(X;Y)$.

Solution of Problem 5.15

Let

$$|x|_X = \|x\|_X + \big\|A(x)\big\|_Y \qquad \forall\, x \in X.$$

Then $|\cdot|_X$ is a norm in X and so by Remark 5.23, norms $|\cdot|_X$ and $\|\cdot\|_X$ are equivalent. Hence there exists $c > 0$ such that

$$|x|_X \leqslant c\|x\|_X \qquad \forall\, x \in X,$$

so

$$\|A(x)\|_Y \leqslant c\|x\|_X \qquad \forall\, x \in X.$$

This proves that $A \in \mathcal{L}(X;Y)$ (see Proposition 5.12).

Solution of Problem 5.16

Let $\{x_n\}_{n\geqslant 1} \subseteq X$ be a Cauchy sequence. Since

$$\|A(x_n) - A(x_m)\|_Y = \|x_n - x_m\|_X \qquad \forall\, n, m \geqslant 1,$$

it follows that $\{A(x_n)\}_{n\geqslant 1} \subseteq Y$ is a Cauchy sequence. Because Y is a Banach space, we can find $y \in Y$ such that

$$A(x_n) \longrightarrow y \quad \text{in } Y.$$

Note that A^{-1} is an isometry too, hence continuous and so

$$x_n \longrightarrow A^{-1}(y) \quad \text{in } X.$$

This proves that X is a Banach space.

Solution of Problem 5.17

"**(a)** $\Longleftrightarrow$ **(c)**": It is clear from the homogeneity of A.

"**(a)** $\Longrightarrow$ **(b)**": Clearly, if A is a surjective isometry, then $A(\overline{B}_1^X) = \overline{B}_1^Y$.

"**(b)** $\Longrightarrow$ **(c)**": Suppose that $A(\overline{B}_1^X) = \overline{B}_1^Y$. Then $A(\partial B_1^X) \subseteq \partial B_1^Y$. Suppose that we can find $u \in \partial B_1^X$ such that $\|A(u)\|_Y = \vartheta < 1$. Then

$$\|\tfrac{u}{\vartheta}\|_X > 1 \quad \text{and} \quad \|A(\tfrac{u}{\vartheta})\|_Y = 1.$$

Let $x \in \overline{B}_1^X$ be such that $A(x) = A(\frac{u}{\vartheta})$ with $x \neq \frac{u}{\vartheta}$. This contradicts the injectivity of A. Therefore $A(\partial B_1^X) = \partial B_1^Y$.

Solution of Problem 5.18

Let

$$D = X \setminus C = \{x \in X : A_n(x) \longrightarrow A(x) \text{ in } Y\}.$$

Due to the linearity of the operators, D is a vector subspace of X. Let $u \in C$. Then for all $\lambda \in \mathbb{R}$ and all $x \in D$, we have that $x + \lambda u \in C$. To see this, note that, if $x + \lambda u \in D$, then $u \in D$ (since D is a vector subspace of X), a contradiction. Hence $x + \frac{1}{n}u \in C$ and so $x \in \overline{C}$. Therefore $D \subseteq \overline{C}$ and so $X = C \cup D \subseteq \overline{C}$, i.e., $X = \overline{C}$, which proves that C is dense in X.

Solution of Problem 5.19

Let $u \in \ker x^*$. Then, since $\|x^*\|_* = 1$, we have

$$\|x - u\| \geqslant \left|\langle x^*, x - u\rangle\right| = \left|\langle x^*, x\rangle\right|,$$

so

$$\operatorname{dist}(x, \ker x^*) \geqslant \left|\langle x^*, x\rangle\right|.$$

On the other hand, for a given $\varepsilon > 0$, we can find $y \in \partial B_1 = \{y \in X : \|y\| = 1\}$ such that

$$\left|\langle x^*, y\rangle\right| \geqslant 1 - \varepsilon.$$

Note that

$$x - \tfrac{\langle x^*, x\rangle}{\langle x^*, y\rangle} y \in \ker x^*.$$

Therefore

$$\operatorname{dist}(x, \ker x^*) \leqslant \tfrac{|\langle x^*, x\rangle|}{|\langle x^*, y\rangle|} \leqslant \tfrac{|\langle x^*, x\rangle|}{1-\varepsilon}.$$

Since $\varepsilon > 0$ is arbitrary, we let $\varepsilon \searrow 0$ and we obtain

$$\operatorname{dist}(x, \ker x^*) \leqslant \left|\langle x^*, x\rangle\right|.$$

Thus finally, we conclude that

$$\operatorname{dist}(x, \ker x^*) = |\langle x^*, x\rangle|.$$

Solution of Problem 5.20

(a) Let $u \in X$. Then

$$\|A(u)\|_\infty = \|u \circ \xi\|_\infty = \max_{0 \leqslant t \leqslant 1} |u(\xi(t))| = \max_{y \in \xi([0,1])} |u(y)|. \quad (5.12)$$

If ξ is surjective, then $\xi([0,1]) = [0,1]$ and so

$$\max_{y \in \xi([0,1])} |u(y)| = \|u\|_\infty,$$

hence $\|A(u)\|_\infty = \|u\|_\infty$ (see (5.12)). On the other hand, if $C = \xi([0,1]) \neq [0,1]$, then C is a compact interval in $[0,1]$ and we can find a nontrivial closed interval $I \subseteq [0,1]$ such that $C \cap I = \emptyset$. Let $\widehat{u} \in X$ be such that $\operatorname{supp} \widehat{u} \subseteq I$. Then, from (5.12), we have $\|A(\widehat{u})\|_\infty = 0$, hence A is not an isometry (in fact it is not even injective).

(b) Let $\xi\colon [0,1] \longrightarrow [0,1]$ be defined by

$$\xi(t) = 4t(1-t) \qquad \forall\, t \in [0,1].$$

Then ξ is surjective and from **(a)**, we have that

$$A(u) = u \circ \xi \qquad \forall\, u \in X$$

is an isometry. Let $h \in R(A)$. Then

$$h(t) = h(1-t) \qquad \forall\, t \in [0,1],$$

which of course is not true for all elements of X. Therefore A is not surjective.

Solution of Problem 5.21

The linear operator is not continuous. To see this let

$$u_n(t) = t^n \qquad \forall\, t \in [0,1],\ n \geqslant 1.$$

Then $\{u_n\}_{n\geqslant 1} \subseteq C^1([0,1])$ and

$$\|u_n'\|_\infty = n \quad \text{and} \quad \|u_n\|_\infty = 1 \qquad \forall\, n \geqslant 1.$$

So, we cannot find $M > 0$ such that

$$\big\|A(u_n)\big\|_\infty \leqslant M\|u_n\|_\infty \qquad \forall\, n \geqslant 1,$$

hence A is not continuous (see Proposition 5.12).

Solution of Problem 5.22

Let $\{u_n\}_{n\geqslant 1} \subseteq V$ be a sequence such that $u_n \longrightarrow u$ in X. Then

$$P(u_n) = u_n \qquad \forall\, n \geqslant 1$$

and

$$P(u_n) \longrightarrow P(u) \quad \text{in } X.$$

So, $u = P(u)$, hence $u \in V$. This proves that V is closed.

Solution of Problem 5.23

No. We know that

$$(l^1)^* = l^\infty = \big\{\widehat{x}^* = \{x_n\}_{n\geqslant 1} : \ \|x\|_{l^\infty} = \sup_{n\geqslant 1} |x_n| < +\infty\big\}.$$

Let $\widehat{u}^* \in l^\infty$ be the constant sequence equal to 1. Then

$$\langle \widehat{u}^*, e_k\rangle = 1 \qquad \forall\, k \geqslant 1$$

and so $\{e_k\}_{k\geqslant 1} \subseteq (\widehat{u}^*)^{-1}(1)$. Since the latter is closed and convex, we infer that $C \subseteq (\widehat{u}^*)^{-1}(1)$ and so $0 \notin C$.

Solution of Problem 5.24

For a given $\varepsilon > 0$, let

$$\begin{aligned} B_{\frac{1}{n}} &= \{y \in Y : \ \|y\|_Y < \tfrac{1}{n}\} \\ D_n &= \big\{x \in X : \ \big|a(x,y)\big| \leqslant \varepsilon \text{ for all } y \in B_{\frac{1}{n}}\big\}. \end{aligned}$$

The continuity of the map $x \longmapsto a(x,y)$ implies that for every $y \in B_{\frac{1}{n}}$, the set $\{x \in X : |a(x,y)| \leqslant \varepsilon\}$ is closed, hence the set D_n being the intersection of closed sets is closed. The continuity of the map $y \longmapsto a(x,y)$ implies that

$$X = \bigcup_{n \geqslant 1} D_n.$$

By the Baire category theorem (see Theorem 1.26), for some integer $n_0 \geqslant 1$, we have

$$\operatorname{int} D_{n_0} \neq \emptyset.$$

So, we can find $\delta > 0$ such that for all $x \in B_\delta^X(0) = \{x \in X : \|x\|_X < \delta\}$ and all $y \in B_{\frac{1}{n_0}}$, we have that $|a(x,y)| \leqslant \varepsilon$, hence a is continuous at $(0,0)$. Now, if $(x_n, y_n) \longrightarrow (x,y)$ in $X \times Y$, then

$$\begin{aligned} & |a(x_n,y_n) - a(x,y)| \\ = & |a(x_n - x, y_n - y) + a(x_n, y) + a(x, y_n) - 2a(x,y)| \\ \leqslant & |a(x_n - x, y_n - y)| + |a(x_n - x, y)| + |a(x, y_n - y)| \longrightarrow 0, \end{aligned}$$

so a is continuous everywhere.

Solution of Problem 5.25

By Problem 5.12, it suffices to show that, if a sequence $\{y_n\}_{n \geqslant 1} \subseteq Y$ is such that

$$\sum_{n \geqslant 1} \|y_n\|_Y < +\infty,$$

then the series $\sum\limits_{n \geqslant 1} y_n$ converges in Y. Let $x_0 \in X$ be such that $\|x_0\|_X = 1$. By Corollary 5.26, we can find $x_0^* \in X^*$ such that

$$\langle x_0^*, x_0 \rangle = \|x_0\|_X^2 = 1.$$

Let $\{A_n\}_{n \geqslant 1} \subseteq \mathcal{L}(X;Y)$ be defined by

$$A_n(x) = \langle x_0^*, x \rangle \, y_n \qquad \forall\, x \in X,\ n \geqslant 1.$$

Then

$$\|A_n\|_{\mathcal{L}} \leqslant \|x_0^*\|_{X^*} \|y_n\|_Y \qquad \forall\, n \geqslant 1$$

and so $\sum_{n\geqslant 1} \|A_n\|_{\mathcal{L}} < +\infty$. Since by hypothesis $\big(\mathcal{L}(X;Y), \|\cdot\|_{\mathcal{L}}\big)$ is a Banach space, by Problem 5.11,

$$\sum_{n\geqslant 1} A_n \xrightarrow{\|\cdot\|_{\mathcal{L}}} A \in \mathcal{L}(X;Y).$$

Then

$$\Big\|\sum_{n=1}^{m} A_n(x_0) - A(x_0)\Big\|_Y \leqslant \Big\|\sum_{n=1}^{m} A_n - A\Big\|_{\mathcal{L}} \|x_0\|_X \longrightarrow 0 \quad \text{as } m \to +\infty,$$

so

$$\sum_{n=1}^{m} y_n = \langle x_0^*, x_0 \rangle \sum_{n=1}^{m} y_n \longrightarrow A(x_0) \quad \text{in } Y \quad \text{as } m \to +\infty$$

and thus $\sum_{n\geqslant 1} y_n$ converges in Y. So, by Problem 5.12, Y is a Banach space.

Solution of Problem 5.26
No. To see this, let $X = L^\infty(0,1)$ and let $A \in \mathcal{L}(X)$ be defined by

$$A(u) = \chi_{[0,\frac{1}{2}]} u \qquad \forall\, u \in L^\infty(0,1).$$

Let $u_n = n\chi_{(\frac{1}{2},1]} \in L^\infty(0,1)$ for $n \geqslant 1$. Then $\|u_n\|_\infty = n$ and so $\|u_n\|_\infty \longrightarrow +\infty$. On the other hand

$$A(u_n) = \chi_{[0,\frac{1}{2}]} n\chi_{(\frac{1}{2},1]} = 0 \qquad \forall\, n \geqslant 1.$$

Solution of Problem 5.27
By hypothesis

$$X = \bigcup_{n\geqslant 1} N(A^n).$$

Invoking the Baire category theorem (see Theorem 1.26), we can find $n_0 \geqslant 1$ such that $\operatorname{int} N(A^{n_0}) \neq \emptyset$. Then Problem 5.4 implies that $X = N(A^{n_0})$ and so $A^{n_0} = 0$.

Solution of Problem 5.28

For $n, m \geqslant 1$ and $x \in X$, we have

$$\begin{aligned} \|(A_n - A)(x)\|_Y &= \|(A_n - A_m)(x) + (A_m - A)(x)\|_Y \\ &\leqslant \|A_n - A_m\|_{\mathcal{L}} \|x\|_X + \|(A_m - A)(x)\|_Y. \end{aligned}$$

For a given $\varepsilon > 0$, we can find $n_0 = n_0(\varepsilon) \geqslant 1$ such that

$$\|A_n - A_m\|_{\mathcal{L}} \leqslant \tfrac{\varepsilon}{2} \qquad \forall\, n, m \geqslant n_0$$

(recall that $\{A_n\}_{n \geqslant 1}$ is $\|\cdot\|_{\mathcal{L}}$-Cauchy). So, we have

$$\|(A_n - A)(x)\|_Y \leqslant \tfrac{\varepsilon}{2}\|x\|_X + \|(A_m - A)(x)\|_Y \qquad \forall\, x \in X,\ n, m \geqslant n_0.$$

Since $A(x_n) \longrightarrow A(x)$ in Y, we can find $\widehat{n}_0(\varepsilon, x) \geqslant n_0$ such that

$$\|(A_m - A)(x)\|_Y \leqslant \tfrac{\varepsilon}{2} \qquad \forall\, m \geqslant n_0.$$

Therefore,

$$\|(A_n - A)(x)\| \leqslant \tfrac{\varepsilon}{2}\|x\| + \tfrac{\varepsilon}{2} \qquad \forall\, x \in X,\ n \geqslant n_0,$$

so

$$\|A_n - A\|_{\mathcal{L}} \leqslant \varepsilon \qquad \forall\, n \geqslant n_0$$

and thus $A_n \xrightarrow{\|\cdot\|_{\mathcal{L}}} A$.

Solution of Problem 5.29

Arguing by contradiction, suppose that no such y exists. Then the set $\overline{A(B_1^X)}$ is absorbing (see Definition 5.7(a)), hence $\overline{A(B_1^X)}$ is a neighbourhood of the origin in Y, which means that A is surjective, a contradiction.

Solution of Problem 5.30

Since X is a Banach space and A is an isomorphism onto $A(X)$, it follows that $A(X)$ is complete. Therefore $A(X)$ is closed in Y. Since by hypothesis it is also dense in Y, we conclude that $A(X) = Y$, i.e., A is surjective.

Solution of Problem 5.31

"$\Longrightarrow$": Clearly the continuity of A implies that $\ker A$ is closed.

"$\Longleftarrow$": Suppose that $\dim A(X) = n$ and let $\{e_k\}_{k=1}^n$ be a basis of $A(X)$. Then for every $x \in X$, we have

$$A(x) = \sum_{k=1}^{n} \lambda_k(x) e_k,$$

with $\{\lambda_k\}_{k=1}^n$ being linear functionals on X (the ***coordinate functionals***). For every $k \in \{1, \ldots, n\}$, let $x_k \in X$ be such that $A(x_k) = e_k$ and let us set $X_n = \operatorname{span}\{x_1, \ldots, x_n\}$. We have

$$X = X_n \oplus \ker A.$$

Since by hypothesis $\ker A$ is a closed linear subspace of X, $\operatorname{dist}(x, \ker A)$ is a norm for X_n and the finite dimensionality of X_n implies that it is equivalent to the norm $\|\cdot\|$ which X_n inherits from X. Hence we can find $c > 0$ such that

$$\|x\| \leqslant c \operatorname{dist}(x, X_n) \qquad \forall\, x \in X_n.$$

The finite dimensionality of X_n implies the continuity of λ_k on X_k for every $k \in \{1, \ldots, n\}$. So, we can find $M_k > 0$ such that

$$|\lambda_k(x)| \leqslant M_k \|x\| \qquad \forall\, x \in X_n.$$

Let $u \in X$. Then $u = x + y$ with $x \in X_n$, $y \in \ker A$ and

$$\begin{aligned} |\lambda_k(u)| &= |\lambda_k(x)| \leqslant M_k\|x\| \leqslant cM_k \mathrm{dist}(x, \ker A) \\ &\leqslant cM_k\|x+y\| = cM_k\|u\|, \end{aligned}$$

so λ_k is continuous on X for every $k \in \{1, \ldots, n\}$ and so $A \subset \mathcal{L}(X;Y)$.

Solution of Problem 5.32

Let $V_1 = V \oplus \mathbb{R}x$ and let $h\colon V_1 \longrightarrow \mathbb{R}$ be the linear functional, defined by

$$h(u + \lambda x) = \lambda \mathrm{dist}(x, V).$$

Evidently $h(x) = \mathrm{dist}(x, V)$ and $h(u) = 0$ for all $u \in V$. Also, we have

$$|h(u + \lambda x)| = |\lambda| \mathrm{dist}(x, V) \leqslant |\lambda| \big\|x - \big(-\tfrac{u}{\lambda}\big)\big\| = \|u + \lambda x\|,$$

so

$$h \in V_1^* \quad \text{and} \quad \|h\|_* \leqslant 1.$$

In addition, we have

$$\mathrm{dist}(x, V) = |h(x - u)| \leqslant \|h\|_* \|x - u\| \qquad \forall\, u \in V,$$

so

$$\mathrm{dist}(x, V) \leqslant \|h\|_* \mathrm{dist}(x, V)$$

and thus

$$\|h\|_* \geqslant 1.$$

We infer that $\|h\|_* = 1$.

Finally, by the Hahn–Banach theorem (see Theorem 5.24), we can find $x^* \in X^*$ such that $x^*|_{V_1} = h$ and $\|x^*\|_* = \|h\|_* = 1$.

Solution of Problem 5.33

If h is continuous, then the set $N(h) = h^{-1}(\{0\})$ is closed. Conversely, suppose that the set $N(h)$ is closed. If $N(h) = X$, then $h = 0$. If $N(h) \neq X$, let $u \in X \setminus N(h)$ be such that $h(u) = 1$. For $x \in X$, we have $x - h(x)u \in N(h)$, hence $x = y + h(x)u$ for some $y \in N(h)$. Suppose that $x = y' + \lambda u$ for some $y' \in N(h)$ and $\lambda \in \mathbb{R}$. Then $h(x) = \lambda$ (recall that $h(u) = 1$) and so $y = y'$. Therefore, $X = N(h) \oplus \mathbb{R}u$. Using Problem 5.32, we can find $x^* \in X^*$ such that $\|x^*\|_* = 1$, $x^*|_{N(h)} = 0$ and $\langle x^*, u\rangle = 1$. For all $x \in X$, we have $x = y + h(x)u$ with $y \in N(h)$ and so $\langle x^*, x\rangle = h(x)$, hence $x^* = h$ and so h is continuous.

If $N(h)$ is not closed, then h is not continuous and since $\overline{N(h)}$ is a proper vector subspace of X, we have $h \neq 0$. Suppose that $\overline{N(h)} \neq X$. Then for some $u \in X \setminus \overline{N(h)}$, $h(u) = 1$ and for some $x^* \in X^*$, we have $x^*|_{N(h)} = 0$, $\langle x^*, u\rangle = 1$. As above, we show that $x^* = h$, hence x^* is not continuous, a contradiction. So, $N(h)$ is dense in X.

Solution of Problem 5.34

Let $\{u_n\}_{n\geqslant 1} \subseteq \operatorname{span}\{V, x_0\}$ and suppose that $u_n \longrightarrow u$ in X. We have

$$u_n = y_n + \xi_n x_0, \quad \text{with } y_n \in V,\ \xi_n \in \mathbb{R} \qquad \forall\, n \geqslant 1.$$

By Problem 5.32, for every $n \geqslant 1$, we can find $x_n^* \in X^*$, with $\|x_n^*\|_* = 1$ such that

$$x_n^*|_V = 0 \quad \text{and} \quad \langle x_n^*, x_0\rangle = 1.$$

So, we have

$$\|u_n\| \geqslant \big|\langle x_n^*, u_n\rangle\big| = |\xi_n|\big|\langle x_n^*, x_0\rangle\big| = |\xi_n| \qquad \forall\, n \geqslant 1$$

and thus the sequence $\{\xi_n\}_{n\geqslant 1} \subseteq \mathbb{R}$ is bounded.

Hence, by passing to a suitable subsequence if necessary, we may assume that $\xi_n \longrightarrow \xi$. Then

$$u_n - \xi_n x_0 \longrightarrow u - \xi x_0 \quad \text{in } X,$$

hence $y_n \longrightarrow u - \xi x_0 \in V$. Therefore $u = y + \xi x_0$ with $y \in V$, $\xi \in \mathbb{R}$, which proves that the set $\operatorname{span}\{V, x_0\}$ is closed.

Solution of Problem 5.35

Let $f, g\colon X \longrightarrow \mathbb{R}$ be continuous, linear functionals which extend h and

$$\|f\| = \|g\| = \|h\|$$

(i.e., they are Hahn–Banach extensions of h; see Theorem 5.24). Let $\lambda \in (0,1)$ and let us set

$$\widehat{h}_\lambda = \lambda f + (1-\lambda) g \colon X \longrightarrow \mathbb{R}.$$

This functional $\widehat{h}_\lambda$ is continuous, linear. Also, if $x \in V$, then

$$\widehat{h}_\lambda(x) = \lambda f(x) + (1-\lambda) g(x) = \lambda h(x) + (1-\lambda) h(x) = h(x)$$

(recall that $f|_V = h$, $g|_V = h$). So, $\widehat{h}_\lambda$ extends h.

For every $x \in X$, we have

$$\left|\widehat{h}_\lambda(x)\right| \leqslant \lambda\left|f(x)\right| + (1-\lambda)\left|g(x)\right| \leqslant \left(\lambda\|f\| + (1-\lambda)\|g\|\right)\|x\| = \|h\|\,\|x\|$$

(recall that $\|f\| = \|g\| = \|h\|$). Therefore,

$$\|\widehat{h}_\lambda\| \leqslant \|h\| \qquad \forall\, \lambda \in [0,1].$$

Since $\widehat{h}_\lambda$ extends h we also have

$$\|\widehat{h}_\lambda\| \geqslant \|h\| \qquad \forall\, \lambda \in [0,1].$$

We conclude that

$$\|\widehat{h}_\lambda\| = \|h\| \qquad \forall\, \lambda \in [0,1].$$

Solution of Problem 5.36

By hypothesis $\left\|u^*\big|_{\ker x^*}\right\|_* \leqslant \varepsilon$ and by the Hahn–Banach theorem (see Corollary 5.25), we can find $y^* \in X^*$ such that

$$y^*\big|_{\ker x^*} = u^*\big|_{\ker x^*} \quad \text{and} \quad \|y^*\|_* \leqslant \varepsilon.$$

We have

$$(u^* - y^*)(x) = 0 \qquad \forall\, x \in \ker x^*,$$

so

$$u^* - y^* = \lambda x^* \quad \text{for some } \lambda \in \mathbb{R}.$$

Then

$$\big|1-|\lambda|\big| \;=\; \big|\|u^*\|_* - \|u^*-y^*\|_*\big| \;\leqslant\; \|y^*\|_* \;\leqslant\; \varepsilon.$$

If $\lambda > 0$, then we have

$$\|x^*-u^*\|_* \;=\; \big\|(1-\lambda)x^*-y^*\big\|_* \;\leqslant\; |1-\lambda| + \|y^*\|_* \;\leqslant\; 2\varepsilon.$$

If $\lambda < 0$, then we have

$$\|x^*+u^*\|_* \;=\; \big\|(1+\lambda)x^*+y^*\big\|_* \;\leqslant\; |1+\lambda| + \|y^*\|_* \;\leqslant\; 2\varepsilon.$$

Solution of Problem 5.37

From Proposition 5.37, we know that $V^\perp$ and $Y^\perp$ are both closed. Next we show that $V^\perp \cap Y^\perp = \{0\}$. To this end, let $u^* \in V^\perp \cap Y^\perp$. If $x \in X$, then since $X = V \oplus Y$, we can write (in a unique way) $x = v + y$, with $v \in V$, $y \in Y$. We have

$$\langle u^*, x\rangle \;=\; \langle u^*, v\rangle + \langle u^*, y\rangle \;=\; 0 \qquad \forall\, x \in X,$$

so

$$u^* \;=\; 0, \quad \text{hence } V^\perp \cap Y^\perp = \{0\}.$$

Finally, we show that $X^* = V^\perp + Y^\perp$. To this end, let $x^* \in X^*$ and let $\mathrm{proj}_V, \mathrm{proj}_Y \in \mathcal{L}(X)$ be the projection operators, defined by

$$\mathrm{proj}_V(x) = v, \quad \mathrm{proj}_Y(x) = y \qquad \forall\, x = v + y,\ v \in V,\ y \in Y.$$

Then $u^* \circ \mathrm{proj}_V \in X^*$ and for all $y \in Y$, we have $\mathrm{proj}_V(y) = 0$ and so $u^* \circ \mathrm{proj}_V \in Y^\perp$. Similarly $u^* \circ \mathrm{proj}_Y \in V^\perp$ and $u^* = u^* \circ \mathrm{proj}_V + u^* \circ \mathrm{proj}_Y$.

Solution of Problem 5.38

Let $id\colon (X, \|\cdot\|_\infty) \longrightarrow (X, \|\cdot\|)$ be the identity operator. Suppose that $u_n \xrightarrow{\|\cdot\|_\infty} u$ and $u_n \xrightarrow{\|\cdot\|} y$. Then from the two convergences, we have

$$u_n(t) \longrightarrow u(t) \quad \text{and} \quad u_n(t) \longrightarrow y(t) \qquad \forall\, t \in [0,1],$$

so

$$u(y) = y(t) \qquad \forall\, t \in [0,1],$$

i.e., $u = y$. Invoking the closed graph theorem (see Theorem 5.50), we infer that id is continuous. So, we can find $c > 0$ such that

$$\|u\| \leqslant c\|u\|_\infty \qquad \forall\, u \in X = C\big([0,1]\big).$$

Finally an application of Corollary 5.49 establishes the equivalence of the two norms $\|\cdot\|_\infty$ and $\|\cdot\|$.

Solution of Problem 5.39

Clearly S is linear. Let $\big\{\big(x_n, S(x_n)\big)\big\}_{n\geqslant 1} \subseteq \operatorname{Gr} S$ be a sequence such that $x_n \longrightarrow x$ in X and $S(x_n) \longrightarrow v$ in V. Then

$$A(x_n) = B\big(S(x_n)\big) \qquad \forall\, n \geqslant 1$$

and so passing to the limit as $n \to +\infty$, we have $A(x) = B(v)$, hence $S(x) = v$. Therefore $\operatorname{Gr} S$ is closed and so by the closed graph theorem (see Theorem 5.50), we have that $S \in \mathcal{L}(X;V)$.

Solution of Problem 5.40

From Corollary 5.42, we know that $A \in \mathcal{L}(X;Y)$ and there exists $M > 0$ such that

$$\|A_n\|_{\mathcal{L}} \leqslant M \qquad \forall\, n \geqslant 1.$$

We have

$$\begin{aligned} \big\|A_n(x_n) - A(x)\big\|_Y &\leqslant \big\|A_n(x_n) - A_n(x)\big\|_Y + \big\|A_n(x) - A(x)\big\|_Y \\ &\leqslant M\|x_n - x\|_X + \big\|A_n(x) - A(x)\big\|_Y \longrightarrow 0, \end{aligned}$$

so $A_n(x_n) \longrightarrow A(x)$ in Y.

Solution of Problem 5.41

Let $\{x_n\}_{n\geqslant 1} \subseteq X$ be a sequence such that

$$x_n \longrightarrow x \quad \text{in } X \quad \text{and} \quad A(x_n) \longrightarrow x^* \quad \text{in } X^*.$$

By hypothesis, we have

$$\langle A(x_n)-A(u), x_n-u\rangle \;=\; \langle A(x_n-u), x_n-u\rangle \;\geqslant\; 0 \qquad \forall\, n \geqslant 1,\; u \in X,$$

so

$$\langle x^* - A(u), x - u\rangle \;\geqslant\; 0 \qquad \forall\, u \in X.$$

In the last inequality, we choose $u = x + \lambda h$ for $\lambda > 0$, $h \in X$. We obtain

$$-\lambda\langle x^* - A(x), h\rangle + \lambda^2\langle A(h), h\rangle \;\geqslant\; 0,$$

so

$$\langle x^* - A(x), h\rangle - \lambda\langle A(h), h\rangle \;\leqslant\; 0.$$

Let $\lambda \searrow 0$. Then

$$\langle x^* - A(x), h\rangle \;\leqslant\; 0 \qquad \forall\, h \in X,$$

hence $x^* = A(x)$. So, $\operatorname{Gr} A$ is closed and invoking the closed graph theorem (see Theorem 5.50), we conclude that $A \in \mathcal{L}(X;Y)$.

Solution of Problem 5.42

Let $\{x_n\}_{n\geqslant 1} \subseteq X$ be a sequence such that

$$x_n \longrightarrow x \quad \text{in } X \quad \text{and} \quad A(x_n) \longrightarrow x^* \quad \text{in } X^*.$$

By hypothesis, we have

$$\langle A(x_n), u\rangle \;=\; \langle A(u), x_n\rangle \qquad \forall\, u \in X,\; n \geqslant 1,$$

so

$$\langle x^*, u\rangle \;=\; \langle A(u), x\rangle \;=\; \langle A(x), u\rangle \qquad \forall\, u \in X$$

and thus

$$\langle x^* - A(x), u\rangle \;=\; 0 \qquad \forall\, u \in X,$$

i.e., $x^* = A(x)$. Therefore A has a closed graph and so invoking the closed graph theorem (see Theorem 5.50), we conclude that $A \in \mathcal{L}(X;X^*)$.

Solution of Problem 5.43

(a) "$\Longrightarrow$": Note that

$$N(A) + C = A^{-1}\big(A(C)\big).$$

Since $A(C)$ is closed and $A \in \mathcal{L}(X;Y)$, we have that the set $A^{-1}\big(A(C)\big)$ is closed in X. Therefore the set $N(A) + C$ is closed in X.

"$\Longleftarrow$": Let $N(A) + C$ be closed in X. Since A is surjective, we have that

$$A\big((N(A) + C)^c\big) = A(C)^c.$$

But by the open mapping theorem (see Theorem 5.47), we have that the set $A\big((N(A) + C)^c\big)$ is open and so the set $A(C)$ is closed.

(b) Since $\dim N(A) < +\infty$, it follows that the set $N(A) + V$ is closed (see Remark 5.53) and so by part **(a)**, we infer that the set $A(V)$ is closed in Y.

Solution of Problem 5.44

From the open mapping theorem (see Theorem 5.47), we can find $\varepsilon > 0$ such that $\varepsilon B_1^Y \subseteq A(B_1^X)$, where $B_1^X = \big\{x \in X : \ \|x\|_X < 1\big\}$ and $B_1^Y \big\{y \in Y : \ \|y\|_Y < 1\big\}$. Therefore for every $y \in Y$ with $\|y\| = \delta < \varepsilon$, we can find $x \in B_1^X$ such that $y = A(x)$. Let $M = \frac{1}{\delta} > 0$. This is the desired $M > 0$.

Solution of Problem 5.45

"$\Longrightarrow$": Since the set $\mathcal{Y}$ is equicontinuous, for a given $\varepsilon > 0$, we can find $\delta = \delta(\varepsilon) > 0$ such that

$$\big\|A(x)\big\|_Y \leqslant \varepsilon \qquad \forall\, A \in \mathcal{Y},\ \|x\|_X \leqslant \delta,$$

so

$$\big\|A(\delta \tfrac{x}{\|x\|})\big\|_Y \leqslant \varepsilon \qquad \forall\, A \in \mathcal{Y},\ x \in X$$

and thus

$$\big\|A(x)\big\|_Y \leqslant \tfrac{\varepsilon}{\delta}\|x\|_X = M_1\|x\|_X \qquad \forall\, A \in \mathcal{Y},\ x \in X,$$

with $M_1 = \frac{\varepsilon}{\delta}$.

So, by the Banach–Steinhaus theorem (see Corollary 5.40), we can find $M > 0$ such that

$$\|A\|_{\mathcal{L}} \leqslant M \qquad \forall\, A \in \mathcal{Y}.$$

"$\Longleftarrow$": By hypothesis, we have

$$\big\|A(x)\big\|_Y \leqslant M\|x\|_X \qquad \forall\, x \in X.$$

So, for a given $\varepsilon > 0$, we choose $\delta = \delta(\varepsilon) = \frac{\varepsilon}{M} > 0$ and we have

$$\big\|A(x)\big\|_Y \leqslant \varepsilon \qquad \forall\, A \in \mathcal{Y},\ \|x\|_X \leqslant \delta,$$

so $\mathcal{Y}$ is equicontinuous.

Solution of Problem 5.46

We know that $\big(\mathcal{L}(X), \|\cdot\|_{\mathcal{L}}\big)$ is a Banach space (see Proposition 5.14). Since $\|A\|_{\mathcal{L}} < 1$, the real series $\sum\limits_{n\geqslant 0} \|A\|_{\mathcal{L}}^n$ converges. Because

$$\|A^n\|_{\mathcal{L}} \leqslant \|A\|_{\mathcal{L}}^n \qquad \forall\, n \geqslant 0,$$

the real series $\sum\limits_{n\geqslant 0} \|A^n\|_{\mathcal{L}}$ also converges. Hence, by Problem 5.11, we have

$$S = \sum_{n\geqslant 0} A^n \in \mathcal{L}(X).$$

Let us set

$$S_m = \sum_{n=0}^{m} A^n \in \mathcal{L}(X).$$

Then

$$S_m \longrightarrow S \quad \text{in } \mathcal{L}(X).$$

We have

$$\big\|(id - A)S_m - id\big\| = \|A^{m+1}\|_{\mathcal{L}} \leqslant \|A\|_{\mathcal{L}}^{m+1}.$$

Since $\|A\|_{\mathcal{L}} < 1$, we have

$$\lim_{m\to+\infty} (id - A)S_m = id,$$

so

$$(id - A)S = id.$$

Similarly, we show that $S(id - A) = id$. Therefore $(id - A)^{-1} \in \mathcal{L}(X)$ and $(id - A)^{-1} = S = \sum\limits_{n\geqslant 0} A^n$.

Solution of Problem 5.47

Since $A_n \longrightarrow A$ in $\mathcal{L}(X)$, we can find $n_0 \geqslant 1$ such that

$$\|A_n - A\|_{\mathcal{L}} < 1 \qquad \forall\, n \geqslant n_0.$$

Then, for all $n \geqslant n_0$, we have

$$\big\|id - A_n^{-1}A\big\|_{\mathcal{L}} = \big\|A_n^{-1}(A_n - A)\big\|_{\mathcal{L}} \leqslant \|A_n^{-1}\|_{\mathcal{L}}\|A_n - A\|_{\mathcal{L}} < \|A_n^{-1}\|_{\mathcal{L}} < 1,$$

so $A_{n_0}^{-1}A$ is an isomorphism (see Problem 5.46) and thus $A = A_{n_0}A_{n_0}^{-1}A$ is an isomorphism too.

Solution of Problem 5.48

For a normed space V, let

$$c_0(V) = \big\{\widehat{v} = \{v_n\}_{n\geqslant 1} : v_n \in V,\ v_n \longrightarrow 0 \text{ in } V\big\}.$$

Furnished with the norm

$$\|\widehat{v}\|_\infty = \sup_{n\geqslant 1} \|v_n\| < +\infty,$$

the space $c_0(V)$ becomes a normed space (it is a "vector-valued" version of the classical Banach space c_0). It is straightforward to check that if V is a Banach space, then so is $\big(c_0(V), \|\cdot\|_\infty\big)$.

For every $n \geqslant 1$, let $S_n\colon c_0(X) \longrightarrow c_0(Y)$, be defined by

$$S_n(\widehat{x}) = \big(A_1(x_1), A_2(x_2), \ldots, A_n(x_n), 0, \ldots\big).$$

Also, let $S\colon c_0(X) \longrightarrow c_0(Y)$ be defined by

$$S(\widehat{x}) = \big\{A_n(x_n)\big\}_{n\geqslant 1}.$$

Then hypothesis of the problem implies that S is well defined. Clearly

$$S_n \in \mathcal{L}\big(c_0(X), c_0(Y)\big).$$

We have

$$\big\|S(\widehat{x}) - S_n(\widehat{x})\big\|_\infty = \sup_{k\geqslant n+1} \big\|A_k(x_k)\big\|_Y \quad \text{and} \quad \lim_{n\rightarrow+\infty} \sup_{k\geqslant n+1} \big\|A_k(x_k)\big\|_Y = 0$$

(by hypothesis), so

$$S_n(\widehat{x}) \longrightarrow S(\widehat{x}) \qquad \forall\, \widehat{x} \in c_0(X),$$

thus $S \in \mathcal{L}\big(c_0(X); c_0(Y)\big)$ (see Corollary 5.42) and

$$\|S\|_{\mathcal{L}} = \sup\big\{\big\|S(\widehat{x})\big\|_\infty : \ \|\widehat{x}\|_\infty \leqslant 1\big\} < +\infty.$$

For $n \geqslant 1$ and $u \in X$, let $\widehat{x}_u = \big(0, \ldots, 0, u, 0, \ldots\big)$ (where u is at the nth position). Then $\widehat{x}_u \in c_0(X)$ and we have

$$\big\|S(\widehat{x}_u)\big\|_\infty \leqslant \|S\|_{\mathcal{L}} \|\widehat{x}_u\|_\infty = \|S\|_{\mathcal{L}} \|u\|_X,$$

so

$$\big\|A_n(u)\big\|_Y \leqslant \|S\|_{\mathcal{L}} \|u\|_X$$

and thus

$$\|A_n\|_{\mathcal{L}} \leqslant \|S\|_{\mathcal{L}} \qquad \forall\, n \geqslant 1.$$

Solution of Problem 5.49
"$\Longrightarrow$": Let $A\colon V \times Y \longrightarrow V + Y$ be defined by

$$A(v, y) = v + y.$$

Evidently A is linear, continuous and bijective. So, by the Banach theorem (see Theorem 5.48), A is an isomorphism. Thus, there is a bounded linear projection

$$\operatorname{proj}_V : V + Y \longrightarrow V,$$

defined by

$$\operatorname{proj}_V(v + y) = v.$$

Then, for every $v \in \partial B_1^V$ and every $y \in \partial B_1^Y$, we have

$$1 = \|v\| = \|\operatorname{proj}_V(v - y)\| \leqslant \|\operatorname{proj}_V\|_{\mathcal{L}}\|v - y\|,$$

so

$$\frac{1}{\|\operatorname{proj}_V\|_{\mathcal{L}}} \leqslant \operatorname{dist}(\partial B_1^V, \partial B_1^Y)$$

(since $v \in \partial B_1^V$ and $y \in \partial B_1^Y$ were arbitrary).

"$\Longleftarrow$": We proceed by contradiction. So, suppose that $V + Y$ is not closed. Then from the first part of the solution, there is no $c > 0$ such that

$$c\|v - y\| \geqslant \|v\| \qquad \forall\, v \in V,\ y \in Y.$$

So, for every $n \geqslant 1$ we can find $v_n \in V$ and $y_n \in Y$ such that

$$\|v_n\| = 1 \quad \text{and} \quad \|v_n + y_n\| \leqslant \tfrac{1}{n},$$

thus

$$\big|1 - \|y_n\|\big| = \big|\|v_n\| - \|y_n\|\big| \leqslant \|v_n + y_n\| \leqslant \tfrac{1}{n}$$

and so $\|y_n\| \longrightarrow 1$. We have

$$\begin{aligned}\Big\|v_n + \tfrac{y_n}{\|y_n\|}\Big\| &\leqslant \|v_n + y_n\| + \Big\|\tfrac{y_n}{\|y_n\|} - y_n\Big\| \\ &\leqslant \tfrac{1}{n} + \|y_n\|\Big|1 - \tfrac{1}{\|y_n\|}\Big| \qquad \forall\, n \geqslant 1,\end{aligned}$$

thus

$$\Big\|v_n + \tfrac{y_n}{\|y_n\|}\Big\| \longrightarrow 0,$$

i.e., $\eta = 0$, a contradiction.

Solution of Problem 5.50

"$\Longleftarrow$": Evidently, if A and S are invertible, then $(AS)^{-1} = S^{-1}A^{-1} \in \mathcal{L}(X)$.

"$\Longrightarrow$": Suppose that AS is invertible. Then we show that A and S are both injective. Indeed, if for $x \neq 0$ and $S(x) = 0$, then

$$(AS)(x) = A(S(x)) = A(0) = 0$$

and so AS is not injective, a contradiction. So S is injective. Similarly, if for some $x \neq 0$ we have $A(x) = 0$ (recall that $SA = AS$). If $A(X) \neq X$, then AS is not surjective, a contradiction. Similarly, if $S(X) \neq X$, since $AS = SA$. Therefore, invoking the Banach theorem (see Theorem 5.48), we have that $(AS)^{-1} \in \mathcal{L}(X)$.

Solution of Problem 5.51

We argue by contradiction. So, suppose that the convergence is not uniform on C. Then we can find $\varepsilon > 0$, a subsequence of $\{A_n\}_{n \geqslant 1} \subseteq \mathcal{L}(X;Y)$ (denoted for simplicity by the same index) and $x_n \in C$ for $n \geqslant 1$ such that

$$\left\|A_n(x_n) - A(x_n)\right\|_Y \geqslant \varepsilon \qquad \forall\, n \geqslant 1. \tag{5.13}$$

By passing to a further subsequence if necessary, we may assume that

$$x_n \longrightarrow x \quad \text{in } X.$$

By Corollary 5.40, we can find $M > 0$ such that

$$\|A_n\|_{\mathcal{L}} \leqslant M \qquad \forall\, n \geqslant 1, \quad \|A\|_{\mathcal{L}} \leqslant M. \tag{5.14}$$

Then, we have

$$\begin{aligned}
\left\|(A_n - A)(x_n)\right\|_Y &\leqslant \left\|(A_n - A)(x)\right\|_Y + \left\|(A_n - A)(x_n - x)\right\|_Y \\
&\leqslant \left\|(A_n - A)(x)\right\|_Y + \left\|A_n - A\right\|_{\mathcal{L}} \|x_n - x\|_X \\
&\leqslant \left\|(A_n - A)(x)\right\|_Y + 2M\|x_n - x\|_X,
\end{aligned}$$

so

$$\left\|A_n(x_n) - A(x_n)\right\|_Y \longrightarrow 0 \quad \text{as } n \to +\infty,$$

a contradiction to (5.13). Therefore, the convergence is uniform on C.

Solution of Problem 5.52

"(a) $\Longrightarrow$ (b)": Let $V = R(S)$. Note that

$$\|y\|_Y = \big\|A(S(y))\big\|_Y \leqslant \|A\|_{\mathcal{L}}\big\|S(y)\big\|_X \qquad \forall\, y \in Y,$$

so

$$\tfrac{1}{\|A\|_{\mathcal{L}}}\|y\|_Y \leqslant \big\|S(y)\big\|_Y \qquad \forall\, y \in Y$$

and thus $R(S) = V \subseteq X$ is a closed vector subspace.

Let $W = N(A)$ and let $x \in V \cap W$. Then $x = S(y)$ for some $y \in Y$ and so

$$y = A\big(S(y)\big) = 0,$$

so $x = 0$. Therefore $V \cap W = \{0\}$.

From the Banach theorem (see Theorem 5.48), we know that $\widehat{A}\colon X/_W \longrightarrow Y$ is an isomorphism and for every $y \in Y$, we have that $y \in A\big(S(y)\big)$. Hence $X = V + W$. Therefore $X = V \oplus W$ (see Definition 5.52).

"(b) $\Longrightarrow$ (a)": Let V be the topological complement of $W = N(A)$. Let $\mathrm{proj}_V \in \mathcal{L}(X)$ be the projection onto V. Let $y \in Y$. Then, since $R(A) = Y$, we can find $x \in X$ such that $A(x) = y$. Let $S(y) = \mathrm{proj}_V(x)$. Clearly S is well defined (i.e., independent of the choice of $x \in A^{-1}(y)$). Moreover, using the isomorphism $\widehat{A}\colon X/_W \longrightarrow Y$, we see that $S \in \mathcal{L}(Y;X)$ and $AS = id_Y$.

Solution of Problem 5.53

(a) Let $z^* \in Z^*$. Then for every $x \in X$, we have

$$\begin{aligned}\big\langle (S\circ A)^*(z^*), x\big\rangle_X &= \big\langle z^*, S\big(A(x)\big)\big\rangle_Z \\ &= \big\langle S^*(z^*), A(x)\big\rangle_Y \\ &= \big\langle (A^*\circ S^*)(z^*), x\big\rangle_X,\end{aligned}$$

so

$$(S\circ A)^* = A^*\circ S^*.$$

(b) From the Banach theorem (see Theorem 5.48), we have that $A^{-1} \in \mathcal{L}(Y;X)$. We have

$$AA^{-1} = id_Y \quad \text{and} \quad A^{-1}A = id_X,$$

so

$$(A^{-1})^* A^* = id_{Y^*} \quad \text{and} \quad A^*(A^{-1})^* = id_{X^*}$$

(see part **(a)**), hence

$$(A^*)^{-1} = (A^{-1})^*.$$

Solution of Problem 5.54

"$\Longrightarrow$": Follows from Definition 5.33.

"$\Longleftarrow$": Suppose that $D(A^*) = Y^*$. Then, for every $y^* \in Y^*$, we have

$$\sup\left\{ \langle y^*, A(x)\rangle_Y : \|x\| \leqslant 1\right\} = \|y^* \circ A\|,$$

so the set $\{A(x) : \|x\| \leqslant 1\}$ is w-bounded. Invoking the uniform boundedness principle (see Corollary 5.40), we see that we can find $M > 0$ such that

$$\|x\| \leqslant 1 \quad \Longrightarrow \quad \big\|A(x)\big\| \leqslant M,$$

so $A \in \mathcal{L}(X;Y)$.

Solution of Problem 5.55

First we show that ∂B_1 is w-dense in $\overline{B}_1$, i.e.,

$$\overline{\partial B_1}^w = \overline{B}_1.$$

So, let $x_0 \in X$ be such that $\|x_0\| < 1$. Let $U \in \mathcal{N}_w(x_0)$ (the filter of weak neighbourhoods of x_0). We may assume that

$$U = \left\{x \in X : \big|\langle x_k^*, x - x_0\rangle\big| < \varepsilon \text{ for all } k \in \{1,\ldots,m\}\right\},$$

where $\varepsilon > 0$ and $x_k^* \in X^*$ for all $k \in \{1, \ldots, m\}$. Consider the map $\xi \colon X \longrightarrow \mathbb{R}^m$, defined by

$$\xi(x) = \left(\langle x_1^*, x \rangle, \ldots, \langle x_m^*, x \rangle \right).$$

This map is not injective or otherwise $\dim X = m$, a contradiction. So, we can find $u_0 \neq 0$ such that $\xi(u_0) = 0$, hence

$$\langle x_k^*, u_0 \rangle = 0 \qquad \forall\, k \in \{1, \ldots, m\}.$$

Then

$$x_0 + \lambda u_0 \in U \qquad \forall\, \lambda \in \mathbb{R}.$$

Let us set

$$\vartheta(\lambda) = \|x_0 + \lambda u_0\| \qquad \forall\, \lambda \in \mathbb{R}.$$

Then $\vartheta(0) = \|x_0\| < 1$ and since $\vartheta(\lambda) \longrightarrow +\infty$ as $\lambda \to +\infty$, we can find $\lambda_0 > 0$ such that $\vartheta(\lambda_0) = 1$. This means that $x_0 + \lambda_0 u_0 \in \partial B_1 \cap U$, hence $x_0 \in \overline{\partial B_1}^w$. Since $x_0 \in B_1 = \{x \in X : \|x\| < 1\}$ is arbitrary, we conclude that $\overline{\partial B_1}^w = \overline{B}_1$ (see the Mazur theorem; Theorem 5.58). So, we have proved that ∂B_1 is w-dense in $\overline{B}_1$.

Next we show that ∂B_1 is G_δ in $(\overline{B}_1, w)$. Note that

$$\partial B_1 = \bigcap_{n \geqslant 1} U_n,$$

where

$$U_n = \left\{x \in \overline{B}_1 : \|x\| > 1 - \tfrac{1}{n}\right\} \qquad \forall\, n \geqslant 1.$$

By the weak lower semicontinuity of the norm functional (see Proposition 5.56(c)), U_n is open in $(\overline{B}_1, w)$, so, we conclude that ∂B_1 is G_δ in $(\overline{B}_1, w)$.

Solution of Problem 5.56

Suppose that X_w is metrizable and let ϱ be a metric on X generating the weak topology. Since $0 \in \overline{\partial B_1}^w$, we can find $x_n \in \partial(nB_1)$ such that $\varrho(x_n, 0) < \frac{1}{n}$. Then $x_n \xrightarrow{w} 0$ in X and so the sequence $\{x_n\}_{n \geqslant 1} \subseteq X$ is bounded, a contradiction.

Solution of Problem 5.57

Let $x^* \in X^*$ be such that $\|x^*\|_* = 1$ and

$$\ker x^* = \operatorname{span}(x_1, \ldots, x_{m-1})$$

(see Corollary 5.31). Also, let $u^* \in X^*$ be such that

$$\langle u^*, x_k \rangle = 1 \qquad \forall\, k \in \{1, \ldots, m-1\}.$$

Since ∂B_1 is compact, we have

$$C = \big\{x \in \partial B_1 : \langle x^*, x \rangle = 1\big\} \neq \emptyset.$$

Let us choose $x_m \in C$ such that

$$\langle u^*, x_m \rangle = \inf_{x \in C} \langle u^*, x \rangle.$$

Then

$$\|x_m - x_k\| \geqslant \langle x^*, x_m - x_k \rangle = \langle x^*, x_m \rangle = 1 \qquad \forall\, k \in \{1, \ldots, m-1\}.$$

Note that

$$\langle u^*, x_m - x_k \rangle = \langle u^*, x_m \rangle - 1 < \langle u^*, x_m \rangle \qquad \forall\, k \in \{1, \ldots, m-1\}$$

and so $x_m - x_k \notin C$, hence

$$\|x_m - x_k\| > 1 \qquad \forall\, k \in \{1, \ldots, m-1\}.$$

Solution of Problem 5.58

Since by hypothesis $\{x_n\}_{n \geqslant 1}$ is a norm Cauchy sequence, for a given $\varepsilon > 0$, we can find $n_0 = n_0(\varepsilon) \geqslant 1$ such that

$$x_n \in x_m + \overline{B}_\varepsilon \qquad \forall\, n \geqslant m \geqslant n_0.$$

The set $x_m + \overline{B}_\varepsilon$ is closed, convex, so by the Mazur theorem (see Theorem 5.58), it is w-closed. Passing to the limit as $n \to +\infty$ and since by hypothesis

$$x_n \xrightarrow{w} 0 \quad \text{in } X,$$

we have

$$0 \in x_m + \overline{B}_\varepsilon \qquad \forall\, m \geqslant n_0,$$

so

$$x_m \in \overline{B}_\varepsilon \qquad \forall\, m \geqslant n_0$$

and thus $x_n \longrightarrow 0$ in X.

Solution of Problem 5.59
We know that

$$X = \bigcup_{n\geqslant 1} n\overline{B}_1 \quad \text{and} \quad w\text{-int}\,(n\overline{B}_1) = \emptyset.$$

So, X_w is a countable union of nowhere dense sets, thus it is of the first Baire category (see Definition 1.25).

Solution of Problem 5.60
Since $\{x_n\}_{n\geqslant 1} \subseteq C$ and the set C is compact, we can find a subsequence $\{x_{n_k}\}_{k\geqslant 1}$ of the sequence $\{x_n\}_{n\geqslant 1}$ such that $x_{n_k} \longrightarrow y$ in X for some $y \in X$. We know that $x_{n_k} \xrightarrow{w} x$ in X. Therefore $y = x$. Hence every subsequence of $\{x_n\}_{n\geqslant 1}$ has a further subsequence that norm converges to x in X. Thus by the Urysohn criterion (see Problem 1.3), we conclude that the original sequence norm converges to x in X.

Solution of Problem 5.61
Let $\{x_\alpha\}_{\alpha\in J}$ be a net in X such that $x_\alpha \xrightarrow{w} x$ in X. Then

$$(y^* \circ A)(x_\alpha) \longrightarrow (y^* \circ A)(x) \qquad \forall\, y^* \in Y^*,$$

so

$$\langle y^*, A(x_\alpha)\rangle_Y \longrightarrow \langle y^*, A(x)\rangle_Y \qquad \forall\, y^* \in Y^*$$

and thus

$$A(x_\alpha) \xrightarrow{w} A(x).$$

Therefore $A \in \mathcal{L}(X_w; Y_w)$ and invoking Proposition 5.61, we conclude that $A \in \mathcal{L}(X; Y)$.

Solution of Problem 5.62

"$\Longrightarrow$": From Proposition 5.61, we have $A \in \mathcal{L}(X_w; Y_w)$ (X_w being the Banach space X endowed with the weak topology). In particular then $A \in \mathcal{L}(X; Y_w)$.

Alternatively, recall that the weak topology is smaller than norm topology. Therefore directly from the definition of continuity, we have that if $A \in \mathcal{L}(X; Y)$ then $A \in \mathcal{L}(X; Y_w)$.

"$\Longleftarrow$": Let $x_n \longrightarrow x$ in X and $A(x_n) \longrightarrow y$ in X. Since by hypothesis $A \in \mathcal{L}(X; Y_w)$, we have that $A(x_n) \xrightarrow{w} A(x)$ in Y, hence $y = A(x)$ and so $\operatorname{Gr} A$ is closed in $X \times Y$. By the closed graph theorem (see Theorem 5.50), we have that $A \in \mathcal{L}(X; Y)$.

Solution of Problem 5.63

Let $\{x_k^*\}_{k=1}^n \subseteq X^*$ and $\varepsilon > 0$. For $i\colon V \longrightarrow X$ being the identity (inclusion) map, let $v_k^* = x_k^* \circ i \in V^*$ for all $k \in \{1, \ldots, n\}$. We have

$$\begin{aligned} U\big(0; \{v_k^*\}_{k=1}^m; \varepsilon\big) &= \big\{v \in V : \ \big|\langle v_k^*, v\rangle_V\big| < \varepsilon \text{ for all } k = 1, \ldots, n\big\} \\ &= \big\{v \in V : \ \big|\langle x_k^*, v\rangle_X\big| < \varepsilon \text{ for all } k = 1, \ldots, n\big\} \\ &= V \cap \big\{x \in X : \ \big|\langle x_k^*, x\rangle_X\big| < \varepsilon \text{ for all } k = 1, \ldots, n\big\} \\ &= V \cap \widehat{U}\big(0; \{x_k^*\}_{k=1}^m; \varepsilon\big) \end{aligned}$$

(see Remark 5.55). Next, let $\{v_k^*\}_{k=1}^n \subseteq V^*$ and $\varepsilon > 0$. Let $x_k^* \in X^*$ be the extension of $v_k^* \in V^*$ on all of X for all $k \in \{1, \ldots, n\}$ (see Corollary 5.25). Then reasoning as above, we show

$$V \cap \widehat{U}\big(0; \{x_k^*\}_{k=1}^m; \varepsilon\big) \;=\; U\big(0; \{v_k^*\}_{k=1}^m; \varepsilon\big).$$

Therefore, the two topologies coincide on V.

Solution of Problem 5.64

Let $\{x_\alpha\}_{\alpha \in J} \subseteq C + E$ be a net and assume that

$$x_\alpha \xrightarrow{w} x \quad \text{in } X.$$

We have $x_\alpha = c_\alpha + e_\alpha$ with $\{c_\alpha\}_{\alpha \in J} \subseteq C$ and $\{e_\alpha\}_{\alpha \in J} \subseteq E$. Since E is w-compact, we can find a subnet $\{e_\beta\}_{\beta \in I}$ such that $e_\beta \xrightarrow{w} e \in E$ in X. Then

$$c_\beta \;=\; x_\beta - e_\beta \xrightarrow{w} x - e \;=\; c \in C \quad \text{in } X.$$

So, $x = c + e$ with $c \in C$, $e \in E$, hence $x \in C + E$ and we conclude that $C + E$ is w-closed.

Solution of Problem 5.65

Let

$$X_0 \;=\; \big\{\widehat{x} = \{x_n\}_{n \geqslant 1} : \; x_n \neq 0 \text{ only for a finite number of } n\text{'s}\big\}.$$

We furnish X_0 with the $\|\cdot\|_{l^1}$-norm. Clearly $\big(X_0, \|\cdot\|_{l^1}\big)$ is a noncomplete normed space. Let $\xi_n > 0$ be a sequence such that $\xi_n \longrightarrow \infty$. We consider the sequence $\{\widehat{x}_n^*\}_{n \geqslant 1} \subseteq X_0^*$, defined by

$$\langle \widehat{x}_n^*, \widehat{x} \rangle \;=\; x_n \qquad \forall\, n \geqslant 1,\ \widehat{x} = \{x_n\}_{n \geqslant 1}.$$

Let $C^* = \{0\} \cup \big\{\xi_n \widehat{x}_n^* : \; n \geqslant 1\big\} \subseteq X_0^*$. Note that

$$\big\|\xi_n \widehat{x}_n^*\big\|_{X_0^*} \;=\; \xi_n \longrightarrow +\infty \quad \text{as } n \to +\infty,$$

so the set $C^* \subseteq X_0^*$ is unbounded. On the other hand, for every $\widehat{x} \in X_0$, we have

$$\langle \xi_n x_n^*, \widehat{x} \rangle = \xi_n x_n \longrightarrow 0 \quad \text{as } n \to +\infty$$

so the set is w^*-bounded.

Solution of Problem 5.66

"$\Longrightarrow$": Follows from Proposition 5.65(c).

"$\Longleftarrow$": A norm $|\cdot|$ on X which is the predual norm of $|\cdot|_*$ is given by

$$|x| = \sup\big\{ \langle x^*, x \rangle : |x^*|_* \leqslant 1 \big\} = \sup_{x^* \neq 0} \tfrac{|\langle x^*, x \rangle|}{|x^*|_*}.$$

Then, we have

$$\begin{aligned} \sup\big\{ \big| \langle x^*, x \rangle \big| \ |x| \leqslant 1 \big\} &= \sup_{x \neq 0} \tfrac{|\langle x^*, x \rangle|}{|x|} = \sup_{x \neq 0} \frac{|\langle x^*, x \rangle|}{\sup\limits_{u^* \neq 0} \frac{|\langle u^*, x \rangle|}{|u^*|_*}} \\ &\leqslant \sup_{x \neq 0} \tfrac{|\langle x^*, x \rangle| |x^*|_*}{|\langle x^*, x \rangle|} = |x^*|_*. \end{aligned}$$

Let

$$\xi(x^*) = \sup\big\{ \big| \langle x^*, x \rangle \big| : |x| \leqslant 1 \big\}.$$

Suppose that for some $x^* \in X^*$ we have

$$\xi(x^*) < |x^*|_*,$$

so

$$\lambda \xi(x^*) < |x^*|_*,$$

for some $\lambda > 1$. Note that

$$\tfrac{\lambda x^*}{|x^*|_*} \notin \overline{B}_1^{|\cdot|_*} = \big\{ v^* \in X^* : |v|_* \leqslant 1 \big\}.$$

Also, from the choice of λ, for every $x \in X$, we have

$$\lambda \tfrac{|\langle x^*, x \rangle|}{|x^*|_*} < \sup_{u^* \neq 0} \tfrac{|\langle u^*, x \rangle|}{|u^*|_*} = \sup\big\{ \big| \langle u^*, x \rangle \big| : |u^*|_* \leqslant 1 \big\}.$$

So, $\frac{x^*}{|x^*|_*} \in \overline{\overline{B}_1^{|\cdot|_*}}^{w^*}$, hence $\overline{B}_1^{|\cdot|_*}$ is not w^*-closed which means that $|\cdot|_*$ is not w^*-lower semicontinuous, a contradiction.

Solution of Problem 5.67

(a) Let $y \in \operatorname{ext} A(C)$. Let $C_0 = C \cap A^{-1}(y)$. Since $A \in \mathcal{L}(X_w; Y_w)$ (see Proposition 5.61), we see that $C_0 \subseteq X$ is w-compact and so, we can find $u_0 \in \operatorname{ext} C_0$ such that $A(u_0) = y$. Let $u_0 = \frac{1}{2}(x+z)$ with $x, z \in C$. Then

$$A(u_0) = y = \tfrac{1}{2}\big(A(x) + A(z)\big).$$

Since $y \in \operatorname{ext} A(C)$, we must have

$$A(x) = A(z) = A(u_0),$$

hence $x, z \in C_0$. Since $u_0 \in \operatorname{ext} C_0$, we must have $x = z$. Therefore $u_0 \in \operatorname{ext} C$ and so we conclude that $\operatorname{ext} A(C) \subseteq A(\operatorname{ext} C)$.

(b) Let $x \in C \setminus \operatorname{ext} C$. Then $x = \frac{1}{2}(u+z)$ with $u, z \in C$, $u \neq z$. Since

$$A(x) = \frac{1}{2}\big(A(u) + A(z)\big)$$

and $A(u) \neq A(z)$ (A being an isomorphism),

$$A(C \setminus \operatorname{ext} C) \subseteq A(C) \setminus \operatorname{ext} A(C).$$

Since A^{-1} exists (A being an isomorphism), we also have opposite inclusion. Therefore $A(\operatorname{ext} C) = \operatorname{ext} A(C)$.

Solution of Problem 5.68

"$\Longrightarrow$": We assume that X is finite dimensional with $m = \dim X$. Let $\{e_k\}_{k=1}^m \subseteq X$ be a basis of X. For $x \in X$, we have

$$x = \sum_{k=1}^m \lambda_k e_k$$

and then let us set

$$h_k(x) = \lambda_k \qquad \forall\, k \in \{1, \dots, n\}.$$

Then $\{h_k\}_{k=1}^n \subseteq X^*$ (see Problem 5.15), they are linearly independent and span X^*. Therefore $\dim X^* = \dim X = m < +\infty$.

"$\Longleftarrow$": Suppose that X^* is finite dimensional. Then from the previous part of the solution, we infer that X^{**} is finite dimensional too and $\dim X^{**} = \dim X^*$. Let $j\colon X \longrightarrow X^{**}$ be the canonical embedding (see Definition 5.68). We know that j is an isometric isomorphism. Moreover, from the Goldstine theorem (see Theorem 5.70) and since X^{**} is finite dimensional, we have that $j(X) = X^{**}$ and so $\dim X = \dim X^{**} < \infty$.

Solution of Problem 5.69

"$\Longrightarrow$": Let $V = \overline{\operatorname{span}}^{w^*} D$ and suppose that $X^* \neq V$. Then by the strong separation theorem (see Theorem 5.29), we can find $x \in (X^*_{w^*})^* = X$ such that $x|_V = 0$ and $x \neq 0$. Then

$$\langle x^*, x \rangle = 0 \qquad \forall\, x^* \in D$$

and so D is a non-separating family, a contradiction.

"$\Longleftarrow$": Suppose that $D \subseteq X^*$ does not separate points. Then we can find $x \in X \setminus \{0\}$ such that

$$\langle x^*, x \rangle = 0 \qquad \forall\, x^* \in D.$$

Then

$$\langle x^*, x \rangle = 0 \qquad \forall\, x^* \in \overline{\operatorname{span}}^{w^*} D$$

and so we conclude that $\overline{\operatorname{span}}^{w^*} D \neq X$.

Solution of Problem 5.70

"$\Longrightarrow$": We argue by contradiction. So, suppose that

$$\overline{A^*(Y^*)}^{w^*} \neq X^*.$$

Then, by Corollary 5.31, we can find $x \in (X^*_{w^*})^* = X$, $x \neq 0$ such that

$$x|_{\overline{A^*(Y^*)}^{w^*}} = 0.$$

By definition, we have

$$0 = \langle A^*(y^*), x \rangle_X = \langle y^*, A(x) \rangle_Y \qquad \forall\, y^* \in Y^*,$$

so $A(x) = 0$, hence $x = 0$ since A is injective, a contradiction.

"$\Longleftarrow$": Suppose that $A(x) = 0$. Then

$$\langle y^*, A(x) \rangle_Y = 0 \qquad \forall\, y^* \in Y^*,$$

hence

$$\langle A^*(y^*), x \rangle_X = 0 \qquad \forall\, y^* \in Y^*.$$

Since by hypothesis $\overline{A^*(Y^*)}^{w^*} = X^*$, we infer that $x = 0$. Therefore A is injective.

Solution of Problem 5.71

Let X be a nonreflexive Banach space and let

$$\overline{B}_1 = \{x \in X : \|x\| \leqslant 1\}.$$

Then $\overline{B}_1$ is closed, convex in X^{**}, but by the Goldstine theorem (see Theorem 5.70), it is not w^*-closed. So, the Mazur theorem is not valid for the w^*-topology.

Solution of Problem 5.72

Let X be a normed space and let $j\colon X \longrightarrow X^{**}$ be the canonical embedding. We know that j is an isomorphism onto $V = j(X) \subseteq X^{**}$. Let

$$\widehat{X} = \overline{V}.$$

Then $\widehat{X}$ is a closed subspace of X^{**} which is a Banach space and V is dense in $\widehat{X}$. Since V is isometrically isomorphic to X, we conclude that X is dense in a Banach space.

Solution of Problem 5.73

Let

$$K = \overline{B}_1^* = \{x^* \in X^* : \|x^*\| \leqslant 1\}$$

furnished with the relative weak* topology. By the Alaoglu theorem (see Theorem 5.66), this is a compact topological space. Recall that using the canonical embedding, we can view X as a vector subspace of X^{**}. Then every $x \in X$ belongs in $C(K)$ and

$$\begin{aligned}\|x\| &= \sup\{|\langle x^*, x\rangle| : \|x^*\|_* \leqslant 1\} \\ &= \sup\{|x(x^*)| : x^* \in K\} = \|x\|_{C(K)}.\end{aligned}$$

So, X is isometrically isomorphic to a subspace of $C(K)$.

Solution of Problem 5.74

By hypothesis $A(\overline{B}_1^X)$ is relatively w-compact in Y. Viewing $A(\overline{B}_1^X)$ as a subset of Y^{**}, we have

$$\overline{A(\overline{B}_1^X)}^{w^*} = \overline{A(\overline{B}_1^X)}^{w},$$

since w-compact sets in Y are w^*-compact sets in Y^{**}. Since the set $A(\overline{B}_1^X)$ is convex, by the Mazur theorem (see Theorem 5.58), the last equality becomes

$$\overline{A(\overline{B}_1^X)}^{w^*} = \overline{A(\overline{B}_1^X)} \subseteq Y.$$

But by the Goldstine theorem (see Theorem 5.70), $\overline{B}_1^X$ is w^*-dense in $\overline{B}_1^{X^{**}}$, while it is easy to see that $A^{**} \in \mathcal{L}(X^{**}_{w^*}; Y^{**}_{w^*})$. Hence

$$A^{**}(\overline{B}_1^{X^{**}}) \subseteq A(\overline{B}_1^X) \subseteq Y,$$

so

$$A^{**}(X^{**}) \subseteq Y.$$

Solution of Problem 5.75

Let $\{v_n^*\}_{n\geqslant 1} \subseteq V^\perp$ be a sequence such that $\|u^* - v_n^*\|_* \searrow m^\perp$ as $n \to +\infty$ (see Definition 5.36(a)). Then the sequence $\{v_n^*\}_{n\geqslant 1} \subseteq V^\perp$ is bounded. So, by the Alaoglu theorem (see Theorem 5.66), the sequence

$\{v_n^*\}_{n\geqslant 1}$ is relatively w^*-compact. So, we can find a subnet $\{v_\alpha^*\}_{\alpha\in J}$ of $\{v_n^*\}_{n\geqslant 1}$ such that

$$v_\alpha^* \xrightarrow{w^*} \widehat{v}^* \quad \text{in } X^*$$

and $\widehat{v}^* \in V^\perp$. From the w^*-lower semicontinuity of the norm function (see Proposition 5.65(c)), we have

$$\|u^* - \widehat{v}^*\|_* \leqslant \liminf_{\alpha\in J} \|u^* - v_\alpha^*\|_* = m^\perp \quad \text{and} \quad \widehat{v}^* \in V^\perp,$$

so

$$\|u^* - \widehat{v}^*\|_* = m^\perp.$$

Solution of Problem 5.76

Recall that the w^*-topology of X^{**} restricted on $X \subseteq X^{**}$ equals the w-topology of X. We view C as a subset of X^{**}. If we can show that $\overline{C}^{w^*} \subseteq X$, then clearly we are done.

By hypothesis, we have

$$\overline{C}^{w^*} \subseteq \overline{C_\varepsilon + \varepsilon \overline{B}_1}^{w^*} \subseteq \overline{C_\varepsilon}^{w^*} + \varepsilon \overline{\overline{B}_1}^{w^*} = C_\varepsilon + \varepsilon \overline{B}_1^{**}$$

(by the Goldstein theorem; see Theorem 5.70), so

$$\overline{C}^{w^*} \subseteq \bigcap_{\varepsilon>0} (C_\varepsilon + \varepsilon \overline{B}_1^{**}) = C_\varepsilon \subseteq X.$$

Solution of Problem 5.77

Let X be an infinite dimensional subspace of l^1 and suppose that X is reflexive. Then $\overline{B}_1^X = \{u \in X : \|u\|_{l^1} \leqslant 1\}$ is w-compact (see Theorem 5.73). Then from the Schur property (see Remark 5.57) and the Eberlein–Smulian theorem (see Theorem 5.78), we have that $\overline{B}_1^X$ is compact which in turn implies that X is finite dimensional, a contradiction.

Solution of Problem 5.78

From Proposition 5.37(a), we know that $V^{\perp}$ is a closed vector subspace of X^*, hence it is reflexive. Since $\left(X/_V\right)^*$ is isomorphic to $V^{\perp}$ (see Proposition 5.38), we infer that $\left(X/_V\right)^*$ is reflexive. Invoking Theorem 5.76, we conclude that $X/_V$ is reflexive.

Solution of Problem 5.79

Let $\{u_n^*\}_{n\geqslant 1} \subseteq C$ be a sequence such that

$$\|x^* - u_n^*\|_* \searrow \operatorname{dist}(x^*, C).$$

Evidently $\{u_n^*\}_{n\geqslant 1}$ is a bounded sequence and so by the Alaoglu theorem (see Theorem 5.66), $\{u_n^*\}_{n\geqslant 1}$ is relatively w^*-compact. So, we can find a subnet $\{u_\alpha^*\}_{\alpha\in J}$ of $\{u_n^*\}_{n\geqslant 1}$ such that $u_\alpha^* \xrightarrow{w^*} u^*$ in X^*. Since C is w^*-closed, we have $u^* \in C$. Also, from the w^*-lower semicontinuity of the dual norm (see Proposition 5.65(c)), we have

$$\|x^* - u^*\|_* \leqslant \liminf_{\alpha\in J} \|x^* - u_\alpha^*\|_* = \operatorname{dist}(x^*, C) \quad \text{and} \quad u^* \in C,$$

so

$$\|x^* - u^*\|_* = \operatorname{dist}(x^*, C).$$

Solution of Problem 5.80

"$\Longrightarrow$": Since C is w-closed (by the Mazur theorem; see Theorem 5.58) and X is reflexive, this implication is a particular case of Problem 5.79.

"$\Longleftarrow$": Suppose that X is not reflexive. Then by Theorem 5.76, we can find $x^* \in X^*$, with $\|x^*\|_* = 1$ which does not attain its norm. Then

$$\operatorname{dist}\big(0, (x^*)^{-1}(\{1\})\big) = 1 < \|x\| \qquad \forall\, x \in (x^*)^{-1}(\{1\}).$$

but the set $(x^*)^{-1}(\{1\})$ is closed, convex, hence by hypothesis proximinal, which contradicts the above inequality.

Solution of Problem 5.81
Let $\widehat{u} \in X$ be such that $\widehat{\lambda} = \varphi(\widehat{u}) < +\infty$ and consider the set

$$\widehat{C} = \{x \in X : \varphi(x) \leqslant \widehat{\lambda}\}.$$

The convexity and lower semicontinuity of φ imply that the set $\widehat{C}$ is convex, closed, hence by the Mazur theorem (see Theorem 5.58), $\widehat{C}$ is w-closed. Moreover, the coercivity hypothesis on φ implies that $\widehat{C}$ is bounded. Therefore, from the reflexivity of X, we have that the set $\widehat{C}$ is w-compact (see Theorem 5.73). The functional $\varphi\colon X \longrightarrow \overline{\mathbb{R}} = \mathbb{R} \cup \{+\infty\}$ being convex, lower semicontinuous, is weakly lower semicontinuous (see Corollary 5.60). Hence by Theorem 2.86, we can find $x_0 \in \widehat{C}$ such that

$$\varphi(x_0) = \inf_{x \in \widehat{C}} \varphi(x).$$

If $x \in X \setminus \widehat{C}$, then $\varphi(x) > \lambda_0$ and so

$$\varphi(x_0) = \inf_{x \in X} \varphi(x).$$

Solution of Problem 5.82
From Proposition 5.61, we know that $A \in \mathcal{L}(X_w; Y_w)$, while from Theorem 5.73, we have that $\overline{B}_1^X$ is w-compact. So, the set $A(\overline{B}_1^X) \subseteq Y$ is w-compact, hence w-closed. Therefore the set $A(\overline{B}_1^X)$ is closed.

Solution of Problem 5.83
The operator $\widehat{A}\colon X/_{N(A)} \longrightarrow Y$, defined by $\widehat{A}([x]) = A(x)$ is an isomorphism. Therefore Y is isomorphic to $X/_{N(A)}$. But since X is reflexive and $N(A) \subseteq X$ is a closed vector subspace (recall that

$A \in \mathcal{L}(X;Y)$), from Problem 5.78, we have that the space $X/_{N(A)}$ is reflexive. Then by Theorem 5.76, we conclude that Y is reflexive.

Solution of Problem 5.84

"(a) $\Longrightarrow$ (b)": Let $\{C_n\}_{n\geqslant 1}$ be a sequence of nonempty, closed, convex and bounded subsets of X. For every $n \geqslant 1$, the set C_n is w-closed (by the Mazur theorem; see Theorem 5.58) and bounded. Then the reflexivity of X implies that for every $n \geqslant 1$, the set C_n is w-compact (see Theorem 5.73). Since the sequence $\{C_n\}_{n\geqslant 1}$ is decreasing, $C_n \subseteq C_1$ for all $n \geqslant 1$ and the family has the finite intersection property. Then Theorem 2.81 implies that

$$\bigcap_{n\geqslant 1} C_n \neq \emptyset.$$

"(b) $\Longrightarrow$ (a)": Let $x^* \in X^* \setminus \{0\}$ and let

$$C_n = \left\{x \in X : \|x\| \leqslant 1, \langle x^*, x\rangle \geqslant \|x^*\|_* - \tfrac{1}{n}\right\}.$$

Evidently $\{C_n\}_{n\geqslant 1}$ is a decreasing sequence of nonempty, closed, convex and bounded subsets of X. Hence by hypothesis, we have $\bigcap\limits_{n\geqslant 1} C_n \neq \emptyset$. Let $x \in \bigcap\limits_{n\geqslant 1} C_n$. Then

$$\|x\| \leqslant 1 \quad \text{and} \quad \langle x^*, x\rangle \geqslant \|x^*\|_* - \tfrac{1}{n} \qquad \forall\, n \geqslant 1,$$

so

$$\|x\| \leqslant 1 \quad \text{and} \quad \langle x^*, x\rangle = \|x^*\|_*.$$

Since $x^* \in X^* \setminus \{0\}$ is arbitrary, from Theorem 5.76, we infer that X is reflexive.

Solution of Problem 5.85

Let $V = A(X)$ and suppose that it is a closed vector subspace of Y. Since Y is reflexive, V is reflexive too, while $A\colon X \longrightarrow V$ is an isomorphism (see the Banach theorem; Theorem 5.48). Then by Theorem 5.76, X is reflexive, a contradiction.

Solution of Problem 5.86

No. Indeed, if $A(l^1) \subseteq l^2$ is closed vector subspace of l^2, then $V = A(l^1)$ is reflexive (see Proposition 5.77 and recall that l^2 is a Hilbert space, thus reflexive). From the Banach theorem (see Theorem 5.48), we have that $A\colon l^1 \longrightarrow V$ is an isomorphism and so Theorem 5.76 implies that l^1 is reflexive, a contradiction (the result also follows from Problem 5.85).

Solution of Problem 5.87

Let $\{x_n\}_{n\geqslant 1} \subseteq X$ be a sequence such that for every $x^* \in X^*$, $\lim\limits_{n\to+\infty} \langle x^*, x_n\rangle$ exists. Then, by Corollary 5.43, the sequence $\{x_n\}_{n\geqslant 1}$ is bounded. Since X is reflexive, Corollary 5.75 and the Eberlein–Smulian theorem (see Theorem 5.78) imply that there exists a subsequence $\{x_{n_k}\}_{k\geqslant 1}$ of $\{x_n\}_{n\geqslant 1}$ such that $x_{n_k} \xrightarrow{w} x$ in X. Hence

$$\lim_{n\to+\infty} \langle x^*, x_n\rangle = \lim_{k\to+\infty} \langle x^*, x_{n_k}\rangle = \langle x^*, x\rangle \qquad \forall\, x^* \in X^*,$$

so $x_n \xrightarrow{w} x$ in X.

Solution of Problem 5.88

First, we show that

$$\inf_{x\in C} \big\|x - f(x)\big\| = 0.$$

To this end, we fix $u \in C$ and for $\varepsilon \in (0,1)$, we consider the map

$$f_\varepsilon(x) = \varepsilon u + (1-\varepsilon) f(x) \qquad \forall\, x \in C.$$

Since C is convex, we have $f_\varepsilon\colon C \longrightarrow C$. Note that

$$\big\|f_\varepsilon(x) - f_\varepsilon(y)\big\| \leqslant (1-\varepsilon)\|x-y\| \qquad \forall\, x, y \in C.$$

Invoking the Banach fixed point theorem (see Theorem 1.49), we can find a unique $x_\varepsilon \in C$ such that $f(x_\varepsilon) = x_\varepsilon$. We have

$$\begin{aligned} \|x_\varepsilon - f(x_\varepsilon)\| &= \|\varepsilon u + (1-\varepsilon) f(x_\varepsilon) - f(x_\varepsilon)\| \\ &= \varepsilon \|u - f(x_\varepsilon)\| \leqslant \varepsilon \operatorname{diam} C, \end{aligned}$$

so

$$\|x_\varepsilon - f(x_\varepsilon)\| \longrightarrow 0 \quad \text{as } \varepsilon \searrow 0$$

(recall that the set C is bounded). This proves that $\inf\limits_{x \in C} \|x - f(x)\| = 0$. We can find a sequence $\{x_n\}_{n \geqslant 1} \subseteq C$ such that

$$\|x_n - f(x_n)\| \longrightarrow 0.$$

Since the set C is w-compact and $\{x_n\}_{n \geqslant 1} \subseteq C$, by the Eberlein–Smulian theorem (see Theorem 5.78), we may assume that

$$x_n \xrightarrow{w} x \in C.$$

We have

$$\begin{aligned} & \|x_n - f(x)\|^2 - \|x_n - x\|^2 \\ = \; & \big(\|x_n - f(x_n)\| - \|x_n - x\|\big)\big(\|x_n - f(x)\| + \|x_n - x\|\big) \\ \leqslant \; & \|x_n - f(x_n)\| \big(\|x_n - f(x)\| + \|x_n - x\|\big) \longrightarrow 0. \end{aligned}$$

On the other hand, we have

$$\begin{aligned} & \|x_n - f(x)\|^2 = \|x_n - x + x - f(x)\|^2 \\ = \; & \big((x_n - x) + (x - f(x)),\ (x_n - x) + (x - f(x))\big)_H \\ = \; & \|x_n - x\|^2 + \|x - f(x)\|^2 + 2\big(x_n - x,\ x - f(x)\big)_H, \end{aligned}$$

so

$$\|x_n - f(x)\|^2 - \|x_n - x\|^2 \longrightarrow \|x - f(x)\|^2$$

(recall that $x_n \xrightarrow{w} x$ in X). Thus, we conclude that

$$\|x - f(x)\| = 0$$

and so $x = f(x)$.

Solution of Problem 5.89
We know that

$$\|A\|_{\mathcal{L}} = \sup\{\|A(x)\|_V : \|x\|_H = 1\}.$$

We can find a sequence $\{x_n\}_{n\geqslant 1} \subseteq \partial B_1^H$ such that

$$\|A(x_n)\|_V \longrightarrow \|A\|_{\mathcal{L}}.$$

By the Eberlein theorem (see Theorem 5.78) and by passing to a suitable subsequence if necessary, we may assume that $x_n \xrightarrow{w} x$ in H, $x \in \overline{B}_1^H$. We have

$$\|A(x_n) - A(x)\|_V^2 \leqslant \|A\|_{\mathcal{L}}^2 \|x_n - x\|_H^2. \tag{5.15}$$

Note that

$$\begin{aligned}\|x_n - x\|_H^2 &= (x_n - x, x_n - x)_H = \|x_n\|_H^2 + \|x\|_H^2 - 2(x_n, x)_H \\ &= 1 + \|x\|^2 - 2(x_n, x)_H,\end{aligned}$$

so

$$\lim_{n\to+\infty} \|x_n - x\|_H^2 = 1 - \|x\|_H^2.$$

Also, we have

$$\begin{aligned}\|A(x_n) - A(x)\|_V^2 &= \big(A(x_n) - A(x), A(x_n) - A(x)\big)_V \\ &= \|A(x_n)\|_V^2 + \|A(x)\|_V^2 - 2\big(A(x_n), A(x)\big)_V.\end{aligned}$$

From Proposition 5.61, we know that $A \in \mathcal{L}(H_w; V_w)$ and so

$$A(x_n) \xrightarrow{w} A(x) \text{ in } V.$$

So, if in the last equality, we pass to the limit as $n \to +\infty$, then

$$\lim_{n\to+\infty} \|A(x_n) - A(x)\|_V^2 = \|A\|_{\mathcal{L}}^2 - \|A(x)\|_V^2.$$

Returning to (5.15) and passing to the limit as $n \to +\infty$, we obtain

$$\|A\|_{\mathcal{L}}^2 - \|A(x)\|_V^2 \leqslant \|A\|_{\mathcal{L}}^2 - \|A\|_{\mathcal{L}}^2 \|x\|_H^2,$$

so

$$\|A\|_{\mathcal{L}} \|x\|_H \leqslant \|A(x)\|_V \quad \text{with } \|x\|_H \leqslant 1$$

and thus

$$\|A(x)\|_V = \|A\|_{\mathcal{L}} \|x\|_H.$$

If $x \neq 0$, then

$$\big\|A\big(\tfrac{x}{\|x\|}\big)\big\|_V = \|A\|_{\mathcal{L}}$$

and since $\left\|\frac{x}{\|x\|}\right\|_H = 1$, we contradict our hypothesis. Therefore $x = 0$ and so $x_n \xrightarrow{w} 0$.

Solution of Problem 5.90

First we show that $\mathcal{F}$ is single-valued. Let $x_1^*, x_2^* \in \mathcal{F}(x)$. Then

$$\langle x_k^*, x\rangle \;=\; \|x\|^2 \;=\; \|x_k^*\|_*^2 \quad \text{for } k = 1, 2.$$

So, we have

$$2\|x_1^*\|_*\|x\| \;\leqslant\; \|x_1^*\|_*^2 + \|x\|^2 \;=\; \langle x_1^* + x_2^*, x\rangle \;\leqslant\; \|x_1^* + x_2^*\|_*\|x\|,$$

thus

$$\|x_1^*\|_* \;\leqslant\; \tfrac{1}{2}\|x_1^* + x_2^*\|_*.$$

Since $\|x_1^*\|_* = \|x_2^*\|_*$ and X^* is strictly convex, it follows that $x_1^* = x_2^*$, i.e., $\mathcal{F}$ is single-valued.

Let $x_n \longrightarrow X$ in X. Then $\big\|\mathcal{F}(x_n)\big\| = \|x_n\| \longrightarrow \|x\|$. Note that the sequence $\{\mathcal{F}(x_n)\}_{n\geqslant 1} \subseteq X^*$ is bounded and X^* is reflexive (since X is). So, we may assume (at least for a subsequence) that

$$\mathcal{F}(x_n) \;\xrightarrow{w}\; x^* \quad \text{in } X$$

(see Theorem 5.73 and the Eberlein–Smulian theorem (see Theorem 5.78)). We have

$$\langle x^*, h\rangle \;=\; \lim_{n\to+\infty} \big\langle \mathcal{F}(x_n), h\big\rangle \;\leqslant\; \lim_{n\to+\infty} \|x_n\|\,\|h\| \;=\; \|x\|\,\|h\| \qquad \forall\, h \in X$$

and

$$\langle x^*, x\rangle \;=\; \lim_{n\to+\infty} \big\langle \mathcal{F}(x_n), x_n\big\rangle \;=\; \lim_{n\to+\infty} \|x_n\|^2 \;=\; \|x\|^2.$$

Therefore

$$\|x^*\|_* \;=\; \|x\|,$$

i.e., $x^* = \mathcal{F}(x)$.

Then, by the Urysohn criterion (see Problem 1.3), for the original sequence, we have

$$\mathcal{F}(x_n) \xrightarrow{w} \mathcal{F}(x) \quad \text{in } X,$$

so $\mathcal{F}$ is sequentially continuous from X into X^*_w.

Finally, let $x, u \in X$. Then

$$\begin{aligned} \langle \mathcal{F}(x) - \mathcal{F}(u), x - u \rangle &= \langle \mathcal{F}(x), x \rangle - \langle \mathcal{F}(x), u \rangle - \langle \mathcal{F}(u), x \rangle + \langle \mathcal{F}(u), u \rangle \\ &\geqslant \|x\|^2 - 2\|x\|\,\|u\| + \|u\|^2 = \big(\|x\| - \|u\|\big)^2 \geqslant 0, \end{aligned}$$

so $\mathcal{F}$ is monotone.

Solution of Problem 5.91

(a) No. First let

$$X = c_0 = \big\{\widehat{x} = \{x_n\}_{n\geqslant 1} : x_n \longrightarrow 0\big\}$$

furnished with the norm

$$\|\widehat{x}\|_\infty = \sup_{n\geqslant 1} |x_n| \qquad \forall\, \widehat{x} \in X.$$

Let $\widehat{x}^k = \big\{x_n^k = \frac{1}{n}\delta_{nk}\big\}_{n\geqslant 1}$ for all $k \geqslant 1$ and let us set $C = \big\{\widehat{x}^k\big\}_{k\geqslant 1} \cup \{0\}$. Evidently $\widehat{x}^k \longrightarrow 0$ as $k \to +\infty$ and so the set C is compact. But

$$\widehat{x} = \sum_{k\geqslant 1} \tfrac{1}{2^k}\widehat{x}^k \in \overline{\operatorname{conv}}\, C, \quad \widehat{x} \notin \operatorname{conv} C.$$

Therefore, the set $\operatorname{conv} C$ is not closed (in particular then $\operatorname{conv} C$ is not compact).

(b) No. Next, let

$$X = \big\{\widehat{x} = \{x_n\}_{n\geqslant 1} : x_n \neq 0 \text{ only for a finite number of } n\big\}.$$

We furnish X with the supremum norm

$$\|\widehat{x}\|_\infty = \max_{n\geqslant 1} |x_n|.$$

Note that $\big(X, \|\cdot\|_X\big)$ is not a Banach space. Let C be as above. Then C is compact in X, but

$$\widehat{x} = \sum_{k\geqslant 1} \tfrac{1}{2^k}\widehat{x}^k \notin X$$

and so $\widehat{x} \notin \overline{\operatorname{conv}}\, C$. Let

$$u_n = \sum_{k=1}^{n-1} \tfrac{1}{2^k}\widehat{x}^k + \tfrac{\widehat{x}^n}{2^{n-1}} \qquad \forall\, n \geqslant 1.$$

Then $\{u_n\}_{n\geqslant 1} \subseteq \operatorname{conv} C$ is a Cauchy sequence, but it is not convergent in $\overline{\operatorname{conv}}\, C$. This proves that $\overline{\operatorname{conv}}\, C$ is not compact in X (hence in Theorem 5.86 completeness of X is crucial).

Solution of Problem 5.92

We have

$$x = \sum_{k=1}^m \lambda_k x_n \quad \text{with } x_1, \ldots, x_m \in C \quad \text{and} \quad \lambda_1, \ldots, \lambda_m \geqslant 0,$$
$$\sum_{k=1}^m \lambda_k = 1.$$

We assume that this representation of x is so chosen that x cannot be written as a convex combination of fewer than m vectors from C. This implies that all vectors $\{x_k\}_{k=1}^m$ are distinct and $\lambda_1, \ldots, \lambda_m > 0$. We will show that $m \leqslant n+1$. Arguing by contradiction, suppose that $m > n+1$. Then $\{x_k\}_{k=1}^m$ must be affinely dependent and so we can find $t_1, \ldots, t_m \in \mathbb{R}$ not all zero such that

$$\sum_{k=1}^m t_k x_k = 0 \quad \text{and} \quad \sum_{k=1}^m t_k = 0.$$

Let $r > 0$ be such that $\lambda_k + t_k r \geqslant 0$ for all $k \in \{1, \ldots, m\}$ and at least one of them is zero. Such a choice of $r > 0$ is possible, since all λ_k are positive and at least one of t_k is negative. Then

$$x = \sum_{k=1}^m (\lambda_k + t_k r) x_k.$$

If we drop the zero coefficients, we have a convex combination of fewer that m vectors in C, which contradicts the minimality of m. Therefore $m \leqslant n+1$.

Solution of Problem 5.93

Let $n = \dim X$ and let

$$E = \{\widehat{\lambda} = \{\lambda_k\}_{k=1}^{n+1} \in \mathbb{R}_+^{n+1} : \sum_{k=1}^{n+1} \lambda_k = 1\}$$

and consider the map $h\colon E \times X^{n+1} \longrightarrow X$, defined by

$$h(\widehat{\lambda}, x_1, \ldots, x_{n+1}) = \sum_{k=1}^{n+1} \lambda_k x_k.$$

Evidently h is continuous, while E and C^{n+1} are compact sets. Hence $h(E, C^{n+1}) \subseteq X$ is compact. But according to Problem 5.92, $h(E, C^{n+1}) = \operatorname{conv} C$. Therefore $\operatorname{conv} C$ is compact.

Solution of Problem 5.94

Let U be a convex neighbourhood of the origin. Since C is totally bounded, we can find a finite set $F \subseteq X$ such that

$$C \subseteq F + \frac{1}{2}U$$

(see Definition 5.7(c)). From Problem 5.93, we have that $\operatorname{conv} F$ is compact. Let $x \in \operatorname{conv} C$. Then

$$x = \sum_{k=1}^{m} \lambda_k x_k \quad \text{with } \{x_k\}_{k=1}^{m} \subseteq C,\ \lambda_1, \ldots, \lambda_m \geqslant 0,\ \sum_{k=1}^{m} \lambda_k = 1.$$

For each $k \in \{1, \dots, m\}$, there exists $y_k \in F$ such that $x_k \in y_k + \frac{1}{2}U$, then

$$x = \sum_{k=1}^{m} \lambda_k (x_k - y_k) + \sum_{k=1}^{m} \lambda_k y_k, \quad \sum_{k=1}^{m} \lambda_k (x_k - y_k) \in U$$

(since U is convex) and

$$\sum_{k=1}^{m} \lambda_k y_k \in \operatorname{conv} F.$$

Therefore

$$\operatorname{conv} C \subseteq \operatorname{conv} F + \tfrac{1}{2}U.$$

But $\operatorname{conv} F$ is compact. So, we can find a finite set $F_0 \subseteq \operatorname{conv} F$ such that

$$\operatorname{conv} F \subseteq F_0 + \frac{1}{2}U.$$

So, finally we have $\operatorname{conv} C \subseteq F_0 + U$, which shows that $\operatorname{conv} C$ is totally bounded.

Solution of Problem 5.95

Since C is compact, it is totally bounded and so by Problem 5.94, the set $\operatorname{conv} C$ is totally bounded. Because X is a metric space (see Definition 5.5(a)), it follows that $\overline{\operatorname{conv}}\, C$ is compact.

Solution of Problem 5.96

Evidently X^* with the norm topology is a Polish space (see Definition 2.150). Also

$$X^*_{w^*} = \bigcup_{n \geqslant 1} (n\overline{B}^*_1, w^*),$$

where $\overline{B}^*_1 = \{x^* \in X^* : \ \|x^*\|_* \leqslant 1\}$. Since X is separable (X^* being separable), by Theorems 5.66 and 5.86(a), $(n\overline{B}^*_1, w^*)$, the set $n\overline{B}^*_1$ with the relative w^*-topology is a compact metric space, hence also a Polish space. Then

$$X^*_{w^*} = \bigcup_{n \geqslant 1} (n\overline{B}^*_1, w^*)$$

is a Souslin space (see Definition 2.156 and Proposition 2.159). But then invoking Corollary 4.36, we conclude that

$$\mathcal{B}(X^*) = \mathcal{B}(X^*_{w^*}).$$

Solution of Problem 5.97

Let

$$l^\infty = \{\widehat{x} = \{x_n\}_{n\geqslant 1} : \|\widehat{x}\|_{l^\infty} = \sup_{n\geqslant 1} |x_n| < +\infty\}.$$

We know that l^∞ is a nonseparable Banach space (see Problem 1.18) with the $\|\cdot\|_\infty$-norm. Let $\{u_n\}_{n\geqslant 1} \subseteq C_b((0,1))$ be a sequence such that

$$\operatorname{supp} u_n \subseteq \left[\tfrac{1}{n+1}, \tfrac{1}{n}\right] \quad \text{and} \quad \|u_n\|_\infty = 1 \qquad \forall\, n \geqslant 1.$$

Let $A\colon l^\infty \longrightarrow C_b((0,1))$ be defined by

$$A(\widehat{x}) = \sum_{n\geqslant 1} x_n u_n.$$

Evidently $A \in \mathcal{L}(l^\infty; C_b((0,1)))$ and it is injective. So, $C_b((0,1))$ contains an isometric copy of l^∞ which is nonseparable, hence $C_b((0,1))$ cannot be separable.

Solution of Problem 5.98

We consider the nonseparable Banach space $X = l^\infty$ (see Problem 1.18). For $n \geqslant 1$, let $u_n\colon l^\infty \longrightarrow \mathbb{R}$ be defined by

$$u_n(\widehat{x}) = x_n \qquad \forall\, \widehat{x} = \{x_n\}_{n\geqslant 1} \in l^\infty.$$

Clearly,

$$u_n \in (l^\infty)^* \quad \text{and} \quad \|u_n\|_{(l^\infty)^*} = 1 \qquad \forall\, n \geqslant 1.$$

By the Alaoglu theorem (see Theorem 5.66), the unit ball in $(l^\infty)^*$ is w^*-compact. Suppose for the moment that the Eberlein–Smulian theorem was valid in the w^*-topology. Then we can find a w^*-convergent subsequence $\{u_{n_k}\}_{k\geqslant 1}$ of $\{u_n\}_{n\geqslant 1}$. Hence $\{x_{n_k}\}_{k\geqslant 1}$ is convergent for every $\widehat{x} = \{x_n\}_{n\geqslant 1} \in l^\infty$, which of course is not true. Therefore the sequence $\{u_n\}_{n\geqslant 1}$ cannot have any w^*-convergent subsequence and so the Eberlein–Smulian theorem is not true for the w^*-topology.

Solution of Problem 5.99

By the Alaoglu theorem (see Theorem 5.66) and Theorem 5.85(a), the dual unit ball $\overline{B}_1^* = \{x^* \in X^* : \|x^*\|_* \leqslant 1\}$ furnished with the w^*-topology is a compact metric space. So, we can find a sequence $\{u_k^*\}_{k\geqslant 1} \subseteq \overline{B}_1^*$ such that $\overline{B}_1^* = \overline{\{u_k^*\}_{k\geqslant 1}}^{w^*}$. Let $A\colon X \longrightarrow l^\infty$ be defined by

$$A(x) = \{ \langle u_k^*, x\rangle \}_{k\geqslant 1} \qquad \forall\, x \in X.$$

Then

$$\|x\| = \sup_{k\geqslant 1} \big| \langle u_k^*, x\rangle \big| = \big\|A(x)\big\|_{l^\infty}$$

and so A is a linear isometry into l^∞.

Solution of Problem 5.100

Let $\{x_n\}_{n\geqslant 1} \subseteq C$ be a sequence and let $V = \overline{\operatorname{span}}\, \{x_n\}_{n\geqslant 1}$. Then V is a separable, closed subspace of X. So, by hypothesis the set $C\cap V$ is w-compact. Since $\{x_n\}_{n\geqslant 1} \subseteq C \cap V$, from the Eberlein–Smulian theorem (see Theorem 5.78), we can find a subsequence $\{x_{n_k}\}_{k\geqslant 1}$ of the sequence $\{x_n\}_{n\geqslant 1}$ such that $x_{n_k} \xrightarrow{w} x$ in V, hence in X too. We have that $x \in C\cap V$ and so once again the Eberlein–Smulian theorem implies that the set C is w-compact.

Solution of Problem 5.101

Since by hypothesis, X is w-separable, we can find a countable set $D \subseteq X$ such that $\overline{D}^w = X$. Consider the set

$$E = \operatorname{span}_{\mathbb{Q}} D$$

(the set of linear combinations of elements of D with $\mathbb{Q}$-coefficients). Then the set E is countable and norm dense in $V = \operatorname{span} D$. From the

Mazur theorem (see Theorem 5.58), we have that $X = \overline{V} = \overline{V}^w$ and so $\overline{E} = X$, which proves the separability of X.

Solution of Problem 5.102

Since X is separable, we can find a sequence $\{x_n\}_{n\geqslant 1} \subseteq \partial B_1 = \{x \in X : \|x\| = 1\}$ such that $\partial B_1 = \overline{\{x_n\}_{n\geqslant 1}}$. By Corollary 5.26, we can find a sequence $\{x_n^*\}_{n\geqslant 1} \subseteq \partial B_1^*$ such that

$$\langle x_n^*, x_n \rangle \;=\; 1 \qquad \forall\, n \geqslant 1.$$

If $x \in \partial B_1$, then we can find an index $n_0 \geqslant 1$ such that $\|x - x_{n_0}\| < \frac{1}{2}$. Then, we have

$$\left\langle x_{n_0}^*, x_{n_0} - x \right\rangle \;\leqslant\; \|x_{n_0} - x\| \;<\; \tfrac{1}{2},$$

so

$$\tfrac{1}{2} \;\leqslant\; \left\langle x_{n_0}^*, x \right\rangle.$$

Because $X = \bigcup\limits_{r>0} r\partial B_1$, we conclude that the sequence $\{x_n^*\}_{n\geqslant 1} \subseteq \partial B_1^*$ separates points in X.

Solution of Problem 5.103

Recall that $0 \in \overline{\partial B_1^X}^w$, where $\partial B_1^X = \{x \in X : \|x\|_{l^1} = 1\}$ (see Problem 5.55). Suppose that X^* is separable. Then by Theorem 5.85(b), the set $(\overline{B}_1^X, w)$ is metrizable. So, we can find a sequence $\{x_n\}_{n\geqslant 1} \subseteq \partial B_1^X$ such that $x_n \xrightarrow{w} 0$ in l^1. Since l^1 has the Schur property (see Remark 5.57), we have that $x_n \longrightarrow 0$ in l^1, a contradiction.

Solution of Problem 5.104

Let $\overline{B}_1^* = \{x^* \in X^* : \|x^*\|_* \leqslant 1\}$ and let $(\overline{B}_1^*, w^*)$ denote the set $\overline{B}_1^*$ equipped with the relative w^*-topology. From the Alaoglu theorem (see Theorem 5.66) and from Theorem 5.85(a), we know that the set $(\overline{B}_1^*, w^*)$ is compact and metrizable, hence it is a Polish space (see Definition 2.150). Then

$$X^*_{w^*} = \bigcup_{n\geqslant 1} n(\overline{B}^*_1, w^*)$$

is a Souslin space (see Definition 2.156 and Proposition 2.159(c)).

Solution of Problem 5.105

Let X be a separable Banach space and let $\{x_n\}_{n\geqslant 1}$ be a sequence dense in $\overline{B}_1 = \{x \in X : \|x\| \leqslant 1\}$. Let $A\colon l^1 \longrightarrow X$ be defined by

$$A(\widehat{\lambda}) = \sum_{n\geqslant 1} \lambda_n x_n \qquad \forall\, \widehat{\lambda} = \{\lambda_n\}_{n\geqslant 1} \in l^1.$$

We have

$$\|A(\widehat{\lambda})\| \leqslant \sum_{n\geqslant 1} |\lambda_n| \|x_n\| \leqslant \sum_{n\geqslant 1} |\lambda_n| = \|\widehat{\lambda}\|_{l^1},$$

so $A \in \mathcal{L}(l^1; X)$.

We claim that A is surjective. Due to the linearity of A, it suffices to show that it is onto $\overline{B}_1$. So, let $x \in \overline{B}_1$. We choose $x_{n_1} \in \{x_n\}_{n\geqslant 1}$ such that $\|x - x_{n_1}\| < \frac{1}{2}$ and then we pick $x_{n_2} \in \{x_n\}_{n\geqslant 1}$ such that $\|2(x - x_{n_1}) - x_{n_2}\| < \frac{1}{2}$ (so also $\|x - (x_{n_1} + \frac{1}{2} x_{n_2})\| < \frac{1}{2^2}$) and $n_2 > n_1$. Suppose that $x_{n_1}, \ldots, x_{n_k}$ have been chosen with $n_1 < n_2 < \ldots < n_k$. We choose $x_{n_{k+1}} \in \{x_n\}_{n\geqslant 1}$ such that

$$\Big\|x - \sum_{i=1}^{k+1} \tfrac{1}{2^{i-1}} x_{n_i}\Big\| < \tfrac{1}{2^{k+1}} \quad \text{and} \quad n_{k+1} > n_k.$$

Let $\lambda_n = 0$, if $n \neq n_k$ for all $k \geqslant 1$ and $\lambda_n = \frac{1}{2^{k-1}}$ if $n = n_k$. Then $\widehat{\lambda} = \{\lambda_n\}_{n\geqslant 1} \in l^1$ and $A(\widehat{\lambda}) = x$. If $N = N(A)$, then $\widehat{A}\colon l^1/_N \longrightarrow X$, defined by

$$\widehat{A}([\widehat{\lambda}]) = A(\widehat{\lambda})$$

is the desired isomorphism.

Solution of Problem 5.106

"$\Longrightarrow$": Let $K = \overline{C}$ and let us proceed indirectly. So, suppose that the sequence $\{x_n^*\}_{n\geqslant 1} \subseteq X^*$ does not w^*-converge to 0 uniformly in K. Then we can find $\varepsilon > 0$ and $\{x_n\}_{n\geqslant 1} \subseteq K$ such that

$$\left|\langle x_n^*, x_n\rangle\right| \geqslant \varepsilon \qquad \forall\, n \geqslant 1.$$

Since K is compact, by passing to a suitable subsequence if necessary, we may assume that $x_n \longrightarrow x \in K$. Also, since $x_n^* \xrightarrow{w^*} 0$, we can find $M > 0$ such that

$$\|x_n^*\|_* \leqslant M \qquad \forall\, n \geqslant 1.$$

Let $n_0 \geqslant 1$ be such that

$$\|x_n - x\| \leqslant \tfrac{\varepsilon}{2M} \qquad \forall\, n \geqslant n_0.$$

Then

$$\begin{aligned}\left|\langle x_n^*, x\rangle\right| &\geqslant \left|\langle x_n^*, x_n\rangle\right| - \left|\langle x_n^*, x_n - x\rangle\right| \\ &\geqslant \varepsilon - \|x_n^*\|_*\|x_n - x\| \geqslant \tfrac{\varepsilon}{2} \qquad \forall\, n \geqslant n_0,\end{aligned}$$

which contradicts the fact that $x_n^* \xrightarrow{w^*} 0$.

"$\Longleftarrow$": Let $K = \left(\overline{B}_1^*, w^*\right)$, where $\overline{B}_1^* = \left\{x^* \in X^* : \ \|x^*\|_* \leqslant 1\right\}$ and w^* denotes the relative w^*-topology on $\overline{B}_1^*$. By the Alaoglu theorem (see Theorem 5.66) and of Theorem 5.85(a), the set K is compact metrizable. From Problem 5.73 (see its solution), we know that there exists an isometric isomorphism A of X into $C(K)$. Suppose that C is not bounded. Then we can find a sequence $\{x_n\}_{n\geqslant 1} \subseteq C$ such that $\|x_n\| > n$ for all $n \geqslant 1$. Let $u_n^* \in X^*$, $\|u_n^*\|_* = 1$ be such that

$$\left|\langle u_n^*, x_n\rangle\right| > n \qquad \forall\, n \geqslant 1.$$

We set $v_n^* = \frac{u_n^*}{\sqrt{n}}$. Then $\|v_n^*\|_* \longrightarrow 0$ and in particular $v_n^* \xrightarrow{w^*} 0$, but $\left|\langle v_n^*, x_n\rangle\right| > \sqrt{n}$ for all $n \geqslant 1$, contradicting our hypothesis. So, C is norm bounded. Hence, we can find $M > 0$ such that $\|x\| \leqslant M$ for all $x \in C$. Then

$$\|A(x)\| \leqslant \|A\|_{\mathcal{L}}\|x\| \leqslant M\|A\|_{\mathcal{L}} \qquad \forall\, x \in C,$$

so $A(C) \subseteq C(K)$ is equibounded.

We will show that the set $A(C)$ is also equicontinuous. Let $x_n^* \longrightarrow x^*$ in $K = \overline{C}$ and arguing by contradiction, suppose that we can find $\varepsilon > 0$ and a sequence $\{x_n\}_{n\geqslant 1} \subseteq C$ such that

$$\big|\langle x_n^* - x^*, x_n\rangle\big| \geqslant \varepsilon \qquad \forall\, n \geqslant 1.$$

Since $x_n^* - x^* \stackrel{w^*}{\longrightarrow} 0$ in X^*, this contradicts our hypothesis. So, $A(C) \subseteq C(K)$ is equicontinuous. By the Arzela–Ascoli theorem (see Theorem 2.181), the set $A(C)$ is relatively compact in $C(K)$ and since A is an isometric isomorphism, C is relatively compact in X.

Solution of Problem 5.107

Clearly the set C is also w^*-compact and $id\colon (C,w) \longrightarrow (C,w^*)$ is a continuous bijection, hence a homeomorphism. From Theorem 5.85(a), (C,w^*) is compact metrizable, hence separable. Then so is the homeomorphic space (C,w). Let $D \subseteq C$ be a countable set such that $\overline{D}^w = C$. Let $V_0 = \operatorname{span}_{\mathbb{Q}} D$ ($\operatorname{span}_{\mathbb{Q}}$ denoting rational linear combinations). Then V_0 is countable and $\overline{V}_0 = V = \overline{\operatorname{span}}\, D$. Hence V is norm separable, closed vector subspace of X^*. Note that $C \subseteq V$ and so C is norm separable.

Solution of Problem 5.108

First suppose that X^* is separable. Setting $\overline{B}_1 = \{x \in X : \|x\| \leqslant 1\}$, from Theorem 5.85(b), we know that $(\overline{B}_1, w)$ (w denoting the relative weak topology on $\overline{B}_1$) is metrizable. Moreover, from Problem 5.55, we know that $\overline{\partial B_1}^w = \overline{B}_1$. So, we can find a sequence $\{x_n\}_{n\geqslant 1} \subseteq \partial B_1$ such that $x_n \stackrel{w}{\longrightarrow} x$ in X.

Now suppose that X is reflexive. Since X is infinite dimensional, we can find $\{u_n\}_{n\geqslant 1} \subseteq X$ linearly independent. Let $V = \overline{\operatorname{span}}\,\{u_n\}_{n\geqslant 1}$.

Then V is infinite dimensional, separable reflexive and so we are back to the previous case.

Solution of Problem 5.109

Evidently V is separable. Let $\{v_n\}_{n\geqslant 1}$ be dense in V and let $E_k = \operatorname{span}\{v_n\}_{n=1}^k$. Then $\{E_k\}_{k\geqslant 1}$ is an increasing sequence of finite dimensional vector subspaces of V and the set $\bigcup\limits_{k\geqslant 1} E_k$ is dense in H. Let $e_i \in E_i$ be such that $\|e_i\| = 1$. If $E_2 \neq E_1$, then we can find $e_2 \in E_2$ such that $\{e_1, e_2\}$ is an orthonormal basis of E_2. Continuing this construction, we produce an orthonormal basis $\{e_n\}_{n\geqslant 1} \subseteq \bigcup\limits_{k\geqslant 1} E_n \subseteq V$ of H.

Solution of Problem 5.110

Let $[x], [y] \in X/_V$ be such that

$$\big\|[x]\big\| \leqslant 1, \quad \big\|[y]\big\| \leqslant 1, \quad \text{and} \quad \big\|[x]-[y]\big\| \geqslant \varepsilon.$$

Since X is reflexive, we can find $u_1, u_2 \in X$ such that

$$\|x-u_1\| \leqslant 1 \quad \text{and} \quad \|x-u_2\| \leqslant 1$$

(see Problem 5.80). Also, we have

$$\big\|(x-y)-u\big\| \geqslant \varepsilon \qquad \forall\, u \in V$$

(see Definition 5.18). The uniform convexity of X implies that

$$\tfrac{1}{2}\big\|(x-u_1)+(y-u_2)\big\| < 1-\delta$$

(see Definition 5.88), so $\frac{1}{2}\big|[x]+[y]\big| < 1-\delta$ which proves that $X/_V$ is uniformly convex (see Definition 5.88).

Solution of Problem 5.111

By Remark 5.94, it suffices to show that the parallelogram law fails for the supremum norm $\|\cdot\|_\infty$. To this end, let

$$u_1(t) = \min\{0,t\} \quad \text{and} \quad u_2(t) = t \qquad \forall\, t \in [-1,1].$$

Then $u_1, u_2 \in C([-1,1])$. We have $\|u_1\|_\infty = \|u_2\|_\infty = 1$, $\|u_1+u_2\|_\infty = 2$ and $\|u_1 - u_2\| = 1$. So, the parallelogram law fails.

Solution of Problem 5.112

Let $\xi > 0$ be such that $\xi M^2 < 1$. By the Egorov theorem (see Theorem 3.76), we can find a Lebesgue measurable subset D of T such that $|T \setminus D| < \xi$ (by $|\cdot|$ we denote the Lebesgue measure on $\mathbb{R}$) and $\lambda_n u_n \rightrightarrows 0$ on D. Then

$$\begin{aligned}\lambda_n^2 &= \int_0^1 \big(\lambda_n u_n(t)\big)^2\, dt = \int_D \big(\lambda_n u_n(t)\big)^2\, dt + \int_{T\setminus D} \big(\lambda_n u_n(t)\big)^2\, dt \\ &\leqslant \sup_D (\lambda_n u_n)^2 + M^2\lambda_n^2\xi \qquad \forall\, n \geqslant 1,\end{aligned}$$

so

$$(1 - M^2\xi)\lambda_n^2 \leqslant \sup_D (\lambda_n u_n)^2 \longrightarrow 0 \quad \text{as } n \to +\infty$$

and thus $\lambda_n \longrightarrow 0$.

Solution of Problem 5.113

By replacing x_n with $x_n - x$ for $n \geqslant 1$, if necessary, we may assume that $x = 0$. Let $n_1 = 1$ and choose integer $n_2 > 1$ such that

$$\big|\,(x_{n_1}, x_{n_2})\,\big| \leqslant 1$$

(such a choice is possible since $x_{n_k} \xrightarrow{w} 0$ in H). Having chosen $n_1 < n_2 < \ldots < n_k$, we choose $n_{k+1} > n_k$ such that

$$\big(x_{n_i}, x_{n_{k+1}}\big)_H \leqslant \tfrac{1}{k} \qquad \forall\, i \in \{1, \ldots, k\}.$$

Since $x_n \xrightarrow{w} 0$, we can find $M > 0$ such that

$$\|x_n\| \leqslant M \qquad \forall\, n \geqslant 1.$$

Therefore

$$\begin{aligned}\Big\|\tfrac{1}{k}\sum_{i=1}^{k}x_{n_i}\Big\|^2 &= \tfrac{1}{k^2}\Big(\sum_{i=1}^{k}x_{n_i},\sum_{j=1}^{k}x_{n_j}\Big) = \tfrac{1}{k^2}\sum_{i=1}^{k}\sum_{j=1}^{k}(x_{n_i},x_{n_j})\\ &= \tfrac{1}{k^2}\Big(\sum_{i=1}^{k}\|x_{n_i}\|^2+2\sum_{i=1}^{k-1}\sum_{j=i+1}^{k}(x_{n_i},x_{n_j})\Big)\\ &\leqslant \tfrac{1}{k^2}\Big(kM^2+2\sum_{i=1}^{k-1}\sum_{j=i+1}^{k}\tfrac{1}{j-1}\Big)\\ &= \tfrac{kM^2+2k}{k^2} = \tfrac{M^2+2}{k} \longrightarrow 0 \quad \text{as } k\to+\infty.\end{aligned}$$

Solution of Problem 5.114

Let $(\cdot,\cdot)_A : H\times H \longrightarrow \mathbb{R}$ be defined by

$$(x,u)_A = \big(A(x),u\big)_H \qquad \forall\, x,u\in H.$$

Clearly, this is a semi-inner product and from the Cauchy–Schwarz inequality (see Remark 5.94), we have

$$\big|\,(x,u)_A\,\big|^2 \leqslant (x,x)_A\,(u,u)_A\,,$$

so

$$\big|\big(A(x),u\big)_H\big|^2 \leqslant \big(A(x),x\big)_H\big(A(u),u\big)_H.$$

Let $u=A(x)$. Then

$$\begin{aligned}\big\|A(x)\big\|^4 &= \big(A(x),A(x)\big)_H^2 \leqslant \big(A(x),x\big)_H\big(A^2(x),A(x)\big)_H\\ &\leqslant \big(A(x),x\big)_H\big\|A^2(x)\big\|\,\big\|A(x)\big\| \leqslant \big(A(x),x\big)_H\big\|A\big\|_{\mathcal{L}}\big\|A(x)\big\|^2,\end{aligned}$$

so

$$\big\|A(x)\big\|^2 \leqslant \|A\|_{\mathcal{L}}\big(A(x),x\big)_H \qquad \forall\, x\in H.$$

Solution of Problem 5.115

Without any loss of generality, we assume that $\{A_n\}_{n\geqslant 1}$ is increasing (i.e., $A_n \leqslant A_{n+1}$ for all $n \geqslant 1$). The reasoning is similar, if the sequence is decreasing. For $m \geqslant n$ and using Problem 5.114, we have

$$\|A_m(x) - A_n(x)\|^2 \leqslant 2M\big(A_m(x) - A_n(x), x\big)_H \qquad \forall\, x \in H.$$

Note that the sequence $\{(A_n(x), x)\}_{n\geqslant 1}$ is bounded and nondecreasing, hence convergent. Therefore, from the last inequality, it follows that

$$A(x) = \lim_{n\to+\infty} A_n(x) \quad \text{exists for all } x \in H.$$

Then $A \in \mathcal{L}(H)$ (see Corollary 5.42) and clearly it is self-adjoint.

Solution of Problem 5.116

(a) For every $x \in H$, we have

$$\big(A^{2n}(x), x\big) = \big(A^n(x), A^n(x)\big) = \|A^n(x)\|^2 \geqslant 0$$

(since for every $m \geqslant 1$, we have that A^m is self-adjoint), so $A^{2n} \geqslant 0$.

(b) For every $x \in H$, we have

$$\big(A^{2n+1}(x), x\big) = \big(A(A^n(x)), A^n(x)\big) \geqslant 0$$

(since $A \geqslant 0$), so $A^{2n+1} \geqslant 0$.

Solution of Problem 5.117

Consider the identity operator

$$id\colon \big(C([0,1]), \|\cdot\|_\infty\big) \longrightarrow \big(L^2([0,1]), \|\cdot\|_2\big).$$

Since $\|\cdot\|_2 \leqslant \|\cdot\|_\infty$, it follows that id is continuous. Moreover, since V is closed with respect to the norm $\|\cdot\|_2$, it follows that $(V, \|\cdot\|_\infty)$ is closed (inverse image under a continuous map of a closed set). Therefore $(V, \|\cdot\|_2)$ and $(V, \|\cdot\|_\infty)$ are Banach spaces and so the Banach theorem (see Theorem 5.48) implies that $id\colon \big(V, \|\cdot\|_\infty\big) \longrightarrow \big(V, \|\cdot\|_2\big)$ is an isomorphism. Therefore, we can find $c > 1$ such that

$$\|u\|_2 \leqslant \|u\|_\infty \leqslant c\|u\|_2 \qquad \forall\, u \in V.$$

Let $\{u_n\}_{k=1}^n$ be an orthonormal sequence in V ($n \leqslant \dim V$). If $\lambda_1, \lambda_2, \ldots, \lambda_n \in \mathbb{R}$, from the Parseval identity on $\operatorname{span}\{u_k\}_{k=1}^n$ (see Theorem 5.105(b)), we have

$$\Big\|\sum_{k-1}^{n} \lambda_k u_k\Big\|_2^2 = \sum_{k=1}^{n} \lambda_k^2.$$

Thus, we have

$$\Big|\sum_{k-1}^{n} \lambda_k u_k(t)\Big| = c\Big(\sum_{k=1}^{n} \lambda_k^2\Big)^{\frac{1}{2}} \qquad \forall\, t \in [0,1].$$

Let $\lambda_k = u_k(t)$ for every $k = 1, \ldots, n$, $t \in [0,1]$. Then, we have

$$\sum_{k-1}^{n} u_k(t)^2 \leqslant c\Big(\sum_{k=1}^{n} u_k(t)^2\Big)^{\frac{1}{2}},$$

so

$$\sum_{k=1}^{n} u_k(t)^2 \leqslant c^2 \qquad \forall\, t \in [0,1].$$

Integrating both sides over $[0,1]$ and recalling that $\|u_n\|_2 = 1$ for all $n \geqslant 1$, we obtain

$$n \leqslant c^2,$$

so $\dim V \leqslant c^2$, i.e., V is finite dimensional.

Solution of Problem 5.118

Let

$$H = l^2 = \Big\{\widehat{x} = \{x_n\}_{n\geqslant 1} :\ \|\widehat{x}\|_2 = \Big(\sum_{n\geqslant 1} x_n^2\Big)^{\frac{1}{2}} < +\infty\Big\}$$

and let

$$V = \big\{\widehat{x} \in l^2 :\ x_n = 0 \text{ for all but only a finite number of } n\text{'s}\big\}.$$

We know that $H = l^2$ is a Hilbert space and $V \subseteq H$ is a dense vector subspace of H. For $n \geqslant 1$, let $A_n : V \longrightarrow H$ be defined by

$$A_n(\widehat{x}) = \{nx_n\delta_{kn}\}_{n\geqslant 1} \qquad \forall\, \widehat{x} = \{x_n\}_{n\geqslant 1} \in V.$$

Note that

$$A_n(\widehat{x}) \longrightarrow 0 \quad \text{in } H \qquad \forall\, \widehat{x} \in V.$$

So,

$$\text{the sequence } S(\widehat{x}) = \{A_n(\widehat{x})\}_{n\geqslant 1} \text{ is bounded} \qquad \forall\, \widehat{x} \in V.$$

It is clear from the definition of A_n that $\|A_n\|_{\mathcal{L}} \leqslant n$. Also, if $e_n = \{\delta_{kn}\}_{k\geqslant 1}$, then $A_n(e_n) = n$ and so $\|A_n\|_{\mathcal{L}} \geqslant n$, hence $\|A_n\|_{\mathcal{L}} = n$ and so the sequence $\{A_n\}_{n\geqslant 1} \subseteq \mathcal{L}(V;H)$ is not bounded, although it is pointwise bounded.

Solution of Problem 5.119

From the Riesz–Fréchet representation theorem (see Theorem 5.100) and the Banach–Steinhaus theorem (see Corollary 5.40), we have that

$$\sup_{n\geqslant 1} \|A_n(x)\| < +\infty \qquad \forall\, x \in H.$$

Using once more Corollary 5.40, we infer that

$$\sup_{n\geqslant 1} \|A_n\|_{\mathcal{L}} < +\infty.$$

Solution of Problem 5.120

Let $x \in H$. Then from the Parseval equality (see Theorem 5.105(b)), we have

$$\|x\|^2 = \sum_{n\geqslant 1} |(x, e_n)_H|^2,$$

so

$$(x, e_n)_H \longrightarrow 0 \qquad \forall\, x \in H.$$

From the Riesz–Fréchet theorem (see Theorem 5.100), we have

$$e_n \xrightarrow{w} 0 \quad \text{in } H.$$

Solution of Problem 5.121

Let $u = \operatorname{proj}_C(x)$ and let $\xi\colon \mathbb{R} \longrightarrow H$ be defined by

$$\xi(t) = tx + (1-t)u \qquad \forall\, t \in \mathbb{R}.$$

Evidently ξ is continuous and $\xi(0) = u$. Arguing by contradiction, suppose that $u \in \operatorname{int} C$. Then $\xi(0) \in \operatorname{int} C$ and by the continuity of ξ, we can find $\delta \in (0,1)$ small such that

$$\xi(t) \in C \qquad \forall\, t \in [-\delta, \delta].$$

Hence

$$\xi(\delta) = \delta x + (1-\delta)u \in C,$$

so

$$\operatorname{dist}(x, C) \leqslant \big\| x - \big(\delta x + (1-\delta)u\big)\big\| = (1-\delta)\|x-u\|,$$

thus

$$\|x-u\| \leqslant (1-\delta)\|x-u\|,$$

a contradiction since $\delta \in (0,1)$.

So, $u \notin \operatorname{int} C$ and since $u \in C$, we infer that $u \in \partial C$. Moreover, $\operatorname{dist}(x, C) = \operatorname{dist}(x, \partial C)$.

Solution of Problem 5.122

Let $x \in H$. By Theorem 5.97, we can find a unique $u_n \in C_n$ such that

$$m_n = \|x - u_n\| = \operatorname{dist}(x, C_n) \qquad \forall\, n \geqslant 1.$$

Since the sequence $\{C_n\}_{n\geqslant 1}$ is decreasing, the sequence $\{m_n\}_{n\geqslant 1}$ is increasing and bounded above by $m = \operatorname{dist}(x, C)$. So, we have $m_n \nearrow \widehat{m}$. From the parallelogram law (see Remark 5.94), we have

$$\big\|x - \tfrac{u_n+u_m}{2}\big\|^2 + \big\|\tfrac{u_n-u_m}{2}\big\|^2 = \tfrac{1}{2}\big(\|x-u_n\|^2 + \|x-u_m\|^2\big) \qquad \forall\, m \geqslant n \geqslant 1,$$

so

$$\|u_n - u_m\|^2 \leqslant 2\big(m_n^2 - m_m^2\big) \qquad \forall\, m \geqslant n$$

and thus $u_n \longrightarrow u$ in H. Evidently $u \in C_n$ for all $n \geqslant 1$ (recall that the sequence $\{C_n\}_{n\geqslant 1}$ is decreasing) and so $u \in C$. We have

$$m_n = \|x - u_n\| \leqslant \|x - y\| \qquad \forall\, y \in C_n,\ n \geqslant 1,$$

so

$$\|x - u_n\| \leqslant \|x - y\| \qquad \forall\, y \in C$$

and thus

$$\|x - u\| \leqslant \|x - y\| \qquad \forall\, y \in C,$$

so

$$\widehat{m} = m = \operatorname{dist}(x, C) = \|x - u\|, \quad \text{i.e., } \operatorname{proj}_C(x) = u.$$

Finally, we conclude that $\operatorname{proj}_{C_n} \longrightarrow \operatorname{proj}_C$.

Solution of Problem 5.123
We know that

$$\|A^*A\|_{\mathcal{L}} \leqslant \|A^*\|_{\mathcal{L}}\|A\|_{\mathcal{L}} = \|A\|_{\mathcal{L}}^2$$

(see Proposition 5.110). Also, for all $x \in H$, we have

$$\big\|A(x)\big\|^2 = \big(A(x), A(x)\big)_H = \big(A^*A(x), x\big)_H \leqslant \|A^*A\|_{\mathcal{L}}\|x\|^2,$$

so

$$\|A\|_{\mathcal{L}}^2 \leqslant \|A^*A\|_{\mathcal{L}}$$

and thus

$$\|A\|_{\mathcal{L}}^2 = \|A^*A\|_{\mathcal{L}}.$$

Solution of Problem 5.124
Since A is normal (see Definition 5.108(c)), we have

$$\big(AA^*(u), u\big)_H = \big(A^*A(u), u\big)_H \qquad \forall\, u \in H,$$

so

$$\big\|A(u)\big\| = \big\|A^*(u)\big\| \qquad \forall\, u \in H.$$

Then, using the last equality with $u = A(x)$, for every $x \in H$, we have

$$\begin{aligned} \|A(x)\|^2 &= \big(A(x), A(x)\big)_H = \big(A^*A(u), u\big)_H \\ &\leqslant \|A^*A(x)\|\,\|x\| \\ &= \|A^2(x)\|\,\|x\|. \end{aligned}$$

Solution of Problem 5.125

Let $\{x_n\}_{n\geqslant 1} \subseteq \overline{B}_1 = \big\{x \in H : \|x\| \leqslant 1\big\}$ be a sequence. From Problem 5.124, we have

$$\begin{aligned} \|A(x_n) - A(x_m)\|^2 &\leqslant \|A^2(x_n) - A^2(x_m)\|\,\|x_n - x_m\| \\ &\leqslant 2\|A^2(x_n) - A^2(x_m)\| \qquad \forall\, n, m \geqslant 1. \end{aligned}$$

So, if $\big\{A^2(x_n)\big\}_{n\geqslant 1}$ is a Cauchy sequence, then $\big\{A(n_n)\big\}_{n\geqslant 1}$ is a Cauchy sequence too. Because A^2 is compact, by passing to a subsequence if necessary, we have that the sequence $\big\{A^2(x_n)\big\}_{n\geqslant 1}$ is convergent. Then the sequence $\big\{A(x_n)\big\}_{n\geqslant 1}$ is convergent (as a Cauchy sequence), which proves that A is compact.

Solution of Problem 5.126

"**(b)** $\iff$ **(c)**": Follows from the polarization identity (see Remark 5.94).

"**(a)** $\Longrightarrow$ **(c)**": Suppose that $A \in \mathcal{L}(H)$ is unitary. Then

$$\big(A(x), A(u)\big)_H = \big(A^*A(x), u\big)_H = (x, u)_H \qquad \forall\, x, u \in H.$$

"**(c)** $\Longrightarrow$ **(a)**": Suppose that A preserves the inner product. Then, we have

$$\big(A^*A(x), u\big)_H = (x, u)_H \qquad \forall\, x, u \in H$$

so

$$A^*A(x) = x \qquad \forall\, x \in H.$$

Since A is surjective and an isometry, $A^{-1} \in \mathcal{L}(H)$ (see the Banach theorem; Theorem 5.48). So $A^{-1} = A^*$, i.e., A is unitary.

Solution of Problem 5.127

Let λ_1, λ_2 be two distinct eigenvalues of A and let $u_1, u_2 \in H \setminus \{0\}$ be corresponding eigenvectors. We have

$$A(u_1) = \lambda_1 u_1 \quad \text{and} \quad A(u_2) = \lambda_2 u_2.$$

Then

$$\big(A(u_1), u_2\big)_H = \lambda_1 (u_1, u_2)_H$$

and

$$\big(A(u_1), u_2\big)_H = \big(u_1, A(u_2)\big)_H = \big(u_1, \lambda_2 u_2\big)_H = \lambda_2 (u_1, u_2)_H .$$

Therefore

$$(\lambda_1 - \lambda_2)(u_1, u_2)_H = 0,$$

so $(u_1, u_2)_H = 0$ (since $\lambda_1 \neq \lambda_2$).

Solution of Problem 5.128

"(a) $\Longrightarrow$ (b)": From Theorem 5.97, we know that

$$u \in C \quad \text{and} \quad (x - u, c - u)_H \leqslant 0 \qquad \forall\, c \in C. \tag{5.16}$$

Note that $0 \in C$ and $2u \in C$. Therefore, from (5.16), we have

$$(x - u, -u)_H \leqslant 0 \quad \text{and} \quad (x - u, u) \leqslant 0,$$

so

$$(x - u, u)_H = 0,$$

i.e., $x - u \perp u$.

So, from (5.16), we have

$$(x - u, c)_H \leqslant 0 \qquad \forall\, c \in C.$$

"**(b)** $\Longrightarrow$ **(a)**": Since $x - u \perp u$, we have that $(x-u,u)_H = 0$. Hence

$$(x-u, c-u)_H \leqslant 0 \quad \text{and} \quad u \in C,$$

so $u = \operatorname{proj}_C(x)$ (see Theorem 5.97).

Solution of Problem 5.129

Since S is uniformly continuous (see Proposition 5.12), it admits an extension $\widehat{S} \in \mathcal{L}(\overline{V};X)$ (see Theorem 1.47). Let $\widehat{p} = \operatorname{proj}_{\overline{V}} \in \mathcal{L}(H)$ be the orthogonal projection onto $\overline{V}$ and let us set $A = \widehat{S} \circ \widehat{p}$. Then $A \in \mathcal{L}(H;X)$ and $A(x) \in \widehat{S}(\widehat{p}(x)) = \widehat{S}(x)$ for all $x \in V$. We have

$$\|A\|_{\mathcal{L}} \leqslant \|\widehat{S}\|_{\mathcal{L}}\|\widehat{p}\|_{\mathcal{L}} = \|\widehat{S}\|_{\mathcal{L}} = \|S\|_{\mathcal{L}}.$$

On the other hand, since A is an extension of S on H, we have

$$\|A\|_{\mathcal{L}} \geqslant \|S\|_{\mathcal{L}},$$

so

$$\|A\|_{\mathcal{L}} = \|S\|_{\mathcal{L}}.$$

Solution of Problem 5.130

Let H be any infinite dimensional Hilbert space. Let $\{e_n\}_{n\geqslant 1}$ be an orthonormal sequence and let $\{\lambda_n\}_{n\geqslant 1} \subseteq [1,+\infty)$ be such that

$$\sum_{n\geqslant 1} \frac{1}{\lambda_n^2} = +\infty \quad \text{and} \quad \lambda_n \nearrow +\infty$$

(for example, we can choose $\lambda_n = \sqrt{n}$). Let $C = \{\lambda_n e_n : n \geqslant 1\}_{n\geqslant 1}$. We have

$$\|\lambda_n e_n - \lambda_m e_m\|^2 \geqslant \sqrt{2} \qquad \forall\, n,m \geqslant 1,\ n \neq m.$$

So, every Cauchy sequence in C is eventually constant, hence C is closed.

Let $\{u_n\}_{n\geqslant 1} \subseteq C$ be a weakly convergent sequence. Then it is bounded and so, since $\lambda_n \nearrow +\infty$, the sequence must be eventually constant, hence the set C is sequentially weakly closed.

Let U be a weak neighbourhood of 0. We may assume that

$$U = \{x \in H : \left|(x, y_k)_H\right| < \varepsilon \text{ for all } k = 1, \ldots, m\},$$

for some $\varepsilon > 0$ and $\{y_k\}_{y_k}^m \subseteq H$. Let

$$\vartheta_n = \sum_{k=1}^m \left|(y_k, e_n)_H\right|.$$

We have

$$\{\left|(y_k, e_n)_H\right|\}_{n\geqslant 1} \subseteq l^2 \qquad \forall\, k \in \{1, \ldots, m\},$$

so

$$\sum_{n\geqslant 1} \vartheta_n^2 < +\infty = \sum_{n\geqslant 1} \tfrac{1}{\lambda_n^2}.$$

So, the set

$$D = \{n \geqslant 1 : \vartheta_n < \tfrac{\varepsilon}{\lambda_n}\}$$

is infinite. For every $k \in \{1, \ldots, m\}$ and every $n \in D$, we have

$$\left|(y_k, \lambda_n e_n)_H\right| = \lambda_n\left|(y_k, e_n)_H\right| \leqslant \lambda_n \vartheta_n < \varepsilon,$$

so $\{\lambda_n e_n\}_{n\in D} \subseteq U$. Thus $0 \in \overline{C}^w$ and so the set C is not weakly closed.

Solution of Problem 5.131

For every $x^* \in X^*$ and $x \in X$, we have

$$\begin{aligned} \langle (P^*)^2(x^*), x\rangle &= \langle P^*(x^*), P(x)\rangle = \langle x^*, P^2(x)\rangle \\ &= \langle x^*, P(x)\rangle = \langle P^*(x^*), x\rangle, \end{aligned}$$

so $(P^*)^2 = P^*$, i.e., P^* is a projection in X^*.

Solution of Problem 5.132
Since $P \in \mathcal{L}(H)$ is a projection, we have $P^2 = P$. Therefore, for every $x \in H$, we have

$$\big(P(x), x\big) \;=\; \big(P^2(x), x\big) \;=\; \big(P(x), P(x)\big) \;=\; \big\|P(x)\big\|^2 \;\geqslant\; 0$$

(since P is self-adjoint being an orthogonal projection; see Problem 5.114), so $P \geqslant 0$.

Also, since P is an orthogonal projection, we have

$$H \;=\; \ker P \oplus P(H), \quad \ker P \perp P(H), \quad \ker P = (id - P)(H).$$

Hence

$$\|x\|^2 \;=\; \big\|P(x)\big\|^2 + \big\|(id - P)(x)\big\|^2,$$

so

$$\big\|P(x)\big\|^2 \;\leqslant\; \|x\|^2 \qquad \forall\, x \in H,$$

therefore $P \leqslant id$.

Solution of Problem 5.133
For every $t \in [0,1]$, we have $\big\|tx + (1-t)u\big\| \leqslant 1$. Let $y = \frac{x+y}{2}$. Then $\|y\| \;=\; \frac{1}{2}\|x + u\| \;=\; \frac{1}{2}\big(\|x\| + \|u\|\big) \;=\; 1$. Suppose that there exists $v \in tx + (1-t)y$ with $t \in (0,1)$ such that $\|v\| < 1$. The vector y belongs in the segment which contains x and v. Let $\lambda \in (0,1)$ be such that $y = \lambda v + (1-\lambda)x$. We have

$$\|y\| \;\leqslant\; \lambda\|v\| + (1-\lambda)\|x\| \;<\; 1,$$

a contradiction.

Solution of Problem 5.134
Let

$$|x| \;=\; \sup\Big\{\Big\|\sum_{n=1}^{m} \lambda_n e_n\Big\| \;:\; m \in \mathbb{N}\Big\}.$$

It is easy to check that $|\cdot|$ is a norm on X and $(X, |\cdot|)$ is a Banach space. We have

$$\|x\| \;=\; \Big\|\sum_{n\geqslant 1} \lambda_n e_n\Big\| \;\leqslant\; \sup\Big\{\Big\|\sum_{n=1}^{m} \lambda_n e_n\Big\| \;:\; m \in \mathbb{N}\Big\} \;=\; |x|.$$

Invoking Corollary 5.49, we have that $\|\cdot\|$ and $|\cdot|$ are equivalent norms and so there exists $c > 0$ such that

$$|x| = \sup\Big\{\Big\|\sum_{n=1}^{m}\lambda_n e_n\Big\| : m \in \mathbb{N}\Big\} \leqslant c\|x\| = c\Big\|\sum_{n\geqslant 1}\lambda_n e_n\Big\|$$

for all $x = \sum\limits_{n\geqslant 1}\lambda_n e_n$. Evidently $c \geqslant 1$.

Solution of Problem 5.135

"$\Longrightarrow$": We assume that P is an orthogonal projection onto a closed vector subspace V. If $u \in H$, then $u = v + y$ with $v \in V$ and $y \in Y$. We have

$$P\big(P(u)\big) = P(v) = v = P(u),$$

so $P^2 = P$.

Also, if $\widehat{u} = \widehat{v} + \widehat{y}$, with $\widehat{v} \in V$, $\widehat{y} \in V^{\perp}$, then

$$\big(P(u), \widehat{u}\big)_H = (v, \widehat{u})_H = (u, \widehat{v})_H = \big(u, P(\widehat{u})\big)_H,$$

so $P = P^*$, i.e., P is self-adjoint. "$\Longleftarrow$": Let $V = R(P)$. Note that $V = N(id - P)$ and so V is closed in H. Also, from Corollary 5.121, we have

$$V^{\perp} = N(P).$$

Therefore, for a given $v \in V$ and $y \in V^{\perp}$, we have

$$P(v + y) = P(v) + P(y) = P(v) = v,$$

so P is the orthogonal projection onto V.

Solution of Problem 5.136

Let $A \colon X \longrightarrow Y^*$ be the linear operator, defined by

$$\big\langle A(x), y\big\rangle_Y = a(x, y) \qquad \forall\, y \in Y.$$

Then, from hypothesis **(i)**, we have

$$\left|\left\langle A(x), y\right\rangle_Y\right| \leqslant M\|x\|_X\|y\|_Y,$$

so

$$\left\|A(x)\right\|_Y \leqslant M\|x\|_X,$$

i.e., $A \in \mathcal{L}(X;Y^*)$. Similarly, $\widehat{A} \in \mathcal{L}(Y;X^*)$, where

$$\left\langle \widehat{A}(y), x\right\rangle_X = a(x,y) \qquad \forall\, x \in X.$$

Hypotheses **(ii)** and **(iii)** imply that

$$\left\|\widehat{A}(y)\right\|_{X^*} \geqslant c_1\|y\|_Y \qquad \forall\, y \in Y$$

and

$$\left\|A(x)\right\|_{Y^*} \geqslant c_2\|x\|_X \qquad \forall\, x \in X.$$

By Theorems 5.124 and 5.125 and the above estimates, we have that

$$A \text{ and } \widehat{A} \text{ are both bijective.}$$

Therefore, there exist unique $x_0 \in X$ and $y_0 \in Y$ such that

$$A(x_0) = y^* \quad \text{and} \quad \widehat{A}(y_0) = x^*,$$

so

$$\left\langle A(x_0), y\right\rangle_Y = a(x_0, y) = \langle y^*, y\rangle_Y \qquad \forall\, y \in Y$$

and

$$\left\langle \widehat{A}(y_0), x\right\rangle_X = a(x, y_0) = \langle x^*, x\rangle_X \qquad \forall\, x \in X.$$

Solution of Problem 5.137
First note that the fact that

$$\left\langle A(x), x\right\rangle \geqslant c\|x\|^2 \qquad \forall\, x \in H,$$

implies that $N(A) = \{0\}$ and so A is injective. From the last estimate, we have

$$\left\langle x, A^*(x)\right\rangle \geqslant c\|x\|^2,$$

so

$$\|A^*(x)\| \geqslant c\|x\|$$

and thus

$$N(A^*) = \{0\}.$$

From Corollary 5.121(d), we know that

$$\overline{R(A)} = N(A^*)^{\perp}.$$

So, $\overline{R(A)} = H$ and we infer that $R(A)$ is dense in the Hilbert space H. Since

$$\|A(x)\| \geqslant c\|x\| \qquad \forall\, x \in H,$$

it follows that $R(A)$ is also closed in H. Therefore, $R(A) = H$, i.e., A is surjective, hence a bijection. Invoking the Banach theorem (see Theorem 5.48), we conclude that A is an isomorphism.

Solution of Problem 5.138

"**(a) $\Longrightarrow$ (b)**": Since A is an isomorphism, $R(A) = Y$. Also

$$\|x\|_X = \|A^{-1}(A(x))\|_X \leqslant \|A^{-1}\|_{\mathcal{L}}\|A(x)\|_Y \qquad \forall\, x \in X,$$

so

$$c\|x\|_X \leqslant \|A(x)\|_Y \qquad \forall\, x \in X$$

and with $c = \|A^{-1}\|_{\mathcal{L}}^{-1}$.

"**(b) $\Longrightarrow$ (a)**": From Theorem 5.125, we know that $R(A)$ is closed. Since by hypothesis, the set $R(A)$ is dense in Y, it follows that $R(A) = Y$. If $x \in N(A)$, then $A(x) = 0$ and so

$$0 = \|A(x)\| \geqslant c\|x\|,$$

hence $x = 0$. Therefore, A is a bijection and so the Banach theorem (see Theorem 5.48) implies that A is an isomorphism.

Solution of Problem 5.139

No. Indeed, suppose that we could find surjection $A \in \mathcal{L}(l^2; l^1)$. Invoking Theorem 5.124, we can find $c > 0$ such that

$$\|\widehat{u}^*\|_{l^\infty} \leqslant c\big\|A^*(\widehat{u}^*)\big\|_{l^2} \qquad \forall\, \widehat{u}^* \in l^\infty$$

(since $A^* \in \mathcal{L}(l^\infty, l^2)$ and A^* is injective).

Therefore, a closed vector subspace of l^2 is isomorphic to l^∞. But this cannot happen since l^2 is separable and l^∞ is nonseparable.

Solution of Problem 5.140

Let $A \in \mathcal{D}(X;Y)$. By Theorem 5.125, we can find $c > 0$ such that

$$\big\|A(x)\big\|_Y \geqslant c\|x\|_X \qquad \forall\, x \in X.$$

Let $S \in \mathcal{L}(X;Y)$ be such that

$$\|A \quad S\|_{\mathcal{L}} < \tfrac{c}{2}.$$

Then for all $x \in X$, we have

$$\begin{aligned}\big\|S(x)\big\|_Y &= \big\|(S-A)(x) + A(x)\big\|_Y \\ &\geqslant \big\|A(x)\big\|_Y - \big\|(S-A)(x)\big\|_Y \\ &\geqslant c\|x\|_X - \|S-A\|_{\mathcal{L}}\|x\|_X \\ &\geqslant \tfrac{c}{2}\|x\|_X,\end{aligned}$$

so $S \in \mathcal{D}(X;Y)$ (by Theorem 5.125). This proves that $\mathcal{D}(X;Y)$ is an open subset of $\mathcal{L}(X;Y)$.

Solution of Problem 5.141

Let $\vartheta\colon \mathcal{L}(X;Y) \longrightarrow \mathcal{L}(Y^*;X^*)$ be defined by

$$\vartheta(A) = A^* \qquad \forall\, A \in \mathcal{L}(X;Y).$$

Evidently ϑ is linear and an isometry (see Proposition 5.34). Moreover, by Theorem 5.124, we have

$$\mathcal{Y}(X;Y) = \vartheta^{-1}(\mathcal{D}(Y^*;X^*))$$

(see Proposition 5.34). From Problem 5.140, we know that $\mathcal{D}(Y^*;X^*)$ is open. Hence $\mathcal{Y}(X;Y)$ is open in $\mathcal{L}(X;Y)$.

Solution of Problem 5.142

By hypothesis, we have

$$\big(A(x),x\big)_H \geqslant \|x\|^2$$

and so by Theorem 5.125, we have that $N(A) = \{0\}$ and $R(A)$ is closed. Also, we have

$$\big(x,A^*(x)\big)_H \geqslant \|x\|^2 \qquad \forall\, x\in H.$$

Invoking Theorem 5.124, we infer that A is surjective, i.e., $R(A)=H$. Therefore, we conclude that A is an isomorphism.

Solution of Problem 5.143

Let

$$\widehat{C} = \big\{x\in X:\ \langle x^*,x\rangle \leqslant \sigma(x^*,C) \text{ for all } x^*\in X^*\big\}.$$

Note that the function $x^*\longmapsto\sigma(x^*,C)$ being the supremum of linear, w^*-continuous functional, is sublinear and w^*-upper semicontinuous (see Definition 5.136). Therefore the set $\widehat{C}$ is closed, convex and clearly $C\subseteq\widehat{C}$, hence $\overline{\operatorname{conv}}\,C\subseteq\widehat{C}$.

Now, suppose that $y\notin\overline{\operatorname{conv}}\,C$. By the strong separation theorem (see Theorem 5.29), we can find $\widehat{x}^*\in X^*\setminus\{0\}$ and $\varepsilon>0$ such that

$$\langle\widehat{x}^*,u\rangle \leqslant \langle\widehat{x}^*,y\rangle-\varepsilon \qquad \forall\, u\in C,$$

so

$$\sigma(\widehat{x}^*,C) < \langle x^*,y\rangle,$$

i.e., $y \notin \widehat{C}$. Therefore, we infer that

$$X \setminus \overline{\operatorname{conv}} C \subseteq X \setminus \widehat{C},$$

hence $\widehat{C} \subseteq \overline{\operatorname{conv}} C$. We conclude that $\widehat{C} = \overline{\operatorname{conv}} C$.

Solution of Problem 5.144

Let $x, u \in X$, $x \neq u$. From the strong separation theorem (see Theorem 5.29), we can find $v^* \in \overline{B}_1^*$, $v^* \neq 0$ and $\varepsilon > 0$ such that

$$|\langle v^*, x-u \rangle| \geqslant \varepsilon.$$

From Corollary 5.130, we know that $v^* \in \overline{\operatorname{conv}}^{w^*} \operatorname{ext} \overline{B}_1^*$. So, for a given $\delta \in \left(0, \frac{\varepsilon}{4}\right)$, we can find $\{w_k^*\}_{k=1}^n \subseteq \operatorname{ext} \overline{B}_1^*$ such that

$$\Big|\Big\langle v^* - \sum_{k=1}^n \lambda_k w_k^*, x \Big\rangle\Big| < \delta \quad \text{and} \quad \Big|\Big\langle v^* - \sum_{k=1}^n \lambda_k w_k^*, u \Big\rangle\Big| < \delta,$$

with $\lambda_1, \lambda_2, \ldots, \lambda_k \geqslant 0$ and $\sum\limits_{n=1}^{k} \lambda_k = 1$. Then

$$\Big|\Big\langle \sum_{k=1}^n \lambda_k w_k^*, x-u \Big\rangle\Big| > \tfrac{\varepsilon}{2},$$

and so for some $k_0 \in \{1, \ldots, n\}$, we have

$$\langle e^*, x-u \rangle \neq 0,$$

with $e^* = \lambda_{k_0} w_{k_0}^*$.

Solution of Problem 5.145

Arguing by contradiction, suppose that A is surjective. Then by the open mapping theorem (see Theorem 5.47), we can find $\varepsilon > 0$ such that

$$\varepsilon B_1^X \subseteq A(B_1^X) \subseteq \overline{A(B_1^X)}$$

and $\overline{A(B_1^X)}$ is a compact subset of X (since $A \in \mathcal{L}_c(X)$), where $B_1^X = \{x \in X : \|x\|_X < 1\}$. So, the set $\overline{B}_1^X$ is compact, hence X is finite dimensional (see Proposition 5.9(a)), a contradiction to our hypothesis that X is infinite dimensional.

Solution of Problem 5.146

From Proposition 5.61, we have $A \in \mathcal{L}(X_w; l_w^1)$. If $\overline{B}_1^X = \{x \in X : \|x\|_X \leqslant 1\}$, then since X is reflexive, $\overline{B}_1^X$ is w-compact (see Theorem 5.73). Hence $A(\overline{B}_1^X)$ is w-compact in l^1. Let $\{\widehat{u}_n\}_{n\geqslant 1} \subseteq A(\overline{B}_1^X)$. Then, by passing to a suitable subsequence if necessary, we may assume that $\widehat{u}_n \xrightarrow{w} \widehat{u} \in A(\overline{B}_1^X)$ in l^1. By the Schur property (see Remark 5.57), we have that $\widehat{u}_n \longrightarrow \widehat{u}$, hence $A(\overline{B}_1^X)$ is norm compact in l^1 and so $A \in \mathcal{L}_c(X; l^1)$.

Solution of Problem 5.147

Since $A \in \mathcal{L}_c(X;Y)$, the set $\overline{A(\overline{B}_1^X)} \subseteq Y$ is compact, hence separable. Note that

$$A(X) = \bigcup_{n\geqslant 1} n\overline{A(\overline{B}_1^X)},$$

so the set $A(X)$ is a separable vector subspace of Y.

Solution of Problem 5.148

Suppose that $x_n \xrightarrow{w} x$ in X. Then we can find $r > 0$ such that

$$\|x_n\|_X \leqslant r \qquad \forall\, n \geqslant 1.$$

Since $A \in \mathcal{L}_c(X : Y)$, we have that the set $\overline{A(\overline{B}_r)} = C$ is compact (with $\overline{B}_r = \{x \in X : \|x\| \leqslant r\}$). Let C denote the set with norm topology and let (C, w) be the set with the weak topology. We consider the identity operator $id: C \longrightarrow (C, w)$. Evidently id is a continuous bijection and because C is compact, it is a homeomorphism. Hence on C, the norm and the weak topologies coincide. From Proposition 5.61, we have that $A(x_n) \xrightarrow{w} A(x)$ in Y, $\{A(x_n)\}_{n\geqslant 1} \subseteq C$ and $A(x) \in C$. Therefore, $A(x_n) \longrightarrow A(x)$ in Y. Thus A is completely continuous.

Solution of Problem 5.149

"$\Longrightarrow$": This was established in Problem 5.148.

"$\Longleftarrow$": Let $\overline{B}_1^X = \{x \in X : \|x\|_X \leqslant 1\}$ and let $\{y_n\}_{n\geqslant 1} \subseteq A(\overline{B}_1^X)$ be a sequence. Then $y_n = A(x_n)$, with $\{x_n\}_{n\geqslant 1} \subseteq \overline{B}_1^X$. Since X is reflexive, we may assume (at least for a subsequence) that

$$x_n \xrightarrow{w} x \in \overline{B}_1^X \quad \text{in } X,$$

so

$$y_n = A(x_n) \xrightarrow{w} A(x) = y \quad \text{in } Y$$

(since $A \in \mathcal{L}(X_w; Y_w)$). Invoking the Eberlein–Smulian theorem (see Theorem 5.78), we conclude that the set $A(\overline{B}_1^X)$ is w-compact, hence $A \in \mathcal{L}_c(X;Y)$.

Solution of Problem 5.150

"$\Longrightarrow$": Let X_w be the Banach space X furnished with the weak topology and let us consider the function $\varphi\colon X_w \longrightarrow \mathbb{R}$, defined by

$$\varphi(x) = \|A(x)\|_Y \qquad \forall\, x \in X.$$

From Problem 5.148, we know that φ is sequentially continuous. Also due to the reflexivity of X (see Theorem 5.73) and the Eberline-Smulian theorem (see Theorem 5.78), the closed unit ball $\overline{B}_1^X = \{x \in X : \|x\|_X \leqslant 1\}$ is sequentially w-compact. So, we can find $x_0 \in \overline{B}_1^X$ such that

$$\|A(x_0)\|_Y = \sup\{\|A(x)\|_Y : \|x\|_X \leqslant 1\} = \|A\|_{\mathcal{L}}.$$

"$\Longleftarrow$": Let $x^* \in X^* \setminus \{0\}$ and $y \in Y$, $\|y\|_Y = 1$. Let $A \in \mathcal{L}(X;Y)$ be defined by

$$A(x) = \langle x^*, x\rangle\, y \qquad \forall\, x \in X.$$

Then $A \in \mathcal{L}_f(X;Y)$ (i.e., A is of rank 1), hence $A \in \mathcal{L}_c(X;Y)$ (see Remark 5.139). Then by hypothesis, we can find $x \in \overline{B}_1^X$ such that $\|A\|_{\mathcal{L}} = \|A(x)\|_Y$. Hence $\|x^*\|_* = |\langle x^*, x\rangle|$ and so x^* is norm attaining which, by Theorem 5.76, implies that X is reflexive.

Solution of Problem 5.151

Yes. According to the Mazur theorem (see Theorem 5.58), it suffices to show that the set C is norm closed. So, let $\{u_n\}_{n\geqslant 1} \subseteq L^1(0,1)$ be a sequence such that $u_n \longrightarrow u$ in $L^1(0,1)$. By Propositions 3.132 and 3.131 and by passing to a suitable subsequence if necessary, we may assume that $u_n(t) \longrightarrow u(t)$ for almost all $t \in [0,1]$. Since $u_n(t) \geqslant 1$ for almost all $t \in [0,1]$, we infer that $u(t) \geqslant 1$ for almost all $t \in [0,1]$ and so $u \in C$. This proves that the set C is w-closed in $L^1(0,1)$.

Solution of Problem 5.152

We argue by contradiction. So, suppose that we can find $\varepsilon > 0$ and a sequence $\{x_n\}_{n\geqslant 1} \subseteq X$ such that

$$\|x_n\|_X = 1 \quad \text{and} \quad \varepsilon + n|x_n|_X < \|A(x_n)\|_Y \qquad \forall\, n \geqslant 1. \tag{5.17}$$

Since X is reflexive, the set $\overline{B}_1^{\|\cdot\|_X} = \{x \in X : \|x\|_X \leqslant 1\}$ is w-compact (see Theorem 5.73) and by the Eberlein–Smulian theorem (see Theorem 5.78), it is sequentially w-compact. So, by passing to a suitable subsequence if necessary, we may assume that

$$x_n \xrightarrow{w} x \quad \text{in } (X, \|\cdot\|_X). \tag{5.18}$$

Because the norm $\|\cdot\|_X$ is stronger than $|\cdot|_X$, the identity map

$$i\colon (X\|\cdot\|_X) \longrightarrow (X, |\cdot|_X)$$

is continuous, hence weakly continuous too (see Proposition 5.61). Therefore, we have

$$x_n \xrightarrow{w} x \quad \text{in } (X, |\cdot|_X) \tag{5.19}$$

(see (5.18)). Moreover, since $A \in \mathcal{L}_c(X;Y)$, from Problem 5.148 and (5.18), we have

$$A(x_n) \longrightarrow A(x) \quad \text{in } Y. \tag{5.20}$$

Returning to (5.17) and using (5.19) and (5.20), we infer that $|x_n|_X \longrightarrow 0$. Hence $x = 0$ (see (5.19)) and so $A(x_n) \longrightarrow 0$ in Y (see (5.20)). On the other hand, from (5.17), we have

$$\varepsilon \leqslant \|A(x_n)\|_Y \qquad \forall\, n \geqslant 1,$$

a contradiction.

Solution of Problem 5.153

Arguing by contradiction, suppose that $0 \notin \overline{A(\partial B_1)}$. Then

$$\inf\left\{\|A(x)\| : \|x\| = 1\right\} = c > 0,$$

so

$$\|A(x)\| \geqslant c\|x\| \qquad \forall\, x \in X.$$

By Theorem 5.125, the last inequality implies that $R(A)$ is closed. Let $Y = R(A)$. Then Y is an infinite dimensional Banach space and $A \in \mathcal{L}_c(X;Y)$ is bijective, hence an isomorphism (see the Banach theorem; Theorem 5.48), a contradiction to the compactness of A (see Remark 5.139).

Solution of Problem 5.154

Since $A \in \mathcal{L}_c(X;Y)$, we have that the set $\overline{A(\overline{B}_1^X)}$ is compact in Y (where $\overline{B}_1^X = \{x \in X : \|x\|_X \leqslant 1\}$). Hence the set $A(\overline{B}_1^X)$ is separable. Note that

$$\bigcup_{n\geqslant 1} nA(\overline{B}_1^X) = \bigcup_{n\geqslant 1} A(n\overline{B}_1^X) = A\Big(\bigcup_{n\geqslant 1} n\overline{B}_1^X\Big) = A(X) = R(A).$$

Therefore

$$R(A) = Y$$

is separable.

Solution of Problem 5.155

Let $\overline{B}_1^X = \{x \in X : \|x\| \leqslant 1\}$ and $K = \overline{A(\overline{B}_1^X)}$. Since $A \in \mathcal{L}_c(X;H)$, $K \subseteq H$ is compact. So, for a given $\varepsilon > 0$, we can find a finite set $\{h_1, \ldots, h_{N(\varepsilon)}\} \subseteq H$ such that

$$K \subseteq \bigcup_{k=1}^{N(\varepsilon)} B_\varepsilon(h_k)$$

(where $B_\varepsilon(h) = \{y \in H : \|y - h\|_H < \varepsilon\}$ for all $h \in H$). Let

$$Y = \operatorname{span}\{h_1, \ldots, h_{N(\varepsilon)}\} \subseteq H$$

and let $\operatorname{proj}_Y : H \longrightarrow Y$ be the orthogonal projection onto Y (it exists since Y is finite dimensional). Then $\operatorname{proj}_Y \circ A \in \mathcal{L}_f(X;H)$.

Let $x \in \overline{B}_1^X$. Then, we can find $k_0 \in \{1, \ldots, N(\varepsilon)\}$ such that

$$\|A(x) - h_{k_0}\|_H < \varepsilon.$$

We have

$$\begin{aligned} \|\operatorname{proj}_Y(A(x)) - h_{k_0}\|_H &= \|\operatorname{proj}_Y(A(x)) - \operatorname{proj}_Y(h_{k_0})\|_H \\ &\leqslant \|\operatorname{proj}_Y\|_{\mathcal{L}} \|A(x) - h_{k_0}\|_H \\ &= \|A(x) - h_{k_0}\| < \varepsilon \end{aligned}$$

(recall that $\|\operatorname{proj}_Y\|_{\mathcal{L}} = 1$), so

$$\|\operatorname{proj}_Y(A(x)) - A(x)\|_H < 2\varepsilon \qquad \forall\, x \in \overline{B}_1^X,$$

thus

$$\|\operatorname{proj}_Y \circ A - A\|_{\mathcal{L}} < 2\varepsilon \quad \text{and} \quad \operatorname{proj}_Y \circ A \in \mathcal{L}_f(X;Y).$$

Solution of Problem 5.156

Consider the bilinear form $\widehat{a}\colon X \times X \longrightarrow \mathbb{R}$, defined by

$$\widehat{a}(x,y) = a(x,y) + c_2\,(x,y)_H \qquad \forall\, x,y \in X.$$

Clearly $\widehat{a}$ is continuous and from hypothesis, we have

$$\widehat{a}(x,x) \geqslant c_1\|x\|_X^2.$$

Also, the map $x \longmapsto (u_0, x)_H$ is a continuous linear functional on X. Then by the Lax–Milgram theorem (see Theorem 5.114), there exists a unique $v \in X$ such that

$$\widehat{a}(v, x) = (u_0, x)_H \qquad \forall\, x \in X.$$

Let $A\colon H \longrightarrow X$ be defined by

$$A(u_0) = v \qquad \forall\, u_0 \in H.$$

Then clearly $A \in \mathcal{L}(H; X)$ and exploiting the fact that X is embedded compactly into H, we have $A \in \mathcal{L}_c(H)$.

Note that $x_0 \in X$ is a solution of our problem (i.e., $a(x_0, x) = (u_0, x)_H$ for all $x \in X$) if and only if

$$\widehat{a}(x_0, x) = a(x_0, x) + c_2\,(x_0, x)_H = (u_0 + c_2 x_0, x)_H \qquad \forall\, x \in X.$$

Therefore, $x_0 \in X$ solves our problem if and only if $x_0 = A(u_0 + c_2 x_0)$. It follows that

$$w - c_2 A(w) = u_0 \quad \text{with } w = u_0 + c_2 x_0. \tag{5.21}$$

So, our problem is uniquely solvable if and only if (5.21) is unique solvable. By the Fredholm alternative theorem (see Theorem 5.147):

$$id - c_2 A \text{ is injective if and only if } id - c_2 A \text{ is surjective.} \tag{5.22}$$

Suppose that $(id - c_2 A)(w) = 0$. Then $w = A(c_2 w)$ and so $w \in X$ and

$$a(w, x) + c_2\,(w, x)_H = (c_2 w, x)_H \qquad \forall\, x \in H.$$

Thus

$$a(w, x) = 0 \qquad \forall\, x \in H,$$

so

$$a(w, w) = 0,$$

hence by hypothesis $w = 0$. So $id - c_2 A$ is injective, hence surjective too (see (5.22)).

Therefore problem (5.21) is solvable and by (5.22) the solution is unique. Thus we conclude that our problem is uniquely solvable.

Solution of Problem 5.157

Let $\{e_k\}_{k\geqslant 1}$ be the Schauder basis and let p_n be the corresponding projections on span $\{e_k\}_{k=1}^n$. For every $x \in X$, we have $p_n(x) \longrightarrow x$ in X as $n \to +\infty$. Let $A \in \mathcal{L}_c(X)$. Then

$$\|p_n \circ A - A\|_{\mathcal{L}} = \sup_{\|x\|\leqslant 1} \big\|(p_n - id)\big(A(x)\big)\big\| \longrightarrow 0 \quad \text{as } n \to +\infty$$

(see Problem 5.51 and recall that $\overline{A(\overline{B}_1)}$ is compact since $A \in \mathcal{L}_c(X)$).

Solution of Problem 5.158

Let $A \in \mathcal{L}_f(X)$. Since $R(A)$ is finite dimensional and by hypothesis X is infinite dimensional, A is not injective (see the Banach theorem; Theorem 5.48) and so we can find $x \in N(A)$ with $\|x\| = 1$. We have

$$\big\|(id - A)(x)\big\| = 1,$$

so

$$\|id - A\|_{\mathcal{L}} \geqslant 1 \qquad \forall\, A \in \mathcal{L}_f(X)$$

and thus $\operatorname{dist}\big(id, \mathcal{L}_f(X)\big) \geqslant 1$. On the other hand $\operatorname{dist}\big(id, \mathcal{L}_f(X)\big) \leqslant \|id\|_{\mathcal{L}} = 1$, hence

$$\operatorname{dist}\big(id, \mathcal{L}_f(X)\big) = 1.$$

Solution of Problem 5.159

No. To see that, let us take any $K \in \mathcal{L}_c(H)$ such that $AK = KA$. Then

$$AK(e_n) = KA(e_n) = K(e_{n+1}) \qquad \forall\, n \geqslant 1,$$

so

$$\big\|K(e_n)\big\| = \big\|K(e_{n+1})\big\| \qquad \forall\, n \geqslant 1$$

(since A is an isomorphism), so

$$\|K(e_n)\| = \|K(e_1)\| \qquad \forall\, n \geqslant 1.$$

From Problem 5.120, we know that $e_n \xrightarrow{w} 0$ in H. Since $K \in \mathcal{L}_c(H)$, from Problem 5.148, we have that $K(e_n) \longrightarrow 0$ in H. Therefore from the last equality, we infer that

$$K(e_n) = 0 \qquad \forall\, n \geqslant 1,$$

so $K \equiv 0$. Since A commutes with itself, if A is compact, then by the above argument we would have that $A \equiv 0$, a contradiction.

Solution of Problem 5.160

Let $A \in \mathcal{L}_c(X)$ and let $\overline{B}_1 = \{u \in X : \|u\|_\infty \leqslant 1\}$. We know that $A(\overline{B}_1)$ is relatively compact in X. By the Arzela–Ascoli theorem (see Theorem 2.181), $A(\overline{B}_1)$ is equicontinuous. So, for a given $\varepsilon > 0$, we can find $\delta = \delta(\varepsilon) > 0$ such that

$$|t-s| \leqslant \delta \implies |v(t)-v(s)| \leqslant \varepsilon \qquad \forall\, v \in A(\overline{B}_1).$$

Let $m \geqslant 1$ and let $\{t_i\}_{i=1}^m \subseteq (0,1)$ be such that $\{(t_i-\delta, t_i+\delta)\}_{i=1}^m$ is an open cover of $[0,1]$. We can find a continuous partition of unity $\{\vartheta_i\}_{i=1}^m$ subordinate to this cover (see Theorem 2.147). Let $A_\varepsilon \in \mathcal{L}(X)$ be defined by

$$A_\varepsilon(u) = \sum_{i=1}^m A(u)(t_i)\vartheta_i \qquad \forall\, u \in X.$$

Evidently $A_\varepsilon \in \mathcal{L}_f(X)$. For every $u \in \overline{B}_1$ and $t \in [0,1]$, we have

$$\begin{aligned} |(A(u)-A_\varepsilon(u))(t)| &= \Big|\sum_{i=1}^m \big(A(u)(t)-A(u)(t_i)\big)\vartheta_i(t)\Big| \\ &\leqslant \sum_{i=1}^m \varepsilon\vartheta_i(t) = \varepsilon, \end{aligned}$$

so

$$\big\|A(u) - A_\varepsilon(u)\big\|_\infty \leqslant \varepsilon \qquad \forall\, u \in \overline{B}_1,$$

thus

$$\|A - A_\varepsilon\|_{\mathcal{L}} \leqslant \varepsilon.$$

Solution of Problem 5.161

Let $\widehat{u}_n = \{\delta_{kn}\}_{k\geqslant 1}$. Then $\{\widehat{u}_n\}_{n\geqslant 1} \subseteq l^p$ and $\|\widehat{u}_n\|_{l^p} = 1$. We may assume (at least for a subsequence) that

$$\widehat{u}_n \xrightarrow{w} \widehat{u}.$$

For every $\widehat{u}^* \in l^{p'} = (l^p)^*$ (with $\frac{1}{p} + \frac{1}{p'} = 1$), we have

$$\langle \widehat{u}^*, \widehat{u}_n\rangle_{l^p} = u_n \longrightarrow 0 \quad \text{as } n \to +\infty,$$

so

$$\widehat{u}_n \xrightarrow{w} 0 \quad \text{in } l^p,$$

i.e., $\widehat{u} = 0$.

If the embedding was compact, then from Problem 5.148, we would have $\widehat{u}_n \longrightarrow 0$ in l^q, a contradiction, since $\|\widehat{u}_n\|_{l^q} = 1$ for all $n \geqslant 1$.

Solution of Problem 5.162

"$\Longrightarrow$": We know that $X/_{N(A)}$ is a Banach space and $\widehat{A} \in \mathcal{L}\big(X/_{N(A)}; Y\big)$. Let $\widehat{B}$ be a bounded set in $X/_{N(A)}$ and let $[x] \in \widehat{B}$. Choose $u \in [x]$ such that $\|u\| \leqslant 2\big|[x]\big|$. We denote it by $u\big([x]\big)$. Let $\widetilde{B} = \big\{u\big([x]\big) :\ [x] \in B\big\}$. We have

$$A(\widetilde{B}) = \widehat{A}(\widehat{B}) \quad \text{and} \quad \widetilde{B} \text{ is bounded.}$$

Since $A \in \mathcal{L}_c(X;Y)$, $A(\widetilde{B}) = \widehat{A}(\widehat{B}) \subseteq Y$ is relatively compact, hence $\widehat{A} \in \mathcal{L}_c\big(X/_{N(A)}; Y\big)$.

"$\Longleftarrow$": Let $p\colon X \longrightarrow X/_{N(A)}$ be the canonical "projection" $p(x) = [x]$. Then

$$A = \widehat{A} \circ p,$$

so $A \in \mathcal{L}_c(X;Y)$ since $\widehat{A} \in \mathcal{L}_c(X/_{N(A)};Y)$ and $p \in \mathcal{L}(X;X/_{N(A)})$.

Solution of Problem 5.163

For every integer $m \geqslant 1$, we consider the finite rank operator K_m, defined by

$$K_m\big(\{x_n\}_{n\geqslant 1}\big) = (\lambda_1 x_1, \ldots, \lambda_m x_m, 0, \ldots).$$

Evidently, for every $m \geqslant 1$, K_m is linear and continuous. Since $\lambda_n \longrightarrow 0$, for a given $\varepsilon > 0$, we can find an integer $n_0 \geqslant 1$ such that

$$|\lambda_n| \leqslant \varepsilon \qquad \forall\, n \geqslant n_0.$$

Then for $m \geqslant n_0$, we have

$$\big\|(A - K_m)(\{x_n\}_{n\geqslant 1})\big\|_{l^2} = \Big(\sum_{k\geqslant m+1} |\lambda_k|^2 |x_k|^2\Big)^{\frac{1}{2}} \leqslant \varepsilon \sum_{k\geqslant m+1} |x_k|^2,$$

so

$$\|A - K_m\|_{\mathcal{L}} \leqslant \varepsilon \qquad \forall\, m \geqslant n_0.$$

Thus, A is the limit of finite rank operators, hence it is compact (see Proposition 5.141).

Solution of Problem 5.164

From Theorem 5.144, we know that $(id - A)(X) \subseteq X$ is closed, hence a Banach space. Suppose that $id - A$ does not have a continuous inverse. Then for every $n \geqslant 1$, we can find $x_n \in X$ such that

$$\big\|(id - A)(x_n)\big\| < \tfrac{1}{n}\|x_n\|.$$

Let $y_n = \frac{x_n}{\|x_n\|}$ for all $n \geqslant 1$ and since A is compact, we can find a subsequence $\{y_{n_k}\}_{k\geqslant 1}$ of $\{y_n\}_{n\geqslant 1}$ such that $y_{n_k} \longrightarrow y$ in X. We have

$$\left\|(id - A)(y_{n_k})\right\| < \frac{1}{n_k} \qquad \forall\, k \geqslant 1,$$

so

$$A(y) = y, \quad \|y\| = 1$$

and thus $id - A$ is not injective, a contradiction.

Solution of Problem 5.165

(a) First suppose that $p \in [1, +\infty)$ and let $e_n = \{\delta_{kn}\}_{k\geqslant 1} \in l^p$ for $n \geqslant 1$ (with δ_{kn} being the Kronecker symbol). Then $\{e_n\}_{n\geqslant 1}$ is a Schauder basis for l^p (see Remark 5.107). So, for every $\widehat{x} = \{x_n\}_{n\geqslant 1} \in l^p$, we have

$$\widehat{x} = \lim_{m\to+\infty} \sum_{n=1}^{m} x_n e_n \quad \text{in } l^p.$$

Note that $A(e_1) = 0$ and $A(e_n) = e_{n-1}$ for all $n \geqslant 2$.

For $\widehat{x} = \{x_n\}_{n\geqslant 1} \in l^p$, we have

$$\left\|A(\widehat{x})\right\|_{l^p}^p = \sum_{n\geqslant 2} |x_n|^p \leqslant \sum_{n\geqslant 1} |x_n|^2 = \|\widehat{x}\|_{l^p}^p,$$

so

$$A \in \mathcal{L}(l^p) \quad \text{and} \quad \|A\|_{\mathcal{L}} \leqslant 1.$$

On the other hand

$$1 = \|e_1\|_{l^p} = \left\|A(e_2)\right\|_{l^p} \leqslant \|A\|_{\mathcal{L}}$$

and so finally $\|A\|_{\mathcal{L}} = 1$.

If $p = \infty$, then

$$\left\|A(\widehat{x})\right\|_{l^\infty} = \sup_{n\geqslant 2} |x_n| \leqslant \sup_{n\geqslant 1} |x_n| = \|\widehat{x}\|_{l^p}$$

and so $\|A\|_{\mathcal{L}} \leqslant 1$. But

$$1 = \|e_1\|_{l^\infty} = \left\|A(e_2)\right\|_{l^\infty} \leqslant \|A\|_{\mathcal{L}}$$

and so, again $\|A\|_{\mathcal{L}} = 1$.

(b) From Proposition 5.151, we know that

$$\sigma(A) \subseteq \{\lambda \in \mathbb{C} : |\lambda| \leqslant \|A\|_{\mathcal{L}}\} = \{\lambda \in \mathbb{C} : |\lambda| \leqslant 1\}.$$

If $|\lambda| < 1$, then $\widehat{x}_\lambda = \{\lambda^{n-1}\}_{n\geqslant 1} \in l^p$ and

$$(A - \lambda id)\widehat{x}_\lambda = \{\lambda^n\}_{n\geqslant 1} - \lambda\{\lambda^{n-1}\}_{n\geqslant 1} = 0,$$

so

$$\{\lambda \in \mathbb{C} : |\lambda| < 1\} \subseteq \sigma_p(A) \subseteq \sigma(A) \subseteq \{\lambda \in \mathbb{C} : |\lambda| \leqslant 1\}$$

(see Definition 5.153). Since $\sigma(A)$ is closed (see Proposition 5.151), from the last inclusions, we infer that

$$\sigma(A) = \{\lambda \in \mathbb{C} : |\lambda| \leqslant 1\}.$$

If $1 \leqslant p < +\infty$ and $\lambda \in \sigma_p(A)$, then

$$(A - \lambda id)(\widehat{x}) = 0 \quad \text{with } \widehat{x} \in l^p,\ \widehat{x} \neq 0.$$

Hence

$$x_n = \lambda^{n-1} x_1 \implies \widehat{x} = x_1 \widehat{x}_\lambda,$$

so

$$\sigma_p(A) = \{\lambda \in \mathbb{C} : |\lambda| < 1\}$$

(since $1 \leqslant p < +\infty$).

Note that for every $\lambda \in \mathbb{C}$, with $|\lambda| < 1$, the eigenspace corresponding to $\lambda \in \mathbb{C}$ is one dimensional (i.e., λ is simple) and it is generated by $\widehat{x}_\lambda$.

If $p = +\infty$, then it is clear from the above argument that

$$\sigma(A) = \sigma_p(A) = \{\lambda \in \mathbb{C} : |\lambda| \leqslant 1\}.$$

Solution of Problem 5.166

Let $\lambda = a + i\beta$ with $\beta \neq 0$. For every $x \in H$, we have

$$\begin{aligned} & \big(A(x) - \lambda x, x\big)_H - \big(x, A(x) - \lambda x\big)_H \\ = \; & \big(A(x), x\big)_H - \lambda\|x\|^2 + \overline{\lambda}\|x\|^2 - \big(x, A(x)\big)_H \\ = \; & (\overline{\lambda} - \lambda)\|x\|^2 \; = \; -2i\beta\|x\|^2 \end{aligned}$$

(since A is self-adjoint).

So, for every $x \in H$, we have

$$\begin{aligned} 2|\beta|\|x\|^2 \; &= \; \big|\big(A(x) - \lambda x, x\big)_H - \big(x, A(x) - \lambda x\big)_H\big| \\ &\leqslant \; \big|\big(A(x) - \lambda x, x\big)_H\big| + \big|\big(x, A(x) - \lambda x\big)_H\big| \\ &\leqslant \; \big\|A(x) - \lambda x\big\| \, \|x\| + \|x\| \, \big\|A(x) - \lambda x\big\| \\ &= \; 2\big\|A(x) - \lambda x\big\| \, \|x\|, \end{aligned}$$

thus

$$2\beta\|x\| \; \leqslant \; \big\|A(x) - \lambda x\big\|.$$

Then, from Proposition 5.159, we conclude that $\lambda \notin \sigma(A)$. Therefore, $\sigma(A) \subseteq \mathbb{R}$.

Solution of Problem 5.167

Clearly $A \neq 0$. For every $x \in X$, we have

$$A^2(x) \; = \; \langle x_0^*, x\rangle \, A(x_0) \; = \; \langle x_0^*, x\rangle \, \langle x_0^*, x_0\rangle \, x_0 \; = \; \langle x_0^*, x_0\rangle \, A(x),$$

so

$$A^2 \; = \; \langle x_0^*, x_0\rangle \, A.$$

Therefore A is a projection (i.e., $A^2 = A$) if and only if $\langle x_0^*, x_0\rangle = 1$.

Evidently $A \in \mathcal{L}_f(X)$ (in fact it is of rank 1). So, by Theorem 5.155, we have

$$\sigma(A) \; = \; \{0\} \cup \sigma_p(A).$$

An eigenvalue $\lambda \in \sigma_p(A)$ has an eigenspace $E(\lambda) \subseteq R(A) = \mathbb{R}x_0$. Therefore $\lambda = \langle x_0^*, x_0\rangle$ and so finally $\sigma(A) = \{0, \langle x_0^*, x_0\rangle\}$.

Solution of Problem 5.168
We have

$$\begin{aligned} \|A\|_{\mathcal{L}}^2 &= \sup\big\{\big(A(x),A(x)\big) : \|x\| \leqslant 1\big\} \\ &= \sup\big\{\big(A^*A(x),x\big) : \|x\| \leqslant 1\big\} \\ &= \sup\big\{\big(A^2(x),x\big) : \|x\| \leqslant 1\big\} = \|A^2\|_{\mathcal{L}} \end{aligned}$$

(since A and A^2 are self-adjoint and using Remark 5.158).

For every $m \geqslant 1$, $A^m \in \mathcal{L}(H)$ is self-adjoint. So, from the previous part, we have

$$\|A^{2^m}\|_{\mathcal{L}} = \|A\|_{\mathcal{L}}^{2^m}.$$

Now, let $1 \leqslant n \leqslant 2^m$. Then

$$\begin{aligned} \|A\|_{\mathcal{L}}^{2^m} &= \|A^{2^m}\|_{\mathcal{L}} = \|A^n A^{2^m-n}\|_{\mathcal{L}} \leqslant \|A^n\|_{\mathcal{L}}\|A\|_{\mathcal{L}}^{2^m-n} \\ &\leqslant \|A\|_{\mathcal{L}}^n\|A\|_{\mathcal{L}}^{2^m-n} = \|A\|_{\mathcal{L}}^{2^m}, \end{aligned}$$

so

$$\|A^n\|_{\mathcal{L}}\|A\|_{\mathcal{L}}^{2^m-n} = \|A\|^{2^m}$$

and thus $\|A^n\|_{\mathcal{L}} = \|A\|_{\mathcal{L}}^n$.

Solution of Problem 5.169
First we show that A is well defined. So, let $u \in X$ and let $t, s \in [0,1]$, $s \leqslant t$. Then

$$\begin{aligned} \big|A(u)(t) - A(u)(s)\big| &= \Big|\int_0^s \big(k(t,\tau)-k(s,\tau)\big)u(\tau)\,d\tau \\ &\quad + \int_s^t k(t,\tau)u(\tau)\,d\tau\Big| \\ &\leqslant \|u\|_\infty \int_0^s \big|k(t,\tau)-k(s,\tau)\big|\,d\tau + M\|u\|_\infty|t-s|, \end{aligned}$$

so

$$\big(A(u)(t) - A(u)(s)\big) \longrightarrow 0 \quad \text{as } t \to s,$$

i.e., $A(u)(\cdot) \in X$.

We have

$$\big|A(u)(t)\big| \leqslant M\|u\|_\infty t, \tag{5.23}$$

so

$$\big\|A(u)\big\|_\infty \leqslant M\|u\|_\infty$$

and thus

$$\|A\|_{\mathcal{L}} \leqslant M.$$

Next, using induction, we will show that

$$\big|A^n(u)(t)\big| \leqslant \tfrac{M^n}{n!}\|u\|_\infty t^n \qquad \forall\, n \geqslant 1. \tag{5.24}$$

For $n = 1$, (5.24) is true (see (5.23)). Suppose that (5.23) is true for $n \geqslant 1$. We have

$$\begin{aligned}\big|A^{n+1}(u)(t)\big| &= \Big|\int_0^t k(t,s)A^n(u)(s)\,ds\Big| \leqslant M\int_0^t \big|A^n(u)(s)\big|\,ds\\ &\leqslant \tfrac{M^{n+1}}{n!}\|u\|_\infty \int_0^t s^n\,ds = \tfrac{M^{n+1}}{(n+1)!}\|u\|_\infty t^{n+1}\end{aligned}$$

(by the induction hypothesis), which proves (5.24). From (5.24), it follows that

$$\big\|A^n(u)\big\|_\infty \leqslant \tfrac{M^n}{n!}\|u\|_\infty,$$

so

$$\|A^n\|_{\mathcal{L}}^{\frac{1}{n}} \leqslant \tfrac{M}{(n!)^{\frac{1}{n}}}.$$

Recall that $(n!)^{\frac{1}{n}} \longrightarrow +\infty$. So,

$$\lim_{n\to+\infty} \|A^n\|_{\mathcal{L}}^{\frac{1}{n}} = 0.$$

Then from Remark 5.150, we infer that $\sigma(A) = \{0\}$.

Solution of Problem 5.170

Suppose that $\lambda \notin \sigma(A)$. Then $(A - \lambda id)^{-1} \in \mathcal{L}(X)$ and so we can find $c > 0$ such that

$$\big\|(A - \lambda id)(x)\big\| \geqslant c\|x\| \qquad \forall\, x \in X.$$

Let $x = x_n$. Then

$$\big\|(A - \lambda id)(x_n)\big\| \geqslant c \qquad \forall\, n \geqslant 1,$$

a contradiction to the hypothesis that $A(x_n) - \lambda x_n \longrightarrow 0$ in X.

Solution of Problem 5.171

Let $\{x_n\}_{n\geqslant 1} \subseteq X$ be a sequence such that

$$\|x_n\| = 1 \qquad \forall\, n \geqslant 1 \quad \text{and} \quad \big|(A(x_n), x_n)\big| \longrightarrow \|A\|_{\mathcal{L}}.$$

By passing to a subsequence if necessary, we have $\big(A(x_n), x_n\big) \longrightarrow \lambda$ with $\lambda = \pm\|A\|_{\mathcal{L}}$. Then

$$\begin{aligned} 0 \leqslant \big\|A(x_n) - \lambda x_n\big\|^2 &= \big\|A(x_n)\big\|^2 - 2\lambda\big(A(x_n), x_n\big) + \lambda^2\|x_n\|^2 \\ &\leqslant 2\lambda^2 - 2\lambda\big(A(x_n), x_n\big) \longrightarrow 0 \quad \text{as } n \to +\infty, \end{aligned}$$

so $\lambda \in \sigma(A)$ (see Problem 5.170).

Solution of Problem 5.172

"**(a)** $\Longrightarrow$ **(b)**": Let $x, u \in X \setminus \{0\}$ and suppose that $\|x + u\| = \|x\| + \|u\|$. Without any loss of generality, we may assume that $0 < \|x\| \leqslant \|u\|$. We have

$$\begin{aligned} \Big\|\tfrac{x}{\|x\|} + \tfrac{u}{\|u\|}\Big\| &\geqslant \Big\|\tfrac{x}{\|x\|} + \tfrac{u}{\|x\|}\Big\| - \Big\|\tfrac{u}{\|x\|} - \tfrac{u}{\|u\|}\Big\| \\ &= \tfrac{1}{\|x\|}\|x + u\| - \|u\|\big(\tfrac{1}{\|x\|} - \frac{1}{\|u\|}\big) \\ &= \tfrac{1}{\|x\|}\big(\|x\| + \|u\|\big) - \|u\|\big(\tfrac{1}{\|x\|} - \tfrac{1}{\|u\|}\big) = 2, \end{aligned}$$

so

$$\Big\|\tfrac{x}{\|x\|} + \tfrac{u}{\|u\|}\Big\| = 2.$$

This by the strict convexity of $\|\cdot\|$, implies that $\frac{x}{\|x\|} = \frac{u}{\|u\|}$, hence $x = \lambda u$, with $\lambda = \frac{\|x\|}{\|u\|} > 0$.

"(b) $\Longrightarrow$ (c)": Let $x, u \in X$ and suppose that $2\|x\|^2 + 2\|u\|^2 = \|x+u\|^2$. Evidently $x = 0$ if and only if $u = 0$. So, we may assume that $x, u \in X \setminus \{0\}$. Then

$$\begin{aligned} 0 &= 2\|x\|^2 + 2\|u\|^2 - \|x+u\|^2 \geqslant 2\|x\|^2 + 2\|u\|^2 - \big(\|x\| + \|u\|\big)^2 \\ &= \big(\|x\| - \|u\|\big)^2, \end{aligned}$$

so

$$\|x\| = \|u\| \quad \text{and} \quad \|x+u\| = \|x\| + \|y\|,$$

thus $x = y$.

"(c) $\Longrightarrow$ (a)": Let $x, u \in X$ with $\|x\| = \|u\| = 1$ and $\|x+u\| = 2$. Then

$$2\|x\|^2 + 2\|u\|^2 = \|x+u\|^2,$$

so $x = u$ and so $\|\cdot\|$ is strictly convex (see Remark 5.169).

Solution of Problem 5.173

First, we check the Gâteaux differentiability. Let $\widehat{x} = \{x_n\}_{n\geqslant 1} \in l^1$ be such that $x_n = 0$ for some $n \geqslant 1$ and let $e_n = \{\delta_{kn}\}_{k\geqslant 1}$. Then

$$\|\widehat{x} + \lambda e_n\|_{l^1} - \|\widehat{x}\|_{l^1} = |\lambda|,$$

so

$$\tfrac{\|\widehat{x}+\lambda e_n\|_{l^1} - \|\widehat{x}\|_{l^1}}{\lambda} = \tfrac{|\lambda|}{\lambda}.$$

But $\lim\limits_{\lambda\to 0} \frac{|\lambda|}{\lambda}$ does not exist. So, $\|\cdot\|_{l^1}$ is not Gâteaux differentiable at $\widehat{x} = \{x_n\}_{n\geqslant 1} \in l^1$ if $x_n = 0$ for some $n \geqslant 1$.

Next, let $\widehat{x} = \{x_n\}_{n\geqslant 1} \in l^1$ with $x_n \neq 0$ for all $n \geqslant 1$. Let $\widehat{h} = \{h_n\}_{n\geqslant 1} \in l^1$. For a given $\varepsilon > 0$, we can find $m \geqslant 1$ such that

$$\sum_{n>m} |h_n| < \tfrac{\varepsilon}{2}.$$

If $\delta > 0$ is small and $|\lambda| \leqslant \delta$, then

$$\operatorname{sgn}(x_n + \lambda h_n) = \operatorname{sgn} x_n \qquad \forall\, n \in \{1, \ldots, m\}.$$

We have

$$\begin{aligned}
&\Big|\tfrac{1}{\lambda}\big(\|\widehat{x}+\lambda\widehat{h}\|_{l^1}-\|\widehat{x}\|_{l^1}\big)-\sum_{n\geqslant 1}h_n\operatorname{sgn}x_n\Big| \\
\leqslant\ &\Big|\sum_{n=1}^{m}\tfrac{1}{\lambda}\big(|x_n+\lambda h_n|-|x_n|-\lambda h_n\operatorname{sgn}x_n\big)\Big|+2\sum_{n>m}|h_n| \ <\ \varepsilon \qquad \forall\,|\lambda|<\delta,
\end{aligned}$$

so $\|\cdot\|_{l^1}$ is Gâteaux differentiable at $\widehat{x}$.

Now we check the Fréchet differentiability of the l^1-norm. Since Fréchet differentiability at a point implies Gâteaux differentiability at that point, from the first part of the solution, we see that it suffices to consider the points $\widehat{x} = \{x_n\}_{n\geqslant 1} \in l^1$ such that $x_n \neq 0$ for all $n \geqslant 1$. Let $\widehat{h}^m = \{h_n^m\}_{n\geqslant 1} = \big(0,\ldots,0,-2x_m,-2x_{m+1},-2x_{m+2},\ldots\big)$ for $m \geqslant 1$. Then $\|\widehat{h}^m\|_{l^1} \longrightarrow 0$ as $n \to +\infty$. From the first part of the solution, we know that the Gâteaux derivative at $\widehat{x}$ is $\{\operatorname{sgn} x_n\}_{n\geqslant 1}$. So, this is the only candidate for Fréchet derivative (see Remark 5.164). Then

$$\Big|\|\widehat{x}+\widehat{h}^m\|_{l^1}-\|\widehat{x}\|_{l^1}-\sum_{n\geqslant 1}h_n^m\operatorname{sgn}x_n\Big| \ -\ \Big|\sum_{n\geqslant m}\big(\ \ 2|x_n|\big)\Big| \ =\ 2\|\widehat{h}^m\|_{l^1},$$

so $\|\cdot\|_{l^1}$ is not Fréchet differentiable at $\widehat{x}$.

Solution of Problem 5.174

Clearly, we can assume that $x_n \neq 0$ for all $n \geqslant 1$ and $x \neq 0$. Then $y_n = \frac{x_n}{\|x_n\|}$ for $n \geqslant 1$ and $y = \frac{x}{\|x\|}$. Then

$$\|y_n\| \ =\ \|y\| \ =\ 1 \qquad \forall\, n \geqslant 1.$$

Evidently $y_n \xrightarrow{w} y$ in X and we have

$$\begin{aligned}
2 \ =\ 2\|y\| \ &\leqslant\ \liminf_{n\to+\infty}\|y_n+y\| \ \leqslant\ \limsup\|y_n+y\| \\
&\leqslant\ \|y\| + \lim_{n\to+\infty}\|y_n\| \ =\ 2,
\end{aligned}$$

so $\|y_n+y\| \longrightarrow 2$. Since X is locally uniformly convex, we have $\|y_n-y\| \longrightarrow 0$ (see Definition 5.168(b)).

Solution of Problem 5.175
Let $\{u_n\}_{n\geqslant 1} \subseteq \partial B_1$ be such that $\langle x^*, u_n\rangle \longrightarrow \|x^*\|_*$. Then

$$2 \geqslant \|u_n + u_m\| \geqslant \tfrac{1}{\|x^*\|_*}\langle x^*, u_n + u_m\rangle \longrightarrow 2 \quad \text{as } n, m \to +\infty,$$

so

$$\|u_n - u_m\| \longrightarrow 0 \quad \text{as } n, m \to +\infty$$

(due to uniform convexity). Hence $\{u_n\}_{n\geqslant 1} \subseteq \partial B_1$ is a Cauchy sequence and so

$$u_n \longrightarrow u \in \partial B_1.$$

Then

$$\langle x^*, u\rangle = \|x^*\|_* \quad \text{and} \quad \|u\| = 1.$$

Due to the uniform convexity of X, this u is unique.

Solution of Problem 5.176
"**(a)** $\Longrightarrow$ **(b)**": For $\lambda > 0$, we have

$$\tfrac{\varphi(x_0+\lambda h)-\varphi(x_0)}{\lambda} - \tfrac{\varphi(x_0-\lambda h)-\varphi(x_0)}{-\lambda} = \tfrac{\varphi(x_0+\lambda h)+\varphi(x_0-\lambda h)-2\varphi(x_0)}{\lambda}. \tag{5.25}$$

Let

$$\begin{aligned} \varphi'_+(x_0)(h) &= \lim_{\lambda\to 0^+} \tfrac{\varphi(x_0+\lambda h)-\varphi(x_0)}{\lambda}, \\ \varphi'_-(x_0)(h) &= \lim_{\lambda\to 0^+} \tfrac{\varphi(x_0-\lambda h)-\varphi(x_0)}{-\lambda}. \end{aligned}$$

Since φ is Fréchet differentiable at x_0., we have

$$\varphi'_+(x_0)(\cdot) = \varphi'_-(x_0)(\cdot)$$

and the two limits are uniform in $h \in \partial B_1$. So, if in (5.25) we let $\lambda \searrow 0$, then we have **(b)**.
"**(b)** $\Longrightarrow$ **(a)**": If **(b)** is true, then from (5.25), we have

$$\varphi'_+(x_0)(\cdot) = \varphi'_-(x_0)(\cdot).$$

Hence

$$\varphi'(x_0)(\cdot) = \varphi'_+(x_0)(\cdot) = \varphi'_-(x_0)(\cdot)$$

is a linear functional. We need to show that it is bounded (i.e., $\varphi'(x_0) \in X^*$). Since by hypothesis φ is continuous at x_0, then we can find $M > 0$ and $\delta > 0$ such that

$$\varphi(x) \leqslant M \qquad \forall\, x \in \overline{B}_\delta(x_0)$$

(where $\overline{B}_\delta(x_0) = \{x \in X : \|x - x_0\| \leqslant \delta\}$). The convexity of φ implies that

$$\tfrac{\varphi(x_0+\lambda h)-\varphi(x_0)}{\lambda} \leqslant \tfrac{M-\varphi(x_0)}{\delta} \qquad \forall\, |\lambda| \leqslant \delta,\ h \in \partial B_1,$$

so

$$\langle \varphi'(x_0), h \rangle \leqslant \tfrac{M-\varphi(x_0)}{\delta} \qquad \forall\, h \in \partial B_1.$$

Solution of Problem 5.177

"(a) $\Longrightarrow$ (b)": Since $\|\cdot\|$ is Fréchet differentiable at $x \in \partial B_1$, by Problem 5.176, for a given $\varepsilon > 0$, there exists $\delta = \delta(\varepsilon) > 0$ such that

$$\|x+h\| + \|x-h\| \leqslant 2 + \varepsilon\|h\| \qquad \forall\, \|h\| \leqslant \delta.$$

Let $\{x_n^*\}_{n\geqslant 1}, \{u_n^*\}_{n\geqslant 1} \subseteq \partial B_1^*$ be two sequences such that

$$\lim_{n\to+\infty} \langle x_n^*, x\rangle = \lim_{n\to+\infty} \langle u_n^*, x\rangle = 1.$$

We can find $n_0 \geqslant 1$ such that

$$\big|\langle x_n^*, x\rangle - 1\big| < \varepsilon\delta, \quad \big|\langle u_n^*, x\rangle - 1\big| < \varepsilon\delta \qquad \forall\, n \geqslant n_0.$$

Then for all $n \geqslant n_0$ and for $\|h\| \leqslant \delta$, we have

$$\begin{aligned}
\langle x_n^* - u_n^*, h\rangle &= \langle x_n^*, x+h\rangle + \langle u_n^*, x-h\rangle - \langle x_n^*, x\rangle - \langle u_n^*, x\rangle \\
&\leqslant \|x+h\| + \|x-h\| - \langle x_n^*, x\rangle - \langle u_n^*, x\rangle \\
&\leqslant 2 + \varepsilon\|h\| - \langle x_n^*, x\rangle - \langle u_n^*, x\rangle \\
&\leqslant \big|1 - \langle x_n^*, x\rangle\big| + \big|1 - \langle u_n^*, u\rangle\big| + \varepsilon\|h\| \\
&\leqslant 3\varepsilon\delta,
\end{aligned}$$

so

$$\begin{aligned}
\|x_n^* - u_n^*\|_* &= \sup\big\{\langle x_n^* - u_n^*, h\rangle : \|h\| = 1\big\} \\
&= \sup\big\{\tfrac{\langle x_n^* - u_n^*, \delta h\rangle}{\delta} : \|h\| = 1\big\} \\
&\leqslant 3\varepsilon \qquad \forall\, n \geqslant n_0
\end{aligned}$$

and thus

$$\|x_n^* - u_n^*\|_* \longrightarrow 0.$$

"**(b) $\Longrightarrow$ (a)**": Suppose that $\|\cdot\|$ is not Fréchet differentiable at x. By Problem 5.176, we can find $\varepsilon > 0$ and a sequence $h_n \longrightarrow 0$ in X such that

$$\|x + h_n\| + \|x - h_n\| \geqslant 2 + \varepsilon\|h_n\|. \tag{5.26}$$

For every $n \geqslant 1$, we choose two sequences $\{x_n^*\}_{n\geqslant 1}, \{u_n^*\}_{n\geqslant 1} \subseteq \partial B_1^*$ such that

$$\langle x_n^*, x + h_n\rangle = \|x+h_n\| \quad \text{and} \quad \langle u_n^*, x - h_n\rangle = \|x-h_n\| \qquad \forall\, n \geqslant 1$$

(see Corollary 5.26). We have

$$\big|\langle x_n^*, h_n\rangle\big| \leqslant \|h_n\| \quad \text{and} \quad \big|\|x + h_n\| - \|x\|\big| \leqslant \|h_n\| \qquad \forall\, n \geqslant 1,$$

so

$$\langle x_n^*, h_n\rangle \longrightarrow 0 \quad \text{and} \quad \|x + h_n\| \longrightarrow \|x\| = 1$$

and thus

$$\begin{aligned}\lim_{n\to+\infty} \langle x_n^*, x\rangle &= \lim_{n\to+\infty} \big(\langle x_n^*, x + h_n\rangle - \langle x_n^*, h_n\rangle\big)\\ &= \lim_{n\to+\infty} \big(\|x + h_n\| - \langle x_n^*, h_n\rangle\big) = 1.\end{aligned}$$

Similarly, we show that

$$\lim_{n\to+\infty} \langle u_n^*, x\rangle = 1.$$

Then

$$\begin{aligned}\langle x_n^* - u_n^*, h_n\rangle &= \langle x_n^*, x + h_n\rangle + \langle u_n^*, x - h_n\rangle - \langle x_n^* + u_n^*, x\rangle\\ &\geqslant \|x + h_n\| + \|x - h_n\| - 2\\ &\geqslant \varepsilon\|h_n\|\end{aligned}$$

(see (5.26)), so

$$\|x_n^* - u_n^*\|_* \geqslant \varepsilon \qquad \forall\, n \geqslant 1,$$

a contradiction.

Solution of Problem 5.178

Let $x \in \partial B_1$ and choose $x^* \in \partial B_1^*$ such that $\langle x^*, x\rangle = 1$ (see Corollary 5.26). Suppose that $\{x_n^*\}_{n\geqslant 1}, \{u_n^*\}_{n\geqslant 1} \subseteq \partial B_1^*$ are two sequences satisfying

$$\langle x_n^*, x\rangle \longrightarrow 1 \quad \text{and} \quad \langle u_n^*, x\rangle \longrightarrow 1$$

Then

$$2 \geqslant \|x_n^* + u_n^*\|_* \geqslant \langle x_n^* + u_n^*, x\rangle \longrightarrow 2 \quad \text{as } n \to +\infty,$$

so

$$\lim_{n\to+\infty} \left(2\|x_n^*\|_*^2 + 2\|u_n^*\|_*^2 - \|x_n^* + u_n^*\|_*^2\right) = 0$$

and thus

$$\|x_n^* - u_n^*\|_* \longrightarrow 0$$

due to the fact that $\|\cdot\|_*$ is locally uniformly convex (see Definition 5.168(b)). By Problem 5.177, $\|\cdot\|$ is Fréchet differentiable.

Solution of Problem 5.179

By Theorem 5.190, we can find a countable set $D \subseteq E$ such that $x \in \overline{D}^w$. Let $V = \overline{\operatorname{span}} D$. Then V is a separable Banach space and $\overline{D}^w \subseteq V \cap C$, hence $\overline{D}^w$ is w-compact. Invoking Theorem 5.85(c), we have that $(\overline{D}^w, w)$ is compact metrizable. So, we can find a sequence $\{x_n\}_{n\geqslant 1} \subseteq D$ such that $x_n \xrightarrow{w} x$ in X.

Solution of Problem 5.180

From Theorem 5.85(a), we know that $\overline{B}_1^* = \{x^* \in X^* : \|x^*\|_* \leqslant 1\}$ furnished with the w^*-topology is metrizable. The set $C^* \cap \overline{B}_1^*$ is sequentially w^*-closed and metrizable for the w^*-topology. So, the set $C^* \cap \overline{B}_1^*$ is w^*-closed. Invoking the Banach–Diedonné theorem (see Theorem 5.189), we have that the set C^* is w^*-closed.

Alternative Solution

By Theorem 5.85(a), the w^*-topology on $\overline{B}_1^* = \{x^* \in X^* : \|x^*\|_* \leqslant 1\}$ is metrizable. In metric spaces closedness is sequentially determined. Hence, since the set $C \cap (n\overline{B}_1^*)$ is sequentially w^*-closed (in $n\overline{B}_1^*$), it is w^*-closed in $n\overline{B}_1^*$. But the latter is w^*-closed in X^*. Therefore, we conclude that the set C is w^*-closed.

Bibliography

[1] Beauzamy, B.: Introduction to Banach Spaces and Their Geometry. Notas de Matemática, vol. 86. North-Holland Publishing Co., Amsterdam (1985)

[2] Brézis, H.: Functional Analysis, Sobolev Spaces and Partial Differential Equations. Springer, New York (2011)

[3] Brown, A., Pearcy, C.: Introduction to Operator Theory I: Elements of Functional Analysis. Graduate Texts in Mathematics, vol. 55. Springer, New York (1977)

[4] Day, M.M.: Normed Linear Spaces. Ergebnisse der Mathematik und ihrer Grenzgebiete, vol. 21. Springer, Berlin (1973)

[5] Diestel, J.: Sequences and Series in Banach Spaces. Graduate Texts in Mathematics, vol. 92. Springer, New York (1984)

[6] Dunford, N.: Schwartz, J.T.: Linear Operators. I. General Theory. Pure and Applied Mathematics, vol. 7. Wiley, New York (1958)

[7] Enflo, P.: A Counterexample to the approximation problem in Banach spaces. Acta Math. **130**, 309–317 (1973)

[8] Fabian, M.J., Habala, P., Hájek, P., Montesinos Santalucía, V., Pelant, J., Zizler, V.: Functional Analysis and Infinite Dimensional Geometry. CMS Books in Mathematics, vol. 8. Springer, New York (2001)

[9] James, R.C.: A non-reflexive Banach space isometric to its second conjugate space. Proc. Natl. Acad. Sci. U.S.A. **37**, 174–177 (1951)

[10] Lax, P.D.: Functional Analysis. Pure and Applied Mathematics. Wiley-Interscience, New York (2002)

[11] Megginson, R.E.: An Introduction to Banach Space Theory. Graduate Studies in Mathematics, vol. 183. Springer, New York (1998)

[12] Reed, M., Simon, B.: Methods of Modern Mathematical Physics. I: Functional Analysis. Academic, New York (1972)

[13] Rudin, W.: Functional Analysis. International Series in Pure and Applied Mathematics. McGraw-Hill Book Co., New York (1991)

[14] Yosida, K.: Functional Analysis. Die Grundlehren der Mathematischen Wissenschaften, vol. 123. Springer, Berlin (1978)

Other Problem Books

[1] Aksoy, A., Khamsi, M.: A Problem Book in Real Analysis. Problem Books in Mathematics. Springer, New York (2009)

[2] Aliprantis, C.D., Burkinshaw, O.: Problems in Real Analysis: A Workbook with Solutions. Academic, San Diego (1999)

[3] Costara, C., Popa, D.: Exercises in Functional Analysis. Kluwer Texts in the Mathematical Sciences, vol. 26. Kluwer, Dordrecht (2003)

[4] de Souza, P.N., Silva, J.-N.: Berkeley Problems in Mathematics. Problem Books in Mathematics. Springer, New York (2004)

[5] Gelbaum, B.R.: Problems in Real and Complex Analysis. Problem Books in Mathematics. Springer, New York (1992)

[6] Gelbaum, B.R., Olmsted, J.M.H.: Theorems and Counterexamples in Mathematics. Problem Books in Mathematics. Springer, New York (1990)

[7] George, C.: Exercises in Integration. Problem Books in Mathematics. Springer, New York (1984)

[8] Kaczor, W.J., Nowak, M.T.: Problems in Mathematical Analysis I: Real Numbers, Sequences and Series. Student Mathematical Library, vol. 4. American Mathematical Society, Providence (2000)

[9] Kaczor, W.J., Nowak, M.T.: Problems in Mathematical Analysis II: Continuity and Differentiation. Student Mathematical Library, vol. 12. American Mathematical Society, Providence (2001)

L. Gasiński and N.S. Papageorgiou, *Exercises in Analysis: Part 1*, Problem Books in Mathematics, DOI 10.1007/978-3-319-06176-4,

[10] Kirillov, A.A., Gvishiani, A.D.: Theorems and Problems in Functional Analysis. Problem Books in Mathematics. Springer, New York (1982)

[11] Ta-Tsien, L. (ed.): Problems and Solutions in Mathematics. Major American Universities Ph.D. Qualifying Questions and Solutions - Mathematics. World Scientific, Hackensack (2003)

List of Symbols

Symbol	Page	Meaning
d_X	p. 1	metric on space X
$\operatorname{diam} C$	p. 4	diameter of set C
$\operatorname{dist}(x, A)$	p. 4	distance of x from set A
$\operatorname{dist}(A, B)$	p. 4	distance of sets A and B
$\operatorname{int} E$	p. 6	interior of set E
$\overline{E}$	p. 6	closure of set E
∂E	p. 6	boundary of set E
$\omega_f(x)$	p. 11	oscillation of function f at point x
$C(X)$	p. 20	space of continuous functions
$\operatorname{supp} f$	p. 26	support of function f
$\operatorname{Gr} f$	p. 31	graph of function f
$\operatorname{epi} f$	p. 31	epigraph of function f
$\liminf\limits_{n\to+\infty} E_n$	p. 31	lower Kuratowski limit of sets $\{E_n\}_{n\geqslant 1}$
$\limsup\limits_{n\to+\infty} E_n$	p. 32	upper Kuratowski limit of sets $\{E_n\}_{n\geqslant 1}$
$P_f(X)$	p. 32	collection of nonempty closed subsets of X
$A_n \xrightarrow{h} A$	p. 32	convergence in Hausdorff metric space $(P_f(X), h)$
$\mathcal{N}(x)$	p. 194	family of neighbourhoods of x
$\operatorname{int} A$	p. 195	interior of set A

L. Gasiński and N.S. Papageorgiou, *Exercises in Analysis: Part 1*,
Problem Books in Mathematics, DOI 10.1007/978-3-319-06176-4,

$\overset{\circ}{A}$	p. 195	interior of set A
$\operatorname{ext} A$	p. 195	exterior of set A
∂A	p. 195	boundary of set A
$\operatorname{bd} A$	p. 195	boundary of set A
A'	p. 196	derived set of A
$\overline{A}$	p. 196	closure of set A
$\operatorname{cl} A$	p. 196	closure of set A
$\tau(\mathcal{A})$	p. 198	topology generated by $\mathcal{A}$
$\omega_f(x)$	p. 203	oscillation of function f at point x
$\mathbb{R}^*$	p. 203	$\mathbb{R}^* = \mathbb{R} \cup \{\pm\infty\}$
$\overline{f}$	p. 207	relaxed function of $f\colon X \longrightarrow \mathbb{R}^*$
$\overline{\mathbb{R}}$	p. 208	$\overline{\mathbb{R}} = \mathbb{R} \cup \{+\infty\}$
$\overline{\mathbb{R}}^*$	p. 208	$\overline{\mathbb{R}}^* = \mathbb{R} \cup \{-\infty\}$
$\{U_i\}_{i\in I} \prec \{V_j\}_{j\in J}$	p. 224	$\{U_i\}_{i\in I}$ is a refinement of $\{V_j\}_{j\in J}$
$\operatorname{supp} f$	p. 225	support of function f
$\operatorname{Gr} F$	p. 230	graph of multifunction $F\colon X \longrightarrow 2^Y$
$f_0 \overset{V}{\simeq} f_1$	p. 234	f_0 is homotopic to f_1 relative to V
$f \simeq 0$	p. 234	f is nullhomotopic
$f \simeq_p g$	p. 234	paths f and g are path homotopic
$(X,C) \simeq ((Y,D))$	p. 235	homotopy equivalent pairs
$\pi_1(X,x_0)$	p. 238	set of equivalence classes of homotopy relation between loops
σ_n	p. 245	n-simplex
$\dim K$	p. 246	dimension of geometric simplicial complex K
$\|K\|$	p. 246	polyhedron of K
$N_K(x)$	p. 247	simplicial neighbourhood of x
$Lk_K(x)$	p. 247	link of x
$st_K(\sigma)$	p. 247	star of simplex σ

$C_k(K)$	p. 248	group of (oriented) k-chains
∂_k	p. 249	boundary operator
$Z_k(K)$	p. 249	kernel of boundary operator $\partial_k : C_k(K) \longrightarrow C_{k-1}(C)$
$B_k(K)$	p. 250	image of boundary operator $\partial_{k+1} : C_{k+1}(K) \longrightarrow C_k(K)$
$H_k(K)$	p. 250	kth (simplicial) homology group
$\widetilde{H}_0(K)$	p. 250	reduced homology group of K in dimension 0
$C_k(K, L)$	p. 251	group of relative chains of $K \bmod L$
$Z_k(K, L)$	p. 251	group of relative k-cycles
$B_k(K, L)$	p. 251	group of relative k-boundaries
$H_k(K, L)$	p. 251	kth relative (simplicial) homology group
Δ_n	p. 252	standard n-simplex
$S_n(X)$	p. 252	nth singular chain group
f_*	p. 252	homomorphism induced by f
$C_n(X, A)$	p. 255	relative n-singular chain group of $X \bmod A$
$C^n(X; G)$	p. 262	set of all singular n-cochains
$\#X$	p. 266	cardinality of set X
$\overline{\mathbb{R}}_+$	p. 267	$\overline{\mathbb{R}}_+ = [0, +\infty]$
$\operatorname{cont} f$	p. 268	$\operatorname{cont} f = \{x \in X : f \text{ is continuous at } x\}$
$C(X)$	p. 271	space of all continuous functions $f : X \longrightarrow \mathbb{R}$
$C_b(X)$	p. 271	space of all bounded continuous functions $f : X \longrightarrow \mathbb{R}$
$\sigma(\mathcal{Y})$	p. 405	σ-algebra generated by $\mathcal{Y}$
$\mathcal{B}(X)$	p. 405	Borel σ-algebra
$\lambda(\mathcal{F})$	p. 406	λ-class generated by $\mathcal{F}$

$m(\mathcal{F})$	p. 406 monotone class generated by $\mathcal{F}$
$\mathcal{L}(\mathbb{R}^N)$	p. 412 σ-algebra of Lebesgue measurable sets
$\Sigma \otimes \Sigma'$	p. 416 product σ-algebra
D_x	p. 416 x-section
D_y	p. 416 y-section
μf^{-1}	p. 419 image measure (or measure induced by function f)
$\mu \circ f^{-1}$	p. 419 image measure (or measure induced by function f)
χ_A	p. 421 characteristic function (or indicator function)
$\mathcal{L}^1(\mu)$	p. 428 set of all $\mathbb{R}^*$-valued, μ-integrable functions
f_x	p. 437 x-section of function $f\colon \Omega \times \Omega \longrightarrow \mathbb{R}$
f_y	p. 437 y-section of function $f\colon \Omega \times \Omega \longrightarrow \mathbb{R}$
$f_n \xrightarrow{\mu} f$	p. 443 convergence in measure
$L^0(\Omega)$	p. 444 vector space of all equivalence classes of measurable functions
μ^+	p. 448 positive part of signed measure μ
μ^-	p. 448 negative part of signed measure μ
$\|\mu\|$	p. 448 total variation of signed measure μ
$\\|\mu\\|$	p. 448 total variation norm of signed measure μ
$\nu \ll \mu$	p. 449 ν is absolutely continuous with respect to μ
$L^1_{\text{loc}}(\mathbb{R}^N)$	p. 450 space of locally integrable functions on $\mathbb{R}^N$
$E^{\Sigma_0} f$	p. 451 conditional expectation of f with respect to Σ_0
Σ_σ	p. 454 collection of all sets prior to stopping time σ

$f_n \xrightarrow{au} f$	p. 491	f_n converges to f almost uniformly
$\mathcal{B}(X)$	p. 633	Borel σ-algebra in X
$\mathcal{B}a(X)$	p. 633	Baire σ-algebra in X
$\operatorname{supp} \mu$	p. 636	support of measure μ
δ_x	p. 636	Dirac measure at x
$C_c(X)$	p. 637	space of continuous functions $f\colon X \longrightarrow \mathbb{R}$ which have compact supports
$C_0(X)$	p. 637	space of continuous functions $f\colon X \longrightarrow \mathbb{R}$ which vanish at infinity
$M_b(X)$	p. 638	space of signed Radon (tight) Borel measures
$M_1^+(X)$	p. 638	probability measures on $\mathcal{B}(X)$
$UC_b(X)$	p. 639	set of all bounded and uniformly continuous function $f\colon X \longrightarrow \mathbb{R}$
$\widehat{\Sigma}$	p. 642	universal completion of Σ
$\mathcal{B}_u(X)$	p. 642	universal σ-algebra of X
$\operatorname{dom} F$	p. 643	domain of F
$P_f(X)$	p. 643	$P_f(X) = \{A \subseteq X :$ A is nonempty and closed$\}$
$P_k(X)$	p. 643	$P_k(X) = \{A \subseteq X :$ A is nonempty and compact$\}$
$\widehat{P}_f(X)$	p. 643	$\widehat{P}_f(X) = P_f(X) \cup \{\emptyset\}$
$P_{fc}(X)$	p. 643	$P_{fc}(X) = \{A \in P_f(X) :\ A$ is convex$\}$
$P_{kc}(X)$	p. 643	$P_{kc}(X) = \{A \in P_k(X) :\ A$ is convex$\}$
$P_{wkc}(X)$	p. 643	$P_{wkc}(X) = \{A \subseteq X :\ A$ is nonempty, w-compact and convex$\}$
$L^p(\Omega)^*$	p. 648	dual of Banach space $L^p(\Omega)$
$\operatorname{Var} u$	p. 660	total variation of function u
$BV(T)$	p. 660	space of functions of bounded variation
$BV_{\text{loc}}(T)$	p. 660	space of functions locally of bour' variation

$\mathrm{Var}_T u$ p. 660 total variation of function u on T

$BV(T;\mathbb{R}^N)$ p. 660 space of functions of bounded variation

$BV_{\mathrm{loc}}(T;\mathbb{R}^N)$ p. 660 space of functions locally of bounded variation

$V,\ V_{t_0},\ V_{t_0,u}$ p. 661 indefinite variation of u

$N_u(\cdot;A)$ p. 663 Banach indicatrix of u on A

$N_u(\cdot)$ p. 663 Banach indicatrix of u on X

$AC(T)$ p. 664 space of absolutely continuous functions $u\colon T \longrightarrow \mathbb{R}$

$AC_{\mathrm{loc}}(T)$ p. 664 space of locally absolutely continuous functions $u\colon T \longrightarrow \mathbb{R}$

$AC(T;\mathbb{R}^N)$ p. 664 space of absolutely continuous functions $u\colon T \longrightarrow \mathbb{R}^N$

$AC_{\mathrm{loc}}(T;\mathbb{R}^N)$ p. 664 space of locally absolutely continuous functions $u\colon T \longrightarrow \mathbb{R}^N$

$BV(\Omega)$ p. 668 space of functions of bounded variation

$BV_{\mathrm{loc}}(\Omega)$ p. 668 space of functions which are locally of bounded variation

$V(u;\Omega)$ p. 669 variation of function u in set Ω

$\partial^* A$ p. 670 reduced boundary of set A

$n(x,A)$ p. 672 measure-theoretic outer normal of A at x

$\partial_* A$ p. 672 measure-theoretic boundary of A

$H^s(A)$ p. 673 s-dimensional Hausdorff measure of A

$\mu_n \xrightarrow{v} \mu$ p. 692 sequence of Radon measures $\{\mu_n\}_{n\geqslant 1}$ converges vaguely to Radon measure μ

$\overline{A}_V$ p. 773 closure of A relative to V

$\partial_V A$ p. 773 boundary of A relative to V

$\mathrm{Disc}\,(f)$ p. 810 discontinuity set of f

$\|\cdot\|_X$ p. 836 norm in X

p_C	p. 837	Minkowski functional (or gauge functional)
$A^{\perp}$	p. 843	annihilator of A in X^*
${}^{\perp}C$	p. 843	annihilator of C in X
$D(A)$	p. 861	domain of operator A
$R(A)$	p. 861	range of operator A
$\operatorname{Gr} A$	p. 861	graph of operator A
$N(A)$	p. 861	kernel of operator A
A^*	p. 861	adjoint of operator A
$\operatorname{ext} C$	p. 863	set of all extreme points of C
$S_A(x^*, \eta)$	p. 864	slice of set A
$\sigma(x^*, A)$	p. 865	support function of set A
$\mathcal{L}_c(X; Y)$	p. 865	space of compact operators from X into Y
$\mathcal{L}_f(X; Y)$	p. 865	space of finite rank operators from X into Y
$i(A)$	p. 866	index of operator A
$\sigma(A)$	p. 867	spectrum of operator A
$\varrho(A)$	p. 867	resolvent of operator A
$\sigma_p(A)$	p. 868	point spectrum of operator A

Index

(Σ, Σ')-measurable function, 418
F_σ-set, 16, 197
G_δ-set, 16, 197
H-elliptic form, 859
H^1-rectifiable set, 674
L^1-metric, 3
N-dimensional Lebesgue measure, 412
N-function, 468
T_0 space, 194
T_1 space, 194
T_2 space, 194
T_3 space, 194
T_4 space, 194
α-Hölder continuous function, 439
ε-chain, 26
ε-net, 18
λ-class, 405
μ-a.e., 422
μ-almost everywhere, 422
μ-equivalent sets, 446
μ-integrable function, 428
μ-null set, 409
π-λ theorem, 406
π-class, 405
inf-compact function, 692
σ-additive
  function, 407
  set function, 408
σ-algebra, 403
  generated by functions, 419
  relative, 406
σ-compact
  metric space, 20
  space, 216
σ-field, 403
σ-finite
  measure space, 408
  signed measure, 448
σ-ring, 403
σ-subadditive set function, 408
c-topology, 230
i-face function, 253
k-boundaries, 250
k-chain, 248
k-contraction, 13
k-cycles, 249
k-space, 275
kth homology group, 250
m-face, 245
n-cochain
  singular, 262
n-simplex, 245
  singular, 252
  standard, 252
nth homotopy group, 241
nth relative singular homology group of $X \bmod A$, 256

L. Gasiński and N.S. Papageorgiou, *Exercises in Analysis: Part 1*, Problem Books in Mathematics, DOI 10.1007/978-3-319-06176-4,

nth singular
  chain group, 252
  homology group, 253
s-dimensional Hausdorff measure, 673
w^*-topology, 708
x-section, 416
y-section, 416

a.e., 422
absolutely
  continuous
    function, 649, 664
    measures, 449
  convergent series, 877
absorbing set, 836
accumulation point, 6, 196
additive set function, 407
adjoint operator, 842, 859, 861
admissible class of spaces, 259
affinely independent vectors, 245
Alaoglu theorem, 849
Alexander duality theorem, 263
Alexandrov
  one-point compactification, 216
  theorem, 16
algebra, 403
almost
  convex set, 874
  everywhere, 422
  uniform convergence, 491
analytic set, 227
annihilator, 843
approximately continuous function, 703
approximation
  simplicial, 247
arcwise-connected space, 221
area formula, 667, 675
Arzela–Ascoli theorem, 21, 232
Asplund space, 872
asymptotically nonexpansive function, 40
atom, 414
augmentation function, 250
axiom
  dimension, 260
  exactness, 260
  excision, 260
  homotopy, 260

backward shift operator, 899
Baire
  σ-algebra, 633
  category theorem, 9
  function, 634
  measurable function, 634
  metric space, 20
  space, 225
balanced set, 836
ball
  closed, 6
  open, 3
Banach
  fixed point theorem, 14
  indicatrix, 663
  space, 836
  theorem, 845
Banach–Dieudnné theorem, 875
Banach–Steinhaustheorem, 843
Banach–Zaretsky theorem, 665
barycentric coordinates, 245
base, 7
  local, 7
Basel inequality, 858
basis, 198
  local, 198, 846
  neighbourhood, 198

orthonormal, 857
Schauder, 858
Besicovitch covering theorem, 653
best approximation point, 874
Betti number, 254, 256
Bolzano theorem, 22
Bolzano-Weierstrass property, 215
Borel
σ-algebra, 405, 633
function, 634, 641
isomorphic spaces, 641
measurable function, 420, 634
measure, 417
set, 405
space, 641
Borel–Cantelli lemma, 462
boundary, 6, 195
measure-theoretic, 672
operator, 253, 255
point, 195
reduced, 670
relative, 773
boundary operator, 249
bounded
set, 3, 691, 837
signed Radon measure, 638
variation, 660, 668
box topology, 211
brackets
duality, 262
Brezis–Lieb lemma, 433

canonical embedding, 849
Cantor
function, 425
intersection theorem, 9
set, 413
Cantor–Bendixson theorem, 35
Carathéodory
function, 675
theorem, 411
Cauchy
in measure sequence, 442
sequence, 4
Cauchy–Schwarz inequality, 854
Cauchy–Schwarz–Bunyakowski inequality, 434
chain
elementary, 248
rule, 666
change
of variable formula, 439, 440, 675
of variables, 667
characteristic function, 421
Chebyshev
inequality, 430, 435
set, 874
circle representative of a loop, 240
Clarkson inequality, 435, 854
class
σMK, 647
of spaces
admissible, 259
clopen set, 22, 194
closed
ball, 6
function, 11, 208
graph theorem, 845
operator, 861
relation, 271
set, 5, 193
closure, 6, 196
relative, 773
cluster point, 6, 196
coarea formula, 675

coboundary operator, 262
cocountable set, 405
coercive
  form, 859
  function, 217
cofinite set, 404
cohomology theory, 261
compact
  metric space, 17
  operator, 865
  set, 213
  space, 213
compact-open topology, 230
compactification, 216
  Alexandrov, 216
complement
  topological, 845
complementary subspaces, 845
complete
  measure space, 410
  metric space, 4
completely
  continuous operator, 865
  regular space, 271
completion, 14
  of a measure, 410
  of a measure space, 410
  universal, 642
component
  connected, 218
  path-connected, 25, 222
composition function, 231
conditional expectation, 451
conjugate
  exponents, 649
  subgroup, 289
connected
  component, 23, 218
  set, 22, 217
  space, 217
connectivity relation, 25
continuous
  convergence, 233
  form, 859
  function, 9, 202
  multifunction, 229
  selector, 230
  set function, 650
contractible space, 235
contraction, 13
convergence, 4
  almost uniform, 491
  continuous, 233
  in measure, 443
  in probability, 443
  pointwise, 16
  uniform, 16
  uniform on compacta, 232
convergent series, 877
convex
  body, 876
  function, 433
  set, 229
convexity theorem, 451
coordinates
  barycentric, 245
countably
  additive set function, 408
  compact set, 215
  generated σ-algebra, 407
  separated, 686
cover, 652
  fine, 652
  locally finite, 26
  open, 8, 199
  Vitali, 652
covering
  function, 242

space, 242
universal, 244
theorem, 653
curve
simple, 706
cut
pair, 220
point, 220

Darboux property, 59
decomposition
Hahn, 447
Jordan, 448
Lebesgue, 450
theorem, 454
definiteness property, 854
deformation
retract, 236
rctraction, 236
deleted neighbourhood, 194
dense set, 7, 196, 852
densely defined operator, 861
density
lower, 703
of a set at a point, 704
upper, 703
derivative
Radon–Nikodym, 449
with respect to measure, 654
derived set, 196
diameter, 4
differentiability
set, 656
with respect to measure, 654
differentiation theorem, 655
dimension, 246
axiom, 260
Hausdorff, 674
Dini theorem, 49, 274
Dirac measure, 636
directed set, 200
disconnected
set, 22
space, 217
disconnection, 217
discontinuous function, 202
discrete
metric, 2
space, 194
topology, 194
dispersed space, 228
dispersible space, 228
distance, 1
domain, 643
dominated convergence theorem, 431, 452
Doob
L^p martingale inequality, 456
decomposition theorem, 454
first martingale inequality, 456
upcrossing inequality, 457
doubling measure, 685
dual
norm equivalent, 885
space, 838
duality
brackets, 262
map, 841
Dunford–Pettis theorem, 651
Dynkin
class, 405
theorem, 406

Eberlein–Smulian theorem, 851
ecart, 1
Egorov theorem, 423
eigenfunction, 868
eigenspace, 868
eigenvalue, 868

elementary chain, 248
embedding
  canonical, 849
epigraph of function, 31
equality
  Parseval, 858
equibounded set, 882
equicoercive functions, 272
equicontinuous
  family, 21
  set, 232
    functions, 650
equiintegrable family, 441
equimeasurable functions, 687
equivalent
  dual norm, 885
  measures, 450
  metric spaces, 14
  metrics, 14
  sets, 446
Euclidean metric, 2
evaluation function, 231
exactness axiom, 260
excision
  axiom, 260
  property, 258
excisive pair, 258
exponential function, 240
exponents
  conjugate, 649
extension, 10
  theorem, 12
exterior, 195
  point, 195
extreme
  point, 863
  set, 864
family
  closed under proper differences, 461
  equicontinuous, 21
  equiintegrable, 441
  locally finite, 203
  neighbourhood finite, 203
  pointwise bounded, 21
  regular, 657
  separating, 209
  total, 209
  uniformly
    bounded, 21
    equicontinuous, 21
    integrable, 441
Fatou lemma, 427, 431, 452
field, 403
final topology, 212
finale point, 220
fine cover, 652
finite
  intersection property, 18, 213
  measure space, 408
  perimeter, 668
  rank operator, 865
  signed measure, 448
finitely additive set function, 407
first
  category
    set, 8
  countability, 8
  countable space, 198
fixed point, 14
form
  H-elliptic, 859
  coercive, 859

continuous, 859
symmetric, 859
formula
area formula, 667, 675
change of variable formula, 675
coarea, 675
integration by parts, 667
Fréchet
compact set, 215
differentiable function, 870
differentiable norm, 871
space, 836
Fréchet–Jordan–von Neumann theorem, 854
Fredholm
alternative theorem, 867
operator, 866
Fubini theorem on series of monotone functions, 659
Fubini–Tonelli theorem, 437
function
μ-integrable, 428
μ-quasi integrable, 428
inf-compact, 692
σ-additive, 407, 408
σ-subadditive, 408
k-contraction, 13
absolutely continuous, 649, 664
additive, 407
approximately continuous, 703
asymptotically nonexpansive, 40
Baire, 634
measurable, 634
Borel, 634, 641
measurable, 420
Borel measurable, 634
Cantor, 425
Carathéodory, 675
closed, 11, 208
coercive, 217
composition, 231
continuous, 9, 202
at a point, 202
contraction, 13
convex, 433
countably additive, 408
covering, 242
discontinuous, 202
epigraph, 31
evaluation, 231
exponential, 240
finitely additive, 407
Fréchet differentiable, 870
Gâteaux differentiable, 870
graph, 31
Hardy–Littlewood, 450
identification, 212
integrable, 428
Lipschitz continuous, 13
locally
absolutely continuous, 664
integrable, 450
Lipschitz, 13
of bounded variation, 660
lower semicontinuous, 43, 204
measurable, 418
monotone, 407
nonexpansive, 13
of bounded variation, 660, 668
of local bounded variation, 668
open, 11, 208
oscillation, 11, 203

proper, 19
quasi integrable, 428
relaxed, 207
saltus, 667
sequentially
coercive, 217
continuous, 205
lower semicontinuous, 205
upper semicontinuous, 205
simplicial, 246
singular, 649, 667
strictly convex, 433
subadditive, 408
superadditive, 408
uniformly continuous, 13
universally measurable, 642
upper semicontinuous, 204
Vitali continuous, 650
functional
gauge, 837
locally bounded, 277
Minkowski, 837
sublinear, 840
functions
equicoercive, 272
equicontinuous, 232
equimeasurable, 687
uniformly ν-absolutely continuous, 650
Vitali equicontinuous, 650
fundamental group, 238

Gâteaux differentiable
function, 870
norm, 871
gauge functional, 837
generated topology, 198
geometric simplicial complex, 246
dimension, 246
Goldstine theorem, 849
graph, 861
measurable multifunction, 643
of function, 31
of multifunction, 230
group
fundamental, 238
of relative chains, 251

Hölder
continuous function, 439
inequality, 434
Hahn
decomposition, 447
decomposition theorem, 447
Hahn–Banach theorem, 840
Hanner inequality, 436
Hardy–Littlewood function, 450
Hardy–Littlewood–Wiener maximal theorem, 451
Hausdorff
dimension, 674
measure, 673
metric space, 32
space, 194
Heine–Borel theorem, 214
Helly
first theorem, 663
selection theorem, 663
Hilbert space, 855
homeomorphic
metric spaces, 15
spaces, 11, 208
homeomorphism, 11, 208
local, 208
homology
group
reduced, 250
sequence
reduced exact, 258

singular, 253
theory, 259
homotopic loops, 237
homotopy
axiom, 260
class, 234
equivalent pairs, 235
relative, 234
relevance, 234
types, 235

identification
function, 212
points, 291
space, 212
topology, 212
identity
polarization, 854
image measure, 419
indefinite variation, 661
index, 866
indicator function, 421
indicatrix, 663
indiscrete topology, 193
induced topology, 197
inequality
Basel, 858
Cauchy–Schwarz, 854
Cauchy–Schwarz–Bunyakowski, 434
Chebyshev, 430, 435
Clarkson, 435, 854
Doob
L^p martingale, 456
first martingale, 456
upcrossing, 457
Hölder, 434
Hanner, 436
interpolation, 435
isodiametric, 675
Jensen, 434, 453
Markov, 429, 435
Minkowski, 434
initial
point, 220
topology, 209
inner
product, 854
regular measure, 635
with respect to compact sets, 635
regular set, 679
with respect to compact sets, 679
integrable function, 428
integral, 426, 428
integration by parts formula, 667
interior, 6, 195
point, 6, 195
intermediate value theorem, 22
interpolation inequality, 435
intersection property, 18
isodiametric inequality, 675
isolated point, 6, 196
isometric spaces, 12, 839
isometry, 12, 839
isomorphic
sets, 641
spaces, 641, 839
isomorphism, 641, 839

Jacobi transformation theorem, 440
James theorem, 850
Jensen inequality, 434, 453

Jordan decomposition, 448
  theorem, 448

Kadec–Klee property, 853, 901
kernel, 838
Kolmogorov space, 194
Krein–Milman theorem, 863
Krein-Rutman theorem, 870
Kuratowski
  limit
    lower, 31
    upper, 32
  measure of noncompactness, 55
  theorem, 642
Kuratowski–Ryll Nardzewski selection theorem, 644
Kuratowski–Ulam theorem, 270

lattice property, 429
Lax–Milgram theorem, 860
Lebesgue
  decomposition, 450
    theorem, 450
  differentiation theorem, 655
  dominated convergence theorem, 431
  measurable set, 412
  measure, 412
  monotone convergence theorem, 430
  number, 55
  point, 656
  set, 656
  theorem on monotone functions, 658
Lebesgue–Stieltjes measure, 418
lemma
  Borel–Cantelli lemma, 462
  Brezis–Lieb lemma, 433
  Fatou lemma, 427, 431, 452
  Riesz lemma, 840
  shrinking lemma, 27
  Urysohn lemma, 12, 223
lift, 242
limit, 4, 195
  point, 6, 196
Lindelöf
  metric space, 8
  space, 199
linear operator
  weakly continuous, 848
link, 247
Lipschitz
  continuous function, 13
  equivalent metric spaces, 15
local
  base, 7
  basis, 198, 846
  bounded variation, 668
  homeomorphism, 208
  maximum, 32
  minimum, 32
  strict
    maximum, 32
    minimum, 32
localizable space, 415
locally
  absolutely continuous function, 664
  bounded functional, 277
  compact
    metric space, 20
    space, 215
  connected
    metric space, 23
    space, 219
  convex space, 836

finite
cover, 26
family, 203
perimeter, 668
integrable function, 450
Lipschitz function, 13
path-connected
metric space, 25
space, 221
uniformly convex
norm, 871
space, 871
loop
based at a point, 237
loops
homotopic, 237
lower
density, 703
Kuratowski limit, 31
semicontinuous
function, 43, 204
multifunction, 229
lsc multifunction, 229
Lusin
N-property, 665
space, 229
theorem, 423
Lyapunov convexity theorem, 451

map
duality, 841
Markov inequality, 429, 435
martingale, 453
inequality, 456
regular, 459
maximum
local, 32
Mayer–Vietoris sequence, 258
Mazur theorem, 847
meager set, 8
measurable
function, 418
multifunction, 643
space, 407
countably separated, 686
measure, 407
Borel, 417
doubling, 685
induced by a function, 419
inner regular, 635
with respect to compact sets, 635
Lebesgue, 412
Lebesgue–Stieltjes, 418
mutually singular, 449
nonatomic, 414
of noncompactness, 55
outer regular, 635
purely atomic, 414
Radon, 635
regular, 635
separable, 446
signed, 407
singular, 449
support, 636
tight, 635
uniformly distributed, 682
measure space, 408
σ-finite, 408
complete, 410
finite, 408
localizable, 415
probability, 408
semifinite, 408
measure-theoretic
boundary, 672
outer normal, 671

measures
  absolutely continuous, 449
  equivalent, 450
metric, 1
  L^1, 3
  discrete, 2
  Euklidian, 2
  Hausdorff, 32
  projection, 855, 874
  supremum, 3
  topology, 5
  uniform, 3
metric space, 1
  σ-compact, 20
  Baire, 20
  compact, 17
  complete, 4
  Lindelöf, 8
  locally
    compact, 20
    connected, 23
    path-connected, 25
  of property S, 56
  path-connected, 24
  separable, 8
  topologically complete, 15
  totally disconnected, 23
  well-chained, 26
metric spaces
  equivalent, 14
  homeomorphic, 15
  Lipschitz equivalent, 15
  topologically
    equivalent, 14
    homeomorphic, 15
  uniformly equivalent, 15
metrically separated sets, 682
metrics
  equivalent, 14
  topologically equivalent, 14
metrizable topological space, 194
Michael selection theorem, 229
Milman–Pettis theorem, 853
minimum
  local, 32
Minkowski
  functional, 837
  inequality, 434
monotone
  class, 406
    theorem, 406
  convergence theorem, 427, 430, 452
  sequence, 892
  set function, 407
multifunction
  continuous, 229
  graph, 230
    measurable, 643
  lower semicontinuous, 229
  lsc, 229
  measurable, 643
  upper semicontinuous, 229
  usc, 229
  Vietoris continuous, 229
mutually singular
  measure, 449
  signed measure, 449

negative
  part of signed measure, 448
  set, 447
neighbourhood, 5
  basis, 198
  deleted, 194
  finite family, 203
  of ε-continuity, 232
  of a point, 194

of a set, 194
simplicial, 247
net, 200
convergence, 200
frequently in a set, 200
Nikodym theorem, 650
nilpotent operator, 880
nonatomic measure, 414
nonexpansive function, 13
norm, 836
Fréchet differentiable, 871
Gâteaux differentiable, 871
locally uniformly convex, 871
quotient, 839
rotund, 871
strictly convex, 871
normable space, 836
normal
operator, 859
space, 194
normed space, 836
nowhere dense set, 196
null set, 409
nullhomotopic, 234
nullspace, 838
number
Lebesgue, 55

open
ball, 3
cover, 8, 199
function, 11, 208
mapping theorem, 845
relation, 271
set, 5, 193
subcover, 8
operator
adjoint, 842, 859, 861
backward shift operator, 899
boundary, 253, 255
closed, 861
coboundary, 262
compact, 865
completely continuous, 865
densely defined, 861
eigenvalue, 868
finite rank, 865
Fredholm, 866
index, 866
nilpotent, 880
normal, 859
positive, 892
resolvent, 867
self-adjoint, 859
spectrum, 867
unbounded, 861
unitary, 859
weakly
compact, 886
continuous, 848
opposite orientation, 248
ordering
partial, 200
orientation, 248
opposite, 248
oriented simplex, 248
orthogonal
projection, 855
spaces, 857
orthonormal
basis, 857
set, 857
oscillation function, 11, 203
outer
measure, 410
normal measure-theoretic, 671
regular measure, 635

pair
  excisive, 258
  triangulated, 247
pairs
  homotopy equivalent, 235
paracompact space, 224
parallelogram law, 854
Parseval equality, 858
partial ordering, 200
partition, 660
  of unity, 26
partition of unity, 225
  subordinated to a covering, 225
path, 220
  homotopic paths, 234
  homotopy, 234
path-connected
  component, 25, 222
  metric space, 24
  space, 221
path-connectivity relation, 26
paths path homotopic, 234
perfect set, 6
perfectly normal space, 223
pivot space, 857
point
  extreme, 863
  isolated, 196
  of best approximation, 874
  of density, 656
  spectrum, 868
pointed topological space, 233
pointwise
  bounded family, 21
  convergence, 16
polarization identity, 854
Polish space, 226
polyhedron, 246
Portmanteau theorem, 639
positive
  operator, 892
  part of signed measure, 448
  set, 447
precise refinement, 224
principle
  uniform boundedness, 843
probability
  distribution, 419
  measure space, 408
product
  σ-algebra, 416
  measure, 416
  space, 211
  topology, 211
Prohorov theorem, 640
projection, 29, 855
  metric, 874
  orthogonal, 855
proper
  differences, 461
  function, 19
property
  S, 56
  Bolzano-Weierstrass, 215
  Darboux, 59
  definiteness, 854
  excision, 258
  finite intersection, 18, 213
  Kadec–Klee, 853, 901
  lattice, 429
  Lusin, 665
  Schur, 847
  separation, 640
proximinal set, 887

pseudometric, 1
    space, 1
purely
    H^1-rectifiable set, 674
    atomic measure, 414

quasi integrable function, 428
quotient
    norm, 839
    normed space, 839
    space, 213
    topology, 213

Radon measure, 635
Radon–Nikodym
    derivative, 449
    theorem, 449
random variable, 419
range, 861
reduced
    boundary, 670
    exact homology sequence, 258
    homology group, 250
    singular cohomology groups, 263
refinement, 27, 224
    precise, 224
reflexive space, 850
regular
    family, 657
    martingale, 459
    measure, 635
    space, 194
relation
    closed, 271
    open, 271
relative
    σ-algebra, 406
    n-singular chain group of $X \bmod A$, 255
    nth singular cohomology group, 262
    boundary, 773
    closure, 773
    homotopy, 234
    singular
        n-boundaries of $X \bmod A$, 255
        n-cycles of $X \bmod A$, 255
    topology, 197
relatively compact set, 216
relaxed function, 207
renorming theorem, 875
representation theorem, 856
resolvent, 867
restriction, 10
retract, 236
    deformation, 236
    strong deformation, 237
retraction, 236
    deformation, 236
    strong deformation, 237
reverse
    martingale, 458
    submartingale, 458
    supermartingale, 458
Riesz
    lemma, 840
    representation theorem, 638, 648
Riesz–Fréchet representation theorem, 856
ring, 403
rotund norm, 871

saltus function, 667
saturated set, 212
scalar product, 854

Schauder basis, 858
Schur property, 847
Scorza–Dragoni theorem, 676
second
  category
    set, 8
  countable set, 7
  countable space, 198
selection theorem, 663
selector, 230
self-adjoint operator, 859
semifinite measure space, 408
semiinner product, 854
semimetric, 1
  space, 1
seminorm, 837
semiring, 403
separable
  σ-algebra, 407
  measure, 446
  metric space, 8
  space, 196, 852
separate points, 407, 885, 890
separated space, 194
separating family, 209
separation, 22
  property, 640
  theorem, 842
sequence
  Cauchy, 4
  Cauchy in measure, 442
  convergent
    in measure, 443
    in probability, 443
    vaguely, 692
  limit, 4
  Mayer–Vietoris, 258
  monotone, 892
  weak Cauchy, 888
sequence of measures
  weakly convergent, 651
sequentially
  closed set, 206
  coercive function, 217
  compact set, 18, 215
  continuous function, 205
  lower semicontinuous
    function, 205
  open set, 206
  upper semicontinuous
    function, 205
series
  absolutely convergent, 877
  convergent, 877
set
  F_σ, 16, 197
  G_δ, 16, 197
  H^1-rectifiable, 674
  f-saturated, 212
  absorbing, 836
  almost convex, 874
  analytic, 227
  balanced, 836
  Borel, 405
  bounded, 3, 691, 837
  Cantor, 413
  Chebyshev, 874
  clopen, 22, 194
  closed, 5, 193
  cocountable, 405
  cofinite, 404
  compact, 213
  connected, 22, 217
  convex, 229
  countably compact, 215
  dense, 7, 196, 852
  directed, 200
  disconnected, 22

equibounded, 882
equicontinuous, 232
evenly covered by p, 241
everywhere dense, 7
extreme, 864
Fréchet compact, 215
inner regular, 679
with respect to compact sets, 679
Lebesgue measurable, 412
meager, 8
negative, 447
nowhere dense, 8, 196
of finite perimeter, 668
of first category, 8
of locally finite perimeter, 668
of second category, 8
open, 5, 193
orthonormal, 857
perfect, 6
positive, 447
proximinal, 887
purely H^1-rectifiable, 674
relatively compact, 216
second countable, 7
sequentially
closed, 206
compact, 18, 215
open, 206
totally bounded, 18, 837
set function
Vitali continuous, 650
set functions
uniformly ν-absolutely continuous, 650
Vitali equicontinuous, 650
sets
isomorphic, 641
metrically separated, 682
shrinking lemma, 27
signed measure, 407, 447
σ-finite, 448
finite, 448
mutually singular, 449
simple
curve, 706
function, 421
simplex
dimension, 245
oriented, 248
vertices, 245
simplicial
approximation, 247
function, 246
of simplicial pairs, 246
neighbourhood, 247
simply connected space, 239
singular
(n)-boundaries, 253
(n)-cycles, 253
n-chain, 252
n-cochain, 262
n-simplex, 252
function, 649, 667
homology, 253
measure, 449
slice of a set, 864
Souslin space, 227
space
T_0, 194
T_1, 194
T_2, 194
T_3, 194
T_4, 194
σ-compact, 20, 216
simply connected, 239
arcwise-connected, 221

Asplund, 872
Banach, 836
Borel, 641
compact, 17, 213
complete, 4
completely regular, 271
connected, 217
contractible, 235
covering, 242
disconnected, 217
discrete, 194
dispersed, 228
dispersible, 228
dual, 838
first countable, 198
Fréchet, 836
Hausdorff, 194
Hilbert, 855
identification, 212
Kolmogorov, 194
Lindelöf, 8, 199
localizable, 415
locally
compact, 20, 215
connected, 23, 219
convex, 836
path-connected, 25, 221
uniformly convex, 871
Lusin, 229
metric, 1
metrizable, 194
normable, 836
normal, 194
normed, 836
paracompact, 224
path-connected, 24, 221
perfectly normal, 223
pivot, 857
pointed, 233
Polish, 226
product, 211
pseudometric, 1
quotient, 213
normed, 839
reflexive, 850
regular, 194
second countable, 198
semimetric, 1
separable, 8, 196, 852
separated, 194
Souslin, 227
strictly convex, 871
strongly Lindelöf, 228
Tietze, 194
topological, 193
topologically complete, 15
totally disconnected, 23, 219
triangulated, 247
uniformly
convex, 853
tight, 640
Vietoris, 194
WCG, 873
weak Asplund, 872
weakly
compactly generated, 873
sequentially complete, 888
well-chained, 26
spaces
Borel isomorphic, 641
equivalent, 14
homeomorphic, 11, 15, 208
isometric, 12, 839
isomorphic, 641, 839
Lipschitz equivalent, 15
orthogonal, 857

topologically homeomorphic, 15
uniformly equivalent, 15
spectrum, 867
Stampacchia theorem, 860
standard
n-simplex, 252
representation, 421
star, 247
stopping time, 454
theorem, 455
strict
local maximum, 32
local minimum, 32
strictly convex
function, 433
norm, 871
space, 871
strong
deformation
retract, 237
retraction, 237
separation theorem, 842
topology, 212
stronger topology, 194
strongly Lindelöf space, 228
subadditive set function, 408
subbasis, 198
subcomplex, 246
subcover, 199
open, 8
subgroup
conjugate, 289
sublinear functional, 840
submartingale, 453
convergence theorem, 457
subnet, 201
subpair, 233
subpolyhedron, 246
subspace topology, 6, 197
superadditive set function, 408
supermartingale, 453
support, 26, 225
of a measure, 636
supremum metric, 3
suspension space, 291
symmetric
form, 859

theorem
π-λ theorem, 406
Alaoglu theorem, 849
Alexander duality theorem, 263
Alexandrov theorem, 16
area formula, 667, 675
Arzela–Ascoli theorem, 21, 232
Baire category theorem, 9
Banach
fixed point theorem, 14
theorem, 845
Banach–Dieudnné theorem, 875
Banach–Steinhaus theorem, 843
Banach–Zaretsky theorem, 665
Besicovitch covering theorem, 653
Bolzano theorem, 22
Borel–Cantelli lemma, 462
Brezis–Lieb lemma, 433
Cantor intersection theorem, 9
Cantor–Bendixson, 35

Carathéodory theorem, 411
Cauchy–Schwarz–Bunyakowski inequality, 434
chain rule, 666
change
  of variable formula, 439, 440, 675
  of variables, 667
Chebyshev inequality, 435
Clarkson inequality, 435
closed graph theorem, 845
coarea formula, 675
convexity theorem, 451
covering theorem, 653
differentiation theorem, 655
Dini theorem, 49, 274
dominated convergence theorem, 452
Doob
  L^p martingale inequality, 456
  decomposition theorem, 454
  first martingale inequality, 456
  upcrossing inequality, 457
Dunford–Pettis theorem, 651
Dynkin theorem, 406
Eberlein–Smulian theorem, 851
Egorov theorem, 423
excision property, 258
Fatou lemma, 431, 452
Fréchet–Jordan–von Neumann, 854
Fredholm alternative theorem, 867
Fubini theorem on series of monotone functions, 659
Fubini–Tonelli theorem, 437
Goldstine theorem, 849
Hölder inequality, 434
Hahn decomposition theorem, 447
Hahn–Banach theorem, 840
Hanner inequality, 436
Hardy–Littlewood–Wiener maximal theorem, 451
Heine–Borel theorem, 214
Helly
  first theorem, 663
  selection theorem, 663
integration by parts formula, 667
intermediate value theorem, 22
interpolation inequality, 435
isodiametric inequality, 675
Jacobi transformation theorem, 440
James theorem, 850
Jensen inequality, 434, 453
Jordan decomposition theorem, 448
Krein–Milman theorem, 863
Krein-Rutman theorem, 870
Kuratowski theorem, 642
Kuratowski–Ryll Nardzewski selection theorem, 644
Kuratowski–Ulam theorem, 270
Lax–Milgram theorem, 860
Lebesgue
  decomposition theorem, 450
  differentiation theorem, 655

dominated convergence theorem, 431
monotone convergence theorem, 430
theorem on monotone functions, 658
Lusin theorem, 423
Lyapunov convexity theorem, 451
Mazur theorem, 847
Michael selection theorem, 229
Milman–Pettis theorem, 853
Minkowski inequality, 434
monotone
class theorem, 406
convergence theorem, 452
Nikodym theorem, 650
on monotone functions, 658
open mapping theorem, 845
Portmanteau theorem, 639
Prohorov theorem, 640
Radon–Nikodym theorem, 449
Riesz
lemma, 840
representation theorem, 638, 648
Riesz–Fréchet representation theorem, 856
Scorza–Dragoni theorem, 676
selection theorem, 663
separation property, 640
shrinking lemma, 27
Stampacchia theorem, 860
stopping time theorem, 455
strong separation theorem, 842
submartingale convergence theorem, 457
Tichonov theorem, 215
Tietze extension theorem, 12, 223
Troyanski renorming theorem, 875
uniform boundedness principle, 843
Urysohn
lemma, 12, 223
metrization theorem, 224
Van Kampen theorem, 245
Vitali
covering theorem, 653
theorem, 443
Vitali–Hahn–Saks theorem, 449
weak separation theorem, 842
Weierstrass theorem, 19
Yankov–von Neumann–Aumann
projection theorem, 648
selection theorem, 646
Yosida–Hewitt theorem, 649
theory
cohomology, 261
homology, 259
Tichonov theorem, 215
Tietze
extension theorem, 12, 223
space, 194
tight measure, 635
topological
complement, 845
pair, 233
space, 193

metrizable, 194
vector space, 835
topologically
complete metric space, 15
equivalent metric spaces, 14
equivalent metrics, 14
homeomorphic metric spaces, 15
topology, 5
box, 211
compact-open, 230
discrete, 194
final, 212
identification, 212
indiscrete, 193
induced by metric, 194
initial, 209
product, 211
quotient, 213
strong, 212
stronger, 194
upper limit, 199
Vietoris, 286
weak, 209, 638, 846
weaker, 194
total
family, 209
variation, 448, 660
totally
bounded set, 18, 837
disconnected
metric space, 23
space, 219
trace, 406
translation operator, 835
triangle inequality, 1, 429
triangulated
pair, 247
space, 247
triangulation, 247
Troyanski renorming theorem, 875

unbounded linear operator, 861
uniform
boundedness principle, 843
convergence, 16
metric, 3
uniformly
ν-absolutely continuous functions, 650
bounded family, 21
continuous function, 13
convex space, 853
distributed measure, 682
equicontinuous family, 21
equivalent metric spaces, 15
integrable family, 441
tight space, 640
unitary operator, 859
universal
σ-algebra, 642
completion, 642
covering space, 244
universally measurable function, 642
upcrossing, 457
upper
density, 703
Kuratowski limit, 32
limit topology, 199
semicontinuous
function, 204
multifunction, 229
Urysohn
criterion for convergence, 33
lemma, 12, 223
metrization theorem, 224
usc multifunction, 229

vague topology, 708
vaguely convergent sequence, 692
Van Kampen theorem, 245
variation, 669
vector space
  topological, 835
vertices, 245
Vietoris
  continuous multifunction, 229
  space, 194
  topology, 286
Vietoris continuous multifunction, 229
Vitali
  continuous set function, 650
  cover, 652
  covering theorem, 653
  equicontinuous set functions, 650
  theorem, 443
Vitali–Hahn–Saks theorem, 449

WCG space, 873
weak
  Asplund space, 872
  Cauchy sequence, 888
  separation, 842
    theorem, 842
  topology, 209, 638, 846
weaker topology, 194
weakly
  compact operator, 886
  compactly generated space, 873
  continuous linear operator, 848
  convergent sequence of measures, 651
  sequentially complete space, 888
wedge product, 395
Weierstrass theorem, 19
well-chained metric space, 26

Yankov–von Neumann–Aumann
  projection theorem, 648
  selection theorem, 646
Yosida–Hewitt theorem, 649

Printed in the USA
CPSIA information can be obtained
at www.ICGtesting.com
CBHW071536151124
17467CB00008B/511

9 783319 355351